BOXING
MEDICAL ASPECTS

Dedicated to

LLUIS BARRAQUER-BORDAS, Barcelona

A.H. ROBERTS, London

and in memory of

J.A.N. CORSELLIS

JAMES HAMILTON DOGGART

F.R. FERGUSON

JULIUS HALLERVORDEN

KAREL HENNER

OTAKAR JAROS

ERNST JOKL

HARRISON MARTLAND

C. MAWDSLEY

KARL T. NEUBÜRGER

KARL SELLIER

MIROSLAV SERCL

J.D. SPILLANE

GEORG STRASSMANN

LADY EDITH SUMMERSKILL

BOXING
MEDICAL ASPECTS

FRIEDRICH UNTERHARNSCHEIDT

AND

JULIA TAYLOR-UNTERHARNSCHEIDT

ACADEMIC PRESS

An imprint of Elsevier Science

Amsterdam • Boston • London • New York • Oxford • Paris
San Diego • San Francisco • Singapore • Sydney • Tokyo

This book is printed on acid-free paper

Academic Press
An Imprint of Elsevier Science
84 Theobald's Road, London WC1X 8RR, UK
http://www.academicpress.com

Academic Press
An Imprint of Elsevier Science
525 B Street, Suite 1900 San Diego, California 92101-4495, USA
http://www.academicpress.com

ISBN 0-12-709130-0

Library of Congress Catalog Number: applied for

A catalogue record for this book is available from the British Library

Typeset and printed by PlantijnCasparie Heerhugowaard, J. Duikerweg 14, 1703 DH Heerhugowaard, The Netherlands
Cover Design by Chris van Egmond, PlantijnCasparie Heerhugowaard, The Netherlands

02 03 04 05 06 07 9 8 7 6 5 4 3 2 1

Table of contents

Preface

There is probably no more controversial, volatile and incendiary topic than the question of the relative merits and risks of boxing. Every time a boxer is killed in the ring, public debate is renewed. Governmental bodies and athletic organizations are continually discussing ways to make boxing safer for participants or even whether it should be banned altogether.

There is a general understanding among most physicians that boxing can cause debilitating or fatal injuries. Hence the expression 'punch drunk', in common usage for more than a century. However, that is where consensus ends. After agreeing that boxing can cause death or serious injuries, opinion is divided between two equally committed and vocal camps. There are those who support boxing and argue that it has positive benefits for society, countered by those who believe that the health risks outweigh the possible benefits of boxing.

Yet in spite of continuing debate, boxing, both professional and amateur, is a growing activity, continually taking in new groups of participants. In many countries today women are boxing professionally, and young girls and boys are entertaining spectators by exchanging blows in the ring. Boxing classes are tremendously popular in the United States, as the latest fitness fad.

It is important to note from the beginning that references to boxing and boxing injuries throughout this work will pertain only to those activities involving delivery of direct blows making bodily contact. In gyms throughout the world many people participate in 'recreational boxing', where little or no bodily contact takes place. There may be no head contact at all. These participants shadow box or strike punching bags or special mitts on the hands of a partner. Sometimes they spar in a manner similar to American touch football, where light contact is substituted for tackling or applying hard force to other players. The touch is intended to be more symbolic than disabling.

It is possible to engage in some of these boxing activities with no more risk than that found in other recreational sports. Although when sparring partners begin to apply force behind their blows, the potential for injuries will rise accordingly, even in a friendly round.

Unfortunately it is not possible to draw a hard line between amateur and professional boxing in terms of risk of harm. Amateur boxing matches are more like professional boxing than the low contact exercises discussed above. Even where children are concerned, forceful contact is risky, especially to the head or neck. As will be seen in subsequent chapters, even relatively light blows to the head, especially if repetitive, can lead to brain damage that may not be obvious for many years after the match.

When the dangers of boxing are questioned, in either scientific or journalistic publications, supporters fervently plead for the *'noble art of self-defense,' the 'sweet science,'* or *'fist-fencing'* said to be wholesome for body and mind. There is no shortage of defenders of boxing as a beneficial sport, even among those who most clearly demonstrate the permanent debilitating effects.

Boxing advocates often refer to statistical accounts claiming that boxing is less dangerous than a number of other sports. For example, it has been claimed that boxing injuries have a frequency of about 2 %, which is lower than in soccer (2.3 %), handball (2.8 %) and heavy athletics (3.1 %). With a ratio of fatal incidents of 0.13/1000 boxing would seem to be a safer sport than American football (0.3/1000), motor racing (0.7/1000), scuba diving (1.1/1000), mountain climbing ((5.1/1000), hang gliding (5.6/1000), parachuting (12.3/1000) or steeple chase riding (12.8/1000) (CUNEO and RUSSO 1984).

Unfortunately there is little information available on how these often quoted studies were conducted. A *Report of the Medical Council on Scientific Affairs of the American Medical Association* (1980) emphasized that the reliability and validity of the statistics cannot be established. Among other problems, the number of participants and times at risk in each sport are not stated, making comparison among sports difficult.

Perhaps the most serious defect of statistical studies showing low accident rates for boxing is that permanent brain damage is excluded. The studies usually consider only the more obvious but less significant injuries, such as fractures of bones, bruises, distortions, hematomas, etc. Late or permanent damage in the central nervous or visual system, unaccounted for in the statistics, is considered by many to be the most serious and debilitating risk of boxing.

Fatalities aside, a few defenders of boxing fervently deny that it can cause brain damage. They say that they have never seen any related clinical syndrome, and that 'punch drunkenness' does not exist. They claim that lesions in the brains of boxers are the result of other activities or diseases, as alcoholism, venereal diseases, different forms of dementia, etc., not related to boxing.

Just as fervent as advocates are the detractors of both amateur and professional boxing. There are those who oppose boxing on philosophical grounds as well as those who believe that injuries from boxing are underestimated. Some believe that witnessing boxing events is harmful to society in general and particularly so to children. Among those concerned about the degree of risk in boxing, most notably, are many physicians and leading medical associations. Many such organizations have stated positions in writing, based primarily on medical findings. In view of the substantial risk of injury in boxing, they have called for a complete legal prohibition against it.

From the time of the earliest medical publications, boxing was condemned. It was called "the only legal way of killing... The moment has come that boxing should no longer be officially encouraged, but generally prohibited... A way of acting that in everyday life is punished by prison, hard labor, and even death, is rewarded in boxing with publicity, gold medals, and money" (JOKL 1962). "A person who seeks to justify boxing as a 'noble art of self-defense' seems to be blindfolded in the use of an inaccurate phrase," noted PÖSCHL and KRIEGER (1963).

CRITCHLEY said "boxing is different from other athletic endeavors in that points are scored only by violent contact, i.e. connecting blows. The sole objective is to punch the opponent in order to disable him. Thus boxing causes injury by intention rather than by accident. Causing injury is at the very heart of boxing, as opposed to being incidental to fair play. Matches are unequivocally won by flooring the opponent so that he cannot rise or continue fighting. (Boxing injuries) are not so much regretted as regrettable" (CRITCHLEY 1957).

WINTERSTEIN had written in 1937, that boxing presented a medical experiment for the study of traumatic brain injuries.

LUNDBERG, the former editor of the *Journal of the American Medical Association* formulated 1983:

"Boxing seems to me less sport than is cockfighting, boxing is an obscenity... Boxing, as a throwback to uncivilized man, should not be sanctioned by any civilized society."

VAN ALLEN noted 1983:

"How strange that, in this climate of preoccupation with health and physical fitness and with near-hysterical concern for every conceivable deleterious factor in the environment, so few raise their voices against boxing... Is now not the time to suppress exposure of this fragment of our savagery by the mass media and leave boxing to those who enjoy privately staged dogfights."

EBBING wrote 1988:

"Boxing is everything else but an athletic sport. It is not a sport at all... Celebrated is an obscene male picture, false bravery and empty heroism."

National medical associations in the United States, Great Britain, Canada, and Australia, to name a few, have condemned boxing. To date no medical associations have spoken out in support of boxing as a safe activity. Many physicians find it difficult to accept boxing, as it is the only sport in which the sole objective is to use physical force, in the form of direct blows, to render the adversary defenseless and knock him out of his senses. ... Or the sole objective is to disable the opponent by inflicting blows sufficient to make him drop to the floor, unconscious or at least unable to get up.

The position assumed by medical associations should not be easily accepted simply because it has been published. However, it is fair to say that the physical risk of harm from boxing can only be reasonably evaluated in light of data found in medical publications over the last four decades. During that time more has been learned about the origins and significance of closed head injuries than throughout the rest of history altogether.

My purpose here is not to argue for or against the continuation of boxing, but to lay out the literature and the body of scientific knowledge that are necessary to provide a meaningful background for the ensuing debate. I wanted to provide a comprehensive resource for those who are involved in regulating boxing and those who participate directly, as well as for the medical and scientific communities. It is impossible to make sound decisions about the risks of boxing, or safety measures, without understanding what the specific risks entail. For the sake of completeness I have carefully quoted case histories and research, more extensively than in other scientific publications, in order to provide complete information with which each reader can form his or her own well considered opinion.

There is no doubt that many a discussion on boxing has generated heated arguments that have deteriorated into irrelevant quarreling. Some of the material in this volume will be controversial. I have tried to list, quote, and discuss alternative theories with as much objectivity as possible, but in doing so, felt free to employ a mild form of irony in reflecting on some of the more extreme opinions voiced.

'Sua fata habent libelli!' The translation of this Latin phrase originates from *'De litteris, syllabis, metris' ('About litteries, syllables, metrics')* by Terentianus MAURUS, who lived during the end of the third century A.D.). That is to say, the impulse to write, the concept, the process of writing, in other words its whole history, is very specific for each book written. This is the eighth book I have written, and it calls for a particular introduction, by way of the preface – which I am writing after the book is finished.

I think it necessary to say a few words about my professional background. I am a Board certified neurologist and psychiatrist and neuropathologist. As a psychiatrist I consider myself somatically oriented; I do not practice analytical psychiatry. As a neuropsychiatrist and neuropathologist I see myself as a morphologically oriented clinician.

I have worked as a medical scientist for 40 years in close collaboration with physicists and engineers, and have thus remained deeply entrenched in the field of biomedical engineering without having specific training or a degree in it.

Between the early 1970s and 1987 I was chief of the Department of Neuropathology of the Detachment of the Naval Aerospace Medical Research Laboratory – the later Naval Biodynamics Laboratory – in New Orleans, Louisiana.

Prior to this monographic treatment of the entire subject, I wrote a number of smaller papers, several book chapters, and gave numerous lectures on the historical and medical aspects of boxing. The result of my continuing ruminations over head trauma and boxing injuries is this comprehensive, perhaps Teutonic compilation of the material. The charitably inclined reader, hopefully, will excuse the seemingly excessive thoroughness; it was deemed necessary, since there are many old misapprehensions about boxing injuries that are still repeated today.

In order to achieve a better understanding of the entire topic, as expressed in the words of Johann Wolfgang von GOETHE: "Ein Gegenstand, wohl beschaut, schließt ein neues Organ in uns auf." ("An object, well studied, opens a new organ in us"), a synopsis is necessary. That is to say, we suddenly see and thoroughly understand a subject only after close inspection, reflection, and research on it. One of my objectives has been to set forth the literature, in one place, as a reference source for anyone interested in boxing, as well as for those who study head trauma.

Regardless of how one feels about boxing as a sport, there is unquestionably a substantial body of medical literature describing and discussing known injuries that are seen after participation in boxing. The literature is pervasive, appearing in journals of many different medical specialties and in different languages.

Pathologist Harrison MARTLAND, of New York, published a paper entitled *'Punch drunk'* in 1928. He described the neurological and psychopathological permanent damage of the central nervous system in boxers in classic form. He referred to "a peculiar condition occurring among prize fighters." Many cases remain, according to MARTLAND, mild in nature and do not progress beyond this point. In others a very distinct dragging of the leg may develop and with this there is a general slowing down in muscular movements, a peculiar mental attitude characterized by hesitancy in the head, necessitating withdrawal from the ring... Later on, in severe cases, there may develop a peculiar tilting of the head, a marked dragging of one or both legs, a staggering, propulsive gait with the facial characteristics of Parkinson syndrome or a backward swaying of the body, tremors, vertigo and deafness. Finally, marked mental deterioration may set in necessitating commitment to an asylum. This classic paper will be discussed in detail on p. 288.

The first comprehensive treatment of the medical aspects of boxing was presented by Ernst JOKL, in 1941. JOKL summarized the permanent brain damage and acute pathomorphological findings known at the time. Reports of the pathomorphological permanent brain damage in boxers did not yet exist.

The English neurologist McDONALD CRITCHLEY, in a series of publications, referred to the severe neurological and psychopathological brain damage in boxers (1935, 1937, 1957, 1960, 1965).

LA CAVA described the permanent neurological and psychopathological findings in boxers in a series of papers (1949, 1950, 1951, 1953, 1954, 1963, 1964, and 1973).

The Chairman of the Department of Neurology of the University of Prague, Karel HENNER, published a series of scientific papers (1955, 1956, 1963, and 1966) about the neurological damage in boxers. HENNER was critical of boxing in his time, due to the severe damage it inflicted on the central nervous system. In 1956 he wrote:

"Boxing is a dangerous sport, even when the present rules are strictly observed... It must be born in mind that traumatically induced unconsciousness is an absolutely normal and legitimate feature of this form of sport. Both criminal and civil law punish even small injuries to the head and provide for compensation for the victim, and rightly so. In boxing, the contestant who succeeds in 'putting his opponent to sleep' is held in esteem, fêted and set up as an example. This cynical euphemism means that the aim of boxing is to pummel your opponent's head so cleverly and brutally that he falls unconscious. If he falls and rises again, he is given a series of terrible blows until he finally remains motionless. There is the moral damage to the spectators. This leads to over-admiration of brute force and exerts an unhealthy influence on young, immature onlookers and may even lead to sadism."

TICHY (1997) reminded us of HENNER's legacy, on the occasion of his one hundredth birthday, in his warning of the neurological effects of boxing as it was practiced in his time. It was to his credit that the protective helmet was introduced. He also commented on the inadequacy of medical examinations carried out at the time.

Further contributions that dealt with the clinico-neurological aspects of the encephalopathy of the boxer were published by SERCL (1955, 1969), SERCL and JAROS (1956, 1957, 1963, 1968, and 1974).

In 1956 the English politician, Lady Edith SUMMERSKILL, wrote a book with the title *'The*

Ignoble Art,' warning of serious brain and eye injuries in boxers.

The English neurologist ROBERTS published a book entitled *'Brain Damage in Boxers,'* in 1969. The author evaluated a group of 224 boxers from a group of 16,781 professional boxers regulated by the *British Boxing Board of Control* from 1929 to 1956. Of the 224, 37 had permanent brain injuries that could be directly attributed to boxing.

The first thorough and splendid description of the chronic pathomorphological findings in the brain of a boxer was written by BRANDENBURG and HALLERVORDEN. In 1954 they published a paper entitled *'Dementia pugilistica mit anatomischem Befund' ('Dementia pugilistica with Anatomical Findings').*

Further publications dealing with chronic morphological aspects of the encephalopathy of boxers are from GRAHMANN and ULE (1957), PAYNE (1958), NEUBÜRGER et al. *(1959),* CONSTANTINIDES and TISSOT (1967) and CORSELLIS et al. (1967).

The English neuropathologist CORSELLIS and his collaborators published a classic study about neuropathological findings in 15 former boxers. The brain damage was often complex and differed in its severity; no permanent brain damage could be seen in only one case out of 15. 11 of 13 cases exhibited the most striking finding, where widening and fenestration of the septum pellucidum was observed. Identical findings were published earlier by SPILLANE (1962) in pneumencephalograms of boxers. Additionally CORSELLIS et al. found atrophies in the cerebellum with glial proliferation, which especially affected the folia. There was a marked loss of Purkinje cells at the lower surface of the cerebellum. The substantia nigra had a loss of pigmented neurons, especially in its lateral parts. The neurons of the cerebral cortex revealed a disseminated loss of nerve cells and Alzheimer's fibrillary tangles without presence of senile plaques.

In the early 1960's the influence of linear and rotational acceleration and deceleration forces, respectively, on the heads and brains of animals was examined (UNTERHARNSCHEIDT and SELLIER 1962, UNTERHARNSCHEIDT and HIGGINS 1969). One of the most important questions was to find the threshold for producing a cerebral concussion (Commotio cerebri) and delineate it from the threshold for production of primary traumatic lesions of the brain. One observation was especially significant for boxers, with wide ranging ramifications. The application of so-called subconcussive intensities of impact, after a single application, produced neither disturbances of behavior nor morphologic alterations in the brain. However, after repeated applications of the same lower intensity of force, clear and severe permanent brain damage was produced. The pathomorphological findings were published in detail by UNTERHARNSCHEIDT (1963).

The permanent brain damage in the animal model, especially the severe disseminated loss of neurons in the cerebral cortex, in the Ammons' horn formation, and in the cerebellum, were identical with the corresponding permanent brain damage found in boxers. The experimental setup in the animal model resembles the boxer who receives numerous repeated blows against his head and brain during his active training and career. This observation first prompted me to begin collecting and evaluating the medical literature about boxing.

Especially in the last decade, a plethora of new clinical data and morphological findings in amateur and professional boxers has emerged. Renowned medical organizations in different countries have analyzed the degree of brain damage in boxers that can be attributed to boxing and whether boxing should continue to be permitted.

In collaboration with my late friend Karl SELLIER – the noted physicist and physician – the physics of impact was thoroughly examined (SELLIER and UNTERHARNSCHEIDT 1963, UNTERHARNSCHEIDT and SELLIER 1966, 1970, 1971, UNTERHARNSCHEIDT 1975). The results of physical, clinical and morphological examinations from these experiments and those of other authors appeared in a series of publications, reviews and chapters in books on sports medicine (UNTERHARNSCHEIDT 1970, 1972, 1975,

1985, 1993, 1994, UNTERHARNSCHEIDT and SELLIER 1978).

The time is now ripe to publish a comprehensive presentation of the traumatic lesions of the central nervous system in boxers. The health advantages and disadvantages of boxing must be discussed in the light of currently available medical findings. The scientist and physician in today's society is not only entitled, but obliged to report these facts and conclusions, as they are directly relevant to the public interest and current debate.

New diagnostic tools and techniques have been introduced into clinical neurology which allow us to reach better working knowledge and diagnosis of the somatic processes of the brain. These techniques, especially computed axial tomography (CAT) and magnetic resonance imaging (MRI), have been used to achieve a better understanding of traumatic lesions of the CNS, and traumatic encephalopathy of the boxer. These findings are included and discussed extensively below.

A short summary of the physics of impact and the acute and chronic clinical and pathomorphological findings of the central nervous system of boxers is presented. Fundamental experiments using animal models were conducted, bearing similarity to the experience of boxers. These findings are quoted in summary, followed by a discussion on their application to boxers and central nervous system.

Occasionally boxing advocates argue that protracted investigations on the health value of boxing need to be undertaken before reaching any conclusions. They call for long term studies on individual boxers. Suffice it to say here that long term studies have never been considered necessary in order to reach conclusions about head injuries from other sources of trauma, yet conclusions have been accepted and are relied upon every day. It will be within my scope to demonstrate both that reliable information has existed for a long time and that it constitutes a sufficient basis on which to draw relevant conclusions.

While it is not my intention to support or oppose boxing as a sport, my own research and experience with head trauma and boxing injuries compels me, as a physician and a scientist, to let the findings be known, so that prospective boxers and in some cases their parents, can make an informed decision to accept the risks. Boxers who deliver and receive blows should understand what takes place in the brain and exactly what risks they are accepting. Every youth who boxes for recreational purposes, as well as every parent, needs to know that in addition to the risk of immediate harm there are certain long range effects to consider. It is important also for the spectator to understand the long-range ramifications of the spectacle in the ring. I lack all missionaric zeal and elan to prohibit boxing, but am of the opinion that this topic is of immense interest to the public and that it is especially important for youngsters and others to know what risks they are consenting to accept when they step into the ring.

Some remarks seem necessary in regard to the presentation of the scientific material in this book. Medical case histories will be treated in an unusually broad form, more so than in other medical publications. The case histories of the clinical and pathomorphological findings are presented in extreme detail. This is for two reasons: Once the case histories have to be evaluated under different aspects, to be summarily presented, which has not been undertaken before. On the other side, defenders and advocates of boxing utter and assert, again and again, that no or only a little literature about the permanent brain damage of boxers exists, and when some data were published, then these were not convincing findings. This publication will show convincingly that the clinical and pathomorphological findings are completely sufficient to demonstrate that boxing, amateur and professional, as it is performed today, produces permanent brain damage, and has to be condemned from a medical standpoint. Should defenders and proponents of boxing in future use the argument that adequate medical literature does not exist, so is the answer to such statements, that they are not familiar with the medical literature.

Writing about boxing raises an interesting aspect, which was discussed by EDMONDS (1973) who noted:

"For the professional historian, athletics has long been considered a subject beyond the scope, or perhaps better, beneath the contempt of academic concern. Indeed, sports have usually been relegated to Clio's junk-heap and have been viewed as part of what historian Roland BARTHOFF calls 'pots and pans' history. Those who have written accounts of athletic heroes have generally consisted of professional sportswriters, or perhaps athletes themselves, composing with the help of ghost writers, inspirational autobiographies which are usually sold at little tables during Fellowship of Christian Athletes conventions. I would, however, agree with philosopher Paul WEISS, who noted that the study of the ideals and practice of athletes can inform the scholar about some of the basic values of a given society. In fact, an estimated eight million Americans participate in some form of athletics each year, while far more are spectators at sporting events."

EDMONDS continued:

"Of all sports, prizefighting has come under the most sustained criticism. Boxing conjures up the image of sleazy characters hanging around small gymnasiums, eagerly seeking fights to fix and boxers to buy. At worst, boxing is a brutal sport, its critics argue, a sport which causes physical harm, brain damage, and often death. But such charges, no matter how valid, should not deter the scholar from examining the sport and its participant. If moral stench be allowed to inhibit scholarship, then no right minded historian would go near such subjects as crime, perversion, or the fate of six million Jews in Europe."

Amateur boxing is an acknowledged sport at modern Olympic games, though some have questioned whether boxing is compatible with the Olympic ideal. The writings of Baron de COUBERTIN have been included to elaborate on this debate.

While writing about the medical aspects of boxing, I realized that such a text cannot be satisfactorily written without being thoroughly familiar with the development of boxing in ancient cultures and its subsequent history. Since my schooling was based on the humanities, focusing on the old classical languages – Latin and Greek – I could not resist looking into the development of boxing in ancient cultures. From there I read on, into the revival of boxing in the seventeenth century England and the Regency epoch and then to the development of boxing in Colonial America and the United States. The informative background discussion grew until it encompassed an entire volume of its own. During these years of work the project became a family affair, and a separate volume on the legal aspects of boxing was added (Julia TAYLOR UNTERHARNSCHEIDT and Friedrich UNTERHARNSCHEIDT).

In the historical volume of this work, cultural-sociological aspects of boxing in different civilizations and in different eras are presented and examined in detail. The relative value or unworthiness of boxing, in its many different forms, as viewed by different classes in society, will be examined. Statements from contemporaneous literature and the press are quoted, summed up or described.

The volume on the legal aspects of boxing includes discussions on the regulatory schema, legal problems and relevant case law on liability.

Introduction

Prerequisite for the study of the acute and chronic clinical and morphological injuries of boxers is a solid and informed knowledge of the traumatic lesions of the brain. The pursuit of neurotraumatology requires an intimate and thorough knowlcdgc of biomechanics.

This treatise begins with an introductory chapter on impact mechanics, followed by a short synopsis of the neuropathology of head injuries, particularly closed head injuries.

Impact Mechanics of Head and Brain Injuries

The most important features of an impact event are the amount of force delivered and the duration of the impact event. Careful analysis has shown that biological tissue and structures are damaged by the relative movement between two mass elements. Also important is whether the force on the biological system can be considered *static* or *dynamic*.

Two physical quantities play a role in blunt impact on a head: the velocity, v, and the acceleration, a. Velocity is equal to the movement in space per unit of time:

$$v = \frac{\Delta s}{\Delta t}$$

expressed in m/s or km/h (1 m/s = 3,6 km/h).

The acceleration indicates how velocity changes in a unit of time:

$$a = \frac{\Delta v}{\Delta t}$$

A given change of velocity v can produce a large a, when the time interval is very small.

An impact in a very brief time interval can produce high acceleration. The velocity, v_a, in connection with the mass of the striking body, m_a, indicates the so-called momentum, $m_a \cdot v_a$. By the theorem of momentum this is equal to the impulse, I, imparted by the applied force.

To understand traumatic brain damage due to collisions, one must discern two cases: *translational trauma* and *rotational trauma* (Fig. 1). The best known example of a rotational trauma is the uppercut or hook. The direction of the blow does not go through the center of gravity of the head, so the cranium will not only be accelerated in translation, but it will also accelerate in a rotatory fashion. The brain is unable to follow the quick rotation of the skull immediately, so the inner surface of the cranium slides against the brain surface. The brain will pull at its suspensions; the *connecting* or *bridging veins* will be stretched and may rupture. One may think of this as a nodding motion. Since the medulla oblongata and the adjoining white substance are fixed in place, such a rotation also causes a folding of different parts of the brain.

That these brain motions have occurred in a rotation trauma may be concluded from the kind of lesions found and the rupture of the connective or bridging veins. The folding of single parts of the brain is a working hypothesis, to be sure.

The most frequent violence suffered by the skull

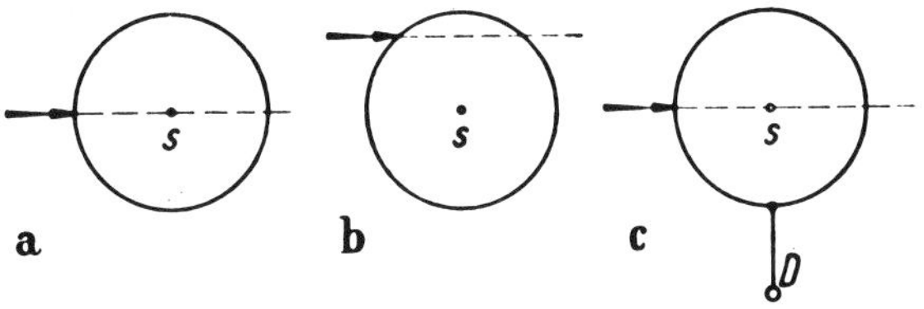

Fig. 1. Possibilities of the head to experience a rotation (and thus a rotation of the brain relative to the skull) in a collision. a. Collision axis through the center of gravity 'S'. Translatory, but no rotatory movement. b. Collision axis misses the center of gravity 'S', consequently translatory and rotatory movement about 'S'. Model for an 'uppercut'. c. A body may rotate about D to which it is fixed by a pole. Even when the collision axis goes through 'S', translation and rotation will occur, but the latter will be small. Model for head and cervical cord. In a collision of short duration only the translatory movement will count. (From SELLIER, UNTERHARNSCHEIDT, 1963).

and its contents is translation trauma, i. e., the head will be accelerated without much rotation. At higher accelerations this results in the so-called 'contrecoup' lesions of the brain. The term contrecoup is as incorrect as is the term point of contrecoup, since the brain does not collide with the antipole, but tries to get away from it. This type of lesion initiated several theories because of its seemingly paradox location, opposite to the point of impact.

According to our experiments (SELLIER and UNTERHARNSCHEIDT 1963), the facts may be explained as follows: In a collision between the head and a solid wall, the skull will be abruptly decelerated, but inertia keeps the brain moving (Fig. 2). It is easily seen that this will produce an overpressure at the surface of the brain at the site of impact. At the side opposite to the impact point (at the 'antipole'), an under-pressure or reduced pressure will be produced near the brain surface (and in the cerebrospinal fluid), since there the brain is moving away from the skull. In a sphere with ideally rigid walls, i. e., with nonflexible walls, the under-pressure may easily be calculated, also the overpressure; except for the sign, over- and under-pressure are equal (Fig. 3). Over- and under-pressure are proportional to the acceleration, the distance between collision pole and antipole, and the specific weight of the brain. The pressure gradually

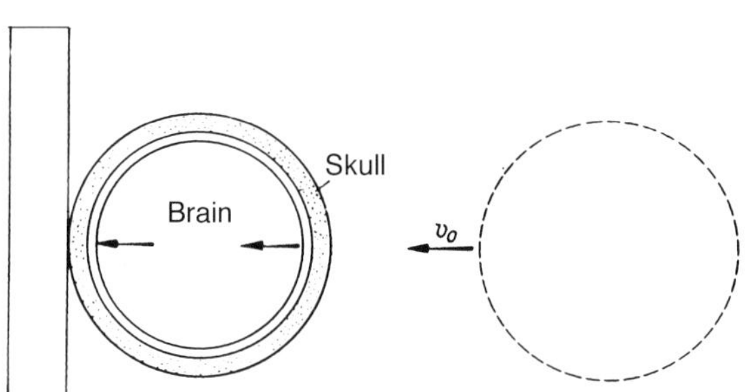

Fig. 2. Colliding with a rigid wall, the skull will be quickly decelerated while the brain is moving on, because of inertia, and is pressed against the cranium. This causes overpresseure to develop at the impact point (left arrow), and underpressuere (negative pressure, right arrow) at the antipole. Physically, it is irrelevant which of the collision partners is moving, the wall or the head, because the relative movement will count. (From SELLIER, UNTERHARNSCHEIDT, 1963).

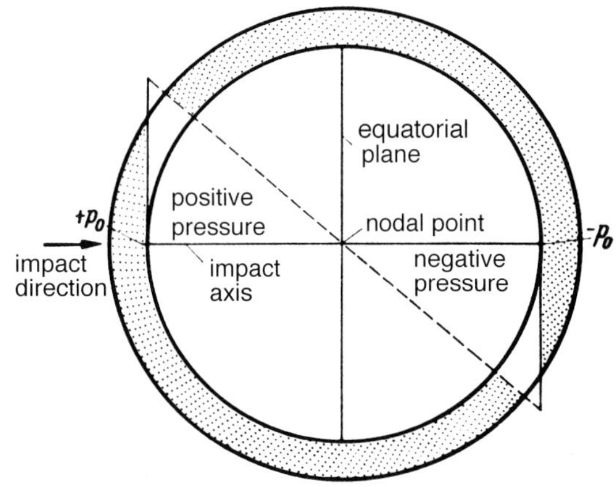

Fig. 3. Schematic diagram of the pressure transient over the collision axis, maximum and minimum values at the margin to the right and left, zero at the nodal point. (From SELLIER, UNTERHARNSCHEIDT, 1963).

decreases in going from the collision pole to the antipole, going through zero in the center of the sphere (if this has rigid walls). These considerations can be supported by experiments.

A rigid sphere with strong walls (Fig. 4) was equipped with two pressure sensors (pressure transducers, piezo-electrical principle) at the collision point and another at the antipole, which could be moved relative to the axis (SELLIER and UNTERHARNSCHEIDT 1963). The results are shown in Fig. 5. The positions of the two sensors and the corresponding pressure distributions are

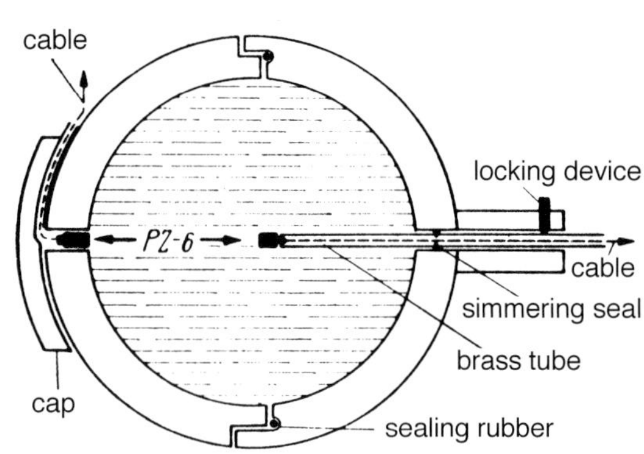

Fig. 4. Model of the skull as used for measuring the pressure distribution. Vessel: Trogamid B plastic. At the left, screened point of impact with a pressure sensor PZ 6. The pressure sensor at the antipole (to the right) may be moved near to the impact pole. (From SELLIER, UNTERHARNSCHEIDT, 1963).

designated by *a* and *b*. At the collision pole, there is an overpressure of about the same magnitude. There follows an oscillation originated by vibrations of the wall of the sphere. When the sensor, *b*, was located in the center of the sphere (see Fig. 5, second graph), it registered practically no pressure, while the course of the pressure at point *a* is similar to that in the first diagram. In the third graph, the sensor was about halfway between the center and the antipole. The pressure distributions for *a* and *b* are very similar, but the amplitudes at *a* are about twice those at *b*. This means that the pressure decreases linearly from the collision pole to the antipole, the zero point being at the center of the sphere. This is valid only for a sphere with ideally rigid walls.

From physical laws, it follows that in a liquid there cannot be a larger under-pressure than – 1 atm, since gas bubbles will suddenly be formed

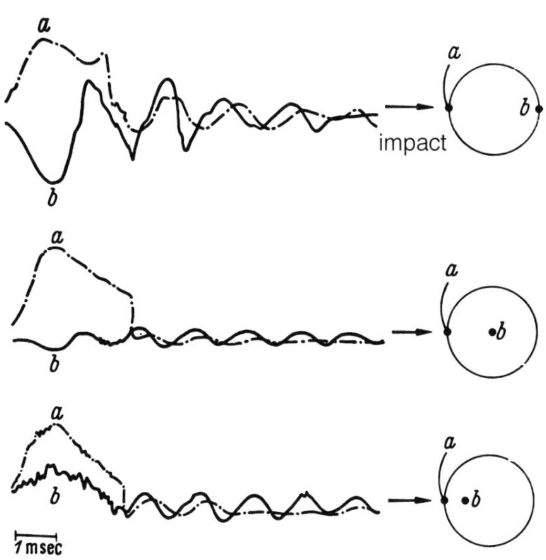

Fig 5. Experimental pressure transients in a water-filled hollow sphere with almost rigid walls for different positions of the antipole pressure sensor (at b). Curve a gives the pressure transient at the impact pole, curve b that at the antipole. The maximum amplitude at a is 3.0 atm, at b 2.2 atm. The time scale is given on the abscissa. The absolute presseure at the impact pole is greater than at the antipole because the walls are not ideally rigid and will be slightly deformed during the collision; also, the rise of the transient will be flattened a bit. The relative positions of the pressure sensors are given to the right of the diagram. (From SELLIER, UNTERHARNSCHEIDT, 1963).

(the liquid will begin to boil), which will destroy the water-containing tissue. The question is which physical quantity – the over- or the under-pressure – is causing brain damage? Experiments by EBBECKE give evidence that nervous tissue will stand a high overpressure for a long time without damage. So there is no doubt that it will stand it for a short time. Convincing evidence for the harmlessness of a higher overpressure to brain substance is given by bullets penetrating the skull. According to our measurements, a bullet of 7,65 mm (0.30 in.) penetrating the head produces a pressure of about 40 atm at the brain surface, for a short period. The tremendous overpressure will not cause any damage to the cortex of the 'contrecoup' type. The overpressures occurring with a head trauma are very much lower. The under-pressure must therefore be responsible for the injury: at a certain magnitude, gas bubbles will be produced, which cause the capillaries to burst *(known in engineering as 'cavitation').* Imagine that these capillaries were planned to take an inner pressure of a few mm/Hg. An under-pressure of 1 atm is equivalent to an inner pressure of 1 atm. This great under-pressure therefore will certainly destroy these tiny vessels by producing gas bubbles. These may also develop in the tissue, which will be inflated whereby its structure will be destroyed. Since during a blow under-pressure lasts only a very brief time, a few ms, these gas bubbles will not be very large and will collapse immediately after cessation of the collision, and similarly, of the under-pressure. The occurrence of these gas bubbles is no hypothetical assumption; it was demonstrated by slow motion pictures of a dummy skull filled with water (GROSS 1958). A detailled description is presented in the chapter on contrecoup injuries . A fixed head will show no 'contrecoup' lesions at the brain. If, for example, a hammer would fall and strike a head rested on the floor, or leaning to a wall, one would not find this type of lesion in the brain. This fact is a prime support to the under-pressure (suction) theory.

These considerations are valid for a sphere with ideally rigid walls. However, a bony shell in no case is ideally rigid; it will yield in a collision, i. e., the sagittal diameter will decrease. Thereby the movement of the brain surface relative to the inner wall of the skull is reduced, and as a consequence the under-pressure produced by a given acceleration of the head will be less than calculated (Figs. 6 and 7). In other words, the acceleration required to produce the so-called contrecoup lesions must be higher than calculated. Then, how high must the acceleration be, theoretically? The lower limit, derived from the specific weight of the brain tissue and the sagittal diameter, is 120 g, approximately. In practice it will be higher, by the factor of about 2, because of the flexible cranium – i. e., in the vicinity of 250 g, so-called contrecoup lesions may be expected.

Theoretical considerations suggest that the smaller the skull is, the higher the critical acceleration necessary to provoke lesions will be. Also, the area where the under-pressure of about – 1 atm will be effective should be a segment of the sphere adjoining to the antipole. The more severe the violence is, the more the volume of this segment will be increased (i. e., its depths will be increased.

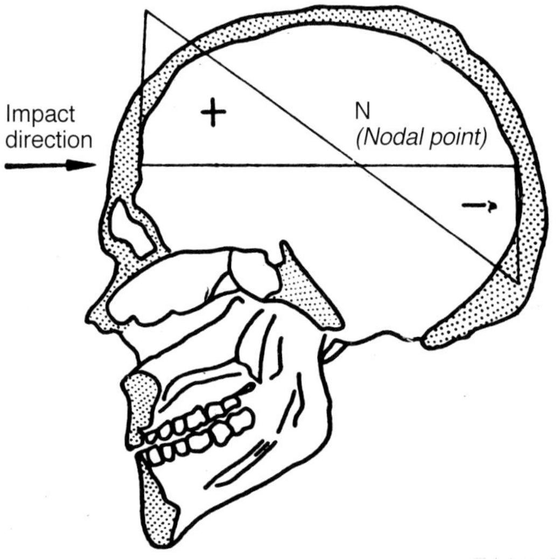

Fig. 6. Schematic diagram of the pressure distribution in the cranium during a head-on impact. The distribution is not symmetrical about the nodal point because the skull deforms during the impact. (From SELLIER, UNTERHARNSCHEIDT, 1963).

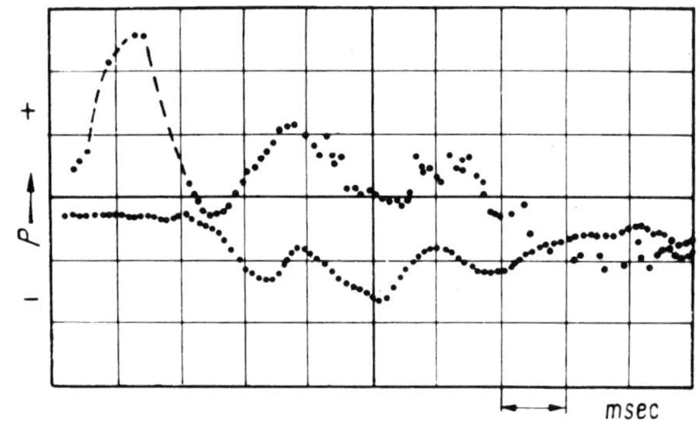

Fig. 7. Pressure measurements with a macerated skull. Head-on impact by a 'concussion gun'. *Upper curve:* pressure transient at the impact pole; *lower curve:* pressure transient at the antipole. The negative pressure at the antipole was produced more than 0.2 ms later, since then skull was deformed in the collision; its amplitude is therefore smaller than that of the positive pressure. An identical pressure scale was used for both curves. (From SELLIER, UNTERHARNSCHEIDT, 1963).

While at small accelerations only the summits of the gyri are destroyed by cavitation, at higher acceleration the sulci will be affected also. The primary (physical) damage will extend to segment-shaped structures of the cortex and possibly to the white substance. The typical picture of a wedge-shaped lesion can be explained by anatomy only.

In a collision, the cranium will not behave like a homogenous hollow sphere, which would be deformed to an ellipsoid. This anticipated deformation will be influenced greatly by the *bony sutures,* which cannot transduce bending stresses as well as bones. That is why in a severe collision a bone will be rather deformed at the point of impact, and the bending stresses will only weakly be transduced to the neighboring bones, so that the reactive forces from the adjacent bone to the impacted bone may be neglected – contrary to the events during the impact to a homogeneous hollow sphere.

As an example, consider the impact of a ball (as a blunt object of small mass) to the occiput. With an ideal hollow sphere, strains (distensions) will occur in the outer surface in the vicinity of the impact point, just at the circle-shaped boundary between the undeformed sphere and the indentation at the impact pole (Fig. 8 a). But if this area is sort of hinged to the inner parts of the sphere (so that

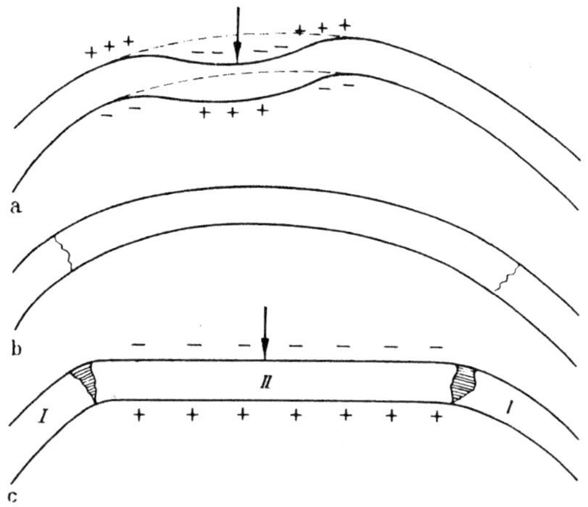

Fig. 8. Deformation of the skull with and without sutures. a. Violence to a homogenous hollow sphere. Strains (+) occur in a circular region around the impact. Bending forces are transmitted to the deformed area form adjoining areas. b. Undeformed hollow sphere with sutures. c. Deformation similar to a. The sutures yield like hinges. No bending forces are transduced from the neighboring areas (I, II). The entire area within the sutures (II) is uniformly bent in ('lid of a pot'). (From SELLIER, UNTERHARNSCHEIDT, 1963).

only weak bending forces may be transmitted), the area of impact will immediately be bent as a whole, i. e., up to its pliable boundaries (Figs. 8 b, c). Then stresses (shortenings) will occur in the outer layer, because the sutures will not transduce bending forces from the boundaries.

Classification of Head and Brain Injuries

Open and Closed Head and Brain Injuries

In exploring the nature of head and brain injuries and the development of after or 'late' effects, the simple classification of *'open'* and *'closed' brain injuries* provides a solid principle of order. The basic criterion is the *involvement of the dura mater*. If *the dura mater* has been *penetrated*, the injury is termed an *open brain injury*. Where the *dura mater is* intact, there is a *closed brain injury*. Fracture of the skull does not necessarily imply an open brain injury since fractures can occur without rupture of the dura mater.

The *morphological alterations* produced by sharp and blunt blows to the head are divided into *primary* and *secondary traumatic alterations. Primary traumatic* or *mechanical injuries* are a direct consequence of the mechanical effects of the traumatic force. They originate during the impact and are focal or frequently multilocular. *Secondary traumatic lesions* are produced by circulatory disorders. They appear subsequent to the impact as hypoxic or ischemic cell alterations, partial or total necroses, diapedetic hemorrhages, or edema.

Primary and Secondary Traumatic Lesions

The mechanisms producing traumatic brain damage can be conceptually divided into *blunt* and *sharp impact forces.*

It is necessary to distinguish between impacts on the *freely movable head* and one that is *fixed.* DENNY-BROWN termed impact forces on a freely movable head *'acceleration concussion,'* and impact forces on one that is fixed *'compression concussion.'*

A head may be considered fixed not only when the head is lying, for example, against a hard foundation, but also when the head is impacted while in a normal posture, by a small object from above (Figs. 9 and b)

SPATZ separates the *traumatic, passive lesions,* produced in the moment of impact, from the *post-traumatic, reactive alterations,* which occur after variable intervals, due to brain edema or circulatory disturbances.

Blunt Non-perforating Impacts

Closed brain injuries are usually caused by *blunt impacts*, where the force is distributed over a large area of the head. The skull may be fractured and the victim rendered unconscious, but the dura mater remains intact. Consequently there is no brain wound and no danger of infection. The principal primary traumatic alterations are so-called cortical contusions, especially near the opposite side of the brain, which are sometimes still termed contrecoup lesions.

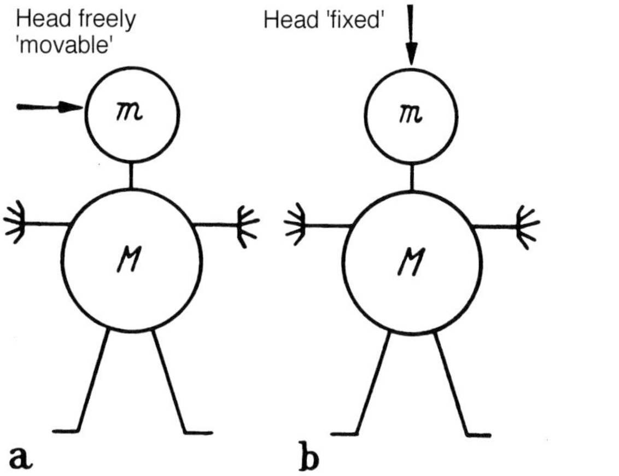

Fig. 9. Different axes of impact. a. Frontal, occipital, temporal impact. The head may be considered unsupported during the collision (lasting only a few ms); the spinal column has no influence on the event. Acceleration trauma. b. Impact in the direction of the axis of the spinal column. The head is 'dynamically fixed' by the spinal column and the mass of the body. (From SELLIER, UNTERHARNSCHEIDT, 1963).

Frequent *complications* are *epidural* and *subdural hemorrhages* followed by *cerebral compression. Central traumatic lesions,* such as hemorrhages of the ependyma of the walls of the lateral ventricles, third ventricle, the aqueduct of Sylvii, fourth ventricle, and the corpus callosum are sometimes found as well. Hemorrhages of the brainstem may also occur.

Brain edema resulting from circulatory disturbances can be sufficiently severe to produce a number of secondary traumatic alterations. Partial and total tears in the walls of extracerebral vessels may lead to the formation of *dissecting aneurysms and hemorrhages.*

A *blunt impact to the head* may produce brain injury by inducing stress waves in the brain, by the stress waves it induces in the brain or by the acceleration which it imparts to the head, or by a combination of both. In general, it has been found that the force of an impact on the head has such a slow *'rise time'* due to the cushioning effect of the skin and scalp, that the action produced is essentially quasi-static in nature. As a result, the stress waves generated by the impact are not important. Thus the injuries which result from a blunt impact must be attributable to both the accelerations imparted to the head and to the inertial forces which must be developed if each part of the system is to accelerate at the same rate.

Sharp Perforating Impacts

Impacts by sharp objects such as missiles, bullets, shell fragments, tools, propellers, etc. may penetrate the skull and dura mater and cause brain tissue alterations in varying degrees. These primary traumatic lesions resulting from mechanical destruction of brain tissue are collectively referred to as *brain wounds.*

There is a strong possibility of infection of the leptomeninges and brain when the dura mater is opened and the brain communicates with the external environment. The possibility of infection determines the clinical and morphological picture, in contrast with the closed brain injury.

Brain injuries in boxing consist largely of closed head injuries. Open head injuries are not common, but they are possible, and sometimes they are intentionally inflicted. When former heavy weight boxer Mike TYSON stated in an interview that he wanted to hit his opponent in the face and smash the bones into his brain, he expressed a good understanding of how to inflict a devastating open head injury. TYSON is quoted, saying: "I try to catch my opponent on the tip of his nose because I try to punch the bone onto his brain." TYSON made this statement after a boxing match in which he had broken Jesse FERGUSON's nose. When a well executed blow strikes the face directly, the relatively fragile craniofacial bones can fracture, allowing the sharp fragments to puncture the dura mater and brain.

Some related publications are quoted:

FINNEY et al. (1964) in their neurological survey of comminuted sub-frontal fractures, found that in frontal trauma, the cranium gave way to fracture lines along the floor of the frontal fossa, often with small bony fragmentation and dural lacerations. The authors concluded "that the simplest of such wounds must be considered compound, as the bony fragments come into contact with the air containing sinus."
ALONSO (1970) described an unusual case of nasal injury, wherein the nasal bones were propelled into the base of the skull. Sinus x-ray showed fragments of bone in the frontal si-

nuses. A fracture was noted in the right infra-orbital rim. The crista galli was fractured.
ALONSO continued to say that the fracture lines often extended to the adjacent ethmoidal, cribriform plate complex.

Injury-Producing Mechanisms

Translational and Rotational Acceleration (Fig. 10 a and b)

An analysis of the effects of trauma must consider *two kinds of impact – central* and *oblique.* In the first case, the projected impact axis passes through the skull's center of gravity. The resulting head motion is simply a *translational (linear) acceleration,* exemplified by a straight blow into the face (Figs. 11 and 12). In the second case, however, the axis is not centered and the resulting motion is a *combined translational and rotational acceleration of the head.* This occurs in swinging blows to the head, uppercuts (Fig. 13), or hooks. The degree to which both motions participate in the acceleration depends on the oblique nature of the blow. For example, tangential impact, such as an uppercut to the chin, produces practically pure rotational acceleration. Initially the skull alone begins to rotate, while the brain, by reason of its inertia, does not follow the motion immediately. At that point it pulls

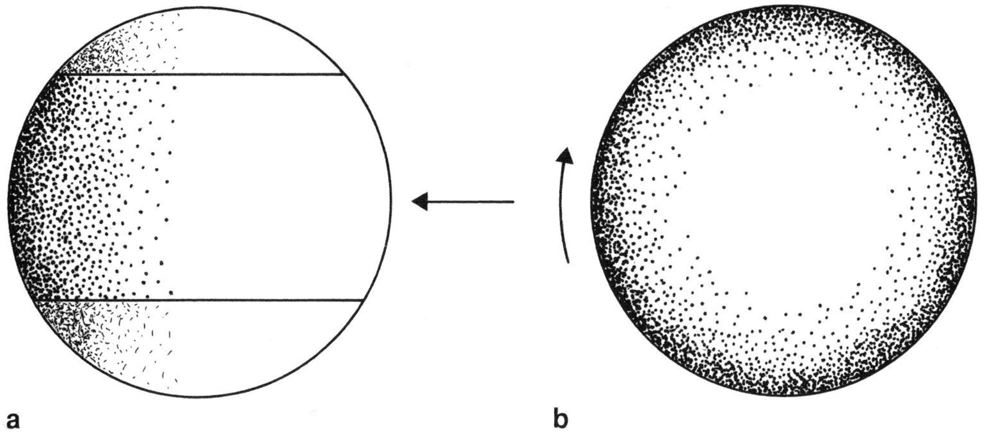

Figs. 10 a and b. Schematic representation of a rigid sphere subjected to (a) translational (linear) acceleration and (b) rotational acceleration. The two schematic representations show the topography of the primary traumatic lesions due to (a) translational (linear) acceleration, and (b) rotational acceleration.
a. The primary traumatic alterations at a broad blunt impact of force and course of the impact axis through the center, or near the center of the system produce primary traumatic tissue alterations at the side opposite to the impact area ('lesions par contrecoup'). These lesions are not genuine cortical contusions, but have to be termed so-called cortical contusions, since the fluid in a model and the brain, too, does not strike at the inner wall of the shell or the interior of the skull, but recedes from it. The arrangement of the primary traumatic lesions is cylindrically symmetric.
b. At a tangential impact of the force a rotational acceleration of the model or the head occurs. The interior wall of the shell or the interior of the skull are accelerated, while the brain stays due to its inertia behind. Thereby shifts develop between model wall and the fluid respective skull and brain. The primary traumatic lesions show at rotational acceleration a radially symmetric arrangement. We may not speak here of cortical contusions or so-called cortical contusions, but venorhectic, occasionally capillo- and arteriorhectic hemorrhages located predominantly in the more superficial layers of the cerebral cortex. The hemorrhages consistently follow vessel systems running vertical to the cortical surface. The characteristic pinpoint hemorrhages indicative of cortical contusions in the first stage cannot be seen. Thus, not only is there a difference in quality between primary traumatic hemorrhages associated with angular acceleration and the so-called cortical contusions found in translational trauma (with the impact axis running through the center of gravity of the system), but there is also a difference in the pattern of distribution of the primary traumatic cortical lesions encountered in angular and translational acceleration. These tissue alterations are located close to the midline, not in a cylindrycally symmetrical pattern, but radially symmetric pattern, as predicted by injury mechaics.
These tissue alterations are identical with those, described by UNTERHARNSCHEIDT and HIGGINS (1969) using animal models with the Head Acceleration Device II (HAD – II) with nondeforming angular acceleration of the head.

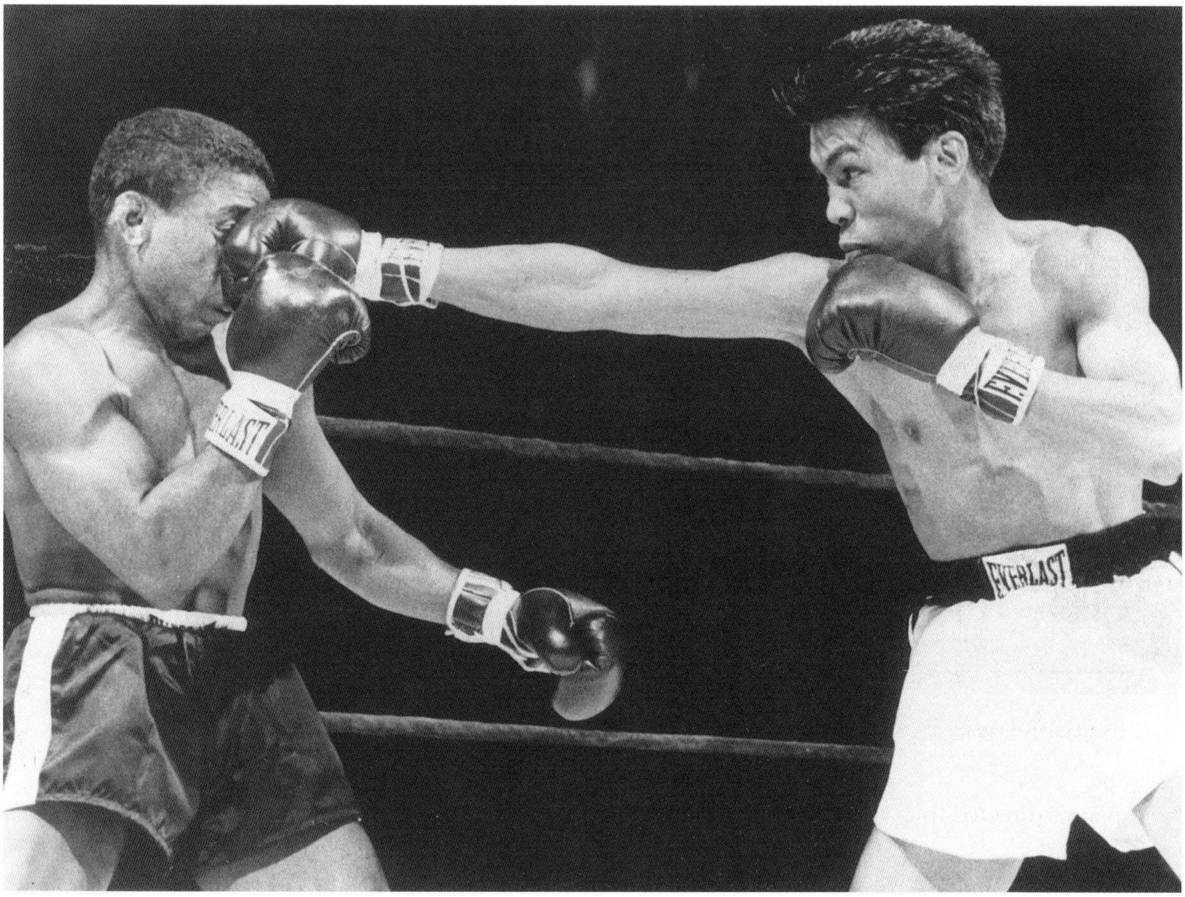

Fig. 11. Professional featherweight bout between Miguel BARRIOS (left) and Gabriel ELORDE (right) on November 9, 1956 at Madison Square Garden, New York. Photo: Frank MASTRO.

against its suspensions. Such relative displacement of skull and brain distends the *bridging veins* between the *brain* and *superior sagittal sinus,* sometimes tearing them. The skull always transmits the effects of force to the brain; the impact on the brain depends on the specific nature of the acceleration.

Translational acceleration sets the skull in motion before the brain can overcome its inertia. It follows that, at the impact point, the distance between the inner surface of the skull and that of the brain is first reduced and then increased as pressure develops during the event. The distance between skull and brain becomes greater at the point opposite the impact, the antipole, causing a decrease in pressure or rarefication. The pressure gradient thus produced in the brain has been found to be approximately linear.

In boxing, the majority of blows hit the head either straight on, slightly obliquely into the face, or from below. The soft tissue of the face reduces acceleration to some extent, by reason of its compressibility. Blows directed to the unprotected chin or the temporal bone generate greater acceleration of the head than those to the face.

Translational (Linear) Acceleration
(Figs. 14 a and b and 15 a)

The simplest action, from an analytical standpoint, develops if the head is given pure translational (linear) acceleration. Such an acceleration is produced if the head is free of all constraints and the force of

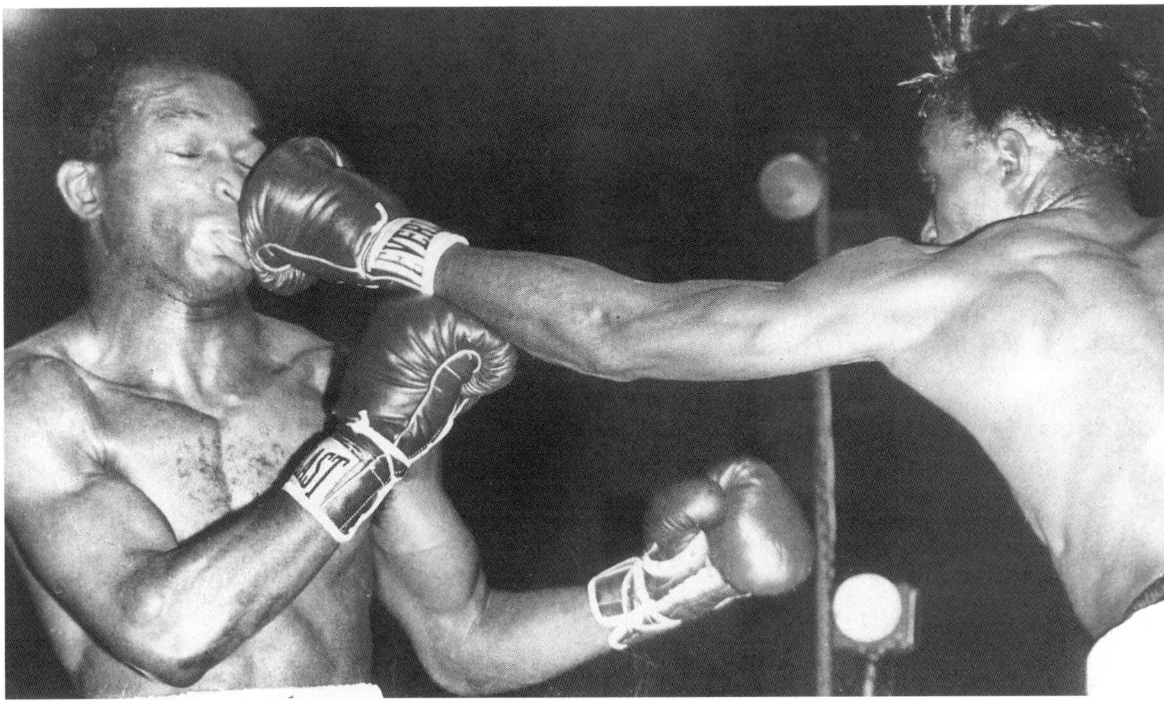

Fig. 12. Professional lightheavyweight fight between Wallace SMITH (left) and Joe BROWN (right). Photo: U.P.I.

the impact is directed through the center of gravity of the head, as illustrated in Fig. 1 a. The brain is comparatively uniform in density. All of the elements of the brain, nerve tissue, blood and cerebrospinal fluid having approximately the same density as water. The brain is encased in a skull that is very rigid in comparison to the rigidity of the brain. The entire system, for present analytical purposes, can be regarded as a thin-walled spherical shell filled with water. It may readily be shown that if such a skull is given a translational acceleration, a pressure gradient will develop across the skull, in the direction of the acceleration, with the maximum pressure occurring at the point of impact and the minimum at the point opposite the impact, the '*antipole*'. The pressure at any single point is essentially hydrostatic, i. e., no shearing stresses develop. At the impact point the pressure is positive (above ambient) while at the antipole the pressure is negative (below ambient). There will always be some intermediate point along the continuum where no pressure change has occurred. This is referred to as the 'nodal' point (see Figs. 2 and 3). With an incompressible fluid and an inextensible skull the *nodal point* is at the center of the shell. If the fluid is compressible and the shell extensible, the pressure variation is still linear, but the position of the nodal point shifts.

Positive pressure causes no damage, so far as we now know. However, negative pressure does cause injury if at any point it drops below the vapor pressure, producing cavitation. Cavitation and the injuries it produces are described in detail in a later section.

Cavitation produces the injuries that were classified as contrecoup injuries in the 18th century. Physicians noticed long ago that in some cases, an impact to the head produced damage somewhere on the opposite side of the brain. These injuries were discussed and debated in the medical literature during the 18th century (CHOPART, GRIMA, LOUIS, SABOURAUT, SAUCEROTTE). At that time it was assumed that the brain must be battered against the inside of the

Fig. 13. Professional featherweight bout between 'Little Red' LOPEZ and Jorge REYES, who was knocked out in the seventh round. The photo shows the face of REYES fractions of a second after a right 'uppercut' of LOPEZ landed.

skull opposite the impact point, so they were called *contre-coup lesions*. The term is still used, although today it is generally understood to be technically imprecise.

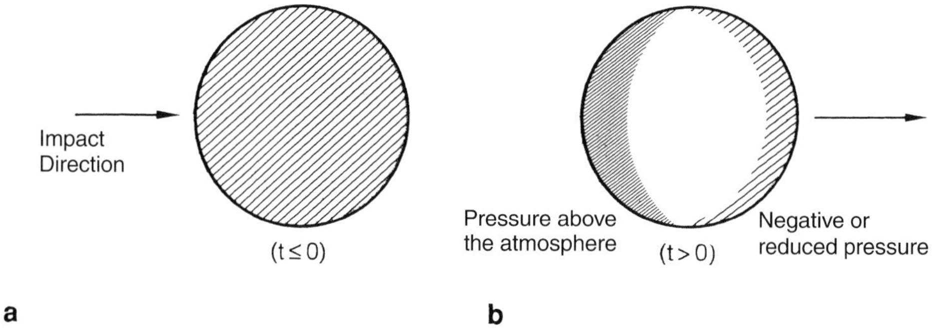

Fig. 14. Schematic representation of a rigid sphere accelerated from left to right. a. System at rest. b. Positive pressure at the impact point, reduced pressure at the antipole. (From UNTERHARNSCHEIDT, 1993).

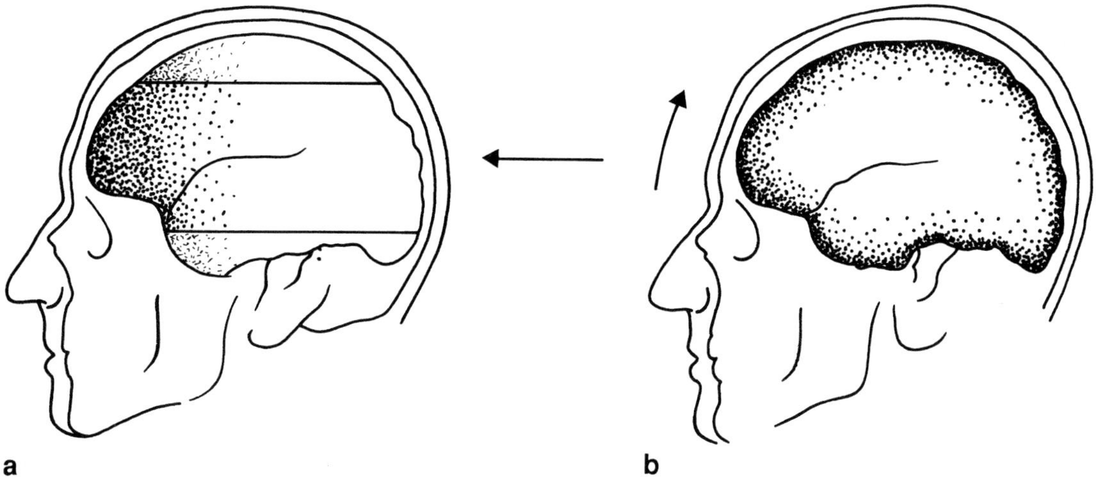

Fig. 15 a and b. The two schematic representations show the topography of the primary traumatic lesions due to (a) translational (linear) and (b) rotational acceleration.
a. The primary traumatic tissue alterations at a blunt impact of the force and course of the impact axis through the center or near the center of gravity of the head produce primary traumatic lesions at the site opposite to the impact area ('lesions par contrecoup'). These lesions are not genuine cortical contusions, but have to be termed so-called cortical contusions, since the fluid in a model or the brain tissue, too, do not strike at the inner wall of the shell or the interior of the skull, but recede from it. The primary traumatic alterations develop in the zone of negative or reduced pressure.
b. At tangential impact of the force a rotational acceleration of the model or the head occurs. The skull is accelerated while the brain first due to its inertia stays behind, thereby developing shifts between the skull and the brain. It develop tears of bridging veins (here not shown) which lead to subdural hemorrhages and primary traumatic tissue alterations occur which may be termed 'gliding contusions'. The term cortical contusion for these primary traumatic alterations of the brain is incorrect, they are not cortical contusions, but tear-offs and tear-outs of vessels of the cerebral cortex. (From UNTERHARNSCHEIDT, 1993).

Rotational Acceleration

If the force of a blunt impact is not directed through the center of gravity of the head, as in Figs. 15 b, 16 a and b, or if the head is constrained to rotate about some fixed point, as shown in Fig. 1 c, rotational acceleration is imparted in addition to translational acceleration, see also Fig. 10 b. For a purely rotational acceleration to develop, the impact would have to be in the form of a couple, or pure moment. Such impacts are improbable, but it is helpful to consider the reaction to a purely rotational acceleration imparted to the head. It may be compared to the reaction of a rigid container filled with water after an angular acceleration. The container moves and the water remains essentially at rest. If a more viscous liquid is substituted for the water, the relative motion between the container and the liquid is be reduced. The force necessary to accelerate the fluid is developed as a result of shearing action between the container and the fluid, as well as between adjacent layers of fluid. Obviously, there are conditions that can produce very large shearing deformations.

The brain and skull form a much more complicated system than the above-described liquid filled container model. There are blood vessels and

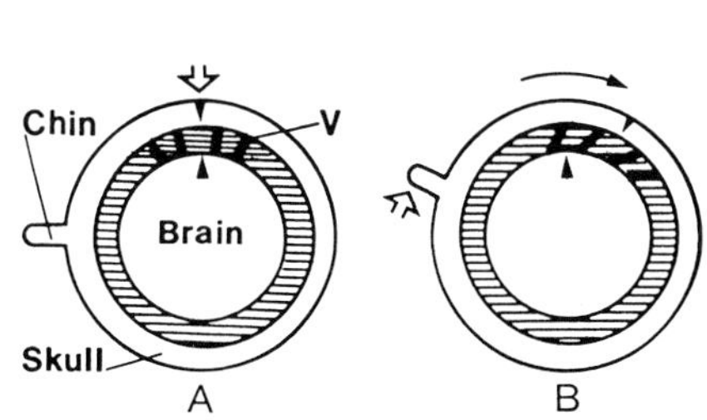

Fig. 16. a. Schematic representation of the effect of rotational acceleration of the head upon the brain. b. Due to inertia, the brain stays initially behind the rotational momentum of the skull. Thus a strain is placed on the bridging veins (V) that may disrupt them. (From SELLIER, UNTERHARNSCHEIDT, 1963).

nerves passing into the brain from openings in the skull, there are protuberances from the skull, and there are membranes and other supporting structures around the brain that help to maintain its position relative to the skull. Relative motion between the brain and skull and between the different parts of the brain must exist before sufficient force is developed to accelerate the brain along with the skull. During the early phase of relative motion the weaker elements of the system will be injured. Blood vessels will be torn, axons may be torn, and lesions will be produced at points of greatest stress throughout the brain. Detailed discussions of the various types of injuries found are presented in subsequent sections.

Combined Linear Acceleration and Rotational Forces

In reality one type of acceleration rarely occurs to the total exclusion of the other. The head is constrained by its support system, so that it must rotate about a point in space. Consequently, even a force directed through the center of gravity will produce some degree of angular acceleration along with the linear translation. In some impacts, one type of acceleration may predominate initially, while the other type predominates later. Both types of acceleration may be present with comparable magnitudes throughout the impact. Whatever the combination of features, some superposition of effects probably occurs as well. That is, any injuries that might be produced by the translational acceleration acting alone will also be produced if the translation is accompanied by rotational acceleration, and vice versa. Direct evidence on this point is not available, but comparisons of the pathomorphology of head injuries produced by pure rotational acceleration and those produced by more general types of impacts seem to support this assumption.

Impression versus Acceleration Trauma (Table 1)

Pure impression trauma (Figs. 17 a – c) and *pure acceleration trauma* (compare Figs. 14, 15 a) are the two extremes on a continuum of physically possible events. *Pure impression trauma*, where no acceleration of the head is involved, occurs mainly at the impact point, as a consequence of deformation of the plastic skull. The skull is first forced inward, developing pressure on the brain. Then the *skull re-*

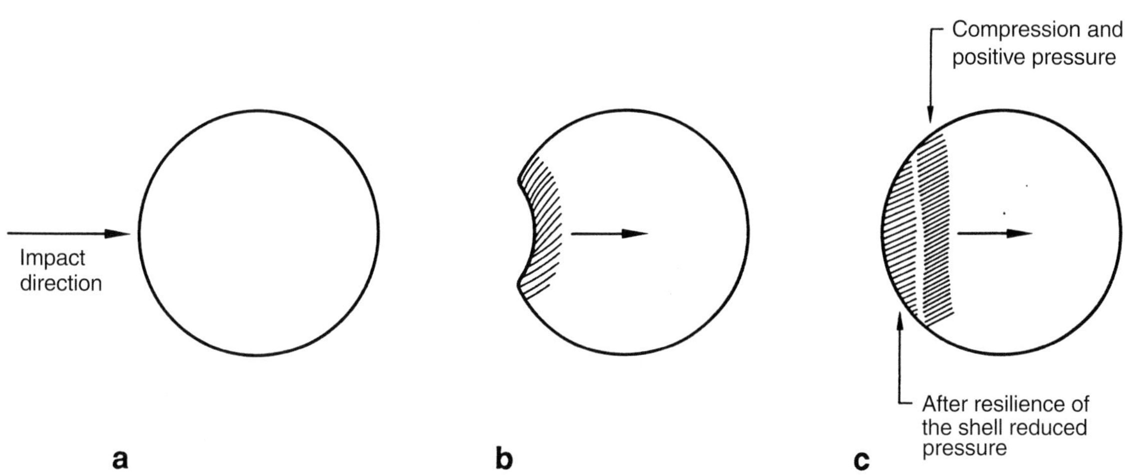

Fig. 17 a – c. Impression trauma.
a. Impact direction from left to right. Mechanism of cavitation due to concentrated impact. b. Impact force applied at the rigid shell causes local deformation (indentation) and compression and positive pressure in the sphere. c. After the force application ends, the local deformation responds to its normal position and causes a local reduction in pressure, fluids near the impact point produce localized foci of 'contusion', impacting object has low mass and high velocity. This accounts for the complete absence or only negligible presence of acceleration trauma. (From UNTERHARNSCHEIDT, 1993).

TABLE 1
Impression and acceleration trauma. (From SELLIER, UNTERHARNSCHEIDT, 1963).

	IMPRESSION TRAUMA	ACCELERATION TRAUMA
Object involved in collision	Small mass, load; cross section	Great mass, load; cross section
Acceleration of whole head	Practically zero	High
Impact energy	Exhausted at impact point	Transmitted to entire skull
Bones	Local indentation, fracture = bending fracture	Possible deformation of entire skull, indirect fracture
Brain injury	At impact pole, with high intensities also at antipole	At antipole, with high intensities also at impact pole
Mechanism	Limited indentation by projected object, rebound of bone, local negative pressure	Acceleration of skull as a whole, deceleration of brain as a whole, positive pressure at impact pole, negative pressure at antipole

turns, elastically, to its original configuration and perhaps a little beyond for a brief instant. The *elastic rebound of the skull* produces a *rarefication zone near* the point of impact that propagates across the brain. Lesions are produced at the impact site but not elsewhere. Impression trauma results if the impacting object is small, if the impacted part of the head is relatively soft and the duration of the impact short, or if the head is in a fixed position.

Pure acceleration impacts cause little or no flattening of the skull and produce *lesions mainly at the counter-pole (the so-called contrecoup effect).* This type of injury usually involves impact by a large object on a rigid part of the skull, where the blow has a relatively long duration. Pure acceleration-type injuries occur only if the head is freely movable.

Between the two extremes, all sorts of intermediate combinations are possible, where both acceleration and impression trauma occur simultaneously. Which type predominates depends critically on the specific characteristics of the impact and the structural characteristics of the affected area of the head. Unfortunately this important fact has been ignored completely in medical literature. There the tendency has been to summarize clinical and pathomorphological findings into one statistical account irrespective of whether the injuries were produced predominantly by acceleration or impression. If clinical and morphological findings are to be satisfactorily analyzed, some qualitative classification of the distribution of cerebral lesions, as a function of impact characteristics, (i. e., predominantly impression or predominantly acceleration) must be used.

Deformation of the Boxer's Skull Due to Severe Boxing Blows (Fig. 18)

Boxing photos capturing blows to the face or head, such as those included here, show that the skull is considerably deformed by the massive impact. Analysis of collected action photos shows that instantaneous skull deformation occurs with regularity in boxing matches, though it is too brief to observe directly. The press photos also show that several different bones of the skull are usually involved in the deformation.

We discussed impression trauma earlier, where a localized object hits the skull and produces a local indentation at the point of contact. The blows impose massive force on the convex surface of the skull, which is then broadly deformed.

Hard blows to the head of a boxer could argu-

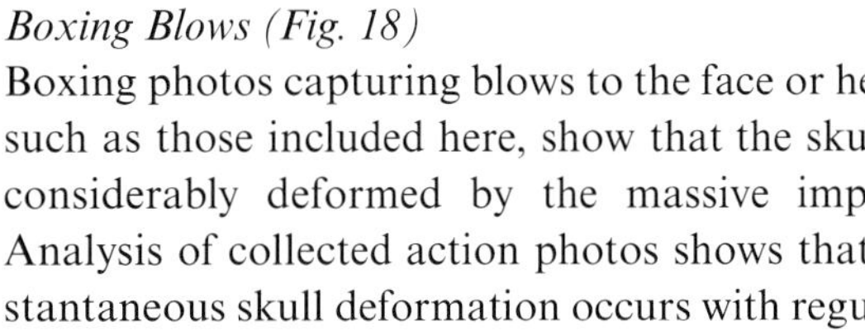

Fig. 18 a – c. For the computation of the deformation of the skull and the time interval of the rise time of the pressure at the point of impact and beginning of decrease of the pressure at the point opposite the point of impact. a. Before, the impact and antipole have the same velocity v_o ; b. during the impact at the impact pole $v = 0$, at the antipole is $v = v_o$; c. The point opposite the impact point is also decelerated to $v = 0$. It is moved from the begin of the impact the distance s. (From SELLIER, UNTERHARNSCHEIDT, 1963).

ably be included within the category of acceleration or deceleration trauma, as they are also characterized by broad impact to the head. However, by definition, acceleration or deceleration trauma can be caused by many different objects, or by materials of different mass. For example, a kick with a shoe, or a bare fist can accelerate a head. During an automobile collision, a head can be decelerated violently by contact with the interior parts of the passenger compartment or the windshield. In contrast, the boxing blow to the face is always the result of contact between the gloved fist and various parts of the head, though, admittedly, with varying intensities.

The action photos of the moment of impact and the reaction to the blow, a fraction of a second after contact, show conclusively that both the viscero-cranium and the encephalo-cranium are visibly distorted or markedly deformed. The deformability of the skull is especially clear when the photos are viewed under low magnification. Close study of the photos strongly suggests that massive head impacts in boxing cause an effect similar to that already described in the chapter on non-linear acceleration.

Clinical Diagnoses

Contusio Capitis

Contusio capitis ('Schädelprellung') describes a clinical picture after blunt impact against the head, without unconsciousness and loss of tone. The intensity of the kinetic energy is under the threshold necessary to produce a cerebral concussion (commotio cerebri). It is a clinical syndrome, because there are no functional or pathomorphologic alterations in the Central Nervous System.

The cause of a contusio capitis is most often knocks on doorframes or panels, against pendent objects, or falls. They are often the result of impacts against the head in traffic, house and sports accidents. In this regard the unfortunate term 'bagatell trauma' is sometimes used. In circumscribed impacts against the head a subgaleal hematoma may develop. A contusio capitis can be combined with cuts and bruises of the viscero-cranium or cerebro-cranium.

A blunt impact against the head producing the acute stage of a clinical picture of contusio capitis can be combined with a fracture of the skull and an epidural hemorrhage. A contusio capitis can be followed by a chronic subdural hematoma after several weeks.

Contusio capitis is a frequent occurrence in boxing, where numerous blows are directed against the head in each round. The obvious results are bruises and swellings of the skin of the head, which is *'puffed'* with hemorrhages in clear view. The more dangerous sequel is the less obvious permanent brain damage resulting from repeated blunt impacts. See the next chapter on subconcussion or subscommotio cerebri.

The Term Subcommotio Cerebri or Subconcussion

One of the most surprising findings in our research is that repeated subconcussive injuries cause permanent brain damage, without loss of consciousness at any time of impact. This was an unexpected revelation, almost counterintuitive, since subconcussive injuries do not appear to be as destructive as more dramatic events with loss of consciousness. The ramifications of this point are especially significant with respect to boxing, as it is easy to seriously underestimate the consequences of repeated head blows, especially over a long career.

Pathomorphological findings on human brains and experimental data from animal models (UNTERHARNSCHEIDT 1958, 1963, 1975, 1993, 1994) show that repeated impacts of subconcussive intensity ultimately result in a severe, permanent brain damage due to secondary, circulatory tissue alteration *(summation effect).* A typical example is the obvious permanent brain damage of ex-boxers, so well known that the word 'punch drunk' is commonly used to describe them. (UNTERHARNSCHEIDT 1970, 1972, 1973, 1975, 1985, 1994), UNTERHARNSCHEIDT and SELLIER 1971, ROBERTS 1969, CORSELLIS et al 1973). Another example is the severe permanent brain damage of jockeys after frequent falls (FOSTER et al 1976), and epileptics after many seizures and falls (NEUMANN 1959).

A single subconcussive impact does not create this kind of brain damage. Severe and permanent brain damage is the cumulative effect after receiving multiple subconcussive blows to the head. *Thus, we coined the matching syndrome subcommotio cerebri.* Frequent contusiones capitis lead to a severe and permanent brain damage due to secondary traumatic processes, which occur after an interval. It is then no longer a clinical syndrome but a morphological one.

Repeated Subconcussive Impacts

These impacts are a subclass of the blunt non-perforating impacts. Their distinguishing feature is that they produce no apparent immediate effects. It is likely, however, that each blow to the head causes circulatory disturbances resulting in the destruction of vulnerable neurons, particularly the large neurons in the cerebral cortex and Ammon's horn formation. They also cause the loss of Purkinje cells in the cerebellum with consecutive glial proliferation. The result is glial scar tissue. The damage due to a single blow of such impact intensity can sometimes be completely harmless, but the cumulative effect of repeated impacts clearly leads to irreversible cell alterations and permanent brain damage. As a consequence, after many such blows, which a boxer has to suffer during his active career, his mental activity is affected. The punch-drunk ex-prize fighter is a typical example of the eventual result of many subconcussive and concussive blows to the head. Such brain damage is sufficient to render a boxer ineffective in the ring, but is often unnoticed by the examining physician who gives a clear bill of health before a match. A diminished sense of timing or balance can be catastrophic in the ring, as happened in the last fights of MUHAMMAD ALI, who fought against his personal doctor's advice.

Commotio cerebri (cerebral concussion)

Commotio cerebri (cerebral concussion) is a *clinical diagnosis describing* the immediate post-traumatic state of unconsciousness and loss of muscle tone after a blunt impact to the freely movable head. The impacting force must reach a certain threshold to produce a commotio cerebri (cerebral concussion). Physically it does not matter whether the object is accelerated or decelerated; the relative movement is the important factor. The concussion syndrome is the result of an acceleration or deceleration event and must be considered an inertia phenomenon. The deformity of the skull due to different age influences the threshold, younger skulls being more deformable than older ones.

A commotio cerebri (cerebral concussion) can be produced from any impact direction. Physical analysis reveals, however, that in the lateral, bitemporal impact direction the effective kinetic energy must be double that of the sagittal impact direction. This is because the lateral diameter of the skull is only half of the sagittal diameter.

The duration of unconsciousness has been used as a criterion for the severity of the clinical picture; this practice can be traced back to SCHÖNBAUER and BRUNNER (1928) and SYMONDS (1962). LOEW (1950) made an important objection to using duration of unconsciousness as an indicator of injury severity: "Since only about 14% of all cases are unconscious for more than two hours, there is no way to judge the severity of the brain damage by that means for most cases." Duration of unconsciousness should only be seen as one symptom among others and not as the sole criterion for the severity of the injury. RICKER and DÖRING made the same observation in 1955. This condition is not associated with any detectable macroscopic or microscopic lesions of the brain. The whole process is *traceless* as noted by SPATZ (1950).

Concussion

The English term *'concussion'* has different meanings. 'Concussion' signifies the essential process of an impact. It can, thus, result in different tissue alterations, also primary traumatic lesions. The other usage of 'cerebral concussion' is identical with commotio cerebri, the sequel (result?) of a force. In this case the term is identical with a clinical diagnosis. Here the term 'cerebral concussion' is always identical with commotio cerebri, a preferable term.

Types of Injuries of the Brain and its Enveloping Structures

Epidural (Extradural) Hemorrhages and Hematomas

An *epidural hemorrhage,* or hematoma, is a collection of blood that extends between the inner table of the skull and the dura mater. The dura mater forms the periosteum and if undamaged it adheres tightly to the bone, thereby limiting the space between skull and dura mater into which hemorrhaging blood can flow.

This lesion is almost always due to arterial bleeding as a result of rupture of the middle meningeal artery (Fig. 19), which arises from the external carotid artery and runs between the dura mater and the inner table of the skull in the temporal fossa. These hemorrhages are caused mainly by blunt blows of low or medium intensity, such as in a fall or fist fight (Fig. 20). Not much intensity is necessary to injure the temporal region, as both the temporal bone and lower portion of the parietal bone are thin. Penetrating injuries, usually of low intensity, are another cause of extradural hemorrhage. In both open and closed head injuries, tearing of the middle meningeal artery most frequently produces the hemorrhage. The anterior or posterior meningeal arteries may also be involved.

Rarely is a hemorrhage produced by venous bleeding or from an injury to a major sinus (superior sagittal or transversal sinus).

Although the vessel lesion occurs simultaneously with the impact, clinical signs appear only after several (4 to 12) hours. Obviously even the most attentive ring physician could not rule out the possi-

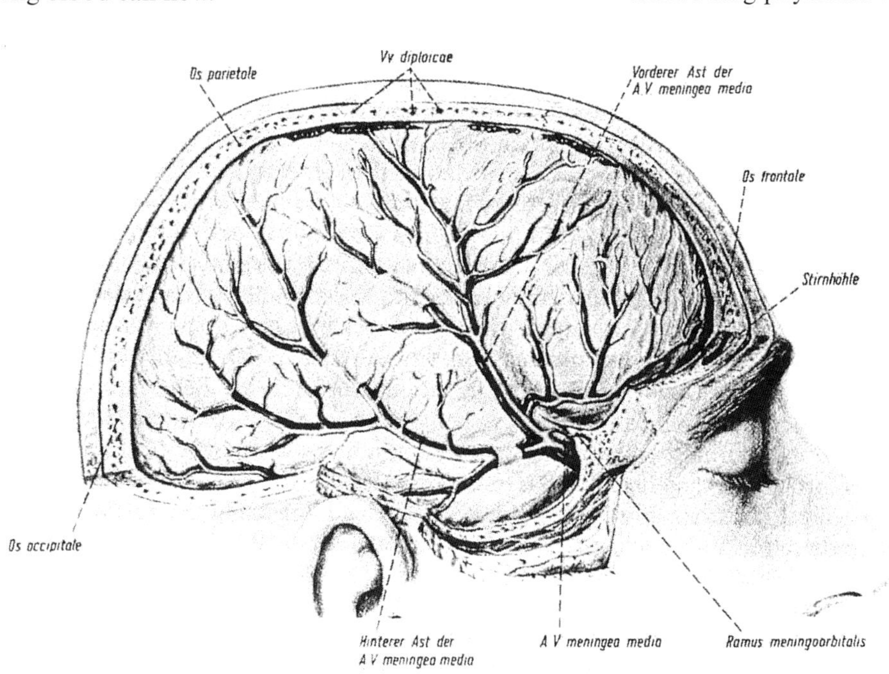

Fig. 19. Lateral view of the right cerebral hemisphere. Representation of the right anterior meningeal artery and its branches. (From KESSEL et al., 1969).

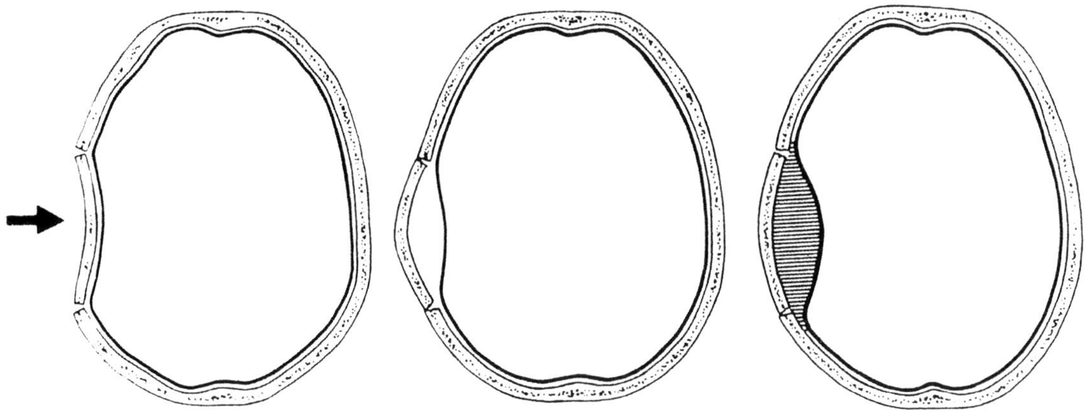

Fig. 20. Genesis of an epidural hematoma according to Charles Bell (1816). (After LEOPOLD 1977, from UNTERHARNSCHEIDT, 1993).

bility of an epidural hemorrhage or hematoma in a ringside examination during or immediately after the fight.

As the first symptom the befallen side reveals a constriction of the pupil due to an irritation of parasympathic pupilloconstrictive fibers. This is very important for the ring physician to recognize and diagnose. This phase lasts only a short time. About four hours after the injury it can no longer be seen. After the short-lived contraction of the homolateral pupil follows mostly a slow dilatation, which can develop into a full dilatation and a complete fixed pupil (*Fig.* 21). The state of a maximally dilated fixed pupil may be present already after one hour after the injury.

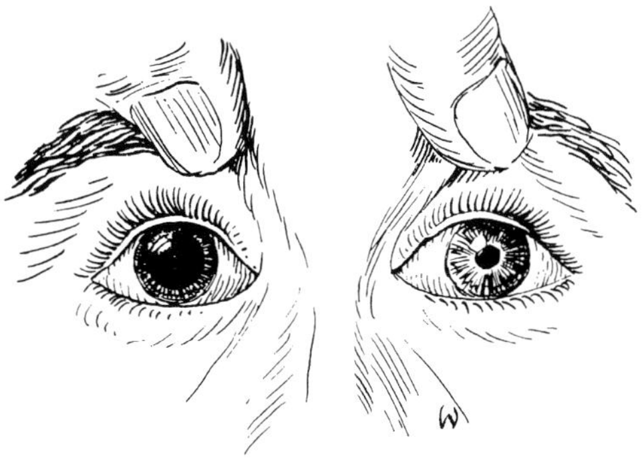

Fig. 21. Dilated right fixed pupil at epidural hematoma (From KESSEL et al., 1969).

Develops the hemorrhage slower, several hours, six or eight, exceptionally twelve hours or more may elapse before the homolateral pupil is dilated and completely fixed. An onesided dilated pupil is an alarming sign, because it points toward a developing temporal hernia of the brain. Waiting is very dangerous, immediate intervention is absolutely necessary.

Unconsciousness after an impact to the head is normally followed by a period in which the patient is conscious, the so-called '*lucid interval*'. However, in the following hours, continuous bleeding and enlargement of the hematoma leads to a progressive impairment of the state of consciousness. Prompt and rapid clinical diagnosis and surgical removal of the epidural hematoma are critical, because of the otherwise rapidly deteriorating clinical picture (Figs. 22 a and b, 23, 24). If not removed, the circumscribed hematoma, a space-occupying disc-like mass, produces a local indentation and compression of the brain. Once this occurs, edema and necrosis of the underlying brain tissue will follow.

Subdural Hemorrhages and Hematomas

A *subdural hemorrhage is* defined as an accumulation of blood between the dural and arachnoid membranes. The subdural hemorrhage is much more common than an epidural hematoma, occur-

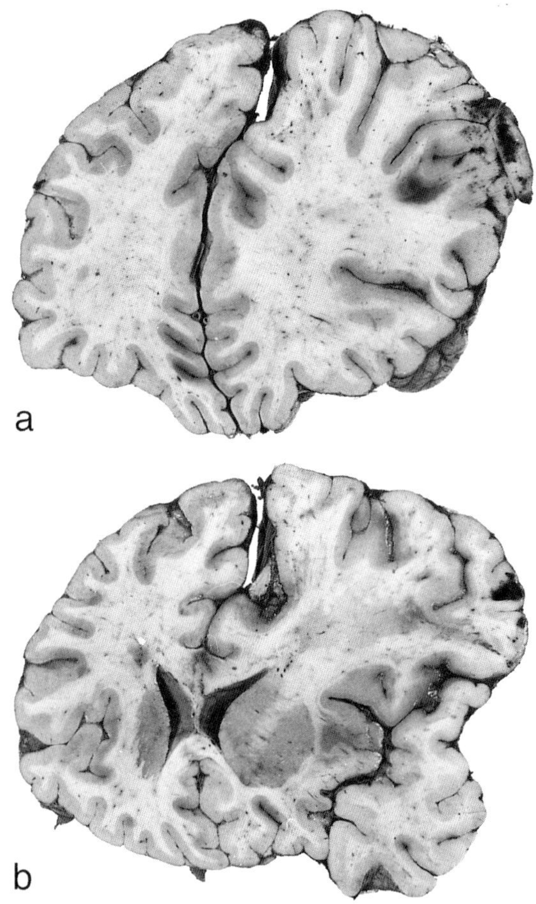

Fig. 22 a and b. Man cerebrum. Head injury five days before exitus, skull fracture, bilateral epidural hemorrhages, at the right side at the frontal convexity subdural hematoma of 2 mm thickness. Massive edema of the right cerebral hemisphere with shifting of the midline structures to the contralateral side. Displacement of midline structures of medial parts of the right cerebral hemisphere to the contralateral side under the free edge of the falx cerebri. (From UNTERHARNSCHEIDT, 1993).

ring approximately ten times more frequently. Subdural hemorrhages are nearly always of venous origin, whereas epidural hemorrhages are usually caused by arterial bleeds.

Subdural hemorrhages are caused by both blunt and sharp impacts to the head. The sharp impact penetrates the dura mater, producing an open brain injury. From this brain wound, blood can spread through disrupted brain tissue and form a subdural and intracranial hemorrhage. A subdural hemorrhage can also result from injury of a venous sinus, the meningeal artery, or vessels of the dura mater.

Subdural hemorrhages develop from blunt impacts when the impact axis does not run through the center of gravity of the head. In such case the relative rotation of the skull against the brain ruptures the bridging or connecting veins (Fig. 25 a and b), which then drain into the superior sagittal sinus.

Rupture of the bridging veins is usually regarded as the principal cause of subdural hemorrhages, unless other complicating injuries are found to have been responsible. Occasionally, subdural hemorrhages of arterial origin can be seen.

Large subdural hemorrhages, called *subdural hematomas,* are *space-occupying lesions; they* bring about *cerebral compression* (Fig. 26), leading to additional circulatory brain damage, such as necrosis, hemorrhagic necrosis, etc. (Figs. 27, 28, 29).

The hematoma is organized similarly to a thrombus or embolus, from the uppermost of the enveloping structures, the dura mater, and finally results in a cicatrix. A *subchronic* or *chronic subdural hematoma* is a slowly expanding intracranial space-occupying lesion.

Subdural hematomas can be *acute, subacute or chronic.* GURDJIAN and WEBSTER's (1958) classification is based on the premise that "(1) when bleeding is so severe as to produce signs and symptoms almost immediately after injury (up to 3 days), then the hematoma is *acute;* (2) when the bleeding is less severe and is tolerated for a week or more (four to fifteen days), the hematoma is *subacute;* (3) when the initial hemorrhage is not associated with obvious cerebral trauma and produces no clinical signs for several weeks, or when there is severe cerebral trauma, so that the patient is severely ill from the beginning but whose improvement is interrupted by renewed signs of an intracranial dynamic disease, the lesion is *chronic* (over sixteen days)."

Subarachnoid Hemorrhages

The sources of *hemorrhages in the subarachnoid space are* its arterial and venous vessels and branches of the cerebral vessels. If these are damaged through any of several possible mechanisms,

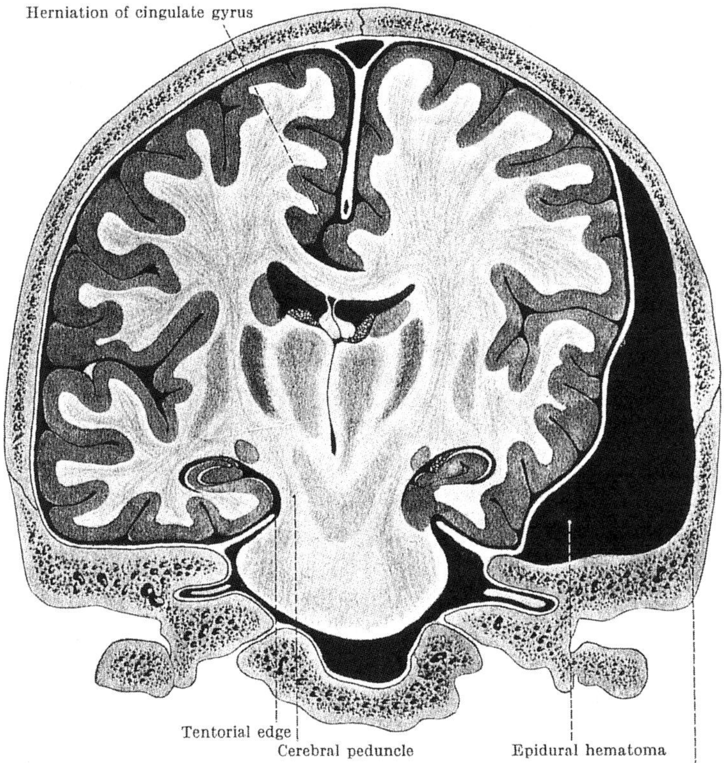

Fig. 23. Anatomical demonstration of epidural hematoma in a coronal section. Observe: This section runs in an oblique fashion from anterior superior to posterior inferior to demonstrate the compression on the cerebral peduncle on the side opposite the hematoma. (From KEMPE, 1985).

blood will mix with cerebrospinal fluid. The typical result is a hemorrhage with poor demarcation (Fig. 30). In cases of severe impact, subarachnoid hemorrhages can spread over the entire surface of both cerebral hemispheres. Lesser hemorrhages can be completely absorbed, except for small quantities of blood pigment. Larger hemorrhages transform by stages of resorption and organization into fibrosis and scar tissue.

Subarachnoid hemorrhages are almost always present where cortical contusions are found. They develop instantly with the impact and are simultaneous with the rhectic hemorrhages[1] found in the cerebral cortex and subcortical white matter.

1 (Rhexis, Greek, means rupture. Thus a rhectic hemorrhage is one caused by rupture of the vessel wall due to the mechanical force. The distribution of arteriorhectic, venorhectic and capillorhectic hemorrhages indicates the involvement of arteries, veins or capillaries in the rupture.)

So-called Cortical Contusions

Cortical contusions ('Rindenprellungsherde') are common, specifically traumatic alterations that are caused by blunt impacts on a freely movable head. They appear at the moment of the impact without delay. Hugo SPATZ has described different stages of the so-called cortical contusions in three stages, as follows: (1) *Hemorrhage and necrosis*, (2) *resorption and organization*, and (3) *final* or *defect stage*. A detailed description of the pathomorphology is presented in a later section.

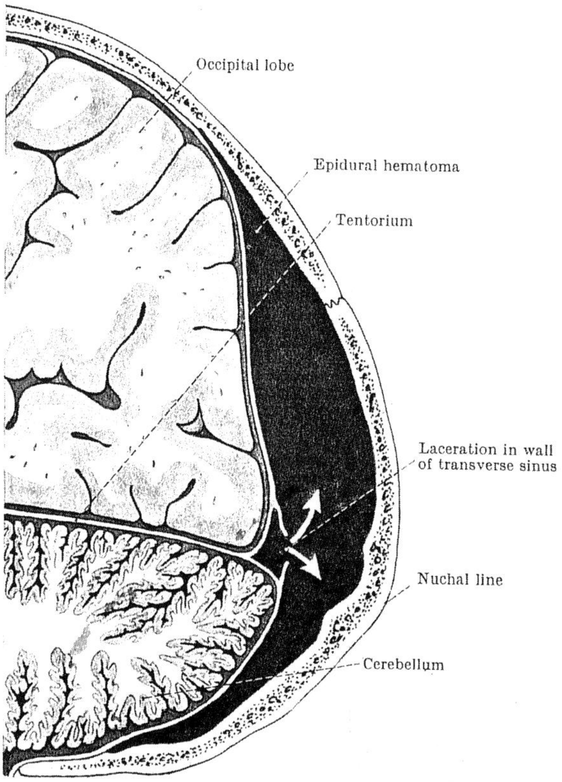

Fig. 24. Anatomical demonstration of extradural hematoma over occipital and suboccipital area due to laceration of transverse sinus (parasagittal section 2 cm from the midline. (From KEMPE, 1985).

Cortical contusions ('Rindenprellungsherde') are specific traumatic alterations caused by the mechanical force of blunt impact. Initially, secondary tissue alterations caused by subsequent circulatory disturbances of the brain after blunt impact were not distinguished from the primary cortical contusions, leading to an over-broad category of 'so-called' cortical contusions. The French term *'plaques jaunes'* even included the final stage lesions appearing after mass hemorrhages or necroses along with primary and secondary injuries. E. KAUFMANN (1922) believed that the same tissue alterations were sequelae of so-called cortical contusions and of necroses of the cerebral cortex due to thrombotic or embolic occlusions. MARBURG (1926) also did not consider the tissue alterations of the so-called cortical contusions to be specifically traumatic.

Hugo SPATZ and his school, from 1929 on, quoting prior publications of BENDA (1921), emphatically accentuated that primary traumatic alterations of the cerebral cortex, the so-called cortical contusions, in all stages, especially in the final stage, are well characterized and can be certainly differentiated from circulatory, reactive tissue alterations. SPATZ was able to show that the *'etat vermoulu' was* not a sequel of tissue necrosis, especially of scleratheromatous origin, but that it is the final or defect stage of the so-called cortical contusion.

Predilection areas of cortical contusions are on the prominent areas of the cerebral hemispheres, the summits of the convolutions. These traumatic alterations are more frequently localized at the antipole than at the impact area of a blunt force (MARTLAND and BELING 1929, SPATZ 1936, KALBFLEISCH 1940, PETERS 1943, BLOOMQUIST and COURVILLE 1947, WELTE 1948, RIPPERGER and UNTERHARNSCHEIDT (1970), UNTERHARNSCHEIDT 1993). The direction and intensity of the impacting force is a controlling factor in the type and extent.

The areas of transition from the base of the brain to its convexity are the predilection sites of the so-called cortical contusions. These are the third frontal gyrus in the frontal area and the second and third temporal gyrus for the temporal area. The first temporal gyrus is less frequently affected. So-called cortical contusions are also found at the transition zone between the base of the cerebral hemispheres and the convexity in the occipital lobes. They are found less often than in the fronto-temporal area. At the occipital pole they can be isolated. They are seldom found on the inferior plane of the occipital lobes.

At the base of the cerebral hemispheres, the anterior cranial fossa and the corresponding orbital area of the frontal lobes are often involved. The area adjacent to the orbital roofs is often affected and, less often, the area medial to the gyri recti, in

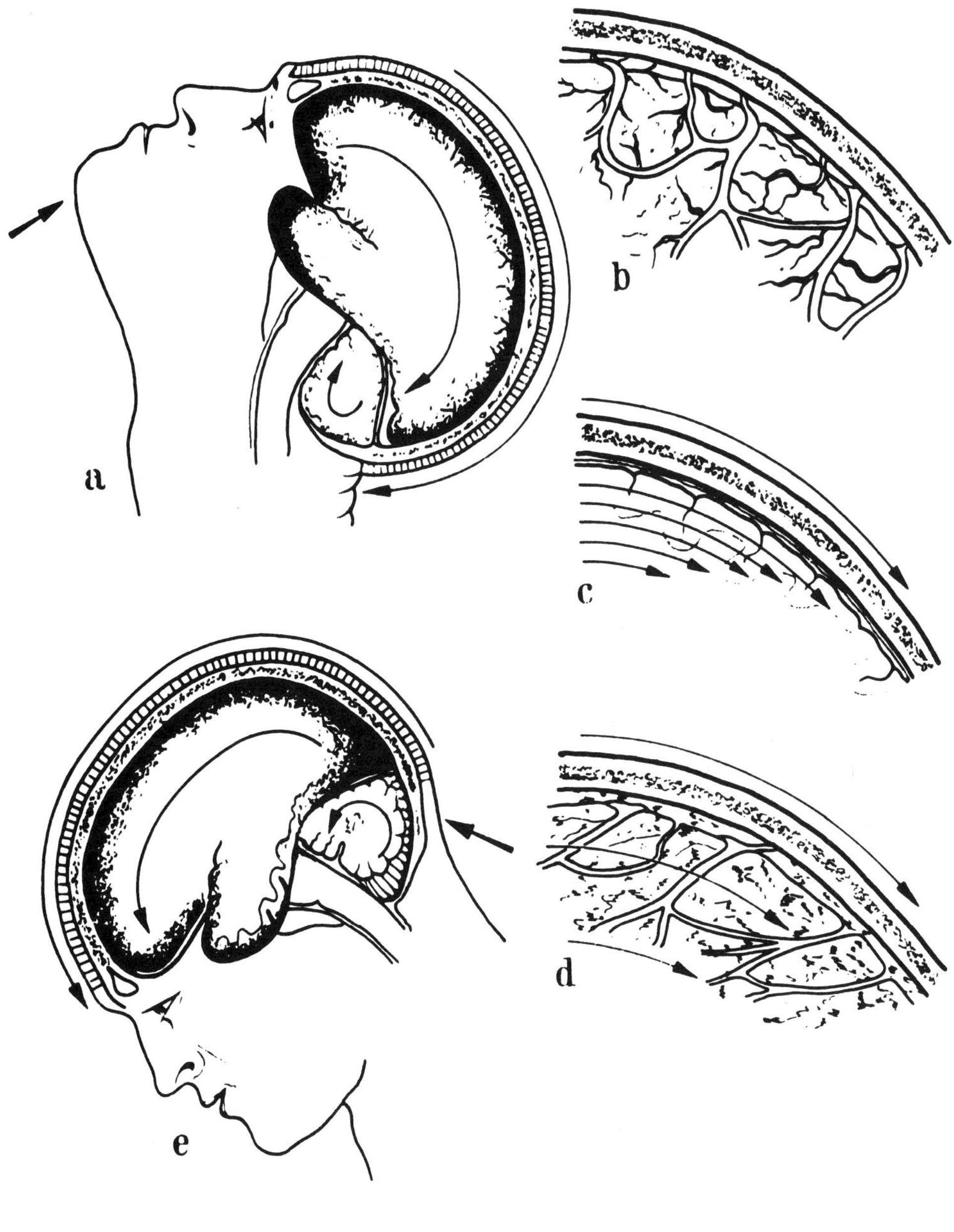

Fig. 25 a – e. Injury of the brain by rotation. Whenever the head is struck by a force not directed along that line which passes through the centre of gravity of the head and the occipito-atlantal joint (the fulcrum), it is set into rotation. The skull necessarily takes the first impact of the blow and moves before the brain. Then the brain is secundarily set into motion by the skull and particularly by projecting bony prominences and dural septa. Since the brain is soft and not rigid, it rapidly becomes deformed. In closed injuries it is the shearing forces associated with deformity that cause the maximum damage to the cerebral tissues and tear the cerebral arteries and veins. When a patient is struck on the chin the head is thrown backwards.
b, c, d Showing the resulting deformity of the brain in relation to the vault of the skull.
e, If a patient falls backwards and strikes his head on the ground, the head will be knocked forwards into an anterior rotation, the shearing forces in this case being in the opposite direction. Probably, in most accidents the head is set into violent rotation about different axes at different phases of the infliction of the violenence. (From ROWBOTHAM GF, 1964). The arrows in the brain must have the opposite direction!

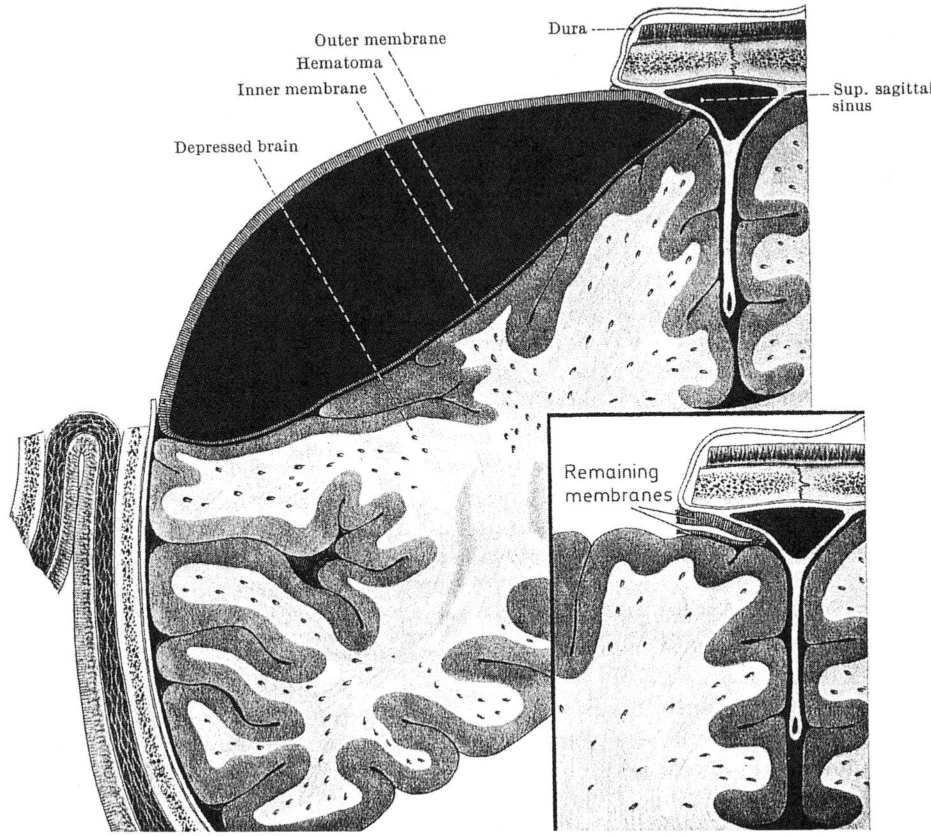

Fig. 26. Chronic subdural hematoma coronal section. Observe: Insert shows part of the membranes to be left in place. (From KEMPE, 1982).

whose area bulbus and tractus olfactorius are situated. The frontal poles are often affected as well.

Another area where the so-called cortical contusions are commonly found is in the middle cranial fossa, at the temporal pole and the areas around the inferior parts of the temporal lobes. Here the third and fourth temporal gyri (gyrus fusiformis or collateralis) are especially involved, as well as the uncus gyri hippocampi, whilst the deeply localized Ammons' horn area is not involved.

SPATZ called special attention to the fact that in all of his cases an extensive area of the base of the cerebral hemispheres remained completely free of all so-called cortical contusions. It was the surface of the basal ganglia and the brain stem, from the optic chiasm to the olives of the medulla oblongata.

The convexity of the cerebral hemispheres reveals cortical contusions, *so-called schizogyri,* only in rare cases.

Summarily, one can conclude that cortical contusions reveal a typical distribution that is dependent on the impact axis. Consequently one might expect that the always-recurring localization of the foci is also associated with certain well-known and typical clinical pictures, but that is not the case. SPATZ noted that usually clinicians do not take the contrecoup effect into account, considering the convexity of the cerebral hemispheres but not the base. Frequently they failed to identify the area of the impacting force, an indicator of impact direction details.

The classification of injuries into six types would direct the attention to the contrecoup area. Impact directions from left and right are intentionally dis-

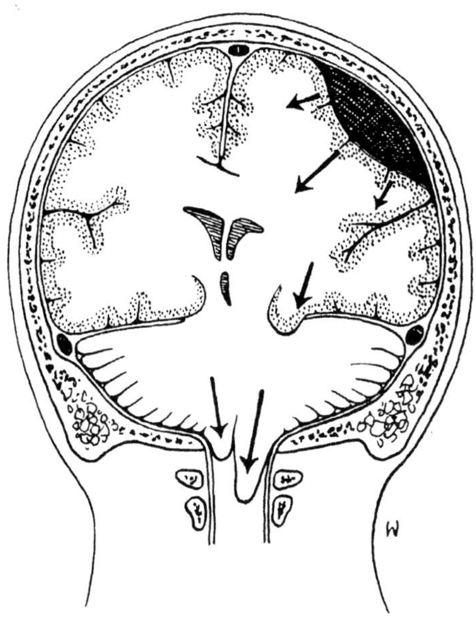

Fig. 27. Drawing of the final stage of an acute subdural hemorrhage. After development of a temporal pressure cone (arrow) a pressure cone and incarceration of both cerebellar tonsils (arrows) into the spinal canal has developed. (From KESSEL et al., 1969).

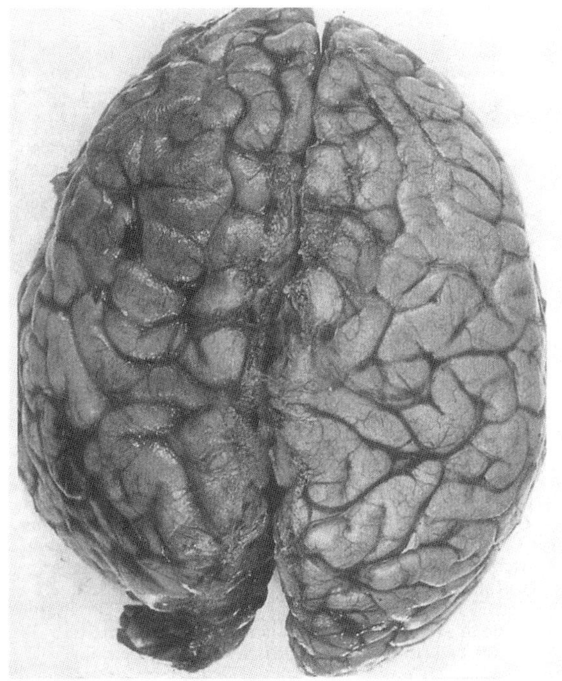

Fig. 28. Man. Closed head injury. Subdural hemorrhage left. There is an extensive swelling of both cerebral hemispheres with flattened summits and narrowed sulci. The left cerebral hemisphere shows a brownish-red discoloration of the lobuli, which indicates beginning hemorrhagic-necrotic tissue alterations. Macrophoto. (From UNTERHARNSCHEIDT, 1993)

tinguished for an important reason. Due to the predominance of contrecoup effects, impacts from the right affect the left side of the brain and vice versa. The functionally more important (dominant) left cerebral hemisphere is more affected by an impact from the right (type 4), while at type 3, the less critical right cerebral hemisphere is affected. Thus, aphasic disorders can be explained by the location of the tissue damage on the left, resulting from an impact from the right.

The commonly found type 1 and 5 cortical contusions of the bulbus and tractus olfactorius are the cause for parosmias, hyposmias and anosmias.

Cortical contusions are not tissue alterations of a uniform mechanical genesis. Different accident mechanisms result in different tissue alterations that are collectively termed cortical contusions. The following analysis of these tissue alterations will show that some of these tissue lesions are genuine contusions, but that the majority of these tissue alterations are not the result of direct contusions of the surface of the brain. Particularly the so-called cortical contusions at the point opposite the impact pole, the antipole – the essential lesions par contrecoup, cannot be explained by a direct contusion of the cerebral cortex against the skull.

The group of genuine contusions of the cortex consists of: (1) The direct contusions of the cortex beneath an impression fracture, (2) the ring-like contusions around the point of entrance of a bullet, (3) cortical contusions at the external ricochet shot, and (4) cortical contusions at the inner ricochet shot.

In addition to the four groups of cortical contusions, there are two other forms of traumatic tissue alterations that are often erroneously included with the so-called cortical contusions. They can be distinguished unequivocally by their pathogenesis, namely: (5) the *'gliding contusions'* and (6) the *so-*

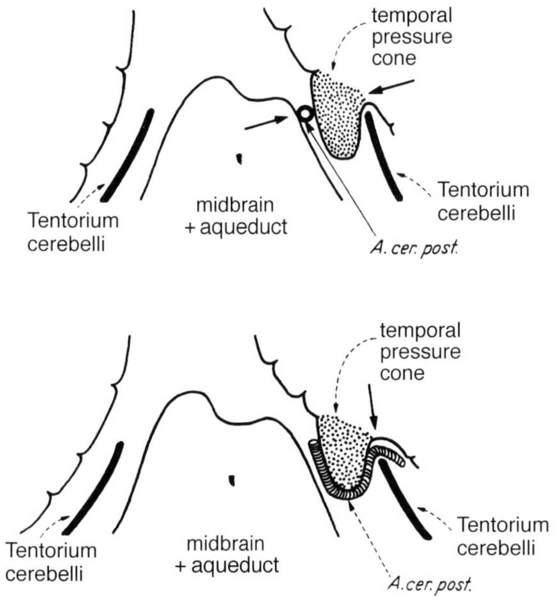

Fig. 29 a and b. Mode of origin of hemorrhagic infarctions in the occipital lobe at 'temporal pressure cone'. It is not decisive that the posterior cerebral artery is flatly pressed against the cerebral cortex. Important is rather more that the artery is dragged over the sharp edge of the tentorium cerebelli and is stretched in its longitudinal axis, lower drawing. (From ZÜLCH, 1959).

called cortical contusions 'par contrecoup,' which forms the larger group.

The traumatic alterations of the cerebral cortex – sometimes the subcortical white matter is also involved, are described below.

(1) Traumatic alterations of the cerebral cortex beneath a circumscribed impression fracture.

Following a circumscribed direct impact at the skullcap, with consecutive impression fracture, the impressed bony skull fragments develop traumatic cortical lesions. When bone fragments are displaced into the depths they also appear in the subcortical white matter. Hereby, it is about a directly cortical contusion of the cerebral cortex. Even when the imprimatur is lifted up immediately after the injury, the broken-in and displaced bony fragments execute not only momentary, but a persistent pressure on the brain tissue. Despite the fact that this is a direct cortical contusion, one can rightfully attribute it to the longer lasting pressure, a *traumatic necrosis* or a *contact necrosis* (Fig. 31).

(2) *Cortical contusions* in a ring around the point of entrance of a bullet. These do not occur in boxing.

(3) *Cortical contusions* at external and internal ricochet shots do not occur in boxing.

(4) The *'gliding contusions'* develop after rotational accelerations of the head (Fig. 32). While the bony skull with the affixed dura mater experiences a rotational acceleration, the brain remains behind, due to its inertia. During these gliding movements between skull and dura mater on the one side and the surface of the brain of the other side, bridging veins are over-stretched and ultimately torn. Vessels of the cerebral cortex are pulled out and torn off. These are not genuine cortical contusions, but the result of pulled out or torn off vessels of the cerebral cortex. These tissue alterations in humans were described by VOIGT and LÖWENHIELM (1974) and VOIGT et al. (1977), just as in monkeys due to non-deforming angular acceleration of the head using *Head Acceleration Device (HAD – II)* (UNTERHARNSCHEIDT and HIGGINS (1969). The essential morphological distinction between so-called cortical contusions due to linear acceleration and those of gliding contusions due to rotational acceleration were described in detail by UNTERHARNSCHEIDT and HIGGINS (1969).

(5) So-called cortical contusions par contrecoup

These traumatic lesions are the most frequent traumatic alterations of all groups discussed. This term is a misnomer, since they are not contusions of the cortex at all. Rather, they occur in zones of reduced pressure.

Impact Directions

The location and extent of a cortical contusions is highly dependent on the direction from which the impact is delivered. A system for identifying the impact direction has been suggested by SPATZ (1950) (Fig. 33).

The system consists of six principal directions as follows:

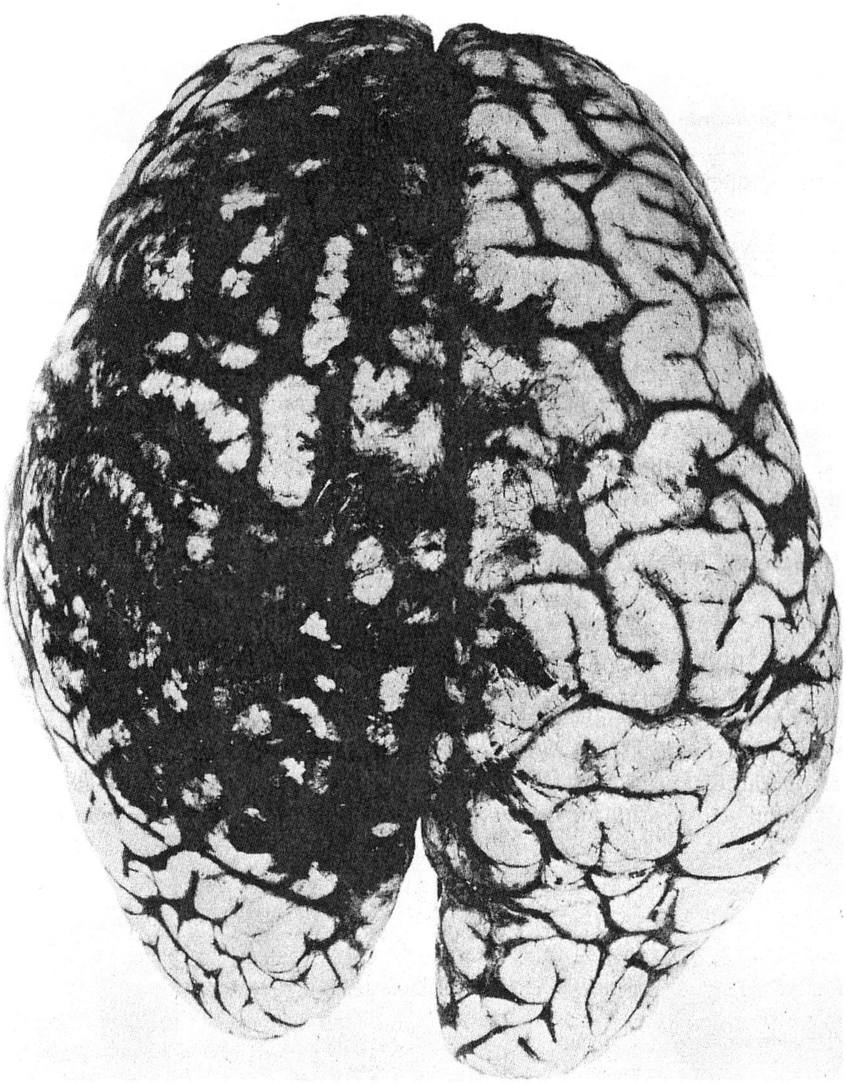

Fig. 30. Man. Closed head injury. Subarachnoid hemorrhage, mostly over the left cerebral hemisphere. Macrophoto. (From UNTERHARNSCHEIDT, RIPPERGER, 1970).

Type I – Impact from behind. Impact from behind frequently results in bursting fractures of the occipital bone (Fig. 34 a). The most consistent finding (97.2% of the cases) (WELTE 1948) is injury of the brain tissue across the impact point (Figs. 34 b and c). The clinical signs are prolonged disturbance of consciousness and motor activity.

Type II – Impact from the front. The frontal impact may be directly into the face or higher, on the forehead. The blow, if it comes on the face, is cushioned by the relatively weak structures of that part of the anatomy. The human face acts as shock absorber for the brain. This is the main direction of impacts in accidents, and in general the impact is quite severe by any criteria that may be applicable. In a high percentage of cases cortical lesions are produced in the immediate vicinity of the impact. In relatively few cases cortical contusions as seen on the brain opposite the point of impact.

The low resistance to fracturing of the relatively thin bones of the anterior cerebral fossa, i. e., of the orbital roofs and ethmoid sinuses, frequently results in compression fractures of the anterior cerebral fossa with accompanying perforation of the dura mater. As a consequence, open injuries with infection of the brain wound starting at the sinuses are seen quite often.

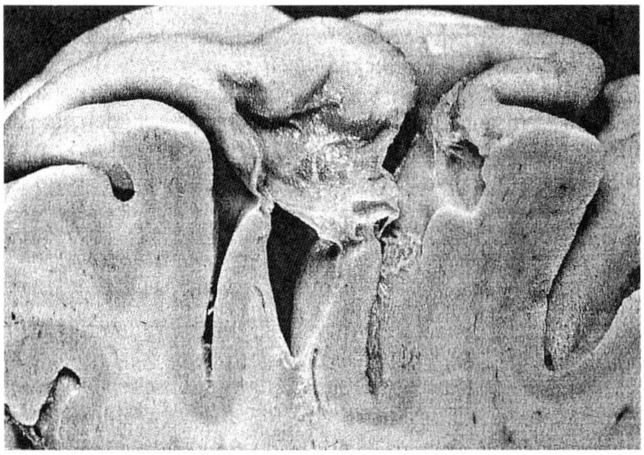

Fig. 31. Man. Impression fractures of the calotte of the skull. So-called cortical contusions in the third stage (defect stage) with cystic defect. Cerebral cortex and subcortical white matter are affected. This form of cortical contusions situated at the point of impac are called also schizogyri. Macrophoto. (From UNTERHARNSCHEIDT, 1993).

A *subclassification, type II b,* has been used because of the differences between injuries produced by blows to the face and those which strike higher, near the line of the scalp and at a slight angle from above. As a result of such a blow, lesions are found in the frontal and temporal parts of the cerebrum. Some investigators believe that this type of blow displaces the brainstem. In our opinion this action can occur only if the blow is so severe as to produce considerable deformation of the skull and therefore of the brain.

Type III – Impact from the left. These impacts produce fractures at the impact point, and in two-thirds of observed cases cortical contusions occurred at the opposite side (WELTE 1948).

Type IV – Impact from the right. The effects produced by these impacts correspond to those for type III. It should be noted that lateral impacts are not frequent because the shoulders absorb a large part of the impact forces that might otherwise have been applied to the head. In addition anatomical particularities reduce the frequency of cortical contusions in lateral portions of the cerebral hemispheres. It has been shown analytically that the positive and negative pressures developed in an ellipsoidal water-filled shell are less for a given acceleration in the direction of the minor axis than they are if the acceleration is in the direction of the major axis (SELLIER and UNTERHARNSCHEIDT 1963, KOPECKY 1968).

Type V – Impact from above. This impact differs from the other types in that the direction is not perpendicular to the spinal column. The motion of the head in a direction parallel to the spinal column is more restricted than it is in the perpendicular direction. There is little acceleration of the head produced by this impact force. Instead compression in the direction of the spinal column is produced. This type of impact seldom occurs as a blunt broadly surfaced blow. It occurs more frequently in accidents involving falling objects with small mass and a small surface. Accidents of this type cause almost no acceleration damage because the energy is absorbed by local indentation of the skull, and since there is very little acceleration, no lesions are produced in the antipole area.

One type of damage at the cerebral convexity, the *so-called schizogyri* (MITTELBACH 1929, RIEDERER von PAAR 1936) which have been observed, is actually a cortical contusion in the defect stage, or more correctly, the result of a cavitation phenomenon at the impact point. These tissue damages are located near the impact point and are believed to be the re-

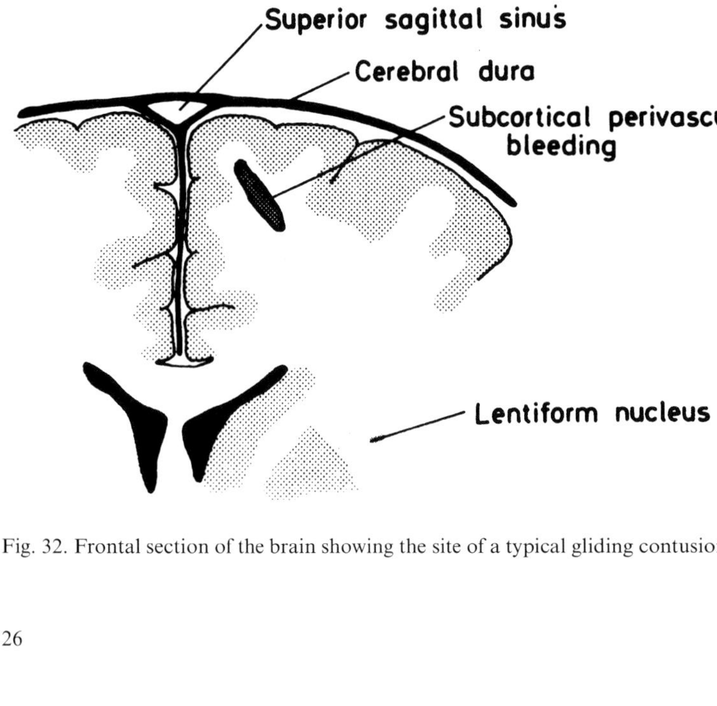

Fig. 32. Frontal section of the brain showing the site of a typical gliding contusion hemorrhage. (From LÖWENHIELM, 1975).

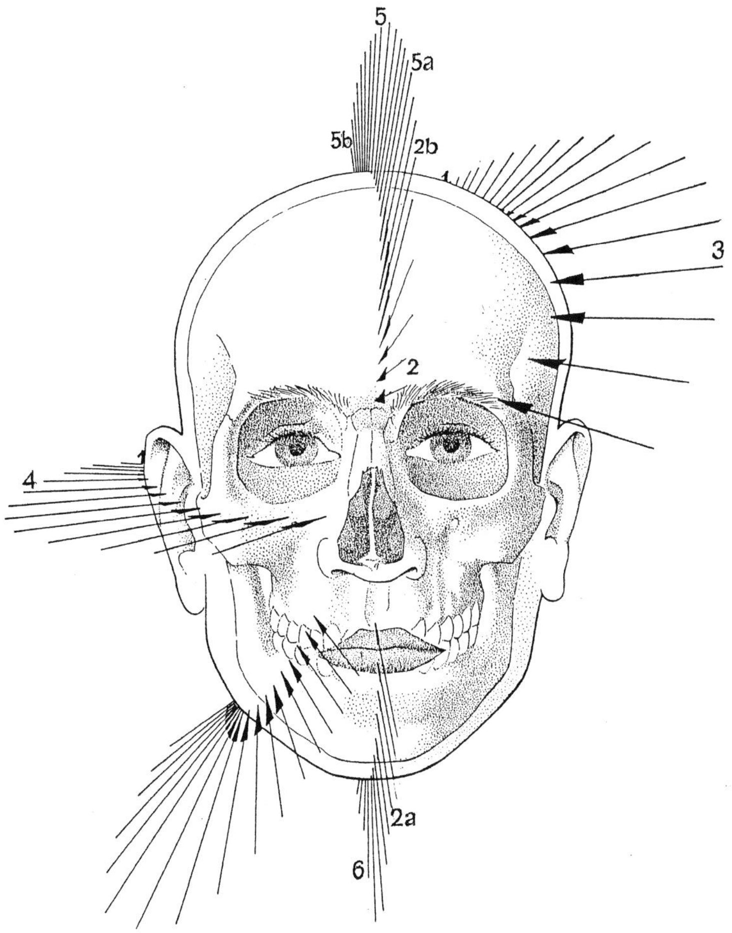

Fig. 33. Main impact directions of direct impact at the head. Impact direction 3 is from left to right, and impact direction 4 is from right to left, inpacy direction 1 is from behind. (From UNTERHARNSCHEIDT, 1993).

sult of a local deformation of the skull. They are the result of an impact direction from above – type V.

Type VI – Impact from below. This type of impact can affect the head only through the spinal column. The vertebral column acts as a shock absorber to reduce the intensity of the impact forces to the head, and the acceleration it produces. The low acceleration forces are not sufficient to produce cortical contusions at the cerebral convexity.

Statistical analyses do not distinguish between acceleration and impression traumas. Instead the frequencies of injuries at the impact area and the antipole are usually lumped together. If a distinction were made it would be seen that the acceleration traumas are located mainly at the antipole, and the impression traumas at the impact area.

The Different Stages of the so-called Cortical Contusions

SPATZ has described the development of cortical contusions in three stages as follows: (1) *Hemorrhage and necrosis,* (2) *resorption and organization,* and (3) *final or defect stage* (SPATZ 1936, 1950),

First stage: hemorrhages and necrosis. At the moment of impact rhectic hemorrhages occur, in which the vessels are cuffed by erythrocytes and the surrounding brain tissue is displaced (Fig. 35). These hemorrhages affect mainly the summits of the gyri, with the smaller hemorrhages being restricted to the cortex and the larger ones reaching as deep as the subcortical white matter. SPATZ has pointed out the remarkable finding that the tissue continuity is usually fully preserved in

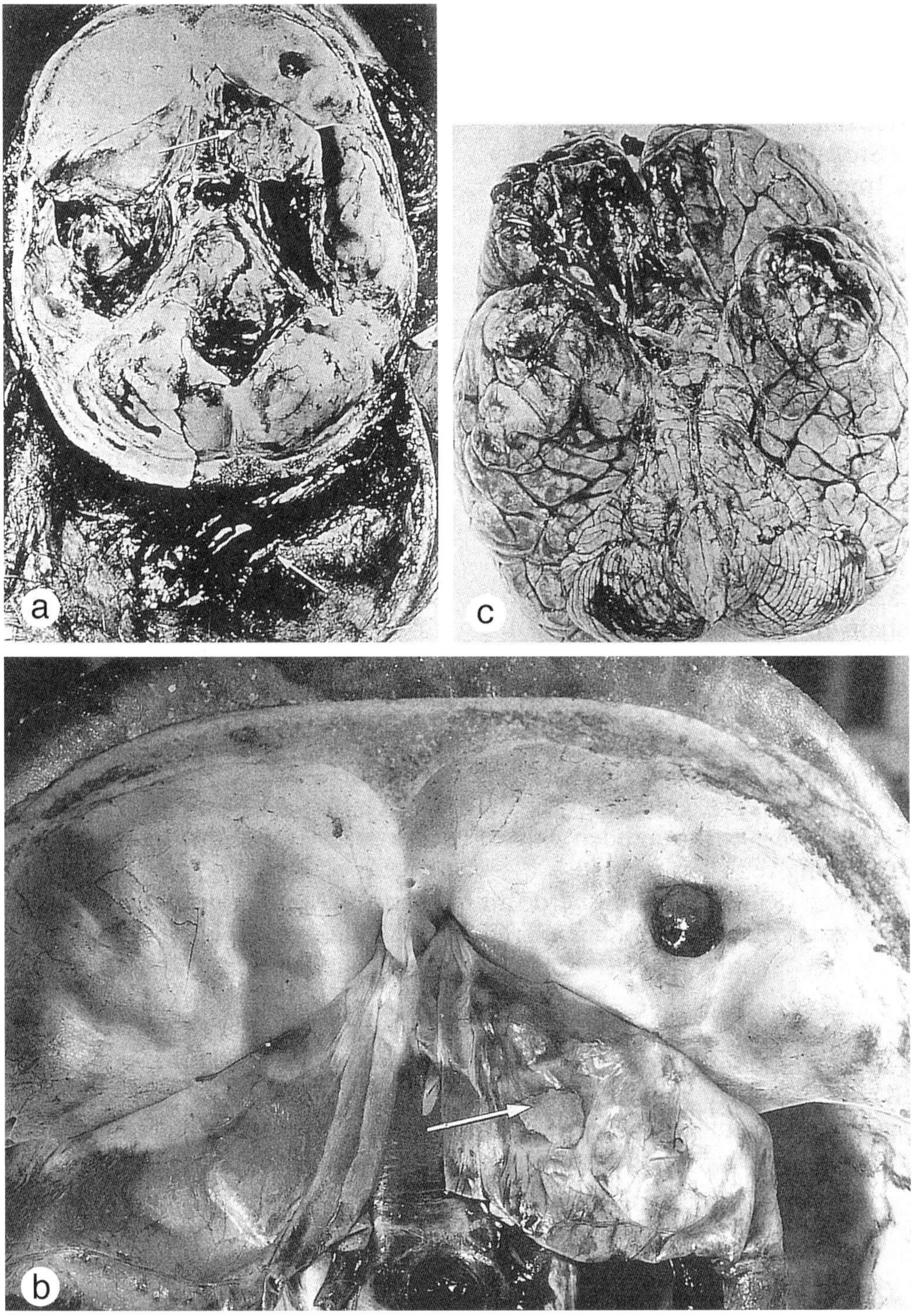
a
c
b

the earliest stages of the cortical contusion and that tissue lacerations are rare. Subarachnoid hemorrhages are a consistent feature with rhectic hemorrhages (Fig. 36).
After a certain period of survival, additional tissue alterations develop, which are complicated by phenomena of glial reaction or proliferation. Prior to the far reaching investigations of SPATZ and his school, and the more recent studies in human cases, the distinction between immediate and delayed tissue alterations was neglected.
On *gross examination*, subarachnoid, and sometimes subdural, hemorrhages are visible. They are caused by the bursting of small vessel branches on the summits of the gyri, with blood collecting initially in the sulci of the convolutions (Fig. 37).
On *microscopic examination*, the vessels are cuffed by erythrocytes (Fig. 38). When death occurs immediately, tears in the vessel walls can be found. The cyto-architecture is affected only inasmuch as the hemorrhages are occasionally densely spaced. In the area of perivascular hemorrhage the cell body of the neurons is shrunken and elongated, Nissl or tigroid substance is seldom clearly discernible (Fig. 39 a and b). The nucleus is deformed and frequently diffusely stained. The hemorrhage usually shows a rather pale stain using the Nissl technique, whereby the tissue appears somewhat lighter, simulating *'paling'* (SPATZ), characteristic of necrosis. There is no firm evidence of glial changes in this early stage of the process (Fig. 40 a – c).
Second stage: resorption and organization. If the impact has been survived for several hours, the effused blood pigment is grossly visible as a diffusion halo around each hemorrhage. After twelve hours survival, the outlines of the foci become visible. Then necrosis sets in, and the foci assume the typical trough or wedge shape (Fig. 41). Necrosis is complete in the foci, partial in adjacent tissue, with a predominant loss of nerve cells. Starting from the margin, a marked proliferation of the vascular connective tissue takes place (Fig. 42). The network of proliferating vessels contains abundant numbers of macrophages and gitter cells that gradually dissolve after four or five weeks. The destroyed brain tissue (debris) is carried off into nearby vessels by *phagocytic cells* from the proliferation zone that develops around the necrosis. This process is called *'mobile disintegration of the tissue'.*
Third stage: final or defect stage. In this stage the lesions represent a cystic defect filled with cerebrospinal fluid (Fig. 43). This is the final aspect of *total necrosis* (Fig. 44) - brain tissue lesions in which parenchyma, neurons, and glial elements have suffered irreversible damage (Fig. 45).
In *partial necrosis*, only the more vulnerable neurons are affected while the more resistant glial elements and the mesodermal elements of the vessels remain unaffected and develop progressive alterations. The proliferation involves mainly astroglial elements. *The final product is* a *so-called glial and/or mesodermal scar.* Partial necroses are usually found next to cortical contusions. It is important to distinguish between partial and total necroses since each form results in a different defect.
Clinically, it is important to know whether these permanent traumatic lesions are located in silent zones, or in areas which produce psychopathological or neurological deficits. In general, the defect stages of the so-called cortical contusions produce little scar tissue and no dural adhesions (Fig. 46) so that focal epilepsy develops less often than with open brain injuries.
For detailed neuropathological studies of the brain after closed head injury, see SELLIER and UNTERHARNSCHEIDT (1963), UNTERHARNSCHEIDT and RIPPERGER (1970), UNTERHARNSCHEIDT (1993, 1994). For neurological and neurosurgical aspects see GURDJIAN and WEBSTER (1958), EVANS (1963), GURDJIAN (1975), COOPER (1987).

Schizogyri (Figs. 47 a – c)

The *schizogyri* were described by MITTELBACH and SPATZ (1927), MITTELBACH (1929), and RIEDERER von PAAR (1936). According to

Fig. 34 a – c. Man. Closed head injury. Individual was hit into the face, fell.backward and hit the pavement of a road with his occiput. Deceleration about 200 G.
a. Hemorrhages between galea aponeurotica and skull (occipital region, lower arrow). Extensive bursting fractures at the occiput in the area of the posterior cerebral fossa. Opposite of the impact pole 'impression fracture' of the right orbital roof due to reduced intracranial pressure (upper arrow). Macrophoto.
b . Close up of (a). The dura mater in the anterior cranial fossa is removed and turned over. The broken out bone fragment is due to the suction occurring in this area sucked out of the orbital roof to the inside of the bone formation. Adipose tissue from the right orbita is likewise sucked inside the anterior cerebral fossa. Macrophoto.
c. Basal view of the brain of the patient. Localized subarachnoid hemorrhages at the entire base of the brain, especially at the orbital gyri, the lower part of the temporal lobes, the temporal poles and the lower cerebellar hemipheres, so-called cortical contusions in the first stage, which are more pronounced at the opposite site of the impact area. Traumatic lesions of both olfactori bulbs. The so-called cortical contusion are located at the orbital gyri, right more pronounced than left, at the frontal pole, both temporal poles and the lower surface of both temporal lobes. (We are indebted for this photo series to Professor Dr. Otto PROKOP, former Chairman of the Department of Forensic Pathology, Humboldt University, Berlin, from SELLIER, UNTERHARNSCHEIDT, 1963).

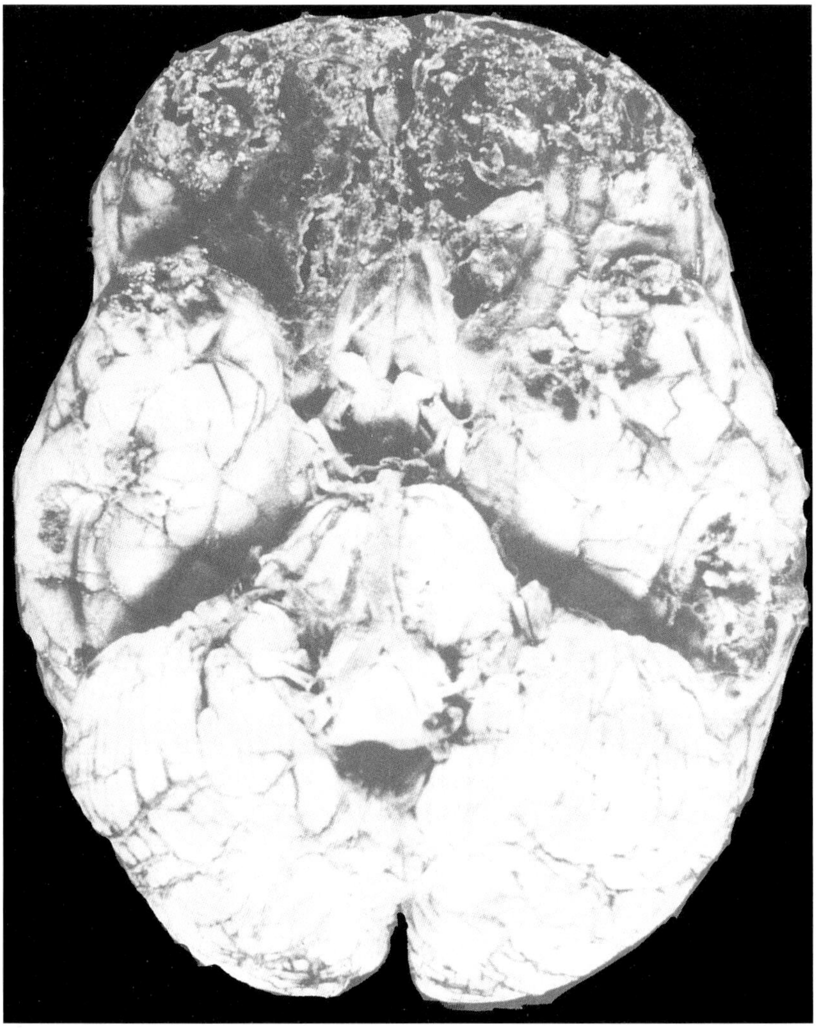

Fig. 35. Human. Cerebrum. Basal parts of both frontal lobes show massive so-called cortical contusions in the first stage. Both fila olfactoria are destroyed. Both temporal poles are affeceted, too. 'Lesions par contrecoup' due to fall at the occiput. Macrophoto.

these authors, schizogyri are localized at the convexity of the cerebral hemispheres, especially in the posterior third of the first and second frontal gyri, as well as at the anterior central lobe, in patients who have suffered a closed head injury. Fractures of the skull are usually not found. However, there are exceptions.

There is no longer any doubt about the traumatic origin of schizogyri. In our opinion, their origin is not uniform; they are usually sequelae of either an impression trauma or a contre-coup lesion. In some cases they may be the final stages of 'gliding contusions.'

Contrecoup Effects

These effects, which are observed in the antipole region of the impact (hence the term contrecoup), were first described in the second half of the 18th century by French authors (LOUIS, CHOPART, GRIMA, SABOURAUT, SAUCEROTTE).

The term contrecoup was first used in 1766, at which time a series of papers or memoirs on the

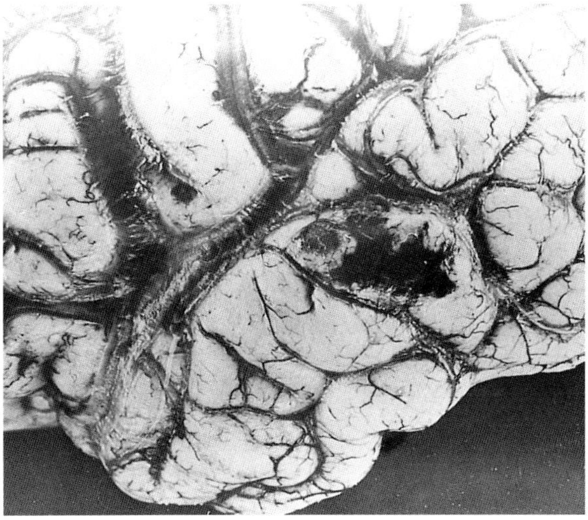

Fig. 36. Cortical contusions in the first stage at the temporal lobe. Macrophoto.

subject was presented be several members of the Academie Royale de Chirurgie in Paris.

PUDENZ and SHELDEN (1946) published a paper on *'The Lucite Calvarium – A Method for Direct Observation of the Brain.'*

Since 1766, many explanations of the mechanism of coup and contrecoup brain injuries have beeen presented in the literature.

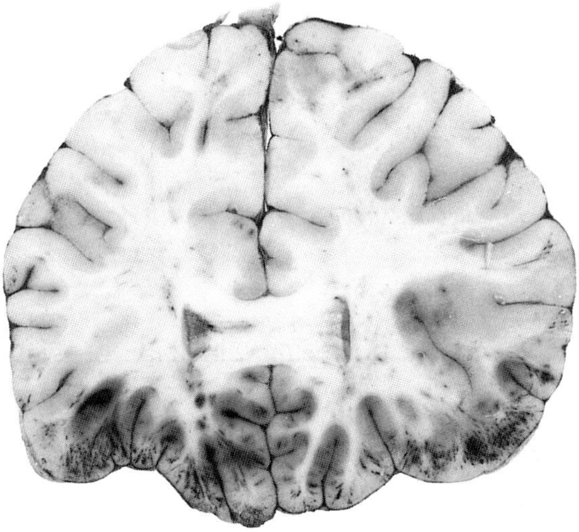

Fig. 37. Man. Cerebrum. 10 year old patient. Traffic accident, fracture of the base of the skull, subdural hematoma. Coronal section through anterior parts of the frontal lobes. Extensive so-called cortical contusions in the first stage, essentially limited to the cerebral cortex of basal parts of both frontal lobes. Macrophoto. (From UNTERHARNSCHEIDT, 1993).

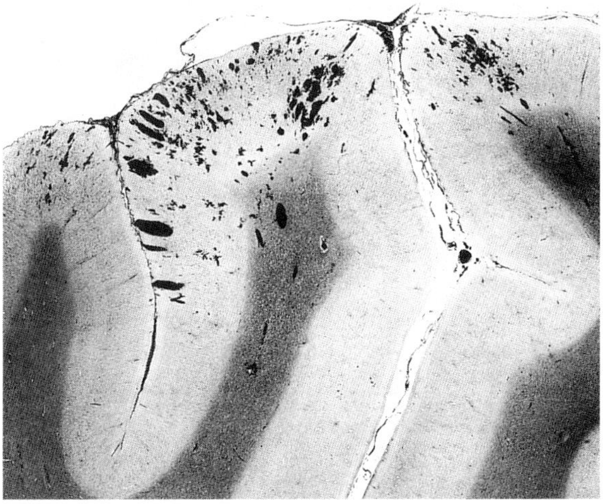

Fig. 38. Man. Cerebrum. So-called cortical contusions in the first stage, especially in the cerebral cortex, in the middle lobulus extending also into the subcortical white substance. The sulci are not affected. The Heidenhain technique was chosen, not to show demyelinations, which are not to be expected in this early phase, but because this technique staines fresh erythrocytes distincly black. Heidenhain technique, 6:1.

PUDENZ and SHELDEN published these theories, which may be classified under six separate mechanisms (Fig. 48), explain brain damage on the basis of (a) vibrations transmitted from the skull to the underlying brain; (b) pasage of waves of force throught the brain; (c) brain displacement at the pole opposite the site of impact; (d) deformation of the skull; (e) pressure gradients set up within the brain as a result of the blow; (f) rotation of the cerebral mass within the confines of the skull.

" (1) High-speed cinematography has beeen used to observe the effcts of subconcussive blows and gunshot wound of the head on the brains of macaque monkeys fitted with lucite calvaria."

" (2) Blows on the head cause swirling, rotating movements of the brain within the cranial cavity."

" (3) The patterns of convolutional motion are determined by the direction in which the head is displaced by the blow."

" (4) Midline frontal and occipital blows displace the head through a sagittal arc and produce convolutional gliding through the same arc."

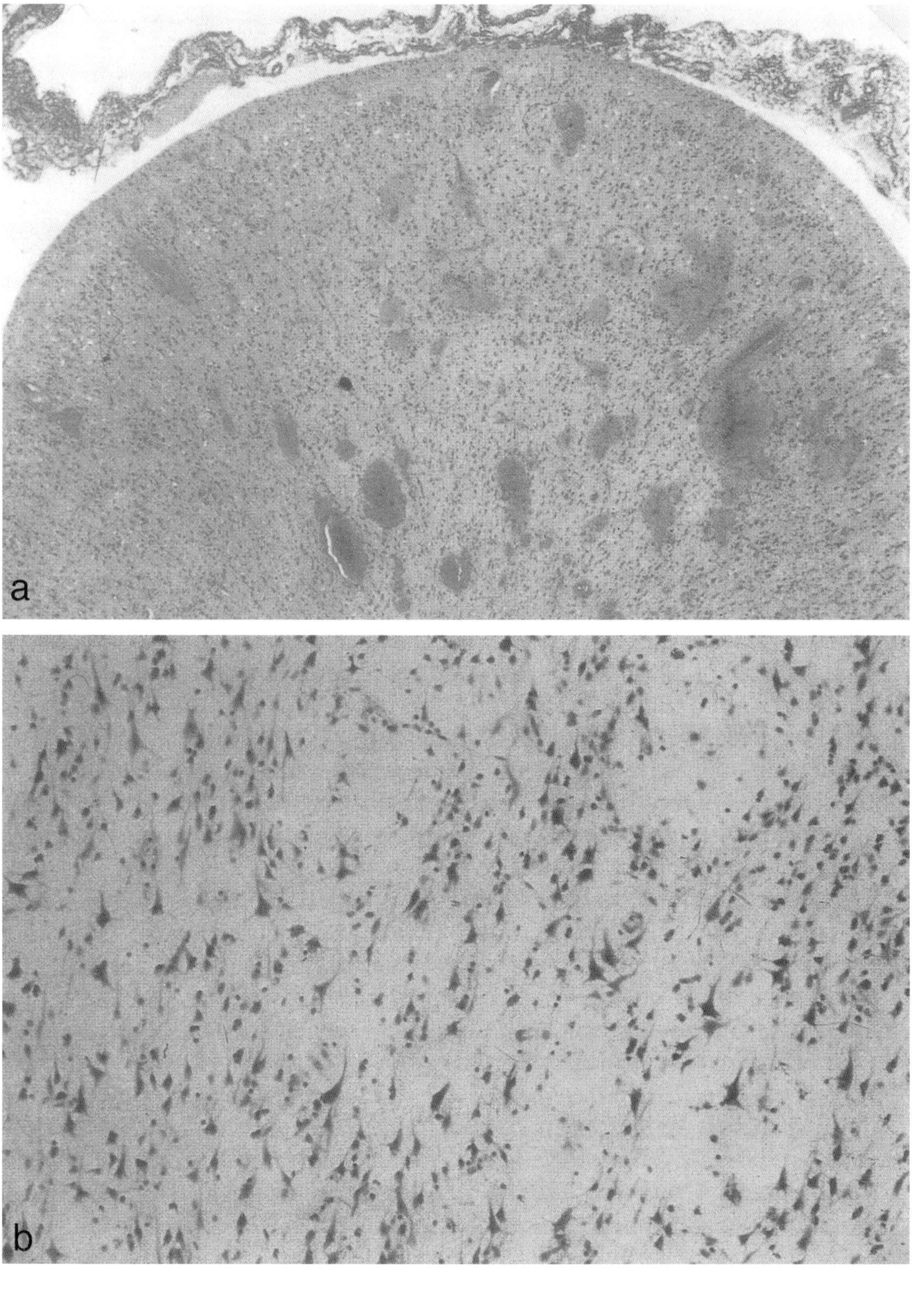
a
b

" (5) In contrast, temporal and parietal blows rotate the head through a combination of sagittal, horizontal and coronal planes and cause a complicated type of brain movement in which convolutional glide in the sagittal and horizontal planes is striking. Coronal rotation of the brain was not visualized distinctly."

GROSS (1958) begins his pioneering paper that the basic purpose of this research effort has been to investigate the dynamics of the head and its contents when subject to impact. The actual experimentation involved was limited to the study of the dynamics of fluid-filled glass flasks simulating the human head.

GROSS continues:

"Let us now take the case in which the glass tube is sealed at both ends and completely filled with a fluid as shown in Fig. 49. Neglecting radial expansion of the tube and assuming the compressibility of the fluid to be greater that the compressibility of glass, we conclude that the fluid column will be accelerated by both pressure from the impacted end of the tube and tension from the opposite end of the tube."

GROSS continues:

"Although the human skull is much more complex than the simple case stated above, there is reason to believe that distribution of pressure upon impact of the unrestrained head is of the same nature. For lack of substantial evidence, let us assume that the distribution of pressure within the unrestrained skull upon blunt impact graduates from a positive pressure on the impacted site, as shown in Fig. 50. This differs from the case in which the head is restrained and the forces are of a crushing nature."

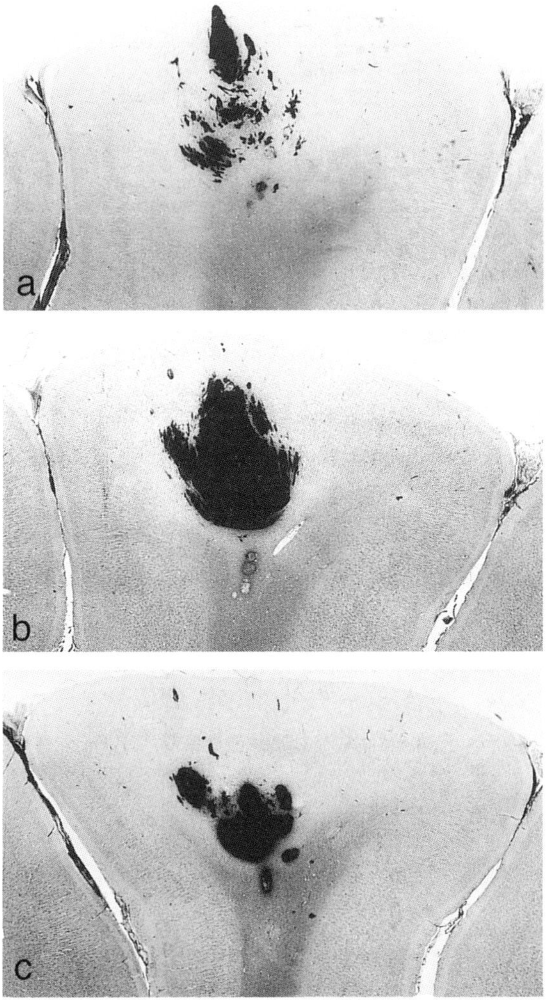

Fig. 40 a - c. Man. Cerebrum. Closed head injury. Serial sections through a cortical contusion in the first stage. Shown are the 1., 48., and 88. sections. Hematoxylin-Eosin technique, 4:1. (From SELLIER, UNTERHARNSCHEIDT, 1963).

Fig. 39 a and b. Man. Closed head injury. So-called cortical contusions in the first stage, multiple rhectic hemorrhages at a summit of a lobulus.

a. Shinkage and hyperchromatosis of neurons, especially in the neighborhood of hemorrhages. A glial proliferation has not began yet. The leptomeninges above the so-called cortical contusions are intact, there exists only a moderate subarachnoid hemorrhage. Van Gieson technique, 30:1.

b. Section of an area of Fig. a of the same patient. Small rhectic hemorrhages, which extend into the surrounding brain tissue, with distinct shrinkage and hyperchromatosis of neurons. Beginning halo formations around single neurons. Many neurons are already completely destroyed. The stratification of the cerebral cortex is lost. A glial reaction has not begun yet. Nissl technique, 80:1. (From UNTERHARNSCHEIDT, 1993).

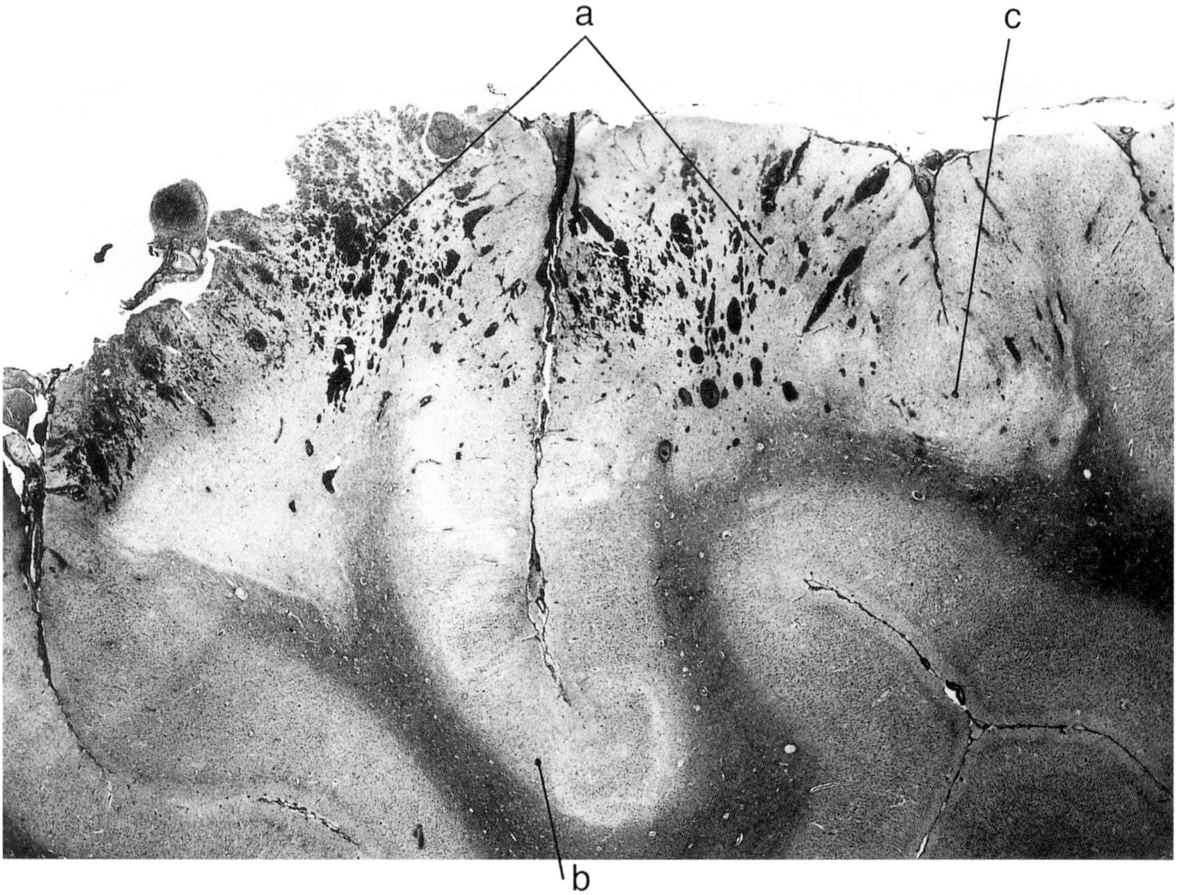

Fig. 41. Man. Closed head injury. Fleabite-like hemorrhages at the summits of gyri. Cortical contusion in the frist stage at (a). The profound sulcus at (b) presents no primary but secundary tissue alterations. The more shallow sulcus at (c), located within the area of critical pressure, exhibits both primary and secondary tissue alterations. Van Gieson technique, 5.5:1. (From SELLIER, UNTERHARNSCHEIDT, 1963).

In discussing the mechanism of tissue damage GROSS noted that in the various theories that are based on fluctuations of pressure it is generally concluded that it is the negative phase of the pressure cycle that is causing the tissue damage. It has been most difficult to visualize just how the application of a uniform field of negative pressure could produce brain damage of a focal nature. GROSS continues that in the various theories that are based on fluctuations of pressure it is generally concluded that it is the negative phase of the pressure cycle that is causing the tissue damage. It has been most difficult to visualize, according to GROSS, just how the application of a uniform field of negative pressure could produce brain damage of a focal nature.

GROSS writes that if we assume formation of cavities within the brain caused by acceleration of the head, upon cessation of the accelerative force the cavities will collapse immediately because of the ambient pressure and the surface tension of the fluid.

GROSS noted that the mechanical violence of collapsing cavities has long been known to the engineering profession. The first practical problems of cavitation came with the introduction of rotary

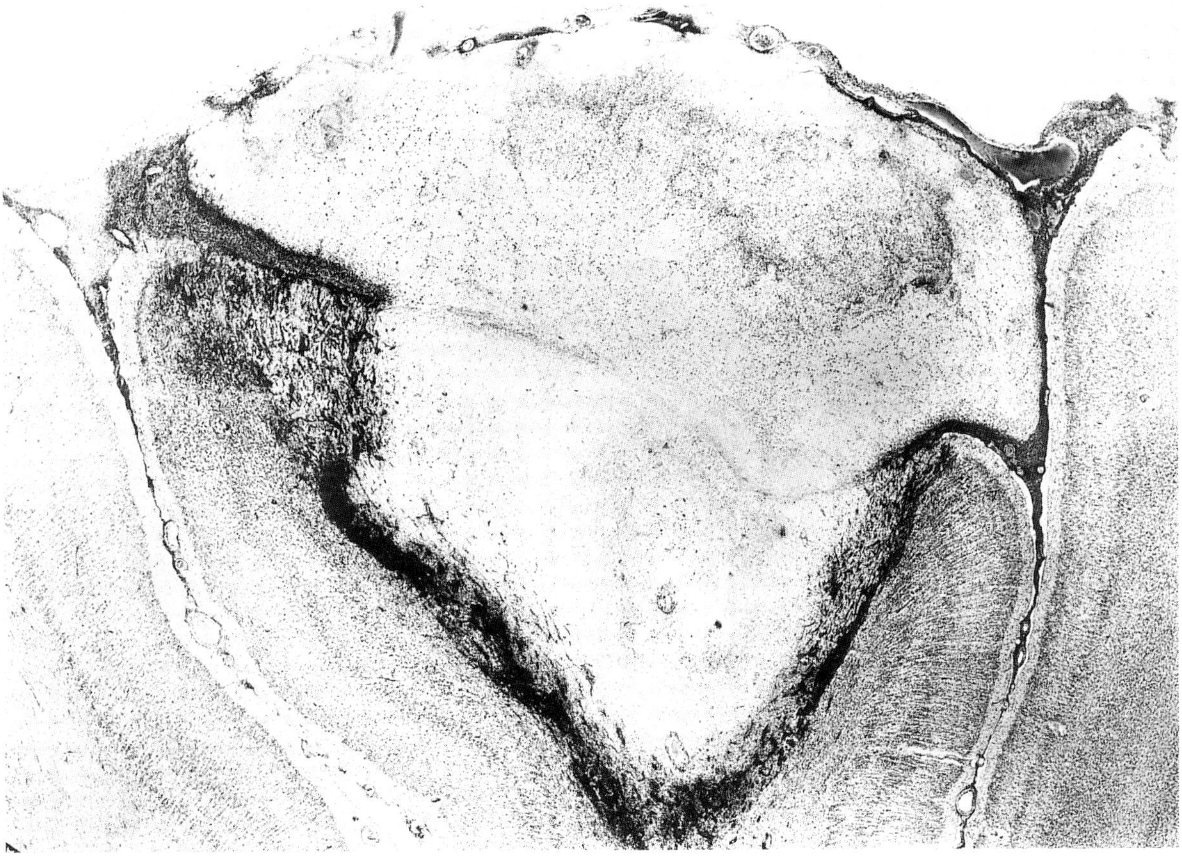

Fig. 42. Man. Closed head injury. Wedge-shaped cortical contusion in cortex and subcortical matter. A considerable proliferation of vascular connective tissue is observed. Nissl technique, 20:1. (From SELLIER, UNTERHARNSCHEIDT, 1963).

hydraulic apparatus, such as ship's propellers and zentrifugal pumps. If, according to GROSS, a propeller or pump rotor is driven too hard, cavities will form on the low pressure side of the blades because of tension failure of the fluid. The violent collapse of these cavities causes the metal to become spongy and wear away. This process is known as cavitation erosion and has been the subject of much research. According to GROSS, it is generally agreed that the erosion is caused by the impacting force of the collapsing cavity exceeding the strength of the metal or at least the fatigue strength. GROSS believed the impacting pressure produced by the collapsing cavities is due to be in the order of 30.000 atmospheres.

GROSS continues:

"The mechanism of brain concussion being proposed is based on the premise that the various forms of focal brain damage are the result of violent collapse of cavities formed within the brain."

Contrecoup Cavitation

GROSS writes that on the basis of the assumption made in relation to the distribution of pressure, a plausible explanation for the phenomenon of contrecoup becomes apparent. If the zone of negative pressure, which is opposite the point of impact, exceeds the sum of the cohesive pressure and the ambient pressure, cavitation will occur as shown in Fig. 51. The location of the cavities will be dependent on the actual distribution of pressure and the location of the gas nuclei. The maximum size of the cavities will be dependent on the magni-

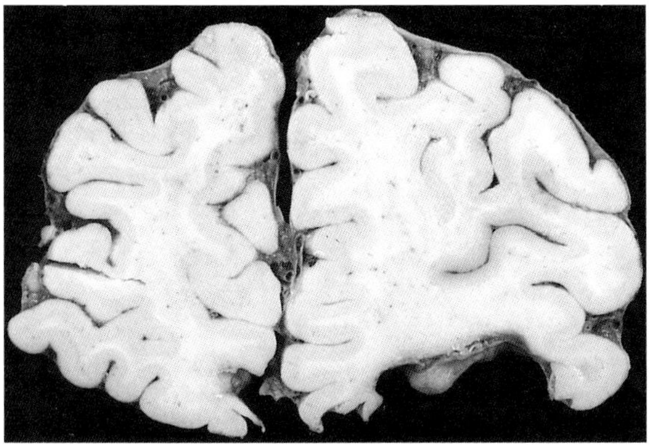

Fig. 43. Man. Cerebrum. 74 year old patient, since seven years needing care due to arteriosclerosis. No further data available. Time of accident or fall unknown. Coronal section through anterior parts of both frontal lobes. The entire basal cortex and subcortical white matter of the right frontal lobe reveals an extensive sharp-edged bowl-shaped cystic defect, a so-called cortical contusion in the third stage. Macrophoto. (From UNTERHARNSCHEIDT, 1993).

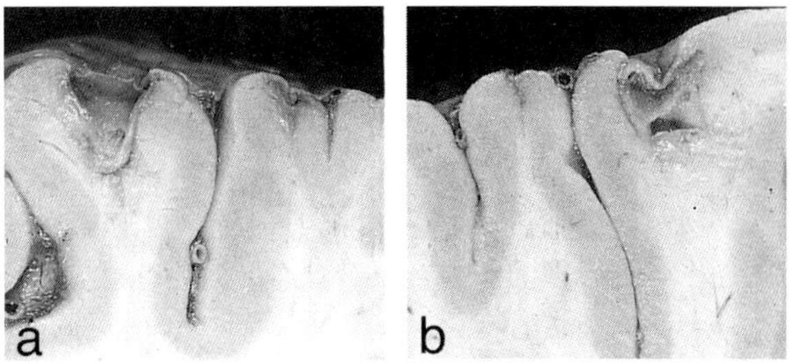

Fig. 44. Man. Cerebrum. Closed head injury. On four neighboring summits of gyri are wedge-shaped tissue defects visible, which extend into the subcortical white matter. Over the defect are intact but thickened leptomeninges visible. So-called cortical contusions in the third stage. Macrophoto. (From UNTERHARNSCHEIDT, 1993).

tude of the negative pressure and its duration above the threshold of cavitation. Once the negative pressure falls below the threshold necessary to support the cavities, the sequence of collapse begins.

GROSS emphasizes that it is this violent collapse of these cavities (Fig. 52) with the associated impacting pressures that produce the physiological damage recognized as contrecoup. If, according to GROSS, a larger number of small cavities are formed, the damage will show as a number of small petechial hemorrhages where the individual cavities formed and collapsed; however, if the cavitaion is localized at one point on the surface of the brain, the characteristic conical contusion with the apex extending into the brain will result.

Coup Cavitation

Damage of a local nature occurring in the area

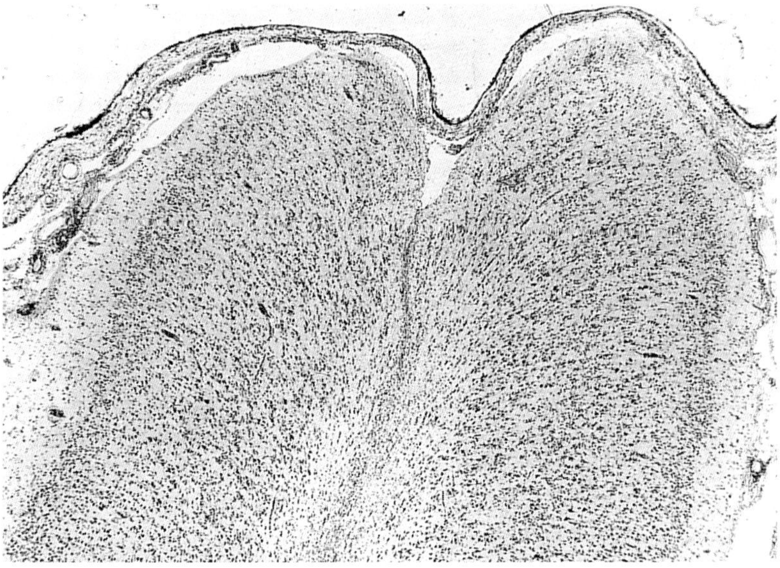

Fig. 45. Man. Cerebrum. Closed head injury. Wedge-shaped cyctic defect covered by thickened leptomeninges, so-called schizogyri, cortical contusions at the third stage. Macrophoto. (From MITTELBACH, 1929).

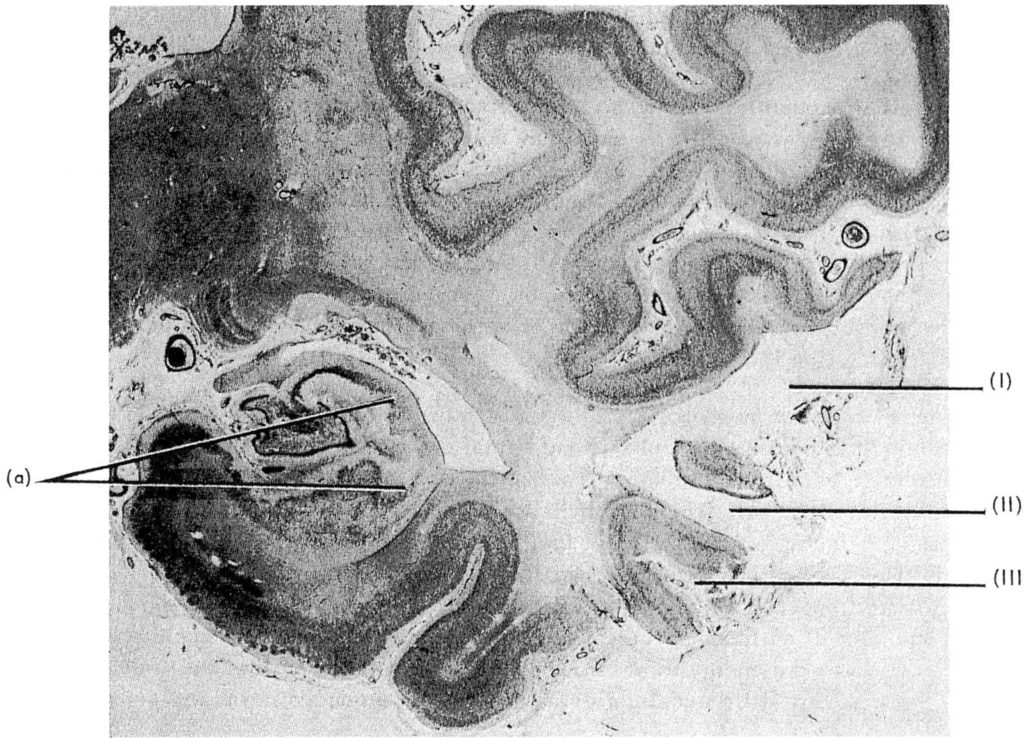

Fig. 46. Man. Closed head injury with fracture of the base of the skull, suffered 36 years before death occurred. Extensive so-called cortical contusions in the third or defect stage. Located in the frist, second, and third temporal lobe of the right cerebral hemisphere. Total loss of neurons in Sommer's sector (h_1) at (a). Nissl technique

immediately surrounding the blow can be explained, according to GROSS, in the following manner: A sharp nonpenetrating blow to the skull (Fig. 53 a – d) will produce the distribution of negative-positive pressure discussed above as long as the force of impact and acceleration of the head are maintained. Upon cessation of the impacting force there is no longer the acceleration of the head essential to support this distribution of pressure.

GROSS explains the mechanism further that at the instant of the collapse of the field of pressure because of cessation of the acceleration, the skull is left locally depressed as a result of the force of the blow. If fracture does not occur, the elasticity of the skull will cause a very rapid recovery of shape at the point of depression. The elastic forces in the skull returning the depression to its original shape can be sufficiently high to cause localized negative pressure sufficient to cause tension failure within the fluid. Coup injury, therefore, frequently shows not only the contusions associated with direct impact but also shows the petechial hemorrhages caused by the subsequent coup cavitation.

Experimental Findings

GROSS used asymmetrical glass flasks in an attempt to obtain a greater understanding of the coup-contrecoup cavitation phenomena, simulating the dimensions of the human head were fabricated for use in impact threshold experiments. The unique feature of these flask is the off-center location of the foramen [magnum] in the human skull.

GROSS subjected the asymetrical flasks to impact under two conditions: (1) with the neck up and the fluid partially up the neck, and (2) with the neck down and the flask completely filled with fluid, with a rubber diaphragm sealing the neck opening. The flasks were filled with distilled water

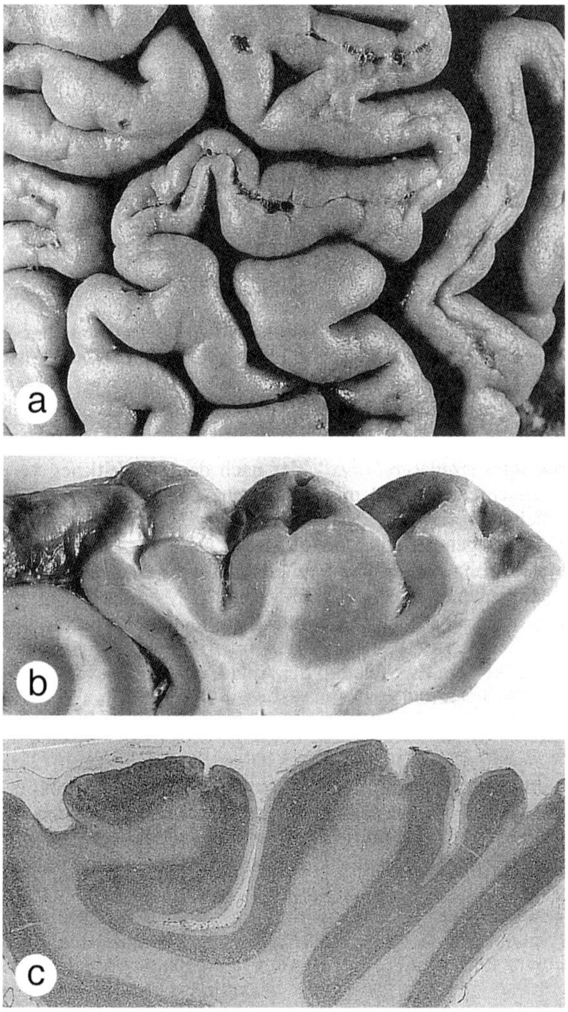

Fig. 47 a – c. Man. Closed head injury. a. Schizogyri at the summits of gyri, run parallel to it. Macrophoto. b. The cutting plane through the summits of the lobuli shows wedge- and trough-shaped defects of the cerebral cortex, which extend here and there into the subcortical white matter. Macrophoto. c. shows the wedge- and trough-shaped clear cut tissue defects in he cerebral cortex and subcortical substance. Accident 28 years ago, details unknown. Nissl technique, 25:1. (From UNTERHARNSCHEIDT, 1993).

without special treatment. Impact was produced with a rubber mallet.

GROSS obtained evidence of cavitation by the following means: (1) visual (high speed motion pictures); (2) audible (cavitation produces a sharp metallic 'clink'); and (3) accelerometer ring (flash vibration due to cavitation collapse).

GROSS attached an accelerometer to the flask at a point opposite the point of impact for the flask impact test. This installation not only enabled the determination of the maximum acceleration of the flask due to the impact, but gave a clear indication of the occurrence of cavitation. A typical acceleration record is shown in Fig. 54. The initial position of the trace shows the acceleration-time history of the flask resulting from impact with a rubber mallet. From this portion of the record (point A) maximum acceleration data is obtained. Immediately following this portion of the record there will or will not occur (point B) a high frequency burst, depending upon whether or not cavitation occurs. This indication is produced by high frequency vibration of the flask being excited by the collapse of the cavities formed.

Fig. 54 in GROSS' paper presents data obtained with the flask filled with mineral oil. Figs. 55 and b in GROSS's paper present the data obtained with the flask filled with distilled water. The difference between the g thresholds obtained is mainly due to the difference between the specific gravities of the two fluids.

GROSS continues:

"From the pevious contrecoup cavitation thresholds formula developed, we can now take the above data and compute the coefficient 'K' for the above test configurations."

$$K = \frac{Pt}{\gamma GL}$$

"For the experiments with mineral oil

$$K = \frac{14.7}{(0.0306)(6.40)(161)} = .465$$

$P_t = 14.7$
$L = 6.40$
$\gamma = .0306$"

"For the experiments with water

$$K = \frac{14.7}{(.0361)(6.50)(121)} = .517$$

$P_t = 14.7$
$L = 6.50$
$\gamma = .0361$"

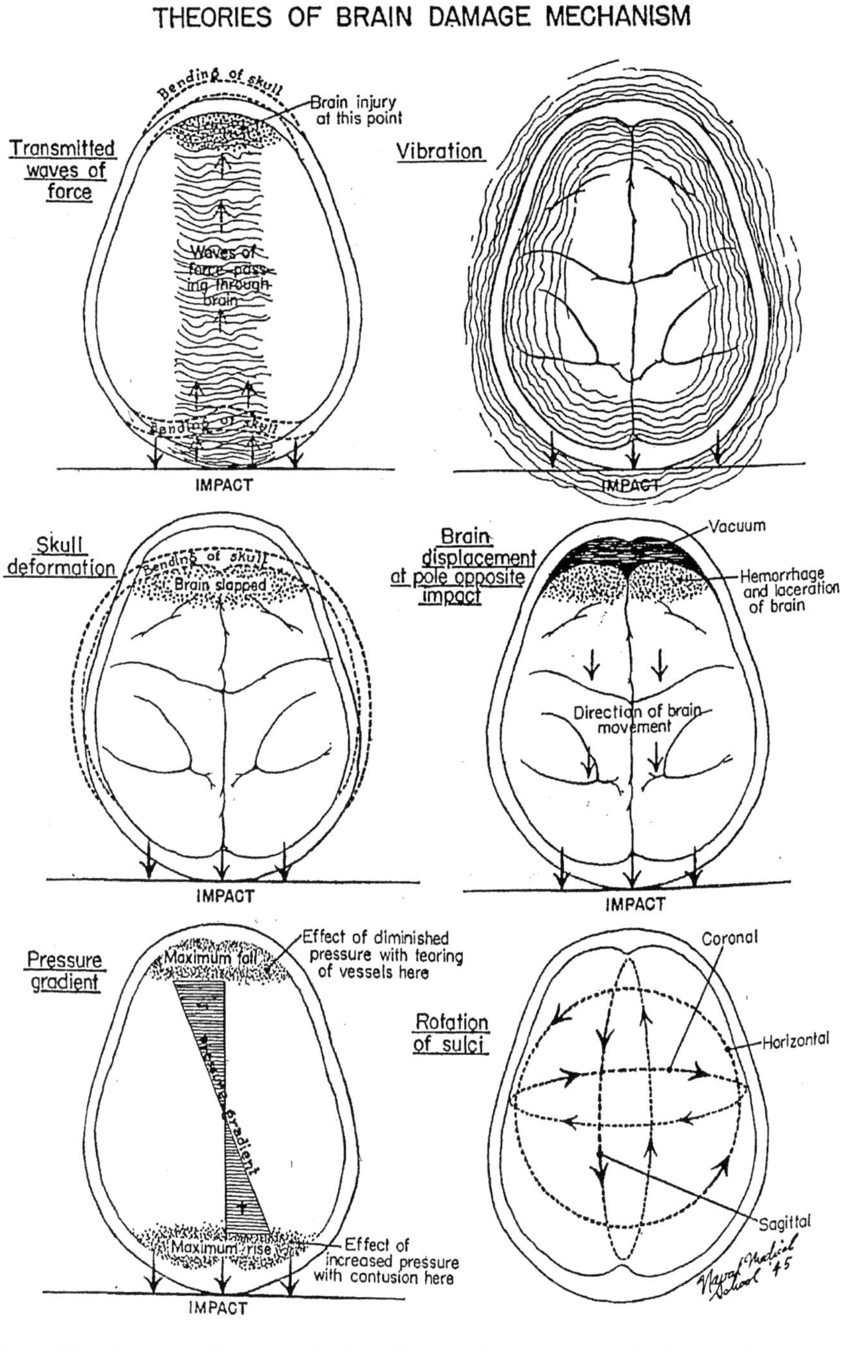

Fig. 48. Theories concerning the mechanism of coup and contrecoup cerebral damage. The skull deformation shown in the diagram representing the theory of transmitted waves of force has not been described in all of the writings on this theory but is included for the sake of simplicity. (From PUDENZ, SHELDEN, 1946).

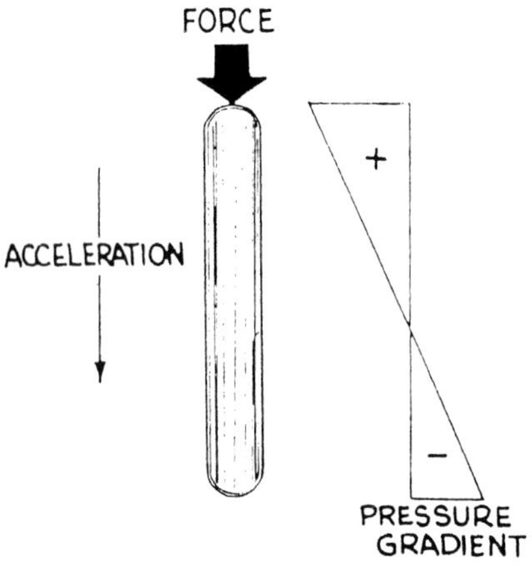

Fig. 49. Tensile stress gradient developed in the fluid contained in a closed tube because of acceleration produced by impact to one end. (From GROSS, 1958).

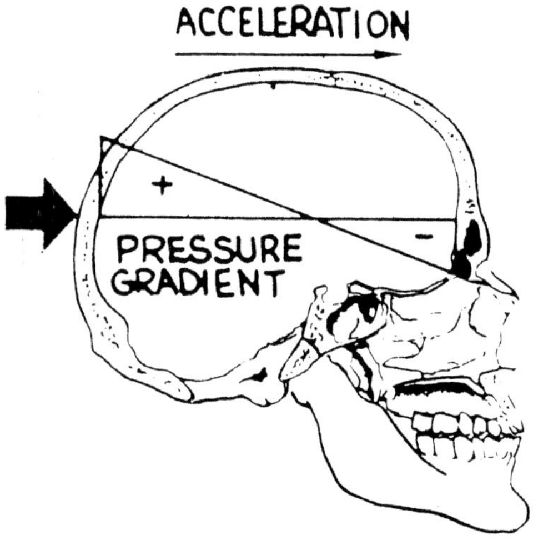

Fig. 50. Tensile stress gradient developed in the fluid of the brain because of acceleration produced by impact. (From GROSS, 1958).

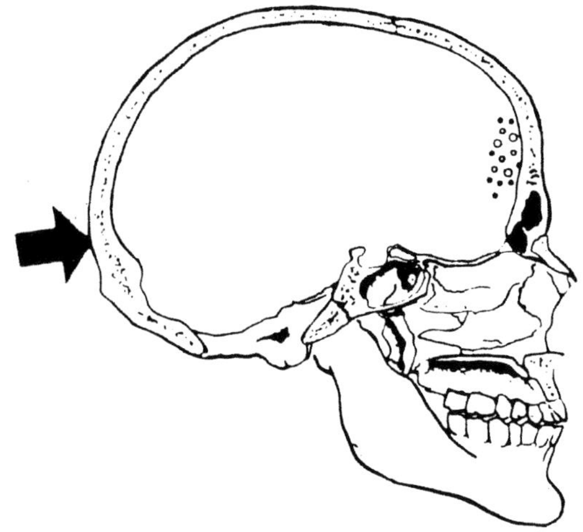

Fig. 51. Contrecoup cavitation occurring at a point opposite the impact because of tension failure of the fluid. (From GROSS, 1958).

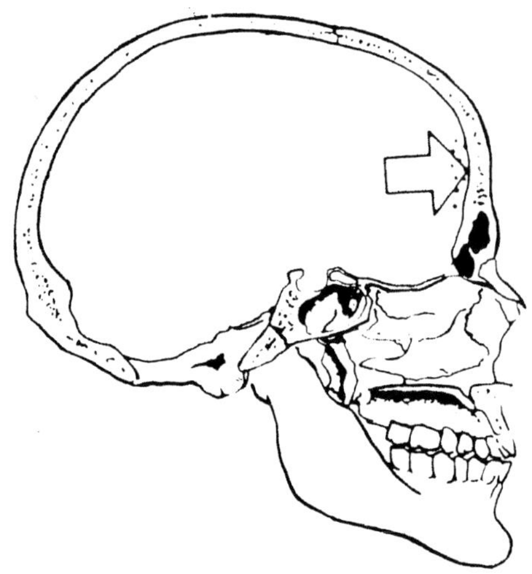

Fig. 52. Contrecoup brain injury of a focal nature resulting from the violent collapse of the cavities formed during the interval of acceleration. (From GROSS, 1958).

GROSS summarises:

"An analysis made of the resonance-cavitation theory of brain concussion, evaluating the various physical factors that relate to the impact threshold of brain concussion. A formula for computing the impact threshold for contrecoup cavitation requires the obtaining of 'K' factors by impact research on human cadavers. Instrumentation techniques were developed for use in obtaining threshold acceleration cavitation data on human cadavers. Impact experimetnts were conducted on glass simulated skulls to demonstrate the technique of determining 'K' factors."

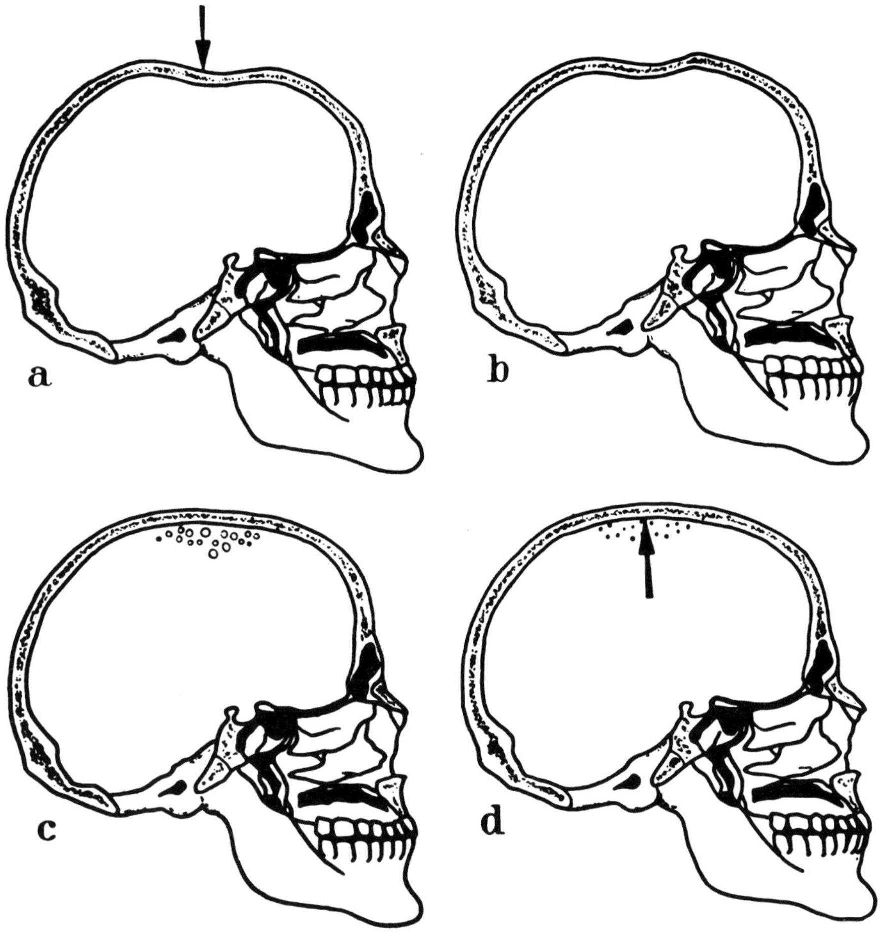

Fig. 53. Mechanism of coup cavitation. (a) Impact to top of skull causes local deformation of the skull inward. (b) Cessation of the impacting force leaves the skull locally deformed in a highly stressed condition. (c) Snap-back of the deformed skull to its normal position results in localized cavitation. (d) Collapse of the cavities formed produces the punctate contusions associated with coup injury. (From GROSS, 1958).

Shock Waves

In medical literature almost every pressure pulse propagating through body tissue is referred to as a 'shock wave.' Shock waves, as they are defined in mechanics, never occur in brain injuries because the short-strong impacts required to produce them do not occur and if they did the injuries produced would be so massive that death would result, practically instantaneously.

Central Traumatic Brain Lesions

The description of the pathomorphology of the traumatic hemorrhages near the ventricular system (subependymary and hemorrhages in inferior parts of the corpus callosum) must be introduced with some biomechanical remarks.

Compression of a shell causes a change of volume in the impact direction. Three logical variations are possible (Figs. 56 a – c):

(1) The large diameter *a is* in the impact direc-

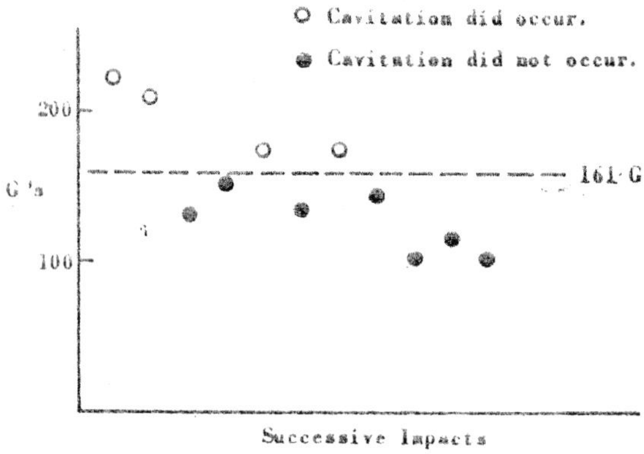

Fig. 54. Maximum g data for oil-filled asymmetrical flask. (From GROSS, 1958).

tion, (2) both diameters are equal, and (3) the small diameter *b* is identical with the impact axis.

In case (1) *a* becomes smaller, *b* becomes larger. The result is increased volume.

In case (2) the volume remains constant.

Case (3) produces a deformation in the direction of the small axis and volume is reduced.

Both effects, the deformation of the shell and the changes in volume have consequences for brain tissue. Examining case 1, it is evident that with the shortening of *a* an increase of *b* occurs. The brain tissue is compressed in the direction of the impact axis, with transversal dilation. Resulting movements cause considerable dislocation of tissues, especially in the middle area (Figs. 57 a and b).

The brain is not homogenous. Oblong, fluid filled cavities pass through it in the direction of axis *a*. These are the lateral ventricles. Since the lateral ventricles can only extend laterally when impacted in the direction of axis *a*, the volume increases. However, because there is insufficient cerebrospinal fluid to follow the movement in such a short time span, pressure in the ventricular space is reduced for an instant. Such movements and resulting pressure reduction may cause hemorrhages in the ventricular walls *(subependymary hemorrhages)*.

Isolated central hemorrhages are caused by a compression/expansion mechanism, which takes maximal effect around the center of the brain, with negligible effects in the periphery.

Case 2 will not be treated.

Case 3 is identical with an impact in the bitemporal axis. A decrease in the volume occurs, as previously discussed.

The *subependymary hemorrhages in the ventricle walls* cannot be explained by the 'normal' reduced pressure theory alone, where maximal effects occur on the surface of the brain. In this case, the effects of skull deformation must also be considered.

Traumatic lesions developing in *central parts of the brain* after a single sagittal blow to the head occur when the impact axis is identical with the sagittal diameter of the head. In this type of impact the bitemporal axis of the skull is lengthened and the

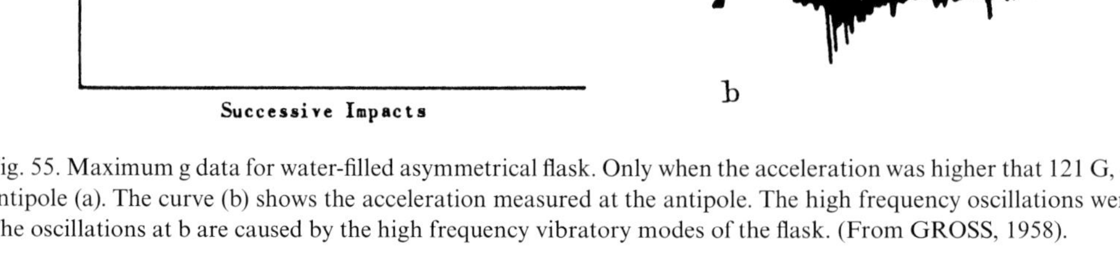

Fig. 55. Maximum g data for water-filled asymmetrical flask. Only when the acceleration was higher that 121 G, cavitation occurred at the antipole (a). The curve (b) shows the acceleration measured at the antipole. The high frequency oscillations were caused by the collapse. The oscillations at b are caused by the high frequency vibratory modes of the flask. (From GROSS, 1958).

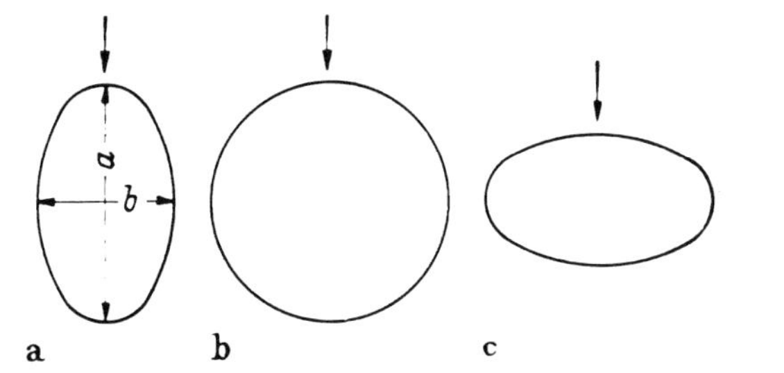

Fig. 56 a - c. Volume change of different skulls during a collision. a. Ellipsoid collision in the direction of the long axis: axis *a* will be shortened, *b* lenghtened, volume increased. b. Sphere deformed to ellipsoid : no volume change. c. Ellipsoid collision in the direction of the short axis, volume decrease. As an extreme example, consider a water-filled cup. (From SELLIER, UNTERHARNSCHEIDT, 1963).

ventricles are dilated. Since the time of impact is too short for the cerebrospinal fluid to flow in and fill the suddenly enlarged ventricular space, pressure reduction occurs in the ventricular system and cavitation may result, as in the contrecoup effect. We have referred to this phenomenon as been the *'central effect of cavitation'* (SELLIER and UNTERHARNSCHEIDT 1963). The ventricular walls (Fig. 58) and inferior portions of the corpus callosum located next to the ventricles are particu-

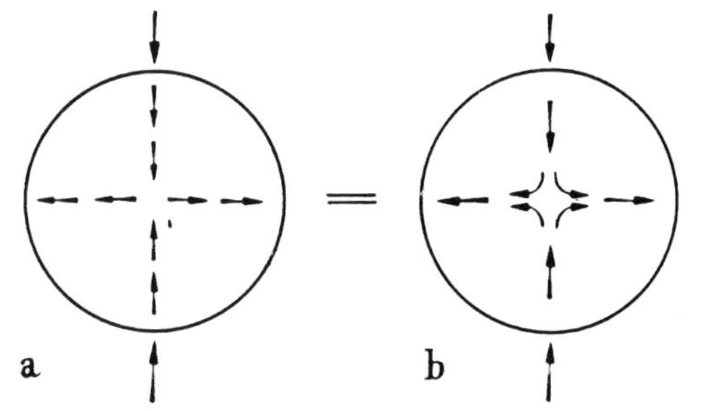

Figs. 57 a and b. Model for the possible stresses and strains in the brain substance following lateral deformation by collision or compression. Portions of the brain located along the collision axis *a* will move to the center (arrows); those parallel to a vertical axis will move away from the center (arrows). The resulting displacement is shown in *b*. Thus destructive shearing strains will occur in the center of the brain. (From SELLIER, UNTERHARNSCHEIDT, 1963).

larly subject to the effects of reduced pressure (Figs. 59 a – d). Thus, these anatomical structures are the principal sites of primary traumatic hemorrhages.

Rhectic hemorrhages have been observed mostly on the inferior surface of the corpus callosum. The ependymal hemorrhages are the result of cavitation, as previously described. Occasionally the

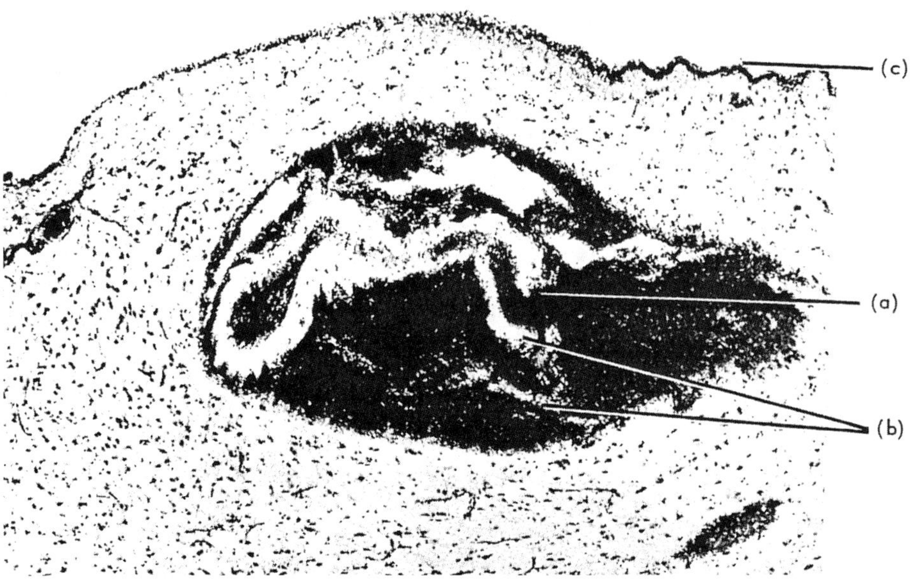

Fig. 58. Man. Cerebrum. Closed head injury. (a) Rupture of a venous wall in the subependymal white substance of the lateral ventricle, at (b) splintered vessel wall. (c) Intact spendymal cell layer. Masson-Goldner staining technique.

ependyma contains scattered erythrocytes (Fig. 60), or microhemorrhages erupt into the ventricular lumen and form a thin film of blood on the walls. Laminar splits and ruptures of vessel walls are also found in the basal ganglia and in extracerebral vessels (Figs. 61 a and b).

In cases of longer periods of survival, the typical sites of the primary traumatic, rhectic hemorrhages in the direct vicinity of the ventricles are not affected. Here the injury pattern is different. Rhectic hemorrhages are not only found in the walls of the lateral ventricles and inferior parts of the corpus callosum, but ruptures of the ependymal layer are also found. Parts of the ependyma with subependymal white matter can be detached from the tissue and can be seen free 'swimming' or 'floating' in the hemorrhages which invaded the ventricular lumen. The detached parts of brain tissue from areas near the ventricular wall are visible in hemorrhages in the ventricles, completely surrounded by them. These pictures are not rare. In the literature these lesions have hardly been given notice.

General Aspects of the Biomechanics of Boxing

The mechanics of impact in boxing are treated comprehensively by SELLIER and UNTERHARNSCHEIDT (1963), UNTERHARNSCHEIDT (1963, 1975, 1985), and UNTERHARNSCHEIDT and SELLIER (1966, 1970, 1971), UNTERHARNSCHEIDT and RIPPERGER (1970). and ATHA et al. (1985).

The force of a blow to the head is related to the velocity of the fist and the masses involved. It further depends on the mechanical properties of the colliding bodies, namely glove and head. On impact, both the gloved fist and the head experience force; the head is accelerated and the fist is decelerated. During contact, the force of the impact deforms both head and fist.

A better knowledge of the process of collision between two given bodies with elastic surfaces may be obtained using the following analogous model: the masses of the colliding partners are represented by two rigid spheres of corresponding mass, connected by a spring. The spring is chosen to possess the stiffness that is required to achieve a realistic collision time. The stiffness c of a spring is expressed in kp/cm, indicating the weight (kp) required to compress the spring by 1 cm: the higher the value of c, the stiffer the spring. Often several springs participate in the collision. For example, in a blow dealt with the gloved fist, one would count the effects of the compressible lining of the glove, the tissue overlying the bony skull, the compressibility of the wrist and possibly that of the elbow, as behaving in a spring-like manner. With regard to our problem, the compressibility of the glove is so much greater than that of all other elastic elements in the system, such as the skin and bones, that it can be ignored.

The accurate assessment of the striking mass is also a problem, since it is composed of the masses of the glove, hand, forearm and upper arm, and to a certain degree, that of the thorax. Moreover, the total mass involved in a blow differs not only with the individual but also with technique, depending on how much body weight is thrown into the blow.

Measurement of Head Acceleration in Boxers

Structural damage to the human brain in boxing can be the result of *two main causes*: (1) *The direct impact of a single blow*, and (2) the *repeated* and *cumulated blows* extending over an entire boxing match of three rounds in amateurs and 12 to 15 rounds in professionals.

Model experiments, carried out by SCHMID (1970), using 8, 10, and 12 oz. gloves, reduced blow intensity by 40%, in comparison with blows by the naked fist.

Measurements of the physical properties of a boxer's punch were carried out by SCHMID (1970), BAGREEV and TRAHIMOVITCH (1981), JOCH et al. (1981), SELLIER (1971), UNTERHARNSCHEIDT and SELLIER (1965, 1967, 1970, 1971), ATHA (1985), UNTERHARNSCHEIDT (1992).

ATHA et al. (1985) designed an instrumented padded target mass suspended as a ballistic pendulum to determine the mechanical properties of a boxing punch. Data were gathered from a world ranked British professional heavyweight, Frank BRUNO, who delivered a blow to the above-mentioned pendulum. The target was a cylindrical metal mass of 7 kg (16 lb.), that is roughly equivalent to the head and associated neck segments of a heavyweight opponent. The punches were filmed by ATHA et al., in color, using a high quality news film or in black and white. A 16 mm, variable shutter Bolex camera was used, at a nominal 64 Hz as well as a Hycam rotating prism camera operating at either 400 or 1.500 Hz. A millisecond timer and a fiducial system were included in the field of view, to permit the timing and scaling of events. Within 0.1 s of the start, the punch had traveled 0.49 m and attained a velocity of 8.9 m/s on impact. The peak force was 4096 N (0.4 ton) on impact and was attained within 14 ms of contact. This represents a blow to the human head of up to 6320 N (0.63 ton). The impulse accelerated the target head at the rate of 520 m/s 2 (53 g). For comparison, a padded wooden mallet with a mass of 6 kg (13 lbs.), swung at 20 mph, would deliver an equivalent blow.

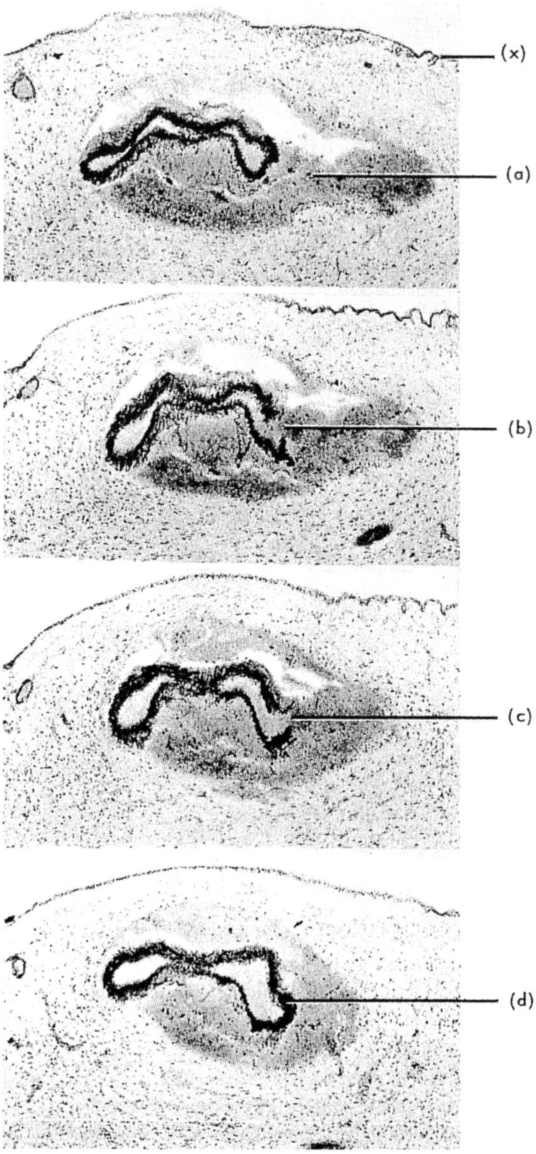

Fig. 59 a - d. Man. Cerebrum. Closed head injury. Serial sections of a subependymal venous hemorrhage has caused a considerable reduction of the vessel lumen; the vessel wall was left intact. (b) Broad rupture of the vessel wall, with lamellar splintering of the wall. (c) The tear is less broad. (d) The vessel wall is no longer torn, but the henmorrhage extends further along the vessel wall and reduces the volumen. Van Gieson staining technique.

Table 2 shows the results of the analysis.

ATHA et al. drawing attention to the shortness of the time taken to deliver the punch from first ex-

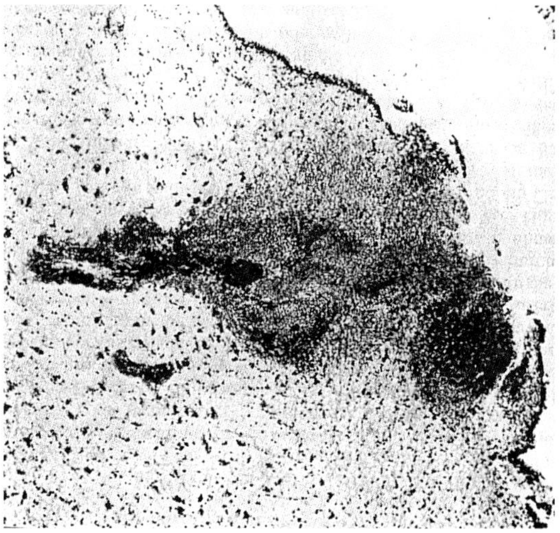

Fig. 60. Man. Cerebrum. Closed head injury. (a) Ruptured vein. (b) Eruption of the ependymal hemorrhage through an ependymal defect into the ventricle system. (c) Intact ependyma. Van Gieson staining technique.

tension of elbow to peak contact. The authors noted:

"If basic visual reaction time is taken as 140 ms the chances of an opponent dodging a blow he sees coming are clearly slim. His protection will instead depend primarily on the adoption of successful random evasion strategies or anticipation based on earlier subtler cues than those associated with the start of punch delivery."

The peak force recorded by ATHA et al. was the equivalent of a blow to the unpadded human face ranging up to 0.63 tons. This peak force was in excess of that required to fracture facial bones (NEWMAN 1988).

ATHA et al. continued:

"Of course the duration of the contact was brief, but it still lasted long enough to generate an impulse of lifting a 100 kg opponent some 8 mm off the ground and accelerating the target head at 520 m/s^2 (53 g), well beyond safety margins quoted by JOHNSON et al. (1983)."

ROY et al. (1984) examined the influence of bandages on the strength of impact of punches in boxing. In boxing, a punch can be directed to the head or the torso of the opponent; however, because of the gravity of the ensuing injuries, the limit of toler-

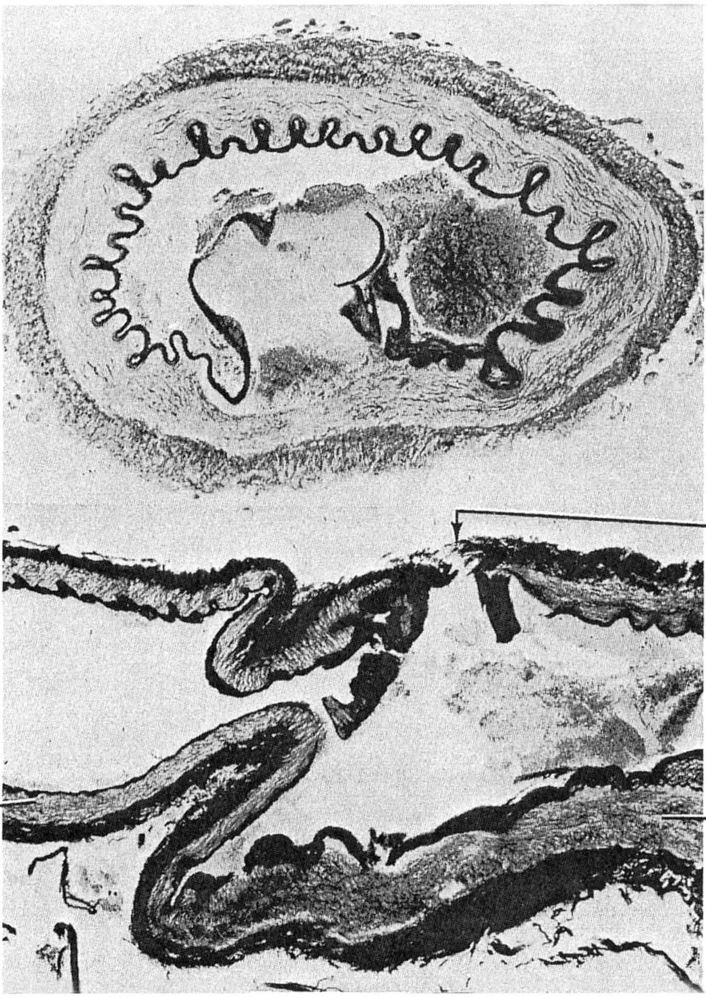

Figs. 61 a and b. Man. Primary traumatic longitudinal and transversal tears of the basilar artery. (a) Frontal collision impact, type II a, according to Spatz). Immediate death. Fractures of the calvarium and cranial base, fracture of the fifth thoracic vertebra, so-called cortical contusions, primary traumatic brainstem hemorrhages. Longitudinal tear of elastic bundle of basilar artery, with dissecting hemorrhage in the vessel wall. Van Gieson Elastica staining technique. (b) Pedestrian. Frontal collision with automobile going at 60 km/h. He is thrown on the hood which he struck with the upper part of the frontal bone, while the cervical spine was hyperextended (impact type II b, according to Spatz). Immediate death. Fracture of the base of the skull, fracture of the third cervical vertebra, beside other fractures. Subarachnoid hemorrhages, so-called cortical contusions, primary traumatic hemorrhages in the basal genglia and brainstem. Transversal tear of the basilar artery caused by tearing off arteria circumferens longa. Van Gieson Elastica staining technique.

ance of the head and face to repeated impacts is particularly important to keep in mind. The objective of the author's research was to compare the impact forces, which could be attained barehanded versus those obtained with different types of bandages, made up of gauze and diachylon.

For the purpose of this study, 22 boxers were selected. The impact force was measured with a Kistler force plate, covered with a synthetic mattress. A system of photoelectric cells was used to measure the velocity of the punch. Each boxer had to impact the target with ten punches while his hand was covered with one of four types of bandages as well as barehanded. This study has shown that the bandaged hand significantly increases the impact force of a punch. The force increases in relation to the thickness of the bandage. The use of additional pieces of diachylon has a greater influence than gauze. Therefore, it can be concluded that bandaging increases the rigidity of the hand and facilitates the transfer of the force in comparison to a similar punch without a bandage or with a thinner one.

Relationship between Weight of Boxing Gloves and Severity of Impact

When attempting to quantitatively assess the violence exerted on the head during boxing matches, the properties of boxing gloves (of 6 to 16 oz) and of various types of ring canvas were examined with regard to their influence on the impact. An 8 oz glove of 228 g consists of 114 g leather and 114 g padding.

UNTERHARNSCHEIDT and SELLIER (1970) employed a simple experimental device which consisted of two pendulums representing the striking fist and the unrestrained head. The first pendulum was a metal tube carrying at its free end a wooden hand covered with the test glove. The second pendulum carried a wooden block with plain surface; the side facing the gloved pendulum, was covered by a piece of wooden fabric about 1.5 mm thick, and the opposite side carried an accelerometer (Fig. 62). Naturally, the magnitude and the velocity of the impacting masses are important factors in the successful simulation of real fight conditions. One of the author's first concerns, however, was to obtain comparable values of the violence generated with boxing gloves of different weights.

TABLE 2
Some properties of a professional boxer's punch. (From ATHA et al., 1985).

Variable	Value
Duration of trunk advance before start	201 ms
Velocity of trunk and fist before start	1.0 m/s (2.2 mph)
Time taken from start of punch to impact	100 ms
Distance punch travels from start to impact	490 mm
Distance of follow through	110 mm
Peak acceleration of fist before contact	90 m/s^2
Velocity of fist at impact	8.9 m/s (20 mph)
Residual velocity after impact	1.8 m/s
Rise time of force to peak	14 ms
Rise time of acceleration to peak	14 ms
Peak contact force	4096 N (0.4 ton)
Estimated maximum force to human head	6320 N (0.63 ton)
Impulse transmitted	41 Ns
Acceleration induced in head mass model	520 m/s^2 (53 g)

1 N = 0.255 lb; 1 m/s = 2.24 mph; 1 g = 9.81 m/s^2

To measure the intensity of the blows in an actual bout, two physical education students, unskilled in boxing, fought for ten minutes using 12 oz. gloves, with accelerometers bandaged to their heads. The measurements obtained indicated that 21 blows accelerated the head by 0 to 5 g, 12 blows by 6-10 g, 3 blows by 11-15 g, 3 blows by 16-20 g, and 2 blows by 21-25 g. The first group of 21 blows included some defensive movements of the head that had the same effect as blows (Fig. 63). The results correspond to those obtained for 16 oz. gloves in the model experiments. If the students had worn the 6 oz. gloves commonly used in professional contests, these values probably would have been at least twice as high. Had the subjects been professional boxers, capable of delivering more effective blows, the values would certainly have been higher. If this many blows hit the head in such a 10-minute fight, the equivalent of little more than 3 rounds, one must assume that the number of blows in a professional bout of 12 to 15 rounds would be at least several times greater. (UNTERHARNSCHEIDT and SELLIER 1970).

The velocity of the glove achieved with a glove of 6 oz was found to be about 2.7 times greater than that achieved with a 16 oz glove. Strangely, enough, two supposedly identical gloves of 8 oz produced different acceleration values that were well outside our other experimental data. Repeated measurements showed the same deviation. Further, it was observed that on the average, higher accelerations were registered from the last blow of a series of three or four consecutive blows. This indicated that the material slightly 'faded' or retained its deformation, an effect that may be observed in regular fights (UNTERHARNSCHEIDT and SELLIER 1970).

The impressive intensity of the boxing blow is

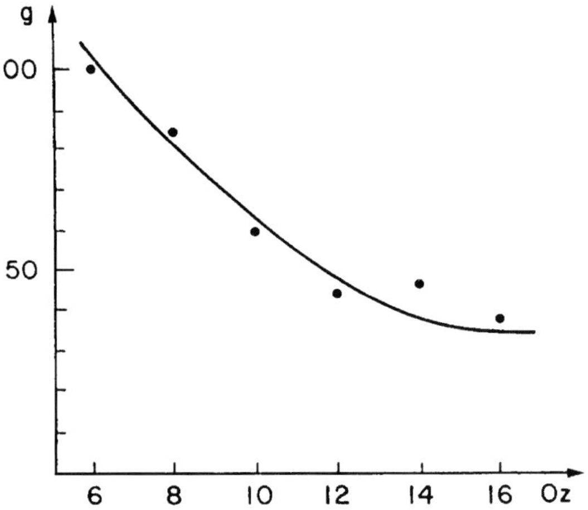

Fig. 62. This table shows the measurements of acceleration using boxing gloves of different weights (ordinate, acceleration expressed in g; abscissa: ounces of gloves). In serial testing, the gloves' material showed 'fatigue' or 'fading' of its elastic qualities, and consequently the acceleration measurements rose. (From UNTERHARNSCHEIDT, SELLIER, 1973).

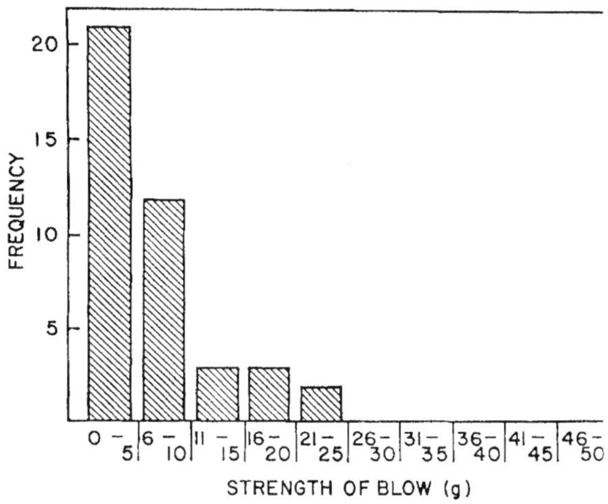

Fig. 63. Statistical data on the frequency of blows of a given strength, in 5-g increments. (From UNTERHARNSCHEIDT and SELLIER, 1971).

not only recorded in the measurements of damage (both clinical and morphological) to boxers but is directly reflected in the photographs of the moment a boxer is struck. The tremendous force of the blow deforms and displaces the soft parts of the face, deforming the bony structures too. The selection of press photographs demonstrates the effects of deformation, acceleration and inertia on facial features, which, at times, are hardly recognizable. It must be obvious that blows of such distorting power do indeed damage the brain. Although it has been argued that like effects are only seen among professional boxers, the blows dealt by amateurs are generally damaging enough to produce similar results. Blows of such intensity as those depicted are by no means rare events, as a random search through the illustrative material of press archives will prove.

Influence of Headguards and Canvas

The role of headguards and bandaging in the decrease of the violence can be readily assessed (Fig. 64). The influence of the head guard with its rather large elastic constant is small; then the utilized glove is soft (higher number of ounces) and, consequently, its elastic constant small. The effectiveness of the head protection grows in relation to the progressing rigidity of the glove, i. e., with the utilization of lighter gloves. This rule applies also to bandaging and the like.

To achieve optimal head protection in a fall, the head must be decelerated by the canvas over the longest possible distance; the violence of the impact is reduced as the stopping distance is gradually increased (Fig. 65). The medical requirement, however, cannot find practical application, because a canvas of more than a certain thickness and compressibility would impede the boxer's agility. With every step the fighter would sink into the canvas and before long would be fatigued without having performed, simply by 'being on his toes.' Furthermore, if the thickness of the canvas is limited for practical reasons, the softest canvas is not also the most suitable one, as will be shown. The head should be fully decelerated and the fall energy completely absorbed when the point of maximum compression of the canvas is reached. A highly compressible canvas, however, decelerates the head but little, with the result that at the point of maximum compression of the canvas, the head still possesses considerable fall energy. This energy is fi-

Fig. 64 Tamy Mauriello (left), the Bronx fighter who meets British boxer Bruce Woodcock in Madison Square Garden, New York, on May 17, 1946, has a sparring bout with Cleo Everett in his training quarters. Mauriello usually trains without a headguard, but during the last week he wears one to prevent any injuries which could possibly bring about cancellation of the fight. The photo shows two types of head gears; the left head gear should protect additionally nose and chin. In training soft gloves are used, so that an attenuation of the intensity of the boxing blows is barely to expect, they are used above all to protect the face from lacerated wounds. Photo: U.P.I.

nally abolished over a very short stopping distance by the hard floor underneath the canvas (concrete, wood), with the effect that the head sustains considerable force in the impact. On the other hand, if the canvas is rather hard the head is stopped before it has traversed the whole energy-absorbing thickness of the canvas. The stopping distance then is shorter than necessary.

The compressibility of the ideal canvas of a given practicable thickness is such that the head is stopped when the maximum compressibilty of the canvas is reached. In this case, the forces acting on the head are minimal. It follows that the efficiency of the canvas is related to the fall energy of the head, that is, to the height of the fall. The quality of the canvas, therefore, must be chosen with respect to the average fall height. This amounts to considerable less that the average body size, since a stiff fall of the upright body may generally be considered beyond probability.

The authors' experiments confirmed their view that highly compressible material, like latex foam rubber, was inadequate to the purpose because it produced high deceleration values, whereas felt

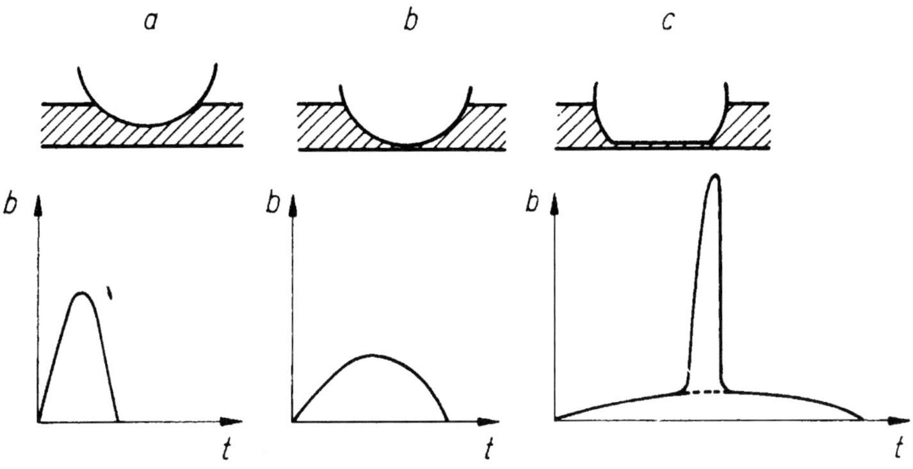

Fig. 65. Schematic presentation of the influence of the ring canvas (hatched, top) on the impact of the head. The acceleration curves are below. (*a*) Canvas too hard. The head is rapidly decelerated, reaching a relatively high peak deceleration. (*b*) Canvast right. The head is completely decelerated if the canvas is maximally compressed, producing relatively little deceleration. (*c*) Canvas too soft. The head is not decelerated by the canvas layer, and impacts against the hard foundation produce extremely high deceleration. (From UNTERHARNSCHEIDT, SELLIER, 1971).

proved to be very suitable because it possessed the most adequate rigidity.

LÖFGREN (1967) commented on the canvas of the boxing ring: "In rule 1.c) of the 1955 edition of *Articles of Association and Rules for International Meetings, European Championships and International Tournaments of the Association Internationale de Boxe Amateur (A.I.B.A.)*, it is stated: "The floor shall be covered with felt not less than half an inch (1.5 cm) and not more than three-quarters of an inch thick (1.9 cm)." To this rule, regarding the material was added in the 1958 edition:

"... rubber or other suitable approved material having the same quality of elasticity. These rules have been introduced since severe injuries to the skull and brain in boxing seem to occur mostly when the boxer falls backwards from a blow to the head. The addition concerning the material was caused by the fact that new materials, such as sorbo rubber, are being used in many places since felt was insufficient to lessen the effect of blows to the head in falls. This fact had been established wholly empirically. From a medical traumatological point of view opinions may vary as to the suitability of the regulation concerning the thickness of the undercover ("... not less than half an inch (1.5 cm) and not more than three-quarters of an inch (1.9 cm)." Rather than stating the metric thickness of under-covers of varying quality with the sole function of affording the most adequate protection against injuries to the head, in my opinion, a rule should be accepted stating that the under-cover must be of a kind which does not obstruct free foot work. It seems evident that an under-cover with a thickness of 1.5-1.9 cm, made from some "... suitable approved material having the same quality of elasticity..." does not, or at least not without some kind of special construction, offer anything like the efficient protection aimed at.

LÖFGREN carried out experiments on trauma-preventing under-covers for the floor of boxing rings. "The oscillographic examination of its different constructions of foam, plastic and wallboard compared with felt as an under-cover for the floor in the boxing ring proved that the first lessened the force of the impact by about 3/4 - 3/5. In this respect the felt cover was hardly better than a floor with vinyl resin covering. These tests show that an efficient protection against severe brain and skull injuries of a boxer when he falls backwards from a blow to the head seems possible. It is therefore our duty to make common use of the most efficient under-covers."

Pneumatic Boxing Gloves

In 1967, LÖFGREN introduced a *pneumatic boxing glove,* of which he wrote:

"It is not perfect and does not fulfill all demands which can be set on an ideal glove. Yet it performs its preventive task surprisingly well and because of these great medical advantages its disadvantages weigh very little in the scales."

According to LÖFGREN "The denomination pneumatic glove reveals that the usual horsehair, straw, or fiber padding of the standard glove has been substituted by a specially constructed filling which is filled with air through a tube coupled to the padding. The pressure in the glove may be read by means of a manometer. This is not necessary, however, as the pressure in the glove, within a wide scope of variation, and already at low pressure, reduces the accelerating power of the blow to such an extent that the pressure may be adjusted to that found most suitable by the boxer himself. These statements are illustrated by the graph following further on in the text. The impact of the blow given with the pneumatic glove is not so hard on the combatant as with the standard boxing glove. Moreover, it does not lose its firmness and grow limp. The padding does not get raveled, nor does it shift out of place, as does the filling of the usual type of boxing glove. Consequently, the glove retains all its good qualities even after use over a long period. Small boxing clubs with limited financial resources would thus be able to provide their members with first rate gloves and the use of defective ones avoided. The price of the pneumatic glove exceeds that of the standard glove by about one-fifth only. The pneumatic glove is somewhat heavier than the one commonly used in the European countries – which is another advantage, to my mind. The weight of the new glove is 13-16 oz., while the standard glove does not usually exceed 12 oz." The standard boxing glove as required by the rules usually weighs 8 oz.

The tests were carried out, according to LÖFGREN, as follows: By means of a pendulum movement a constant velocity was given to the glove which was fitted onto a wooden model of a fist. At this velocity the glove met a mass, also fixed onto a pendulum to which an induction vibration indicator was attached. Two series of test were carried out:

(a) In the first series the mass of the fist pendulum was about 20 kilograms and its velocity about 12 meters per second.

(b) In the second series, the mass was about 6 kilograms and the corresponding velocity about meters per second. The electric current excited in the indicator by the blow was led into an oscilloscope. The extent of the acceleration given to the indicator by the blow could be judged from the coefficient of the angle of the rising part of the curve shown on the screen of the oscilloscope.

In all measurements only the relative acceleration was observed. The acceleration values obtained in cases (a) and (b), by delivering blows with a standard glove, were considered as a basis for comparison.

Thumbless or Thumb-restrained Boxing Gloves

A new boxing glove developed by Jhoon RHEE (a karate expert) has among other features *no separated thumb* (1982). The application on file at the U.S. Patent and Trademark Offices describes, according to KOENIG (1982) a spongy synthetic shell shaped like a gourd and laced at the tapered end. RHEE's goal in the construction of the new boxing glove was to eliminate the protrusion of the thumb that can injure the eye. The boxer using the new glove forms a fist by gripping a flexible crossbar inside its simple chamber. The leather of the conventional boxing glove is replaced by vinyl, the horsehair by foam rubber. KOENIG described the appearance of the new thumbless glove: "It looks like something dropped prematurely off of a rubber-boot production line." The new type of thumbless boxing glove is developed to eliminate the danger of 'thumbing' of the eyes.

According to KOENIG, RHEE has made one concession to tradition: the laces on his gloves. Eventually, however, he plans to use Velcro fasteners that won't tear across facial cuts. 'Thumbing' of eyes is mainly blamed for a detachment of the retina. It has yet to be seen whether retinal detachment is the only or main result of 'thumbing' the eyes; a heavy blow against the eye region or the head properly applied without any 'thumbing' may also produce retinal detachment.

A thumbless or thumb-restrained boxing glove does not allow the thumb to extend into the 'hitchhiker's position.'

Discussion of Some Ideas and Terms used in Boxing

The term *'ring'* is used for the roped square in which a boxing match takes place. A roped ring was first mentioned in the *London Prize Ring Rules of 1838.* During the time of 'bare knuckle fist fights' the rings were 20 to 24 feet square.

Southpaw

A *'southpaw'* is a boxer with a stance in which he keeps his right arm and right foot forward. His left hand is generally the better one. Orthodox opponents of southpaws are often confused when they fight against a southpaw; they have to move to the right instead to the left to find an opening.

To *'throw a fight'* or a *'fixed fight'*, means that a pugilist loses a fight on purpose. A well known 'fixed fight' was that of Bill STEVENS, who was beaten in 17 minutes by George MEGGS in March 1761. There was a rumor that Jack SLACK had given him £ 50 to lose his combat against

MEGGS. There exists a slang expression, a boxer 'takes a dive.' Bill DARTS had been bribed with £ 100 to lose his fight against Peter CORCORAN in May 1771.

'Fouls' in boxing, either in amateur or professional boxing, include blows beneath the belt, on the back of the head and nape of the neck ('rabbit punches'), kidney blows, hitting with an open glove, the wrist or elbow. A boxer may not duck below his waistline. A 'butt' with the head is considered a foul. 'Butts' can be produced by boxers, who intentionally ignore the rules, or they can occur accidentally. It is considered a foul to hit a boxer who is down or on his knees.

So-called 'Long Counts:' There are some important boxing matches known, in which a knocked down boxer had been allowed more that 10 seconds to rise.

The most well-known of these *'long counts' took* place in the second heavyweight fight between Gene TUNNEY and Jack DEMPSEY. TUNNEY was leading all the rounds until the seventh when DEMPSEY drove TUNNEY against the ropes and knocked him down with two hooks to the jaw. DEMPSEY thought that this was the end of the fight. But DEMPSEY remained close to TUNNEY, more or less standing above him and hesitated to move into a neutral corner. The count had begun, but the referee ignored it until DEMPSEY had moved into a neutral corner. The referee then started his count, in compliance with the rules. TUNNEY got up at '9' and fought defensively to recover for the remainder of the round.

In the opinion of the official timekeeper TUNNEY had been on the canvas for 14 seconds, but there were eyewitnesses who said it was more than 15 seconds. There are two possible versions of what took place. Many had doubts, that TUNNEY would have been able to get up before the count of '10'. On the other hand, one could argue that TUNNEY deliberately stayed down on the canvas as long as possible, taking advantage of the time.

Another kind of *'long count'* occurred in the heavyweight fight between Jack DEMPSEY and Luis Angel FIRPO. After FIRPO had knocked and pushed him through the ropes in the first round, there were eyewitnesses who were convinced that DEMPSEY was out of the ring for more than ten seconds.

The Longest Boxing Matches with and without Gloves

The *longest boxing match with gloves* was fought at the *Olympic Club* in *New Orleans on* April 6, 1893 between Andy BOWEN and Jack BURKE. This boxing match was declared 'no contest' after 110 rounds extending over 7 hours 19 minutes.

Some of the longest bouts on record are, according to GOLESWORTHY (1960) (G = gloves, K = knuckles):

6 hrs. 40 min. Dan NEEDHAM v. Patsy KERRIGAN, San Francisco, February 27, 1890, G.

6 hrs. 15 min. James KELLY v. Jonathan SMITH, Melbourne, Australia, October 19, 1856, K.

6 hrs. 3 min. Mike MADDEN v. Bill HAYS, Edenbridge, Kent, July 17, 1849, K.

5 hrs. 45 min. Mike MADDEN v. Jack GRANT, Woking, Surrey, December 12, 1848, K.

5 hrs. 8 min. Harry SHARPE v. Frank CROSBY, Nameski, Ilinois, February 2, 1892, G.

5 hrs. 3 min. William SHERIFF v. Jack WELSH, Philadelphia, April 4, 1884, K

4 hrs. 30 min. Jack JONES v. Patsy TUNNEY, Cheshire, 1825, K.

4 hrs. 20 min. Jack FITZPATRICK v. James O'NEAL, Berwick, Maine, December 4, 1860, K.

3 hrs. 49 min. Michael DONOVAN v. W. C. McCLELLAN, San Francisco, August 18, 1879, K.

3 hrs. 40 min. Jem MACE v. Joe COBURN, St. Louis, November 30, 1871, K.

3 hrs. 30 min. Benjamin BRAIN v. William HOOPER, Newery, Berks, August 30, 1790, K.

3 hrs. 20 min. Ned PRICE v. Joe COBURN, Stoneham, Mass., May 1, 1856, K.

3 hrs. 16 min. James 'Deaf' BURKE v. Simon BYRNE, St. Albans, Herts, May 30, 1833, K. longest championship fight ever.

3 hrs. 12 min. Joe JEANNETTE v. Sammy McVEY, Paris, Apl., April 17, 1909.

3 hrs. 10 min. J. L. SULLIVAN v. Charlie MITCHELL, Chantilly, France, March 10, 1888, K.

3 hrs. 8 min. Tom SAYERS v. Harry PAULSON, Appledore, Kent, January 29, 1856, K.

3 hrs. 2 min. Aaron JONES v. Tom SAYERS, Canvey Island, June 26, 1857, K.

3 hrs. James 'Deaf' BURKE v. Bill FITZMAURICE, Harpenden Common, Herts, June 9, 1829, K.

So-called "No Decision Bouts"

Before the *FRAWLEY Law* became effective in the State of New York in 1911, boxing could be carried out only in clubs and witnessed by members, but they were considered exhibition bouts in which no decisions were announced. No pugilist could win such a contest unless he was able to knock out his opponent. In this time newspaper reporters decided by a majority vote a 'popular decision,' but this was not generally accepted. The FRAWLEY Law was followed in 1920 by the *WALKER Law.*

The so-called Second Impact Syndrome

CANTU and VOY (1995) discuss the *'second impact syndrome'* which occurs when an athlete who has sustained an initial head injury, most often a concussion, sustains a second head injury before symptoms associated with the first have fully recovered. These authors reported six cases, of which five were boxers.

The first case, a high school hockey player will not be quoted here.

Case 2: A 17 year old Golden Glove boxer fought a difficult semifinal bout in a regional tournament. During the two day rest period before the final bout, he complained of a severe headache; his mother stated he had literally 'eaten' two bottles of aspirin. On the morning of the final bout he was in an automobile accident and struck his head forcefully. He felt nauseated on the way to the finals bout that evening, and his coach twice had to pull the car over to the side of the road so he boxer could vomit. During the pre-competition examination the physician asked if there were any problems: The boy and his coach answered 'none.' After an uneventful first round and minor head blows during the second round, the boxer collapsed in the ring. Shortly thereafter he became decerebrate with fixed dilated pupils.
He was intubated and taken to the hospital where, after he received intravenous mannitol and dexamethasone, his condition remained unchanged. Emergency computed tomography (CT) scan showed massive brain swelling and a small left frontal subdural hematoma. After conferring with his parents, surgeons performed an emergency craniotomy. After removing the subdural hematoma, surgeons observed massive brain herniation out of the craniotomy defect and performed a decompressive cranioectomy. Despite intensive treatment, the boxer's condition did not improve. Two postcraniotomy electroencephalograms were flat, and the patient died of cardiac arrest one week after surgery.

This unfortunate young man had already participated in a 'difficult semifinal bout' in a regional Golden Gloves tournament. During the two day rest period before the final bout, he complained about severe headache, his mother stated that he had literally 'eaten' two bottles of aspirin. There is no doubt that this boxer suffered from the sequelae of a subdural hemorrhage, more precisely of an acute subdural hemorrhage. Immediate medical treatment should have been initiated and the patient admitted to a hospital and the acute subdural hemorrhage removed.

On the morning of the final bout the patient was in an automobile accident, he struck his head forcefully. He felt nauseated on the way to the final bout that evening and his coach had to pull the car over to the side of the road so the patient could vomit. During the pre competition examination the physician asked of there were any 'problems' the boy and the coach answered 'none.' After an uneventful first round and minor head blows during the second round, the boxer collapsed in the ring. Shortly thereafter he became decerebrate with fixed dilated pupils.

The blows are described as 'minor,' but one must take into consideration that punches were directed at a boxer with a pre-existing acute subdural hemorrhage. Boxing blows against a head with an acute subdural hemorrhage must have a more devastating effect than usual. Indeed, in this case they were certainly devastating.

This story is incomprehensible and staggering. Already after the first bout, which was described as 'difficult,' the boxer complained about a severe headache. These headaches must have been severe, because his mother made the very unusual statement, that he had literally "eaten two bottles of aspirin."

As if this weren't enough, on the morning of the final bout the patient was involved in a car accident and struck his head forcefully. He felt nauseated on the way to the final bout that evening and the coach had to pull the car over to the side of the road so the boxer could vomit. The impact suffered in this car accident probably intensified the hemorrhage. Here again, a physical examination should have revealed the existence of a subdural hemorrhage. The size of the pupils is not mentioned.

The boxer collapsed after receiving only minor blows during the second round and became decerebrate shortly thereafter, with fixed dilated pupils. The 'minor' blows landed on the head of a boxer who had passed a physical and entered the ring with an acute subdural hemorrhage. A CT scan in the hospital revealed a small frontal subdural hemorrhage and massive brain swelling. The subdural hemorrhage was described here as a subdural hematoma, or space-occupying lesion. The surgeons performed an emergency craniotomy, removed the hematoma and observed a massive herniation of the cerebrum. The postoperative EEG's were flat and the boxer died of a cardiac arrest one week after surgery. The pathomorphological findings indicate without doubt, that the traumatic lesions in the brain were not recent, but the result of the boxing blows received in the first boxing match, not due to the car accident, since massive subjective complaints existed before the car accident. This sad case reveals that coaches are not familiar with possible complications that can develop and he lied to the examining physician about injuries. The examining physician did not bother to check the size of the pupils, so his examination can only described as superficial and inadequate.

Case 3: This boxer was knocked down in a bout during a tournament and complained of a headache on the next day. The following day he was allowed to box again after a limited pre-competition examination in which he reported a 'slight headache.' He collapsed in the second round after receiving minor blows. He then had a seizure and became decerebrate with fixed pupils.
At the hospital, an emergency CT scan revealed a small subdural hematoma and massive brain swelling. The blood clot was removed surgically and the swollen brain was observed to herniate through the skull opening. The post-surgical intracranial pressure was impossible to control, and the patient died six days later.

There are no statements concerning the date of the last boxing match before the two mentioned in the report. This tragic case reveals, as do some of the others, the relative violent trauma from multiple boxing matches within a short period of time, in tournaments, and also in Olympic Games. A boxer complaining about a headache after a bout, or a series of bouts, should not be allowed to participate in another boxing match immediately. Participation should be allowed only after a comprehensive medical examination with supplementary tests and diagnostic procedures.

Case 4: A 17 year old amateur boxer fought poorly in his first bout in an international tournament. He received several standing '8' counts after being stunned by head blows. Afterwards, he was obsessed with boxing again and (in retrospect) exhibited inappropriate obsessive behavior. Six hours later after another boxer's bout was cancelled, he boxed again, encouraged by his coach. He again performed poorly, taking several severe hits that included two standing '8' counts, and the referee stopped the bout in the second round. A few minutes later while the referee was announcing the winner and holding the contestants' hands, the injured boxer collapsed in the center of the ring. He immediately lapsed into a coma with fixed dilated pupils.

Despite medical treatment that included decompression craniotomy, he died the next day. The autopsy showed severe cerebral edema with bilateral uncal herniation.

This 17 year old amateur boxer was an inferior pugilist in this international tournament and received several standing '8' counts after being stunned by head blows. There are no commentaries about subjective complaints. A thorough physical examination would have been appropriate; instead he boxed again six hours later. The commentary about 'obsessive behavior' is silly. It is the job of a coach to say 'no' to such behavior. Not only did the coach fail to prevent the boxer from continuing to box after he had performed poorly in a first bout, receiving four standing '8' counts, but he encouraged the boxer to continue in an additional bout. The negligence of the coach, whose duty was to act in the interest of his boxer, could only be described as outrageous or criminal. Not surprisingly, in the following match the boxer again performed poorly. He was severely beaten, taking serious blows against his head. He suffered two more standing '8' counts before the referee stopped the bout in the second round. A few minutes later the boxer collapsed in the ring and lapsed into a coma with fixed pupils.

Despite a decompression craniotomy the boxer died the next day. The autopsy revealed a severe brain edema with bilateral uncal herniations of both temporal lobes.

The brain edema was extreme and pronounced because there was bilateral uncal herniation. No histological data or histological evaluation are presented for this case.

Case 5: A 21 year old college student prepared for a weekend fraternity 'smoker' - an unofficial boxing match between two inexperienced boxers in which allegedly few rules are followed. He trained with an experienced boxer who inflicted head blows and other hits that were of considerable force. Afterwards, the college student developed headache and saw a physician at the student health center. He downplayed the severity of the pain and failed to keep a follow-up appointment. Two days later he sparred again in a training session without coaching or corner supervision. After two minutes of a light workout he collapsed in the ring, lapsed into a coma with fixed dilated pupils, became apneic, and died before medical attention could be obtained.

This 21 year old college student was a very inexperienced boxer who prepared for an unofficial boxing match for a weekend fraternity fistfight, a so-called smoker. He trained with an experienced boxer who landed severe blows on his head and punished him severely. Afterwards, he developed a headache and saw a physician at the student health center. He dissimulated the severity of the pain and failed to keep a follow-up appointment.

Two days later he sparred again in a training session without coaching and supervision. After two minutes of a light workout he collapsed in the ring, lapsed into a coma with fixed dilated pupils, became apneic, and died before medical attention could be obtained.

There are no data of an autopsy presented, but it is possible that there was an acute subdural hematoma or a massive hemorrhagic infarction of parts of the brain.

Case 6: A 24 year old novice boxer fought an extremely hard bout with a ranked boxer and received many forceful blows. During three rounds he received four standing '8' counts – the fourth terminated the fight. He left the ring unassisted and returned to the dressing room. Within minutes he collapsed into deep coma.
He was taken to the hospital, but despite intensive medical treatment with ventilator, mannitol, and corticosteroids; he died shortly thereafter. Autopsy revealed massively increased intracranial pressure from edema and vascular congestion and herniation that caused midbrain necrosis. No subdural hematoma was present.

In this match a novice boxer fought an extremely hard bout against a ranked boxer. He took many forceful blows and was severely punished. During three rounds he had four standing '8' counts – the fourth terminated the boxing match. He was able to return unassisted to the dressing room, where he collapsed into a deep coma within minutes.

Despite intensive medical treatment with ventilator, mannitol and corticosteroids, he died shortly thereafter. The autopsy showed a massive brain edema, herniation of the cerebrum at the tentorium cerebelli and a midbrain necrosis. There was no subdural hematoma present.

The case demonstrates that severe blows against the head with four standing '8' counts produced a very massive brain edema which lead to a herniation of inferior parts of the temporal lobes at the tentorium cerebelli and produced a traumatic brainstem syndrome. This case demonstrates that a severe life-threatening brain edema may develop without the presence of a space-occupying subdural hematoma. This happened perhaps, too, in case 5. This case demonstrates very convincingly, too, how dangerous boxing fights can be when a novice is paired with a very experienced boxer. Coaches should prevent such bouts from taking place.

The term '*second impact syndrome*' introduced by CANTU should not be used for all of these cases, which were published by CANTU and VOY (1995). There is no doubt that such a syndrome exists, but only when two single impacts occur. Boxing differs from other sports, where a pair of impacts may clearly be seen, because the object of boxing is to inflict a series of blows to the head as rapidly as possible. In boxing, a single pair of blows would be rare, so such a term is of limited use. A preferable term would be *'multiple impact syndrome'*, since in these cases dozens of hard blows were delivered. In many boxing matches hundreds of blows are landed against the head, so that one would have to refer to *'hundred impact syndrome'* or *' thousand impact syndrome.'*

The severity of the blows is evident from the fact that several standing '8' counts were given; the boxer was repeatedly out on his feet, or too dazed to box. These cases show the treacherous nature of a standing '8' count convincingly. The boxer only partially recovers during the eight seconds, after which he is forced to stand and defend in a dazed or groggy state. His defensive capabilities are rendered ineffective. Three or four standing '8' counts in a round or a match demonstrate that the boxer was unable to recover during the pause, which instead of protecting him, may have served only to prolong the now one-sided fight.

We have argued previously that although the knock out punch is recognized as dangerous, boxers regularly face even more devastating risks of harm in the ring. In fact, the boxer who receives a knock out blow early in a match is spared further injury from successive blows, at least for the immediate future. Research and empirical data confirm that it is far more dangerous for a boxer to receive a pummeling, especially after he has become dazed and less capable of avoiding head blows. At no time is a boxer in greater danger than after he has received a standing '8' count, for during his subsequent incapacitation the opponent has the time and opportunity to inflict his worst. Left with impaired balance, timing, judgment, coordination and or vision, in short, all of the abilities necessary for self preservation in the ring, the dazed boxer is virtually a defenseless target. The cases above, as well as many other case histories, clearly demonstrate that this is when the most devastating injuries occur. Indeed, in the 6 cases of CANTU and VOY (1995) the injuries were fatal.

The most dangerous moments in boxing occur when a boxer has been beaten into a state of grogginess, but remains on his feet. His senses are incapacitated and he is unable to defend at his usual ability. Without intervention, the natural course of events is that his opponent will give him an undefended pummeling until the bell rings or until he falls to the ground. During these remaining moments of the fight, his compromised brain is susceptible to serious injury or death from blows that would ordinarily be survivable. In addition to receiving more and better landed head blows because he is unable to avoid them, the dazed boxer is taking them on his already compromised brain, which, as we have seen, enhances the injuries from each successive blow.

Paradoxically, in this situation, the rules provide little protection for the boxer. A referee has the authority to stop the fight, but when to do so is a matter of his discretion. The referee is in a difficult position, if not a conflict of interest, because typically this is the most intense, exciting, frenzied point of the match. Both the fans and the industry may feel disappointed or cheated if deprived of the ultimate conclusion of the fight. Referees who stop fights early know that they may suffer a backlash of disapproval by not being chosen to officiate in future matches. Unquestionably there is little time to make a considered decision in the heat of the moment.

The rules usually require the post concussion boxer to wait for a specified time interval before accepting another match, often a thirty day delay. This is done to allow for the fullest possible recovery. However, the dazed boxer after an '8' count, when he is most vulnerable and in desperate need of institutionalized protections is without help. If one were to ask what single change would make boxing safer from the perspective of our experience, it would be to find some way of protecting the injured or incapacitated boxer from further (enhanced)head injuries. Once these injuries are amassed, there is little that medical intervention can do to help the boxer or prevent death.

There is a good argument, that a knock out is not the main danger for a boxer, because that saves him, occurring in an early round, from more punishment. It is much more dangerous for a boxer to get one or several standing '8' counts. The rule to stop a boxing match only after the third standing '8' count in one round is not sufficient to protect a boxer, who is dazed and groggy. Furthermore, in general, the rest period of one minute between two rounds is to short to guarantee a recovery of a boxer who had, for example, two standing '8' counts in the preceding round to protect him from additional standing '8' counts in the following rounds. Each of these so-called protective measures are not a protection for a boxer, but allow that he receives additional numerous repeated blows which produce a permanent brain damage.

A boxer who has received several standing '8' counts in a boxing match is more severely injured than a boxer who is knocked out in one of the early rounds of a boxing match. Protective measures protect the knocked out fighter, he may not participate in boxing matches for a certain time interval, the boxer who received one or several standing '8' counts is not protected, despite the fact that he needs more and longer protection.

CANTU and VOY ask, what do these cases have in common?

Of the cases describing boxers, five patients out of six had a cerebral concussion with residual cerebral symptoms. All were said to have lapsed into a coma with brainstem involvement from herniation of the cerebrum at the uncus gyri hippocampi secondary to massively increased intracranial pressure. Though subdural hematomas were present in half of the cases, the pathomorphological alterations occurred with what appeared to be minor head trauma and were refractory to medical treatment.

CANTU and VOY summarized:

> "In short, the findings of the six case studies are consistent with 'second impact syndrome.'

In preceding paragraphs it was emphasized that there were not relatively minor head traumas, but severe ones. After a standing '8' count, a boxer has been severely punished, is dazed and groggy, and only partially capable to defend himself. The standing '8' count does not protect him from further punishment. The authors continue: "Again, we have stressed the point, that the term 'second impact syndrome' is not only inadequate, but incorrect and untrue. There were not two impacts, but they were multiple in the sequential event of a boxing match. One can here only speak of a 'multiple impact syndrome.' The term 'second impact syndrome' should not be used in boxing, where it could be used only in a case, in which a boxer is hit by a single severe blow at the begin of a bout, which knocked him down and he hits then the canvas in a second impact, in so-called a deceleration trauma.

CANTU and VOY continued:

> "Again, the typical pattern of 'second impact syndrome' starts with postconcussion symptoms, including visual, motor, or sensory changes and difficulty with thought and memory. Before these symptoms resolve – which may take days or weeks – the athlete returns to competition and receives a second blow to the head... The second blow may be remarkably minor, perhaps involving a blow to the chest, side, or back that can snap the athlete's head and impact accelerative forces to the brain. The athlete may appear stunned, but usually does not lose consciousness and often completes the play. The athlete usually remains on his or her feet for seconds or minutes, but seems dazed, similar to someone suffering from a grade 1 concussion without loss of consciousness. Often, the athlete remains on the playing field and sometimes walks off under his or her own power."

We cannot agree with the authors statement, that "the typical pattern of 'second impact syndrome' starts with postconcussion symptoms..." At least in the boxers' case histories there was a series of significant head blows followed by a state of dazed grogginess. All of them received one or more standing '8' counts. Some of the boxers were unable to stay in competition, as evidenced by the fact that bouts were stopped. In other words, the officials declared technical knock outs. The so-called 'second impact' consisted of not one, but a series of blows. The effects of the blows, however strong, were not "remarkably minor". They landed on the heads of injured boxers who were unable to get out of the way. Subdural hematomas were hammered and punched with devastating effect. The boxers did not always remain "on their playing field," or on their feet in the ring. Their bouts were concluded when they collapsed.

Again, the *'second impact syndrome'* can be seen in some sports, we do not deny it. But this term should not be used in boxing, where often the blows are so fast and numerous that they are difficult to count. *'Multiple impact syndrome,'* or *'repetitive impact syndrome'* would be more accurate terms.

CANTU and VOY continued:

> "In addition, educating athletes and their parents about concussion symptoms and the risk of a 'second impact syndrome' is vital, their decisions directly influence a safe return to play."

We would add that the injured boxers should never lie to the ring physicians who perform the pre-competition examinations. The coaches should not support the boxers to lie, nor should they recommend that injured athletes participate in further boxing. The coaches must be better educated if they will be responsible for making informed decisions about when it is safe for a boxer to continue. We would like to believe that if the coaches in the preceding case histories had been better informed of risks and symptoms themselves, they would have prevented the tragic results to athletes in their charge.

The prefight physical examinations performed on these boxers were insufficient to reveal problems that would have been obvious to a careful, competent ring physician. The physicians in these cases missed the very signs for which they should have been looking. If needless maiming and deaths of athletes are to be prevented, the quality and extent of pre-fight physicals must be improved.

Prize Ring Rules

Under the old Prize Ring rules a round ended every time one of the participants went down. The record for the most knock downs in a boxing match must be the same, according to GOLESWORTHY (1960), as for the record number of rounds – 276. That was the total number of knock downs when Jack JONES beat Patsy TUNNEY at Cheshire, England, in 1825.

Under the Queensberry Rules the record probably belongs to the long match between 'Battling' NELSON and Christy WILLIAMS at Hot Springs, Arkansas, on December 26, 1902. NELSON knocked WILLIAMS down 41 times before knocking him out for the full count in the seventeenth round. NELSON himself was knocked down seven times.
The record for scoring the most knock downs in a world heavyweight title fight with boxing gloves probably belongs, according to GOLESWORTHY (1960) to Max BAER, who knocked down the giant Italian Primo CARNERA twelve times in eleven rounds, when winning the title in 1934.

Some other contests with a large number of knock downs are, according to GOLESWORTHY:

Joe GRIM 'The Iron Man,' an Italian cruiserweight who settled in America and earned a reputation as one of the toughest men in the ring was down 17 times in a fight with Bob FITZSIMMONS at Philadelphia in 1903, and at least 17 times – some say 30 – when he met Jack JOHNSON nearly two years later.
Jack DEMPSEY floored Jess WILLARD seven times in the first round of their world title fight, July 4, 1919. The 'Manassa Mauler' also put Luis Angel FIRPO down nine times in two rounds in their famous fight of 1923.
Bruce WOODCOCK put Jock PORTER down nine times before the referee intervened in the third round, November 27, 1945.
Charlie HILL was floored ten times by Bobby NEILL before the referee intervened in the ninth round of their British featherweight title contest, April 13, 1959.
Floyd PATTERSON put Pete RADEMACHER down seven times when winning by a sixth round knock out, August 22, 1957.
The record for a world title fight at any weight goes to Vic TOWEEL of South Africa. When defending his bantamweight crown against Danny O'SULLIVAN in Johannesburg, December 2, 1950, he had the Englishman down 14 times before forcing him to retire in the tenth round.
When scoring a first round knock out, Jock McAVOY floored Eddie RISKO six times in 2 minutes 28 seconds actual fighting, New York, December 29, 1935.
Peter KIERNAN, British bantamweight champion put challenger Frankie WILLIAMS (Birkenhead) down eleven times before the referee stopped the fight in the seventh round, January 28, 1953.
Olle TANBERG floored Jock PORTER 21 times before the referee intervened in the sevent round, Stockholm, January 8, 1946.
Benny LYNCH scored ten knock downs over Jackie BROWN (Manchester) to win the world flyweight title, September 9, 1935. BROWN retired in the second round.
Pat KERRY (Derry) floored Johnny O'CALLAGHAM (Dagenham) 14 times in a featherweight contest before the referee stopped the fight in the fourth round, Belfast, September 20, 1958.

The following rules exist:

Every time a fighter is knocked down, it is mandatory for a boxer to take a count of '8.' Since 1963 it is generally accepted rule that the referee should continue at the end of any round, except the last, until the boxer gets up or is counted out.

In most states of the United States the referee should stop a boxing match immediately a boxer has been knocked down three times in any one round.

Knocked out of the Ring

Cases in which a boxer is knocked out of the ring have occurred many times in major fights. The most famous one was the fight between the American heavyweight Jack DEMPSEY and Luis Angel FIRPO at the New York Giants' baseball ground on September 14. 1923. Jack DEMPSEY knocked down the giant Argentinian seven times but once FIRPO knocked DEMPSEY through the ring ropes landing in the press section of the first row. DEMPSEY was able to scramble back into the ring, some visitors say he was pushed back by some spectators, but FIRPO was not able to knock DEMPSEY out in this round. DEMPSEY landed a knock out blow in the second round. The fight had lasted 3 minutes 37 seconds.

Another world heavyweight champion was knocked out of the ring but was able to enter the ring again was Joe LOUIS. Buddy BAER knocked LOUIS through the ring ropes in the first round of their boxing match at Washington, May 23, 1941. LOUIS was very fast back in the ring at the count of '4.' LOUIS won the bout when BAER was disqualified in the seventh round.

Vic TOWEEL, the world bantamweight champion, was knocked out of the ring in the first round of the boxing match with Jimmy CARRUTHERS, on November 15, 1952, at Johannesburg, South Africa. He was able to get back into the ring but was knocked out in 2 minutes 19 seconds.

Freddie MILLS knocked Len HARVEY out of the ring in the second round at the Tottenham Hotspur football ground, on June 20, 1942. HARVEY was not able to beat the count and lost his British and Empire lightheavyweight titles.

In the heavyweight boxing match between Jack JOHNSON and middleweight Stanley KETCHEL at Colma, California, on October 16, 1909, KETCHEL surprised JOHNSON in the twelfth round and knocked him down. JOHNSON jumped up and knocked KETCHEL out with a right to the chin. JOHNSON had already left the arena before KETCHEL had fully recovered. Two of the teeth of KETCHEL were found embedded in JOHNSON's boxing glove.

Double Knock Outs

Ad WOLGAST, the lightheavyweight champion and the challenger Joe RIVERS connected simultaneously in the thirteenth round, at Vernon, California, on July 14, 1912. Both boxers dropped to the canvas. It appeared that neither would be able to get up, but the referee Jack WELCH pulled WOLGAST up to his feet while counting RIVERS out.

Fewest Knock Outs

William 'Gorilla' JONES who had won the middleweight title in 1931, had an active career which extended over 16 years, during which he participated in aboutt 200 boxing matches, was never knocked out.

The profesional boxer Len HARVEY was counted out only once in 418 professional fights, and that occurred in his last fight when he was knocked out by Freddie MILLS.

The 'Iron Man' Joe GRIM was knocked out only three times in more than 300 fights. In a boxing match against Jack JOHNSON in July 1905, he was knocked down 17 times, in a six round bout, but JOHNSON was unable to knock him out.

If a contestant is knocked down he must get up unassisted within ten seconeds, his opponent has to retire to the farthest neutral corner. A boxer is considered down even when he is on one foot or both feet.

When ordered by the referee to 'break,' both boxers shall immediately take one step back before recommencing to box. A boxer shall not strike back or attempt to strike his opponent on a 'break.'

The referee shall have the power to disqualify a contestant for any of the following acts: (1) Hitting below the belt, (2) using a 'pivot blow,' (3) hitting on the back of the head or neck (nape), (4) kidney punching, (5) hitting with the open glove, the inside or butt of the head, or with the wrist or the elbow, (6) holding, butting or careless use of the head, shouldering, wrestling or roughing, (7) not trying, (8) persistently ducking below the waist-line, (9) intentional falling without receiving a blow, (10) failing to 'break' when so ordered, or striking or attempting to strike an opponent on the 'break', (11) for deliberately striking an opponent when he is dropping to the floor or when he is down, (12) for ungentlemanly conduct which he may deem foul.

A contestand disqualified for any cause whatever shall not be entitled to any prize or renumeration.

The referee shall have the power to stop the contest if in his opinion a contestant is outclassed or accidentally disabled.

The referee shall decide (1) any question not specified for in these rules, (2) the interpretation of any of these rules on matters arising.

Punches

NOVICH (1963) noted:

"Boxing developes physical fitness, provides wholesome recreation, satisfies psychological need and, in the case of professionals, affords them a career."

NOVICH continued:

"But in those head injuries which cause varying degrees of unconsciousness which should caution the physician to use extreme conservation. For instance, in the recent LISTON – PATTERSON fiasco, from where I saw it, PATTERSON didn't suffer too unduly from LISTON's blows; just enough to go to the floor and be counted out. If the clinical examination and follow-up examination show no neurological sequelae with or without the EEG, two monts of honest living and training should be enough to clear up the cobwebs in his brain. Two months of inactivity guarantee that if there are any residuals they will present themselves."

This passage points to a fundamental misconception about 'effective' blows in boxing. KAPLAN and BROWDER (1954), for example, wrote:

"Blows on the chin or side of the head in the anterior temporal region seemingly produce the most knock outs. Relatively few blows in a single contest are truly effective. By 'effective blow' is meant one that produces some alteration in the conscious state."

KAPLAN had asked Jack DEMPSEY about the number of 'effective' head blows sustained by a boxer in a single fight. His answer was that possibly one in a 1.000 hit the mark. One must remember that when a fighter like Jack DEMPSEY in his prime struck a person effectively, that is solidly on the head and with the person not being set for it, unconsciousness usually resulted. Yet Jack DEMPSEY had fought many ten, twelve or fifteen round bouts without knocking out his opponent."

In other words, an 'effective' blow is only that sort of punch which produces immediate unconsciousness as the result of a cerebral concussion. A cerebral concussion of similar intensity to a boxing blow, but of another origin, usually occurs as a single event for individuals as the result of an accident. Physicians and laymen take such injuries quite seriouly, even if only transient symptoms occur. Most patients consider themselves hurt and not rarely file insurance claims. But the opinion that only a punch against the head, which produces a cerebral concussion is an 'effective' punch is incorrect. A serious of hammering blows against head and body whose intensity is not high enough to reach the threshold in which a cerebral concussion occurs contributes to the development of severe permanent brain damage. This can be deduced from the morphological findings in boxer brains and from transfer of data from animal models dealing with repeated impacts of subconcussive strength.

It is interesting to quote boxers about the effect of a hard blow to the head:

James BRADDOCK after his fight agaisnt Joe LOUIS said:

"The left jab was someone jammed an electric bulb in your face and busted it," the right was worse: "Like someone nailed you with a crowbar. I thought half my head was blowed off. I figure he caved it in. I felt it after he hit me and I could't even feel it was there."

Randall (Tex) COBB, on whether he'd be willing to fight Larry HOLMES again after taking a beating from him in 1982 *(Sports Illustrated, 1984):* I don't think his hands could take the abuse."

The lightweight Roberto DURAN, nicknamed 'Hands of Stone,' knocked Ray LAMPKIN in the sixth round with a left hook down.

The 27year old challenger tried to get on his feet, but fell back and was counted out. He had to be carried on a stretcher out of the ring and was unconscious for a period of 30 minutes.

"DURAN, 23, said after the sixth defense of his World Boxing Association title that it was a difficult fight for him because he was not in good shape."
"'If I had been in normal shape LAMPKIN would not have lasted six rounds and he would have gone to the morgue instead of the hospital,' DURAN said.

COHANE (1963) writes about Gene TUNNEY's left jab:

As a fighter, TUNNEY was unspectacular, but demoniacally efficient. He was a manufactured boxer – only his spiritual assets were natural... From Benny LEONARD, the lightweight maestro, TUNNEY borrowed the trick of shoving his left jab in a slightly upward plane, to sharpen the cutting edge. This method helped make a mess of DEMPSEY's eyes. TUNNEY's arsenal included also a fine right to the head, straight as a string, a vicious left hook to the liver, and a hurtful right under the heart."

Peter HELLER (1973) reported about the professional boxing fight between Carmen BASILIO and Tony De MARCO, which took place on June 10, 1955 in Boston. The referee stopped the fight in round twelve and awarded a t. k. o. to BASILIO.
BASILIO said:

"I got hit on the point of the chin. It was a left hook that hit the right point of my chin. What happens is it pulls your yawbone socket from the right side and jams into the left side and the nerve there paralyzed the whole left side of my body, especially my leg. My left knee buckled and I almost went down, but when I got back to my corner the botton of my foot felt like it had needles about six inches high and I just kept stamping my foot on the floor, trying to bring it back. And by the time the bell rang for the eight round it was all right."

Whether the explanantion of the effect of the left hook, given by Carmen BASILIO, is correct, we will not further discuss. It is in my opinion more probable that the powerful blow delivered by Tony De MARCO against the right side of his head had an immediate effect on his brain, producing a transient weakness and paresthesia of the left, contralateral extremity.

It seems to me more probable that the motoric and sensible cortex of the right cerebral hemisphere, was irritated by the force of the powerful blow which produced a disturbance of the left lower extremity.

Blows against the Side of the Rib

Paul GALLICO described a punch, for which the lightheavyweight boxer Paul BERLENBACH was famous:

"But he had a magic, this boy. It was a numbing, paralyzing body punch that caused his opponent suddenly to crumple up as though they had been struck by a .44 slug. It was a slow punch, too. None of your quicker-than-the-eye Joe LOUIS drives that are so swift that they blurr on camera plates. Sometimes Paul seemed merely to be dropping his right hand to the side of an opponent, in the manner in which one places a friendly hand on someone's shoulder. Sometimes he would hit opponent on the chin, but more often he just hit him in the side and caved in his ribs. Often it would happen in a clinch and you would't see the blow at all. His opponent would suddenly wilt, go limp, collapse to the floor, and stay there."

So-called Cork-screw Punch

The so-called *'cork-screw punch'* is not an illegal punch, it was allegedly introduced by the professional middleweight boxer 'Kid' McCOY. Turning of the boxing glove in the moment of impact at the skin of an opponent produces often lesions of the skin and the subcutaneous tissue. This invention brought McCOY the nickname 'The Corkscrew Kid.' (COLLINS).

Effect of a Blow under the Heart

About the effect of a *blow under the heart* wrote Gene TUNNEY (1922):

In the spring of 1929, while coming North on the train after promoting the SHARKEY – STRIBLING fight at Miami Beach, DEMPSEY went into the compartment of Roy HOWARD and Ray LONG, the publishers. After talking a little while about the Chicago fight and, in particular, that seventh round, DEMPSEY exclaimed:"

"'The right-hand punch under the heart that TUNNEY hit me when he got off the floor in the seventh round was the hardest blow I ever have received. It was not a question on my mind of being knocked out – I thought I was going to die. I could not get my breath. A second rubbed away the congestion around my heart when I came back to my corner; but for that I would not have been able to come out for the eight round.' Gus WILSON, his worshipping trainer, has told me the same story."

Solar Plexus Blows

FITZSIMMONS is considered the boxer who introduced the *'solar plexus blow.'*

TOPEROFF (1987) noted:

"It seems that the vulnerable area just below the chest and above the stomach, when struck properly with force, simply knocks the wind out of an opponent, causing him to deflate."

Punches into the Stomach Region

Terrific beating in the stomach may produce nausea and wretching (Fig. 66).

Professional heavyweight fight between Rocky MARCIANO and the British heavyweight Don COCKELL on May 16, 1955. COCKELL is shown wretching as a result of the terrific beating in the stomach he received at the hands of heavyweight Rocky MARCIANO. MARCIANO won the fight by a TKO in 54 seconds of the ninth round.

Fig. 66. Professional heavyweight fight between Rocky MARCIANO and Don COCKELL on May 16, 1955. Photo: Herb SCHARFMAN.

Illegal Punches

Boxing regulations were developed to make boxing less dangerous for the participants. Rules commonly *prohibit blows below the belt, rabbit punches, hitting while in a clinch,* among other especially debilitating moves. However, prohibition has not stopped boxers from using illegal blows. In fact, a science of how to get the better of the opponent using subtle or disguised illegal moves has developed, and is even discussed in printed publications on boxing technique. Some of the popular questionable strategies advocated in print are described below.

Jay C. 'Champ' THOMAS (1997) who had been an active boxer for many years, wrote a book: *'Boxings Dirty Tricks and Outlaw Killer Punches.'* He began his career as boxer in 1923, he is a veteran of allegedly 10.000 bouts. THOMAS only used these tricks and punches, as he writes, to defend himself against unscrupulous fighters, and he cautioned against employing them recklessly.

THOMAS wrote:

"Boxing is brutal, Maybe it's too brutal. I've seen boxers blinded. I've seen many 'clean-type' boxers annililated by unethical opponents. I've watched the 'dirty-tricks artists' completely destroy the young and sports-minded fighters. And most of the youngsters didn't even know they were on the receiving end of a croocked attack."

THOMAS brings an example:

Once he was handling a pretty fair heavyweight prospect in an important fight. He was in excellent physical condition.

THOMAS wrote:

"My fighter won the initial round by a comfortable margin. He fairly radiated confidence when he answered the bell for the second round. But the picture changed quickly. My guy jabbed a left and bounced the target's head backward a foot or more. That was the last good punch my man threw. He was a sad case when the round ended and he wobbled back to his corner."

"'He gouged my eyes,' Bobby moaned. 'He hit my balls. His elbow zapped into my mouth. The referee is helping him. I couldn't even complain. What can I do?'"

"That was the first time and last time I ever felt that I let a fighter down. Since then I always taught my boxers enough dirty tricks fo 'foul' on even terms with the dirty fighters. When a bum started the dirty-trick business with one of my boxers... that bum bit off more than he can chew. You see, any fighter I trained and/or taught soon became an expert in dirty trick fighting. But they seldom employed unfair tactics until the opposition provoked it. Example: Fouling Freddie launches a dirty attack, and my athletes retaliates with a double dose of the same. Then, for good measure and to assure Fouling Freddie that he has picked on the wrong guy, my fighter fires off an assortment of unethical punches."

"One boxer I handled used to counter a dirty trick with a solid wallop to the groin. Then he'd mutter sarcastically, 'If you hit me with another dirty punch... I may decide to get rough.' That usually stopped the dirty fighter. He probably figured that anything rougher than the crusher to his groin would be too much to handle. Anyway, the naughty punks got religion after Joey gave him his special punch."

"Getting back to the dude I let down... I knew I didn't have the time to teach him to fight dirtier than the slob he was fighting. But right then and there I vowed that I'd never send a man into a battle until I'd taught him how to handle all emergencies. And I consider a dirty-fighting opponent an emergency that calls for drastic counter-tactics."

"I'm a firm believer in sportsmanship in boxing, in all sports, and in every area in the battle of life. I dislike war and unfairness in sports or business, but I'm realistic and know that war, unfairness and unscrupulous people will always be a part of the world. Therefore, a country should have the firepower to cope with war, and individuals should have the firepower to cope with war, and individuals should have the know-how to handle unpleasant situations. In fact, a country should have superior firepower to discourage potential military enemies. And a boxer should have the firepower to defeat any fistic opponent."

"I can't do much to assure that my country or any country has the military punch to win, but I could help my boxing protégés attain the knowledge they needed in order to win. This book, hopefully, will give aspiring young boxers enough tricks to make any foul-minded opponent afraid to try his disreputable tactics."

"My advice to all boxers, beginners or veteran pros, is to learn the dirty tricks I teach in this section of the book. Perfect the tricks, invent a few more of your own, but hold them back until your opponent attempts to take advantage of you. Then make him sorry he ever tried a foul blow on you."

"So... I won't waste any more time with a preamble. I'll rush to the next page and start filling your bag with the punches and moves which will give you superiority over any and all dirty-trick fighters."

Blows with an Open Boxing Glove ('Lacing Punches')

Professional heavyweight fight between Joe LOUIS and Tami MAURIELLO on September 18, 1946 in Yankee Stadium, New York. At the opening of the first round MAURIELLO rushed out of his corner and landed with a right hand that shuck and staggered LOUIS and sent him backwards into the ropes. LOUIS used his open right boxing glove to 'smear' MAURIELLO's face (Figs. 67 a to d). and knocked him out in 2 minutes, 9 seconds. An alert referee should have recognised this rude breach of the rules and this foul and should have warned LOUIS and after repetition disqualified.

THOMAS (1987) wrote:

"Employing boxing-glove laces to injure a foe is probably boxing's most common dirty trick. Laces can be used repeatedly without being detected by the victim, the officials, or the victim's seconds."

"Any good pro can make a lacing punch look natural and legal. And you can't find many pros who haven't 'ripped' a few opponents. They often cut their opponents to ribbons, and are seldom forced to pay for their fouling tactics. That is... they don't pay unless they meet one of their students. Then they get paid in 'triplicate' for their brutal deeds."

"When I was a youngster I had extremely fair skin, and I bruised easily. I was barely 17 when I fell victim to a real pro lacer. Even though I had already begun to learn dirty tricks and keep them handy to use in case I had to 'pay' an evil opponent for giving me the business, the dude was so slick that I didn't know I was getting the business from him."

"I called on all my reserve strength, guts and staying power and emerged victorious. Later in the dressing room, I learned that I had been on the receiving end of a concentrated lacing attack. The mirror told me that I had tangled with a buzz saw."

"'Hey, THOMAS', said a battered old pug, eyeing my sagging frame on the bench, 'the guy from El Centro marked you like a farmer marks a pig.'"

"'I didn't even know he was marking me,' I mustered through my cut lips. Then I glanced at my handler and asked, "Why didn't you stop JUAREZ from fouling me?"

"'He didn't foul you,' mine handler said. 'So he heeled you with the laces... so what? It's routine. You shoulda put laces to him.'"

"It was one of the times I rode a freight train to town, hurried to the arena, and accepted a fight for that night. I hadn't trained, and I was weak from hunger. And the promoter had given me a total stranger as a second."

"I glared at the handler and said, 'Why'd you let him do it?'"

"'Punk' the man snapped. 'You ain't nothing but a railroad bum. The promoter paid me a buck to second you. All that buys is the water bottle between the rounds. Why worry? You won.'"

""When I think of that incident I get mad at my self for not bestowing that 'house second' with a barrage of punches. I should've given him a few lumps. He might have more respect for the next 'railroad bum' he was asked to mishandle.'"

"That lacing incident did 'inspire' me to learn how to out-lash the rippers. I made it a point to watch the man who had done me in. I soon learned most of his cute lacing tricks. And, as with any punch or move I've learned, I've cultivated and developed the lacing trick to the point of perfection."

'Lacing Punch' to Rip Open the Brow and Produce a Quick T K O

THOMAS (1997) wrote:

"This trick is used so often and so successfully that most referees seem to consider it a legitimate punch. They seldom warn or penalize those guilty of using it. Newspaper men, even if they notice it, brush it off as an accident. And is it isn't used too often, the victim's handlers don't even complain."

"Study the punch closely. The expert tricksters make it look like a legitimate punch that is slightly short of the target. The lace, like a knife, slices across the bony part of the eyebrow. It cuts a deep gash, and the spurt of blood causes the referee to stop the fight."

"Cuts about the eyes have terminated hundreds of fights in early rounds and presented victory to the culprits who did the damage. And you'd be surprised at the few complaints lodged by the victims or their backers. They simply don't know what to look for. Even if they do know what to look for, even if they know what happened, the victims and their handlers seldom find a referee and/or ring officials who will

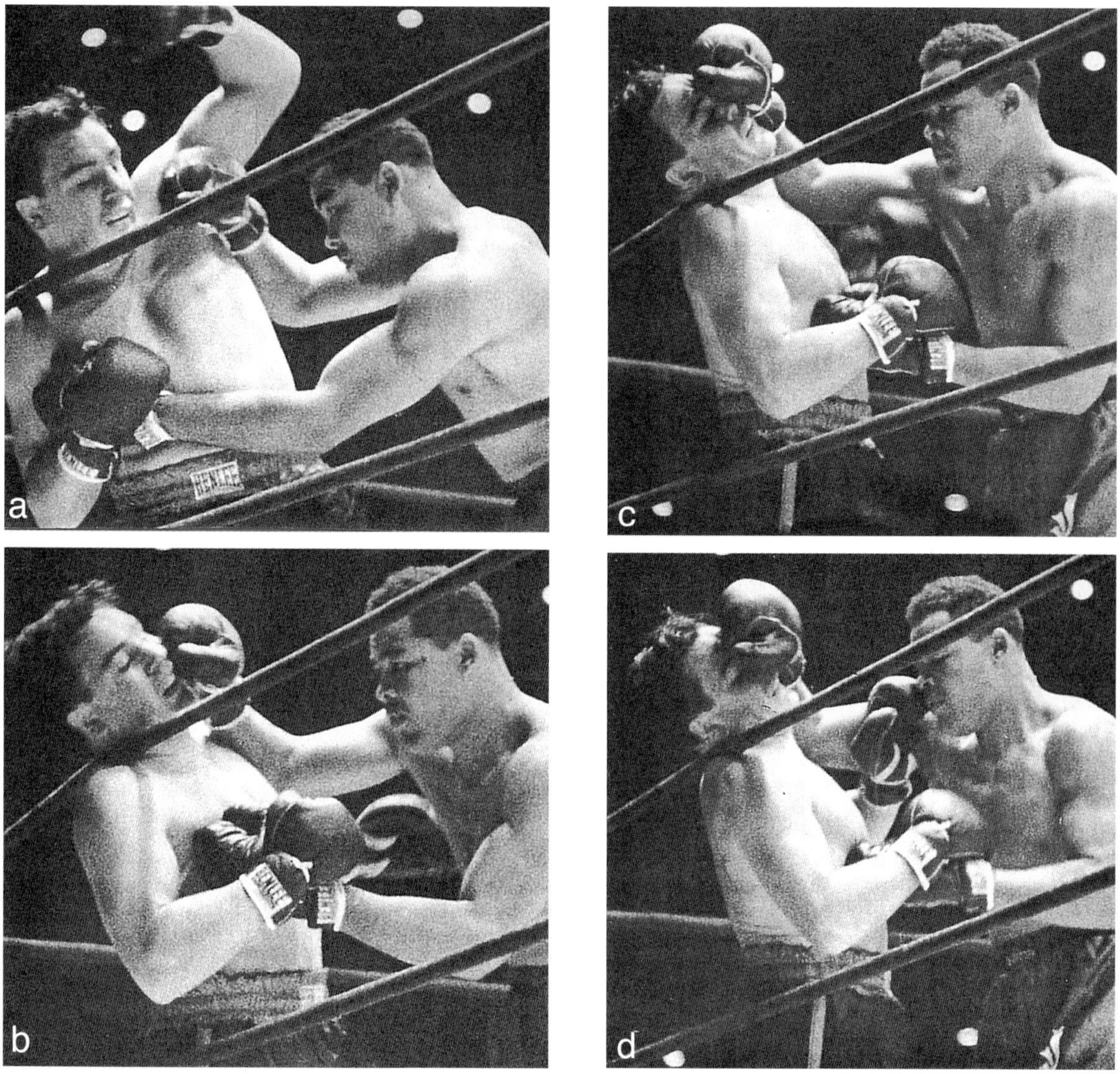

Fig. 67 a – d. Professional heavyweight fight between Tami MAURIELLO and Joe LOIUS on September 18, 1946.

listen to their complaints. So I consider my method the best defense against dirty trickster.

"Learn more brutality and tricks and be able to use them better than any other and all the 'dirty trick' fighting opponents."

"The best way to make this lacer and all lacers pay off for you? Fire a right or left straight, hook, upward drive, make it look like a fair punch. Keep the palm toward your opponent and make contact on the vulnerable target with the heel portion of your glove. This, of course, puts the laces on the spot you want to rip. Inject all the speed you can muster, put it in the shot... and have an early victory."

"Lace-slasher punches can come from any direction."

'Lace-slasher' to Injure the Nose

"Imagine some punch hitting that same nose, with the punch carrying a thousand pounds of force and zipping at a speed of thirty miles per hour."

"I actually saw one like this in a Minneapolis gymnasium. The victim's nose was in the same position, and the culprit's landed in the angle shown. But, unfortunately for the victim, the nose wasn't that way for more then a thousandth of a second. When I helped him up off the ring mat, his nose was on his right cheek, and shattered bone protruded through the skin that had been the bridge of his nose. The bloody sight caused me to vomit... but the unlucky victim never recovered. That was in a gym, but is was the most damage I've seen a lace-slasher do."

'Lace Punch' to the Mouth

This one can completely wreck a mouth... by removing teeth, putting any number of stitches in the lips, and causing blood to gush like water spurting from a fountain. In addition, the nose can be broken with the same punch."

"As most fights are stopped because of eye cuts, it indicates that the lace-punchers concentrate on the brow area. But the nose and mouth are almost as vulnerable. Don't ignore them. I've employed the lace-slasher punches with devastating results. Since the night I was inducted into the 'dirty-trick victim's club,' I've managed to stop the slashers with a better slashing attack of my own. To put it bluntly... I've used the gloves' laces to out-gun the rats who tried to cut me up."

"I made the lace punches my ounce of prevention. When a boxer fouled me, I laced him in a dangerous and vulnerable target area. Then I said, 'That's my sample.' As a rule, he got the message and didn't try another illegal punch. So... I went ahead and defeated him with overwhelming boxing firepower."

"Lacing punches can be delivered with either hand. In fact, I often employed a buzzsaw-type lacing attack. Of course, I usually reserved the buzzsaw blitz for second offenders. Example: If they fouled me a couple of times, and I responded with a warning shot, and they tried to out-dirty-trick me... then I started with the buzzsaw and followed it with the full treatment. And, in dozens of alley-type, pier-six brawls, I never lost."

Lace punches are easy to learn and develop. The heavy punching bag is a good target to practice on. The punches can be cultivated nicely on the big bag."

"Caution! These punches are to be used for defensive purposes only. Let the other fighter be the aggressor... then you're justified in countering with your dangerous shots."

Eye Thumping and Gouging

Pete AXTHELM *(Dark Night of the Fiery Warriors, Newsweek, November 21, 1983, pp. 109-110).*

The middleweight champion Marvin HAGLER had fought and beat Roberto DURAN in a 15-round title fight at Cesars Palace, in Las Vegas, Nevada.

AXTHELM wrote:

Sixth round: "Then he [DURAN] began to counter in the little ways that DURAN watchers have loved and hated for so long. No longer blessed with his hands of stone, he used his still-hard head to butt his rival and raise a welt. He appeared to get a thumb in HAGLER's left eye, followed by enough short rights to open a profuse cut. When the punches didn't quite connect, he once held Mavin's head in place as he took aim at the target. And in the clinches, his head was always scraping the eye. No, it wasn't nice or pretty or particularly admirable. But it was, after too long an absence, the real DURAN."

THOMAS (1997) wrote:

"Thumb his way to victory? Just what do I mean?

"Eye-gouging is a heel trick generally associated with professional wrestling. But, unknown to most boxing fans and many ring officials and even the blinded victims, the eyes are prime targets for many slick boxing villains. And those slick ring villains can literally blind opponents without drawing the slightest warning from the ring officials. At least the wrestling villains are playing, and they do get warnings from the referees."

"How do the boxers do it?"

"It's not too difficult to blind your opponent temporarily or permanently when you make his eyes your special target. In fact, it's very easy. Modern boxing gloves are made to order for the dirty tricksters who like to blind their opponents. No matter how you try to clamp your thumb under your curled fingers to give your fist natural protection, the glove's thumb sticks out like something hanging from an overloaded suitcase you've tried to repack."

"With a little practice you can 'thread a needle' with that protruding thumb. It's dead end to stab it into the eyes of your boxing adversary. And once a victim gets the thumb he's at a disadvantage. After all, if a fighter can't see... how can he fight."

"My first experience with the 'eye rippers' was in bootleg amateur bouts in southern California in 1935. The pros often came from Los Angeles and, pretending to be amateurs, booked themselves on the amateur programs; Naturally, they had a field day fighting the less-experienced amateurs."

"You may wonder why a pro would bother to 'waste his time' performing on an amateur program. Well, let me remind you that it was in the middle of the Great Depression. Many pro fighters, especially the preliminary boxers, fought opponents for purses of ten to twenty dollars. The bootleg promoters, who knew the invading boxers were veteran pros, gave them the same amount to fight on under amateur cards. Thus, the pros were paid to have their fun at our expense. It also gave them an opportunity to keep in condition. And, last but not least, they sharpened their dirty tricks on us luckless amateurs."

"One night I was fighting a veteran pro from San Diego. He thumbed my right eye with his left thumb a dozen times in the first round. The eye was closed when I went to my corner. My left eye had't caught a thumb, but it was watering badly and I couldn't see through the blur."

"Fat Eddie did the booking for my team. He often had a half-dozen boxers fighting on different programs, and he had a number of trainers working for him. My trainer for that night was a character who liked to be called 'DILLINGER.' He had been tagged with the name because he bragged to one and all that he had met the infamous outlaw. Rumors flowed (probably started by the man himself) that DILLINGER was as tough or tougher than the real John DILLINGER."

"When DILLINGER saw my eye, he cursed, then went to the referee and spoke a few words. The referee swaggered over to my opponent (who called himself Al HERNANDEZ for the fight) and talked heatedly for a few seconds. He was still talking and arguing when the bell sounded for the next round."

"I started out of my corner, but DILLINGER held my arm. Then the ref went to the announcer. The announcer climbed up in the ring and yelled for attention. He informed the crowd that 'HERNANDEZ' had been disqualified for giving false information."

"On the way back to Yuma I asked DILLINGER what he had said to the referee. He replied, 'I told the bastard to stop the fight and declare you the winner. He started to give me back talk but I told him I'd kill him and the phony HERNANDEZ if he didn't do as I ordered.'"

"Notice how the punch appears to be a long left jab with the glove open and the thumb protruding. The fist should be closed, but with the thumb sticking out as it is done."

"The jabbing thumber is extremely easy to use. With a bit of practice, any fighter can become an expert at it. You can literally annihilate an opponent. What's more... the referee, the handlers, the spectators, and many victims can't catch the culprit."

"Many tricky maneuvers can be executed in a clinch. You can either push yourself away from your opponent or push him away from you. In either case you can do a great deal of damage... even end the fight with one or more of these eye gougers."

"I stiffen my fingers and stab them straight into my opponent's eyes, his solar plexus, or his mouth. One reason is that the stiffened fingers give you about four inches more reach, and you are in better position to guard against a counterattack."

"Incidentally, I call this one 'the spear shot,' as it has speared me out of many tight spots. To strengthen my fingers I often did pushups on the tips of my fingers. I stopped one opponent in a navy bout when I used the 'spear' to stab his solar plexus."

"This one shot has 'probed' openings for me and helped set the stage for many victories. You will do well to develop it."

Blows with the Elbow

After missing with a hook boxer intentionally or accidentally using their elbow to hit their opponents on the head.

Professional heavyweight fight between Rocky MARCIANO and Joe LOUIS on October 26, 1951, at Madison Square Garden, New York. MARCIANO who knocked out LOUIS in the eigth round, has his right elbow struck into LOUIS left eye after he had missed with a right hook. (Fig. 68).

THOMAS (1997) wrote:

"During Henry ARMSTRONG's heyday, I saw the little 'triple champion' display his talents several times. I was surprised at the number of dirty tricks he had developed as part of his style of fighting. He did as much damage with his head and elbows as he did with his fists."

"I wondered why the ring officials didn't stop ARMSTRONG's bouts and award the decisions to his opponents. The least a ref should've done was to warn Henry and/or take a round from him. Of course, awarding one or more rounds to ARMSTRONG's opponent wouldn't have helped much. Few ARMSTRONG foes escaped being knocked out." Supporters of the triple champion defended him and insisted that he wasn't fouling victims. 'Hell no!' they said. 'Ol' Hank is just fighting his style. He's just too good for the other fighters. Those bums duck right into his elbows and head. Ain't Henry's fault.'"

ARMSTRONG's great success in the ring started a trend. Dozens of boxers tried to copy his style. Just as the fighters who attempted to ape MUHAMMAD ALI failed, those wanting to copy Henry ARMSTRONG never came close to being second ARMSTRONGs. But they all did one thing... they fouled and illegally whipped hundreds of opponents."

"The ring officials shouldn't have permitted such a ring mayhem. Those officials were just as guilty as the fighters who butted and elbowed their way to hundreds of illegal victories."

"Being of the old school that teaches you 'if you can't lick 'em, join 'em.' I did just that. I started a personal crusade against the butters and elbowers. With my simple logic, I figured that if I could be rougher, meaner, and better than my opponent, I would win all my bouts."

"I was boxing in a tough bootleg amateur circuit in Arizona, California , and Mexico. Just anything short of a machine-gun attack was permitted in those ring bouts. Death was common in that boxing territory. To survive, you had to be rough, mean, and a bit crazy."

"ARMSTRONG hadn't yet won a title when I first saw him in action. I was impressed by his relentless attack and by the

Fig. 68. Heayvyweight fight between Rocky MARCIANO and Joe LOUIS on October 26, 1951, at Madison Square Garden, New York. MARCIANO who knocked out LOUIS in the eighth round, has his elbow struck into LOUIS left eye after he had missed with a right hook.

way he used his head, shoulders, elbows, and head. And when the club's coach told me that I was fighting an ARMSTRONG-type foe, I sought advice. How could I counter and/or survive the flying elbows and butting head?"

"The club in which I was a member was called *The Cactus Boxing Club*. Fat Eddie (I never knew his last name) was the owner/boss of the club. He managed about fifty boxers on a regular basis, and he had almost as many ex-fighters hanging around begging for as chance to 'train' boxers for the boss. Each 'trainer' had his own idea of how to 'train' boxers. As a result, one's educational progress was in a constant state of confusion."

"My favorite trainers in that club were DILLINGER and a black dude called 'The Preacher.' They always tried to help the men assigned to them. When I learned I was fighting a Henry ARMSTRONG-type, I asked the Preacher how I should cope with the man. Preach was in a hurry to leave with a team going in the opposite direction from where I was fighting, but he delayed the trip fifteen minutes to give me a few pointers. That fifteen minutes probably saved my scalp."

"My bobbing opponent came at me headfirst, his fists end elbows flying like a windmill out of control. He managed to hit me with all his goodies before I caught him coming in. My elbow slammed into his mouth as he came charging toward me. It was the only counter-foul I needed. It won the fight for me."

"You make good use of your elbows, you fire first a left hook, deliberately miss, but follow through with your elbow and crash it into your foe's jaw. Or you miss deliberately with both your fists and elbow, then reverse your elbow and slam it into your opponent's exposed jaw."

"Any of the punches demonstrated in this chapter can end a fight for you. But there are pitfalls. You can break your elbow on the top of a hard head. And that's exactly what I did in one fight. I chipped a bone in my left elbow. I did win the fight, but I wonder if the victory was worth the pain I gave myself. Now, many years later, my elbow still bothers me at times. Of course, I keep telling myself that if I hadn't stopped the guy, he might've done killing damage with his elbows and butting head."

"Most of the time I've reserved my elbow shots for the bobbing-butting-elbowing cuties. A few times, though, I've chastised dudes for other rule violations by slipping them a dose of elbow poison."

"In one bout, when my foe thumbed my eye, I countered with an elbow hook to his mouth. The shattering crunch ripped out two of his teeth. The referee took one look and stopped the fight."

"You should practice the elbow shots and keep them honed and sharp... but save them for the would-be tough guys."

Head Butting

Head butting can occur accidental or deliberate.

Heavyweight fight between Rocky MARCIANO and the English heavyweight Don COCKELL on May 16, 1955, in San Francisco, California (Fig. 69). COCKELL was the inferior boxer and was severely punished in the first eight rounds by the powerful and hammering blows and the unfair and illegal fighting of MARCIANO. The English radio commentator Eamon ANDREWS noted:

"MARCIANO is one of the toughest champions who ever rubbed a foot in resin, but he has never read the rule books. He played a different sport from the one COCKELL was taught. He butted unmercifully, he hit with the elbow, he hit low. A British referee would have sent him to his corner after three rounds."

The catalogue of fouls included kidney blows in the first round, low blows in the third round, in the fourth round opened a butt a wound on the forehead of COCKELL, in the fifth and sixth round again low blows, and he hit three times after the gong at the end of the round, in the ninth round occurred again a butt and a low blow and in the same round COCKELL was hit after he was lying on the canvas. The referee Frankie BROWN admonished or warned MARCIANO not a single time.

This photo shows an illegal butt of MARCIANO. The referee stopped the bout in the ninth round, MARCIANO won by a technical knock out.

The 'Nodder'

THOMAS (1997) wrote:

"The 'nodder" is one of boxing's slickest and sneakiest dirty tricks. I've seen it employed by smart boxers, and I've seen it win dozens of fights. But I have never seen a referee give a

Fig. 69. Professional heavyweight fight between Rocky MARCIANO and Don COCKELL on May 16, 1955.

warning or take a round from a fighter who used it. The nodder is readily accepted by ring officials, sportswriters, and spectators as an accidental butt. It's considered just ' one of those things' that happens in a prizefight. A sort of hazard of the game."

"Don't believe it!"

Most head-butting incidents are planned. And the nodder is probably the best and most dangerous of all the head-butting 'accidents.' It looks so natural that your opponent and his handlers let it pass as unavoidable."

"One boxer I knew used the nodder often that we dubbed him "The Billygoat' and a few other names. But he was an ingrate, and disapproved of all the nicknames. He liked to respond to our jeers by saying. 'Look, fellas, I ain't no billygoat. I'm a very polite boxer. Like when the other guy misses with a left, I let him know that I think he's smart. I nod respectfully. It's not my fault if he wants to kill the top of my head. And it's not my fault if he's so eager that his rushing face crashes my head so hard it breaks his face. Who can blame me?"

"Apparently no one blamed him. All the referees let him get away with his polite nods during the two years I knew him. Since then, I've watched others employ a version of the nodder many times."

"Few of the other had Billygoat's skill, but most of them ended their fights with the nodder. However, I consider Billygoat's nodder the best headbutt I 've ever seen. In fact, I've borrowed it a few times, and it has always worked for me."

"One boxer is starting the nodder to 'slide' his straight left past the head of his opponent. It's a deliberate miss."

"To get the most out of the nodder, you should start your jab viciously, deliberately miss, and then step forward and bow your head. Of course, to get the maximum out of the nodder you must make all the moves – the jab, the nod, and the step forward – simultaneously."

"Even the sharp boxers fall for the nodder. When the classy boxer sees that your left is missing, he will attempt to counter. And, as he's coming forward and your nodding head is plowing in, the inevitable crash is a real knock out finisher."

"Every referee buys the nodder as a natural accident. They seem to think it's one of those things that can't be helped. A regular hazard of the fight game. I've seen more than a hundred fights terminated with the nodder. I've seen noses broken, eyebrows gashed, lips almost completely destroyed, cheek bone busted, and jaws broken when the nodder experts go into action."

"You can be an expert user of the nod trick, and it doesn't take much practice. Of course, I suggest that you not practice with a sparring partner. Work on a heavy punching bag, take your boxing stance, fire a left jab and miss, nod and step forward. Let the top of your head collide with the bag. After a few practice runs you can swing the bag and then butt it as it comes toward you."

"In closing, I must warn you that the nodder can, at times, be very painful to you. I've seen more then one dude get his scalp gashed from the shattered teeth of his victim. Of course, most of the victims catch ten times more hell than the nodding boxer does."

"I don't like to use any tactic unless I can hurt the other guy and not get hurt myself. Of course, an important victory is often worth the chance you must take."

Blows below the Waistline

Blows below the waistline are illegal. A blow *'below the belt'* is an euphemistic term for direct injury to the testicles, they are extremely painful.

The referee has to admonish and warn a boxer who commits such a foul.

Professional welterweight fight between champion Emile GRIFFITH and Jorge FERNANDEZ on December 8, 1969, at Las Vegas, Nevada. FERNANDEZ writhes on the canvas (Fig. 70) as referee Harry KRAUSE stands over him after going down from a hard blow under the waistline thrown by GRIFFITH in the ninth round. GRIFFITH was awarded the fight for being ahead on points at the time.

The so-called 'Rabbit Punch'

The referee has to warn a boxer who commits such an illegal blow and has to disquality him immediately if such a blow is repeated. Everybody who watches boxing fights will attest that 'rabbit punches' are relatively frequent and are very often not penalised.

The former professional heavyweight Jack SAVAGE, who retired from his active career in 1918, discussed in 1936 the so-called 'rabbit punch' (*Old Fighter, The Saturday Evening Post, 208, March 7, 1936*):

"One night in 1914 I invented the 'rabbit punch'. I was fighting Jack ROWED, a big, powerful man, at Brown's Gymnasium. He fought in a crouch and with his head thrust forward on his neck as he threw a punch. Reflex action brought my right hand down hard, as tough wielding a hammer, to the base of the skull. He dropped as though polaxed. Out for a quarter of an hour. Bill BROWN was refereeing:

"I had discovered a new punch – one that, not even FITZSIMMONS had ever known. I felt pretty good. I tried it

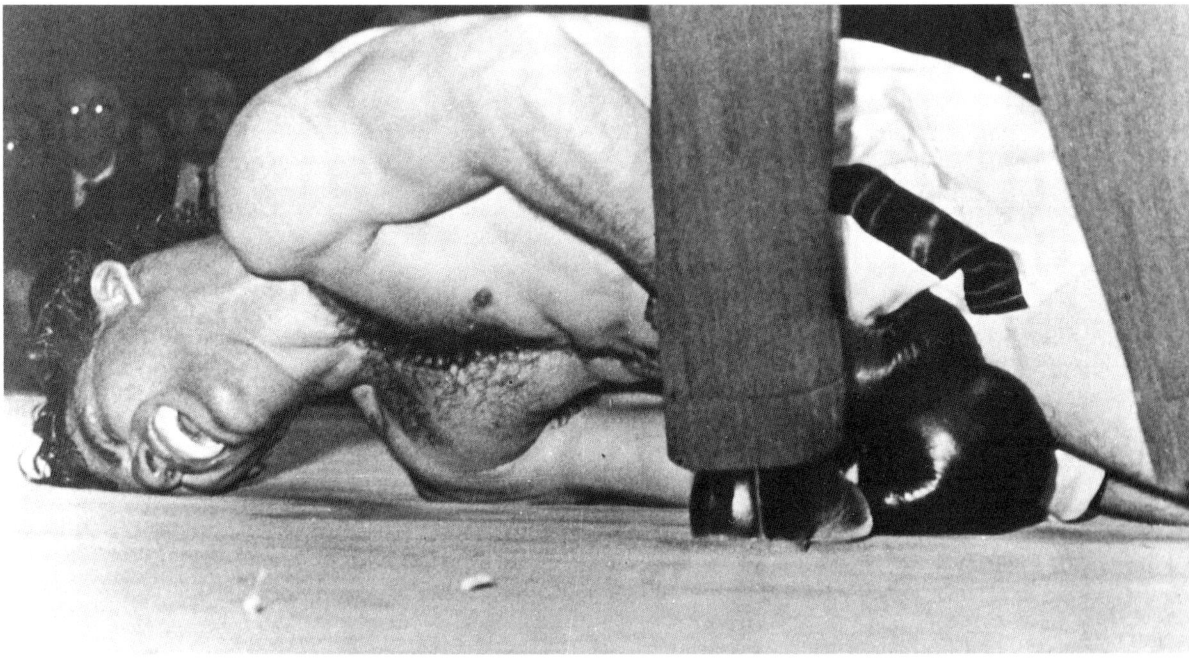

Fig. 70. Professional welterweight fight between Jorge FERNANDEZ of Argentina and Emile GRIFFITH on December 8, 1962 in Las Vegas, Nevada. Photo: A.P.

out on a number of times in shadow boxing. I used it again, two fights later, on Jom DONOVAN. And the same thing happened. DONOVAN went down for the full count in the first minute of the first round. And again Bill BROWN was the third man in the ring. He finished counting DONOVAN out and the turned to me."

"'Jim.' He said gravely, 'I'd throw that punch away if I were you. If you keep on using it, you're going to be arrested some day for murder.'"

"'I threw it away, I never used it again, I taught it to Jess WILLARD, and he used it a couple of times, without effect, on Jack JOHNSON. A rule banning it was put into effect.'"

TUNNEY wrote 1932 about the 'rabbit punch:'

"This rabbit blow is the most dangerous blow in boxing. It was barred originally through the action of Jack DEMPSEY in his fight with Bill BRENNAN in Madison Square Garden. After BRENNAN had started to the canvas from a body blow, DEMPSEY hit him a fearful blow on the back of the neck on his way down. This left BRENNAN prostrate on the canvas."

TUNNEY continued:

"The rabbit blow is particularly dangerous and damaging because it is directed at the point where the base of the skull and the cervical vertebra meet – the lower brainstem, called the medulla oblongata. This nerve center is actually the center of life. It controls respiration, heart action, and locomotion. Many doctors are agreed that it is the twisting of the brainstem by a blow or jolt that causes the so-called knock out. No man can stand eighty or ninety rabbit blows from the fists of a heavy hitter like DEMPSEY, as I did in each fight, without feeling the effects. It brings a deadening headache which seems to rest in the base of the skull. There is a numbness which reacts on the whole nervous system."

In this area of the nape of the neck, the region where that base of the skull and upper cervical vertebrae meet, the lower brainstem, the medulla oblongata, and the upper cervical spinal cord are located. This area can be termed a life center, because it controls respiration, heart action and locomotion. Blows against this region are therefore on the one side very effective and on the other very dangerous. Many authors emphasized that a blow at this region could lead to a shifting of the brainstem.

The professional fight between Jörg EIPEL and Alain MARION (for details see p. 449), in which EIPEL suffered a severe brain injury with following death, MARION delivered in the first six rounds 18 times 'rabbit punches' and the following rounds a total of 21. The referee warned MARION very late in the fight.

Professional Boxing Fights in which Participants committed Different Fouls of which some were photographically documented

COHANE (1963) described the boxing match between Gene TUNNEY and Harry GREB:

"Few warriors would have stomached the full fifteen rounds of TUNNEY's only professional defeat. He was defending the light-heavyweight title against Harry GREB, a cruel, ceaseless mixer, whose disregard of rules made DEMPSEY look like a ladyfinger. GREB fractured TUNNEY's nose twice, opened cuts under his chin with head butts, and practically punched him into a pulp."

COHANE continued:

"The blood TUNNEY swallowed, the adrenalin chloride with which his seconds tried staunch his bleedings, and the rum and orange juice which they plied him – all combined to nauseate him, but he fought on. Only when he tried to climb the steps to his dressing room did he collapse."

The sports journalist W. O. McGEEHAN termed Harry GREB's style as 'the manly art of modified murder.'

The sports journalist Bert SUGAR (1984) wrote about him:

"The GREB style was dictated by his personal estimate of what the opponent and the referee would hold still for. Somewhat lacking the social amenities of the Queensberry rules, GREB fought as an early-day terrorist, without rules, inhibitions or prisoners taken, his opponents more often than not feeling as if they had to fight him off, rather than actually fight against him."

Harry GREB, 'The Pittsburgh Windmill' who died at age 32 participated in 294 boxing bouts – alone in the year 1919 in 44 fights – became blind in one eye in a bout with Kid NORFOLK on August 29, 1921. He was able to keep his blindness secret until his death about five years later.

The sports journalist Bert SUGAR (1984) wrote about Jack DEMPSEY:

"He used every possible means at his disposal to win, his definition of survival less a breaking of the rules than a testing of their elasticity – hitting low, after the bell, behind the head, while a man was on the way down, and even while he was on the way up. 'Hell,' he said, 'it's a case of protecting yourself at all times.'"

Importance of Animal Models for the Interpretation of Boxing Injuries

Investigations in which uncontrolled and controlled mechanical forces applied to the heads of animals are correlated with resulting morphological alterations are few. The first of these was conducted by Alfons JAKOB in 1913. Further studies were conducted by PETERS (1943), WINDLE et al. (1944), WINDLE and GROAT (1945), GROAT et al. (1945), DENNY-BROWN (1945), GROAT and SIMMONS (1950), UNTERHARNSCHEIDT (1958, 1963, 1983, 1986, 1994), and FRIEDE (1960). Injuries caused by translational and rotational motion of the unrestrained head are characterized by typical morphological alterations, which are predictable for each type of impact. Predictable patterns have been observed through analysis of the results of impact. Translational and rotational motion are most frequently observed in combined form in boxing, but may be seen individually in the straight blow (see Figs. 11 and 12) and the hook (or uppercut) (see Fig. 13) respectively.

Any broadly surfaced impact imparts to the head a force by which it is accelerated. If the impact is directed at the center of mass of a freely movable rigid object, the resulting motion is a translational acceleration. If the impact is directed eccentrically, the result is a combined translational and rotational acceleration. The magnitude of the rotational acceleration is related to the degree of eccentricity of the acting force. The magnitude of the translational acceleration is related to the distance between the point of fixation and the center of gravity of the head. The distinction between the two types of acceleration is important in view of the different tissue alterations they initiate in the brain. Pure translational acceleration creates pressure gradients, while rotational acceleration produces rotation of the skull relative to the brain. Both processes are the effects of mass inertia of the brain.

It can be expected, according to the physical analysis of translational and rotational accelerations with exactly measured inputs, that different mechanisms produce different pattern of tissue lesions.

Experiments with different animal species, which employed translational and rotational accelerations with exactly measured inputs are summarized and the morphological alterations in respect to distribution and quality are discussed.

The inertia effect is fundamental for all injuries caused by mechanical forces. The effect can be studied on a rigid, freely movable spherical shell filled with an incompressible, nonviscid and homogenous fluid.

After a central impact, the shell moves only in translation, carrying with it the contained liquid. While the blow lasts and the sphere moves, its contents are compressed at the impact side, in consequence of the inertia of the liquid particles, but are rarefied (expanded) at the side opposite to the impact pole (the 'counterpole'). This is the basic pressure distribution picture due to an acceleration motion of the skull. Of major importance is the dependence of the pressure distribution on the time variation and duration of the blow. If the impact is not centrally directed, the skull receives an angular momentum in addition to the acceleration. The farther from the center of rotation the particle of interest is, the greater the acceleration suffered, and thus the greater the potential for injury. Therefore, points near the center of rotation are likely to be damaged only if the acceleration is severe. Rotational traumas involve considerable tensile forces between the accelerated skull and the inert brain.

When investigating brain damage caused by boxing, interest is focussed on the subconcussive dose, as it has been observed that a single application does not create immediate clinical symptoms (specifically no loss of consciousness or in animals unresponsiveness). By contrast, the concussive dose produces immediate unconsciousness of several seconds' or minutes' duration, although after a single application no additional clinical symptoms are observed. Both levels of impact are too low to cause primary traumatic lesions, such as hemorrhages, between the enveloping structures of the brain, in the brain, or so-called cortical contusions, etc. Higher intensities – those which are capable of producing primary traumatic brain lesions – will not be considered here, as they do not have great significance for the boxer. However, boxing blows and falls in the ring do produce lesser primary traumatic lesions, including superficial 'cortical contusions.'

Models with Translational Acceleration

It was observed during the experiments conducted by UNTERHARNSCHEIDT (1958, 1963, 1965, 1970) and SELLIER and UNTERHARNSCHEIDT (1963) on translational acceleration, that behavior and histological findings remain unchanged after the application of a single subconcussive blow. Repeated applications however, produced severe permanent brain damage due to circulatory disorders.

'Repeated applications' here means blows that were delivered at intervals ranging from 5 to 20 seconds, simulating closely the situation in the boxing ring when a groggy contender takes an increased number of rapidly succeeding blows. It is especially noteworthy that animals examined immediately, or a few days after, the experiment presented no histopathological findings, and that severe brain lesions were seen only after prolonged survival times (from several days to weeks).

A *single impact of subcommotio or subconcussive strength,* at a speed of 7.1 m/s, corresponding to an acceleration of 205 g, imparted to the free head of a cat, caused neither behavioral nor histological changes in the central nervous system; whereas repeated impacts of the same intensity, without causing primary traumatic tissue lesions, did produce secondary traumatic alterations due to circulatory disturbances. Lesions of the cerebellum included disseminated or total loss of Purkinje cells – especially at the summits of the lobuli of the vermis –, they were associated with proliferation of Bergmann's glia, thinning of the granular cell layer combined with glial reaction, and glial proliferation in the striae medullares and cerebellar white substance (Figs. 71, 72 a and b, 73 a – e, 74 a and b). By comparison, the pathomorphological alterations in the cerebrum were less severe; they consisted of disseminated ischemic nerve cells and a moderate glial proliferation in the white substance.

Impacts of commotio or concussive strength, that is, producing the clinical symptoms of commotio cerebri (unresponsiveness in cats) have a velocity of 8.3 – 9.4 m/s, corresponding to an acceleration of 280 – 400 g. After one such impact, the morphological alterations which account for the clinical symptoms of commotio cerebri or cerebral concussion prove to be traceless, according to SPATZ,

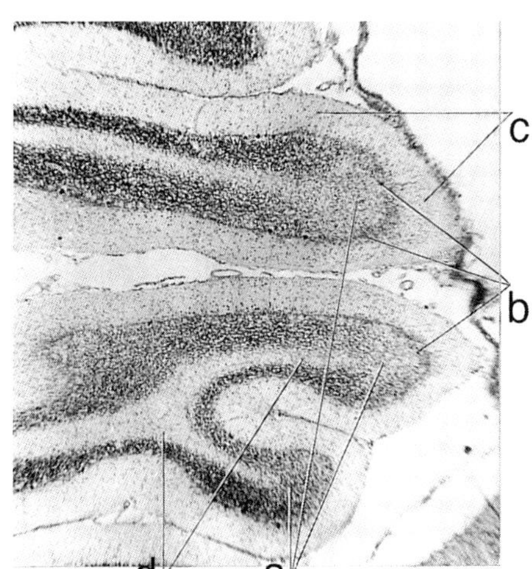

Fig. 71. Cat. Cerebellum. Pronounced reduction of the granular cell layer at the summits of the cerebellar vermis at a. Loss of Purkinje cells in rows and proliferation of Bergmann cells in rows at b. Astro- and microglial proliferation in the molecular cell layer at c and in the white substance at d. Nissl technique, 20:1.

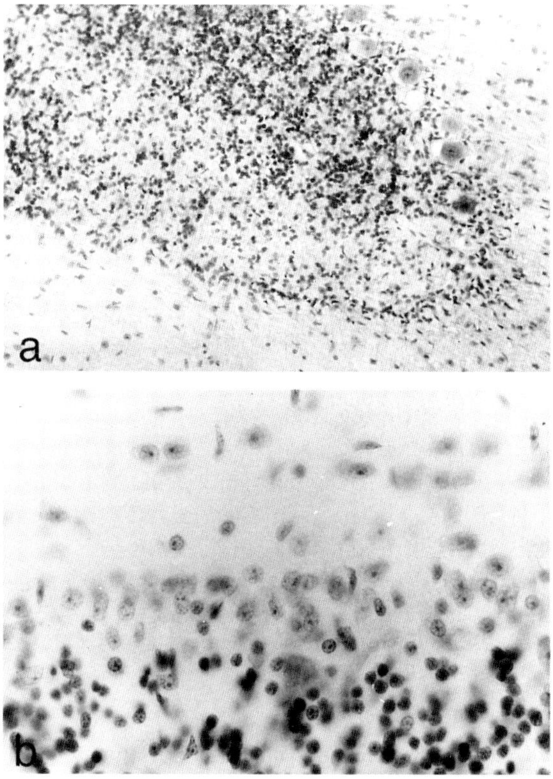

Figs. 72 a and b. Cat. Complete loss of Purkinje cells and proliferation of Bergmann glia. Microglial and astroglial proliferation in the molecular cell layer. Moderate loss of cells in the granular cell layer. Nissl techniques, 320:1.

with the methods of investigation used today. We found, in particular, no evidence for glial cell proliferation. There are no staining methods available for light microscopy that would detect any possible alterations.

However, after repeated impacts of equal intensity and at intervals of 1 –2 days, the result was apparent – severe permanent brain damage; the cerebral cortex showed, beside disseminated ischemic nerve cells, extensive focal and pseudolaminary necroses of the parenchyma and loss of nerve cells (Figs. 75, 76, 77, 78 a and b, 79) in various parts of the Ammon's horn formation. The loss of neurons was either disseminated or occurred as the complete destruction of a sector of the band (Figs. 80 a and b, 81 a and b). Tissue alterations in the cerebellum, although less intense, corresponded in quality to those caused by successive impacts of subcommotio or subconcussive strength.

These findings lead to the unfortunate but inescapable conclusion that impacts of intensities that do not cause noticeable tissue alterations when applied single, but that repeated impacts of lesser intensity (as employed in these experiments), when applied in close succession, caused brain damage of greater severity than that produced by the higher impact intensities applied at longer intervals (several days to one week). A sustained permanent brain injury can therefore result from secondary tissue lesions alone, with no primary traumatic alterations present at all. It follows that impact intensity and time interval are significant pathoplastic factors in relation to the extent of the permanent brain damage that is produced by blunt blows.

Spinal Cord The spinal cord shows extensive demyelinations in the pyramid tracts (Figs. 82 a and b, 83).

Considerable primary traumatic lesions are produced by impacts with a velocity of 10.5 m/s or more, corresponding approximately to more than 400 g. In all instances, there were subarachnoid and subdural hemorrhages, cortical contusions at the counterpole and, at the impact pole, single in-

Figs. 73 a – e. Cat, cerebellum. a. Summit of a convolution showing complete loss of Purkinje cells and formation of glial shrubs, reduction of the granular cell layer with survival of the more resistent cells of the Golgi type, and proliferation of astroglia and microglia in the molecular cell layer, Nissl technique, 128:1. b. Cat. Cerebellum. Serial loss of Purkinje cells is associated with considerable glial shrub formation (proliferated microglia). There is disseminated proliferation of astroglia and microglia in the molecular cell layer. Nissl technique, 135:1. c. Rabbit. Cerebellum. Control animal. Purkinje cells, granular cell and molecular cell layers are unremarkable. Nissl technique, 320:1. d. Cat. Cerebellum. The Purkinje cells contain vacuolated cytoplasm. The nuclei are no longer clearly discernible. There is beginning proliferation of local glia. Nissl technique, 320:1. e. Cat. Cerebellum. Shrunken and hyperchromatic Purkinje cells with beginning reaction of local glial elements. The granular cell layer is unremarkable. Nissl technique, 320:1. (From UNTERHARNSCHEIDT, 1975).

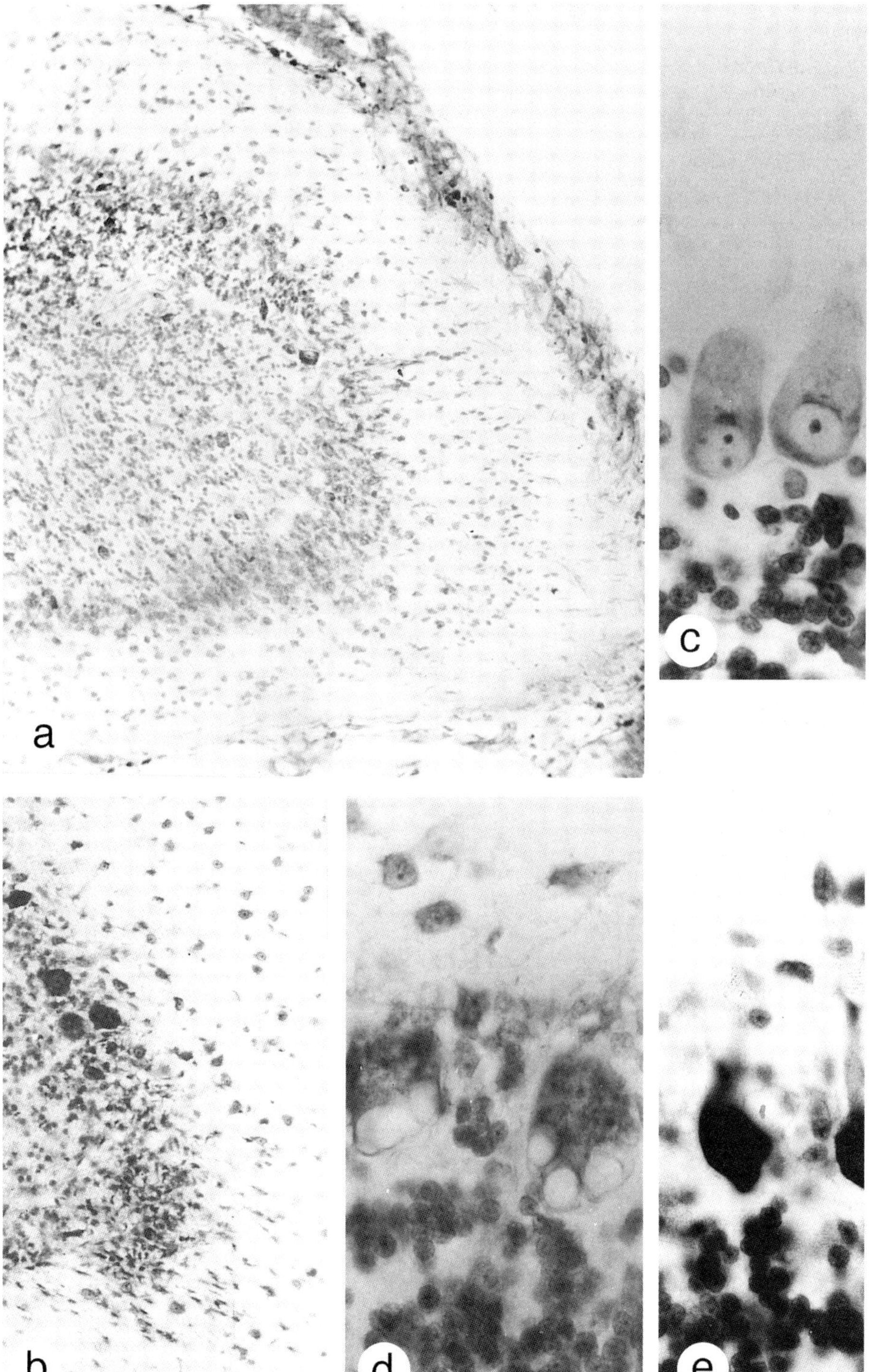
a
b
c
d
e

tracerebral hemorrhages and traumatic necroses. Velocities of the impacting instrument of 17.2 and 18.3 m/s are fatal to a cat.

Rotational Accelerations

In the initial phase of a rotational trauma (as in the uppercut or hook), the accelerated skull moves ahead while the brain is delayed by its inertia. This produces considerable strain and shearing forces. It was hypothesized that by comparison to the translational trauma, the rotational impact would produce a different pattern of lesions. This was proven to be correct. Unlike the cylindrically symmetric location of translational impact lesions, these rotational impact lesions were located radially symmetrically along the midline of the brain. Findings included subdural hemorrhages caused by tearing of the bridging veins, subarachnoid hemorrhages situated mainly close to the midline, and straining and tearing of vessels in cranial nerves, in superficial layers of the cerebral cortex (and, with increasing impact intensity, in deeper portions of the brain). In addition, there were vascular tears due to strain in the medulla oblongata and the entire spinal cord, located far from the impact location but along the axis of the strain.

These experiments have direct significance for boxing injuries. The experimental results throw light on the pathogenesis of subdural hemorrhages, which alone account for more than 75% of ring incidents and thus occupy first place in the cause of boxer deaths.

Non-deforming Rotational Acceleration of the Head using Head Acceleration Device II (HAD-II) (Fig. 84)

The effects of rotational acceleration on the CNS were not investigated previously. The first experimental study with measured input to study the pathomorphology of rotational acceleration in an animal model was carried out by UNTERHARNSCHEIDT and HIGGINS (1969). Adult female squirrel monkeys (Samirii sciureus) were used. The kind and location of traumatic lesions related to this type of trauma were studied. Surviving animals were sacrificed five days after the experimental injury. Histological examinations were performed without any foreknowledge of the acceleration endured, and only after the pathological findings were recorded was a correlation with mechanical inputs attempted.

The experiments were conducted using a machine designed for rotational acceleration inputs to the head. It consists of a helmet in which the monkey's head can be secured with plaster of Paris. The path taken by the head is limited to a 45° arc by the geometry of the machine (Fig. 84). The helmet is driven by a rod that follows an eccentric pattern, machined in a flywheel, giving a triangular waveform. A motor-driven transmission and tachometer provide exact control of flywheel speed, and the rotational acceleration of the head.

Because of technical difficulties in direct measurement of the very high angular accelerations available, the rotational acceleration was calculated from the linear acceleration of the follower rod and the known geometry of the machine. Details of the calculations used and the expected small errors resulting from this method of measurement may be found in HIGGINS and SCHMALL (1967) and HIGGINS et al. (1966).

The machine is capable of producing only non-deforming blows. Fig. 85 shows kinematic arrangement of the machine. The point F on the diagram corresponds to the ear canals for the plaster-imbedded monkey head. The axis of rotation for the monkey is somewhere in the lower cervical spine, probably in the neighborhood of C_7/T_1. The position of the head within the helmet is fixed by ear and canal pins and a maxillary bite bar used temporarily during the imbedding procedure. (Pins and bar are removed after the plaster hardens).

The head acceleration machine produces a waveform consisting of a pair of triangles – one representing acceleration, the other deceleration. Only the peak positive acceleration is reported here. However, a complete description of this waveform for each monkey tested can be found in a previous report (HIGGINS et al. 1966). From a knowledge of the mechanical properties of the aluminum and steel structure and measured properties of plaster of Paris, it was concluded that there was little or no attenuation of the waveform delivered to the monkey head by the rod and helmet.

On the day of the test, a locally acclimated monkey was selected at random from the colony, weighed, examined and anesthetized with sodium methohexital by the intravenous route. The head was shaved and electroencephalogram leads attached. The scalp was then covered with a rubber condom and the helmet, bite bar and ear pins placed with the monkey secured in a restraining chair.

Plaster of Paris was then poured over the head and allowed to set, pins and bar were removed, leads were attached, and the

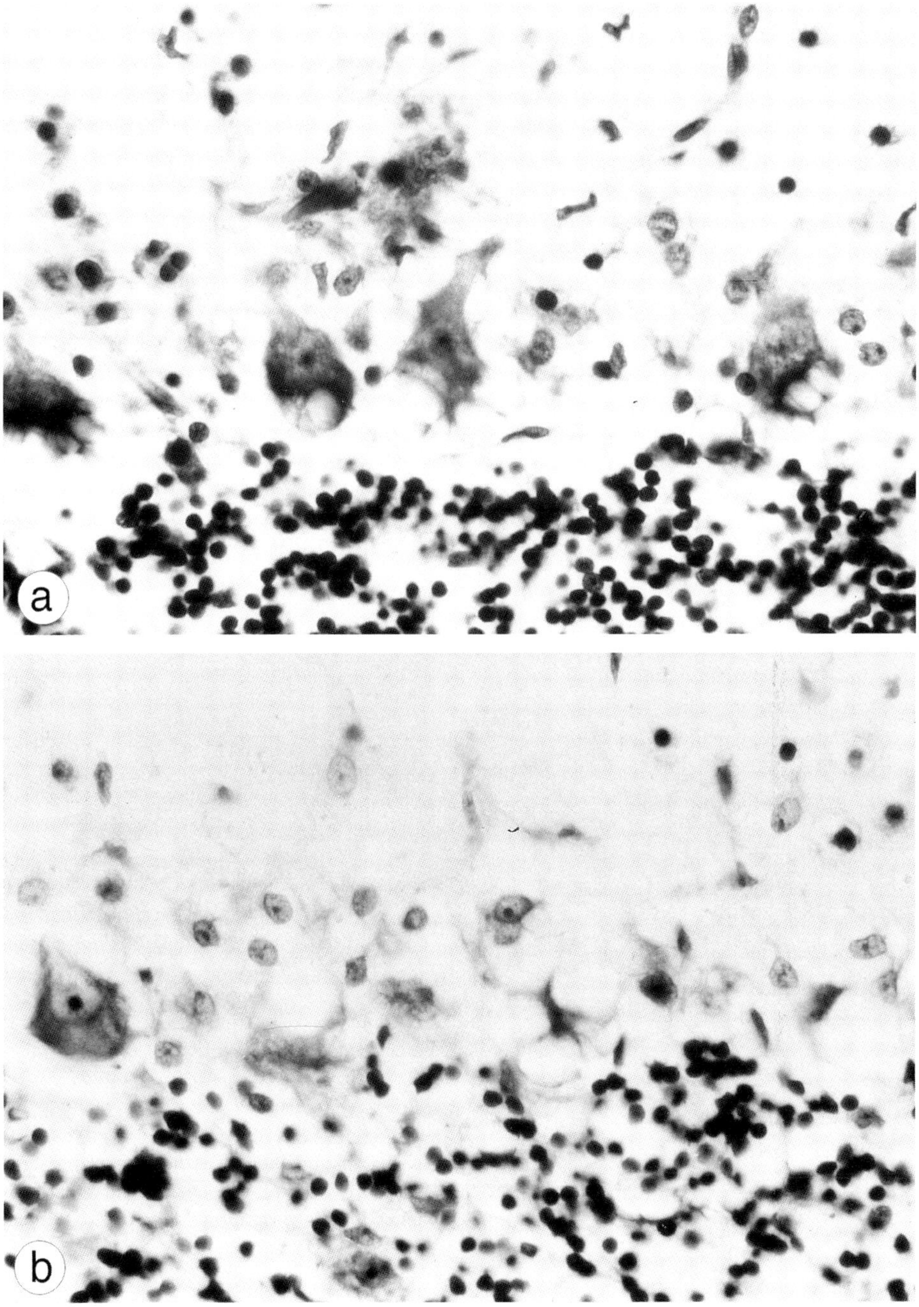

Figs. 74 a and b. Cat. Cerebellum. Extremely altered Purkinje cells, partly with vacuoles. The nucleus is severely altered. Proliferation of Bergmann glial elements, microglia and astroglia. Formation of symplastic cell formations. Nissl technique, 320:1.

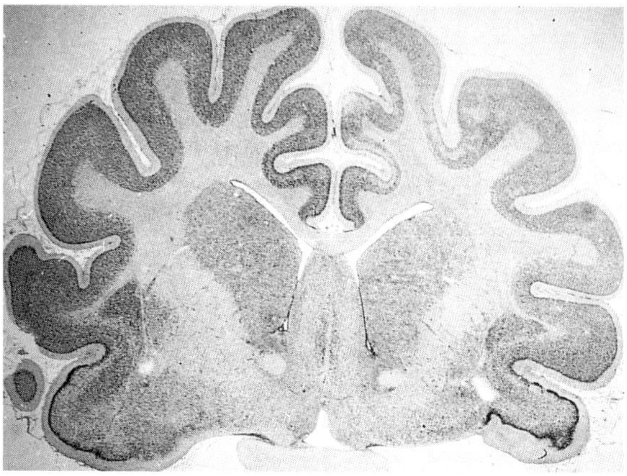

Fig. 75. Cat. Cerebrum. Extensive spotty and pseudolaminary elective parenchyma necroses, mainly in the right cerebral hemisphere. The loss of neurons is clearly discernible with this low magnifiacation. Nissl technique, 2:1.

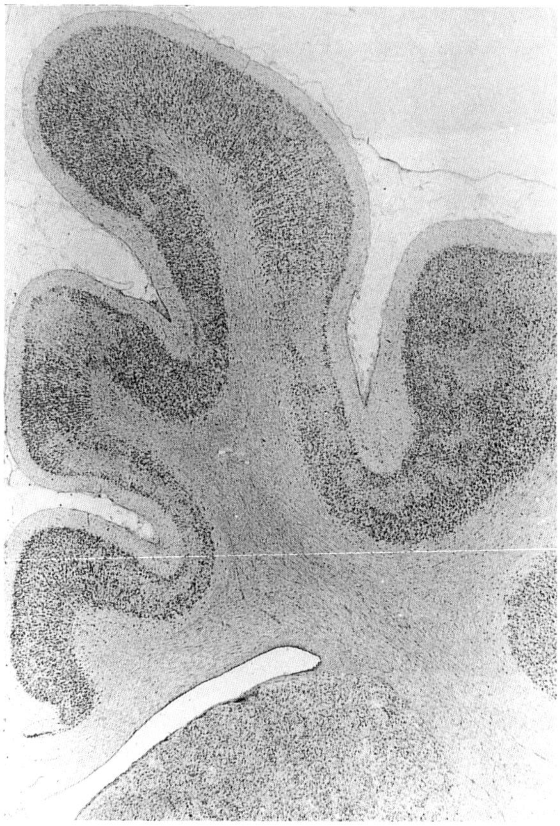

Fig. 76. Cat. Cerebrum. Extensive spotty and pseudolaminary elective parenchyma necroses in the cerebral cortex, proliferation of glial elements, especially astrogia. Nissl technique, 5:1. (From UNTERHARNSCHEIDT, 1963).

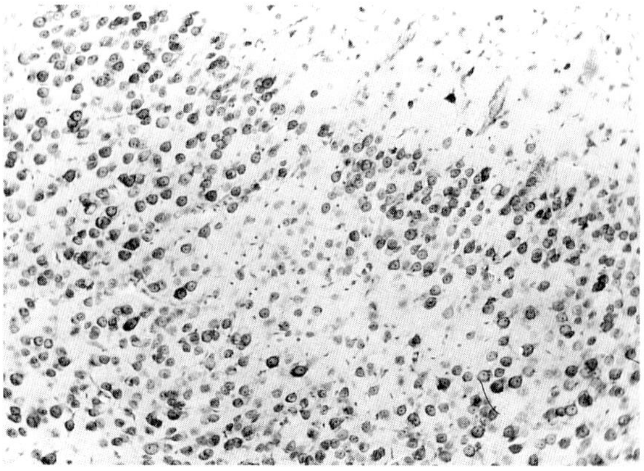

Fig. 77. Cat. Cerebrum. Circumscribed elective parenchyma necrosis in layers of the cerebral cortex. Numerous neurons with swollen nuclei and at the margin located nuclei. Proliferation of microglial elements in the molecular cell layer. Nissl technique, 80:1. (From UNTERHARNSCHEIDT, 1963).

helmet attached to the maschine, the electronic recording system calibrations recorded and the motor started. A brief physical examination was then performed. Anesthetic level was adjusted so that purposive responses to stimuli were present.
The machine was fired at the predetermined acceleration level. Once the acceleration level required for a lethal blow was determined as well as that for essentially no observable clinical effect, an experimental design was formulated to determine the 50% level for concussion. Each monkey received only a single blow (with the exception of monkey 49-2 (A-67-24), which received multiple blows).
The squirrel monkeys were autopsied immediately following death. Brain and spinal cord were removed in toto and fixed in 10% formalin. The brain was cut in coronal sections, midbrain, pons, medualla oblongata, and cerebellum were removed applying the midbrain section as described by SPATZ, then cut in coronal sections and embedded in celloidin.
The entire spinal cord at the level of the nerve roots of each cervical, thoracic and lumbar segment was cut in blocks and embedded in paraffin, then cut stepwise in serial sections. The thickness of the celloidin sections was 20 μ, that of the paraffin sections 9 μ. The histological sections of the cerebrum, midbrain, pons, medulla oblongata, and cerebellum were stained with the Nissl method and with hematoxylin-eosin; those of the spinal cord were also stained with the Van Gieson and Masson-Goldner method.

Results

Twenty-five monkeys were subjected to the controlled acceleration environment. Three additional monkeys were acclimated to the colony and then sacrificed as controls.

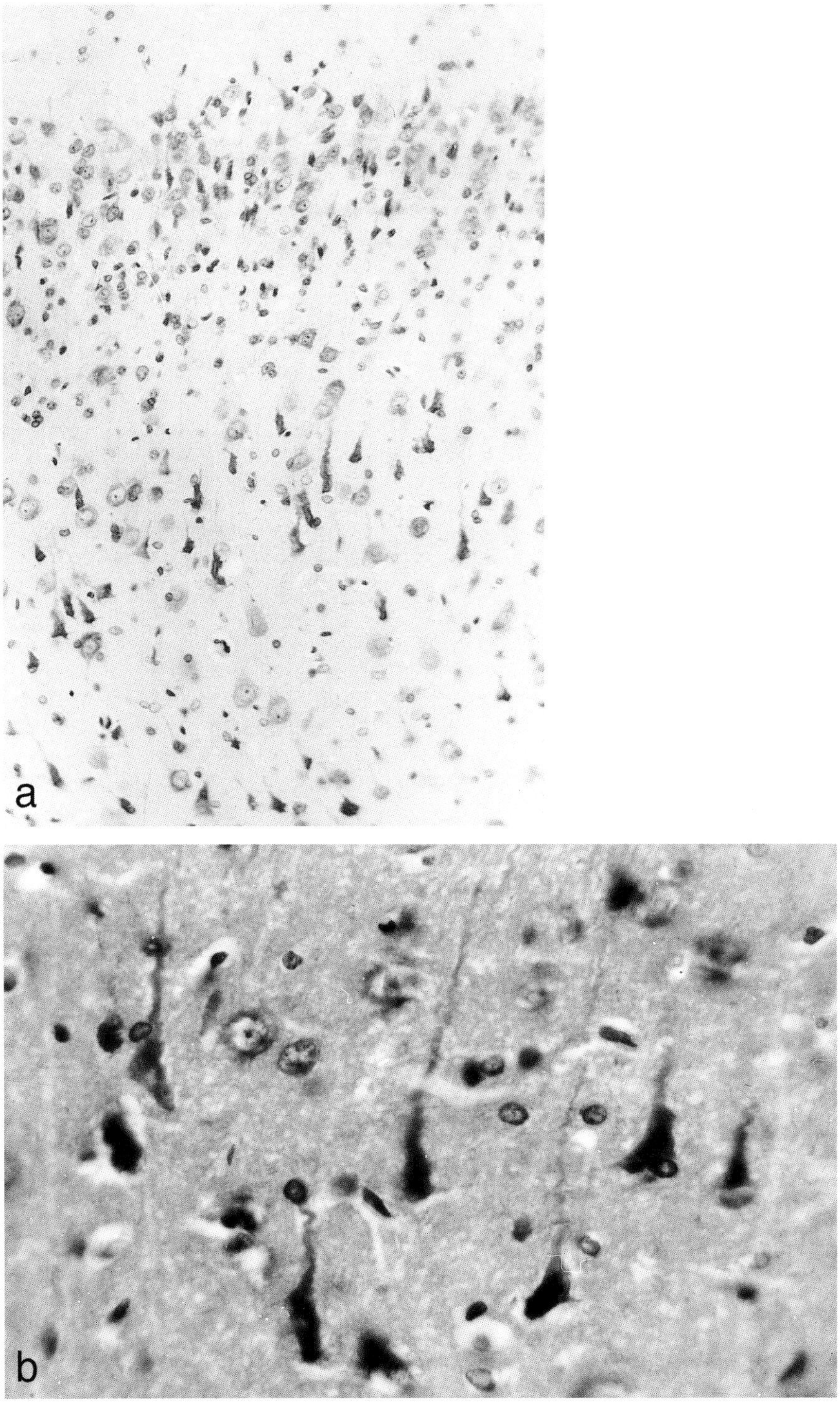

Figs. 78 a and b. a. Rabbit. Cerebrum. Ischemic neurons with pink stained corkscrew like processes. Beside neurons with considerable swollen nuclei and marginal cytoplasm around the nucleus. Nissl technique, 128:1. b. Close up of Fig. a. Ischemic neurons with pink stained corkscrew like processes and shrinkage and hyperchromatosis. Nissl technique, 320:1.

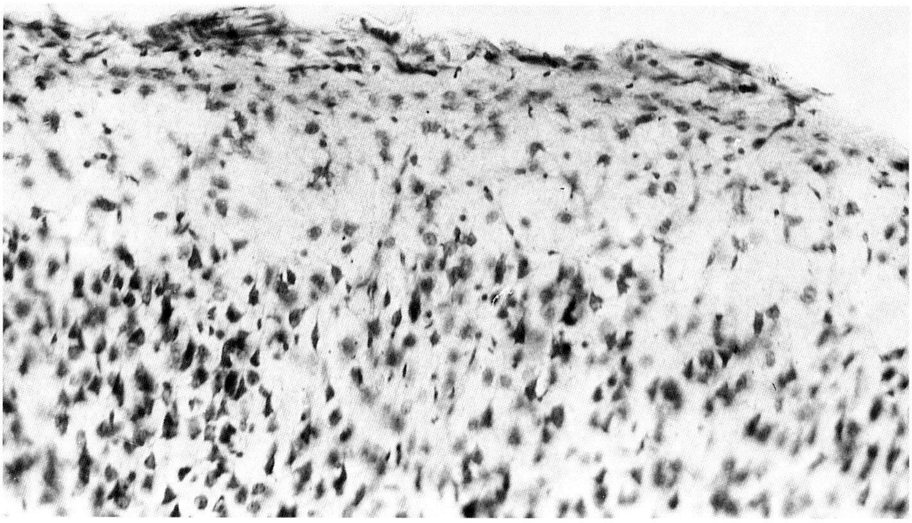

Fig. 79. Cat. Cerebrum. Thickening of leptomeninges. Massive proliferation of astroglial and microglial elements in the molecular cell layer. Loss of neurons in the second and third cortical layers, with hyperchromatosis and astroglial and microglial reaction. The stratification of the cortex is lost. Nissl technique, 128:1.

The range of available rotational accelerations was found to produce a continuum of clinical effects from no observable signs through concussion to death. Table 3 shows the monkeys tested, levels of peak angular acceleration, and the major clinical findings. Results of the statistical study of 50% concussion levels have been reported by HIGGINS and SCHMALL (1967).
Three animals died immediately after the acceleration, and the remaining 22 were sacrificed at five days. At the time of autopsy, the gross pathological changes were noted and are presented in Table 4. These findings are with the dura mater intact. Table 5 summarises the morphological findings.

Kinematic Aspects of Head Injury

In any generalized motion of the head, the motion can be described as a translation of the center of mass of the head (Fig. 86) and a rotation about the center of mass. It can be shown on kinematic grounds alone (OELKER et al. 1966) that the head-brain system would be more sensitive to the rotational component than to the translational. In fact, the farther from the center of rotation the particle of interest is, the greater the potential for injury. Thus, a point near the center of rotation is likely to be damaged only if acceleration is severe.

Rotational traumas involve considerable tensile forces between the accelerated head and the inert brain. It can be expected, according to the physical analysis of translational and rotational trauma, that different mechanisms produce different patterns of lesions. In the histological evaluation special attention was given to the effect of rotation of the various parts of the spinal cord – an aspect that had been neglected thus far.
Very few groups have studied rotational acceleration. None of these have controlled the path taken by the head, or the waveform of acceleration used. The experimental designs closest to our own are those in which the head is caused to move by inertia forces resulting from whole body acceleration (MARTINEZ et al. 1968). However, in our setup, the head is moved while encased in a plaster-filled helmet. Hence, the primary input is to the head and not to the whole body. Nonetheless, we have demonstrated significant lesions in the spinal cord.

Mechano-Pathologic Correlation: Immediate Death

Three of 24 animals died spontaneously (A-67-13 after 8 minutes, A-67-2 after 16 minutes, and A-67-1 after 20 minutes). They had suffered the highest rotational acceleration, with 3.27, 3,63 and 3.86 x 10^5 rad/s^2. Subdural (A-67-2) and subarachnoid (A-67-13) hemorrhages predominated. Additionally, there were rhectic hemorrhages of the cerebral hemispheres →

Figs 80 a and b. Cat. (a) Ammons' horn formation. At the left side segmental loss of neurons in a row. Nissl technique, 5:1. (b) Segmental loss of neurons in Sommer's sector (h_1) with marked microglial proliferatiobn Nissl technique, 20:1. (From UNTERHARNSCHEIDT, 1963)

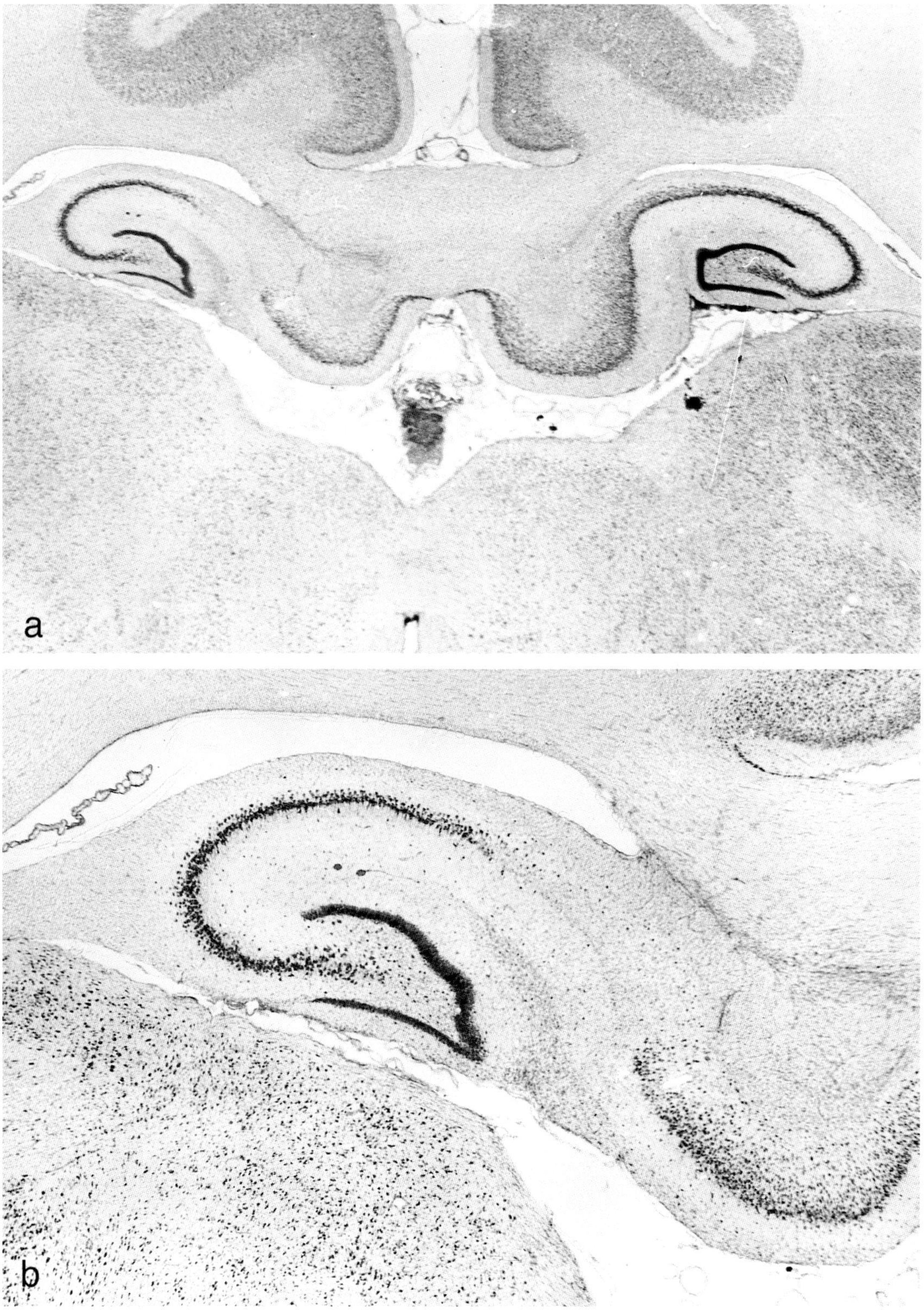
a
b

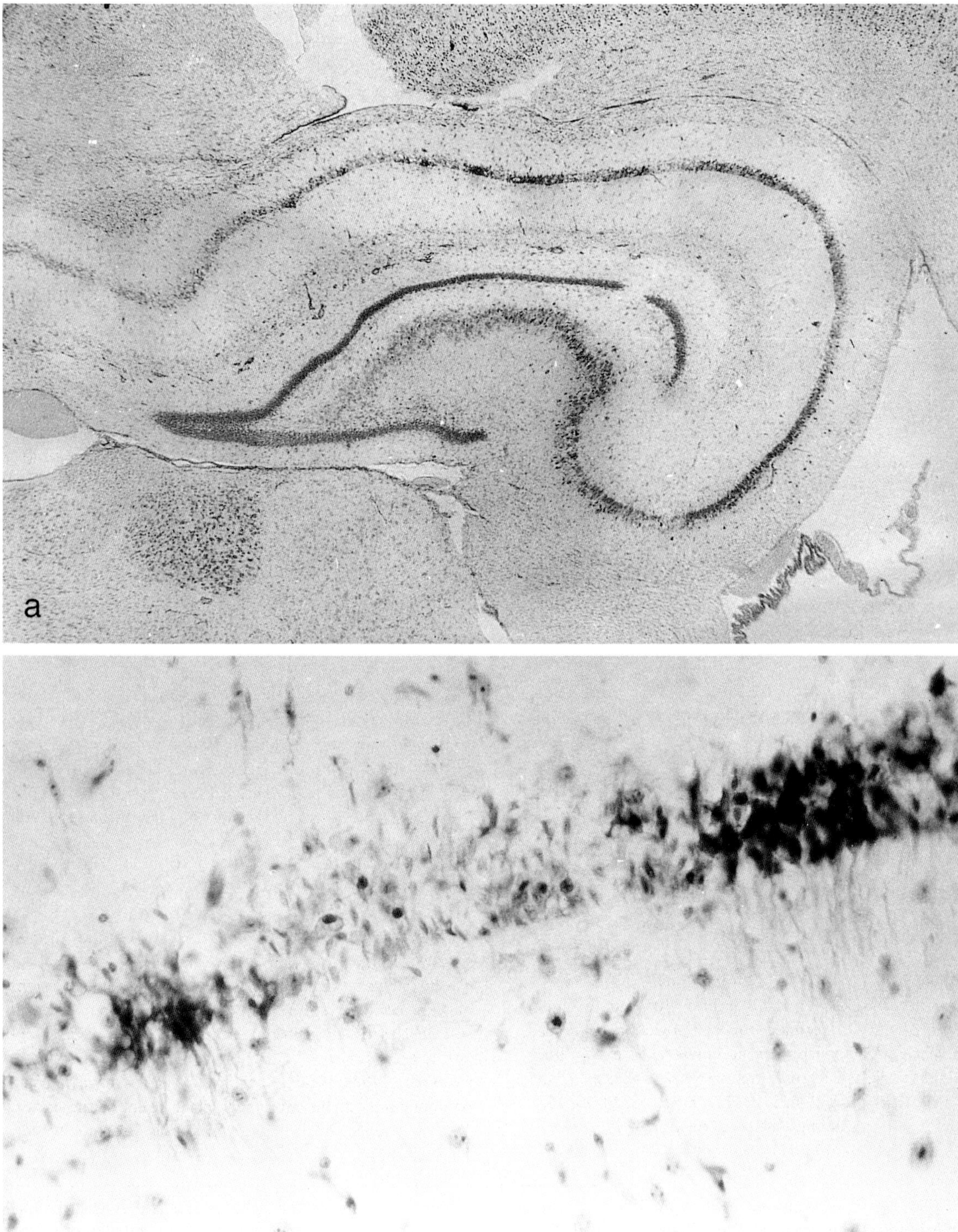
a
b

near the midline fissure (A-67-13), rhectic hemorrhages at the base of the frontal lobes and/or temporal lobes, near the midline (A-67-13, 1, 2), and at the occipital lobes, near the midline (A-67-1). These lesions consisted of rhectic hemorrhages, mostly venous, of the outer layers of the cortex, and capillary hemorrhages in the same areas. Only these animals showed additional hemorrhages in the more central parts of the brain. For example, in the hippocampus formation (A-67-13, 1) and around the third ventricle (A-67-1, 2) venorhectic or capillary hemorrhages due to overstretching of vessels in these areas were found.

Rhectic hemorrhages as the name suggests, are caused by tearing of the vessel wall by mechanical force. They are subdivided into arteriorhectic, arteriolorhectic, capillarorhectic, and venorhectic hemorrhages. Occasionally, a massive hemorrhage covering the wall makes it impossible to ascertain the origin of the bleeding. *Diapedetic hemorrhages,* on the other hand, develop over a period of time, blood cells and –plasma penetrate the vessel walls in a process called *diapedesis.*

Mechano-Pathologic Correlation: Surviving Animals

The remaining 21 monkeys were sacrificed after five days. It was obvious that the degree of primary traumatic alterations increased with the angular acceleration.

The lesions included subdural hemorrhages, subarachnoid hemorrhages, tearing and avulsion of veins in superficial cortical layers leading to rhectic hemorrhages and partial and total traumatic necroses, as well as rhectic hemorrhages in cranial nerves.

Cerebrum (Figs. 87 a and b, 88 a and b)

Five animals (A-67-7, 22, 17, 11, 5) were exposed to the lowest rotational accelerations (1.01 to 1.50 x 10^5 rad/s^2) and showed no primary or secondary morphological alterations in the cerebrum. These were present, however, in ten of thirteen animals subjected to rotational accelerations from 1.50 to 1.97 x 10^5 rad/s^2 (A-67-6, 8, 23, 21, 18, 15, 16, 20, 23, 10, 14, 19, and 9). The findings included subarachnoid hemorrhages in A-67, 23, 21, and 25, combined in one instance (A-67-21) with primary traumatic hemorrhages in the N. oculomotorius, and tears and avulsions largely of veins and capillaries in superficial layers of the cerebral cortex in eight animals (A-67-1, 6, 8, 15, 10, 14, 19, 9). The subarachnoid hemorrhages were found in the lower range of angular acceleration, while the primary traumatic hemorrhages of the cerebral cortex, which must be considered the more severe lesions, were always seen in animals exposed to the higher accelerations. Thus, the four monkeys (A-67-10, 14, 19, 9 – 1.75, 1.88, 1.91, 1.97 x 10^5 rad/$s^{2)}$ subjected to the highest angular accelerations suffered primary traumatic lesions of the cerebral cortex. This fact indicated that with increasing intensity of the angular acceleration, the primary traumatic alterations extended from the periphery to central portions of the brain and primary traumatic lesions are added to the initially present subarachnoid hemorrhages in the periphery of the system. This change in the pattern of distribution was foreseen in the analysis of injury mechanics.

The highest angular accelerations administered, i. e., 2.05 – 3.86 × 10^5 rad/s^2, in four of six monkeys, caused severe primary traumatic hemorrhages in the cerebral cortex (A-67-12, 3, 4, 13, 1, 2 – 2.05, 2.54, 2.84, 3.27, 3.63, 3.86 × 10^5 rad/$s^{2)}$), and in three animals with intensities of 3.27 × 10^5 rad/s^2 and more, additional primary traumatic lesions in the white substance (located even closer to the central pivot). Angular accelerations of more than 3.0 × 10^5 rad/s^2 were not survived (A-67-13, 1, 2). Although the number of animals in the test is relatively small, the relation between intensity and tissue lesions is quite obvious.

The direct correlation between primary traumatic alterations in the brain and magnitude of angular acceleration can be demonstrated equally well when the 24 test animals listed in Table 3 are divided into two groups and arranged in ascending order of angular acceleration. The first group of 12 monkeys (A-67-7 to A-67-16 – 1.01 to 1,74 × 10^5 rad/$s^{2)}$ showed primary traumatic alterations in the cerebrum in five animals. By contrast, the second group, exposed to higher accelerations (A-67-20 to A-67-2 – 1.74 to 3.86 × 10^5 rad/s^2) showed similar alterations in nine of twelve animals.

The histological examination of the primary traumatic lesions in the cerebral cortex reveals that these lesions do not present so-called cortical contusions, a finding that is clearly supported by photographs. Rather they present venorhectic, occasionally capillo- and arteriorhectic, hemorrhages located predominantly in the more superficial layers of the cortex. There is evidence of tearing of vessel walls. Also, the hemor-

Figs. 81 a and b. Rabbit. (a) Ammons' horn formation. Circumscribed loss of neurons in the Ammons' horn band in h_3, h_2 and h_1. Only small islands of intact neurons at a, especially in the so-called dorsal resistent part of Spielmeyer (h_2). Beginning microglial proliferation. Nissl technique, 20: 1. (b) Rabbit. Ammons' horn formation. Close up of Fig. a. Two smaller islands with still intact neurons. In-between extensive loss of neurons. Proliferation of microglia and astroglia. Nissl technique, 128:1. (From UNTERHARNSCHEIDT, 1963).

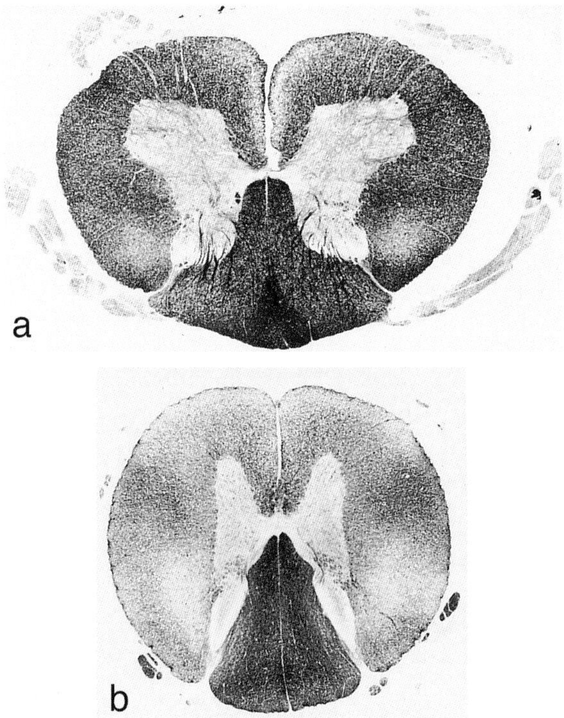

Figs. 82 a and b. Cat. Spinal cord, cervical region. Considerable demyelination in the anterior pyramidal tracts and in the lateral pyramidal tracts. Additionally differently distinct demyelinations in different parts of the white matter of the anterior and lateral tracts. Moderate demyelination in Burdach's tract, none in Goll's tract. Heidenhain technique, 9.5:1 (From UNTERHARNSCHEIDT, 1963).

rhages consistently follow vessel systems running vertical to the cortical surface and are thus draining vessels. The characteristic pinpoint hemorrhages indicative of cortical contusions in the first stage cannot be seen. Thus, not only is there a difference in quality between primary traumatic cortical hemorrhage associated with angular acceleration and the so-called cortical contusions found in translational trauma (with the impact axis running through the center of gravity of the system), but there is also a difference in the pattern of distribution of the primary traumatic cortical lesions encountered in angular and translational acceleration. These alterations are located close to the midline, not in a cylindrically symmetric pattern but radially symmetric, as predicted by the injury mechanics.

These findings present, in our opinion, substantial evidence against HOLBOURN's concept that cortical contusions are due to rotational motion. We attribute great importance to angular rotation,

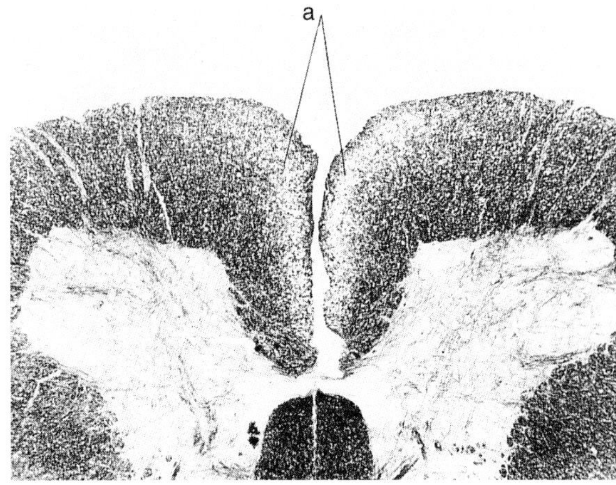

Fig. 83. Cat. Spinal cord, cervical region. Close up of Fig. 77 a. Considerable demyelination in both anterior pyramidal tracts at a, less pronounced and of different intensity in other areals of the anterior and lateral tracts. Heidenhain technique, 27:1. (From UNTERHARNSCHEIDT, 1963).

in such cases as the uppercut in boxing, or hooks, when tearing of bridging veins causes subdural hemorrhage. But we also advance a number of facts indicating that the quality and distribution of tissue lesions after translational and angular accelerations are essentially different when all of the possibilities of histological examinations are brought to bear.

Hippocampus Formation

Four of 24 animals exhibited morphological alterations in the hippocampus formation (A-67-10, 14, 13, 1 – 1.75, 1.87, 3.27, 3.63 est. × 10^5 rad/s^2)[2]. These belonged to the higher intensity group. Two of these (A-67-10, 14) also had secondary alterations due to circulatory disturbances that were located adjacent to the primary traumatic tissue alterations in the frontal lobes. Perifocal edema and its sequelae seem to be at the origin of these secondary traumatic, reactive lesions. They consist of hyperchromatosis and shrinkage of neurons in the Spielmeyer dorsal resistant sector (h2) and Sommer's sector (h1).

There is early evidence of cell loss, however there is no clear glial reaction as yet. It can, nevertheless, be expected after a sufficiently long period of survival. These secondary traumatic lesions were not found in animals that remained free from primary traumatic lesions. (Admittedly, the survival time of five days was rather short). Two animals (A-67-13, 1) that died soon after the experiment exhibited rhectic hemorrhages in the Ammons' horn band. There were, of course, no reactive glial cell proliferations so shortly after the experiment.

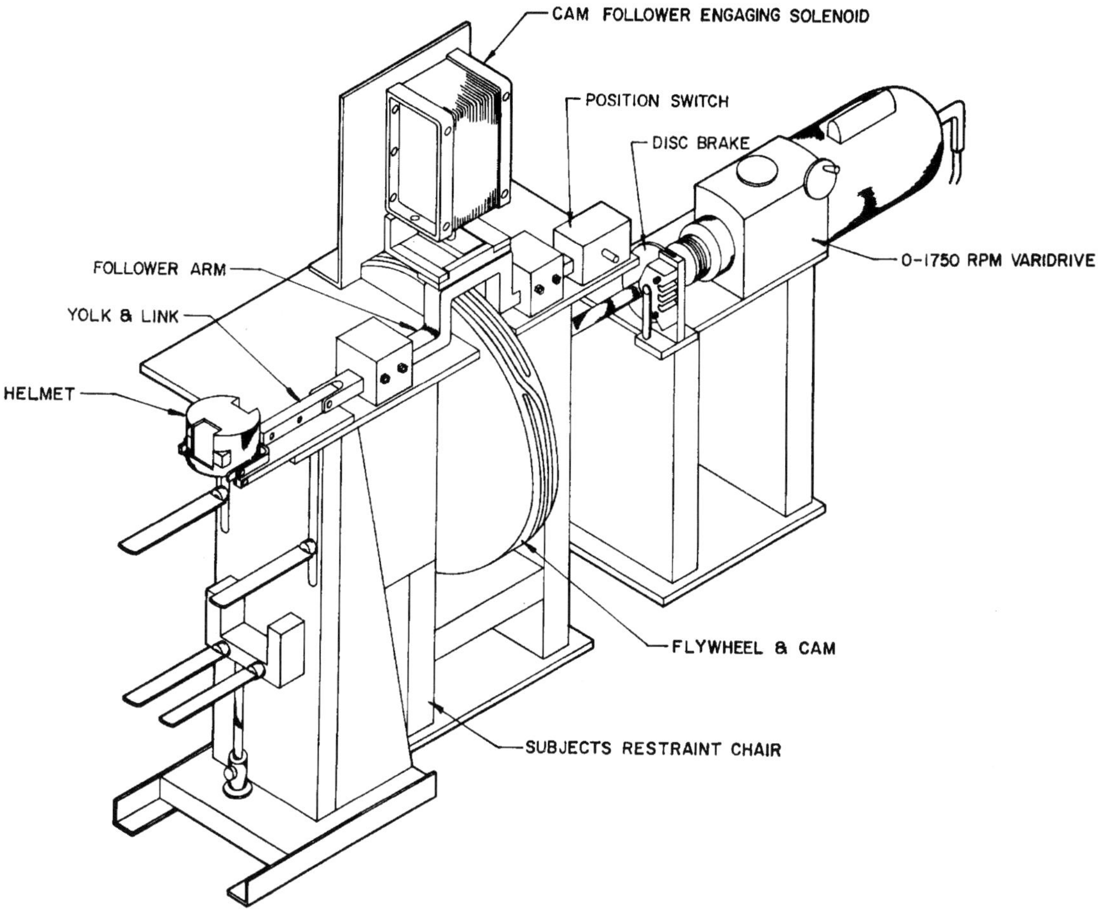

Fig. 84. The experiments were conducted using a maschine designed for rotational acceleration inputs to the head. It consists of a helmet in which the monkey's head can be secured with plaster of Paris. The path taken by the head is limited to a 45° arc by the geometry of the maschine. The helmet is driven by a rod which follows an eccentric maschined flywheel, giving a triangular waveform. A motor-driven transmission and tachometer provide exact control of flywheel speed and, thus, the rotational acceleration of the head.
Because of technical difficulties in direct measurement of the very high angular acceleration available, the rotational acceleration was calculated from the linear acceleration of the follower rod and the known geometry of the maschine. Details of the calculations used and the expected small errors resulting from this method of measurement may be found in Higgins and Schmall (1967) and Higgins et al. (1966). This maschine is capable of producing only non-deforming blows. C From UNTERHARNSCHEIDT, HIGGINS, 1969).

Primary versus Secondary Tissue Alterations

It is important to distinguish *primary* from *secondary tissue alterations.* The former, also called *direct* or *mechanical alterations appear* instantly and are the immediate consequence of a lesioning force. The latter, by contrast, are the indirect result of the impact and are based on circulatory disorders, among others, and as such take hours or days to develop. SPATZ named the two categories *passive* and *reactive tissue lesions.* The quality and pattern of distribution of primary lesions is dependent on the direction, intensity, and duration of the impact. If the victim survives, the area and vicinity of primary traumatic lesions will naturally develop secondary, organizational and resorptive processes that eventually form either *glial* or *mesodermal-glial scar tissue,* or *cystic tissue defects.* The differentiation of primary and secondary traumatic lesions is by no means a purely academic question, but enables us to understand the causation of the lesions

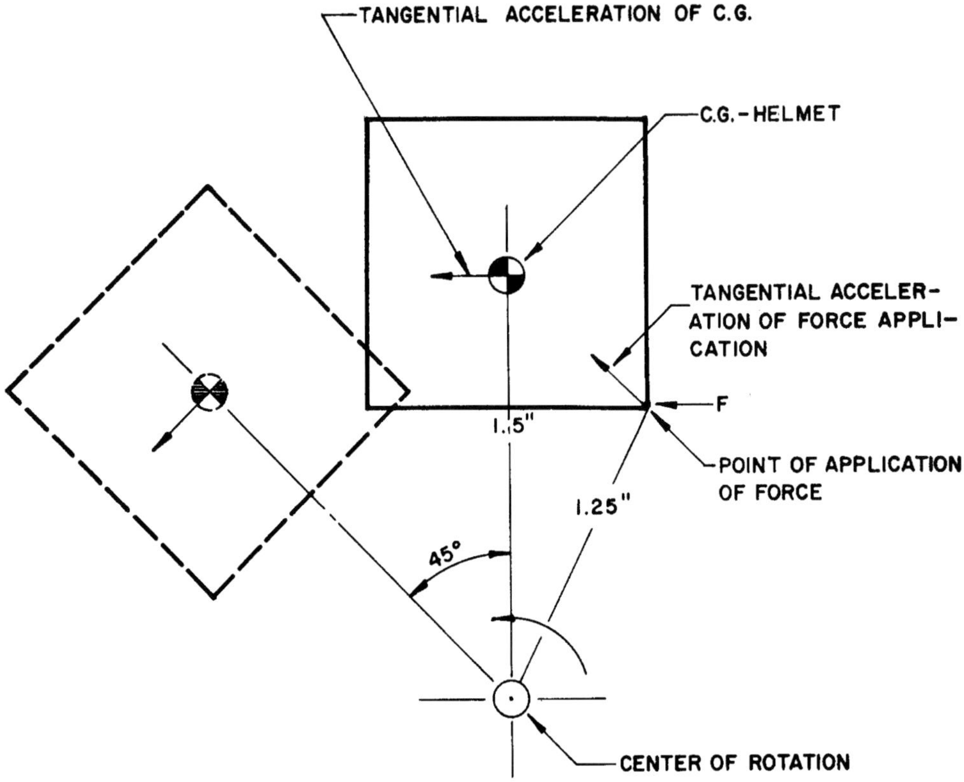

Fig. 85. Motion of the head restraint system (From UNTERHARNSCHEIDT, HIGGINS, 1969).

with regard to the mechanics involved, such enabling the concept of protective devices which are to eliminate or reduce these lesions.

Cranial Nerves, Midbrain, Pons, Medulla Oblongata, Cerebellum

Three animals (A-67-21, 12, 13) showed primary traumatic lesions in cranial nerves, including the N. oculomotorius, optic chiasm and N. vestibuli.
These findings are equally well explained by vascular tears after strain due to angular motion of the skull relative to the brain. Only very few descriptions of these lesions have been made in human pathology.
Primary traumatic lesions in midbrain, pons, medulla oblongata, and cerebellum, too, are clearly related to the intensity of the angular acceleration, as Table 3 indicates. They were found in three animals of the low intensity group, as compared to seven animals in the high intensity group.

Spinal Cord (Fig. 89)

With the exception of a few animals, the spinal cord always showed small rhectic hemorrhages in various segments. The rhectic hemorrhages of capillaries and small veins are found more frequently disseminated in the grey substance, caused by the longitudinal and transverse stretching of ascending and descending spinal branches of spinal cord arteries, veins, and capillaries. The primary traumatic alterations in the spinal cord affect the cervical, thoracic, as well as the lumbar-sacral regions. The severity of these lesions, too, is related to the magnitude of angular acceleration and increases with intensity. The correspondence is not as obvious as it is with the cerebrum, cerebellum, midbrain, pons, and medulla oblongata. These spinal cord findings are not fatal, and the test animals showed, indeed, no clinical signs.
These small hemorrhages are difficult to discern when stained with Nissl technique. After three days of survival, they hardly showed any mesodermal glial reaction. After five days of survival, unlike the more severe rhectic hemorrhages in the cerebrum and brainstem.
Two monkeys (A-67-22, 3 – 1,0 and 2.54×10^5 rad/s^2) presented subdural hematomas in the cauda equina. The relation

TABLE 3
Clinical findings. (From UNTERHARNSCHEIDT, HIGGINS, 1969).

Autopsy number	α, 10^5 rad/sec^2	Heart rate	Electro-cardio-gram	Respira-tory rate	Gasp	Electro-encepha-logram	Corne-al reflex	Direct light reflex	Con-sensual light reflex	Pupil size	Extra ocular move-ments	Audito-ry re-sponse	Biting	Tongue move-ments	With-drawal	Mic-turition	Con-cussion	Duration of con-cussion in minutes
A-67-7	1.01	+	+	...	...	...	...	...	...	...	+	...	...	...	...	...	...	...
A-67-22	1.11	...	...	+	...	...	...	...	...	...	...	...	...	...	...	...	...	...
A-67-17	1.40	+	...	...	+	...	...	+	+	...	...	+	+	+	...	...	+	24
A-67-11	1.48	+	+	+	+	+	...	+	+	...	...	+	+	+	+	...	+	11
A-67-24	1.50	...	...	...	...	...	...	...	...	...	...	...	+	+	...	...	...	...
A-67-5	1.50*	...	...	+	...	...	...	+	+	+	+	+	+	+	+	...	+	16
A-67-6	1.50*	...	...	+	...	...	...	+	+	+	+	...	...	...	...	...	...	...
A-67-8	1.53	+	+	+	+	...	...	+	+	...	...	...	...	...	...	...	+	9
A-67-23	1.54*	...	...	+	...	...	...	...	...	...	...	...	...	...	...	...	...	...
A-67-21	1.54	...	...	...	+	...	+	+	+	...	+	...	+	+	+	...	+	1
A-67-18	1.56	+	+	+	...	...	...	...	...	...	...	...	...	...	+	...	...	...
A-67-15	1.59	+	+	+	...	...	...	...	...	...	...	...	+	+	+	...	+	5
A-67-16	1.74	+	...	+	...	...	...	...	+	...	...	...	...	...	...	...	...	...
A-67-20	1.74	+	+	+	...	+	...	...	...	+	...	...	...	+	...	...	...	...
A-67-25	1.74	...	...	+	...	...	...	...	...	...	...	...	...	...	...	...	...	...
A-67-10	1.75	+	+	...	...	...	...	+	+	+	...	+	...	...	+	...	+	5
A-67-14	1.87	+	+	...	...	+	...	+	+	...	...	...	...	+	+	...	+	8–10
A-67-19	1.91	...	...	+	+	+	+	+	+	...	+	...	...	...	+	...	...	...
A-67-9	1.97	+	+	...	...	...	...	+	+	+	...	...	...	+	+	...	+	10
A-67-12	2.05	...	...	+	+	...	...	+	+	+	...	...	+	+	+	...	+	15
A-67-3	2.54	...	...	+	...	+	...	+	+	+	...	...	+	+	...	...	+	12
A-67-4	2.84	+	...	+	...	+	+	+	+	...	...	...	+	+	+	+	+	13
A-67-13	3.27	...	+	+	...	+	+	+	+	+	...	+	...	...	+	...	+	Lethal
A-67-1	3.63*	+	+	+	...	...	+	+	+	...	+	+	+	+	+	...	+	Lethal
A-67-2	3.86	+	+	+	†	+	+	+	+	+	...	...	+	...	+	...	+	Lethal

+ indicates a change following blow.
α = peak angular acceleration.
* = estimated value.
† = EEG channel inoperative.

to the magnitude of angular acceleration could not be clearly established.

Three *control animals* (A-68-1, 2, 3) showed no remarkable morphological alterations.

As we have mentioned above, the pattern and quality of histological lesions was different in rotational traumas. We expected and found subdural hemorrhages due to tearing of bridging or connecting veins between the cerebral hemispheres and dura mater, subarachnoid hemorrhages predominantly in the midline, straining and tearing of vessels in cranial nerves, in superficial cortical layers, and with stronger intensities, also in deeper parts of the cerebrum. We expected vessel ruptures due to strain in the spinal cord far from the impact site, but along the principal axis of stress. That hyperflexion and hyperextension can produce changes in position of

TABLE 4
Gross autopsy findings. (From UNTERHARNSCHEIDT, HIGGINS, 1969).

Monkey number	Autopsy number	Peak angular acceleration (10^5 rad/sec^2)	Pos. phase duration (msec)	Total duration (msec)	Concussion	Gross autopsy findings
9	A-67-7	1.01	10.32	25.8	−	No gross evidence of damage
138c-165c-24	A-67-22	1.11	4.32	10.8	−	No gross evidence of damage
145-172-19	A-67-17	1.40	3.52	8.8	+	No gross evidence of damage
154c-181c-12	A-67-11	1.48	3.20	8.0	+	No gross evidence of damage
116-7	A-67-5	1.50 est.	...	...	+	No gross evidence of damage
48c-8	A-67-6	1.50 est.	...	...	−	No gross evidence of damage
144c-75c-10	A-67-8	1.53	3.12	7.80	+	No gross evidence of damage
50c-25	A-67-23	1.54 est.	...	...	−	No gross evidence of damage
6c-23	A-67-21	1.54	3.52	8.80	+	Hemorrhage over upper aspect of right frontal lobe
82-20	A-67-18	1.56	3.76	9.40	±	No gross evidence of damage
51c-16	A-67-15	1.59	3.36	8.40	+	Hemorrhage at base of brain and left frontal lobe
52c-18	A-67-16	1.74	3.76	9.40	−	No gross evidence of damage
150-177-22	A-67-20	1.74	3.68	9.20	−	No gross evidence of damage
146c-173c-26	A-67-25	1.74	3.84	9.60	−	Possible hemorrhage in posterior cranial fossa around tentorium cerebelli
149c-11	A-67-10	1.75	3.04	7.60	+	No gross evidence of damage
83c-15	A-67-14	1.87	3.20	8.00	+	Bilateral hemorrhage in area of superior sagittal sinus
155c-182c-21	A-67-19	1.91	3.20	8.00	−	No gross evidence of damage
152c-179c-6	A-67-9	1.97	3.12	7.80	+	Old hemorrhage on right temporal lobe
47c-13	A-67-12	2.05	3.36	8.40	+	No gross evidence of damage
4	A-67-3	2.54	2.88	7.20	+	Biparietal and occipital hemorrhage
163-5	A-67-4	2.84	No timing data		+	Right parietal and bilateral frontal hemorrhage
153c-18c-17	A-67-13	3.27	2.40	6.00	+ Lethal	Bilateral hemorrhage on occipital crest; osseous bleeding over entire skull; subdural hemorrhage over frontal poles
46c-1	A-67-1	3.63 est.	...	...	+ Lethal	Biparietal hemorrhage
137c-164-3	A-67-2	3.86	2.40	6.00	+ Lethal	Frontal lobes show generalized superficial hemorrhage; vessels on parietal lobes on bleeding (appears to be venous)
(49-2	A-67-24	1.50	Multiple runs. Pathology not reported here)			

Observations reported are correlated with monkey number, (positive) peak angular acceleration, duration of positive phase of the dual triangular waveform, and total duration of the acceleration.

TABLE 5
Rotational acceleration (squirrel monkeys; saimiri sciureus). (From UNTERHARNSCHEIDT, HIGGNS, 1969).

Rotational acceleration in rad/sec^2	Morphological findings
$1.01–1.50 \times 10^5$ rad/sec^2	No subdural, no subarachnoid hemorrhages. No lesions in cerebrum, cerebellum, midbrain, and pons.
$1.53–2.84 \times 10^5$ rad/sec^2	Subdural and subarachnoid hemorrhages. Rhectic hemorrhages in cranial nerves. Primary traumatic hemorrhages in superficial layers of the cerebral cortex, near the midline. Single rhectic hemorrhages in the medulla oblongata.
$3.27–3.86 \times 10^5$ rad/sec^2	Massive subdural and subarachnoid hemorrhages. Severe primary traumatic hemorrhages in the cerebral cortex, extending into the central white matter. Small rhectic hemorrhages in the hippocampus formation. Rhectic hemorrhages in cerebellum, midbrain, pons, and medulla.

With the exception of a few animals, the spinal cord showed small rhectic hemorrhages in various segments. The severity of these lesions, too, was related to the magnitude of angular acceleration and increased with intensity. The correspondence was not as obvious as it was with the cerebrum, cerebellum, midbrain, pons, and medulla. These spinal-cord findings were not fatal, and the test animals showed, indeed, no clinical signs. (UNTERHARNSCHEIDT and HIGGINS, 1969.)

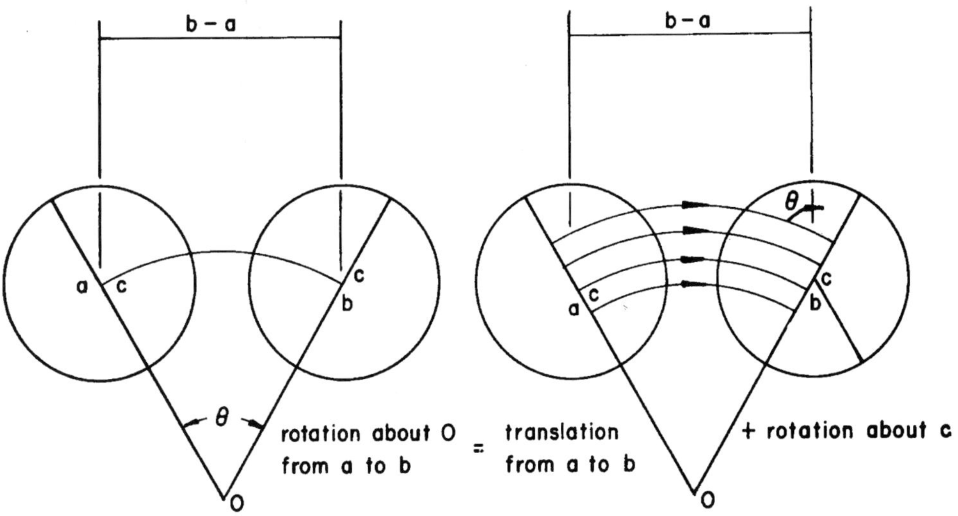

Fig. 86. Equivalence of arbitrary displacement. (From UNTERHARNSCHEIDT, HIGGINS, 1969).

the spinal cord has already been pointed out by Alf BREIG (1960) on cadavers.

Although the kinematic equations for determination of levels of acceleration at any point in the head are simple for the rigid body case, solution of an equation that would adequately predict these locations of histological lesions is beyond the scope of this contribution, or indeed any currently available theoretical treatment. Furthermore, the number of monkeys studied at higher acceleration levels is too small for an explicit cataloguing of expected pathologic findings.

Conclusions

The experiments have shown that a *non-deforming rotational acceleration delivered* to the head through a controlled path is capable of producing lesions not only in the brain, but through the entire length of the spinal cord, extending to the cauda equina.

In clinical circumstances in which humans are subjected to severe rotational acceleration of the

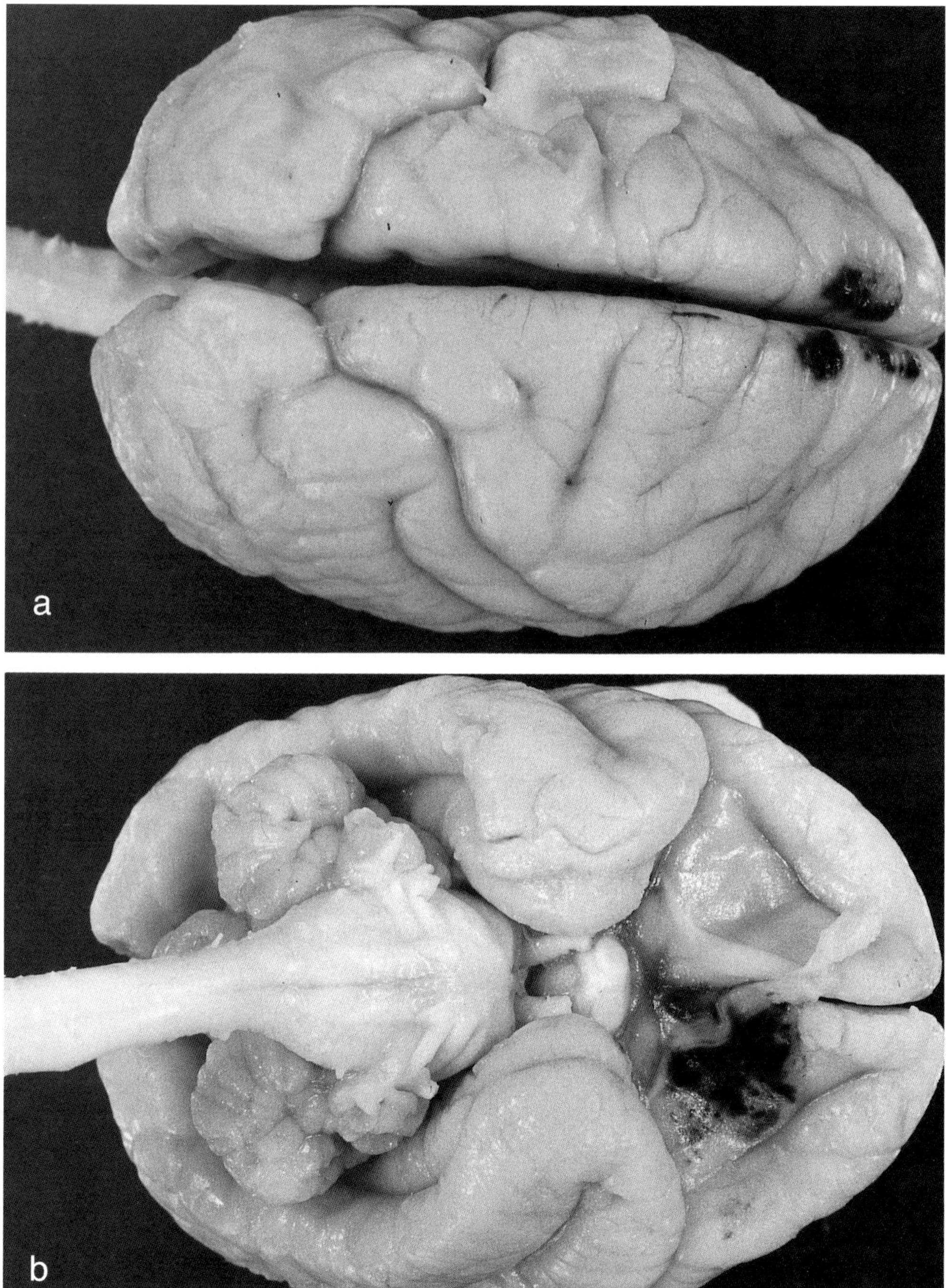

Figs. 87 a and b. Squirrel monkey, no. 14. Rotational acceleration of the head $187 \times 10^5 \times rad/s^2$. a shows small focal subarachnoid hemorrhages (and histologically rather superficial rhectic hemorrhages due to vessel rupture after strain) near the midline on the pallium and in central parts of the interhemispheric fissure. Macrophoto. b shows subarachnoid hemorrhages (and histologically rather superficial rhectic hemorrhages due to vessel rupture after strain) in basal portions of the left frontal lobe. The left olfactory tract also reveals hemorrhages. Macrophoto. (From UNTERHARNSCHEIDT, HIGGINS, 1968).

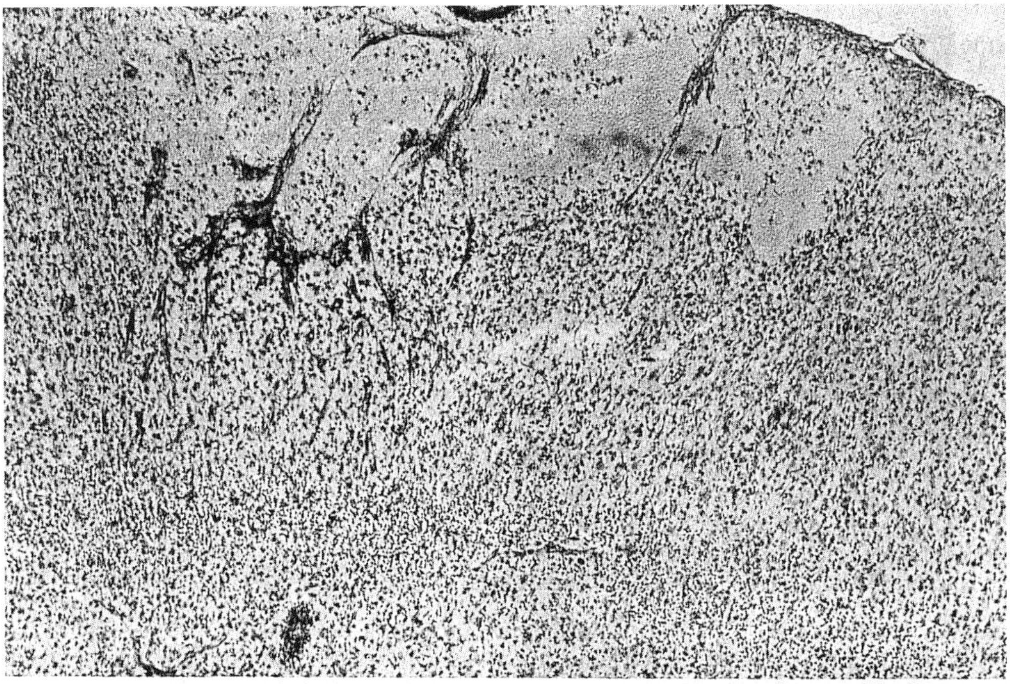

a

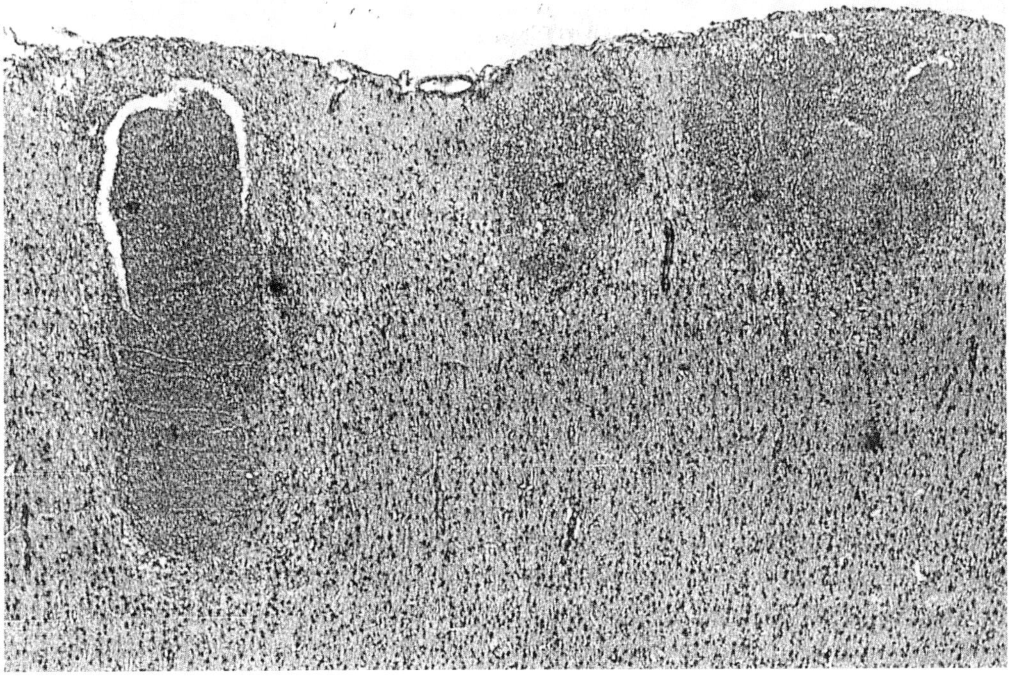

b

Figs. 88 a and b. a. Squirrel monkey, no. 8. Rotational acceleration of the head 1.53 x 10^5 rad/ s^2. Occipital region. Large confluent hemorrhages in superficial layers of the cerebral cortex, due to venous ruptures. Astroglial and microglial reaction in the surrounding brain tissue, with partial necroses. Beginning mobile disintegration by macrophages. Beginning proliferation of capillaries and connective tissue at the edge of the lesions. Disseminated loss of neurons in the vicinity is combined with glial reaction. Nissl technique, 16:1. b. Squirrel monkey, no 14. Rotational acceleration of the head 1.87 x 10^5 /s^2. Frontal region, Venorhectic hemorrhages with blood diffusing into the surrounding tissue. In the these areas the neurons are shrunken, sometimes encircled by a halo. There is a focal loss of neurons combined with glial cell reaction. Hematoxylin-Eosin technique, 16:1 (From UNTERHARNSCHEIDT, HIGGINS, 1968).

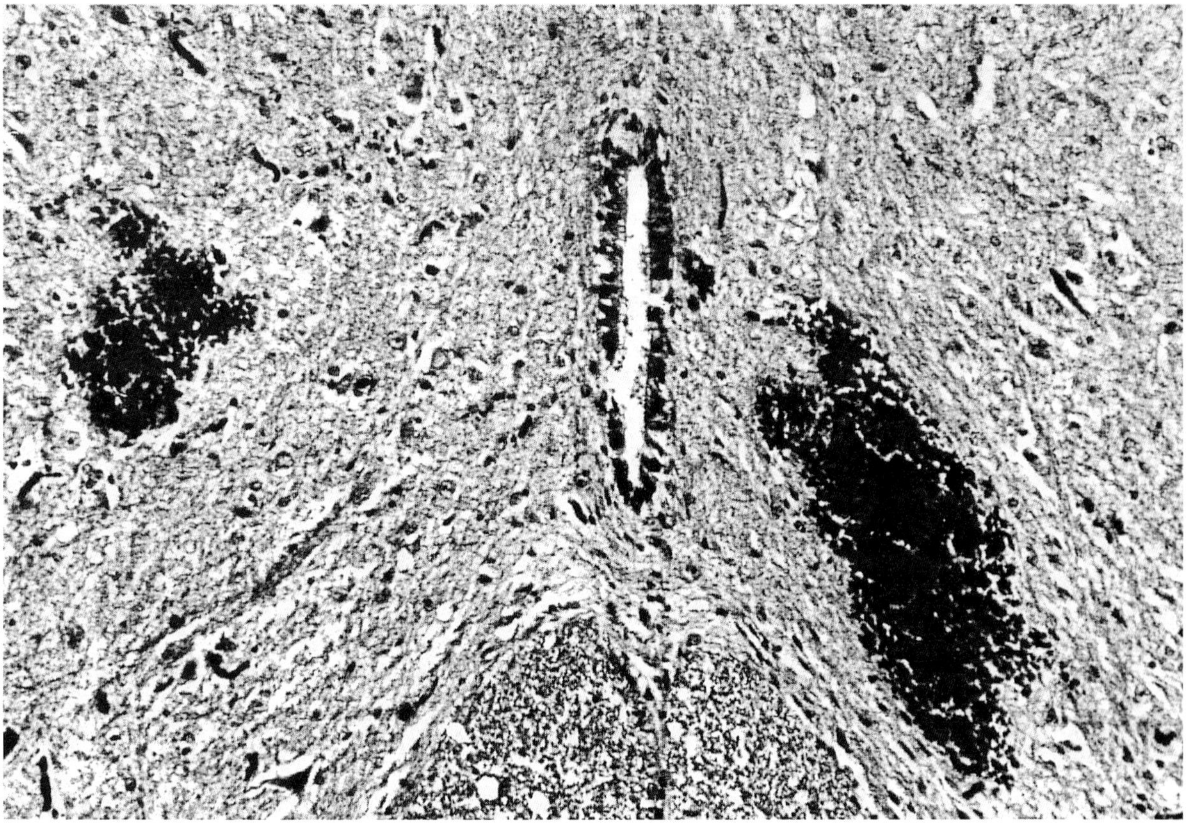

Fig 89. Squirrel monkey, no. 6. Rotational acceleration of the head 1.50 x 10^5 rad/s^2 estimated. Masson-Goldner technique, 40:1. (From UNTERHARNSCHEIDT, HIGGINS, 1969).

head, the spinal cord as well as the brain should be examined histopathologically.

In summary, not only does a qualitative difference exist between the primary traumatic cortical contusions found in translational injuries, but there are also different patterns of distribution for the primary traumatic lesions encountered in both types of acceleration. Inasmuch as these lesions are arranged in a cylindrically symmetric pattern after translational acceleration, as compared to a radially symmetric pattern located close to the midline after rotational acceleration.

Except for the question of location, these considerations seem to be valid also for the interpretation of findings in the spinal cord, although correlations are not as patently manifest here as they are in the brain. Nevertheless, the relation between severity of primary traumatic lesions and magnitude of acceleration is evident throughout the entire central nervous system.

Injuries Apart from CNS Damage

Facial Injuries

Facial injuries occur frequently in amateur and professional boxing because the face is all that lies between the gloved fist and its destination, the head and brain. The existence of a laceration is self-evident. Coexisting injuries to the *facial nerve and separation of the parotid duct* must be considered.

The most frequent injuries of the face of boxers are: (1) *Contusions*, (2) *abrasions*, and (3) *lacerations of the face*, (2) contusions, abrasions or lacerations around the eye, (3) *lacerations* of the lips, (4) *injuries of the ear, in their final stage so-called 'cauliflower ears,'* (5) *fractures* of the *mandible*, (6) *lesions of the teeth or loss of them,* and (7) *injuries* and *fractures* of the *nose.*

Injuries of the face occur in amateur and professional boxers frequently, since these anatomical structures are again and again hit by the boxing gloves of the opponent. The skin is swollen (*'puffed'*) and circumscribed hemorrhages in the edematous tissue can develop. At the same time injuries of the *facial nerve* and *disconnection of the ductus paroticus* can exist. Application of ice, as-

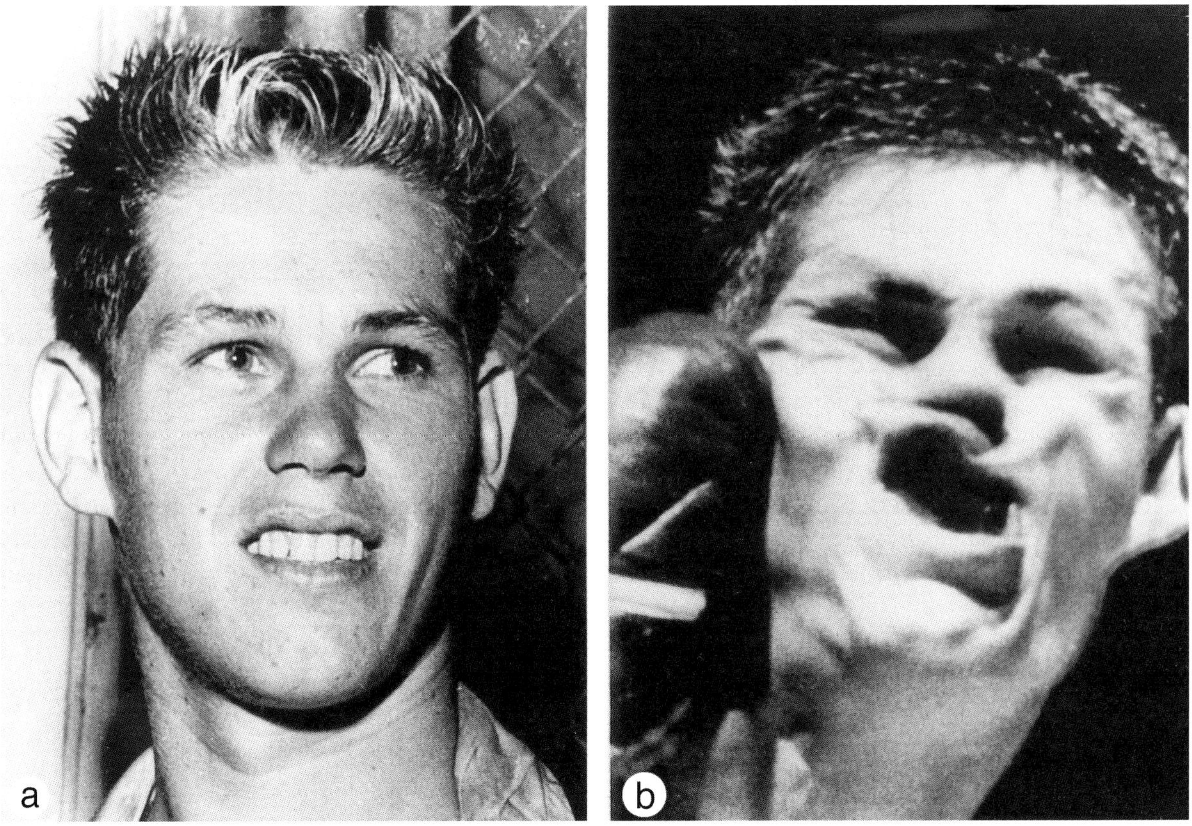

Figs. 90a and b. Professional lightweight fight between Mark TESSMAN (Houston) and Ronnie WILLIAMS (Fort Worth). Photo: A.P.

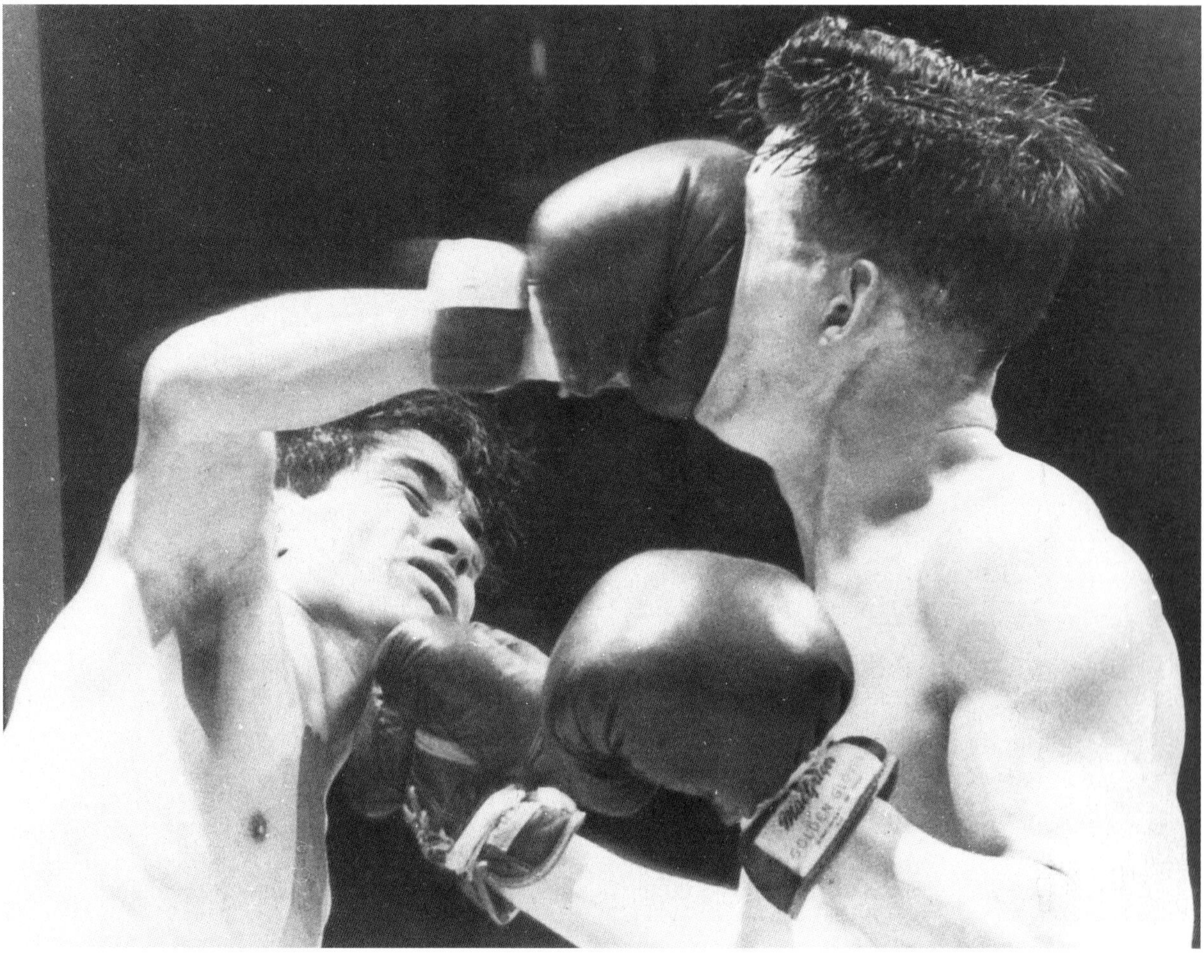

Fig. 91. Flyweight Golden Glove bout between Christopher RAFTER (right) and Paz MOORE (left) on February 26, 1958. Photo: U.P.I.

tringents, or raw meat have no influence on the swelling.

Selection of Action Photos of Press Photographers in Professional Fights

The action photos of press photographers were divided into two groups, the first shows the actual effects of a single hard blow on the features of a boxer and the second group shows the more chronic sequelae they leave on the face of the pugilist after a knock out, a technical knock out or after the end of the fight. Both groups cannot be completely separated because they overlap in some areas.

Deformation and Distortion of the Faces of Pugilists after a Single Hard Blow

The potential intensity of boxing blows to the head and their clinical and morphological consequences are not only suggested by our experimental measurements of head accelerations, but action photos of boxers immediately after a hard blow to the face show the intensity and deformation in several ways.

The dreadful force of the blow deforms and displaces the human features. Inspection of press photographers' action photos of boxers' heads, taken at the moment of impact or milliseconds later, reveal that through deformation, distortion, acceler-

Fig. 92. Professional middleweight fight between Willie VAUGHAN and Ralph 'Tiger' JONES. Photo: U.P.I.

ation, and inertia, single features are pushed against each other and so displaced that occasionally it is difficult to identify them. It is not necessary to be a physician to comprehend that blows causing such terrible distortion of human features must have a damaging effect on the much more vulnerable brain tissue too. The human countenance as shock absorber! We do not know to which extent the hammer hard blows broadly impacting the head deform the skull and therefore involve the brain; we must expect direct brain damage due to deformation of the skull.

The photos of bloody, swollen, 'puffy' and battered faces of boxers whose fights were stopped, who were knocked out, or those who finished, for example, a twelve round draw fight, speak for themselves. It is important to bear in mind the realistic number of hard face and head blows in a single ten or twelve round professional fight; they often number in the hundreds. The very delicate brain tissue is even more vulnerable to permanent injury from hard boxing blows than the face. When head and face distortion occurs, the brain is directly injured by the mechanical forces and indirectly injured later, by circulatory disturbances.

Selection of a Series of Action Photos of the Faces of Professional Boxers in the Moment of Impact:

Professional lightweight fight between Mark

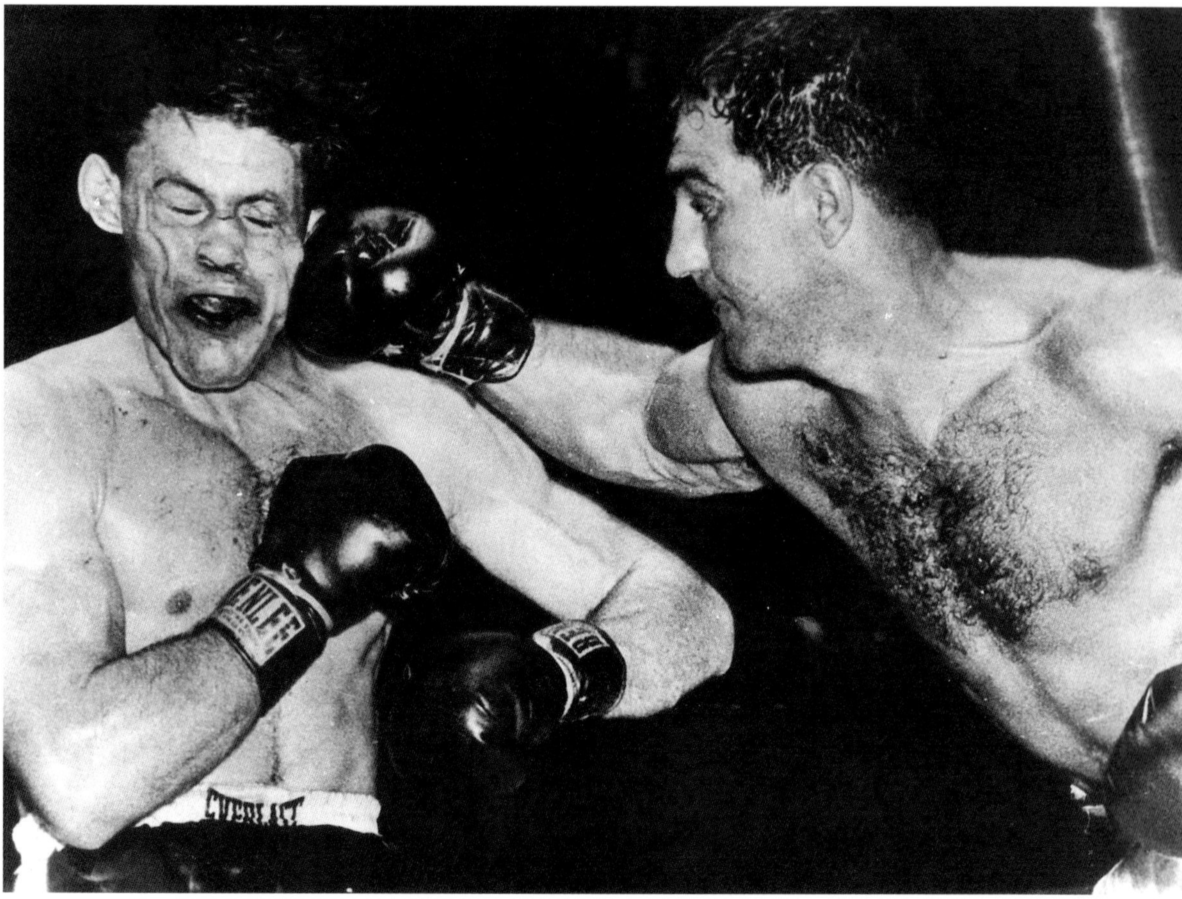

Fig. 93. Second professional heavyweight fight between Rocky MARCIANO and Roland LA STARZA on September 24, 1953 at the Polo Grounds, New York. Photo: Jack FRANK, New York Tribune.

TESSMAN (Houston) and Ronnie WILLIAMS (Fort Worth) (Fig. 90 a and b). A hard left of TESSMAN deforms and distorts the face of WILLIAMS, who has been hit on the right side of his face. The photo demonstrates how parts of the skin and subcutaneous tissue are displaced by the considerable kinetic energy. The eyes are closed, the mouth wide open. It also reveals that while the features of the face are affected by the direct impact of the blow, the left ear is bent forward due to its inertia. The photo clearly demonstrates the tremendous intensity of a blow by a lightweight! The picture on the left shows WILLIAMS' face before the fight.

Flyweight Golden Glove bout between Christopher RAFTER (right) and Paz MOORE (left) (Fig. 91). RAFTER's face is hit by a powerful right from MOORE, while connecting with a right to the chin of MOORE. The left side and probably parts of the median structures of RAFTER's face have taken a concave form due to the enormous force of the impacting boxing glove. MOORE won the 3-round decision.

The photo shows the moment in which a hammering right of 'Tiger' JONES connects with the face of Willie VAUGHAN (Fig. 92). The tremendous kinetic energy of the blow produces a grotesque displacement and flattening of the face, especially of the left side. Eyelids, nose, upper and lower lips are completely deformed and distorted.

Fig. 94. Professional heavyweight championship fight between Rocky MARCIANO (left) and Roland LA STARZA (left). Photo: KEYSTONE.

The tremendous impact has accelerated the head, while the right ear lags behind due to its inertia. The human face is acting here as a shock absorber!

Second heavyweight fight between Rocky MARCIANO (right) and Roland LA STARZA. The photo shows the deformed and distorted face of LA STARZA milliseconds after MARCIANO connected with a hammering right (Fig. 93). In the eleventh round the weary LA STARZA was knocked through the ropes onto the ring apron. He got to his knee at '5,' was up at '9,' but referee Ruby GOLDSTEIN had seen enough and stopped the bout. LA STARZA, who had taken MARCIANO to a very close decision in 1950, in the second match was again able to win for six rounds with smart boxing coupled with MARCIANO's repeated fouling. Ruby GOLDSTEIN, the referee, penalized him in the

Fig. 95. Professional title fight in the middleweight between Gene FULLMER (right) and Florentino FERNANDEZ (left) on August 5, 1961 in Ogden, Utah.

sixth round, which moved LA STARZA into a clear point lead. Journalists reported that MARCIANO responded in the seventh round with an assault of sustained ferocity seldom seen in the ring. For the next three and a half rounds MARCIANO punched, battered and clubbed at LA STARZA's head and body.

The heavyweight championship fight between Rocky MARCIANO (left) and Roland LA STARZA (right). In the eleventh round MARCIANO connected with a flurry of hard blows to the already defenseless LA STARZA, who is hanging on the ropes. MARCIANO is starting a hook with his right, which knocked down LA STARZA (Fig. 94); referee Ruby GOLDSTEIN stopped the bout in this round.

Description of the fist fight between Rocky MARCIANO and Roland LA STARZA

LA STARZA had won 37 straight fights since he had turned professional in July 1947. In this fight MARCIANO was the underdog.

Fig. 96. Professional lightheavyweight fight between Archie MOORE (right) and Jimmy SLADE (left) on February 27, 1952 at the St. Louis Stadium. Photo: U.P.I.

The fight took place on March 24, 1950 at Madison Square Garden in New York. At the beginning LA STARZA was outboxing MARCIANO, but was unable to hit him effectively. After three rounds the spectators began to jeer both boxers.
In the fourth round MARCIANO became aggressive, throwing hard punches. He landed a right hook at LA STARZA's head, who went down. He got up to one knee. When the count reached '8' the bell range and the round was over. In the fifth round LA STARZA used his jabs and stayed at long range. LA STARZA was the better boxer in the sixth and seventh rounds and scored points.
In the eighth round MARCIANO became very aggressive and threw a low punch. The round was given to LA STARZA on a foul.
The ninth and tenth rounds saw furious fighting. LA STARZA continued to box very well and scored with his jabs. Both fighters got a standing ovation from the spectators when the fight ended.
The decision of the judges were split, one voted 5-5 and one even for MARCIANO, the other 5-4 and one even for LA STARZA. Referee Jack WATSON scored 5 rounds for MARCIANO and 5 rounds for LA STARZA.
Ring announcer Johnny ADDIE: "in points he scores six points for LA STARZA and nine for MARCIANO".
There were 13,658 spectators in the Madison Square Garden; they had paid $ 53.723.
It was a tremendous setback for LA STARZA in his ring career. Only two of his ten next fights were held in New York, and not in Madison Square Garden. LA STARZA won all but two of his seventeen fights after the defeat by MARCIANO. MARCIANO was heavyweight champion by Christmas 1952.
A second fight took place on September 24, 1953; MARCIANO knocked out LA STARZA in round 11.
Rocky MARCIANO had 49 fights, won 49.

Fig. 97. Professional heavyweight fight between Carl 'Bobo' OLSON and Ralph 'Tiger' JONES in Chicago. Photo: d.p.a.

Professional title fight in middleweight between Gene FULLMER (right) and Florentino FERNANDEZ (left) (Fig. 95). FULLMER winces as FERNANDEZ lands a left on the right side of FULLMER's head behind the right ear, at the same time FULLMER connects with a right to the body in the eight round of their title fight.

Gene FULLMER had 64 fights, 55 won, 6 lost, 3 drawn.

Professional fight in light heavyweight between Archie MOORE (right) and Jimmy SLADE (left). MOORE delivers a right in the seventh round that causes Jimmy SLADE to wince (Fig. 96). SLADE's features are distorted, his eyes closed. MOORE won the fight by a unanimous decision.

Archie MOORE had 232 fights, 195 won, 27 lost, 9 drawn, 1 no contest.

Fig. 97. Professional middleweight fight between

Fig. 98. Professional welterweight title bout between Emile GRIFFITH (right) and Gaspar ORTEGA (left) on June 3, 1961 in Los Angeles. Photo: A.P.

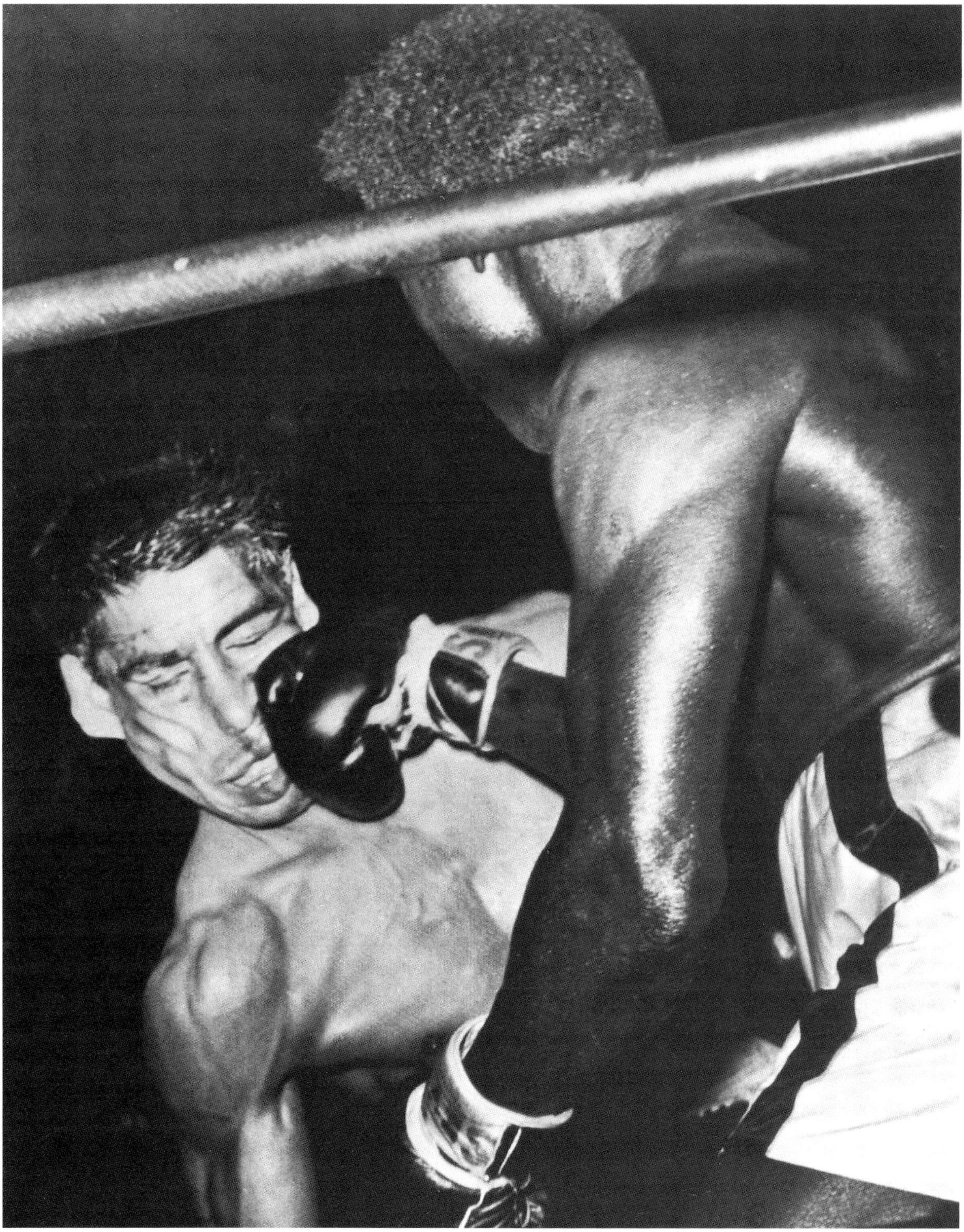

Fig. 99. Professional lightweight fight between Kenny LANE and Ludwig LIGHTBURN on September 19, 1956 at Miami, Florida. Photo: U.P.I.

Carl 'Bobo' OLSON and Ralph 'Tiger' JONES. OLSON connects with a hammering right hand to the head of JONES. The impact of the blow has distorted JONES' face. OLSON won the fight by a decision.

Welterweight title bout between Emile GRIFFITH (right) and Gaspar ORTEGA (left). A right hand punch thrown by champion Emile GRIFFITH lands flush on the left side of the face of Gaspar ORTEGA in the waning moments of the eleventh round, contorting and deforming his features. ORTEGA's eyes are closed, his right ear remains inert (Fig. 98). Referee Tommy HART stopped the fight in the next round to spare ORTEGA further punishment.

Professional lightweight fight between Kenny LANE and Ludwig LIGHTBURN (left). LIGHTBURN takes a vicious right across the jaw in the end phase of this bout (Fig. 99). His eyes are closed; sprays of sweat splash from his head. LANE won by decision.

The photo shows a scene of the light-heavyweight bout between the British light heavyweight champion Chic CALDERWOOD and the American boxer VAN CLAY (Fig. 100). CALDERWOOD connects with a straight right to the face of VAN CLAY, who is stunned by the blow. His eyes are closed. CALDERWOOD lost the fight by points.

The photo shows another scene of the fight be-

Fig. 100. Professional lightheavyweight fight between the British lightheavyweight champion Chic CALDERWOOD (right) and the American boxer VAN CLAY on December 11, 1962 at the Empire Pool, Wembley, England. Photo: KEYSTONE.

tween Chic CALDERWOOD (left) and VAN CLAY (right) (Fig. 101). It was a bloody ring battle; both boxers are marked and bleeding.

First heavyweight title bout between Rocky MARCIANO and Ezzard CHARLES (Fig. 102). This fight is considered one of the most grueling ring battles for the heavyweight championship. With blood streaming from a cut above MARCIANO's right eye, he managed a roaring finish and was awarded the title by a close decision. The photo shows the face of CHARLES fractions of a second after MARCIANO connected with a hard blow to the left side of CHARLES face, which led to a gross deformation and distortion of CHARLES features.

Ezzard CHARLES had 122 fights, 96 won, 25 lost, 1 drawn.

In the professional heavyweight fight between Jimmy ELLIS (right) and Leotis MARTIN (left) a right hook of ELLIS connects with the chin of MARTIN (Fig.103), with considerable deformation of his features. MARTIN's eyes are closed, his mouth is distorted.

Professional lightweight fight between title holder Willie PASTRANO (left) and Gregorio PERALTA (right) (Fig. 104). PASTRANO connects with a stinging left on PERALTA's head in the fourth round, who is stunned, his eyes are closed. PASTRANO won the fight in the sixth round by a technical knock out.

Fig. 101. Another scene of the previously shown fight between Chic CALDERWOOD (left) and the American boxer VAN CLAY (right). Photo: KEYSTONE.

Professional heavyweight fight between Cassius CLAY and Doug JONES. CLAY lashes out with his left to the chin of Doug JONES and contorts his face in the tenth and final round (Fig. 105). JONES eyes are closed. CLAY had prophesized that he would knock out his opponent, but JONES remained of his feet for ten rounds. CLAY had said before the fight: "JONES likes to mix, so I'll let it go to six. If he talks jive, I'll will cut it to five. And if he talks some more, I'll cut it to four." CLAY won a unanimous decision, which received an unpopular reaction in the jammed arena.

Cassius CLAY (MOHAMMAD ALI) had 56 fights, 51 won, 5 lost, 37 knock outs.

Professional title fight in welterweight between Carmen BASILIO (right) and 'Sugar' Ray ROBINSON (left). In the photo (Fig. 106), BASILIO connects with a smashing right to ROBINSON's jaw in the tenth round of their bout. He is stunned and his eyes are closed. The impact propels a spray of sweat from ROBINSON's head and hair. BASILIO won the championship by a split decision.

Carmen BASILIO had 79 fights, 56 won, 16 lost, 7 drawn.
Ray ROBINSON had 202 fights, 175 won, 19 lost, 6 drawn, 2 no contest.

Second heavyweight fight between MUHAMMAD ALI (right) and Ken NORTON (left). The photo shows the effect of a hard right on

Fig. 102. First professional heavyweight fight between Rocky MARCIANO (right) and Ezzard CHARLES (left) on June 17, 1954 at Yankee Stadium, New York.

NORTON's face just milliseconds after the blow connected (Fig. 107). A spray of sweat demonstrates the severity of the impact. NORTON grimaces under the effect of the blow; his eyes are closed.

Professional light heavyweight fight between Harold JOHNSON (right) and Eddie COTTON (left) (Fig. 108). The face of COTTON, already swollen around the left eye is hit by a hard right from JOHNSON. His head is turned to the right and distorted by the force of the blow, his eyes are closed. JOHNSON retained his championship by a split decision after their fifteen round bout. The commentary for this press photo reads: 'Block buster connection'.

Professional lightweight fight between Ray PORTILLA and Carlos ORTIZ (Fig. 109). The photo shows a moment in which PORTILLA neglected his guard. ORTIZ with a stinging blow connects and distorts PORTILLA's features. Both eyes are closed, the nose bent to the right, the mouth is open. ORTIZ won the eight round bout with a close decision.

Professional featherweight fight between Sandy SADDLER (left) and Willie PEP (right). PEP staggered SADDLER with a hard right to the jaw (Fig. 110), and then landed thirteen blows without a return. The photo shows SADDLER's face milliseconds after the impact of a hard blow and reveals the extensive lateral deviation of the corkscrew turned nose and closed eyes. PEP could not continue to fight in the eighth round because of a shoulder injury.

Championship fight in middleweight between Joey GIARDELLO and Rubin CARTER. GIARDELLO winces as CARTER bounces off a

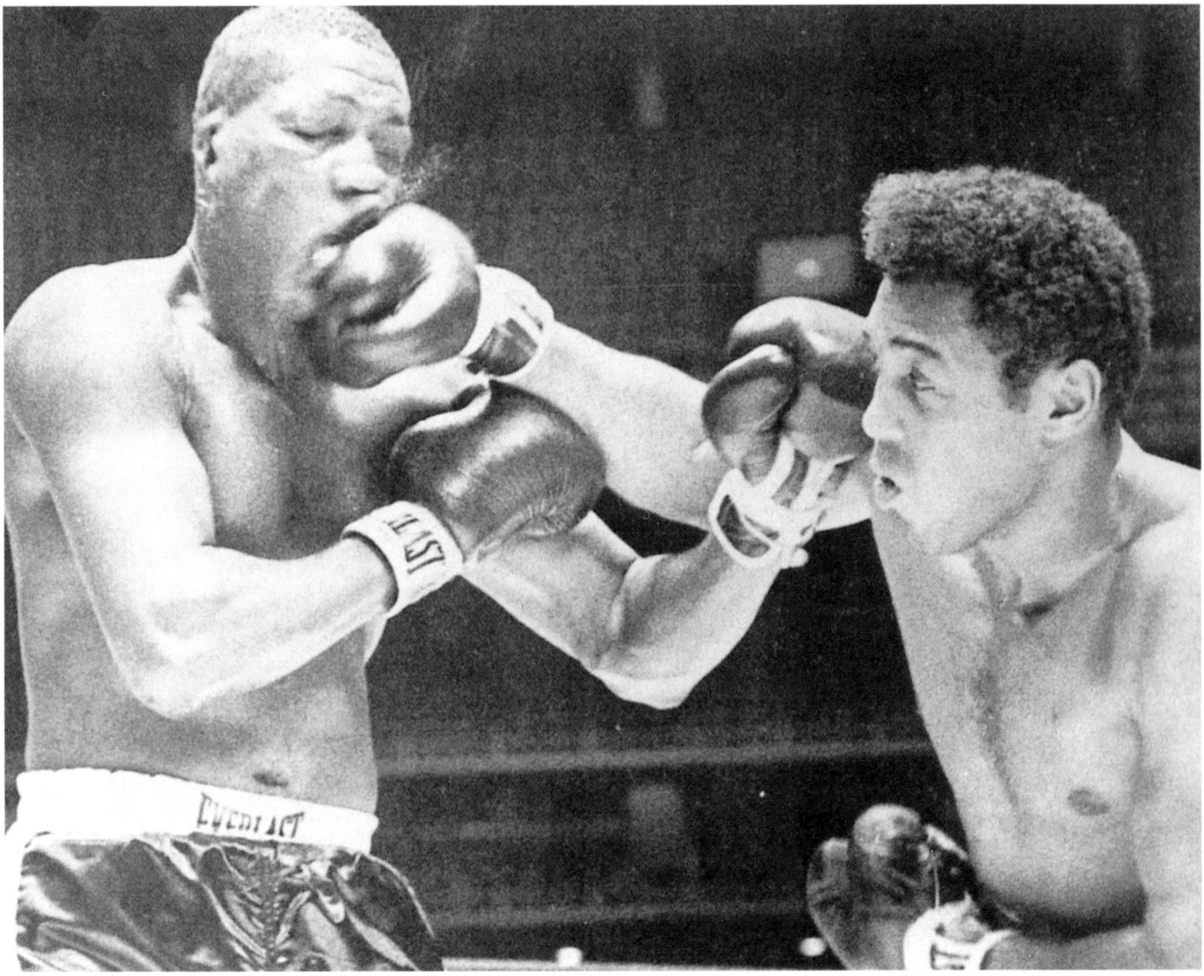
Fig. 103. Professional heavyweight fight between Jimmy ELLIS (right) and Leotis MARTIN (left).

hard right to his head in the fourteenth round (Fig. 111). The comment to this press photo reads: "The receiver was the winner." GIARDELLO, whose face shows hemorrhages of his left lower eyelid and lateral left eye region, was awarded a decision over CARTER, although CARTER carried the fight throughout most of the rounds.

Professional fight in welterweight between the Argentinean boxer Jorge FERNANDEZ (right) and Isaac LOGAN (Fig. 112). FERNANDEZ connects with a sharp right to the left side of LOGAN's face, distorting his features and turning his head abruptly to the right side by the force of the blow. A shower of sweat is flying from LOGAN's abruptly accelerated head. FERNANDEZ won a unanimous decision.

Professional boxing fight in middleweight between Florentino FERNANDEZ, the 'Cuban Ox,' and Johnny FEATHERMAN (Fig. 113). A left hand of FERNANDEZ, which connects with the right side of FEATHERMAN's jaw and neck, knocking his face out of shape. The intensity of the blow can be conjectured from the extent of the deformation of his face. The fight was stopped in the fifth round, after FERNANDEZ had opened deep cuts over FEATHERMAN's eyes. It was the 38th victory for the Cuban, 31 of these by a knock out.

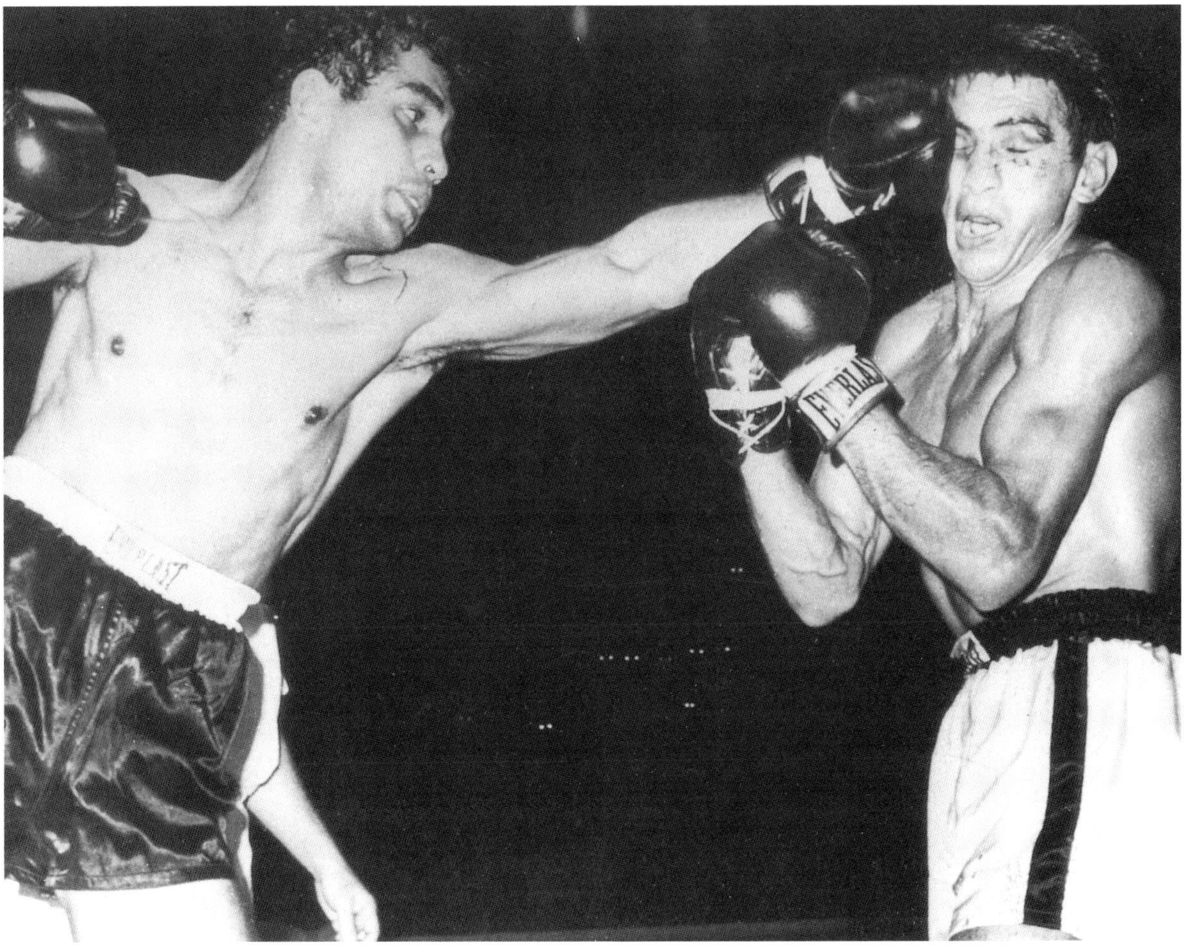

Fig. 104. Professional lightheavyweight fight between title holder Willie PASTRANO (left) and Gregorio PERALTA (right) on April 11, 1964, New Orleans, Photo: A.P.

Beaten-up Faces of Professional Boxers

The photo shows the face of Jerry QUARRY after his technical knock out by Joe FRAZIER (Fig. 114). The wound under his right eye required eight stitches . QUARRY's beaten up face is swollen, both eyes are nearly closed.

The photo shows a scene from the fight between Rocky MARCIANO and Joe WALCOTT. MARCIANO is landing a hard right on lateral parts of WALCOTT's chin and lateral parts of his neck (Fig. 115 a). The photo demonstrates the enormous deformation and distortion generated by the powerful impact.

Fig. 115 b. This photo is a close-up of Fig. 115 a. MARCIANO connects with a hard blow on left lateral parts of WALCOTT's chin and lateral parts of his neck. WALCOTT's face is severely distorted and displaced. While the head is accelerated by the direct impact, parts of his right cheek are following fractions of a second later due to its inertia; also the right ear remains inert. Blows against the head of such intensity also deform the skull and injure the brain.

Rocky MARCIANO won the heavyweight championship from Jersey Joe WALCOTT on September 23, 1952 at Philadelphia's Municipal Stadium by a knock out in the thirteenth

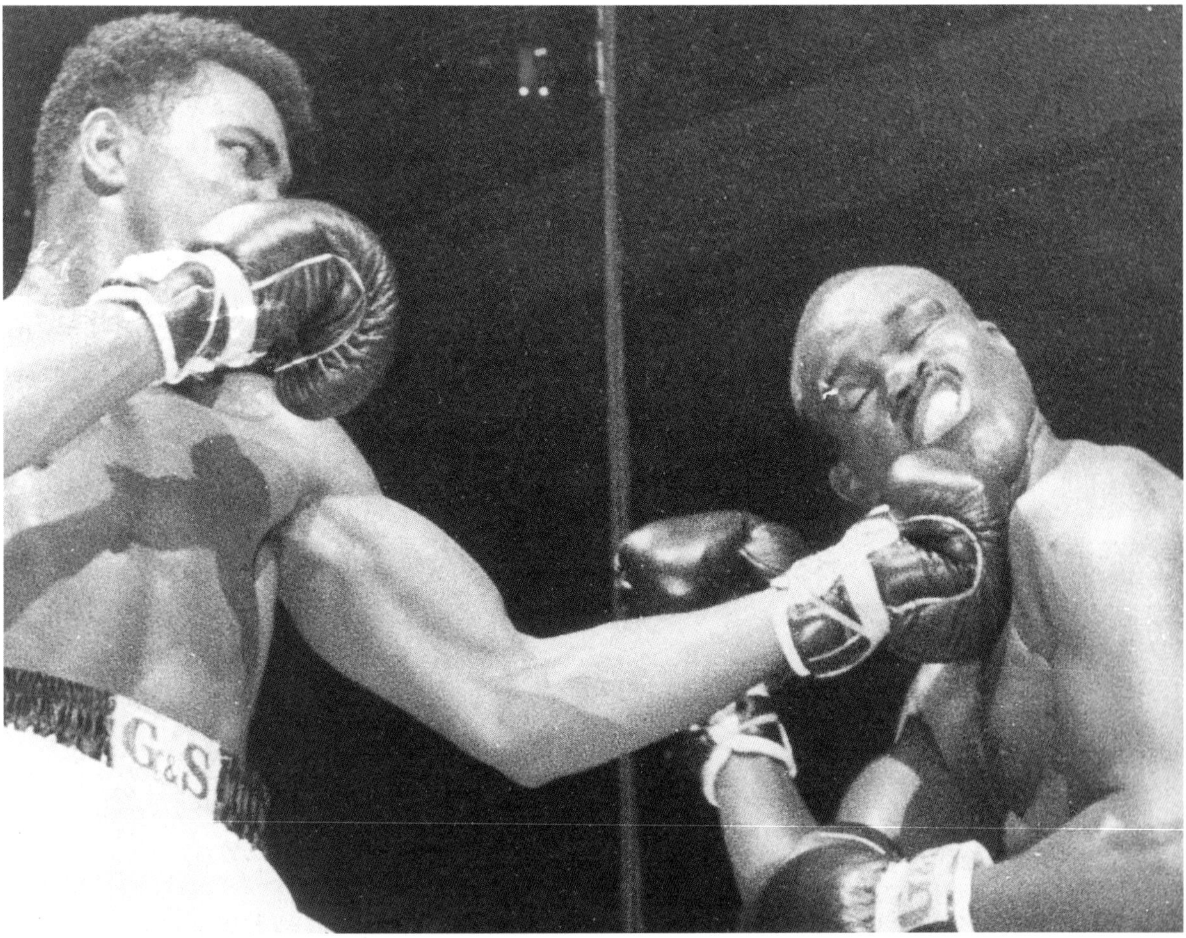

Fig. 105. Professional heavyweight fight between Cassius CLAY (left) and Doug JONES (right) on March 13, 1963 in Madison Square Garden, New York. Photo: A.P.

round. He defended his title six times before he retired undefeated from the ring in April 1956. MARCIANO's record was 49-0, with 43 knock outs. He knocked out 88% of his opponents, in comparison to 76% of Joe LOUIS.

Jersey Joe WALCOTT had 72 fights, 53 won, 18 lost, 1 drawn.

Description of the fight between Rocky MARCIANO and Jersey Joe WALCOTT

In the first round both fighters landed with hard lefts and rights. WALCOTT connected with a left hook to the jaw of MARCIANO that knocked him down. He later said: "I wasn't hurt by the punch, not even dazed." He quickly got on his knees. WALCOTT continued punching MARCIANO, blood ran from his lips. At the end of the first round MARCIANO's left eye was swollen.

MARCIANO pressed the attack, but WALCOTT was 2" taller and ten pounds heavier and had a 7" reach advantage. WALCOTT was the superior boxer.

Shortly before the third round ended WALCOTT landed a hard blow on MARCIANO's head, but MARCIANO continued the attack. MARCIANO landed a right, left and another left combination that hurt WALCOTT. MARCIANO landed some hard punches but WALCOTT showed no effect. They battered each other. It was WALCOTT who hit more often and harder.

In the sixth round MARCIANO's and WALCOTT's heads collided. The top of MARCIANO's head connected with WALCOTT's bone above the left eye. Blood began to flow from a deep gash in MARCIANO's forehead. And the flesh above of WALCOTT's eye was torn open. MARCIANO tried to make WALCOTT's wound the target of his blows but blood

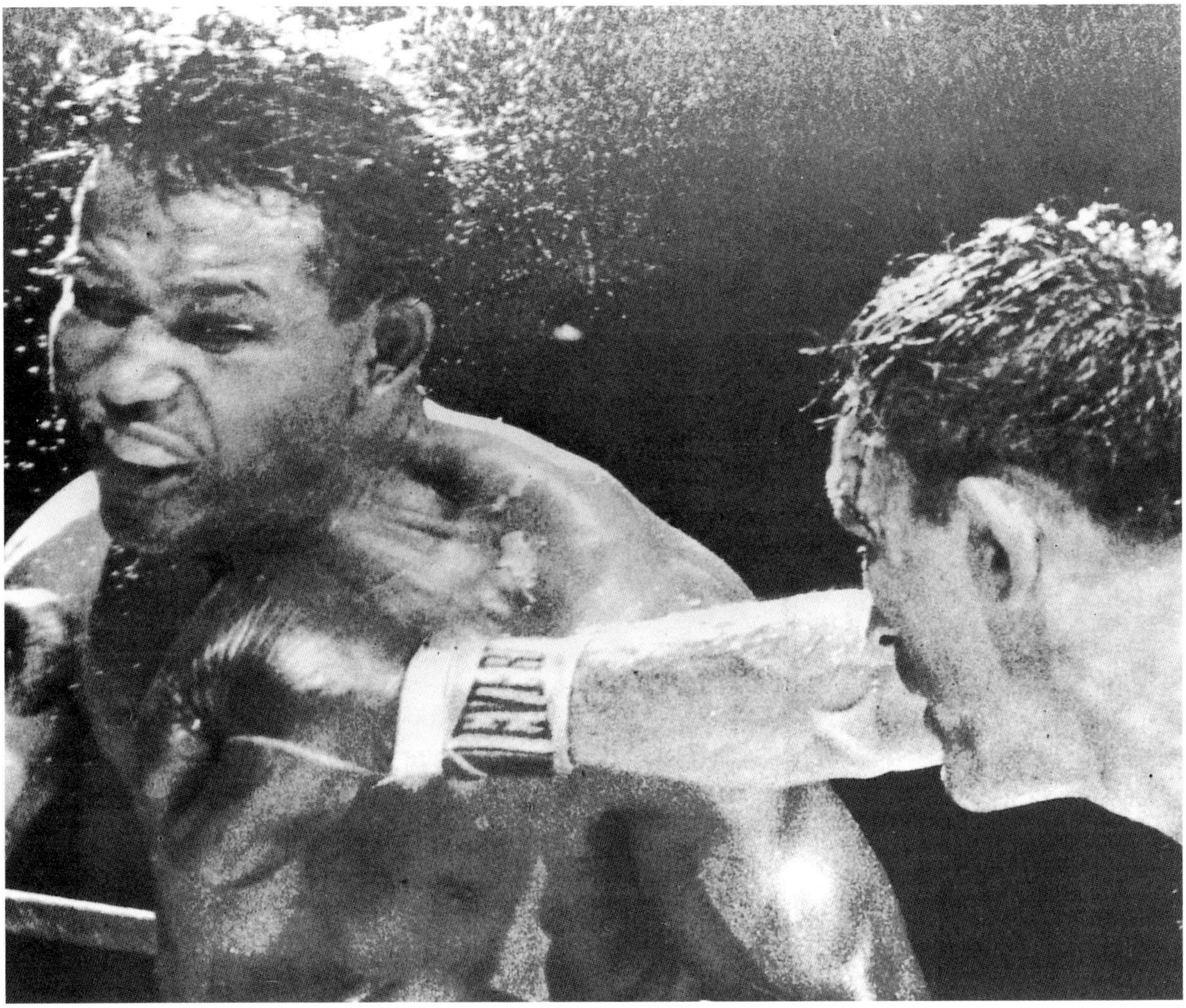

Fig. 106. Professional title fight in welterweight between Carmen BASILIO (right) and 'Sugar' Ray ROBINSON (left) on September 23, 1957 at Yankee Stadium, New York.

run down into his own eyes, he could not see WALCOTT clearly enough.

MARCIANO attacked relentlessly.

In the eleventh round WALCOTT connected with a right to MARCIANO's heart region. The impact of this punch stopped MARCIANO and WALCOTT landed a series of left and right blows to his head. A gash opened above MARCIANO's right eye. The bell saved MARCIANO. WALCOTT later said that he thought he could finish MARCIANO in the next, the twelfth round.

Both boxers began the thirteenth round each trying to knock out the other. WALCOTT was ahead on points. Now MARCIANO smashed a left to the body of WALCOTT who backed off. Then MARCIANO connected with a short right that distorted WALCOTT's face. A left hook landed on his head, WALCOTT sank down to one knee, his left arm around the middle rope, finally his head lying on the canvas. The sports journalist LIEBLING described it: "WALCOTT flowed down like floor out of a chute."

Dr. NARDIELLO needed fifteen stitches to close the wounds on MARCIANO's scalp, above the bridge of his nose and the corner of his left eye.

Professional heavyweight fight between Rocky MARCIANO and Joe LOUIS on October 26, 1951 at Madison Square Garden, New York.

The photo shows the end phase of the fight. MARCIANO connected with a hammering right to LOUIS head in the eight round (Fig. 116).

Fig. 107. Second professional heavyweight fight between MUHAMMAD ALI (right) and Ken NORTON (left) on September 28, 1976.

LOUIS eyes are already closed and his legs buckle. LOUIS was counted out. This bout lead to a title for MARCIANO.

This photo shows the end of the bout a few seconds later, from another line of sight. MARCIANO pounded the former champion and had him in a helpless state. LOUIS was knocked through the ropes (Fig. 117). Referee Ruby GOLDSTEIN, with his back to the camera, stopped the fight.

Joe LOUIS had 67 fights, 64 won, 3 lost.

Description of the fight between Rocky MARCIANO and Joe LOUIS

The fight took place on October 26, 1951. LOUIS was 37, he desperately needed money. LOUIS was 6'2", his weight was 212 pounds, MARCIANO 5'10", his weight was 187 pounds. LOUIS had a 9" reach advantage. 17.241 spectators were at the Garden and several millions watching on national television.

Fig. 108. Professional lightheavyweight fight between Harold JOHNSON and Eddie COTTON on August 30, 1961, in Seattle, Washington. Photo: A.P.

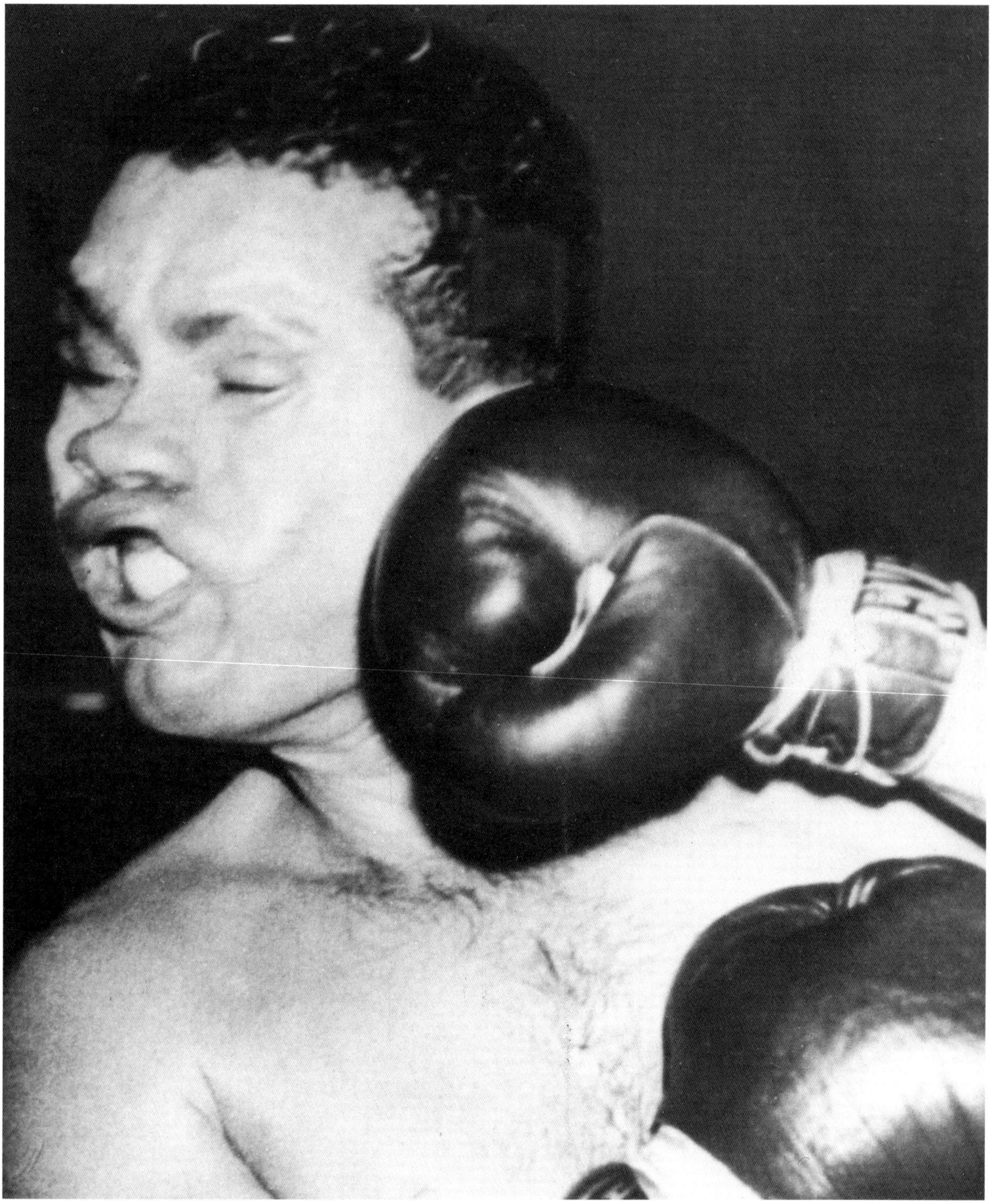

Fig. 109. Professional welterweight fight between Ray PORTILLA and Carlos ORTIZ. Photo: d.p.a.

Fig. 110. Professional bantamweight fight between Sandy SADDLER (left) and Willie PEP (right) on September 26, 1951 in Yankee Stadium, New York.

From the beginning MARCIANO was the aggressor, LOUIS used his left jab to keep him away. In the fifth round LOUIS' left cheek was swollen and he had a mouse under his eye. A trickle of blood run from MARCIANO's nostrils, he was slightly ahead.

At the end of the sixth round LOUIS seemed weary. In the seventh round, MARCIANO continued to keep the pressure on LOUIS, who became slower. Referee Ruby GOLDSTEIN had scored four rounds for MARCIANO and two for LOUIS and one even, Judge Joe ANGELO had MARCIANO heading 5-2, and judge Harold BARNES MARCIANO 4-3.

In the eighth round MARCIANO drove LOUIS toward the ropes and a short left hook brought LOUIS to the canvas. LOUIS lying flat on his back, rolled over and rose to one knee. LOUIS took a count of '8' and began to get up. MARCIANO forced LOUIS again to the ropes and landed two hammering left hooks. LOUIS eyes were shut, his arms were hanging at the side of his body. LOUIS began to sag to the canvas and was hit by a right to the jaw. LOUIS fell backward through the ropes, one leg remained in the ring. Referee GOLDSTEIN began to count, but when he reached three or four waved his arms, signaling that the fight was over. LOUIS had to be assisted in order to avoid falling into the press row. This fight ended LOUIS active career. LOUIS al-

Fig. 111. Professional championship middleweight fight between Joe GIARDELLO (left) and Rubin CARTER (right) in Philadelphia. Photo: A.P.

Fig. 112. Professional welterweight fight between the Argentinian boxer Jorge FERNANDEZ (right) and Isaak LOGAN, Cuba, (left) on August 19, 1961 in Madison Square Garden, New York. Photo: A.P.

legedly said that MARCIANO hit harder than SCHMELING.

Professional heavyweight fight between Rocky MARCIANO and Archie MOORE on September 21, 1955 at Yankee Stadium, New York.

Archie MOORE, battered and dazed, tries in vain to get up as he hangs onto the rope, while referee Harry KESSLER administers the count (Fig. 118). Victor and still champ, Rocky MARCIANO, walks away after delivering the knock out in the ninth round.

COHANE (1963) gave the following description of the professional heavyweight bout between Rocky MARCIANO and Archie MOORE:

"When he faced MARCIANO, MOORE was thirty-six, with more than 150 fights in 19 years behind him. In December, 1952, when he won the light-heavyweight championship from Joey MAXIM, Archie was thirty-two. While still an accomplished pro, against MARCIANO and MAXIM, MOORE was well past his prime. He had been made to wait too long by fate and fortune; for one thing, when Archie was at his best the negro fighter was not encouraged toward the top, as he has been now for several years."

Professional heavyweight fight between Joe LOUIS and Lee SAVOLD on June 15, 1957 at Madison Square Garden, New York (Fig. 119). Lee SAVOLD was brutally beaten up. The photo shows SAVOLD severely beaten with a bloody and swollen face collapsing after a hard blow of LOUIS (Fig. 120).

The professional middleweight bout between

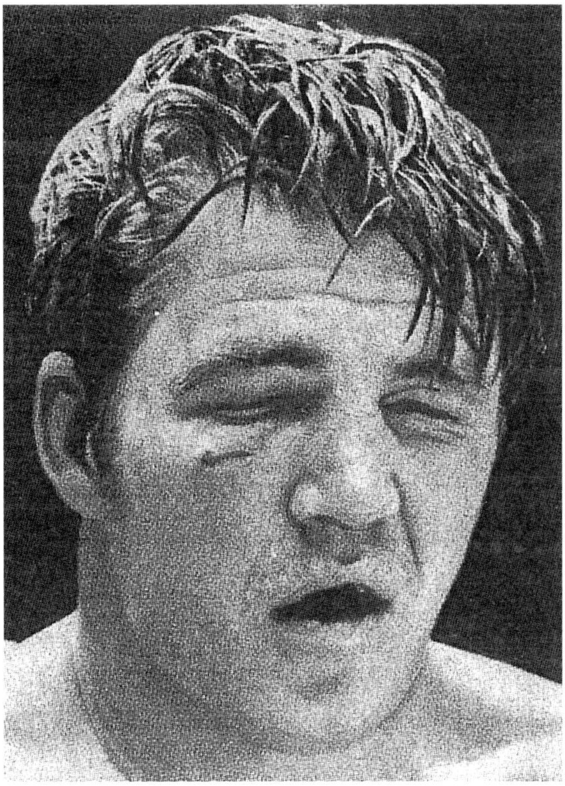

Fig. 114. Professional heavyweight fight between Joe FRAZIER and Jerry QUARRY in 1969. FRAZIER won the fight by a technical knock out. Photo: Globe.

Hans STRETZ and Gerhard HECHT was judged a 'draw' after a bloody 12-round fight. In the eighth round STRETZ sinks 'groggy' to the canvas after HECHT landed a hard right. His eyes are closed (Fig. 121). There is massive bleeding from his mouth.

Professional featherweight fight between 'Harlem' Tommy MURPHY (left) and Abe ATTELL (right) on March 9, 1912 in Daly City, California (Fig. 122). It was one of the bloodiest fights ever. Both boxers were covered with blood during the 20-round fight. The referee declared MURPHY the winner.

Sixth professional boxing fight between 'Sugar' Ray ROBINSON and Jake LA MOTTA on February 14, 1951, in Chicago (Fig. 123). LA MOTTA leans beaten and defeated on the ropes after the referee stopped the fight in the thirteenth round. Both eyes are tightly closed.

Fig. 115 a. Professional heavyweight championship fight between Rocky MARCIANO and Joe WALCOTT on September 24, 1952. Photo: d.p.a.

Second professional boxing match between 'Sugar' Ray ROBINSON and Carmen BASILIO in 1958 in Chicago (Fig. 124). ROBINSON won the fight by a decision. From the sixth round on, BASILIO had to take severe punishment. The battered and puffy face shows the result of the numerous blows, his left eye is tightly closed.

Professional fight in middleweight between Peter MÜLLER and Paul ROUX on June 6, 1961, in Cologne, Germany (Fig. 125). The photo shows the bloody and swollen face of Peter MÜLLER in his corner, in severe traumatic shock after he was defeated on points.

Professional heavyweight fight between Joe FRAZIER and George CHUVALO on July 1967, in Madison Square Garden, New York (Fig. 126). The photo taken, after the fight ended, reflects the severe punishment and numerous blows against the head of CHUVALO – it is beaten in a bloody pulp. Both eyelids are swollen and nearly closed.

Fig. 113. Professional middleweight fight between Florentino FERNANDEZ, the 'Cuban Ox,' and Johnny FEATHERMAN at Miami Beach. Photo: A.P.

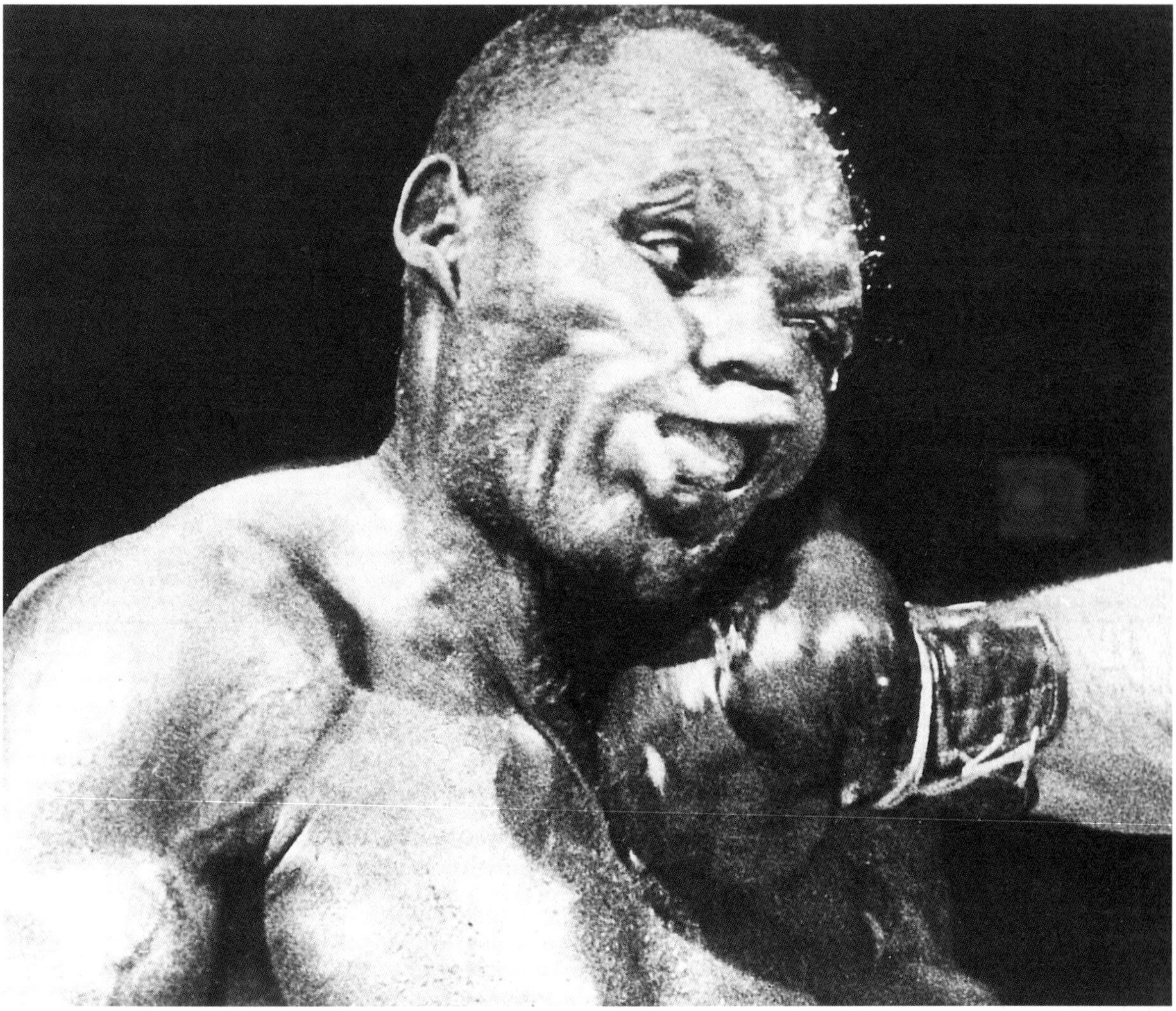

Fig. 115 b. Close-up of the precedimg photo of the professional championship fight between Rocky MARCIANO and Joe WALCOTT on September 24, 1952. Photo: d.p.a.

Joe FRAZIER (with Phil BERGER) writes about the fight with George CHUVALO:

"In the meantime, Medison Square Garden had offered me fifty thousand dollars to fight George CHUVALO that July. CHUVALO was a fighter with a reputation for taking punches without showing any visible effect. The Human Shock Absorber of the heavyweight division."

"Over the years he had fought the best men in the division – fighters like CLAY, TERRELL, PATTERSON, BONAVENA, FOLLEY, Mike De JOHN, and Doug JONES. And while most of them had beaten CHUVALO, in sixty-two bouts none of them had been able to knock the twenty-nine-year-old Canadian down or knock him out. Trying to knock down CHUVALO, it was said, was trying to knock down Canada'a Parliament Building in Ottawa. It just couldn't be done."

"Well, the night I fought CHUVALO I put a crimp in the reputation he had for being undestructible. From the opening bell, I began tattooing that big-jawed face with left hooks. Almost immediately, small cuts opened under CHUVALO's left eye and under his right eye."

"By the second round, the swelling around CHUVALO's eyes grew worse and his face was smeared with blood. But give him credit, CHUVALO hung in there. In the third round, in a kind of last-ditch effort, he went toe to toe and landed a few good licks, rousing the crowd of 13.984 paying customers."

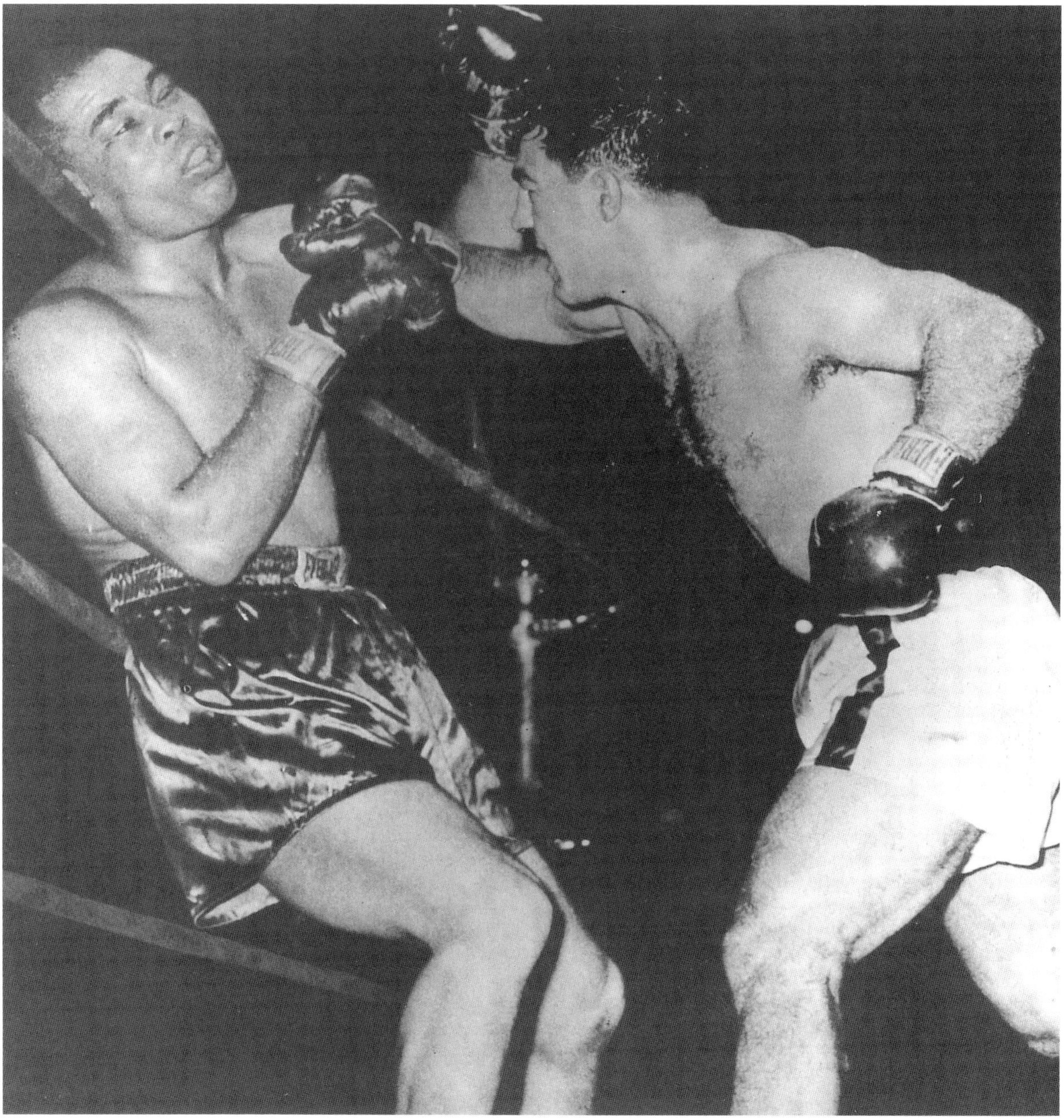

Fig. 116. Professional heavyweight fight between Rocky MARCIANO (right) and Joe LOUIS (left) on October 26, 1951.

"By now, the cuts and bruises were running purple. CHUVALO looked bad enough to prompt Dr. Edwin CAMPBELL, the New York State Athletic Commission physician, to examine him between the third and fourth rounds. CAMPBELL let the fight go on, but it wouldn't last long."

"At the start of the fourth round, I walked out and hit CHUVALO with a left to his already-closed right eye. Suddenly he turned his back to me and groped his way to his corner, like a man feeling his way through a pitch-black room."

"It turned out a blood vessel leading to his eyelid had erupted. 'My eyball felt like a grapefruit,' CHUVALO would say after-

Fig. 117. Professional heavyweight fight between Rocky MARCIANO and Joe LOUIS on October 26, 1951. This photo was taken immediately after Fig. 116.

ward, 'The pain was excruciating. I thought my eye was going to come out.'"

"The referee, Johnny COLAN, stopped the fight."

"Days later, CHUVALO, complaining of pain and double vision, ended in a Toronto hospital to have surgery for a bone fracture under his right eye."

Joe FRAZIER had 37 fights, 32 won, 4 lost, 1 drawn.

Professional fight in light-heavyweight between Gerhard HECHT (left) and Herbert KLEINWÄCHTER (right) (Fig. 127). The faces of both boxers are severely marked, they are swollen and bloody, the eyes are partially or completely closed due to hemorrhages in the eyelids.

TOPOROFF (1987) reported on severe beatings he had seen in professional boxing:

"The worst beating I ever saw close up at the old Garden was on December 10, 1965, in a rematch between two heavyweights named James J. – surnamed BEATTIE and WOODY."

"Jim BEATTIE was the sacrificial lamb, seemingly chosen for his role by destiny. A huge six-foot-nine, 235-pound, rawboned Minnesota farm boy with a deceptively successful amateur career behind him – his sheer size had defeated most of his opponents – BEATTIE had gotten his job as a professional fighter through *The New York Times.* He had answered an ad promising up to $ 150,000 a year for the right sort of young man willing to try to become heavyweight champion of the world."

"The idea of advertising for a big white galoot who could be retooled into a heavyweight champ, thereby making everyone involved in the project rich, was the brainstorm of a New York restaurateur and some friends with disposable cash. The 'white' made success more likely, since in the heavyweight class the word attached itself to 'Hope,' and the combination often produced significant revenue. Rarely a champion."

For an on-the-job trainee, BEATTIE started well enough as a pro. He'd lost just once in thirteen tries; then WOODY beat him by a decision in the Garden. BEATTIE then strung together five consecutive knockouts – they were mostly tomato cans – and came back to the Garden for the rematch with WOODY. There was a point midway through the third round when I had BEATTIE still ahead on my card and thought he might beat his shorter, well-muscled black opponent. Perhaps it was weariness that reached him first, or the hooks WOODY was digging into the big blond's exposed ribs. Or both. However it was, the beating became monstrous."

"A number of things conspired to make it so. WOODY put his head on big Jim's chest and stayed so close that BEATTIE had absolutely no punching room, all the while digging to the long, exposed flanks. When the big man dropped his hands and doubled over, WOODY whipped lacerating punches to the pink face. The blood then splattered was all BEATTIE's. I'd worn a red sweater."

"Because BEATTIE was so incredibly game and WOODY was punishing but not powerful enough to put him away, the

Fig. 118. Professional heavyweight championship fight between Rocky MARCIANO and Archie MOORE on September 21, 1955 in New York. Photo: d.p.a.

equipoise of damage shifted by slim degrees, making the accumulation seem less terrible than it was. In addition, BEATTIE was so huge, it might have appeared to the referee he had more of himself to offer up than a normal-sized heavyweight. I could never prove it, but I had the distinct feeling that the crowd wanted to see BEATTIE punished precisely because he was so big. Lilliputian revenge."

"Frankly, I didn't see the specific blow that put him down and out in the seventh round, but felt the boards beneath the ring rattle when he spun helplessly and fell face forward to the canvas. After they got him up and into his robe – white with green trim, if memory serves – he ran down the ring steps alongside me. I looked up into his blinding kliegs above the ring and saw a blood-smeared face haloed in light; his eyes were dazed, the mouth hung open. I thought he'd topple over onto me and was glad when he went past."

"But BEATTIE walked up the aisle past the turn to the dressing room area. Alone, James J. BEATTIE staggered up the aisle right toward the end arena. Patrons were booing at him and pointing comically. I ran after him, caught him by the arm, and turned him toward his dressing room. James J. BEATTIE was senseless. "

"Later, as I stood in his dressing room, his eye was not merely swollen, it looked as though it might drop off. A gaping slash lined the brow. There was hardly a square inch of face that was not ripped raw or cut. Had Jim BEATTIE stayed around New York, he'd have given Chuc WEPNER, the celebrated 'Bayonne Bleeder' and real life model for Rocky BALBOA, a run for his money as blood letting champ. I broke my long tradition of asking losers foolish questions that night with BEATTIE: "What in the world,' I said, 'are you doing in the business?"

Fig. 119. Professional heavyweight fight between Joe LOUIS and the British boxer Lee SAVOLD on July 1951 at Madison Square Garden, New York.

"James J. BEATTIE didn't bat a bloody eye: 'I'm still sharpening my skills!' Three minutes earlier he hadn't been sharp enough to find his dressing room."

Injuries of the Eye Lids and Eye Brows

The skin around the eye lids, especially the upper lids, possesses an abundant blood supply. It is thin and easily displaced. *Periorbital hematomas* ('*black eyes*') or *hemorrhage into the upper eyelids* can become so large and voluminous, that an eye can be closed completely, so that the visual faculty is completely abolished. An ophthalmologist is then unable to see and examine the fundus. The injuries are the results of lesions of blood vessels. A boxing blow can press the vessel against the *more or less sharp edge of the superior margin of the orbita* and in that way injure and open it. Tangentially directed blows may displace the blood vessel rich tissue of this region, rupturing it.

Fig. 120. Lee Savold's bruised and swollen face reflects the severe punishment he took from the ex-champion Joe LOUIS in their Madison Square Garden bout. Louis knocked him out in the sixth round of the match. Photo: U.P.I.

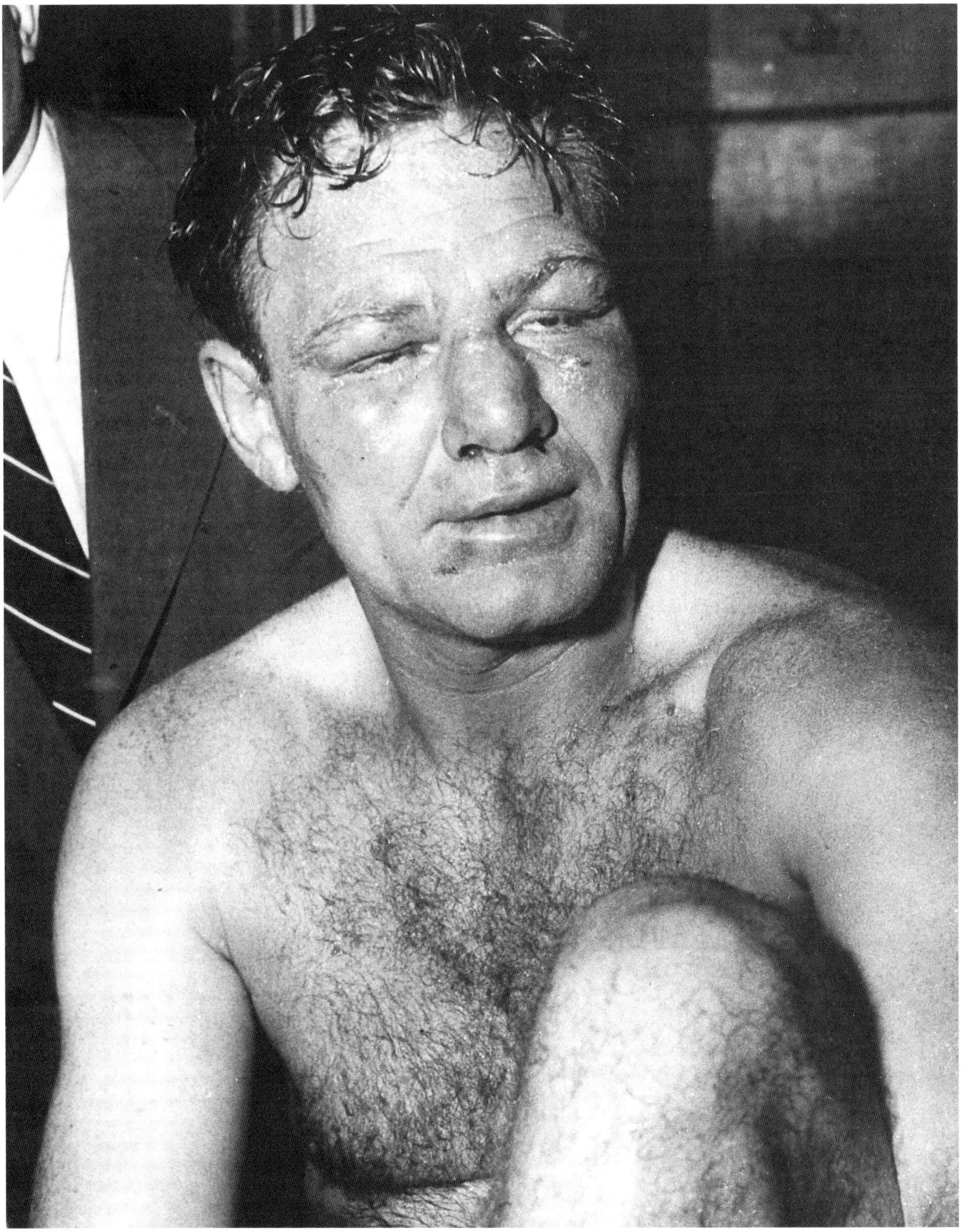

Fig. 121. Professional middleweight championship fight between Hans STRETZ and Gerhard HECHT, Germany. Photo: SCHIRMER.

Fig. 122. Professional fight between Abe ATTELL (right) and 'Harlem' Tommy MURPHY on March 9, 1912 in Daly City, California.

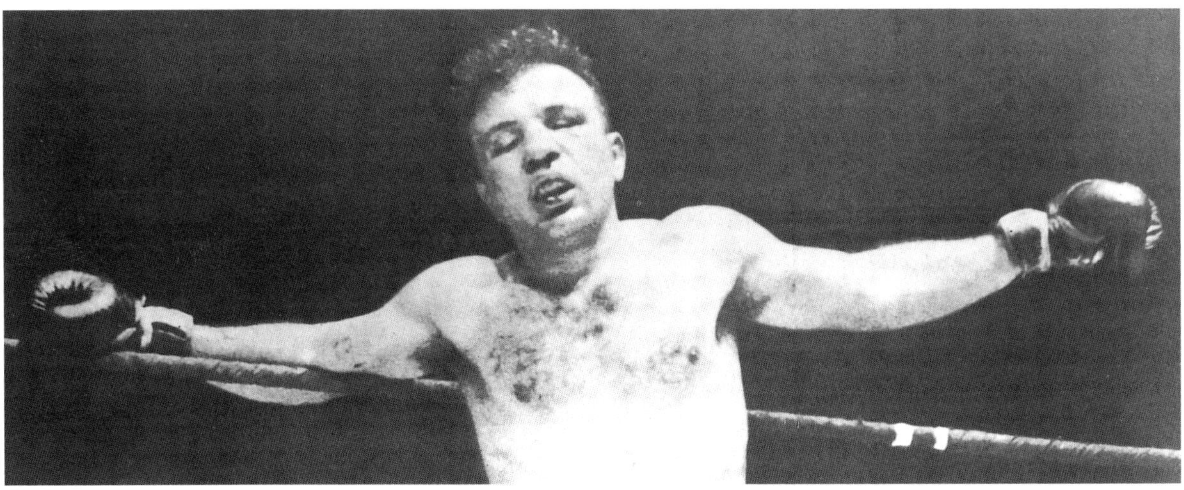

Fig. 123. Sixth professional lightheavyweight fight between 'Sugar' Ray ROBINSON and Jake LA MOTTA on February 14, 1951 in Chicago. LA MOTTA leans beaten and defeated on the ropes after the referee stopped the fight in the thirteenth round.

McCOWN (1959) reported that there were 1.010 cases of wounds of the soft tissue of the face over a seven year period. These consisted of contusions, abrasions and lacerations, and there was no attempt made to differentiate between them because there were elements of each type in all of the wounds. In most of them the contusion seemed to predominate, particularly in wounds about the

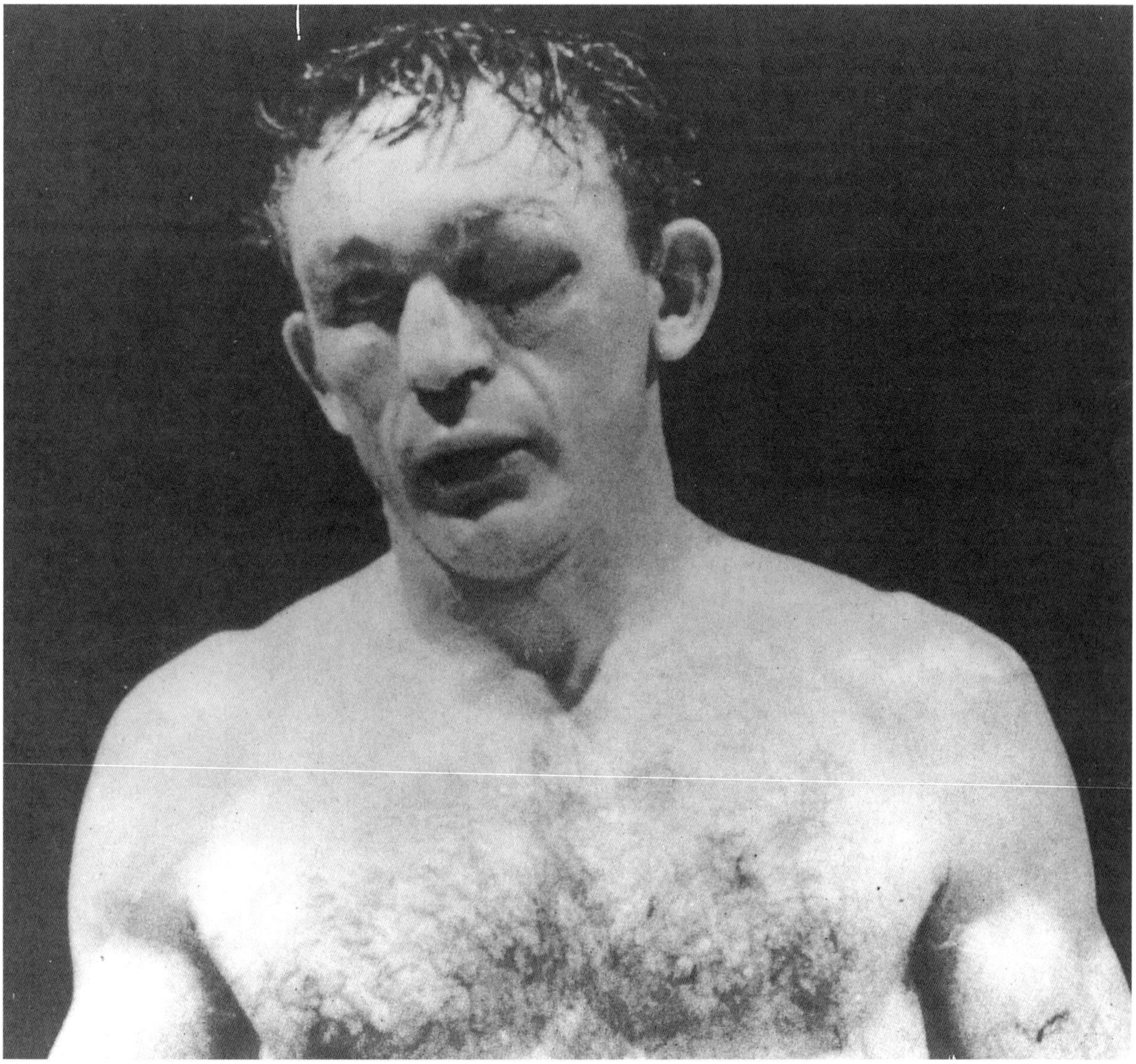

Fig. 124. Second professional boxing match between 'Sugar' Ray ROBINSON and Carmen BASILIO in Chicago in 1958. The photo shows BASILIO's battered and puffy face.

cheek and jaw; whereas in wounds over the underlying bony prominences such as the frontal bone, supraorbital ridge, malar eminence, zygomatic arch or chin, lacerations predominate. In wounds about the eyelids and cheek, hematomas are common.

McCOWN continues to report that 575 bouts had to be stopped because of extensive injuries to the soft tissue. It was also significant that many of

Fig. 125. Professional middleweight fight between Peter MÜLLER and Paul ROUX on June 6, 1961, in Cologne, Germany. MÜLLER lost the fight by points, is beaten and in a severe traumatic shock. Photo: A.P.

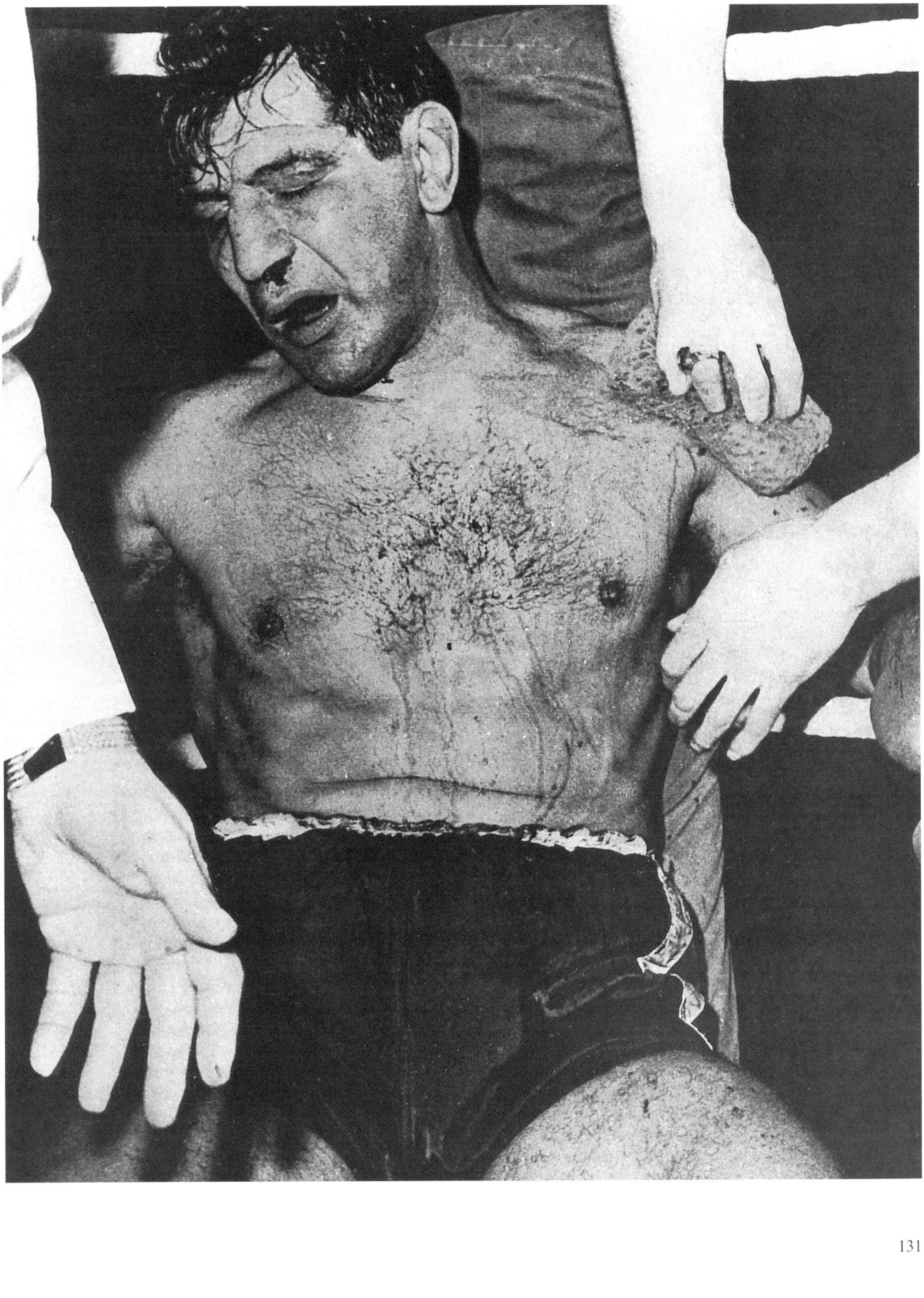

Fig. 126. Professional heavyweight fight between Joe FRAZIER and George CHUVALO on July 19, 1967 at Madison Square Garden, New York. The photo, taken after the fight ended, reflects the severe punishment of CHUVALO.

Fig. 127. Professional lightheavyweight fight between Gerhard HECHT (left) and Herbert KLEINWÄCHTER (right). The fight ended in a draw. Photo: Helmuth RUDOLPH.

the lacerations and contusions were over the exact sites of previous similar injuries. The immediate treatment of these lacerations at ringside is far from satisfactory. He mentioned especially that Monsels solution, silver nitrate and adrenalin, was banned. However, other equally obnoxious solutions, greases and ointments are applied such as zinc oxide ointment, bismuth subgallate and Vaseline. Thus, the wounds are not only contaminated, but their immediate treatment and definitive care is rendered difficult. Under such conditions, the results achieved leave much to be desired. All lacerations should be treated by thorough irrigation with copious amount of saline solution, adequate but not unnecessary débridement, avoidance of chemicals or antiseptics and meticulous layer closure using 5-0 plain interrupted catgut sutures in the deep layers and fine no. 5-0 silk or horsehair sutures for the skin. These sutures should be removed early to minimize formation of cicatrix.

McCOWN continues to say that since all wounds heal by cicatrization one should minimize it in these boxers, because a dense broad thick scar is an easy target to reopen, and the density of the scar determines the vulnerability to opening when subjected to trauma. Therefore, a fine linear cicatrix will protect and perform much better than the broad, dense, deep variety resulting from poor care.

Ideally, boxers who sustain lacerated wounds

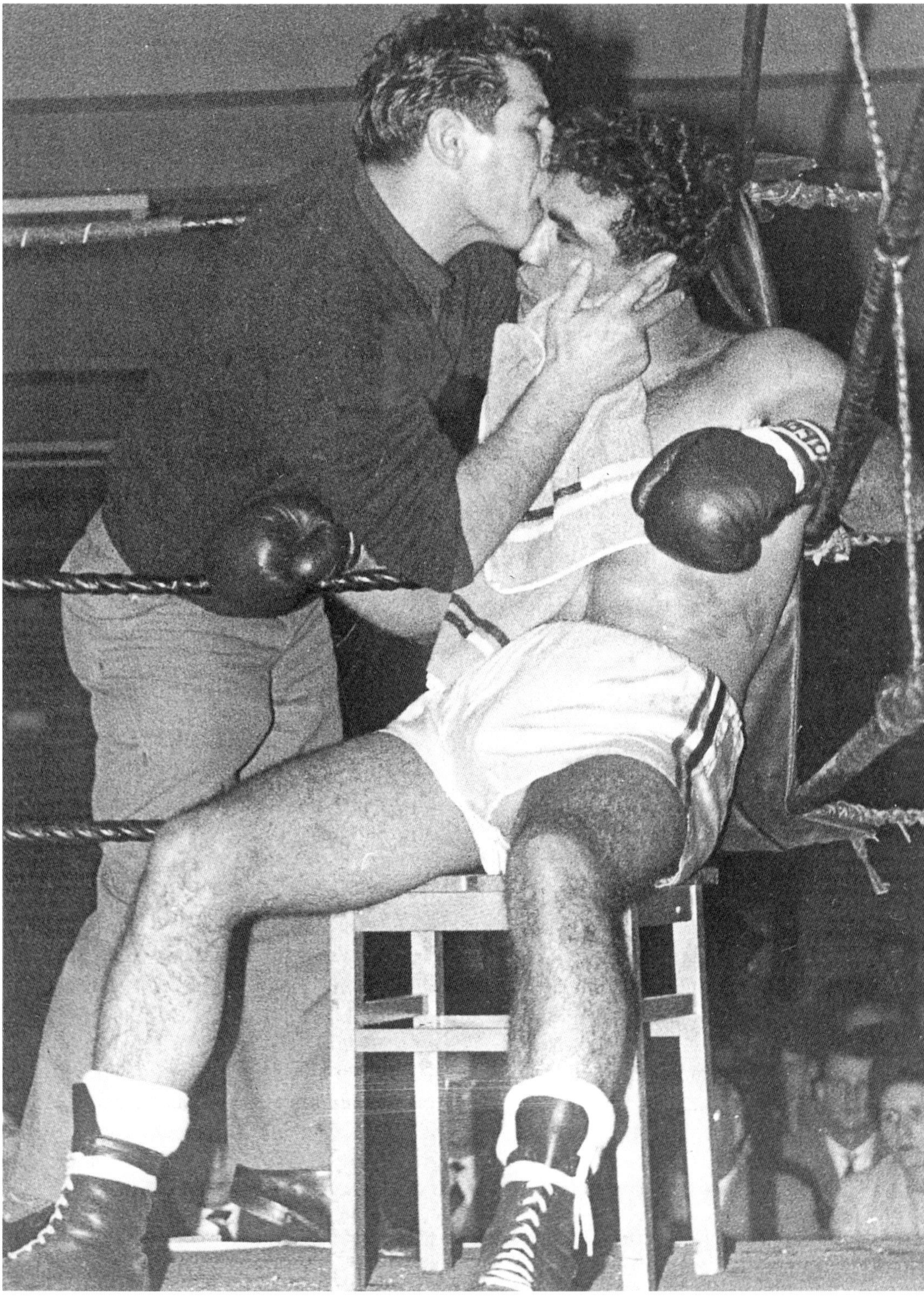

should not resume boxing before three months of healing time without reinjury. As a practical matter, this time limit is very difficult to put into practice. Many boxers believe that a scar is merely a cosmetic alteration, not worth the time lost to full recovery.

Craniofacial lesions in boxers were described by BOIGEY (1920), PALAZZI (1925), FRACKELTON (1959) and BORGOGNA et al. (1994), with reference to the different types of boxing blows and their typical lesions.

It is surprising how few papers are published on craniofacial lesion in boxers.

In the so-called *uppercut* against the mandible, this structure presses against the maxilla with great intensity. The upper teeth can be damaged severely. Fracture of the mandible and maxilla are possible complications. The effect of an uppercut is less severe when the boxer keeps his mandible in occlusion with maxilla, so that both parts form a block. PALAZZI saw a boxer who was hit by an uppercut and succumbed a few hours later in hospital due to a fracture of the base of the skull.

Swinging blows which connect with the horizontal ramus of the mandible may lead to sudden dislocations of the mandible to the contralateral side, which can be complicated by ruptured ligaments, especially the spheno-maxillar ligament and the capsular ligament.

A *straight blow* against the tip of the chin produces an abrupt retropulsion of the mandible against the dorsal wall of the cavum articulare glenoidale. Such blows can fracture the anterior wall of the meatus auditivus, often causing hemorrhaging from the ear.

A *blow straight against the mouth* impacts against the frontal areas of maxilla and mandible simultaneously. This will damage the incisors, with possible loss of one or several teeth.

Common Facial Injuries from Boxing

The *most common facial injuries* are: (1) *Bruises, abrasions, or cuts*; (2) *bruises, lacerations, or cuts around the eye, eyebrows or eyelids*; (3) *cuts of the lips*; (4) *'cauliflower ears,'* (5) *fractures of the mandible*; (6) *injuries to* and *loss of teeth*; and (7) *injuries* and *fractures of the nose*.

Bruises, Abrasions and Cuts of the Face

(1) The general *bruises and abrasions of the face* can be the result of multiple repetitive impacts. The skin may be swollen ('puffy') or there may be 'lumps' of swollen tissue due to edema and hemorrhages.

(2) The skin around the eyelids has an abundant blood supply, is thin and slides easily. *Periorbital hematomas* (*'black eyes'*) may reach so large a dimension that the eye is completely closed. They are the result of injuries to blood vessels, which can be severed by a blow by which the vessel is 'cut' by the more or less sharp edge of the orbital rim or ruptured due to the displaceable thin skin in this area. Some individuals are especially predisposed to these types of injuries. Application of ice bags or the administration of astringents or raw meat will not influence the swelling at all.

Injuries to the Periorbita and Orbit: Possible boxing injuries to the periorbita and orbit include, according to GIOVINAZZO et al. (1995), lid and soft tissue lacerations, injuries to the extraocular muscles, and fractures to the orbit rim and floor.

Lid and Soft Tissue Lacerations: Lacerations to the lid and soft tissue, are, according to GIOVINAZZO et al., commonplace. Lid lacerations involving the lid margin should be repaired by an ophthalmologist or a plastic surgeon well versed in lid anatomy. Improper suturing of lid margin lacerations can result, as the authors write,

Fig. 128. Professional middleweight fight between SCHILLING and the Syrian Mohammad SAHIB on March 21, 1961 in Rüsselsheim, Germany. SAHIB's right eye lid and brow were injured. Peter MÜLLER, his second, tried instead of using a swab of cotton, to suck out the hemorrhage in the eyelid. Photo: d.p.a.

BERG
BERG

Fig. 130. Professional lightweight bout between Danny Jo PEREZ (right) and Franky RYFF (left) on March 11, 1955 in Madison Square Garden, New York. Photo: Harry WAGER, International News Photo.

in scarring of the lid margin, with subsequent corneal irritation and breakdown. The physician should assume that substantial periorbital trauma is associated with intraocular injury and should consult an ophthalmologist. Damage to the levator aponeurosis may result in ptosis.

Damage to Extraocular Muscles: Injury to these muscles may result, according to GIOVINAZZO et al., in motility problems and diplopia. The most common muscle injured in boxers is the superior oblique, probably from damage to the trochlea. "When motility disturbances are identified, injuries have occurred to the third, fourth, or sixth cranial nerve intracranially from closed head trauma." Orbital fractures, edema, and hematoma may also restrict extraocular ocular muscle motion.

Fractures of the Orbital Rim and Orbital Floor: Such fractures do not require repair, according to GIOVINAZZO et al., unless there is associated facial flattening, enophthalmos, or diplopia. Here an evaluation of the eye by an ophthalmologist is essential, as fractures of the orbit and orbital floor

Fig. 129. Professional boxing match between Peter MÜLLER, Cologne, Germany, (right) and Paul ROUX (left) on June 6, 1961 in Cologne, Germany. Photo: SCHIRMER.

are commonly associated with severe and potentially blinding intraocular injury.

GIOVINAZZO et al. comment:

"Ophthalmologists who actually care for the injured boxers realize that several factors prevent accurate counts of eye injuries due to boxing:"

"(1) Boxers tend not to be seen in hospital emergency rooms, as eye injuries are the accepted result of the sport and are usually 'toughed out' with little or no treatment."

"(2) Blinding injuries most often affect one eye, and the boxer will frequently hide the defect for fear of being disqualified from the sport."

"(3) Other blinding injuries, such as glaucoma from angle recession, may occur years after retirement from the sport and a correlation between the blindness and boxing will not be made – and if made, it will not be reported to any central monitoring agency."

Certain boxers are known to be especially vulnerable to this type of injury, because the upper ridge of the orbita is sharp-cornered and acts like a knife when hit.

Chymotrypsin has been used for reduction of the hematomas and swelling. Other authors could not confirm the allegedly good results reported by BLONSTEIN (1964).

Many cuts around the eye are the result of a clash or head butting by the opponent, and can be avoided by a scrupulous and attentive referee, to some degree. If a boxer has an eye cut and the referee does not stop the fight, the injured area is subjected to more immediate blows, exacerbating the damage. Since attacking weakened areas is an accepted boxing strategy, repeated blows to cut eyes are predictable.

Cuts around the eyes must be treated by a physician, with great care, in order to achieve first intention healing. Ideally, they should be treated in an emergency room or hospital, with good lighting and aseptic conditions. *Untrained cornermen, or 'cut men,'* should not attempt to treat these injuries between rounds. Such emergency treatment often consists of daubing at wounds with a dirty towel draped the neck of the second. Quack remedies in the ring are legendary, including covering skin cuts with carpenters' glue, or rubbing adrenalin into open cuts to stop the bleeding. Adrenalin has been applied in dosages sufficient to affect the boxer's pulse, blood pressure, etc, as reported in *Scientific American* (1952).

ROLAUF (1937, quoted by JOKL, 1941) described the following experience

"In the *Albert Hall* in *London,* I once watched a boxer whose left eye became totally closed by a large hemorrhage into the eyelids, caused by punches. Consequently he was unable to aim properly, and his punches often missed their objective by more than a foot. He seemed to be doomed, when during the next interval, his second took a knife, made a small incision underneath the brow and sucked the blood out of the wound with his mouth. The swelling immediately subsided, the boxer was able to see again, and won the fight."

Fig. 128 shows a similar example. Professional middleweight fight between SCHILLING and SAHIB. SAHIB's right eyebrow developed a massive hematoma after one of SCHILLING's blows. SAHIB's handler, the professional boxer Peter MÜLLER (Cologne) tries, instead of using a cotton swab, to suck out the wound. SAHIB lost by a knock out. Here, a second is trying to suck out the hematoma of the right upper eyelid with his mouth. In spite of it, the injured boxer lost the fight by a knock out. Clearly such treatment falls beneath all modern standards of medical care and would not even be acceptable under battlefield conditions.

Peter MÜLLER is profusely bleeding from a laceration of his left temple. ROUX suffered a cut of his right eyelid (Fig. 129). Both hemorrhages do not influence the visual acuity. ROUX won the bout by a decision. The professional boxer Paul ROUX, left, with a wound beneath his right eye from which blood could not run over the eye (Fig. 129). Nonetheless, the swelling connective tissue

Fig. 131. Professional heavyweight fight between Cassius CLAY and the British heavyweight champion Henry COOPER on June 18, 1963 in Wembley Stadium, London. CLAY was knocked down in the fourth round.

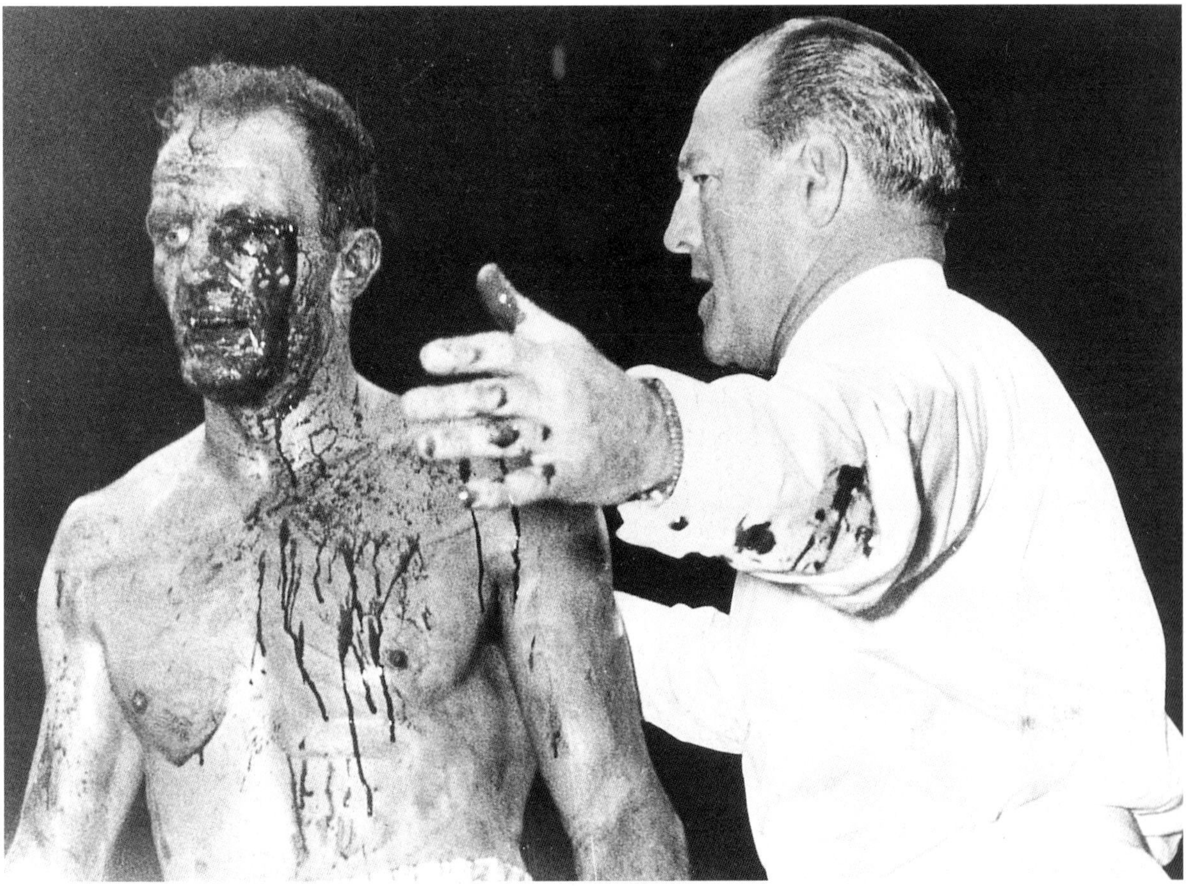

Fig. 132. This photo shows a scene from the fifth round of the same boxing match. Tommy LITTLE, the referee, stopped the boxing match. CLAY won by a technical knock out. Photo: A.P.

and extension of the hemorrhage cause loss of vision.

The professional boxer Danny Lo PEREZ (right) has severe hemorrhages in the upper and lower lid of his left eye (Fig. 130). The original legend of this press photo is interesting. It said that his opponent, Franky RYFF (left) continued with blows on the injured eye, and the hemorrhage enlarged. He won the bout on a decision. Neither the referee nor the ring physician stopped the bout. It is well known that an injured area of the face is a favorite target of boxers. The object is to defeat the opponent by capitalizing on an especially vulnerable area. Commentaries of sport reporters and testimony of boxers confirm this.

The next two photos show scenes from the professional boxing match between Cassius CLAY and English champion Henry COOPER on June 18, 1963, in Wembley Stadium in London. COOPER knocked CLAY down in the fourth round (Fig. 131), who was saved by the bell. The break between the two rounds was timed at 90 seconds, as Angelo DUNDEE, CLAY's manager, insisted replacing a boxing glove CLAY had split open on his fall on the canvas. By this maneuver CLAY won additional time to recover from the knock down. DUNDEE many years later admitted that he had aggravated a tear in the boxing glove of CLAY, which was replaced.

This photo shows a scene form the fifth round of this boxing match (Fig. 132). Tommy LITTLE, the referee, stopped the boxing match, when

Fig. 133. Professional heavyweight fight between Cassius CLAY and Sonny LISTON on February 25, 1964 in Miami, Florida. Photo: A.P.

COOPER's eye region was injured and was heavily bleeding. COOPER is told by referee LITTLE, splattered with blood from COOPER, that the fight is over. COOPER lost this fight by a technical knock out.

A left by CLAY connects with LISTON's right eye, in the third round of the title fight. Blood trickles from a wound under LISTON's left eye (Fig. 133). LISTON refused to continue, as the seventh round was to begin, due to eye cuts and an alleged injury to his left shoulder. Up to this time the referee and the judges had called the bout a draw.

The end of the professional fight between Jose NAPOLES and Billy BACKUS, on June 4, 1971, in Los Angeles (Fig. 134). Referee Dick YOUNG stopped the fight in the eighth round. The effect of the numerous blows of NAPOLES can be seen on BACKUS' face. His right eye is completely closed. His face battered and swollen and covered with blood. His nostrils are bleeding. There is a cut on lateral parts of the left eyebrow of BACKUS.

The end of the professional boxing match between COOKES and NAPOLES on April 17, 1969 in Los Angeles (Fig. 135). The photo shows the beaten-up, swollen and battered face of

Fig. 134. Professional title fight between José NAPOLES and Billy BACKUS on June 4, 1971 in Los Angeles, California. Referee Dick YOUNG stopped the fight in the eighth round. The effects of NAPOLES' blows can be seen on BACKUS' face.

COOKES with a tightly closed right eye and a bleeding nose and mouth. The referee stopped the fight in the thirteenth round.

The fight is over and lost (Fig. 136). The French professional boxer Gaston MATHIEU is sitting in his corner, blood running of his battered face from a laceration of his left eye region. MATHIEU was severely punished and was severely beaten by the Danish boxer Christian CHRISTENSEN.

Profssional boxer Stefan REDL, of Germany, holds his stamd im the third round of a 15-round welterweight bout against Frank IPPOLITTO, of the Bronx, on June 3, 1957 at St. Nicholas Arena, New York. REDL won the bout by a split dcision and scored his 15 th straight victory and remained undefeated. The facial hemorrhage (Fig. 137) originates lateral of his right eye and does not impede his visual acuity.

Again, a boxer whose face is so severely beaten-up and swollen must have damage to his much more vulnerable brain tissue. The clinical and morphological brain damage will be manifest after an interval. There is no doubt that any individual can have such tissue alterations of his face without severe damage to the brain tissue.

Only in some of the fights where they had the authority to do so, did the referee or ring physician stop the fight. In all other cases, boxers finished the match with considerably decreased visual faculty. Without both eyes a boxer is unable to see

Fig. 135. Professional welterweight fight between José NAPOLES and Curtis COKES on April 17, 1969 in Los Angeles, California. The photo shows the battered face of COKES.

blows coming from the side of the injured eye and cannot defend himself effectively. As a practical matter, he will be unable to avoid more inevitable, severe head blows, and blows to the damaged eye. Since most boxers are right handed, the left eye is more frequently and more severely damaged.

MILLARD (1963), who discussed the closure of boxing lacerations between rounds, noted that the problem of the laceration during a contest was of particular interest. In his opinion, "too often a doctor is forced to stop a contest and the referee pronounces a technical knockout against a lacerated fighter, who may be well ahead in points at the time. This is unfair and a paradox, since the majority of lacerations are caused by head butting, which is considered a foul and an illegal blow. Victory awarded to the butter not only encourages, but rewards, the act of butting.

MILLARD continued:

"Many promising fighters have given up their quest, and many a title fight has been lost because of an inconsequential laceration. Lacerations are so common that surgical procedures have been attempted, to smooth the sharp edge of bone on the supraorbital ridge, and even to pad it with a dermolipomatous graft, with questionable success (GILLIES and MILLARD 1957). The answer could be immediate closure of the laceration during a one minute corner rest period" (MILLARD 1963).

MILLARD is of the opinion that it is possible to close the laceration of the eyelid-eyebrow region with one or more sutures during a one-minute corner rest period and, if all other factors are favorable, the contest could be allowed to continue.

MILLARD produced lacerations 3 cm in length on skin over bony prominences of a laboratory dog that was placed under general anesthesia. A special Mersilene No. 4-0 suture prepared by Ethicon on a traumatic cutting needle was used to place a figure-of-eight stitch, knotted snugly with 4 ties. This effectively closed the wound and checked the hemorrhage. A stopwatch timed a single stitch in 35 seconds, a double stitch in 45 seconds – well within the 60 seconds limit of the corner rest period. Each suture line was then tested with strong straight and glancing blows by a boxer with boxing gloves. The wounds remained closed and the sutures in position without tears or breaks. The sutures were removed in the usual five days, and all wounds healed well.

MILLARD presented the proposition in an open meeting of the *Miami Boxing Commission on* July 2, 1962. The *Commission* unanimously agreed to allow an immediate suturing of a laceration if their ring physicians approved. At a special meeting the proposition was presented to the managers and contestants for the next program. The majority of the managers accepted; several declined. The treatment was to be made available only to those requesting it.

During the third round of a semifinal welterweight bout a boxer received a 2.5 cm (1 inch) laceration below the left brow which was gaping enough to threaten a technical knock out. During the next rest period the wound was cleaned by MILLARD with benzalkonium chloride, closed snugly with

a figure-of-eight Mersilene suture, and the brow was greased with Neosporin ointment. The fighter was able to answer the bell and continued the bout, receiving several blows to the sutured area. Eventually the fight was stopped for reasons not associated with the laceration or the suturing. Examination of the sutured wound after the contest was encouraging according to this author. Four days later the sutures were removed revealing an exceptionally well healed wound. MILLARD stresses the point that not all lacerations are suitable for between-rounds suturing. "Yet it would seem that this immediate suturing should be available to amateurs in the National or Olympic quests for glory or in important professional fights where not only pride but the purse is important. Possibly this suturing would be best limited strictly to world title fights. Equal treatment should be available to both fighters."

"Although to date no suture has torn out, untied, or broken, the possibility must be considered. Should a disruption occur, it would depend on the judgement of the doctor whether a second try would be justified during the contest" (MILLARD 1963).

A normally unclean dressing room is not the adequate place for suturing wounds, as shown in Fig. 138. VILLEMAIN, making his American debut, took a bloody beating but was not floored once. He lost a unanimous decision after a 10-round bout. Dr. Vincent NARDIELLO, the New York boxing commission physician, stitches the wounds of Robert VILLEMAIN in the dressing room of Madison Square Garden in New York.

Suturing should always be done in an emergency room or hospital with good lighting and sterile conditions. Exceptions may include suturing in *life-threatening situations* such as battlefield environments or in similar situations in civilian emergencies. Whether this procedure is available in 'National or Olympic quests for glory', in 'professional fights where not only pride but the purse is important', or is "possibly... limited to world title fights" is irrelevant where priority must be given to basic universally accepted principles of medicine. To accept and engage in substandard risky medical practices for the sake of enhanced prestige or financial gain would be a complete departure from the goals and objectives of modern medicine. For these reasons, the practice of hasty suturing in and around the boxing ring must be condemned on medical grounds.

Examples of Professional Fights in which Boxers suffered Extensive Injuries of the Eyelids

The following photos reveal extensive injuries of the eyelids of professional boxers.

The Korean boxer SOO HWAN HONG won the boxing fight against TAYLOR by a point decision. Fig. 139 shows TAYLOR after the end of the 15-round boxing match. His right eye is so tightly closed that he cannot see at all. An extensive periorbital swelling and edema extends into the periorbital tissue and the skin of the right forehead region. The lips are battered and the nose is swollen.

Such a massive lesion of his eye is not only a problem for the boxer during the fight, because he is not able to see incoming blows, but also an ophthalmologist cannot see the interior of the eye, for example the fundus. Examination of the fundus is an important part of the boxing examination, but it is impossible when the eye is swollen tightly shut.

A left hook of 'Sugar' Ray ROBINSON in the fifth round resulted in the injury of the left eyelid of Carmen BASILIO. His left eye is 'puffed up.' BASILIO lost his title due to a split decision. BASILIO's left eye is completely closed due to a monstrous massive hematoma of the upper eyelid completely closed (Fig. 140). The injured boxer was unable to see or evade blows coming from the right side. The referee should have stopped the fight because of the disabling eye injury.

Carmen BASILIO had won the championship of the world by a majority decision from 'Sugar' Ray ROBINSON on September 23, 1957 at Yankee Stadium, New York, in the presence of 38.072 spectators. It was BASILIO's fifty-second victory in the boxing ring and his greatest triumph in 71 professional fights. He had won the championship in the wel-

Fig. 136. Professional fight between the Danish boxer Christian CHRISTENSEN and the French boxer Gaston MATHIEU. The French professional boxer Gaston MATHIEU is sitting in his corner, blood running of his beaten-up and swollen face. Photo: KEYSTONE.

terweight two times and won the championship in the middleweight.

Subsequent examples of professional boxing matches show how extensive injuries of the eye region and eyelids can be. There are lacerated wounds of the eyelids and extensive hemorrhages in one upper eyelid, which produced such marked swelling and hematoma that the eye was tightly closed and vision was impossible. Only some of these professional fights were stopped, the others continued, regardless of blindness or risk of permanent injury. They ended in knockouts, seldom in favor of the boxer whose eye was injured.

In the first professional fight between Harry GREB and Gene TUNNEY, GREB fractured TUNNEY's nose, in two places. In the first round, TUNNEY received a severe hemorrhage, pouring blood from a wound above his left eye. TUNNEY's face was beaten to a bloody pulp to such an extent that referee Kid McPARTLAND had to interrupt the fight frequently, to remove coagulated blood from the boxing gloves of GREB (COLLINS 1990).

The sports journalist W.O. McGEEHAN termed Harry GREB's style as 'the manly art of modified murder.'

The sports journalist Bert SUGAR (1984) wrote about him:

"The GREB style was dictated by his personal estimate of what the opponent and the referee would hold still for. Somewhat lacking the social amenities of the Queensberry rules, GREB fought as an early-day terrorist,without rules, inhibitions or prisoners taken, his opponents more often than not feeling as if they had to fight him off, rather than actually fight against him."
Harry GREB, 'The Pittsburgh Windmill' who died at age 32 participated in 294 boxing bouts – alone in the year 1919 in 44 fights – became blind in one eye in a bout with Kid NORFOLK on August 29, 1921. He was able to keep his blindness secret until his death about five years later.

FLEISCHER et al. (1975) reported:

In Chicago Stadium on July 16, 1947 ZALE and GRAZIANO fought until the sixth round, when GRAZIANO "let loose a barrage of lefts and rights and ZALE went down against the ropes, forcing the referee to stop the slaughter." Rocky GRAZIANO, with a completely open battered left eyelid and a near tightly closed right eye, was the champion. GRAZIANO wrote in his autobiography: "This was no boxing match. It was a war, and if there wasn't a referee, one of the two of us would have wound up dead."

PLIMPTON (1977) reported on the heavyweight championship fight between Rocky MARCIANO and Archie MOORE in the Polo Grounds in New York:

"Archie was in the eighth round in terrible difficulties. Ring physician Dr. NARDIELLO looked at him in his corner. He was battered and one eye was tightly closed. Dr. NARDIELLO bent forward, to have a better look, and heard MOORE whisper: 'Don't stop it, Doc; let me try once more with a desperado!' The ring physician let him continue with the fight, and MARCIANO knocked out MOORE in the next round."

The heavyweight 'Bonecrusher' SMITH, who during his career as amateur boxer had the nickname 'Sominex' - a sleeping medication, which could be obtained without a prescription ('over the counter'), fought later as professional boxer against Larry HOLMES.

HOLMES landed a short right in the eleventh round, which ripped open SMITH's left eyelid. HOLMES continued to 'work on the cut,' in other words, again and again he hit the same area to enlarge, tear, and cause hemorrhage. In the twelfth round referee Davey PEARL interrupted the fight and asked the ring physician, Dr. Daniel ROMEO, to look at the tear. ROMEO informed SMITH that the wound was considerable and asked him whether the fight should be stopped. "Call it off," answered SMITH. The tear required 20 sutures.

Injuries of the Lips

(3) *Lip injuries* are frequent in boxing, but *cuts of the lips* are relatively rare, although they do occur despite properly fitted gum-shields. Extensive inju-

Fig. 137. Professional welterweight fight between Stefan REDL of Germany, and Frank IPPOLITTO on June 3, 1957 at St. Nicholas Arena in New York. Photo: UPI.

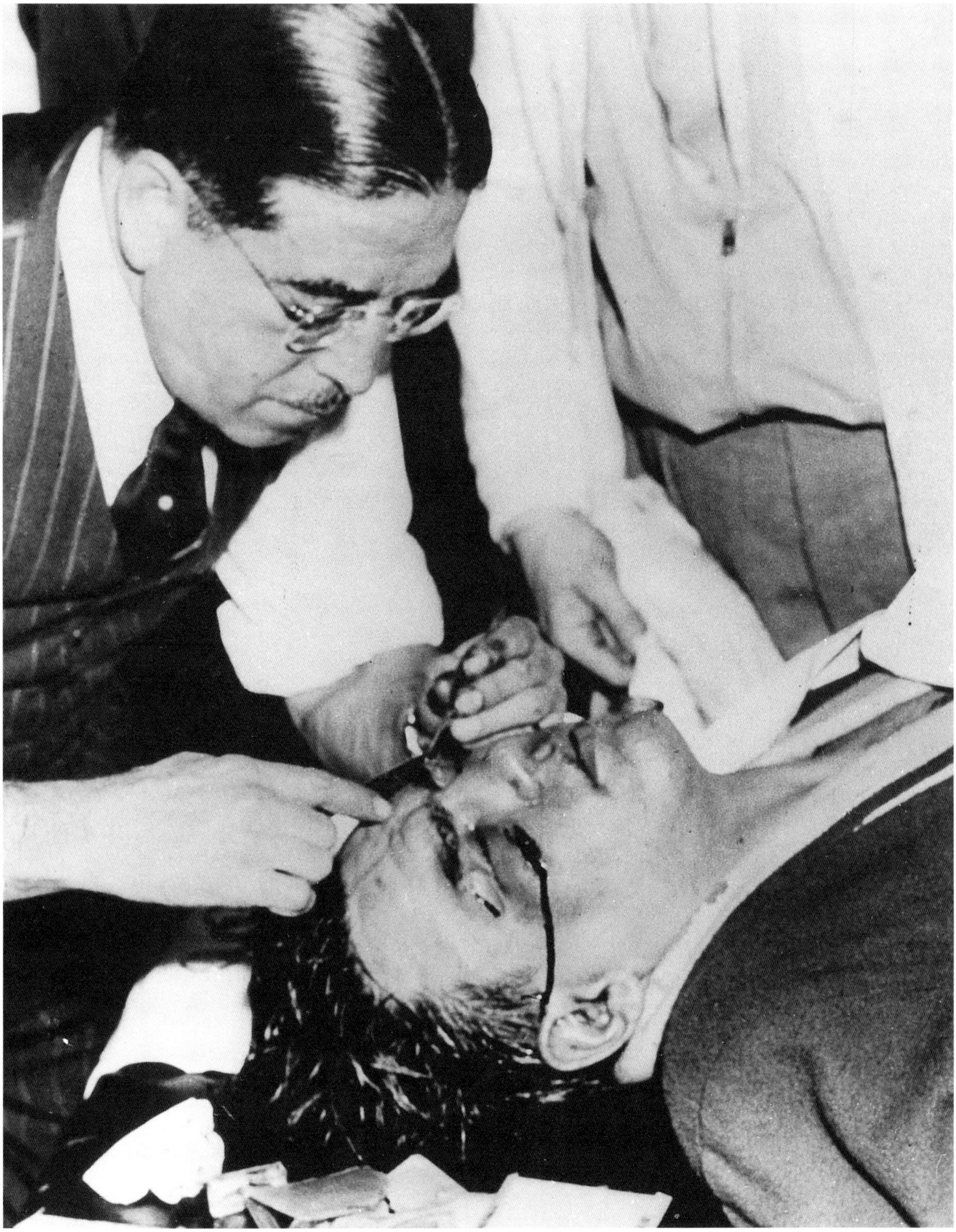

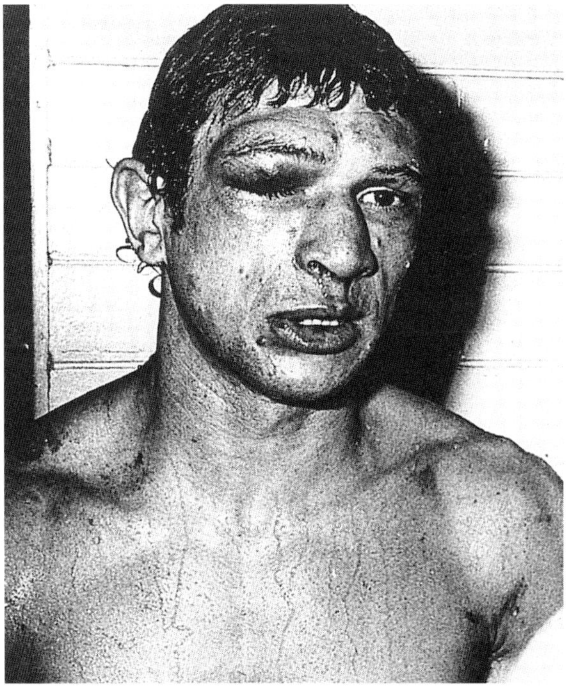

Fig. 139. Professional bantamweight fight between Arnold TAYLOR of South Africa (shown in the photo) and the Korean boxer SOO HWAN HONG on July 3, !974 in Durban, South Africa. The photo shows TAYLOR after the end of the 15-round fight.

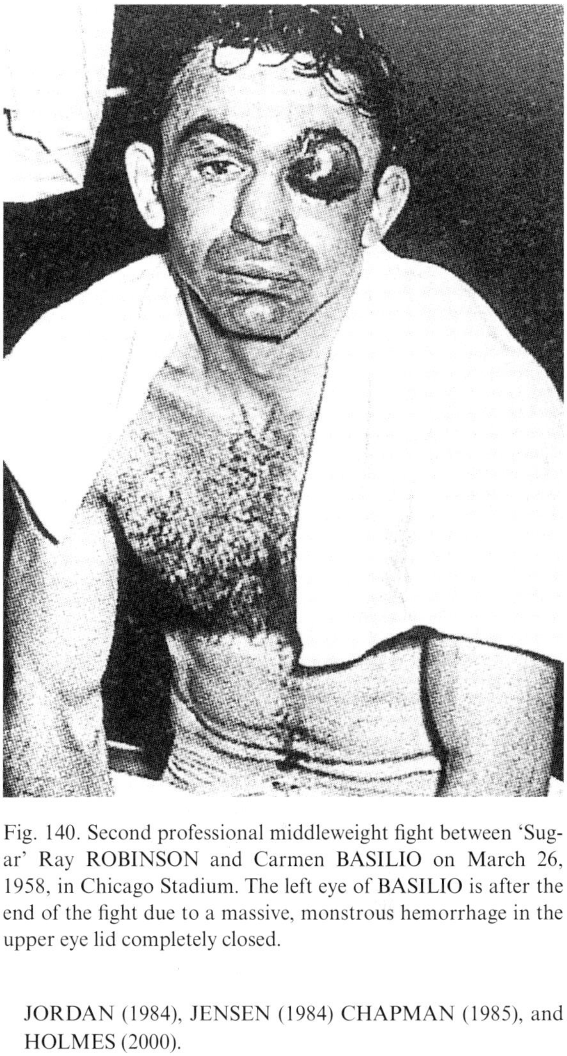

Fig. 140. Second professional middleweight fight between 'Sugar' Ray ROBINSON and Carmen BASILIO on March 26, 1958, in Chicago Stadium. The left eye of BASILIO is after the end of the fight due to a massive, monstrous hemorrhage in the upper eye lid completely closed.

ries and larger cuts of the lip must be sutured with fine stitches using local anesthesia in an emergency room or a hospital.

PALAZZI (1925) saw an injury in a boxer who was hit against the lower teeth. As he had his lower lip flexed inside at the time, the punched-out teeth fell through the perforated wound of the lower lip.

Mouth protectors and *mouthguards* for boxers were discussed and described by GÖHLE (1953), BROCK (1953,1954), RAABE (1958), STENGER et al. (1964), HEINTZ (1968) CLEGG (1969, 1979), MIKOLEIT (1965), WEHNER and HENDERSON (1965), HICKEY et al. (1967), PICOZZI (1975), DAVIES et al. (1977), TURNER (1977), HUGHSON (1980), LUKE et al. (1982), UPSON (1982), BOLLMANN et al. (1973), BASISKE and von SCHWANEWERDE (1973), DOMBROVSKI (1974), JORDAN (1984), JENSEN (1984) CHAPMAN (1985), and HOLMES (2000).

Mouthguards are considered by most authorities to be an essential part of equipment for players in any body-contact sport, especially the combative type, e. g. boxing, rugby, American Football.

CHAPMAN (1985) discussed orofacial injuries and the use of mouthguards. According to CHAPMAN, mouthguards provide dental protection, especially of the upper anterior teeth. They

Fig. 138. The physician of the New York State Athletic Commission, Dr. Vincent NARDIELLO, stitches the wounds of Robert VILLEMAIN in the dressing room of Madison Square Garden, New York, on January 7, 1949, after the French middleweight lost an unanimous 10-round decision to Steve BELLOISE of New York. VILLEMAIN making his American debut, took a bloody beating without being floored once. Photo: U.P.I.

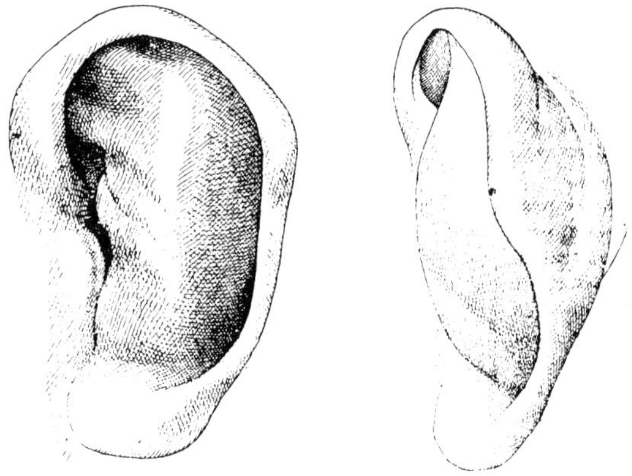

Fig. 141. Drawing of a fresh leftsided othematoma (48 hours). Left view shows the othematoma from lateral, and the right view from posterior (From BRAINE, and RAVINA, 1923).

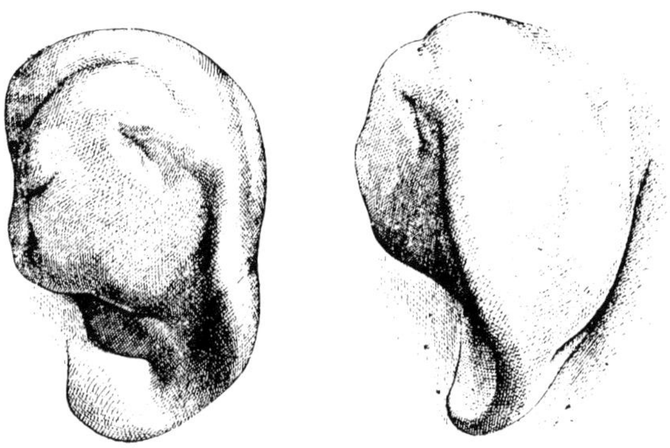

Fig. 142. Drawing of an old othematoma (15 days old). Left view from lateral and right view from posterior, which developed into a 'cauliflower ear' ('Blumenkohlohr'). (From BRAINE, and RAVINA, 1923).

also provide soft tissue protection around the mouth, a reduction in the risk of mandibular fracture, and reduction of concussion force from a blow to the mandible (CLEGG 1969, UPSON 1982, and DAVIES, et al. 1977). Of the 28 players interviewed, only 7 (15%) wore mouthguards. The most common reason given for not using mouthguards was difficulty with breathing. Today, boxers generally use mouthguards.

Injuries of the Tongue

(4) *Injuries of the tongue* are frequently due to bites.

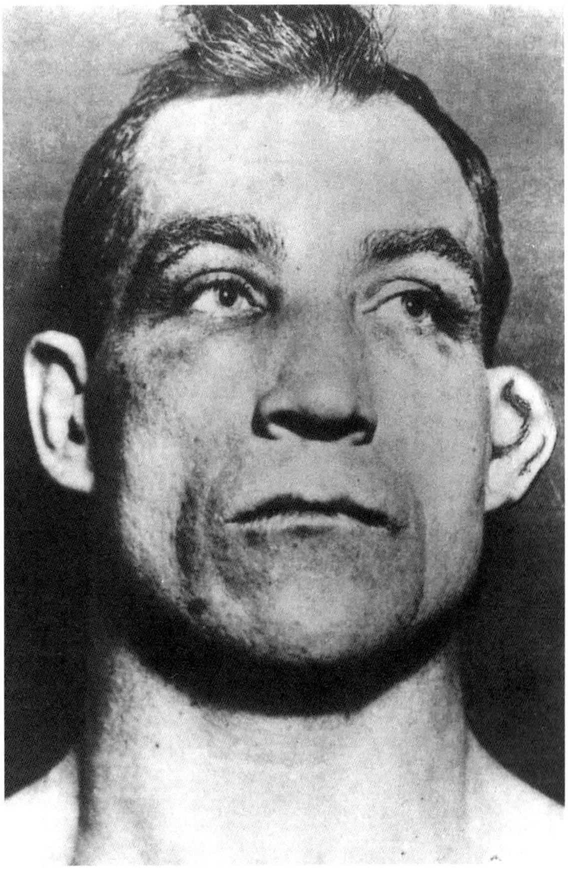

Fig. 143. Typical left 'cauliflower ear' or 'tin ear' of Tom SHARKEY.

Injuries of the Ear

(5) *Boxing injuries of the external ear* were described by LYDSTON (1907), CZERNER (1957), EADE (1964), and TEMPLER and RENNER (1990).

'*Cauliflower ears*,' resulting in the deformation of the auricle, are relatively frequent in boxers. Typical ear injuries are the result of a tangential blow against the ear with the skin sliding across the perichondrium (*Decollement traumatique*), or fracture of the cartilage of the ear which may occur when the auricle is folded against the temporal bone by a blow. In such a case a hemorrhage develops on both sides of the cartilage (Fig. 141) and blood collects on both sides of the cartilage of the ear. This leads to an isolation of the cartilage from the connective tissue, which is important for its metabo-

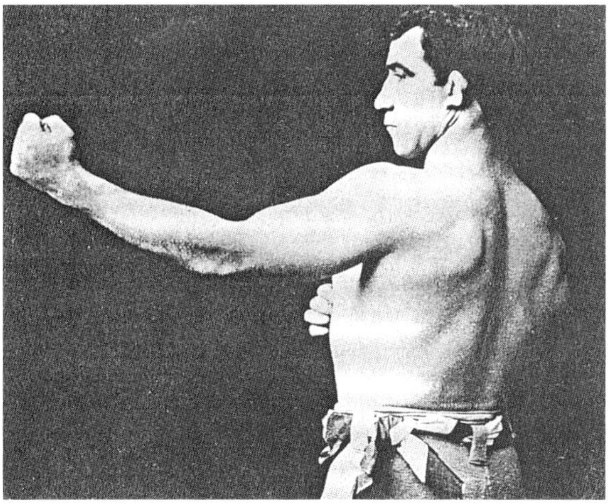

Fig. 144. Another photo of Tom SHARKEY, who fought CORBETT and FITZSIMMONS but was defeated by JEFFRIES for the heavyweight championship in 1899.

Fig. 145. Another boxer with a massive 'cauliflower ear' was Johnny BASHAM. His left ear stands stiffly laterally out and is markedly thickened. (From Corri E, Gloves and the Man; the Romance of the Ring, 1933, Fig. 34, p. 64).

lism. First an *othematoma* exists, then when the damage is not corrected, a *necrotizing chondritis* (Fig. 142) follows. The so-called *'cauliflower ear'* is the result. The auricle is grossly distorted and thickened, and the concha and auditory meatus are filled with hyperplastic cartilaginous tissue masses of almost the density of bone. The process can be diagnosed as *hyperplastic chondritis* or *perichondritis*. The treatment consists, in early phases, of aspiration of the fluid from both sides of the cartilage and the prevention of new fluid deposits by compresses. The so-called 'cauliflower ear' or 'tin ear' is the final end stage.

The left ear of the professional boxer, Tom SHARKEY, shows a typical 'cauliflower ear' or 'tin ear' (Fig. 143). These injuries are seen more frequently on the left ear, since about 85% of boxers are right handed, therefore landing more blows on the left side of the head of their opponents. The left ear of SHARKEY stands out stiffly, is enlarged at least to twice its normal size and four times its natural thickness. These chronic conditions are a combination of both repeated thumping and consistent neglect. A 'cauliflower ear' or 'tin ear' is a valuable indicator of the amount of punishment a boxer had to take during his active career.

The same 'cauliflower ear' or 'tin ear' of Tom SHARKEY is shown in Fig. 144, from a lateral view.

Another boxer with a marked 'cauliflower ear' was Johnny BASHAM. His stiff left ear stands out laterally and is markedly thickened (Fig. 145).

TEMPLER and RENNER (1990) summarized and discussed the modern techniques of treating external ear injuries. These injuries occur in people of all ages, but predominate in active people such as wrestlers, boxers, and bike riders. The types and extent of injuries are a function of the force causing the injury. Shearing forces of moderate intensity cause hematoma formation, whereas greater force causes lacerations or even amputation. Sharp objects cause lacerations determined by the force, direction, and point of impact. People participating in high-risk activities should wear protective headgear. The goal of treatment is to restore the normal contours while preventing infection. Hematoma results in disfigurement by organization or chondritis. Evacuation and pressure dressings using sterile technique correct the condition. Simple lacerations are closed under aseptic technique using either skin-to-skin sutures only or sutures of the skin combined with intercartilage sutures. Extensive and complex lacerations require meticulous care to match all fragments and prevent infection or loss of tissue. Bare cartilage must be covered with vascularized tissue. Chondritis is the most feared complication of injury or surgery of the pinna. It is an aggressive process requiring prompt removal of pus and necrotic cartilage. Exteriorization and removal of all cartilage is effective but disfiguring. Removal of only affected cartilage and constant irrigation with antibiotic solutions is effective but requires prolonged hospitalization. Iontophoresis of antibiotics into the auricle may be effective and conserve tissue. Traumatic deformities are corrected with composite grafts from the opposite ear, costal cartilage, and local pedicled flaps.

LYDSTON (1907) published the case history of a 33 year old retired professional boxer, who was the hero of a number of 'bruising battles.' In several of these, he stated that his left ear was beaten into a shapeless mass. On each occasion the traumatism produced deformity, until finally the auricle was greatly distorted and thickened, the concha and auditory meatus filled with cartilaginous tumors of almost the density of bone, and the auditory canal was completely occluded by a tumor the size of a large hazelnut. The average thickness of the auricle was about three-fourths of an inch. The author operated at five sessions. Each operation consisted of a subcutaneous resection of the cartilaginous overgrowth.

PAMPUS and GROTE (1956) performed EEG recordings in 190 active and 17 non-active amateur boxers – these results will be discussed on p. 415. They found 'cauliflower ears' in 11 cases, in more or less marked form, thereof 10 times left and one time on the right side.

That the left ear is more frequently involved, is not surprising, it can be deduced from the fact, that 85% of all boxers are right hander and that right blows are more frequent than left blows.

Fractures of the Mandible

(6) *Fracture of the mandible* is an allegedly relatively rare occurrence in boxing, but may occur at the condyle neck or at the angle of the mandible. The injuries were described by KOHLRAUSCH (1921), DIETRICH (bilateral fracture), and MANDL (1925). They are easy to diagnose; common signs are: severe pain, dental malocclusion, trismus, etc. A specialist should treat these injuries.

GOTTLIEB and GOTTLIEB (1951) reported eleven cases of fracture of the mandible sustained by amateur boxers. Typically a *'boxing fracture' of the mandible occurs in the region of the left angle* and is caused by a blow with the right hand, a so-called *'haymaker.'* Only one of the boxers had used a rubber bite-block, which is considered to be relatively good protection against mandibular fracture.

BLONSTEIN (1965) mentioned two fractured jaws in 6.000 amateur boxing contests. BAILEY (1965) discussed his findings and remarked that he had no conflicting figures. He basically agreed with BLONSTEIN's opinion. However, the previously mentioned figures from GOTTLIEB and GOTTLIEB (1951) are substantially higher. Such a large discrepancy in the reported number of mandibular fractures in amateur boxers raises questions about the thoroughness and accuracy of BLONSTEIN's published data. The noted English ophthalmologist DOGGART expressed similar doubts about the number of eye injuries reported by BLONSTEIN (comp. p. 147).

Injuries to and Loss of Teeth

(7) *Injuries to and loss of teeth* occur despite properly fitted gum-shields. A specialist also must treat these. The literature on dental injuries in boxing is rare (KLUGE 1966, PALATELLA 1969).

Injuries and Fractures of the Nose

(8) Case histories of nose injuries and fractures in boxers were published by SELTZER (1950), ZORZOLI (1951), BOLANOS CACHO (1964), ELLIS (1965), PAKOWSKI (1969), CHAPMAN (1985), CROCE and CROCE (1987), and LAVORGNA et al. (1995).

Injuries and fractures of the nose occur frequently. They can be identified by displacement and crepitus on palpitation. The inferior half or three-fifths of the septum is unprotected by the nasal bones and is, according to ELLIS (1965), vulnerable to direct trauma. Laterally applied force may dislocate the septum from its seat on the maxilla, pushing it to one side. Sometimes the cartilage cracks along the line of curvature into which it is bent by the blow, and occasionally the ethmoidal plate will also snap across vertically. In the same paper, ELLIS states that more direct frontal violence may produce a number of different deformities all based on the production of crumbling injuries. The bony portion of the septum, chiefly the ethmoid and vomer, snaps across in varying places depending on the direction and force of the blow, while the cartilage may buckle or break if its elasticity is exceeded. The fragments sometimes slide alongside one another producing a thickened septum, whilst at other times they become angulated.

When the blow has been sufficiently strong to break the nasal bones, the septum is invariably fractured at the same time and the fragments are dragged into positions governed by the final positions of the nasal bones.

The net result of these various injuries ranges, according to ELLIS from virtually nothing on the one hand to severe cosmetic deformity on the other. Sometimes, however, when neither the appearance nor the function are affected, epistaxis may later occur with the slightest trauma.

ELLIS also mentions compound fractures, which can occur when the nose strikes against the rope-supporting poles, the floor of the ring, or against the opponent's head.

SELTZER (1950) reported that he had seen fractures of the perpendicular plate of the ethmoid, with penetration of the cubiform plate and injury to the brain substance, on at least two occasions.

In the beginning, the signs and symptoms of fresh injury to the external nose are bleeding from one or both nostrils. This is the result of injury to the nasal mucosa with tearing of the smaller blood vessels. Hemorrhaging ranges from slight to profuse. There is early swelling of the mucosa, with edema, and possible hematoma, both of which can cause nasal obstruction. The accumulation of blood in the nasal cavities can also obstruct the nasal passages. If the septal cartilage is injured it may deviate to one side or the other, filling the passageway and obstructing respiration. Soon after such injuries to the nose, the loose tissue about the eyes may become infiltrated with blood, giving the familiar blackened appearance of one or both eyes.

Veteran boxers with 50 bouts or more seldom bleed or bleed only a little from the nose. Repeated injuries destroy the septum nasi and normal nasal lining, replacing them with dense scars of connective tissue so that vascularity is reduced or the vessel walls are thickened and less vulnerable. There is some individual variation in this effect, so that some blood vessels can remain thin and elastic.

In rare cases there may be no external deformity in early injury to the boxer's nose, but on examination with the speculum the various sign are plainly evident.

After repeated fights, the tip of the nose always appears rounded and bulbous as a result of blows on the lower lateral cartilage, by which they may be repeatedly fractured (especially about the angle).

SELTZER concludes his excellent review with the statement:

"An important point for the surgeon before performing a reconstructive operation of the nose of a professional boxer is to have it conclusively understood that his pugilistic life must be given up if he wishes to preserve a normal appearing nose, since the operation cannot be successfully repeated and a nose, once reconstructed, is particularly vulnerable."

PAKOVSKI (1964) examined 450 amateur boxers from 16 to 36 years of age, and divided them into 5 groups depending upon the length of time they had practiced boxing. In 194

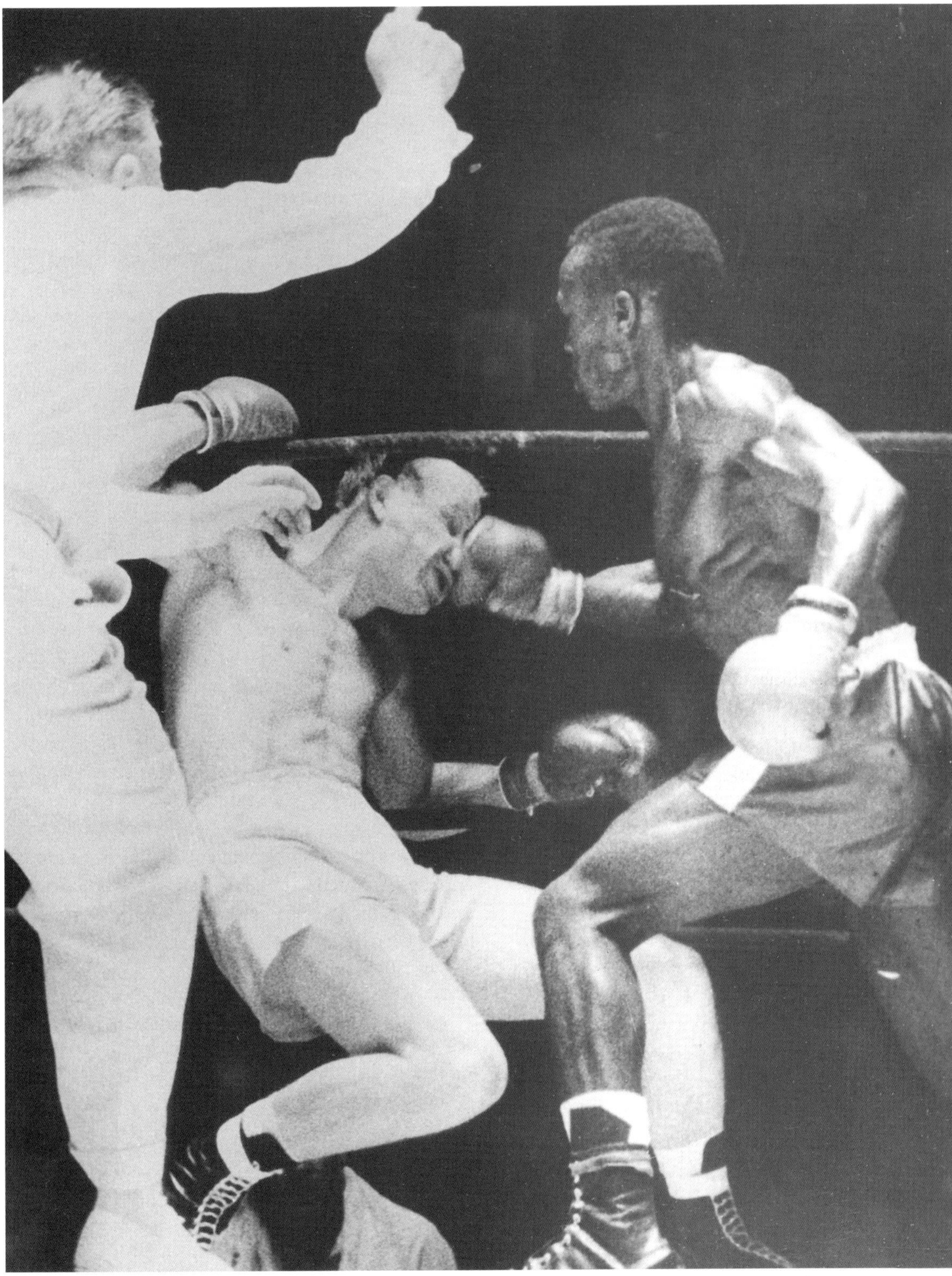

boxers (43.1%) he found a crooked nasal septum making the patency of the nasal passages deficient. This was seen bilaterally in 36 (8.4%). In 152 boxers (33.7%) examination with an otoscope showed that the tympanic membranes were drawn in. In 64 of these 152 boxers the patency of the Eustachian tubes was checked, and in all cases examined there existed a marked deficiency or decrease in size. Deformation of the external part of the nose, in the form of so-called '*boxer's nose*', was found in 43 (9.6%) of 450 boxers. The percentage of boxers with the above-mentioned changes showed a close dependence upon the length of time they had been boxing; the percentage was the highest in those who had boxed the longest.

It should be remembered with regard to facial injuries that in general boxers suffer not only an altered visage, but also, what is more dangerous, an irreversibly altered brain and mind.

After the preceding findings on face injuries in boxers, it would be helpful to consider a debate between several eminent physicians, published in the book '*Medical Aspects of Boxing*,' edited by Bass et al. (1965).

ELLIS (1965) commented:

"I have very little to say about the emergency treatment of nasal injuries for I do not believe that ringside measures are often successful. Lacerations of the nares or free epistaxes are reasons for stopping the bout."

"Lacerations are best cleaned and covered with a sterile dressing and the boxer sent to where the wound can be accurately sutured."

"Speaking of injuries to a young boxer," ELLIS says, "If impairment of the cosmetic appearance and function are minimal, there is a case for no treatment at all, especially if the boxer is at the beginning of his boxing career and even more if he shows promise and is likely to do well competitively. This is because of the weakened state of the nose (in most cases) after surgery, and because repeated trauma may nullify the effect of surgery altogether."

BAILEY (1965) presented the following comment:

"Prevention of boxing injuries is desirable so long as the purpose of the sport is not lost. However much devotees decry the necessity for knock downs and knock outs in boxing, violent physical contact – its avoidance where possible and its endurance where necessary – are the heart and soul of boxing. This, of itself, rules out the use of complete protection as used by American footballers. Protection of some tissues is accepted, and gum-shields and protective mouth pieces should be worn."

BAILEY continued:

"Boxing gloves that present a supple unbroken surface reduce the risk of facial abrasions and lacerations."

"Amateur treatment by cornermen should be discouraged. In the early rounds of a professional bout, the opponent will play upon a cut over the eye or a severe bruise round the eye until such time as the eye is closed or covered by a sheet of blood. By the time the referee stops the fight – with or without reference to a doctor – a skin-deep laceration which could, with proper treatment, heal with a minimum of scarring becomes a contused, deep, infected gash plastered with quack remedies by the 'cut-man.' This may be the source of recurrent trouble for the rest of the boxer's career."

BAILEY commented on various substances used to prevent and treat haematoma and edema:

"BLONSTEIN used prophylactic *streptokinase* and *streptodornase* in one set, and dummy tablets in another set of boxers and compared the results. He formed the impression that there were fewer hematomas and contusions in the treated series than in the control group. Streptokinase is reputed to act by hastening the removal of the fibrin barrier that forms round any tissue damage. Whether or not it does this, it is impossible that it should prevent the formation of a haematoma or contusion, as these are purely dependent on the mechanical tearing of a blood vessel. It may hasten the resolution of a haematoma, although objective proof of this is very hard to obtain."

BAILEY continued:

"Fracture of the mandible is a rare occurrence only likely to occur in the late stages of a hard-punching professional bout – the more so where one of the contenders is on his last legs with open mouth and slack jaw."

BAILEY summarized:

Fig. 146. Professional featherweight fight between the British title holder Rafia KING (right) and the French boxer Gracieux LAMPERTI, France (left), on October 19, 1961, at Palais des Sports, Paris, France. Photo: A.P.

Fig. 147. Face of the boxer at the left (unidentified) is hit by a very forceful right of his opponent. Photo: d.p.a.

"Sports injuries in general deserve expert attention... in boxing, we would see a reduction in the incidence and recurrence rate of facial injuries by using only the best equipment, by strict refereeing, and by ready access to expert medical treatment."

An article *(Manly Art, Scientific American, 187, October 1952, p. 46) noted:*

"To keep prizefighters operative round by round, their seconds often resort to emergency treatment. The therapeutic regimen may include: daubing the wounds with the dirty towel that the second carries draped around his neck, sealing the edges of torn skin with carpenter's glue, and rubbing adrenalin into open cuts to stop bleeding, sometimes using enough to affect the fighter's pulse, blood pressure and respiratory rate. These practices were deplored by physicians attending a recent medical symposium sponsored by the *New York State Athletic Commission*. The symposium concluded that national medical standards should be established for the ring, and that ringside doctors should have the authority to protect injured boxers against the ministrations of their handlers."

"A Pennsylvania Athletic Commissioner sounded, however, a note of caution about 'medical standards.' 'You must know who you are dealing with – they are rough and though,' said John 'Ox' DA GROSA. Requiring too much expensive equipment and too many medical tests might force young fighters to become ' slaves to the moneylenders and managers.' Their indebtedness might then tempt the boxers to agree to a 'tank job.'"

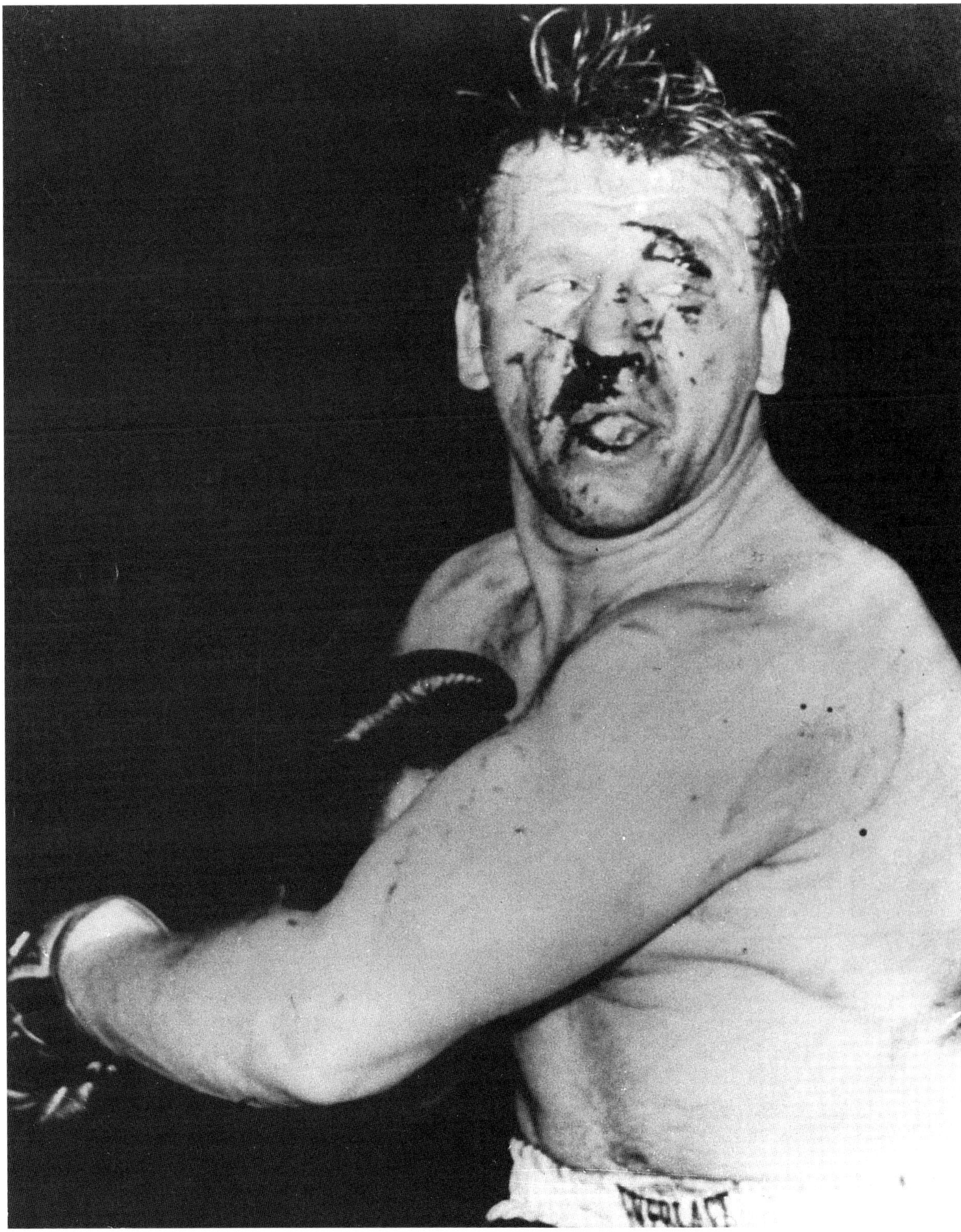

Fig. 148. The professional boxer Freddie BESHORE bleeding from nose, mouth and left eye. Photo: UPI.

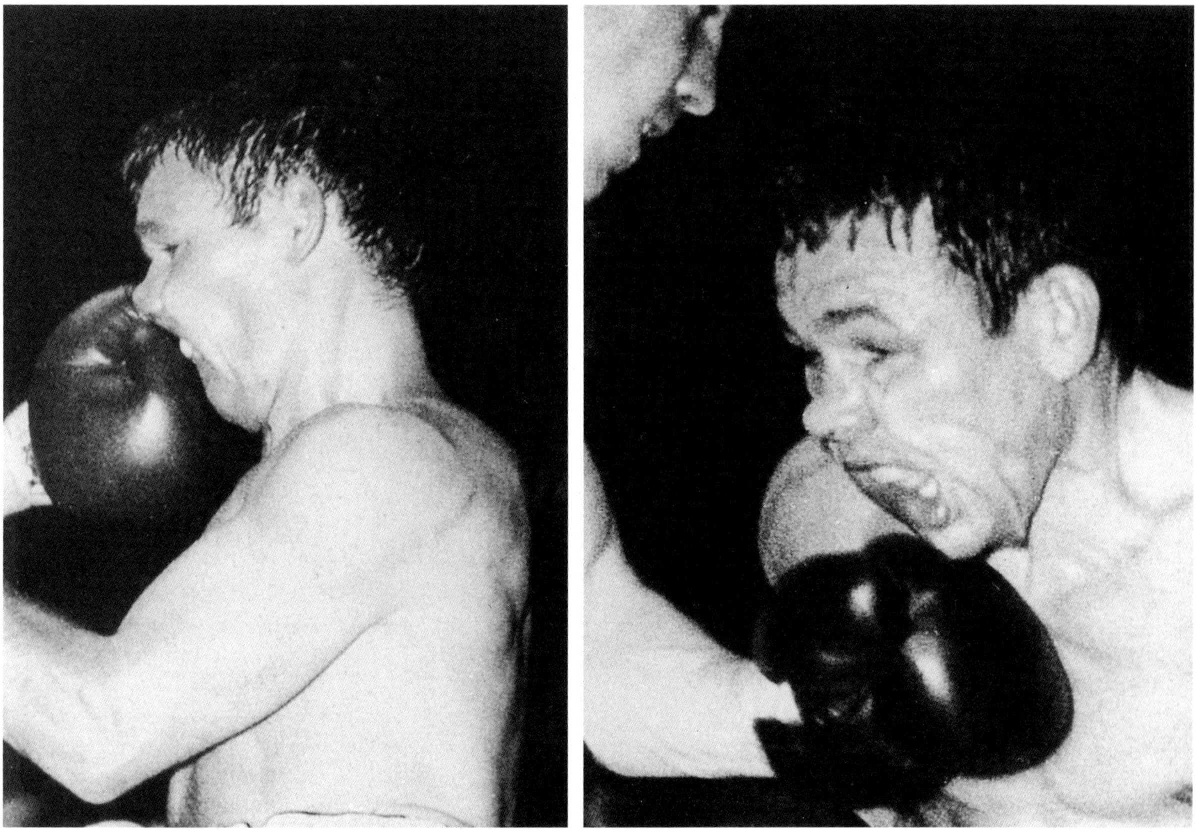

Figs. 149 a and b. Professional lightweight bout between the British champion Dave CHARNLEY and the American boxer Keene LANE. Photo: d.p.a.

Press Photos of Professional Boxers with Different Injuries of the Nose

The following action photos of professional boxers show different injuries of the nose. Severe hemorrhage from the nose can occur along with other injuries of the face, eyelids, and mouth.

Shortly before the end of the sixth round of a ten-round fight the English title holder Rafia KING connected with a hard right to nose and face of Gracieux LAMPERTI (France) (Fig. 146). The fight was stopped by the referee.

Punches against lateral areas of the nose can produce compression and lateral shifting of facial structures. The face of the boxer (unidentified) at left is hit by a very forceful right of his opponent. Nose and mouth are deformed, the nose compressed (Fig. 147).

The professional heavyweight Freddie BESHORE bleeding from nose, mouth and left eyelid (Fig. 148).

Professional lightweight bout between the American boxer Kenny LANE and the British champion Dave CHARNLEY. Kenny LANE connects with a solid left to the face of Dave CHARNLEY, which produces a marked deformation and distortion of his face (Fig. 149 a and b). Commentary of the sports writer: "Boom, that was a real nose-bender." The left photo shows the moment of impact and the right fractions of seconds later. The effect

Fig. 150. Profesional heavyweight fight between Rocky MARCIANO and Keene SIMMONS on January 29, 1951 in Providence, Rhode Island. Photo: David LAMONTAGNE.

of the blow is visible on CHARNLEY's face. CHARNLEY won the fight.

The face of professional heavyweight fighter Keene SIMMONS in Fig. 150 shows a powerful right blow inflicted by heavyweight champion Rocky MARCIANO. In the fifth round KEENE's face is clearly distorted and deformed. The left of his lower lip is displaced sidewise and downward, and there is bleeding from mouth and nostrils. Both eyes are closed by the impact of the blow. His nose, which is not directly hit, is indirectly deformed. MARCIANO knocked SIMMONS out in the eighth round.

Drawing of the nose of a Spanish champion boxer. Old fractures of the cartilage are present (Fig. 151).

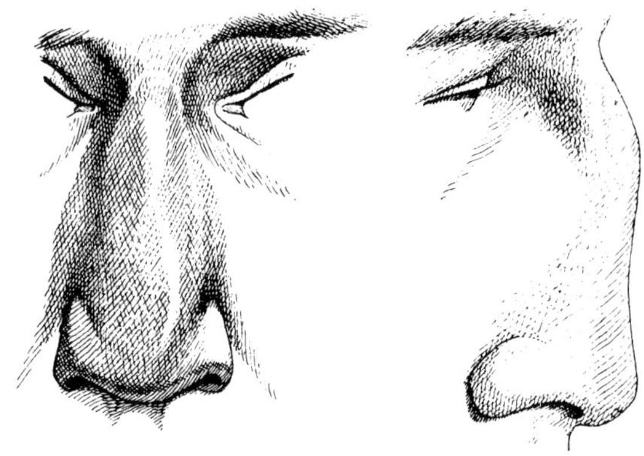

Fig. 151. Drawing of the nose of a Spanish champion boxer. Old fractures of the nasal cartilage exist. (From BRAINE, and RAVINA, 1923).

Discussion of Facial Injuries

"In the modern professional prize ring one sometimes reads sickening accounts of ploys designed to win contests. Apparently the clashing of skull against eyebrow in a deliberate attempt to split it open is part and parcel of the game" (ELLIS 1965).

The previous discussion on facial injuries focused attention on: (1) The use of preparations for treating/preventing hematoma and edema formation; (2) the influence of age on injury and the speed of recovery; and (3) any proposals for legislation concerning the best treatment of boxers with facial injuries, most notably a fractured septum or badly cut eye.

The *first point* drew directly conflicting statements from BLONSTEIN, who introduced the treatments mentioned above to the UK, and from BAILEY, whose statements concerning BLONSTEIN's eminently favorable statistics are much less acidic on paper than their content implies.

He noted:

"With regard to BLONSTEIN's comments on the use of *Varidase,* there is no method of stopping the bruise from developing unless there is some already present deficiency in the blood- clotting system of the body and you cannot clot broken blood-vessels. You cannot make it disappear sooner until the tissue is healed. As far as causing the resolution of hematomas is concerned there is really nothing that is effective."

The general consensus on the *second point* seemed to be that the tissues of the young would perhaps bend further before they break than those of older fighters, but beyond this there is little difference between boxers of different ages. Recovery, though perhaps slightly aided – in the case of fractures – by the fact that the bony tissues are still growing, takes just as long in the young as in the older boxer. The idea of not repairing an injury until the boxer was somewhat older – apparently practiced in some parts of the boxing world – was dismissed.

As for the *third point,* concerning legislation for the best treatment of boxers with facial injuries, ELLIS believes "that the bout should be stopped immediately once the eye is cut and the cut should not be 'doctored' by the seconds."

BAILEY stated, "With regard to cuts there is no question that you have to adopt elaborate measures. You have simply got to say that without exception a fight will be stopped when a cut appears over the eye." We agree that BAILEY's conclusion is appropriate if protection against unnecessary injury and exacerbation of injuries are priorities.

Eye Injuries

Cases of eye injuries in boxers were presented and discussed by SEDAN (1932), DURANDY (1935), FAVORY (1937), CANDIAN (1937), ZENNEK (1938), VÖLKEL (1943), BOSHOFF and JOKL (1948), FAVORY and SEDAN (1951), DOGGART (1951, 1955, 1965), ALBAUGH (1952), GLEES (1954), BOEHMIG (1954), KLENKA (1955), AICHMAIR (1956, 1957), ALJAMO (1957), LÖHLEIN (1959), SEGAL (1959), HUGONNIER (1964), RUGG-GUNN (1965), BUSCH (1965), KLAR (1966), BLAGOJEVICH and ERAK (1966), PAUL (1966), VENCO et al. (1966), DODEN and STARK (1974), PALMER et al. (1976), COLOSI and YANOFF (1977), WHITESON (1981), BARASCU (1985), MAGUIRE and BENSON (1986), GIOVINAZZO et al. (1987). CARTER and PARKE (1987), TOCZOLOWSKI et al. (1991), McLEOD (1992), LEACH et al. (1992), WEDRICH et al. (1993), VADALA et al. (1997).

As these quotations illustrate, the incidence and severity of eye injuries due to boxing blows is a much-debated and emotionally charged subject. The proponents and opponents hold diametrically opposed views, and both insist fervently that they are right and that the others are misinformed (that is to say, wrong).

"Although retinal detachment is rightly dreaded as one of the most devastating ocular hazards of the ring, it is only one among a formidable array" (DOGGART 1965).

"Serious eye injuries can occur in boxing. Many of you have seen thousands of contests and know that they are so rare as to be negligible and cannot be regarded as a serious hazard in boxing" (GRAHAM 1965).

Case Histories of Eye Injuries in Boxers

We begin our discussion of eye injuries in boxers with case histories of injured boxers, before we present the summaries of these lesions as presented by different ophthalmologists.

The case histories presented by BOSHOFF and JOKL (1948) vividly depict these injuries:

Case 1: An 18 year old boxer received a blow to the right eye during a boxing lesson at school. Almost immediately, vision in this eye became blurred.

Ophthalmologic examination conducted on the same day revealed a hyphema filling two thirds of the anterior chamber. The blood was absorbed within eight days, but the pupil was flattened in its upper part. A peripheral tear in the iris, extending from 11 to 2 o'clock, was evident.

Case 2: This case of a youth aged 17 is essentially similar to case 1.

Case 3: An 18 year old boxer noticed that vision in his left eye was blurred after a boxing match.

Ophthalmological examination on the same day revealed visual acuity of 6/12 and small tears on the lateral edge of the left pupil, which was semi-dilated. *Gonioscopic examination* (Goldmann) revealed a small, fresh area of cyclodialysis of traumatic origin, situated opposite 2 o'clock, in the angle of the anterior chamber. During the subsequent 14 days the tear closed gradually.

Case 4: A 23 year old man received a severe blow to his left eye while boxing. He subsequently observed that the lower part of the visual field was blacked out. Four weeks later a peripheral area of traumatic choroiditis of recent origin was found in the upper part of the left fundus. Five months later the diagnosis of detachment of the retina, extending from 11 to 2 o'clock, was made. Two attempts at surgical treatment failed; and when he was last seen, a year after the boxing injury, a cataract was beginning to form.

Case 5: A 30 year old man was struck by a severe blow on his left eye during a boxing match. Within a month the whole left retina became detached. An operation was performed in Switzerland, with unsatisfactory results. Two years after the injury the retina was still detached and lying in loose folds. No tear could be located.

Case 6: A 16 year old youth was hit in the right eye during a boxing match. At an *ophthalmologic examination 14* days later, a vertical white area of retinal edema was seen. Three weeks later a choroidal tear was evident in the interior temporal part of the fundus. Two years later the diagnosis of a flat, circumscribed retinal detachment was made. Surgical reposition was successfully carried out. The retina has since remained in position, for 18 months.

Cases 7 and 8: Two additional cases of choroidal tear in boys aged 14 and 15 years, respectively, were essentially similar to the preceding cases. In both, the tear was in the macular region, and visual acuity was as poor as 6/24 and 6/36, respectively. No retinal detachment appeared in the course of more than nine months.

Case 9: A 39 year old teacher was demonstrating dodging in the boxing ring when his opponent hit him in the right eye. Within an hour he could not see with that eye. An extensive hemorrhage in the vitreous was observed; this cleared within a year, though at this stage visual acuity was still reduced to 6/18. Widespread macular changes were present. General examination revealed early arteriosclerosis and slight hypertension.

Case 10: During a boxing bout, a 17 year old youth received a blow on the left eye, with the result that vision in this eye became blurred about an hour after he received the blow.

At *ophthalmological examination* on the following day it was evident that he had unilateral keratoconus and that the blow had ruptured Descemet's membrane, with resultant edema and irritation of the eye. This condition lasted four weeks. The end result was slight scarring below the apex of the corneal cone.

GLEES (1954) published the case history of a 44 year old professional boxer who was treated in a psychopathic hospital for a paranoid psychosis. We refer to the publication of GELLER (1953) who presented the neuropsychiatric aspects of this patient.

Marked thickening of the eyebrow, in this area numerous fine scars of the skin. Upper eyelids markedly thickened. The left upper and lower eyelids are closer together than the right. Bilateral trochlear paresis was found. Both pupils no longer round, slow and uneven reaction to light, a little better convergence. Bilateral severed ruptures of the sphincter and iridodonesis. Bilateral subluxation of the lens. Water clefts in anterior parts of the lens. Right temporo-caudal at the margin of the pupil pinhead sized props of the corpus vitreum into the anterior chamber, corporea vitrea otherwise unremarkable. Visual acuity bilaterally after adjustment of a myopic inverse astigmatism decreased to 6/12.

GLEES commented that several boxing-related eye injuries exist. Alterations of the eyebrows and eyelids are only of cosmetic interest, whereas all the other injuries have to be considered more or less severe. The bilateral trochlear paresis can be explained as possible local damage to the trochlea, as frequently seen homolateral in boxers, but also the result of a central injury, i. e. a hemorrhage in the area of the velum medullare.

LÖHLEIN (1954) published case histories of eye injuries in two boxers.

(*Case 1*) reported on a 26 year old professional boxer who received a blow against his left eye on October 5, 1952. No unconsciousness, but hemorrhage in the left upper eyelid. After a few days a 'shadow' developed over the injured eye, which extended to the right and which increased rapidly in size.

An *ophthalmological examination* on October 17, 1952 revealed unremarkable findings of the right eye. The visual field of the left eye nasal inferior into the fixation point was to a large extent abolished. The corpus vitreum revealed a few black erythrocytes. The retina revealed a wavelike detachment between 1100 and 1700 o'clock from the periphery to near the macula. On October 18, 1952, diathermy operation, undisturbed recovery. Follow-up on January 23, 1953 – about three months later – shows a reattached retina. In the following three months, there were no complications occurred and no subjective complaints.

LÖHLEIN discussed the question of whether the boxer should take up boxing again, since the boxer himself wanted to do so. He mentioned that a redetachment could not be excluded with surety, but since in this case the process of healing was uncomplicated and no subjective complaints remained, he consented to boxing.

A few weeks after the boxer had taken up his training again, he returned with the complaint that for the last week there seemed to be a veil before his left eye. He had nearly complete visual acuity, but there was a wavelike detachment of the retina, completely apart from the area of the first detachment. The detachment increased and finally covered the area of the macula with rapid loss of vision and visual field. The attempt to reattach this with a second diathermia was successful only in some areas. There remained a marked detachment causing poor eyesight and limited visual field. The patient was forced to give up boxing.

LÖHLEIN came then to the conclusion that a boxer who has suffered a detachment of the retina must give up boxing immediately. LÖHLEIN made the final statement that no one should become a professional boxer until he had finished the training for another profession so that he could have another career.

Case 2: Trochlear paresis caused by boxing blow. A 32 year old professional boxer received a heavy blow against the head on August 1, 1948, and immediately had double vision. This improved in the next few weeks, but no further recovery after November 1948.
The *ophthalmological examination* on January 5, 1949, revealed visual acuity right and left after correction of a mixed astigmatism of 6 dioptries 5/7.5 p. Anterior part of the eye unremarkable. Pupils both the same size. Operation March 18, 1949, retro-positioning of rectus inferior muscle of the left eye. Patient remarked improvement, but this did not last. One month after the operation, the double vision was as bad as before. December 1949, nine months later, rectus inferior muscle was again sewn back into place. Uneventful recovery, double vision no longer existed looking down and to the left, only minimal looking down and to the right. Patient took up boxing again, but recognized that his ability to react had decreased. He gave up boxing.

FAVORY and SEDAN (1952) reported two case histories of boxers who, after heavy blows against the head, revealed constriction of vessels of the retina with paling of the pupils and transient blindness. In a similar case FAVORY and SEDAN described a very marked vasoconstriction of the arteries of the retina; the pressure was so low that it could not be measured.

FAVORY (1937) published three case histories of boxers with eye injuries.

Case 1: In 1934, a flyweight boxer consulted the ophthalmologist for treatment of a wound of the left eye one day after a hard fight. The boxer did not recognize any symptoms during the fight. The first symptoms appeared the following night and worsened rapidly. His manager decided to bring him immediately for an ophthalmological examination. His eyes were watering and he was very sensitive to light.
The *ophthalmological examination* revealed a small oval ulcus immediately under the center of the cornea, measuring 2 mm in diameter. The sensitivity was reduced in the center of the ulcus, but was normal in the rest of the cornea.
Case 2: This boxer suffered a severe knock out – vomiting and headache followed. There was an enormous hematoma of the eyelids, so that it was impossible to examine the eye. A few days later a hypema was seen which filled the anterior chamber. It was completely resorbed after two weeks. An iridodialysis could be seen between 7 and 10 o'clock. The iris was mutilated between 7 and 11 o'clock. The lens oscillated in a pendulum-like manner.
Case 3: A 23 old flyweight boxer was seen on March 24, 1935. He had received a series of hard blows against his head, especially against the left eye region, in a bout the evening before. There was a conjunctival ecchymosis. He complained about a disturbance that occurred only while reading. The left pupil shows only a slight mydriasis.

A number of ophthalmologists have decried boxing from their standpoint as doctors. The great amount of experience represented by these physicians lends credence to their statements.

In 1952, ALBAUGH reported on his experience as a consulting physician to the *Athletic Commission of the State of California* for the previous six years. This included 154 injured or allegedly injured eyes in 92 cases.

He noted that:

"Although similarities exist between the types of eye injury resulting from boxing and those resulting from other occupations, some important differences must be noted. Eye injuries resulting from boxing are confined to a definite age group. Boxers, for example, are licensed in California only between the ages of 18 and 36. Exceptions are extremely rare."

ALBAUGH noted:

"Damage to the eye is almost always the result of a direct blow upon the eyeball, and is usually severe enough to cause profound pathologic changes."

He also noted that:

"because of fear of disqualification, many severe injuries are not reported until so long after they occur that healing processes have become well established and it is much too late for adequate constructive treatment. In some cases the injured men retire from fighting before asking for help."
"One of the tragic features of eye injuries sustained in boxing is that all too often they are bilateral."
"Of the 154 injured eyes, 38 had cataracts and in 10 instances this was bilateral. Other structures besides the lens frequently were damaged. In the only operative case, the results were excellent – 20/20 vision (corrected). In four the cataracts cleared spontaneously (taking from one month to two years), and in the remaining thirty-three cases no follow-up was possible.
Of the 154 injured eyes, 29 showed retinal separation, and in two this was bilateral. "Generally the period between injury and examination was a long one, and in only five of these cases was operation considered advisable. Of the five patients operated on, only one obtained a reattachment. In the hands of all ophthalmologic surgeons in this vicinity, the results of

reattachment operations have been so poor that the trustees of the *Boxers and Wrestlers Benefit Fund* have been consistently hesitant to authorize such surgical therapy and its attendant expense. They have been willing to have operation attempted only when the prognosis appeared unusually favorable."

"Nearly all eyes showed dis-insertion of the retina, with or without other holes. In all of them there was profound disturbance of the vitreous, as evidenced by many opacities, both large and small, present for a long time after the initial injury."

"Statistically, at least, boxers who sustain eye injuries make a poor showing for successful medical rehabilitation. Furthermore, the follow-up is embarrassingly poor."

"As a class, boxers seem to be quite blind to the fact that their career must of necessity end after a pitifully few years."

Three years later, DOGGART (1955) wrote specifically describing his great experience with eye injuries resulting from boxing.

Classification of Eye Injuries by DOGGART

According to DOGGART (1955): Derangement of the visual apparatus from boxing may conveniently be divided into: (1) *Ocular damage;* (2) *injuries to neighboring structures, including the ocular adnexa,* and (3) *lesions of the visual pathways and other parts of the brain.*

Ocular Damage

DOGGART noted:

"One important asset to a boxer is a pair of well-formed orbital margins, but his eyes are nevertheless liable to be damaged by blows falling tangentially upon the cornea from a downward and outward direction, where the orbital cavity is at its shallowest. Moreover, the thumb of a boxing glove can easily impinge upon the eyes even if they be deeply set; hence boxing is particularly dangerous for myopes, because their eyes are not only more prominent but more vulnerable than those of the average emmetrope."

"*Lesions of the Outer Eye: Conjunctival hemorrhages* are not in themselves of much importance, but they may, of course, be accompanied by damage to other structures. Any or all of the corneal layers may be injured. Small abrasions may heal uneventfully, but larger ones are liable to infection; and the resulting ulcers will sometimes leave dense opacity in their train. Edema and striate opacity of the substantia propria have been repeatedly observed, and Descemet's membrane may be folded or ruptured. *Scleral rupture* is another serious possibility to be borne in mind, because it can at first be masked by conjuctival hemorrhage, and will commonly be associated with damage to the inner eye."

"*Lesions of the Inner Eye:* These occur in bewildering variety. *Hyphema* often arises from *iris hemorrhage* and will at first preclude detailed inspection of the fundus. The iris itself may display single or multiple ruptures of the pupillary margin, with or without implication of the sphincter muscle; or its ciliary margin may be partly torn across to produce an iridodialysis. Occasionally the whole peripheral attachment of the iris gives way, and the retracted ball of uveal tissue is later seen lying free in the anterior chamber. Interference with the ciliary body is presumably the explanation for that persistent hypotony which has been detected in some boxers, and sometimes an insidious iridocyclitis supervenes. The worst cases advance relentlessly to a state of *phthisis bulbi.* Occasionally a lesion of the ciliary body is limited to its muscle; whence follows paralysis of accommodation. The *lens* may be *subluxated* or *totally dislocated,* and *cataract* is not uncommon. In this connection, it must be remembered that the full effect of lens concussion may be delayed for months or years. Eyes that appear to be undamaged may have sustained stresses and damage that will manifest after a time lag. *Cataract* can also emerge as a complication of cyclitis in an eye damaged by boxing."

"*Hemorrhage into the Vitreous*: Hemorrhages into the vitreous may be derived from the anterior or the posterior segment of the globe or from both. In all except very severe cases, absorption tends at first to be complete, but repeated hemorrhage encourages the deposition of those fibrous strands, which may herald *retinitis proliferans*."

"*Retinal detachment* is one of the most dreaded consequences of boxing, and a number of bilateral cases are recorded in the literature. Here it must be emphasized that the chances of permanent reposition by surgical treatment are much less than they are in other types of juvenile detachment. Among those who have testified to this effect is ALBAUGH (1952). Operation was considered feasible in only five out of his series of 29 retinal detachments due to boxing, and even among these five relatively favorable cases, only one retina was successfully reattached. *Cystic and pigmentary changes at the macula* are another tragic outcome of blows upon the eyes, and *retinal hemorrhages* are commoner than many people realize."

"*Choroidal hemorrhages* are even more destructive, and almost certainly produce irregularly shaped, mottled, raised areas of degeneration so conspicuous in the seasoned pugilist. Whenever we see a bruiser with a pair of *cauliflower ears*, one must be on the lookout for *cauliflower choroids*. These multiple fundus lesions account for many a boxer's defective visual acuity. Then again, a choroidal rupture may involve the macula. *Optic atrophy* is another possible fundus finding. *Avulsion of the optic nerve* can produce *complete atrophy* and so can *hemorrhage into the nerve sheaths. Partial consecutive atrophy* is a sequel of repeated traumatic hemorrhage, especially from the choroid."

"*Recurrent congestive glaucoma* is another potential danger for a man exposed to repeated blows upon the eyes, and this condition may produce a blind, painful eye. But it is abundantly clear that for every troublesome eye requiring removal, several remain as nonfunctioning, though painless, vestiges, and in many others vision is permanently curtailed."

"Injuries to Neighboring Structures, including the Ocular Adnexa:

"*Diffuse thickening of the bone along the orbital margins* is not uncommonly evident in ring veterans. *Localized hyperostoses* have also been found encroaching upon the orbital cavity. Any of the *orbital bones can* be *fractured,* especially the ethmoid, maxilla, malar, and lacrimal bones. *Orbital emphysema* is apt to arise when the accessory air cells, especially those of the ethmoid, are involved in a fracture; *obstruction of the naso-lacrimal duct is* one of the complications of a broken nose. The *lacrimal gland* thus suffers damage. Another possible source of deranged lacrimal function is injury to a canaliculus. *Orbital emphysema is* apt to arise when the accessory air cells, especially those of the ethmoid, are involved in a fracture. Obstruction of the naso-lacrimal duct is one of the complications of a broken nose."

STALLARD (1951) reported severance of the left superior and inferior lacrimal canaliculi from a punch landed in amateur army boxing. *Conjunctivodacryocystostomy* had to be performed.

Bruising of the eyelids constitutes an ordinary '*black eye*', and is not usually serious, but sometimes ptosis occurs from damage to the levator palpebrae superioris.

Any of the extra-ocular muscles may be implicated, but the superior and inferior obliques are the most directly vulnerable because they so nearly approach the anterior margin of the orbit. Even in the absence of direct muscular injury, e. g., laceration, the action of the recti muscles, as well as that of the obliques, may be impeded by connective tissue developing after repeated hemorrhage into the orbit. Most of the interstices between the orbital structures are filled with fat, a tissue slow to absorb effused fluid; and absorption tends to be progressively slowed if these orbital hemorrhages keep on recurring. Accordingly, limitation of movement in one or more directions, and sometimes in all directions, has been registered in many boxers. Improvement of ocular motility through surgery can be frustrated by adhesions that obscure anatomical landmarks normally encountered in this cavity.

Divergent squint is often manifest when one eye has been rendered blind or grossly defective; and high degrees of *exophoria* have been recorded even among those with equal vision in the two eyes. In some countries boxers suffering from diplopia are banned from official matches, but we must remember that the ocular muscle balance can be seriously deranged even in a patient not complaining of double vision. It has been playfully suggested that diplopia may stimulate a boxer to punch harder in the hope of knocking out two opponents with one blow, but WHITING's (1954) patient, Bill SOFTLY, was assuredly handicapped by reduplication of his images. "Yes," he said, "I see two chaps in the ring and I hit the one that isn't there and the one that is there hits me."

A number of years later, in 1963, the *British Association of Sports and Medicine* held a conference on the *Medical Aspects of Boxing*, publication edited by BASS et al. (1965).

The two speakers at the session on eye injuries were DOGGART (1965) and RUGG-GUNN (1965).

Additional Literature on Eye Injuries

RUGG-GUNN (1965) said that, "Every ophthalmic surgeon has large numbers of cases (of detached retina) under his care... and modern methods of operation have become very successful." However, he adds that, "The occasional examples due to boxing differ from what we may call the ordinary types. The latter usually begin as a tear in the upper periphery of the retina, and gradually the detachments work down to the inferior periphery. In the case of the true boxing type, apparently the result of a series of injuries acquired over a long period of time, ultimately a collection of minute cysts develop in the macular area, and may lead finally to a central detachment. This rare type is less amenable to operative procedures than the ordinary type."

RUGG-GUNN summarized:

"In conclusion, I wish to say that my association with boxing is entirely with amateurs. It happened, however, that when appointed *Surgeon to the Western Ophthalmic Hospital*, a boxing school for professionals was round the corner, and I saw many cases of injuries to the eyes, including some detachments of the retina."

"The risk of injury to the eye in boxing certainly seems to be small, but when injury does occur it is often serious."

DOGGART (1965) presented a paper at this conference in which he argued forcefully, eloquently, and with great emotion, against boxing from an ophthalmologist's viewpoint. His main points were: (1) Surgery for detached retinas was less successful in boxers than in other patients because of the other damage that was usually present; (2) boxers whose eyes are damaged tend to leave the ring as soon as this happens, and do not usually go to boxing doctors for treatment. He expounded, "I make this suggestion to all those of you who seek the truth. If you want to know the facts about serious damage sustained in the ring, put your questions to neurologists and ophthalmologists rather than to people who can bear to watch the young bashing each other into unconsciousness." (3) medical controls over boxing do not affect the central problem of eye injuries: "JOKL has pointed out that in its reaction to injury, living tissue is not influenced by the monetary arrangements between the boxers' managers. No amount of regular medical inspection will alter the fact that the retina is flimsier than a piece of tissue paper," and (4) "the infliction of severe, permanent, progressive damage to the brain from boxing has been repeatedly established, and I will not presume to encroach upon ground that neurologists will cover..."

DOGGART also quoted a letter he had received from the late Professor WEVE, of Utrecht, The Netherlands, a world authority on detachment of the retina. WEVE had treated no less than four world champions for detachment of the retina, among them was one who had detachments in both eyes within a period of three years, both unquestionably traumatic cases. WEVE mentioned that not so long ago he once met with some difficulties because he refused to certify a well-known boxer from South Africa so that he could enter the ring again, after having a traumatic detachment treated and healed. After his experience with boxing injuries, WEVE disapproved of boxing, and supported DOGGERT's position that it should not be continued as a competitive sport. WEVE's views were emphatically endorsed by the late LARSSON, another principal authority on retinal detachment.

One of the most striking elements of the discussion concerning eye *injuries (British Association of Sports and Medicine, Medical Aspects of Boxing)* was the fact that a number of those present, beginning with DOGGART, spoke not only of eye injuries but of the dangers of boxing. There are several possible explanations for this: (1) This was the first session of the conference, and the attendants wanted to air their views; (2) DOGGART, ever a crusader on the side of abolishing boxing, used his paper as a forum for his wider views and so provoked rebuttal; or (3) there is some intrinsic link between eye injuries and the more grave dangers of boxing. Nothing is more likely than that all three are correct.

DOGGART's paper is quoted at length below for several reasons: First, it is largely upon this that the rest of the discussion hangs; second, he makes (and refutes) most of the main points in defense of boxing; and third, his eloquent and heartfelt argument is a good indication of the extent to which this subject may affect those who oppose boxing.

DOGGART wrote:

"At a critical stage in the battle of Copenhagen, NELSON heard that his superior officer was signaling him to withdraw. NELSON, who had lost the sight of one eye in an earlier battle, put up a telescope to his blind eye, and quietly remarked that he saw no such signal. This device of *facultative Nelsonism,* if we can agree to call it so, has proved a godsend for many a clinician seeking to establish the value of a certain treatment, or to research workers bent upon indicting one particular factor as the main cause of a disease. Equipped with facultative Nelsonism they can suppress, belittle or ignore all data which do not advance their theses."

DOGGART continued:

"Some of the medical men dedicated to boxing seem to carry this facultative Nelsonism to the verge of ostrichism. They blandly maintain that injuries sustained in the ring are rare, trivial and transient. They enlarge on medical precautions taken, especially at amateur fights, and they go on to recite the toll of damage sustained by players of ball games. A flying ping pong ball perhaps scratches a player's cornea. 'Ah,' say the boxing enthusiasts, wagging their heads in mock solemnity, 'he should have stuck to a safe sport like ours.'"

"Let us remember, however, that many of the people whose eyes or brains were badly injured at boxing took warning and left the ring before it was too late. They are soon forgotten by the boxing devotees and they do not seek treatment from boxing physicians as a rule. Therefore I make this suggestion to all those of you who seek the truth. If you want to know the facts about serious damage sustained in the ring, put your questions to neurologists and ophthalmologists rather than to people who can bear to watch the young bashing each other into unconsciousness."

DOGGART summarized the ocular damage:

"Although retinal detachment is rightly dreaded as one of the most devastating ocular hazards of the ring, it is only one among a formidable array. Do not seek information about these from the boxing doctors. Look at the relevant ophthalmic literature, in which you will find a fearful catalogue of lacrimal derangement, paralytic diplopia, iritis, cataract, glaucoma, vitreous opacity, choroido-retinal degeneration and optic atrophy. The normal macula, upon which are focused the clear images of objects under direct gaze, is a particularly vulnerable portion of the ocular fundus, and I have examined many boxing victims in whom diminished visual acuity of one or both eyes was amply explained by choroido-retinal pigmentary degeneration, scarring and atrophy in this macular region."

"Do not believe the people who pretend that amateurs are exempt from these disasters. JOKL has rightly pointed out that in its reaction to injury living tissue is not influenced by the monetary arrangements between the boxer's managers. No amount of regular medical inspection will alter the fact that the retina is flimsier than a piece of tissue paper."

"Eye doctors are, of course, well aware that damage to the eyes themselves is not the only explanation for lowered visual efficiency. The eyes cannot interpret their own images. Interpretation depends upon the intricate ramifications of the visual pathways and upon the linkage of these pathways with other parts of the brain, that very organ which is the main target for a knock out punch. The infliction of severe, permanent, progressive damage to the brain from boxing has been repeatedly established, and I will not presume to encroach upon ground that neurologists will cover tomorrow, but I have thought it right to stress the close relationship of the eyes with the central nervous system. The proper timing of punches depends upon the state of the central nervous system even more than it does upon a healthy pair of eyes."

DOGGART quoted now excerpts from circulating newspapers:

"Here is a typical boxing word-picture from one of our widely circulating newspapers, 'Near the end of the sixth round there came a clash of heads and a bad cut opened over HOWARD's left eye. Skilled hands repaired that cut in the interval but it only needed two or three left jabs early in the seventh round to open it again.'"
Here is another, 'An injury to McATEER's left eye in the second round caused most of the trouble. It was not possible for me to see definitely whether it was a punch or a butt with the head that led to the wound.'
"'HUMEZ is a rough customer and he was twice warned in the first two rounds to keep his punches up and for the questionable use of his head in charging in to the attack... A trickle of blood ran down McATEER's cheek from the second round on and the eye became a target for many of HUMEZ's best punches. Before the end it was like a half-opened oyster, and it was plain that he could not focus properly through it.'"

When a boxer has an injury of the eye region and the fight is not interrupted by the referee or ring physician, the injured area becomes a preferred target of the opponent, who tries to enlarge the tissue damage with more blows.

Here is another example from a newspaper:

I hesitated for awhile where to include the second professional fight between the lightweights Livingstone BRAMBLE and Ray 'Boom Boom' MANCINI. MANCINI's face was severely bruised and battered, but the most severe damage was the cut over the left eye, because blood running down over MANCINI's left eye impaired his vision in such a way that he was blind on this eye and could therefore not see the punches coming from the left. Therefore, the injuries are filed under injuries of the eyelids, though the entire face was lacerated and bloody.

PUTNAM commented on the fight between Livingstone BRAMBLE and 'Boom Boom' MANCINI *(PUTNAM P: Bloody. Beaten, but unbowed. Sports Illustrated, Vol. 62, No. 8, February 25, 1985, pp. 14-17):*

"'MANCINI is an arrogant fool, and I hope he gets seriously hurt in this fight,' said BRAMBLE, now on a fiery roll."
"'How seriously?' asked a stunned listener."
"BRAMBLE weighed the question, and then backed off. 'Seriously enough for the referee to count to 10,' he said."
"While MANCINI elected not to join verbally with BRAMBLE before the fight, WOLF showed no such reluctance. 'BRAMBLE is the most despicable person I ever met in my life,' WOLF told Jerry IZENBERG of the *Newark Star Ledger*. 'When he called Ray a murderer [for having been the victor in the 1982 bout that led to the death of DUK KOO KIM in Buffalo, I wanted to kill him. If he were crossing a street against the light, and I was in a car and the light was in my favor, I wouldn't stop. In fact, if he were dying in bed, and I had the pill to save him, I'd flush it down the toilet.'"
"If they'd fought this one in a saloon, the cops would have stopped it. Then they would have arrested Livingstone BRAMBLE. The other brawler, Boom Boom MANCINI, would have gone free: The guy with the face that looks as if it has just been hammered with a brick always gets a pass."
"But the authorities let this bout go the distance, and even the courage of a beaten MANCINI couldn't overshadow the shame the fight racket visited on Reno Saturday night."
"The three judges compounded matters by indicating that MANCINI had been in the hunt, that BRAMBLE had escaped with his WBA lightweight title by the thinnest of margins. Although they followed different routes, at the end, Dave MORETTI, Jimmy RONDEAU and Edward LEVINE all had it figured that BRAMBLE had won by a single point. One had only to look at MANCINI's swollen and battered face, his slammed-shut left eye and the blood pouring from long rips above that eye and beside the right one to wonder what fight the judges had watched."

The bout was telecast by HBO: the blows were counted by a computer. According to the computer, BRAMBLE landed 674 punches to MANCINI's 381.

This was the rematch of a fight in Buffalo, which was stopped in the fourteenth round by the referee. MANCINI lost his title by a technical knock out to BRAMBLE.

PUTNAM continued:

> "As in the first fight, MANCINI was cut in the first round, this time over his left eye. By the fourth, the eye began to swell badly. In the fifth, BRAMBLE ripped a ditch beside MANCINI's right eye."
> "By the seventh round, referee Mills LANE was anxiously studying MANCINI's tortured features. BRAMBLE was making a curious fight: He'd rattle MANCINI's head with wicked volleys, and then he'd back off, seemingly content to let his defense frustrate the former champ."
> "In the eighth, LANE called a cease-fire, sent BRAMBLE to a neutral corner and led, MANCINI, his left eye now all but closed, over to Dr. Charles FILIPPINI, the ring physician, 'I'm all right,' MANCINI said. LANE had a good idea, but FILIPPINI sent MANCINI back to face the grinning BRAMBLE. MANCINI finished the fight strong, although LANE called on Dr. FILIPPINI once more in the fifteenth round."

"MANCINI embraced BRAMBLE at the close: 'You're the champion,' he said."

Twenty stitches were necessary to repair the damage.

Twenty stitches repaired the damage of the skin and subcutaneous tissue of the face. There remained only a scar, but the numerous blows – BRAMBLE landed 674 – damaged the much more vulnerable brain tissue. For this damage there is no method of repair; it is permanent damage which can be progressive.

Later MANCINI said:

> "I couldn't see the punches coming. The blood kept going into my eyes."

This statement of MANCINI, often repeated by other boxers, reveals the main problem: it is not the bloody wound which is so dangerous or unaesthetic, but the fact that the running blood impedes the vision of a boxer, he is unable to see and avoid punches coming from the side of the injured eye.

In another newspaper article, the severe tears and lacerations above both eyes of Ray 'Boom Boom' MANCINI were discussed *(Did MANCINI's blood bath go to far? Post-Valentine day massacre. The Houston Post, February 18, 1985, 3 D)*. It said:

> "Livingstone BRAMBLE and Ray 'Boom Boom' MANCINI proved they were tough, throwing an incredible 2.000 punches over 15 brutal rounds."
> "But perhaps the toughest guy at Saturday night's World Boxing Association lightweight championship fight was Dr. Charles FILIPPINO. It was FILIPPINO, the ring physician, who repeatedly examined two huge gashes over MANCINI's eyes and kept sending the brave but outgunned fighter back for more punishment... MANCINI bled heavily from both eyes from the fifth round on as BRAMBLE hammered away with stinging punches. At the end of each round, MANCINI's face was a sheet of blood... Dr. FILIPPINI didn't think the gashes were serious, but other doctors at the Reno hospital did. They performed plastic surgery on MANCINI early Sunday morning and eventually closed the wounds with nearly 75 stitches."
> "'The doctor was definitely swayed by the crowd,' said BRAMBLE's manager, Lou DUVA. 'It was a pro-MANCINI crowd and he was afraid to stop it. But he should have stopped that fight.'"
> "'Those were two of the worst cuts I've ever seen. At the end of the eighth round MANCINI told his guys in the corner that he couldn't see out of his left eye. He was blind. And he was allowed to keep fighting.'"
> "Veteran referee Mills LANE, who was also empowered to halt the fight, seemed willing to let the burden fall on the doctor's shoulders. DUVA said LANE was also a victim of the crowd."
> "'I believe that if the people weren't yelling so much to let the fight continue, Mills LANE would have stopped that fight,' DUVA said. 'He wanted to stop it, but decided to let the doctor make a decision.'"
> "'I mean who are those doctors? Where do they come from? They could be veterinarians for all we know.'"
> "BRAMBLE who appeared at a Sunday morning news conference, said he was also surprised that MANCINI lasted 15 rounds."
> "'I'll tell you, man,' the native Virgin Islander said, 'If I bled like Ray MANCINI did, I would have stopped fighting and come back another day. He was really in bad shape.'"
> "MANCINI said before the fight that another loss to BRAMBLE – he lost his title last June to him – would be the end of his career. He stopped short of announcing his retirement Saturday night, but indicated that his career had ended."
> "'The cuts really hurt me,' he said. 'The last half of my career the cuts have really hurt. I just seem to cut easier than I used to, and I can't keep fooling myself anymore.'"

"'When you're in trouble, you've got to listen to your heart, and that's what I'll do now. I'll go home and get some rest and think it over.'"

The 15-round encounter between Livingstone BRAMBLE and Ray 'Boom Boom' MANCINI was described as a grueling fight. MANCINI's face was beaten up, 'a mess', as one observer noted, they had to get a plastic surgeon."

"In this 15-round lightweight fight for the World Boxing Association (WBA) more than 2.600 punches were thrown and MANCINI got the worst of it."
BRAMBLE landed 674 punches compared with MANCINI's 371, BRAMBLE throw an average of 81 punches a round, most of the hard right jabs to MANCINI's face.

BRADLEY *(The Washington Post, February 18, 1985, p. 7)* reported:

"Most of the damage came from BRAMBLE's jab, which tagged MANCINI in the face 255 times. Although BRAMBLE won the unanimous decision by one point, the bout was hardly that even. BRAMBLE owned most of the middle rounds. MANCINI was early and late. The crowd of 11.758, at times chanting 'Boom Boom,' seemed to be challenging the three judges at ringside to rule against MANCINI, their hero."
"BRAMBLE said the booing helped his game plan. 'If they didn't boo,' he said, 'MANCINI wouldn't have kept punching. He punched himself out early, and I was able to come back with better stuff later on. He didn't hurt this coconut (BRAMBLE's head, author). I could feel his strength, but I know the body shots I threw took it out of him!'"

The description of this 15-round boxing match shows convincingly that the number of blows is not so low as KAPLAN and BROWDER (1954) believed. A total of 2.600 punches were thrown in this 15-round bout. That is an average of 173 per round, or 87 per boxer. These punches did not all land. BRAMBLE landed 674 punches in the 15 rounds, which is about 45 per round. BRAMBLE said: 'A lot of times I hit MANCINI in the body.'

Let us assume that only half of these, about 22 per round, landed. These were not blows of concussive strength, since no knock down occurred, but numerous blows of a subconcussive strength that repeatedly landed on his head over 15-rounds, in a time interval of 45 minutes. Such a barrage of impacts against head and brain cannot be endured without damage to the brain. During a fight where each boxer receives numerous subconcussive blows to the head, brain edema will have begun to develop before the fight is over. The brain edema causes damaging permanent alterations, so that both boxers sustain injury, regardless of the outcome of the fight. Thus the winner may lose more than he has gained from the fight, since the purse cannot be said to offset the effects of permanent brain damage.

"'A lot of times I hit Ray MANCINI in the body and he looked like he wanted to quit,' BRAMBLE said yesterday. I had him hurt in the eighth and thought maybe the ref or the ring doctor would stop the fight, there was so much blood.'"

J.C. OATES noted 1985:

"... the startling picture of Ray 'Boom Boom' MANCINI after his second losing fight with Livingstone BRAMBLE, for instance, when MANCINI's face was hideously battered (photographs in *Sports Illustrated* and elsewhere were gory, almost pornographic); the much reprinted photograph of the defeated Thomas HEARNS being carried to his corner in the arms of an enormous black man in formal attire – the 'Hit-Man' from Detroit now helpless, only semiconscious, looking precisely like a black Christ taken from the cross. These are powerful, haunting, unsettling images, cruelly beautiful, very much bound up with the primitive appeal of the sport."

A letter to the editor *(Sports Illustrated, Vol. 62, No. 11, March 18, 1985, pp. 86 + 88 reads:*

"Sir
Sports Illustrated (SI) has long been the national sports conscience, decrying scandal in college athletics, drug abuse in professional sports, illegalities in horse riding, etc., but when is SI going to take a stand against the brutality of professional boxing?
Some SI readers were morally outraged by your photographs of beautiful women in revealing swimsuits, but I wonder how many mustered any indignation at seeing the bloody, distorted face of Ray MANCINI *(Bloody, Beaten, Unbowed, Feb. 25)*. Thank you for not putting that face on the cover.
Ken CHAMLEO, Brevard, NC."

We continue with DOGGART's comments:

"When I hear medically qualified people proclaim that the chances of serious eye injury in boxing are remote, I know only too well that they are not likely to be convinced by argument. They have been given the opportunity to study the anatomy and physiology of our special sense organs, among which the eyes are clearly the most vulnerable; but so blind is their devotion to boxing that they airily dismiss the likelihood of havoc from a volley of punches directed upon the eyes."

TUNNEY (1932) reported on a boxer's loss of eyesight following a punch over the left eye. He wrote:

"While in Hollywood making the 'greatest serial ever produced' - *'The Fighting Marine'* - I attended a fight show one night. Gene DELMONT, my friend who seconded me in some A. E. F. matches in France, was in the star bout. Gene went very well for the first three rounds, winning rather easily. At the beginning of the fourth he stepped into a terrific right-hand punch that seemed to land on his forehead. There was an instant reversal of form. Gene backed away. He clinched. He hung on. The referee had a hard time breaking them. The crowd started razzing. Gene was stalling. The fifth round was a repetition of the fourth. DELMONT's opponent started getting to him in the sixth, pummeling him rather severely. The audience started booing. The eighth, ninth and tenth rounds were test of endurance for little DELMONT. Leaving the ring, Gene got an unmerciful booing and razzing. The brave ones in the audience called him profane names."

"I believed something was wrong, so I went back to the dressing room. Gene was weeping; the names he had been called as he left the ring had hurt. I asked him what happened in the fourth round. And this was his answer: 'I got a punch over the left eye that didn't hurt much. I couldn't see with it. When I got back to my corner, I told one of my seconds that the eye-lash was in my eye. I asked him to open the eye and turn the lash out.'"

"'Your eye is open,' he said."

"'Stop kidding me,' I said. 'I can't see a thing. Open it.'"

"'I tell you it is open,' he insisted."

"'Like a death sentence I realized I was blind in that eye.'"

DOGGART continued:

"We are told that boxing promotes patience and courage, and it is only fair to admit that there is some force in this argument. There are, however, many forms of sport which not only inculcate pluck and endurance, but also allow the participants to triumph without deliberately damaging a fellow mortal."

"'Yes,'" they will say, "'but isn't a boxer favorably equipped to deal with thugs and bullies?'" Not at all. A thug is not likely to stand in a good light waiting to be hit on the jaw. He is more likely to attack you with a knife from behind. Incidentally, we know that a man skilled in wrestling, judo or jiu-jitsu can make short work of a boxer."

"Boxing is hailed as an excellent form of physical culture, but there are many athletic pursuits which develop the body in better proportion. Long years in the ring are all too likely to produce a simian, top-heavy appearance, from over-development of the upper half of the body. Another excuse put forward by boxing fans is that they chiefly enjoy the adroit footwork and skilled defense mechanism of an expert boxer. No doubt these discriminatory enthusiasts exist, but Lt.-Col. FARTHING, a former sparring partner to Joe BECKETT has stressed the less savory aspect of those fights. In his view the majority of persons who attend professional boxing matches are not interested in a skillful boxer, who by adroit movements avoids punishment and wins his fights on points decisions."

"All they clamor for," he says, "is a man with a 'killer punch' and they wait with avid and sadistic anticipation for the moment when it is successfully brought into action and his opponent is battered into submission or sinks unconscious to the floor of the ring."

BOARDMAN caught RYFF with a flurry of punches that ended with a right cross to the jaw that flattened RYFF in 0.46 of the ninth round. Referee Barnie FELIX counts RYFF out. The crowd in the background cheers BOARDMAN's victory. This photo is presented to show the jubilation of the spectators (Fig. 152). A cerebral concussion is celebrated.

DOGGART wrote:

"It may be true, as boxing devotees are fond of proclaiming, that more people have been killed on the football field than in the ring, but surely the total number of football players vastly exceeds that of boxers. Nearly all sports are attended by hazards, and hazards cannot altogether be eliminated, even if we were to wrap people up in cotton wool

Fig. 152. Professional ligtweight fight between Larry BOARDMAN and Franky RYFF on January 6, 1956 at Madison Square Garden, New York. Photo: Frank MASTRO.

and put them in a glass case. I wonder what you will think about this specious argument recently put forward by a medical man in the lay press. He said: 'Boxing is not the only sport where one opponent deliberately tries to hurt the other.' He goes on to speak of '... kicking, elbowing, scratching, gouging and even biting that goes on in the rugger scrum, the ankle-tapping and punching at soccer, and the deliberate attempt to injure one's opponent at hockey, polo and lacrosse.' Alas, there are foul players in nearly every kind of sport, and in this imperfect world many bad things will escape the notice of a referee. Nevertheless, the man who wrote that letter must know quite well that all the nasty things which he lists are contrary to rules, and will be penalized if they are detected by an honest referee. We should not judge a game by the disreputable tricks of its less scrupulous players. What a howl of indignation he would let forth if boxing were condemned on the score of the deliberate head-butter, kidney-punchers, rabbit-punchers and hitters-below-the-belt. Serious and even fatal injuries will from time to time arise in many different sports, even without intentional foul play, but the deliberate disabling of your opponent is forbidden in ball-games, running contests and most other forms of sport. In boxing the main point is to disable your opponent, and the surest way to do that, if you can catch his chin unguarded, is to concuss him with a blow which may damage him for life or strike him dead."

MANNING has recently pointed out that since 1945 an average of 18 fighters, professional and amateur, have been punched into their graves every year. 'The medical case against boxing,' he suggests, 'is overwhelming. It remains largely acceptable in Britain and the United States only because ballyhoo for the next big fight distracts public attention from the sadism of the one before.' MANNING is a layman to medicine, but we who have been shown the central nervous system and the sense organs in all their beautiful intricacy should be all the more eager to oppose a pastime which wrecks Nature's artistry."

The discussion following both papers was lively, to say the least, and a classic example of completely conflicting information.

LYLE (1965) said, "Detachment of the retina may be a condition which can be successfully remedied by means of operation, but cases due to boxing injury are much more difficult to deal with and results are very rarely so successful as cases of detachment of the retina due to other causes."

LYLE continued:

"The point I would like to make is that repeated trauma to the eye, just as to the brain, gives rise to a form of concussion. This condition sets up fibrosis in the retina, giving rise eventually to gross defects of vision. When I was serving with the R.A.F. in the war, it was my duty to examine a number of professional boxers who wished to be admitted as P. T. instructors. As far as my memory serves me, I think the whole lot had evidence of detachment of the retina. This condition was so severe that it grossly affected a large percentage of these men and their physical ability and most were unfit to be admitted to the service. So many of them were not really aware of their defect; they had come to accept that their sight was not too good and could not be treated by any means at all."

LYLE noted

"In my experience there are a number of very serious injuries which affect the eye due to boxing and I think we must accept these facts. Some of us are likely to come across more cases than others and there are many ophthalmic surgeons who may not have the experience."

"As a surgeon at *Moorfield's,* I would say that we do get a number of serious cases of eye injuries due to boxing."

However, McDOUGALL disagreed completely:

> "In the past ten years I have not seen one boxer with an eye injury that much affected him. Three or four retinal haemorrhages have occurred which resolved completely with adequate treatment."

McDOUGALL continued:

"Mr. DOGGART gave a very good outline of all the injuries to which the boxer is susceptible, but not only due to boxing. His remarks were rather emotional and when he states that in other games had the referee seen the breach he would have penalized the man, it is no consolation to the man that had the referee seen the incident his opponent would have been penalized. One feels that the referee must have detached retina in both eyes to allow the sort of conduct that all of us see in some games."

> "As a previous amateur boxer myself and as one who takes extreme care and interest in boxing in amateur circles, to my mind the dangers to the eyes have been grossly overstressed as far as amateur boxing is concerned."

SAVILLE (1965) also discussed the findings of DOGGART and came to the following conclusions:

> "At the onset I would like to say I have every respect for Mr. DOGGART's skill as an ophthalmic surgeon, but I must say I have never listened to a more unscientific, emotional and biased talk than his. I was hoping to hear some detached scientific evidence of the sort given by our second ophthalmic speaker. Whilst having every respect for Mr. DOGGART's opinion on normal ophthalmic matters, I cannot accept what he said. The figures he quoted – do they come from England? As far as we are concerned, who control professional boxing in this country, we do not know them and I am sure they do not come from this country. There are some countries which have very little control (these countries should have control) where these things happen; they certainly do not happen here. To say that the non-boxing doctor or ophthalmologist or neurologist sees all the cases and we do not is, to my mind, a lot of nonsense. It is not just a question of examining a boxer before a fight – these men are followed up and we are in close contact with them. If there are all these cases, surely we must come across them?"

SAVILLE continued:

"I note Mr. DOGGART made great play on Nelsonism – I would say he is a case of reversed Nelsonism, if there is such a thing. I would also like to add that in this country each boxer is examined at least once a year, apart from the normal pre-fight medical examinations, when we look particularly for eye injuries. If there is any lowering of the standard at all the boxer is suspended. We had a case only yesterday where a very well known boxer whose eyes were lower than our standard (6-9 in one eye and 6-8 in the other) had his license withdrawn. All boxers' medical sheets are looked at very closely, and over a period of years it is rare that we get any real deterioration in the very short space they continue boxing. In the case of the ten boxers mentioned, if my memory serves me correctly, they were mainly (9 out of 10) boxers of the pre-control period before the *Board of Control* came into being with its very strict medical regulations... Today has convinced me more than ever that we cannot find objectivity in certain types of doctors."

This somewhat personal diatribe was followed by the opinions of BLONSTEIN (1965).

"I am speaking as a *senior medical officer* of the *Amateur Boxing Association* and the first point I would like to take up is Mr. DOGGART's reference to retinal detachment which he said was on the whole remote. As far as amateur boxing goes it is extremely remote. I have been vetting [sic] amateur boxers for 34 years and during that time we have only had two cases of detached retina. I personally have seen three cases of detached retina in a very short time amongst Scottish rugby players."

"As for the other serious injuries he mentioned I say emphatically we do not get these serious eye injuries in amateur boxing. I personally have never seen any. Mr. DOGGART talked about football deaths as compared to boxing deaths... however deplorable these deaths are, the deaths resulting from boxing compare very favorably with such a widely-played sport as football, and even cricket... Mr. DOGGART said the main object of boxing was to knock your opponent unconscious. The object of boxing is to defeat your opponent by out-pointing him, the main object is not to knock your opponent unconscious, although it does occur."

DOGGART, undaunted, responded:

"What I would say when replying to most of the people in favour of boxing I have already said in the course of my paper. I do think perhaps Dr. BLONSTEIN is a little confused because he quoted me as mentioning detachment as a remote possibility. I did not suggest that retinal detachment was a remote possibility and I drew attention to its special dangers in boxing. I have heard Dr. BLONSTEIN's reassuring remarks for many years; he keeps up a stream in support of his pastime. Some of his reassurances have worn a little thin so far as I am concerned. I have often heard them contradicted. I happen to have with me a letter from the *British Medical Journal* written by a doctor in reply to an article in which Dr. BLONSTEIN made the usual reassuring remarks about the rarity or absence of injuries. This was a man who had attended as medical officer over 2.000 tournaments, and disagrees with Dr. BLONSTEIN's conclusions."

"Dr. BLONSTEIN's reassurances have worn a bit thin as far as I am concerned, but I was not so optimistic as to suppose I could convince him by any arguments of mine."

In fact, although the feelings of the participants ran high, there was little concrete evidence either way. It does seem, though, that the bulk of such evidence as was offered lies with the opponents of the sport. SAVILLE, who denies the figures quoted by DOGGART, does not offer any reason for his refusal to accept them, but instead proceeds to deliver a diatribe against those who oppose boxing. The statement of McDOUGALL – that in ten years he had not seen one case – comes into perspective only when considered in the light of ALBAUGH's paper. Both may well be telling the truth, but simply because Dr. McDOUGALL does not see the patients does not prove that they do not exist. Obviously the dangers would seem to McDOUGALL to be 'grossly overstressed' – they did not seem so to ALBAUGH.

Obviously, some of the preceding pages had little to do, specifically, with eye injuries. However, they do serve to prove that eye injuries (with the concomitant possibilities of brain damage) are among the more serious of the injuries due to boxing, and that the great majority of ophthalmologists recognize that boxing is a large, specific source of damage to the eye.

COLOSI and VANOFF (1977) published the

case of a 28 year old boxer who had sustained a *rhegmatogenous retinal detachment that* was repaired successfully.

Repeated blunt trauma caused a second detachment two years later. Three attempts at repair by scleral buckling with diathermy and drainage of subretinal fluid were unsuccessful, and the eye was enucleated.

Pathological examination showed internal erosion of the intrascleral silicone implant into the vitreous compartment. Conjunctival-like epithelium lined the intrascleral bed of the implant and extended through the internal erosion onto intraocular structures, and this conjunctival epithelium was traced onto the external surface of the eye. Repeated applications of diathermy, combined with drainage of subretinal fluid, had weakened the sclera so that the silicone implant eroded internally and epithelium grew into the eye.

Reforms for the prevention of Eye Injuries in Boxing
ELKINGTON (1985) reported on *'Boxing and the Eye: Results of a Questionnaire'* that:

"In June 1982 a resolution to campaign for the abolition of professional boxing was approved at the *Annual Representative Meeting of the British Medical Association*. Subsequently the Council of the Association asked its Board of Science to look into the scientific aspects of this issue and the Board, in their turn, set up a Working Party to give detailed advice."

"Members of the Working Party set about this task, requesting and reviewing the literature and sifting information from many interested bodies and hearing oral evidence from those who wished to give it. One specific aspect was injuries to the eye."

"The literature contains a plethora of case reports about boxers with eye injuries. However, there are relatively few reports of surveys carried out on boxers. Mc COWN (1959) reporting an American study, stated that over a five year period, of the 9.871 boxers studied, sixteen had suffered eye damage; eleven had corneal injuries and five had retinal detachments. In a further report from the USA PALMER (1976) found that of fifty-five retired boxers, nine had evidence of past contusion of the angle of the anterior chamber. However, despite an average follow-up period of thirty-three years, not one had developed glaucoma – a well recognised complication of such angle recession."

"As far as the United Kingdom is concerned DOGGART (1951) in his paper 'Fisticuffs and the Eye' set out the many ways in which the eye and its adnexae may be damaged as a result of boxing. Almost all the recognised consequences of blunt trauma are listed. The survey carried out by ROBERTS (1969) into the prevalence of traumatic encephalopathy amongst 224 ex-professional boxers, gives some useful information on the state of their eyes. Five boxers had double vision as a consequence of a palsy of the extraocular muscle. Two had retinal detachment. However, one of these boxers said the detachment stemmed from a blow to the eye in the childhood. Another had a pigmentary disturbance of the macula of one eye which was thought to be due to trauma. The precise impaired vision of several others with defective sight was not determined. When assessing this study account should be taken of the fact that the subjects were not examined by an ophthalmologist and although the interior structures of the eye were routinely examined the pupil was dilated prior to examination in only two cases. Boxing injuries occurring in Army personnel between 1969 and 1980 were reported by OELMAN and co-workers (1983). Ten men were discharged from the Service for medical reasons which resulted from damage sustained in the ring. Two head injured eyes; one having a detached retina and the other an injury to an optic nerve."

"In the written evidence given to the Working Party by Medical Officers of the Amateur Boxing Association (ABA) it was reported that no serios eye injuries had been seen in the past five years. When given oral evidence the Medical spokesman for the ABA said that the Association had no system of following up boxers, although they were currently working hard on setting up such a system."

"As far as professional boxing was concerned, the British Boxing Board of Control declined to give either written or oral evidence to the Working Party. The Board's Senior Medical Officer (WHITESON 1981) when writing three years ago about retinal detachments amongst professional boxers' commented "I can think of only ten cases since 1945...' This is a rate of one case every three or four years."

"Enquiries made amongst eye surgeons suggested that boxers with eye injuries were seen only rarely. Nevertheless, the statements that there had been no serious eye injuries amongst amateur boxers during the past five years and that retinal detachments were seen only every three or four years amongst professional boxers did not accord with the experience of some surgeons questioned. In the circumstances the Working Party felt it right to seek information from as many ophthalmologists as possible and to this end a questionnaire was circulated to the entire United Kingdom membership of the Ophthalmological Society of the United Kingdom (OSUK). The inevitable limitations of this approach were recognised from the outset. Provisional figures, based on questionnaires returned by January 3, 1984 were published in the Reports of the Working Party. The study has been concluded and the final figures are now presented."

Material and Methods

"Copies of the questionnaire (Fig. 153) were sent to the 717 members of the OSUK on December 8, 1983. Details of the membership of the OSUK and the way in which members in various grades replied are given in Table 6. A total of 361 replies were received by May 1, 1984 when the study was concluded. The reply rate was 50.35 per cent. Seven replies were excluded for the reasons shown in Table 7. Of the remaining 354 replies, 188 (53.12 per cent) were from ophthalmologists who did not recall having seen a boxer with an eye injury whereas 166 (46.89 per cent) were from those who could remember seeing a boxers with such damage. The 166 surgeons who had seen eye injuries gave details about 210 boxers. The status of the boxers (professional, amateurs, status unknown) and whether or not their injury had become apparent in the last five years in shown in Table 8. The types of injury of the 210 boxers in the three status categories are set out in Table 9. Of the 166 respondents who had seen boxing injuries, 130 (78.31 per cent) gave details of only one boxer. The remaining 36 (21.69 per cent) reported details about two or more boxers. One surgeon gave details of thirteen injured boxers seen in the past four years. Seven of these had retinal detachments, two had retinal tears and two had a cataract in one eye. The 166 surgeons thought that they had seen, in addition, a total of 104 boxers about whom they could not give specific details. Of these 166 surgeons, 31 (18.67 per cent) indicated that they had seen boxers in question at a centre other than the one at which they held their definitive appointment or, if retired, where they held their appointment. A further five surgeons failed to name a centre at which they currently work. Of the 354 respondents, 35 (9.88 per cent) made additional comments."

Discussion

"Any questionnaire is necessarily a compromise. Ask too much, and it will not be completed. Ask too little, and one learns no more than one knows already. The questionnaire on which this study was based was no exception. It was reassuring, however, that the overall replay rate was so high (50.35 per cent) and that amongst the practising surgeons, who examine and treat those with serious eye conditions, the reply rate was even higher (64.41 per cent) (Table 6)."

"The information given by two surgeons was excluded because they were not sure that boxing was the cause of two boxers' bilateral cataracts. One boxer was in his twenties and the other in his forties. It was interesting to learn from the comments by several surgeons that they though that boxers develop cataracts earlier than do the general population. This is also the view of one medial officer of the ABA to whom I have spoken. Of the two surgeons who declined to take part in the study, one did so because he felt a prospective study would prove more informative and the other wrote that he disapproved of a Working Party such as ours being set up if the outcome of its findings was that boxing was banned."

"Those replying to the questionnaire could not always recall whether a boxer had beeen a professional or an amateur and this led to the use of 'unknown' categories in Table 8, 9 and 10."

"The possibility that one boxer's injury might be reported by more than one surgeon was appreciated from the beginning and this led to the inclusion in the questionnaire of Question 2: 'At which Eye Unit did you see the patient?' It was obvious from several of the returns that there was indeed over reporting, namely that two or more surgeons were describing the same boxer. However, of the 166 surgeons who had seen injured boxers, only 31

Southampton Eye Hospital

Southampton and South West Hampshire Health Authority

WILTON AVENUE
SOUTHAMPTON
SO9 4XW

Telephone 38883
Appointment 25681

ARE/SEE

Dear Colleague,
Please forgive my approaching you in this rather impersonal way. But if you read on I think you will understand the reason.
As you may know the British Medical Association has set up a Working Party to look into the Medical aspects of Boxing. As a member of this group I have been asked to find out about eye injuries amongst boxers.
An epidemiological study is not currently feasible. Thus I am writing to all Ophthalmologists in the United Kingdom asking for their help in advising the Working Party on this topic.
I should be most grateful, therefore, if you would answer the following few questions and return this sheet in the attached stamped addressed envelope. An early reply would be very much appreciated.
Yours sincerely,
A. R. Elkington, FRCS
Consultant Ophthalmic Surgeon.

Your Name: Address:
(Please print)

1. Do you recall ever seeing a boxer with an injury to an eye or adjacent structures? Yes/No
2. If the answer to Question 1 is Yes:
   a. Was the Boxer Professional/Amateur/Status unknown
      (Please delete as appropriate)
   b. Briefly, what was the extent of the injury?
   c. At which Eye Unit did you see the patient?
   d. Was the injury incurred within the last 5 years? Yes/No
   e. Have you treated more than one boxer? Yes/No
      If so—how many?

Should you have any other information that you would like me to pass on to the Working Party please write your comments on the reverse side of this sheet.

Fig. 153. Questionaire of Southamptom and South West Hampshire Health Authority. (From ELKINGTON, 1985).

(18.67 per cent) had seen them at a centre other than that at which they now work, or from which they have retired. The remaining 135 surgeons (81.33 per cent) wrote about boxers seen in their own units. (If all the replies of the five respondents who did not write down the name of the unit are assumed to relate to a different centre than the percentages are respectively.) These figures, taken in conjunction with the finding that of the 166 surgeons, 130 (78.31 per cent) had seen only one

TABLE 6
Details of 354 replies from the membership of the Ophthalmological Society of the United Kingdom. (From ELKINGTON, 1985).

Number of members circularised		*Number of members who replied that they:* *had seen an injured boxer (a)*	*had not seen an injured boxer (b)*	*Subtotal a+b=10*
Consultants (practising)	380	115	113	228
Consultants (retired)	76	23	18	41
Clinical assistants	210	7	17	24
Senior registrars	27	5	15	20
Registrars	12	0	0	0
Irish members	12	0	4	4
'Unidentified'		16	21	37
Total	717	166	188	354

TABLE 7
Reasons for excluding the replies of surgeons. (From ELKINGTON, 1985).

Reason	*No. of surgeons*
Uncertainty as to aetiology of 2 boxers' bilateral cataracts	2
Boxer with bilateral retinal detachments – from Nigeria	1
Boxer with a retinal detachment – seen in th U.S.A.	1
Uncertainty about cause of angle recession	1
Unwilling to take part in the study	2
Total	7

boxer, suggest that over reporting does not introduce a major discrepancy into the figures."

"Such 'over reporting' is likely to have been far exceeded by 'under reporting' for the following reasons:"
"(i) The study relied upon the memory of those approached. There was no reason for surgeons to 'make up' cases, but they may well have forgotten them."
"(ii) A boxer may have presented as a patient with an eye disorder may not have been ascribed to boxing."
"(iii) The 166 surgeons who gave specific details about 210 boxers, also recalled seeing a further 104 boxers whose injuries they could not remember in detail."
"(iv) Almost half (49.65 per cent) members of the OSUK did not return the circular."

"The ABA are surely right to be evolving a system whereby their boxers are followed up for, as shown in Table 8, at least twenty-two injured amateur boxers have been seen within the past five years. And of these 22, 13 had retinal detachments."

"The Britieh Board of Boxing control should do the same, if they have not done so already, for rather than there being only one or two professional boxers presenting with retinal detachments in the last five years, as they have suggested, at least eighteen have been identified (Table 10). And this is likely to be an under estimate for the reasons outlined above."

"The frequency with which boxers sustain retinal detachments as opposed to other ocular injuries has beeen one of most interesting findings of this study. Table 9 shows that of the 210 reported injuries 118 (56.10 per cent) were of detachments. The rates for the three boxing categories were respectively: Professional 63.44 per cent: Amateur 43.66 per cent and 'Unknown' 60.87 per cent."

"The incidence of retinal detachment amongst the general population is rare (1 per 10.000 patients per year). Most of these patients are elderly. Many are either myopic or are aphakic. In contrast boxers who develop detachments are relatively young and the routine assessment of their unaided visual acuity should have prevented those who are appreciably myopic from having entered the ring. Moreover the causative tears in the retina were frequently reported to be of a type that is characteristically seen after trauma to the eye. Several boxers evidently presented with giant retinal tears, which are seen only very rarely amongst patients with detachments drawn from the general population. Despite the great advances in retinal detachment surgery in recent years, separation of the retina is still a very serious condition, and there is a very natural tendency amongst surgeons to equate surgical success with anatomical reapposition of the retina rather than with the return of a good acuity."

TABLE 8
Status of 210 injures boxers. (From ELKINGTON, 1985).

	Injury more than 5 years ago	*Injury less than 5 years ago*	*Total*
Professional	68	25	93
Amateur	49	22	71
Unknown	28	18	46
Total	145	65	210

TABLE 9
Details of the types of 210 injures boxers. (From ELKINGTON, 1985).

	Retinal detachment	*Lid laceration*	*Corneal damage*	*Hyphaema*	*Traumatic cataract*	*Vitreous haemorrhage*	*Retinal tear*	*Macular damage*	*Choroidal tear*	*Fracture of orbital wall*	*Total*
Professional	59	12	1	3	3	0	3	5	1	6	93
Amateur	31	17	0	6	5	2	1	4	0	5	71
Status unknown	28	4	0	3	5	0	2	2	0	2	46
Total	118	33	1	12	13	2	6	11	1	13	210

TABLE 10
Time of onset of retinal detachment of 118 boxers. (From ELKINGTON, 1985).

	Onset less than 5 years ago	*Onset more than 5 years ago*	*Total*
Professional	18	41	59
Amateur	13	18	31
Unknown	7	21	28
Total	38	80	118

"The findings of this survey can bring no comfort to anyone who boxes or, indeed, to those members of the medical professional advise them. The fact that apparently advisers have no data on ocular injuries amongst boxers must be of concern. For to quote a pertinent adage 'Absence of evidence is not evidence of absence'. There is clearly a need, as recommended by the Working Party to collect comprehensive relevant medical information about all boxers and ex-boxers in a systematic way."

Additional literature on Eye Injuries MAGUIRE and BENSON (1986) begin their study on retinal injury and detachment in boxers with the statement that the recent surge in morbidity and mortality associated with the sport of boxing has reactivated the debate on the ethics of modern pugilism.

They noted:

"Official policy recommendations of the American Medical Association, supporting widespread curtailment and /or abolition of the sport, have highlighted the increased public and media attention given to this controversial subject."
MAGUIRE and BENSON, quoting WHITESON, *senior medical officer* of the *British Board of Control and cochairman* of the *World Boxing Council, reported* that ocular injuries, except adnexal lacerations, were rare in boxing and that "retinal detachment... is not a common occurrence in boxing." He could recall only ten cases since 1945.

MAGUIRE and BENSON recently, however, became aware that they were treating a large number of professional and amateur boxers with retinal injuries. Prompted both by the debate and recent evidence showing that many boxers have brain damage, these authors reviewed the *Will Eye Hospital, Philadelphia*, experience over a 31-month interval.

MAGUIRE and BENSON evaluated and treated a total of nine boxers with ten involved eyes from April 1983 to October 1985 (Table 11). Eight patients had a retinal detachment. One of these patients had a traumatic retinal dialysis in the fellow eye but had not yet developed a retinal detachment. The ninth patient had a traumatic tear in the retina but did not have a retinal detachment. Their average age was 26 years.

Two boxers were amateurs and seven were professional. Of the seven professional fighters three were champions of divisions while three others were highly ranked within their weight class. All had strong antecedent histories of trauma to the face and head. Each individual underwent a complete ophthalmologic examination, including history, visual acuity, applanation, tonometry, slit-lamp examination, and indirect

TABLE 11
A summary of individual patient data. (From MAGUIRE, BENSON, 1986).

Patient No./Age, yr	Preoperative Visual Acuity	Symptoms	Follow-Up Visual Acuity	Result
1/28	20/80	Decreased visual acuity	20/200	Reattached
2/29	20/30	Decreased visual acuity	20/30	Reattached
3/28	20/400	Photopsias	20/400	Reattached
4*/24	20/30	Unknown	20/25	Reattached
	20/40	Routine examination findings	20/25	Reattached
5/24	20/15	Decreased visual acuity	20/40	Reattached
6/26	20/20	Decreased visual acuity	Hand motion	Reattached
7/26	20/20	Visual field loss	20/25	Reattached
8/26	Hand motion	Decreased visual acuity	Hand motion	Detached
9/23	Hand motion	Decreased visual acuity	20/70	Reattached

* Patient 4 had both eyes affected

ophthalmoscopy. All affected eyes required operative intervention for their retinal tears and detachments.

Of the ten eyes treated, six were left eyes and four were right eyes. The best corrected visual acuity was 20/20 or better in three eyes (30%), and 20/25 to 20/40 in three eyes (30%), and 20/80 or worse in four eyes (40%).

Eight eyes (80%) had *rhegmatogenous retinal detachments,* while two were noted to have a retinal tear without evidence of subretinal fluid. Of the eight retinal detachments, six were secondary to a retinal dialysis, which is a linear tear along the vitreous base. Three detachments were secondary to giant tears, which are tears longer than 90 degrees. In four cases the macula was detached, resulting in visual acuities of 20/30 and 20/80 and movement perception (two patients).

Of the eight patients with retinal detachment, seven presented with decreased visual acuity. The eighth patient with retinal detachment noted peripheral visual field loss without decreased visual acuity. Only two patients reported *photopsias (light flashes).* One other experienced *entopsias (floaters).* For all of the pathological findings in the group, the retinal damage was detected on routine examination in only one case. This patient had a retinal dialysis without subretinal fluid. Considering this fact, it is likely that both groups in the debate were correct. BLONSTEIN's group of physicians probably did not see many eye injuries related to boxing. However, DOGGART must have seen some of the injuries that were not detectable on routine examination. His insistence that serious eye injuries are common in boxing is based on subsequent, more detailed examinations after injuries became problematic. *From this one may conclude that, initially, eye injuries are not easily detected in generalized examinations, and that a specialist must be consulted promptly in order to obtain the best possible results.* It is important to consider the potential for serious eye injuries as a significant risk of boxing, and not merely incidental or rare.

All retinal tears were, according the MAGUIRE and BENSON, clearly caused by blunt trauma to the eye. Six (60%) were located supero-nasally, three (30%) were inferonasally, and one was supero-temporally. In most cases there were other signs of vitreo-retinal trauma. Five patients had a definitive posterior vitreous detachment, three had a significant vitreous hemorrhage, and two had an avulsed vitreous base, which is known to be pathognonomic of blunt trauma. In only two eyes were there no associated vitreo-retinal findings. Four eyes had a traumatic cataract and one eye had a subluxed lens.

All affected eyes underwent surgery. The patients with retinal detachment were initially treated with a scleral buckling procedure and cryotherapy. Two patients underwent vitrectomy at the same time, because they had giant tears. The two eyes that had retinal tears without subretinal fluid were treated with cryotherapy alone. Four of the eight patients with retinal detachments required a second operation. Three had a re-accumulation of subretinal fluid due to insufficient closing of the retinal break and later, one had infection around his scleral buckle. The retina was successfully reattached in all patients except one who developed severe proliferative vitreo-retinopathy.

To study the risks of boxing further, MAGUIRE and BENSON, reviewed *Wills Eye Hospital* cases of r*hegmatogenous retinal detachment in* patients between the ages 20 and 30 years from July 1, 1883, to June 30, 1985. The authors found 127 cases, 70 of which had a history of ocular trauma. Nineteen (27%) of the 70 had a history of a blow to the eye by a fist. Of the 19, six were boxers. Thus, in a 24-months period, 5% (6/127) of rhegmatogenous retinal detachments, found in patients in the third decade of life, were in boxers.

According to MAGUIRE and BENSON there are three mechanisms by which blunt trauma can damage the eyes and associated structures: coup, contrecoup, and equatorial expansion. Coup injury refers to local damage at the site of impact and includes injuries such as lid lacerations, corneal abrasions, subconjunctival hemorrhage, local retinal necrosis, and choroidal rupture and hemorrhage.

According to MAGUIRE and BENSON there are two postulated mechanisms for peripheral reti-

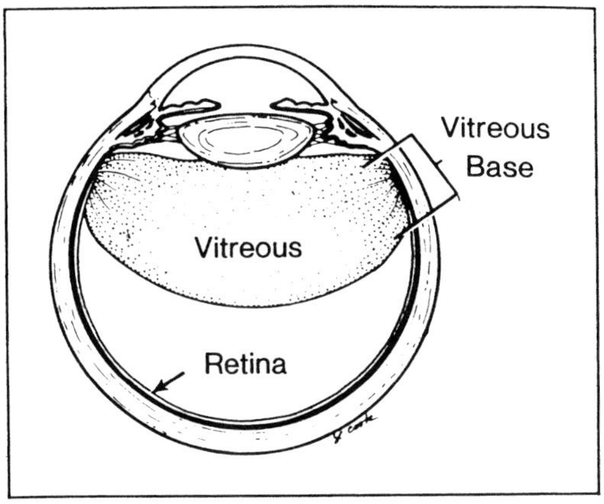

Fig. 154. Vitreous gel normally fills posterior segment of globe; its firmest attachment to peripheral retina is in area of ora serrata. This is vitreous base. In this eye, posterior vitreous detachment is shown. (From MAGUIRE, and BENSON, 1986).

nal tears following blunt trauma. Most tears are related to strong adhesions between the vitreous gel and the retina. The vitreous contains large numbers of collagen fibers that insert into the retina near its most anterior border, the ora serrata. The area where these fibers insert is called the 'vitreous base' (Fig. 154). Elsewhere, the vitreous can usually separate relatively easily from the retina if severe traction is applied. The vitreous is relatively elastic; however, when the eye is struck, as by a fist, the eye must compress rapidly along its antero-posterior axis. Since the globe has a fixed volume it must therefore expand in its equatorial plane or it will rupture. If the vitreous body does not have time to stretch, severe traction is placed on the area of the vitreous base, tearing it along its anterior or poste-

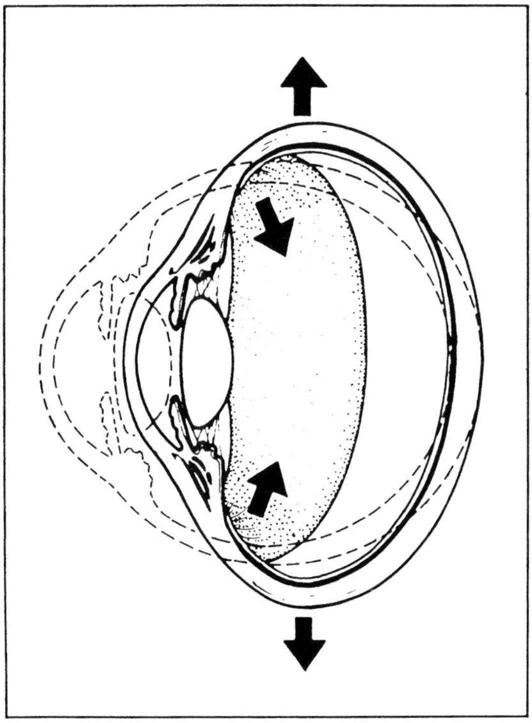

Fig. 155. Blow causes rapid compression along eye's anteroposterior axis. Resultant expansion along equatorial plane leads to traction at vitreous base. (From MAGUIRE, and BENSON, 1986).

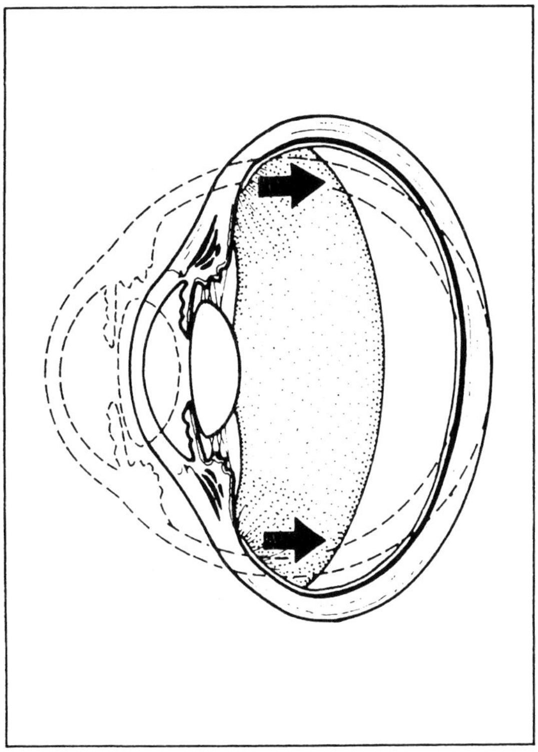

Fig. 156. Blow to eye directly pushes back on vitreous base, shearing it from wall of eye, thereby tearing retina. (From MAGUIRE, and BENSON, 1986).

rior margin (Fig. 155). Such tears are linear and are called 'dialyses'.

John SCOTT (oral communication to MAGUIRE and BENSON, June 1982) proposed a second explanation for tears at this location. He proposed that the blow to the eye directly pushed back on the vitreous base, shearing it from the wall of the eye and thereby tearing the retina (Fig. 156).

MAGUIRE and BENSON reported the surprising observation that most of the fighters were totally asymptomatic until visual acuity decreased noticeably. According to them, this is entirely consistent with the inherent characteristics of traumatic tears and detachment. Many detachments go, according to these authors, undetected for months or years, their slow advance documented only by multiple subretinal pigmented lines (demarcation lines), which are believed to take three months to form. MAGUIRE and BENSON noted that the anterior location of many of these tears, along with their initial lack of symptoms, causes them to be easily overlooked on routine examination.

MAGUIRE and BENSON quote the paper by DODEN and STARK (1974), who studied 247 patients suffering from significant trauma to the head but not to the eye. None had a retinal tear. MAGUIRE and BENSON concluded that the retinal tears in their group of boxers were caused by direct blows to the eye.

GIOVINAZZO et al. (1987) reported on a study conducted with the cooperation with the *New York State Athletic Commission*. Boxers applying for a new or yearlong renewal license were sequentially referred over for a *complete dilated ocular examination* over a two year period. Examinations of 74 boxers were conducted at the *Sports Vision Institute of the Manhattan Eye, Ear, and Throat Hospital*. At least one ocular injury was found in 66% of boxers. Vision-threatening injuries were defined as significant damage to the angle, lens, macula, or peripheral retina, and they were found in 58% of boxers. Nineteen percent of boxers had angle abnormalities. Nineteen percent of boxers had pathologic cataracts, over 70% of these were posterior subcapsular. Six boxers had macular lesions. A total of 24% of boxers had retinal tears. Attempts were made to identify risk factors in boxing that might be predictive for ocular injury. Variables included age, weight division, left- or right- handedness, total number of losses, and total number of bouts. There was a significant correlation between the total number of bouts and losses, and the percentage of retinal tears. College varsity athletes were selected as control. Significant differences were found between boxers and controls for the total number of injuries, total vision-threatening injuries, and the number of retinal tears.

The ages of the 74 examined boxers ranged from 16 to 42 years (mean age, 25 years). The distribution by race indicated that 45 (60%) of the boxers were black, 16 (22%) were Hispanic, 10 (14%) were white, one was Asian and one was Middle Eastern (Table 12). The years of boxing had a range of 1 to 24 (mean, 9 years) (Table 13). The number of amateur bouts ranged from 0 to 365 (mean, 51). Total bouts ranged from 3 to 373 (mean 61). Total losses ranged from 0 to 41 (mean 8). Fifty-eight boxers were right-handed and fought with a right dominant stance. Twelve fought left-handed and three fought variably right- and left-handed, switching their stances as the situation dictated. In the standard ring stance, the nondominant hand is held high and toward the opponent. The eye on the nondominant side is then closer to the opponent, but it is perhaps better protected by the nondominant hand held directly in front. None of the young athletes reported using drugs of any kind, including anabolic steroids.

Refractive Error: Results of examination of the boxer's *refractive states* showed 21 (28%) boxers to

TABLE 12
Demographic characteristics. (From GIOVINAZZO et al., 1987).

Characteristics Race	No. of patients (%)
Black	45 (60)
Hispanic	16 (22)
White	10 (14)
Other	2 (3)
Refractive error	
Myopia	21 (28)
Hyperopia	1 (1)
Emmetropia	52 (70)

TABLE 13
Career statistics. (From GIOVINAZZO et al., 1987).

	Range	Mean
Years boxing	1–24	9
Amateurs bouts	0–365	51
Total bouts	3–373	61
Total losses	0–41	8

be myopic, one (1%) hyperopic, and 52 (70%) boxers were emmetropic (Table 12).

External Examination: Significant pathologic findings in the *external and muscle examination* were found in nine (12%) boxers (Table 14). The presence of ptosis greater than 2 mm was found in three (4%) boxers. There were scars present along the superior orbital ridges of many boxers but only one so prominent as to cause *extensive madarosis and distortion of the eyebrow*. There were four motility disturbances.

Conjunctiva and Cornea: An asymptomatic, unilateral injection of the conjunctiva was noted in five (3%) patients.

Iris: Numerous types of iris type pathology were noted (Table 14). Sphincter tears were not uncommon, occurring in seven (9%) boxers. One patient was noted to have *early rubeosis at the pupillary margin*. This was the boxer with the longest standing retinal detachment.

Lens: The number of boxers with *cataracts* was 14 (19%). Ten (7%) of these were posterior subcapsular cataracts (Table 14). Only one posterior subcapsular cataract had vision reduced to the 20/40 level. Three boxers had bilateral posterior subcapsular cataracts. Two (3%) boxers had anterior subcapsular and nuclear sclerotic changes.

Angle: *Angle abnormalities* were noted in 14 (19%) boxers; six (8%) had bilateral changes (Table 14). Angle recession was seen in seven (9%) boxers.

Macula: Six patients had *increased pigmentation in the macular region* significant enough to be classified as pathologic (Table 14). Two patients had *macular cysts* with no evidence of hole formation. The visual acuity was 20/25 in each of these eyes. One patient had a macular hole that was diagnosed by deep white deposits as well as transmission of choroidal fluorescenec on fluorescein angiography. The vision in this eye was reduced to 20/200.

Peripheral Retina: A total of 18 boxers (24%) had significant *retinal breaks or frank detachments* (Table 14). Three (4%) boxers had atrophic holes smaller than our standard photograph. These may be considered by some observers as pathologic, but did not meet the criteria established for the study. Operculated tears were found in six (8%) boxers. All these eyes had evidence of vitreous traction. Multiple breaks were identified in three eyes. Horseshoe tears or flap tears were identified in 12 (16%) boxers. One tear was associated with lattice degeneration. Multiple breaks with either laser or cryopexy. One of these patients chose to stop boxing after treatment.

TABLE 14
Type and frequency of injuries. (From GIOVINAZZO et al., 1987).

Injuries	No. of Boxers (%)	No. of Eyes
External		
Ptosis	3 (4)	3
Motility	4 (5)	4
Decreased sensation	1 (1)	1
Scars	1 (1)	1
Iris		
Sphincter tears	7 (9)	1
Rubeosis iridis	1 (1)	8
Angle		
Recession	7 (9)	8
Anterior synechiae	3 (4)	4
Pigment dispersion	4 (5)	5
Lens-cataract		
Posterior subcapsular	10 (14)	13
Anterior subcapsular	2 (3)	2
Nuclear sclerosis	2 (3)	2
Macula		
Hole	1 (1)	1
Cyst	2 (3)	2
RPE degeneration	3 (4)	6
Peripheral Retina		
Atrophic hole	3 (4)	4
Operculated tear	6 (8)	9*
Horseshoe tear	12 (16)	12†
Detachment	2 (3)	2

RPE = retinal pigment epithelium.
* Three eyes with multiple tears.
† Eight eyes with multiple tears.

Retinal detachments were noted in two patients. One patient had a chronic detachment with rubeosis iridis. His vision was decreased to 20/200. He presented for a medical examination in an attempt to resume his boxing career. He refused treatment. A second patient had 20/25 vision, but a large retinal dialysis superiorly and a localized detachment. He was reattached with a scleral buckling operation, but lost to follow-up after two months. Lattice degeneration was noted in five (3%) of the eyes, but only two were noted to have retinal breaks.

Controls: Fifteen subjects were examined, and no pathologic lesions were found in the anterior segment. A pigmentary maculopathy was present in one control patient (Table 15). Peripheral retinal examination showed two control subjects with lattice degeneration.

Statistical Analysis

Both univariate and multivariate analyses were performed to evaluate which risk factors might be independently predictive of an ocular complication of boxing. Injuries to the structures that are potentially vision threatening, specifically the angle, lens, macula, and peripheral retina, were also individually evaluated, looking for possible risk factors in boxers and controls.

Chi-square analysis comparing injuries in study boxers and controls showed the frequency of boxer injuries to be significantly increased for total injuries, and for the total visually threatening injuries to the angle, lens, macula, and peripheral retina. The number of retinal tears was also significantly greater in boxers when compared to controls (Table 15).

TABLE 15
Comparison of injuries between boxers and controls. (From GIOVINAZZO et al., 1987).

Injuries	Boxers (n = 74)	Controls (n = 15)	Significance Level
Total ocular injuries	66%	7%	$P < 0.001$
Vision-threatening injuries (angle, lens, macula, and peripheral retina)	58%	7%	$P < 0.001$
Retinal tears	24%	0%	$P < 0.05$

Univariate analyses were also performed evaluating the variables of age, weight division, total bouts, total number of losses, years of boxing, and handedness as risk factors for the occurrence of boxing injuries. Using the Student's *t* test, the only significant correlation was between the presence of retinal tears and total losses multivariate analysis was performed to eliminate any confounding influences among the potential risk factors. Using a multiple logistic regression equation, the potential risk factors of age, weight division, handedness, years of boxing, total number of bouts, and total number of losses were evaluated against vision-threatening injuries to the angle, lens, macula, and peripheral retina.

Significant correlations were found between both the total number of bouts and total number of losses and the prevalence of retinal tears. The relationship is almost linear between losses and reti-

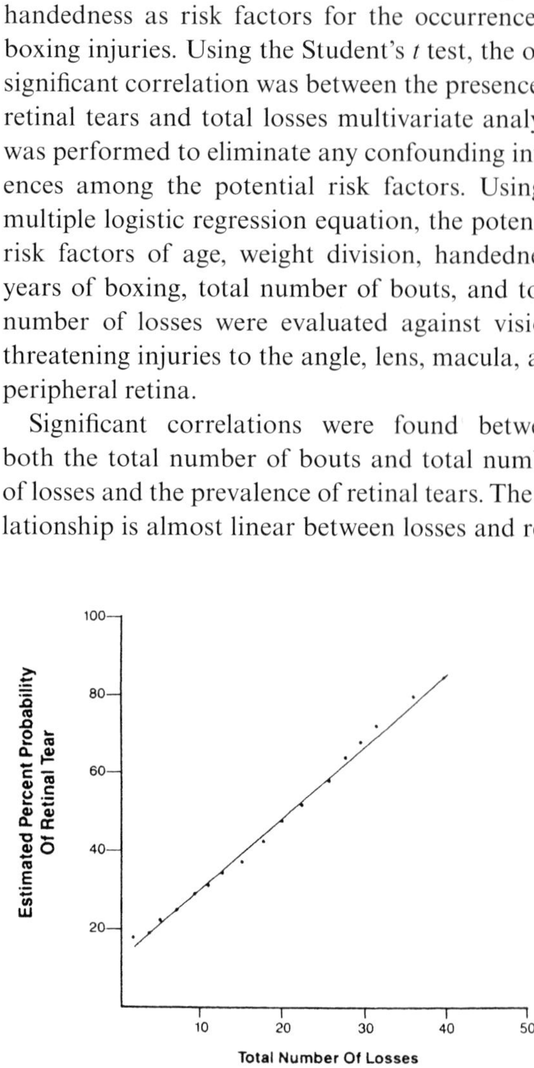

Fig. 157. Probability of retinal tears from total number of losses. Probability is based on multiple logistic regression analysis, R = 16 %. Regression coefficient was significant at P = 0.01. (From GIOVINAZZO et al., 1987).

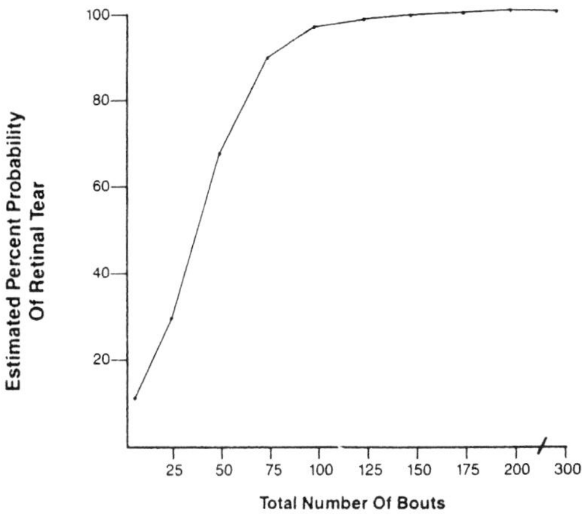

Fig. 158. Probability of retinal tears predicted from total number of bouts. Probability based on multiple logistic regression analysis, R = 40 %. Regression coefficient was significant at P = 0.01. (From GIOVINAZZO et al., 1987).

nal tears with a 20% prevalence of retinal tears after five losses (Fig. 157). After 75 bouts, the prevalence of retinal tears in the study boxers was over 90% (Fig. 158).

GIOVINAZZO et al. noted that boxing is a sport that by design inflicts acute physical injuries on its participants. They continue: "It is the only professional sport in which winning is predicated on intentional physical damage to one's opponent."

Ocular injuries from boxing have, according to the authors, been noted in the sports medicine and ophthalmologic literature, but with differing conclusions. Most reports are based on the personal observations of doctors who have clinical experience in treating individual boxers.

The study by GIOVINAZZO et al. is the first to study a prospective series of boxers on a large scale to determine the ocular complications of boxing. It is also unique in its design, in that it has evaluated the asymptomatic boxer who is still active in his profession. This study shows an extremely high prevalence of ocular injuries that can result from blunt trauma in boxing. Over half (52%) of all eyes and two-thirds (66%) of the boxers examined had pathologic changes.

There was a high (19%) frequency of angle abnormality in boxers ranging from recession and tears in the ciliary body to peripheral anterior synechiae and increased angle pigmentation. Although no patient had an elevated intraocular pressure, glaucomatous disc changes, or visual field defects, the potential for glaucoma exists in these boxers. PALMER et al. (1976) examined 55 former boxers and found a similarly high degree of abnormalities, but no instances of glaucoma. The reported prevalence of chronic glaucoma after a traumatic angle recession ranges from 7 to 9%. The onset of glaucoma may not occur until ten or more years after the injury. Many patients who have glaucoma from trauma have angle recession of at least 180° and perhaps an underlying tendency towards the development of chronic open angle glaucoma. A long-term, longitudinal study of former boxers would be helpful to establish the frequency of late onset glaucoma in eyes with angle injuries.

The overall prevalence of cataracts was 19%. This is an extremely high frequency of cataracts in a population with a mean age of 25. No cataracts were present in the control population. Over 70% of these cataracts were posterior subcapsular in nature, a finding that has heretofore not been fully recognized and reported. Cataract formation secondary to drug use was considered, but no boxer gave a history of drug use, including the use of steroids. Boxers are not inclined to use steroids because maintaining a minimum weight is of great importance to remain within a specific weight division. Only one cataract in this series of boxers was sufficiently severe to produce a visual decline to 20/40. However, it is not unreasonable to assume that there will be a future visual decline because of the progressive nature of posterior subcapsular changes. In young (age, 25-35 years) symptomatic boxers examined by the authors outside of this study, cataracts have been observed with visual acuity loss to the level of 20/200.

Several different traumatic manifestations were noted in the *macula* including *macular cysts, macular hole,* and *hyperplastic-atrophic pigment epitheli-*

opathy. Choroidal ruptures and *secondary neovascularized degeneration of disciform scarring* was not seen in these asymptomatic boxers. These manifestations have been seen by the authors in boxers not examined in this study.

A finding of considerable significance was the extremely high number of *retinal tears. Peripheral retinal tears* were found in 24% of boxers. There were no retinal tears found in the control population of varsity athletes. This finding suggests, according to GIOVINAZZO et al. that boxers are at high risk for retinal detachments developing.

The frequency of retinal tears correlated with the total number of bouts and the total number of losses of a boxer. This is probably, a reflection of the number of concussive blows to the eye, according to the authors. As a boxer suffers more blows, the tendency for retinal tears to develop increases. Fig. 157 of GIOVINAZZO et al. shows that the relationship between retinal tears and losses is almost linear. " The relationship between the total number of bouts and retinal tears rises steeply after six bouts (Fig. 158). After 100 bouts, nearly all of the study boxers had suffered retinal tears."

These findings have, according to GIOVINAZZO et al. implications for both the time and method for examining boxers. Because a significant percentage of injuries cannot be detected in standard exams, *every examination of a boxer should include a dilated vitreoretinal examination.* One of the study boxers best illustrates the importance of a dilated examination. The boxer, who was asymptomatic, was found to have a visual acuity of 20/25 in the right eye and 20/20 in the left eye. After dilation, a large supernasal retinal dialysis was observed in the right eye with a localized detachment. It was important to note that the macula was not involved in this detachment. This boxer was scheduled to fight four days after his examination. The fight was cancelled and the boxer's retina was repaired. GIOVINAZZO et al. conclude, that an examination limited to visual acuity testing and to an undiluted view of the fundus would not have detected this serious, vision-threatening injury.

An important question for ophthalmologists, according to GIOVINAZZO et al., concerns the risk for a boxer who has had a retinal detachment repaired and who wants to resume his boxing career. The medical advisory board of the *New York State and California Athletic Commission recommends* that no such boxer be permitted a license to box. The *Medical Council of Australia* has recommended that retinal detachments be a contraindication to a boxing career *(Proceedings, 101st Session of the National Health and Medical Research Council, Brisbane, Australia, The Health Aspects of Boxing, June 1986).*

GIOVINAZZO et al. noted that there is, however, no data available regarding the specific issue of reattachment after trauma either from boxing or from other contusive injury. Suggestive evidence can be found in the series of MAGUIRE and BENSON (1986) which indicated that a tendency for reattachment does exist in boxers. Four of their eight boxers with retinal detachments required a re-operation even though none of these eight individuals returned to boxing. GIOVINAZZO et al. know of three boxers who have experienced a reattachment of the retina after returning to the ring.

The macular complications of boxing offer, according to GIOVINAZZO et al., fewer prospects for treatment. Cystic and pigmentary changes, choroidal rupture and in particular, macular hole formation may be associated with acute or delayed, severe, and irreversible loss of central vision. As with angle recession, evaluation of the potential long-term visual loss from these macular changes should be determined by a longitudinal study of boxers.

One fortunate consideration in the study of GIOVINAZZO et al. was that many of the vision-threatening injuries were detected at a treatable stage. For example, most retinal tears were not associated with a retinal detachment. Other injuries such as angle recessions that may eventuate in glaucoma and posterior subcapsular cataracts are also treatable.

Also, the potential for visual decline exists in all of the boxers with vision-threatening injuries. Most subjects in this study had relatively good visu-

TABLE 16
Risk factors predictive of retianl tears in boxers. (From GIOVINAZZO et al., 1987).

Fisk Factors	Peripheral Tears
Age	NS
Weight division	NS
Total bouts	$P < 0.01$
Total losses	$P < 0.03$
Handedness	NS
No. of years boxing	NS

NS = not significant.

TABLE 17
Visual acuity. (From GIOVINAZZO et al., 1987).

Visual Acuity	
20/200	Retinal detachment
20/200	Macular hole
20/80	Amblyopia
20/40	Cataract
20/30	Cataract
20/25	6 boxers, 9 eyes

al acuity (Table 16) at the time of examination. Of the 74 boxers (Table 17), only two had visual loss to the 20/200 level. One boxer had ignored and denied treatment for a long-standing retinal detachment. Another boxer, a –10.00 myope, had a macular hole. A third boxer was found to be amblyopic with 20/80 vision. It was surprising, indeed alarming, to learn that two boxers were actually engaged in their sport with a visual acuity equivalent to legal blindness in one eye. It is possible that these boxers did not recognize their severe visual loss. Denial of their visual impairment may also be a factor because they were seeking re-licensure to box.

One fortunate consideration noted in the study was that many of the vision-threatening injuries were detected at a treatable stage. For example, most retinal tears were not associated with a retinal detachment. Other injuries such as angle recessions that may eventuate in glaucoma and posterior subcapsular cataracts are also treatable. This observation emphasizes the importance of an adequate and frequent ocular examination to detect severe but treatable complications.

The study of GIOVINAZZO et al. established that an extremely high prevalence of ocular injuries in boxers exists. The overall sample size, 74 boxers is, according to the authors, somewhat small and the selection methods, performed by the *New York State Athletic Commission,* were not under the examining ophthalmologist's strict control.

GIOVINAZZO et al. summarized that it is clear that participation in boxing results in a large number of ocular injuries. The effects of these injuries can be quite variable and are best exemplified by the story of two United States Olympic gold medal winners, the two 'Sugar' Rays. Sugar Ray LEONARD, Olympic gold medal winner in 1976, turned professional and became the welterweight champion. He allegedly earned several million dollars by the age of 26. Although he suffered a retinal detachment, he has been successfully treated. 'Sugar' Ray SEALES, Olympic gold medal winner in 1972, suffered bilateral retinal detachments. After more than 400 bouts and seven vitreo-retinal surgeries, he is now bilaterally blind and relative poor. GIOVINAZZO et al. observed that these two individuals illustrate the wide range and potentially devastating social and physical effects of boxing.

Damaged eyes have ended the careers of the professional boxers Archie SEXTON, Andy NEWTON, George ANGELO, Bruce WOODCOCK and Jack PETERSEN.

Pat VALENTINO, a former heavyweight contender, revealed, according to BARNES (1958), that he had undergone an eye operation nine months before his fight with Ezzard CHARLES in 1949, and was partially blind at that time. He also admitted that when he lost to Joe LOUIS two months later he was completely blind in the right eye.
The professional boxers 'Kid' GAVILAN, Henry ARMSTRONG, Jimmy CARTER, Frankie GENERO, 'Speedy' DUDO and Vince DUNDEE were partial or total blind. Damaged eyes have ended the careers of fighters Archie SEXTON, Andy NEWTON, George ANGELO, Bruce WOODCOCK and Jack PETERSEN.

GIOVINAZZO et al. (1995) presented a review of

eye injuries in boxers. This review is based on extensive personal experience and thorough knowledge of the literature. Along with the summary of the English ophthalmologist DOGGART (1955), this is a major contribution to the literature of eye injuries in boxers. The review of GIOVINAZZO et al. is therefore discussed in detail.

GIOVINAZZO et al. (1995) begin their review of eye injuries in boxers with the statement:

"Boxing is a preventable cause of blindness."

GIOVINAZZO et al. discuss in general and in particular:

Injuries to the Eye and the Ocular Contents

Possible boxing injuries to the eye are, according to GIOVINAZZO et al., ruptured or perforated globe, hyphema, cataracts, retinal tears or detachments, and macular injuries.

Ruptured or Perforated Globe: Although thumbing (which transmits the greatest possible force to the globe) has resulted, according to GIOVINAZZO et al., in one incident of a ruptured globe, a normal eye would usually not be perforated or ruptured by the blunt forces involved in the sport.

Hyphema: Hyphema, or bleeding into the anterior chamber is, according to GIOVINAZZO et al. common and is due to tearing of vessels at the iris root. Often associated with hyphema are pigment dispersion and recession and scarring (anterior synechiae) of the iridocorneal angle and the trabecular meshwork; these may result in glaucoma and loss of vision years after the injury.

Cataracts: This opacity of the lens resulting in decreased vision is, according to GIOVINAZZO et al. common, most often posterior subcapsular in boxers, and is often treated by surgery to remove the opaque lens and insert a plastic intraocular lens implant. The authors continue to write that a return to boxing after cataract surgery should be discouraged.

Retinal Tears and Detachment: Blows may cause, according to GIOVINAZZO et al., either tears in the retina or a complete detachment. Although retinal tears and retinal detachments are undeniably linked, as GIOVINAZZO et al. note, the difference in prognosis and severity justifies the separation. While the State of California prohibits any boxer from obtaining a license if he has suffered a retinal detachment or a retinal tear, GIOVINAZZO et al. believe that many boxers with successfully repaired retinal detachments may continue their careers. They go on to conclude that a treated tear should almost never compel a boxer to retire.

Retinal Detachment is caused, according to GIOVINAZZO et al., by retinal tears and leaking of liquefied vitreous through the tear into the potential space between the retina and the retinal pigment epithelium. Large defects of the peripheral visual field may be present, as well as loss of central vision from the detachment of the central retina, the macula, as GIOVINAZZO et al. note.

Macula Injuries: Compression of the globe can cause, according to GIOVINAZZO et al., a rupture of the choroid and create an underlying scar that reduces central vision.

Injuries to the Central Visual Pathways: The optic nerves, optic chiasm, optic tracts, and visual cortex are, according to GIOVINAZZO et al., susceptible to permanent damage from repeated blows to the head. Cortical blindness may result from a sharp blow to the head (such as the impact from falling onto the canvas at the time of a knock out).

GIOVINAZZO et al. (1995) treat Olympic and professional boxing separately, because they are dissimilar sports.

The authors noted that unfortunately very little data are available concerning eye injuries to Olympic boxers. The only study available is one conducted by Dr. David W. SIME (personal communication to GIOVINAZZO et al. 1995). Dr. SIME examined the eyes of 13 Olympic boxers in 1984. Of the 13, three had retinal holes or tears, probably as a result of boxing, and one had an unrelated amblyopic eye to 20/400. U.S. Olympic boxers have not have any regular mandated eye examination program, nor was there any follow-up on the meeting of the U.S. Olympic Committee task force on head injuries in amateur boxing held on August 20, 1983.

In regard to the instructional program at the U.S. Military Academy at West Point, GIOVINAZZO et al. write that it is fashioned after the Olympic program. Although the total injury rate for the Academy's program seems low (less that 4% injuries in 2.100 cadets who received boxing instruction between 1983 and 1985), the incidence of eye injuries is impossible to evaluate, since no asymptomatic participants had the benefit of an adequate ophthalmological examination for this study (WELCH et al. 1986). Colonel John RIFILE is now conducting a study at West Point to determine the incidence of ocular problems in the boxing program there.

CARTER and PARKE (1987) published the case of unusual retinal tears in an amateur boxer.

> A 30 year old man presented with poor vision in his left eye. He had boxed for ten years in an amateur boxing association. No eye examination had been done prior to or during his brief boxing career. He did not wear headgear because his boxing coach wanted him to 'feel the punches.' Six months prior to his initial evaluation, he noted temporal visual field loss in his left eye progressing to loss of all formed vision. He was unaware of field loss in the right eye.
> On presentation, his best corrected visual acuity was 20/40 OD, and he had motion in the left eye. There was a marked afferent pupillary defect in the left eye. Examination of visual fields revealed central and infero-nasal fields remaining in the right eye. Intraocular pressures were 51 mm Hg OD and 5 mm Hg OS. Gonioscopy revealed recessed angles in both eyes.
> *Ophthalmoscopic examination* revealed multiple retinal tears in both eyes. In the right eye, there were radial tears extending anteriorly from the disc and an inferior circumferential 90° giant tear. The retina was detached in the supero-nasal and inferior quadrants anterior to the arcades. There was a total retinal detachment in the left eye with an enormous tear through the fovea and a large 60° tear extending posterior to the equator in the infero-nasal quadrant. Epi-retinal membranes were present in a pattern of C3 proliferative vitreo-retinopathy.

This case demonstrates unusually severe retinal trauma in an amateur boxer, not only by the multiplicity of large posterior tears, but also by the tear transversing the macula of the left eye where the retina is thickest.

The coach's advice to forgo headgear in order 'to feel the punches' reflects appalling ignorance and professional incompetence for one in a position of leadership and trust. Boxing coaches must be better educated in physiology and the specific risks of boxing trauma. With such poorly informed coaches, it is no wonder that boxers sometimes want to take additional punishment voluntarily, to become accustomed to it, as if the head can somehow become resistant to injury through repetitive blows.

ENZENAUER and MAULDIN (1989) examined boxing-related ocular injuries in the *United States Army, 1980 to 1985*. The authors noted that:

> "Boxing is a controversial sport in which neurological and ocular injuries can be particularly devastating."

Currently, competitive boxing is promoted throughout the military. Ocular injuries attributable to military boxing were studied as a subset of a larger study of boxing-related injuries.

The authors reviewed records of hospitalizations for boxing related injuries in the *US Army Medical Treatment Facilities (MTF), worldwide,* between January 1, 1980 and December 31, 1985. The subset of patients with ocular injuries was considered separately.

During the six-year period under consideration (1980 through 1985), there were 401 admissions for boxing-related injuries among active duty sol-

TABLE 18
Ocular injuries among 401 boxing-related injuries in the US Army, 1980 to 1985. (From ENZENAUER, MAULDIN, 1989).

	No.	Patients
Primary ocular diagnosis	14	(3.5%)
Secondary ocular diagnosis	7	(2.0%)
Total	22	(5.5%)

(From R.W. Enzenauer and R.J. Mauldin 1989)

TABLE 19
Ocular injuries other than primary admission diagnosis. (From ENZENAUER, MAULDIN, 1989).

Associates with:
Cerebral laceration/contusion
Contusion head/neck
Intracranial injury (2 cases)
Fracture of malar bones
Fracture of facial bones
Headache

(From R.W. Enzenauer and R.J. Mauldin 1989)

diers and cadets. More that 5% of patients (22) had some ocular injury at the time of admission, and almost 4% (15) had an ocular injury as their primary cause for admission (Table 18). Seven patients (2%) had ocular injury associated with a more major injury, which was the primary cause for admission. When the ocular injury was not the primary admitting diagnosis, it was associated with cerebral laceration/contusion, contusion of the head/neck, intracranial injury, fractured malar bones, and headache (Table 19).

Table 20 of the paper of ENZENAUER and MAULDIN shows the number of cases by *length of hospitalization ('bed days') and* by *length of disability ('sick days').*

For ocular injuries, the mean length of hospital stay was 8.1 days and the mean length of disability was 17.8 days. They were both significantly larger than the average 'bed days' and 'days' for all other boxing-related injuries.

ENZENAUER and MAULDIN noted that the association of boxing and ocular injury is not new.

> "In 1955 DOGGART eloquently described the 'impact of boxing upon the visual apparatus,' ending with an impassioned plea that boxing would soon 'be as obsolete as galley slaves or the chain gang.'" The findings of DOGGART are quoted in detail on p. 164.

The authors quoted several physicians who have alleged that serious eye injuries are uncommon among boxers. WHITESON (1992) noted that ocular injuries, other than simple lid/brow lacerations, are rare. Dr. Edwin CAMPBELL, of the New *York State Athletic Commission stated:*

TABLE 20
Number of admissions, inpatient 'bed days,' and disability 'sick days' for ocular injuries from boxing, US Army, 1980 to 1985.

	ICD	No.	Bed Days		Sick days	
	Code	Admissions	Total	Average	Total	Average
Total injuries – all sites	–	401	2,056	5.1	3,581	8.9
Total eye injuries	–	22	179	8.1	391	17.8
Eye – Primary Diagnosis	–	15	153	10.2	353	23.5
Open wounds, ocular adnexa	8708	1	2	2.0	1	2.0
Contusion of eye, unspecified	9219	1	1	1.0	1	1.0
Contusion of orbital tissues	9212	2	3	1.5	3	1.5
Optic nerve injury	9500	1	2	2.0	2	2.0
Superficial injury to cornea	9181	3	6	2.0	6	2.0
Laceration of eye, unspecified	9714	1	2	2.0	2	2.0
Contusion of eyeball	9213	1	6	6.0	6	6.0
Contusion of eyelids	9211	1	1	1.0	1	1.0
Disorders of globe/blindness	3696	1	124	124.0	330	330.0
Fracture of orbital floor	8026	3	6	2.0	13	4.3
Eye – Secondary Diagnosis	–	7	26	3.7	38	5.4
Contusion of orbital tissues	9212	3	12	4.0	12	4.0
Corneal ulcer	3702	1	7	7.0	1	7.0
Fracture of orbital floor	8026	1	2	2.0	14	14.0
Disorders of orbit NEC	3768	1	1	1.0	1	1.0
Avulsion of eye	8713	1	4	4.0	4	4.0

(From R.W. Enzenauer and R.J. Mauldin 1989)

"since we began using thumbless gloves in 1981, we haven't had a series of eye injury."

ENZENAUER and MAULDIN wrote

"Such naive statements only serve to lull the boxers, promoters, and fans into a false sense of security that the threat to vision among boxers is small, when in fact, entirely the opposite is true. Serious eye injuries are not uncommon in boxing."

ENZENAUER and MAULDIN continue that the continued military promotion of competitive boxing needs to be addressed.

"Proponents of boxing in the armed forces argue that boxing develops physical courage while teaching the boxer how to surmount fear. These alleged benefits of boxing need to be weighed against the definitive medical risks associated with participation."

ENZENAUER and MAULDIN continued:

"A proposal for a prospective study of boxing-related ocular injuries has been submitted to the *command and staff at the United States Military Academy,* though eventual approval is unlikely. Boxing is as much an integral part of cadet life as physical hazing was a quarter of a century ago" (ENZENAUER and MAULDIN).

TOCZOLOWSKI et al. (1991) performed *ophthalmological examinations* in 139 boxers. They showed that the injuries concerned most frequently the ocular adnexa and not the eye globe itself. Only two contestants who practiced boxing for more than ten years exhibited peripheral opacities of the lens and pale foci of pigment at the fundus periphery connected with a deepening of the filtration angle; this could have been caused by contusion of the eye. The introduction of headgear was clearly shown to reduce the frequency of eye injuries in boxing.

WEDRICH et al. (1993) examined a group of 25 active, asymptomatic, amateur boxers to evaluate the nature and incidence of ocular pathologic conditions in amateur boxing. An approximately age-matched group of 25 men who were not boxers was used as a control group. The number of fights varied from 1 to 220 (mean 29.4) and the number of spar rounds from 200 to 600 (mean 192.5). Visual acuity was 20/20 in all subjects of both groups, except for one boxer with keratoconus. Intraocular pressure was below 20 mm Hg in all eyes. In 19 (76%) boxers, pathologic anatomic findings were attributed to contusion trauma. Lesions of the anterior eye segment included injuries of the lid in three eyes (12%), angle abnormalities in five (20%), and slight lens opacities in five (20%). Posterior vitreous detachment was observed in three (12%) boxers and peripheral retinal scars were seen in 15 (60%). In 6 (24%) eyes, retinal tears or atrophic holes were detected. In the seven (28%) boxers with lesions of the anterior eye segment, the posterior segment was also affected in six (85%). In the control group, an atrophic hole was found in one (4%) patient with no other ocular pathologic findings.

The *American Academy of Ophthalmology* has adopted guidelines for reform to prevent ocular injuries in boxing, which are presented in the next paragraph:

Reforms for the Prevention of Eye Injuries in Boxing Policy:

"The American Academy of Ophthalmology recognizes the high incidence of ocular injuries in boxing and the lack of appropriate guidelines and regulations to prevent these injuries and protect participants in the sport. Severe reform measures with strict enforcement are necessary to reduce ocular injuries, promote early diagnosis and treatment, and prevent visual disability in the sport of boxing."

Background:

"Serious eye injuries, such as retinal detachment, have long been known to occur in boxing (1, 2). Recent research has shown that even asymptomatic boxers, those without any visual complaints, have a high incidence of injuries that are potentially visually threatening (3). Over half of the boxers examined had injuries that could cause a loss of vision, either acutely or insidiously."

"Many of these injuries, when detected at an early stage, can be effectively treated before the onset of visual loss. This emphasizes the importance of frequent and thorough eye examinations for all boxers. In addition, examinations, including a dilated fundus examination, should be performed whenever ocular injury may have occurred."

"The goal of frequent examinations is not only to diagnose and treat early injuries, but also to detect those boxers who have already suffered visual loss and to prevent any further risk to their vision by continued participation in the sport."

Guidelines:

Examination of Boxers

"There should be specific requirements regarding ocular examinations of boxers as follows:

An initial complete eye examination should be required before licensure for either amateur or professional boxing. This initial examination should include visual acuity, visual fields, slit-lamp biomicroscopy, intraocular pressure measurement, gonioscopy, and a dilated vitreoretinal examination including indirect ophthalmoscopy with scleral depression by an ophthalmologist."

"(b) A repeat, complete examination should be performed after one year, six bouts, two losses (GIOVINAZZO et al. 1987), the stopping of a fight after an eye injury, or at the discretion of the ringside physician."

"(2) A mandatory, temporary suspension from sparring or boxing should be required for specific ocular pathology. When a retinal tear is identified and treated, a minimum 30-day suspension from the ring should be enforced. When a retinal detachment is treated and if a boxer is to resume his career, a minimum of 60 days mandatory suspension should be enforced. Suspensions may be individualized in consultation with the medical advisory board of each state athletic commission."

"(3) The minimal visual requirements for boxing should be:

A corrected visual acuity of 20/40 or better in each case.

(b) A full central visual field acuity of not less that 30m degrees in each eye."

"(4) An ophthalmologist should serve on each state and medical boxing advisory board to advise on a boxer's eligibility for licensure."

Safety Equipment

"(5) There should be adopted by state and local licensing agencies adequate safety equipment to minimize ocular injuries. The thumb of most regulation boxing gloves is small enough to penetrate the orbital rim. It is an important cause of blunt ocular trauma sustained in the boxing ring. The injury can be eliminated by the use of the thumbless boxing glove. The use of the thumbless boxing glove should be mandatory not only in all amateur and professional bouts, but also in all sparring matches."

Regulation

"(6) There should be established in the United States a 'National Registry of Boxers' for all amateur and professional boxers including sparring mates. A computer-based, central registry of this kind would be effective in recording licensed bouts and specifically noting technical knock outs, knock outs, and significant ocular injuries. When a bout is stopped because of an ocular injury, proper records and documentation should insure adequate examinations of injured boxers."

"(7) There should be a program of training and re-certification of ringside physicians in the proper identification of serious eye injuries sustained in the ring. The diagnosis of serious eye injuries and specific criteria for suspension of a bout should be emphasized during this training period. Seminars should be planned and conducted in cooperation with the American Association of Ringside Physicians and the American Academy of Ophthalmology."

"(8) There should be adopted by state and local licensing agencies a uniform code of safety requiring that a boxing match shall be stopped upon the occurrence of specified ocular symptoms or injuries such as visual field loss and blood in the anterior or posterior segment of the eye."

Summary:

"A high incidence of ocular injuries occurs in boxers, but many of these injuries are treatable if detected early. A complete, dilated eye examination by an ophthalmologist is necessary to diagnose ocular injuries before they cause visual loss. Uniform regulations, enacted on a nationwide basis, mandating required eye examinations and safety equipment, will reduce the frequency and severity of ocular injuries in boxing."

Footnotes:

(1) JH DOGGART: 'The impact of boxing on the visual apparatus.' Arch Ophthalmol 54:161-169, 1955

(2) JI MAGUIRE and WE BENSON: 'Retinal injury and detachment in boxers.' J Am Med Ass 255:2451-2453, 1986

(3) VJ GIOVINAZZO, LA YANNUZZI, JA SORENSON, DJ DELROWE and EA CAMBELL: 'The ocular complications of boxing.' Ophthalmology 94:587-596, 1987

Developed by: Eye Safety and Sports, Ophthalmology Committee

Approved by: Board of Directors, June 23, 1990

Eye Injury suffered by President Theodore ROOSEVELT

Theodore ROOSEVELT boxed as a lightweight at Harvard University.

ISENBERG (1988) noted:

"Young ROOSEVELT never distinguished himself athletically while wearing the crimson; he was not a natural and had little interest in organized sport. Nevertheless he possessed a flaming competitive urge that would help seal a bond of friendship with SULLIVAN in later life, as well as with other prize fighters such as Mike DONOVAN, Battling NELSON, and Bob FITZSIMMONS. For years T. R. strongly supported either professional or amateur boxing as long as the sport was free of corruption, and as police commissioner of New York he heartily approved of an effort to get boxing clubs established in the city" (ISENBERG 1988).

Theodore ROOSEVELT commented on boxing when he was *Civil Service Commissioner*, in 1890:

"The people who attend it and make a hero of the prize fighter, and, – excepting boys who go for fun and don't know any better – to a very great extent, men who hover on the borderlines of criminality; and those, are not speedily brutalized, and are never rendered more manly. They form as ignoble a body as do the hundred frequenters of rat-pit and cockpit. The prize fighters and his fellow professional athletes of the same ilk are, together with their patrons in every rank of life, the very worst foes with whom the cause of general athletic development has to contend."

Despite of these statements, Theodore ROOSEVELT actively participated later as governor and president, sparring with 'Professors of Boxing' and former professionals. One of his regular sparring partners was Mike DONOVAN, former 'bare knuckle' middleweight champion of the world, who later became a boxing instructor in New York. DONOVAN became his sparring partner when ROOSEVELT was Governor of the State of New York and later president of the United States. In one of the sparring matches ROOSEVELT lost nearly the visual acuity of one eye. (DONOVAN, M., *The ROOSEVELT That I Know*, New York, 1909, Chapter 1; WAYNE, A., (ed.) *The Autobiography of Theodore ROOSEVELT*, Cond. Ed. New York, 1958, pp. 30-31)

Hearing Impairments

While there is a vast body of literature concerning the results of single head injuries, only a few papers have been published on the effects of repeated head injuries and the effects on the hearing organ. While head injuries usually occur as isolated accidents in the average person's life, boxers are continually subjected to numerous repeated head blows, spanning the years of their involvement in boxing. Often, during the last years of a boxing career, slowed responses reduce the ability to defend against head blows, so that head trauma is even more frequent. Damage to the hearing apparatus is of particular interest to otologists, since the hearing losses in these cases are caused by repeated traumatic head injuries.

Case Histories of Hearing Loss in Boxers

Cases of vestibular abnormalities and hearing loss in boxers were published and discussed by De KLEYN (1941), CZERNER (1957), OTTLITZ (1962), WILKE and OEKEN (1964), PUXEDDU et al. (1964), PAKOVSKI (1965), MONTANARO (1965), BRETKOWSKI (1968), PAULSEN and HUNDHAUSEN (1971), MOSER et al. (1980), SOUTSU (1988) and KREJCOVA and CZERNY (1989).

Disturbances in hearing were first reported by De KLEYN (1941). The majority of the boxers whom he examined revealed disturbances of hearing at the 4.000 Hz frequency.

CZERNER (1957) reported impairment of hearing as late damage in 75% of all examined boxers.

OTTLITZ (1962) carried out *cochleo-vestibular examinations* in four groups of individuals: (1) Twenty-six adolescent amateur boxers; (2) 50 adolescents of the same age group as in (1) who did not participate in boxing; (3) 20 adult amateur boxers; and (4) 21 adult wrestlers.

In the group of adolescent boxers there were no noticeable pathological findings; there existed no difference between groups 1 and 2.

In the adult group of amateur boxers six audiograms revealed typical curves of the acoustical trauma type, indicating impairment of hearing on the upper range with a dip at C^5.

WILKE and OEKEN (1964) attempted to interpret the otological aspects of a cerebral knock out, since some considered a *commotio labyrinthii* sufficient to produce a cerebral knock out. Immediately after the knock out, each of the boxers was carried into an examination room adjacent to the ring, and examinations of the cochlear and vestibular organs were performed. The results indicate, according to WILKE and OEKEN, that a commotio labyrinthii does not play a role in the pathophysiology of a cerebral knock out. A direct pressure effect of the labyrinthic fluid for the genesis of a cerebral knock out must be rejected, according to these.

MONTANARO (1965) examined 54 amateur and professional boxers, who had participated in at least 20 and not more than 120 bouts. Each of them had suffered one knock out, and there were pathological findings in 54 cases.

He described *three categories:* (1) In 26 cases the audiogram revealed a unilateral reception deficiency with a decrease of 4.000 Hz; (2) in 12 cases the audiogram showed a bilateral depression in the curves; and (3) in 16 cases there were various different disturbances.

Twenty of the examined boxers showed symptoms of organic brain damage.

Impaired hearing after blunt injuries of the head without skull fractures was first demonstrated using audiometric curves after World War II.

PAKOVSKI (1965) conducted audiometric examinations on 117 amateur boxers who were divided into *four groups, depending* on the *length of their boxing careers*. The results showed a disturbed sound conduction at frequencies of 128, 256, and 512 Hz, which was associated with an obstruction of the Eustachian tubes. The disturbance of sound perception at the frequency of 4.096 Hz was due to the sequelae of concussions of the internal ear, as the result of blows to the head. The author came to the conclusion that boxing must be considered a dangerous sport in regard to the state of the organ of hearing.

Series of PAULSEN and HUNDHAUSEN

PAULSEN and HUNDHAUSEN (1971) examined 43 boxers audio-metrically; 35 were amateur, the remaining eight were professional boxers. Twenty-five of them were between 20 and 30 years of age, eleven were under 20 years, and seven over 30 years. Apart from the question of the frequency and extent of these hearing losses, their reversibility and any typical features of the characteristic audiometric curves of the hearing pattern, the injuries to the hearing function are interesting because they are the result of repeated head trauma.

The boxers in this study were grouped by weight into the following classes: (1) *Feather-lightweight* (up to 61 kg); (2) *half welter-welterweight* (from 62 to 67 kg); (3) *light middleweight-middleweight* (68 to 72 kg); (4) *light-heavyweight* (73 to 80 kg), and (5) *heavyweight* (above 80 kg).
The number of bouts ranged from one to 400. Thirteen boxers were involved in between one and eight bouts, twelve boxers between ten and 20 bouts, nine boxers between 30 and 90 bouts, six boxers between 100 and 200 bouts, and three boxers between 300 and 400 bouts.
Threshold audiometric recordings were taken of all the boxers, as well as *"above threshold" audiometric curves*.
The greatest difficulty for all subjects, according to PAULSEN and HUNDHAUSEN, was convincing the boxers to submit to an audiometric examination. More than the boxers, however, the trainers and the representatives of the *Amateur Boxing Association of the State of Schleswig-Holstein, Germany*, rejected audiometric examinations. They argued that the yearly medical routine examination should be sufficient, and that medical publications concerning the sport of boxing were considered detrimental, especially with regard to young recruits.
It was only with great difficulty that, finally, 43 boxers could be examined. Only those boxers whose hearing impairments could be related only to boxing were examined.

The audiometric findings were evaluated using the following criteria:

(1) Extent of hearing impairment in different weight classes. Body weight influences the potential force of a blow. A heavyweight boxer has greater striking force potential than a flyweight does. Since boxing opponents are matched by weight class, the average striking force will increase along with the weight class.
(2) The extent of hearing impairment in relation to the total number of bouts. The longer a boxer is active, the more frequently he is hit by his opponent against the head. It has to be taken into consideration that the boxer is not only subjected to blows against his head in matches and meets, but also during training and while involved in sparring bouts. Each boxer is subjected to repeated blows against his head in every week of his career, since boxers normally train at least twice a week.

Audiometric Findings in the Different Weight Classes

1) *Feather-lightweight* (up to 61 kg – 8 boxers)
Two boxers in this class revealed no hearing losses; the remaining six exhibited hearing losses in the frequencies between 1.000 and 8.000 Hz (Fig. 159).
(2) *Half-welter – welterweight* (62 to 67 kg – 13 boxers)
Six audiograms were normal, but the remaining seven showed hearing losses in the high frequencies. In the mean curves a slight hearing impairment occurred in the higher frequencies (Fig. 160).
(3) *Light middle – middleweight* (68 to 72 kg – 10 boxers)
One boxer had a normal audiogram; in two cases there was a bilateral, continuous decrease in the high frequencies. The curve of a boxer with 300 amateur and 98 professional bouts dipped on the left side at 1.000 Hz, and on the right side at 1.500 Hz to 70 dB hearing impairment and revealed a dip at C5 with 50 dB hearing loss. The remaining seven boxers revealed hearing impairments in the middle and high frequencies. In the mean curve

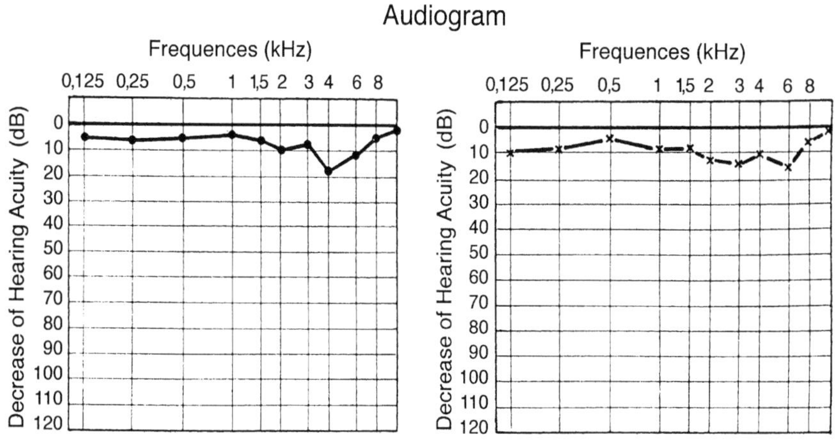

Fig. 159. Audiogram. (From PAULSEN, and HUNDHAUSEN, 1971).

(Fig. 161), slight hearing impairments can be seen in the middle and higher frequencies.

(4) *Light-heavyweight* (73 to 80 kg – 6 boxers)

This group also revealed hearing losses in the middle and higher frequencies. In three cases the left ear was more involved than the right. The mean curve revealed a trough-shaped hearing loss above 2.000 Hz, with accentuation of the frequencies between 3.000 and 4.000 Hz (Fig. 162).

(5) *Heavyweight* (above 80 kg – 6 boxers)

The curves of the six boxers in this class revealed depressions between 3.000 and 6.000 Hz in two cases. The remaining four audiograms showed a continuous hearing impairment above 1.000 or 1.500 Hz. This is shown in the mean curve as well (Fig. 163).

Threshold Audiometric Findings of Boxers Related to the Total Number of Bouts

(1) *1 to 9 bouts* – 13 boxers

This group was composed of four boxers in

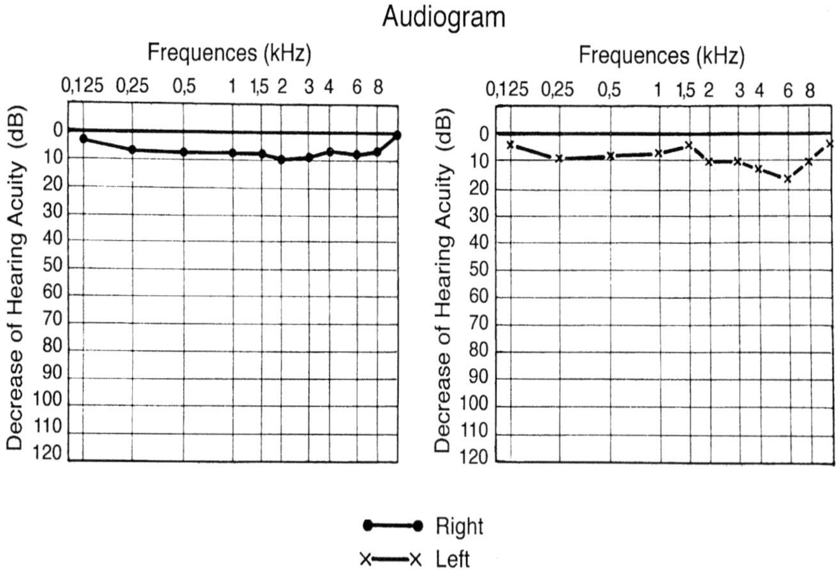

Fig. 160. Audiogram. (From PAULSEN, and HUNDHAUSEN, 1971).

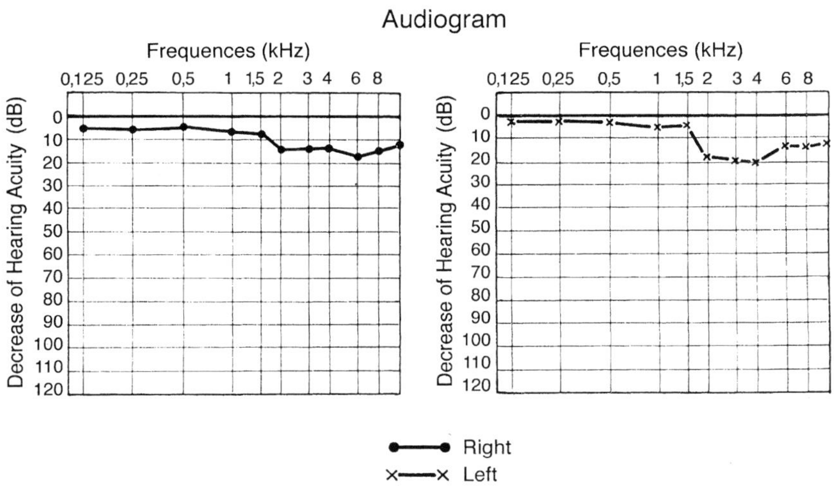

Fig. 161. Audiogram. (From PAULSEN, and HUNDHAUSEN, 1971).

weight class one (WC 1), six in WC 2, two in WC 3, and one in WC 4. The audiogram of seven boxers revealed no hearing impairments with more than 10 dB. The remaining boxers showed in their audiograms only a slight hearing loss in the higher frequencies. This can be seen in the mean curve (Fig. 164) as a flat trough between 1.500 and 8.000 Hz.

(2) *10 to 29 bouts* – 12 boxers

This group was made up of two boxers of WC 1, four of WC 2, two of WC 3, two of WC 4, and two of WC 5. Two boxers revealed practically no hearing loss. The audiogram of the remaining boxers revealed a very slight hearing loss in the lower frequencies, on the right side more than on the left, and slight but definite hearing impairments in the form of troughs between 1.500 and 12.000 Hz. In single cases isolated dips can be seen. The mean curve (Fig. 165) revealed a slight but definite hearing loss in trough form between 1.500 and 12.000 Hz, mainly in the higher frequencies.

(3) *30 to 99 bouts* – 9 boxers

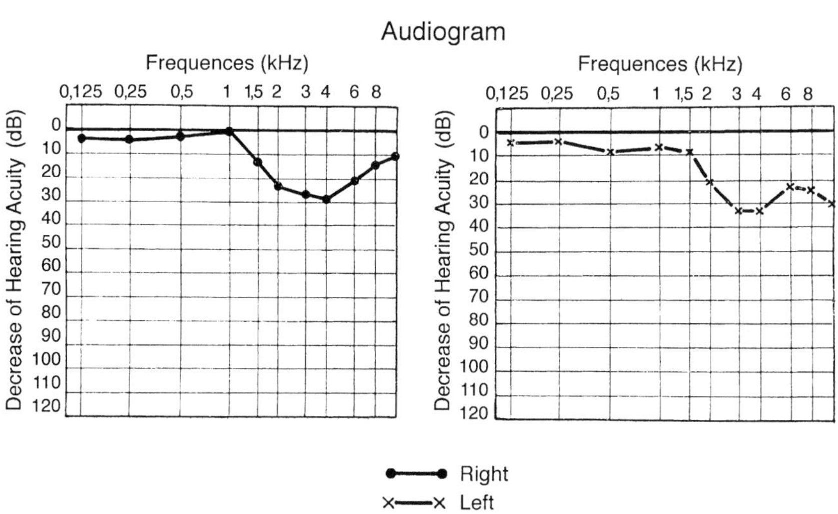

Fig. 162. Audiogram. (From PAULSEN, and HUNDHAUSEN, 1971).

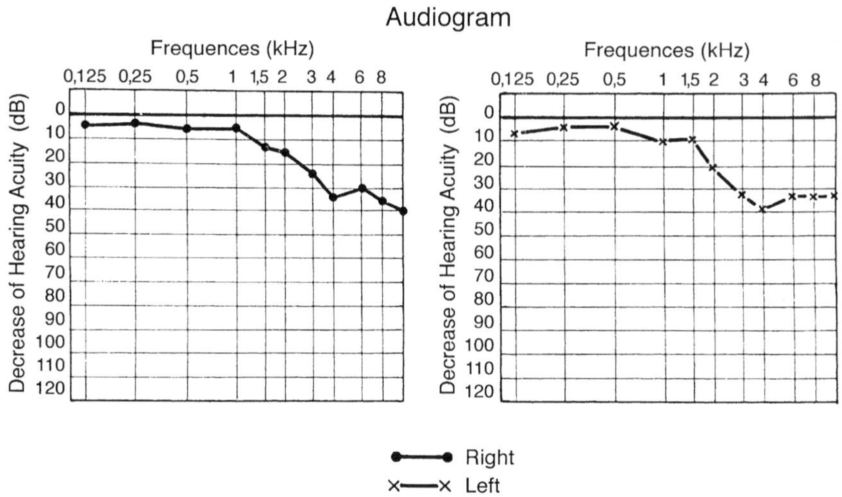

Fig. 163. Audiogram. (From PAULSEN, and HUNDHAUSEN, 1971).

There were two boxers in the group of WC 1, two of WC 2, 3 of WC 3, and two in WC 5. In every case there was a hearing impairment of more than 10 dB. Altogether, the hearing loss is more pronounced. In one case, there was a continuous dip in the high frequencies, in another case a steep dip at 1.500 Hz with 75 dB hearing loss at 3.000 Hz. Single audiograms had dips in other flat troughs. The mean curve is characterized by slight to medium hearing losses in the middle but mainly in the higher frequencies. Bilaterally a slight trough at C5 is visible. The hearing impairment on the left side is more pronounced than on the right (Fig. 166).

(4) *100 to 200 bouts* – 6 boxers

This group consisted of one boxer in WC 2, two in WC 3, two in WC 4, and one in WC 5. All audiograms revealed hearing losses ranging in severity from light to medium. With the exception of one boxer the left side was more severely damaged than the right side. One boxer who participated as an amateur for eleven years, with 141 bouts, revealed a deep, trough-shaped hearing loss at C5 on the

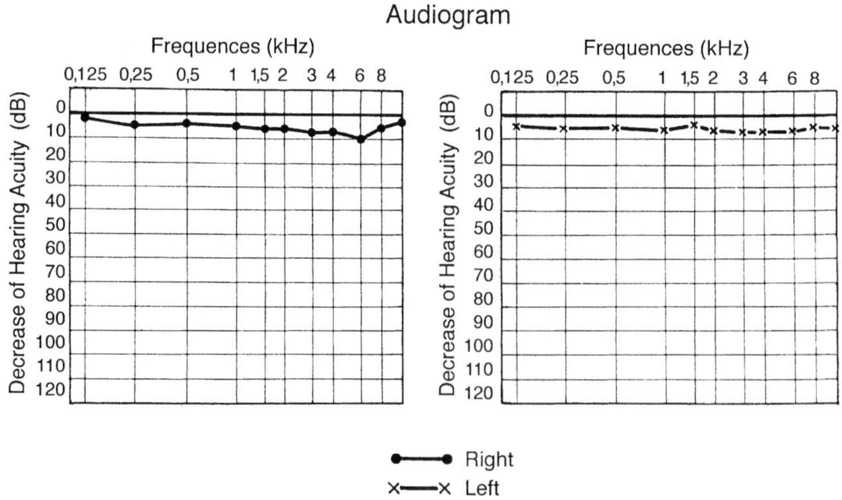

Fig. 164. Audiogram. (From PAULSEN, and HUNDHAUSEN, 1971).

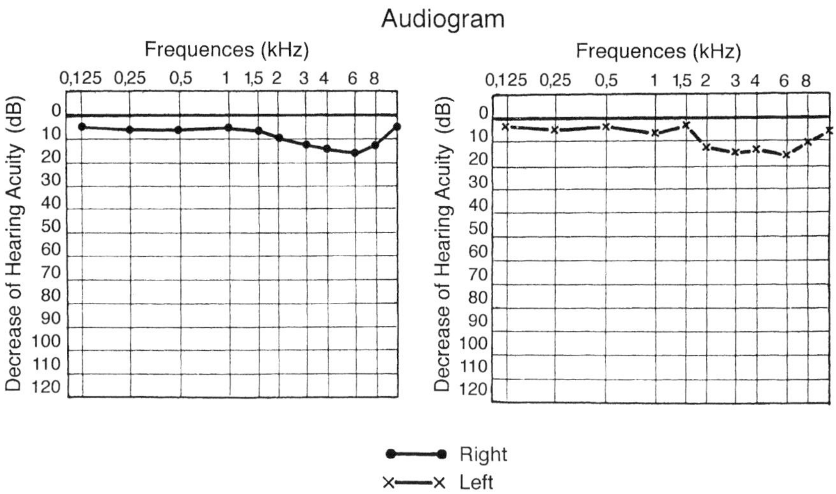

Fig. 165. Audiogram. (From PAULSEN, and HUNDHAUSEN, 1971).

right side, on the left side at fis4. The mean curves showed on the right side a flat trough-shaped hearing loss in the middle and especially in the higher frequencies, with slight C5 dip, and on the left side a continuous hearing loss with accentuation in the higher frequencies and slight fis5 dip (Fig. 167).

(5) *300 to 400 bouts* – 3 boxers

All the boxers of this group revealed a marked hearing loss; each of these boxers is a professional, who has boxed for many years. In two cases the hearing impairment was bilateral. In the third case (Fig. 168) the left side was slightly more damaged than the right side. In each of these cases there is a continuous hearing loss, extending from the middle to the higher frequencies. In every case there are dips and troughs, which are most pronounced at C5. The mean curve revealed bilaterally a marked hearing loss in the middle and especially in the higher frequencies. The hearing loss has already begun at 500 Hz. Bilaterally, there was a slight C5 dip, which was more pronounced on the left side than on the right.

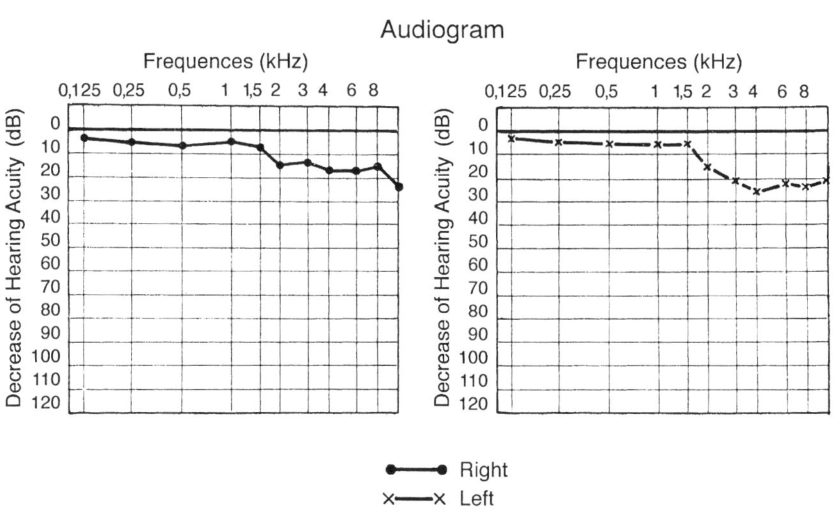

Fig. 166. Audiogram. (From PAULSEN, and HUNDHAUSEN, 1971).

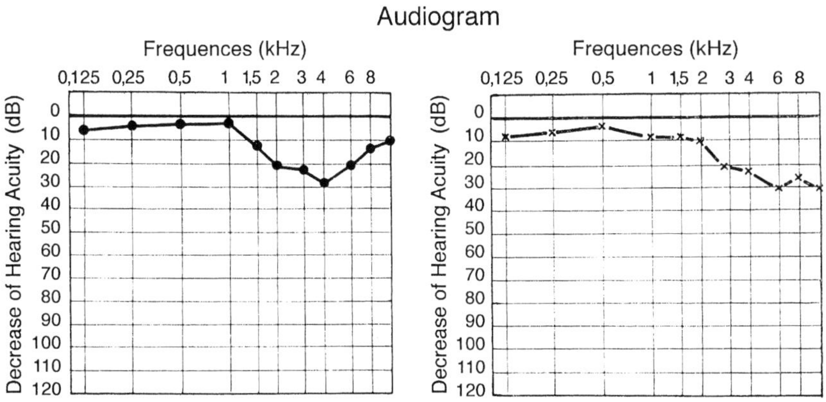

Fig. 167. Audiogram. (From PAULSEN, and HUNDHAUSEN, 1971).

Extent of the Hearing Loss in the Different Weight Classes

The findings demonstrate that 76.6% of the boxers had a pathological audiogram.

Two of eight boxers in the first weight class have normal hearing. In the second class, there is a hearing loss of less than 10 dB in seven out of 13 cases. In the third class (light middle – middleweight), only one out of 10 boxers had no hearing loss. In the fourth and fifth weight classes (light heavyweight and heavyweight) all boxers revealed hearing losses.

In the lighter weight classes up to 67 kg, no boxers show a steep decline (dip) in the high frequencies. Troughs exist normally between 2.000 and 6.000 Hz, with a maximum of 4.000 Hz.

Two boxers in the third class showed high tone dips in their audiograms, with slight C5 dips (Fig. 169). The remaining classes revealed depressions and troughs between 2.000 and 6.000 Hz. The troughs in heavier boxers are wider than they are for boxers lower weight classes.

Depressions and troughs in the middle and higher frequencies also occur in the light heavyweights. In two cases, depressions in the higher frequencies are visible along with the troughs at 3.000 and 2.000 Hz (Fig. 170).

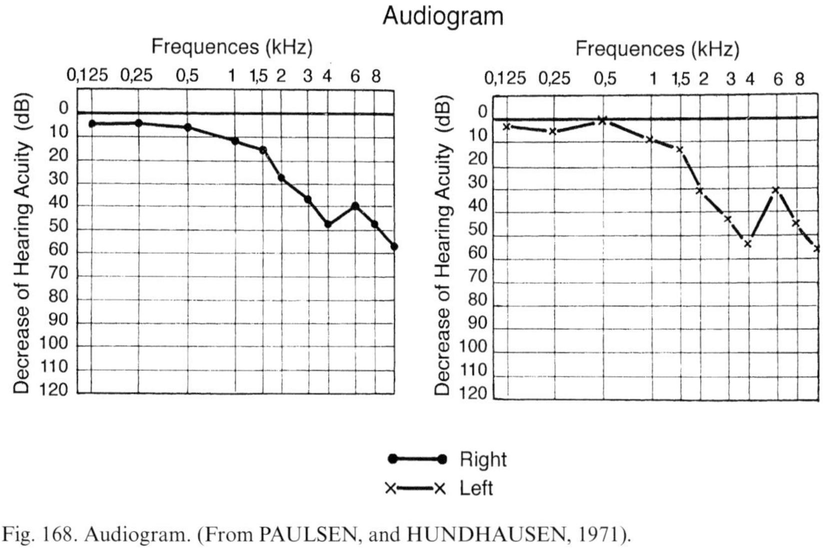

Fig. 168. Audiogram. (From PAULSEN, and HUNDHAUSEN, 1971).

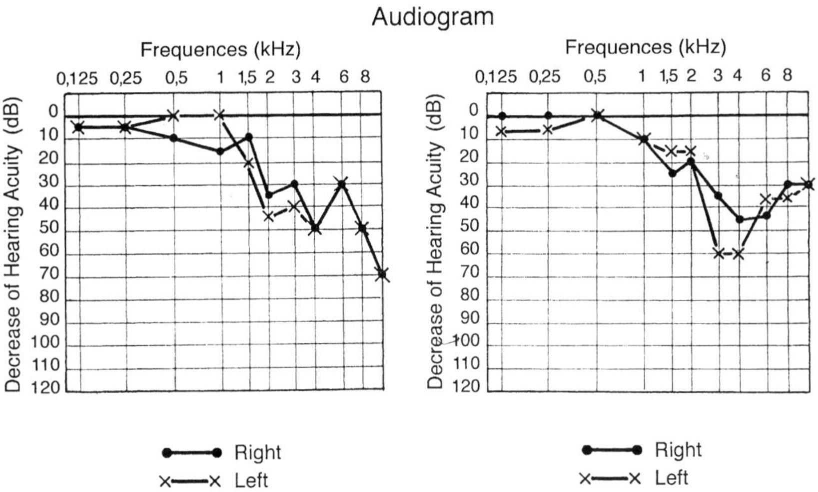

Figs. 169 and 170. Audiograms. (From PAULSEN, and HUNDHAUSEN, 1971).

In four out of six audiograms of heavyweights, there were both trough-shaped hearing losses in the middle and higher frequencies and depressions in the higher frequencies (Fig. 171).

Extent of the Hearing Loss in Regard to the Number of Bouts

The hearing impairment grows with the increasing number of bouts. Boxers with less than ten fights have practically normal hearing, with only minimal hearing losses in the high frequencies. In boxers with up to 30 fights hearing impairments can be recognized by depressions and troughs in the middle and especially in the higher frequencies. These hearing losses can be interpreted as minimal, despite the fact that they are more severe than in boxers with a lower number of fights.

Boxers with 30 to 99 fights reveal, usually in the middle and high frequencies, trough-shaped and sometimes continuous hearing losses. The mean curve reveals bilaterally flat troughs. The hearing

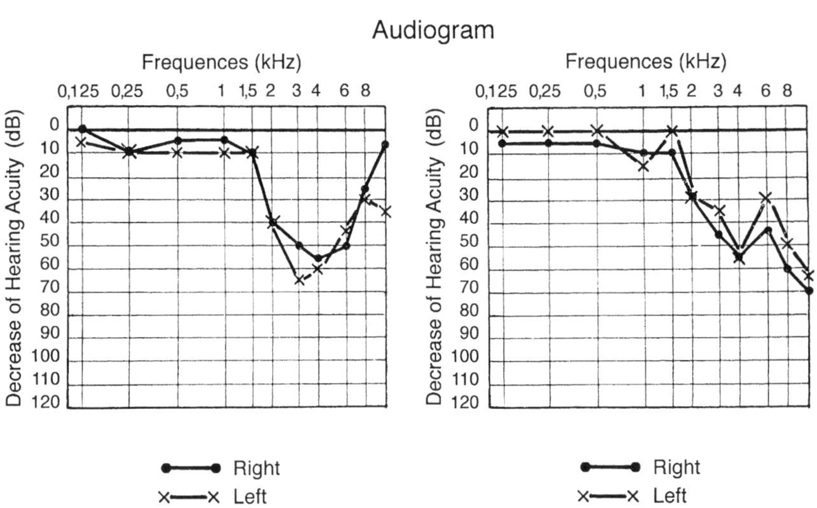

Figs. 171 and 172. Audiograms. (From PAULSEN, and HUNDHAUSEN, 1971).

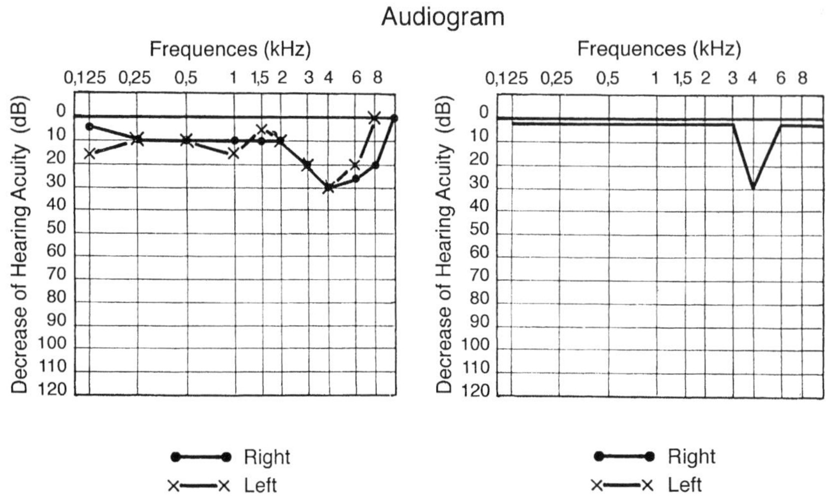

Figs. 173 and 174. Audiograms. (From PAULSEN, abd HUNDHAUSEN, 1971).

impairments in boxers with 100 to 200 bouts are more severe. In the middle and higher frequencies, these boxers have trough-shaped or continuous hearing losses. The mean curve shows bilaterally a slight C5 dip. The most severe hearing impairment appears in boxers with more than 300 bouts (Figs. 172, 173, and 174). The hearing deficit is especially severe in the middle frequencies, and hearing is severely impaired.

In summary, it can be said that with an increasing number of bouts, the hearing impairment increases (but it should be mentioned that in single cases the hearing losses can be minimal).

The extent of the hearing impairment after blunt impacts to the head is in general related to two factors, namely the intensity and number of impacts. PAULSEN and HUNDHAUSEN took both factors into consideration, by evaluating the audiograms as they related on the one side to the weight classes of the boxers, and on the other side to the number of fights. The results show convincingly that with an increasing intensity of impacts (higher weight class) and an increasing number of impacts (number of fights) there is an increase in hearing impairment. The number of fights seems to be of greater importance.

Classification of Hearing Acuity after Head Injury

Despite the small number of cases examined, the audiograms can be classified into *three types of curves:*

Curve type A (Fig. 174) shows slight hearing impairments in the high frequency of the type of a dip at C5 (Fig. 176). This dip can appear at higher or lower frequencies. In this area flat trough-shaped hearing impairments can be seen, which do not appear in excessive noise trauma. This type of curve may possibly be interpreted as specific for a status after repeated head impacts.

Curve type B (Fig. 175) – reveals a wider trough-shaped hearing deficit with a maximum of around 3.000 – 4.000 Hz. In single cases this trough-shaped hearing deficiency may appear in higher frequencies.

Curve type C (Fig. 176) – is characterized by continuous hearing loss beginning in the lower part of the middle frequency, and severe damage in higher frequencies combined with a slight C5 dip.

If one compares these three curve types with the corresponding curves of hearing impairment due to sound and noise trauma, it is impossible then to decide, using the audiogram, whether the cause for the hearing impairments is a mechanical trauma, a noise trauma, or chronic exposure to exces-

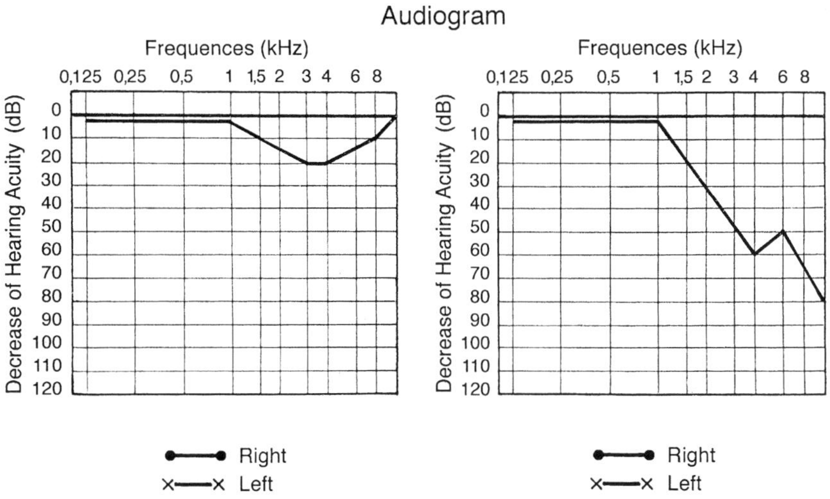

Figs. 175 and 176. Audiograms. (From PAULSEN, and HUNDHAUSEN, 1971)

sive noise. The ear reacts to different types of the same noxious event in the same way. As a rule, slight injury leads to alterations and hearing loss in the high frequencies, especially around 4.000 Hz. More serious injuries produce more severe and marked alterations in the form of trough-shaped hearing losses that can extend into the middle frequencies. Severe injuries, or injuries resulting from repeated impacts over a longer period, lead to a high-tone loss that finally extends into large parts of the middle frequencies as well.

In general, the classification of the types of hearing impairment due to boxing is not as simple as the classification due to noise or explosion trauma.

Relationship between Localization of Impact and Hearing Impairments

PAULSEN and HUNDHAUSEN quote the literature and finally conclude that the nearer the impact occurs to the ear, the heavier is the hearing loss. The authors quoted the papers of KOBAYASHI (1925), KOCH (1933), WAGEMANN (1956/1958), JUDICIBUS (1957), KAWAMURA (1958), and GLASENAPP (1962), and concluded that if the occipital and temporal areas are hit, hearing losses will frequently result.

The boxer is hit mainly against the frontal parts of his face and on the forehead, but also against lateral parts of the temple and the ear. Blows against the occiput are against the rules, but they are dished out from time to time.

The results of PAULSEN and HUNDHAUSEN reveal a very interesting finding. In 12 cases, the left ear showed a more severe hearing deficit than the right one. The explanation should be that the vast majority of all boxers are right-handed, the left hand leads, and the right hand hits. The most severe blows will be delivered by the right hand against the left side of the head of the opponent. In this regard the data in the literature are divided. PIQUET et al. (1953/1954), DESPONS (1955/1956), and JUDICIBUS (1957) find, in head injuries, hearing losses on both sides to the same extent. On the other hand, ZANGE (1915), CASTELLO (1955), BOURGEOIS (1955), PAGNO et al. (1956) and KAWAMURA (1957) more frequently describe unilaterally severe hearing impairments.

Above-threshold Audiometry: The Localization of Hearing due to the Results of above-threshold Audiometry

Twenty boxers out of 43 had an above-threshold audiometry. In 17 cases this demonstrated an above-threshold disturbance, and in three there was a retro-labyrinthine lesion. All three cases were professional boxers who had been involved in a large number of fights, and all had severe hearing losses.

Hearing Impairment as an Occupational Hazard in Professional Boxers

In the seventh Edition of the *Occupational Hazards Manuel in the Federal Republic of Germany* (1969), damages due to repeated blunt impacts against the head and hearing impairments of the kind that can only be present in professional boxing are not mentioned. Theoretically it is possible to register the damage due to professional boxing as an occupational hazard, since each illness due to noxious events of the professional activity has to be recognized as an occupational disease. Practically, this is not the case, because professional boxing does not have a *trade organization ('Berufsgenossenschaft') to* provide insurance for them.

The authors summarize their findings, that few instances of damage to the hearing in boxers have been reported. Apart from the question of the frequency and extent of hearing losses, their reversibility and any typical features of the characteristic audiometric curves of the hearing pattern, the injuries to the hearing function are of particular interest to the otologist because the hearing losses involved in these cases are caused by *repeated injuries to the skull.*

While head injuries are usually an isolated accident in the average person's life. The boxer is subject to multiple and repeated head injuries over a period of some years. In this sport or occupation there are persons whose avowed purpose in life is to strike their opponents with the gloved fist so often and so hard as to cause them temporarily to loose their senses. This is known as the knock out, though it is also possible to win on points. Only with boxers is it possible to ascertain the effect of repeated head injuries on the hearing function, and the hearing organ. The course and extent of hearing disturbances after repeated head injuries are still unknown.

PAULSEN and HUNDHAUSEN carried out audiometric examinations in 43 boxers of all weight classes, from featherweights to heavyweights.

The authors obtained the following results:

(1) Hearing losses occur regularly in training bouts.

(2) The extent of the hearing loss depends in the main on two factors. The weight class of the boxer as a measure of the severity of the head trauma and the number of fights he has undertaken as a measure of the frequency of head injuries received. The weight category plays a smaller part here. A heavyweight usually has more severe hearing losses than a featherweight does, because the greater weight of his opponents enables them to deliver stronger punches. Heavyweights also tend to have a considerably greater number of fights behind them than boxers in the lighter weight categories.

(3) The course of the hearing loss after repeated blows to the head (slight to medium head injuries) is first evidenced by a dip at c5; later a trough appears in the higher frequencies, extending also into the middle frequencies. Finally, large portion of the hearing in the high frequencies and reaching into the middle frequencies is lost altogether.

The shape of the curve is very similar to curves obtained after repeated noise trauma and chronic exposure to excessive noise. The hearing thus reacts to different sorts of external damage in a similar manner.

The following particular results were obtained. Out of 43 there were 35 amateurs and eight professional boxers. 33 boxers (76.7%) showed hearing losses and 10 boxers had practically normal hearing, though the latter had only a few fights behind them. The course of the hearing loss in different weight categories corresponds to the pattern indicated in paragraph 3 above.

The question of recovery of the hearing after a fight remains unanswered, as all the boxers were only examined once. In the history, rushing noises, whistling and deafness after one fight were regularly denied.

In twelve cases the left ear showed greater hearing loss than the right ear. This finding is explained by the greater incidence of boxing blows to the left ear, as most boxers are right handed.

Twenty out of 43 boxers had above threshold audiometry. In 17 cases this demonstrated an above threshold disturbance and in three there was a retrolabyrinthine lesion. All these three cases were professional boxers who had been involved in a large number of fights and they all had severe hearing losses.

As boxing regularly leads to hearing losses and even more importantly, to general cerebral damage, all boxers should at least wear protective helmets. To prohibit the sport of boxing altogether would, of course, also be quite effective, as has been seen in Sweden.

SOUTSU (1988) studied ABRs of boxers, who had received repeated impact to the head and were often concussed, immediately after bouts or training, in order to reveal the influence of knock out or repeated mild head injuries.

Forty-six male professional boxers, ranging from 18 to 28 years old, were examined at the stadium and at the boxing gymnasium. Eighty-five recordings were done, since some boxers were examined more than once. Twenty-seven cases of pre-exertion group (resting: five, pre-sparring: ten, pre-training: twelve) were compared with post-bout group (35 cases). The I-V IPL (inter-peak latency) of pre-exertion group was 4.30 +/- 0.18 msec and that of post-bout group was 4.02 +/- 0.15 msec. The latter showed significantly shorter ($p < 0.01$) IPL. Among 35 cases of post-bout group the I-V IPL of knocked out group (14 cases) was 4.04 +/- 0.11 msec, and that of non-knocked-out group (21 cases) was 4.01 +/- msec 0.17 msec. There was no significant difference. The I-V IPL of pre-sparring state was 4.24 +/- 0.15 msec, and that of the post-sparring state was 4.10 +/- 0.13 msec. The difference was significant between these two groups ($p < 0.01$). After training without impacts to the head, I-V IPLs were shortened in all twelve cases. IPL: of pre-training state was 4.34 +/- 0.22 msec. and that of post-training state was 4.19 +/- 0.22 msec. This also showed significant difference ($p < 0.01$). On one boxer, the authors recorded ABRs during disturbance of consciousness. He was knocked out by a blow to the face in the first round, and was completely unconscious. ABRs were recorded immediately after he had been laid on the bed.

KREJCOVA and CERNY (1989) reported that vestibulogical examination of boxers revealed that 80% had vestibular abnormalities. These were obvious but well compensated, so that they remained hidden in everyday life and appeared only on laboratory examination or under stress. Regarding the origin of vestibulo-ocular reflex, abnormalities in boxing contribute three etiopathogenetic factors: The first is the direct traumatization of the vestibular apparatus by a direct blow. The second is the diffuse damage of the brainstem white matter and of vestibular nuclei. The third factor is the traumatization of the cervical spine via the traumatization of the proprio-receptive systems, cervical sympathetic and the adjacent vascular mechanisms. One-third of the abnormalities is caused by direct labyrinthine traumatization. The findings of simultaneous EEG, OKN and ETT abnormalities in one-half of the subjects suggest that there are probably also supra-tentorial lesions caused by the above-mentioned mechanisms.

Injuries to the Hand, Wrist, Muscles and Peripheral Nerves

Case histories of boxers' injuries of the hand and wrist were published by DELAPCHIER (1909), LENOIR (1909), LEAKY (1909), BRAINE and RAVINA (1923), HELLGREN (1933), KLAUSS (1934), LA CAVA (1954), ISELIN and DUCROQUET (1948, 1949). WAXMAN and GESHELM (1948), MEYRAN (1951), LA CAVA (1952), ISELIN (1960), PIJADE (1961), LOYARTE and MAZAN (1963), FARROW (1965), PARRY (1965), FRANKE (1965), LAROSE and SIK (1968), ZEMINA (1980), KARPOW (1983), PORTER et al. (1988), POSNER and AMBROSE (1989), SIM (1990), MAITRA and SEN (1990), CAVANAGH (1992) and KANERVA (1998).

An understanding of the fist position inside of a boxing glove is a necessary prerequisite to study of hand injuries caused by boxing.

FARROW (1965) observed that the gloved hand as it is in boxing bears no resemblance whatsoever to the clenched fist of bare fisted fighting. He x-rayed some willing boxers in their gloves with fists clenched as tight as possible, and the metacarpo-phalangeal joints were only flexed to about 50-60 degrees, with the fingers slightly open and the thumb almost straight and abducted from the side. This is largely due to the fact that the hand has to grip the palm of the glove and is therefore more open. Due to the splinting effect of the thumb encasement and the fact that the padding on the side causes the thumb to be a little on one side, the thumb, if it is not held close into the side of the glove, is placed in a position of maximum vulnerability and is therefore most liable for injury. The distortion of the metacarpo-phalangeal joint is the most common hand injury of boxers.

The following drawings show convincingly a properly clenched fist and some of the most frequently occurring fractures of the hand.

A properly clenched fist is shown in Fig. 177. The four points delineate the dorsal plane of the hand (metacarpal 2 – 5) ('c'est le marteau vulnerant'). The crosses mark the most common sites of fracture (From BRAINE J, and RAVINA A, 1923). Compare the comment of FARROW above.

A one month old fracture of the base of the 1st metacarpal (Bennett's fracture) is shown in Fig. 178. (From BRAINE J, and RAVINA A, 1923).

A one month old fracture of the base of the 1st metacarpal is shown in Fig. 179. (From BRAINE J, and RAVINA A, 1923).

Fig. 180 reveals an old fracture of the head of the 1st metacarpal, known as the *'professional stigma of the pugilist'.* (From BRAINE J, and RAVINA A, 1923).

A recent fracture of the body of the 2nd metacarpal and an old fracture of the head of the 1st metacarpal is shown in Fig. 181. (From BRAINE J, and RAVINA A, 1923).

Fig. 182 shows the hand of Georges CARPENTIER, French heavyweight champion. Repeated fractures of the head of the 1st metacarpal suffered in the second round of his fight against Jack DEMPSEY in the United States. Note the considerable raccourcissement of the 1st metacar-

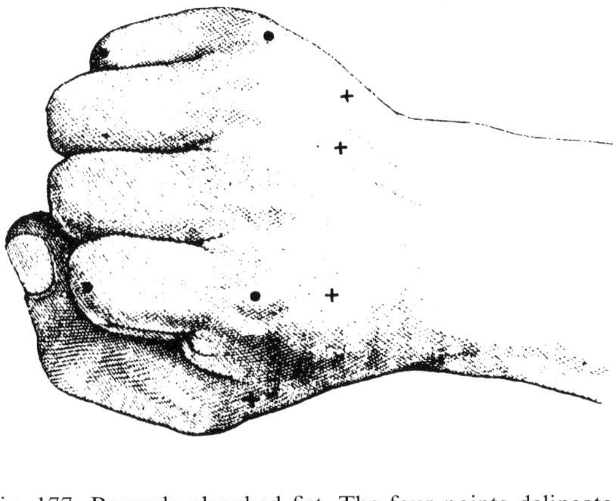

Fig. 177. Properly clenched fist. The four points delineate the dorsal plane of the hand (metacarpals 2 – 5) ('c'est le marteau vulnerant'). The crosses mark the most common sites of fracture. (From BRAINE J, and RAVINA A, 1923).

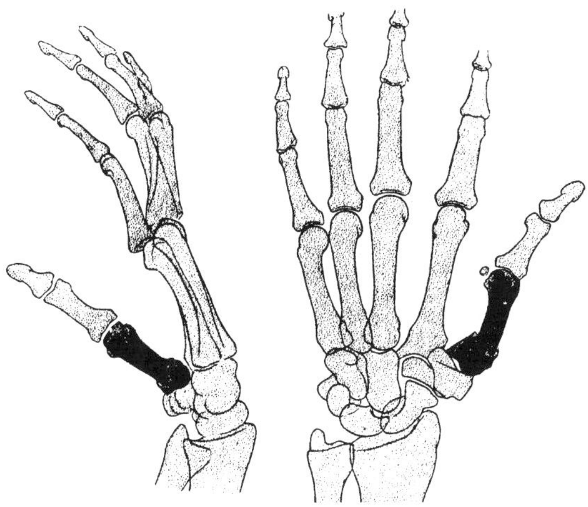

Fig. 178. One month old fracture of the base of the first metacarpal (Bennett' fracture). (From BRAINE J, and RAVINA A, 1923).

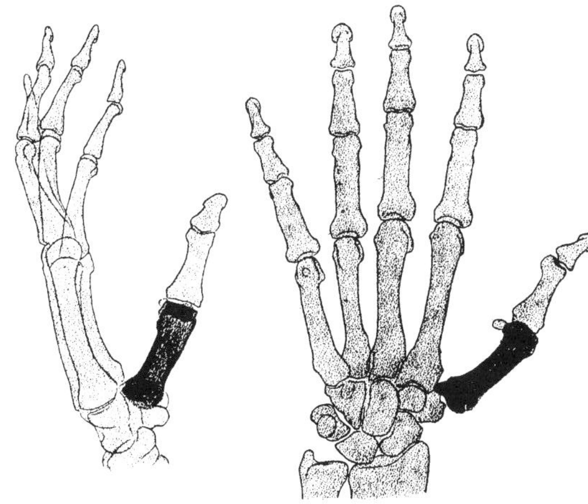

Fig. 180. Old fracture of the head of the first metacarpal, known as the 'professional stigma.' (From BRAINE J, and RAVINA A, 1923).

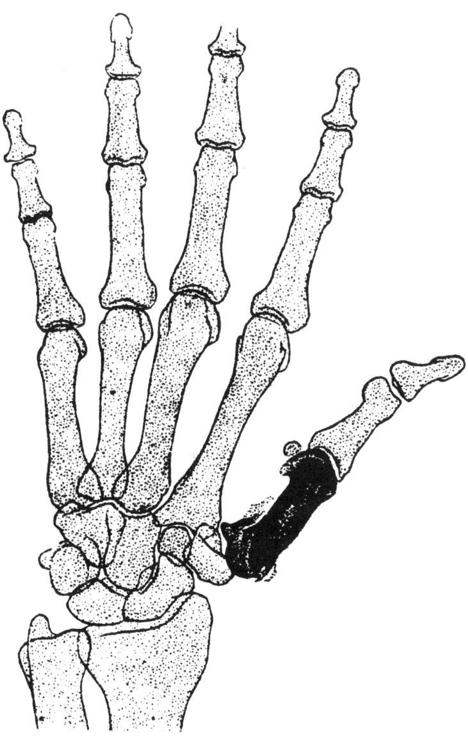

Fig. 179. One month old fracture of the base of the first metacarpal. (From BRAINE J, and RAVINA A, 1923).

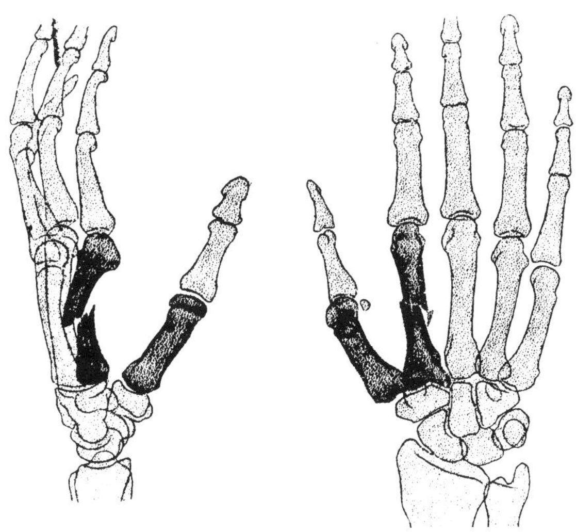

Fig. 181. Recent fracture of the body of the second metacarpal; old fracture of the head of the first metacarpal. (From BRAINE J, and RAVINA A, 1923).

pal at the epaisseur of its base. (From BRAINE J, and RAVINA A, 1923)

Fig. 183 shows a fracture of the main body of the 2nd metacarpal, 2 months old. Gross callus formation with dorsal angulation. Old fracture of the

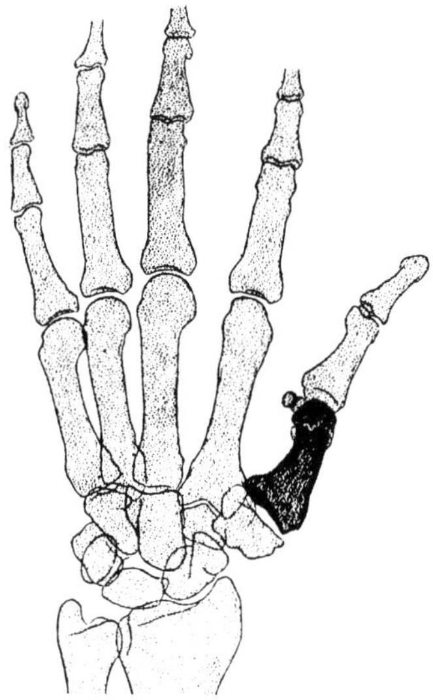

Fig. 182. Hand of Georges CARPENTIER, French heavy-weight champion. Repeated fracture of the head of the first metacarpal suffered in the second round of the fight against Jack DEMPSEY, in America. Note the considerable shortening of the first metacarpal. At the thickness of its base. (From BRAINE J, and RAVINA A, 1923).

head of the 1st metacarpal (From BRAINE J, and RAVINA A, 1923).

Georges Carpentier had 109 fights, 88 won, 14 lost, 6 drawn, 1 no contest.

A statistical analysis of hand and arm injuries in boxers was presented by LA CAVA (1952). The author evaluated 3.000 accidents covering the years 1946-1951. A total of 591 injuries occurred in which the recovery took more than eight days. Injuries to the arms totaled 375 (63%) cases: fractures occurred in 183 cases, distortions and ruptures of muscles in 95 cases, and contusions in 87 cases.

The fractures occurred at the following areas: (1) Metacarpal I (86 cases = 46.99%); (2) metacarpal II (43 cases = 23.49%); (3) metacarpal III (10 cases = 5.46%); (4) metacarpal IV (3 cases = 1.63%); (5) metacarpal V (8 cases = 4.37%); (6) os naviculare (4 cases = 2.18%); (7) os pyramidale (1 case = 0.54%); (8) fingers – not specified (10 cases = 5.48%); and (9) hand – not specified (18 cases = 9.85%).

LA CAVA concluded that he could demonstrate

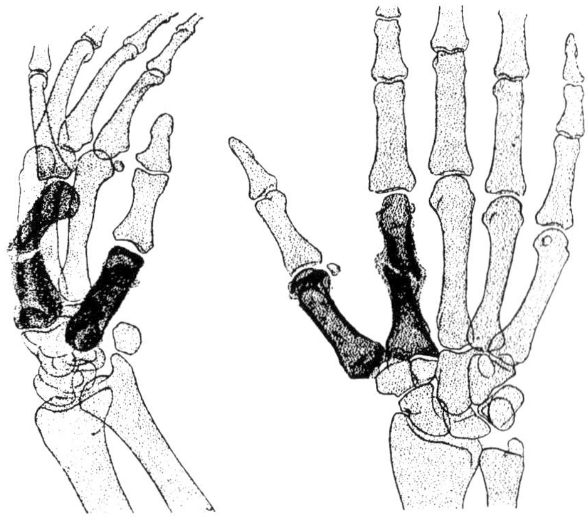

Fig. 183. Fracture of the main body of the second metacarpal, 2 months old. Gross callus formation with dorsal angulation. Old fracture of the head of the first metacarpal. (From BRAINE J, and RAVINA A, 1923).

that the fractures of the metacarpal I and II are most frequent hand injuries in boxing, in contrast to the statistics of DELAPCHIER (1909). LA CAVA explained the frequent fractures as being the result of the thumb being in the wrong position while hitting.

Hand and wrist injuries occur frequently in boxers. ISELIN (1960) reported that out of 105 case histories shown to him by DUCROQUET, Surgeon of the *French Boxing Federation*, 52 revealed, not fractures, but symptoms of carpo-metacarpal arthrosis (compare ISELIN and DUCROQUET 1948, 1949). This extremely painful arthrosis is caused, provoked or maintained by hitting. The end result is a permanent metacarpal lump, which never disappears, remaining as a professional stigma of the pugilist. This statement contradicts that of FARROW (1965), who reported that injuries to the hand and wrist are not common in boxing.

McCOWN (1959) published findings on professional boxers during a seven year period organized by the *New York State Athletic Commission:*

There were 16 fractures of the nasal bone, 14 of the metacarpals (ten of which involved the right hand), and three of the phalanges. There were four fractures of the jaw, three of the mandible and one of the maxilla. The disability of the fractured hands was approximately three months, while the disability as a result of fractures of the jaw was six months.

McCOWN also reported on the combined bone and soft tissue injuries from the same seven year period.

There were 18 cases of *acute* and *chronic traumatic synovitis* usually involving the *second* and *third metacarpal phalangeal joints*. This injury is rather characteristic and has aptly been paraphrased *boxer's knuckle. It consists of swelling of the skin, subcutaneous tissues and inflammation of the synovia.* Occasionally, a small amount of fluid may be present in the joint and slight to moderate or marked swelling occurs. McCOWN wrote that he had no experience with the operative procedures. In those cases of extensive involvement he advised hot soaks, occasional aspiration of fluid and injection of .5 cc of hydrocortisone into the joint, and periods of rest along with appropriate physiotherapy. Most of them responded. Some, however, became chronic and interfered with effective punching.

Most of these injuries involve the metacarpophalangeal joint of the thumb, then follow the heads of the metacarpalia 2 to 5, then the distortions of the hand joint and the fractures of the navicular bone.

The *most common fractures* are: (1) *Fracture of the head of the first metacarpal;* (2) *impacted fracture of the head of the first metacarpal;* (3) *fracture of the body of the second metacarpal combined with old fracture of the head of the first metacarpal;* (4) *fresh fracture of the body of the second metacarpal combined with old fracture of the head of the first metacarpal;* and (5) *fracture of the base of the first metacarpal,* the so-called *Bennett's fracture dislocation.* (So-called *'stove-in knuckles' are angulated fractures of the neck of the metacarpal).*

Among 523 injuries to professional and amateur boxers and wrestlers treated in a five year period by WAXMAN and GESHELM (1948), 22 were injuries to the knuckles of 21 boxers, all in their twenties. Twelve of the knuckle injuries were diagnosed as *simple contusions* and complete recovery followed a few days of treatment with hot soaks. In ten instances a *traumatic bursitis* of the *metacarpophalangeal joints of the hand* was diagnosed. The diagnosis of *bursitis* was confirmed by aspiration of bloody or gelatinous fluid from the affected bursa. Review of the literature by the authors indicated that this is a relatively rare disease. Conservative treatment in two of the cases resulted in clinical cures; in the remaining three the bursal sac was excited and clinical cure followed.

The authors published two case histories:

Case 1: A 22 year old boxer injured the middle knuckle, right hand, on April 4, 1945. There was swelling and tenderness over the knuckle. On several occasions fluid was aspirated from the bursa. *Conservative treatment* brought no beneficial results. The *sac was excised* on April 24, 1945, and the patient was discharged on June 19, 1945, clinically cured. There has been no further trouble referable to this knuckle.

Case 2: The patient, a pugilist, aged 23, injured the left middle knuckle on March 8, 1946. A *roentgenogram* on March 11 was negative for fracture. Clinically there was tenderness and thickening of the soft tissues over the metacarpo-phalangeal joint. The patient complained of pain on impact. One half cc. of amber fluid was aspirated. Adhesive strapping, hot soaks and aspiration brought no relief and on April 12 the *bursal sac was excised.* Seven weeks later the joint was in excellent condition and the patient resumed boxing.

In mid-November of the same year, this patient injured the knuckle of the little finger of the right hand. No fracture was shown by roentgenogram. The diagnosis was *bursitis of the metacarpo-phalangeal joint, right little finger,* and 1/2 cc of amber fluid was aspirated. Soft tissue over the joint was thickened, and there was pain on impact, and tenderness to touch. No improvement resulted from hot soaks, physical therapy and aspiration, and the *bursal sac was excised.*

A *histological report* of the *excised bursa* follows, "Reveals portion of a cystic structure composed of dense collagenous stromal wall. There are many blood vessels and slight round cell infiltration. Diagnosis: Bursa."

Four months after the operation the knuckle was apparently sound and the patient resumed boxing.

HELLGREN (1933) reported on a 22 year old bantamweight boxer who injured his right hand.

X-ray examination carried out immediately after the bout revealed no fracture. The patient had noticed that the knuckle of the right index finger had come to occupy a lower level in relation to other knuckles of the same hand. On the dorsum of the hand just below the wrist a painful swelling arose. The injury was regarded as a distortion of the articulation between the multangulum minus and the II metacarpal bone with disruption of the corners of the latter, probably as a result of the dorsal ligaments.

GLADDEN (1957) remarked that the literature contains many articles on the impairment of function of the tendons of the hand, but there is a paucity of material on the pathology of injured knuckles in which tendons are involved as a result of repeated direct trauma. There is even less information available as to the surgical treatment of these knuckles that will permit prizefighters to continue in their profession. Many terms have been used in the description of these hands: *'brittle,' 'water on the joint,'* and *'bursitis,'* amongst others. The disability is usually progressive.

GLADDEN surgically explored a number of metacarpo-phalangeal joints after a single trauma. There was marked swelling of soft tissue over the affected joints, and x-rays revealed impacted fractures of the neck of metacarpal bones. There was palmar angulation of the distal fragment in all instances. The edema of the soft tissue and capsule subsided following intra-medullary fixation of fracture and immobilization. None of the four patients treated by GLADDEN had any further swelling, pain, or any specific discomfort after their discharge.

GLADDEN presented the following classification:

Type I: Thickening of soft tissue (including the tendon and capsule) over the distal end of the metacarpal bone, with no evidence of a tear.
Type II: Thickening of soft tissue, tendon and capsule over the affected joint, with tear of the more superficial portion of the soft tissue.
Type III: Thickening of soft tissue, tendon and capsule over the affected joint, with tear of soft tissue and tendon.
Type IV: Thickening of soft tissue, tendon and capsule with tear of these structures; the defect extends into the joint space.

According to GLADDEN it is possible to treat a Type I injury conservatively. It is also possible to treat an early Type II injury with a small defect in tendon sheath as a Type I injury. In the latter instance, the swelling will continue to persist to some degree and some discomfort will follow punches on bony prominences such as the head, elbow, and shoulder.

In all instances where a diamond-shaped tear exists in Type II, III and IV injuries, surgery offers the best prognosis.

WOLFF (1928) published a *fracture of the ulna* in a 21 year old amateur boxer. Another case was reported by REFSHAUGE (1963).

In a discussion about the medical aspects of amateur boxing in 1965, PARRY reported that he had analyzed all the cases of hand injuries due to sports that he had treated during the past fifteen years and found that few of them had come from boxing. FARROW (1965), participating in the same conference, remarked that boxing injuries to the hand and wrist are not common. He continued to say that when they do occur, the pain, swelling, and deformity are not outstanding features, and because of this, he indicated that hand injuries are very likely to go unnoticed.

WHITING et al. (1988) used two synchronized cameras to film four experienced boxers while they threw a series of punches at a practice bag. Three-dimensional (3D) coordinates of each boxer's shoulder, elbow, wrist, and glove were used to estimate linear and angular kinematics of the upper extremity. Average velocities of contact ranged from 5.9 to 8.2 m/s with peak velocities of 6.6 to 12.5 m/s reached 8 to 21 ms prior to hand/glove contact with the bag. Significant differences in shoulder and wrist velocities, elbow angle excursions, and elbow angular velocities were seen when comparing hooks and jabs. Few differences were evident when comparing the kinematics of gloved versus barehanded punches. Results are significant in providing kinematic data characteristic of experienced performers, which may be used in a kinetic model of punch impact and its relationship to potential injury mechanisms.

POSNER and AMBROSE (1989) examined injuries to the dorsal aspect of the metacarpo-phalangeal joint of a finger, which usually damages the sagittal fibers of the extensor tendon mechanism. Rupture of the underlying joint capsule, a far more serious injury, has received scant attention in the medical literature. The problem is often unrecognized because the extensor tendon may remain in its normal midline position. Eight dorsal capsular ruptures are described in six patients who had closed direct trauma to either the knuckle or their index finger or long fingers. Most of the injuries occurred in professional boxers.

SIM (1990) examined dislocation of the thumb saddle joint as a rare sports injury. His report deals with four dislocations of the carpo-metacarpal joint of the thumb during the period 1957-1983 as an injury caused by sports. Three dislocations could be found in boxing, one was caused by a fall from a horizontal bar. All of them were closed: two times the dislocation was to the dorsal, one to the dorso-ulnar and one to the dorso-radial side. Closed dislocations do not require operative treatment. Reduction and immobilization for four weeks in a plaster splint with inclusion of the thumb were sufficient, recurrence of the dislocation did not occur. Correct bandage and the use of effective boxing gloves are of great importance.

JORDAN et al. (1990) conducted a survey of boxing injuries at the *US Olympic Training Center (USOTC)* from January 1977 through August 1987. The purpose was to determine the types of acute and chronic injuries sustained by amateurs, during training, sparring, and competition. During the study period 447 injuries occurred. (The total number of boxers was unobtainable). Most injuries were to the upper extremity (32.9%), followed by the lower extremity (23.9%), head and face (including eyes, ears, nose, and mouth/throat (20.5%) and back (6.9%). Cerebral injuries accounted for only 6.5% of all injuries, probably boxing at the *USOTC* is primarily noncompetitive and more blows to the head occur during competition than during sparring and training.

A total of 447 injuries were recorded during the investigation period (Table 21).

Upper extremity: The hand and wrist accounted for 62% of all upper extremity injuries (Table 22). Of the 35 metacarpal injuries (the most common hand injury), 20 were contusions and six were fractures. Fourteen of the finger injuries were either contusions or sprains. There were two fractures of the proximal phalanx and one avulsion fracture of the middle phalanx. Other finger injuries included one tendon arthritis, extensor tendinitis, and metacarpophalangeal synovitis.

The 18 *thumb injuries* included a Bennett's fracture, one avulsion fracture of the proximal phalanx, and an avulsion fracture of the metacarpal. Thirteen thumb injuries were sprains, strains, or contusions. Nine of the 13 wrist injuries were sprains or strains. One fracture of the radius was noted, but there were no fractures of the scaphoid.

Of the 32 *shoulder injuries,* 12 involved the rotator cuff, six were deltoid strains, and four involved the long head of the biceps. Eight of the 16 elbow injuries resulted from hyperextension.

Lower Extremity: Most of the lower extremity injuries were to the knee (Table 23), and most of these involved the patella (11) or the iliotibial band (9). Seventeen of the 27 ankle injuries were sprains. Injuries to the leg, foot, hip/groin, and the tigh were tentatively less common.

Hand Nose Injuries and Face: – the most common

TABLE 21
Types of boxing injuries at the US Olympic Training Center during a 10-year period. (From JORDAN B D, et al., 1990).

Injury Site	No.	% of Total Injuries
Upper extremity	147	32.9
Lower extremity	107	23.9
Head and face (includes eye, ear, nose, and mouth/throat)	92	20.6
Back	31	6.9
Cerebral	29	6.5
Cervical spine, brachial plexus	23	5.1
Chest	17	3.8
Kidney	1	0.2
Total	447	100.0

TABLE 22
Upper extremity injuries at the US Olympic Traning Center during a 10-year period. (From JORDAN B D, et al., 1990).

Injury Site	No.
Hand and wrist	92
Metacarpal	35
Finger	26
Thumb	18
Wrist	13
Shoulder	32
Elbow	16
Upper arm	4
Forearm	3
Total	147

TABLE 23
Lower extremity injuries at the US Olympic Training Center during a 10-year period. (From JORDAN B D, et al., 1990).

Injury Site	No.
Knee	36
Ankle	27
Leg (calf, tibia, fibula)	19
Foot	12
Hip/groin	7
Thigh	6
Total	107

TABLE 24
Head and face injuries at the US Olympic Training Center during a 10-year period. (From JORDAN B D, et al., 1990).

Injury	No.
Nose	34
Eye	23
Contusion	7
Abrasion (not specified)	4
Other (not specified)	4
Corneal abrasion	3
Retinal detachment	1
Orbital injury	1
Periorbital hematoma	1
Hyphema	1
Subconjunctival hemorrhage	1
Mouth/throat	13
Face laceration/contusion	8
Jaw	7
Ear	5
Head	2
Total	92

head and face injury – included primary epistaxis (16) and contusions (13), with three nose fractures and two lacerations (Table 24). Regarding *eye injuries* - the second most common head and face injury – it is of interest that only one case of retinal detachment was recorded. *Jaw injuries* included five temporomandibular joint sprains and two contusions. No fractures were recorded. *Ear injuries* included the traumatic perforations of the tympanic membrane. Eight of the mouth/throat injuries were oral lacerations.

Back: Twenty of the *back injuries* involved the lumbar spine; 11 involved the thoracic spine. Twenty-five of the back injuries were strains or sprains, five were muscle sprains, and one was a contusion.

Cerebral: Surprisingly, only 29 cerebral injuries occurred, 28 were *concussions,* and one was diagnosed as a *traumatic headache* (for all practical purposes, this injury could be categorized as post-concussion syndrome). The authors were unable to determine the neurologic manifestations of the concussions, with the method of data extraction the authors used. However, 26 of the concussions were recorded as mild.

Cervical spine, brachial plexus: Most of the 23 *cervical spine* and *brachial plexus injuries* were strains of the neck or the trapezius muscle. Twenty-one cervical spine and two brachial plexus injuries were observed.

Chest: The 17 *chest injuries* included 11 contusions (to the rib, chest, or sternum) and two *rib fractures.*

Kidney: Only one *kidney contusion* was recorded.

Cerebral injuries accounted for only 29 (6.5%) of the 447 injuries reviewed during the period investigated. This relative low percentage is, according to JORDAN et al., similar to that reported by WELCH et al. (1986) for the USMA institutional boxing program. These authors noted a total of 294 moderate cerebral injuries, 22 (7.4%) of which were attributed to blunt trauma to the head. Although WELCH et al. did not include whether their definition of head injury was the clinical equivalent of cerebral concussion, they did state that no head injuries resulted in neurologic deficits.

JORDAN et al. stress that the low frequency of cerebral injuries among boxers at the *USOTC* and cadets at the *USMA* is probably attributable to the fact that boxing at these institutions is primarily noncompetitive – it consists mainly of sparring and instruction. JORDAN et al. are quoting WELCH et al. (1986) who reported a head injury rate of 2.38 per 1.000 hours of competitive boxing vs. 0.76 per 1.000 hours of instructional boxing.

JORDAN et al. noted that the frequency of cerebral injuries at the *USOTC* is considerably lower than reported for the 1981 and 1982 *USA/Amateur Boxing Federation National Championships.* During these events 48 matches were stopped because of a blow to the head. Assuming that the boxers who received the head blows sustained cerebral injury, 48 of the 100 boxing injuries incurred at these events were cerebral. This high number may reflect, according to JORDAN et al. an increased injury rate in competition compared with sparring and training. The authors mentioned that most of these boxers were fully alert, and only a few lost consciousness or had amnesia. Although 48% of the total injuries were head blows, the rate of head blows was 4.38 per 100 exposured.

JORDAN et al. noted that the two year survey of acute injuries among professional boxers in *New York State* (JORDAN and CAMPBELL 1988) showed a total of 376 injuries, of which 262 (69.7%) were craniocerebral. This relatively high proportion of craniocerebral injuries, compared with the low percentage of cerebral injuries noted at the *USOTC* (6.5%), is attributable to the following factors:

"The definition of craniocerebral injury in the New York survey was any technical knock out (TKO) or knock out (KO) resulting from a head blow. Accordingly, all TKOs and KOs were considered craniocerebral injuries unless otherwise specified. Since not all TKOs and KOs necessarily result in a concussion, injuries could be classified as craniocerebal even though there was no neurologic dysfunction, and therefore these injuries could be overrepresented."

"The New York survey included only acute injuries incurred during competition. Therefore, chronic injuries sustained during sparring and training were not assessed. If one assumes that chronic injuries and injuries sustained during sparring and training are less likely to be craniocerebral, then excluding these would elevate the percentage of craniocerebral injury."

"The higher percentage of craniocerebral injuries reported in the New York survey may reflect the fact that the boxers were professionals rather than amateurs."

"Upper extremity injuries appear to be fairly common in boxing, accounting for 32.9% of injuries in this study. Upper extremity injuries incurred at two amateur boxing national championships and for 45.6% of all injuries evaluated at the *USMAS* during a 2-year period. However, they were relatively uncommon (2.4% of all injuries) among the professional boxers in New York State."

"The results of this survey and others indicate that eye injuries make up a rather small percentage of all injuries. Only 23 (5%) of the 447 injuries in the present study involved the eye. ESTWANIK et al. (1984) found only three eye injuries (none serious), and WELCH et al. (1986) encountered no eye injuries. Eye injuries accounted for about 2.1% of all injuries sustained by the New York professional boxers."

"The major advantage of tabulating the boxing injuries at the *USOTC* is that it reflects the distribution of acute as well as chronic injuries sustained during sparring, training and competition. Injuries that occur during training are more likely to include chronic overuse injuries to the extremities. Overuse injuries to the lower extremity (e. g., the knee and ankle) may be associated with jogging and jumping rope. Comparable overuse training injuries to the upper extremity (primarily the shoulder and elbow) are associated with repeated delivery of punches during noncontact boxing drills."

"By comparison, the traumatic injuries incurred

during sparring and competitive boxing might include more facial lacerations, eye injuries, and cerebral concussions. Traumatic injuries to the hand, however, can occur in sparring, training, and competitive boxing. Repetitive pounding of the heavy punching bag with the wrapped and/or gloved fist can result in significant hand injury during training."

The authors noted that the methodological issues listed below can influence the distribution of boxing injuries encountered in a survey:

"The population studied may consist of amateur or professional boxers. For example, one might expect a lower frequency of facial lacerations among amateurs because of protective headgear these boxers wear."

"In view of these considerations, comparing various surveys can be difficult because different methods may have been used."

The authors close their discussion: A major limitation of their investigation is that they presented specific injuries as a percentage of the total number of injuries. They did not calculate the number of injuries per number of boxers because the total population base of amateur boxers who attended the *USOTC* was unobtainable. Despite this limitation, however, the distribution of various boxing injuries can provide useful information regarding the relative frequency of injuries within a defined population.

CAVANAGH (1992) noted that *fractures of the trapezium* are rare, representing about 3.5% of all fractures of the carpus (BOGESKOV et al. 1966). Of the reported fractures only 20% have been vertical.

A *vertical fracture* of this bone was sustained by a 28 year old professional boxer, the British cruiser weight champion, during the course of a fight. Radiographs and physical examination failed to show evidence of subluxation of the first metacarpal and a short period of immobilization in a plaster cast was therefore advised. He declined, however, in view of his imminent Commonwealth title fight.

CAVANAGH emphasizes that the mechanism of injury is controversial. KÖHLER (1928) wrote that isolated fractures of the trapezium were particularly rare and almost impossible to reproduce experimentally. KINDLE (1910) who performed cadaveric experiments wrote that the trapezium was more likely to fracture when the thumb was abducted and the first metacarpal strongly flexed. ROCKWOOD and GREEN (1984) wrote, "fracture through the articular surface of the trapezium is produced by the base of the first metacarpal being driven into the articular surface of the adducted thumb."

COHEN et al. (1990) discussed in their contribution *athlete's nodules* and their treatment by surgical excision. Athlete's nodules are connective tissue nevi of the collagen type (collagenomas) which appear as thick dermal masses at sites of chronic trauma. They have been described as occurring on the dorsum of the feet, knees, and knuckles of surfers, boxers and marble players. Recurrent minor blunt trauma and pressure are etiological factors in this condition.

Rib Fractures in Boxers

Rib fractures in *boxing* are probably not rare but they are rarely described in the literature.

A *fracture of a rib in boxing was* seen by REFSHAUGE (1963).

A *spontaneous pneumothorax as* the result of boxing was published by REFSHAUGE (1963).

Traumatic Pneumothorax

Traumatic pneumothorax in association with rib fracture and spontaneous pneumothorax are both recognised conditions (BAISCH 1965, WILSON 1970, CHESTER et al. 1982, GETZ et al. 1983). Pneumothorax may occur after blunt abdominal trauma when the diagnosis may be difficult radiologically but is suggested by rib fractures (WALL et al. 1983).

BELHAM and ADLER (1985) reported a case of pneumothorax following abdominal trauma in a boxer with minimal symptoms and no evidence of a rib fracture:

A previously healthy, non-smoking, twenty-three year old male professional boxer was struck severely in the left loin during a bout. He won his contest after a further three rounds but then complained of some poorly located pleuritic left chest pain since the delivery of the blow but distinct from the discomfort at the site of impact. At that time there were no abnormal physical signs and he was referred to the Accident Department of The Middlesex Hospital. On his arrival his only persisting symptom was left loin discomfort.

He was a fit well-looking young man without dyspnoea or cyanosis. There was mild tenderness of the left loin and slight, poorly localised, left chest discomfort on maximal inspiration. Physical examination in general and of the chest in particular was otherwise unremarkable, even no reassessment following chest radiography. There was a complete absence of rib tenderness or pain on springing the chest. He was mildly pyrexial – 37.5°C – and urine examination showed traces of both protein and blood.

Chest X-ray showed a small apical and lateral pneumothorax with no evidence of rib fracture.

Management: He was admitted to hospital for observation and remained well. His pain, proteinuria and haematuria, presumably due to minor renal haemorrhage, settled within twenty-four and forty-eight hours. Serial chest X-rays at twenty-four and forty-eight hours showed minimal resolution of his pneumothorax and no new evidence of rib fracture. He was discharged for out-patient follow-up and his pneumothorax subsequently resolved spontaneously and he remained well.

The authors discuss, that closed traumatic pneumothorax typically occurs in association with rib fracture when these diagnoses may be made easily both clinically and radiologically. The authors draw attention to this case as illustrating the ease with which pneumothorax may be overlooked when symptoms are mild and transient and the usual physical signs absent.

In this case, according to the authors, the definite history of pleuritic pain, although mild and transient, alone led to the diagnosis.

The injury should be, according to BELHAM and ADLER, considered particularly in boxers and others subjected to direct blows, even where these do not involve the chest, and be excluded radiologically if there has been pleuritic pain despite the absence of abnormal physical signs.

The authors summarise: "The demonstration of this injury is clearly important not only for its immediate management but also for its implications in advising the timing of a return to body contat sport."

Muscle Ruptures in Boxers

Ten cases of *muscle ruptures in boxers* were reported by BORCHERS and TONTSCHEFF (1932). The ruptures affected the pectoralis major, the anterior brachial, the biceps, and the abdominal muscles on the right side.

Another case of a ruptured biceps muscle was reported by REFSHAUGE 1963).

KRAUS (1934) published the case history of a 23 year old boxer, who had been hit on the chin while sparring three days earlier. At the same moment when he tried to land a straight left, he was hit by a left counter, thrown by his opponent. The patient missed with his left, and at the same time he felt a lightning-like pain in the left upper extremity, which was weak ever after.

The *clinical examination* revealed a stretching of the roots C_5-C_7 left, causing a primary weakness of the serratus muscle.

KRAUS concluded that a stretching of the brachial plexus occurred at the same moment as the patient missed with his blow and was hit by the counter of his opponent.

Paralysis of the right anterior serratus muscle due to paralysis of the long thoracic nerve was described by GUTTMANN (1937) in a 28 year old boxer. During a boxing match he had suddenly felt a lightning pain in the right shoulder.

WOODS (1957) described the *dislocation of the shoulder joint in a boxing match.*

In the course of a light heavyweight bout between *R.A.F. College, Cranwell,* and *Cambridge University, a* cadet was seen suddenly holding his left elbow with his right glove, completely disabled. The round ended almost immediately, and he sank into his corner in great pain.

It is often difficult to see some decisive incidents in the ring (hence multiple judges as well as a referee), and in this instance WOODS was 'stymied' by the referee and saw nothing.

As he climbed up to investigate, a colleague in the front row helpfully hazarded the opinion that the man had dislocated his shoulder. On examination this 'spot' diagnosis was seen to be perfectly correct with the classical flattening of the deltoid region and the bulge of the head of the humerus anteromedially and subcoracoid. It was not the moment or the place for bothering the patient about the detailed mechanism whereby this extraordinary happening had come about. WOODS was puzzled because in boxing, the left upper limb is mainly concerned with the safe piston-like movements of leading and guarding, in which no considerable abduction is involved. It would have been much more explicable in the case of the right shoulder-joint, where wide abduction occurs in swinging a right hook to the jaw. Moreover, this was not a recurrence. The man had never dislocated his shoulder before, and so lacked the notorious tendency to recurrence. WOODS wrote to the patient and to the S.M.O. at Cranwell R.A.F. College, asking if they could throw any further light on the dislocation, and I also questioned the opponent.

The cadet replied: "It was very difficult, as I didn't actually feel it happen at all. The only sensation was of not being able to raise my left arm, and then I thought it must be cramp. I didn't realize the shoulder was dislocated until I returned to my corner at the end of the round." He continued, "I have asked independently all the members of the team who were watching, and... it appears that... I attempted a left hook, and while my arm was fully extended (my punch having missed) he hit my arm with a right-hand punch."

The opponent confirmed this missed hook, and stated that he himself had a very sore right ulna from the violent impact of his own upthrusting forearm blow in warding off the said hook.

WOODS concluded:

"It would appear that the patient's left shoulder-joint was relaxed, with flaccid muscular control, at the moment when his arm was widely abducted and the blow missed. Simultaneously the Cambridge man violently forced the arm much further upwards, bruising his own forearm, and levering the head of the humerus downwards out of its shallow socket and through the capsule of the joint."

The stability of the shoulder joint, with its shallow glenoid cavity providing little or no support to hold the head of the humerus, depends on the tension of the tendons passing under, over, or through the joint and the elastic compressing pull of the muscles acting upon it. The dislocation happened at the moment of misfire, when all these structures were relaxed.

WOODS (1949) pointed out that he had never seen this accident in the boxing ring before, during thirty years at the Cambridge ringside, nor had he heard of it elsewhere. It would seem that primary dislocation of the shoulder-joint in the boxing-ring is an extremely rare accident, comparable in rarity with dislocation of the hip joint at 'rugger'.

Another case of a *dislocated shoulder,* which resulted during infighting, was seen by REFSHAUGE (1963).

Injuries to Internal Organs

Cardiac Injuries

The *heart and pericardium,* lying close behind the anterior chest wall, are very vulnerable to blows in the region of the precordium. Traumatic injuries of the heart while boxing are often reported. JOKL (1941) wrote that professional boxers have found that the effects of well-placed hooks to the heart last for a considerable period, usually for the rest of the fight. He also considers that chest injuries in boxers are very much more common than is generally assumed.

A blow against the thorax near the heart can produce a sudden loss of muscle tone, so that the boxer goes down, and unconsciousness may well occur *(heart knock out).*

TUNNEY (1932) mentioned details about a blow that he landed on DEMPSEY's heart region:

"In the spring of 1929, while coming north on the train after promoting the SHARKEY - STRIBLING fight at Miami Beach, DEMPSEY went into the compartment of Roy HOWARD and Ray LONG, the publishers. After talking a little about the Chicago fight and, in particular, that seventh round, DEMPSEY exclaimed: 'The right-hand punch under the heart that TUNNEY hit me when he got off the floor in the seventh round was the hardest blow I have ever received. It was not a question in my mind of being knocked out I thought I was going to die. I could not get my breath. A second rubbed away the congestion around my heart when I came back to my corner; but for that I would not have been able to come out for the eighth round.' Gus WILSON his worshipping trainer, has told the same story."

KÜLBS (1937) reported the case history of a young boxer who survived such an injury, but fatal injuries have been described in a few cases (von BRAMANN 1927, DEUTSCH 1928, JANKOVITCH 1935).

Such cases are clinically diagnosed as *commotio cordis.* The blow against the precordial area of the thorax can be considered a blunt impact, which initiates coronary spasms. These spasms have, in regard to their localization and duration, effects on the myocardium. The arterial blood pressure decreases.

According to SLAPAK (1956), the following clinical symptoms appear: A drop in blood pressure, vertigo, and unconsciousness, parallel with changes in the ECG. In mild cases extrasystolies or bradycardies with sinus rhythm occur, and occasionally auricular fibrillation. In severe cases atrioventricular conduction defects of all types appear. In cases with a complete block, an Adams-Stokes syndrome can develop and the patient can succumb to such an attack. The further complication of irreversible ventricular fibrillation may also develop. Cases of commotio cordis have no relation to the results of a blow against the solar plexus.

JOKL (1941) reported the case history of a ten year old boxer who had received several blows to the thorax and abdomen and died immediately after the fight. The *autopsy* revealed a coronary sclerosis and coronary thrombosis.

PROHIC (1959) reported the case history of an 18 year old amateur boxer who died shortly after receiving a blow against the precordial region.

The *autopsy revealed* a suffocative asphyxiation. The blood reached the bronchi from a small rupture of the lung tissue.

MUNSCHECK (1964) reported on an 18 year old all round athlete, who was in good health and had no subjective complaints. He participated in a sparring match using 12-ounce gloves. He was hit at the end of the fourth round, receiving a relatively light (according to his trainer) counter blow against the left thorax. He immediately discontinued the fight and went to the ropes where he collapsed. He was dead on arrival at the hospital. The *autopsy* revealed severely stenosed coronary arteries. The left and right descending rami showed massive occlusions by atheromatous plaques. The other vessels were unremarkable.

BURGER and WELSCH (1965) described two reflex deaths in the ring:

Case 1: A 20 old boxer received, during a training bout, a blow against the upper abdomen above the beltline. Immediate unconsciousness followed. He died before a physician could arrive, in about 20 minutes' time.

Autopsy findings revealed heart and other internal organs unremarkable. *Diagnosis*: Reflex-death due to severe mechanical irritation of the solar plexus.

Case 2: A 20 year old boxer received, in a training bout, a severe blow against the pre-cordial region with immediate nausea. Bout stopped. Increased unconsciousness with bluish discoloration of face. Attending physician, arriving 10 minutes later, declared him dead. Heart and coronary vessels were unremarkable. *Diagnosis*: Contusio cordis.

DEUTSCH (1928) published the case history of a 26 year old professional boxer who had taken up boxing at 19.

He fought for two years as an amateur, and turned professional at 21. He had participated in a large number of fights and was the champion in the light-heavyweight class. He was known for training very rigorously and did a lot of running. He had never complained of heart problems.

His last fight should have lasted eight rounds against an opponent with whom he had previously fought to a draw. Before the fight, he was pale and irritable. In the third round he received a hook against the precordial area without having previously taken any heavy punishment. He dropped like a stone and was counted out. Immediate death followed.

Autopsy revealed a large hypertrophic heart. Coronary vessels, aorta and other organs were unremarkable. *Diagnosis*: Reflex death due to severe blow to the pre-cordial region.

A 1925 editorial in the *Journal of the American Medical Association* reports on an inquest held on the body of a boxer, aged 21, who died suddenly after a contest. The evidence showed that the length of each round was two minutes of boxing, with one-minute intervals. At the end of the fifth round his opponent was half a point in front. The boxing had been fair on the part of both participants. In the sixth round they had been boxing half a minute when the referee stopped the fight. The decedent appeared to hang on to his opponent. At that point the fight seemed tame, an exhibition sort of affair. All through it was a very fair contest. The decedent looked surprised at his stopping the fight, but shook hands with his opponent and proceeded to his corner, walking across the ring. As he did so he was seen to hold his side, to sink gradually to the floor, and to roll over. The referee went to him and massaged his heart, but he showed no signs of life. There was no breaking of any of the ordinary rules of boxing in the match and no signs of anger on the part of either boxer. There was no knock down blow in the match.

Dr. R.M. BRONTE, the government pathologist, made a *necropsy*. He found two or three little bruises on the body of an external character and of no particular moment. The skull and brain were normal and the heart weighed 11 ounces (312 g), the normal weight. Death he considered due to reflex inhibition of the heart as the result of a blow or blows. He knew of the similar death of a man who had had a knock in the same spot by another man's elbow. Similar deaths had also occurred in the football field. The coroner said that in view of the evidence it was difficult to see how a charge of manslaughter could be brought, and the jury returned a verdict of accidental death.

It is impossible to or one cannot over emphasize the point that in many kinds and cases of boxing injuries, there is often a marked discrepancy between the severity of the morphological alterations in a given organ and the overall physical fitness of the subject. That is to say, often, serious boxing injuries do not show obvious symptoms?

VAJDA (1960) examined 68 amateur boxers aged from 20 to 24 years, 56 after fights and 12 after sparring matches. The distribution of the 68 boxers according to the localization of the blows received was as follows: (1) *Head,* 22 cases, five of them knock outs; (2) *epigastrium,* 5; (3) *chest* 23, four of them knock outs; and (4) *no blows received,* 6; total 56 cases. An ECG was recorded in each case not later than 10 to 20 minutes after the fight. Two hours later a second ECG was recorded on the subjects who had shown some change.

The boxers were divided into *two groups. Group (1)* consisted of those 23 fighters who had received chest blows and the 12 subjects who had been their sparring partners. The remaining 38 fighters, who had suffered no blows, or no blows to the chest, were classified as *Group (2).*

There were eleven cases of *electro-cardiographic changes* in *Group 1* (31.4%). Three of these were knock outs from chest blows. There was also a fourth knock out in this group: a boxer received a series of blows to the chest in each of two consecutive days and was counted out on the second day, although repeated electro-cardiography failed to detect any pathological signs. Of the 12 trainees in Group 1, only 3 showed ECG changes. Although these men had exchanged blows aimed almost entirely at the chest, they had put less force behind their punches in sparring than they would have done in a contest.

Electro-cardiography revealed changes in four cases (12.4%) from Group 2; one of them had been knocked out, and three had received several series of blows to the head.

Deviations mostly presented themselves in the ST interval and the T wave; in one case extrasystolia was observed. Especially marked were the changes in a 21 year old boxer who had been knocked out by a blow against the precordium. An ECG taken ten minutes after the event revealed a deep Q_1 (3mm), negative-positive T_1, flat T_2, high R wave in V_1, and deep and wide Q and negative T in V_2.

Similar changes were recorded in another 22 year old boxer, who in a contest had gone down on two occasions from punches to the chest for, 2 or 3 seconds, and was counted out on the third: in leads II and III the ST segment was expressed, T_2 flat, T_3 negative; two hours later all reading were again normal.

Of the eleven subjects with ECG changes, six showed injury to the posterior and four to the anterior wall of the heart, and one exhibited disturbances of rhythm. Normalization always took less than 24 hours.

With a view to establishing the extent to which ECG changes were due to physical strain and direct chest injury respectively, records taken after sparring were compared with those taken after contest fights of each of a number of boxers who had suffered no chest blows.

In view of the evidence, VAJDA concluded, the ECG changes were probably due more to the chest injury itself than to the physical and nervous strain during the fight. This was supported by the fact that in several of the cases, ECG changes were recorded in consequence of blows to the chest the boxer had received in sparring, while in the same subject no such changes appeared after fighting in a contest when he had suffered no chest blows.

PARSONS-SMITH and WILLIAMS (1949) presented an interesting observation of a patient who was a very robust, well-developed schoolboy aged 15 1/2, who weighed 72.12 kg.

He had lived in Tanganyika until he was five, and whilst there he had experienced several attacks of fever. He had otherwise been healthy, and in September of 1946, March of 1947, and March of 1948, had been found fit in routine school medical examinations. There was no history suggesting rheumatic fever.

The *first incident* occurred about one year before the present illness. In March, 1947, when he was boxing at school in a competition, the fight had to be stopped in the first round as he was in severe pain following concentrated punishment to the region of his heart. This pain, which was dull and aching in character, continued for the next week or so. It was aggravated by exercise, and the patient remembered that he had to stop in a cross-country run and walk home because of the severe dyspnea and pain in his left chest, produced by exercise. After this there was no recurrence of the pain, and for the next year he played various school games, including football, with some distinction and without restriction.

He next took part in competitive boxing when he had to fight the boy who had beaten him so thoroughly the previous year. Again he received severe punishment to the left chest, and at the end of the first round he complained to his seconds of dull and sickening pain over his heart. The fight was stopped in the second round, as the boy was in agony. He thinks that he received only one glancing blow to the left side of his head, and he was in no way dazed or confused. The dull pain persisted, and some hours later he noticed a sharp pain, which from his description was pleuritic in type, over his precordium. He slept badly that night, but the next morning the pain was less, though he felt generally ill and very stiff. He took things quietly, as he found that exercise brought on the pain. Two days later he had improved considerably but was still afraid to exert himself. During that afternoon he was sitting watching the boxing finals, when he became aware that his left leg had suddenly become dead below the hip and that he felt very faint. Two hours later his arm became affected, and shortly afterward he had a general convulsion followed by confusion and headache. No further convulsion occurred, and during the next few days his condition was stationary.

He was *admitted to hospital* five days after the injury. The boy was alert, intelligent, and cooperative. The skull was normal in shape and was not tender. Neck movements were full and there was no limitation of straight leg raising. The cranial nerves were perfectly normal, and in particular there was no papilloedema and no field defect. Speech was normal. On the affected left side the leg was very spastic, there was ankle clonus, tendon jerks were increased, and there was an extensor plantar response. Voluntary movements and coordination were considerably impaired in the leg. The arm was similarly affected, but to a much smaller degree. Sensation was normal to full testing. The pulse rate was 80/min and was irregular. The blood pressure was 130/70 mm/Hg. There was no clinical enlargement of the heart. The heart sounds were distant and difficult to hear. A soft systolic murmur was heard over the mitral area, and a basal pericardial rub. Blood was normal. Blood Wassermann and Kahn reactions were negative.

Roentgenograms of the skull, heart, ribs, and lung fields all showed no abnormality.
The *EEG was* abnormal. There was a low-voltage alpha rhythm of 10 c/s that underlay a generalized symmetrical abnormally which consisted of runs of high voltage 1 c/s and medium voltage 3-4 c/s. There was no evidence of focal abnormality nor of one side being worse than the other. An *ECG taken* at the same time was also abnormal. The curves showed evidence of partial heart block due to two distinct lesions. First there was an occasional omission of auricular systole, though the auricular beats were for the most part regular. This would indicate some partial sino-auricular block. Secondly, the P-R interval was prolonged and variable. It varied from 0.2 - 0.4 seconds in length, which would indicate partial disturbance of conduction of the auriculo-ventricular system. The ventricular rate was 84/min. The terminal deflections in the three limb leads were normal.
He was kept in bed and given occupational therapy, with resultant improvement in the power of his left upper limb after a week, though the leg was still very spastic. The cardiac arrhythmia persisted, but the heart sounds improved in quality. The systolic murmur increased in intensity; it became harsh and occupied almost the entire systolic period. It was audible all over the praecordium, and was maximal in the left third and fourth interspaces 6.25 cm. from the mid-sternal line. The pericardial rub was heard only during the first two days in hospital.
An *EEG taken* one week later showed a considerable change. Whereas there were generalized slow waves in the record taken on April 5, these persisted in the left hemisphere and there was suppression of all waves in the right parietal region.
A *second EEG* two weeks later showed a slight improvement. A low-voltage dominant frequency of 12 c/s had returned. There were more 6 c/s waves on both sides, and runs of 3 c/s were seen which were more evident on the right than on the left. The records were still far from normal, but showed slight and gradual improvement.
Further ECG recordings were made, and these showed very little change. An intramuscular injection of 1.3 mg. of atropine sulphate caused an increase of pulse rate to 120/min, the rhythm became normal and the P-R interval shortened to 0.16 s.
The hemiplegia continued to improve so that he was fit to be discharged from hospital six weeks after admission. Since then his convalescence has been controlled and uneventful.
An *ECG* taken four weeks later was reported to show partial heart block of varying character without the usual Wenckebach's periods. The auricular and ventricular complexes appeared normal, but the auricular rate was not regular. It appeared, therefore, that there was a disturbance of sino-auricular activity as well as A-V conduction. Nearly five months after the injury, examination showed that improvement still continued. There were only residual signs of the hemiplegia, which caused him very little disability. The *EEG* showed a further slight improvement, the records from the hemispheres being symmetrical. The heart was not enlarged, the sounds were of good quality, and no murmurs were heard. The pulse was still irregular, rate 78/min. The *ECG still* showed sinus and A-V block, but the P-R interval was shorter than had been seen previously and varied between 0.18 and 0.22 second. He was fully ambulant and was due to start on a sedentary engineering course of instruction.

BELLOTTI et al. (1992) described a case of a cardiac trauma observed in a professional boxer after a world championship boxing match.

A 32 year old professional boxer was referred to the authors with a diagnosis of traumatic detachment of the retina two days after a world championship boxing match. He was asymptomatic and the only remarkable findings were multiple ecchymotic injuries of the face. During his hospitalization, while waiting for a possible surgical intervention for repair of the detachment of the retina, he underwent routine chest-x-rays, electrocardiogram and serum blood examinations.
The *electrocardiogram,* in absence of symptoms, revealed unexpected abnormalities in the precordial leads, with a negative T wave in V1 and a two to four mm ST – segment elevation of V2 - V3. To assess the evolution of the myocardial injury, serial cardiography and 2-dimensional echo-cardiography, and one radionuclide angiography at rest were performed. The electro-cardiographic abnormalities returned to normal during one month. Technetium pyrophosphate radionucleotide angiography was normal with homogenous cardiac uptake distribution. Cardiac enzymes revealed a slight increase in total creatinine phosphokinase and MB fraction with peaks of 297 and 39 nV/liter, respectively, (creatinine phosphokinase normal laboratory range, less than 160 nV/liter).

The authors concluded that contusion of the heart is a common and sometimes lethal sequelae of blunt trauma to the chest. It has been well documented after motor vehicle accidents and only rarely during sport activities. The contusion of the heart may occur without serious external injury, and may go unnoticed clinically. With regard to the time interval between the heart contusion and electro-cardiographic changes, ROTHSTEIN reported that the onset of the abnormality is frequently delayed for 48 hours (1983), and it may appear as many as five days after the injury, usually resolving over a period of 30 days.

COUSTEAU (1993) noted that the case of myocardial contusion after a boxing match by BELLOTTI et al. (1992) was not absolutely conclusive. He wrote, that although this diagnosis may be the right one, there is another one, not discussed by the authors. The *athletic heart syndrome* that can mimic the abnormalities seen on the electrocardiogram and that disappeared after one month of rest. There was no significant deviation of the cardiac enzymes, according to COUSTEAU and no alteration of the echocardiogram. The author asked: "Did this athlete have an electrocardiogram immediately before his match."

BELLOTTI answered that it is well known that intensive athletic training can be associated with alterations of the resting electrocardiogram (HUSTON et al. 1985). In the case reported by BELLOTTI et al., the electrocardiogram recorded shortly before the boxing match (not shown in the text) had a normal pattern of repolarization. Therefore, according to BELLOTTI, the ST elevation resembling transmural ischemia that could be observed in this athlete after the boxing match is likely to be an expression of myocardial contusion rather than the consequence of athletic training.

Acute Renal Damage

AMELAR and SOLOMON (1954) studied the changes that occur in a man's urine after he has engaged in a boxing match. It is necessary to consider not only the factor of trauma but also that of strenuous exercise, for it is with both factors that we are dealing, and the effects of the latter should not be confused with those due to the constant infliction of blows on the kidney region during a fight. That urinary changes do occur following muscular exercise is a fact well chronicled in the world's physiological literature.

Following exercise there is a decrease in the urine volume. The renal plasma flow and filtration rate have been shown to be markedly reduced during exercise. If the exercise is severe, it may be accompanied by albuminuria. Following exercise there is a decrease in the chloride concentration in the urine and an increase in the urinary acidity, ammonia and phosphate.

AMELAR and SOLOMON examined 139 professional boxers; in 103 of these cases, a comparison was made of urinary findings before and after the bout.

A freshly voided specimen of urine was collected from each boxer on the day of the fight. While the bout was in progress, an attempted estimate was made of the number and severity of blows received in the region of the kidneys, and a note made of the number of rounds the fight lasted. Back in the dressing room approximately five to twenty minutes after the fight, another voided specimen was collected from each boxer.

The authors found that 46% of the fighter's urine specimens changed from clear before the fight to turbid immediately afterwards. The urinary specific gravity increased in 80% of the cases. Traces of acetone were manifest in 14%. The urinary PH decreased in 30%. Sugar was spilled in the urine of 9% of the boxers. Albuminuria, which was not present before the bout, was present in 68% of the boxers after the fight.

Microscopically red blood cells were found in significant pathological amounts in 73% of the boxers after fighting, and 26% were found to have casts of granular or hyalin variety in their urine. In each case, when casts were present, red cells were also present microscopically.

Small numbers of crystals, mostly calcium oxide, were seen in the urine of 20% of the boxers.

The data obtained from urinalysis were broken down further, in an attempt to correlate the severity of post fight urinary changes. Several factors were examined, including the number of rounds fought, estimated severity of blows to the kidney region, and body weight and over all protective muscle mass of the boxer. The only correlation found was the number of rounds fought.

For example, 65% of those who fought *one to six rounds* had *hematuria,* but it was present in 80% of those fighting *seven* to *twelve rounds.* The number of red cells present was much greater in the latter group, four of whom had *total gross hematuria.* Similarly, *albuminuria* was present in 60% of those who fought 1-6 rounds and in 87% of those who fought 7-12 rounds.

AMELAR and SOLOMON concluded that their data proved the obvious. In summary they found that, in addition to the effects of exercise, there is also a separate acute traumatic reaction observed in the urine of boxers, immediately after a fight. *The severity of urinary changes correlates with the duration of the fight, which, in turn, determines the amount of renal trauma.*

The fighters experience acute trauma to the kidneys evidenced by the finding of a high percentage of albumin, casts and red blood cells in the urine. Exercise alone will not produce these changes.

The authors conclude that it was not in the scope of their paper to arrive at any conclusion as to the long range effects of acute renal trauma experienced during the course of repeated boxing matches. The question of the effect of repeated trauma in the production of scarring in the kidneys and eventual diminution of renal function should be studied. *The authors close their discussion with the statement that a boxer should not be allowed to return to the ring until his urinary findings have returned to normal.*

KLEIMAN (1958) studied the renal and vesicle functions of 764 professional boxers. A urologic history was obtained from each subject, and altogether 2.200 specimens of urine were obtained after bouts. Only 1.518 urine specimens were obtained immediately after bouts, because 550 boxers were unable to urinate for several hours after fighting. Comparisons were also made between 132 substitutes who were prepared to box but not called upon to do so, and pugilists who did box.

Significant hematuria was found in 27% of the post-bout specimens of urine; 30 specimens were salmon pink, eight were bright red, and four were dark red. *Roentgenographic studies* revealed structural abnormalities of the kidneys in many cases of hematuria, and suggested that such defects render the kidneys more susceptible to the trauma of boxing (as demonstrated by the high incidence of recurrent post-bout hematuria).

FLOOD (1960) studied the occurrence of albuminuria and hematuria in 35 boxers at an Army boxing tournament. *Albuminuria* occurred in 15% and *hematuria* in 28%. These findings in amateur boxers are in agreement with the statistical incidence in professional boxers reported above, but the incidence of albuminuria in this study was much lower than in the larger series dealing with professional fighters (15% versus 68%).

It has been well established that self-limited, benign changes frequently occur in the urine following exercise. This study was undertaken by HOPPES et al. (1991) to evaluate potential changes in the urine of amateur boxers. Demographic data, a boxing history, a complete medical history, and a physical examination were obtained for each subject. Prefight and postfight urine were examined for comparison. The data were subjected to statistical analysis in order to evaluate the significance of mean changes by using an independent one-way analysis of variance. The authors used the least significant difference test designed by FISHER. Results of this study revealed no clinically significant effect on urine variables other than changes normally associated with strenuous exercise of other types. However, the long-term effect of boxing on renal functions was not explored. Therefore, the authors recommend that all boxers undergo a urine evaluation as a routine part of their ongoing physical examination.

ABARBANEL et al. (1990) examined *sports hematuria.* They noted that strenuous exercise makes extraordinary demands. The transition from rest to intensive physical activity can cause pathological changes in various organs, particularly in the urinary tract. Hematuria (microscopic or macroscopic) is one of the abnormalities commonly found after sports activity. This phenomenon can occur in noncontact sports (such as rowing, running and swimming) as well as in contact sports (boxing, football and so forth). The pathophysiology can be either traumatic or nontraumatic. Renal trauma and/or bladder injury due to repeated impact of the posterior bladder wall against the bladder base can cause vascular lesions and consequently hematuria. There are two mechanisms of nontraumatic injury.

(1) Vasoconstriction of the splanchnic and renal vessels occurs during exercise in order that blood can be redistributed to the contracting skeletal muscles. This causes hypoxic damage to the nephron and results in increased glomerular permeabil-

ity favoring increased excretion of erythrocytes and protein into the urine.

(2) A relatively more marked constriction of the efferent glomerular arterioli results in an increased filtration pressure, which favors increased excretion of protein and red blood cells into the urine. It must be noted that sports hematuria differs from other conditions that may cause reddish discoloration of the urine due to physical exercise, such as *march hemoglobinuria* and *exercise myoglobinuria*. In the latter two abnormalities there is excretion of hemoglobin and myoglobin molecules in the urine and not whole blood or intact red blood cells. Sports hematuria usually has a benign self-limited course. However, coexisting urinary tract pathological conditions should be excluded carefully.

HOPPES et al. (1991) examined the renal response to boxing and changes in the urine in amateur boxers. According to the authors it has been well established that self-limited, benign changes frequently occur in the urine following exercise. This study was undertaken to evaluate potential changes in the urine of amateur boxers.

Demographic data, a boxing history, a complete medical history, and a physical examination were obtained for each subject. Prefight and postfight urine specimens were examined for comparison. The data were subjected to statistical analysis in order to evaluate the significance of mean changes by using an independent one-way analysis of variance. The authors used the least significant difference test designed by FISHER.

Results of this study revealed no clinically significant effect on urine variables other than changes normally associated with comparable strenuous exercise of other types. However, the long-term effect of boxing on renal function was not explored. Therefore, the authors recommend that all boxers undergo a urine evaluation as a routine part of their ongoing physical examinations.

Damage to the Central Nervous System

Consciousness

A discussion about the different disorders of unconsciousness must begin with definitions of consiousness.

William JAMES (1893) wrote that everybody knew what consciousness was until he tried to define it, and no definition that has been given is completely satisfactory.

Purdon MARTIN (1949) uses the word to mean a sense of awareness of self and the environment, well knowing that this definition begs many questions.

The author continues:

"Allowing for marginal overlap, we can divide the states of consciousness in which we observe our patients into four groups:"

(1) The *normal condition,* in which the person is responsive and reacts to psychological stimuli and indicates by his behavior and conversation that he has the same awareness of himself and his environment as ourselves;

(2) *sleep,* a condition of inactivity from which he can be roused and resume his normal state;

(3) a *state of unconsciousness* in which he appears to be deeply asleep but cannot be roused to normal consciousness;

(4) a *much less common state,* or group of states, in which the patient is awake and active but gives evidence that he is not completely, if at all, aware of himself and his doings, and of which he has no recollection afterwards; that are the states presented by certain *epileptic equivalents, postepileptic automatisms,* and a *posttraumatic excitement.*

The *normal state of consciousness* fluctuates in the course of the day within a wide range, from intense attentiveness to drowsiness or to deep abstraction of though with greatly reduced awareness of self and of much of the environment. "It must vary from one person to another," noted MARTIN, "for the stolid and stupid person cannot have the same average degree of awareness as the alert, intelligent, and cultural person who is aware of so many more things with such greater intensity."

Sleep is according to MARTIN a buffer between consciousness and unconsciousness. A person in sleep, if not disturbed, gives little evidence of being aware of himself and his environment.

The outstanding fact about the unconscious person is that he cannot be roused to consciousness.

Is there any part of the brain with whose activity consciousness is particularly associated? GALEN considered the lateral ventricles as seat of consciousness, later the pineal body of the corpus striatum were mentioned. Since Hughling JACKSON and FERRER it is generally accepted that consciousnes depends on the activity of the cerebral cortex.

FOG (1958) noted that our understanding of the mechanism of consciousness can be prompted by studying unconsciousness. The relationship of consciousness to unconsciousness is not merely that of positive to negative.

FOG noted:

"Most recent observations which have enriched our insight into the mechanism and morphological correlates of consciousness have been collected by experiments on lower animals and only partly on man."

The term 'consciousness' is ambiguous: One meaning is the perceptiveness and reactivity of the nervous syetem; and the other is the content of the conscious mind.

Awareness is the basic unspecific function underlying conscious activities as a whole. It is presumably connected with the activator system of the reticular formation of the brainstem, and its tone is maintained both by afferent external and proprioceptive stimuli, as well as by impulses descending from the cortex, and, to some degree, by humoral regulations parallel to chemical regulation of the respiratory and vasomotor centres.

FOG continues:

"The physiological basis of the psychological content of consciousness is unknown but must presumably be sought primarily in the cortex."

"The complicated ramifications of the conscious conduct make it impossible to fix and define consciousness. As Hughling JACKSON pointed out, we are differently conscious (in all respects) from one moment to the next."

"Purdon MARTIN (1949) stated that what we call consciousness is unconditionally connected with cortical activity."

Ritchie RUSSELL (1971) wrote:

The effects of head injury on the mechanism of remembering are both of academic and practical importance. The process of being fully orientated and capable of correlating one event with the next requires that the sequence of events be remembered at least for a considerable time: for this happen it seems necessary to assume a high grade of normal behaviour by the brain."

Different Knock out Mechanisms

Clinically, boxing injuries can be grouped into *acute* and *chronic*. The most notable example of the *acute* group is the *knock out.*

One must distinguish between a *knock out sui generis* and a *technical knock out.* In a *knock out sui generis* a boxer is disabled by a blow or a series of blows and is unable to get to his feet unassisted for a period of ten seconds to resume fighting. Moreover, when he has been knocked down and the bell has rung before the end of the ten count, he is often revived and put back into the ring after one minute to continue the fight.

"The 'knock out' – so dramatic a spectacle in professional contests – entails the deliberate and violent production of a state of motor hypotonus and helplessness coupled with a severe – if short-lived disturbance of consciousness," so wrote CRITCHLEY in 1957. The victim collapses heavily (loss of tone) and remains immobile or inert for a period of seconds or minutes. In falling the head may forcibly strike a ring rope, a ring post or the canvas, hereby producing a second and another direct impact to the head (deceleration trauma) which of itself may complicate the clinical picture.

CRITCHLEY continues:

"If the boxer is rendered so helpless as to be unable to regain the erect stance within ten seconds, he is 'counted out.' After a knock out blow most boxers are able to leave the ring within a minute or two, though ordinarily with the help of their seconds. Longer periods of unconsciousness are by no means unknown, however."

"When a boxer is 'knocked out' it does not mean, as it's commonly thought, that he has been knocked unconscious, or even incapacitated; it means rather poetically that he has been knocked out of Time. (The referee's dramatic count of ten constitutes a metaphysical parenthesis of a kind through which the fallen boxer must penetrate if he hopes to continue in TIME.) There are in a sense two dimensions of Time abruptly operant: while the standing boxer is *in time* the fallen boxer is *out of time.* Counted out, he is counted 'dead' – in symbolic mimicry of the sport's ancient tradition in which he would very likely be dead" (OATES 1987).

It can be indubitable agreed to NOONAN *(Boxing and the Brain. The New York Times Magazine, June 12, 1983):*

> "The knock out is the heart of boxing. It is what every boxer wants to do to his opponent in every fight – as soon as possible. While boxing's defenders have often claimed that knock outs are relatively infrequent, an examination of recent records indicates otherwise. The top ten ranked heavyweight contenders in 1981, for instance, had 149 knock outs listed among their 206 ring victories, according to *The 1982 Ring Record Book,* Larry HOLMES has won 43 fights as a professional and 30 of them are listed as knock outs. And 'Sugar' Ray LEONARD won 23 of his 33 pro victories by knock outs."

CRITCHLEY wrote 1937 that;

> "The 'knock out' is the most acute manifestation of cerebral dysfunction and the 'punch drunk state' is the most chronic among the disorders."

The classical or ordinary cerebral knock out is the result of a blow where the point of impact is the 'button,' that is the point of the chin. The impact is transmitted to the base of the skull. But there is no doubt, that many other types of head blows lead to a knock out. There exists a lengthy discussion about the source of loss of consciousness and tone. The exact mechanism of loss of consciousness and tone is not known.

LA CAVA (1952) is of the opinion that a hard blow to the mandible involves some nerve structures in the mandibular joint. The irritation produces allegedly an abrupt drop in blood pressure. This theory is not proven.
FLINT (1920) believed the knock out is the result of a distortion of the jaw which mechanically impinges upon the carotid artery and so produces a temporary cerebral anemia. This theory has to be refuted for two reasons. First the jaw cannot impinge upon the carotid artery and second a cerebral anemia cannot produce an immediate unconsciousness and loss of tone, at least 8 to 10 seconds were necessary.
PFEIFFER (1922) believed that a short mechanical obstruction of the jugular vein produces a knock out. This theory must be refuted, since a temporary obstruction of the jugular vein cannot produce an immediate unconsciousness and a loss of tone.
HERING (1927) and following him ASK-UPMARK (1935) have pointed out that the carotid sinus reflex plays a part in it, but this probably concerns only a special and rare type, as discussed elsewhere.
WINTERSTEIN (1937) holds that the knock out syndrome is due to acute pressure on the medulla oblongata.

A *boxer who is knocked out* may lie on the canvas in a variety of postures: flat on his back Figs. 184, 185, 188, completely prone with his head on the canvas or with his head lying on his arms (186, 187.

As the result of a hard left hook PORTUGUEZ is lying on his back unconscious on the ring canvas (Fig. 184). The ring canvas around his head shows splashes of water which were probably administered to 'revive' him.

Rocky GRAZIANO was knocked out by Tony ZALE by a crushing right to his chin in the third round (Fig. 185). The knocked out boxer is lying unconscious on the ring canvas on his back. His trainer empties a sponge filled with water at his face in an attempt to regain consciousness.

Such methods, administered by a physician, a nurse or paramedics at the site of an accident or in an emergency room, for treatment of an unconscious patient suffering a cerebral concussion (commotio cerebri) would be immediately involved in a malpractice suit and win it easily for such type of treatment. Hundredth and thousands of spectators who are watching the scene seem to find this therapeutic procedure apt and proper, but would probably vehemently protest, when they were treated for a cerebral concussion in this fashion.

Marcel PIGOU was knocked out in the second round in the fight against Florentino FERNANDEZ (Fig. 186). The boxer lies unconscious flat on his face.

Richard GRUPE landed a powerful blow on the chin of his opponent Werner VOLLMER, who is lying unconscious with closed eyes on the ring canvas (Fig. 187).

In a sparring bout Gene FULLMER knocked out his sparring partner Eddie HICKMAN with a powerful right (Fig. 188). FULLMER's trainer administers oxygen to HICKMAN. FULLMER, the former middleweight champion, is preparing to meet Dick TIGER at the Los Angeles Center.

The professional boxer Julian KEENE lies twenty minutes after being knocked out unconscious on his back in his dirty and filthy dressing room (Fig. 189). I wonder that nobody thought about it to transport him to a hospital.

When a boxer tries to rise after regaining consciousness he is in a dazed state, he staggers and lurches while attempting to gain or keep his equilibrium. He may fall down to the canvas or ropes for a second time due to his disturbed equilibrium.

Fig. 184. Professional middleweight fight between European title holder Gustav SCHOLZ and Tuzo PORTUGUEZ (Costa Rica) in Munich, Germany. SCHOLZ chased PORTUGUEZ around the ring and hit him during the first few minutes with a heavy left hook which knocked him out immeddiately.

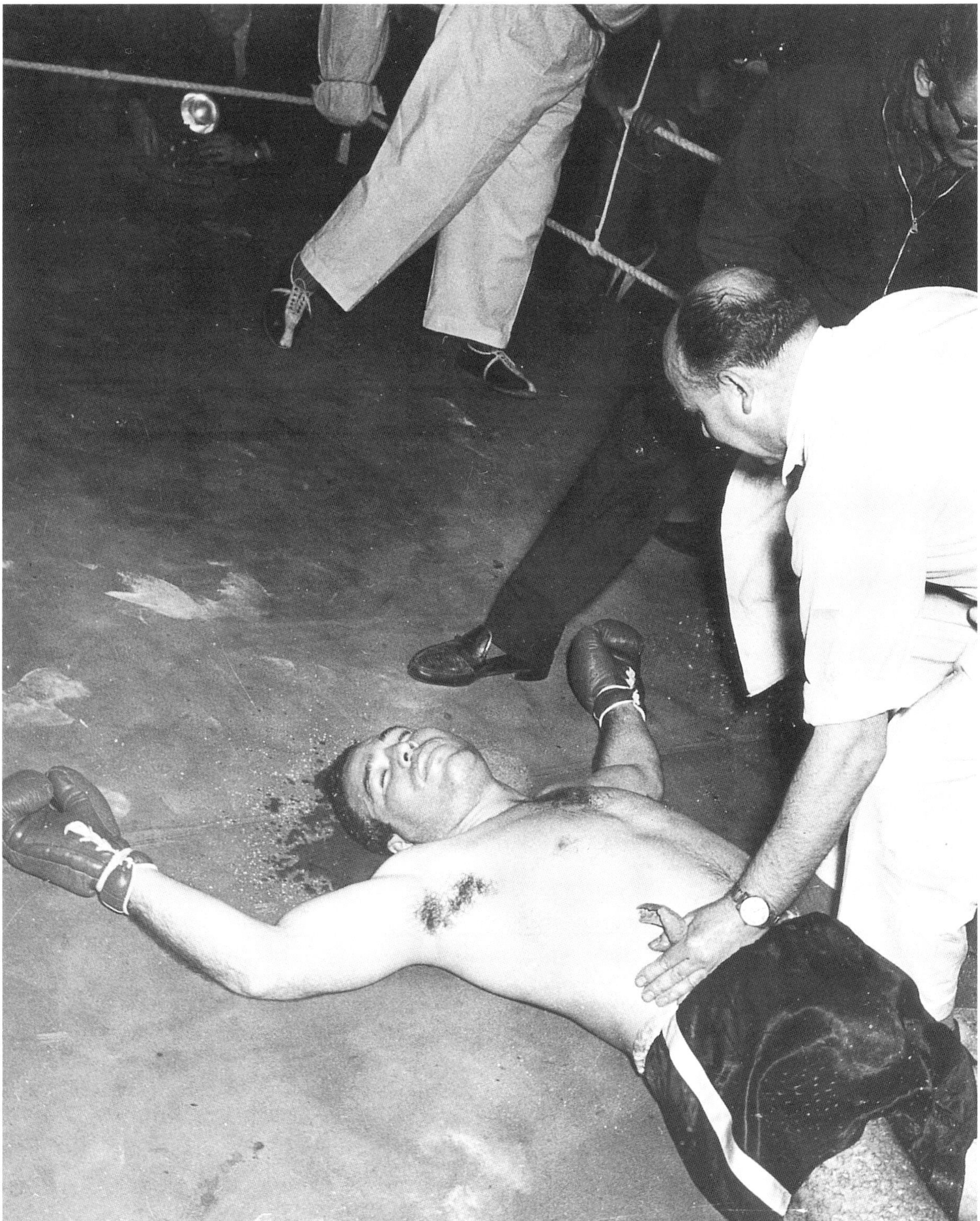

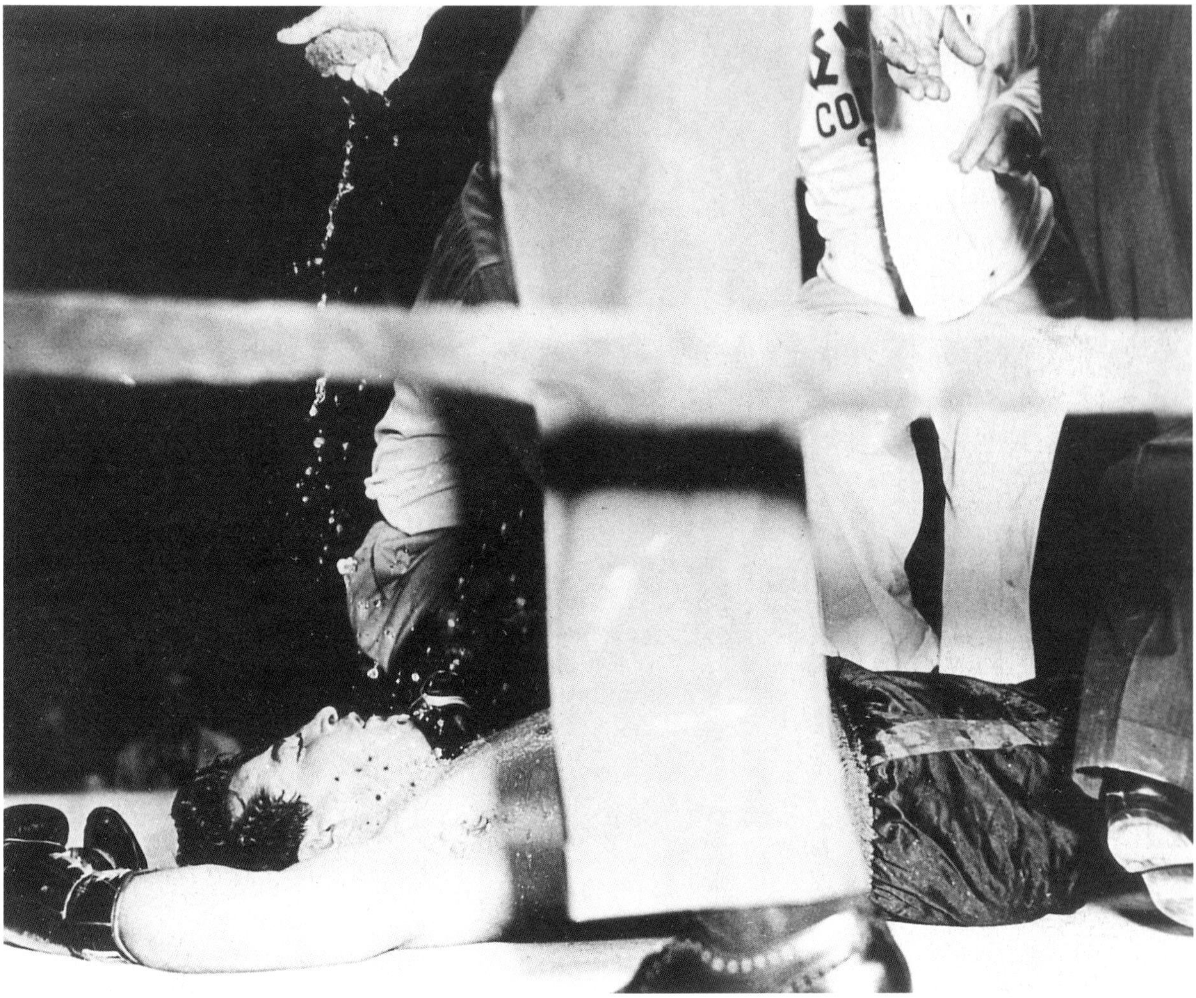

Fig. 185. Professional title fight between Tony ZALE and Rocky GRAZIANO at Newark Stadium. Rocky GRAZIANO is lying unconscious on his back on the canvas while his trainer squeeses a wet sponge at his face, to 'revive' him.

A *technical knock out is* declared when the fight is stopped by the referee due to an apparent injury or the defenselessness of one participant, as determined by the referee. Severe cuts of the face, one contestant completely outclassed by the opponent or excessive fatigue may be reasons for the referee to stop the bout and declare a technical knock out.

Within the definition of the *knock out sui generis* a distinction must be made between a *cerebral knock out* and a *body knock out.*

A *cerebral knock out* consists of: (1) *cerebral concussion (commotio cerebri), with immediate unconsciousness* and *loss of tone*; (2) cases where a series of repeated blows to the head cause a *'pummeling knock out'* ('Verhämmerungs knock out'); (3) *loss of posture* along with *amnesia due to anemia of the brain;* (4) *loss of posture combined with unconsciousness* due to an *oculo-cardiac reflex mechanism* (5) a knock down, *without unconsciousness,* when the fallen boxer is unable to stand up and continue fighting within the ten-count; (6) *several forms of cerebral reflex knock out* in which the mechanism is debatable; and (7) the cases involving *compression* of *arteries leading to the brain,* especially the carotid arteries.

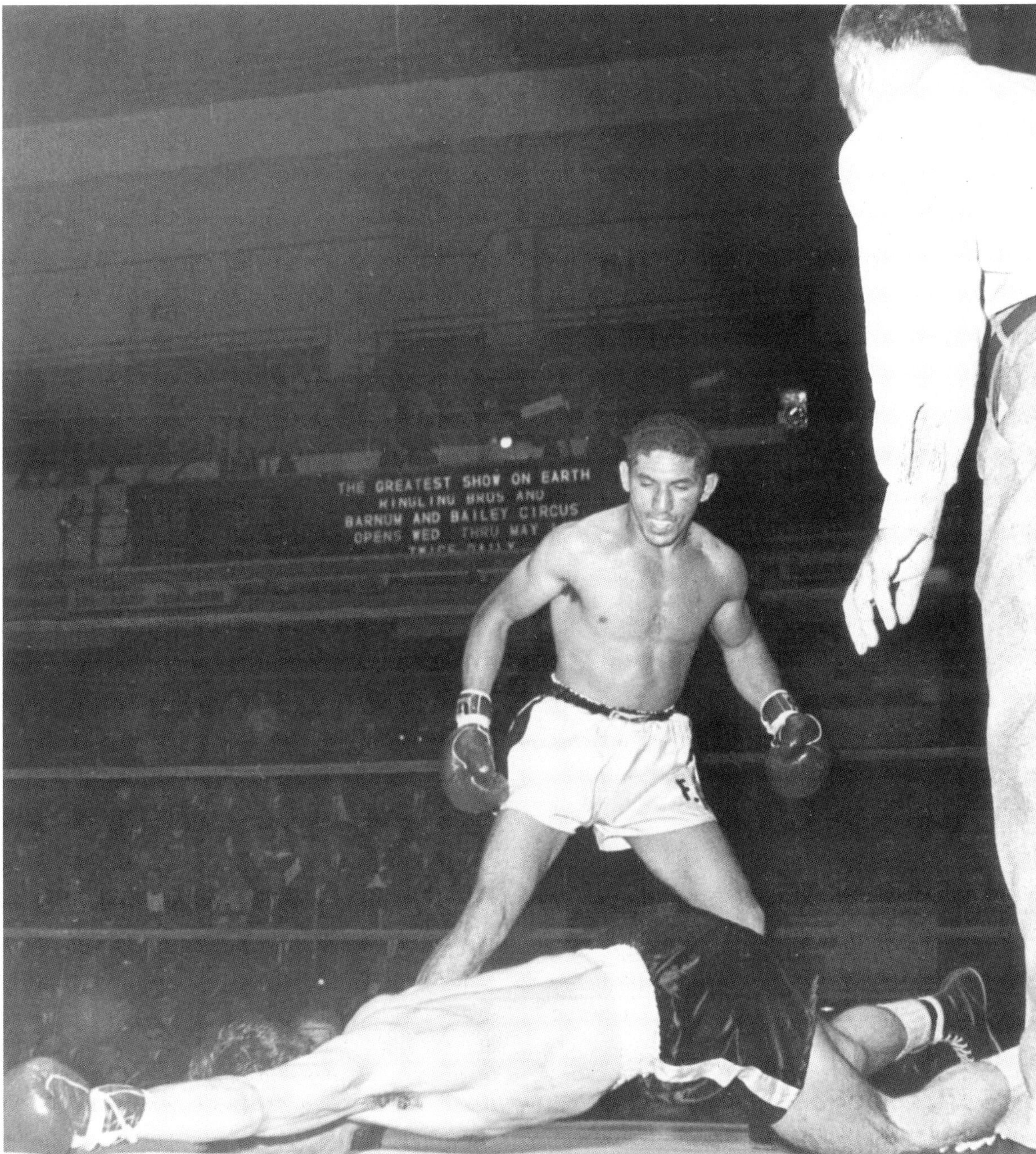

Fig. 186. Professional boxing fight between Florentino FERNANDEZ and Marcel PIGOU on March 23, 1961 at Madison Square Garden, New York. FERNANDEZ knocked out PIGOU in the seceond round of their bout. PIGOU lies flat on his face. The referee is Al BERL. Photo: A.P.

Commotio cerebri (Cerebral Concussion)

A heavy blow may accelerate the head of a boxer sufficiently to produce the *symptoms* of a *cerebral concussion* or *commotio cerebri*, thereby precipitat-

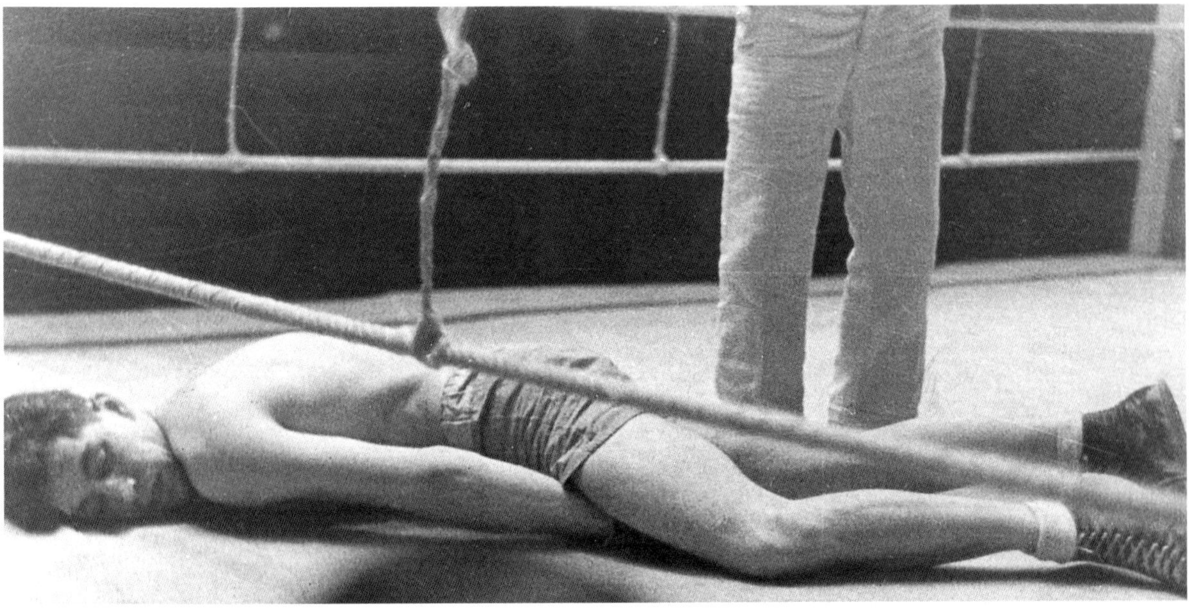

Fig. 187. Professional boxing fight between Richard GRUPE, Hamburg, and Werner VOLLMER, Magdeburg, on December 16, 1946 in the Sport Palace in Hamburg, Germany. GRUPE hit his opponent in the second roud on the chin and knocked him out. VOLLMER is shown lying unconscious on the canvas with closed eyes.

ing *immediate unconsciousness* and *loss of muscle tone.* The intensity of the impacting glove has, in such a case, been high enough to accelerate the head to the threshold of the concussive syndrome. The unconsciousness and the loss of posture occur *instantly*, within a fraction of a second after the impact, and the boxer collapses on the spot; he may land on his back, in a prone position, or even draped over the ropes (for details see page 234). *While falling,* the head may heavily strike the ropes and/or the canvas, thereby producing a *second direct impact to the head,* which may complicate and increase the severity of the clinical picture.

A punch on the *'button'*, as boxers call the point of the chin, is considered especially effective as a knock out blow. It is said of some boxers that they have a so-called *'glass chin'* e. g. they are said to react to a blow against the chin with a knock out, while others with a so-called *'cast iron jar'* can take severe punishment without obvious effects.

The reasons for the special effectiveness of blows against the point of the chin are not clear. It is possible that the transfer of kinetic energy through the chin to the skull is extraordinarily efficient. Alternatively, a blow to the point of the chin might trigger one of the reflex mechanisms leading to unconsciousness (discussed later under (6).

SERCL and JAROS who have studied medical aspects of boxing extensively noted in 1962:

"The pathophysiology of the mechanisms of cerebral concussion is very complex and has been thoroughly studied experimentally in animals. In man, the study of pathophysiologic mechanisms of cerebral concussion meets with objective difficulties consisting in inaccurate history of the injuries and usually very incomplete data concerning the proper mechanisms of the injury... A boxing match gives us the opportunity to study the mechanisms of cerebral concussion. When we observe matches we can see the hits, we can record them exactly and compare them with the subsequent neurological findings. Even photographs and films are often at our disposal which enable us to be certain of a facts."

KAPLAN (quoted from McCOWN (1959) stated that, "a blow from a human being with a

Fig. 188. Gene FULLMER's trainer administers oxygen to FULLMER's sparring partner Eddie HICKMAN, after FULLMER had knocked him out with a powerful right during a training bout on February 18, 1963. Photo: A.P.

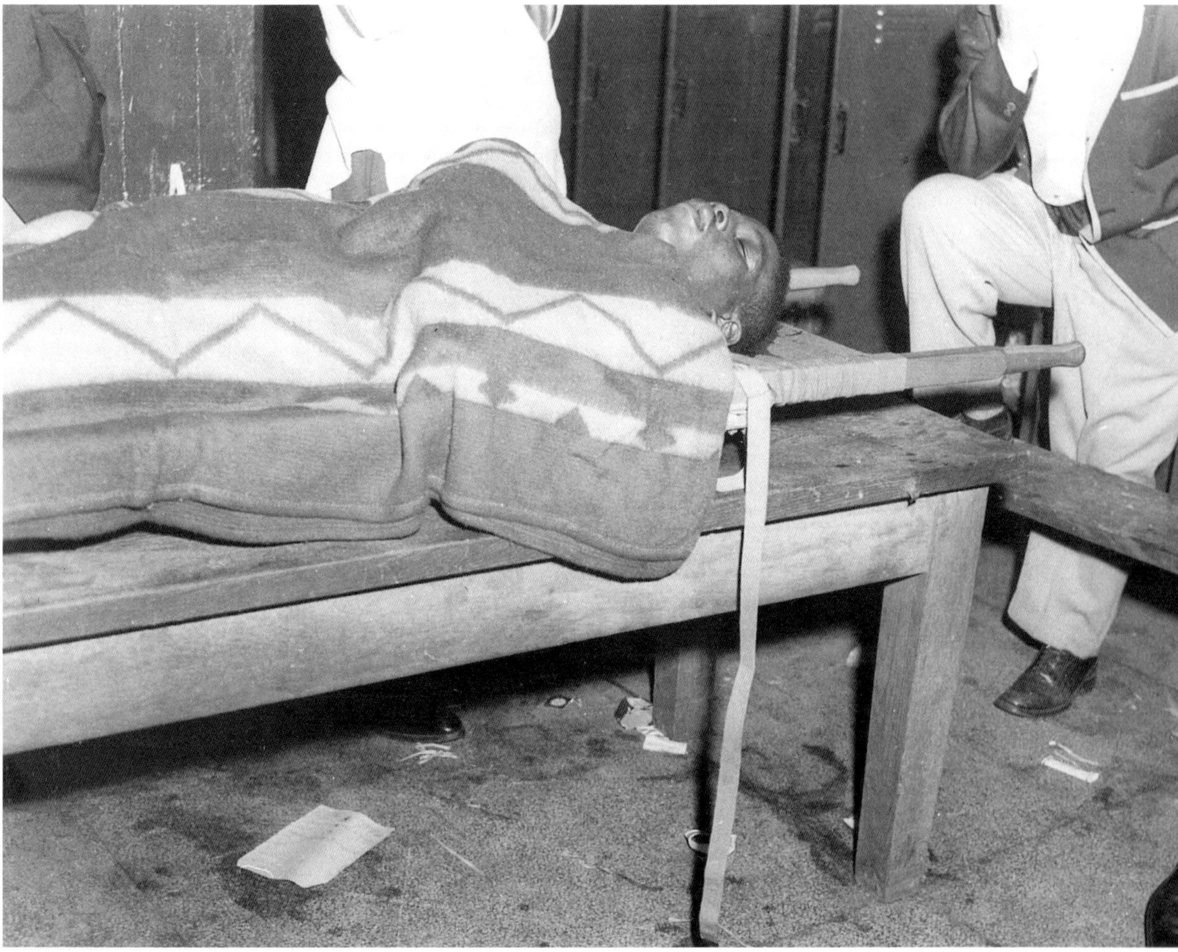

Fig. 189. The professional boxer Julian KEENE lies twenty minutes after being knocked out in his dressing room, still unconcious on a stretcher. U.P.I.

padded gloved fist probably is not forceful enough to produce any direct damage to the brain. Most dazed and unconscious states are due to an excessive inflow of stimuli from many sources."

This statement of KAPLAN must be considered untenable. There is no doubt that a heavy blow of a skilled boxer can accelerate the head of an opponent, that the threshold for a commotio cerebri (cerebral concussion) is reached. It is not necessary that the blow is delivered by a heavyweight, the kinetic energy by a blow delivered by a lightweight can be sufficient.

DIXON (1962) wrote:

"The number of theories which have been advanced to explain cerebral concussion gives a fair estimate of our ignorance about its real mechanism."

Anemia of the brain cannot be used to explain this type of cerebral concussion (commotio cerebri), since such a process would require at least 8 to 10 seconds.

A cerebral concussion may be also produced by a sudden fall, in which the boxer is hit considerably above his body's center of gravity and is literally knocked off his feet, hitting his occiput in a rapid deceleration trauma on the canvas. For details see pp. 25 and 28.

Various theories have been put forward as a

physiologic explanation for concussion. Some authors have theorized that the force of the impact has a direct effect on the neurons producing a paralysis (MILLER 1927, DENNY-BROWN and RUSSELL 1941). A certain physiologic threshold must be reached in which the neurons are transiently not functioning. Other authors are of the opinion that the blow acts on neurons producing a neuronal excitation followed by depression of function parallel to the occurrence of seizures (WALKER et al. 1944).

BRAIN noted in 1958:

> "There are perhaps no more complex problems in neurology than those which arise in connection with consciousness, and for topics more plentiful been confusion."

Ritchie RUSSELL (1971) wrote:

> "My conception of the mechanics of concussion is, therefore, that at the moment of injury the whole of the brain tissue undergoes molecular disturbances within the nerve elements (probably in the myelin sheath) which brings about an interruption in the conducting functions of the nerve-cell processes, and leads to the instantaneous loss of consciousness."
> DENNY-BROWN and RUSSELL (1941) concluded from their animal experiments on head injuries: "Subthreshold blows often stimulate the vagoglosso-pharyngeal system and in this way result in depression of cardiac, vasomotor and respiratory function for ten to thirty seconds, occasionally longer. This is a possible explanation for the 'knock out' blow in boxing." Again, this reflex mechanism cannot explain the immediate unconsciousness, as it would require several seconds. TROTTER (1932) described cerebral concussion as "an essentially transient state due to head injury which is of instantaneous onset, manifests widespread symptoms of a purely paralytic kind, does not as such comprise any evidence of a structural cerebral injury and is always followed by amnesia for the actual moment of the accident."

In neuropathology, cerebral concussion (Commotio cerebri) is defined as a morphologically traceless (at least, using light microscopy) reversible functional disorder, which is therefore only clinically defined (SPATZ 1936, UNTERHARNSCHEIDT 1963, 1993, 1994).

As mentioned above, it is not so rare that a boxer is *'hanging draped' over the upper* or *middle rope* of the ring. He may be only 'groggy' or completely unconscious, and therefore, technically knocked out. Nat FLEISCHER (1958) revealed such a case:

> "Can a man, standing on his feet, be unconscious and be counted out by the referee? It has been done, and I've seen it. I saw a lightweight championship change hands, May 28, 1917, in such a bout. Benny LEONARD had cornered champion Freddy WELSH. WELSH was unsteady on his feet and desperately trying to hold himself up. He slipped one arm over the highest rope, and braced himself just as LEONARD landed a hard punch to the jaw. WELSH started to fall toward the floor to get a moment's rest. Benny, always a very shrewd fighter, started a quick flurry of punches which knocked WELSH cross the top rope. Kid McPARTLAND, a very good referee, made LEONARD step back. The champion was out, hanging on the ropes, his feet on the canvas. McPARTLAND counted the final 10 over him, and as convincing proof that poor WELSH was absolutely unconscious, he slid off the ropes at the end of the count and fell in a limp heap on the canvas."

Boxers Hanging Draped Over Ring Ropes

A series of action photos of professional boxers is presented hanging draped over ropes.

Jack NELSON hangs in the ropes while the referee is counting (Fig. 190). It is practical impossible to deceide whether NELSON is unconscious, or 'groggy,' and severely dazed. Would it not be better for the referee to help the severely injured boxer to get out of the ropes instead to continue with counting him out? Ten seconds are here lost to help a severely injured individual.

Already in the first round RESCH was knocked down to '9' and was in the following round the inferior boxer. At the begin of the fifth round RESCH was able to connect fully, collapsed then after a flurry of hard blows by MÜLLER and was counted out after the third knock down by the referee, who had to pull him then from the ropes. RESCH is hanging draped, anterior parts of his throat lying on the rope, completely unconscious over the rope and referee THOMPSON is counting to '10,' before he lifted him from the ropes (Fig. 191). In the background the jubilant victorious Peter MÜLLER. Referees should not continue to count, if a boxer is hanging draped, dazed or unconscious in the

Fig. 190. Professional boxer Jack NELSON hangs beaten, defenseless, 'groggy,' perhaps unconscious over the middle ring rope. Photo: U.P.I.

ring ropes, but should stop the fight immediately and help the injured boxer.

OATES (1985) wrote:

> "In a recent heavyweight fight in Buffalo, 220-pound Tim WITHERSPOON repeatedly struck his 260-pound opponent James BROAD, caught him in the ropes, while the referee looked on without acting – though a number of spectators called for the fight to be stopped."

TUNNEY (1932) described a knock out of an opponent in his second bout with K.O. SULLIVAN, a southpaw:

> "Putting both hands before his face, he shuffled in until he got within hitting distance. I was waiting. As he pulled his left hand down from his face to swing it at my jaw, I shot out from my chest a straight right hand that did not travel more than 9 inches (23 cm). It landed right between his eyes. It turned him completely around. With his back to me, he dived head-first into the canvas. There he lay face down until he was carried to his corner, which was just one minute and a half after the contest had started."

Fig. 191. Professional middleweight fight between Peter MÜLLER, Cologne, Germany, and Max RESCH, Magdeburg, Germany, October 25, 1958 at the Frankfurter Festhalle, Germany. Photo: d.p.a.

BER

William TAAFFEE (1985) wrote about the welterweight championship fight between Milton McCRORY and Donald CURRY:

> "When *Home Box Office'* cameras caught the dreadful cross eyed star of Milton McCRORY after his near extinction at the hands of Donald CURRY last Friday night, viewers had to ask themselves two questions: (1) Should TV be showing this brutality in the first place, and (2) should HBO be zooming in for chilling closeups of McCRORY in his extremity? The answer is yes on each count. As long as boxing is legal in America, TV has a right, even a documentary duty, to be there. And once there, TV ought to show boxing for what it is, blood, blue language and all. There aren't any virgins in the ring, nor should be there in boxing's living rooms."

Ritchie RUSSELL (1971) wrote:

"The recovery from an uncomplicated knock out never passes through a stage of delirium. When the knock out blow does not lead to an interruption of the fight, the men report retrospectively that their sensations during the following rounds were more or less vague. This is sometimes described as a 'narrowing of the visual field.' People say that they saw only 'the white skin of their opponent,' or ' saw his face only: everything else was black.' Very commonly a mild depersonalization exists, the feeling of a kind of detachment from the person's own body and its activity."

"After a complete knock out the subjects describe the beginning of regaining consciousness in various ways. Sometimes the first thing they hear is the roar of the crowd, the counting of the referee and the tapping of their seconds on the floor, shouts from their corner to get up: the drive to get up before being counted out is reported by many as the strongest and the first self-observed impulse. Others who have been more severely knocked out and come round only in the corner report the sensation of cold water and manipulation of their seconds to bring them round as the first subjective recollection of recovering consciousness. Among the severe cases one hears of boxers who did not regain consciousness until they found themselves under the shower."

"During recovery they are slow in response: they are not correctly orientated, but their orientation is never disturbed by hallucinations: mistakes can be generally accounted for by the amnesic period. In their self-description as quoted, one finds only descriptions of their vague and undifferentiated sensory perceptions but never any of hallucinations, not even of illusion or misinterpretation."

"Summarising, it may be said that after knock outs, states of dimmed and altered consciousness can be observed, but delirium is never seen, either as independent reaction or a transitional stage in either of the other reaction types. Therefore, it is likely that the mechanism underlying delirium is different from that of dimming and alteration of consciousness, and it is possible that the latter two are produced in more than one way."

CANTU (1986) had developed a practical scheme for grading the severity of concussion.

According to the author there are three grades;

> *Grade 1* is the most mild concussion, which occurs without loss of unconsciousness. The only neurologic deficit is a brief period of posttraumatic amnesia that may be fleeting or may last up to 30 minutes.
> *Grade 2* is a moderate concussion, usually involves a brief period of unconsciousness not exceeding five minutes. It may also occur rarely with no loss of consciousness but with a protracted period of posttraumatic amnesia lasting over 30 minutes but less than 24 hours.
> *Grade 3* is a severe concussion, occurs with a more protracted period of unconsciousness lasting over five minutes. Rarely it may occur without unconsciousness but with a sustained period of posttraumatic amnesia lasting longer than 24 hours.

This grading of concussion is in my opinion neither helpful nor significant. In all three grades an unconsciousness may be missing. *Grade 1* I term a subconcussion or subcommotio cerebri, an unconsciousness is missing. One can use the term the individual is 'dinged.'

Grade 2 is considered a moderate concussion usually involves a brief period of unconsciousness, not exceeding five minutes. This should be termed commotio cerebri or cerebral concussion, in cases in which an unconsciousness exists. When it occurs without loss of consciousness it has to be termed a subcommotio cerebri or subconcussion.

Grade 3 is a commotio cerebri or a cerebral concussion if it occurs with unconsciousness. If it occurs without unconsciousness it has to be termed a subcommotio cerebri or subconcussion. The inclusion of periods of amnesia with different length in a all three grades is in my opinion confusing. A traumatic amnesia consists of a period in which the individual seems to perform purposeful activities, sometimes covering hours or days in a dream-like state. MUNRO (1938) said of such a patient: "He knows that he has been unconscious, but his friends do not." We agree with the content of this statement, but object to the use of the term 'unconscious.' Patients in this state ('twilight,' 'fugue,' 'dream state,' 'traumatischer Dämmerzustand') are by definition amnesic but not unconscious. These traumatic amnesias are also called 'traumatic fugue,' or 'dream states,' They occur frequently in boxers. They are discussed in a separate chapter.

'Pummeling knock out' ('Verhämmerungs – knock out')

Occasionally after a *series of blows to the head,* a boxer is 'groggy.' His movements are then slower than normal, less certain, and his sensorium is impaired. The muscle tone, most obviously that of the neck, is reduced and the head moves like a pendulum under the blows of his opponent. It therefore sustains higher accelerations and, consequently, more severe effects. Unless a sensible referee stops the bout it will end with the boxer's being hammered into a knock out state or with his no longer being able to instigate voluntary motor activity. States of 'grogginess,' in which consciousness is reduced, are not rare according to the testimony of boxers and descriptions in the literature. The boxer is *'out on his feet'"* and is *unable to defend himself.* His arms hang loosely down at his sides, and he runs a very high risk of being subsequently hit and severely injured by a series of blows to the head and neck. Finally, he sacks down to the canvas. This is known as a *'pummeling knockout' ('Verhämmerungs-knock out').*

Loss of Posture and Amnesia due to Anemia of the Brain

Loss of posture with *unconsciousness* and *amnesia due to anemia* of the *brain* is an event in which at least several seconds elapse before all symptoms become evident. This is a *delayed clinical syndrome,* developing only after an *interval.* It is often seen in boxing, when the boxer goes down some seconds after being hit.

Nat FLEISCHER (1958) reported such a case:

> "I also recall the strange ending to the New York debut of Babe HERMAN, flashy young Californian, whom Jack KEARNS managed during the early 1920's.
> "HERMAN's opponent was Irish Johnny CURTIN, colorful Jersey City slugger, and the scrap was held at Charley Dressrick's Pioneer Club."
> "For five rounds it was HERMAN's brilliant boxing and sharpshooting against CURTIN's rugged two-handed aggressiveness."
> "The Baby nailed Irish Johnny on the chin with a vicious right just as the bell clanged the end of the fifth. Beyond jolting him as it landed, the punch seemed to have no serious effect on CURTIN. Gingerly, Irish Johnny turned to walk to his corner. He had taken about two steps when, suddenly, the full force of that wallop caught up with him, and he pitched forward on his face."
> "Despite the frenzied manipulations of his handlers during the minute's rest, Irish Johnny was still out and unable to answer the gong for the sixth."

Oculo-cardiac Reflex Mechanisms (ASCHNER-DAGNINI Reflex)

Bernhard ASCHNER (1883-1960) in Vienna and Guiseppe DAGNINI in Bologna independently described a slowing down of the pulse after compression of the eye bulb in 1908. This phenomenon initially gained lively interest in the literature. It was called the *oculo-cardiac reflex* or later *ASCHNER-DAGNINI reflex.* The literature in recent decades is scarce, however, in spite of improvements in clinical determination of heart and circulatory statistics.

If the *optic nerve* is *irritated mechanically,* impulses are aroused in the nerve and these are conceived as stars, flashes of light, etc; however, a slowdown of cardiac activity or a cardiac arrest may occur when the *eye* and *surrounding areas* are hit (*oculo-cardiac reflex*). These incidences can be included in the *syndromes with syncopal attacks*; they do not belong to epilepsy.

Detailed findings about this reflex can be found in the papers of PETZETAKIS (1914), MOUGEOT and DUVERGER (1917), BARRE and CRUSEM (1923), DANIELOPULO et al. (1932), LAIGNEL-LAVASTINE (1924), VAN DOREN (1925/1926), GILLESPIE et al. (1926), SÖDERBERGH (1926), and SZOKOLNIK (1929).

The *reflex arc* (Fig. 192) consists of: (1) *the centripetal limb,* the *first branch of the trigeminal nerve.* The *receptors* are in the *retrobulbar tissue of the eye,* because even after enucleation of the bulbus oculi the reflex can be triggered by pressure on the retro-orbital tissue (BERNARDS 1960). Previous theories of impulse conduction by chemical substances (IMACHI 1939) over the fasciculus opticus (ASCHNER) were subsequently disproven.

The *oculocardiac reflex* is of *theoretical as well as considerable practical interest,* not only for ophthalmologists, but also for sports medicine practitioners interested in brain damage from boxing. A

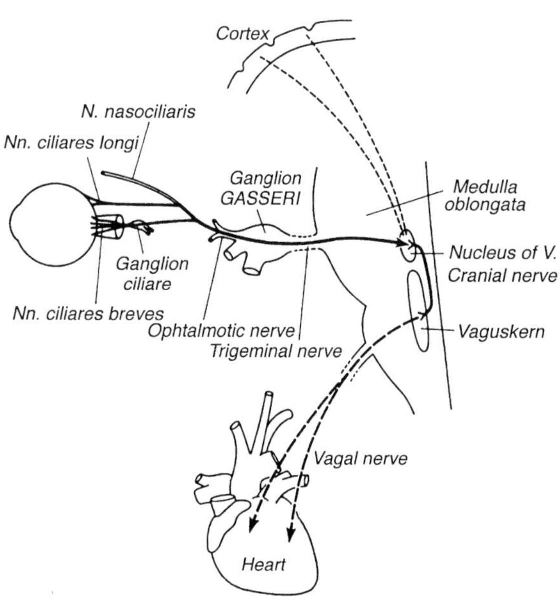

Fig. 192. Reflex arc of the Aschner-Dagnini reflex. (From HOLLWICH F, et al., 1965).

hard blow against the eye may lead to *bradycardia, decrease of the respiratory rate* and *vomiting* (MICHAUD 1938, PLANTEN 1958, 1961, 1963). The after-effects of this reflex can extend for hours and can be misdiagnosed as a concussion syndrome (commotio cerebri) or shock, etc.

HOLLWICH et al. (1965) performed examinations on 100 test objects using an EEG. The authors used compressions of 50, 100, 150 and 200 g. Beyond it was the by many authors recommended digital pressure utilized (STYTSCHINSKI 1931, LAIGNEL-LAVASTINE et al. 1939, PASTEUR et al. 1948, TARDIEU and TARDIEU 1948, TINEL (1948). A distinct dependence of the reflex response could be observed in regard to the amount of pressure used, whereby age differences exist, as was already reported by ASCHNER (1908) and PUMPLUN and STYTSCHINSKI (1938). The reflex response was dependent on the degree of pressure applied, with distinct differences determined by the age of the subject (Figs. 193 and 194). With *digital pressure,* HOLLWICH et al. (1965) were able to get a *positive reflex* in 80% of the test subjects, a *paradox response (tachycardia)* in 6%, and a *negative reflex* in 8%. A positive reaction was found more often in women who reacted more frequently positive (94%) than men (80%). In the age group between 20 and 30 years in every patient the reflex could be produced. More than 60 year old individuals revealed in only 81% a positive reaction. Intensive compression of the eye bulb produced extensive reactions more frequently than slight compression (Fig. 195). A typical reflex answer in shown in Fig. 196.

Dangerous sequels of pressure on the bulbs can develop as slackening of the sinus rhythms and of the conductive system, extreme bradycardia up to

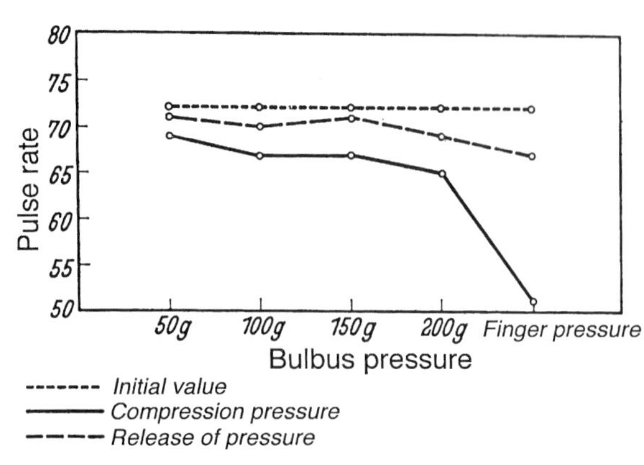

Fig. 193. Age group 20 – 30 years. (From HOLLWICH F, et al., 1965).

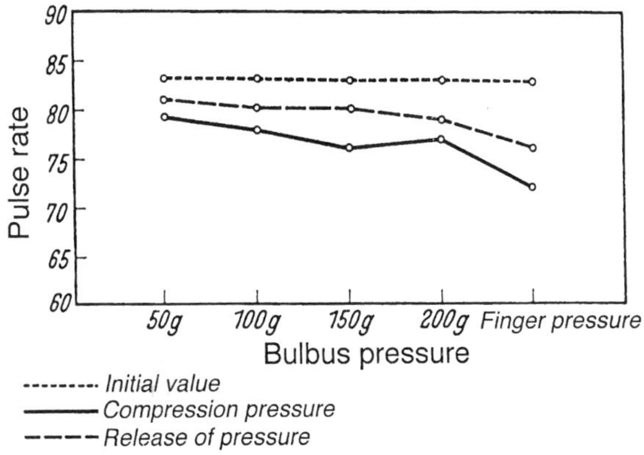

Fig. 194. Age group 50 – 60 years. (From HOLLWICH F, et al., 1965).

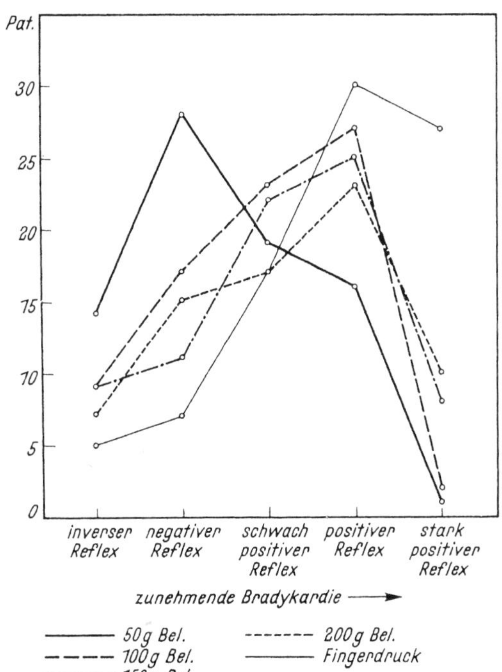

Fig. 195. Elicitation of the refeflex in dependence of the compression. (From HOLLWICH F, et al. 1965).

chamber automatic or heart arrest, and following cerebral seizures due to a cerebral ischemia (BALABANE 1935, PRUCHE 1933, YOVANOVITCH and MILANOVITCH 1938).

Pull on the external eye muscles in strabotomy and *enucleation of the eye bulb can trigger the ASCHNER-DAGNINI reflex* (HOLLWICH 1955, 1964, SORENSEN and GILMORE 1956, KIRSCH et al. 1957, PLANTEN 1958, 1961, 1963, RHODE et al. 1958, MENDELBLATT et al. 1962). Severe general symptoms with circulatory collapse, due to a compression bandage of the eye after an enucleatio bulbi was reported by TREIGNY (1940).

The ophthalmologists are familiar with the oculo-cardiac reflex, since they have for diagnostic or therapeutic reasons quite frequently to carry out pressure on the bulbus oculi (HOLLWICH et al. 1965).

We follow these authors:

(1) For the softening of the eye bulb before extraction of a cataract some ophthalmologists carry out preoperatively through the eyelids a digital compression of the eye bulb for a period of 5 – 10 minutes. Is the finger replaced by a weight, which acts onto the bulbus, it has to amount to 100 to 150 g at least, to achieve the desired hypotonic effect (ARRUGA 1959). This compression pressure is, however, not sufficient to induce the oculo-cardiac reflex.

(2) In measurements of the arterial pressure in the eye, using the dynamometer of H. K. MÜLLER, with pressure of a spring on a stamp which compresses the eye on the sclera near the temple side, until the central retinal artery pulse (diastolic pressure) or anemic (systolic pressure). To that end often a compression of more than 100 g is necessary. In many hundreds of hypotonic patients up to 320 g pressure was exerted (WEIGELIN and LOBSTEIN 1962).

(3) Pull on the external eye muscles in a strabotomy or enucleatio bulbi the ASCHNER-DAGNINI reflex is produced. According to PLANTEN (1958, 1961, and 1963) pull on the eye muscles leads to the same effect as pressure on the retrobulbar tissue.

(4) A severe contusio bulbi may lead to bradycar-

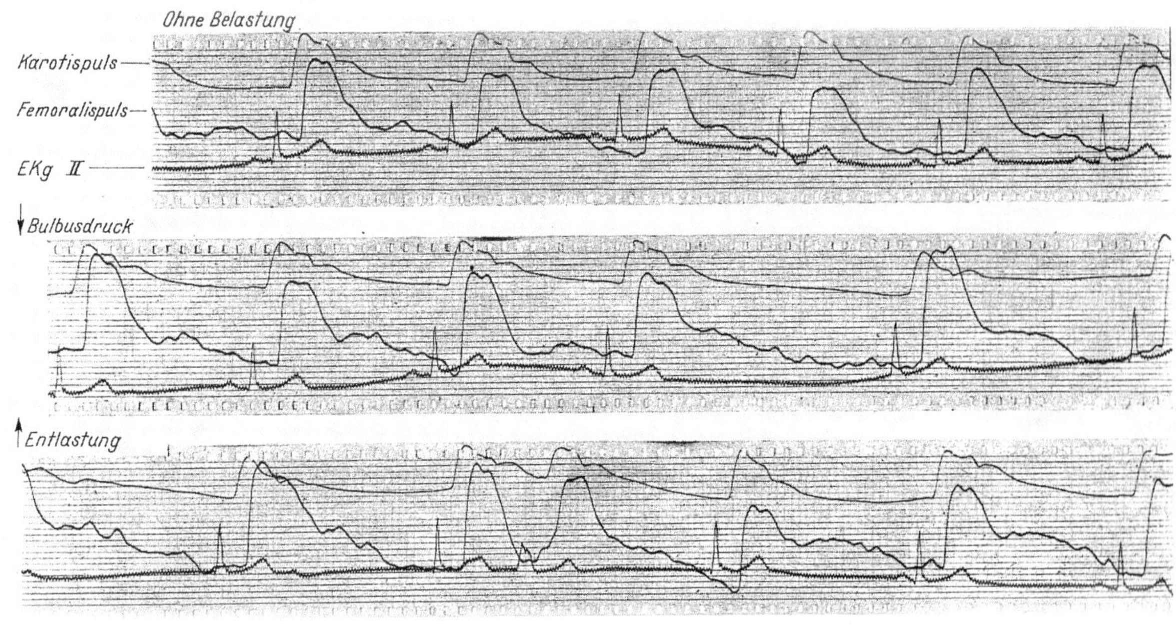

Fig. 196. Typical EEG example of a oculocardiac reflex. F. K., male, 22 years, pulse frequency 54/min. After bulbus compression slowed down sinus rhythm (1 – 4 seconds), chamber automatic after the 4th heart action. After release continuation of the nodal rhythm. Appearance of a extrasystole, restoration of the sinus rhythm. (From HOLLWICH F, et al., 1965).

dia, a retardation of respiration, vomiting and other severe general manifestations (MICHAUD 1938, PLANTEN 1958, 1961, 1963), which can extend for hours and can lead to diagnostic misinterpretations (shock, cerebral concussion, commotio cerebri) due to a long after-effect of the reflex.

The oculo-cardiac reflex can be interrupted by anesthesia of branches of the trigeminal nerve, which can be achieved by a retrobulbar injection of Novocaine.

STÖRTEBECKER (1953) described an observation of a oculo-cardiac reflex in a 25year old, who had received a blow against his left eye, though not in boxing. Within a few hours he developed an extreme bradycardia with reduction of pulse frequency down to 36/min, without any other cerebral symptoms. Pressure on the bulbus oculi of the uninjured eye changed the bradycardia almost to a asystolia, findings which were recorded electro-cardiographically.

I am convinced that the oculo-cardiac reflex plays a very important role in boxing, and occurs frequently, despite the fact that except of the short communication of JIMENES ESPINOSA and ESPINOSA IBARRA. To my knowledge there are no other publications about this reflex in boxers.

JIMENEZ ESPINOSA and ESPINOSA IBARRA (1960) examined ten boxers, who had suffered a knock out. They recommended, in boxers, who had suffered a knock out, to record immediately an EEG, to be able to induce oculo-cardiac and carotid sinus reflexes during the EEG recording.

Case 1: A 28 year old amateur, a boxer since the age of 20 had boxed 80 times in the past year and lost four fights, the last one (in August 1957), when he was knocked out with a blow to his neck, was followed by a normal neurological examination. *EEG* was normal. He was very sensitive to Metrazol producing a syncopal state.

Case 2: A 24 year old amateur, a boxer since the age of 20 had to his credit 68 victories and 20 defeats. He lost his last two fights by knock outs. *Neurological examination* was negative. Oculo-cardiac reflex slowed his heart and made him faint."
Case 3: A 19 year old boxer had a history of six victories and six defeats, the last two defeats being knock outs. Thirty-two hours after the last knock out the oculo-cardiac reflex showed a 16-second systole."

The authors concluded that EEG's should be conducted on boxers shortly after they have experienced multiple knock outs, and oculo-cardiac and carotid sinus massage reflex should be carried out during the EEG recording.

Knock down with Loss of Muscle Tone
In some instances after a *single heavy blow* or a *series of blows against the head,* a boxer is *knocked down* due to *temporary loss of muscle tone,* but is not unconscious; he is unable to move his extremities for a certain time. The mechanism of this *traumatic atony* is not known, nor do we know how frequently this type of knock out occurs. This knock down is frequently the result of blows against the body.

Different Forms of Cerebral Reflex Knock outs.
Each of the authors below has advanced his own *theory* regarding a *different knock out mechanism* .

SHERRINGTON (1906) proposed the theory that *stimulation of the labyrinth results* in a loss of tone of vestibular origin, and that the boxer would collapse. "The knock out blow, where the lower jaw conveys concussion to the otocyst, reduces in a moment a vigorous athlete to an unstrung bulk of flesh whose weight alone determines its attitudes, if indeed a reactionless mass can be described as possessing an attitude at all." SHERRINGTON does not, however, offer an explanation for the instantaneous unconsciousness.
SOMEN (1913 /1914) championed the theory that a blow against the boxer's *lower jaw,* which is considered especially effective in causing a knock out, produces dizziness through the *impact* of the *jaw* against the *inner ear*. In his opinion, the *mechanical irritation of the labyrinth could* be an explanation of the knock out.
According to SERCL and JAROS (1962), a hard blow on the *jaw may* trigger a *reflex mechanism*. The *mandibular joint* is richly *innervated* by fibers of the *trigeminal nerve* as can be seen in preparations of HROMADA (quoted after SERCL and JAROS 1962). These are the *articular branches* of the *auriculo-temporal nerve*, the *masseteric nerve,* and the *deep posterior nerves* and inconstant innervation from the *facial nerve.* A blow on the jaw triggers an excessively intensive pathologic excitation there propagates through the *trigeminal nerve* into the *reticular formation* of the brainstem, and causes a rapid loss of the tone of the antigravity muscles and a state of unconsciousness.

This is an interesting theory, but literature about this reflex is non existent. There are no additional anatomical studies on the mandibular joint and its richly innervated fibers of the trigeminal nerve and there are no clinical papers about this reflex.

GOVONS (1968) carried out interesting experiments in an effort to investigate posture, as it is the only sign of concussion susceptible to objective analysis. His experiments revealed that all higher animals possess the fundamental capability of assuming an integrated posture. They are able to maintain a stance with the head raised, in the face of unpredictable unbalancing forces. Sustaining attitudes and postures is an active process, like movement, requiring coordination and integration of a great number of reflexes. Normal maintenance of posture requires constant input, in the form of messages from receptors, particularly proprio-receptors (including the labyrinth). Based on the information received, a continuous stream of impulses is transmitted to the musculature, commanding adjustments necessary to maintain appropriate posture and movement under the circumstances. Through reflex mechanisms, the nervous system responds to unbalancing forces by producing counter forces of appropriate direction and magnitude to maintain equilibrium. Maintaining balance under changing conditions is a complex state. In the event that equilibrium cannot be maintained, gravity exerts its effects. The nervous system continues to relay messages, even during complete loss of balance, when defensive reflex mechanisms are brought into play, in order to break the force of the fall.
The head is of special importance and assumes a dominant role and in maintenance of posture. The head is the leading segment of the body, containing, in addition to the brain, the labyrinth, the retina, and the olfactory organs. The important proprio-receptors of the neck are also a part of the system. A blow to the head disturbs its position relative to the rest of the body. It is inconceivable that the postural reflex mechanism will fail to react. The problem is to determine whether the brain first reacts to the violence to the head, as a result of messages pouring in from receptors, or as a result of the force acting directly on central neurons. It is also important to discover whether paralysis after concussive blows is instantaneous, or whether, in the brief interval between the blow and

the total loss of body posture, there is an intense period of reflex activity.

GOVONS studied the knock down blows to six boxers using moving pictures taken at 32 frames per second. There was a time interval of about 30 milliseconds between frames. The frame capturing the instant that the glove the first contacted the jaw was labeled zero.

Of six boxing events studied, five knock down incidents were the result blows to the left jaw, rotating the head to the right. The remaining knock down followed a blow to the right jaw, rotating the head to the left. In every case, rotation of the head followed the direction of the blow was marked and characterized the knock down punch. Extension of the head in varying degrees occurred in four boxers, and was absent in two. The head attained maximum rotation approximately 60 milliseconds after the blow; by that time the gloved fist had cleared the jaw.

Instead of falling in the direction of the punch, all boxers reacted, that is, they generated forces which were opposite in direction to the applied force. In all boxers the head counter-rotated; in four, where no further body contact between the two opponents occurred, the counter-rotation of the head continued and approached 180 degrees. This counter-rotation of the head was followed by a counter-rotation of the body until balance was lost. The end result was that the boxer fell in a direction opposite to the direction of the blow. In two boxers, the counter-rotation could not be completed because of body contact between the contenders, and the boxer fell at the point of interference by this force.

Upon contact, the boxer began his descent to the canvass, under the force of gravity. During the fall, he displayed reflex activity. In frames where the head position could be followed clearly, it turned abruptly, to gaze in the direction of the fall. The head turn was followed by a drop to the knee. The boxer then fell to the gloved fist or the elbow. The head was always the last part of the body to reach the canvas. From the time of the punch, to the complete loss of body posture, less than 0.8 seconds elapsed. If the head were a freely falling object, falling six feet, it would take 0.6 seconds to reach the ground. Considering the fact that the descent of the head did not begin until after an interval of about 0.3 seconds from the point of contact, and that the fall was impeded by the extremities and muscle turns of the neck, the rate of the fall was too rapid to be a passive motion. For these reasons, it appeared that the boxer literally threw himself to the ground, as an over-reaction to the punch.

GOVONS concluded that the fall followed a flurry of intense reflex activity, and, as such, could not possibly be the result of central neuron paralysis. Since a reflex is an impulsive response to sensory information input. From the demonstration of predictable reflex patterns, one can conclude that the force of the blow is first detected by receptors that are sensitive to sudden head movement. These are probably proprio-receptors, including the muscles spindles, the Golgi tendon organs, Pacinian corpuscles, and labyrinths. The stimulus is the impulsive movement of the head, which in the case of the boxer is mainly rotation. Presumably the afferent impulses are fired into the brainstem and upper cervical cord through a synchronous discharge of the proprio-receptors. The loss of posture reflex represents only a portion of the response to the stimulus.

This theory championed by GOVONS is not proven. There exists no additional confirming literature.

Body Knock outs

The *body knock out* is the second classification under the heading of *knock out sui generis*. The body knock out includes all forms that are the result of : (1) A *carotid sinus syndrome*; (2) *injuries* to the *larynx* and *glottis*; (3) *blows* against the *precordium*; (4) blows which provoke a *vagovagal reflex*, (Goltz experiment); and (5) *blows* to the *coeliac plexus*.

The different syndromes are described in detail in the following chapter

Mechanical Irritation of the Carotid Sinus

Mechanical compression of the carotid artery can produce *bradycardia* and even *asystolia of the heart*. This is not new information, in fact, physicians have known for several centuries. The terminus carotis derives from the Greek word (karos), which means 'deep sleep,' 'stupor,' 'unconsciousness,' (the verb (kareuein) means 'stupefy,' 'deaden.' The anatomist VESAL termed the artery 'Arteria soporifica,' the 'sleep-bringing' artery.

PARRY (1799) described this condition in a half-sided, paralyzed admiral. The pressure of a single finger on his carotid artery produced immediate reduction of the pulse rate to between 15 and 20/min.

KUSSMAUL and TENNER (1857) reported additional observations. They found that compression on the carotid bifurcation produced unconsciousness with convulsive arm movements, although these authors attributed their patient's unconsciousness to insufficient blood supply to the brain resulting from direct clamping of the blood supply to the brain.

The next authors, the English physician WALLER (1862) and the German physiologist LANDOIS (1865), observed a distinct, recognizable slowing down of the pulse in certain patients, due to finger pressure on the carotid artery. They, too, were unable to explain the phenomenon.

The Chairman of the Department of Physiology at the University of Jena, Johann Nepomuk CZERMAK (1828-1873) reported a self-observation – he was 38 years old in 1866, – containing a clear description of the hypersensitive carotid sinus:

> "When I locate the carotid artery at the anterior margin of the sternocleidoid muscle and if I a little above this point exert a pressure with my finger, the systolic interval of my heart becomes longer, the frequency of my pulse decreases."

CZERMAK believed the phenomenon to be the result of direct mechanical irritation of the vagal nerve. He spoke of a *'Vagusdruck' a ('vagal compression experiment')*. Luigi CONCATO 1872/73) (quoted from BÜTTNER (1957) and FRANKE (1963) explained that a mechanical compression of the carotid bifurcation in eleven of his patients with atheromatous lesions of the vessel wall of the carotid arteries produced a slowing down, or even a pause, in the pulse.

CONCATO, too, considered an irritation of the vagal nerve. CZERMAK (1865) presented the correct explanation, that the mechanical compression of the carotid artery due to the carotid sinus reflex initiated a stimulus of the heart inhibitory branches of the vagal nerve. HEINECKE et al. (1908), WENCKEBACH (1914), HERING (1924, 1928), KOCH (1924) presented additional observations.

HERING (1928) furnished interesting related hints when he reported that cases of death due to hanging and the unconsciousness in a knock out in boxing could be explained by blows to the carotid artery as a result of a carotid sinus reflex. ROSKAM (1930) and WEISS and BAKER (1933) presented further studies. These authors introduced the group of three different types of this reflex: namely: (1) a *cardiac*, (2) *depressoric*, and (3) the (not secured) *cerebral*.

The publications of SCHARF and BOYD (1935), LENEGRE (1938), and ROSSIER (1939), brought further progress. They made electro-cardiographic recordings before the syndrome was initiated.

Hans FRANKE (1963) presented the last essential, comprehensive and classic monographic treatment of spontaneously occurring carotid sinus syndrome and the hyperactive reflex response in experimental irritation. FRANKE, the leading authority in this domain, stated emphatically that the genuine, spontaneously occurring, carotid sinus syndrome must be separated from the general hyperactive reflex reaction due to experimental excitation of a sensible carotid sinus. French authors (WARTER et al. 1959) followed FRANKE's lead, distinguishing the *'syndrome d'hyperréflectiviteé' spontanée,'* as an separate syndrome, distinctly separated from the *'syndrome d' hyperréflectivité provoquée.'* The two syndromes are, according to FRANKE, closely related to each other, but not identical. The reason given is the following: The spontaneous carotid sinus syndrome conforms to the special clinical form of an unusually hyperactive carotid sinus in a diseased state. It is caused by the patient's own behavior, on particular occasions. In a routine examination, compression of the carotid sinus can evoke symptoms that never occur spontaneously (GREELEY 1955). These cases involve a hyperactive carotid sinus reflex, but no genuine carotid sinus syndrome (FRANKE 1963).

FRANKE and his collaborators (1963) tested the *exovasal irritation of the carotid sinus* in nearly 3.900 individuals of both sexes and different age groups. I will follow the description of these authors:

(1) More than 70% of the examined individuals and practical all juveniles younger than 20 years of age reveal no substantial reaction or, at the most, an unimportant bradycardia, in case single extrasystolies, without involvement of the blood pressure, identical with the first group of SIGLER.
(2) In about 10% of the probands the heart slows down about 20 beats per minute, the blood pressure decreases about 10, maximal 20 mm/Hg, the cardiac arrest less than two seconds, this second grade according to SIGLER produces no clinical signs.
(3) In another 10% of the probands the heart frequency decreases about 30 to 50%, the asystolia lasts longer than two seconds, the blood pressure decreases at about 30 mm/Hg, it

constitutes a beginning pathological reflex (third reflex grade according to SIGLER).
(4) In the remaining group cardiac arrest occurs between 3 – 5 seconds, and longer, as well as a decrease of the blood pressure of 50 mm/Hg and lower (fourth reflex grade according to SIGLER).

According to FRANKE (1963), these objective signs correspond with subjective symptoms, such as vertigo, paresthesias, and visual disturbances. The consciousness of a patient lying down is affected by an asystolia of more than 7 – 9 seconds, of the sitting patients of more than 5 seconds.

Arteriosclerosis increases reflex activity, and is present to some degree by the age of 35. For this reason, only younger patients are good candidates for such studies. Standardized techniques must be used, and only studies using comparable, standardized techniques should be compared with each other. Following these considerations, FRANKE (1948) has tested the degree of the carotid sinus reflex in more than 200 students without any circulatory disease (age until 25 years). The author could find only a moderate sinus bradycardia (up to 20% lower than the starting point) without any electrocardiographic deficits. The decrease of the blood pressure in a healthy juvenile is seldom more than 20 mm/Hg.

FRANKE (1963) summarizes his observations as follows, that the *cardiac type* exists then, when with standardized test methods the reflex induced asystolia amounts to more than 2 to 3 seconds

The *depressoric type* of hyperactive reflex occurs with a reflex related decrease of the blood pressure to 50 mm/Hg or lower.

The spontaneous carotid sinus syndrome produces brisk movements of the patient's head, according to FRANKE (1963). Certain situations such as sudden turning, excessive bending backward, and contraction of muscles in sudden neck movements (FRANCESCHETTI and DURET 1949), cause irritation or stretching of HERING's reflex apparatus. Contrary to popular opinion, these syndromes are by no means uncommon.

The reflex provoking mechanism, the so-called *'trigger mechanism' of the carotid sinus syndrome* can vary and as FRANKE rightly emphasized, occasionally very dramatic. In older literature, the starched shirt collar was of more importance than the softer shirt collars worn today. Common trigger mechanisms include brusque bending of the head backward, as at the barber or dentist, or movements of the neck and head while driving a car. By 1914 WENCKEBACH had already reported the case of a priest, who noticed vertigo while preaching at the pulpit. The pressure of his stiff, narrow shirt collar produced cardiac arrest, as WENCKEBACH correctly concluded.
ROSKAM (1930) reported the case of a 53 year old patient with spontaneous carotid sinus syncope. He always suffered cardiac attacks when his wife buttoned his tight collar or bound his tie.
Trigger mechanisms provoking carotid sinus syndrome attacks in predisposed persons were compiled by FRANKE. They included carrying of corn sacks on the shoulder, tying fabric tightly around the neck by a barber (GROSSMANN 1956), fruit gathering on ladders (FRANKE 1963), looking at upper advertisements on an advertisement pillar or a movie screen from the front row (KUNERT 1960), looking upward at a circling airplane (LAFON and LAFON 1957), difficult defecation in older patients with an arteriosclerosis (LAFON and LAFON 1957), exertion while putting on shoes (MARTELL), and painting a house ceiling (FREEDBERG and SLOAN 1937).

Frequency of the Different Forms of the Carotid Sinus Reflexes

There is a great disparity among the reported data in the literature on different reflex types.

WEISS et al. (1933, 1936) looked at 56 cases of carotid sinus syncope. They found that 19 individuals (34%) had the heart inhibiting form, 3 (5%) had the purely depressoric form, and 34 subjects (60%) a vaso-depressoric and at 12 the purely cerebral form. In the large series of FRANKE (1963) totaling 550 syndrome cases, is the vagal-cardiac form with 431 = 78.3% of the observations the most frequent one. Only 10% of the patients had the pure vessel type syndrome, and the other 12% showed the combined vagal-cardiac/vaso-depressoric types.

The Reflex Pathway of the Different Types

The *reflex pathway* of the different forms was commented on by FRANKE (1963) (Fig. 197). I am

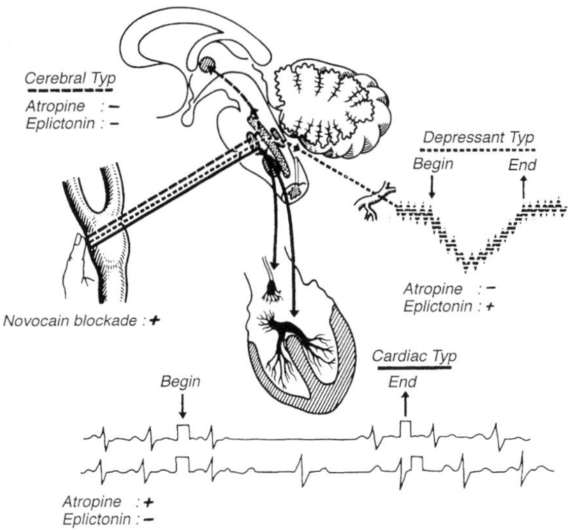

Fig. 197. Reflex path of the different types of the carotid sinus syndrome and of the hyperactive carotid sinus reflex. (From FRANKE H, 1963).

following this author: while the afferent limb, from the carotid sinus to the medulla oblongata extending centripetal reflex limb in all types of the carotid sinus syndrome is the same, are different efferent, centrifugal reflex limbs known:

(1) A *vagal-cardial* or *heart inhibiting type* with asystolia of the heart lasting more than two seconds. The stroke volume of the heart is decreased, with secondary hypotension (Fig. 198). Thereby it develops a hypoxia of the brain with following unconsciousness, sometimes, too, convulsions of the type of a Morgagni-Adams-Stokes attack.

(2) *Depressoric type with a primary vasodilatation and decrease of the blood pressure in the arteries.* It can by all means at missing inhibition of the heart occur as a vagovasal attack in a hyperactive carotid sinus.

A *primary central type* was described by the previous authors, which affects without changes in the blood supply of the brain isolated vegetative center of consciousness in the hypothalamus,

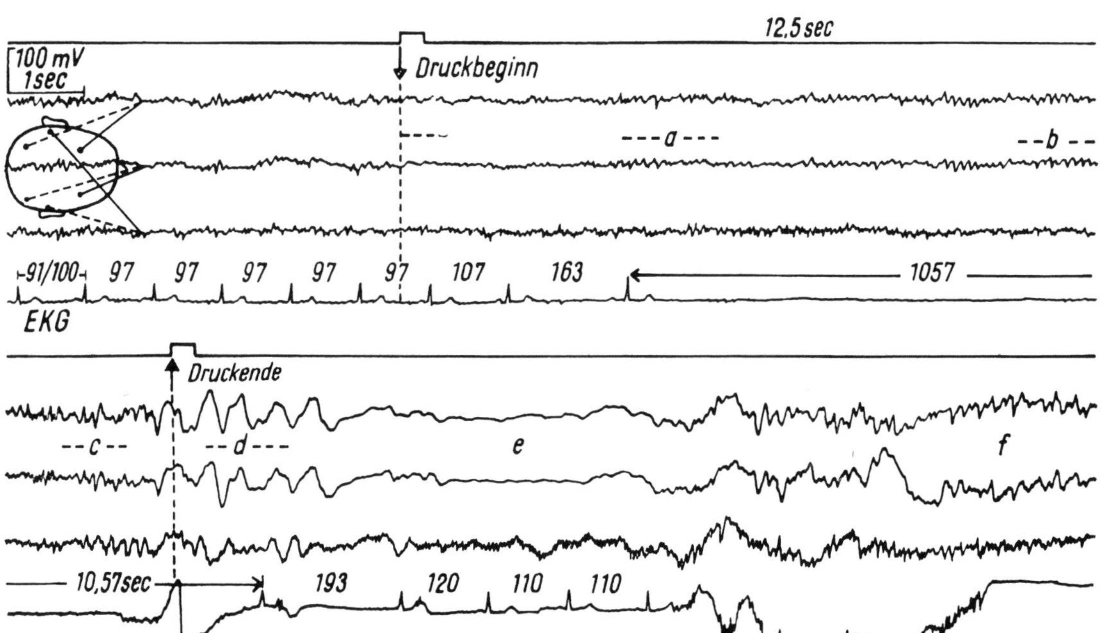

Fig. 198. EEG- and ECG situation at a patient with an hyperactive carotid sinus reflex of cardiac type. The curves link up directly. (From FRANKE H, 1963).

whereby unconsciousness and respiratory arrest occur.

According to FRANKE (1963) the bifurcation of the carotid artery is in most healthy adults slightly bulbous dilated (Fig. 199). The patients with a carotid sinus syndrome or a hyperactive reflex, especially those in the higher age groups, have a distinct arteriosclerotic ectasis of the vessel, up to plum size, an ectatic-aneurysmatic widening (SUNDER-PLASSMANN 1930). Preferentially spontaneous aneurysms develop, since the vessel wall in the area is physiologically thinned (FRANKE 1963).

The *centripetal trajectory of the carotid sinus reflex* extends to the in distal parts of the fossa rhomboides located group of nerve cells of the N. alare cinerea, in which the reflex is directed into two directions for the centrifugal trajectory.

The *cardiac type* of hyperactive carotid sinus reflex or carotid sinus syndrome, the reflex arc proceeds through the Nucleus dorso-ventralis of the vagal nerve and exits the central nervous system over his posterior roots and extends via the Nn. depressores of the vagal nerve to the heart. In some of the probands, FRANKE was able to demonstrate that compression of the right carotid artery produces a slow-down of the heart frequency respective an asystolia, the reflex arc of a compression of the left carotid artery follows the left vagal nerve and the conduction of impulses from the auricle to the chamber of the heart. Thus a left-sided compression of the carotid artery occurs in patients with a hyperactive carotid sinus reflex. Commonly found are varying degrees of auricle/chamber disorders (FRANKE 1963).

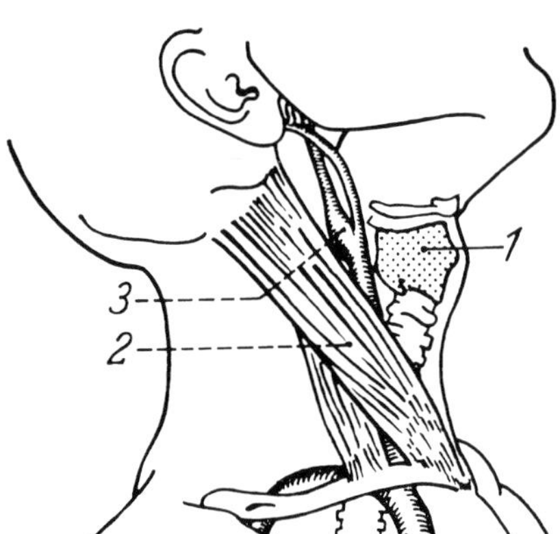

Fig. 199. Topography of the carotid sinus in the anterior neck triangle (modified after SALOMON). 1. Thyroid cartilage (Adam's apple), 2. sternocleidomastoid muscle, 3. carotid sinus. (From FRANKE H, 1963).

The *primary depressoric type* of the hypersensitive carotid sinus syndrome has the same afferent reflex path, which is then switched over to the nearby vasomotoric center. Sympathetic fibers of the sympathicus lead centrifugal dilatotrophic irritations to the peripheral nerves. The result is a decrease of the arterial blood pressure.

At the *central type,* the afferent reflex pathway allegedly extends via the reticular substance as the reflex center via the midbrain to the cerebrum, with the result of a centrally induced unconsciousness, with participation of heart and blood pressure. BRAUCH (1954) alluded rightly, that the exact course of the intracerebral reflex pathway of the controversial central reflex type is not known with certainty.

The existence of the carotid sinus syndrome of the central type remains controversial today, while there is no doubt about the validity of the vagal-heart inhibitory and the vaso-depressoric forms of the carotid sinus syndrome.

Carotid sinus syncope of the central type was first described by WEISS et al. (1936). According to these authors, an imitation of the carotid sinus leads to unconsciousness after 3 – 4 seconds, reflectively without changes in blood pressure and heart actions. The latent period is therefore considerably shorter than in other kinds of carotid sinus syncope.

The symptomatology of the central type was described on account of reports from literature as follows: weakness, vertigo, deafness, itching of the extremities, paleness of the face with following unconsciousness, thereby hyperpnea and convulsions.

Claims in the literature about the frequency of central type carotid sinus syndrome are unusually controversial. COPE et al. (1947) consider the central type to be the most frequent, while ENGEL (1959) described this type as existing in only in a few cases. There are authors who reject the central type completely, such as BROSER (1958), GURDJIAN et al. (1958), and FRANKE (1963).

ENGEL (1950) described two cases of the cerebral form of a carotid sinus syncope, in which EEG recordings were recorded. They were young men in the age of 28 and 36 years. Mechanical occlusion of each single carotid artery did not produce clinical symptoms or ECG alterations. After mechanical occlusion of both carotid arteries developed after 10 seconds headache and bilateral slowing down of the EEG curves (Fig. 200). The mechanical irritation at each carotid bifurcation on the lying patient produced within ten seconds contralateral symptoms of deafness, itching, clonic seizures and a slowing down of the homolateral EEG potentials (Fig. 201). FRANKE (1963) called special attention that ENGEL did not describe significant changes of pulse and blood pressure in his cases, strict critics demand to prove the existence of the cerebral type a continuous blood pressure recording, which was not carried out.

WEBSTER et al. (1955) and GURDJIAN et al. (1958) advanced the most important objection to the cerebral classification of carotid sinus syndrome. The mechanical influencing of the blood supply of the brain in the case of existing atheromatous alterations of the carotid arteries initiated a syndrome, which was before described as the cerebral type of a hypersensitive carotid sinus. These authors used a group of patients with partial or complete occlusion of the internal carotid arteries by arteriosclerotic processes to demonstrate the point. Some of the patients described by WEISS et al. (1936) are explained by a purely mechanical blood supply disturbance, due to previous atheromatous damage.

The pathologically enhanced carotid sinus reflex is determined by the starting rhythms of the heart and its momentary susceptibility, according to FRANKE (1963). The hyperactive cardiac reflex in the various cardiac insufficiencies lasts at least a few seconds. Accordingly, cardiac arrest manifests for a duration of up to 12 seconds, by complete asystolia of the auricles and ventricles. In rare cases even in distinct electro-cardiographic disturbances of the potentials of the QRS group, of the ST – tract and the T – spikees.

FRANKE (1963) surmised that irritation of the right carotid sinus produces a greater effect on the sinus knot (with sinus bradycardia and sinu-auricular block), while left carotid sinus stimulation more negatively affect dromotop on the AV – knot (with partial to complete AV – block). This pattern of cardiac reflex responses is attributable to the distribution of the vagal terminations on each side of the heart.

A sinu-auricular block with six seconds of asystolia of the entire heart using a fivefold registered recording is shown in Fig. 202. The left-sided carotid irritation produced only minimal delayed action of the auriculo-ventricular transfer time, without a manifest asystolia; accordingly, there was no hyperreactive reflex on the left side.

Symptomatology of the Provoked Hyperactive Carotid Sinus Reflex

The results of 150 patients with a provoked hyperactive carotid sinus reflex were presented by FRANKE (1963), in summary, in a synoptic diagram of symptomatology, shown in Fig. 203. Accordingly, the feedback of the hyperactive carotid sinus reflex of the cardiac type depends essentially on the duration of the cardiac arrest. FRANKE noted:

"In the first *phase* are despite the beginning asystolia of the ventricles of the heart the complexion of the face, the consciousness and the cerebral electro-encephalographic activity during a latent period of 3 – 4 seconds due to the yet sufficient oxygen reserve unchanged. Afterwards *(2. phase)* the patients complain while the blood pressure is falling about increasing vertigo as in the spontaneous syndrome. The complexion of the face becomes pale and the skin becomes cooler. In the EEG develops after a beginning leveling off as sign of a beginning transportative cerebral hypoxia a slight activation of the potentials with beta waves. In the *3rd phase,* the reflex asystolia, the patients experience increased duration of cardiac arrest (7 to 13 seconds) and, at sinking arterial blood pressure of less than 40 mm, deep unconsciousness. The patient looks looks pale as death due to his pallid skin. His skin feels cold.

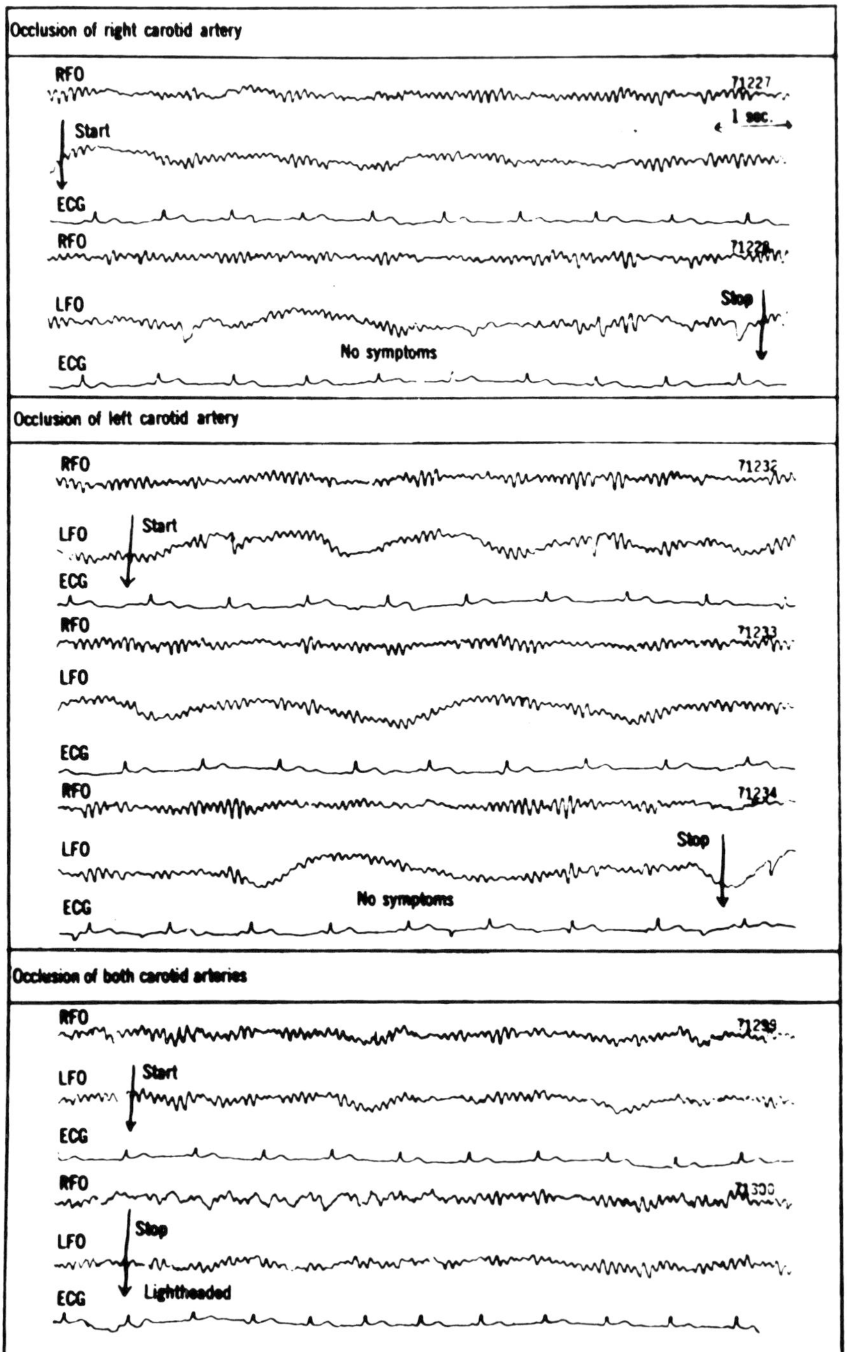
Occlusion of right carotid artery
RFO
71227
Start
1 sec.
ECG
RFO
71228
LFO
Stop
No symptoms
ECG
Occlusion of left carotid artery
RFO
71232
LFO
Start
ECG
RFO
71233
LFO
ECG
RFO
71234
LFO
Stop
No symptoms
ECG
Occlusion of both carotid arteries
RFO
71299
LFO
Start
ECG
RFO
71300
LFO
Stop
ECG
Lightheaded

In this moment of the manifest syncope with deepened respiration, and in some cases appearing Bell's eye phenomenon, the pressure on the carotid artery should be interrupted. With the lessening of the finger pressure at the carotid bifurcation, the heart autonomy recovers after shorter (8-10 second) chamber asystolias fast in 8 – 15 seconds. After cardiac arrest of more that 12 seconds duration, the face pales and then a sudden flash develops. Sometimes slight Morgagni-Adams-Stokes attacks of short duration occur, accompanied by transient electrical convulsive paroxysms.

Sequels of Blows at the Carotid Sinus in Boxing

HERING had already in 1927 alluded that it was well known to boxers, that blows at the carotid sinus could lead to a knock out. Moreover FRANKE (1963) remarked that one cannot speak of a knock out blow with unconsciousness, especially in young individuals, from a hypersensitive carotid sinus syndrome, since the stimulant in these cases would be physiologically abnormal. The peculiarity in boxing blows against the neck region, that the syndrome here is not induced by the pressure of a finger, but that massive blows broadly act on the neck region, not only single, but by increasing length of the boxing match repeatedly whereby especially some or a single blow would be particularly hard. Here one must emphasize that the side is important, as it was detailed described on p. 246. We can assume that about 85% of boxers are right-handed, as is true of the entire population. For the population of all boxers, blows more frequently hit the left side of the neck than the right side. Apart from the short allusions in the literature by HERING (1927) and FRANKE (1963) there are surprisingly, to my knowledge, otherwise no hints about the sequelae of blows to the carotid sinus in boxing in the literature.

BRACK (1934) as well as FERRARO (1937) reported about fatal incidents in the boxing ring, in which boxers received blows against lateral parts of the neck and whose autopsies revealed not any pathological findings.

(1) *Mechanical irritation* of the *carotid sinus* may provoke one of the three *forms* of *carotid sinus syndrome*. Unconsciousness is the result of a depressive, cardiac, or central type of carotid sinus syndrome.

In the first two forms mentioned, the drop in blood pressure in the bradycardia or asystoly leads, respectively, to hypoxemia or anoxemia of the brain, after which unconsciousness may follow a few seconds later. In the third form, immediate unconsciousness with loss of posture occurs reflexively. This form should be considered a cerebral knock out.

HERING (1927) had called attention to a network of receptor nerves at the bifurcation of the internal and external carotid artery; he called these structures 'carotid sinus.' Mechanical irritation of these receptors leads to changes in respiration, blood circulation and muscle tone.

BOIGEY (1938) called special attention to the fact that the right side of the neck is more sensitive to blows than the left in regard to the carotid sinus reflex.

Injuries of Larynx and Glottis due to Blunt Blows to the Neck

Blunt blows to the neck also produce injuries to the *larynx and glottis*. A boxer hit on the throat may stagger and gasp for air, possibly resulting in a technical knock out, if called by the referee.

Von SURY (1922) reported on a 22 year old amateur boxer who received a blow against the area of the left mandible and neck in the second round of a sparring fight. The boxer collapsed after an interval of about two to three minutes and immediate death occurred. The autopsy revealed a marked edema and swelling of the vocal cords. There was an extensive hemorrhage in and around the neck musculature, which resulted in an acute glottis edema.

Fig. 200. Case of ENGEL: Cerebral type of a hyperactive carotid sinus reflex. After mechanical compression of the carotid artery no reaction. After mechanical occlusion of both carotid arteries after 10 seconds slowing of the homolateral brain potentials. (After ENGEL GL, from FRANKE H, 1963).

G.E.: Age 29 - Carotid Sinus Reflex Hypersensitivity, Cerebral Type

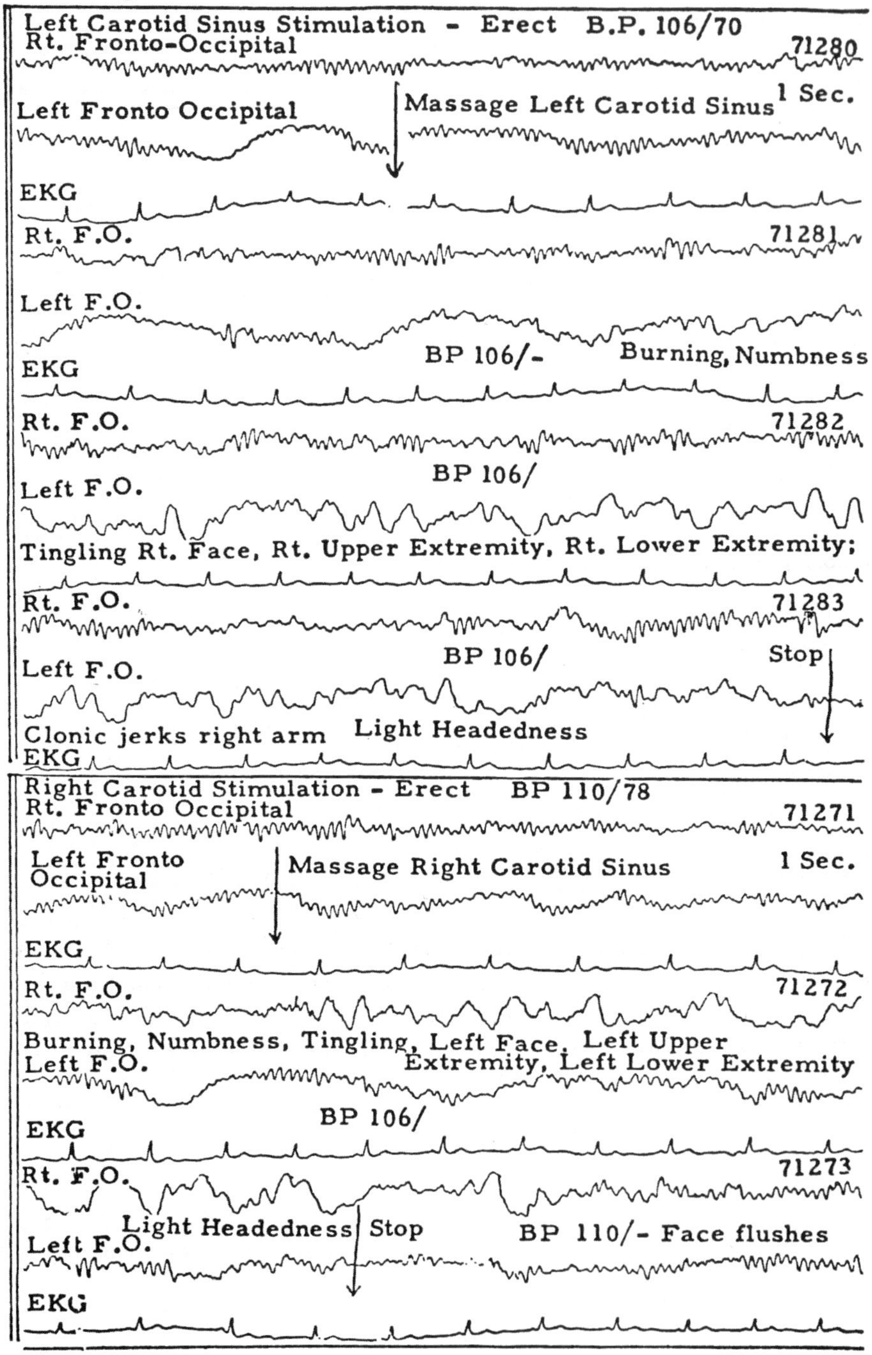

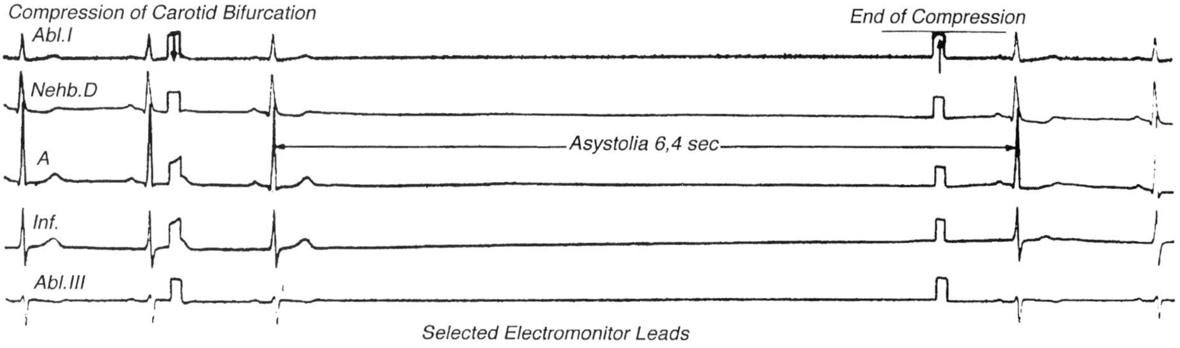

Fig. 202. Carotid type of hyperactive carotid sinus reflex. Sinu-auriular block with an asystole of the chambers of 6,4 seconds. (From FRANKE H, 1963).

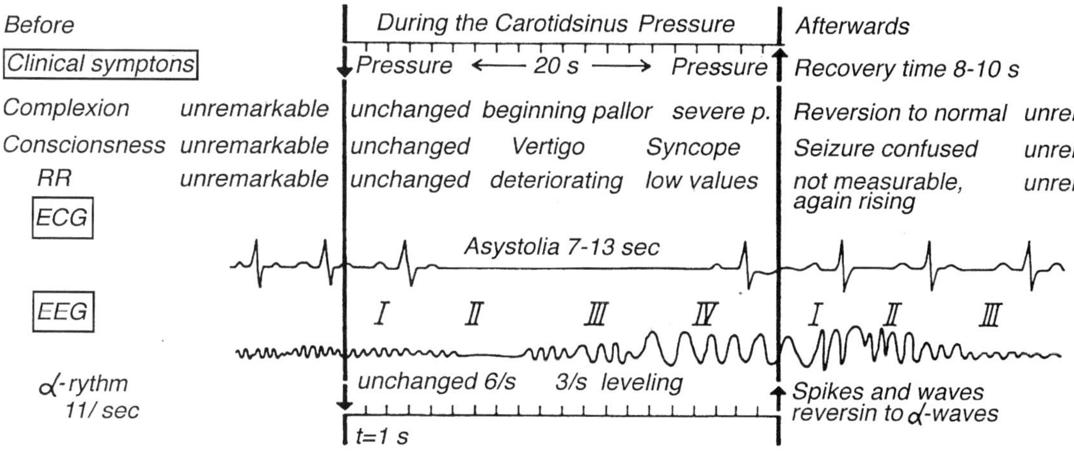

Fig. 203. Synoptic scheme of the symptomatology of the hyperactive carotid sinus reflex. (From FRANKE H, 1963).

Gene TUNNEY (1932) described a boxing blow against the Adam's apple [larynx]:

> "That first blow won the fight for me. Jack [DEMPSEY] was very ineffective after that. He hit me but one hard blow during the whole contest; it was a terrific left hook in the sixth round, landing on my Adam's apple. The cartilage was pushed into my throat and lacerated the mucous membrane on the side. I cought blood and was hoarse for several days."

Press photographs of boxers hit against the throat reveal that these injuries can easily incapacitate (Fig. 204a-c). The author remembers a professional heavyweight boxer whose speech was severely restricted for several weeks after a blow against the throat.

Hyoid Fracture

BPKIC et al. (1986) described ab extremely rare fracture of the hyoid bone and the larynx in a boxer

---

Fig. 201. G. E., age 29, carotid sinus reflex hypersensitivity, cerebral type. Unilateral excitation of the carotid sinus leads to retardation of the homolateral brain potentials. (After ENGEL G L, 1959, from FRANKE H, 1963)

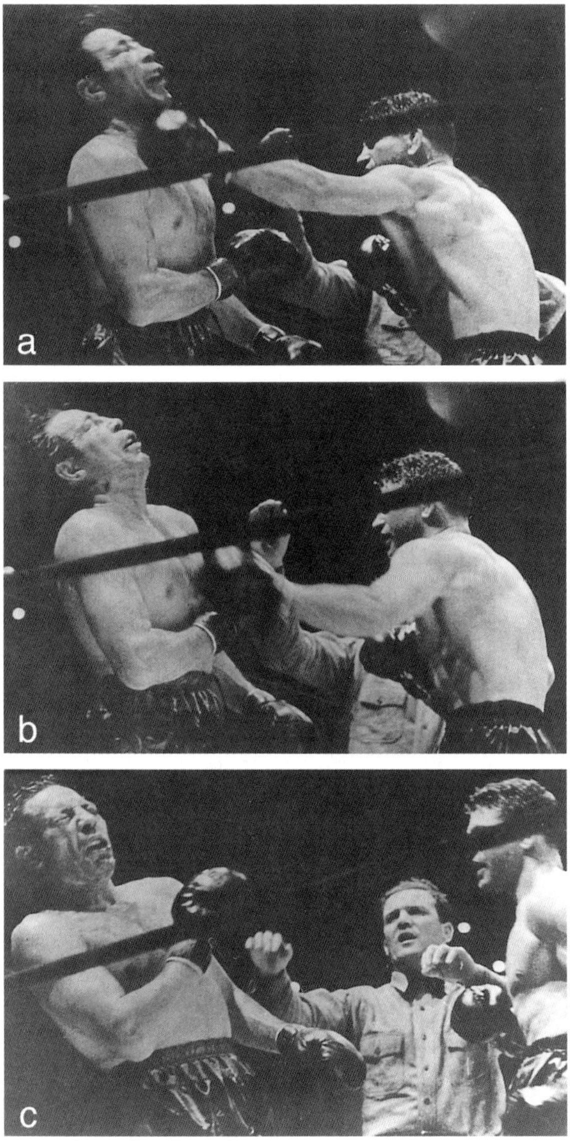

Fig. 204 a – c. This photo series shows a blow against anterior parts of the neck, the larynx (upper photo) and its effect on the boxer, whose agonising features, closed eyes and his bearing – the left arm hangs down, the right fist seeks support at the upper ring rope – the effect of the received blow and the complete defenselessness of the hurt boxer. The referee seems in this case to be sensible and to stop the fight. Such blows against the larynx occur more frequently as generally assumed. Such blows against the larynx are in boxing not illegal.

which occurred during a boxing match. The patient refused an operation.

Blows against the Precordium

Blows against the precordium may produce *alterations in cardiac actions,* resulting in a knock out. There exists a 'heart knock out,' or commotio or contusio cordis, which is sometimes fatal. These disturbances are discussed in the section on Cardiac Injuries.

Blunt Blows to Thorax and Abdomen

GOLTZ (1863) carried out experiments on frogs which are referred to as the *'Goltz's Klopfversuch.'* The author cut small windows into the frog's thorax carefully avoiding injuring the pericardium. By striking against the abdomen of the frog he initiated bradycardia and ultimately cardiac arrest in diastole. Shortly before a respiratory arrest occurred, he reported that the muscular tone decreased markedly. If both vagal nerves were transected, no reaction occurred, but if only one vagal nerve was cut the experiment could be carried out. After destruction of brain and spinal cord the experiment could not be reproduced, but with resection of the brain alone, with medulla oblongata and spinal cord left intact, the experiment produced positive results.

A *vagovagal syncope* consists, according to KOPALD et al. 1964, of hypotension and loss of consciousness associated with cardiac slowing or standstill.

The afferent or centripetal impulses originate in the sensory endings of the vagal nerve. These inhibit the vasomotor and cardioaccelerator centers while exciting the cardioinhibitory center. Efferent or centrifugal vagal impulses may produce marked bradycardia or varying degrees of heart block. The resulting hypotension leads to decreased cerebral blood flow and results in syncope.

Vagovagal syncope must differentialdiagnostically distinguished from *vasovagal syncope,* in which a sharp decline in blood pressure is secondary to a loss of vasomotor tone resulting from a variety of stimuli, as changes in posture, pain, fright, etc. Sometimes reflex tachycardias associated with the hypotension, is the only cardiac change seen. In a vagovagal syncope occurs a cardiac slowing or arrest.

Blows to the Coeliac Plexus

Blows to the *coeliac plexus* (the old *solar plexus)* may lead to extensive reactions and effects on the *circulatory system.* A hard punch into a boxer's abdomen may cause nausea and vomiting, and a shock-like state. This is associated with a pain contorted face and writhing weakness, low blood pressure, generally slow and sometimes irregular pulse, marked pallor, profuse sweating, syncopal attack with unconsciousness, widely dilated pupils, occlusion of nasal passages, possible lacrimation, and if consistent, the shock from the pain may lead to decrease in urinary output.

The *coeliac plexus,* which *contains many ganglia,* is the largest of the vegetative nervous system and thus was at one time called the *cerebrum abdominale.* This plexus rests upon the abdominal aorta and extends into the area of the branches of the coeliac artery, cranial mesenteric artery and renal artery; cranially it connects with the thoracic aortic plexus and caudally with the abdominal aortic plexus.

The *coeliac ganglion* is a paired crescent-shaped ganglion that reaches laterally into the neighborhood of the adrenal gland. It can form a ring-shaped ganglion around the coeliac artery, thus the old name solar plexus. The ganglion contains the following roots; the splanchnic nerves, the vagal nerves (especially the right one after it has branched out from the gastric rami with the plexus), the right phrenic nerve, and several direct branches from the lowest thoracic and the first two lumbar ganglia. From these derive further 'secondary' plexus, which surround the arteries of the abdominal organs. These ganglia, often containing ganglion cells, are named according to the arteries they accompany. There are direct branches extending to the adrenal glands, pancreas, and stomach. It is worthwhile to note the extensive connection of the plexus with the adrenal glands.

This demonstrates that the coeliac plexus supplies the sympathetic innervation for all organs and blood vessels, for the vasoconstriction and the smooth musculature of the abdominal organs, and for the secreting fibers and sensoric fibers of the organs. Blows or pressure at the coeliac plexus, via afferent fibers (especially extending into the splanchnic nerves), produce an increase in arterial pressure in the peripheral vessels, vasoconstriction of vessels in the lungs (dyspnoe), an increase of the cardiac rate, changes in the blood chemistry due to action of the smooth musculature of the spleen, increase in adrenalin, excretion by the secretory fibers, and influence on the vessel tone finally leading to a massive vagal reflex.

The *'solar plexus knock out'* is not ordinarily accompanied by unconsciousness. JOKL (1941) thought that this injury is identical with the 'Klopfversuch' studied by GOLTZ (1863). This is incorrect, there exist two different mechanisms.

Charles DICKENS described in 'David COPPERFIELD' a *'bare knuckle fight'* between a *young gentleman* and a *young butcher,* the terror of the youth of Canterburry. The description of the 'bare knuckle fight' reveals the disturbances of consciousness of the young gentleman in a literally form.

DICKENS wrote:

"The shade of a young butcher rises, like the apparition of an armed head in MacBeth. Who is this young butcher? He is the terror of the youth of Canterbury. There is a vague belief abroad, that the beef suet with which he anoints his hair gives him unnatural strength and that he is a match for a man. He is a broad-faced, bull-necked young butcher, with rough red cheeks, an ill-conditioned mind, and an injurious tongue. His main use of the tongue is to disparage Doctor STRONG's young gentlemen. He says, publicly, that if they want anything he'll give it 'em. He names individuals among them (myself included), whom he could undertake to settle with one hand, and the other tied behind him. He waylays the smaller boys to punch their unprotected heads, and calls challenges after me in the open streets. For these sufficient reasons I resolve to fight the butcher."

"It is a sommer evening, down in a green hollow, at the corner of a wall. I meet the butcher by appointment. I am attended by a select body of our boys; the butcher, by two other butchers, a young publican, and a sweep. The preliminaries are adjusted, and the butcher and myself stand face to face. In a moment the butcher lights ten thousand candles out of my left eyebrow. In another moment, I don't know where the wall, or where I am, or where anybody is. I hardly know which is myself and which is the butcher, we are always in such a tangle and tussle, knocking about upon the trodden grass. Sometimes I see the butcher, bloody but confident; sometimes I see nothing, and sit grasping on my second's knee; sometimes I go in at the butcher madly, and cut my knuckles open against his face, without appearing to discompose him at all. At last I awake, very queer about the head, as from a giddy sleep, and see the butcher walking off, congratulated by the two other butchers and the sweep and publican, and putting on his coat as he goes; from which I augur, justly, that the victory is his."

"I am taken home in a sad plight, and I have beef-steaks put to my eyes, and am rubbed with vinegar and brandy, and find a great white puffy place bursting out of my upper lip, which swells immoderately. For three or four days I remain at home, a very ill-looking subject, with a green shade over my eyes; and I should be very dull, but that Agnes is a sister to me, and condoles me, and reads to me, and makes the time light and happy. Agnes has much confidence completely, always; I tell her all about the butcher, and the wrongs he has heaped upon me; she thinks I couldn't have done otherwise than fight the butcher, while she shrinks and trembles at my having fought him."

In some instances such a disturbance of consciousness of a longer duration is observed. In a boxer, who was knocked out in the first minute of the first-round, his recollection began only 36 hours later, when he found himself sitting in a train and wondering where he was travelling. His behavior was apparently unexceptional to lay observers. When he was examined he had a slurred speech and all reflexes were very active. He revealed a slight unsteadiness in standing and walking. He dragged one leg slightly.

In this case unconsciousness developed immediately after a knock out. The patient then developed a traumatic amnesia, that covered a period of 36 hours, in which he found himself sitting in a train. He must have bought the train ticket himself. His behavior must have been inconspicuous, because nobody discovered the traumatic amnesia. MUNRO (1938) noted of such a patient: "He knows that he has been unconscious, but his friends do not." We object to the use of the term unconsciousness, the traumatic amnesia developed from a status of unconsciousness. There are no data about the duration of the unconsciousness from the knock out in this case.

Amnesias

A *classification of amnesias* was presented by VERJAAL (1939):

Amnesia, retrograde amnesia and *anterograde amnesia.* The author examined 43 patients and the following details were taken into consideration:

(1) The type of trauma, (2) the location of trauma, (3) whether the accident occurred unexpected, or whether the patient recognized it in advance, (4) the duration of unconsciousness, (5) the duration in which after the accident psychic disturbances were manifest, (6) the extent of the retrograde amnesia after the end of the unconsciousness, (7) the time in which the retrograde amnesia ended, (8) the extent of the retrograde amnesia after no psychic anomalies were any longer present, (9) whether recollection of the accident existed or not, (10) the duration of the anterograde amnesia, and (11) the extent of the retrograde amnesia and the anterograde amnesia at the reexamination.

According to VERJAAL here exist three main theories:

(1) The amnesia has a somatic foundation, it results by an impairmant of those parts of the cerebrum, probably cerebral cortex, which play an important role in the preserving of recollection.

(2) The disturbance of memory, or paramnesia, is genuine psychogenic. This theory relies mainly on the practical experience, that many forms of loss of memory can be eliminated by hypnosis.

(3) the loss of memory is a psychological manifestation, it is based on repression mechanisms. This theory is defended by many psychoanalysts.

These are only the main theories.

VERJAAL (1939) noted that those cases of loss of memory are best suited for an analysis in which the disturbance developed due to a head injury.

Naturally it was not possible to answer all questions satisfactory, many data had to be found out by questioning witnesses and family members.

Twenty-eight of the 43 examined patients had in the first examination an amnesia, in 17 of these 28 patients existed moreover a distinct retrograde amnesia. In several patients the disturbance of memory disappeared during the observation period, that finally still 19 patients revealed an amnesia, in whose only in four cases a distinct and in eleven cases only a very slight retrograde amnesia existed. At the end of the treatment was the retrograde amnesia always less extensive than at the first examination. This recession of the retrograde amnesia during the observation period resulted in a different way:

(1) At the begin the amnesia changed often quite

considerable, it was now and then more or less extensive, shortly afterwards it was missing completely.

(2) The retrograde amnesia can in such a way decline, that the recollection returns in the course of hours or days of events, which seemed in the beginning to be forgotten.

The retrograde amnesia can remain unchanged for some time and then very suddenly disappear.

The retrograde amnesia can subside to a certain time before the accident, which never can be remembered. It remains a gap, which never disappeares.

Retrograde Amnesia

Ritchie RUSSELL (1971) defined *retrograde amnesia* as the "interval preceding the injury, of which the patient has no recollection." In it is the interval preceding between a single heavy blow or series of blows of which the pugilist has no recollection. It is not necessary that the boxers suffered a knock out.

SCHACTER and CROVITZ (1977) summarized that:

> "... the single greatest deficiency in the literature concerned with retrograde amnesia following closed head trauma is the lack of appropriate methods which permit quantitative study of the phenomenon."

While CRITCHLEY in 1937 noted that a retrograde amnesia is permanent, published RUSSELL (1935), RUSSELL and NATHAN (1946) and BENSON and GESCHWIND (1967) convincing cases of shrinking retrograde amnesia.

RUSSELL (1935) described the phenomenon of *shrinking retrograde amnesia* in a patient who had suffered a closed head injury – not due to boxing – who revealed a retrograde amnesia of nine years duration during the transition from coma to resolution of posttraumtic amnesia. He finally was able, over a period of ten weeks to gradually recollect events in the remote past beginning with the earliest memories and finally was able to recall everything within a few minutes of his motorcycle accident.

These findings were later confirmed by RUSSELL and NATHAN (1946) in other cases; they reported that in cases of retrograde amnesia, memories were recalled and recovered in a chronological order beginning with those from the distant past with later recollecting those of more recent events. These authors found that shrinkage of retrograde amnesia usually went parallel with resolution of posttraumatic amnesia; but in single cases gaps in the past remained in oriented patients.

A very well documented case of *shrinking retrograde amnesia* was described by BENSON and GESCHWIND (1967) in a patient who had a right temporal subgaleal hematoma of undetermined etiology, who was "stuperous but not really unconscious."

We are quoting in detail:

> During the early phase of recovery, a retrograde amnesia of two years' duration was exhibited; responses to questions concerning address, occupation, and family life corresponded to the patient's life two years prior to injury (e. g. he insisted that he was living in Washington, D.C., although he had moved to Boston two years before). After a period of three months, recall of remote information steadily improved until the patient was left with a residual gap of 24 hours prior to admission. The authors conjectured that the patient's probable alcoholic intoxication shortly before injury extended the period of permanent retrograde amnesia. They contrasted their brief residual retrograde amnesia after head injury with the permanent one in metabolic Korsakoff's disease, which may span periods as long as several decades. The shinkage of retrograde amnesia was interpreted as evidence for gradual restoration of retrieval of memories from long-term storage; posttraumatic amnesia was interpreted as reflecting impaired consolidation of new information, which undoubtedly was consolidated prior to injury, constitues strong evidence for a retrieval hypothesis of retrograde amnesia, there is no reason to doubt that impaired retrieval also contributes to deficient learning and memore after head injury.

REISNER and SCHERZER introduced the term *'isolated traumatic amnesia' ('isolierte traumatische Amnesie').* According to ARDITO et al. (1974) such a traumatic amnesia can occur completely isolated but also be combined with a retrograde amnesia. These authors noted that these amnesic states are not discussed in papers which examine the traumatic encephalopathy of the boxer.

BENDER (1956) described short *'transient amnestic episodes,'* which in general extend from a few hours to one or two days, in which the patient seems to act normal but in reality is disoriented and confused.

Anterograde Amnesia ('traumatischer Dämmerzustand')

The so-called *anterograde amnesia,* i. e., a gap of memory for a period after a knock out, covering not only the unconscious state but a time when to all outward appearance the boxer behaved normally, suggests that here again a boxer is in a state of 'altered consciousness,' which can clear up after minutes, hours or days. (The German term is *'traumatischer Dämmerzustand').*

Each trauma to the head can cause a cerebral concussion (commotio cerebri), a contusio cerebri or a traumatic amnesia ('traumatischer Dämmerzustand'). Multiple preceding injuries of the brain due to mechanical trauma of the head seem to be an important factor for these traumatic amnesias. But a single severe blow can cause a traumatic amnesia, too.

Frequently forensid problems result after traumatic amnesia, for example after traffic accidents. The affected person begins with a relatively normal and orderly impression, but attracts attention later due to tactless actions, irritated annoyance, aggressive behavior and disregard for social manners. Conflicts situations show up inevitably. Hit-and-run driving after automobile accidents in a traumatic amnesia were described.

Traumatic amnesias ('traumatische Dämmerzustände') are very common in boxers. Almost every boxer has reported that he had one at least once, many report about multiple. The affected persons, despite a disorientation in time and space, make to direct aquaintances a completely orderly impression. Traumatic amnesias of different length, affective disturbances, confusion and delirious disturbances are in single cases observed. BODECHTEL and R. SPATZ (1971) noted that these symptoms are similar to a slight cerebral concussion. We do not agree with this statement, since the clinical symptoms of both entities are so different. In traumatic amnesias unconsciousness is missing. We quote here only the papers on traumatic amnesias in boxing in which they occurred (LIVET 1922, DARTWELL 1925, BLITMAN 1928, MARTLAND 1928, TUNNEY 1932, CRITCHLEY 1937, WINTERSTEIN 1937, HERZOG 1938, LA CAVA 1949, LARSSON et al. 1954, BERGLEITER and JOKL 1956, Nat FLEISCHER 1958, MAWDSLEY and FERGUSON 1963, UNTERHARNSCHEIDT and SELLIER 1971, ARDITO 1974).

Such a traumatic amnesia can develop without cerebral concussion (commotio cerebri) with an immediate loss of consciousness and muscle tone. It is worth mentioning here that a heavy boxing blow or a series of blows can produce the same syndrome in a boxer without having been knocked down or knocked out.

When the knock out blow does not lead to an interruption of the fight, the men report retrospectively that their sensation during the following rounds were more or less vague. This is sometimes described as a *narrowing of the visual field.* People say that they saw 'the white skin of their opponent,' or 'saw his face only: everything else was black.' Very commonly a mild depersonalization exists, the feeling of a kind of detachment from the person's own body and its activity. After a complete knock out the subjects describe the beginning of regaining consciousness in various ways. Sometimes the first thing they hear is the roar of the crowd, the counting of the referee and the tapping of their seconds on the floor, or shouts from their corner as strongest and the first self-observed impulse. Others who have been more severely knocked out and come round only in the corner report the sensation of cold water and the manipulations of their seconds to bring them around as the first subjective recollection of recovering consciousness. Among the severe cases one hears of boxers who did not regain consciousness until they found themselves under the shower.

We advice the readers of the chapters about the clinical and morphological case histories of this book to read them in the original very attentively and thoroughly to learn not only about the miserable and shameful state of health of ex-boxers, but also about the frequency and duration of traumatic amnesias. Lack of space does not allow a thorough analysis of this topic.

During recovery they are slow in response. They are not correctly oriented, but their orientation is never disturbed by hallucinations. Mistakes can generally be accounted for by the amnesic period. In their self-description as quoted, one finds only descriptions of vague and undifferentiated sensory perceptions, but never any hallucinations, not even of illusions or misinterpretations.

In a *cerebral concussion,* different states of altered consciousness may occur: (1) *Amnesia;* (2) *retrograde amnesia*; and (3) *anterograde amnesia,* (*'traumatische Dämmerzustände'*). The latter may occur after a severe single blow or after a series of blows against the head without producing unconsciousness.

(1) The *main clinical symptom of cerebral concussion* is a *simple gap in memory* – an *amnesic period* – commencing at the moment of the impact against the head. The *period of unconsciousness* normally covers an *interval* of *seconds* to *minutes,* or (in some cases) *hours.* The return of consciousness normally covers a more or less brief period, but in some cases it may extend over several days (as described below under fugue states). The first symptoms after recovering from unconsciousness are described by a boxer in this way: he remembers the sensation of cold water his seconds splashed on his face, manipulations by his seconds or by the referee, the roar of the spectators, the counting of the referee, and shouts from his cornermen to get up and continue the fight.

(2) In *some cases* of *retrograde amnesia,* there is not only *amnesia* or a *gap in memory* concerning the *blow,* but also a *shorter or longer period of time preceding the blow has disappeared from memory.* This is a so-called *retrograde or retroactive memory loss.* Boxers forget not only the knock out blow but also preceding rounds, how they got into the ring, or events that occurred well before the match.

(3) In addition to retrograde amnesia, there exists an *anterograde loss of memory.* In this type, the *amnestic phase covers a period following the apparent recovery of consciousness.*

The return from a state of total unconsciousness and the renewal of motor activity do not permit us to conclude that lucid consciousness has been completely restored.

The *'twilight,' 'fugue' or 'dream states'* (SCHILDER 1934) resemble conditions that occur in epileptics. They are, according to BOWMAN and BLAU (1960), characterized by the performance of purposeful activities, sometimes covering a period of several days, in a half conscious dream-like state very much like somnabulism. The patients may appear entirely rational and perform adequate acts in this condition, yet there is a complete amnesia or loss of memory concerning these occurrences. Wounded soldiers in this state may, after a short period of unconsciousness, return to their positions; other patients act and talk in a senseless manner, and wander around aimlessly. Extreme furor with violence, similar to the epileptic excitement, has also been described.

It has to be stressed that groggy boxers or boxers in a traumatic amnesia may commit fouls without being responsible for their actions due to their state of altered consciousness.

The acts performed in such a traumatic amnesia appear purposeful. In most cases a partial or complete amnesia for this condition remains. MUNRO (1938) said of such a patient, "He knows that he has been unconscious, but his friends do not." We agree with the content of the statement, but object to the use of the term 'unconscious'. Patients in this traumatic amnesia ('traumatische Dämmerzustände') are by definition amnestic but not unconscious.

Amnesia may at first be complete, but later the amnestic period may become shorter and the boxer may reveal some more or less vague islands of memory during this period. This is not based on the reports given to him by other persons, but represents a later 'recovery' of primarily lost information.

Different and sometimes not very precise names are used to describe these conditions: *semiconscious state, partial consciousness, state of altered consciousness,* etc. These terms should not be used, since an adequate and widely accepted nomenclature (which has been used above) already exists.

The best known story of such a state is that told by Gene TUNNEY of himself. He suddenly realized after a sparring fight that he had slipped into a twilight state and consequently broke off and quit boxing for fear of becoming insane.

> "One day while boxing with a sparring partner, Frank MUSKIE, we bumped heads. The part of my skull that is the thinnest, near the temple, struck the toughest part of his, the top. I was terribly dazed. As I straightened up, a long hard swing landed on my jaw. Without going down or staggering, I lost all consciousness of what I was doing and instinctively proceeded to knock MUSKIE out. Another sparring partner, Eddie EGAN entered the ring; we boxed three rounds. I have no recollection of this nor have I any recollection of anything that occurred until next morning when I was awakened in my little cabin by the water's edge, wondering who I was and what I was doing there... As I lay in this awful state of returning consciousness, I became greatly frightened. Gradually my name came to me. That I was a pugilist soon followed, then the thought of being a champion – impossible, unbelievable. I must have had a long dream. Gradually, the realization came that I had not been dreaming. I rose and asked guarded questions. I wanted to know all about the events of the day before. For three days I could not remember the names of my most intimate acquaintances. I had to stop training. I did not leave my cabin except to eat or take a short walk. On these occasions all seemed queer. I was unable to orientate myself. The sensation I had was as though hot water had been poured through a hole in my skull and flowed slowly over my brain to my eyes, leaving a hot film."
> "There were three newspaper men at the camp reporting my activities. They had to be deceived. This story was too sensational to permit it to get out."
> "I confided my condition to none but Eddie EGAN. He was keeping a diary. It is amusing to read his notes of those few goofy days."
> "After returning to normal I decided that any sport in which such accidents could occur was dangerous. I realized I had had a concussion. The first seed of retirement was sown then. The possibility of becoming punch-drunk haunted me for weeks."

Boxing enthusiasts may object, suggesting that in this case the twilight state was the result of an illegal head butt, but descriptions of twilight or fugue state resulting from boxing blows against the head are numerous in the literature of both professional and amateur boxers.

Fred DARTNELL ('Long MELFORD') (1925) called in this context to notice the Australian boxer Fred RAY, who participated in England in a boxing match against Fred DYER, who knocked him down with a hard right. RAY himself reported about this knock down:

> "It amused me to hear people saying afterwards how game I was to fight after this. Honestly, it was fully a round before I regained my senses. I knew nothing of being dropped by DYER until I was seated in my corner. Then I asked my second what had happened. After being rattled like this, a boxer is a mere automaton, for the next round or so. In such a punch a faintharted man may fight back gamely, yet if he possessed his complete senses he would probably quit."

In another example, featherweight contender Harry BLITMAN describes his fight with Connie MACK which took place September 10, 1928 at Shiba Park (now Connie MACK Stadium) in Philadelphia. BLITMAN, who spent eight years in the ring, winning most of his bouts but taking some severe beatings, stopped boxing after the following incident:

> "Just before the bell at the end of the second round, BASS caught me on the jaw with a thunderous right, I was dropped for the count of seven when the round ended.
> "For the first two minutes of the third, following instructions in a sort of animal way, I was able to make BASS miss repeatedly. I was pumping uppercuts against his chin, but not hard enough to do any real damage."
> "The fourth was my round. I smashed my left into Benny's mouth and nose time after time. I had him dazed and reeling for two minutes. Yet I could not put him away, and he came back like the great fighter he was."
> "The fifth round was all BASS. It was a case of a good right hand. I don't like to think about the punishment I took in

that round or the fury I felt with myself because my timing was off and I couldn't seem to land."
"It was all over at 1:04 in the sixth round. With the exception of the 10 seconds it required to count me out, Benny was punching me at will. I got up at the count of '8' on the last knock down, but I couldn't stand and slumped to the canvas. I tried to push myself up by my finger tips, but I couldn't do that either."
"All that came later. The morning after the BASS fight, HOFF sent two of his boys to my hotel. Since the guarantee was in my name – Boo Boo was having some financial trouble in the bursting Florida land boom – I was told to go by the old Franklin Trust Company at 15th and Chestnut streets and pick up the purse – in cash. The left side of my face was badly swollen. My body, particularly my stomach and ribs, was a corset of pain. But what really alarmed me was that my hands wouldn't behave in the job of putting on my socks. And when I stood up on one leg to slide the other into my trousers, I fell on the floor.
"One of the gunmen who helped me up laughed and said, 'That was some going-over you got last night!'"

LIVET (1922), WINTERSTEIN (1937), HERZOG (1938), and BERGLEITER and JOKL (1956) mentioned reports on traumatic amnesia ('traumatische Dämmerzustände') in boxers.

LIVET (1922) published the following case history:

A 30 year old professional boxer received a butt to the head from his opponent, in the first round of a fight. He mentioned the foul to his opponent, who apologized. In the break after the first round, the boxer felt well and was fully oriented. From the beginning of the second round, the boxer was not able to remember the fight. At first he continued to box in a normal fashion, but after a few steps he fell down, without having been hit. The referee counted to '5' and '6', the boxer heard '7', '8' and '9' and tried to get to his feet. At this point he slipped into an amnestic phase again.
When the second round ended, the boxer walked into the wrong corner of the ring and had to be brought to his own. He seemed to recover completely. The third round began, the boxer attacked his opponent and hit him with a series of severe blows, but was then knocked out by a powerful blow from his opponent. He was carried to his corner, and his gloves were removed. He left the ring on his own, and walked to his dressing room. Here his twilight state ended and he asked "Where am I?" The answer was, "You are here to box." He recognized that his hands were covered with bandages and answered, "Yes, that is right."

After a free interval of about two minutes after the head butt he became unconscious and later amnestic. He had an amnestic period from the break after the first round and can remember only a few seconds, hearing the referee counting, then recovery from the amnestic state after about another ten minutes.

WINTERSTEIN (1937) reported on five boxers:

The last thing one boxer remembers is the bell at the end of the fourth round; then everything went blank. He next remembers arriving home that evening and seeing his mother; later he hears from his friends that he won the fight, his opponent being disqualified for hitting him after the bell. His friends report that he was picked up by the referee after having been on the floor a few seconds; and that without any assistance he left the ring, washed and dressed himself, fetched his money from the manager, went to the tram and took four tickets for himself and for his three friends.
Another boxer was knocked down for a short count in the first round. He next remembers standing under the shower and his father telling him that he won the fight in the third round on a technical knock out, having fought like a fighter and not at all like the scientific boxer he usually was.
In a *third case,* full recollection began in the train the morning after the fight in which he was knocked out in the first round. He inquired what town he was going to.
In yet *another case* a boxer caught a heavy punch in the first round but did not go down. He next remembers sitting in his corner, the roar of the crowd suddenly becoming audible, and seeing a sign going up marked 'seventh round,' whereas he was under the impression that it was only the second round; he won his fight on points.
As a *last example* there is the case of a boxer who remembers the fight only up to the fifth round. Recollection came back in the dressing room, his first thought being, "Did I fetch my money?" In spite of many indirect questions he was unable that evening to find out from his friends what the result of the fight was. As he did not wish to appear stupid by asking directly, he had to wait for the next morning's paper to read of his victory on points.

WINTERSTEIN reported, that nearly half the boxers he examined described such instances of anterograde amnesia. Many were deeply impressed by this experiment. The boxer is in a *'state of altered consciousness'* or *'posttraumatic twilight state.'*

HERZOG (1938) published three examples of knock outs in boxers; clinically they could be considered cerebral concussions with unusually long unconscious periods combined with a complete amnestic period afterwards.

Case 1: Knock out with immediate unconsciousness of about three to four minutes duration. No vomiting. A traumatic twilight or dreamstate of about three to four hours' duration developed after regaining consciousness. The boxer seemed to be orderly, was able to speak coherently, and was adjusted to the situation, but was exceptionally disoriented. In the beginning of the twilight state he mistook the situation which led to slips. The boxer later had a complete amnesia for this period.
Case 2: Professional boxer who was knocked out by an uppercut. Immediate unconsciousness for a period of about 30 minutes. No nausea, no vomiting. About two hours after the knock out the boxer was fully oriented. Complete amnesia for this period. The twilight state, with disturbances in orientation and in which the patient was confused, reverted completely.
Case 3: Is similar to case two. The unconsciousness was only of 15 minutes duration, and the following twilight state lasted an hour.

STEIN (1951) reported about a boxer, who had to take one week after a severe knock out in an other match in the first round only slight blows against his head and was counted out, he watched the two following boxing matches. In the dressing room, he became very suddenly unconscious. A leftsided hemiplegia developed. At an operative intervention which was performed under suspicion of an epidural hemorrhage, no bleeding was found. The symptoms slowly receded and two months later he was discharged after retrogression of all clinical symptoms. For the duration of the fight and the following five days existed a traumatic amnesia or 'fugue sate.'

This observation with a five days 'state of altered consciousness' or 'posttraumatic twilight state' is the longest described in a pugilist.

BERGLEITER and JOKL (1956) published three case histories of boxers who had experienced dream or fugue states.

Case 1: A 20 old amateur boxer, in the third round of a boxing match, 'walked' into a straight left from his opponent. A handler remarked: "That punch hit!" Friends noticed after the bout, that the boxer was unable to remember details of the bout. He was completely disoriented and did not recognize his girlfriend, who expected him. He had a fit of excitation and tried to beat up his girlfriend, but finally went to a pub with her. Late in the evening he went to a house where he had lived a long time ago, and was surprised not to be able to open the lock. It is not known how he got to his apartment. On the following morning he was unable to remember the events of the last evening, the boxing match, the argument with his girlfriend, the visit to the pub, and his attempt to enter his old apartment. The neurological examination three days later was unremarkable.
Case 2: A boxer was knocked down in the first round for a count of '4'. He continued the fight until his opponent was disqualified in the twelfth round for a low blow. In the dressing room, he could not remember the fight, and asked who had won. He was able to remember only three short episodes. He remembered sitting in his corner during a break and thinking he was in a restaurant. Then he could remember seeing scaffolding in a corner of the arena and asking himself what it was there for. Finally, he remembered being under the cold shower after the fight. There had been no unconsciousness during the fight. In the dressing room he was unable to find his shoes and his suit. After dressing he seemed helpless and asked: "How do I get home?" He knew his first name but not his last name. At home he slept to the next forenoon, when his mother woke him up to remind him of his work. He could not remember what kind of work he was doing or where he was employed. He complained about headaches.
Case 3: A former British heavyweight champion gave the following report. He later committed suicide:
"My manager and I kept the fact that I had severe disturbances in my consciousness and memory during my last fights absolutely secret. But now, after I finished my career, I can tell you that after the fight my manager locked the doors to the dressing room so that he was able to tell me what had happened during the bout, so that I could tell the newspapermen waiting outside something sensible. During the interview he whispered to me details about the fight in 'Africaans' and 'Zulu,' so that I talked no nonsense. I have heard much about brain damage in boxers, and I know that many boxers are able to contribute to that from their own experiences. I find these injuries terrible, especially when they accumulate."

Nat FLEISCHER (1958) contributed two cases he had seen at ringside:

"In December 15, 1947, I saw a comical performance by Mexican welterweight Nava ESPARZA, in a bout with Brooklyn's Rocco ROSSANO at the St. Nicholas Arena in New York City.
"Twenty-six seconds after the fight began, ESPARZA was counted out in the quickest knockout in the history of the St. Nick Club. After the '10' count had been tolled, Nava was helped out of the ring. He climbed down clear-eyed, apparently in complete control of himself. As he got to the bottom step he reached down and picked up his own water bucket."
"Quickly reversing his movements, he climbed back into the ring with the bucket. What for? To await the next fighter. Later

in the dressing room he snapped back into focus, and was surprised to find out that while dazed, he had attempted to become a second."

"Some years ago two husky lightweights, Hugh ROSS of Kentucky and Tom KELLY of New York were pitted against each other in a 10-round bout. When the final gong sounded, ROSS was all in, on the verge of collapsing. As he fell into his chair, grasping the ropes, he turned to his second: 'What round was I knocked out in?' he inquired."

"Seconds later the announcer was hailing ROSS the winner, but the groggy fighter didn't realize it. In the dressing room he couldn't be convinced that he hadn't been kayoed. He thought his handlers were spoofing."

LARSSON et al. (1954) carried out investigations in conjunction with the *Swedish Amateur Junior Championships* in 1950 and 1951 and with the *Swedish Championships* in 1951. This study included 268 participants altogether.

The authors concentrated their investigation on junior amateur boxers: altogether 75 boxers were examined, in connection with 102 matches. The age of the boxers ranged between 17 and 25 years with few exceptions. Thirty had fought less than 20 contest bouts; 21 had boxed less than three years and 15 more than six years. According to their own report, 13 had never been knocked down previously, and only 17 had been knocked down more than three times.

On the whole, the junior matches appeared harder than the adult matches did. They were generally fought with youthful drive and lack of caution.

In the entire study of 75 boxers, 35 had been knocked down, with 14 knock outs.

Of the 44 boxers in the group examined before and after the matches, 13 had been knocked down, with 5 knock outs. Clinical observations were made during and after the match, and the cases with the severest symptoms were:

(1) Several knock downs and knock out. Believes he lay prone when referee stopped match (lay supine). Foregoing round gong rang while he was down; does not remember this.

(2) Several knock downs and knock out. No recollection of latter part of match; does not know he was knocked down or that he was knocked out. 'Wakened' sitting in examination room. Disorientation in time.

(6) Knock out. Amnesia for match. Disorientation in time. Euphoric. Amnestic period of twenty minutes duration. Headache afterwards.

(7) Hard head blows, match stopped. Believes he was in center of ring when match stopped (hung on ropes). Examination five minutes later showed disorientation in time; completely lucid some minutes later.

(15) Knock down. Does not remember match. Disorientation in time. Amnesia for about 20 minutes.

(20) Several knock downs and knock out. Amnesia for knock out and a few seconds before and after.

(25) Several knock downs. Believes that match was even. Stated a moment later that he had been knocked down, but remembered no details.

(30) Several knock downs. Remembers nothing of match. "What year is it? 1943, 1952... no, 1950." Sits contentedly and wonders where he is. Amnesia for twenty-five minutes.

(32) Several knock downs and knock out. Does not remember knock downs.

(56) Knock down twice. No recollection of knock down. Disorientation in time. Euphoric. Lucid after 25 minutes.

(64) Knock downs. Confused memory of match. Disorientation in time. Completely lucid after five minutes.

(66) Knock downs. Confused memory of match and does not know definitely that he lost the match. Disorientation in time. Lucid after 20 minutes.

(72) Match stopped due to hard hard blows to the head. Recalls only entering the ring, 'wakened' subsequently in examination room. Appeared hazy. Amnesia for at least five minutes.

(74) Several knock downs. Confused memory of match. Lost on points, but believes he was knocked out. Does not recall coming for examination. Completely lucid after one hour.

The most important results summarized above show that pronounced periods of amnesia and disorientation occurred in fourteen boxers. Superficially, these boxers appeared entirely recovered from their knock down, but by testing their mental functions the actual condition was revealed.

Subjects experiencing brief amnesia or slight disorientation:

(4) No knock down. Thinks he lay on floor when the referee stopped the match (stood upright with hanging arms).

(13) Knocked down once. Thinks he was able to continue match when it was stopped by referee (stood then with back to opponent). Slight disorientation; subjectively hazy.

(14) No knock down. Thinks he stood in the middle of the ring when the match was stopped (stood against the ropes in the corner).

(48) Knocked down once. Blow to left ear, count to '8'. Doubtful memory of remaining seconds of match. Deafness and ringing in his left ear for half an hour.
(52) No knockdown. Mentally hazy, 'cannot see clearly.' Appears to think sluggishly. Uncertain orientation in time. Lucid after half an hour.
(55) No knockdown, but reported severe effect of blows to the head. Disorientation in time.
(57) Knocked down once. Still mentally affected after count to '9', so the match was stopped. Well oriented at examination.
(60) Knocked down twice. Swaying and staggering for a while after the knock out. Recalled the knock down at examination, but reported that he had not been able to follow the count of the referee. Slight headache all day, worse after the match.
(62) Knock out at the second knock down. Does not remember clearly where he was when the match stopped. Some disorientation in time.
(68) Knocked down twice. Has difficulty in understanding things said after match. Subjectively hazy at examination and 'cannot see clearly.' Complete recovery after one hour.
(70) Knocked out at the second knock down. Thinks he stood up when the referee stopped match (lay on the floor).
(73) Knocked down once. Thinks he fell to his knees in knock down (lay flat). Headache both before and after match, vomited ten minutes after match.

According to LARSSON et al., an amnestic period with confusion seems to be characteristic of most knock outs. The boxers typically described the events in this manner: They did not recall any particularly hard blows; after a while they began to hear the referee's count and to notice the shouting of the public, but little else of what went on around them. Later their thoughts would begin to clear. During observation of the matches the boxers typically appeared to be confused, especially after a knock out, but also sometimes in connection with a knock down or a hard blow.

The boxers might report that they saw stars or that everything went black or dim momentarily following a hard blow, but it was never a question of a definite blackout. Nor was there any notable vertigo, although the slight unsteadiness and sense of weakness that sometimes occurred could be difficult to analyze, and the authors never found 'loss of tone' (resembling the inability to move – with preserved consciousness – occurring in the cataleptic attacks of narcolepsy).

A number of boxers exhibited amnesia after a knock down, covering periods of the matches during which they more or less effectively continued to box. The most pronounced example was case 15. This boxer, who was knocked down early in the match, won the bout notwithstanding amnesia for the entire match and for a period of 20 minutes thereafter.

In this condition, with evident confusion and disorientation, the boxers were still capable of automatic and rather effective action. The motor functions seem to be retained as well as, remarkably enough, the ability to follow the changing situations during the match. That such a condition may exist in the slight confusion after concussion of the brain is well known.

MAWDSLEY and FERGUSON (1963) reported posttraumatic amnesia in professional boxers. At the end of his active career, boxer number one had amnesia for 12 hours after a hard bout. Boxer number 8 was confused for hours after bouts he had lost, and boxer 9 had an amnestic period of six hours after a technical knock out.

But two cases of severe amnesias ('traumatischen Dämmerzustände') in professional fights which had a decisive influence on the outcome of heavyweight fights will be discussed in detail. The first case is the first heavyweight fight between Max SCHMELING and Joe LOUIS in 1936 and the second one is the heavyweight fight between Floyd PATTERSON and Ingemar JOHANSSON in 1969.

A case in which a traumatic amnesia occurred was the first heavyweight fight between Max SCHMELING and Joe LOUIS on June 19, 1936 at Yankee Stadium, New York. After a knock down in the fourth round due to a hard blow by SCHMELING (Fig. 205), LOUIS seemed to be befuddled, struck his opponent low and after the bell. Referee Arthur DONOVAN warned LOUIS (Fig. 206). There are no reports about LOUIS traumatic amnesia. In the twelth round LOUIS stood groggy and defenseless with his left arm hanging down as a powerful right of SCHMELING landed on his jaw (Figs. 207 a – c). LOUIS dropped to his

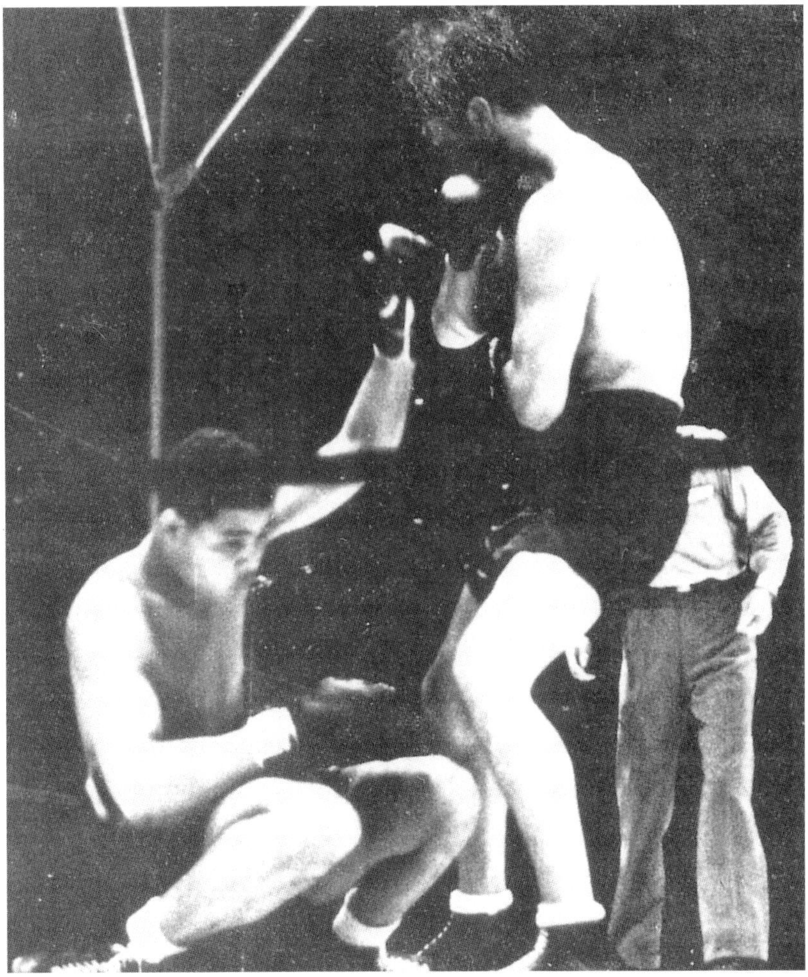

Fig. 205. First heavyweight fight between Max SCHMELING (right) and Joe LOUIS (left) on June 19, 1936 at Yankee Stadium, New York. LOUIS had won 27 consecutive fights, all with four exceptions, by a knock out, before his last fight with SCHMELING. LOUIS is about to hit the canvass for the first time in the fourth round of their first meeting. SCHMELING an 8–1 underdog, scored a 12-round knock out of LOUIS.

knees on the ropes (Figs. 208 and 209), fell then backwards to the canvas and was counted out.

Another typical case of a posttraumatic amnesia is the first professional title fight between Ingemar JOHANSSON, Sweden, and Floyd PATTERSON, on June 26. 1969 in Yankee Stadium, New York. In the first two rounds PATTERSON seemed to have the fight under control. JOHANSSON had not landed a hard blow in these rounds. PATTERSON connected with a hard right on JOHANSSON's mouth and nose (Figs. 210 and 211).

At the begin of the third round JOHANSSON connected with a powerful left hook and a following right, PATTERSON was knocked down to '9' (Fig. 212), he was so dazed and disoriented that he walked into a neutral corner. He said later that he had thought JOHANSSON had been knocked down and the referee had started counting. It existed a treumatic amnesia as the result of the combination of the two hard blows which had him knocked down. Another photo shows another scene of this fight. PATTERSON rubs his nose, defenseless, with his left arm hanging down, and an

Fig. 206. First professional heavyweight fight between Max SCHMELING and Joe LOUIS. After LOUIS first knock down in the fourth round LOUIS seemed to be in an amnestic state ('traumatischer Dämmerzustand'). LOUIS obviously befuddled, struck his opponent low and after the bell. Here referee Arthur DONOVAN warns LOUIS.

easy target for JOHANSSON's left (Fig. 213). PATTERSON suffered seven knock downs in this round before referee Ruby GOLDSTEIN signaled at 2:02 min that it was all over (Fig. 214).

GOLDSTEIN (1959) commented on this fight:

"The unwritten code of the ring demands that the heavyweight champion, above all others, must be allowed to continue to the end. This code is intended for the protection of a champion who may seem to be on the loosing end of a title bout. As long as he can stand and see, even if he can't keep his hands up, he is usually allowed to continue, on the theory that he deserves even a dying chance to cling to the title."

GOLDSTEIN continued:

"In the Floyd PATTERSON – Ingemar JOHANSSON fight in June, 1959, I gave PATTERSON every chance to recover from the knock down he suffered the first time JOHANSSON reached him with a straight right hand. Of the six that followed, four resulted from glancing blows, as he still was feeling that first one in the legs. To assure myself that he was not badly hurt, I once asked him if he was o. k? He said he was and his eyes looked all right, but the next time he was toppled, I called a halt."

Interestingly, in their rematch were victorious. Traumatic amnesias ('traumatische Dämmerzustände') are the results of accidents which occur in horseback riding and automobile accidents.

A traumatic amnesia was described by BISMARCK. After a fall from his own horse, he mounted his groom's horse and rode home. There he mistakenly reported that his groom had an accident and he did not recognize his dogs. He became en-

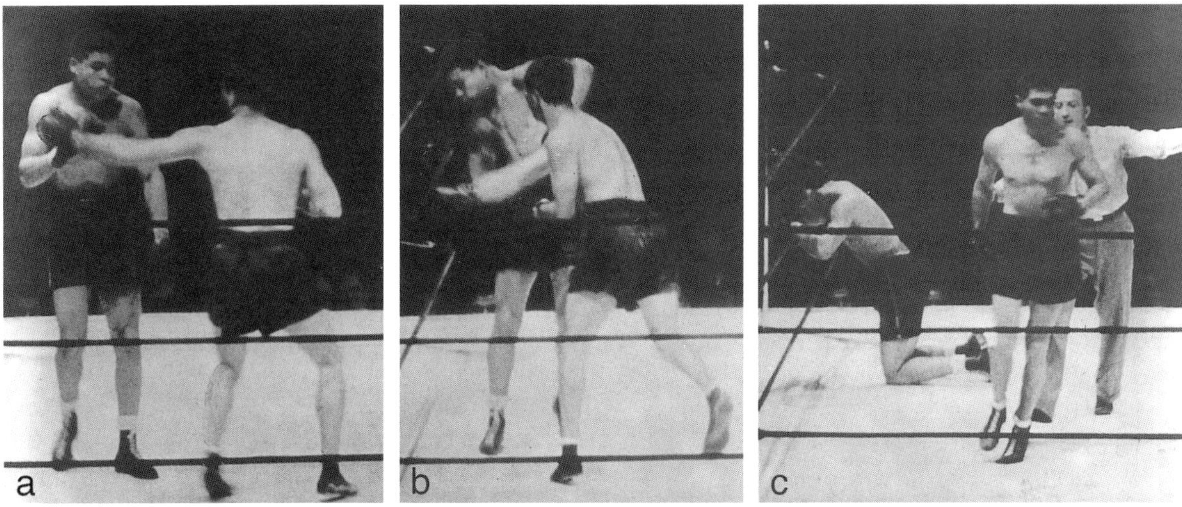

Figs. 207 a – c. In the twelfth round LOUIS (left) stands defenseless and groggy with his left arm hanging down (left photo) as a powerful right of SCHMELING lands on his jaw. The middle photo was taken shortly after the right had connected. LOUIS dropped to a praying position on the ropes (right photo). Referee Arthur DONOVAN directs SCHMELING to a neutral corner. The left side of LOUIS' face was so swollen that it appeared he had a double cheek. His left eye was completely closed and his lips were cut and bleeding. SCHMELING had scored his first knock down in the fourth round. (See photo 205).

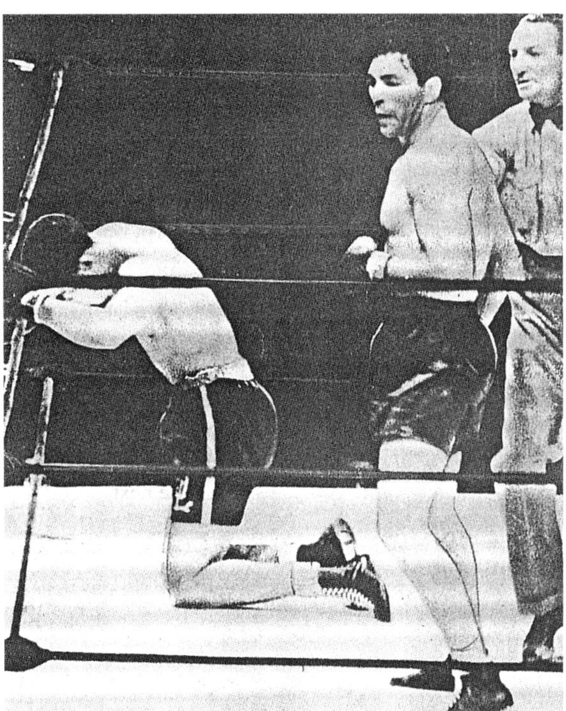

Fig. 208. Close-up of the end of the fight. LOUIS is beaten and groggy in a 'praying position' on the ring ropes. Referee Arthur DONOVAN is directing SCHMELING to a neutral corner. It is a phase some fractions of a second earlier than Fig. 127 c.

raged when the joyous dogs greeted him with howling. Subsequently he dropped off into a deep sleep. As he wakened up at the next morning there existed a sharp gap in his memory. All together BISMARCK fell about 50 times from a horse, so that a summation effect could have occurred due to many small head injuries.

Another traumatic amnesia occurred after a hit-and-run driving accident.

BODECHTEL and R. SPATZ published the following case history of a hit-and-run driving accident:

A 40 year old man caused at 6:30 h in the morning in the center of Munich a rear-end collision, as he overlooked a red traffic light. By it he dashed with his forehead against an anterior crossbeam of his car. There existed no unconsciousness. In his history were three cerebral concussions mentioned. After the impact he left his car and offered the victim of the accident DM 500.- for the case that he would not inform thr police. As the victim of the accident subsisted at a report of the accident, the causer of the accident drove off with his car suddenly, whereby a passer-by, who wanted to hinder him to leave the site of the accident, was dragged along for a long stretch. Following he parked his car in a hotel garage and traveled to his father in another city, went there to bed and woke up at evening after a deep sleep with a complete amnesia for the entire accident.

Fig. 209. This photo, taken from another line of sight, shows LOUIS in a 'praying position,' dazed and groggy on the ropes. Referee Arthur DONOVAN gestures SCHMELING to step back and then directs him to a neutral corner. LOUIS, finally, slumped off the ropes, fell backward to the canvas and was counted out. A great ring upset had occurred.

Fig. 210. First professional heavyweight fight between Ingemar JOHANSSON, Sweden (left), and Floyd PATTERSON on June 26, 1969 in Yankee Stadium, New York. The scene shows action in the first two rounds. PATTERSON connects with a right on JOHANSSON's mouth and nose. PATTERSON seemed to have the fight under control. JOHANSSON had not landed a hard blow in these rounds. Photo: U.P.I.

Knock out as Result of Foul or Illegal Punches

In addition, a knockout may be the result of *foul* or *illegal punches,* such as a *chopping blow against the back of the neck ('rabbit punch').* A blow *'below the belt'* (which CRITCHLEY (1957) calls a euphemistic term for direct injury to the testicles), is apt to produce much pain, shock, and occasionally unconsciousness, or punches against the kidneys with resulting hematuria or renal lacerations.

'Dreamy States'

The term *'dreamy state'* was introduced by Hughling JACKSON (1888). In these disturbances of consciousness no unconsciousness develops. The involved individuals have the impression, that they experience already former experienced or witnessed events or long familiar situations again. The patient may have perhaps a temporary sensation of oddness, strangeness or a familiarity with a situation or a landscape. Obviously there are similarities to the *de vecu* or *des avu states.* During this period a certain cognisent insight remains. The patient believes to see a scene from his past. Important is the indication that the individual is able to form an opinion and that he recognizes the unreality of the scene. JACKSON used also the term 'mental diplopia'. Frequently such a situation is combined with sensations from the gustatory or olfactory system, seldom on the contrary with acustic, optic or vestibular perceptions. The symptoms of the 'dreamy state' were 1938 from GIBBS et al. as 'psychomotor attacks' described. After such a ' dreamy state' is over, the patient can recall the cortical functions undisturbed and he has control over his coordination.

Fig. 211. This photo shows a close-up of the previous photo. Catching a punch flash in the face, Sweden's JOHANSSON is jolted by a right thrown by PATTERSON in the second round. Note the marked deformation and distortion of JOHANNSON's face. Photo: UPI.

G & S
SPORTING GOODS

Fig. 212. At the begin of third round JOHANSSON connected with as left hook and a following right, the so-called 'Ingo's Bruge.' PATTERSON was knocked down to '9', he was so dazed that he went into a neutral corner. He said later that he had thought JOHANSSON had been knocked down and the referee had started counting. It existed probably a traumatic amnesia ('traumatischer Dämmerzustand'). PATTERSON was knocked down in this round additional six times. After the seventh knock down referee Ruby GOLDSTEIN stopped the bout. Photo: U.P.I.

'Groggy' States

The colloquial English expression *'groggy'* implies that one is *under the influence of alcohol, or 'staggering and stupified with drink.'* The term 'groggy' derives from the English name for a strong mixture of alcoholic spirits and water, 'grog.'

The British Royal Navy had an old custom of distributing rum, served *'neat'* (undiluted), to all officers and seaman, up until the mid eighteenth century. In 1740, British Admiral Edward 'Old Grog' VERNON ordered all officers and seaman to drink their daily dose of rum, to prevent scurvy. The rum did not solve the scurvy problem, until later, when sugar and lemon were added, offering some protection. Also, the rum was no longer 'neat', but was diluted with water. Admiral VERNON probably instituted these changes because, after

Fig. 213. This photo shows another scene of the previous heavyweight title fight. PATTERSON rubs his nose, probably in a traumatic amnesia ('traumatischer Dämmerzustand'), defenseless and an easy target for JOHANSSON's left. We mentioned already that PATTERSON was so confused that he went to the neutral corner, because he thought that the Swede was knocked down and counted out. Photo: U.P.I.

Fig. 214. PATTERSON suffered seven knock downs in this round before referee Ruby GOLDSTEIN signals at 2:03 min. that it is all over after PATTERSON had hit the deck for the seventh time in this round. JOHANSSON became the first heavyweight champion in Sweden's history. Photo: U.P.I.

drinking their undiluted rum all at once, some of the seamen became acutely drunk.

Admiral VERNON was called 'Old Grog' because of a *'grosgrain coat',* which he was known to wear in foul weather. British seamen called the watered down rum 'old grog drink,' which was eventually shortened simply to 'grog drink,' and 'grog.' A person who had consumed too much of the grog drink was referred as being 'groggy.' The book of Captain James PACK *"Nelson's Blood: The Story of Naval Rum"* is worth reading in this context.

The Virginia home of George WASHINGTON, Mount Vernon, was named after Admiral Edward VERNON. George WASHINGTON inherited the land from his elder half brother Lawrence, who once served under 'Old Grog.'

The term 'groggy' eventually fell into general use. It is still used, especially with reference to boxing, although the connection with alcohol is no longer necessary. In boxing, 'groggy' refers to the cold sober boxer after one or more blows to the head. The 'groggy' boxer resembles the drunken seaman in that he is unsteady on his feet, his equilibrium is disturbed. He is teetering and staggering to and fro on his feet. His speed and reaction time are slow. The causal blow(s) may occur rapidly, or may extend over several rounds.

It is possible for a boxer to become 'groggy' after a single hard blow (not resulting in a knock down, normally this condition would end after a short interval, or the boxer 'cleared up,' in boxers' jargon. Sometimes in a match a boxer will shake his head several times, to 'clear' his head.

Repeated blows to the head can also produce grogginess, during one round, or over several rounds of a match. Frequently a boxer will suffer several hard head blows, making him 'groggy.' His

ability to defend himself is reduced, or he becomes completely defenseless, creating an opening for his opponent to land a series of punishing blows. The head is the favored target in boxing, and in this situation, it is unprotected. The groggy boxer's sensorium and equilibrium are confused, and his responsiveness is slow. If he is able to throw a punch, it is off target, late. Boxers say the 'timing' is off, or that the boxer is 'befuddled' at the end of the round, barely on his feet. He staggers unsteadily, his legs working at odds with each other ('rubberlegged'). Such disturbances in coordination originate in the cerebellum.

Such a condition presents serious dangers for the boxer's brain, and is the result of a summation phenomenon. The extent of the damage depends not only on the blows themselves, but their proximity in time to one another. Ten head blows extended over time have a different effect than the same ten blows delivered in rapid succession, repeated or cumulative.

The presence of a groggy condition presents an opportunity for the experienced and sensible referee (or the ring physician, if he is empowered to act, to stop the bout immediately and prevent serious additional injury from an undefended and defenseless beating. In such a case *a technical knock out* would be granted in favor of the boxer who is now deprived of his opportunity to commit unbridled mayhem before a bloodthirsty crowd.

A groggy state can resolve during a bout or the interval between rounds, when consciousness returns and the boxer 'clears up.' One cannot conclude, however, that the boxer is 'cured' and able bodied. The groggy state is neither benign nor traceless for the brain. Groggy states, especially in professional boxers, are the injuries that later produce comprehensive, severe, often progressive, and permanent clinical and pathomorphological brain damage.

Exhaustively questioned, boxers report that they have repeatedly been groggy, they deny the degree of incapacitation, or they dissimulate. Sometimes, they simply don't know about their past experience, as grogginess can produce an amnestic state.

Groggy states are distinct from punch drunkenness, although both conditions can produce an inebriated appearance. A groggy state may be measured in terms of minutes or hours, but under no circumstances can one speak of it as being cured. Experiencing groggy states is the cause of the permanent condition called punch drunkeness.

The blows that cause groggy states also produce the pathomorphological changes, which, over time, manifest in the permanently disabling traumatic encephalopathy of the boxer or punch drunk condition.

After a single hammering blow or series of hard blows a boxer may be dazed or stunned, he is stupefied, beaten-up, bewildered or befuddled, he is rubber-legged. These are more colloquial terms instead of 'groggy.' The brightness of his consciousness is reduced, his ability to defend himself properly is reduced, he has to take, therefore, more punishment.

DE JOHN (left) with battered face, is stunned and dazed by a heavy right from BESMANOFF (right) (Fig. 215). Despite this punishment DE JOHN's smashing uppercuts brought him through this bout with an unanimous decision over BESMANOFF.

Fig. 215. Professional heavyweight fight between Mike DE JOHN and Will BESMANOFF on November 28, 1958 at Madison Square Garden, New York. Photo: U.P.I.

Acute Clinical Findings

Survived Pathologic Injuries

Survived Subarachnoid Hemorrhages

Several histories of boxers who survived punches to the head resulting in *traumatic subarachnoid hemorrhages* have been published.

DOLEMAN (1941) reported such a case history in one boxer:

A Naval rating, aged 20, was taking part in a service boxing contest. A decision of a 'technical knock out' was given against him in the second round after he had received a certain amount of head punishment and was knocked down twice for short counts. It took the boxer a few hours to recover from his 'indisposition.' For the following three days he attended to his duties. On the fourth day he reported sick because of severe pains across the small of the back, radiating towards the posterior iliac spine. There was no headache, but his temperature was slightly elevated. Pulse rate was 72/min. On extension of the legs, there was some pain along the course of the sciatic nerves. No abnormal signs of the central nervous system were discovered.

On the fifth day following the boxing fight the young man complained of acute frontal headache. His temperature was 100° F and he was in a stuporous state, reluctant to converse. Twice he vomited. The cranial nerves appeared normal, positive Kernig's sign was present in both legs and there was mild neck rigidity. The right knee-jerk was absent; the left could be elicited with reinforcement. *Lumbar puncture* revealed cerebrospinal fluid under pressure and deeply bloodstained. Wassermann reaction was negative. Lumbar puncture was repeated on the next day and 20 ccm of fluid were removed. Pressure had fallen and the fluid was now less colored. The headache disappeared in a few days, but the stiffness of the legs and neck continued for a week. X-ray of skull showed no evidence of bony injury.

The patient returned to duty six weeks later, and has shown no further effects from the injury.

In accordance with the classification laid down by SYMONDS, this case was diagnosed as one of 'early delayed traumatic subarachnoid hemorrhage.'

Survived Subdural Hemorrhages and Hematomas

There have been descriptions of boxers who survived *subdural hemorrhages* and *hematomas* after evacuation of the hematoma.

KÖNIG (1938):

Twenty-two year old amateur boxer who was knocked out by a blow against his head. He was unconscious for about 15 minutes. Immediately after regaining consciousness, complaints about severe headache. On admission the patient was numb but oriented. Complaints of severe headache. Retrograde amnesia. Pulse 80/min, regular. Neurologically unremarkable. No abnormalities of the pupils. Three hours after admission, vomiting. An uncomplicated cerebral concussion was diagnosed. Two days later pulse 54/min. Lumbar puncture showed sanguinolent CSF, pressure 400 mm/H_2O. On the next day increase of headaches, complaints about reduced hearing in his left ear. Hematotympanon left. The neurological examination reveals no right patellar reflex. Neck rigidity. On the fifth day deterioration with apathy and motoric restlessness. On osteoplastic trepanation of the left temporal bone, a *subdural hematoma is removed.* Increase of pulse to 80/min. On the next day subjective and objective improvement, especially of his headaches. Discharged three weeks later without complaints.

KRAYENBÜHL and NOTO (1940) published the case history of a *chronic subdural hematoma* after a boxing match.

Case: Thirty-six year old business man was hit on the chin in a 'friendly' boxing match. He did not become unconscious, but was stunned. Since then severe frontal headache. About five weeks later repeated vomiting and paresis of his right foot. Difficulties in finding words; bladder incontinence.

On *admission* a right-sided mimic facial paresis. Slight weakness of the right arm and paresis of the right foot. Spasticity of the right leg. Right patellar and Achilles reflexes increased, and positive Babinski right. Amnestic-aphasic disturbances. Disoriented as to time, euphoric. The left-sided carotid angiogram revealed a downward shift of the left middle cerebral artery. A glioblastoma multiforme was suspected. A subdural hematoma over the left frontal lobe was removed. Fast recovery. Discharge without complaints.

In summary, six weeks after a boxing blow against the chin a *chronic subdural hematoma* was removed. The hematoma was completely autolytic and only a beginning fibrous organization in the form of a delicate membrane could be detected.

ROGERS (1941) published the case history of a 20 year old amateur boxer with a *chronic subdural hematoma:*

A Marine, aged 20, was knocked out in the boxing ring. On regaining consciousness a few minutes later, he had severe headache and vomited. However, he did guard duty that night and went out on a route march next day. Since headache accompanied by pain in the back continued, the patient was admitted to hospital with neck rigidity, raised temperature (99° F) and a slow pulse (48 to 50/min). Both optic discs were blurred; the tendon reflexes were normal and the plantar responses were flexor. Leucocytes numbered 10,000 per c/mm. X-ray examination of the skull was negative and the cerebrospinal fluid pressure was 250 mm of H_20.

Five days later the patient 'was inclined to be fatuous.' The left knee-jerk was now brisker than the right, and the plantar response extensor. High papilloedema with exudates and retinal hemorrhages were present in both optic discs. An EEG-record was obtained at this stage. There was a very low potential discharge at varying frequencies in the right fronto-temporal region, most definite in the neighborhood of the Sylvian point. Nine days later the patient was free from headache and declined operation. The highly choked discs, however, were still present, and because of these and the slight left pyramidal signs and the EEG findings, he was thought to have a chronic subdural haematoma overlying the right cerebral hemisphere. Five weeks after the damaging knock out, a right subtemporal craniotomy was performed. When the dura mater was opened a blood cyst was exposed, occupying the Sylvian fissure and overlying a large area of the cortex. Part of the cyst wall was removed and its contents were evacuated. The cerebrum, which was flattened and depressed beneath the cyst, now gradually expanded. At an examination six weeks later the optic discs were found normal and the patient was able to return to duty.

HALL et al. (1952) reported the case history of another amateur boxer, who showed immediate improvement after evacuation of a *subdural hematoma.*

The 24 year old white male laborer was training for a local Golden Glove tournament when he was struck on the right temple with a moderate blow that brought on temporary dizziness. He continued the bout and noted nothing more than a slight headache that persisted. That evening he noted nausea, vomiting, and increased severity of the headache with the additional symptoms of drowsiness and short periods of being irrational.

On the third day the patient was *admitted to a hospital* in a semiconscious condition with a pulse of 56/min, respiration 16/min, rectal temperature 99° F, and blood pressure 140/80 mm/Hg. His *physical examination* was otherwise unremarkable. *Neurological examination* revealed nuchal rigidity, bilateral choking of the disk, and hyperactive reflexes, but no localizing sign. Spinal puncture revealed a cloudy blood-tinged spinal fluid. The pressure was 500 mm/H_2O, on entering, and withdrawing pressure 144 mm/H_2O.

The patient improved following this spinal puncture. X-rays of the skull and cervical spine were normal. He showed gradual improvement with daily spinal punctures, until the sixth hospital day when he suddenly went into a coma. Blood pressure increased to 170/80 mm/Hg, and respiration to 24/min. The pulse remained unchanged. The first sign of localization occurred with dilation of the right pupil. Pulse and respiration gradually became irregular and the patient developed generalized tonic convulsions.

A large amount of dark, thick blood containing small clots was forcibly expelled from the subdural space during surgery. The patient's condition improved immediately.

McCOWN (1959) published the case history of a boxer who developed a *subdural hematoma:*

A 22 year old lightweight boxer had 50 fights since September of 1952. He won 43 with five losses, of which two of the most recent were by technical knock outs and three were draws. This boxer was taking a severe beating about the head in the eighth round. He went to his corner unaided, complained of headache; twitching of muscles of arms and face developed and he became unconscious. A resuscitator was used at ringside and consciousness returned in about ten minutes. He walked back to his dressing room unaided and immediate hospitalization was advised. He refused and returned to his home in Springfield, Massachusetts. Three days later he was seen by a neurosurgeon and at that time complained of severe headaches and was hospitalized. *EEG was* performed and was negative for abnormalities.

Neurological examination revealed no localized signs but showed evidence of choking of the disc and papilledema. The next day a lumbar puncture was performed. The spinal fluid was xanthochronous and showed markedly increased pressure 380 mm/H_2O. On July 29, three days later, his headaches increased, as well as choking of the discs, and he showed bilateral Babinski reflex.

Cerebral arteriography was performed and revealed a right-side lesion. Tentative diagnosis of a right-sided subdural hematoma was made.

Surgery was performed on the same day, during which a large right-sided subdural hematoma was evacuated. The immediate postoperative course was satisfactory. The patient's lethargy disappeared, as well as headaches and other neurologic signs. He was discharged ten days later with no residual neurological signs or symptoms, and was retired from boxing because of severe craniocerebral trauma and subdural hemorrhage.

Comments: This boxer's record reveals that he probably fought too frequently during the previous year and in the two previous bouts which he lost by technical knock outs; he sustained considerable punishment to the head. In addition, the *EEG taken* pre-bout revealed beginning evidence of deterioration and paroxysmal dysrhythmia on hyperventilation with a slightly slow record anteriorly.

LECHOWSKI and JEDLINSKI (1965) reported on *four young amateur boxers* with *subdural hematomas* who showed immediate improvement after evacuation of the hematoma.

Case 1: The 18 year old patient came to the clinic with the idea that he had a brain tumor. For two weeks he had had violent headaches and vomiting. There was increased intracranial pressure, and paresis of the right side.

After the hematoma was removed, the patient's condition improved immediately. Several days before the headaches and vomiting had started, the patient had received several blows to the head during a boxing fight. The patient left the clinic without complaint four weeks after the operation.

Case 2: A 22 year old patient who, ten days before being admitted to the clinic, was hit on the external area of the left eye during a boxing training bout. He was in great pain and was nauseous, and these symptoms increased in the hospital. *Examination* revealed increased intracranial pressure. The diagnosis was *subdural hematoma.* A craniotomy was performed on both sides, and a liquid hematoma found on the left side. The patient left the clinic without any complaints after several days.

Case 3: A 23 year old patient who received a blow to the base of the nose while boxing and was unconscious for a moment. He did not fall down and was not counted out by the referee, and was able to finish the fight. Two hours after the fight, however, a violent headache appeared and several hours later he was unconscious. He was admitted to the clinic 14 days after the trauma with symptoms of increased intracranial pressure, right paresis and subtentorial herniation. The diagnosis was *subdural hematoma,* and a cranial puncture was performed. A liquid hematoma was removed from the right side. The patient left the clinic in good condition after three weeks.

Case 4: A 22 year old boxer who received a blow in the face during training and fell down in the ring, but did not lose consciousness. Next day an increasing spastic paresis appeared on the left side and there was divergent strabismus of the left eye, headaches, and hypersomnia. In the evening of the same day he was admitted to the hospital. The diagnosis was *subdural hematoma.* A cranial puncture was performed on both sides. On the left side a liquid subdural hematoma was found (volume about 50 ml). After removal of the hematoma, immediate improvement was noted. He left the clinic on his own after three weeks, and the divergent stabismus of the left eye and paresis of the left hand and leg disappeared completely after six months.

MORIYASU et al. (1969) reported the case histories of two Japanese boxers with *acute subdural hematomas* that were surgically resolved.

SAWAUCHI et al. (1996) published a paper about *acute subdural hematoma* and *professional boxing. Acute subdural hematoma,* the most common acute brain injury in boxing, accounts for 75% of all acute brain injuries in boxing and is the leading cause of boxing fatalities. The aim of this study was to evaluate acute subdural hematoma caused by professional boxing, by analyzing the content of bouts, the level of consciousness on admission, CT scan, and outcome three months after admission.

This series is for didactic reasons included under the summary 'Survived Subdural Hemorrhages and Hematomas' despite the fact that one of the 15 professional boxers died.

Fifteen boxers who had suffered from *acute subdural hematoma* were classified into three groups according to patterns in loss of consciousness.

Transient unconsciousness (Transient type): Boxers who had returned to alertness within an hour from the time of injury.

Lucid interval type: Neurological deteroriation appeared with a lucid interval from ten minutes to an hour afterward.

Deterioration of consciousness type (deterioration type): A state of unconsciousness appeared and worsened a few minutes after knock out.

Analyzing the number of rounds in bouts indicated that a hematoma developed most frequently in bouts of ten rounds. All of the authors' subjects presented *subdural hematomas without cerebral contusions on CT scan.* With regard to hematoma location, nine involved the left side, three were on the right, two were suboccipital and one was interhemispheric.

The *transient type* (Table 25) was found in seven patients who had GCS scores of 14, 16 on admission. Since CT scans revealed thin subdural hematomas, with or without mild midline shift, conservative therapy was carried out in all patients. All patients had a good recovery.

Five lucid interval type patients with admission GCS scores of 4, 6 and 7, demonstrated thicker hematomas (compared with the transient type findings) with significant midline shift on CT scan. All patients required surgery. Outcome of this type was good recovery (n = 2), moderate disability (n = 1), persistent vegetative state (n = 1), death (n = 1).

Three *patients of deterioration type* had GCS scores of 5, 6. Because of subdural hematoma with remarkable midline shift on CT scan, all patients underwent surgery. Outcome was good recovery (n = 1), moderate disability (n = 1), persistent vegetative state (n = 1).

Overall outcome (Table 26) was good recovery 66.7%, moderate disability 13.3%, persistent vegetative state 13.3%, with a GCS score of less than eight exhibited good recovery 37.5%, moderate disability 25%, persistent vegetative state 25%, and death 12.5%. CT scan of lucid interval and deterioration type showed a tendency to show thick subdural hematoma and remarkable midline shift compared to transient type. Outcomes of lucid interval and deterioration type were generally worse than those of the transient type. This result suggests that the influence of repeated head injury and diffuse brain injury might make a difference between these groups. Repeated head injury means that at least one series of several successive impacts to the head were inflicted.

TABLE 25
Summary of 15 cases of acute subdural hematoma caused by professional boxing. (From SAWAUCHI S, et al., 1996).

		Bout				CT			Therapy	
No.	Age (y.o.)	Round	Result	Down before final round	GCS	Location of SDH (mm)	Thickness of SDH (mm)	Midline shift (mm)	Op Postop therapy	GOS
[Transient unconsciousness type]										
1	22	10	Win by judge	+	15	IH	6	0		GR
2	23	10	Lose by KO	+	15	Lt	3	3		GR
3	26	8	Lose by KO	+	15	SO	5	0		GR
4	22	6	Lose by KO	+	14	Lt	3	3		GR
5	29	4	Lose by judge	+	14	Lt	5	5		GR
6	28	12	Lose by KO	–	14	SO	5	0		GR
7	21	10	Lose by KO	+	14	Lt	9	12		GR
[Lucid interval type]										
8	27	6	Lose by KO	–	7	Lt	12	15	DC	MD
9	24	10	Lose by KO	+	7	Lt	15	18	HIT	GR
10	17	4	Win by judge	+	6	Lt	–	–	DC	PVS
11	21	4	Lose by KO	+	6	Rt	12	15	HIT BT	GR
12	32	10	Lose by judge	–	4	Rt	15	20	HIT BT, HT	D
[Deterioration of consciousness type]										
13	28	10	Lose by KO	+	6	Rt	10	18	HIT BT	GR
14	19	10	Lose by KO	+	5	Lt	9	15	DC BT	MD
15	28	4	Lose by KO	+	5	Lt	–	–	DC	PVS

GCS: Glasgow coma scale, SDH: Subdural hematoma, GOS: Glasgow outcome scale, IH: Interhemisphere, SO: suboccipit, DC: Decompressive craniectomy, HIT: Hematoma irrigation with trephination, BT: Barbiturate therapy, HT: Hypothermia, GR: Good recovery, MD: Moderate disability, PVS: Persistent vegetative state, D: Death.

TABLE 26
Outcomes of acute subdural hematome caused by professional boxing. (From SAWAUCHI S, et al., 1996).

	GR	MD	PVS	D	Total
Overall outcome (%)	66.7	13.3	13.3	6.7	100
Lucid interval	37.5	25.0	25.0	12.5	100
Deterioration type (%)					

Fractures of the Cervical Spine and Rupture of Vertebral Arteries in Boxing

A severe blow against the head can result in acceleration of neck and head in different vector directions, depending on the nature of the blow. These accelerations of neck and head can be the result of not only a translation trauma due to a straight blow, but also the result of a tangential blow producing rotational acceleration of the head due to the force of an uppercut or a hook.

Head and neck accelerations can result not only in the +Gx, but also in the –Gx and Gy vector directions. The different vector directions of the impacting force were described by UNTERHARNSCHEIDT in 1993. Head impacts from the front can cause considerable hyperextension of the cervical spine, usually with an extreme retroflexion of the head. Such action has been directly observed in cases where witnesses report that a blow had the effect of pulling the legs out from under the body.

At first the boxer falls backward with the cervical spine hyperextended, receiving an acceleration trauma due to the impact of the fist. In the second phase of the blow a marked deceleration trauma may occur, when the head of the sometimes unconscious boxer strikes against the lower ring ropes, or worse, against the canvas. In this phase the head, usually the occiput, rapidly decelerates to zero velocity. This is the classic mechanism of a deceleration trauma, compare p. 25. In cases where the head or cervical spine hit the ropes or canvas, additional damage can result due to fractures of the cervical vertebrae and, possibly, injury of the cervical spinal cord.

Such observations were published by CZEKALOWA (1969), SERCL and JAROS (1969) with complaints for just one hour, KEWALRAMANI and KRAUS (1981), STRANO and MARAIS (1983), and JORDAN et al. (1988).

CZEKALOWA (1969) published a case history of a compound fracture of the fifth cervical vertebra after a boxing injury, with late spinal cord complications.

The patient was a 30 year old miner. He was admitted to a neurological department of a hospital with complaints of strong lumbar pain, spasticity, and involuntary movements of the lower extremities, affecting also the upper extremities in a lesser degree.

"Three weeks before admission to the hospital he had temperatures up to 38° C and an exacerbation of the previous symptoms. He had been first treated in an outpatient clinic with the diagnosis myelitis.

The patient had been a professional light heavyweight boxer in the past, and on April 30, 1957, during a professional fight, had suffered a blow to the head after which he was unconscious for about ten minutes. Upon recovering consciousness, he experienced pain and an increasing edema of the neck for about two weeks. A fracture of the fifth cervical vertebra was diagnosed and paralysis of the upper and lower extremities were diagnosed at a local hospital. He was placed in a plaster of Paris cast for six months. He began to get up after nine months. All symptoms receded and after two years he had recovered to full physical fitness.

Abnormalities noted at the physical examination included muscular weakness in the lower extremities, especially on the right side, bilateral foot clonus, bilateral positive Rossolimo signs and a slight dragging of both feet. The muscular tone was increased in the lower extremities.

The *roentgenological examination of the cervical spine* revealed a wedge-shaped body of C_5 vertebra, with a large beak on the anterosuperior margin. The upper limiting lamina of the C_5 vertebra showed an infraction in its anterior segment. The lateral view revealed a wide fissure in the middle of the vertebral body on its inferior surface. The posterior part of the vertebra was displaced backward, and the posteroinferior margin projected into the spinal canal to a depth of about 4 mm. The base of the spinous process of the vertebra was displaced 2 mm in the posterior direction. The *anteroposterior tomogram* showed a vertical lucid fracture of the body of the vertebra near the right base of the pedicle.

A beak-shaped deformity was observed on the anteroinferior margin of the body of the C_4 vertebra.

The roentgenogram revealed a transverse fracture of the body of the vertebra with displacement of its posterior part toward the lumen of the spinal canal and infraction of the right base of the pedicle, with upward displacement.

SERCL and JAROS (1969) reported that they found three spinal cord injuries among 2.153 amateur boxers who were followed up over a period of years. All were the result of backward falls in the boxing ring. The boxers had slight transitory parapareses of the lower extremities and sensory disturbances for all qualities below the injured spinal segment.

According to SERCL and JAROS the prognosis is usually good; in all three boxers the clinical picture normalized itself within 24 hours. Repeating similar spinal cord injuries may result in lasting spinal cord damage.

The authors noted that despite the rarity of such injuries in their data, it is possible that additional injuries escaped the attention of sports physicians and were not reported.

KEWALRAMANI and KRAUS (1981) reported the case history of a boxer who suffered a fracture of the sixth cervical vertebra, with a resulting quadriplegia.

STRAUS and MARAIS (1983) reported a fracture of the cervical spine in a boxer.

JORDAN et al (1988) described the case history of a 22-year old professional boxer, who suffered a brain contusion associated with an uncal herniation and a cervical spine fracture, in the ring.

During a fight, the professional boxer received a major concussive blow that immediately rendered him unconscious. As the boxer fell, his neck struck the bottom rope and he hyperextended, then hyperflexed, his neck. At that time, the boxer was unresponsive to verbal and noxious stimuli, and his right pupil was dilated.

The boxer's neck was stabilized with a cervical collar. Then he was hyperventilated and was immediately transported to the New York Hospital via a waiting ambulance. Upon arrival in the hospital, he was nasally intubated and treated with mannitol. Notable findings of a neurological examination at the time were extreme agitation, with thrashing movement of all four extremities, and a sluggishly reactive right pupil. The patient also exhibited diffuse hyperreflexia and positive Babinski's signs bilaterally.

A *computed tomographic (CT) scan* showed a right to left middle shift of the frontal horns of the lateral ventricles, without associated hemorrhage or edema. In addition, the CT scan revealed a small subdural hematoma involving the right posterior convexity and interhemispheric tissue. The extent of the subdural hematoma was better defined on the *magnetic resonance imaging (MRI) scan* obtained five days after injury. A *cervical CT scan* revealed a nondisplaced fracture through the left lateral mass of the third cervical vertebra.

Three days after admission, the patient awoke and was able to engage in full conversation. Follow-up cervical x-rays in flexion and extension showed a stable cervical spine, so the cervical collar was removed.

Ten days after admission, the boxer was discharged in stable condition. The only neurologic defect included minimal left-sided hyperreflexia and equivocal Babinski's signs bilaterally.

JORDAN et al. commented that the exact pathophysiologic mechanism of cerebral injury in this case was difficult to explain. Although the MRI scan revealed a small subdural hematoma, it did not appear to account for the extent of universal intracranial pressure and midline shift. For this reason, the boxer was diagnosed as having a cerebral contusion. The CT scans did not show a cerebral contusion. Tissue alterations associated with an uncal herniation are not cerebral contusions, but the tentorium cerebelli produces an incisura in the pons; the tissue degeneration that follows is termed a *'tissue necrosis' ('Gewebsnekrose')*.

PLANT and BUTT (1993) reviewed an historic case of a boxing death (c. 1913), since the cause of death given at that time has since been deemed incorrect. The cause of death is now hypothesized to have been a tear in the right vertebral artery, associated with hyperextension and rotation of the neck.

Court documents and the *autopsy report* (provided in evidence) were used to reassess the extent of the fatal head injury. The documentation suggested a different cause of death that the authors believe is more appropriate than that given at the time (1913).

The case under review involves a former heavyweight boxing champion. The feature bout on the card was for May 24, 1913 in Calgary, Canada, was a title defence between the champion, Luther McCARTY, and challenger Arthur PELKEY. During the latter part of the first round, PELKEY landed minor blows to the head of McCARTY while the two men were engaged in a clinch. As they backed away, the champion turned to his corner directly behind him and nodded to his trainer, apparently to indicate that he was not affected by the last exchange. As he redirected his attention to his opponent, McCARTY went down very slow and rolled over (Fig. 216). (Testimony from the witness taken from the 'Verbatim Report of Proceedings at the Inquisition into the death of Luther McCARTY, May 26, 1913, Calgary, A B'). The champion was dead, and Arthur PELKEY was charged with his murder.

Fig. 216. The photo shows the tragic ending of Luther McCARTY. He was carried from the ring and laid on straw on the floor of the arena. McCARTY died without regaining consciousness. Manager Billy McCARNEY can be seen in the background with both hands on his head. The referee of the fatal bout was Ed SMITH.

Transcripts from PELKEY's murder trial and from coroner's inquisition into the death reveal, according to PLANT and BUTT, two critical events. Testimony by Tommy BURNS, the arena manager, indicated that approximately ten seconds had lapsed from the time of the last physical contact until collapse. Furthermore, this same witness, as well as four doctors and a cameraman, had testified that the blows delivered by PELKEY were seemingly not violent enough to cause death.

Autopsy Report and Evidence in Court

A full autopsy was performed by Dr. H. H. MOSHIER. The only signs of external trauma were confined to the head and neck region with some 'slight swelling and purplish discoloration...' of the right angle of the mandible. Gross internal examination revealed normal abdominal and thoracic organs, except for some adhesions associated with the right lung. Examination of the central nervous sytem revealed '... scalp and cranium were normal. The meninges over the cortex of the brain were normal. The brain substance itself was normal. After removing the scull cap, blood commenced flowing from the base of the brain and a small clot was present in the right occipital region, below the cerebellum. In addition to the clotted blood, a large quantity of unclotted blood poured from the spinal canal.'

There was a *subluxation* of the fourth cervical vertebrae noted, which reportedly impinged on the spinal cord. No site of rupture was detected in the major vessels supplying the brain and no fractures were noted in the skull.

From his examination, Dr. MOSHIER determined the cause of death to be 'subluxation of the fourth cervical vertebrae and the resulting hemorrhage into the spinal canal and the base of the brain. This would by pressure destroy the vital centres in the medulla oblongata... '

Key elements drawn from Dr. MOSHIER's evidence at the inquisition showed that, in his opinion, the fatal hemorrhage did not result from the decedent falling to the canvas. He stated that there was no 'dilation of the ventricles in the abdomen (sic) or the chambers of the heart' to suggest a fainting episode. In addition, Dr. MOSHIER testified that he did not believe that the last exchange of blows could have caused the trauma. This was based on his belief that, given the level of the physical activity, McCARTY could not have survived for the 10 s reported by witnesses if a fatal blow had been delivered.

Anatomical Considerations

The relevant arteries involved in *traumatic basal subarachnoid hemorrhage (TBSAH)* can be divided, according to PLANT and BUTT, into two groups: extra- and intracranial. In the former group, SIMONSEN (1983) reports that 50% of cases were due to vertebral artery rupture. Similarly, HARLAND et al. (1983) showed that 70% of cases reviewed had ruptures in the vertebral artery. Other extracranial rupture sites were also located in the carotid artery (DRAGON et al. 1981, DAVIS and ZIMMERMAN 1983, DOWLING and CURRY 1988). In the latter group, intracranial arteries were identified as the source: cerebral arteries (DOWLING and CURRY 1988, SIMONSEN 1984), cerebellar arteries (DOWLING and CURRY (1988), and the basilar artery (SIMONSEN (1984).

The *posterior inferior cerebellar artery (PICA)* and the *vertebral artery* are of particular interest in this case, according the PLANT and BUTT. The location of the small clot uncovered by the postmortem examination was just below the right cerebellar hemisphere, the area supplied by the right PICA. The origin of the PICA is just superior to the entrance of the vertebral artery into the cranial vault (Fig. 217).

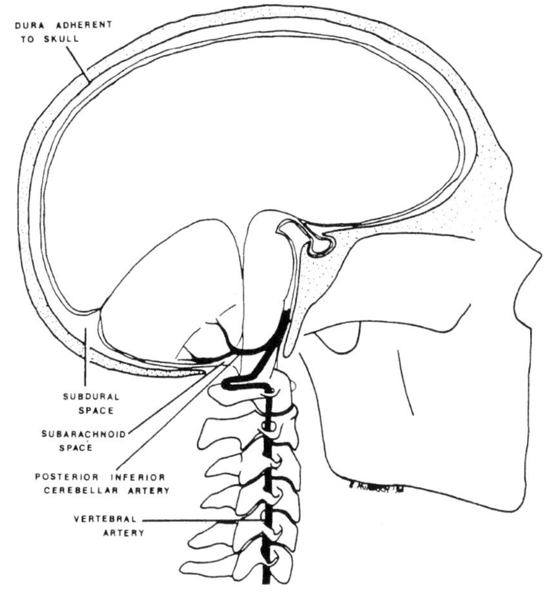

Fig. 217. Relevant arteries in traumatic basal subarachnoid hemorrhage. The origin of the posterior inferior cerebellar artery is just superior to the entrance of the vertebral artery into the cranial vault. (From PLANT, BUTT, 1993).

The vertebral artery has been described by CONTOSTAVLOS (1971) as consisting of three sections. The artery starts as the first branch of the subclavian artery and continues cephalad through the foramen transversarium of cervical vertebrae 6 – 1. The third section begins at the foramen transversarium of the atlas and follows a posterior path around the lateral mass of the atlas into the cranial cavity (Fig. 217). As the artery enters the foramen magnum, it is contained within the subarachnoid space for a short distance, in the cerebellomedullary cistern.

Tension of the vertebral artery, as in hyperextension of the head following a blow to the face has been shown to be a mechanism of tearing, according to PLANT and BUTT, (DAVIS and ZIMMERMAN (1983). Stress would be transmitted along the longitudinal axis of the vertebral artery until the change of direction at the lateral mass. At this point, the direction of the artery is posterior. Stress is now perpendicular to the longitudinal axis of the artery. A concomitant shift in skeletal and vascular configuration due to hyperextension places the vertebral arteries at risk of tearing, as PLANT and BUTT noted. Additional stress involved in rotating the neck increase the tension of this configuration.

Discussion

PLANT and BUTT discuss first the autopsy report of Dr. MOSHIER:

At the time of the postmortem examination, Dr. MOSHIER found blood flowing freely from the base of the brain and the spinal canal, including a subdural hemorrhage. In addition, he found a small clot located in the right occipital region below the cerebellum. His interpretation of these findings was an undefined hemorrhage that caused '... pressure destroy(ing) the vital centres of the medulla oblongata...' A rupture site causing the fatal hemorrhage was not identified; however, there was mention of the venous plexus surrounding the vertebral artery as a possible source. The cause of the rupture, as determined by Dr. MOSHIER, was subluxation of the fourth cervical vertebrae.

PLANT and BUTT comment:

"In our opinion, this is not an adequate description of the cause of death. We believe that there was a TBSAH involving a laceration of the vertebral artery in the superior portion of the spinal canal. The mechanism of the fatal rupture is hypothesized to be a combination of the final blow to the head causing subluxation of C_4, with simultaneous rotation of the neck when the decedent turned to face his corner. The additional stress placed on the vertebral artery as the head was turned extended the vascular injury from the blow."

The authors noted:

"The suggestion of rupture of a major vessel due to rotation or hyperextension of the head and neck is well documented."

Details of the discussion should be studied in the original paper.

PLANT and BULL summarize that in the light of current literature, the fatal injury sustained by McCARTY, as it was described, meets the criteria of TBSAH. The fatal subarachnoid hemorrhage sustained by McCARTY was the result of a discrete injury isolated from any other cerebral injury. PLANT and BUTT speculate that at the time of last physical contact, there was an incomplete tear of the vertebral artery. This injury was extended when the decedent turned his head in an attempt to look at his corner.

According to PLANT and BUTT, the jury deliberated for two hours on June 24, 1913 and decided that PELKEY was not guilty of manslaughter, but that this bout met the definition of a prize fight. Acts which were otherwise prohibited by criminal law were immunized from prosecution if they occurred in a 'prize fight', so this finding of fact protected PELKEY against conviction. This led to authorities charging Tommy BURNS, the arena manager, with manslaughter. Three months later, a letter from the Deputy Attorney General persuaded the prosecution to exonerate Mr. BURNS of the charge, ending the affair with no further legal action taken.

Injuries to the Carotid Arteries due to Blows against the Neck

The range of effects of boxing blows against the *lateral region* of the *neck* embraces still another possibility of acute damage, e.g. the *compression* and the *injury of the walls of arteries leading to the brain*, especially in the *carotid arteries*. A thrombotic occlusion of the carotid artery may result. Both knock downs and knock outs are possible (in a knock down the boxer is able to continue the fight after ten seconds; in a knock out he is not). *Ruptures of internal layers* of the *vessel wall with intralamellar hemorrhages* and *thrombosis* have been described, which can lead to reduced cerebral blood supply and eventually (due to necrosis in their supporting areas) to *hemipareses*.

Blows to the lateral side of the neck may trigger a carotid sinus syndrome.

These blows are completely legal in boxing. They occur quite frequenly in boxing matches.

Only two examples are presented, although many press photos are available.

A right from Teddy WRIGHT (right) lands on the left neck region of his opponent Phil MOYER (left) and displaces the soft tissue of the neck (Fig. 218). Such blows may result in a thrombotic occlusion of the carotid artery or trigger a carotid sinus reflex. WRIGHT won the fight after ten rounds on points.

'Kid' GALIVAN connects with a hard right on Gil TURNER's left neck (Fig. 219). TURNER was knocked out in the eleventh round.

FLINT (1920) advanced the theory that if the lower jaw is hit with great force, the whole jaw jerks backward as a unit and comes into contact with the carotid arteries. Blood circulation is temporarily interrupted by this action, the brain becomes anemic, and unconsciousness sets in. As soon as circulation is restored, consciousness returns. This theory is an improbable explanation of cerebral concussion, in our opinion.

The cases of ROSEGAY (1956), HOCKADAY (1959), MURPHY and MILLER (1959),

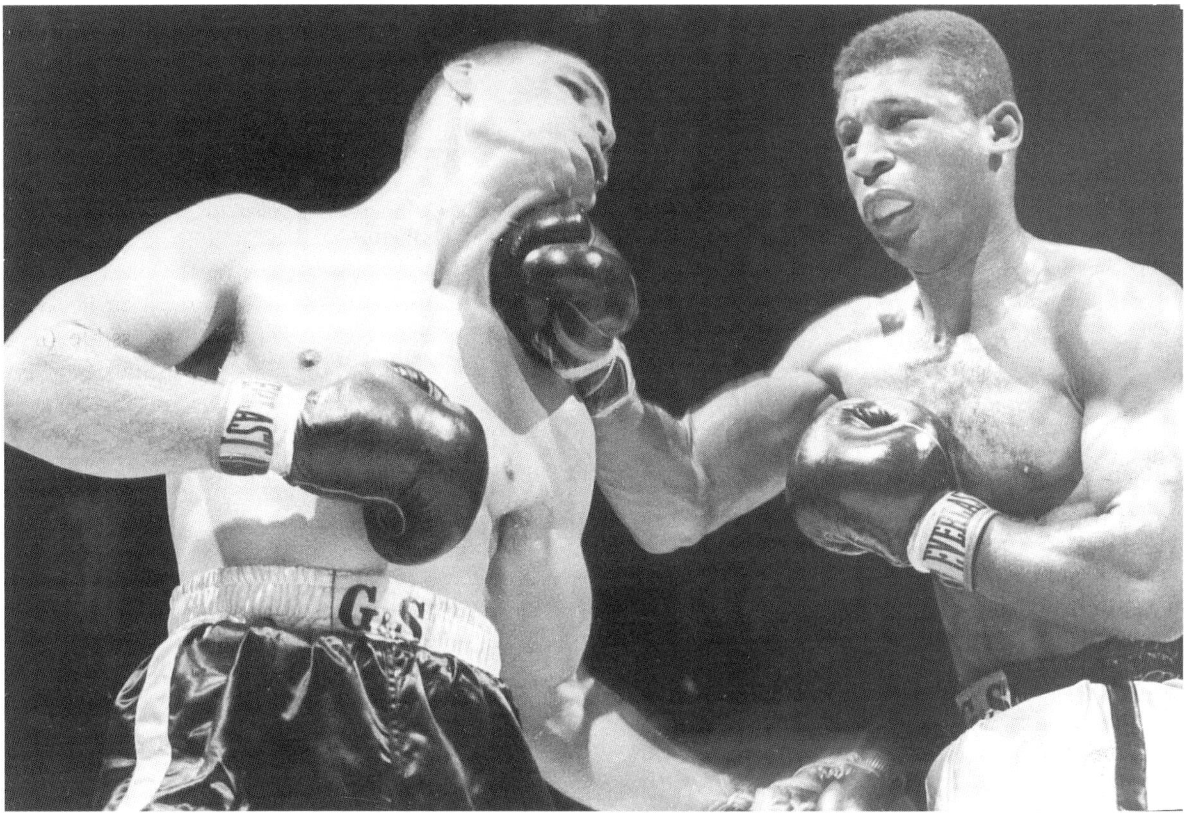

Fig. 218. Professional welterweight fight between Teddy WRIGHT (right) and Phil MOYER (left) on May 26, 1962 in Madison Square Garden, New York. A right from WRIGHT lands on the left neck region of his opponent MOYER and displaces the soft tissue of the neck. Such blows may result in a thrombotic occlusion of the carotid artery or trigger a carotid sinus reflex. WRIGHT won the fight after ten rounds on points. Photo: U.P.I.

STRASSMANN and HELPERN (1968) revealed traumatic occlusions of the carotid arteries.

ROSEGAY (1956) described a 29 year old sergeant who was knocked out. When he arrived at a dispensary and recovered consciousness, the patient noted numbness of the left arm and leg. During the course of the day this weakness progressed. Strength began to return in the lower left extremity, to a limited degree, in a few days. A right carotid arteriogram done 18 days later disclosed an *occlusion* of the *internal carotid artery* just distal to the origin of the posterior communicating artery. The patient made a gradual limited recovery and became ambulatory, but walked with a spastic, hemiparetic gait. There was little return of function in the hand. There was no delay between the trauma and the onset of neurological deficit in this case.

HOCKADAY (1959) reported the case history of a boxer with an occlusion of the left internal carotid artery. A 21 year old professional boxer lost a fight on points but did not receive particularly severe punishment or any noticeable injury to his neck. About three-quarters of an hour later, he vomited and then rapidly developed motor aphasia and a right hemiparesis. An *arteriography* showed the obstruction to be high in the left internal carotid, at the base of the skull. Recovery in the right leg and in speech began 10 days after the fight while six weeks later. When he came to Oxford for rehabilitation, there was only mild nominal dysphasia, slow speech, a moderate right facial weakness, severe spastic paralysis of the arm, a moderate spastic paresis of the leg (but not severe enough to need a toe- spring), and a mild left hemianaesthesia, but no visual loss. Three years later, he is completely independent but there is still almost complete lack of movement in the right forearm and hand. Interestingly, in view of the transitory aphasia, he is a left-handed man with a history of left-handedness in the family.

MURPHY and MILLER (1959), who reported 21 cases of carotid insufficiency, described a 16 year old Golden Glove

Fig. 219. Professional fight between 'Kid' GAVILAN and Gil TURNER on July 7, 1952 in Philadelphia. GAVILAN connects with a hard right on TURNER's left neck and knocking him out in the eleventh round.

boxer who was admitted one day after having been defeated on points in a boxing bout. He was not knocked down or rendered unconscious and went home without complaint. The next morning, however, he was unable to get out of bed and could not use his right side. He became progressively stuporous. The family was able to arouse him at intervals during the day, but he was unable to speak. *Examination* showed right hemiplegia and aphasia without evidence of increased intracranial pressure. A *leftsided carotid arteriogram* revealed an incomplete occlusion of the internal carotid artery at the bifurcation in the neck. The left carotid artery was explored, and a subintimal clot found just above the bifurcation (which had completely occluded the artery by the time of surgery). It was evacuated and retrograde flow was obtained. The artery was closed. There was no immediate postoperative change in the patient's condition. By the eighteenth postoperative day he was alert and eating fairly well, but there was only minimal improvement in his hemiparesis. At that time a *postoperative left carotid arteriogram* was performed and revealed the artery in the neck to be patent. However, there was complete occlusion of the middle cerebral artery, presumably from a preoperative embolus. Twenty-one months later his family physician reported that the boy had impaired but useful speech and could walk with a spastic gait, but showed no significant improvement in the arm. The authors concluded that there is no reason to believe he benefited from the operation.

PAMPUS (1964) is of the opinion that a severe blow may press the carotid artery so intensely against the lateral processes of the cervical vertebrae that it will cause mechanical damage to the vessel wall, with the resulting sequelae. In cases where the Circulus arteriosus cerebri provides sufficient collateral circulation, the clinical findings can be more or less reversible.

STRASSMANN and HELPERN (1968) discussed a correlation between a knock out and a right-sided thrombosis of the carotid artery:

A 30 year old professional boxer was knocked out and remained unconscious for about 30 minutes. The boxer recuperated so that he was able to travel from Philadelphia, where the fight had been, to his home in New York. He developed a headache, syncopes, and an increasing left side hemiparesis. The *angiography* showed an occlusion of the right internal carotid artery. A craniotomy was performed, because a subdural hematoma was suspected; this revealed a swelling of the brain. The patient died 24 hours later, comatose, three weeks after the knock out.

The *autopsy* showed thrombi in the intracranial part of the right internal carotid artery and the middle cerebral artery, with hemorrhagic necrosis of the right cerebral hemisphere, especially the basal ganglia and the temporal lobe. The hemorrhages in the pons were more recent, therefore they must have been secondary to the trauma.

Traumatic Thrombotic Occlusion of the Basilar Artery

KLAGES (1969) described a 16 year old male apprentice with a traumatic thrombosis of the basilar artery.

The 16 year old amateur boxer was trained in boxing by his father since his seventh year of life. As an amateur boxer he participated in more than 60 fights, though allegedly he suffered no knock outs.

Approximately three months before his death, he lost an amateur boxing fight against a far superior opponent, but was not knocked down. Subsequently he won four more additional boxing matches by points. In this entire time he allegedly received only one effective blow against the head, which had 'shocked' him for a short period of time. Before his last fight, in which he won a junior championship, he had had to lose a significant amount of weight and occasionally felt a little exhausted.

Six days after the championship fight, the patient complained for the first time about inconveniences. During the lunch break he became suddenly unconscious. At *admission* to the hospital he was in a shock, was somnolent and had to vomit. His condition deteriorated progressively. He developed *neurological signs* to left side of the brain and disturbances of eye movements developed. Two exploratory trepanations did not reveal hemorrhages. He died eleven days after the onset of the sudden illness, 17 days after his last match.

The *medico-legal autopsy* revealed the following findings: Increased intracranial pressure and tonsillar herniation of the cerebellum. The examination of the brain (Prof. COLMANT) did not reveal posttraumatic alterations at the surface of the cerebral hemispheres, but numerous focal soft-

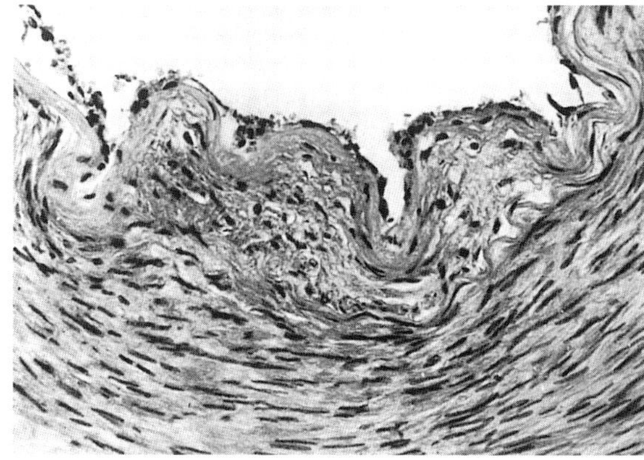

Fig. 220. 16 year old amateur boxer. Basilar artery. Intima with collagenous fiber propagation and fragmentation of the elastica interna. Hematoxylin-Eosin technique, 240:1. (From KLAGES, 1969).

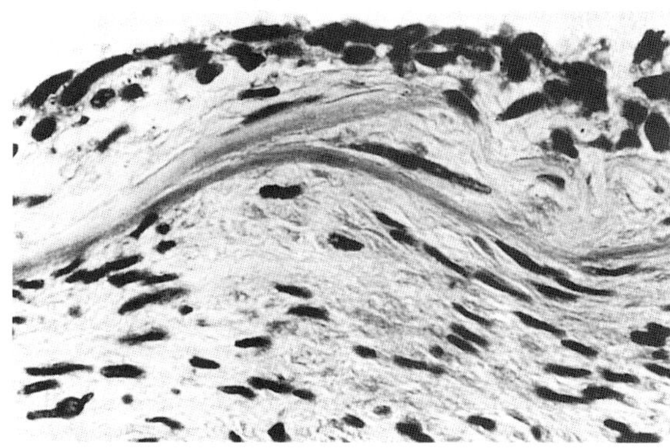

Fig. 221. 16 year old amateur boxer. Basilar artery. Higher magnification. Hematoxylin-Eosin technique, 350:1. (From KLAGES, 1959).

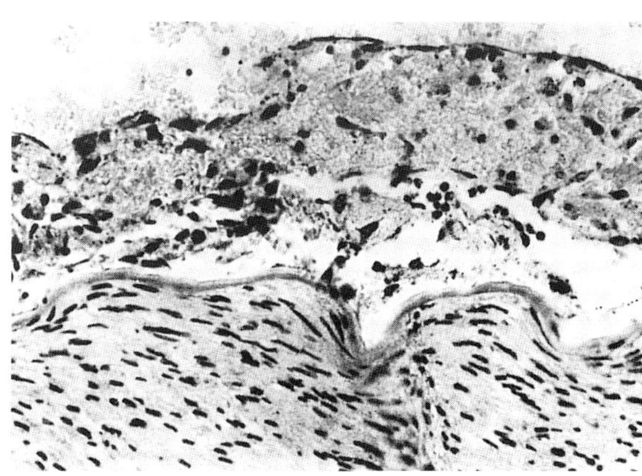

Fig. 222. 16 year old amateur boxer. Basilar artery with partially organized bleeding of the intima. Hematoxylin-Eosin technique, 240:1. From KLAGES, 1959).

enings in thalamus and the substantia nigra bilaterally, also paramedian at the dorsal surface of both cerebellar hemispheres. *Microscopically,* in the margins of the softening a wall of compound granular cells was found with proliferation of capillaries. In the Purkinje cell layer of the cerebellum extensive ischemic cell alterations, elective parenchyma necroses and extensive loss of fibers.
Cause of the multiple softenings was a fresh thrombus, which was partly already in an organization stage, a thrombotic occlusion of the basilar artery at the rostral pons (Figs. 220 – 222). Serial sections revealed subendothelial hemorrhages in an organization stage.

KLAGES sunmarized that there were no hemorrhages of the coverings of the brain, no cortical contusions and no central alterations in the brainstem. The cause of death was the extensive softening in the basal ganglia, the midbrain, the pons and the cerebellum, as result of a thrombotic occlusion of the basilar artery. A process of the vessel wall, which could have explained a spontaneous thrombus, could not be found.

Fig. 223. The professional boxer Jonny SHKON falling backward at full stretch. Photo: U.P.I.

Fig. 224. Professional middleweight fight for the German Championship between Gustav SCHOLZ (left) and Peter MÜLLER (right) on June 19, 1957, in the Sports Palace, Berlin. Photo: KEYSTONE.

Backward Falls of Boxers and Possible Consequences

The force of a boxing blow may accelerate the boxer, and especially his head and the boxer himself in such a way that he falls backward at full stretch, hitting his occiput on the canvas (Fig. 223). Such a mechanism produces a dangerous deceleration trauma. The boxer has already received a severe blow to the head, a so-called acceleration trauma, which may result in unconsciousness. The unconscious boxer receives a second impact when his head impacts the canvas, causing a deceleration trauma. We have already mentioned, see Fig. 34 a – c, that in such a deceleration trauma with impact on asphalt, for example, g forces of about 200 g may result.

Fig. 223 shows the professional boxer Jonny SHKON falling backward at full stretch after being hit. The impact of his head on the canvas would be additional to the blow against his head while boxing on his feet. Only the heels have contact with the canvas as the boxer falls backward.

Fig. 224 shows the professional middleweight fight for the German Championship between Gustav SCHOLZ (left) and Peter MÜLLER (right) on June 19, 1957, in the Sport Palace, Berlin. The title holder Peter MÜLLER was knocked out in the third round of a scheduled 12 round bout. MÜLLER falls backward like a felled tree.

Fig. 225. Professional middleweight fight between Peter MÜLLER, Cologne, Germany, and ex-champion Victorio OLIVERA, Spain, on February 8, 1957 in Hamburg, Germany. Photo: d.p.a.

Fig. 226. Professional fight between Johnny BRATTON and Robert EARL on January 19, 1949. BRATTON connected with a right to the jaw of EARL. The struck boxer falls 'stiff' backwards, only his heels are in contact with the canvas. Photo: Chicago Sun Times.

Fig. 225 show the professional middleweight fight between Peter MÜLLER, Cologne, Germany, and ex-champion Victorio OLIVERA (Spain) on February 8, 1957 in Hamburg, Germany. MÜLLER won the fight in the sixth round by a knock out, after OLIVERA was knocked down several times. The photo shows the extreme rotational acceleration of OLIVERA's head and hyperextension of the cervical spine due to a powerful hook from MÜLLER. The photo shows that in this case, fortunately, the knees of OLIVERA buckle and the backward movement is changed in a fall at his knees.

Fig. 226 shows the professional fight between Johnny BRATTON (left) and Robert EARL (right). BRATTON landed a right on the jaw of EARL. The blow literally knocked EARL stiff. EARL's eyes are closed. Only his heels have contact with the canvas, since there are no ring ropes near enough to catch the fall. EARL will have a second severe impact on the canvas, a deceleration trauma.

Chronic Clinical Findings

MARTLAND's Pioneering Publication on the Punch Drunk Syndrome

Harrison MARTLAND (1928) – more than 70 years ago – in a classic study presented the first clinical description of the psychopathological and neurological features of the punch drunk syndrome. MARTLAND paid special attention to the psychopathological features of this syndrome, he reported that of 23 punch drunk boxers known to him, four were permanently in mental institutions. MARTLAND begins his paper with the statement, "that, unfortunately there is no previous record of any statistics compiled by competent medical authorities as to either the existence or the nonexistence or incidence of the condition known as punch drunk. There are no previous reports to my knowledge of its symptomatology, progress or end-results. We are placed in the position of accepting a series of objective symptoms described to us by laymen. There is undoubted proof that for years fighters, fight-promoters and the sporting world have recognized and talked about this condition."

MARTLAND continued:

> "One sporting writer of note has recently stated that punch drunk was greatly exaggerated and that he had consulted eminent neurologists who assured him that such condition did not exist."

MARTLAND noted however, that the opinion of shrewd layman, many of whom are making a living by observing the physical fitness, actions and characteristics of the professional fighter, is perhaps more substantial than the opinion of medical experts.

MARTLAND noted:

"A fight promoter whose ability to judge the physical condition of fighters is unquestionable has given me the names of twenty-three fighters whom he considers punch drunk (Table 27). Many of these men are scattered over the country and I have been unable to ascertain their exact condition at the present time, especially those who are in asylums. I have examined five of these men and they present the clinical pictures as described. As an illustration I will report one case of advanced parkinsonian syndrome due to punch drunk."

> *Case 2:* History. N. E., aged 38, born in the United States, started to fight in 1906 when 16 years of age. He stopped fighting in 1913, when 23 years of age, because of a tremor in his left hand and an unsteadiness on his legs. During his period of fighting he was a professional featherweight and soon became a top notcher. He had fought such men as Charlie GRIFFIN, Jack BRITTON, K. O. BROWN, Tommy O'-TOOLE, Harry STONE, Kid BURNS, Kid TUTS, Johnny BAKER, Teddy MALONEY and Tommy LANG. He had been knocked out twice; once when 22 years of age he was out for an hour. He was never sick and seldom drank. On account of symptoms of tremor and unsteadiness, he was often wrongly accused of being intoxicated. Since 1913 his condition has slowly progressed until he now resembles a well marked case of paralysis agitans.
>
> *Examination.* – The patient is a well nourished and apparently healthy man. His gait is staggering and propulsive, his facial expression masklike. There is marked stammering and hesitancy in speech. He has a fine tremor in his hands and tongue. The pupils are equal and react to light. The knee kicks are slighly exaggerated. Clonus and Babinski phenomenon are absent. The sensations are normal. The intelligence is normal. He has been under treatment in many clinics for paralysis agitans and told that his fighting did not have anything to do with his present condition. The blood Wassermann reactions have been repeatedly negative. Complete serologic examination of the spinal fluid was entirely negative.

The professional featherweight started to box in 1906 when 16 years of age and stopped fighting in 1913, when 23 years of age. Unfortunately, we do not know in how many fights he participated. We know however that he was a top notcher who fought many ranking list boxers in his weight class.

TABLE 27
List of fighters known by one promoter to be 'punch drunk'. (From MARTLAND, 1928).

No.	Initials	Class	Has Fought	Present Condition
1	B.N.	LHW	Joe Gans	Parkinsonian syndrome
2	J.D.	HW	Weinert, Fulton	Drags leg; bad shape
3	J.T.	LW	Leonard, Kansas, Dundee, Tendler	Drags leg; talks slow
4	B.B.	LW	Walker, Tendler	Punch drunk
5	W.J.	LW	Dundee, Leonard	Punch drunk
6	F.J.	HW	Willard, Weinert	Punch drunk
7	A.W.	LW	....	Asylum
8	B.M.	HW	Moran, Tunney	Asylum
9	J.G.	HW	Sharkey, Jeffries, Fitzsimmons, Johnson	Asylum
10	C.S.	MW	....	Asylum
11	J.C.	....	....	Drags leg; talks slow; thinks slow
12	J.R.	....	....	Punch drunk
13	M.D.	....	....	Punch drunk; almost blind
14	C.C.	....	....	Punch drunk
15	T.S.	....	....	Punch drunk
16	J.S.	....	....	Punch drunk
17	R.S.	....	....	Punch drunk
18	S.M.	....	....	Punch drunk
19	P.J.G.	....	....	Punch drunk
20	T.T.	....	....	Punch drunk
21	B.M.	....	....	Punch drunk
22	J.H.	....	....	Punch drunk
23	D.P.	....	....	Punch drunk

His active career extended only over a period of seven years, he retired when only 23 years old. One should call special attention to the fact that as a featherweight he belonged to the lighter weight classes. This demonstrates convincingly that boxers belonging to lighter weight classes must have considerable punching power, which can be deduced too from the fact that he was knocked out on two occasions. Here is the case of a professional boxer of a lighter weight class who had developed after seven years of boxing a traumatic parkinsonism. This is not a case of punch drunkness, but as already said a traumatic parkinsonism.

But we are greatly indebted to MARTLAND that he presented us with a classic description of punch drunk boxers, because he was able to examine them clinically. His description of the punch drunk state can be described as a classic.

MARTLAND noted:

From a neurological standpoint this case is one of paralysis agitans. It either is traumatic in origin or is a primary, essential for of paralysis agitans of the juvenile type. Epidemic encephalitis is eliminated in this case, since the first symptoms appeared in 1913, three years before epidemic encephalitis was first recorded in France at Bar le Duc, four years before it appeared in Vienna, and six years before cases were making their appearance in America.

The differential diagnosis of MARTLAND has to be seen under the aspect that it was written in the second half of the twenties at a time when it was not general knowledge that traumatic parkinsonism due to boxing is a relatively frequent and generally accepted clinical syndrome.

MARTLAND continued and concluded:

"Very definite anatomic-pathologic proof exists that following a cranial injury death may occur which the autopsy discloses no other lesions but multiple punctate hemorrhages in the deeper structures of the brain. This type of hemorrhage does not occur when the skull if fractured. There is often not even laceration of the scalp. Cortical laceration and hemorrhage is absent or there may be only a slight amount of thin pia-arachnoid bleeding. While this type of cranial injury was known to a few performing medico-legal autopsies, it has never attracted sufficient clinical attention. We are indebted to CASSASA for first describing it and to OSNATO and GILIBERTI for first calling attention to its clinical importance."

(2). "The mechanical theory advanced by CASSASA offers the best explanation of the production of these hemorrhages. Fat embolism may produce similar hemorrhages in the brain which grossly and microscopic are so alike as to be practically indistinguishable. In such cases the anatomic proof of extensive fat embolism must be sought in the other viscera."

(3)."In a series of 309 consecutive cases of cranial injury coming to autopsy, MARTLAND and BELING found this type of hemorrhages in nine cases, or 2.9 per cent. In six of these cases fat embolism was eliminated as the cause of the hemorrhages, and was not proved in the three remaining cases. They first called attention to the frequency of these hemorrhages in the basal nuclei, especially in and near the corpora striata, and to the rarity in the cortex of the cerebrum and below the tentorium cerebelli. They called them 'concussion hemorrhages' because a purely mechanical theory seemed to explain their production. They spoke of them as 'ring hemorrhages' because on microscopic examination they appeared as small rings surrounding the vessels and filling the perivascular space of the Virchow-Robin."

We continue with a later statement of MARTLAND:

(6). "The condition of punch drunk has been described. While most of the evidence supporting the existence of this condition is based at this time on the observations of fight fans, promoters and sporting writers, the fact that nearly one half of the fighters who have stayed in the game long enough develop this condition, either in a mild form or a severe and progressive form which often necessitates commitment to an asylum, warrant this report. The condition can no longer be ignored by the medical profession or the public. It is the duty of our profession to establish the existence or nonexistence of punch drunk by preparing accurate statistical data as to its incidence, careful neurologic examinations of fighters thought to be punch drunk, and careful histologic examinations of the brains of those who have died with symptoms simulationg the parkinsonian syndrome. The late manifestations of punch drunk will be seen chiefly in the neurologic clinics and asylums, and such material will practically fall to the neuropathologist connected with such institutions."

"Punch drunk bears the same relation to multiple concussion hemorrhages as do many of the postconcussion neuroses and psychoses that follow blows or falls on the head."

Some of the sentences of MARTLAND are so important and prophetic that they will be reported:

... "the fact that nearly half of the fighters who have stayed in the game long enough develop this condition [punch drunkness], either in a mild form or a severe and progressive form which often necessitates commitment to an asylum, warrant this report. The condition can no longer be ignored by the medical profession or the public. It is the duty of our profession to establish the existence or nonexistence of punch drunk by preparing accurate statistical data to its incidence, careful neurologic examinations of fighters thought to be punch drunk, and careful histologic examinations of the brains of those who have died with symptoms simulating the parkonsonian syndromes."

It took more than 70 years that the farsighted comments of MARTLAND are generally accepted. If today a few individuals continue to negate the existence of the punch drunk syndrome so one must say of them that they are not familiar with the existing literature, and that MARTLAND more that 70 years ago had a better understanding and a better concept of this condition. There is the urgent plea of MARTLAND that it is the duty of the medical profession to establish the existence of the punch drunk syndrome. His urgency to examine these ex-boxers by neurologists and neuropathologists is met. One feels sorry that MARTLAND is not able to see the accuracy and rightness of his writings.

MARTLAND continued:

"From the studies of CASSASA, OSNATO and GILIBERTI, and MARTLAND and BELING, it would seem that the older theories of cerebral concussion will have to be discarded. We now have the possibility of a definite type of brain injury explaining the various phases and late manifestations following many cases of cranial injuries."

MARTLAND shows two figures, a macrophoto of the horizontal section of the cerebrum and a low power photomicrograph of a section through the corpus striatum in case 1 in which innumerable 'ring hemorrhages' may be noted with their perivascular distribution, of an observation of a case he had published 1927 in collaboration with BELING.

This case is not a boxer, but a man, aged 76, who while going upstairs, stumbled and struck his head. He became unconscious. On admission to the hospital he was in coma. He died thirty hours after admission to the hospital with pulmonary edema, having never regained consciousness.
At autopsy, there was a small amount of hemorrhage in the left temporal muscle and overlying scalp, and laceration of the outer part of the left eyebrow. The skull was not fractured. There was marked edema of the brain, with dilatation of the lateral ventricles. There was no laceration or hemorrhages of the surface of the brain. In the corona radiata of both frontal lobes were multiple, punctate hemorrhages. There were innumerable small hemorrhages less than a pinhead in size in the white matter of the brain over the roofs of both lateral ventricles. There were a few similar areas in both corpora striata and over the ependyma of the third ventricle and in the posterior horns of the lateral ventricles.

To begin with this case is an observation published in a paper of MARTLAND and BELING (1927). The alterations in the white matter of the cerebrum show a hemorrhagic necrosis here described as punctate hemorrhages. The case of the boxer describes only his clinical findings, he did not die. It is objectionable to compare the pathomorphological findings of another case – who was not a boxer – with the clinical observation of the boxer.

Discussion of the so-called Multiple Concussion Hemorrhages

MARTLAND (1928) wrote that the theory of punch drunk assumes that the basic lesions is due to traumatic multiple hemorrhages. He refers to a paper of CASSASA (1924) who reported five autopsies showing what he called *multiple traumatic cerebral hemorrhages.* In all these cases was a history of a single head injury. Three out of five patients were momentarily unconscious at the time of injury. After a lucid interval varying from three to twenty-four hours there developed a period of marked irritability with increase of deep reflexes, which was followed by unconsciousness.

At autopsy, sections of the brain *showed multiple, usually punctate hemorrhages scattered over various parts of the parenchyma of the brain.* Lacerations of the scalp, fractures of the skull, cortical lacerations or hemorrhages except for occasional slight pia-arachnoid hemorrhages were not found.

Microscopic examination showed these punctate hemorrhages to be located around the blood vessels in the perivascular spaces of Virchow-Robin. When large, they broke through into the surrounding parenchyma and often became confluent. CASSASA considered this type of traumatic cerebral hemorrhage as relatively rare. For he had found only these five cases during a period of ten years' work with Otto SCHULTZE, former coroner's physician, and Charles NORRIS, chief medical examiner of New York City.

CASSASA's explanation of the mechanism of concussion depends on the mechanism of the so-called perivascular and perineuronal spaces and the identification of a network of fine fibrils connecting the external wall of the blood vessel with the surrounding brain tissues across the spaces of Virchow-Robin.

CASSASA wrote:

"Sudden overfilling of the perivascular lymph space with cerebrospinal fluid conceivably could produce laceration of a vessel by the tearing of its wall in the neighborhood of such a fibrillar attachment. Otherwise, without such an attachment, the laceration of a vessel surrounded by fluid could not be produced by any pressure exerted through that fluid which would only tend to compress the vessel but not lacerate it. Such an increase of cerebrospinal fluid in one perivascular space could be caused by the cerebrospinal fluid from the surface of the brain being driven into it by pressure exerted by the change of shape under an area of violence is in the direction of flattening and diminution of space for the cerebrospinal fluid in that area. This fluid must find its way out of that area through the various sulci of the brain and in connection therewith such fluid cannot find its way through these channels must find a way into the perivascular lymph spaces in the reverse direction of the normal flow of the cerebrospinal fluid in these channels.

In 1927, OSNATO and GILIBERTI, studied 100 clinical cases of concussion of the brain with or without fracture of the skull and one of CASSASA's cases from a histologic standpoint. They concluded that:

"Anatomic and clinical investigations seem to show definitely that our conception of concussion of the brain must be modified. It is no longer possible to say that 'concussion is an essentially transient state which does not comprise any evidence of structural cerebral injury.' Not only is there actual cerebral injury in cases of concussion but in a few instances complete resolution does not occur, and there is a strong likelihood that secondary degenerative changes develop. When this happens, we have a condition which, clinically at least, resembles some of the reactions seen in encephalitis. We feel, therefore, that the postconcussion neuroses should properly be called cases of traumatic encephalitis."

In 1927, MARTLAND and BELING reported their experience with consecutive cases of traumatic cerebral hemorrhage over a definite period of two years. During this period, 309 consecutive autopsies were analyzed of people dying as the result of cerebral injuries (exclusive of gunshot wounds of the head), occurring in a community of about 800.000 people and representing practically all the deaths due to cranial injury during that period. The number was sufficiently large to give a pretty good idea of the incidence of the various types of traumatic intracranial hemorrhages."

"Extradural hemorrhage was found in twenty-five cases, or 8.2 per cent; cortical laceration and hemorrhage in 254 cases, or 82 per cent, while hemorrhages similar to those described by CASSASA and unassociated with fracture of the skull or other gross surface injury occurred in nine cases, or 2.9 per cent. Subdural and pia-arachnoid hemorrhages and multiple streaky, deep hemorrhages associated with cortical injuries were present in twenty-one cases, or 6.9 per cent."

The theory that punch drunk assumes that according to MARTLAND, CASSASA, OSNATO and GILIBERTI, MARTLAND and BELING, that the basic lesion is due to traumatic multiple punctate hemorrhages. This theory has to be refuted. To begin with, lesions of this type were not described in the pathomorphological findings of the observations of BRANDENBURG and HALLERVORDEN (1954), GRAHMANN and ULE (1957), NEUBÜRGER et al. (1959), COURVILLE (1962), SPILLANE (1962), MAWDSLEY and FERGUSON (1963), QUANDT and SOMMER (1965), BURGER and MINAROVJECH (1966), CONSTANTINIDES and TISSOT (1967), PAYNE (1968), and CORSELLIS et al. (1973), who described the permanent brain damage in ex-boxers.

The description of these tissue alterations, which occur in cases of severe head injuries in non-boxers, and in boxers too, who suffer severe injuries in road accidents and in the ring and survive at least one to two days, must be described as hemorrhagic necroses; they are not the morphological substrate of the punch drunk syndrome. All authors who described them alluding that they are relative rare, the percentage in larger series amounts to a few per cent.

OSNATO and GILIBERTI noted that anatomic and medical investigations seem to show definitely that the conception of concussion of the brain must be modified. They continue, that it is no longer possible to say that 'concussion is an essentially transient state which does not comprise any evidence of structural cerebral injury.' They conclude that not only is there actual cerebral injury in cases of concussion but in a few instances complete resolution does not occur, and there is a strong likelihood that secondary degenerative changes develop. When this happens, according to OSNATO and GILIBERTI, "we have a condition which, clinically at least, resembles some of the reactions seen in encephalitis. We feel, therefore, that the postconcussion neuroses should properly be called cases of traumatic encephalitis.

This interpretation must be vigorously refuted. These tissue alterations are hemorrhagic necroses, which are not the substrate of a cerebral concussion. The autopsy report reveals that the white substance is mainly affected. Commotio cerebri, or cerebral concussion, is a clinical diagnosis, describing the immediate post-traumatic state of unconsciousness and loss of muscle tone after an impact to the freely movable head. The impacting force must reach a certain threshold to produce a cerebral concussion or commotio cerebri. This condition is not associated with any detectable macroscopic or microscopic lesions of the brain using lightmicroscopical techniques. The whole process is *traceless,* as noted by SPATZ.

The terminology was furthermore complicated by COURVILLE (1962, 1964). This author published a paper entitled *"The Mechanism of Boxing Fatalities."* A summary of this case report is presented by COURVILLE:

> "Repeated blows to the head in the last round of a 10-round championship bout. Victim knocked down in last ten seconds of with impact of back of his head against ropes. Lapse into unconscious state about 45 minutes after end of bout. Prompt admission to hospital in deeply comatose state with wide dilatation of pupils and quadriparesis. Death with relatively little change in condition after interval of 72 hours. *Autopsy:* minor terminal subdural, red softening of medial and basilar aspects of right cerebral hemisphere and softening of intermediate portions of corpus callosum. Moderate edema of entire brain but more focalized in upper brain stem, associated with petechiae in midbrain and upper pons, and softening gray matter about cerebral aqueduct. *Microscopic findings* of red softening of cerebral cortex marked also by ischemia (status spongiosus), edema and infiltration by focal hemorrhages of midbrain and upper pons."

COURVILLE describes the pathomorphological findings at autopsy. It existed red softening of medial and basilar aspects of the right cerebral hemisphere and softening of intermediate portions of corpus callosum. The microscopic findings of red softening of cerebral cortex marked also by ischemic (status spongiosus), edema and infiltration by focal hemorrhages of midbrain and upper pons.

The term red softening of COURVILLE is what we called hemorrhagic necrosis. In a following paragraph COURVILLE uses then term 'diffuse contusion,' which he described in the cingulate gyrus and adjoining regions of the medial frontal and basilar temporal areas, characterized by (1) status spongiosus, probably indicative of an ischemia, and (2) hemorrhagic infiltration. COURVILLE concluded rightfully that the presence of status spongiosus indicates that some impairment of circulation was responsible for this change. This is suggested by the fact that zones or areas of status spongiosus are found at the margin of the areas of hemorrhagic infiltration.

We do not agree with COURVILLE, when he writes, that the areas of hemorrhagic infiltration seem to resemble ordinary contusions in that they are particularly associated with a fairly well defined superficial layer of subarachnoid hemorrhage. Unfortunately, COURVILLE termed these alterations *'diffuse contusions.'* These alterations are no contusions, they occur, to begin with, in regions which are free of contusions and the pathomorphological picture does not resemble contusions, but, as COURVILLE first rightfully stated, red softening, or hemorrhagic necrosis. The macrophoto of the base of the brain and the drawing reveal clearly that these lesions are not contusions.

Summarising his findings COURVILLE wrote, that the one feature of this case, that was somewhat unexpected was the unusual extent of the 'diffuse contusions' which extended from beneath the splenium of the corpus callosum forward along the medial aspect of the hemisphere involving practically the entire right cingulate gyrus as well as the midportion of the corpus callosum. The areas described by COURVILLE are involved in hemorrhagic necrosis and are free of contusions.

Again the term *'diffuse contusion'* is incorrect. The lesions in the brain of this boxer who survived about seventy-two hours should be termed hemorrhagic necrosis. COURVILLE quoted another paper of him (1935) where he described 'diffuse cortical contusions' of the occipital lobe. The tissue alterations in the occipital lobe are not cortical contusions but hemorrhagic necroses.

The term 'multiple concussion hemorrhages' is here discussed in an unusually broad and extensive manner, since these tissue lesions are copied indiscriminating since many years and are again and again wrongly quoted in recent publications to be the morphological substrate of the punch drunk syndrome.

Discussion of the Different Diagnoses used

The term *'punch drunk'* in boxers and ex-boxers was first coined by laymen, that is many years before it was accepted by physicians. The pathologist Harrison MARTLAND published 1928 a paper with the title *'Punch drunk.'*

MARTLAND wrote:

> "For some time fight fans and promoters have recognized a peculiar condition occurring among prize fighters which, in ring parlance, they speak of as 'punch drunk.' Fighters in whom the early symptoms are well recognized are said by the fans to be 'cuckoo,' 'goofy,' 'cutting paper dolls,' or 'slug nutty.'"

MARTLAND continues:

> "Punch drunk most often affects fighters of the slugging type, who are usually poor boxers and who take considerable punishment, seeking only to land a knock out blow. It is also common in second rate fighters used for training purposes, who may be knocked down several times a day. Frequently it takes a fighter from one or two hours to recover from a severe blow to the head or jaw. In some cases consciousness may be lost for a considerable period of time."

MARTLAND presents now a brief but classic description of the symptoms:

> "The early symptoms of punch drunk usually appear in the extremities. There may be only an occasional and very slight flopping of one foot or leg in walking noticeable only at intervals; or a slight unsteadiness in gait or uncertainty in equilibrium. These may not seriously interfere with fighting. In fact, many who have only these early symptoms fight extremely well. And the slight staggering may be noticed only as they walk to their corners."

MARTLAND continues:

> "In some cases periods of slight mental confusion may occur as well as distinct slowing of muscular action. The early symptoms of punch drunk are well known to fight fans, and the gallery gods often shout 'cuckoo' at a fighter. I know of one fight that was stopped by the referee because he thought one of the fighters intoxicated."

MARTLAND presents now a detailed description of the 'punch drunk syndrome:

> "Many cases remain mild in nature and do not progress beyond this point. In others a very distinct dragging of the leg may develop and with this there is general slowing down in muscular movements, a peculiar mental attitude characterized by hesitancy in speech, tremors of the hands and nodding movements of the head, necessitating withdrawal from the ring."

MARTLAND continues:

> "Later on, in severe cases, there may develop a peculiar tilting of the head, a dragging of one or both legs, a staggering, propulsive gait with the facial characteristics of the parkinsonian syndrome, or a backward swaying of the body, tremors, vertigo and deafness. Finally, marked mental deterioration may set in necessitating commitment to an asylum."

The syndrome includes, as described, slurred speech, atactic disturbances and an impairment of psychic functions, all symptoms, which simulate an alcoholic intoxication.

MARTLAND described here not only the 'punch drunk syndrome' but also a traumatic parkinsonian syndrome in a boxer. He mentions the occurrence of deafness, or hearing impairment, which was more that 40 years later described in detail by PAULSEN and HUNDHAUSEN (1971).

MILLPAUGH (1937) termed the condition *'dementia pugilistica.'* But not always exists a dementia, this diagnosis is clinically inaccurate in many cases, and is replaced by the diagnosis *'chronic pro-*

gressive traumatic encephalopathy of boxers' (CRITCHLEY 1957). But not all cases are progressive, as SYMONDS (1960) had stated that in some cases the condition may be mild and non-progressive, while BOWMAN and BLAU (1960) note that progression ocurs only for a few years after the retirement of the boxer, the condition thereafter remaining stationary. This statement is not correct There are cases where the progression continues to a extremely severe dementia. LA CAVA (1952) introduced the term *'cumulative traumatic encephalopathy of the boxer'* and emphasized therewith the danger of repeated impacts against the head. COURVILLE (1962) coined the term 'psychopathic deterioration of pugilists' and gave up therefore the generally accepted connotation of psychopathic.

In an Editorial of the *Journal of the American Medical Association* (1969) the term 'Dementia pugilistica' was critisized:

> "The term 'dementia pugilistica,' first introduced by MILLSPAUGH in 1937 to describe the posttraumatic sequelae of boxing, lacks precision and specificity. The complete disorganization of personality implicit in dementia does not occur is all presumed victims of this disorder."

McDonald CRITCHLEY (1937) noted that:

> "... probably the most important neurological consequence of boxing, however, is the so-called punch drunk syndrome or dementia pugilistica."

CRITCHLEY preferred to call this the *'chronic progressive traumatic encephalopathy of boxers,'* a somewhat lengthy term that is both descriptive and explicit. This condition seems to be a late sequel to boxing and its manifestations may not show themelves until after the boxer's career is concluded. Once it appears it is slowly but steadily progressive. The presenting features are diverse, often making in some difficulty in diagnosis although it is, of course, to be suspected when the stigmata of the ring (cauliflower ear, thickened supraorbital margins) are visible.

According to CRITCHLEY (1937) differential diagnosis is between Parkinson's disease, general paralysis of the insane, and a frontally located brain tumor, and in some cases a multiple sclerosis. The patient usually complains of headache, dizziness and disturbances of vision. Intellectual impairmant is seldom obvious to the patient, although relatives will notice changes in personality with euphoria, fatuousness and defective emotional control. There is frequently, according to CRITCHLEY, a correlation between the degree to which the stigmata of a long boxing career, as 'thickened supraorbital ridges,' and 'cauliflower ear,' etc. are manifest and the severity of the dementia. On clinical examination the usual findings are extrapyramidal signs.

CRITCHLEY noted, that this condition is much more common in professionals than in amateurs – though it has been described in amateur boxers – and is more common in whites than negroes. The class of the boxer will frequently determine his liability to this condition since it is much more commonly seen in second-raters who make up for lack of skill by aggressiveness and a tendency to 'slog it out.'

It is the great problem of this condition, we fully agree with CRITCHLEY, that it is extremely difficult to diagnose in the early stages; the commonest early signs include failure to win bouts which the boxer would normally be expected to win and an inaccurate recovery period after the fight. CRITCHLEY noted, that the ability of the punch drunk boxer "to absorb further punishment in the ring is disputed; some seem to have a raised threshold, being able to take progressively more and more punishment before going down, while in others the opposite occurs. These latter are sometimes called to be 'glassy jawed.'

> "So serious is the risk of injury, and so peculiar is boxing," noted CRITCHLEY, "in that it is the only sport in which competitors deliberately set out to inflict injury upon their opponents, that there is ample justification on medical grounds for its discontinuation."

According to CRITCHLEY's experience, a very distinguished neurologist, differentiation is most difficult between cases oc punch drunkenness and disseminated sclerosis. In his own patients that was indeed the previous diagnosis in two cases. Close resemblance may also exist between punch drunkenness and general paralysis of the insane though the negative serology quickly settles the question.

CRITCHLEY continues:

> "There is in some of the reported cases (1935) evidence of disease of the extrapyramidal system. Here then exists an interesting parallel between a traumatic encephalopathy of boxers due to repeated head injuries and the *'traumatic parkinsonism'* developing after a single severe injury to the cranium."

CRITCHLEY noted that this diagnosis is somewhat long, however describing and typical. When such a syndrome would develop it would become slowly and continual progredient. The terms 'punch drunkness' or 'punch drunkenness' ('Schlagtrunkenheit'), 'dementia pugilistica,' 'chronic encephalopathy of the boxer, or 'encephalopathia pugilistica' have to be defined in detail.

The term 'punch drunkness' or 'dementia pugilistica' describes in general the psychopathological pictures of the chronic permanent brain damage of boxers. There are cases in which the 'dementia pugilistica,' the acquired dementia and depravation are placed in the foreground, and in which neurological findings are either nearly completely missing or are only minimal. Most probably slurred speech, an articulatory speech disorder, exists. Such a psychopathological syndrome can progressively develop into a fully developed 'chronic encephalopathy of the boxer ' or an 'encephalopathia pugilistica.'

The 'chronic progressive encephalopathy of the boxer' or 'encephalopathia pugilistica' is a term which describes the neurological and pathomorphological picture of the chronic and permanent brain damage, which can remain stationary, in general, however, takes a progressive course. Such a boxer can additionally show a severe 'dementia pugilistica.' Here the clinical permanent damage is added by the before described psychopathological picture.

In both clinical entities paranoid and/or catatonic psychoses may occur. They have to be termed *symptomatic psychoses,* because there is a causal relationship between boxing and occurrence of the psychosis. Nevertheless in every case must be decided whether the occurrence of a psychosis in a boxer is not a coincidental occurrence. It has to be thought about the possibility whether a latent psychosis or a disposition for a psychosis could have been triggered by multiple and repeated blows against the head. But there is no doubt that symptomatic or traumatic psychoses in boxers exist.

We are not able to distinguish the morphological injury patterns, which are based upon the predominant or complete psychopathological pictures of the 'punchdrunk syndrome' or 'dementia pugilistica' from the 'chronic progressive encephalopathy of the boxer' or the 'encephalopathia pugilistica.' In both cases a diffuse brain damage exists, in which the pyramidal, the extrapyramidal and the cerebellar system is participating. From the pathomorphological picture cannot be decided, why on the one side such a pronounced psychopathological picture exists, and on the other side it is more or less missing. That all these syndromes are the result of structural damage of the brain due to boxing is evident.

Case Histories of Chronic Clinical Findings

PARKER (1934) presented three case histories of boxers:

Case 1: A man aged 24, came to the clinic, because of weakness of the legs and tremor of the hands. There was no family history of a similar disease. He had had influenza in 1922; it was not severe and he was in bed only a few days.

At the age of 15 he had commenced to box, at first as an amateur. He had become a professional pugilist at the age of 16, in 1926. In that year, in the course of a boxing match, he had received a blow over the left temple, which had knocked him down, and blood had run from his left ear. From 1926 to 1931, he had fought as a lightweight and had averaged about two fights a month. He had made no great headway in his profession, for he had lost as often as he had won and he had not improved his technique greatly. He was not a very good boxer, although he was willing to take considerable punishment in the hope of getting a final victory by knocking out his opponent or wearing him down. He had been knocked unconscious many times, but only for short intervals. One time he had finished eight rounds after being knocked down and had no subsequent memory of it. In 1929, he had been knocked down five times, but he had not been knocked unconscious; the referee stopped the fight. The patient usually made little effort to guard himself, and as a consequence he had a flattened nose and 'cauliflower ears.' In 1931, he was told by his manager that the chances of his rising in his profession were slight. As opportunities for good matches were becoming fewer and fewer, he decided, in 1931, to retire from the ring and to become a filling-station attendant. He felt well during that time. In April, 1932, the chance of making money from a bout presented itself, and he re-entered the ring. He was knocked down twice in the third round, once by a blow over the right side of his jaw and another time a blow over the left eye and temple. Immediately after the fight, he felt as usual, but a few minutes later he took a drink of water, which he promptly vomited. He continued to vomit and retch, even on empty stomach, and an hour after the fight, when attempting to go home he reeled and staggered as if drunk. He managed to make his way home, and went to bed, but slept poorly because of nausea. Finally he slept and awakened at 10 a.m. When he attempted to arise he found himself very unsteady and was unable to balance himself. He was able to walk after a fashion, and actually had more difficulty in equilibrium than during the previous night. After consulting a physician, he returned to bed, where he stayed for three days, feeling nauseated and complaining of diplopia. He improved somewhat after this rest, and again he arose, but while the diplopia had disappeared, he was still unsteady, although very much less so than previously. In two weeks he returned to his duty as a filling-station attendant, still slightly ataxic.

Two months later, in June, 1932, the chance for another fight had come to him, and, in spite of his poor balance, he thought to train for it and overcome his difficulty. While in training, his legs seemed to get worse and he found that he tired very readily. He seemed to be fast enough on his feet in the ring, at the beginning, but was unable to endure any considerable length of time, because of a sense of fatigue of his lower extremities and a tendency of them to drag. He had given up the idea of fighting.

From that time up to his admission at the clinic (1934) his condition had changed little. When PARKER first saw him he still had trouble in walking, in that his feet seemed to drag. The right foot was worse than the left and he wore away the toes of his shoes. The sense of a balance was not normal, but slightly better than it had been; yet he frequently staggered, more at times than at others. By conscious attention to walking he would pick up his feet better than he usually did, but he soon relapsed into a dragging gait. His mother complained that he was more nervous since his last fight, a little more forgetful of names and details, and that his hands shook. Under excitement he tended to have a general tremor.

The patient was well developed and muscular, weighing and showing obvious signs of his previous profession in the form of thickened auricles and flattened nose. His systolic blood pressure was 130/76 mm Hg. He walked slowly and unsteadily. His lower limbs were stiff and slow, the right being the worse of the two. At the same time he staggered on, turning suddenly, and was somewhat ataxic. He held the right arm flexed when walking and stiffly at his side, and did not swing like the other. He hopped clumsily on the right foot, but did better with the left. The toes of both shoes were well worn down. There was a coarse, pill-rolling tremor of both hands, more marked on the right. Slight tremor of the head was present, but speech was unaltered. Speed was diminished on the right side, in both the upper and lower extremity, especially for rapid movements like drumming with fingers on a table. All tendon reflexes were greatly exaggerated, but there was no Babinski's sign on plantar stimulation. Romberg's sign was positive, and slight, horizontal nystagmus was present. Facial expression was normal and labile. Urinalysis and examination of the blood disclosed nothing abnormal and the flocculation test for syphilis performed on the blood was negative. X-ray studies of the skull gave no evidence of fracture or other abnormality.

PARKER commented that there is but little doubt that the severe beating received by this pugilist in his last fight was responsible for the chain of circumstances that followed. Although he finished the fight, the symptoms that appeared almost immediately thereafter suggested at least moderately severe injury to the brain, from which he made very far from complete recovery. It might also be suspected that his previous experiences had altered him so that he was an easy victim in his last fight, but of this there is no evidence, and he attributed all his difficulties to his last and futile appearance in the ring. The unbroken chain of cause and effect is very evident in this case, and it is to be noted that in nearly two years the condition had not become worse nor was there much improvement beyond a certain point. Clinically, the picture was a mixed one, suggesting multiple lesions involving the cerebellar, corticospinal, and basal ganglia systems of motor control. No specific nervous syndrome appeared, such as Parkinson's disease, but rather a medley of scattered and incomplete lesions of the brain.

The neurological examination of the first case of PARKER is comprehensive and thorough. When

PARKER writes that no specific nervous syndrome appeared, such as parkinson's disease, but rather a medley of scattered an incomplete lesions of the brain we have to take into consideration that this paper was published in 1934. There existed a pill rolling tremor of both hands, right more than left. There are many papers in which abortive forms of a parkinsonian syndrome in conjunction with other neurological findings in ex-boxers are published. This has to be considered a beginning parkinsonian syndrome, which is accompanied by other neurological findings, as described in the case history.

Case 2: A man, age 30 years, came to the clinic 1928, because of difficulty in walking and in speech. There was no history of any similar condition in his immediate family. He was married and had one child who was living and well. He had had practically no illnesses before the condition concerning which he made his complaint, although he admitted having had minor infections of the respiratory tract, which at no time had necessitated confinement to bed. He denied having had syphilis. He had commenced boxing at sixteen, in 1914, and shortly thereafter he had become a professional pugilist, fighting more or less continuously up to 1921. He had ranked as a second-rate boxer, and although he had demonstrated plenty of courage and had taken considerable punishment, he never had reached professional eminence. He had the usual broken nose and thickened ears of his trade.

On March 4, 1921, he was knocked out in a fight. He was unconscious during the count of '10', and unable to get up a few seconds later. He mentioned that he had received the blow on the left side of the lower jaw and that was sufficient to put him out quickly. He walked out of the ring, however, rode home in a street-car, and, angered by his defeat he proceeded to get drunk, but finally went to bed. The next day he seemed normal, and then had started training for the next fight, which was to take place sixteen days later. During this period of training he noticed that, when he was skipping rope, his right leg tended to catch on the rope, but he took little notice of it at the time. In the subsequent three fights he finished, he had been winner in the first, and he had fought the other two to a draw. The fourth fight was in June 1921, and it proved to be his last. He was knocked down by the first blow of his opponent, which was a moderately hard one. When he attempted to get up he was dizzy, had a severe headache, and promptly gave up the fight. While walking home, he found that the difficulty with his leg became decidedly noticeable: both lower limbs felt numb and they dragged as he walked. He decided, after this, that he was no longer fit to fight in the ring, and became a labourer in a steel mill. During the first month that elapsed after his last fight, and while he was working in the mill, the stiffness and slowness of movement of his lower extremities increased. This was more marked on the right side, for, as time went on, he wore out two soles of the right shoe to one of the left. Walking became increasingly difficult; he noticed that his voice had become nasal, and that articulate speech was becoming more and more indistinct. He continued at the mill about 18 months, when he was discharged because his infirmity had become obvious.

In the latter part of 1923, movements of the right arm became stiff, clumsy and awkward. Spasmodic contractions of the muscles of the right arm, right leg and back, appeared and gradually increased so that in walking he felt a tendency to be pulled backward. He had occasional parietal headaches lasting from one to four hours, and his powers of mental concentration, attention and memory were considerably reduced. Through the subsequent four years, his difficulties slowly and gradually increased. He experienced more and more difficulty in walking and in speech as well as increasing spasms in muscles of the extremities and trunk. At the time of his examination, however, he was still able to walk alone, and he could feed and dress himself.

The patient was well-nourished, weight 59 kg and there were no abnormal physical findings apart from those suggesting disease of the nervous system. Urinalysis, blood count, examination of eye grounds, and the Wassermann test revealed nothing unusual, and the same result was obtained from X-ray studies of the skull and cervical portion of the spinal column. *Neurological examination* gave a complex picture. The man was of unusual muscular development, probably because of his profession and training, and this rendered more conspicuous the various contractions and spasms continuously at play in his muscles. In walking, his gait was unsteady and spastic; the spasticity was more marked on the right than on the left side. More conspicuous, however, were the continuous, irregular spasms of the muscles of the right arm and back. These were sufficiently violent to pull the man backward and to the right with the right arm jerking and his hand slapping his thigh. These contractions were rapid, ever changing and quickly repeated. When the right arm was extended forward, fully supinated, continuous jerking occurred, tending to flex and pronate the arm. When, on the other hand, the arm was drawn behind the trunk in full pronation, these movements ceased. While the patient was seated the muscles of the back were continuously jerking, pulling him backward and to the right. The muscles of the face and neck were quiet. There was no paralysis, but all of the man's movements were slow and clumsy, except when he was running and jumping, which acts he performed, on the whole, better than slow walking. When he was asked to perform rapid movements of the fingers, these movements were slow on the right side, but in the left leg and foot the movements were equally slow and spastic. The tendon reflexes on the left side were exaggerated, and plantar stimulation on the left side produced an extensor reflex. Stroking the lips with a hard object produced a sucking reflex (Oppenheim's sign). Tests of coordination, applied to the upper and lower extremities, did not disclose marked dysmetria or dysenergia, but all tests were interfered with by the irregular, spasmodic contractions. There was no nystagmus. Speech was very indistinct, slow, laboured, grating and nasal in tone, and the breath escaped through the nose while the man was talking apparently because of some weakness of his palate. During animated conversation, he had the tendency to break into a silly guffaw. This, with the mouth wide open and with the tone of voice seemed to suggest the spasmodic laugh seen in pseudobulbar palsy or progressive lenticular degeneration. Mentally the man seemed childish and below par, even when his social standing and previous education were discounted. He was written to and a reply was received dated March 22, 1934. In his letter he stated he was still able to go around, had not become any worse, and, if anything, there was a slight improvement during the six years that had elapsed since his examination at the clinic.

PARKER commented that again, as in case one, this patient was a plucky and aggressive fighter, but of no great eminence in his profession; he was a lightweight, who started to fight when quite young, took plenty of injuries, and his career ended disastrously. However, the clear-cut chain of events existing between cause and effect is not so obvious in this case. All that is known is that after seven years of strenuous fighting and during the course of his career, between one bout and the next, symptoms appeared insidiously, increased in severity, and put an end to his fighting. Even after he had left the ring and tried other pursuits, his difficulties increased steadily up to the time he appeared at the clinic, seven years after his last fight. Argument could be advanced in this case that his trouble was due to a chronic phase of epidemic encephalitis and that the initial episode, in the form of a mild respiratory infection, had passed unnoticed. In Parkinson's syndrome of epidemic encephalitis, the initial episode may be absent or so mild that it is even forgotten, and yet the disease appears and progresses steadily to complete disability. It is, however, begging the question to assume that this patient had epidemic encephalitis, for this is incapable of proof, and it is just as plausible and reasonable to assume that repeated injuries to his brain led to a progressive, bizarre syndrome of disease of the nervous system. Moreover, the clinical picture was exceedingly complicated and did not resemble at least the more usual sequelae of epidemic encephalitis. There was, first, evident dystonia, producing grotesque muscular contractions, but the corticospinal motor system was also affected, as shown by the spasticity and Babinski phenomena. The dysarthria and the spasmodic laugh should be brought to account. Altogether the picture was that of scattered and diffuse lesions of the brain, resembling no specific clinical syndrome. In the final analysis, unsupported by necropsy, it can only be a matter of opinion that this patient's trouble was due to his pugilism, but in view of other reported cases, there remained a strong feeling that it was actually due to repeated boxing blows.

The neurological examination of the second exboxer of PARKER is comprehensive and thorough. The clinical picture is exceedingly complicated and unusual, there are only a few cases of this type described, because of the irregular spasmodic contractions, dystonia, with grotesque muscular contractions.

Case 3: A man, age 28, came to the clinic in 1923, because of stiffness and dragging of his legs in walking, unsteadiness of gait, and indistinct speech. There was no family history of similar difficulty, and the patient had had no previous illness requiring confinement to bed. He admitted having had gonorrhoea several times, and that he had a sore on his penis at the age of seventeen, the nature of which was uncertain.

He had started as a professional pugilist at the age of 18, and had fought steadily up to the time of his admission, i. e. for ten years. He had had as many as three fights in one week, and five fights in one month; in one year he fought 25 times. He had served in the army during the war and had boxed several times. While serving as a mess waiter, just before his discharge in 1919, he noticed that his left hand and foot were clumsy; he spilled glasses of water because the left foot hit the ground awkwardly and the hands shook. He thought nothing of it at the time. On discharge from the army he had gone back to professional boxing immediately, and continued to hold a championship until he lost it because of an adverse decision. During the latter part of 1919, while walking with his friends they jeered him because his left leg hit the ground with a slapping noise, and for the first time he realized that there was something wrong. Nevertheless, this had not interfered with his ability as a boxer at that time. In 1920, episodes of staggering while walking in the street appeared, and his speech became thick, so that he was unjustly accused of being drunk, a number of times. Just before one fight, a ring attendant, after talking to the patient in his dressing room, insisted on calling the manager to stop the fight, because he thought the fighter was intoxicated. The patient won that fight after a terrific battle. Gradually, during 1921 and 1922, the increasing stiffness of the left arm and leg and the unsteady gait became more obvious. He would be slow, awkward and clumsy in the first few rounds, would spar for time, then, becoming heated and loosened up under the influence of exercise and excitement, would finish as well as ever. The left arm would go into clonus, puzzling his opponent by its unexpected and grotesque trepidation, and then the right arm, which never became affected, would do its deadly work. Nevertheless, his disease advanced without remorse; speed and precision were slower in appearing with each bout, and a few days before the patient's first visit to the clinic his career ended in an inglorious fashion. He had been pitted against a boxer whom he easily had beaten before, and had expected to beat again. In the first few seconds of the first round, while still stiff and awkward, he was given a slight push, stumbled and fell, and was unable even to get back on his feet again. While the crowd howled in derision his seconds threw in the towel and his last fight was over.

As a boxer, he had been the aggressive type, fast enough in action, but he had disdained to guard himself much, and had carried the battle into his opponent's quarters continuously. He stated that he hardly ever felt the blows he received, had never been knocked out, but had received many terrible beatings. He won in the end by his superior stamina and ability to stand up under heavy punishment. He never had to go to bed after a fight, in spite of many cuts and bruises received around his face and head. He belonged to a certain school of prizefighters and was the only one affected among his crowd. As a possible explanation was the fact that they were all fast, clever boxers, hard to reach with a blow, and of them all, he was by far the most disfigured and had a particular reputation for being a hard fighter. He had had innumerable Wassermann tests of his blood, all with negative results, and at least several examinations of spinal fluid, not one of which had disclosed evidence of syphilis.

When he was examined in 1923, his disability was obvious. His speech was thick, muffled and hard to understand; there was a rasping labored quality in the voice. His gait was spastic. The spasticity was marked on the left side, and he dragged his legs, wearing out the toe of his left shoe. There was no marked weakness. The grip of the left hand was weaker than that of the right, but the outstanding feature was the slowness and clumsiness of the fingers, hand and arm. In certain positions the left arm went into clonus. Patellar clonus was also present, and there were a few strokes of ankle clonus on the left. Tendon reflexes were enormously exaggerated everywhere, more so on the left, and on that side a Babinski phenomenon could be elicited. The jaw reflex was exaggerated, and sucking reflexes were present on stroking the lips. The man was unable to hop on the left leg, and on performing active movements, such as shadowboxing, he was so stiff and clumsy that he nearly fell, and his left arm was slow and awkward. The right arm and hand seemed but little affected. There were no sensory changes, tremors or nystagmus: facial expression was mobile and pupillary reactions were normal. He seemed below par mentally, and his general intelligence and memory were somewhat reduced. Facial disfigurement was extreme, both ears were thickened and shapeless, his nose was flattened, his eyebrows were thickened and full of scars, his lips were irregular and distorted, and many front teeth were missing. *General physical examination* disclosed nothing more. The Kolmer modification of the Wassermann test, applied to the blood, gave negative results. On examination of the CSF, the Wassermann reaction was negative, the Nonne test was negative, a small lymphocyte was present per cmm, and the colloidal benzonin test gave no curve.

A diagnosis was made of lateral sclerosis of the spinal cord, and some doubt was expressed as to the origin of the condition. Syphilis was suspected because of the history, and a course of antisyphilic treatment was advised on empirical grounds. A very poor prognosis was given, since it was expected that progression of the disease could be expected.

The clinic lost track of him thereafter, until for the purpose of the present investigation, he was written to, and in response he reappeared in 1934, almost exactly eleven years after his first examination. To our surprise we found that time had dealt gently with him. He seemed in no way worse, and beyond acquiring a certain amount of obesity, he presented an identically similar picture to that encountered on his first examination, and was, if anything, somewhat improved, discounting the increased weight and lack of training. His disturbance of speech and gait were as conspicuous as before, but he had not deteriorated any more, physically, and had added to his collection a few more negative Wassermann tests on his blood and spinal fluid. He had been engaged in various occupations, mainly sedentary, had been moderately successful, and seemed at peace with the world.

PARKER commented that the picture here is somewhat different from that encountered in the other two cases. The patient, a high grade pugilist, developed symptoms during a period of four years, coincident with his heaviest fighting. The symptoms slowly progressed up to his final ignominious failure in the ring, and then for eleven years thereafter, his condition remained much the same as it was when he ceased fighting. The picture was also different from that presented in cases one and two: it was mainly that of injury to the corticospinal motor tracts on both sides; hence the original impression of lateral sclerosis. The man's frank admission that he had contracted venereal disease further led to the presumption that he had syphilis, but this was never proved, either serologically or by any clinical finding that was specific for syphilis of the central nervous system. It is reasonable to exclude this disease in its entirety, and more logical to blame the innumerable injuries to the brain received at the height of the patient's career.

An interesting approach was taken by CARROLL (1936):

"I determined to study 'punch-drunk' in its natural habitat, the boxing world. Accordingly, I frequented training quarters, helped examine fighters, and made friends with all sorts of 'pugs.' I observed declining veterans of the ring, talked with their trainers, and drank beer with their managers. I interviewed sports writers, referees and boxing commission physicians. I sought out reputed 'punch-drunks' and studied them, to the fullest extent that their feeling of exuberant good health would permit of a medical examination. My knowledge of the subject, then, is based upon the information given me by laymen whose business is a keen study of boxers, and the incomplete observation, over a period of two years, of many cases of 'punch-drunk'".

"The course of a typical case is somewhat as follows: A sturdy young man trains in a gymnasium as an amateur boxer in order that he may develop himself physically, and master the finer points of pugilism. He becomes a fairly tough opponent in the ring, but he tends to fight rather than box, and so receives some degree of punishment in each of his 30 or 40 short amateur matches. Later, he becomes a professional, meets stiffer opposition, engages in longer fights; but he is young, strong, and is 'making good money!' After a period of approximately four years as a professional boxer, during which time he has engaged in about 30 to 60 bouts, he begins to *'soften up.'* His tolerance to all head blows is reduced. After being struck on the jaw, he remains dazed for a longer period than formerly, and is more likely to be knocked out *('glass jaw')*. After a blow to the head, his legs will be a little shaky, and feel numb. His timing begins to fail and the fighter is no longer as dangerous on the offensive as he had been. His defense, which had been steadily improving with each encounter, becomes also less effective; and the pugilist who previously had been adroit enough to go through a career unmarked, now begins to develop a flat nose and cauliflower ears. Later his knees tend to give way after a head blow, and a slight dragging of the feet may be noticed, although, often, only as the contestant is walking to his corner at the end of a round. The fighter still boasts of feeling fine and capable, but now loses engagements which, formerly, he could have won with ease."

"Meanwhile the boxer's manager and associates begin to notice changes in him. They find that he has developed deterioration in attention, concentration and memory. In the midst of a conversation he may go into a reverie, suddenly change the subject, or ask the same question several times. He is apt to become too sociable, and very voluble. Usually an impediment of speech will develop, and the eyes may become slightly glazed or have a staring expression *(glass eye)*. In short, he may *continuously simulate a person who is just a little drunk.* When these symptoms have once appeared, they will not only persist, but may progress for a period of a year or so when they will usually become stationary. Thus punch-drunk is a self limited rather than a progressive encephalopathy."

"In the severer cases, the voice may become thick, halting and guttural, as though there were something held in the mouth. The gait may become unsteady. In a few instances vision may begin to fail or some degree of deafness may appear. Headache, dizziness and tinnitus, which one might expect as symptoms, are often present for a few hours or days following a fight, but they rarely persist as a part of the syndrome. Mental and personality changes become more marked, and emotional instability becomes apparent. Involuntary habit movements of a boxing nature are common, and are frequently displayed, even years later, when the person is abstractly preoccupied or during periods of emotional stress. A punch-drunk, sitting as a spectator at a fight, may vigorously ape the blows of his friend in the ring, oblivious to the other people about him. Punch-drunks, almost universally, do not realize that there is anything abnormal about themselves or their conduct, and will hotly resent, often with their fists, any application of this stigma to themselves."

"But the majority of boxers do not have this insight. They will continue to fight until they become so ineffectual in the ring that the promoters refuse to book them for any further matches. Some fighters, usually from economic necessity, will then enlist as sparring partners and receive further punishment while assisting in the training of others."

"The type of boxer is a large factor in the likelihood of punch-drunk developing. It is most often seen in the wide open, slugging type of fighter who is out to give as much as he can, and thus usually takes quite a lot in return. It is rarer in clever boxers, except after an extraordinarily long period in the ring. They rely on speed and coordination rather than strength. Such men usually have good defenses, and receive less punishment than crude fighers, but they also seem less able to 'take it.' What would be a knock out blow to their more highly developed nervous systems might be taken without much apparent damage by one of the brute type. The decline of the skilled boxer, when it finally comes, is often acute, and a single severe beating may result in marked decrease in fighting ability. This deterioration is chiefly in loss of speed and coordination. In crude boxers, speed and coordination have not been such important factors, and they continue along on their 'punch' after their finer control is lost, suffering many beating before they are finally 'through.'"

"With reference to the incidence of punch-drunk, competent followers of the sport estimate that about 5% of men who remain in the professional ranks for a period of five years or more will become definitely punch-drunk. About 6% of fighters remaining in the ring for this period will develop mental and emotional changes that are obvious to people who knew them previously. A few deteriorate to the extent that institutionalization is necessary."

CARROLL has presented us here with a classic description of the punch drunk syndrome. The description is so broad and comprehensive that it is impossible to add details. But one statement has to be contradicted, namely that punch drunk is a self limited rather tha a progressive encephalopathy. There are, withot doubt, cases which are progressive.

MILLSPAUGH (1937) adds his own anecdotic observation of a punch-drunk naval officer:

"One morning while awaiting a boat aboard a cruiser I was attracted by the unusual behavior of the officer of the deck. Judging by his boisterous activity (vocal, pantomime, and foot work) one might envisage him directing a deadly engagement, the outcome of which was by no means certain. Inquiry revealed he was a well known athlete while at the Naval Academy. Over a year later I had occasion to see him burst into a room where five medical officers were temporarily assembled. He chose a seat on the windowsill, spouted forth concerning his ailment and abruptly departed in a dash. Upon his departure the doctors variously shook their heads, lightly smiled, regarded one another with questioning gaze and one murmured, 'punch-drunk.'"

This observation by MILLSPAUGH presents a fine portrait of a punch drunk naval officer. It should be read by everybody who discusses boxing as an athletic activity in military academies. Unfortunately, we do not know whether this officer continued to box after he left the Naval Academy.

HERZOG (1938) presented the following case histories:

Case 1: Professional boxer who was stunned by a blow against the right ear and complained in the following days of headache, dizziness, and unsteadiness of gait. Neurological examination revealed a positive Romberg while walking with closed eyes, deviation to the right side. Had trouble speaking, and had reduced perception and memory. The neurological findings were otherwise unremarkable.

His physician advised the boxer to give up boxing, which he agreed to do. He took up a position as a professional coach.

The complaints of this professional boxer began

after receiving a hard blow in a bout. There are no data presented about his pugilistic career. There was a speech disturbance and reduced perception and memory in a boxer with otherwise unremarkable neurological findings.

Case 2: A 24 year old professional boxer began his career at age 15 as an amateur, and turned professional at age 20. He had participated in 80 bouts, 33 as an amateur without being knocked out, and 47 as a professional. Of the latter, he lost six by points or technical knock out. At the age of 22 he participated in a severe bout for the German light heavyweight championship. He received heavy blows against his head and was knocked down to a count of '6'. After this fight, he was nauseous for several days. He did not consult his physician, but interrupted his boxing career for a time because he felt ataxic and had difficulty in walking.
The *neurological examination* was unremarkable, but he had many severe psychopathological abnormalities. These were confirmed by a friend who was also a boxer and had known the patient during his career. The term 'punch drunk' was used. Speech was monotonous and poor in content. Facial expression blank; emotional monotony, except when describing successes of his career, which are described in a verbal manner. Despite the fact that he was not an intelligent person, there was no doubt that a dementia existed. This was confirmed by the boxer's friend.

The amnestic functions were primarily affected; his horizon was narrowed, thinking was poor, and he had difficulties in perception.

This 24 year old professional boxer had participated in 80 bouts, In a fight for a national lightheavyweight championship he was knocked down and severely punished. He had an ataxia and had difficulty in walkimg.

The neurological examination was unremarkable but there existed many psychopathological findings. A friend, who was also a boxer, used the term 'punch drunk'. This professional boxer was 24 years old and had only 80 bouts in a career as an amateur and professional which extended only over a period of seven years.

Case 3: French professional boxer who showed signs comparable to those described by MARTLAND: unsteady gait, ataxic disturbances. His knees buckle, and he makes sharp, abrupt movements of the torso. Follow-up not possible.

The case history is incomplete. There are no data about the number of bouts, the length of his pugilistic career, his weight class and his age. There seems to be mainly a cerebellar picture.

Case 4: German professional boxer, light heavyweight, who was for a long time a sparring partner for well-known national and international boxers. In the training bouts he purposely neglected his own defense, in order to 'accustom himself to heavy punches.' For this reason, neurological and psychopathological findings were severe. Uncoordinated movements of thorax and slight ataxia of both legs. Shuffles feet when he walks. Additionally, disturbances of speech-stammering and stuttering. Reduced amnestic functions, poor memory and perception. Decline in thinking ability.

Details of the profesional career of the professional lightheavyweight boxer are missing, we do not know his age. Remarkable is that he was for a long a sparring partner for well-known national and international professional boxers. This is one of the cases in which a boxer purposely neglected his defense, in order to 'accustome himself to heavy punches.' Such cases are reported from time to time, that boxers assume that taking severe punches against the head can benefit their ability 'to take punches.' Similar statements are made by some coaches, who think it beneficial, when their fighter 'has to feel the punches.' Such an idea is entirely mistaken and erraneous. Such an attitude, taking voluntarily severe punishment causes a severe permanent and irreversible brain damage. It is, therefore, not surprising that neurological and psychopathological findings were severe. The speech was stammering and stuttering and the memory poor.

Case 5: A 33 year old amateur boxer, 100 bouts, no knock outs. According to his trainer, he was a boxer who neglected his own defense in all his bouts, and had to take many hard blows and much punishment against his head. The boxer told the examining physician that he should have given up boxing after seven years due to dizziness and ataxia in both arms and legs. Reversion of the symptoms, but according to his own opinion, occurrence of a severe dementia. Memory (very forgetful) and impeded perception were especially involved. Language was monotonous and slurred, anacoluthic sentences.

Neurological examination was unremarkable. Psychopathologically, he showed poor memory, inability to memorize new material. There were gaps in the patient's memory that he recognized, and did not attempt to fill out by confabulation. Reduced perception.

This 33 year old amateur boxer had 100 bouts. In this boxer there was a neglect of his own defense in all his bouts, he had to take severe punishment. A former amateur boxer had developed a punch drunk syndrome. The neurological picture was unremarkable with the exception of slurred and anacoluthic speech.

Case 6: Sparring partner of ex-European champion (E. P.) in light heavyweight class, who had trained with him before his bout for the championship. Severe neurological and psychopathological findings developed immediately after these sparring bouts. Slight ataxia, intermittent buckling of feet or legs, shuffling of feet.
Psychopathologically severe dementia exists, so marked that patient requires a long time to react to new situations. Sometimes misconstrues a situation completely. Former manager confirms findings.

This lightheavyweight was a sparring partner of an ex-European champion. Sparring partners have to take severe punishment and it is no surprise that severe neurological and psychopathological findings exist. There exists a punch drunk syndrome combined with a neurological deficit.

Cases 7-9: Two sparring partners of German ex-champion Hans BREITENSTRATER and sparring partner of ex-world heavyweight champion Jack DEMPSEY. The three sparring partners in general revealed the same psychiatric and psychopathological disorders as those in cases 5 and 6. The personality changes were so severe in these boxers that they were diagnosed by laymen as being 'punch drunk.'

All three of these sparring partners of first class professional boxers were 'punch drunk,' they resembled the above mentioned cases 5 and 6.

KNOLL (1938) discussed three clinical cases of *boxer's encephalopathy:*

Case 1: A young professional boxer was examined by the author several times before and after fights. He won several difficult bouts, but took severe punishment in each. Despite hard training he was unable to 'in further fights'. Clinically, his speech was slurred and he had a marked ataxia. His 'reflexes' in a bout were reduced; he was not able to see the blows of his opponent and dodge them. His own blows swung wide and missed, and his 'short hard right' did not work any longer. There was a distinct change in his personality.

There are no data about the number of fights in which this boxer participated and his weight class. This young boxer had to take severe punishment in a series of difficult and hard fights. There were neurological findings, ataxia and slurred speech. Distinct changes in his personality were present. One can diagnose this boxer as punch drunk with additional neurological findings.

Case 2: This boxer received a series of severe blows against the head without being knocked down, and later showed a marked susceptibility to additional blows to his head. His speech was monotonous and his facial expression dull; he showed little interest, although he was prolix, and he repeated himself. A dementia was obvious and more pronounced than the neurological findings.

There are no data about his pugilistic career, his age and weight class. The boxer had to take severe punishment in a bout and had to take additionaol blows to the head in consecutive fights. There was a speech disorder. A dementia was diagnosed, a punch drunk syndrome existed with neurological findings.

Case 3: This boxer had to give up his career after seven successful years as a boxer. Dizziness, ataxia in arm and legs and increasing mental deterioration were the main findings. He was known for his ability to 'take punches.' He experienced increasing forgetfulness and his capacity for memory was reduced.

There are no data of his boxing career and his weight class, we learn only that he had a successful career for a period of seven years. He was a boxer who was known for his ability to 'take punches.' One can conclude that he had taken severe punishment in his active career. There existed an ataxia and a mental deterioration was obvious. A beginning punch drunk syndrome existed with ataxic disorders.

GREWEL (1941) reported on a 21 year old heavyweight boxer, who developed clinical symptoms a few weeks after he was knocked out in a bout with an internationally-known heavyweight champion.

Shaking of both arms developed, especially the left one, and a spasmodic torticollis was observed with sudden backwards movements of the head. At the same time his speech changed, he reported double vision, his gait was stomping rather than walking, and swallowing became difficult. At times he was very sleepy and had to go home, to go to bed. The patient had to give up boxing.
A *neurological examination* a year after the onset of the symptoms revealed a rigid face with severely limited facial expression. He rarely blinked. The left pupil was smaller than the right one. The facial nerve on the left was less innervated, especially around the lips. The upper lip was stiff, whistling impossible, and there was marked salivation. The head showed sudden movements backwards, and the chin turned to the right side. There was a tremor of the left arm and hand. The reflexes were increased, but not pathological. The gait was stomping. The patient was dysarthric. It was noted that the patient's character had changed; the formerly well-balanced man became difficult, impatient, distrustful and truculent.
In the following four years, a bilateral ptosis developed, and there was a marked tremor, right > left. The left arm showed weakness. The muscle tone was higher on the right than on the left. Dysdiadochokinese left > right. Changing tremor of the head. The left masseter, inside, was painful. He behaved in a very childish manner, threatening, always complaining, distrustful, 'knows everything better,' and had no insight.

GREWEL diagnosed the syndrome as a *posttraumatic encephalopathy.*

This 21 year old heavyweight boxer developed clinical symptoms after he had been knocked out by an internationally known heavyweight champion. One can conclude that he had to take severe punishment in this fight. An unusual and interesting clinical picture developed, shaking of both arms and a spasmodic torticollis existed. Disturbances of his gait were present. A dysarthria was seen. The neurological findings were progressive, especially the tremor of his head became obvious. A change in his personality developed. There exists a punch drunk syndrome, too.

BOURRAT and MICOLIER reported two case histories of professional boxers (1944):

Case 1: Thirty-four year old boxer was transferred (1940) from an artillery depot to an army psychiatric unit. One brother had a retinal ablation, as did the patient. Normal childhood; had sporadic short absences without loss of tonus or consciousness. Left school at age of ten. Was shoe-black, became an amateur boxer during the first world war, at the age of 15, and later turned professional. Was exempted from military service at age 20 (1916) due to ozena. He had a successful career as a professional boxer, and became European champion. He was defeated in the world champion bout by a black American boxer, but only after a long fight. Up to the age of 28, he participated in numerous fights, allegedly never knocked out.
The above-mentioned absences became more frequent in the last year of his career as a boxer. This was one of the reasons he had to give up boxing. Additionally he had suffered, since age 28, from severe headache and amnestic periods. The headache was described as frontal and in the neck region, mostly during the night, combined with insomnia. He had difficulty speaking coherently, losing track of his sentences. He liked to speak about his bouts in Europe and the United States. His memory is very poor, especially for recent events. Again and again he takes a notebook out of his pocket and takes notes. He has very carefully written down the different grades in the Army and their insignia and stripes. Drafted into the French Army in April 1940, he has difficulty finding his way about. He is restless, and has knitted brows. He has an asymmetric face, with classic scars, especially in the area of the left eyebrow. The left pupil is larger than the right one, but reacts to light and convergence. Normal reflexes. The speech is monotonous but well articulated. The transfer to the psychiatric unit was initiated due to his difficulties in finding his way about in the French Army, in order to exclude a case of simulation. Personally, the patient is concerned because there is no improvement in his condition, despite the fact that he has been in ambulatory and clinical treatment for the last four years.

This 34 year old boxer became an amateur boxer at the age of 15 and later turned professional. He became European champion. A black American boxer defeated him in a bout for the world championship, but only after a hard fight. Up to age 28, after a 13 year long ring career, in which he participated in numerous fights, he was allegedly never knocked out. The exact number of his amateur and professional bouts is not mentioned. In World War II he was transferred from a French artillery depot to an Army psychiatric unit, it was initiated due to his difficulties in finding his way about the French Army, in order to exclude a case of simulation. There exists a dementia in a simple structured personality who left school at the age of ten. This boxer had a retinal ablation. It is not mentioned whether he was released from military service.

Case 2: Begins to box at age of 18 and turns professional at age 20. His career as a boxer extends over a period of 21 years, until he is 39 years old. He became quite famous. The end of his career began in a fight against a black American boxer who knocked him out in the ninth round of a bout. While leaving the ring he collapsed; a long amnestic period followed, and he was unable to remember his trip home from Paris to Lyon. Two weeks later he was matched against a local boxer whom he had defeated very easily before. After a blow against his head he quit without being knocked out or down. The same situation repeats itself one match later, under analogous conditions. After both fights he has an amnestic period. A sports journalist writes about him: "He seemed to be slow (sluggish?)". He quit boxing and began to work in an administrative office; this leads to many difficulties as the result of his poor memory. At the age of forty-two he gives up this job due to a tremor of the right hand. A short time later he falls and is admitted to a hospital. The diagnosis is: fall to the head with following 'hebetude.'
The *examination* reveals a broken nose, cauliflower ears, scars at the eyebrows. No obvious neurological deficit. Slow in all his psychic functions. Forgetful, bad memory, cannot remember dates.

This boxer began to box at the age of 18 and turned professional at age 20. He had an unusual long professional career of 21 years, until he was 39 years old. He was a quite famous boxer. The end of his active career began when in a fight against a black American boxer, he was knocked out in the ninth round. He had a long amnestic period and cannot remember his trip home from Paris to Lyon. This boxer had two more fights against local boxers whom he had defeated very easily before. He had an amnestic period after each of these fights. After giving up his active career he worked in an administrative office which produced many difficulties due to his poor memory. After about two years he had to give up this job due to a tremor of his right hand.

There are no obvious neurological findings, except for a tremor of his right hand. It exists forgetfulness and bad memory.

BOURRAT and MICOLIER stress the point that the patient does not consider that boxing is the cause for all his mental deficits.

McALPINE and PAGE (1949) presented the following case history:

Twenty-four year old boxer who had boxed since the age of 14, had allegedly never received severe punishment in the ring, despite three months' booth boxing, until January 1948, when he was twice knocked out. Amnesia on each occasion did not exceed a few seconds.
February 1948, after losing his temper, his whole body shook and he felt as though he wanted to 'get going' with his fists. This quickly subsided but two days later he noticed slight tremor of the left hand, gradually increasing in severity and, during the next few weeks, spreading up the arm to involve the shoulder and head. Excessive salivation was also noticed during this period, and in March 1948 his speech became slow and indistinct, and his walking unsteady.
All symptoms have been slowly progressive, although the tremor was temporarily improved by moderate doses of belladonna.
The *examination* revealed no evidence of intellectual deterioration. Speech moderately dysarthric. Slight loss of facial expression. Pupils and ocular movements full. No nystagmus. Rhythmic, coarse, static tremor of head, left shoulder, arm and hand, with occasional tremor of right hand. Tremor intensified by emotion but disappeared when limb was supported. On voluntary movement, this coarse tremor is replaced by a less intense and more rapid intention type of tremor. Posture of hand not typical of Parkinsonism; no cogwheel phenomenon. All reflexes were brisk, including jaw jerk. Bilateral extensor plantar responses. No sensory loss. Gait: left arm fixed across body and fails to swing. Mild spastic gait, less apparent when running.
CSF normal, blood Wassermann negative, X-ray of skull normal, *EEG* normal, including a record taken during sleep after seconal medication.

In the following discussion, Russell BRAIN remarked that he had recently seen a very similar syndrome in a professional boxer who exhibited a mixture of torticollis, Parkinsonism and cerebellar ataxia, suggesting that the main damage had fallen upon the striatum and upper mid-brain.

This 24 year old boxer had an active career since the age of 14, allegedly he had never received severe punishment in the ring. But he had boxed for a period of three months in a boxing booth. Tremor of his left hand developed. Gradually increasing in severity, which spread to arm, shoulder and head. It existed an unsteady gait and slow and indistinct speech. All symptoms were severely progressive. The main symptoms were a spastic gait, tremor of the body on the left side and the head. There were signs of an abortive parkinsonian syndrome, combined with a spastic gait and a speech disturbance.

Helen DIMSDALE noted in a discussion remark that it is interesting that this patient complained of excessive salivation, a symptom usually associated with encephalitic Parkinsonism.

RAEVUORI and NAVALLINMAA (1951) reported two case histories:

Case 1: A chauffeur of 33 years. In anamnesis no illnesses to speak of. He has gone in for sports since his youth and boxed for 10 years. Taken part in several matches and been knocked out four times. He has often fought half-muddle-headed after blows on his head. In recent years has become irritable and easily falls out with others. In the war time has had to work in difficult circumstances, to which he has in his own opinion not been able to adapt himself as he should have done. On November 20, 1943, took part in boxing match where he received a blow in his face, so that the nasal bone was fractured. Did not lose consciousness, but was muddle-headed the whole evening. Since this event, has had a headache and felt dizziness, because of which he was taken to a hospital a month afterwards. He slept badly, the nervous system continually sensitive, the patient is easily irritated and loses patience. Other neurological symptoms – including balance disturbances – could not be ascertained. The patient himself complained of his bad memory. When speaking, he sometimes had difficulty in finding words, and when writing in finding capital letters, especially the letter N. Objectively, too, a slight decrease of memory could be ascertained. Intellectually, he proved quite normal. His behaviour was quiet and matter of fact. It is true the patient himself complained of irritability, but was, however, able to control himself outwardly. During a hospital observation of five months there were no changes to speak of in his condition, save that the headache grew less. After leaving the hospital he took up manual work lighter than his former job.

This 33 year old boxer had a career as an active boxer for ten years. He participated in difficult and hard boxing matches and was knocked out four times. He fought often 'muddle-headed.' In a boxing match he received blows in his face with fractures of the nasal bones. The main complaints were bad memory and balance disturbances. Unfortunately, we do not know the total number of bouts in which he participated.

Case 2: A worker of 37 years. Has boxed since he was 17. Has taken part in about 500 matches. Been knocked out twice. Has been known as a man who stands any number of blows. In 1938, received a blow with the fist on his head, after which the left side of the body has been slower than the right one, and the doctor ascertained that he had had a palsy. Has not boxed for four years. During two years the speech has been indistinct and by degrees grown worse. The throat feels stiff and sometimes he gets 'water in the throat' at night. Thinking has become slower than before. Memory grown worse. At times headache. Because of these symptoms he is admitted in a hospital.
Neurologically symptoms of a left sided hemiplegia are verified. When walking the left leg is stiffer than the right one. The movements of the left upper and lower limb are stiffer and clumsier than those of the right one, which is seen also in the diadochokinetic movements. The muscular strength of the left upper and lower limb are weakened. The left side abdominal and cremaster reflexes are weaker than the right side ones. The left side knee and achilles jerks are more active than usual. In Romberg's test he sways perceptibly. In position reflexes the left hand falls lower. Pain sensibility in the left upper limb, in the left half of the neck, head and face, somewhat decreased. Otherwise no sensibility disturbances. About the cranial nerves it may be mentioned that the patient says he sees double pictures when looking to either side and up sidewards. The convergence reaction weakened. The lower and partly the middle branch of the left facial nerve function more faintly than the right one. Uvula wound and withdrawn to the left. The left faucial pillar more contracted and when intoning it places itself higher. The articulation is very indistinct and stiff, like that of an intoxicated person. Slight slurred and hesitant speech. In the liquor, no pathological changes. A slight lability of the pulse appears.
Psychologically the patient makes a primitive impression, clearly seems demented. Thinking and reaction slow and heavy. The expression of the face slack and dull, the look heavy and somnolent. The patient is apathetic, unable to take the initiative, and indifferent. The speech slow and indistinct. Is only with difficulty able to tell about his own life. Memory is perceptibly weakened, so that he in the ten word test is not even at the tenth time able to repeat more than nine words, and after five minutes remember eight words. The intellectual stage defined according to Binet Bobertag 13 years, 4 months. In Kraepelin's addition test the slackening of mental activity proves less than expected so that in ten minute's time the patient does on an average 18.4 sums a minute.

This is a 37 year old ex-boxer who had taken up boxing at age 17 and had taken part in about 500 boxing bouts. He had been knocked out on two occasions. He had to take severe punishment on a number of occasions. After one bout he had a hemiparesis. A clinical diagnosis of a 'palsy' was made. Mental functions are decreased. A left-sided hemiplegia was verified. There exists an articulatory speech disturbance. There was a dementia. In World War II he was released from military service.

The condition of the patient did not change during a hospital observation of more than a month. Because of his illness he was entirely released from military service.

SCHWARZ (1953) published the case history of a 27 year old amateur boxer who was knocked out by a blow against the tip of his chin and hit the canvas with his occiput. Unconsciousness for one and one half hours, later vomiting and headache, which faded away during three weeks of bed rest.
A short time later he was hit on the tip of the chin again, and fell backwards stiff, hitting his occiput on the canvas. Unconsciousness for about 30 minutes. Since then he has participated in 15 bouts, and about two years later during a course of instruction he had been suspended because he had complained about permanent headache after blows against the head.
The *neurological examination* reveals an anisocoria (right pupil more than left), normal reaction to light and normal convergence. Nystagmus with rotatory component, looking sideways. Right side of calvarium sensitive to palpitation. Weakness of lower part of facial nerve on the right side. Tongue deviates slightly to the right side. Reflexes of upper extremity, left more than right. Trömner left positive, left side lower extremity reduced; Mendel-Bechterew left positive, Babinski left suggestive; knee-heel bilaterally ataxic; fingernose test, right misses, left intention tremor. Romberg swaying with tendency to left side; stamping gait.

This 27 year old amateur boxer was knocked out two times, when he hit his occiput on the canvas. He was the first time unconscious for 90 minutes and the second time for 30 minutes. These are two very long periods of unconsciousness. On both occasions there were severe deceleration traumas. It is surprising to learn that he then participated in fifteen more bouts. He was two years later suspended because of complaints of headache.

There are some minor neurological disorders. The psychopathological status is not mentioned. Details of his boxing career are not presented.

Case 2: Boxing career 1939-1948, 176 bouts total. Four of these as a professional. In his whole career he was knocked out three times, once in 1940 due to a blow on the tip of the chin, and in 1948 by a blow against the head. In the last fight, he was knocked out by a blow against his right temple with hemorrhage from the right ear. He was knocked down several times and came to his feet stunned and disoriented, until the referee interrupted the fight and awarded a technical knockout to his opponent.
His wife reported that his memory was bad, and that, unlike his previous behaviour, he is irritable and excitable, sometimes aggressive, and at other times fearful and perplexed.
Neurological examination revealed evident tremor of head and fingers at rest. Gait unsteady, short steps. Left leg drags slightly. Arms slightly flexed without apparent movements in coordination with legs. Face is expressionless. Hyposmia right side; pupils same size and react to light and convergence. Rapid nystagmus in lateral view, weakness in convergence of right eye. Upper extremities: muscle strength bilaterally reduced, reflexes right a little greater than left. Left rigor-like increase in muscle tone. Abdominal reflexes; left negative, right easily fatigued. Lower extremities; muscle strength bilaterally reduced, spastic muscle tone bilaterally achilles clonus. Babinski bilaterally positive; Mendel-Bechterew and Oppenheim right positive. Finger-nose test bilateral deviation with intention tremor. Knee-heel test, left deviation. Romberg-tendency to fall backwards. Speech inarticulate (mumbled) and staccato.
Psychopathological findings: Emotionally unstable, poor memory and difficulty in concentration.

This patient had an active career from 1939 – 1948 – nine years – he had a total of 176 bouts, four of these were professional.

There was tremor of head and fingers, it existed an unsteady gait. The face was expressionless. An articulatory speech disturbance existed. Poor memory and difficulties in concentration. The number of bouts in a nine years period with 176 fights is high. It would have been interesting to have follow-up examinations after five or ten years. There is no doubt that a permanent brain damage existed already at the time of the examination.

SCHWARZ summarized that this case findings are reminiscent of multiple sclerosis. Indeed, this diagnosis was made without knowledge of history. Additionally, there were symptoms that occur in Parkinsonism and are related to disturbances of the basal ganglia.

Case 3: Twenty-five year career as a boxer. About 200 fights, 15 as a professional. No alleged knock outs in this time. He had been frequently groggy and had boxed for many rounds in this state. In some cases, he was fully oriented only hours after the fight. He was known as a boxer who 'could take a punch.' In his last period as a boxer – his last fight was two years ago – he had suffered a continuous headache for nine months over his left forehead. Disturbances of memory, speech clumsy, frequently had to 'look' for words. Sometimes identified objects incorrectly.
Neurological findings showed the left pupil larger than the right, bilaterally normal reactions to light and convergence. Gaze tends towards the left, rapid nystagmus. Slight weakness of the lower facial nerve, cranial nerves otherwise unremarkable. Upper extremities-muscles strength good, no changes in muscle tone, reflexes equal on both sides. Trömner positive left, abdominal reflexes, left weaker than right. Lower extremities, muscle tone unremarkable. Patellar and achilles reflexes increased but equal. Oppenheim positive left, finger-nose test, right showed slight intention tremor. Speech slow, sluggish, hesitant as if looking for right word.
Psychically slow.

This boxer had an active career of twenty-five years, with about 200 fights, 15 as a professional. He was allegedly never knocked out but was frequently groggy for many rounds. He belonged to a group of boxers who were able 'to take a punch.' It can be concluded that this boxer with so many bouts had taken severe punishment.

Complaints about continuous headache are in the foreground. There is an articulatory speech disorder. The neurological examination reveals some minor findings. Disturbances of memory are present.

A follow-up after five or ten years had been necessary, to determine whether there existed a progressive clinical picture.

SCHWARZ comments on this boxer as a so-called *'rubber man'* ('Gummimann') who, after big years as a boxer, is only knocked out in rare cases.

TAYLOR (1953) published the following case history:

A miner aged 26 had boxed on numerous occasions since the age of 15 and as a professional from the age of 20. An ample history, obtained from various sources, failed to reveal evidence of previous mental disorder or any other abnormality, except a minor degree of mental dullness indicated by his school work and Army records. His I. Q. was reported to have been 90 on the Stanford-Binet scale (form L) when he was tested at some unspecified time before leaving school at the age of fourteen.
Two years ago, after becoming cruiser-weight champion of his area, slight changes in character became apparent, shown in the form of egocentricity and of monopolizing the conversation with accounts of the various opponents whom he had knocked out. Soon he was knocked out himself, after only 18 seconds, in the first round, and three months later, in an encounter with a boxer thought by all to be inferior, he took so much punishment that the fight was stopped in the fourth round. His friends were now finding him more quarrelsome and garrulous; he often complained of headaches, and he expressed mild paranoid ideas about his wife that eventually resulted in his admission to a mental hospital a year ago, after he had made threats to kill her with the carving-knife.
He was then found to be slightly confused, was alternately excited or unduly reticent, and had also a slight spatial and temporal disorientation, misidentification of persons around him, confabulation with an associated patchy memory defect for events of the preceding four months and fleeting olfactory hallucinations. Abnormal physical signs were slurred speech, absent abdominal reflexes on the left side and a slightly ataxic gait of cerebellar type. The *CSF* and an X-ray film of the skull were normal. Gradual improvement took place over a period of months, with sedation and rehabilitation, and he was then discharged from hospital and obtained work as an unskilled laborer. Residual signs then were amnesia for events over a period of one or two weeks at the time of his last two fights, and what appeared to be a reduction in intellectual capacity. His I. Q. on the Stanford-Binet scale (form L) was 65. On Raven's progressive matrices and on Kobs's blocks test he was at the five-percentile and on the Mill Hill vocabulary scale at the 10% level. There were now no abnormal neurological signs. The *EEG* showed reduced activity in all leads.

This 26 year old boxer had boxed on numerous occasions since the age of 15 as an amateur and since age 20 as a professional.

Two years earlier, at age 24, after becoming cruiserweight champion of his area, slight changes in his character became apparent. He was knocked out, after only 18 seconds, in the first round, and three months later, in a fight with a boxer considered by all to be inferior, he had to take severe punishment, that the fight was stopped in the fourth round. There is no doubt that he lost two fights in a row, one against an inferior boxer. Both fights had produced amnesias for a longer period. He was admitted to a mental hospital after he had made threats to kill his wive with a karving knife.

An articulatory speech disorder existed. An interesting finding were fleeting olfactory hallucinations. There was a memory defect and a temporal disorientation.

SOEDER and ARNDT (1954) reported five case histories of boxers:

Case 1: Forty-two year old patient, former aircraft mechanic. For the last several years working as an automobile mechanic, but for the last six months no longer employed, no longer able to hold a job because of his 'nerves.'
Had been a good pupil. Active amateur boxer from 1926 to 1942. In 1927 he had been the junior champion of the province, in 1936 an Olympic hopeful. He had had over 300 bouts and, allegedly, no knock outs. He had fallen twice on his occiput and lost consciousness. In 1935 he had had a fracture of the mandible, and in 1950-1951 he had been a boxing trainer. Sixteen years after he had taken up boxing, he had an unsteady gait, a tremor of the arms and legs, slow speech, and reduced manual dexterity. Headache and amnestic disturbances were denied by patient. Mood swings. Had been very thirsty between 1948 and 1950. Libido and potency markedly reduced.
Admission to hospital by factory doctor on suspicion of neurolues. Physician remarked that he had the impression the patient had general paresis. Had talked a lot, been in brawls frequently, pretentious behaviors and at the same time insufficient performance.
Neurological examination revealed a boxer's nose, mask-like face, bilateral nystagmus in lateral position, light reaction of right pupil slow. Hesitant speech, clumsy and slow. Large writing with super-enlarged initial letters (one to two words fill a line). Reflexes of lower extremities increased to pathological extent, Achilles reflex left more than right, clonic. Slow tremor of the head; while excited ataxia of the torso; rolling gait, like a sailor on land.
Psychopathological findings: unusual restlessness, a great deal of pointless, bustling activity accompanying every action. Tries to get into conversation on every occasion. Genially calls other patients 'sports friends.' He manufactured a dice game for the other patients and could not rest until the physician had sufficiently appreciated this clumsy travesty.
"One has to do something for people, get involved in everything." He lost his sense of social distance, speaking familiarly with the director of the hospital. Sometimes irritable. Euphoric mood is in marked contrast to slowness of psychic reactions. Long-winded and ponderous in speech. *Lues reaction* in blood and cerebrospinal fluid negative. *PEG* showed a dilated ventricular system, greater on the left than on the right, with markedly dilated third ventricles. Cerebral atrophy over both frontal lobes. *EEG* marked general disturbances. Paroxysmal waves of high amplitude over frontal areas, focus right-fronto-precentral.

This forty-two year old patient, who had been a good pupil was amateur boxer from 1926 to 1942 for a sixteen year period. He must have been a good boxer because in 1927 he was the junior champion of the province, and in 1936 an Olympic hopeful. The total number of his bouts was over 300, and, allegedly, he had never been knocked out. He continued to be a boxing trainer in 1950 and 1951. The first clinical signs had appeared sixteen years after he had taken up boxing. He had an unsteady gait. Tremor of the arms and legs. Slow speech, and reduced manual dexterity. Interesting is that the referring physician had diagnosed a general paresis. But Lues reactions in blood and cerebrospinal fluid were negative. The face was mask-like. There was an articulatory speech disorder. The gait was rolling like in a sailor. There existed psychopathological findings as restlessness, bustling activity, lost sense of social distance, euphoric mood in marked contrast to slowness of psychic reactions.

The ventricular system was dilated on the left more than on the right and marked cerebral atrophy existed.

The patient had an unusual long active career as amateur boxer and had continued to work as a trainer. The neurological examination revealed moderate disturbances, the psychopathological findings were first considered as typical for a general paresis, which could be excluded.

Case 2: Thirty year old patient, a stoker on a locomotive. For the last six years, a heavyweight boxer. Four knockdowns, last time six weeks before *admission to hospital.* After his last knockdown, he reported headache and nausea for 14 days. The patient had complained for the last one to two years of frequent headaches. Feels agitated.
Neurological examination revealed that his left pupil was wider than the right; there was a weakness of the right lower facial nerve, a gross tremor of the right hand, and increased reflexes without differences with regard to side. Diadochokinesis on the right side reduced. Oppenheim right positive. Overly familiar in the psychic sense, loss of social distance. Pointless,

bustling activity accompanies every action. Easily irritated. Impatient. Mood euphoric (braggart).
Detailed *psychiatric examination* not possible, as patient left hospital on his own after six days. He was admitted for stomatitis. *EEG* low frequency recordings with alpha waves of different frequencies, occasionally changing into beta waves. General frequent lability.

This thirty year old amateur boxer was for six years a heavyweight boxer, who had been knocked out on four occasions. The exact number of his bouts it not given. His complaints of headache and nausea developed after his last knock down two weeks before admission. There were slight to moderate neurological findings. Before the psychopathological examination could be performed, the patient left the hospital on his own after a stay of six days. The patient shows a beginning encephalopathia pugilistica. A follow-up after five or ten years was advisable to establish whether a progressive clinical picture existed.

Case 3: A 28 year old trained auto mechanic. Presently unemployed. Sister had multiple sclerosis. Good pupil in elementary school. Always ambitious. Took up boxing at age 15, at 15 was junior, and at 20 senior champion, of his area. Took up boxing after World War II until 1948. Knocked out in 1944, 1946, and 1948. On the last occasion he sustained an injury to the left ear, since then deafness and tinnitus. He has been a boxing trainer since 1947. He served a short jail term for dealings in opium, and after his release from jail 'nothing else worked.' He had been persuaded by his acquaintances to help them 'clear out' a sporting goods shop, but he had been reported. Shortly before his arrest he had found the one who had reported him and 'took him apart.' He had spent two years in prison. Finally, he wanted to become a 'solid citizen.' During his prison term he had conceived a fool-proof method of winning at roulette. He only needed capital, and his winnings would take care of the rest. ("Can you lend me some money, doctor?") He also intended to bring out a ray-therapy lamp ('put a ray-therapy lamp on the market') and make a lot of money out of that."
Report of family physician who knew the patient from his youth revealed that the patient had become conspicuous by his strategems and restlessness, but had phases in which he was depressed and in which he would brood. Before, he had been a well-balanced and reasonable man. His family had a good reputation locally and was ashamed of this misfit family member.
Neurological findings revealed status after fracture of nose bone, nystagmus to left, combined deafness left, increased reflexes. Radius periostreflex left more than right, achilles reflex left cloni.
Psychopathological: talkative, easily diverted, cannot be held to a single topic. Mood first euphoric, becomes tired during exploration and becomes morose and complains of headache. Later again in good mood. "Earlier I had much money and I will have much money again... when the innovation starts to sell... I could get acquainted with as many women as I want... only have to move a finger... but it would not be worthwhile."
EEG revealed marked general disturbances, and alpha wave reduction, occipital. Distinct focus, right parietal.

This patient, who was a good pupil, began to box at age 15, was at 15 junior, and at 20 senior champion, he must have been a good boxer. He boxed only for three years. The neurological examination revealed only minimal findings. His personality had profoundly changed, he had been a well-balanced and reasonable person, he became conspicuous by his strategems and restlessness, and he had phases of depression. In this ex-boxer a follow-up examination after several years had been interesting to find out whether the process was progressive.

Case 4: Twenty-eight year old patient, postal employee. Went to college but was drafted into military service. He had taken up boxing in school and in the military. He could 'absorb punches.' As a senior, he had 72 bouts. He had been knocked down 17 times; several times he had hit the canvas with his occiput, several times he was briefly unconscious. He gave up boxing in 1953, due to decreased ability (which he related to his fasting in order to make his weight class) on the one hand, and his wife asking him to give it up on the other. He did not finish college, but was employed by the postal service. Complaints of headache and reduced concentration, feels tired and does not sleep well, very thirsty. For the last one to two years, depressed, phlegmatic, sometimes had to cry. Libido non- existent.
Neurological examination revealed left pupil wider than right. Weakness of right lower facial nerve. Reflexes same on both sides.
Psychopathological: Depressed, tired easily, no initiative.
EEG: marked generalized disturbances. Dysrhythmic recording with frequent paroxysm, which consists of theta waves.
After discharge, he comes back two months later without obvious reason. He wanted to find out if anyone could help him. He had been shifted from one position to another in the postal service, away from counter service. He believed he had enemies in the office who didn't like him. He was convinced of his wife's infidelity.

This twenty-eight year old amateur boxer belonged to a group of boxers who could 'absorb punches.' It can be concluded that he took severe punishment. In 72 bouts, he was knocked down seventeen times. It should be called special attention to the fact that he had hit the canvas with his occiput (deceleration trauma), he was several times unconscious.

Neurologically there are only minimal findings. Psychopathologically he is depressed, tires easily and shows no initiative. He developed delusions.

Case 5: A 36 year old. Up until two years ago, he was a professional boxer. Since that time he has been a plumber. He took up boxing at age 14, and had 300 fights with fourteen knock downs. In 1942 and 1949, he suffered longer periods of unconsciousness after knock downs. Since then frequent headache. Twice German champion. Now boxes only occasionally. Was a passable pupil, finished his apprenticeship.
Neurological examination reveals scar tissue in the area of the left eyebrow. Head is held toward the right. Slow reaction to light by the right pupil. Gross tremor of the tongue. Achilles reflex cloni on the right. Rolling gait, like a sailor on land.
Psychopathological: Good mood, arrogant. Reduced concentration, difficult to keep him fixed on his history, difficulty remembering data. "Why is there so much talking? I thought I was here to be examined? Whether I'm cheerful, well that's Rhenish temperament.. Rhenish wine... Well, you know what I mean." Can't do simple calculations, so he says, "you should know, for calculations I have my wife." When asked about the intelligence test, replies, "You do that with somebody else, not with me!"
A co-worker reports, "He has his peculiarities – you have to take that or he explodes immediately. "He is frequently absent from work. Does not tolerate alcohol well – frequently irritated. He forgets everything he does not write down.
EEG revealed low frequency, marked general disturbances, with a marked focus right occipital.

The thirty-six year old boxer had been a professional boxer until two years ago at age thirty-four. He had began his career as amateur boxer at age fourteen and had 300 fights, he suffered fourteen knock outs. This is a long active career. One can accept that he had taken a lot of punishment. After some of the knock outs he had longer periods of unconsciousness. There exist tremor of the tongue, reflex differences and disturbances of gait. The clinical syndrome was a combination of pyramidal, extrapyramidal and cerebellar disorders. He forgets everything what he did not write down.

SOEDER and ARNDT summarized that these are boxers with many bouts and many knock downs. Only Case 1 reveals the full syndrome of encephalopathia traumatica. In other cases, neurological examination demonstrated various pathological neurological findings. Most boxers were euphoric and were hyperactive, sometimes expansive. There were other findings, such as slowness of all psychic functions, early fatigue, iritibility. Four patients had rapid mood swings. Two patients were extremely forgetful, especially Case 5 who had to write down everything. Case 4 revealed a depressed picture with a paranoic attitude towards his own failures.

JOKL (1954) published the following case history:

A boxer, aged 24, who for more than ten years had regularly taken part in fights, 'lost his temper' one day. His whole body shook, and he felt as though he wanted to 'get going' with his fists. Two days later, he noticed a tremor of the left hand which subsequently spread upwards towards arm, shoulder, and head. Excessive salivation occurred and his speech became slow and indistinct, his gait unsteady. All symptoms were progressive. There was no evidence of intellectual deterioration. Some loss of facial expression was noted. Eyes were normal. A coarse, static tremor of head, left shoulder arms and hand were present on the left side, at times also on the right. All reflexes were brisk. No cogwheel phenomenon. The picture was not typical of Parkinsonism. There was sensory loss, the left arm was fixed across the body and failed to swing during walking. Spasticity of gait was less apparent when the patient ran. The *EEG* was normal.

SCHWARZ (1955) reported the following case:

Twenty-five year old patient took up boxing in 1949. Unremarkable personal and family history. He was knocked out in his fifth fight by a hook to the liver. He received heavy blows against his head in a later fight, but he was not knocked down. After all his fights, up to 1951, no main complaints. He started as a light welterweight. In his thirteenth bout, in 1951, he was knocked out in the first round. Shortly before, he had been knocked down after a blow against the head. He had not originally been chosen for this fight, but had jumped in without training, as a substitute. He was transported, unconscious, to a hospital, where he regained consciousness after 24 hours. Amnestic period for the entire fight. Two weeks after admission, typical generalized epileptic seizure, also the week after that. Discharged from the hospital after six weeks, but readmitted three weeks later, because he had had two seizures and several absences at home. He complained of increasing headache, especially in the occiput.

He was described in an *expert neurological opinion,* in April and November of 1951, in a rehabilitation clinic for brain injured. March, 1952, again he had a generalized epileptic seizure; to an increasing extent, general slowing of psychic function, disturbance of memory, difficulty in concentration, anxiety, and remarkable irritability. Complained frequently of paresthesia on left arm and left leg.

He was again put under observation in the department of neurology at a university hospital. General slowness in movements, expressionless face, all psychic functions slow. Difficulty understanding the facts of a case and often needed several explanations before he could understand. Reduced ability to concentrate and early fatigue.

Neurological examination revealed the right pupil wider than the left, and a weakness in the left lower facial nerve. Lower extremity muscle tone higher on the left than on the right. Right lower extremity revealed reduced sensitivity to pain and touch. Finger-nose and knee-heel tests dysmetric, left more than right. Romberg: swaying. *EEG* showed a marked reduction in voltage, right parietal and occipital, and reduction in certain areas of alpha waves. EEG considered pathological, status after closed brain injury. Patient had finished apprenticeship as a carpenter. Worked later in a stoneware factory as an unskilled worker, had to give up several other jobs, and was finally unable to hold any job. He finally was working as a groundsman in a sports club.

The following case history was presented by BERGLEITER and JOKL (1956).

A 37 year old from a healthy family was, as a boy, intelligent, healthy, and had no psychopathological behavior. He had taken up boxing at age 14, made it to the final round of the English championship at age 15, and became an Olympic victor at 17.

He then became a professional boxer and was knocked out two times in this period. After one of these bouts, he had an amnestic period. One day later, severe headache and severe dizziness, blurred and double vision. Reflexes were increased on the left side. Despite this continued boxing and was knocked out in the next bout with a following amnestic period. Patient gave up boxing because he felt helpless in the ring.

Neurological examination revealed a facial paresis in the mid and lower branch of the facial nerve, a positive Romberg, ataxia and dysdiadochokinesis, patellar clonus left. Following period, absences occurred, and a decrease in intellectual functions developed.

Diagnosis: Severe, irreversible, progressive encephalopathy of the boxer.

CRITCHLEY (1957) published 21 case histories of boxers with a traumatic progressive encephalopathy: the symptoms described in these cases are illustrative of the 'punch drunk' state.

The following *clinical notes illustrate* some of the *characteristics of traumatic progressive encephalopathy in boxers:*

Case 1: A punch-drunk syndrome reminiscent of a frontal-lobe tumor.

A professional heavyweight who had been sparring partner to Walter NEUSEL, Tommy FARR, Eddie PHILIPS, Max BAER, Jack HYAMS, and others gave his story, with extreme slowness, and some dysarthria. His actual words were, "I can't think properly and I can't remember things; my speech has been getting bad for 18 months, and for the past two weeks it has been playing hell with me. I can't talk any quicker than I am now. My head goes blank at times. I get pain over my left eye nearly every morning when I wake, but if I rest it goes away, but I feel thick-headed all day long. I get dizzy but I have never lost my senses. The left ear is a bit deaf, and the sight isn't good in the left eye. I can't give my mind to reading; it makes my head worse, I just look at the illustrations."

Case 2: Incipient punch-drunk disability in a first-class heavyweight.

This boxer, who was always liable to headaches after each contest, had found them more persistent in the last few months. Others noticed that he was becoming a little slower, and the speech somewhat thick. On the occasion of his last contest he was knocked out, and headaches thereafter became troublesome. Although the patient maintained that he felt well, he was noticeably dysarthric and slow in cerebration. There was a strong suspicion here of an early punch-drunkenness and he was advised to rest. In this view the *British Boxing Board of Control* concurred, but the *U.S. authorities* dissented. They insisted on his fulfilling his American contracts, but thereafter he rapidly deteriorated both medically and professionally, and the diagnosis of punch-drunkenness was ultimately agreed upon.

Case 3: Mild case of punch-drunkenness with dementia, euphoria, and dysarthria.

This man, aged 61, had fought as a professional lightweight between 1910 and 1926. He became flyweight champion of Essex. Later he went to seek and took on various female boxers, including Madame CARPENTIER, the world's champion lady boxer. When examined, he was complaining of poor powers of concentration and forgetfulness. He would lose the thread in his conversation and had to rely on written memoranda. His mood was extremely euphoric and his speech grossly dysarthric. There was a traumatic arthritis of his thumb-joint with evidence of an old fracture. The *EEG* was flat. The patient asserted that he would never allow his children to box, and said that "it ought to be stopped."

Case 4: Isolated convulsive seizure in a professional boxer? Epilepsy? Early punch-drunkenness.

A man aged 23, who had been a professional lightweight for about four years, had been knocked out only once. Recently, while flying from South Africa, he began to vomit in the aircraft and then lost consciousness and became convulsed. *Clinical examination* was negative, but the *EEG* was abnormal, with bilateral synchronous rhythmic slow-wave discharges suggestive of idiopathic epilepsy. He was advised not to fulfill his engagement to fight the following week; but this warning was not heeded.

Case 5: Delinquency associated with early punch-drunkenness.

A naval rating, aged 29, was sent to the medical officer with the following report from the regulating petty officer, "He has been adrift for two and a half hours... he has lost his respirator in a pub. When spoken to, he belched offensively and without regard. He is a general nuisance, not intelligent, a bad influence. In short, he is evil." When asked what symptoms he complained of, he hooted with laughter and said in a grossly dysarthric fashion, "I've been adrift." Inquiry revealed that he had taken up boxing at thirteen and that he had turned professional. He believed he had won about three-quarters of his 300 odd fights. Latterly, after some of his longer fights, he would be groggy and complain of misty vision.

Case 6: Gross delinquency in a possible mild case of punch-drunkenness.

A featherweight professional boxer aged 27, at one time sparring partner to Seaman WATSON, was seen in 1942. In 1937 he had been charged with inflicting grievous bodily harm, but was acquitted. He entered the service in 1940, promptly went adrift for 24 hours, and then deserted for 14 days. After serving 90 days' detention, he almost at once deserted for 21 days. Later he deserted a third time, and was convicted of theft, going to jail for three months. Thereafter he deserted for seven weeks and married a woman he had met the day previously. He left her immediately after the ceremony and never saw her again. Since then he was charged with a fourth term of desertion. His medical state comprised: Headache, poor memory, a tendency towards losing himself in familiar environments, depression, and intolerance towards alcohol.

Case 7: Recurrent attacks of hysterical blindness associated with a punch-drunk state.

A former bantam-weight champion of Wales and of the west of England aged 27 was knocked out some 60 times in all. He gave up the ring in 1938 after a fight, on account of failing eyesight and hearing. The first attacks of blindness followed a vitriol-throwing incident when he tried to rescue a prostitute from some low joint. He had met her a few days before and had, he thought, persuaded her to go straight, and had given her fare home. Another attack of blindness took place when he narrowly escaped being run down by a lorry. He suddenly lost his eyesight once again in the naval barracks, and he was examined two weeks later in the Royal Navy Hospital, Chatam, showing all the dramatic characteristics of a case of hysterical blindness. His vision suddenly returned, but it disappeared once again when he received an anonymous letter impugning the fidelity of his wife.

Case 8: Striatal type of punch-drunk syndrome.

This man, age 27, began to box at the age of 16 and turned professional. He fought about 200 contests as a bantamweight, sometimes two or three times a week. Later he appeared in boxing booths, for it was becoming increasingly obvious to him that he was deteriorating. About this time – that is, when aged 27 years, he was rejected at the recruiting office, and he realized then that his sight was poor and that his gait was unsteady. Three months later his hands became tremulous. His symptoms increased, up to a point, and then remained stationary.

On examination he was found to be slightly forgetful. He had a nasal type of dysarthria, his face was expressionless, and he tended to dribble. He had a tremor of the outstretched hands, some increase in muscle tonus, and slight but definite ataxia. Tendon-jerks were increased and there was a right extensor response. The gait was titubant, short-steppage, and stiff.

Case 9: Cerebellar type of punch-drunkenness, reminiscent of disseminated sclerosis.

This man, aged 37, was a professional boxer between the ages of 21 and 31. Successful at first as a lightweight, he deteriorated and took on opponents out of his weight class. He did not think he came to much harm, until a particular fight in the Albert Hall – when being distracted by a spectator calling out his name – he was knocked down. He got up after a short count and was floored a second time. This state of affairs went on again and again, and though he finished the fight, he subsequently remembered very little of what happened. His symptoms consisted of progressively increasing tremor and clumsiness of the right upper limb.
Examination revealed a slight dysarthria, bilateral intention tremor, increased tendon-jerks, but flexor plantar responses. Appropriate tests brought to light a mild degree of intellectual deterioration.

Case 10: Striato-cerebellar type of punch drunkenness.

A professional welterweight, aged 27, whose record gradually deteriorated, had been knocked out three times. In 1948 (aged 24) after losing his temper with his fiancée, he started to tremble all over. This subsided except in the left hand, where the tremor persisted with increasing amplitude. After six months it involved the head, and his leg began to drag. His articulation altered. *Examination* revealed a fine tremor of the head, which was held in a curious retracted posture. The speech was dysarthric and staccato. On the left side there was an unusual type of intention tremor, associated with dysdiadochokinesis. The gait showed a steppage abnormality of the left leg. Both plantar responses were extensor in type.

Case 11: Cerebellar type of punch drunkenness in an amateur boxer.

A man aged 39, who had been four times schoolboy champion of Great Britain, began to develop tremor of the arm during the war, on which account he was 'invalided' from the Army. His symptoms increased slightly and he showed an impassive, rather mask-like face; marked dysarthria; titubation of the head, with some hypertrophy of the muscles of the neck; uncoordinated the arms and legs, more so on the left; with intention tremor and dysdiadochokinesis. Originally of superior intelligence, he later showed both an intellectual falling-off and a curious alteration in personality.

Another Group: Illustrative Cases of the 'Groggy state.'

Case 12: This man would feel groggy for days after each contest, with his eyesight getting worse, temporarily, each time. He had particular difficulty in judging distances. This is why he eventually packed up boxing.
Case 13: "I seemed to go dead, and not to feel the punches. It would now take me two or three days to get over each fight." (Ex-welterweight champion of the West of England).
Case 14: Latterly, he took longer and longer to recover from each fight, because of headaches, defective memory and impaired speech. His wife did not approve of his boxing and persuaded him to give it up (ex-lightweight professional boxer with punch drunkenness in which headaches and irritability were conspicuous).
Case 15: A former welterweight champion of Lincolnshire, aged 30, used to recover from his contests promptly and has even boxed on two consecutive days. Latterly, he would remain stiff for some days after each contest. After two years his record deteriorated, especially when he got to London, where he found the style of boxing was different, with more 'in-fighting.' He stopped boxing for four years, but took it up again when he joined the Navy. In his last contest he went down several times in the fourth round, but he finished the fight 'not feeling right,' his 'limbs felt like rubber' and he complained of a thick head and sore neck, and he could not sleep. Early morning headaches followed, with severe postural giddiness, irritability, tearfulness, poor memory, fatigue, and lack of interest.
Examination brought out that his emotivity and a gross dysarthria were evident.
Case 16: A man aged 29, never quick on his feet, he could always 'take it,' and ordinarily 'did twelves' (as he put it, when questioned on the length of the contests he fought). He packed it up because of his headaches. After one fight he became so unsteady on his legs that he was accused of being drunk. Towards the end of his boxing career he would feel queer for about a week after each fight.
Case 17: A man aged 28; two weeks before entering hospital he staggered so much in the ring that the audience howled him down and the promoter declared him drunk. In hospital, he presented a picture somewhat resembling disseminated sclerosis. Three weeks after his discharge from hospital, he fought PAOLINO in Paris for the heavyweight championship of Europe (this was in 1932), but he was knocked out at the first blow, and the crowd stormed the ring.
Case 18: This man, aged 25 had been middle-weight champion of Wales four times and sparring partner to Jack PETERSEN and to Len HARVEY. About six months before coming into *hospital* he received some blows from Jack PETERSEN. He had to fight again two days later, but gave up after three rounds because of weakness in his arms. His next fight was a fortnight later; he gave up after the second round because of weakness and tremulousness of his limbs. His manager thought he was drunk. A doctor told him never to box again. He seemed to improve after a month's convales-

cence in Swansea, but still being unsteady, he went into the National Hospital, where he was found to be punch-drunk.
Case 19: A heavyweight aged 31 was knocked out 'through the ropes' in 1945 and was dazed for the next 24 hours. The right leg was found to be weak immediately after his fight and continued to be so. Within the day or two following this same fight his right hand began to shake: the left hand started to tremble some six months later. He was hospitalized seven years later, a typical severe punch-drunk.
Case 20: A delinquent mild punch-drunk, interviewed in prison, gave up boxing at the age of 20 because he was not feeling himself. He felt drunk, although he had taken nothing in the way of alcohol; his head would be in a swim after each fight.
Case 21: A Royal Navy welterweight, aged 22, matched against a tougher opponent, was struck heavily over the heart in the second round, causing him to feel dizzy or groggy. Shortly after he got a severe blow to the side of the head, flooring him for a count of '9', when he rose and resumed the fight. He could remember every incident vividly up till then; but after that, his memory was a blank. Apparently he finished the second round, and in the third round he received heavy punishment, being knocked down ten times, striking his head on the floor more than once. His next recollection is being on the deck of his ship about an hour later, feeling dizzy and sore about the stomach. Two weeks later he became ill with headaches, vertigo, and vomiting, and was hospitalized. Investigation showed old blood-staining of the *CSF.*

GRAHMANN and ULE (1957) published the following case histories:

Born in 1909, finished elementary school without any difficulties. Seaman for several years, later conductor. Unremarkable history, no venereal diseases, moderate smoking and alcoholic consumption. Good reputation as an employee.

Boxed as amateur from 1930 to 1939 (21 to 30 years old); he had about 100 amateur fights. During the last years of his career he was a trainer for juveniles, often for hours, until completely exhausted in the ring. Allegedly never knocked out, but groggy on several occasions. In June of 1952, accident with bicycle, unconscious for several minutes but was able to go home unaided, no complaints, and took up work again. About a month later he was arrested, because he tried to steal a copper cable from a freight car. "I thought, saw that off, because I am in such economic distress." He accepts the jail term of two months immediately and declared, "Shit, two months is nothing." In the court protocol is the remark that he made "a narrow-minded impression." He was fired from the railway, without a pension, and tried to work as an unskilled laborer on a wharf, but was fired again due to incompetence. "They could not keep me because of my forgetfulness." He became a seaman on a tanker and was known as the 'crazy railway man.' He is used as a rust scraper but was finally dismissed due to his incompetence. From this time on he lived on unemployment benefits.
Admission to hospital. His wife remarks that he struck her as careless and unconcerned during the court procedures. According to her, he is forgetful, he talks more than before, and frequently she is unable to understand what he wants to express. He was careless with money, he had no endurance, he would run around purposelessly, would be boisterous, and then get into a rage. He would complain about headaches frequently and about nausea. On three occasions he had fainted and collapsed.
The *neurological examination* revealed that the right pupil is wider than the left one. Swaying while standing with closed eyes. *EEG*: slight general alterations. The patient is slow, clumsy, easily tires, but bustling with activity, officious. He has no sense of social boundaries, meddles with everything, he is almost always euphoric and only rarely and for short periods morose and impatient. He is overconfident, starts many things but finishes nothing. Self-criticism is missing. Memory is bad; he is inattentive and careless. He omits or doubles letters while writing; the handwriting is disorderly, the paper unclean. He thinks and speaks circumlocuitously, seldom brings a thought to an end, and cannot hold a train of thought while speaking.
A year later, he gives the impression of being demented. He is ethically and esthetically dull.
Case 3: Born in 1925. Unremarkable family history; finished elementary school with good results, apprenticeship as locksmith. After having left school, he joined an amateur boxing club, participated in numerous bouts against national and international opponents; he allegedly suffered no knock outs. During World War II, served in the Navy, continued with boxing. Several times groggy, but allegedly no knock outs. Spent several months as an English prisoner of war. After returning home, his family members were struck by his irritability and fury. He obtained a license as a professional boxer; his career as professional ends after four severe knock outs, in a relatively short time. After the last knock out he became apathetic, indifferent and slovenly. Couldn't hold down a job. February 1951 he became excited, pulled the curtains from the windows, destroyed the radio and put his hand into a burning oven.
Neurological examination revealed *severe dementia; chronic encephalopathy due to boxing* is discussed as the cause. *Transferred to a psychiatric hospital.* Loss of critical faculty, childish euphoria, and marked lack of initiative.
PEG : plump ventricular system and frontal cortical atrophy.
In the summer of 1955, severe state of agitation, in which he insults and threatens his family. Self-inflicted wound of the right hand.
Transferred to psychiatric hospital. Here he presented a *catatonic picture.* Patient lies in bed grimacing with elevated

head. He behaves autistically; from time to time he laughs aloud, without any apparent reason. Pulls out the hair on his head and then for hours sways upper torso back and forth. Thoughts obviously incohesive, impossible to get his attention. Sensible conversation not possible; language completely incoherent.

When asked, "Why did you destroy the radio?" he replied, "It always disturbed me. I had expected dynamite in the radio, because the voices I heard told me that. Then I picked it up and saw there was no cord connected to it."

"Where did the voices come from?"

"From everywhere. I couldn't defend myself against them."

The *neurological examination* was unremarkable with the exception of slurred speech. He remains in a psychiatric hospital.

Case 4: Born in 1911, unremarkable family history. Went to a special school (for low intelligence), tried an apprenticeship as a barber, but had to give up after one year because of insufficient ability. Worked as a messenger boy, in a mortuary, and as a stretcher bearer.

He joined a boxing club at age 12, two years later he was the youth champion of the province of Schleswig-Holstein. Boxed in numerous bouts before and during World War II and continued his career as a boxer after World War II, with varying success.

Allegedly never knocked out, but several times groggy after blows against the head. During the time he boxed, in 1942, friends recognized that his speech became slurred. In 1947 he participated in a series of boxing bouts and took a series of heavy blows against the head, became severely groggy and gave up boxing. The trainer of the boxing club remarked, "They beat him to a pulp." ("Den haben sie matschig gehauen.") He complains of dizziness, his legs seem insecure and weak, as if they were made of rubber. Complains of headache. Somatic and psychic capacities decreased. Finally he became an invalid. Due to his indistinct babbling speech and his wide-legged swaying gait, he is often believed drunk. The children chase him in the street.

Since 1951 he has complained of difficulty in swallowing. His wife thinks he has become very irritable and gets excited about minor things.

During *admission to a neurological hospital,* moderate tremor of the head was noticeable; gait insecure, broad and swaying; speech slurred, face is amimic; innervation of uvula is minimal. All reflexes are greater on the left than on the right. Rossolimo and Mendel-Bechterew are positive, greater right than left. Finger-nose test unremarkable; ankle-knee test shows bilateral ataxia with dysmetria.

Disintegration of his personality. All reactions slowed, ceremonious, awkward. Ability to concentrate reduced, fatigues easily, no spontaneous initiative. Emotionally depressed, affectively incontinent and complains with whining resignation. Somatic and psychic decay.

Egon MÜLLER (1958) reported three case histories:

Case 1: Thirty year old butcher, also an amateur boxer since age 16. Heavyweight champion in 1943, professional boxer from 1945 to 1949. In this time he had at least ten severe bouts, of which he lost five. According to the expert's reports, by 1947-1948 he was already "thoughtless, careless, and boxed without comprehension. He has shown false judgement of his own abilities as a boxer." In 1950, head injury without disturbance of consciousness, due to a traffic accident. About two to three months later, for the first time he had headaches, and noted an increased slowness which was used repeatedly by his employers as a reason to fire him. Slow development of speech and gait disturbances; clumsiness of hands, progressive social depravation. Several follow-ups since 1953 showed early aging and progressive disintegration of his personality with dementia, and especially an inability of the critical faculties.

Neurologically he showed spastic gait, coordination disturbances of the upper extremities, dysarthric but not staccato speech. Progression of neurological disturbances in the following years. Additional findings – nose and maxilla fractures left.

PEG: moderate general dilatation of ventricular space, cerebrospinal fluid unremarkable.

EEG showed changing general alterations and inconstant focal signs.

Carotid angiogram: unremarkable.

Case 2: Forty-five year old patient, employee. Admitted to hospital for expert opinion on an open brain injury from 1941, for which he was receiving an annuity of 70% reduction of earning capacity. He complained of increasing headaches and disturbance of gait, which had not been present in the first years after the injury (he had been active as a training officer). His wife confirmed an increasing insufficiency of performance, mentioned increased irritability, early fatigue, increased perspiration. Family doctor had considered general paresis possible.

Roentgenological examination revealed intracerebral shrapnel fragments.

Neurological examination revealed unsteady gait which was reminiscent of a paraspastic gait disturbance, and severe psychic organic disturbances, such as disturbances of memory functions, difficulty in concentration, increased irritability, loss of critical faculty.

EEG: unremarkable. Further examinations rejected.

Luetic infection can be excluded.

Since the analysis of the war head injury, the course of the process and the neurological findings are not compatible, an additional cerebral process is suspected. This was confirmed, as a detailed history shows that the patient took up boxing in his seventeenth year, was the European champion for several

years, and participated in 412 fights, during which he was frequently groggy.
Case 3: Forty-six year old patient. He had boxed since his early youth, until his 41st year. Five hundred bouts, supposedly with only one knock out and a few groggy states. Now working as a boxing or physical fitness teacher in his own boxing 'stable.' At the age of 42, he suffered a closed head injury due to a motorcycle accident. Slight to moderate brain injury obvious from the first examination and later follow-ups.
Neurological examination revealed anosmia, weakness of convergence, differing radius-periost reflexes, weakness of right side abdominal reflex, weakness of left facial nerve, inconstant Babinski left. These findings could be related to brain contusion as a result of his motorcycle accident, but due to his history, they can be seen as a result of an encephalopathia pugilistica. The progression of the entire process speaks more in favor of the latter. The final decision must remain open, as the patient refused all further examinations or tests.

McCOWN (1959) reported on two boxers.

Case 1: A 24 year old boxer with a poor boxing record of eight losses, three draws and two wins, fought a six round fight at St. Nicholas Arena on October 3, 1952. There was no knock out and he lost the fight by a decision. After returning to his dressing room, chronic seizures of the right upper and lower extremities suddenly developed, and he became unconscious. He was immediately *transported to St. Clare's Hospital* where a *neurological examination* revealed the reflexes to be markedly exaggerated; the pupils were unequal, with the left pupil dilated and right pupil contracted; lumbar puncture disclosed clear spinal fluid but markedly increased pressure, 330 mm H_2O. Patient was unconscious for three hours, after which he became lethargic and drowsy, but responded to questions coherently.
Roentgenological examination of the skull and EEG's were within normal limits. Blood pressure was elevated, 150/90 mm/Hg.

The *diagnosis* was *severe cerebral concussion.* The patient was hospitalized for ten days and *neurological examination* was within normal limits on discharge. Because of the severity of this boxer's injury to the brain, he was retired from boxing.

Case 2 is a subdural hematoma. Surgery was performed and the patient was discharged the days later.

WOLOWSKA (1960) published the following case history:

This ex-amateur boxer is presently an unqualified worker. He was admitted to the hospital with a diagnosis of disseminated sclerosis, because of paralysis in the lower extremities, change in speech, and gait disturbance. His illness began five years ago, with a disturbance of balance. The illness was characterized by remission and exacerbation. In the last year of boxing he appeared drunk, to the point that the physicians suspected him of abusing alcohol. His gait was unsteady, he was irritable, aggressive and over-talkative. Unsteadiness of gait increased, and he was asked to stop boxing. He then started working as an unqualified worker.
Three years ago he began to have memory problems and difficulty with his speech. He was mumbling, his speech was generally unclear, and there was a period during which he was unable to speak at all. Three months before admission he noticed a weakness in his lower extremities, because of which he was *referred to a hospital.* He had never had double vision or bladder dysfunction.
He had been in 70 fights, during which he was never knocked out, as he was a 'hard fighter.' He was always healthy; he had never had venereal disease. He was not having problems passing urine, and his bowel movements were normal. He smoked ten cigarettes per day. He was drinking only a small amount of alcohol, on occasion. He was married but childless. In the last few years he had noticed a decrease in sexual potency, and during the last year he was impotent. According to his wife and close relatives, his illness had changed both his character and intelligence.
His *physical examination* was essentially negative. Blood pressure 130/90 mm/Hg, pulse 72/min.
Neurological examination revealed nystagmus in the lateral gaze. Other cranial nerves were normal. Visual acuity was normal, fundi within the normal limits. Tendon reflexes were normal. Diadochokinesia was less skillfully done on the left side. Sensory nerves were intact. Abdominal reflexes: equal and normal. Achilles clonus and patellar clonus were elicited. Babinski signs, Rossolimo and Mendel-Bechterew were positive, bilaterally. Gait was slightly spastic, and he did not balance his left hand while walking. Overall there was a slowing of all movements.
Psychiatric evaluation revealed a decline in intelligence, memory problems, and lack of interest.
Clinical observation: On the ward, the patient lay passively on his bed. He did not change position often, and did not start conversations with other patients. He had no complaints, no questions, and did not ask for reading material. When asked to read, he was clearly skipping letters and odd words, and not understanding what he was reading. When asked questions, he gave monosyllabic answers. After a few days, when pneumo-encephalography was done, he was more animated, answered questions in full sentences; he was working on the hospital grounds. He tolerated the procedure well.
PEG showed a slightly dilated ventricular system.
EEG record was flat at low amplitude. During the hyperventilation, long waves appeared (for 5 to 6 s), at low amplitude.

In the *CSF* there was 56 mg% of protein. Globulin fraction was increased.
Additional tests, such as sedimentation rate, urinalysis, and CBC, were all within normal limits.

WOLOWSKA summarised:

All of the above symptoms – dementia, personality changes, and disseminated symptoms for cerebellar, pyramidal and extrapyramidal systems – together with a long history of boxing, prompted the diagnosis of boxer's encephalopathy. In confirmation of this diagnosis is the generalized brain atrophy which was found on pneumo-encephalography and EEG. Symptoms would support the diagnosis of disseminated sclerosis, but that diagnosis cannot explain the change of personality and decline of intelligence, lack of characteristic double vision, bladder disturbance, and changes in the fundus.

In contrast to a posttraumatic syndrome, where headache and dizziness seem to be prominent symptoms and a physical examination does not always show an abnormality, personality change and decline of intelligence are quite characteristic of boxer's encephalopathy, as is the complete lack of complaints from the patient. In this situation, the patient had no recognition that anything was wrong with him, and he showed no concern about his symptoms.

BOWMAN and BLAU (1960) presented the case history of a pugilist with traumatic encephalopathy. This classic description covers all salient features of the typical syndrome.

A 28 year old man was *admitted to the hospital for observation,* at the request of his wife. For about the past two years, she had noted that he was becoming gradually more childish in behavior and was occasionally depressed. He wanted to return to professional boxing, but was refused permission by the *Boxing Commissioner,* on account of his poor health. About two weeks previously, he began to cry, became suspicious, apprehensive and fearful. He expressed ideas that people were against him, wanted to poison him and spied on him; that his wife was not the woman he married originally, but that she was a detective, and he called his baby a 'midget baby.' He became excited, yelled in the streets about various people, and was finally arrested.
The past history was incomplete. He completed the seventh grade at school, at the age of 15 years, and then worked as a baker with his father. Later, he became interested in boxing and fought semi-professional fights. He was not successful, but received many severe beatings and was knocked out frequently. He finally had to quit this profession and worked as a laborer until his recent psychotic outbursts. He had married two years previously and had one child 15 months old; his married life was not happy because he was very irritable and fault-finding. He totally abstained from alcohol, and there was no history of venereal disease.
The *physical examination* showed him to be well developed, well nourished and of muscular build. The pupils were normal. There was an absence of the convergence reaction of the eyes. His facial expression was somewhat fixed and rigid, otherwise the neurological examination was normal. His blood pressure was 120/80 mm/Hg and the heart, lungs and abdomen were normal. The blood Wassermann test was negative.
In the *mental examination,* the patient was very dull and presented the superficial impression of a mental defective. At first he was very noisy, disturbed and assaultive. He spoke in a childish immature manner, was circumstantial, disjointed, and frequently lost the direction of his conversation. He appeared confused, was sometimes evasive, and occasionally blocked. His mood was one of indifference, interspersed with frequent silly giggling. He contradicted himself continually. There were some vague ideas of reference, persecution, and infidelity. He had no insight into his condition and was very confident about his boxing abilities; he said that people had accused him of being punch drunk, and that he still trained, by shadow boxing. His orientation was poor. He could not grasp simple concepts, and was able to perform simple calculations only. His memory was poor, especially for recent events.

The *diagnosis of chronic traumatic encephalopathy of pugilists* was made, and he was transferred after a short period of observation to a *state hospital.* A recent report (one and a half years later) indicated that he was unimproved. His usual indifferent dull attitude was occasionally disturbed by periods of excitement and assaultiveness; otherwise there were no significant changes.

SPILLANE (1962) published the case histories of five boxers:

Case 1: A painter and decorator, aged 45, was referred to the Department of Neurology by his practitioner in July, 1960. A former professional boxer since the age of 17, he had retired "because my friends said I was walking unsteady and throwing a leg out, and they said that was a sign of punch-drunk." He had started boxing at the age of 13 and said he had had about 350 bouts and had only once been knocked out. His 'peculiar' gait had developed in an insidious manner about 13 years ago, and for several years he was unaware of it, despite the comments of his wife and friends. He was said to throw out the left leg at times, and some thought he was drunk. He only gradually became aware of his unsteadiness about eight years ago. The symptom fluctuated and was aggravated by alcohol, but he was able to do his work and had never fallen. If he hurries, he catches his left foot on obstacles and frequently stumbles. Slight clumsiness of the fingers, as when tying shoelaces or fastening buttons, has developed in recent years.

The second symptom, again one which was noticed by others for some years before he became aware of it, was of slurring of speech. He found that he has to speak slowly and carefully if he wishes to be understood. The difficulty seems to consist solely in defective articulation; there have been no aphasic symptoms.

Both the unsteadiness and slurring of speech have gradually increased further symptoms made their appearance. There has been urgency and frequency of micturition, with occasional incontinence and bouts of blurring vision. The latter are infrequent and sometimes consist of transient altitudinal hemianopia, "I might be unable to see the top half of your head or the lower part of your body for a few minutes, or sometimes on and off for a day." He had seen several ophthalmologists, who had found no abnormality. There were no complaints of headache, vertigo, diplopia, numbness, or paraesthesiae, and his general health was satisfactory. He complained of dyspepsia and anorexia in 1955 and was investigated in a hospital out-patient department, where it was noted that "there is marked slurring of speech and unsteadiness of gait." There was no family history of neurological or mental disorder.

Examination: General physical examination showed no significant abnormality, except the hallmarks of his profession (broken nose and cauliflower ear). Blood pressure: 110/80 mm/Hg.

The *neurological signs* consisted of slurring dysarthria, a spastic-ataxic type of gait and an extensor plantar reflex (left). There was no ocular or optic abnormality; no nystagmus; optic disks, fundi, and visual fields were normal. Power in his limbs was good, although there was some distal weakness in the left lower limb. Finger and hand movements revealed slight clumsiness. All reflexes were brisk. There was no sensory loss.

He possessed a pleasant frank personality, and there was nothing to suggest any deterioration. He was rather slow, but there was no evidence of intellectual impairment.

Investigation in hospital revealed normal radiographs of skull, chest, and cervical spine. Blood Wassermann Reaction was negative. *Lumbar puncture* produced a fluid under normal pressure; protein 53 mg/100 ml; cells fewer that 5/cmm; and a normal Lange curve and negative Wassermann.

EEG was normal. There was a regular, normally responsive alpha rhythm arising from the occipital lobes, and hyperventilation produced no abnormality.

Lumbar PEG: The lateral ventricles were not enlarged or displaced; the third and fourth ventricles were well outlined and were of normal size and in the midline. The aqueduct was normal. There was no excess of air over the cerebral cortex. On standard antero-posterior and postero-anterior radiographs, the two lateral ventricles were confluent medially. The thin linear shadow produced by the septum pellucidum was absent. In the postero-antero view, the normal separation of the lateral ventricles was seen, but they were confluent anteriorly. An interaxial view confirmed the absence of the septum.

Progress: In the past two years, he has been kept under observation, and the pneumo-encephalogram was recently repeated. Dysarthria and an unsteady gait remain the two chief disabilities. He is euphoric, and psychometric testing indicated that he possesses average intelligence (IQ 106) with little evidence of intellectual deterioration. Further views undertaken on the second *PEG* confirmed the apparent total absence of the septum pellucidum. Antero-posterior tomograms clearly demonstrated this, while the lateral horizontal view, recommended by DYKE and DAVIDOFF (1935) when they reported their first example of absent septum pellucidum diagnosed by pneumo-encephalography, revealed the comma-shaped shadow that they described. It obviously represents the confluent portions of both lateral ventricles at the midline. Finally an antero-posterior view with the head in the horizontal position, and the patient lying on his side, showed that all the air collected in the uppermost ventricle. If a septum separated the ventricles, some of the air should remain trapped in the lower ventricle.

This patient was 45 years of age. His professional career lasted 18 years, from the age of 17 to 35. During this time he had 350 fights, but says he was knocked out on only one occasion. His initial symptom consisted of the insidious development of weakness and ataxia of the left leg. It seems to have been noticed by his friends in the later years of his boxing career and probably had something to do with his retirement. Slurred speech, urinary symptoms, and periodic blurring of vision have since appeared. Although he is slow and euphoric there was no significant evidence of intellectual deterioration; he has continued to work regularly and is not very disabled. However, his speech is very dysarthric, his gait is unsteady, and there is some weakness in the left lower limb with an extensor plantar reflex. There is no nystagmus, no optic or ocular abnormality; the abdominal reflexes are brisk and there is no sensory loss. The disorder has slowly and progressively increased, without remission or exacerbation, for 13 years. *EEG* and *CSF* are normal. The possibility of disseminated sclerosis naturally arose, but when air encephalograms showed that the septum pellucidum was absent, the question of etiology became more obscure.

Case 2: This patient, aged 52, was referred to the Department of Neurology by his practitioner in December, 1960. He retired from the ring when he was 35, after about 300 professional fights, and at that time the only disability he recalled was defective vision of the right eye, the result of injury in one of his fights. As in Case 1, he said he had been knocked out on only one occasion.

His doctor had referred him because of increasing weakness and lack of control of his right leg. The patient stated that this trouble had begun 12 years previously. He recalled a sudden weakness of his right leg while carrying a sack of coal. At first the symptom was intermittent, but over the years it has increased, and he has found that he catches the toes of the right foot on the ground as he walks. He stumbles, rarely falls, and there is no fluctuation of the complaint. During the past 12 years, there have been periods when he experienced paraesthesiae and numbness. His right arm is weak and clumsy at times. He finds difficulty in writing and holding a tea cup. Ten years ago he began to have difficulty in urination. There has been frequency, hesitancy, nocturia, occasional dribbling incontinence, and on one occasion eight years ago, a bout of retention requiring catheterization. There has been no prostatic enlargement. During the past three years his gait has been increasingly slow and unsteady, and he has often been thought to be drunk. He is abstemious, but has found that alcohol aggravates his ataxia. He works regularly and still feels strong. He has no headache, vertigo, or diplopia. He was not aware of any difficulty with speech, and had no complaints suggesting any alteration of personality or intellectual impairment. There was no family history of neurological or mental disorder.

Examination: As in Case 1, general physical examination showed no significant abnormality apart from a characteristically flattened nose and one 'cauliflower ear.' Blood pressure 140/90 mm/ Hg.

The *neurological signs* consisted of slurred speech, unsteady gait, moderate right-sided hemiparesis, and partial right optic atrophy. There was no dysphasia; he dragged his right foot when walking, and all movements were performed slowly and carefully. There was moderate ataxia on formal testing in all limbs, more marked on the right. There was no diplopia; ocular movements were normal, but visual acuity on the right was 6/24, and the optic disk was very pale. Both visual fields showed moderate peripheral contraction. Visual acuity on the left was 6/12. Pupillary reflexes were sluggish on the left and practically absent on the right. The right pupil was larger than the left. The remaining cranial nerves were normal. There was diffuse moderate weakness of the right arm and leg. The arm reflexes were normal and symmetrical. The right knee and ankle jerks were brisker than the left. Both plantar reflexes were flexor. Sensory testing was not very reliable, but there was probably some impairment of superficial sensation in the distal parts of the right limbs: vibration sensation was also impaired on the right side, but joint position sense was apparently completely absent in the right hand and foot.

He was *admitted to hospital*, where *roentgenograms* of his skull and spine were normal. There was slight pneumoconiosis. Wassermann reaction in blood and *CSF* was negative. *Lumbar puncture* revealed a fluid under normal pressure; protein 40 mg/100 ml; cells fewer than 5; normal Lange curve. His memory was unreliable.

Psychometric testing revealed moderate intellectual impairment.

EEG was normal. All activity was of low voltage but alpha rhythm appeared at 8-9 c/sec and was augmented slightly by over breathing.

Lumbar PEG: There was symmetrical enlargement of both lateral ventricles; the third and fourth ventricles were normal. Little air was seen over the cortex. The septum pellucidum could not be identified. On the antero-posterior projection, a faint linear shadow could be seen, which may have been the septum, but on the interaxial projection, which usually gives a good view of the septum, it could not be identified. In the postero-anterior view also, the septum could not be seen, but, unlike case 1, there was no shadow to indicate the confluence of the two lateral ventricles anteriorly.

Progress: He has been kept under observation during the past two years and air encephalography was recently repeated. He is still working, although under some difficulty; he is slower mentally. At times he is anxious or depressed, but on the whole his personality has not significantly been altered.

Psychometry showed that his level of intelligence was dull normal (IQ 81) and that there was significant intellectual deterioration. On the recent air encephalogram the septum could not be seen on the antero-posterior or interaxial views. Tomographic cuts, however, have shown that there is a cavum septi pellucidi with very thin and possibly defective walls; only the lower third of each lamina of the septum could be clearly visualized. In the postero-anterior view it appears as if the lateral ventricles are in communication but the confluence cannot be clearly identified as in case 1. With the patient lying on his side, an antero-posterior horizontal ray projection shows that no air remains in the lowermost ventricle. Lateral horizontal views were not satisfactory, as he fainted during the maneuvre and filling was inadequate; there was no suggestion of the comma-shaped shadow seen in case 1, which DYKE and DAVIDOFF (1935) explained is also present when there is a cavum septi pellucidi whose walls are very thin (so that they cannot be clearly visualized) and also defective, allowing communication between the two ventricles.

The second patient, aged 52, after a boxing career of 20 years during which he had 300 fights and was knocked out on only one occasion, retired at the the age of 35. At that time his only apparent disability was defective vision in the right eye due to injury to his optic nerve. Five years later, at the age of 40, he first experienced weakness of right leg. This has gradually increased and has been accompanied by unsteadiness of gait, slurring of speech, and urinary difficulties. There is moderate intellectual impairment and no significant moderation of personality, and the *EEG* and the *CSF* are normal. *PEG* in 1960 revealed symmetrical enlargement of the lateral ventricles, and the septum pellucidum could not be seen in any of the standard views. It was repeated in 1962, and with the aid of tomography and additional projections it was shown that there was a cavum septi pellucidi whose walls were thin and deficient. Air passed freely from one lateral ventricle to the other.

Case 3: This man, aged 69, retired from the ring at 32 and was found to be suffering from diabetes six years ago. He had had several hundred fights, from the age of 13. For about ten years it had been evident to his friends that his mental faculties were failing. His memory was seriously impaired, and there were episodes of gross amnesia, confusion, and disorientation. He was childish at times, had little insight, and was content and euphoric. He was nearly always smiling and optimistic. Short journeys often resulted in disorientation. On one occasion, while driving a car, he was involved in an accident and sustained a minor head injury, but was confused and disorientated for several days; he had complete amnesia regarding the accident itself.

In 1959 he was admitted to hospital with acute appendicitis, and it was then noted that he could not give a satisfactory history and was confused and disorientated.

Examination: He was found to be confused, disorientated and unsteady, and he complained of headache. Subsequent investigations showed that his diabetes was satisfactorily controlled on insulin (IZS, 60 units daily).

General physical examination showed that he was well preserved. Blood pressure 120/60 mm/Hg.

His speech was clear, but he was unsteady on his feet and stumbled frequently. His pupils were small but active; optic disks were normal. There was slight generalized tremor of his upper limbs at times, but power in all limbs was good. He was ataxic on formal testing; fell on the Romberg test; could not stand on one leg or walk heel-toe fashion. There was no significant reflex or sensory abnormality. Plantar responses were flexor. His peripheral and retinal arteries were not sclerotic.

On *admission,* he did not know what hospital he was in, did not know the day of the week or the year, and was generally childish and petulant in his behavior. He would sulk and complain if left unattended, and would smile and giggle when anyone approached his bedside. His memory of his boxing career was reasonably retained, but places and boxers he could rarely name. He was talkative and happy.

Roentgenograms of his chest showed no cardiac enlargement, although there was some unfolding of his aorta. *Roentgenograms of his skull* showed some calcification of the falx and of both internal carotid arteries. Blood urea was 31 mg/100 ml. No glycosuria. Blood sugar (fasting) 156 mg/100 ml. *Blood* and *CSF* and Wassermann reaction were negative. *Lumbar puncture* revealed a fluid under normal pressure; protein 32 mg/100 ml; cells fewer than 5/cmm; Lange curve normal.

EEG was abnormal. There was normally responsive alpha rhythm at 10 c/s, arising in the post-central areas with short intermittent runs of moderate voltage 2 c/s delta waves bilaterally, predominating in the pre-central leads. This abnormal activity was slightly increased by over breathing.

Lumbar PEG: There was marked symmetrical enlargement of both lateral ventricles with a few pools of air in dilated sulci over the frontal, parietal and temporal lobes. In addition, there was a well outlined cavum septi pellucidi.

The third patient, aged 69, was probably in good health until about ten years ago, but it is difficult to know when mental deterioration actually began. He had been retired from the ring for 30 years and his physical condition was good. He now has progressive dementia, his *EEG* is diffusely abnormal, and *PEG* demonstrates considerable cerebral atrophy and a cavum septi pellucidi.

Case 4: This man, aged 33, had started boxing at the age of thirteen. He said he obtained a professional license at 15 (giving his age as 16). He had had about 200 professional bouts, and retired when he was 28. He said he had won 150, lost 40 and had been knocked out on five occasions. He spent several years in the boxing booths and had 30 to 40 fights a day. He was a wild type but not unpleasant. He served in the Forces, enjoyed the life, had a considerable following, but resented discipline and deserted. He said he acquired and lost a fortune. He had many friends and possessed considerable charm, but his fighting instincts were a nuisance. There was no family history of mental or nervous disorder. He was a powerfully

built man, in excellent physical condition, without any outward abnormality. He admitted that his memory was failing, and that he had 'thumping headaches' when he worked in a bending condition, but he never sought medical advice. He felt his sense of balance was uncertain and that he was now intolerant of alcohol. His speech was quite normal, and I could find no abnormality on physical examination.
Blood Wassermann reaction was negative. *Lumbar puncture* revealed a fluid with raised protein content (60 mg/100 ml.); cells 3; *W.R.* negative; Lange curve normal.
EEG was normal.
Radiographs of skull were normal.
Lumbar PEG: This showed that the ventricular system was normal and a septum pellucidum was present; but it seemed to be absent posteriorly.
A few weeks later encephalography was repeated. On the second examination the septum was clearly seen on the antero-posterior view but not on the postero-anterior view, where the ventricles appeared to be confluent. On the intra-axial view the septum could be seen anteriorly, but not in its posterior part. An antero-posterior view, with the patient lying on his side, showed that all the air collected in the uppermost lateral ventricle; no air remained in the lower ventricle. In several antero-posterior erect tomograms, the septum could always be seen. Lateral horizontal views were normal, except for the presence of a dense shadow within that cast by the lateral ventricles, may conceivably result from a defect in the posterior part of the septum pellucidum. It was present in all horizontal projections in both encephalograms.

The fourth patient, aged 33, is in excellent physical health, there is no abnormality on neurological examination, but he is aggressive and violent, especially when he has been drinking. His EEG is normal, but his cerebrospinal fluid contained 60 mg. of protein per 100 ml. The septum pellucidum is present but apparently deficient posteriorly.

Case 5: A laborer aged 45 entered the Hospital of Neurology in 1962. He started to box when he was 10 and turned professional at 18. Until he retired at the age of 32 he had had approximately 300 fights in various countries. His wife thought that his ill-health began about that time. The first symptom noted was progressive slurring of speech. He had been drinking heavily for a few years, and there were bouts of depression. However, despite the alcoholism there were long periods of abstention, but his family and friends were puzzled because his speech remained slurred. Two pints of beer would often render him incomprehensible. His wife also noticed that nocturnal urinary incontinence, which she had formerly attributed to inebriation, persisted during his periods of sobriety. In the past ten years he has been a patient in a mental hospital on several occasions. During one such admission, about three years ago, it was observed that he was dragging his left leg when he walked and tended to hold his left arm flexed across his body. He had a perforated peptic ulcer in 1951 and a pleural effusion in 1955. In 1957 a partial gastrectomy was performed. At this time an enlarged liver was noted, but liver-function tests were within normal limits. There was no family history of mental or neurological disorder.
Examination: On admission to hospital he was cooperative, and his general physical condition was good. He had been working regularly at a steel mill since August, 1961, and his wife informed me that he had taken no alcohol since then. He was morose and depressed. He had a broken nose and cauliflower ear. His speech was thick and slurred, and his gait unsteady. His pupils were normal. There was nystagmus on right and left lateral gaze. Power in all limbs was good, reflexes were brisk and symmetrical, and the plantar responses were flexor. There was some diminution of sensation (tactile, pinprick, and vibration) in the left leg. Blood pressure 130/80 mm/ Hg. He was slow, his power of concentration and his memory for recent events were poor but he was demented. He was correctly orientated and his behavior was quite normal. A speech therapist formed the view that in addition to his gross dysarthria there was a slight executive dysphasia.
Psychometry revealed a dull normal level of intelligence (IQ 85) with no indication of intellectual deterioration. He attributed his slurred speech, unsteady gait, and alcoholism to boxing, and he thought "professional boxing should be abolished altogether." He recalled one episode during his boxing career of 24 hour amnesia. This included a journey of 50 miles (80 km) to a neighboring town, and eight round contest which he won on points, and the journey home.
Blood Wassermann reaction was negative.
Radiographs of skull were normal.
The *CSF* protein content was raised (60 mg/100 ml); cells 4; normal Lange curve; negative W. R.
EEG was normal.
Lumbar PEG was normal. There was no evidence of myxedema. Liver-function tests were normal, and his liver was not enlarged.
He died suddenly at work of acute myocardial infarction two months after leaving hospital. At *necropsy* there were foci of softening on the left side in the cortex of the parietal temporal, and cerebellar lobes. There was a cystic scar in the anterior limb of the left internal capsule. The meninges were thickened at several sites and the cerebral arteries showed slight to moderate atheroma. The pathological picture did not resemble the restricted form of cerebellar cortical degeneration of alcoholics described by VICTOR et al. (1959) in which the main atrophy is largely confined to the vermis. The lesions were probably the result of vascular disturbance, but what part trauma played could not be determined.

The fifth patient, an alcoholic, had permanent dysarthria and ataxia. At necropsy the only intracranial lesions consisted of small foci of softening in the cortex of the parietal, temporal, and left cerebellar lobes and the left internal capsule.

SERCL and JAROS (1962) studied the mechanisms of cerebral concussion in a series of 1.582 boxers. In total, they observed 1.165 boxing matches. They saw most of the important boxing events, e. g., the *First and Second Spartakiade in Boxing, Army and Air Force Spartakiades in Boxing*, most of the regional boxing matches and, in 1957, the whole *European Championship in Boxing in Prague, Czechoslovakia*, and in 1961, the whole *European Championship in Boxing in Belgrade*, Yugoslavia. In the course of all these boxing matches, the authors examined the boxers neurologically both before and immediately after the fights. During the fights they observed the technique of boxing, and recorded the most important hits bringing about a 'knock out' (k. o.) or groggy state of the fighter. The neurological findings were then compared with the mechanisms of the most important hits. In some cases, mechanisms of the injury were verified by photographs, when necessary. The immediate clinical pictures to the nervous system of boxers were often variegated, as can be seen from Table 28.

Mechanisms of cerebral concussion in boxers are very complex, and in this report SERCL and JAROS wish only to outline the types of the most important mechanisms of lesions of the nervous system of the boxers. We shall not go into details of the single groups.

TABLE 28
Acute findings during 1.165 matches in 427 boxers. (From SERCL, JAROS, 1962).

Reversible within several minutes	336
More permanent up to 24 hours	91
Clinical Pictures	
Derangement of muscular tone	380
Unconsciousness of various grades	112
General muscular weakness	142
Pyramidal symptoms	253
Cerebellar and vestibular signs	319
Extrapyramidal signs	191
Lesions of cranial nerves	7

In the series of SERCL and JAROS of 1.582 boxers, signs of chronic encephalopathy were found in 148 i. e., more than 9%. Table 29 shows the signs of chronic boxer's encephalopathy in 148 boxers.

In evaluating the neurological signs of these chronic findings in boxers, SERCL and JAROS were aware of the fact that mere enumeration of neurological signs may distort both the results and our conclusions. Therefore they compared the findings in boxers with those from a control group of one thousand non-boxing individuals of the same age.

Statistical evaluation of the results by means of the x2 test showed the following results. When SERCL and JAROS considered both series as a whole (Table 30), the pathologic clinical findings prevalent in the boxers was statistically significant, as compared with the control group, with $p < 0.001$.

Individual groups of symptoms were analyzed as well. Extrapyramidal disorders were found more often in boxers than non-boxers, and the difference in frequency was statistically significant, with $p < 0.001$. Vestibulo-cerebellar signs prevailed in boxers, as compared with the control group, but with statistically insignificant frequency, with $p > 0.05$. A number of other miscellaneous signs prevailed in the boxers, as compared with the control group, once again the frequency in boxers was statistically significant, with $p < 0.025$.

SERCL and JAROS also evaluated electro-encephalographic findings in a smaller group of 56 boxers. Analysis of this group shows that boxers have a statistically significant number of pathologic EEG variations, in comparison with results for the control group, with $p > 0.05$.

Based on their studies, SERCL and JAROS concluded: "The statistical evaluation of both the clinical and electro-encephalographic findings shows that pathologic findings in the group of boxers mostly persist, as compared with the control group, the statistical significance ranging from a minimal one to very important significance."

MAWDSLEY and FERGUSON (1963) presented ten case histories:

TABLE 29
Comparison of neurological symptoms. (From SERCL, JAROS, 1962).

Chronic findings in 148 boxers out of 1.582 boxers		*Control group of 1.000 persons aged 18-30 who never boxed*	
Syndromes *Symptoms*		*Syndromes* *Symptoms*	
Extrapyramidal 77		Extrapyramidal 11	
Masking of face	25	Increased nasopalpebral reflex	10
Increased nasopalpebral reflex	45	Increased postular reflexes	6
Increased postular reflexes	45	Puusep's reflex	4
Puusep's reflex	14	Disorder of writing	2
Disorder of writing	29	Decreased synkinesis	8
Decreased synkinesis	6		
Patellar reflex with rapid stop	15		
Vestibulocerebellar 18		Vestibulocerebellar 9	
Nystagmus (1st degree)	15	Nystagmus (1st degree)	7
Hypermetria in upper extr.	11	Hypermetria in upper extr.	5
Hypermetria in lower extr.	8	Hautant sign	6
Hautant sign	9	Romberg sign	5
Romberg sign	20		
Other 53		Other 16	
Abdominal areflexia	10	Abdominal areflexia	6
Pyramidal signs	16	Pyramidal signs	5
Deep reflex asymmetry	35	Deep reflex asymmetry	7
Intellectual changes	5	Intellectual changes	6
Changes of character	3	Changes of character	8
Changes of behaviour	3		

Case 1: A man, now aged 54, began to box professionally at the age of 14. He had 200 fights as a flyweight in a career that lasted until he was 32. He was knocked out four times. Towards the end of his career he was badly beaten on several occasions, and after one bout had a period of amnesia lasting 12 hours. From the age of 30, he noticed slurring of his speech, difficulty in sleeping at night, and increasing lethargy and drowsiness during the day. Shortly after he had given up boxing he became unsteady in his gait and clumsy with his hands; fastening laces and buttons and writing became difficult. His speech and gait slowly deteriorated.

In his youth he had been aggressive, and throughout his boxing career was a heavy drinker. He is still liable to have outbursts of rage and violence but drinks little alcohol, because small amounts aggravate his ataxia. When he was 34 he was discharged from the Army on psychiatric grounds and he has not worked since. Between the ages of 37 and 44 he was admitted to mental hospitals on three occasions after episodes

TABLE 30
Survey of statistical evaluation of findings in series of 1.582 boxers and control group of 1.000 subjects. (From SERCL, JAROS, 1962).

Comparison				*Results of statistical evaluation*
Series of 1.582 boxers		Control series of 1.000 subjects		Pathologic clinical signs in boxers in general prevail statistically significantly with $p < 0,001$ /31,8/
Extrapyramidal signs				Prevalence in boxers as compared with control group statistically significantly with $p < 0,001$ /25,13/
Vestibulocerebellar signs				Prevalence in boxers as compared with control group statistically insignificantly with $p > 0,05$ /0,134/
Other signs				Prevalence in boxers as compared with control group statistically significantly with $p < 0,025$ /6,6/
EEG				In boxers pathologic EEG findings prevail as compared with control group statistically insignificantly with $p > 0,05$ /0,242/
in 56 boxers		in 30 controls		
normal	18	normal	12	
pathologic	38	pathologic	18	
Evaluated by means of χ^2 test.				

of violence. His family noted a progressive deterioration in his memory in the past 20 years. His wife commented, too, on a falling off in his social habits, and on the slovenliness of his dress.

The *clinical examination* revealed a broken nose, a left cauliflower ear, and old metacarpal fractures. There was clinical and psychometric evidence of dementia, with a distinct defect of retentive memory. His speech was slurred; his gait was ataxic, particularly when he turned. His pupils were small; there was restriction of upward gaze and fine nystagmus on lateral deviation of the eyes. Muscular power was excellent, but he showed slight intention tremor in the finger-nose test and was ataxic in the heel-knee test. Fine finger movements were impaired. The tendon reflexes were more brisk on the left side and the left plantar response was equivocally extensor. The *cerebrospinal fluid* was normal.

The *EEG* was within normal limits.

Lumbar PEG showed generalized dilatation of the ventricular system. There was no excess of air over the cerebral cortex, but air outlined the cerebellar folia. There was a cavum septi pellucidi as in Case 5, the outline of the two leaves being very faint. In the antero-posterior view, all the air rapidly collected in the uppermost lateral ventricle when the patient's head was horizontal.

Case 2: A man of 58, had about 500 professional bouts between the ages of 16 and 36 in fly, bantam, and featherweight divisions; in addition, he estimated that he fought 100 times in local booths. At his peak he would often fight three or four times a week and in the booths would take on five opponents in an evening. He was never knocked out but received many beatings, and fights were stopped because of his helpless state. He retired from the ring because his legs were becoming unsteady. People accused him of being drunk, though he is a lifelong teetotaler. The unsteadiness of his gait slowly worsened. When he was about 38, his family and his friends commented on the slurring of his speech, and two years later he noticed tremor of his hands. His disability gradually increased though he managed to work as a road sweeper until ten years ago. Since then he has been unemployed. He has had frequent falls during the past five years and now cannot walk unaided.

The *clinical examination* showed a broken nose, cauliflower ears, and his face was badly marked by the scars of old lacerations. A pleasant garrulous man, clinically showed a gross defect of retentive memory which was confirmed on psychometric testing. His speech was dysarthric and monotonous, his gait distinctly ataxic. The pupils were small; there was restriction of upward gaze, and fine nystagmus on full lateral deviation of the eyes. His facial expression was fixed, and he showed a constant, coarse tremor of the arms and hands, which was more obvious on the right. Tone was increased in his arms and legs. Finger movements were clumsy. Plantar responses were flexor.

The *CSF* was normal.

The *EEG was* abnormal, frequent low and medium voltage theta waves at 4-7 c/s were present symmetrically in all areas. Occasional low to medium voltage delta waves, at 2-3.5 c/s, were seen in temporal leads, symmetrically.

Lumbar PEG showed definite symmetrical dilatation of the lateral ventricles. There was but little air collected over the cerebral or cerebellar hemispheres. There was a cavum septi pellucidi; the two leaves were widely separated and faintly defined. An antero-posterior horizontal view taken with the patient on his side showed all the air in the upper ventricle.

Case 3: A man of 69, boxed as a flyweight from the age of 16. He had about 100 fights; he lost 40, and was knocked out four times. In addition he was a sparring partner to two local champions and said he took a lot of punishment. He gave up boxing, when aged 24, after a blow that impaired the vision in his left eye. At the age of 50, he and his family noticed a deterioration in his memory. This defect was severe for recent events and progressed. On occasion he forgot where to get off the bus on his way to work and would forget in mid-sentence what he wanted to say. He became increasingly depressed and irritable, was apt to fall asleep during the day, and to have occipital, throbbing headaches almost daily.

The *clinical examination* revealed that the patient was a reserved, depressed man with unscarred face and hands. The peripheral vessels showed evidence of arteriosclerosis. Blood pressure was 140/80 mm Hg. Clinical testing showed grossly defective memory, which was confirmed by psychometry. Speech and gait were normal. The left optic disc was abnormally pale. Visual acuity in the left eye was 6/60. There was sustained nystagmus on lateral deviation of the eyes. Both plantar responses were equivocally extensor.

The *CSF was* normal.

The *EEG was* within normal limits.

Lumbar PEG: the lateral ventricles were not dilated; the septum pellucidum was normal; the cerebral sulci were widened, suggesting some cortical atrophy.

Case 4: A man of 57, fought as a flyweight 204 times between the ages of 15 and 24, also acted as a sparring partner and had over a 1.000 bouts in booths. He does not remember being knocked out, though several fights were stopped, to save him from further punishment. He stopped boxing at the age of 24 because he had a grand-mal convulsion. For some years thereafter, he had fits at intervals of a few days; their frequency gradually decreased, but he still has at least one major attack each month. He lost many jobs after fits at work and has been unemployed for the past two years. On four occasions during the past five years, he experienced paranoid hallucinatory states that were accompanied by clouding of consciousness and usually followed a series of grand-mal fits. He became violent at such times, and was admitted to a mental hospital twice. These psychotic states lasted three or four days, and recovery was spontaneous and complete on each occasion. In the past three years, his wife noticed a deterioration in his memory. There was no family history of epilepsy nor of mental illness. He never drank more than an occasional glass of beer and has been a total abstainer for three years.

The *clinical examination* revealed a broken nose. He was pleasant and rather fatuously cheerful. A gross defect of retentive memory was apparent clinically, and *psychometric testing* showed evidence of intellectual deterioration.

The only *abnormal neurological signs* were fine nystagmus on lateral deviation of the eyes and severe impairment of fine hand movements.

The *CSF* contained 50 mg of protein per 100 ml.

The *EEG* was beyond normal limits. Low-voltage theta waves at 4-7 c/s were seen at random in all areas, but were more prominent on the right side and were occasionally focal in the right anterior and mid-temporal region.

Lumbar PEG: The lateral ventricles were greatly dilated. There was no excess of air over the cerebrum or cerebellum. There was a clearly defined cavum septi pellucidi. Antero-posterior views taken with the patient lying on his side with the head horizontal showed all the air collected in the upper lateral ventricle.

Case 5: A 55 year old man fought 100 times as a lightweight professional boxer between the ages of 15 and 21. Thereafter he had many other fights in local booths and acted as a sparring partner until the age of 30. He was knocked out once only, but was beaten often. He remembered, with pride, his ability to survive heavy punishment. When he was 30, tremor of his right hand became manifest, particularly when he lifted a cup to his lips. The tremor slowly increased in amplitude, and his writing steadily deteriorated into illegibility. When he was about 35 his gait became unsteady and his speech became slurred. These symptoms have progressed. Though he was always a querulous person, quick to take offence, his domestic life was fairly placid until about 15 years ago. He then began to show defects of memory and developed paranoid delusions. He was twice admitted to psychiatric units after outbursts of violence.

The *clinical examination* showed a heavily built, plethoric man with a broken nose, old metacarpal fractures, and cauliflower ears. He was morose and irritable. His conversations revolved around his wife's supposed infidelity, his family's enmity, and the ill treatment he had received. There was clinical and psychometric evidence of dementia and gross memory defect. He was grossly dysarthric; his gait was spastic and ataxic; but his muscular development and power were exceptionally good. There was cogwheel rigidity of all four limbs, more obvious on the right side. He had a constant 'pill-rolling' tremor of both hands at rest and gross intention tremor on purposeful movements of the right arm. There was ataxia of the left arm and in the heel-shin tests. Both plantar responses were extensor.

The *CSF* was normal.

The *EEG* showed slight abnormalities: low-medium voltage theta waves at 4-7 c/s in all areas, notably in occipital leads.

Lumbar PEG: The lateral ventricles were abnormally dilated. The cerebral sulci were widened and there was an excess of air in the subarachnoid space overlying the cerebral hemispheres. There was a cavum septi pellucidi with two clearly defined leaves, well separated. With the head horizontal, the patient on his side, antero-posterior view showed air only in the uppermost ventricle.

Case 6: A man of 55, boxed as an amateur between the ages of 12 and 16. He had 90 bouts during this time and later fought 320 times as a professional flyweight, retiring when he was 30. He was knocked out 12 times and often was badly beaten; on a dozen occasions he had long-continued amnesia after fights. At the age of 33, his left arm and hand became weak and unsteady. Five years ago his wife noticed slurring of his speech and unsteadiness of his gait. His writing slowly deteriorated. During the past four years, on seven occasions he lost consciousness for 30 to 45 minutes. He has been unemployed for a year.

The *clinical examination* showed clinical and psychometric evidence of intellectual impairment and memory defect. His facial expression was 'fixed' and Parkinsonian, and his speech was dysarthric. He had an ataxic gait. There was sustained nystagmus on lateral deviation of the eyes and restriction of upward gaze. Tone was increased in the left arm and leg. Fine movements of the left hand were clumsy, and there was gross intention tremor on purposeful movement of the left arm. There were similar, but more slight, signs on the right. There was ataxia in the heel-knee test. Tendon reflexes were more brisk on the left side. Both plantar responses were flexor.

The *CSF* was normal.

The *EEG* showed slight abnormalitites, low to medium voltage theta waves at 4-7 c/s were seen at random in all areas.

Lumbar PEG showed slight dilatation of the lateral ventricles. The cerebral sulci were coarse and wide, and there was an excess of air of the cerebral hemispheres. There was a cavum septi pellucidi; the two leaves were clearly seen and were not widely separated. Air was seen in the lower ventricle when the head was placed horizontally and on one side in the antero-posterior view.

Case 7: A man of 63, fought as a welterweight for ten years. He began as a boy of 16 and said he had boxed nightly until he was 26. He was knocked out once. When he was about 38, a nursing sister drew his attention to his staggering gait. Thereafter many people called him punch-drunk, though he was conscious of little disability until his early fifties. He then noticed unsteadiness of his gait and heaviness of his legs, and found he was unable to run or to walk quickly. He became aware that his speech was slurred, though his wife had commented on this years before. When he was about 50, he lost consciousness for a few minutes. He had no warning and fell to the ground in the street. This was an isolated attack. His unsteadiness of gait and his difficulty with speech slowly worsened. He is now unable to walk unaided. In the past 10 years,

he and his wife had noted a deterioration in his memory. He has never drunk more than an occasional glass of beer.
The *clinical examination* revealed a dysarthric speech, his gait was grossly ataxic, and he was unable to walk without support. Upward gaze was restricted. He had nystagmus on full lateral deviation of the eyes, slower and coarser in amplitude on looking to the right. There was intention tremor of the right hand in the finger-nose test and gross ataxia in each leg when performing the heel-knee tests. An increase in tone of 'clasp-knife' type was found in both legs. The tendon reflexes were exaggerated, particularly on the right side. Both plantar responses were extensor.
Lumbar PEG: A large air-filled subarachnoid space around the cerebellum suggesting advanced cerebellar atrophy was the only abnormal finding.
Case 8: A man of 59 followed his father and brothers into the ring as a professional flyweight, and fought about 600 contests between the ages of 14 and 35. He was knocked out five times, once for an hour. On several occasions he was dazed for hours after being beaten. He gave up boxing because of deterioration in his vision. His only complaint was of poor eyesight. He denied any other disability and it was impossible to obtain a detailed history from him. He had always been a heavy beer drinker, was involved in many brawls, and had convictions for assault. Because of his violent behavior he was admitted to a mental hospital, when he was 40. His wife left him three years ago because of his drunkenness and violence. He has not worked for three years. Whilst unemployed he lived as a tramp, begging drinks and sleeping in public lavatories, until he was admitted to a hospital 18 months ago. During his stay there he made himself a nuisance. He frequently escaped to get drunk and often assaulted his fellow-inmates.
The *clinical examination* revealed that the patient's face was battered and showed the scars of old lacerations and hypertrophied superciliary ridges. His nose was broken; he had a left cauliflower ear, and multiple old metacarpal fractures. Constantly chatting or singing, he was impervious to discipline. There was gross impairment of intellect and memory. His facial expression was fixed. His speech was slow and dysarthric. He walked with an ataxic gait on a broad base, taking short shuffling steps. The optic fundi showed macular degeneration. *Visual acuity* in the right eye was 6/36. The left eye was amblyopic. The pupils were small and reacted sluggishly to light. There was a defect in conjugate upward deviation and in convergence of the eyes. The jaw jerk was brisk. There was rigidity of all four limbs, and a coarse tremor of the hands. Fine movements of the fingers were impaired, and there was slight intention tremor in both hands. The tendon reflexes were brisk, particularly in the arms. Hoffman's sign was positive bilaterally. Both plantar responses were equivocally extensor. The abdominal reflexes were unobtainable.
The *CSF* contained 70 mg of protein per 100 ml.
The *EEG was* abnormal – frequent low to medium voltage theta waves at 4-7 c/s were present in all areas and were variably asymmetrical. There were frequent low to medium voltage delta waves in all areas, more prominently on the left side, in temporal leads.
Lumbar PEG: The lateral ventricles showed considerable symmetrical dilatation. There were abnormal collections of air over the cerebral hemispheres and over the cerebellum. The pattern of the cerebral sulci was coarsened. There was a cavum septi pellucidi, the leaves being faintly defined and well separated. Antero-posterior views, with the head horizontal and on the side, showed no air remaining in the lower lateral ventricle.
Case 9: A metallurgist of 33, had a successful career as a lightweight amateur boxer. He fought about 80 times between the ages of 14 and 26 and was never knocked out, though one of his fights was stopped when he was concussed, and this bout was followed by a period of amnesia lasting six hours. When he was 19 he noticed that his speech, previously precise and clipped, was becoming slurred. This was apparent when he spoke quickly and when he was tired. At this time, too, his memory deteriorated and his handwriting became untidy. In the next two years these symptoms persisted, and he attributed his failure in examinations to the falling off in his memory. Fearing he was becoming punch-drunk, he stopped boxing regularly, and fought only occasionally in the next five years. About the time he stopped boxing completely, he began to have almost constant occipital, throbbing headaches, which latterly came on at weekly intervals. He remained conscious of a speech difficulty, particularly on the telephone, and believed that his memory was faulty. His only neurological abnormality was slight but definite slurring dysarthria.
The *CSF* contained 55 mg of protein per 100 ml. Pandy test showed a faint opalescence.
The *EEG* was probably within normal limits. Occasional low-voltage theta waves at 4-7 c/s were seen at a random in all areas.
Lumbar PEG showed no abnormality.
Case 10: This patient died recently at the age of 51. He began his career as a boy of 14, fighting nightly in local booths. He had 300 bouts as a professional bantam-weight. He retired when he was 30. He was never knocked out, but often lost on points later in his career. At his peak he was very successful, and between the ages of 22 and 26 fought an average of 14 top-class contests each year. When he retired, many people taunted him with being 'punchy.' He had a set, expressionless face and slurring of his speech, and was noticeably clumsy in the use of his hands. His wife noticed a progressive mental impairment from the age of 35. His memory deteriorated, and he lost his last jobs when he was 42, because he could not remember his duties and was unreliable in keeping to a schedule. During the last ten years of his life he spent most of the day sitting apathetically at home. He became slovenly in his dress and habits, wearing the same clothes and underclothes

for weeks on end, unless he was badgered to change. He took no interest in his family's welfare. The lethargy was in striking contrast to the liveliness and alertness described by people who knew him in his twenties. Never a heavy drinker, he was easily rendered ataxic by small amounts of alcohol. He was an amiable man and had no record of aggressive or violent behavior. He was partially impotent from the age of 35, and this was absolute after the age of 47. On three occasions, he lost consciousness for a few seconds. He died from a squamous-cell carcinoma of the floor of his mouth.

The *clinical examination* showed a broken nose and old fractured metacarpals. There was clinical evidence of dementia. He had pronounced slurring dysarthria and an expressionless Parkinsonian like-face. The tendon reflexes were abnormally brisk. The abdominal reflexes could not be elicited. The plantar responses were flexor.

The *EEG was* probably within normal limits, showing intermittent low and medium voltage theta activity at 5-7 c/s in all areas symmetrically.

Macroscopic examination of brain. The *brain weighed* only 1.050 g and there was gross diffuse cortical atrophy, most evident in the frontal lobes. The lateral ventricles were dilated. There was a cavum septi pellucidi. The cavum shows a coronal section through the cerebral hemispheres, cutting the anterior horns of the lateral ventricles in front of the intra-ventricular foramina. The two laminae of the septum are thick and intact here. They remain imperforate anteriorly, where they are attached to the genu and rostrum of the corpus callosum. In the central part of the lateral ventricles just above the foramina of Monro, these two leaves show multiple perforations along their superior margins, near their attachment to the trunk of the corpus callosum. The lattice-like appearance of these perforations is seen in the oblique view. Posteriorly, towards the splenium, the lamellae have almost disappeared. All that remains of their backward continuation are a few strands, arranged in two columns.

MAWDSLEY and FERGUSON summarised that all their patients had been boxers and they showed clinical evidence of neurological disease. The authors stress the point that they did not uncritically assume that their lesions were due to their earlier trade. In each case they looked for evidence of a coindidental lesion unconnected with boxing. A possible alternative diagnosis could be suggested, according to MADSLEY and FERGUSON only in case 3. Here a gap of 25 years of normal life separated the onset of dementia from the end of the patient's boxing career. He is atherosclerotic and his neurological deficit could at least partly be due to cerebrovascular disease.

In the remainder of the patients in the series of MAWDSLEY and FERGUSON progressive disabilities began when they were relatively young, usually at the end of their boxing careers or shortly after retirement. The clinical picture was according to MAWDSLEY and FERGUSON often distinctive and peculiar to this syndrome and was typified in cases 1, 2, 5, 6, and 8. These patients presented an mixture of striatal and cerebellar dysfunction rarely seen in any other disease – though similar signs have been recorded after closed head injuries (KREMER et al. 1947).

Dysarthria, ataxia, and tremors were the principal feature. Evidence of pyramidal tract affection, usually slight, was sometimes present. In case 7 of this series cerebellar defects predominated, striatal features were slight and the addition of severe pyramidal signs gave a superficial resemblance to disseminated sclerosis. In contrast, case 10 was an example of the parkinsonian end of the clinical spectrum.

Epilepsy was in the series of MAWDSLEY and FERGUSON the presenting feature in case 4. The date of onset of the fits and the absence of a family history of convulsions suggested to the authors that this was an example of posttraumatic epilepsy coming on after boxing injuries. There were histories of occasional fits in cases 6, 7, and 10.

There was clinical evidence of dementia in all but one (case 9) of the patients. In seven of these, intellectual deteriorations confirmed by psychometric testing. Personality disorders were manifest in some patients, and three were impotent.

The clinical features shown by the patients of MAWDSLEY and FERGUSON were similar to those described in boxers by earlier writers, notably CRITCHLEY (1957). The authors noted, it seems certain that some boxers sustain chronic neurological lesions as a result of boxing.

Nine of the ten patients of MAWDSLEY and FERGUSON fought in the lighter weight divisions. The authors continue: "It could be that small men are more liable to suffer neurological damage in the ring."

We do not agree with this opinion. The fact that

ex-boxers in the lighter weight divisions are able to land heavy blows which lead to boxer's encephalopathy. We know from analysing the literature that brain damage is more severe in boxers of the light heavyweight and heavyweight divisions, but that the intensity of blows by boxers of the lighter divisions are severe enough to produce permanent brain damage.

We agree with MAWDSLEY and FERGUSON that less skillful boxers are often stated to be more likely to become 'punch drunk.' In the group of the authors the most severely affected patients (cases 2, 5, and 8) had also been less successful. Case 2, the most disabled, was called 'a good curtain-raiser' and was admired for his 'gameness' – i. e. ability to absorb punishment and still remain standing. Champions, however, we agree fully, are not immune, and two included in this series are badly affected. The same findings were reported by CRITCHLEY (1957) and CORSELLIS et al. (1973).

All these men in the series of MAWDSLEY and FERGUSON began to box regularly in their early teens. This early exposure to injury may predispose to later neurological lesions. We fully agree with the statement of the authors that the severity of cerebral affection probably increases with the number of bouts fought.

The *CSF* in the series of MAWDSLEY and FERGUSON was under normal pressure in the nine patients on whom lumbar puncture were performed. In three the fluid contained a slight excess of protein (50 – 70 mg per 100 ml).

EEG recordings from five patients showed an excess of their theta activity, and delta waves were found in two. The EEG abnormalities were not striking, and bore no correlation to the degree of clinical affection, similar non-specific changes were recorded by other authors, as CRITCHLEY (1957). The authors noted that the EEG may be of value in assessing the severity of cerebral injury immediately after a fight. It is of little help in the diagnosis of chronic lesions in boxers.

Nine of the patients in this series of ten had lumbar air-encephalograms, which were abnormal in eight.

In six cases radiographs showed symmetrical dilatation of the lateral ventricles. In three of these the cerebral sulci seemed to be widened, and an excess of air overlay the cerebral hemispheres. Large air-filled spaces also surrounded the cerebellum in two of those three cases.

Air studies of two patients revealed normal ventricular systems; but in one of these cases (case 3) the pattern formed by the cerebral sulci was coarse, and in other pools of air outlined the cerebellar folia (case 7). MAWDSLEY and FERGUSON infered that boxing injuries probably caused the atrophy seen in the brain of case 10 and disclosed by air studies in eight of the other nine patients. Air-encephalograms were normal in an amateur boxer (case 9) whose neurological disability was slight. Radiographs in case 7 showed atrophy of the cerebellum alone, and cerebellar signs were the most prominent clinical features. "There is a correlation between the degree and distribution of cerebral atrophy, demonstrated radiographically, and the severity and type of neurological affection, evidenced clinically."

MAWDSLEY and FERGUSON noted that in six air-encephalograms a cavum septi pellucidi was seen. The degree of separation of the two septal leaves and their radiological definition varied.

In cases 1 and 8 the outlines of the leaves were so faint, that at first sight the septum appeared to be absent. In case 5 the leaves were well defined but thin, and in case 2, though two lamellae were clearly visible, they were interrupted along their length in the antero-posterior view. Case 6 showed relatively thick leaves which were not widely separated, and seemed to be intact. In the remaining five cases the leaves, as well as being separated, were probably perforated, since air passed freely from one lateral ventricle to the other and the 'comma sign' of DYKE and DAVIDOFF (1935) could be seen in lateral views.

MAWDSLEY and FERGUSON noted that a cavum septi pellucidi is not a recognized sequel of acute head injury, nor is it a feature of cerebral atrophy. The authors continue: "In the absence of internal hydrocephalus a cavum septi pellucidi is an uncommon finding and its frequency in ex-boxers must be significant. The conclusion that boxing injuries are responsible for septal abnormalities seems reasonable. We believe that dilatation of the cavum precedes the occasional disappearence of the septum. Changes in the septum pellucidum probably cause no neurological deficit. Radiological demonstration of a cavum, by implicating boxing as its probable cause, is helpful diagnostically."

MAWDSLEY and FERGUSON concluded that

some boxers sustain chronic neurological damage as a result of repeated head injuries in the ring. Clinically these men present a picture of dementia and impairment of motor function, with epilepsy as a less common feature. Boxers so affected are considerably disabled. Eight of the 10 studied had been unemployed for years.

MAWDSLEY and FERGUSON wrote: "The clinical signs of neurological disease are supplemented by the findings on air-encephalography of brain atrophy and abnormalities of the septum pellucidum. The evidence overwhelms legitimate doubts about coincidental causes of this syndrome. Cerebrovascular lesions in ex-boxers of middle age could possibly aggravate pre-existing traumatic damage."

MAWDSLEY and FERGUSON wrote in an addendum that they have examined two more boxers, one showing the typical syndrome of extrapyramidal and cerebellar features, and the other complaining of chronic headache, but with few physical signs. In each, air-encephalography revealed a cavum septi pellucidi.

HUSZAR and Edith KÖRNYEY (1965) published the following case history:

Thirty year old mechanic. Patient began boxing at age 15, active career for 12 years in welterweight or heavier class. About three years before admission to hospital, before the end of his active career, he realized especially when doing delicate work, that he had reduced dexterity in his right hand. According to his statements, the strength of his hand had not weakened, but certain movements were difficult to perform and disturbed by grotesque movements. When trying to move his right hand, he frequently noted a movement of his head to the right. Reduced intensities of complaints, but there was no doubt of the progression of this process. Movements of the head to the right were frequently accompanied by twitches of facial muscle. Increased perspiration, marked forgetfulness, speech clumsy but he reported that this had been so since childhood.

Neurological examination revealed an increase of reflexes on the right side, intentional innervation of the hand, movements of athetotic character occurred, such as head jerks to the right and both halves of the face. Tic-like hyperkinesia. Movements were markedly increased by psychic activity. They were reduced by fixation of head and rest. They ceased completely during sleep. Neurological examination otherwise unremarkable.

Psychopathological: Slowing of thinking, despite his subjective complaints slightly euphoric. Marked disturbance of memory. Repeated examination of *CSF* negative.

EEG frequently repeated and revealed unstable activity without specific or focal disturbances.

PEG revealed moderately dilated lateral ventricles, enlarged third ventricle. Signs compatible with cerebral atrophy (Fig. 227).

A Wilson disease could be excluded by the authors. The amount of copper in blood and urine was normal. The patient had not worked with manganese. The amount of lead in blood and urine were normal.

The authors summarized that a hyperkinetic disturbance of motion existed, of which the etiology was a 12-year boxing activity.

ISHERWOOD et al. (1966) had the opportunity to carry out encephalographic studies on 16 exboxers, all of whom had evidence of neurological disease.

The clinical features of some of these patients were the subject of discussion in a previous communication by MAWDSLEY and FERGUSON (1963).

The main purpose of this paper was to classify the radiological changes that have been observed in association with this type of traumatic encepha-

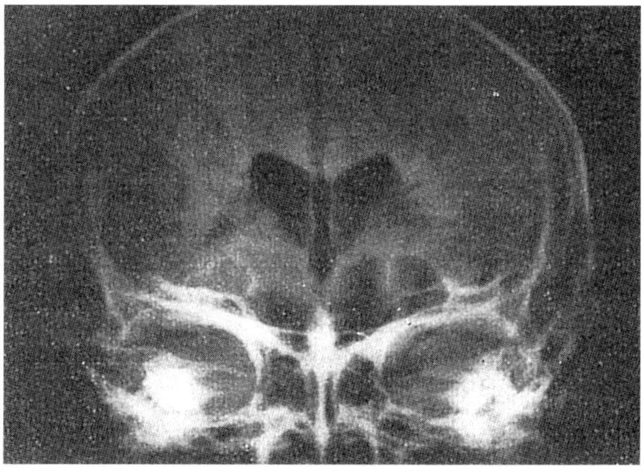

Fig. 227. The pneumencephalogram shows moderately enlarged lateral ventricles. In the midline an enlarged cavum septi pellucidi is visible. The brain reveals furthermore moderate atrophy of the cerebral cortex and the basal cistern. (From HUSZAR, and KÖRNYEY, 1965).

lopathy and to assess their significance and diagnostic value.

On the sixteen patients three had normal encephalograms, whilst the remaining thirteen had a variety of abnormalities (compare Table 31). The abnormal findings were: (1) *Cavum septi pellucidi* in nine cases; (2) *evidence of brain atrophy (generalized ventricular enlargement)* in eleven cases; *cerebral cortical atrophy* in three cases; *cerebellar atrophy* in four cases); and (3) *enlargement of the cistern of the lamina terminalis* in four cases.

The authors found the following alterations in the cavum septi pellucidi: the septum is a thin triangular-shaped membranous partition separating the lateral ventricles, and consisting of two laminae. It is attached above to the corpus callosum, below to the anterior part of the fornix posteriorly and the reflected portion of the corpus callosum anteriorly.

Various abnormalities of the septum have been described in anatomical and autopsy studies. Perforation of the septum is well recognized, and DANDY (1931) considered many such perforations to be acquired. Three types of cavum septi pellucidi have been recorded (Van WAGENEN and AIRD 1943): (1) The *non-communicating type;* (2) *the communicating type with openings into the lateral or third ventricles;* and (3) the *type associated with internal hydrocephalus.*

In this series of cases, in order to demonstrate the integrity of the septum pellucidum or the walls of a cavum septi pellucidi, an antero-posterior film using a horizontal beam, with the patient lying on his side, was taken. Antero-posterior tomography was also carried out in some of the patients. The antero-posterior horizontal view will, in normal circumstances, demonstrate some air trapped in the lowermost ventricle, whereas in a perforated septum, air will rise to the upper ventricle. Some air may of course spill through the foramina of Monro, but in a series of over 300 unselected encephalograms where this procedure was employed, this only occurred in four cases, all of whom had large ventricles. Where linear tomography during encephalography was carried out, transverse movement of the tube, in relation to the patient's head, was found to be essential, in order to avoid linear blur of the septal structures under investigation.

In the present series of 16 cases, nine patients had evidence of a communicating cavum septi pellucidi, an incidence of 56%. If the cases described

TABLE 31
Analysis of the radiological changes observed during encephalography. From MAWDSLEY, FERGUSON, 1963).

Case No.	Cavum septi pellucidi	Brain atrophy			Dilatation of cistern of lamina terminals
		Ventricular enlargement	Cortical atrophy	Cerebellar atrophy	
1	×	×		×	
2		×	×		
3	×	×			
4					
5	×	×		×	
6	×		×		
7	×	×			×
8					
9	×	×			×
10		×			
11					
12		×		×	
13			×	×	×
14	×	×			
15	×	×			×
16	×	×			

by SPILLANE (1962) are considered, the recorded incidence of abnormal communication between the lateral ventricles, with or without a cavum, in boxers, is thirteen out of 21 patients, i. e., 62%.

The natural incidence of a communicating cavum septi pellucidi has never previously been determined accurately, though THIEFFREY et al. (1958) suggested 1%. In the last 300 unselected encephalographies, carried out in the *Manchester Royal Infirmary,* three such cases have been recorded, giving an incidence similar to that suggested by THIEFFREY et al.

Tomography demonstrated the walls of the cavum to be intact anteriorly, but to have multiple perforations behind the foramina of Monro, increasing in number posteriorly. These observations have been confirmed by an autopsy study of the brain of an ex-boxer, where it was noted that the fenestrations increased posteriorly to a point where only a few residual strands of the lamellae remained. A free communication between the posterior ends of the lateral ventricles was thus allowed.

Interestingly, the three cases of cavum septi pellucidi were observed incidentally, none of whom were ex-boxers, and two showed evidence of communication with one lateral ventricle only, and in the third patient, the cavum septi pellucidi communicated with a well-developed cavum Vergae. These variations were not recorded in any of the boxers in the series.

ISHERWOOD et al. concluded, without doubt, that the changes observed in the septum pellucidum in boxers are a significant feature, and traumatic causation has been conclusively proven. The changes may be produced by a sudden elevation in ventricular pressure, with rupture of the floor or walls of the potential cavum septi pellucidi. Later pressure rises might cause dilatation of the cavum and perforation of its walls.

A potential space between the leaves of the septum pellucidum is normally present and the incidence of a non-communicating cavum has been variously recorded from 20% by SCHWIDDE (1952) to 60% by SCHUNK (1963), where the average separation of the leaves was 3 mm.

Dilatation of the cavum and perforation of its walls may precede the occasional disappearance of the septum, as suggested, and as observed in one patient described by SPILLANE (1962). The data also suggest that posterior disappearance of the septum occurs first. None of the boxers without a communicating cavum, however, showed any radiological evidence of widening of the septum.

ISHERWOOD et al. described additional traumatic alterations:

Evidence of brain atrophy. Generalized ventricular enlargement. In ten of the patients, ventricular enlargement was observed. This was recorded initially as a subjective impression.

Cerebral cortical atrophy. In three patients the cerebral sulci were widened.

Cerebellar atrophy. In four patients the subarachnoid space around the cerebellum was markedly increased in size.

Radiological changes indicating brain atrophy, i. e. ventricular enlargement or cortical and cerebellar atrophy, have been observed after head injuries (FRIEDMAN 1932, JACKSON and FLEMINGER 1946, TROLAND et al. 1946).

Direct brain damage due to mechanical linear acceleration and rotation strains, together with localized petechial hemorrhages, gives rise to glial replacement. Such changes may result in local atrophy. It is suggested that the degree and distribution of atrophy on x-rays correlates well with the severity and type of neurological affection evinced clinically.

Enlargement of the cistern of the lamina terminalis. In four of the patients, an unusual dilatation of the cistern of the lamina terminalis was observed, a phenomenon not previously recorded. Three of the patients exhibiting this dilatation also had a cavum septi pellucidi. Tomography in the mid-sagittal plane demonstrated the dilated cistern clearly, together with its relationship to the cavum septi pellucidi. The mechanism of production of this dilatation is uncertain, but the repetitive transient elevation of intracranial pressure brought about by traumatic deformation of the skull may be responsible. It is possible that this mechanism may also result in a rupture of the floor or walls of the cavum septi pellucidi.

GORALSKI and SYPNIEWSKI (1967) reported the case history of a 26 year old electrician, who had been in a ward from March 21 to April 10, 1965.

He was always healthy, and had never had any major illness or any history of head trauma. He had started boxing when he was a student, and had 20 fights in three years. He was knocked out five times and had several so-called technical knock outs. His first knock out took place when he was 19 years old. He had started complaining of headaches in 1963, and had noticed some difficulty in remembering. On March 22, 1964, he was violently knocked out and was unconscious for almost eight minutes. Immediately after the fight he was brought to a hospital. However, because he was feeling good, had no complaints, and physical and neurological examination was essentially negative, he was released home. The next day he went back to his regular work. Three months later he started having feelings of anxiety, was afraid of having cancer and expressed several somatic complaints that were considered hypochondriacal. On June 15, 1964, he was admitted to a hospital again. He was restless, uninhibited, and did not make much sense.
Neurological examination revealed a paralysis of the left facial nerve. He was dismissed after a few days with the diagnosis – status post head injury, and psychosis. However, at home he continued to be restless, paranoid, and not making much sense. He was referred to a psychiatric hospital. He had been there for one month and was discharged, with the diagnosis of status post cerebral concussion.
A few months later he changed. His delusions (mostly paranoid delusions) had increased. He believed someone was spying on him, someone was poisoning his food, etc. At work he behaved bizarrely. For example, he threw his tools into the river, explaining that they needed a longer soaking. When he was seen in the physician's office, he left a brick wrapped in paper on the physician's desk. During the observation period on the neurological ward in the hospital, it was revealed that his pupils were unequal, the left wider than the right and reacting more sluggishly to light. The nasal fold on the left side was smoothed out. The left corner of the mouth sagged and was lower than the normal side. He had no tendon reflexes in the upper extremities. Abdominal reflexes: there was a trace of upper and middle reflexes on the left side, but no others were elicited. The muscle tonus in the upper extremities was slightly increased and the spasticity was pyramidal-type.
Mental status examination: He was oriented as to time and place, displayed good conduct and mood, and was neutral and appropriate in his responses. However, there was a slight psychomotor retardation. He was very critical and had no insight, no feeling that he was sick. He had a lot of delusions, mostly paranoid in nature.
Psychological examination revealed an IQ of 101 on the Wechsler intelligence scale. Abstract thinking was impaired.
EEG was done twice – October 3, 1964, and May 22, 1965. All results were normal.
PEG: The subarachnoid system was only slightly filled up with air. Lateral ventricles were symmetrical but enlarged. The third ventricle was slightly enlarged. The Evans index was 0,29: the *roentgenogram* was suggestive of brain atrophy.
A *second neurological examination* was carried out on May 13, 1966. There was no change in the neurological status. The mental status examination gave no evidence of paranoid delusions. He was planning to return to work. His wife confirmed the improvement of his psychiatric status.

BOUSSELJOT (1969) reported the case histories of *eight amateur boxers:*

Case 1: A 22 year old amateur boxer, who began boxing in 1959. Up to now, *no neurological examination, no EEG recording, and roentgenograms of the skull.* He was knocked out due to head blows several times, had amnestic periods. Sometimes continued to fight while they were occurring. On September 4, 1965, received a blow against the head that produced an amnestic period, which was not recognized by the referee, so the fight continued. Friends reported that he received two more blows against his head, then fell and hit the canvas with his occiput. He was unconscious on admission to the emergency room in a hospital.
The *neurological examination* revealed a numbness of the entire left side of his body, a decrease of muscle strength on the left and positive pyramidal bilaterally. A subdural hematoma was suspected, and a carotid angiography performed revealed signs which excluded a subdural hematoma.
The *EEG* revealed no pathological findings.
Readmission to the hospital from November 5 to December 10, 1965. Numbness of the left hand and the left side of the face. Motoric weakness of the left-sided extremities. Psychic lability (when excited he began to cry).
The *neurological examination* revealed an inconstant hemihypesthesia, left. Slight dysdiadochokinesis, left. Slurred speech, slight weakness of the left lower facial nerve.
EEG: normal recordings.
PEG: moderate internal hydrocephalus, communicating cavum septi pellucidi.
Psychopathologic: A great deal of slow, pointless activity accompanying every action. Mnestic functions decreased, difficulty in concentration. Sudden mood swings.
Diagnosis: Status after cerebral contusion, due to boxing injury with slight left hemiparesis. During a follow up (August, 1966) findings in general the same.

Epicrisis: Twenty-two year old amateur boxer took

up boxing in 1959 without any physical checkup. During his career as a boxer, he was counted out several times, after blows to the head, and he suffered obviously repeated groggy states. In such a groggy state, he received several blows against the head, was knocked out, and had to be transferred to a hospital. Neurological findings revealed half sided, left, neurological symptomatology due to brain contusion, which must be considered, with regard to the follow ups, permanent brain damage.

Case 2: Twenty year old amateur boxer. 1963-1964, motorcycle accident that did not involve the CNS. He has been an amateur boxer since 1960, with 50 bouts. *Before he took up boxing, he did not have a neurological examination, an EEG recording, or a skull x-ray.* He was 'counted' by the referee 30 times after blows to the head. Several times he had a headache, and felt 'wretched;' he also had amnestic periods in some of his bouts. *EEG* recordings in January 1964 and October 1965 revealed repeated general disturbances. During a fight on Sept. 26, 1965, he was hit in the third round by several blows against the head and the referee started counting. After another blow against the head the fight was stopped and his opponent was given a t. k. o. He complained of nausea, and had shivering fits, severe headache, and 'saw everything as through a wall.' *Admission to local emergency room,* from there two days later *transferred to a neurological hospital.*
Neurological examination and *EEG* recordings normal.
PEG revealed a left-sided dilatation of the ventricular system, considered a brain atrophy due to posttraumatic brain edema. Follow up in hospital from December 13 to 17, 1965; Achilles and patellar reflexes, possibly left brisker than right, signs indicative of central irritation of vestibular nerve.
Psychopathologically unremarkable.
Diagnosis: Status after repeated head injury due to boxing, with left sided dilatation of ventricular system.

Epicrisis: Had no neurological examination, EEG recording and shall x-ray before he took up boxing. Twenty year old patient had to be 'counted' 30 times as a result of blows against the head. Inbetween EEG recordings revealed several times slight general alterations. Subjective complaints of headache. On September 26, 1965, bout had to be stopped as result of repeated blows against the head. Clinically developed cerebral concussion. PEG revealed a left-sided ventricular dilation which must be connected with his activities as a boxer.

Case 3: Twenty-three year old amateur boxer with an unremarkable history, especially no epilepsy. Never sick, with the exception of a severe cold in 1957. Boxed as an amateur from 1959-1962, with a total of 16 bouts. In 1961, he suffered a cerebral knock out and was unconscious for five minutes. After consuming alcohol in February, 1964 and January, 1966, sudden episodes of unconsciousness. No details can be given (tendency to dissimulation). The second episode revealed cyanosis, foaming at the mouth and following disorientation.
(In 1967 patient was seen in follow-up after the occurrence of an obvious grand mal seizure, with biting of the tongue).
Admission to a hospital from January 24 to February 3, 1966.
Neurological examination was unremarkable.
Psychopathological examination: unremarkable.
EEG after hyperventilation, occurrence of generalized high voltage spikes and waves over a period of 2-4 s duration.
PEG: dilatation and rounded disfiguration of left lateral ventricle and third ventricle.
Diagnosis: Symptomatic epilepsy, probably of traumatic origin.

Epicrisis: Twenty-three year old patient suffered cerebral knock out in 1961 with five to six minutes unconsciousness. In 1964 and 1966 developed, after alcohol consumption, grand mal seizures. Under consideration of otherwise unremarkable history and findings, a post-traumatic origin (boxing and cerebral knock out) must be considered as overwhelmingly probable.

Case 4: Twenty-five year old amateur boxer. Unremarkable history with the exception of a cerebral concussion, in 1956. Took up boxing in 1963, total of 60 to 70 bouts. Frequently received severe blows against the head. *Before this treatment, no neurological examination was done, no EEG recordings were taken, and no x-rays of the skull were done.* During a bout in February of 1966, he was knocked down. On June 26, 1966, he was knocked out as a result of a blow against his left neck; the boxer was stunned. Swaying and knees were buckling-amnestic period for about five minutes, did not know how he got into the dressing room. *After this event, no neurological examination was done, no EEG recordings taken, and he was not barred from fighting.* On the following days, frequent right-sided headache. Further bouts on July 2 and July 8, 1966. During the last bout he received severe punishment against the head; afterwards, severe headache, which led to admission.
Neurological findings revealed hemihypesthesia, right. Achilles reflexes higher on the right side. Romberg deviation to the right.
EEG normal.

PEG did not reveal any space-occupying lesions.
Diagnosis: Cerebral contusion as result of cerebral knock out June 26, 1966. Admission from February 21 to March 10, 1967, complaints of moderate headache.
Neurological examination revealed inconstant hemihypesthesia, right. Otherwise unremarkable neurological and psychopathological findings.
EEG: unremarkable
Central alterations of vestibular nerve.
Diagnosis: Intermittent disturbances of cerebral circulation due to carotid sinus reflex after boxing blow against left neck.

Epicrisis: Twenty-five year old patient hit June 26, 1966, by blow against the left side of the neck, with short period of numbness, swaying and buckling of the knees, and following amnestic period. Additional bouts on July 2 and July 8, 1966, produced more severe punishment to the head and increased headache. Neurological examination in another hospital revealed slight right-sided disturbances, which were considered the result of a cerebral contusion. Blow against the left side of the neck probably triggered a carotid sinus reflex with intermittent disturbances of cerebral circulation.

Case 5: Twenty-six year old amateur boxer with unremarkable personal and family history. On November 27, 1964, received while training blows against the head with consecutive unconsciousness (details not known).
Admission in surgical department in unconscious condition. Remained somnolent for several days, and had (third and fourth days) bradycardia of about 50/min. After one week, improvement. Subjective complaints of headache, dizziness, and disturbance of vision. *Ophthalmologically* a right-sided hemianopsia was detected. Treatment in neurological department January 11 to March 4, 1965. No complaints with the exception of disturbances in vision. *Ophthalmological* right side homonymous hemianopsia.
Neurological examination shows no pathological findings.
EEG shows slight general disorders; alpha reduction left parieto-occipital.
PEG: bilateral blunt lateral ventricles with marked dilatation in area of left posterior horn and trigonum.
Diagnosis: Boxing injury due to cerebral contusion with cortical contusion in left occipital lobe.
EEG revealed focus left parieto-occipital.
PEG revealed dilatation of posterior horn of left lateral ventricle which can be considered the result of cerebral contusion.

Epicrisis: Twenty-six year old patient received, while in boxing training, a blow against the head with immediate unconsciousness and somnolence for several days. During the same time, a right homonymous hemianopsia developed.

Case 6: Twenty-one year old amateur boxer with unremarkable family history. Cerebral concussion at age four, otherwise unremarkable. Amateur boxer since 1957, 95 bouts; 1963, cerebral knock out.
During bout February 3, 1967, he received blows against the head with following monocle hematoma. Despite headache, dizziness and unconsciousness on two occasions, participated in two more fights before admitted to hospital. He had lost all sense of social boundaries, did not 'keep his distance,' overly familiar, etc. Again, occurrence of periods of unconsciousness. Transferred on February 28, in somnolent status into surgical department. Left pupil wider than right; intracranial hematoma was excluded by x-rays and carotid angiogram.
Diagnosis: Status after head injury with brain edema. Treatment in neurological department, March 3 to March 23, 1967.
Neurological examination reveals spontaneous nystagmus left greater than right, weakness of left lower facial nerve.
Psychopathologically distance loss and loss of critical faculties.
EEG revealed slight dysfunction, possibly more on the right than on the left.
CSF: unremarkable.
PEG: moderate general dilatation of ventricular system.
Diagnosis: Brain damage due to repeated cerebral concussions and contusions.
As a result of recurring periods of unconsciousness and headache, admission to department of neurology of a university hospital April 24 to May 11, 1967.
Neurological examination revealed nystagmus looking to the right; otherwise no pathological findings.
Psychopathologic examination revealed that reactions and responses were dull (college graduate).
EEG showed slight general alterations with indication of diencephalic disturbance.
Diagnosis: Status after repeated head injuries due to boxing with transient neurological and psychopathological findings.

Epicrisis: Twenty year old patient develops monocle hematoma after blows against the head in boxing on February 3, 1967. Despite subjective complaints and periods of unconsciousness, patient participates in two more bouts. Patient admitted in somnolent state to surgical department and later to neurological department. Transient psychopathological, neurological and EEG findings lead to diagnosis of cerebral contusion with recovery to a great extent after several weeks' treatment.

Case 7: Twenty-three year old amateur boxer, 'severely ill' as a small child. *Amateur boxer since 1959, no previous neurological examination, EEG recordings, or skull x-rays.* In 1963, cerebral concussion due to blows against the head with nausea and headache. In 1965, bout was stopped, as, after blows against head, left leg was 'asleep,' and buckled. In practically every bout, received blows against the head, with transient decrease of ability to react ('ringing in his head and like fog before his eyes'). Despite permanent headache, only once (1962) had an EEG been performed.
Admission to neurological department due to permanent headache, treatment from November 18 to December 4, 1965.
Neurological examination unremarkable.
Psychopathological examination revealed moderate decrease of amnestic functions and difficulty in concentration. Simply structured personality, affective labile, easily irritated.
EEG showed alpha reduction left occipital, slight general alterations, left more than right.
PEG revealed rounding of the edges of the ventricular system.
Diagnosis: Status after repeated head injuries due to boxing with neurological residuals.

Epicrisis: No previous neurological examination, EEG recordings, no skull x-rays were performed. Twenty-three year old amateur boxer suffers, in active boxing career extending over six years, repeated severe blows against the head, which lead to increased subjective complaints. As result of existing EEG alterations, psychopathological findings and subjective complaints, cerebral damage indicated.

Case 8: Nineteen year old amateur boxer with unremarkable personal and family history. Amateur boxer since 1963, with 28 bouts. Once 'counted' due to blows against the head. *Up to now, no neurological examination and no x-rays of the skull. Only an EEG recording was done.* On December 9, 1967, fell out of bed in what was obviously an epileptic seizure. House physician observed, on same day second generalized tonic-clonic seizure with terminal sleep. *Admission to neurological clinic,* treatment December 11 to December 22, 1967.
Neurological examination revealed unremarkable findings. Psychopathological examination also unremarkable.
EEG and *CSF* tests refused.
Diagnosis: Epileptic seizures presumable of traumatic origin.

Epicrisis: Nineteen year old amateur boxer develops, after five year long boxing career, two grand mal seizures in one day. Due to otherwise unremarkable history, correlation with his boxing career is probable.

BOUSSELJOT, who reported the case histories of *eight amateur boxers with clinical findings of cerebral damage,* summarized the findings:

The brain damage to the boxers in cases 1 and 2 can be explained by the fact that they were subjected to a series of blows against the head; their reduced defensive reactions either were not or could not be recognized by the referee. Boxer 1 was evidently groggy. Boxer 2 had reported earlier complaints and had had several pathological EEG recordings. The patient continued to box; the findings may be described as the result of a summation effect.

Case 4 participated in a boxing bout one week after a cerebral knock out; case 6 also participated in additional fights, after a blow against his head that had produced a periorbital hematoma and two periods of unconsciousness. In these cases there was not enough recovery time between the bouts, and the insufficient medical supervision before the continuation of the fights seemed to be the main factor in the development of the cerebral damage.

Case 7 also continued boxing, despite the fact that he had had subjective complaints for several years and after a blow to the head he developed transitory hemiparetic disturbances; medical examination and observation were not initiated.

In Case 5 the severe results must be explained both by the fighter's inexperience and by the fact the basic preliminary physical fitness examination was not carried out.

The seizures in cases 3 and 8 are interpreted by the author as being due to brain injuries caused by boxing.

It was evident that practically all eight athletes had no medical examination before they took up boxing, and that even after the development of neurological signs, no neurological examination or EEG recording was initiated.

BOUSSELJOT examined a total of 47 amateur boxers between 1965 and 1967. *In at least 27 of these athletes, no neurological examination or x-rays of the skull had been performed before they took up boxing. In 26 of these athletes, no EEG's had been recorded. Eighteen of these athletes, despite careers as amateur boxers extending over several years and comprising a high number of bouts, had no medical examination performed at all, although 12 of those had received one to three cerebral knock outs, occasionally accompanied by lengthy unconsciousness.*

In ten cases, the recommendations of physicians regarding control examinations, or a decision as to whether boxing should be discontinued, were completely ignored.

The reported experience of BOUSSELJOT contradicts the commentary of SCHWARZ (1965), where he made the assertion that there is no athletic activity in which the screening of the athlete from a medical standpoint is so careful and strict as in amateur boxing.

Many other authors cited above referred to all of the medical information given to them through boxers, their families, and medical records. In relatively few cases were extensive medical tests reported to have been conducted at the beginning of amateur boxing. Perhaps the omission was the result of practical considerations, for example, the novice amateur didn't know in advance that he would pursue boxing for any length of time, or didn't realize the importance of such tests, or couldn't afford to pay for them. Such oversights could be addressed in amateur boxing regulations. If a specified set of medical examinations were to be made mandatory for all prospective participants, it would help identify those with unusually high medical risk factors.

Importance of the Study of ROBERTS (1969)

The most recent study, by ROBERTS (1969), is based on a report on the medical aspects of boxing and the facts provided therein, made by a *Committee of the Royal College of Physicians in London* as a consequence of a public debate in the *House of Lords* in May 1962.

This study has to be considered the most thorough and comprehensive. It has to be considered a classic. Since this publication is of such an importance, we have quoted it due to its importance in a very detailed form.

Two hundred and fifty ex-professional boxers from the population of 16.781 boxers registered between 1929 and 1955, were examined clinically and electro-encephalographically in this investigation. The entire sample of boxers was traced and, with the exception of nine who had emigrated, 16 who died and one who refused, all those traced (224) were examined. EEG's were available in 168 cases and independent accounts in over half the sample.

ROBERTS wrote:

"In 37 cases (17%), there were clinically demonstrable lesions of the central nervous system, which bore a close resemblance to the syndrome previously described in boxers. These followed a relatively stereotyped clinical pattern. At its mildest, this consisted of dysarthria with or without disequilibrium and evidence of symmetrical pyramidal disease, and at its most severe of moderately disabling ataxias, disequilibrium and spasticity, or rigidity and striatal tremor associated with varying degrees of dementia. All grades of disturbed neurological function in cerebellar, pyramidal and extrapyramidal systems were apparent clinically between these two extremes. The occurrence of this syndrome has been analysed with respect to age and occupational exposure, and it was demonstrated that there was an increasing prevalence with increasing exposure... The study has shown the relationship to occupational exposure of an easily recognisable and relatively stereotyped syndrome of disturbed neurological function."

Case 1 (sample no. 34): This man, aged 60, an ex-labourer and store-keeper, had been unemployed for thirteen years and a patient in a mental hospital for seven. There was no family history of neurological or psychiatric disease. He had had no serious injuries or illness in the past, apart from, towards the end of his career, what was probably cellulitis of the hand which he held responsible for the tremor he had been aware of since.

He had started boxing at the age of fifteen and had about fifty amateur contests before his first professional fight at sixteen.

He had fought mostly at middleweight, often every week and usually contests of twelve or fifteen three minute rounds, with a few two-minute round fights. He did not think he had been knocked out but had certainly been knocked down many times. The longest post traumatic amnesia he recalled extended for the last seven rounds of a fight which he won. At the start of his career, he boxed regularly in local booths. Excluding these latter fights he thought he had had about two hundred promoted professional contests. He became area middleweight champion and retired at the age of twenty-nine, after thirteen years in the ring.

His relatives had noticed that his speech was slurred and that he had developed a tremor in his left arm, and his manager that he was limping with his left leg in his last year boxing. He had been forgetful for many years, and the tremor in his left arm and the dragging of his left leg had become more apparent as he grew older. He had held a job as a store-keeper until the firm closed. The death of his wife precipitated a paranoid delusional illness which necessitated his admission, finally for long-term care, to a mental hospital.

Examined he was found to be devoid of insight and euphoric, but it was possible to elicit some poorly systematized paranoid delusions about his wife's relatives, who he felt had dispossessed him of his inheritance. He was demented as judged by the extreme slowness of his thinking and his striking inability to register, during the course of the interview, various pieces of information, including greetings from another boxer whom he had known well, in spite of the fact that these were several times repeated to him. His recall of the details of his boxing career seemed accurate, as far as names and places were concerned, but was hopelessly inaccurate for dates, even those when he was champion. Tested with Raven's Matrices, he scored in the fiftieth percentile, and in the twenty-fifth percentile with the Mill Hill Vocabulary. These tests, used on his initial admission to hospital and repeated seven years later, did not show any decline in the scores, nor had his dementia, as judged clinically and by his behaviour generally, progressed notably in that time.

His speech was slurred, his face expressionless, and there was an increase in tone more marked in the left than the right limbs. All his movements were slow and he was markedly akinetic on attempting to walk, his gait thereafter short-paced and hemiparetic with the left arm partially flexed and the leg intorted and dragging. He was unsteady on his feet and unable to walk tandem fashion, but there was only dubious pyramidal weakness of the left limbs, and no ataxia. There was an intermittent, rhythmical parkinsonian rest tremor of the left limbs. The tendon reflexes were brisker in the left arm than the right, and both plantar responses were extensor. He was normotensive and, apart from a few scattered rhonchi in his chest, general examination was unremarkable. He had the usual stigmata of his career, which were not particularly severe, of broadened nose, left cauliflower ear, and swelling of the dorsum of the wrists as evidence of carpometacarpal arthropathy, less marked on the left since he was right handed.

Investigations included an *EEG,* which was abnormal in that the alpha rhythm, at 7- 8 cps, was ill-organized, and there was persistent delta activity in the right parietal region.

Serological tests for syphilis were negative. A routine blood count, E.S.R., and an examination of the *CSF* were all normal.

Case 2: (sample of no. 54). This man, aged 50, an ex-labourer and merchant seaman, had been unemployed for eight years. His mother had developed diabetes mellitus in middle age, but there was no family history of neurological or psychiatric disease. He had had no serious illnesses in the past and, although he had been injured in several drunken brawls, he had not sustained head injuries outside the ring which had rendered him unconscious.

He had started boxing at a boys' club at the age of twelve, and had had less than fifty amateur contests. One of these had been followed by a post-traumatic amnesia, which lasted until he came to, showering, after the fight. He had his first professional fight at the age of eighteen and about eighty promoted professional contests in the course of the next ten years before he gravitated to travelling with a boxing booth for three years before he retired. He had fought mostly at lightweight, ten and twelve two-minute round contests and denied having been knocked out. Fights had been stopped against him on at least three occasions to prevent further punishment, and he had felt 'hazy' after several fights for at least a day on several occasions. He had attained no great eminence in his career but press records confirmed that he had beaten one former area champion.

His wife had noticed that his speech had been slurred and that he had been unsteady on his feet since his last promoted fight, in which he had received a severe beating. He had been admitted to hospital for observation at the time and had remained unwell for several days. His condition had remained little changed except that he had become increasingly sensitive to alcohol, small quantities apparently making him very drunk, until the gradual development ten years later of a progressively worsening tremor of his limbs. Although his memory did not seem to have deteriorated, he had become increasingly liable to outbursts of uncontrollable temper.

Examined, he was found to be emotionally labile, tearful and laughing alternately during the course of the interview. His memory, as judged by his recall of the details of his career and the development of his symptoms, was not grossly impaired, but formal psychometric tests were not done.

His speech was so slurred as to be almost unintelligible, his face was impassive, and there was a severe asymmetrical parkinsonian tremor of all four limbs, left more than right at rest, increased by emotion and abolished on volition. Tremor also affected his head and trunk, his gait was unsteady and festinant, and his limbs were rigid, more so on the left than on the right. All tendon reflexes were pathologically brisk, there was sustained clonus at knees and ankles, and both plantar responses were extensor. The visual acuity in the left eye was impaired to perception of finger movements by an extensive retinal detachment and conjugate elevation of the eyes was poor. There was irregularity of calibre of the arteries of the fundi in keeping with his hypertension of 210/120 mm /Hg.

Case 3 (sample No. 19). This man, aged 58, had been a skilled craftsman and had built up a small but successful business on his own when he gave up boxing. There was no past history of serious head injuries or illnesses, and no family history of neurological or psychiatric disease. He was a heavy cigarette smoker but drank only an occasional glass of beer.

He had begun boxing in a boys' club at the age of fourteen and had his first professional fight a year later. Over the course of the next fifteen years he had over three hundred promoted professional contests, of which he thought he had lost less than a quarter. Most were as a light heavyweight and were twelve or fifteen two-minute round contests. He had been knocked unconscious once for a few seconds, and the only fights stopped against him were for cut eyes. He recalled one lengthy post-traumatic amnesia extending for the last three rounds of a fight, which he won by a knock out, until he came to, a quarter of an hour later in the dressing room. He admitted to feeling 'hazy' for an hour or so on many occasions after hard fights. He had held no titles but had 'always topped bills' and had evidently been a boxer of some distinction. He had never boxed in a booth, considering it beneath his standing in the profession. He retired at the age of thirty when he began slowing up and losing too many fights.

Both he and his wife had been aware that his speech, for at least the past ten years and perhaps longer, had been 'thick and getting thicker,' and he had considered taking elocution lessons to improve it. Neither were aware of any other disabilities.

Examined, it was apparent that his mentation was extremely slow and that he had difficulty recalling names and places in his boxing career. He scored in the fiftieth percentile on Raven's Matrices, learned his digit span of seven plus one at the fourth repetition, recalled accurately only one of the pair of Binet designs at the first attempt, but both at the second presentation and both accurately after fifteen minutes delay.

His cerebellar dysarthria was so marked as to make his speech almost unintelligible at times, there was a mild degree of ataxia of gait on beginning to walk, on getting out of a chair and on turning, and the left leg was slightly circumducted walking. There was a terminal intention tremor in both arms and all movements were markedly slow. An inconstant rhythmical tremor, apparent at rest and suppressed on volition, was evident in the right arm when he was engaged in tests of intellectual function. The only focal weakness present was in the right triceps, which was wasted and this was associated with a depressed triceps tendon reflex. The left plantar response was extensor. He was normotensive and general examination was otherwise unremarkable. The stigmata of his profession were moderately severe and included a broken nose, thickening of the left supra-orbital margin and a left cauliflower ear.

Investigations were limited to an *EEG,* which was within normal limits.

Case 4 (sample no. 28). This man, aged 48, a dockyard worker, had been an unskilled manual worker in various trades all his life. There was no history of neurological disease, but one sister had had a puerperal depression treated with electroconvulsion. He had contracted gonorrhea in the past, which had been treated with penicillin, but had had no other serious illnesses or injuries. He drank rarely.

He had begun boxing at the age of nine at school, and shortly after this he and his brother had frequently sparred together to provide the kind of light entertainment then popular at some professional promotions. He had had fifty amateur contests before his first professional fight at the age of eighteen. He then had about three hundred professional fights, mostly at bantamweight, during the course of a career, which lasted until the age of thirty, and had at one time had been area champion. He had fought regularly, almost every week during the busiest period of his career, and recalled once having five fights in ten days. He had never been knocked unconscious but had been counted out several times. He recollected several posttraumatic amnesias lasting several rounds. He had had very few fights in booths, feeling it was below his professional standing and 'low stuff.'

His wife had been aware of no alteration in his speech or personality, but he himself thought that his speech had been 'a bit thick' since his last years boxing, and he had been very conscious of the fact, and his children had commented on it, that increasingly over the years, small quantities of alcohol had made him look and feel drunk. When, during the course of the examination, an inconstant parkinsonian tremor of his right arm was drawn to his attention, he claimed it had been present only for a few months. He felt that his memory had deteriorated in recent years, so that he had great difficulty remembering people's names and errands his wife gave him.

Examined, there was no overt intellectual abnormality clinically and he appeared to have good insight into his minor disabilities. He scored in the twenty-fifth percentile on Raven's Matrices and on the synonym selection test of the Mill Hill Vocabulary. He learnt his digit span plus one at the third repetition, but his immediate recall of one of a pair of Binet designs was inaccurate, though after fifteen minutes delay he was able to reproduce faultlessly one of the original designs and his own version of the other.

His speech was slurred, his face expressionless, and there was an inconstant rhythmical parkinsonian tremor of his right arm at rest, abolished on volition. The tendon reflexes were brisker in the right than the left arm but the tone was only marginally increased in the right arm and both plantar responses were flexor. There was no apparent disequilibrium, and his gait was normal. He was normotensive and, apart from clinical evidence of chronic bronchitis, general examination was unremarkable. Investigations were limited to an *electroencephalogram,* which was normal.

Case 5 (sample no. 64). This man, aged 53, had been virtually unemployed until he became a professional boxer and had then lived on his boxing until called up during the war. He was discharged from the services as punch-drunk, had done various labouring jobs for some years thereafter, and been unemployed again for the past nine years. He admitted to luetic infection in his youth, which appeared to have been adequately treated by a course of mercury injections lasting several years. He had once been involved in a road traffic accident in which he had been knocked unconscious for a few seconds. He drank excessively and had done for years. Frequent falls when he was drunk had in recent years caused a number of minor fractures. He had finally been divorced by his wife, because of his increasingly trying behaviour.

He had started boxing in a local gym at the age of fifteen and had had his first promoted professional fight three years later. He had had ninety-six promoted fights in the course on the next ten years, mostly at light heavyweight, and had been knocked unconscious on two occasions, for seconds only at a time, though it had been necessary for him to retire on half a dozen occasions when he was taking a severe beating. He had at one time held the area title.

The records made by the neurologists who had examined him twenty years earlier, during the war, were available. At that time his speech was slurred and he was complaining of headaches, loss of memory and depression. An independent account was also available from his present general practitioner, who had known him as a chronic alcoholic for many years.

Examined, he appeared apathetic, his speech slurred and his mentation slow. His memory for the details of his professional career was very poor, judging by the records he had kept, and he had considerable difficulty grasping news of former colleagues mentioned to him during the course of the interview. No formal psychometric tests were used, but according to his own account of his poor scholastic performance at school, he was constitutionally below average intelligence, though his evident memory difficulties and slow mentation were a certain measure of a degree of dementia superimposed. In addition to his dysarthria, his face was expressionless, he was markedly unsteady on his feet and there was moderate ataxia in the left more than the right arm. There was some spasticity in the pronators of the forearms and wrists, on the left more than the right, the tendon reflexes were brisker in the left limbs, and the left plantar response was extensor. The stigmata of his profession included a broken nose, a left cauliflower ear, and bilateral carpometacarpal arthropathy, none being gross. General examination was precluded by the patient's unwillingness to co-operate further. Investigations were limited to a routine serological test for syphilis, arranged by his general practitioner, which was negative.

Case 6: (sample no. 58). This man, aged 57, a miner all his life apart from a period of three years when he lived entirely on the proceeds of his boxing, had been unemployed for twenty years since a mining accident. He had been buried in a fall of rock, which had fractured lumbar vertebrae and left tibia and fibula, and caused a cauda equina or conus lesion permanently impairing his bladder function. There was a posttraumatic amnesia of about eight hours though he was not unconscious when admitted to hospital, according to the notes made at the time. He had been otherwise well in the past and there was no history of neurological or psychiatric disease in the family.

He started boxing at the age of seventeen, when he was offered a pound for six rounds by someone who noticed him standing watching a fight. In the ten years of his career, he had 110 promoted professional contests, mostly at heavyweight, admitted to several extensive posttraumatic amnesic episodes and to being knocked out on four occasions but for no longer than seconds each time. He had boxed in booths when they came locally during the first two years of his career, mostly fixed fights, and had, for the last five years of his career, been full-time sparring partner to two champion heavyweights. He retired from the ring, at the age of twenty-seven, when his wife and friends told him he was 'getting punchy.'

His wife and a professional colleague, who were interviewed, and the individual himself had been aware that he dribbled and that his speech was slurred and his gait unsteady since his last years in the ring. The wife felt that his speech had remained unchanged, and certainly his dribbling had cleared up, until his mining accident. Since then his speech and gait had progressively deteriorated and his memory, always poor, had become so defective that in recent years he did not know what day it was. The professional colleague, who had known him well, did not think there had been much change in the years immediately following the mining accident, and said it had been well recognized locally that the man had been made punch-drunk by the champion heavyweights he had sparred with for too long. In recent years he had become increasingly irritable and liable to outbursts of uncontrollable temper, and two years earlier had had a severe paranoid delusional illness, making unfounded accusations against his wife. This had required treatment with chlorpromazine and stilboestrol, and he had since been more tractable.

Examined, he was euphoric and evidently demented, but insight was retained and he asserted that he was punch drunk and that "punch-drunk is a terrible thing." His mentation was markedly slow and he was unable to recall any details of his boxing career without considerable prompting. He was disorientated in time and place. He scored in the twenty-fifth percentile in Raven's Matrices and in the synonym selection test of the Mill Hill Vocabulary. His digit span was six but he was quite unable to learn another digit in addition, and although his copy of the Binet designs was accurate, he was unable to recall either.

There was a mild dysphasia, his speech was so slurred as to be occasionally unintelligible, his face was impassive, and there was nystagmus on lateral gaze. His gait was unsteady and high-stepping and there was titubation of head and trunk. All four limbs were ataxic, more so in the right arm and left leg. There was slight spasticity in the pronators of the right forearm and right quadriceps, the tendon reflexes were brisker on that side, and both plantar responses were indefinite extensors. A small area of sacral sensory loss, depressed ankle and absent left knee jerks, together with his urinary incontinence, which necessitated wearing a urinal, were sequelae of the conus or root lesions caused by the mining accident. *General examination* was unremarkable apart from evidence of chronic bronchitis and a blood pressure of 145/105 mm Hg.
Investigations were limited to an *electroencephalogram* which was a rather low voltage ill-organized 9 c/s alpha rhythm, considered to be just within normal limits..
Case 7: (sample no. 63). This man, aged 57, had been a labourer and subsequently a driver for many years until a minor head injury, which did not cause loss of consciousness, incapacitated him with headaches and phobic symptoms, necessitating his retirement. He had been discharged from the services during the war on medical grounds, the details of which were not available. He had otherwise been well in the past and there was no family history of neurological or psychiatric disease.
He had begun boxing at school at the age of twelve and had had about ten amateur contests before his first professional fight at the age of eighteen. He had fought about fifteen to twenty promoted professional contests a year at his busiest, initially at welter and later at middleweight, mostly ten and twelve two-minute rounds, and thought he might have had 150 fights in all before he retired at the age of twenty-six. He denied ever being knocked out or experiencing any memory disturbance during his career. He had had a few fixed fights in booths.
His wife had noticed that for many years that his speech was slurred when he was excited, but thought that this had become much more obvious since his mild head injury, since when she had also noticed that he was unsteady on his feet and tended to drag his left leg.
Examined, apart form his poor recall of the details of his boxing career, there was no overt clinical evidence of intellectual defect. He scored at the fiftieth percentile in Raven's Matrices and in the synonym selection test of the Mill Hill Vocabulary, learned his digit span of five plus one at the second repetition and recalled accurately, both immediately and after fifteen minutes delay, both Binet designs.
His speech was markedly slurred, his gait was unsteady with a mild degree of truncal ataxia and there was an asymmetrical ataxia of the left limbs more than the right. He was generally hypotonic, though the tendon reflexes were brisk, more so on the left than the right. Both plantar responses were flexor. The only weakness present was characteristic of a ruptured supraspinatus tendon. General examination was unremarkable apart from a blood pressure of 145/100 mm /Hg.
Investigations, performed elsewhere after his minor head injury, included an air encephalogram which demonstrated normal ventricles but a septum pellucidum perforated in its mid part. Bilateral carotid arteriograms were normal, and his Wassermann reaction was negative.
An *electroencephalogram* was of low voltage with scanty alpha rhythm, up to 25 microvolts, but otherwise within normal limits.

"There were seven others (sample nos. 4, 14, 29, 67, 133, 169, 216), whose ages ranged from 38 to 58, who presented a similar clinical picture of dysarthria and disequilibrium, with asymmetrical ataxia as evidence of cerebellar lesions, together with pyramidal lesions and immobile facies, but with little or no overt intellectual defect. One of these had had a long standing paranoid delusional illness, for which he had undergone psychiatric treatment, and another was subject to outbursts of uncontrolled temper requiring sedation. The wife of another described a progressive memory impairment and personality change with marked apathy. All were employed. Where independent accounts were available, or the individuals retained insight, there was good evidence that the disabilities had developed in their last years boxing and had been little changed, or become only very slowly more apparent, as they grew older. One had had a air encephalogram elsewhere which demonstrated an absent septum pellucidum but otherwise normal ventricles."

Case 8 (sample no. 23). This man, aged 45, had been a street trader and salesman most of his life and had lately become a sales supervisor. There had been no serious past ill health, but in adolescence he had been involved in a road traffic accident, which had resulted in a head injury causing unconsciousness and a posttraumatic amnesia lasting till the following morning. There was no family history of neurological or psychiatric disease. He drank socially only an occasional beer. He had begun boxing at a youth club gymnasium at the age of eleven, and had had about fifty amateur contests before he turned professional at the age of seventeen. In the next thirteen years he had fought hundred and thirty-seven professional contests, mostly at bantamweight, and at his busiest, as many as twenty a year. He had been knocked out for seconds only on three occasions, but had once felt dazed for an hour afterwards. He had won no titles, but at one time had been rated in the professional press among the first ten in the country at his weight. He had never boxed in a booth and had never worked as a professional sparring partner.

His wife commented that he was forgetful and agreed that his speech was slurred, but did not think there had been any change in the latter or in his equable temperament since she had known him a year before he became a professional boxer. He himself had noted with surprise that his speech was slurred when he had tape recorded himself, but did not think it was due to boxing, and was aware of no other disabilities.

Examined, he seemed mildly euphoric, but there was no clinical suggestion of intellectual impairment, and his memory, as judged by his detailed recall of his career, was good. He scored in the ninety-fifth percentile in Raven's Matrices, and in the fiftieth percentile in the synonym selection test of the Mill Hill Vocabulary. He learned his digit span of six plus one at the third repetition, and his recall of the Binet designs, both immediate and delayed, was accurate.

There was an obvious slurring dysarthria, his face was expressionless, and in repose, a mild inconstant ocular divergence made him appear glassy eyed. All his movements were slow, and repetitive movements of his arms were a little clumsily performed, but there was no definite ataxia. He was unsteady walking tandem and on turning, and his gait was mildly spastic, as were his forearm pronators and both quadriceps, more so in the left limbs than the right. The tendon reflexes were pathologically but symmetrically brisk, there were a few beats of clonus at both ankles, but the plantar responses were flexor and the jaw jerk normal. There was no weakness and no sensory deficit. He was normotensive and general examination was unremarkable. The stigmata of his profession were mild, the only obvious feature being a broken nose.

Investigations included an *electroencephalogram,* which was normal with well developed 10c/s alpha rhythm of good amplitude associated with a little frontal fast activity and occasional theta waves. Routine estimations of his haemoglobin, urea, sugar and liver function tests were all normal, and serological tests for syphilis were negative.

"There were seven other individuals (sample nos. 6, 9, 11, 39, 50, 71, 110) whose ages ranged from 36 to 63 who presented a similar clinical picture of dysarthria, impassive facies, slow and clumsy movements without definite ataxia, and increase in tone, which was more spastic than rigid, associated with pathologically brisk tendon reflexes and unsteadiness of gait. None of those who had psychometric tests achieved similar outstanding percentile scores. Three appeared mildly demented. One of these had been involved in a severe road traffic accident some years earlier, which had caused bilateral subdural haematomata and been associated with a post traumatic amnesia of about two weeks. This seemed likely to have contributed to the intellectual deterioration evidenced by his slow mentation and apathy, although from his wife's account it was clear that this had been present before. There was a family history of mental illness in another, whose paranoid delusions were well systematized in a way characteristic of a chronic paraphrenic illness. Two others were recidivists, and one of these, for whom serological tests for syphilis were not available, had small irregular pupils reacting more briskly on convergence than to light, though there was no other evidence of neurosyphilis."

Case 9 (sample no. 49). This man, aged 58, originally a labourer, had been a driver for thirty years since his retirement from the ring. A philanthropic admirer, who had followed closely his boxing career to the end, had offered him the job, telling him it was pitiful to see what had become of him. He had had no serious illnesses in the past, apart from otitis media as a child, and no severe head injuries. There was no family history of neurological disease, and the only psychiatric history was of a depressive illness in a sister requiring in-patient treatment.

He had begun boxing at the age of thirteen at a working men's club, and had had about thirty amateur contests before he became a professional at the age of nineteen. He had more than two hundred and fifty promoted professional contests, mostly at welterweight, though he had often had to fight heavier men. Most fights had been ten or twelve two-minute rounds, though he thought he had had a number of fifteen three-minute round contests. He had been knocked out on three occasions, but for seconds only though he recalled a few long post-traumatic amnesias lasting for several hours, coming to on each occasion at home or on his way there. He had never won a title but had competed in the eliminator rounds for one. He had never boxed in a booth, considering it 'lowering to a top liner.'

His wife considered him punch drunk, in that his speech was slurred when he was excited, and he dribbled and had done for years. He himself was not aware of this but had noticed that he was unsteady on his feet during his last years boxing and that this had become worse in recent years, so that he often lost his balance. He was aware that his memory was poor.

Examined, there was no impression of intellectual defect clinically, and he was notably insightful and frank. He recalled the details of his boxing career accurately, except for dates. He learnt his digit span of six plus one at the fifth repetition, and his immediate recall of the two Binet designs was accurate, though he was unable to recall one of the figures accurately after fifteen minutes delay. His immediate recall of a flower, a colour and an address were accurate. He scored in the fiftieth percentile in the Raven's Matrices and in the twenty-fifth on the synonym selection test of the Mill Hill Vocabulary.

His speech was slurred and his gait unsteady, particularly on turning. There was slight clumsiness, not amounting to definite ataxia, in all four limbs, but no weakness. Slight spasticity was evident in both quadriceps and the tendon reflexes were pathologically but symmetrically brisk. The left plantar response was an indefinite extensor and the right flexor. He was normotensive, and apart from clinical evidence of severe chronic bronchitis and emphysema general examination was unremarkable.

Investigations were limited to an *electroencephalogram,* which was normal.

"There were five others (sample nos. 1, 40, 53, 111, 224) whose ages ranged from 54 to 63, who presented a similar clinical picture. These in general were similar to, but milder than, those described with the previous case report and included a mild but definite slurring dysarthria, disequilibrium, poverty of facial expression, slight clumsiness of movements, and in all but one, some increase in tone in the limbs and pathologically brisk tendon reflexes. One of these individuals was described by his wife as becoming distressingly apathetic as he grew older, although he was only fifty-five, and another had evidence of a partial palsy of the superior rectus muscle of the left eye. Two were successfully self employed, having built up small businesses for themselves since their retirement from boxing".

Case 10: (sample no. 42). This man, aged 35 at one time a window cleaner and labourer, subsequently took a correspondence course to learn a trade and had then build up a successful small business on his own, when he retired from the ring. There was no past history of serious ill health or injury. A first cousin had committed suicide in middle age, but there was no other history to suggest psychiatric or neurological disease in the family. He did not smoke and drank only an occasional beer.

He had begun boxing at the age of eleven and had had about forty amateur fights before his first promoted professional contest at the age of nineteen. However, he had boxed regularly in booths from the age of fifteen and had continued to do so throughout his career, as many as five four-round contests a night. Although most of these were fixed, it was his opinion that one could still get hurt inadvertently. He had about seventy promoted professional contests, at welterweight, mostly of eight three-minute rounds, and at his busiest only twelve in a year. He was knocked unconscious once, did not think he had been carried from the ring, but recalled that the posttraumatic amnesia lasted till he was in the dressing room. He competed in the eliminator rounds for, but never won, a title, and retired when it became apparent to him that he was not going to get to the top of his profession.

His wife was aware of no change in his amiable and equable temperament in the years she had known him, considered his memory as good as her own and had never seen him unsteady on his feet. She admitted that she had noticed that his speech was slurred at about the time he retired from boxing but did not think there had been any change, for better or worse, since.

Examined, he was undoubtedly jovial but by no means pathologically so, and there was no clinical impression of intellectual defect. His history of his boxing career was accurate and detailed. He learnt his digit span of seven plus one at the fifth repetition. His immediate and delayed recall of the two Binet designs was accurate. He scored at the twenty-fifth percentile in Raven's Matrices, and at the fiftieth in the synonym selection test of the Mill Hill Vocabulary.

The only evidence of neurological lesions was a slight, but quite definite, slurring dysarthria, and an extensor plantar response on the left.

Investigations were limited to an *electroencephalogram,* which was within normal limits, apart from irregular alpha rhythm of low voltage, although often in excess 25 microvolts.

"There were four other individuals (sample nos. 5, 41, 94, 121), whose ages ranged from 47 to 56, who presented a similar clinical picture limited to a mild but definite slurring dysarthria and unilateral or bilateral extensor plantar responses. One of these had presented thirty years earlier to a neurologist with severe dysarthria, unsteadiness of gait, ataxia of all four limbs and a predominantly left-sided pyramidal lesion. The details of the development of this illness were incomplete, but it seems likely that it followed one or a series of hard fights, worsened rapidly over the course of a few days, and gradually cleared up over the succeeding eighteen months, so that he was able to return to boxing. There had been no further episodes of disturbances of neurological function to confirm the original suggested diagnosis of demyelinating disease, and it seems just as likely that this acute brain stem lesion, with almost complete recovery, was due to trauma in the ring."

Case 11: (sample no. 98). This man, aged thirty-nine, had been a clerk before, and a labourer after his boxing career. In recent years he had built up a successful small business in a competitive trade. There was no past history of serious injury or ill health, and the family history was negative for both psychiatric and neurological disease. He had at one time drunk heavily after he had retired from the ring, but for some years had been teetotal.

He had begun boxing at the age of seventeen at a local professionals' gymnasium and had his first professional fight that year. He had about sixty promoted contests, most at heavyweight, and most of six and eight three-minute rounds. He had been knocked out on three occasions, for seconds only, but admitted to feeling groggy for short periods during many fights. He had at one time been area heavyweight champion. In the earlier part of his career he had had a few fights in booths, and in the later part had become full-time sparring partner for several years to two champion heavyweights. He retired from the ring when, on medical grounds, renewal of his license was refused.

An independent account was not available from his wife, but the findings of a neurologist, who examined him when the question of renewal of his license was under consideration fifteen years earlier, were. He had been noted then to be dysarthric and very slow in cerebration. He himself was aware that his memory was appalling, having for several years to write everything down he needed to remember, and commented on the fact that it had always astonished him how slurred his speech was when he heard it on tape recordings.

Examined, it was evident that his memory, judged by his recall of the details of his boxing career, was poor, indeed the area championship he claimed to have held was not the one he had. His mentation was not, however, notably slow, and he scored in the fiftieth percentile in Raven's Matrices and in the twenty-fifth in the synonym selection test of the Mill Hill Vocabulary. He learned his digit span of seven plus one at the second repetition but was unable to recall accurately, either immediately or after fifteen minutes delay, more than one of the pair of Binet designs.

He had a severe slurring dysarthria, but there were no other abnormal physical signs in his central nervous system and he was clearly otherwise extremely fit. The stigmata of his profession were mild and included a broken nasal septum and mild carpometacarpal arthropathy.

The investigations were limited to an *electroencephalogram* which was normal.

"There were three other individuals (sample nos. 30, 118, 178), whose ages ranged from 35 to 52, who presented a similar clinical picture in which the neurological lesions were confined to a moderate or severe slurring dysarthria. One of these was an alcoholic, as judged by his claims to regular excessive consumption and the palpably gross hepatic enlargement and gynecomastia. Another, some years earlier, had a fall which had been followed for several months by paresthesiae provoked by flexing his neck, and the question of demyelinating disease had been raised. Neurological investigations at that time proved normal and included negative serological tests for syphilis and examination of the cerebrospinal fluid."

A detailed discussion of the clinical syndrome of the previous mentioned case histories of ex-boxers is best given in ROBERTS own words. ROBERTS study is so important that we present his findings in detail:

"The clinical evidence in these cases is of lesions of cerebellar, pyramidal and extra-pyramidal systems. In *cases 1* and *2* the extra-pyramidal lesions dominate the clinical picture. In *case 3* the syndrome is predominantly cerebellar, but the tremor affecting the right arm was striatal in character and there was evidence of a pyramidal lesion affecting the left side. *Case 4* presented a mild asymmetrical extra-pyramidal picture with reflex asymmetry indicating a pyramidal lesion of the right limbs more than the left. The history was quite clear on the point of progression in two of these, the second and fourth cases, in that an extra-pyramidal tremor had made its appearance some years later on a background of established dysarthria first noticed during their last years boxing. In the first case described it seemed likely that his disability, due to an extra-pyramidal lesion, had become more obvious as he grew older, but it was difficult to be sure that this indicated something other than the normal degenerative process of ageing. His dementia had clearly progressed over the years, and the development of a paranoid illness in the setting of his dementia was a certain measure of this. In the *third case* it seemed probable since he had considered taking elocution lessons only in recent years, that his dysarthria had worsened, but there was no clear independent account that this was so."

"*Cases 5* and *6* presented a predominantly cerebellar picture with additional evidence of asymmetrical pyramidal lesions and associated dementia. In the one for whom there were available the findings of the neurologist whose diagnosis of punchdrunkenness had invalided him from the services, it seemed that the ataxia and evidence of pyramidal lesions were new developments, though the dysarthria and sluggish mentation were not. In this case it was possible that alcoholism had contributed much to his social decline, and conceivably also to the progression of his cerebellar lesions. Progression had undoubtedly occurred in the other in the years following a serious mining accident, but the head injury at that time was not of the severity seen usually with this degree of disturbance of cerebellar, pyramidal and cortical function, and the subsequent steady progression of these lesions a most unexpected sequel."

"*Case 7* illustrates what is clearly a milder form of predominantly cerebellar and pyramidal disturbance. This case, and the *other seven* who presented a very similar clinical picture, all had a notable poverty of facial expression, although the degree of this varied among them, associated with a striking slowness of movement of the limbs which it was difficult to classify with certainty as due to disturbance of pyramidal or extra-pyramidal function. In general the increase in tone in the limbs was more spastic than rigid, and the tendon reflexes were more frequently pathologically brisk than depressed, so that on the whole this feature of the clinical picture was reminiscent more of pyramidal disease. The evidence for progression in these lesions among this group was poor except in the case in the individual reported in detail whose condition seemed to have worsened, or at least become more noticeable to his relatives, after an apparently trivial head injury; and in the case of one other whose unsteadiness and forgetfulness had, according to his wife, become slowly more disabling over the years despite the fact that his drooling had stopped."

"In *case 8* and *9* it was this poverty and slowness of movement, together with the dysarthria and unsteadiness of gait, that dominated the clinical picture, and the differences between these two individuals, and the twelve bracketed with them, was principally one of degree. Facial impassivity was not marked in all, though very obvious in six, and only one spontaneously mentioned his sialorrhoea. Nor was the character of the dyarthria particularly helpful in providing a distinction between cerebellar, extra-pyramidal or pseudobulbar lesions. It was neither festinant nor syllabic, and although repetitive tongue movements were often slowly and a little clumsily performed protrusion of the tongue was never incomplete enough to suggest a spastic paresis. Indeed, all that could definitely be said about their speech was that it was undisputedly slurred, and half of them had no doubt about this themselves. In only three of these cases was there any definitive evidence to indicate that the condition had progressed since they gave up boxing. In one the progression had been a sequel to a severe head injury in a road traffic accident and there was no suggestion that his condition had remained other than stationary when he had recovered from this. In the remaining two such mild progression in the lesions, and particularly in their forgetfulness, as had occurred, might as well have been due to normal ageing as to any other process."

"The distinction between the lesion demonstrable clinically in *cases 10* and *11,* together with the *seven* referred to after these two reports, was merely one of a pyramidal lesion in addition to dysarthria. There was no suggestion of progression in any of these cases. Indeed, one who presented to a neurologist during his boxing career with clinical evidence suggesting brainstem lesions, had only a mild residual dysarthria and an indefinite left extensor plantar response thirty years later. The interesting of this particular case is considerable. It is difficult to deny that the original diagnosis of brainstem demyelination may still be the correct explanation, but the occurrence within this small sample of two other much more certain cases of disseminated sclerosis, and yet another in whom the possibility of the diagnosis had been raised by neurologists, would indicate a prevalence of this disease among boxers far removed from normal expectation. It seems more probable that this case illustrates more clearly than most in the sample a feature of the encephalopathy which has occasionally been described in earlier case reports, namely that the first evidence of the condition may follow one or a series of particularly hard fights, and that it may then regress if the boxer stops fighting."

"The two clinical syndromes presented by the first thirteen cases *(cases 1–7* and the *six* identified with *case 7),* the one predominantly extra-pyramidal and apparently more clearly progressive, and the other predominantly cerebellar, were the most severe examples of traumatic encephalopathy encountered in this series. The similarity between these and the majority of cases reported in detail previously is close. It is evidently the severity of the disability produced by the encephalopathy which results in their presentation in neurological clinics. The *seventh case,* and *six* of these described with him as having similar clinical evidence of cerebellar and pyramidal lesions, were not greatly disabled by them, though two had been seen in neurological clinics and another by a psychiatrist. These *thirteen cases* seem to be the kind labelled *punch drunk in boxing circles.* Certainly, in eight of these, where it was available, the unsolicited opinion of a professional colleague given during the informal conversation designed to establish rapport, in which the names of boxers from the same area already seen were mentioned, confirmed that they were generally recognised as such. This cannot be said of the other twenty-four cases of traumatic encephalopathy diagnosed here. Although the signs of central nervous system lesions were similar, differing only in degree from those found in the major syndromes, these individuals were on the whole only marginally or in no way disabled. Those with insightful were aware of some disturbance of function, particularly of speech, but it is likely that would fail to provoke much lay comment, nor would the other evidence of the lesions clearly demonstrable clinically. Such cases, with traumatic encephalopathy of this degree, have been reported only by those whose practice during the war included the random selection provided by conscription into the services; and the presentation in these cases seems often to have been due to personality characteristics unsuited to military service as much as to the clinically obvious dysarthria or mild disequilibrium."

ROBERTS continued:

"In this review of the evidence of encephalopathy found in the present sample, and in the cases previously reported, more attention has been paid, intentionally, to the clinical signs which indicate lesions of cerebellar, pyramidal and extra-pyramidal systems, than to the evidence of dementia or personality change. Descriptions of physical signs which are evidence of lesions in these systems are readily comparable from one case report to another. This is not so with description of minor degrees of dementia or minor aberrations of personality, the evidence for which is in itself more difficult to elicit and interpret with certainty. Leaving aside for later consideration the question of dementia and psychiatric disturbance, which undoubtedly occurs in some cases, there are left two major clinical syndromes, explicable on the basis of multiple lesions, the one predominantly cerebellar and the other extra-pyramidal. Often there is evidence of corticospinal disease in addition; indeed, where tremor and rigidity were not evident, the poverty and slowness of movement associated with dysarthria and brisk tendon reflexes, so striking a feature of the condition, were more suggestive of a pseudobulbar than mild degrees of an extra-pyramidal motor disorder. By analogy with what is known of the neuropathology of a number of system degenerations it seems likely that the pathological process in these syndromes is similarly diffuse, the signs of involvement of one system merely being obscured by the predominating clinical evidence of the lesions in another. When there is an adequate independent account, or the individual has sufficient insight, most of these boxers appear to have developed the symptoms of the condition during their last years boxing or after a series of particularly hard fights. It seems that some degree of progression may occur which cannot be accounted for simply on the basis of normal ageing, and occurs more commonly in the extra-pyramidal lesions, but apart from a few isolated exceptions reported previously and in one case in the present series *(case 2),* this progression is not in general characteristic of that seen in system degenerations, or in the commoner presenile dementias. There is a good deal of evidence in the present study to suggest that in most cases the condition remains stationary when the individual has stopped boxing, and indeed there are excellent independent accounts for a few of undoubted improvement after their retirement."

"Considered individually, the few previously reported cases, in which the data are given in sufficient detail to support a causal relationship between the neurological lesions and the boxing career, cannot be held to provide more than anecdotal evidence for the existence of the syndrome. Considered collectively, they may with more justification be held to do this, but even then it might be argued that selective reporting of a chance association has given a false impression of the frequency of the occurrence, among boxers, of an unusual neurological syndrome. However, the composite picture provided by the individual reports is sufficiently similar to the lesions that were demonstrable clinically in the present series of randomly selected individuals to make it unlikely that what has been attributed to boxing in all these cases is merely the fortuitous occurrence of unrelated neurological disease in ex-boxers."

Period of Professional Career

ROBERTS noted that the:

"popularity of the sport in the 1930s and before was such that it was possible for the more able professionals to have fights arranged for them almost every week during some parts of the year. Although it was, according to ROBERTS, exceptional to average more than twenty-five fights over the course of the year, the situation was entirely different by the early 1950s. An initial rise immediately after the end of the war was followed by a steady decline in the number of boxers registering as professionals each year, and the annual number of promoted fights even a successful individual might expect would be nearer a dozen.

ROBERTS noted:

"... that in the earlier years, included in his study, there was not the inclination, as there has been increasingly since the war, on the part of either the referee or the boxer, to stop a fight or concede defeat when one of the pair was taking a beating. The style of contest has been altered to some extent by changes in the length of rounds. Although championship and near-championship fights appear to have extended, before the war, to twelve and fifteen three-minute rounds, many other longer contests were of two-minute rounds."

There appears, according to ROBERTS:

"... to have been little difference in the age at which most started to box, but the younger, post-war group of the sample in general had longer careers in amateur boxing, and many more of those had had in the region of a hundred or more amateur contests, usually three-minute rounds, before they had their first professional fight."

Number of Professional Fights

ROBERTS wrote:

"... that although undoubtedly of greater interest as a measure of occupational exposure, the number of fights and the proportion won or lost were data for which there was no independent check for the whole sample." In calculating the number of fights ROBERTS decided to include only the number of promoted professional contests a boxer had had, since it was virtually impossible to get any clear idea of the number of booth fights he had had; and it was quite pointless to attempt to determine what proportion of these were genuine contests or relative atraumatic entertainment.

In regard to the severity of cerebral trauma and severity of encephalopathy ROBERTS wrote:

"It was not possible to get any reliable data about the number of times any boxer had been knocked out; few considered themselves to have been knocked unconscious if they had merely failed to beat the count after being knocked down and perhaps transiently out, although the fight would have been officially recorded as a knock out. Nor was it possible to determine, with any certainty, what proportion of the total number of fights an individual had won or in which he had taken a severe beating. Knock out appeared to have been an infrequent experience of most of the example, and they were certainly not common in the careers of those for whom the records of each fight and its outcome were available. Few admitted to unconsciousness that lasted for much longer than the count of ten seconds. On the other hand, most recalled vividly and in detail lengthy post traumatic amnesias, and it was possible to analyse the longest, with respect to encephalopathy, as the best available but inexact measure of the most severe head injuries sustained."

The *traumatic amnesias* were analysed by ROBERTS in Table 32 with the overall subdivision of age and the length of the professional career.

ROBERTS continued:

"This later measure of occupational exposure still seem to be the major factor determining the higher frequency of encephalopathy, but there is a suggestion in the oldest group that traumatic amnesias of over an hour, and certainly of over twenty-four hours, increase the frequency further."

In regard to the boxers weight and class ROBERTS noted:

"There remain two further features of the professional careers whose effect on the development of encephalopathy seemed worth studying, both because they might be assumed to be relevant and because the information available was reliable. Clearly the force of impact and the acceleration imparted to the head will vary with the weight of the muscles delivering the blow and steadying the head. Some investigators have suggested that lighter boxers are more liable to encephalopathy, either because they more frequently have to fight heavier men, or because the blows exchanged are inadequate to cause unconsciousness and fights continue longer, resulting in the absorption of larger numbers of subconcussive blows."

ROBERTS continued:

"It will be seen in Table 33 that the reverse is suggested in the present study, that the heavier boxer is more likely to develop encephalopathy, espe-

TABLE 32
Longest post traumatic amnesia related to age and length of professional career. (From ROBERTS, 1969)

Age Examined		50+	49-40	39-30	50+			49-40			39-30			49-30		
Number of Years Professional					10+	9-6	5-3	10+	9-6	5-3	10+	9-6	5-3	10+	9-6	5-3
	Totals															
Total Numbers Examined	224	89	86	49	36	30	23	15	25	46	5	18	26	20	43	72
Post Traumatic Amnesias (of one hour or more)	48	26	13	9	9	10	7	4	5	4	1	3	5	5	8	9
Percentage	21%	29%	15%	18%	25%	33%	30%	27%	20%	9%	20%	17%	19%	25%	19%	13%
Sustained Boxing	36	21	8	7	6	9	6	1	4	3	1	1	5	2	5	8
Sustained Elsewhere	12	5	5	2	3	1	1	3	1	1	0	2	0	3	3	1
Number with Encephalopathy	12	8	3	1	6	2	0	1	2	0	1	0	0	2	2	0
P.T.A. Boxing	7	5	1	1	4	1	0	0	1	0	1	0	0	1	1	0
P.T.A. Elsewhere	5	3	2	0	2	1	0	1	1	0	0	0	0	1	1	0
Percentage with Encephalopathy	25%	31%	23%	11%	67%	20%	0	25%	40%	0	100%	0	0	40%	25%	0
Number of P.T.A.'s of 24 hours or more	7	5	1	1	2	3	0	1	0	0	1	0	0	2	0	0
Number with Encephalopathy	5	4	0	1	2	2	0	0	0	0	1	0	0	1	0	0
Percentage with Encephalopathy	72%	80%	0	100%	100%	67%	0	0	0	0	100%	0	0	50%	0	0

TABLE 33

Boxing weight of those with and without encephalopathy related to age and length of professional career. (From ROBERTS, 1960).

Age Examined		50+	49-40	39-30	50+			49-40			39-30			49-30		
Number of Years Professional					10+	9-6	5-3	10+	9-6	5-3	10+	9-6	5-3	10+	9-6	5-3
	Totals															
Total Numbers Examined	224	89	86	49	36	30	23	15	25	46	5	18	26	20	43	72
Weight																
Heavy, Cruiser, Middle, Welter																
Totals	98	36	32	30	14	13	9	3	10	19	2	12	16	5	22	35
Number with Encephalopathy	22	16	3	3	9	4	3	0	3	0	2	1	0	2	4	0
Percentage with Encephalopathy	22%	44%	9%	10%	64%	31%	33%	0	30%	0	100%	8%	0	40%	18%	0
Weight																
Light, Feather, Bantam, Fly																
Totals	126	53	54	19	22	17	14	12	15	27	3	6	10	15	21	37
Number with Encephalopathy	15	9	3	3	8	1	0	2	1	0	1	1	1	3	2	1
Percentage with Encephalopathy	12%	17%	6%	16%	36%	6%	0	17%	7%	0	33%	17%	10%	20%	10%	3%

cially if he boxed longer or before the war. It may be that the suggestions of higher frequencies among lighter weights reported by others is related to the very much larger numbers of lighter weight boxers fighting professionally before the last war."

ROBERTS continued:

"Previous anecdotal reports have also implied that it is only the second rate boxer who develops the syndrome, the assumption being that the skills of the latter boxers saved them from injury. In the present series it was difficult to grade the abilities of the boxers beyond determining which of them had some time held a championship title. This information was reliable and could be checked. And it seemed a reasonable assumption that those who had held even an area title were on the whole better boxers than those who had not. However, before the war, the practice of awarding area titles does not seem to have been universal, with the result that not all those whose abilities were outstanding were able to claim a title. Seven of the thirteen pre-war champions and one of the six post-war title holders had evidence of encephalopathy."

ROBERTS continued:

"The syndrome described in the case reports in the present study may therefore be considered attributable to boxing on two grounds. First, because although occasionally reminiscent it is not typical of commonly occurring neurological disease and is very similar to that described previously in ex-boxers. Second, because the analysis of prevalence, made possible by the complete tracing of the random sample of ex-boxers who had held professional licenses for three years or more, provides strong evidence that the syndrome occurs more frequently among the most exposed to the occupational trauma under investigation.'

In regard to intellectual functions ROBERTS noted:

"Sixteen of the thirty-seven cases with traumatic encephalopathy showed some degree of intellectual function or personality change. Only two (sample no. 34 and 58) were indisputably demented judging by their ability to live unaided within the community. One was a permanent inmate of a mental hospital, and both had paranoid illnesses complicating their dementias. Nine (sample no. 1, 4, 39, 53, 63, 67, 71, 110 and 224) had severe memory impairment according to independent accounts, confirmed in six by the results of the psychometric tests, and six were to a degree that was sufficiently abnormal to affect their adjustment in the community. In another five (sample nos. 14, 19, 54, 64 and 169) mentation was profoundly slow and memory, judging clinically by their poor anamnesis, was defective suggesting some degree of intellectual deterioration, but independent confirmation in these cases was not available. It seems justifiable to conclude that all these were mildly demented, but, from the two undoubted dements already refered to, only three were unemployed (sample nos. 54, 64, and 224). The one other individual with a clear independent account supporting the psychometric evidence of a profound disturbance of memory, which did not prevent his regular employment, nor, with his wife's help, a relatively normal life within the community, had no other evidence of traumatic encephalopathy and is described in Appendix A (sample no. 137).

The *electroencephalographic findings* are reported in the chapter Encephalography.

The report of ROBERTS is such important, that we include its entire summary:

"Previous reports describing the chronic neurological syndromes found in boxers and attributed to boxing have been reviewed where they were sufficiently detailed to provide a clear picture of both the clinical evidence of an asymmetrical pyramidal lesion and some less clearly defined impairment of the intellectual functions and personality change."

"The method used for sampling at random two hundred and fifty ex-boxers from the population of 16.781 registered between 1929 and 1955, and the method of clinical and electroencephalographic investigation is described. The entire sample of boxers was traced and with the exception of nine who had emigrated, sixteen who had died and one who refused, all were available in one hundred and sixty-eight cases and independent account in over half the sample."

"In thirty-seven cases there were clinically demonstrable lesions of the central nervous system which bore a close resemblance to the syndrome previously described in boxers. These followed a relatively stereotyped clinical pattern. At its mildest this consisted of dysarthria with or without disequilibrium and evidence of asymmetrical pyramidal disease, and at its most severe of moderately disabling ataxias, disequilibrium and spasticity, or rigidity and striatal tremor associated with varying degrees of dementia. All grades of disturbed neurological function in cerebellar, pyramidal and extra-pyramidal systems were apparent clinically between these two extremes. Impairment of intellectual function was disabling to a degree that precluded the unsupported existence in the community of only two cases, and both these had had paranoid psychoses, but as with the other neurological lesions there were indications less clearly demonstrable of all degrees of defective function, particularly of memory. The evidence for progression in these lesions other than that conceivably attributable to normal ageing was poor, except in the case of one individual with a predominantly cerebellar and dementing syndrome, and in the case of three with extra-pyramidal syndromes."

"The occurrence of this syndrome has been analysed with respect to age and occupational exposure, and it was demonstrated that there was an increasing prevalence with increasing exposure, these observations reaching conventional levels of statistical significance. The analysis of a number of other isolated physical signs and symptoms of disturbed neurological function in those not diagnosed as having evidence of traumatic encephalopathy showed similarly their more frequent occurrence among those exposed most to the occupation under investigation. No other systemic disease was distributed among the sample in a way that could not be explained on the basis of chance occurrence or age determined factors."

"Analysis of the electroencephalographic records demonstrated no difference overall between sample and controls, or within the sample when considered with respect to age and occupational exposure. It therefore seems unwise to assume that a single electroencephalogram can be used to support a diagnosis of boxers' encephalopathy."

"The study has shown the relationship to occupational exposure of an easily recognisable and relatively stereotyped syndrome of disturbed neurological function. Seventeen per cent of the sample were affected. Severely disabling degrees of this syndrome were encountered very infrequently, and not all in those professional careers followed the last war. There were indications that much minor clinical evidence of disturbed neurological function, demonstrable among those not diagnosed as having the syndrome of traumatic encephalopathy was also related to their professional careers."

Further Clinical Studies

JEDLINSKI et al. (1971) reported on a series of examinations carried out, between 1961 and 1964, on a group of men aged 21 to 36 years, who had been *amateur boxers* for *periods of five to 17 years* and *had taken part in 100 to 344 bouts.* The study included all amateur boxers in the district of Cracow, Poland, who had fought more than 100 times.

No neurological findings were seen in 23.3%. In 21.7%, only functional symptoms were observed, in the form of nervous irritability. Both of these groups, 45% of the sample, were assessed as neurologically normal. In the remaining 33 boxers (55%), the *following organic symptoms were observed:* (a) *Absent or unequal abdominal reflexes* (18 persons); (b) *indistinct speech* (7 persons); (c) *mild signs of cerebellar uncoordination* (3 persons); (d) *poor mimicry* (5 persons); (e) *distinctly uneven deep reflexes* (3 persons); (f) *oculomotor nerve paresis* (one person); and (g) *focal epilepsy with adversive convulsions* (one person).

In the control groups, no organic symptoms were observed; 10% of the controls showed nervous irritability.

The results of the neurological examinations were classified according to an arbitrary four-degree scale:

0 – negative results of examination (23.3%)
I – functional disorders (21.7%)
II – isolated organic symptoms of slight intensity (33.3%)
III – Pronounced organic symptoms or disease syndromes (21.7%)

The *psychiatric findings* were also *classified into four groups:* The *first group* consisted of boxers in whom no or only slight abnormalities were observed. The *second group* consisted of boxers with neurotic, usually neurasthenic, complaints. Both these groups were considered normal. The *next two groups* included boxers with distinct mental aberrations. In the *third group* were included boxers who, besides headache and vertigo, exhibited irritability, explosive temper and psychomotor retardation. In the *fourth group* , psycho-organic syndromes were evident in the form of beginning dementia.

JEDLINSKI et al. observed an amateur boxing contest lasting nine minutes during the semi-final and final boxing competitions at the *European Championship Tournament* in 1965. Depending on weight, the arithmetic mean of the number of punches was between 60 and 120, of which 82.7% were aimed at the head and 60% attained their aim.

Amateur boxing in Poland, in accordance with sports regulations current in 1971, led to mild neurologic, psychiatric, and electroencephalographic symptoms in about one-half of boxers engaged in this sport for more than five years (after about 150 battles). The symptoms, although clinically not indicating disease, must be assessed as prodromal symptoms of encephalopathy.

Clinical syndromes requiring hospitalization developed in about 10% of amateur boxers. The symptoms may increase after the boxer has stopped fighting.

The *punch drunk state,* or *traumatic encephalopathy of boxer,* is believed widely – and not only by apologists for boxing – to be an occupational hazard of the past. This improvement is attributed to the undoubtedly closer medical supervision of boxers, with shorter fights, better matching of opponents, limitation of the frequency of fights, and a decline of 'booth boxing.' BLONSTEIN, for example, has stated:

"With shorter bouts and modern medical control, the punch-drunk syndrome has not been seen during the last 20 years in amateur boxers."

This view is not supported by empirical evidence

collected during the past 20 years. HARVEY and DAVIS (1974) describe a case of the punch drunk syndrome in a young boxer whose ring career started in 1962, and whose symptoms developed after 72 adult amateur bouts and 16 professional bouts.

The traumatic encephalopathy of boxers presents as a relatively stereotyped clinical picture characterized by evidence of damage to one or more of the pyramidal, extrapyramidal, or cerebellar systems with associated dementia, psychosis, personality change, and social instability in some cases. Symptoms tend to be progressive and may develop during the boxer's career or begin many years after retirement from the ring. Air encephalography may show changes of cerebral atrophy, and a cavum septi pellucidi indicating rupture of its walls. At necropsy, this may be widely fenestrated in posterior parts. Pathological changes include damage to the extrapyramidal pathways, cortical scarring, cell loss of the cerebellum, and neurofibrillary tangles.

A 25 year old middleweight boxer was first seen by HARVEY and DAVIS in November, 1973.

He had previously been in good health, without a personal or family history of psychiatric or neurological disorder. He started boxing at the age of 14, when he would spar once or twice a week and fight club matches once a month. He became a successful schoolboy international. Between the ages of 16 and 21, he fought in about 75 amateur contests, becoming a regional champion and an international champion before turning professional. As an amateur, he lost about ten fights on points, took no severe beatings, was not knocked out, and experienced no episodes of amnesia related to bouts. About once a week during these five years, he was a sparring partner to professional fighters. He and his family deny that he was ever seriously hurt when sparring.

At the age of 21 he became a professional, retiring from the ring 3 1/2 years later in November, 1972. He fought a total of 25 professional bouts, winning 18, losing four 'on points' and drawing once. He was only knocked out once, but another fight was stopped because of the injuries he was receiving. He won his last five fights, one of which was against a boxer who subsequently became British champion.

His family first noticed that his speech was slurred after he had 16 professional bouts. Six months later he started to shuffle slightly when walking, his legs appearing stiff. Despite his success in his last few fights, he was aware of slowness on his feet and of increasing difficulty in avoiding trouble in the ring. He retired from the ring a year before his admission to hospital. During that time, his gait and speech deteriorated. He developed lability of mood, becoming more aggressive and socially unstable. An acute depressive illness with paranoid features then necessitated his admission to a psychiatric unit, where for a few days he received a small dose of chlorpromazine followed by trifluoperazine 5 mg twice daily for about two weeks. A neurological disorder was also diagnosed, and he was later transferred to the National Hospital, at which time he had not taken phenothiazines for four weeks.

He was cooperative but mentally slow. He had a verbal I.Q. of 95 and a performance I.Q. of 88. He was normotensive. Bilateral ptosis was present. He had a symmetrical extrapyramidal disorder characterized by an expressionless face, generalized poverty of movement, cogwheel rigidity in all four limbs, and typically extrapyramidal speech and gait. His tendon reflexes were brisk, with left extensor plantar and variable right plantar response. There was no sensory deficit, and sphincter function was normal.

Roentgenograms of the skull and *cervical spine* were normal, as were the *brain scan* and *EEG. Computerized axial tomography* showed mild ventricular dilatation. This was confirmed by a *lumbar PEG,* which also revealed a cavum septi pellucidi. An intrathecal RIHSA scan showed some ventricular reflux in early films, but normal convexity activity at 24 and 48 hours. He was treated with a combination of levodopa and Carbidopa, which resulted in some improvement in his gait. After discharge he had another episode of disturbed behaviour and had to be readmitted for a period to a psychiatric hospital.

HARVEY and DAVIS concluded that extrapyramidal and pyramidal dysfunction development in this young boxer, coupled with depressive and paranoid mood swings, "leaves us in no doubt that he has the traumatic encephalopathy of boxers." This clinical diagnosis is supported by the radiological demonstration of a cavum septi pellucidi, a characteristic feature of this syndrome, and the changes of early cerebral atrophy.

The predominantly left-sided pyramidal signs accord with the observations of ROBERTS, who, surveying ex-professional boxers fighting between 1929 and 1955, noted a predilection for pyramidal involvement of the left limbs. He attributed this to the fact that the majority of the most powerful blows are delivered from the same direction and thus tend to rotate the brain about the same axis.

Encephalopathy was recorded in several boxers

in the same study, whose careers ended in their mid-twenties, although their professional careers were longer than that of our patient. In individual cases a long career as an amateur or as a sparring partner was thought possibly to be significant. In the patient published by HARVEY and DAVIS, the onset of symptoms was at a relatively early age and at a time in his career when most of his boxing experience had been as an amateur. This suggests that his relatively intensive career as an amateur, coupled with frequent sparring with professional boxers, often heavier and inevitably more experienced, was a major factor in the development of his illness. (Such cases inevitably present evidence that amateur boxers are not categorically immune to neurological injuries from boxing.) The factors that influence the degree and distribution of pathological alterations in professional boxers', such as the number and intensity of matches, must be at play, regardless of the regulatory status of the boxer.

HARVEY and DAVIS concluded that they find it disturbing that the punch-drunk syndrome still occurs despite modern medical control of boxers. On the evidence of the present case, amateur experience can predispose a boxer to the development of a punch drunk state. There are good grounds in particular for recommending closer medical supervision of sparring as well as formal matches.

THOMASSEN et al. et al. (1979) reported that during the period 1955 – 1965, a total of 114 boxers participated in the annual regional amateur boxing championship, Jütland, Denmark.

Twenty-one could not be traced, five had emigrated, two had become professional boxers, and one was disabled from an accident. Of the remaining 83, 70 promised to co-operate in an investigation. As, however, only 61 of these had available boxing start booklets, this group was invited for examination. Of these, 55 completed the examination, but the series was further reduced to 53, as two of the boxers appeared to have had severe head traumata outside the boxing ring.

As a control group, 53 former First Division football players were examined. The controls were selected according to period of football career, age, occupational level and number of possible hazards to the brain (outside football matches) such as traumata, poisonings, diseases, etc.

Among the boxers there were a pair of identical twins, and another two had monozygotic twin brothers, who were later included in the study.

Information about the boxing careers was obtained from the boxers themselves and from their individual start booklets, in which dates, weight, and outcome of all fights as well as quarantines and, to some extent, injuries, were recorded. For each boxer, the length of career, number of fights (total and per year), and average weight class) were noted. The ages at the debut and cessation of the boxing career were registered and so were the number of years passed since first and last fights. Furthermore, the numbers of fights lost and stopped before time in the disfavor of the boxer as well as groggy states and knock outs (k o) were noted. Only defencelessness for 10 seconds caused by blows to the head was reckoned as k o, the severity of a k o was estimated from the duration of posttraumtic amnesia. The incidence of retrograde amnesia and persistent symptoms of concussion after fights were noted. Furthermore, the boxers were questioned about other symptoms in connection with fights, especially amnesia without conciousness and about frequency, length, and cause of quarantines.

In the control group, possible head trauma in connection with football matches were noted, but otherwise they werde not questioned about details of their sports career.

In each case, detailed social and clinical histories were taken, with particular attention to symptoms and causal relationships of brain dysfunction. Anamnestic facts that may influence the result of the examination are called *background variables*. After this, all were subjected to clinical neurological and neuropsychological examination and an electroencephalogram including a period of hyperventilation and photic stimulation was recorded. The individual examination took about three hours.

In the preliminary conversation of the psychological investigation, nearly all boxers expressed the view that they regarded their sport as important for their personal development and social progress.

The neuropsychological examination was initiated with the vocabulary test of Wechsler Adult Intelligence Scale, evaluated according to Danish standardization. The psychological functions were then examined according to the methods of LURIA. The performances were evaluated qualitatively and checked according to the principle of double dissociation, i. e. all functions that include a disturbed factor are affected, while others that do not are preserved.

In addition all subresult were graded quantitatively from 1 to 5 according to the severity of the dysfunctions and subsequently subjected to EDB.

The statistical treatment included profile analysis, multiple and partial regression analysis, and for this was used a statistical package as well as our own programs.

Differences between boxers and control group were investigated for each variable and in each group the correlation between subresults and background variabes was examined, and especially correlations between results and 'occupational exposure' during the boxing career. In the presence of significant relation between a dysfunction and one or more background variables, a attempt was made to correct the results for this influence by means of partial correlation analysis and paired comparisons.

Results

Boxing Career (Table 34)

A total of 56 KO's were recorded, according to THOMASSEN et al. Of these, 41 were judged as slight, eleven as moderate and four as severe; the most severe was followed by 12 hours posttraumatic amnesia. In five cases of KO, there was also retrograde amnesia, most often for one round of the fight. Amnesia without unconsciousness covering from one round to the whole fight had been experienced by six boxers, who apparently had been completely unaffected, and the referee did not interfere. One young boxer had encountered a total of 15 such amnestic episodes during boxing bouts and at the end of his career he also had similar attacks between fights, which made him to stop boxing, following which he had no further attacks.

The usual symptoms of brain concussion, according to THOMASSEN et al. after a fight were rare. Transitory nausea and headache had been experienced by four boxers, and one had stayed in hospital for three days after a KO because of concussion. One hour after winning a fight, another boxer suddenly became unwell with severe headache and nausea; at first, he was in hospital for only three days, but became incapacitated for work, he had to be readmitted for another 42 days with pronounced signs of of postconcussion syndrome.

Of more permanent symptoms, one boxer had had impaired memory and difficulty in concentration; another had had slight dizziness, and a third had been numb in half of his face. All symptoms had diappeared at most one month after the fight.

Background Variables

Figure 228 in the series of THOMASSEN et al. shows that there was a slightly significant deviation of age, the boxers being younger than the control group. A marked difference in schooling and education to the disfavor of the boxers ($P < 0.01$) is seen. While the occupational level is comparable. Nearly all were socially well adjusted.

The football players had been exposed to slightly more head traumata than the boxers (outside the boxing ring), namely 47 against 40. Of these, eight had occurred during football matches. The poisonings mentioned were due mainly to working with organic solvents and were equally represented in the two groups. Neurologically three boxers had

TABLE 34
Information about the boxing careers. (From THOMASSEN et al., 1979).

	Minimum	Maximum	Median	SD
No. of:				
Fights (total)	19	209	76	44
Fights per year	4	22	10	6
Fights lost	5	82	30	16
Fights lost and stopped before time	0	16	6	4
Groggy states	0	>30	5–10	4
Knockouts	0	8	1	2
Lenghts of career (years)	3	16	8	3
Lenghts of quarantine (days)	0	231	30	58
Age at debut of boxing	15	25	16	2
Age at cessation of boxing	18	35	25	4
Years passed since first fight	10	33	18	5
Years passed since last fight	3	19	11	4
Average weight class (kg)	50	93	66	10

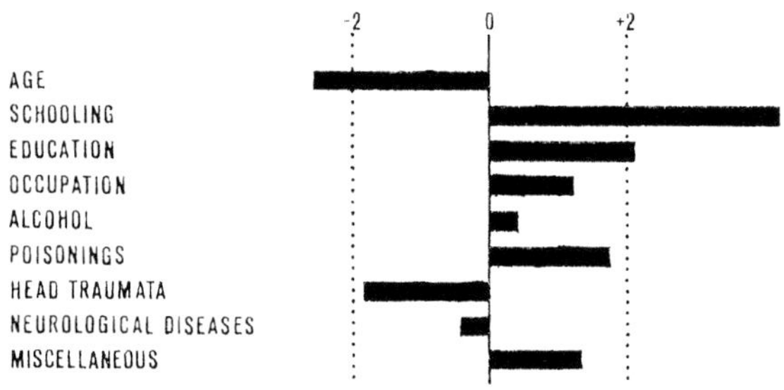

Fig. 228. Background variable, i. e. anamnestic facts, that may influence the results of the examination. Differences between boxers and football players are seen from the length of the colums. A numerical value above 2 indicates a significant difference ($P < 0.05$). A positive deviation means a disadvantage to the boxers and vice versa. (From THOMASSEN, et al., 1979).

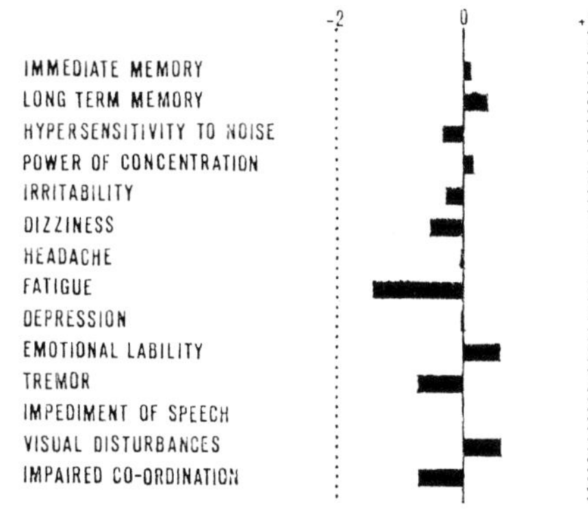

Fig. 229. Complaints – difference between boxers and control group. For key to colums see Fig. 228. (From THOMASSEN, et al., 1979).

had polyradiculitis, meningitis and fits (not clear-cut epilepsy), respectively, and three football players had had the same diseases, all with no or only slight sequelae. 'Miscellaneous' includes such diseases as tetanus, diabetes mellitus, peripheral embolism and severe anemia.

Complaints

From Fig. 229 of the series of THOMASSEN et al. appears that there were no significant differences in the number or degree of complaints in boxers and football players, although there was a slight preponderance in the latter group. A total of 24 boxers and 37 football players had slight complains, especially concerning memory, and concentration. There was no correlataion between complaints and neurological, electroencephalographic or psychological findings, and for the boxers neither with the 'occupational exposure' during the boxing career.

Electroencephalography

The EEG patterns in the series of THOMASSEN et al. did not show any significant differences between boxers and control group (Fig. 230). Normal EEG was found in 66 cases, while 24 had slight and ten had more distinct theta activity diffusely distributed. Six records showed focal slow activity or sharp waves. The EEG dysrhytmias were in 60% predominantly or exclusively located over the left hemisphere in both groups.

EEG dysrhytmias showed no correlation with neurological or psychological findings. In the boxers there was no relation to traumatic boxing experiences.

Neurological Examination

In 34 boxers and 45 football players, the neurological examination showed normal conditions. Positive neurological signs (Fig. 230) were sparse and discreet, e. g. slight nystagmus, slurred speech, clumsiness of fine movements and single brisker tendon reflexes. There were no significant differences between boxers and football players: in particular there was no accumulation of symptoms in individual boxers giving rise to suspicion of incipient boxer encephalopathy.

No correlation was found between neurological and other examination results and especially with regard to traumatic experiences in the boxing careers.

Neuropsychological Investigation

It appears from Fig. 231 of the series of THOMASSEN et al. that the performance of vocabulary test was significantly better that in the control group ($P < 0.01$), but for both boxers and football players the achievements were equal to or above the level of normal intelligence. A number of psychological subfunctions were normal in the members of both groups. Furthermore, a total of 14 boxers and 28 football players showed normal functions in all tests or disturbances in single tests

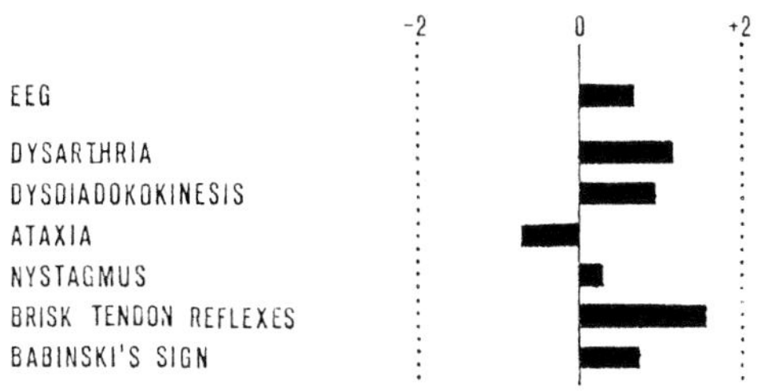

Fig. 230. Electroencephalographic and positive neurological findings – differences between boxers and control group. For key to columns, see Fig. 228. (From THOMASSEN, et al., 1979).

that were insignificant in accordance with the double dissociation principle.

Fig. 231 of the series of THOMASSEN et al. shows that the boxers had more dysfunctions than the control group, and that the differences were significant ($P < 0.05$) in five subtests: (1) Motor function of the left hand, (2) Higher visual functions in intellectual operations in space, (3) Expressive speech, in the repetition of single words, series of words and of short and longer sentences, (4) Pho-

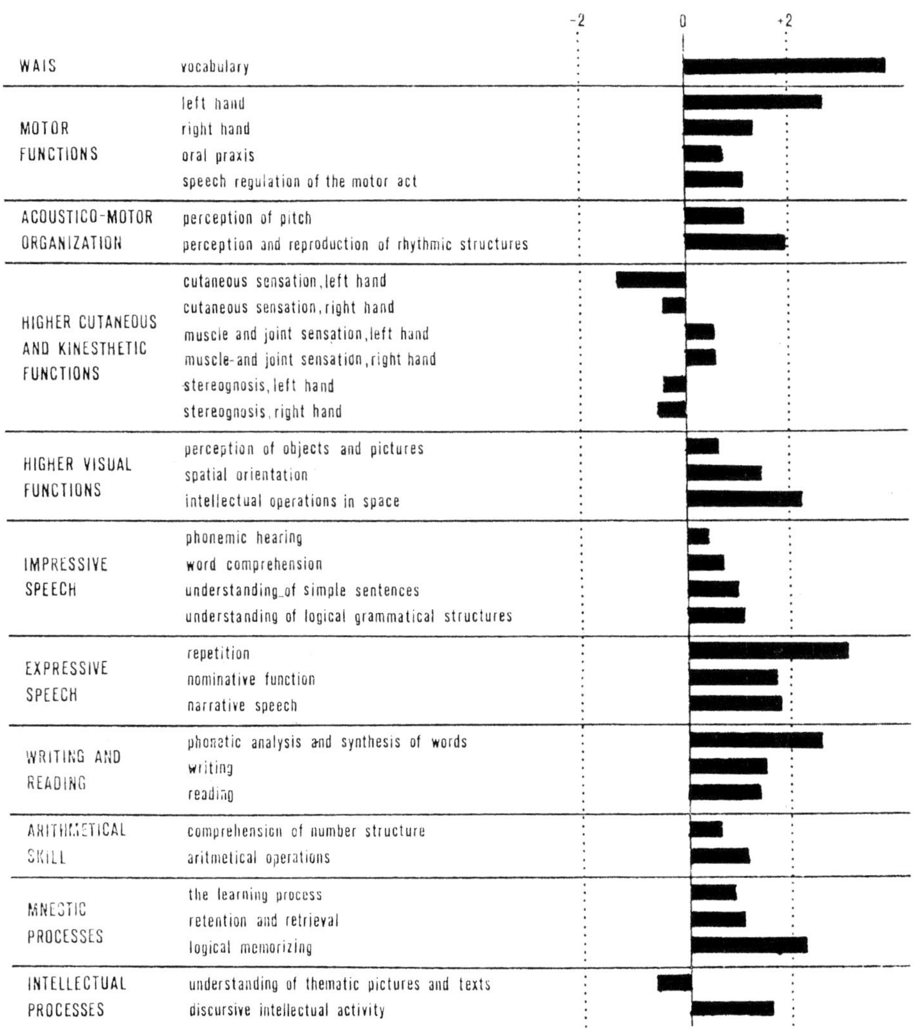

Fig. 231. Neuropsychological investigation – differences between boxers and control group. For key to colums, see Fig. 228. (From THOMASSEN, et al., 1979)

netic analysis of number and synthesis of letters in simple and complex words, (5) Memory, especially logical memorizing, in the testing of which the patient is presented with picture cards corresponding to each word which is to be recalled. By comparing quantitatively the elements of a learning test (10-word test), a slower and more uncompleted learning was found in the boxers ($P < 0.05$).

Discussion

THOMASSEN et al. noted that permanent brain damage has never been diagnosed in sportsmen other than boxers. This clinical picture has especially been observed in old time professional boxers with a career before or during World War II (CRITCHLEY 1937, MAWDSLEY and FERGUSON 1963, ROBERTS 1969). In amateur boxers it has been described, according to THOMASSEN et al., before the introduction of modern safety precautions (MAQUARDSEN 1954, MAWDSLEY and FERGUSON 1963), but after this time only one investigation reports typical encephalopathy in a young British boxer who after an amateur career had 16 professional boxing bouts (HARVEY and DAVIS 1974), while HOLLNAGEL (1974) and RUDIENGAARD (1974) describend atypical cases of uncertain etiology in Danish boxers.

According to THOMASSEN et al. in this study, no neurological signs of the specific syndrome were found in 53 former amateur boxers. Pneumencephalography was therefore not clinically indicated and CAT brain scanning was not yet established. A comparison between the results of the examination of boxers and a control group consisting of 53 former football players showed that complaints, neurological signs and EEG patterns were similar in the two population samples.

In both boxers and football players there was a significant correlation ($P < 0.01$) between vocabulary, psychological dysfunctions and the two background variables, schooling and education. These two background variables differed greatly between the two groups, and therefore a partial correlation analysis was made. The analysis showed coincidence between the psychological functions and the deviations in schooling and education.

Finally, THOMASSEN et al., therefore, made a paired comparison, including 19 pairs, each consisting of one boxer and one football player of the same age, schooling, education and vocabulary. In this comparison, the boxers still showed poorer functions in the same five subtests mentioned above, but now significant only in one test ($P < 0.05$), namely the motor function of the left hand.

Investigation of Twins

Results of the examination of the three pairs of identical twins are seen in Table 35 of THOMASSEN et al. All had been boxers, but with great deviations in number of contests in

TABLE 35
Information of three pairs of identical twins. (From THOMASSEN et al., 1979).

Pair	No.	No. of fights	Complaints	Neurological examination	EEG	Neuropsychological function
1	A	80	None	Normal	Psychomotor variant	Impaired motor function of left hand
	B	140	None	Left-sided Hoffman's reflex	Psychomotor variant	Impaired motor function of left hand
2	A	24	None	Normal	Normal	Impaired logical memory
	B	240	None	Ankle jerk brisker left	4–7 Hz diffusely	Impaired learning
3	A	54	Impaired memory	Normal	Left-sided focal changes	Normal
	B	97	None	Impaired olfactory sense	Left-sided focal changes	Normal

each pair. No correlation between abnormal findings and differences in boxing careers was demonstrable. The neurological signs and psychological dysfunctions were discreet, and the rather distinct EEG dysrhytmias found in pairs 1 and 3 were in each pair identical and probably congenital.

THOMASSEN et al. continue:

"In this study, no neurological signs of the specific syndrome were found in 53 former amateur boxers. Pneumoencephalography was therefore not clinically indicated, and CAT brain scanning was not yet established. A comparison between the results of the examination of boxers and a control group of 53 former football players showed that complaints, neurological signs and EEG patterns were similar in the two population samples. From earlier publications, however, is known that single EEG records are of no diagnostic value in boxing encephalopathy (ROBERTS 1969)."

THOMASSEN et al. conclude:

"The more delicate neuropsychological investigation revealed more dysfunctions in boxers than in controls, but after statistical corrections for deviations in schooling, education, age and vocabulary, the differences between the two groups were significant only for one subtest ($P < 0.05$), which showed slight motor dysfunction of the left hand in boxers. This may be in accordance with earlier reports of predominantly left-sided symptoms in boxers (ROBERTS 1959)."

"It must be emphasized that there were no mutual correlations of abnormal findings or between these differences due to 'occupational exposure' during the boxing career, not even in three pairs of monozygotic twins."

"Furthermore, none of the boxers themselves has noticed embarrassing symptoms and therefore the direct dysfunctions found during the examination affected neither the boxer's normal daily activity nor their social adjustment. Critics may object that the 53 boxers do not constitute a random sample of all former amateur boxers. However, they all had an extensive boxing career with more fights than an average boxer. The frequency of KO's was rather low (1% of fights), but animal experiences have demonstrated that it is the cumulative mild head traumas rather than KO's that cause permanent brain damage (UNTERHARNSCHEIDT 1970). Therefore, injurious effects of this kind should be detectable in this series of former amateur boxers. Since no signs were present, however, this study indicates that modern medical supervision and safety precautions protect amateur boxers against the risk of serious and permanent brain damage."

BRAYNE et al. (1982) noted that creatine kinase isoenzyme BB (CK-BB) is found in high concentrations in the brain. Normally concentrations in blood are undetectable or low. Blood concentrations of CK-BB, were measured in 16 boxers before and after a fight and in 16 track cyclists before and after a race. Blood CK-BB rose to significantly higher concentrations in the boxers than in the cyclists. The number of blows received to the head was estimated in half of the boxers and correlated significantly with the rise in CK-BB. This increase in blood CK-BB may indicate disruption of the blood-brain barrier. This may be one of the mechanisms accounting for brain damage in boxers.

CASSON et al. (1982) performed detailed neurological examinations, EEG and CT scans on ten professional boxers, aged 20 to 31 years, shortly after being knocked out. Intracerebral and subdural hematomas were not suspected or discovered. However, CT scans revealed cerebral atrophy in five of the boxers; this finding was most common in the boxers with the most bouts. The possible relationship of these findings to the chronic encephalopathy of the boxers is discussed by the authors.

The goals of their study were twofold: (1) to determine if CT scanning of the head can contribute to the assessment of boxer's conditions after knock outs and (2) to correlate the results of neurological examination, electro-encephalography, and CT scanning in these fighters.

The authors included in their study ten active professional boxers who were knocked out in bouts or were judged by a ringside physician to have sustained significant head trauma in a technical knock out. According to the authors a 'knock out' occurs when a boxer is knocked to the canvas and is unable to regain erect posture within ten seconds.

The evaluation of the boxers consisted of (1) a formal neurological examination, including a screening mental status examination, (2) an electro-encephalogram performed with an 8-channel Hiltman Medicraft Mark III electroencephalograph, and (3) a non-contrast CT scan of the head performed with a Picker Synerview 600 fourth generation body-scanner. The clinical examination and CT scan were done within one week of the knock out. The EEG was obtained within two weeks of the knock out. The neurological examination, EEG, and CT scan are each performed and interpreted independently.
The boxer's age ranged from 20 to 31 years. All weight classes have been included, with a preponderance of heavier fighters. The total number of professional bouts ranged from two to 52. The authors included fighters of championship caliber as well as mediocre and poor fighters. All the boxers denied alcoholism or drug use. All the subjects of this study had sustained mild head injuries by clinical criteria. In no case did the duration of the posttraumatic amnesia exceed two minutes. One of the ten boxers had been knocked out. One (no. 7) had sustained frequent powerful blows to his head during the two rounds before a fight was stopped, due to his profuse facial bleeding. All the boxers were fully alert and oriented at ringside examination after the fight.

The details of each subject's boxing career and the results of his clinical examination, EEG, and CT scan are depicted in Table 36. Only one boxer (no. 4) had an abnormal neurological examination; he had a mild organic mental syndrome manifested by impairment of recent memory and recall, confusion and dyscalculia. There were no focal neurological deficits. The EEG and CT scan were normal.

Five boxers had definitively a*bnormal CT scans,* three with mild, generalized, cerebral atrophy and two with central cerebral atrophy. One of these boxers also had a cavum septi pellucidi. A generalized atrophy pattern is one in which there is a cortical sulcal prominence and commensurable dilatation of the ventricular system. The central atrophy pattern consisted of dilatation of the ventricular system without evidence of an obstructive lesion and with little or no cortical sulcal prominence. No extracerebral or intracerebral hematomas, contusions, edema, white matter demyelination, infarcts or areas of porencephaly were identified.

Two boxers had definitely abnormal EEG's, both with increased slowing in the theta range anteriorly, bilaterally but asymmetrically. Both had normal clinical examinations. One had a normal CT scan; the other had a central atrophy pattern and a cavum septi pellucidi on the CT scan.

CASSON et al. write, that boxing is a violent sport in which the participants intentionally attempt to knock out their opponent . They continue:

"Although it has been claimed that very few effective head blows are delivered in professional fights (KAPLAN and BROWDER (1954), the alarming incidence of serious head injuries in boxers provides clearcut evidence of the dangers of the sport."

The authors initiated this study to determine if neurological evaluation, electroencephalography and computed tomography might help to prevent some future cerebral injury.

The authors were surprised by the finding of cerebral atrophy on the CT scans of half of these active boxers. Within their group, the age of the fighters did not correlate with the presence of cerebral atrophy. Interestingly, the more proficient fighters had cerebral atrophy more often than the less proficient one. Two of the four ranked boxers in this series had cerebral atrophy on CT. Although the fourth has a normal CT scan, he was the only boxer in the series with an organic mental syndrome. All three of the fighters who had attained champion status during their careers had cerebral atrophy on CT scan and no organic mental syndrome.

The total number of professional bouts also correlated with the CT abnormalities. Of the five fighters with 20 and more fights, four had cerebral atrophy . *The number of fights, not the proficiency of the boxer, appeared to be the critical factor in injury patterns.* Since none of the boxers had been knocked out more than two times in their careers, the cumulative effect of multiple subconcussive head blows is the most likely culprit.

One of the boxers of CASSON et al. (no. 7) had a cavum septi pellucidi, with central cerebral atrophy visible on his CT scan. This former world champion's style of fighting has always been that of a

TABLE 36
Results. (From CASSON, et al., 1982).

Boxer	Age (yr)	Rank, professional stature	Number of pro fights	Record (W-L-T)	Times knocked out	Duration of loss of consciousness in recent knockout	Mental status and neurological examination	CT scan	EEG
1	25	Unranked	22	13-7-2	2	5 seconds	Normal	Mild generalised cerebral atrophy	Normal
2	20	Unranked	2	0-2-0	1	10 seconds	Normal	Normal	Normal
3	31	Ranked top 10 former Olympic champion	34	23-11-0	1	5 seconds	Normal	Moderate central cerebral atrophy	Normal
4	23	Ranked top 10	20	14-6-0	2	5 seconds	Mild organic mental syndrome; no focal deficits	Normal	Normal
5	22	USA champion kickboxer	44	43-1-0	1	5 seconds	Normal	Mild generalised cerebral atrophy	Normal
6	28	Unranked	9	6-3-0	1	10 seconds (post traumatic amnesia 2 minutes)	Normal	Normal	Minimally abnormal anteriorly more on left
7	28	Ranked top 10 former world champion	52	46-5-1	0	None. TKO for facial cuts	Normal	Cavum septi pellucidi Mild central cerebral atrophy	Minimally abnormal anteriorly more on right
8	21	Unranked	4	2-2-0	2	2 seconds	Normal	Mild generalised cerebral atrophy	Normal
9	28	Unranked	12	8-4-0	1	1 second	Normal	Normal	Normal
10	23	Unranked	9	6-3-0	2	10 seconds	Normal	Normal	Normal

'slugger' and he is noted for the great deal of battering punishment that he has received as well as given. The authors summarize:

"We submit that CT findings of cerebral atrophy, with or without a cavum septi pellucidi, might predict the later development of clinical chronic encephalopathy in presently asymptomatic boxers."

The authors referred to a claim sometimes raised in boxing debates that very few effective head blows are actually delivered and received in professional fights (KAPLAN and BROWDER 1954). The alarming incidence of serious head injuries in boxers provides clear evidence of the dangers faced in boxing.

SIRONI et al. (1982) evaluated the possibility of early detection of minimal cerebral damage in professional boxers.

Ten young pugilists were submitted to a *complete neurological examination, EEG recording, and CT-scan.* Neuroradiological data, electrical patterns and neurological aspects were examined in connection with the sporting career of each boxer. *Neurological examination* was normal in all cases. *EEG patterns* were considered normal in three cases, borderline in four cases and pathological in three cases. *CT-scan* was considered normal in four cases, borderline in four and pathological (showing signs of cerebral atrophy) in two cases. A good correlation was found between CT-scan pictures, EEG patterns and the sporting careers of the boxers: the pugilists who suffered knock outs had pathological and/or borderline CT-scan and EEG's. The pugilists who did not experience knock outs during their careers showed a normal CT-scan and EEG, independently of the years of activity and number of matches.

ROSS et al. (1983) published a classic study on *'Boxers – Computed Tomography, EEG, and Neurological Evaluation.'* These authors begin their study with some introductory remarks that we will bring in toto:

'Boxing is the only sport in which the intent is to inflict physical injury. Particularly in professional boxing, the fighter who injures and ultimately causes his opponent to become helpless earns, for some atavistic reason, the plaudits of the fans and therefore the highest monetary recompense."

To investigae the problem of predicting the onset of dementia pugilistica further, these authors examined boxers by *computed tomographic (CT) scan of the brain* and *attempted to correlate their findings with concomitant neurological and EEG examinations.*

The authors mentioned their study in the widely read column of a prominent Cleveland sports writer, and were able to attract 40 ex-boxers for evaluation. All were volunteers, none (except boxers no. 36 and 37) were currently boxing, and all gave informed consent for inclusion in the study.
Thirty-eight boxers had a CT scan, and 24 boxers had a complete neurological examination by a Board-certified neurologist and EEG. Of these 24 boxers, 22 also had a CT scan.
Each boxer filled out a questionnaire indicating present age, age of first and last bout, amateur and professional wins, losses, draws, knock outs, technical knock outs, and specified symptoms. The 24 boxers submitting to a neurological examination had a full neurological and medical history recorded in addition.
The CT scans were evaluated by three radiologists independently; the group had collectively evaluated more than 15.000 scans in clinical practice. Each scan was graded by all three radiologists, on a scale of 0 to 4 regarding sulcal enlargement (i. e., gyral atrophy) and 0 to 4 regarding ventricular enlargement; 0 was normal, 4 most abnormal. Scores of the three radiologists were averaged.

Computed tomographic scan values (Table 37) represent addition of the score for sulcal enlargement plus ventricular enlargement. Thus, 0 would be most normal, 8 most abnormal. Total CT values, as well as separate values, were analyzed statistically.

Statistical analysis and consultation were carried out by means of the Jonckeere-Terpstra test. A p value equal to .05 or less was considered significant.

The data demonstrated a significant relationship between the number of bouts fought and the composite CT score ($p = .0229$) (Table 38). No significant relationship was demonstrable between sulcal enlargement (or gyral atrophy) and the number of bouts fought ($p = .11$) (Table 39). The number of bouts fought was significantly related to the pres-

TABLE 37
Boxers historical data and examination results. (From ROSS et al., 1983).

Boxer/ Age, yr	Amateur W	Amateur L	Amateur D	Professional W	Professional L	Professional D	KO	TKO	S×S	Exam	EEG	CT†
1/49	45	3	0	0	3	4	0	2	+	+	–	0
2/57	8	12	0	0	0	0	4	0	+	–	–	0
3/37	64	6	0	13	2	0	0	8	–	–	–	1
4/57	5‡	9	0	0	0	0	2	1‡	+	+	–	6
5/66	125	10	0	61	4	0	0	2	–	ND	ND	4
6/25	127	20	0	13	0	0	0	0	–	ND	ND	2
7/61	28	4	0	0	0	0	0	0	+	ND	ND	1
8/31	11	1	0	36	9	3	0	0	+	ND	ND	1
9/43	10	2	0	20	14	0	0	4	–	ND	ND	0
10/56	111	10	0	0	0	0	0	0	–	ND	ND	2
11/33	66	13	0	25	7	0	10	2	+	ND	ND	0
12/34	31	1	0	9	0	0	0	0	–	ND	ND	0
13/32	70	10	0	5	0	0	1	0	–	ND	ND	0
14/21	9	6	0	0	0	0	0	2	+	ND	ND	0
15/47	76	4	0	2	0	0	0	0	+	ND	ND	0
16/29	20	4	0	0	0	0	0	0	–	ND	ND	0
17/46	111	11	0	40	8	0	0	3	+	ND	ND	1
18/24	24	7	0	3	5	0	0	2	–	ND	ND	0
19/59	51	3	0	0	0	0	0	0	–	ND	ND	1
20/49	85	12	0	18	6	0	0	0	+	ND	ND	3
21/61	40	0	0	25	20	0	10	6	+	–	–	6
22/66	0	0	0	60	40	. . .	0	1	+	–	–	1
23/50	29	5	0	2	2	0	0	0	–	–	–	0
24/49	76	17	0	6	3	0	5	2	–	–	–	0
25/54	40	3	0	3	0	0	0	0	–	–	–	1
26/65	7	6	0	0	0	0	2	1	+	–	–	2
27/53	50	19	0	2	3	0	0	1	–	–	–	2
28/42	34	7	0	4	0	0	0	0	–	+	–	0
29/70	60	10	0	70	20	0	0	0	–	+	–	3
30/63	35	2	0	0	0	0	0	0	+	–	+	3
31/58	50	20	0	20-30‡	20	0	2	6	0	–	+	1
32/60	84	2	1	0	0	0	1	0	0	–	–	4
33/74	18	7	1	0	0	1	0	1	0	–	–	ND
34/48	105	7	0	15	0	0	0	0	–	–	+	0
35/34	33	6	0	50‡	20‡	1	1	4	+	+	+	4
36/16	125	25	0	0	0	0	0	0	0	–	+	0
37/14	42	9	0	0	0	0	0	1	0	–	+	0
38/73	20	1	0	17	3	0	0	0	0	–	–	0
39/53	20	2	0	0	0	0	1	1	0	–	+	0
40/49	24	6	0	11	5	0	5	6	–	+	–	ND

*W indicates bouts won; L, bouts lost; D, bouts ending in draw; KO, knocked out; TKO, technical knockouts; S × S, clinical symptoms; exam, examination; CT, computed tomography; plus, positive; minus, negative; ND, not done.
†Computed tomogram score graded on scale of 0 to 4, with 0 as normal and 4 most abnormal. Values shown represent the score for sulcal enlargement plus the score for ventricular enlargement; thus, 0 equals normal and 8 most abnormal.
‡Boxer uncertain of exact number. Statistical analysis was not affected, however. See text and Tables.

ence or absence of ventricular enlargement ($p = .027$) (Table 40).

There was not a significant relationship between the presence or absence of neurological symptoms to the number of bouts fought ($p = .29$) (Table 41). Positive findings on neurological examination were not significantly correlated with the number of bouts fought ($p = .514$ (Table 42). Abnormalities found on EEG were significantly correlated with the number of bouts fought ($p = .0582$) (Table 43). There was no significant relationship between knock outs or technical knock outs and symptoms

TABLE 38
Effect of number of bouts on computed tomographic (CT) scan composite score, 38 boxers. (From ROSS et al., 1983).

Group	No. of Bouts	No. (%) of Boxers With CT Score 0	1	2	≥3	Total Boxers in Each Group, No. (%)
1	0-49	10 (67)	2 (13)	1 (7)	2 (13)	15 (39)
2	50-99	4 (40)	3 (30)	1 (10)	2 (20)	10 (26)
3	100-149	3 (38)	2 (25)	1 (12)	2 (25)	8 (22)
4	≥150	1 (20)	1 (20)	1 (20)	2 (40)	5 (13)
Total Boxers		18	8	4	8	38
% of Total Boxers Examined		47	21	11	21	

$Z = 1.99803296585$; $P = .0229$.

TABLE 39
Effect of number of bouts on sulcal enlargement (gyral atrophy). (From ROSS et al., 1983).

Group	No. of Bouts	No. (%) of Boxers With CT Score 0	1	≥2	Total Boxers in Each Group
1	0-49	10 (67)	2 (13)	3 (20)	15
2	50-99	4 (40)	3 (30)	3 (30)	10
3	100-149	3 (37.5)	2 (25)	3 (37.5)	8
4	≥150	2 (40)	2 (40)	1 (20)	5
Total Boxers, No. (%)		19 (50)	9 (24)	10 (26)	38

$Z = 1.22629753566$; $P = .11$. CT indicates computed tomography.

TABLE 40
Effect of number of bouts on ventricular enlargement. (From ROSS et al., 1983).

Group	No. of Bouts	No. (%) of Boxers With CT Score 0	1	≥2	Total Boxers in Each Group
1	0-49	14 (88)	1 (6)	1 (6)	16
2	50-99	7 (78)	0 (0)	2 (12)	9
3	100-149	6 (75)	1 (12.5)	1 (12.5)	8
4	≥150	2 (40)	0 (0)	3 (60)	5
Total Boxers, No. (%)		29 (76)	2 (5)	7 (19)	38

$Z = 1.92257401187$; $P = .027$. CT indicates computed tomography.

(p = .10), neurological signs (p = .16), abnormal EEG (p = .48), composite CT score (p = .62), sulcal enlargement (p = .43) and ventricular enlargement (p = .38). Patients with abnormal findings on CT examination were found to have more frequent neurological symptoms and abnormal neurological findings.

The data of ROSS et al. support the hypothesis that the more one boxes, the more likely he is to have evidence of cerebral atrophy on CT scan as well as

TABLE 41
Effect of number of bouts on clinical symptoms. (From ROSS et al., 1983).

Group	No. of Bouts	No. (%) of Boxers With Symptoms: Negative	Positive	Total Boxers in Each Group
1	0-49	12 (71)	5 (29)	17
2	50-99	6 (60)	4 (40)	10
3	≥100	8 (62)	5 (38)	13
Total Boxers, No. (%)		26 (65)	14 (35)	40

$Z = .546018906233$; $P = .29$.

TABLE 42
Effect of number of bouts on neurological examination. (From ROSS et al., 1983).

Group	No. of Bouts	Examination Results, No. (%): Negative	Positive	Total Boxers in Each Group
1	0-49	8 (73)	3 (27)	11
2	50-99	5 (83)*	1 (17)	6
3	≥100	5 (71)	2 (29)	7
Total Boxers, No. (%)		18 (75)	6 (25)	24

* $Z = .0358$; $P = .514$.

TABLE 43
Effect of number of bouts on EEG. (From ROSS et al., 1983).

Group	No. of Bouts	EEG Results, No. (%): Negative	Positive	Total Boxers in Each Age Group
1	0-49	9 (82)	2 (18)	11
2	50-99	5 (83)	1 (17)	6
3	≥100	3 (43)	4 (57)	7
Total Boxers, No. (%)		17 (71)	7 (29)	24

$Z = 1.570802\ 19856$; $P = .0582$.

an abnormal EEG. Changes of atrophy demonstrable on CT would, according to ROSS et al. show damage that occurred at least months and probably years before. The authors wrote: "Thus, any CT signs of atrophy may be meaningful and would indicate that perhaps the boxer should consider other sports." ROSS et al. continue that the significance of their data is particularly striking in view of the fact that they included a few professional fighters with numerous bouts behind them, one having fought more than 200 bouts. Furthermore, a standard neurological examination (including detailed history but not formal psychometric tests) proved to be a poor tool, in this population at least, for depicting chronic cerebral changes. In the patient population examined, symptoms included headaches, visual problems, and speech difficulty. Neurological findings included tendon reflexes, and loss of pin-prick sensation. The EEG proved to be a better predictor of possible neurological abnormalities ($p = .582$).

The data suggested no relation of cerebral atrophy to the number of knock outs or technical knock outs. The authors did not correlate the results with any estimate of the fighter's standing in his profession (they had no professional champions, but many who claimed to have been amateur or Golden Glove champions), weight class, or size of gloves used.

ROSS et al. summarized their data as showing that the boxers with even a moderate number of bouts may suffer cerebral atrophy (as judged by ventricular dilatation on the CT scan). Furthermore, the correlation of cerebral atrophy and number of bouts is significant. The authors continue: *"Our study thus supports numerous others in the literature showing that boxing is deleterious to the human brain. Computed tomography and EEG of the brain should be considered a part of a regular neurological examination for active boxers, and if possible should be performed before and after each boxing match in order to detect not only the results of acute life-threatening brain trauma, such as subdural hematomas and brain hemorrhage, but the more subtle chronic and debilitating changes of gradual cerebral atrophy."*

CASSON et al. (1984), published another report on brain damage in modern boxers. Eighteen former and active boxers underwent neurological examination, EEG, computed tomographic scan of the brain, and neuropsychological testing.

The subjects were 13 former boxers, two active professional boxers, and three active Golden Glove boxers. Their initial intention was, as they note, to recruit former boxers. They obtained the names and addresses of 29 former boxers who were still associated with the sport in New York (as part time officials, trainers, managers, etc.) from the *New York State Athletic Commission*. The authors selected from their list all the former boxers, who (1) were aged 25 to 60 years, (2) were currently residing in the metropolitan New York area, (3) had not retired from the boxing for medical, neurological, or psychiatric reasons, (4) were retired from boxing at least one year before the study, (5) had no known history of neurological, psychiatric, or serious medical illness, and (6) had no known history of drug or alcohol abuse. The 23 ex-boxers who met these criteria were contacted to participate in a study on the effects of boxing on the brain. Thirteen former professional boxers and three active Golden Glove fighters contacted the authors and volunteered to participate in the study, after learning about it from the manager of their club. None of them had any history of neurological, psychiatric, or medical illness or substance abuse. They were included to increase the number of boxers studied by CASSON et al. (1984). The authors did not attempt to make comparisons between the group of former, active, and amateur boxers included in this study. All 18 pugilists were considered together as one group of modern boxers. Informed consent was given by every subject; they were not paid to participate in the study.

The authors sought no control groups, because they evaluated boxers with standardized tests, which have established norms. According to these authors, standardized neuropsychological tests, computed tomographic (CT) scan of the brain, clinical neurological examination, and electro-encephalography all have clear criteria of normal and abnormal that have been well established and accepted.

The subjects ranged in age from 18 to 60 years, with a mean age of 36 years. Three are college graduates, ten others are high school graduates, and the remaining five had attended junior high school. The mean educational level was grade 12 (Table 44). All the former fighters were currently employed full time in civil service, education, or private industry. None of the subjects had any history of medical, neurological, or psychiatric symptoms or dysfunction. None had any history of drug or alcohol abuse.

All the fighters of CASSON et al. were active in the post World War II era; 15 fought after 1960. The number of total bouts (professional and amateur) ranged up to 249, with an average of 83 bouts (one Golden Glove fighter had been sparring for three months but had not fought in an actual bout). The number of professional bouts ranged up to 128, with a mean of 29 bouts (five subjects had fought only as amateurs). The length of boxing career ranged from three months to 22 years. The group included two former professional champions, three highly ranked professional contenders, and four other amateur champions. Only one former fighter (no. 5) was primarily a 'slugger,' all the others were considered to be scientific, artful pugilists who practiced self-defense.

CASSON et al. obtained a detailed medical history from every subject, with special attention paid to the details of alcohol and other drug use. The boxers underwent the following tests: (1) a noncontrast CT scan of the brain performed with a fourth-generation body scanner, (2) an EEG, (3) a formal neurological examination including mental status, and (4) a battery of neuropsychological tests that included the Trail Making Test, the Digit Symbol Test, the Wechsler Memory Test (including both verbal and visual memory), and the Bender-Gestalt Test (the standard administration and a 5-s recall administration). The results of each were interpreted in-

TABLE 44
Characteristics of sample. (From CASSON et al., 1984).

Subject No.	Age, yr	Educational Level, Grade	Total Bouts (Wins/Losses)	Professional Bouts (Wins/Losses)	Amateur Bouts (Wins/Losses)	Length of Career, yr	Boxing Accomplishments	Times Knocked Out	Episodes of Amnesia
Ex-Boxer									
1	37	9	64 (61/3)	19 (18/1)	45 (43/2)	12	Unranked	0	0
2	42	11	120 (103/17)	23 (12/11)	97 (91/6)	10	Unranked	1	0
3	33	16	28 (20/8)	1 (1/0)	27 (20/7)	13	Unranked	2	0
4	37	12	115 (113/2)	3 (2/1)	112 (111/11)	10	Pan American Games champion	0	0
5	39	8	84 (78/6)	54 (48/6)	30 (30/0)	14	World champion	1	2
6	58	12	106 (90/16)	6 (3/3)	100 (87/13)	8	Unranked	2	2
7	46	12	140 (132/8)	70 (64/6)	70 (68/2)	22	World champion	4	0
8	30	12	77 (58/19)	47 (38/9)	30 (20/10)	15	Ranked top 10, world	0	0
9	37	16	80 (76/4)	0	80 (76/4)	5	National Golden Gloves champion	2	5
10	55	12	7 (4/3)	0	7 (4/3)	5	Unranked	0	0
11	40	16	240 (224/26)	40 (27/13)	200 (187/13)	20	Pan American Games champion	3	4
12	60	10	228 (194/25/9†)	128 (104/15/9†)	100 (90/10)	15	No. 1 contender, world	0	0
13	27	12	70 (62/8)	22 (19/3)	48 (43/5)	6	National Golden Gloves champion	0	0
Active									
14	24	10	75 (67/8)	22 (20/2)	53 (47/6)	8	Ranked top 10, world	0	0
15	25	12	33 (31/2)	1 (1/0)	32 (30/2)	6	Unranked	1	0
Golden Gloves									
16	23	12	21 (10/6/5†)	0	21 (10/6/5†)	4	Unranked	2	1
17	18	12	0	0	0	0.25	Unranked	0	0
18	18	12	12 (9/3)	0	12 (9/3)	1	Unranked	0	0

Weight classes of boxers have not been included to protect the anonymity of the subjects.
† Number of ties.

dependently, without knowledge of the subject's performance on any other examination, I.C.; and neuropsychological tests, o.s.).

CASSON et al. evaluated the computed tomographic scans by measurements of the lateral ventricles according to the method of Huckman et al. (1975) and qualitative assessment of cortical sulcal prominence. Presence or absence of a cavum septi pellucidi was also noted. The following dimensions were measured by the authors: (1) the distance between the most lateral portion of each of the frontal horns, and (2) the width of the lateral ventricles in the region of the caudate nuclei. These measurements were added together to give a single measure of ventricular size.

Each of the neuropsychologial tests used in this study was a standardized test, with established norms.

Eighty-seven percent (13/15) of the boxer and active professional fighters had abnormal results on at least two of the *four tests (neurological, EEG, CT scan, neuropsychological test battery)*. The three active Golden Glove boxers all had normal neurological examination results, EEG, and CT scan.

CT Scan: Eight of the boxers, examined by CASSON et al. had abnormal CT scans. One had cerebral cortical atrophy, two had central cerebral atrophy, and five had generalized cerebral atrophy. A septum cavi pellucidi was noted in three of these eight scans.

Six of the eight fighters with more than 20 professional bouts had abnormal CT scans. Only two of the ten boxers with less than 20 professional bouts had abnormal CT scans (Fisher's exact test, $P = < .10$). Four of the five highly rated or champion professionals had abnormal CT scans (including three with a cavum septi pellucidi). CASSON et al. found that every subject with an abnormal CT scan had an impairment index of greater than 0.50 on the neuropsychological test battery.

EEG: Electroencephalograms were administered by CASSON et al. to 13 of the 18 boxers (the other five refused the test). Seven had abnormal records – five were diffusely abnormal (increased low amplitude theta activity) and two showed a diffuse, moderate abnormality with bilateral accentuation (excessive moderate amplitude theta activity noted diffusely but more prominently in the temporal region). Every subject with an abnormal EEG in the series of CASSON et al. had an impairment index of greater that 0.50 on the neuropsychological test battery.

Neurological Examination (including Mental Status): Five of the boxers of the series of CASSON et al. had abnormal findings on clinical neurological examination. Three subjects had a clinical organic mental syndrome manifested by disorientation, confusion, and memory loss. The former boxer had impaired recent memory without disorientation or confusion. One of the subjects with an organic mental syndrome had a positive right Babinski's sign. One ex-boxer had dysarthria and horizontal nystagmus on lateral gaze in both directions. One ex-boxer had cortical release phenomenon (glabellar, snout, and palmomental), but this cannot be considered definitely abnormal in a 58-year old man without other neurological findings.

Neuropsychological Tests: In the series of CASSON et al. compared with the established norms, every subject had more than one abnormal neuropsychological test score. Fourteen of the 15 former and active professional boxers scored abnormally on the 5-s recall Bender. All 13 who were given the verbal Wechsler Memory test scored in the abnormal range. Eleven of these scored in the abnormal range on the visual Wechsler Memory Test (Table 45). The authors developed an impairment index for each boxer, consisting of the percentage of abnormal neuropsychological test score. Boxers with abnormal CT scans in the series of CASSON et al. had a significantly higher mean impairment index than those with normal CT scan ($t = 0.04$; $P < .001$). Boxers with abnormal EEG's had a higher mean impairment index than those with normal EEG's ($t = 3.44$; $P < .006$). All twelve boxers with impairment indices greater than 0.50 had an abnormal CT scan and/or abnormal EEG, while the six boxers with impairment indices below 0.50 all had abnormal CT scans and EEG's (Table 46). The impairment index in the series of CASSON et al. correlated significantly with the number of professional fights ($r = .48$; $< .05$) and with age ($r = .54$; $P < .02$). The impairment did not

TABLE 45
Summary of neurological and neuropsychological test results. (From CASSON et al., 1984).

Subjects	Neurological Examination	EEG	Computed Tomographic Scan	Trail Making Test	Digit Symbol Test	Wechsler Memory Test: Verbal Immediate	Wechsler Memory Test: Verbal Delayed	Wechsler Memory Test: Visual Immediate	Wechsler Memory Test: Visual Delayed	Bender Gestalt Test: Standard Administration	Bender Gestalt Test: 5-s Administration	Impairment Index
Ex-Boxers												
1	Impaired recent memory	N*	N	N	N	A*	A	A	A	N	A	0.50
2	Mild OMS	A	Cortical cerebral atrophy	A	A	A	A	A	A	A	A	1.00
3	N	N	N	N	N	A	A	N	N	N	A	0.38
4	N	. . .	Generalized cerebral atrophy	N	A	A	A	A	A	A	A	0.88
5	N	N	Central cerebral atrophy, cavum septum pellucidum	A	A	A	A	A	A	N	A	0.88
6	N (Cortical release phenomena)	A	N	N	N	A	A	A	A	N	A	0.75
7	Dysarthria nystagmus	A	Generalized cerebral atrophy, cavum septum pellucidum	A	N	A	A	A	A	A	A	0.88
8	N	A	N	A	A	A	A	A	A	N	A	0.88
9	N	. . .	N	A	N	A	A	N	N	A	A	0.62
10	Mild OMS, right Babinski	. . .	Generalized cerebral atrophy	A	A	A	A	A	A	A	A	1.00
11	N	. . .	N	A	N	. . .	. . .	. . .	. . .	N	A	0.50
12	Mild OMS	. . .	Generalized cerebral atrophy	A	A	. . .	. . .	. . .	. . .	N	A	1.00
13	N		Generalized cerebral atrophy	A	A	A	A	A	A	A	A	1.00
Active												
14	N		Central cerebral atrophy, cavum septum pellucidum	A	N	A	A	A	A	A	A	0.88
15	N	A	N	N	N	A	A	A	A	N	A	0.62
Golden Gloves												
16	N	N	N	. . .	. . .	A	A	N	N	N	N	0.33
17	N	N	N	. . .	. . .	A	A	N	N	N	N	0.33
18	N	N	N	. . .	. . .	A	A	N	N	N	N	0.33

* N indicates normal; A, abnormal; ellipses, not administered.

TABLE 46
Comparison of impairment index. Computed tomography (CT), and EEG results. (From CASSON et al., 1984).

	CT or EEG	
Impairment Index	Normal	Abnormal
Neuropsychological Test Scores		
<0.50	6	0
>0.50	0	12

(From I.R. Casson et al. 1984)

correlate significantly with number of knock outs or episodes of amnesia.

CASSON et al. compared findings in their group of boxers with norms established for the general population. They found unequivocal evidence of organic cerebral dysfunction in 87% (13/15) of the group of ex-boxers and active professional fighters. The other two professional and three active amateur boxers had suggestive evidence of subtle brain injury (i. e. abnormal scores on some neuropsychological tests with normal neurological examination findings, CT scan, and EEG). The authors emphasize that the subjects had no history of alcohol or drug abuse, venerial disease, head injuries outside the boxing ring, or any significant general medical illness. No subjects had any history of abnormal behavior or psychiatric disturbance at the time of the study. CASSON et al. noted the caveat that one must consider the young age of their subjects, their level of education, their overall successful boxing careers, the relatively small number of knock outs, and the fact that they all boxed in the post-World War II era. Inter alia, they all boxed with modern medical controls, regulatory safeguards, and equipment. CASSON et al. noted: "One would not expect to find evidence of organic cerebral dysfunction in a group of men of this age. Thus, the correlation between the impairment index and the number of professional fights is more clinically significant than that for age. This suggests that the longer one's professional boxing career, the greater is the likelihood of brain damage. A cumulative effect of multiple subconcussive blows to the head is a likely etiology."

CASSON et al. found a much higher incidence of brain damage than was reported in earlier studies on boxers. According to these authors, the reasons for this are multiple. Many prior studies relied on only neurological examination, EEG, and, in a few cases, CT scans for study their subjects (MARTLAND 1928, CRITCHLEY 1957, SPILLANE 1962, La CAVA (1963), MAWDSLEY and FERGUSON 1963, ROBERTS 1969, KASTE et al. 1982, ROSS et al. 1983). CASSON et al. stress the point that neuropsychological tests have proved to be sensitive detectors of brain damage. The only two other studies that employed neuropsychological testing investigated, according to CASSON et al., boxers with relatively few fights (KASTE et al. 1982, THOMASSEN et al. 1979). Both groups reported a lower incidence of abnormalities than that found by CASSON et al., whose subjects had more extensive professional careers than in either of the other studies. "Their results complement ours and lend further support to the direct relationship between length of professional boxing career and the presence of brain damage." (CASSON et al.).

Combining the present data of CASSON et al. with their earlier results (CASSON et al. 1982), a cavum septi pellucidi is present on CT scan in 18% of boxers. A cavum septi pellucidi is present on CT scan in 5% of the population (SCHUNK 1963, NAKANO et al. 1981). This is, according to CASSON et al. (1984) a statistically significant difference ($z = 3.90$; $P < .001$). In the pre-CT scan era, pneumo-encephalography of ex-boxers frequently revealed a cavum septi pellucidi (MAWDSLEY and FERGUSON 1963, KASTE et al. 1982). "Septal abnormalities are an integral part of the neuropathological pictures of boxers' encephalopathy (PAYNE 1968, CORSELLIS et al. 1973)." We conclude that a CT scan demonstrating a cavum septi pellucidi in a boxer should not be dismissed as a normal variation but should be considered a sign of encephalopathy.

The subjects presented by CASSON et al. performed particularly poorly in the neuropsychological tests, which measure short-term memory (Wechsler- Memory Test and Bender 5 second recall), and are believed to detect alterations earlier than other diagnostic tests, thus confirming earlier findings that impairment of recent memory is a central feature of the organic mental syndrome of boxers.

The summary of CASSON et al. is of general importance: *"Existing medical controls and safety measures did not prevent chronic brain damage in this group of boxers. There is no reason to believe that future generations of boxers will be any more fortunate."*

JAUHIAINEN et al. (1985) studied apolipoprotein A-I, B, C-II and lipids before and after rapid weight loss schedules. The compared groups were all athletic. Apolipoproteins were determined by electro-immuno assay using apoproteins purified by chromatofocusing column method. Dextran T 10 was shown to increase rocket height in Apo B assay. Over 1% Dextran concentrations gave poor response. The linearity during calibration was from 0.3 to 3.0 Apo B-! Baseline values of Apo A-I in wrestlers, weightlifters, boxers and judocas were slightly higher as compared to 'normal' population; Apo B was clearly reduced (mean value of 690 mg/l). Weight loss was significant in each experimental group, mean value of 4.1 in active exercises, sauna and diuretic groups together. Compared as the whole sportsmen group in passive weight loss (or sauna) and diuretic groups the most pronounced changes were seen to be elevated apoprotein concentrations, whereas weight-loss by active rapid exercise resulted in no apoprotein changes, but instead an increment in HDL cholesterol and decrement in triglycerides, respectively. This study was the first to evaluate baseline values of apoproteins A-I, B, C-II in first class athletes and also the possibility of changes and lipid values in rapid weight-loss practices.

CABANIS et al. (1986) published the results of a prospective study on 52 boxers, 20 – 60 years of age. Thirteen of the subjects were amateurs and 39 professionals. Twelve of the boxers were still active, while the other 40 had retired after about ten years of boxing.

> Magnetic resonance imaging (MRI) at 0.15 T was used, as it provides an opportunity to evaluate the effects of trauma, and the information obtained may be used to identify disabilities among boxers.
> Multi-slices in two planes without gap, were performed using T1 and T2 weighted sequences. The findings were correlated to the number of matches, the number of knock outs, the severity of head trauma, the style of the boxer, the clinical findings, and the findings at EEG. The study revealed a significant correlation between the presence of lesions and the style of the boxer. The changes noted – mainly atrophic – occurred among those with the capacity to withstand blows and continue fighting, or to 'stand punishment.'

DREW et al. (1986) reported, that young, active, licensed boxers (N=19) were found to display a pattern of neuropsychological deficits consistent with the more severe punch-drunk syndrome of years past. These deficits resulted in significantly lower test performance than that of control athletes (N=10) matched for race, age, and level of education. Thests that showed significant differences between groups include subtests of the Quick Neurological Screening test, subtests of the Halstead-Reitan Neuropsychological Test Battery, and the Randt Memory Test. Fifteen of the 19 boxers scored in the impaired range of the Reitan Impairment Index, as compared to two of the ten controls.

SABHARVAL et al. (1987) reported on a neurological, EEG and computed tomography (CT) study on four Indian boxers who had manifestations of chronic traumatic encephalopathy (dementia pugilistica) resulting from boxing.

> Four symptomatic amateur boxer form the basis of this report. All underwent a detailed neuropsychological evaluation. They were subjected to routine blood and urine examination, ECG, STS and x-ray of the skull. EEG recordings were taken in each patient at intervals of six months and CT scanning was done in all.

Results: The ages of the boxers ranged from 34 to 40 years and their boxing careers span over 9–13 years. None of the boxers had suffered head injury outside the boxing ring. All the boxers had more than 100 bouts to their credit. Patient 1 had interrupted his boxing career for two years because of episodes of giddiness. He resumed boxing but had to retire from the sport because of headaches and slurred speech after two years. Patient 3 had suffered five knock outs of mild severity. None had been rendered unconscious nor admitted to hospital with acute aftermaths of boxing. There was no history of alcoholism, drug addiction or exposure coming from any patient. Psychometry revealed low performances in all patients suggestive of organic brain disturbance. Table 47 summarises the clinical findings in the patients.

Investigations: Routine haemogram, urinalysis, bloodbiochemistry, STS and x-rays of the skull were normal in all.

EEG: Abnormalities were detected in all patients with the exception of patient 3. Patient 1 had slow waves of the theta type in the right frontal region whilst the other revealed paroxysms of theta waves and occasionally delta waves symmetrically located.

Scan: Abnormalities were observed in all 4 boxers. Patient 3 and 4 had moderate dilatation of all ventricles, basal cisterns in addition to sulcal widening. 1 had mild dilatation of the lateral ventricles whilst the dilatation of lateral ventricles was moderate in patient 2.

Discussion: Boxing is according to SABHARVAL et al. a collision sport in which the intent is to inflict physical injury. Boxers, whether professional, amateur or sparring partners are at risk to develop neurological aftermaths of repeated multiple head traumata. No reliable test exists to identify boxers at risk for sudden death or impending brain injury (Council on Scientific Affairs, Brain Injury in Boxing, 1983).

The full blown picture of chronic traumatic encephalopathy in boxers involves slowed motor performance, clumsiness, dysarthria, ataxia of gait, rigidity, spasticity, memory deficit, slowness of thought and personality changes. ROBERTS (1969) in an extensive study of 224 former professional boxers with boxing careers of 6 to 9 years, found such changes in 37 of them. Further, clinical evidence of a Parkinsonian syndrome was noteed

TABLE 47
Results of neurological examination. (From SABHARVAL et al., 1987).

Clinical profile	Boxer 1	Boxer 2	Boxer 3	Boxer 4
a) Age (Years)	39	40	34	38
b) Duration of boxing career	9	11	13	10
c) Age at onset of symptoms	27	38	30	30
d) Age at giving up boxing	30	32	29	28
e) Symptoms & signs				
i) Mood changes	–	Irritability	Depressed	–
ii) Forgetfulness	+	+	+	+
iii) Headaches	–	+	+	+
iv) Diplopia	–	+	–	–
v) Slurred speech	+	+	+	+
vi) Ataxia	+	–	+	–
vii) Tremors	–	–	–	–
viii) IQ	55	60	60	55
ix) Memory impairment	+	+	+	+
x) Cranial nerve palsies	–	Paresis of left superior oblique	–	–
xi) Rigidity or spasticity	–	–	–	–
xii) Motor weakness	–	–	–	–
xiii) Motor incoordination (Heel-knee, finger nose)	+	–	+	–
xiv) Brisk reflexes	+	+	+	+
xv) Babinski's sign	–	–	–	–
xvi) Dysdiadokokinesis	+	–	+	–

in 20 of the 52 case reports on boxers that ROBERTS had extracted from literature up to 1969. CORSELLIS et al. (1973) in a thorough histopathological study of the brains of 15 retired boxers (including two former world champions) found mild to moderate enlargement of lateral ventricles and thinning of corpus callosum, widened ventricles, cavum septi pellucidi and fenestrated septal leaves. Scarring was prominent on the inferior surface of the cerebellum with marked loss of Purkinje cells and thinning of granular layer. Cerebral cortical contusions were uncommon and any evidence of previous hemorrhage was lacking. Eleven cases showed loss of pigmented cells in substantia nigra (this was more marked in four cases who had a record of suffering from a parkinsonian syndrome). Neurofibrillary tangles far in excess to elsewhere in the brain and loss of neurons was marked in medial temporal regions. Mammilary bodies were often sunken, small and gliotic.

The opinion that amateur boxers are probably not afflicted by chronic encephalopathy (BORNSTEIN and CLARKE 1954) is in contrast to the findings of SABHARVAL et al., it is in contrast to their report. THOMASSEN et al. (1979) in a study of 53 amateur bozers found no symptoms of the disorder in any of the boxers. Further, the authorts found statistical difference in EEG abnormalities between boxers and an equal number of footballers. On the contrary of SABHARVAL's et al. patients, who were amateur boxers with a reasonable length of boxing careers had clinical manifestations of chronic traumatic encephalopathy in the form of intellectual deterioration, memory impairment, mood changes, ataxia of gait and hyperreflexia. EEG changes in the form of paroxysmal slowing in theta and delta ranges were present in three patients. The value of EEG has been highlighted by authors (BUSSE and SILVERMAN 1952, NESARAJAH et al. 1961). The EEG changes are usually in form of slow waves of theta and delta type. ROSS et al. (1983) found a significant correlation between EEG abnormalities and number of bouts fought.

CT scan abnormalities in the form of generalised atrophy were present in all four patients. These changes have been observed in about half of professional boxers and are related to the numbers of the bouts (ROSS et al. 1983, CASSON et al. 1982). Boxers with even a moderate number of bouts may suffer from cerebral atrophy (as judged by ventricular dilatation on the CT scan).

The report of SABHARVAL et al. highlights the neurological disablement of this collision sport. Boxers, whether professional or amateur, should be under medical survaillance. A frequent neuropsychological assessment, EEG and CT scanning should be performed in them. A medical record highlighting the number of bouts, knock outs, concomitant illnesses, results of clinical, EEG and CT scan evaluation should be made essential, in boxers who are keen and propose to continue being active in this sport. At the earliest abnormality being detected, they should be compulsory retired from this sport.

In a prospective investigation of neurobehavioral functioning in young boxers, published by LEVIN et al. (1987), 13 pugilists and 13 matched control subjects underwent tests of attention, information-processing rate, memory, and visuomotor coordination and speed.

Managers of three boxing clubs in Houston, Texas, assisted in the recruitment of fighters whom they consented to have a long-term commitment to the sport. All of the boxers from two of the clubs were recruited. The manager of the third club recommended three of his boxers, who exhibited a firm commitment to the sport (as reflected by regular training) and who were likely to continue fighting until the time of follow-up evaluation.

The authors obtained a medical and sociological history from each boxer who considered to participate in the study. Exclusion criteria included a history of chronic alcohol or drug abuse, hospitalization for head injuries sustained outside the ring, or other neuropsychiatric disorders. Twelve professional boxers and three amateur boxers volunteered to participate in the study and satisfied the selection criteria. However, one of the professionals dropped out before completing the examination and the authors subsequently deleted the data on one of the amateurs, because he had only three months of organized boxing experience and no fights before entering the study. Table 48 summarizes the experience of the boxers in the ring.

Through newspaper and television advertisements LEVIN et al. recruited a comparison group of 13 young men from the Houston-Galveston area, who regularly participated in an amateur sport other than boxing.

The authors obtained medical and social histories to screen the control candidates and according to the same criteria employed for the boxers. No control subject had a history of hospitalization for head injury or neuropsychiatric disorder. Apart from these exclusionary criteria, the authors selected the 13 men by individually matching them to one of the boxers according to age, years of education, parental socioeconomic background (Table 49), and race.

TABLE 49
Baseline demographic features and reading level of the boxers and control subjects. (From LEVIN et al., 1987).

Feature	Boxers	Control Subjects
age (yrs)	20.5 ± 2.1	20.0 ± 2.2
education (yrs)	11.8 ± 1.5	11.7 ± 1.3
socioeconomic score†	38.2 ± 7.4	37.7 ± 9.7

Values are means ± standard deviations for 13 boxers and 13 control subjects.
† Hollingshead index score of social class based on the father's occupation (or the mother's occupation if information on the father was unvailable).[18] This social position score is the product of multiplying scaled values for occupation and education. Social class I (for instance, a judge) corresponds to a Hollingshead score range of 11–17, Class II = 18–27, Class III = 28–43, Class IV = 40–60, and Class V = 61–77 (for example, an unskilled employee with an eight grade education).

TABLE 48
Amateur and professional experience of the 13 boxers in this series. (From LEVIN et al., 1987).

Boxer No.	Yrs Boxing	Amateur Bouts	Professional Bouts	Amnesic Episodes†	Sparring Days/Mo
1	11	150	2	4	9
2‡	8	130	0	5	12
3§	8	115	5	5	12
4	6	69	—	4	9
5	9	120	3	5	10
6	6	184	5	3	15
7	8	30	8	5	18
8	5	47	7	3	15
9	5	45	10	3	15
10‡	9	100	—	15	15
11	11	95	9	4	15
12	2	14	2	6	15
13‡	7	40	1	1	15

Boxing experience before entry into the study. The 10 subjects who participated in follow-up testing continued to box with two to seven additional bouts since the baseline data were obtained. All boxers had winning records at the outset of the study and at follow-up examination, and none had suffered a knock out in a professional bout. Individual boxing records are not given so as to preserve the anonymity of the boxers. — = Amateur boxer at the time of this study.
† The number of episodes of posttraumatic amnesia without loss of consciousness was determined by retrospective report.
‡ These boxers failed to return for the 6-month neuropsychological examination.
§ This boxer declined to undergo magnetic resonance imaging but returned for 6-month neuropsychological testing.

Statistical Studies

Univariate and multivariate analysis of variance was used to compare the performance of the boxers with that of the control subjects. In view of the disparity in reading level between the boxers and the control athletes (which approached significance) and other individual differences that might predispose an individual to take up boxing, the authors employed a between-groups design rather than a matched-pairs analysis. Table 50 summarizes the baseline results, comparing the boxers and control subjects on each neuropsychological measure. To emphasize the presence of any differences in performance between the two groups, there is no correction in Table 50 for the number of statistical comparisons.

The result of the study of LEVIN et al. disclosed more proficient verbal learning in the control subjects, whereas delayed recall and other measurements of memory did not differ between the two groups. Reaction time was faster in the boxers than in the control groups, but no other differences were significant. The subjects in each group were re-tested six months later and exhibited improve-

TABLE 50
Baseline neuropsychological test performance in boxers and control subjects. (From LEVIN et al., 1987).

Neuropsychological Test	Boxers	Control Subjects	Group Effect F(1.24)	Group Effect p Value
attention/information-processing speed				
PASAT optimal processing rate†	0.40 ± 0.15	0.41 ± 0.11	0.01	0.91
reaction time (msec)				
average reaction time	637.2 ± 82.2	738.5 ± 126.4	8.3	0.008
simple reaction time	489.0 ± 71.8	609.1 ± 139.1		
complex reaction time	785.4 ± 92.5	867.9 ± 113.7		
Continuous Performance Test (CPT) interstimulus interval (msec)				
average interstimulus interval	1006.9 ± 285.3	1039.8 ± 361.9	0.1	0.79
Trial A	1102.9 ± 235.1	1128.3 ± 375.7		
Trial B	883.2 ± 269.2	897.3 ± 306.1		
Trial C	995.1 ± 318.1	991.9 ± 280.7		
Trial D	1046.2 + 318.1	1141.7 ± 484.9		
memory				
verbal memory				
average no. words retrieved	7.5 ± 3.4	10.5 ± 5.1	3.5	0.07
60-minute recall	15.8 ± 2.8	16.2 ± 3.4	0.1	0.76
visual memory				
no. correct	6.8 ± 1.6	7.2 ± 2.0	0.2	0.66
no. errors	4.2 ± 2.0	3.5 ± 2.5	0.6	0.44
Digit Span Scale score	9.9 ± 2.5	11.0 ± 2.6	1.2	0.29
divergent thinking				
verbal fluency (no. of words)	31.5 ± 10.7	38.2 ± 10.9	2.5	0.12
design fluency				
5-min free condition (no. of designs)	14.5 ± 5.0	18.4 ± 9.0	1.9	0.18
4-min fixed condition (no. of designs)	11.6 ± 5.1	13.2 ± 4.9	0.6	0.44
visuomotor speed				
Trail A time (sec)	24.0 ± 7.2	24.5 ± 7.6	0.04	0.85
Trail B time (sec)	71.6 ± 24.6	78.2 ± 40.0	0.3	0.62
Purdue Pegboard Tes				
preferred hand (no. of pegs)	14.8 ± 2.1	15.4 ± 1.8	0.5	0.49
nonpreferred hand (no. of pegs)	13.5 ± 1.9	13.9 ± 1.8	0.3	0.60
Digit Symbol Scale score	9.0 ± 1.4	10.8 ± 2.9	4.4	0.05
reading single words (percentile)	28.1 ± 28.8	46.6 ± 34.9	2.2	0.15

Values are means ± standard deviations for 13 boxers and 13 control subjects.
† PASAT = Paced Auditory Serial Addition Test. Optimal processing rate was obtained for each subject by selecting the trial which yielded the highest processing rate (no. correct responses/total time of presentation).

ment in their neuropsychological performance, as compared to baseline measurement. However, there were no differences in scores between the boxers and the control subjects at the follow-up examination or in the magnitude of improvement from baseline values. Magnetic resonance imaging, which was performed in nine of the boxers, disclosed normal findings.

Results

Table 51 depicts a pattern of generally improved performance in attention/information processing speed on the follow-up examination. Information-processing rate, as measured by the *Paced Auditory Serial Addition Test (PASAT),* significantly improved as compared to the baseline scores. However, there was no significant difference in information-processing rate between the boxers and control subjects, nor was there any differential change in performance between the two groups; that is, there was no interaction of groups and occasions (Table 51).

In contrast to the shorter reaction times dis-

TABLE 51
Baseline and follow-up neuropsychological test performance in serially studied boxers and control subjects. (From LEVIN et al., 1987).

Neuropsychological Test	Baseline Performance		Follow-Up Performance		Group Effect		Occasion Effect		Group × Occasion	
	Boxers	Control Subjects	Boxers	Control Subjects	F(1,18)	p Value	F(1,18)	p Value	F(1,18)	p Value
attention/information-processing										
PASAT optimal processing rate	0.41 ± 0.14	0.43 ± 0.11	0.46 ± 0.18	0.47 ± 0.14	0.1	0.79	4.7	0.05	0.04	0.85
reaction time (msec)										
average reaction time	642.6 ± 85.5	725.0 ± 118.1	645.6 ± 92.4	633.6 ± 99.8	0.9	0.35	8.0	0.01	9.1	0.007
simple	500.4 ± 70.5	618.1 ± 157.0	513.3 ± 86.5	473.3 ± 111.3						
complex	784.7 ± 100.5	831.8 ± 79.1	777.8 ± 98.3	793.9 ± 88.3						
Continuous Performance Test (CPT)										
interstimulus interval (msec)										
average interstimulus interval	987.1 ± 303.7	1078.4 ± 404.8	934.2 ± 327.0	877.5 ± 199.6	0.01	0.89	4.2	0.06	1.5	0.24
Trial A	1069.5 ± 247.5	1149.1 ± 428.1	987.4 ± 228.0	988.1 ± 231.8						
Trial B	863.9 ± 272.2	937.8 ± 337.3	881.0 ± 408.6	819.2 ± 230.6						
Trial C	998.7 ± 349.7	1022.1 ± 305.7	953.4 ± 337.3	820.4 ± 149.4						
Trial D	1016.2 ± 345.2	1204.4 ± 548.0	915.1 ± 334.2	882.3 ± 186.5						
memory										
verbal memory										
average no. words retrieved	7.9 ± 3.7	11.7 ± 3.8	11.0 ± 3.7	12.7 ± 5.2	2.8	0.11	9.3	0.007	2.7	0.12
60-minute recall	16.0 ± 3.1	16.8 ± 3.2	18.2 ± 1.8	18.6 ± 2.1	0.4	0.53	9.2	0.007	0.1	0.77
visual memory										
no. correct	6.8 ± 1.6	7.4 ± 2.2	7.2 ± 1.4	8.3 ± 1.5	1.7	0.21	2.9	0.11	0.4	0.52
no. errors	4.2 ± 2.0	2.8 ± 2.5	3.3 ± 1.9	2.5 ± 2.8	1.5	0.24	1.3	0.26	0.3	0.57
Digit Span Scale score	10.1 ± 2.6	11.3 ± 2.8	10.5 ± 2.4	10.9 ± 2.3	0.6	0.45	0	1.0	0.9	0.34
divergent thinking										
verbal fluency (no. of words)	31.5 ± 10.7	37.0 ± 11.4	24.9 ± 19.6	37.3 ± 12.8	2.9	0.10	1.5	0.23	1.4	0.25
design fluency										
free cond. (no. of designs)	14.4 ± 4.8	19.2 ± 8.9	16.7 ± 11.7	23.6 ± 11.4	2.16	0.16	4.3	0.05	0.42	0.53
fixed cond. (no. of designs)	10.4 ± 5.2	13.8 ± 3.6	11.2 ± 6.8	16.3 ± 4.9	3.68	0.07	4.4	0.05	1.17	0.29
visuomotor speed										
Trail A (sec)	25.0 ± 7.7	24.1 ± 8.5	22.1 ± 5.1	26.3 ± 6.8	0.4	0.56	0.05	0.83	2.6	0.12
Trail B (sec)	71.9 ± 28.0	66.5 ± 38.2	61.1 ± 19.7	58.2 ± 33.4	0.10	0.75	4.4	0.05	0.07	0.79
Purdue Pegboard Test										
preferred hand (no. of pegs)	14.8 ± 2.3	15.5 ± 1.8	15.0 ± 1.8	16.2 ± 1.5	2.0	0.18	0.8	0.39	0.2	0.63
nonpreferred hand (no. of pegs)	13.4 ± 1.6	13.5 ± 1.8	13.7 ± 1.5	14.8 ± 1.3	1.1	0.32	4.6	0.05	1.8	0.20
Digit Symbol Scale score	8.8 ± 1.1	10.8 ± 2.9	11.0 ± 1.5	10.9 ± 2.5	1.7	0.21	3.4	0.08	2.9	0.11
reading single words (percentile)	30.5 ± 31.9	52.3 ± 34.9	31.4 ± 33.1	57.0 ± 36.7	2.5	0.13	0.7	0.41	0.3	0.57

Baseline and follow-up performance values are means ± standard deviations for 10 boxers and 10 control subjects evaluated at each examination time.
PASAT = Paced Auditory Serial Addition Test. Optimal processing rate was obtained for each subject by selecting the trial which yielded the highest processing rate (no. correct responses/total time of presentation).

played by the boxers as compared to the control subjects on the baseline examination, Fig. 232 indicates that the disparity in response latency between the two groups was considerably narrowed at the time of their follow-up examination.

Consistent long-term retrieval on the verbal memory test is plotted against trials for both the baseline and follow-up examination in Fig. 233. In contrast to the trend toward greater acquisition of the word list by the control subjects on the baseline test, the follow-up data show a narrowing of the gap between the groups.

The results of the study of LEVIN et al. differ from previous research (KASTE et al. 1982, ROSS et al. 1983, CASSON et al. 1984). The study of LEVIN et al. was limited to subjects under 30 years of age, who were entering the early stages of a professional career, or who were still amateurs. Moreover, LEVIN et al. stated that all of their participants were actively competitive in the ring at the time of the study, as opposed to previous reports on former boxers. The obvious difference is that boxers are more likely to show the effects of repeated brain injury during the latter stages of their careers. Younger boxers have suffered fewer blows and even when head blows have injured them, the damage is less likely to be detectable.

LEVIN et al. summarize: "Notwithstanding the constraints imposed on extrapolating from this small serially studied group, our findings raise the possibility that young boxers may escape disabling brain injury, provided that their career in the ring

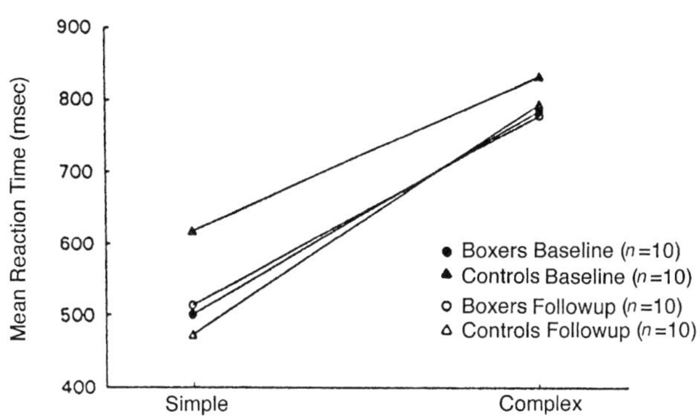

Fig. 232. Graph showing the mean simple and complex reaction times plotted for the boxers and control subjects who were examined at both the baseline and the 6-month follow-up examinations. For a description of simple and complex reaction times see text. (From LEVIN et al., 1987).

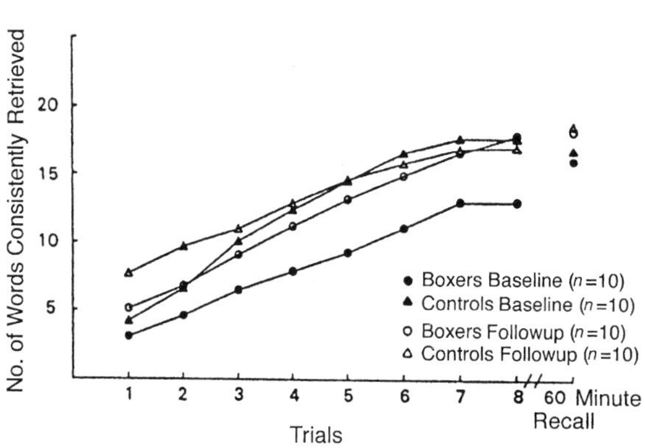

Fig. 233. Graph showing consistent long-term retrieval of words plotted across trials on the modifiwd selective-reminding test given at baseline and at the follow-up examination. For a description of the long-term word-retrieval procedure see text. (From LEVIN et al., 1987).

is limited both in frequency and total duration." While it is likely that limited boxing reduces the extent of injuries, the claim that such injuries can be avoided altogether is highly speculative and perhaps overly optimistic, especially in the absence of long term follow up on individuals with limited boxing experience.

JORDAN (1987) examined the neurological aspects of boxing. He observed that the assessment and prevention of potentially adverse neurological consequences of boxing requires two important considerations. Acute neurological injuries should be distinguished from chronic brain injuries, and the level of competitive boxing (i. e. amateur vs. professional) must also be taken into account. Acute neurological injuries such as concussion, post-concussion syndrome, intracranial hemorrhage, and brain contusion are more readily identified than chronic neurological injuries, because of their immediate devastation of the nervous system. In contrast, chronic neurological injuries differ in their pathophysiological mechanisms, which are exemplified by an insidious onset and progression after the cessation of boxing. Accordingly, the chronic traumatic encephalopathy of boxing poses the most serious neurological threat of boxing. Amateur boxing differs from professional boxing in the duration of fights, rules and regulatory policies, medical evaluation, and protective devices. These factors could produce a differential effect on the risk of injury to the brain The prevention of neurological injuries in boxing requires the integration of proper neurological evaluation by qualified ring-side physicians, the design and utilization of effective protective devices, and the establishment of national regulatory agencies.

ROSS et al (1987) reported about *'Boxing Injuries: Neurologic, Radiologic, and Neuropsychologic Evaluation.'*

ROSS et al begin their paper:

"During the past several years, boxing has undergone both extensive medical and philosophic evaluation. This has resulted in a lively debate concerning the question of whether boxing can be appropriately classified as a sport. Of note, boxing as a sport has been banned in Sweden and Norway. In Great Britain, there is no boxing in the schools, the Royal Air Force, or the Army. The world press has also reacted differently to boxing recently, indicating a heightened awareness of the controversy over this sport."

The authors continue;

"The ultimate purpose of boxing is to defeat one's opponent by injuring him and optimally causing him to become unconscious. Thus the head is the most frequent attacked and injured part of the boxer's body."

Computed tomography has allowed, according to ROSS et al., an in-vivo evaluation of the structural appearance of the brains of boxers that in past years could be evaluated only post mortem.

Using computed tomography, EEG, and neurologic evaluation in the largest study of boxers (Table 52), the authors demonstrated CT abnormalities of atrophy in 19 of 38 boxers (Table 53). CT abnormalities were more frequent in the boxers who had fought more bouts (Table 54). Abnormal EEG's, but not of statistical significance, also were more frequent in boxers with more bouts (Table 55). Neurologic evaluation and clinical symptoms did not correlate with the chronic changes of cerebral atrophy demonstrated on the computed tomography scans (Tables 56 and 57).

ROSS et al. conclude their large series of boxing injuries that boxing is an endeavor that may have to be re-evaluated in the coming years as to whether it should be designated as a sport. They note:

"It is the only 'sport' in which victory is determined by the amount of physical damage done to the opponent."

The authore continue:

"We have presented the largest number of professional and amateur boxers (58) evaluated by various modern diagnostic modalities and have unquestionally demonstrated the deleterious effects of boxing upon the brain."

According to ROSS et al., there have been few, if any, meaningful actions taken by the promoters of boxing to correct the conditions under which

TABLE 52
Boxers' historical data and examination results. (From ROSS et al., 1987).

Boxer/	Amateur			Professional								
(age)	W	L	D	W	L	D	KO	TKO	SXS	EXAM	EEG	CT†
1/49	45	3	0	0	3	4	0	2	+	+	–	0
2/57	8	12	0	0	0	0	4	0	+	–	–	0
3/37	64	6	0	13	2	0	0	8	–	–	–	1
4/57	5‡	9	0	0	0	0	2	1‡	+	+	–	6
5/66	125	10	0	61	4	0	0	2	–	ND	ND	4
6/25	127	20	0	13	0	0	0	0	–	ND	ND	2
7/61	28	4	0	0	0	0	0	0	+	ND	ND	1
8/31	11	1	0	36	9	3	0	0	+	ND	ND	1
9/43	10	2	0	20	14	0	0	4	–	ND	ND	0
10/56	111	10	0	0	0	0	0	0	–	ND	ND	2
11/33	66	13	0	25	7	0	10	2	+	ND	ND	0
12/34	31	1	0	9	0	0	0	0	–	ND	ND	0
13/32	70	10	0	5	0	0	1	0	–	ND	ND	0
14/21	9	6	0	0	0	0	0	2	+	ND	ND	0
15/47	76	4	0	2	0	0	0	0	+	ND	ND	0
16/29	20	4	0	0	0	0	0	0	–	ND	ND	0
17/46	111	11	0	40	8	0	0	3	+	ND	ND	1
18/24	24	7	0	3	5	0	0	2	–	ND	ND	0
19/59	51	3	0	0	0	0	0	0	–	ND	ND	1
20/49	85	12	0	18	6	0	0	0	+	ND	ND	3
21/61	40	0	0	25	20	0	10	6	+	–	–	6
22/66	0	0	0	60	40	...	0	1	+	–	–	1
23/50	29	5	0	2	2	0	0	0	–	–	–	0
24/49	76	17	0	6	3	0	5	2	–	–	–	0
25/54	40	3	0	3	0	0	0	0	–	–	–	1
26/65	7	6	0	0	0	0	2	1	+	–	–	2
27/53	50	19	0	2	3	0	0	1	–	–	–	2
28/42	34	7	0	4	0	0	0	0	–	+	–	0
29/70	60	10	0	70	20	0	0	0	–	+	–	3
30/63	35	2	0	0	0	0	0	0	+	–	+	3
31/58	50	20	0	20-30‡	20	0	2	6	0	–	+	1
32/60	84	2	1	0	0	0	1	0	0	–	–	4
33/74	18	7	1	0	0	1	0	1	0	–	–	ND
34/48	105	7	0	15	0	0	0	0	–	–	+	0
35/34	33	6	0	50‡	20‡	1	1	4	+	+	+	4
36/16	125	25	0	0	0	0	0	0	0	–	+	0
37/14	42	9	0	0	0	0	0	1	0	–	+	0
38/73	20	1	0	17	3	0	0	0	0	–	–	0
39/53	20	2	0	0	0	0	1	1	0	–	+	0
40/49	24	6	0	11	5	0	5	6	–	+	–	ND

* W indicates bouts won; L, bouts lost; D, bouts ending in draw; KO, knocked out; TKO, technical knockouts; S×S, clinical symptoms; exam, examination; CT, computed tomography; plus, positive; minus, negative; ND, not done.

† Computed tomogram score graded on scale of 0 to 4, with 0 as normal and 4 most abnormal. Values shown represent the score for sulcal enlargement plus the score for ventricular enlargement; thus, 0 equals normal and 8 most abnormal.

‡ Boxer uncertain of exact number. Statistical analysis was not affected, however. See text and Tables.

boxers are subjected to physical abuse. Recommendations regarding the creation of a National Board of Boxing to supervise this 'sport' have not been headed.

The authors finish their study with the statement:

"The established medical injuries due to boxing and the lack of any sustained and significant efforts on the part of organized boxing create an atmosphere that is conducive to following the call for the consideration of a ban of boxing."

RYAN (1987) presented a review (1918-1985) of

TABLE 53
Effect of number of bouts on ventricular enlargement. (From ROSS et al., 1987).

Group	No. of Bouts	No. (%) of Boxers With CT Score			Total Boxers in Each Group
		0	1	≥2	
1	0-49	14 (88)	1 (6)	1 (6)	16
2	50-99	7 (78)	0 (0)	2 (12)	9
3	100-149	6 (75)	1 (12.5)	1 (12.5)	8
4	≥150	2 (40)	0 (0)	3 (60)	5
Total Boxers, No. (%)		29 (76)	2 (5)	7 (19)	38

$Z = 1.92257401187$; $P = .027$. CT indicates computed tomography. *From* Ross, RJ, Cole, M, Thompson, JS, Kim, KH: Boxers: Computed Tomography, EEG, and Neurological Evaluation. JAMA 249:211–213, 1983; with permission. Copyright 1983, American Medical Association.

TABLE 54
Effect of number of bouts on computed tomography (CT) scan composite score, 38 boxers. (From ROSS et al. 1987).

Group	No. of Bouts	No. (%) of Boxers With CT Score				Total Boxers in Each Group, No. (%)
		0	1	2	≥3	
1	0-49	10 (67)	2 (13)	1 (7)	2 (13)	15 (39)
2	50-99	4 (40)	3 (30)	1 (10)	2 (20)	10 (26)
3	100-149	3 (38)	2 (25)	1 (12)	2 (25)	8 (22)
4	≥150	1 (20)	1 (20)	1 (20)	2 (40)	5 (13)
Total Boxers		18	8	4	8	38
% of Total Boxers Examined		47	21	11	21	

$Z = 1.99803296585$; $P = .0229$. *From* Ross, RJ, Cole, M, Thompson, JS, Kim, KH: Boxers: Computed Tomography, EEG, and Neurological Evaluation. JAMA 249:211–213, 1983; with permission. Copyright 1983, American Medical Association.

TABLE 55
Effect of number of bouts on EEG. (From ROSS et al., 1987).

Group	No. of Bouts	EEG Results, No. (%)		Total Boxers in Each Age Group
		Negative	*Positive*	
1	0-49	9 (82)	2 (18)	11
2	50-99	5 (83)	1 (17)	6
3	≥100	3 (43)	4 (57)	7
Total Boxers, No. (%)		17 (71)	7 (29)	24

Z1.570802 19856; $P = .0582$. *From* Ross, RJ, Cole, M, Thompson, JS, Kim, KH: Boxers: Computed Tomography, EEG, and Neurological Evaluation. JAMA 249:211–213, 1983; with permission. Copyright 1983, American Medical Association.

the intracranial injuries resulting from boxing. The records available indicate that, when compared with the total number of participants at all levels, the number of fatalities due to intracranial injuries sustained in the ring has been substantial in the past. The frequency of such fatalities has decreased steadily in recent years, owing to regulatory safety

TABLE 56
Effect of number of bouts on clinical symptoms. (From ROSS et al., 1987).

Group	No. of Bouts	No. (%) of Boxers With Symptoms		Total Boxers in Each Group
		Negative	*Positive*	
1	0-49	12 (71)	5 (29)	17
2	50-99	6 (60)	4 (40)	10
3	≥100	8 (62)	5 (38)	13
Total Boxers, No. (%)		26 (65)	14 (35)	40

$Z = .546018906233$; $P = .29$. *From* Ross, RJ, Cole, M, Thompson, JS, Kim, KH: Boxers: Computed Tomography, EEG, and Neurological Evaluation. JAMA 249:211–213, 1983; with permission. Copyright 1983, American Medical Association.

TABLE 57
Effect of number of bouts on neurologic examination. (From ROSS et al., 1987).

Group	No. of Bouts	Examination Results, No. (%)		Total Boxers in Each Group
		Negative	*Positive*	
1	0-49	8 (73)	3 (27)	11
2	50-99	5 (83)	1 (17)	6
3	≥100	5 (71)	2 (29)	7
Total Boxers, No. (%)		18 (75)	6 (25)	24

$Z = .0358$; $P = .514$. *From* Ross, RJ, Cole, M, Thompson, JS, Kim, KH: Boxers: Computed Tomography, EEG, and Neurological Evaluation. JAMA 249:211–213, 1983; with permission. Copyright 1983, American Medical Association.

measures, and increased medical presence at boxing events.

The considerable concern about the long-term effects of repeated brain injury in boxers is adequately justified by the many studies of live boxers and pathologic examinations of brains of former boxers recorded over the years since 1952. These overwhelmingly confirm that there is a significant relationship between the numbers of bouts fought and the presence and severity of chronic encephalopathy. Because it is unlikely that currently adopted protective measures will substantially reduce these hazards, even though they have apparently reduced the number and severity of blows to the head. This might be done by making any blow to the head in boxing a foul and a reason for disqualification. Admittedly, elimination of head blows would constitute a drastic change to the rules and practice of boxing, but the staggering weight of the evidence shows that there is no way to make head trauma safe. The only way to stop head injury is to stop blows to the head. The other alternative, so often brought up, is even more extreme; many support the elimination of boxing altogether.

JORDAN and ZIMMERMAN (1988) reported findings regarding magnetic resonance imaging in amateur boxers.

Nine amateur boxers who competed in the 1985 and 1986 *New York City Golden Glove competitions* were subsequently suspended for medical reasons, secondary to a knock out or excessive head blows (Table 58). All nine underwent detailed neurological examinations and magnetic resonance (MR) scanning. Neurological examination results and MR scans were normal in all nine boxers. Failure to detect abnormalities on the MR scan, by neurological examination, or both, in these amateur boxers may reflect several factors, including a small sample size, the duration between their last bout and neurological evaluation, and the lower exposure to head trauma among some amateur boxers compared with professionals.

JORDAN and CAMPBELL (1988) reviewed over a two year period, all acute boxing injuries among professional boxers statewide (484 the first year, 422 the second year). During the study period, the boxers fought 3.110 rounds and incurred 376 injuries (262 craniocerebral injuries, 114 other injuries), i. e., they incurred 1.2 injuries per 10 rounds fought (0.8 craniocerebral, 0.4 others). Only four boxers required immediate neurological evaluation at a hospital after a fight; one of the four died as a result of bilateral subdural hematomas. Facial lacerations were the most common other type of injury (66 cases), followed by hand and eye injuries (8 cases each). The authors suggest that severe, acute neurological injuries are rare in professional boxing when strict medical supervision is present. However, they caution that their findings should not be used to draw inferences about the development of chronic neurological injuries among professionl boxers.

The authors reviewed all medical injury report forms filed at the New York State Athletic Commission (NYSAC) fom August 1, 1982 through July 31, 1984. The ringside physician routinely completes an injury report form for any professional boxer registered in New York State who either sustains an injury in the ring or is medically suspended. Information on the reports includes the name of the boxer, a description of the injury, the manner in which the injury was sustained, and medical recommendations.

JORDAN and CAMPBELL, after reviewing all the reports, included only acute injuries occurring in the ring during a competitive bout. Chronic injuries, such as traumatic cataracts, retinal detachments, and chronic traumatic encephalopathy (punch drunk syndrome and dementia pugilistica) were excluded. Injuries occurring during training or outside the ring, i. e., not boxing related, were also excluded. Acute sparring injuries were not included because the information is usually unavailable to the NYSAC. In twelve cases, the type of injury was not specified, so these were also not included in the analysis.

According to JORDAN and CAMPBELL injuries were divided into 'craniocerebral injuries' and 'other injuries'. Craniocerebral injuries were defined as any technical knock out (TKO) or knock out (KO) from head blows. A TKO ot KO in which the nature of the injury was not specified the authors assumed was a craniocerebal injury. The decision regarding a TKO or KO from head blows is usually made at the discretion of the referee, but often implies that a boxer sustained head blows that were considered excessive in force and/or quantity; often the boxer may have experienced a transient impairmant in neurological function. Since the referee (not a physician) determines the TKO, which may not always result in a concussion, TKO's from powerful and/or multiple blows to the head were classified as craniocerebral injuries, as opposed to neurological injuries. (The frequency of concussions among those experiencing a TKO is indeterminant from the authors data). Facial lacerations, eye trauma, nasal injuries, or abdominal blows were defined as other injuries in the authores study.

TABLE 58
Results of nine amateur boxers in the 1985 and 1986 New York City Golden Gloves competition. (Fom JORDAN, ZIMMERMAN, 1988).

Boxer No.	Age, y	Bouts Won	Bouts Lost	Wins by Knockouts	Draw	Weight Division, lb	No. of Years Boxing	Age, y Started Boxing	Days of Sparring per Week	Prior Knockouts	Interval Between Last Bout and Magnetic Resonance Imaging
1	16	11	13	5	0	132	3.5	13	2-3	2	1 mo
2	25	3	1	0	0	178	2	22	4	0	1 mo
3	24	2	3	0	1	112	1.5	19	2	0	1 mo
4	25	0	4	0	0	156	3	22	3	0	3 mo
5	25	20	5	7	0	126	2.5	14	4	0	3 mo
6	22	12	3	1	0	132	2	15	1-3	0	3.5 mo
7	24	0	2	0	0	119	1	24	3-4	0	3.5 mo
8	25	24	8	11	0	132	9	15	3-5	0	1 mo
9	26	6	2	4	0	178	2	24	3	0	1 wk

Boxer 4 was medically suspended for one year and lost his last four fights. The other eight boxers were suspended for 90 days.

Results

JORDAN and CAMPBELL reported that from August 1, 1982, to July 31, 1984, 3.110 rounds were fought and 376 injuries incurred (262 craniocerebral and 114 others). These stastistics translate to 1.2 total injuries per ten rounds fought – 0.8 craniocerebral injuries and 0.4 other injuries. Statistics analyzed by year were relatively consistent (Table 59): During the first year, 484 boxers incurred 138 craniocerebral injuries and 53 other injuries, or 2.9 craniocerebral injuries and 1.1 other injuries per ten boxers. During the second year, 422 boxers incurred 124 craniocerabral injuries and 61 other injuries, or 2.9 craniocerebral injuries and 1.4 other injuries per ten boxers.

JORDAN and CAMPBELL wrote that despite the fact that there were relatively more craniocerebral injuries than other injuries, severe neurological dysfunction was rare: Only four boxers required immediate neurological evaluation at the hospital after a fight. Two boxers experienced postconcussion syndrome and the third boxer was semicomatose with amnesia and anisocoria. The fourth boxer sustained bilateral subdural hematomas and was the only boxer who died during the two-year period.

Of the other injuries, Table 60 of the paper of JORDAN and CAMPBELL, facial lacerations comprised 58% (66 cases), followed by injuries to the eye and hand (eight cases each). Other relatively common injuries included abdominal injuries (seven cases). Orbital hematoma or contusions (five cases), and nasal injuries (five cases). Three boxers lost fights as a result of poor athletic conditioning; these injuries were listed as exhaustion injuries. Three boxers incurred possible rib fractures. Ear injuries, abrasions, shoulder injuries, muscle strains, and injuries from low blows were infrequent (one case each).

JORDAN and CAMPBELL note that the overrepresentation of craniocerebral injuries in their study may be attributable to any or all of several factors. First their definition of craniocerebral injury was such that all KO's and TKO's were included, unless otherwise specified, which could result in an overrepresentation. Second, physicians were more likely to file an injury report form for a boxer suffering from a concussion because state law requires that any boxer experimenting a KO or TKO be medically suspended. A suspended

TABLE 59
Acute boxing injuries per year for two years. (From JORDAN, CAMPBELL, 1988).

	Aug 1982 to July 1983	Aug 1983 to July 1984	Mean
Total licensed boxers (906)	484	422	453
Total rounds fought (3,110)	1,636	1,474	1,555
Total craniocerebral injuries (262)	138	124	131
Total other injuries (114)	53	61	57
Total injuries (376)	191	185	188
Head injuries per 10 boxers	2.9	2.9	2.9
Other injuries per 10 boxers	1.1	1.4	1.25
Total injuries per 10 boxers	4.0	4.3	4.15
Craniocerebral injuries per 10 rounds fought	0.8	0.8	0.8
Other injuries per 10 rounds fought	0.3	0.4	0.35
Total injuries per 10 rounds fought	1.1	1.2	1.15

TABLE 60
Noncraniocerebral boxing injuries over a two-year period. (From JORDAN, CAMPBELL, 1988).

Facial lacerations	66
Hand	8
Eye	8
Abdomen	7
Orbital hematoma/contusion	5
Nose	5
Jaw	4
Exhaustion	3
Ribs	3
Ear	1
Shoulder	1
Testicle	1
Abrasion	1
Muscle strain	1
Total	114

boxer is not reinstated until he undergoes a recommended rest period and a neurological evaluation that includes a neurological examination, a computed tomographic scan, and an EEG. Third, since the boxing community is making a conscious effort to make the sport safer, referees have tended to terminate a bout early, before a boxer can incur a major concussion. This practice would increase the number of TKO's recorded as craniocerebral injuries.

JORDAN and CAMPBELL noted that another potential bias for the higher frequency of craniocerebral injuries is the underrepresentation of less serious other injuries. Physicians were less likely to complete an injury report for a boxer suffering from a facial laceration, unless it was severe enough to require sutures, thereby decreasing the number of other injuries. Furthermore, the ringside physician may not detect subtle or minor injuries. Whether these injuries come to the physician's attention is, according to JORDAN and CAMPBELL dependent on whether the boxer or the cameraman volunteers the necessary information. Accordingly, the ringside physician is more likely to be notified only of the serious injuries. This potential bias is limited if the ringside physician evaluates each boxer after he competes.

JORDAN and CAMPBELL noted that another potential methodological limitation of their study is the accuracy of the data obtained from the injury report forms. The information is dependent on the ringside physician. Failure to file an accurate and complete report can influence the data obtained. However, this potential bias is minimized because the NYSAC requires the medical director to check the injury report for accuracy.

The authors summarize, that the present study represents one of the few epidemiological investigations of boxing injuries in a well-defined population. Despite its methodological limitations, the authors believe, that their data provide a reasonable estimate of the frequency and magnitude of acute neurological and non-neurological injuries associated with professional boxing in the setting of strict medical supervision. However, noted the authors, their data should not be used to draw inferences about the development of chronic boxing injuries such as chronic traumatic encephalopathy and should not be used to comment on chronic conditions.

FRIEDMAN (1989) reported on a 55 year old man who was noted by others to be 'slowing down' over the past three years, though he himself was unaware of it. His family is from the Azores; neither his parents, relatives, nor siblings have had any neurological disorder. He had a history of mild poliomyelitis affecting his legs, but no other significant medical problem. He was a highly talented boxer, fighting twenty professional fights in the lightweight division, including an unsuccessful bout for the world championship. When he last fought, he was 22 years old. Upon retiring from the ring he worked with emotionally troubled children and as an artist in oil paintings; a display of his art was exhibited at the Rhode Island School of Design. He has continued to paint, though more slowly. He has worked as a custodian for several years, but has had no known exposure to toxins.

Results of the physical examination were normal. The patient's speech was slow, low in volume, and mildly slurred, and the words were drawn out; he had difficulty repeating short syllables. He was alert, affable, and well oriented but had no insight into his problems. He was somewhat garrulous and apathetic. He could give the months of the year backwards, recall three of three objects at five minutes and draw a clock, but had difficulties with serial '7'. A cartoon drawing of his was clearly above average. Cranial nerves were intact. There was no nystagmus, but he had severe akinesia. Rapid finger tapping was only slightly slow, but loss of amplitude was marked. Heel tapping was very slow on the right and mildly slow on the left. Tone was moderately increased in all joints. Strength was intact. Deep tendon reflexes were brisk, with a mild increase on the left. Babinski's reflex was present on the right. Sensory examination was unremarkable. His gait was abnormal, with a moderate simian stoop. There was a marked tendency to walk on his toes, with diminished arm swing and small steps. Retropulsion was conspicuous. CT scan of the brain showed diffuse atrophy and a cavum septi pellucidi.

Treatment with carbidopa/levodopa was titrated up to 25 mg/250 mg tid, without any effect. Benztropine mesylate slowly increased to 1 mg tid caused mild confusion as the only effect.

Thirty years after he stopped boxing, this patient had a progressive syndrome of mild dementia, akinesia, rigidity, upper motor neuron signs, postural instability, and a slow, slurred hypophonic dysarthria.

FRIEDMAN summarized that this report presents a typical case of Parkinsonism associated with boxing.

JORDAN and ZIMMERMAN (1990) examined 21 boxers (Tables 61 a – c) who underwent both MRI and CT scans. This referred population of boxers ranged in age from 21 to 66 years. Sixteen of these boxers were active professional boxers and there were four active amateur boxers. One boxer was a retired professional.

> Magnetic resonance imaging scans were obtained on a 0.5-T magnetic resonsance image. All patients had axial T1-weighted scans (repetitive time, 500 milliseconds, echo delay, 60 and 120 milliseconds) from base to vertex.

Three boxers (cases 3, 4 and 10) underwent MRI scans to evaluate subtle or equivocal findings seen on CT scans. A suspected hematoma (case 4) and a possible pontine, low-density lesion (case 10) were suspected since MRI scans yielded normal findings. In a third boxer (case 3), MRI was performed because of a possible low-density lesion seen on a CT scan in the left parietal lobe. Although this CT scan had normal findings, other abnormalities not detected on a CT scan were identified on an MRI scan, including right periventricular white-matter hyperintensity (demyelination) and a right thalamic periventricular spinal fluid intensity cyst.

Eleven boxers had both normal findings on CT and MRI scans, and seven boxers (1, 2, 5, 8, 17, 18, and 20) had abnormal findings on both CT and MRI scans. In all cases with abnormal findings, MRI scans were completely delineated and accurately assessed the nature of the lesion(s). Four boxers were noted to have hypodense lesions on a CT scan. In one boxer (case 1) an MRI scan further delineated the lesion as being a dilated sulcus in the left frontal lobe, representating a focal area of atrophy. In the other three boxers with hypodense lesions found on a CT scan (cases 2, 18, and 21), an MRI scan demonstrated cystic lesions of cerebrospinal fluid intensity, most consistent with areas of encephalomalacia. One boxer (case 5) was noted to have an arachnoid cyst of the quadrigeminal plate on a CT and an MRI scan. Boxer 17 was a retired professional boxer whose neurological history was complicated by a vehicular accident, in which he sustained a subdural hematoma that was evacuated by burr holes. The CT scan demonstrated hydrocephalus and atrophy. The MRI scan displayed other abnormalities, including a hyperintense focus in the left frontal lobe consistent with prior hemorrhagic contusions and extensive white-matter hyperintensities. This was indicative of ischemic demyelination or prior deep white-matter injury (shearing) or both. Magnetic resonance imaging also provided additional information in boxer 8 by demonstrating a small subdural hematoma that was not detected on the CT scan.

JORDAN and ZIMMERMAN summarized that the MRI scan seems to be the neuroradiological test of choice in the evaluation of traumatic anatomic lesions of the central nervous system in boxing. The authors continue: "Although MRI is superior to CT in the current level of development, it does not assess the brain function and should not serve as a substitute for the electro-encephalogram, neurological examination, and neuropsychological testing." According to the authors, there were no instances where abnormalities demonstrated on CT scanning were not detected by an MRI scan. However, some abnormalities on MRI scans, were not detected on CT scans.

The supremacy of MRI scanning over CT scanning is, according to JORDAN and

TABLE 61 PART 1
part 1, 2 and 3. Boxers evaluated by computed tomographic (CT) and magnetic resonance imaging (MRI) scans. (From JORDAN, ZIMMERMAN, 1990).

Boxer	Age, y	Boxing Status	Neurological Signs and Symptoms	CT Scan	MRI Scan	Comments
1	27	Professional	Headache, neurological examination had normal findings	Low-density lesion, left frontotemporal area	Dilated sulcus in the left frontal lobe representative of a focal area of atrophy	Boxed 10 y (5 y as an amateur and 5 y as a professional). After winning a difficult fight, developed a headache. Prefight CT scan had normal findings. Suspended from boxing.
2	26	Amateur	Asymptomatic, neurological examination had normal findings	Hypodense lesion, deep left temporal lobe	Small lesion in the left hippocampus of cerebrospinal fluid intensity consistent with a congenital or posttraumatic cyst	Referred for MRI scan because of a CT scan with abnormal findings on prelicensing medical screening; MRI better delineated the lesion as being an arachnoid cyst and not in a parenchymal lesion of the brain.
3	26	Professional	Asymptomatic, neurological examination had normal findings	Slight ventricular dilatation, questionable hypodense lesion in the left parietal lobe	Unusual configuration of the right atrium of the lateral ventricle, right-sided posterior periventricular white-matter changes	Referred for MRI scan because of a questionable hypodense lesion in the left parietal lobe on a CT scan. MRI confirmed that the questionable lesion on the CT scan was an artifact. However, MRI documented white-matter changes not noted on CT and noted an unusual configuration of the right atrium of the lateral ventricle.
4	25	Professional	Asymptomatic, neurological examination had normal findings	Questionable subdural hematoma, left frontal lobe	NL*	Referred for MRI scan because of a small subdural hematoma that could not be ruled out on CT scan.
5	26	Amateur	Occasional headache, neurological examination had normal findings	Arachnoid cyst of quadrigeminal plate	Quadrigeminal plate arachnoid cyst with associated compressor of the colliculi	Noted to have a CT scan with abnormal findings of prelicensing medical screen. Suspended from professional boxing.
6	26	Professional	Amnesia, blurred vision, headache, dizziness, lethargy, neurological examination had normal findings	NL*	NL*	Developed a postconcussion syndrome after winning a 10-round fight. Boxer was amnesic for the last 2 rounds. Symptoms resolved in 2 wk.
7	21	Amateur	Left inferior quadrant anopsia, left-sided paresthesias	NL*	NL*	Immediately after a fight he complained of visual difficulties. Electroencephalogram demonstrated slight alpha asymmetry with slightly lower voltage in right temporal occipital region.

TABLE 61 PART 2
Boxers evaluated by computed tomographic (CT) and magnetic resonance imaging (MRI) scans. (From JORDAN, ZIMMERMAN, 1990).

Boxer	Age, y	Boxing Status	Neurological Signs and Symptoms	CT Scan	MRI Scan	Comments
8	22	Professional	Coma, uncal herniation	Mass effect in the right hemisphere, could not rule out a large isodense subdural hematoma	Right posterior and intrahemispheric falx subdural hematoma, right to left midline shift	Boxer was knocked out in the ring. Also had a cervical spine fracture secondary to striking neck on bottom rope when falling.
9	23	Professional	Asymptomatic, neurological examination with normal findings	NL*	NL*	Boxer was referred for MRI because of abnormal electroencephalographic findings that demonstrated multifocal asymmetrical accentuation.
10	27	Professional	Asymptomatic, neurological examination with normal findings	Ill-defined, low-density lesion within the pons, probable artifact	NL*	Boxer was referred for MRI because of an equivocal CT scan that demonstrated a questionable lesion in the pons. MRI confirmed that it was an artifact.
11	27	Amateur	Asymptomatic, right upper-extremity hyperreflexia at the time of examination	NL*	NL*	Boxer had right upper-extremity hyperreflexia at the time of medical examination when applying for professional license. Electroencephalographic findings were normal.
12	31	Professional	Asymptomatic, neurological examination with normal findings	NL*	NL*	Boxer was referred for MRI because of mild right posterior temporal focal abnormality on electroencephalogram.
13	26	Professional	Asymptomatic, neurological examination with normal findings	NL*	NL*	Boxer was referred for MRI because of a posterior accentuation with right-sided preponderance.
14	28	Professional	Asymptomatic, neurological examination had normal findings	NL*	NL*	Boxer was referred for an MRI because of abnormal electroencephalographic finding.
15	40	Professional	Asymptomatic, neurological examination had normal findings	NL*	NL*	Boxer was referred for an MRI because of abnormal electroencephalographic findings that demonstrated poor organization with sporadic low voltage (4-6 Hz) activity that was mixed with alpha.

TABLE 61 PART 3
Boxers evaluated by computed tomographic (CT) and magnetic resonance imaging (MRI) scans. (From JORDAN, ZIMMERMAN, 1990).

Boxer	Age, y	Boxing Status	Neurological Signs and Symptoms	CT Scan	MRI Scan	Comments
16	20	Professional	Swollen left eye, fever, headache, lethargy	NL* brain, no orbital fracture, periorbital soft-tissue swelling	NL* brain, periorbital soft-tissue swelling	Within 24 h after a fight, boxer developed periorbital cellulitis. MRI and CT performed because of headache and lethargy.
17	66	Retired	Memory loss, stuttered speech, depression	Hydrocephalus	(1) Hyperintensity in left frontal lobe consistent with old region of contusion, (2) Small hyperintensity in left frontal lobe consistent with previous hemorrhage, (3) White-matter punctate hyperintensities, (4) Atrophy	Retired boxer referred for evaluation of decreased memory and question of Parkinson's disease. Prior history of left frontal subdural hematoma secondary to a motor-vehicle accident. Subdural was drained via burr holes.
18	26	Professional	Coma, nonfocal neurological examination, seizure disorder	Small, rounded hypodensity in the ganglionic region on the right	Spherical lesion anterior to the right half of anterior commissure that was of cerebrosphinal fluid intensity, findings consistent with large perivascular space vs lacuna vs encephalomalacia	Collapsed immediately after a fight. Boxer was hospitalized. Initial CT demonstrated subdural hematoma over the right convexity with midline shift and subarachnoid hemorrhage, with blood noted in the region of the tentorium and the posterior aspect of the falx.
19	29	Professional	Asymptomatic, neurological examination had normal findings	NL*	NL*	Boxer was referred for MRI scan because of poor boxing performance. Nine consecutive losses. Permanently suspended. Normal electroencephalographic findings.
20	28	Professional	Asymptomatic, neurological examination had normal findings	NL*	NL*	Routine evaluation. Normal electroencephalographic findings.
21	24	Professional	Asymptomatic	Left frontal lobe hypodensity	Cystic lesion in left frontal lobe, perhaps consistent with posttraumatic cystic encephalomalacia	Abnormal CT findings noted at prelicensing evaluation.

*NL indicates normal.

ZIMMERMAN, a result of several factors that include: (1) Multiplanar scanning capacity, (2) absence of artifact from adjacent bone, and (3) high level of contrast between brain and cerebrospinal fluid on T1-weighted (short repetition time/short echo delay) images.

ROBERTS et al. (1990) examined archival formalin-fixed material from the temporal lobe from 14 of the 15 cases of dementia pugilistica originally reported by CORSELLIS et al. (1973), one additional case of a boxer who died during a fight, three additional cases of professional boxers, and two amateur boxers (Table 62). Each case had previously been investigated, to establish career details, and had undergone extensive neuropathological examination.

The extent of tangle and plaque formation was established by silver and Congo red staining (Table 63). Paraffin embedded 20 μm sections were stained by hematoxylin-eosin and Congo red and by immuno-cytochemical methods (PAP) using a well characterized monoclonal antibody 4 DR 12/2/6 raised to a synthetic peptide corresponding to residues 8 – 17 of a beta-protein amyloid. Immuno-staining was carried out with or without a fifteen minute pre-incubation on 70% formic acid. Sections were incubated with the primary antibody (dilution 1:1000 ascites fluid for one hour) and then with a second layer anti-mouse Ig G peroxidase conjugated antisera (Mies-Yeda 1:100 for 30 minutes). The reaction site was visualized by a final intubation in a 0.05% diaminobenzidine solution.

TABLE 62
Biographical and neuropathological details of cases. (From ROBERTS et al., 1990).

		Details of cases		Neuropathology[1]		
Case	Age	Boxing career (years)	No fights	Cavum septum	Tangles	Plaques
Professional						
1	22	3	3 (died during fight)	+	–	–
2	53	25	>100	+	+	–
3 (5)*	60	18	>100	+	+ +	–
4	61	15	150	+	+ + +	–
5 (7)	62	20	400	+	+ + +	–
6 (3)	63	13	300	+	+ + +	–
7 (1)	63	14	400	+	+ + +	–
8	65	20	>200	+	+	–
9 (11)	67	20	?	+	+	±
10 (10)	67	15	?	+	+ +	±
11 (4)	69	25	600	+	+ +	–
12 (8)	71	23	565	+	+ +	±
13 (9)	72	10	?	+	+ +	–
14 (2)	77	17	700	+	+ + +	–
15 (6)	83	25	500	+	+ +	+ +
Amateur						
16 (13)	57	15	>100	+	+ + +	+ + +
17 (15)	58	5	> 50	±	–	–
18 (14)	61	20	>100	±	+	+
19	68	2	< 10	±	–	–
20[A]	72	2	< 10	–	+ + +	+
AD cases						
N = 20	55–85	N/A	N/A	–	+ + (MIN)	+ + (MIN)
Controls						
N = 20	58–90	N/A	N/A	–	+ (MAX)	+ + (MAX)

[1] Tangle and plaque assessed by routine silver stains.
[*] Case No. in Corsellis *et al.*
[A] Probable Alzheimer's disease.

For comparison, sections from 20 confirmed cases of Alzheimer's disease and 20 age-matched controls were investigated in a similar fashion.

ROBERTS et al. found that all dementia pugilistica cases with substantial tangle formation showed evidence of extensive beta-protein immunoreactive deposits (plaques). The diffuse 'plaques' were not visible with Congo-red or standard silver stains. The degree of beta-protein deposition was comparable to that seen in Alzheimer's disease. The data of ROBERTS et al. show that the present neuropathological description of dementia pugilistica (tangles but no plaques) should be altered to acknowledge the presence of substantial beta-protein deposition (plaques). The molecular markers present in the plaques and tangles of dementia pugilistica are the same as those in Alzheimer's disease. Similarities in clinical symptoms, distribution of pathology and neurochemical deficits also exist. Epidemiological studies have shown that head injury is a risk factor in Alzheimer's disease, according to ROBERTS et al. It is probable that dementia pugilistica and Alzheimer's disease share common pathogenic mechanisms leading to tangle and plaque formation" (ROBERTS et al.).

DALE et al. (1991) obtained brain material from sixteen cases, all of which were retired boxers, except one, who was a young man of 22 who died in the ring. Ten of these had been described by CORSELLIS et al. (1973). The series included both professional and amateur boxers (Table). Frontal and/or temporal lobe tissue was available from each case. Clinical histories indicated that eleven of the sixteen cases suffered from dementia pugilistica (Table 64).

The authors stained sections (20 μm thick) of formalin-fixed paraffin-embedded brain material with (a) an affinity purified rabbit antiserum to ubiquitin or (b) with a monoclonal anti-

TABLE 63
Detection of amyloid plaques. (From ROBERTS et al., 1990).

	β-Amyloid immunostaining		
Case	No Formic Acid	Formic Acid	Silver staining[1]
1	−	−	−
2	−	+ +	−
3	−	+ +	−
4	±	+ + +	−
5	−	+ +	−
6	±	+ + +	−
7	±	+ + +	−
8	+	+ +	−
9	+	+ +	±
10	+	+ + +	±
11	−	+ +	−
12	+	+ + +	±
13	±	+ +	−
14	−	+ + +	−
15	+ +	+ + +	+ +
16	+ +	+ + +	+ + +
17	±	+ +	−
18	+	+	+
19	+	+	−
20	+ + +	+ + +	+ + +
AD Cases	+ + + (MIN)	+ + (MIN)	+ + (MIN)
Controls	+ + (MAX)	+ + (MAX)	+ + (MAX)

[1] von Braunmühls or Bielschowsky (from table 1).

body, BF 10, which labels a phosphorylated epitope of the medium weight neurofilament protein, NF-M.

DALE et al. reported that the brains from all but one of the demented ex-boxers showed some evidence of neurofibrillary tangle formation following staining with BF 10 and also, in most cases, with anti-ubiquitin. BF 10 generally labeled more tangles and with greater intensity than anti-ubiquitin. Of the five non-demented ex-boxers, three showed BF 10 labeling of tangles, but tangles were not visualized with anti-ubiquitin (Table 64). Tangle labeled with BF 10, were less numerous in the non-demented cases than in those who had been demented.

DALE et al. found that no relationship between the degree of neuropathology and the number of years of boxing, the number of fights, or the professional/amateur status of the ex-boxers was discernible. Some cases showed evidence of senile plaque formations, although plaques tended to be few in number and of an 'immature' type. In two cases who were demented (125/71 and 145/74), however, a large number of classical senile plaques and tangles were evident, suggesting Alzheimer's disease.

DALE et al. continued that they have now shown that ubiquitin, which is known to be a component in Alzheimer's tangles, is present in the neurofibrillary tangles of dementia pugilistica. "Ubiquitin is, at present, the only protein consistently detected in inclusions associated with several neurodegenerative diseases including Lewy bodies in Parkinson's disease, anterior horn cell inclusions in motor neuron disease, Pick bodies of Pick's disease in addition to the tangles of Alzheimer's disease and dementia pugilistica" (DALE et al.).

We fully agree with the statement of DALE et al. that the brain damage to boxers results from acceleration and deceleration of the brain within the

TABLE 64
Case details of boxers showing immunolabelling of tangles with anti-ubiquitin BF 10 and silver stain. (From DALE et al., 1991).

Case	Age at death	Boxing Years	Career Fights	P/A[1]	Frontal Lobe Anti-ubiquitin	BF10	Temporal Lobe Anti-ubiquitin	BF10	Bielschowski Temporal Lobe Silver stain
Demented Cases									
237/70 (6)[2]	83	25	500	P	+/–	+/–	NA	NA	+ + +
258/72	61	8	150	P	–	+ +	NA	NA	NA
155/71 (3)	63	13	300	P	+/–	+	+ + +	+ + +	+ + +
257/72	61	15	150	P	+/–	+ +	+ +	+ +	+ + +
158/74	65	20	>200	P	+/–	+	+ +	+ + +	+/–
124/71 (7)	62	20	400	P	NA	NA	+ +	+ + +	+ + +
119/71 (2)	77	17	700	P	+	+	+ + +	NA	+ + +
47/71 (4)	69	25	600	P	–	–	NA	NA	+ +
65/73	68	20	Many	A	+/–	+/–	NA	NA	+ +
125/71[3] (13)	57	15	>100	A	NA	NA	+ + +	+ + +	+ + +
145/74[3]	72	2	< 10	A	+ + +	+ + +	+ + +	+ + +	+ + +
Non-Demented Cases									
126/71 (12)	91	?	?	P	NA	NA	–	+	+
48/72[4]	22	2	3	P	–	–	NA	NA	–
153/71 (11)	67	20	?	P	–	+	NA	NA	–
120/71 (15)	58	5	50	A	–	–	–	–	–
122/71 (14)	61	20	100	A	–	+/–	–	+	+

[1]Professional/Amateur
[2]Case number in Corsellis *et al.* (1973)
[3]Probable Alzheimer's disease
[4]Died during fight

? Not known
NA Not Available
– unstained
+/– weakly stained
+ to + + + mild to strongly stained.

skull. Although the effects of repeated and numerous subconcussive and concussive blows are also responsible. We do not agree, however, with the statement that particularly rotational acceleration, which allegedly results in the stretching of neurons and blood vessels by sheering forces, produces brain damage in boxers. We cannot agree, because two different injury patterns are produced by translational (linear) and rotational acceleration of the brain; both occur in boxing; for details see pp. 7 and 76.

DALE et al. summarized that it is likely that tangle formation in dementia pugilistica occurs as a direct result of these damaging forces on neuronal processes. We would ask: why only on the neuronal processes, why not also on the neurons per se? It had been suggested that head trauma may be a factor in the development of Alzheimer's disease, although genetic susceptibility and environmental toxins currently appear to be more favored as putative causal factors. "It is plausible that tangle formation in Alzheimer's disease and dementia pugilistica share many common features" (DALE et al.).

JORDAN et al. (1992) performed computed tomography in 338 active professional boxers.

CT scans were abnormal in 25 boxers (7%). The most common CT abnormality was brain atrophy (22 cases). Focal lesions of low attenuation consisted with posttraumatic encephalomalacia were noted in only three boxers. Boxers with abnormal CT scans did not differ from those with borderline or normal CT scans in regard to age, win-loss record, number of bouts or history of an abnormal electroencephalogram. Thirty-seven boxers with borderline CT scans (49%) and 17 with abnormal CT scans (68%) reported a previous technical knock out (TKO) or a knock out (KO), compared with only 89 (37%) of the 238 boxers with normal CT scans ($P < .01$). Brain atrophy was noted more frequently in boxers with a large cavum septi pellucidi (CSP) than in those with a small or no CSP ($P < .05$). Boxers with abnormal or borderline CT scans who experienced a TKO or KO were slightly older than those with normal CT scans and a history of a TKO or K O ($P < .05$).

Three hundred thirty-eight active professional boxers in the series of JORDAN et al. underwent CT. All subjects either were licensed to box professionally in New York State or were applying for a license. They ranged in age from 17 to 46 years. A CT scan that was retrievable was obtained in each boxer, and the results of their annual physical examination were recorded in their medical chart at the New York State Athletic Commisssion.

Results

All CT Abnormalities

Of the 338 CT scans 25 (7%) were considered abnormal (Table 65). The most common CT abnormality was ventricular atrophy (eight cases) followed by diffuse atrophy (six cases), sulcal atrophy (five cases), and posterior fossal atrophy (three cases). Three boxers had focal lesions of low attenuation, consistent with posttraumatic encephalomalacia involving the cerebral hemispheres. The vast majority of the boxers had either normal (238 cases) or borderline (75 cases) CT scans. Of the 25 boxers with abnormal CT scans, seven (28%) also had a CSP.

The characteristics of the boxers – classifed according to CT findings – are presented by JORDAN et al. in Table 66. The three groups were similar in terms of mean age, win-loss record, and number of bouts. One hundred forty-three boxers (42%) had experienced a TKO or KO. Thirty-seven boxers with borderline CT scans (49%) and 17 with abnormal CT scans (68%) reported a TKO or KO, compared with only 89 (37%) of the 238 boxersa with normal CT scans ($P < .01$). Of the 143 boxers of the series of JORDAN et al., who had experienced a TKO or KO (Table 67), 89 (62%) had a normal CT scan and 54 (38%) had a borderline or abnormal CT scan. Average age of boxers with a history of a TKO or KO and abnormal or borderline CT scans was 27.7 years, compared with 26.1 years for those with a history of a TKO or KO and normal CT scans. The difference was statistically significant ($P < .05$).

Cavum septi pellucidi (CSP)

Forty-five boxers (13%) had a CSP (39 small, six large) (Table 68). Boxers with a small CSP tended

TABLE 65
CT findings in 338 active professional boxers. (From JORDAN et al., 1992).

Findings	No. of Patients
Normal	238
Borderline	75
Abnormal	25
Ventricular atrophy	8
Diffuse atrophy	6
Sulcal atrophy	5
Posterior fossal atrophy	3
Focal lesions of low attenuation	3

TABLE 66
Characteristics of 338 active professional boxers according to CT findings. (From JORDAN et al., 1992).

Type of CT Scan	No. of Patients	Mean Age (y)	Mean No. of Wins	Mean No. of Losses	Mean Total No. of Bouts	History of TKO or KO*†	Abnormal EEG*
Normal	238	24.7	8.2	2.7	11.3	89 (37)	10 (4)
Borderline	75	26.6	10.4	3.5	14.3	37 (49)	7 (9)
Abnormal	25	26.5	10.3	3.6	14.4	17 (68)	1 (4)
All	338	25.2	8.8	2.9	12.2	143 (42)	18 (5)

*Numbers are numbers of patients; numbers in parentheses are percentages.
† This distribution of history of experiencing a TKO or KO, according to CT findings, was statistically significant ($P < 0.01$).

TABLE 67
Characteristics of 143 boxers who experienced a TKO or KO, according to CT findings. (From JORDAN et al., 1992).

Subject Group	No. of Patients	Mean Age (y)*	Mean No. of Wins	Mean No. of Losses	Mean Total No. of Bouts	Patients with Abnormal EEG	Patients with CSP
TKO or KO with normal CT Scan	89	26.1	10.3	4.5	15.4	3	10
TKO or KO with abnormal or borderline CT scan	54	27.7	12.2	4.7	17.6	3	12

This age difference between the two groups was statistically significant ($P < 0.05$)

to have had more bouts (including wins and losses), but this difference did not reach statistical significance. No association between size of the CSP and frequency of an abnormal EEG or history of a TKO or KO was noted. However, a positive association between frequency of cerebral atrophy and size of the CSP was observed. Seventeen boxers with a small CSP (44%) and four with a large CSP (67%) had cerebral atrophy (all types), compared with 78 without a CSP who had cerebral atrophy (27%) ($P = < .05$). This association between CSP and atrophy was independent of the history of a TKO or KO.

EEG

Of the 338 boxers of the series of JORDAN et al., 335 had an EEG, 18 (5%) of which were abnormal (Table 69). Boxers with abnormal or borderline EEGs tended to have had more bouts than did

TABLE 68
Characteristics of 338 active professional boxers according to severity of CSP. (From JORDAN et al., 1992).

Type of CSP	No. of Patients	Mean Age (y)	Mean No. of Wins	Mean No. of Losses	Mean Total No. of Bouts	History of TKO or KO*	Abnormal EEG*	Cerebral Atrophy*†
Absent	293	25.0	8.2	2.8	11.5	121 (41)	15 (5)	78 (27)
Small	39	26.9	13.5	4.0	17.2	19 (49)	3 (8)	17 (44)
Large	6	26.7	6.2	3.8	10.3	3 (50)	0 (0)	4 (67)
All	338	25.2	8.8	2.9	12.2	143 (42)	18 (5)	99 (29)

*Numbers are numbers of patients; numbers in parentheses are percentages.
†This distribution of size of the CSP, according to the presence of cerebral atrophy (borderline and moderate to severe), was statistically significant ($P < 0.05$).

boxers with normal EEG's; however, this difference did not reach statistical significance. Twenty-three (92%) of the 25 boxers with abnormal CT scans had normal EEGs. There was no statistical significant difference in the frequency of abnormal EEGs among boxers with normal, borderline, or abnormal CT scans.

JORDAN et al. summarize that the prevalence of abnormal CT scans in an unselected sample of 338 active professional boxers was relative low (7%) compared with that reported in the literature. In the series of JORDAN et al. 68% of boxers with abnormal CT scans experienced a TKO or KO. The authors continue: "Whether central nervous system abnormalities, as reflected by an abnormal CT scan, predispose a boxer to a TKO or KO or whether a TKO or KO results in brain injury that is manifested by an abnormal CT scan remains to be determined.

Approximately 13% of the boxers in the series of JORDAN et al. had evidence of a cavum septi pellucidi (CSP). Although the majority (87%) of these CSP were small, the prevalence of CSP appears to be increased in boxers.

In boxers with a history of a TKO or KO in the series of JORDAN et al., these with an abnormal or borderline CT scan tended to be older than those with normal CT scans. One possible explanation for this finding, so argue JORDAN et al., may be that older boxers are more susceptible to the detrimental effects of a TKO or KO, as reflected by the CT scan. Alternatively, older boxers with a history of a TKO or KO, may have a higher prevalence of cerebral trauma by virtue of longer career experience and thus more time for cerebral atrophy to develop. Support of this hypothesis is provided by the finding that the older knocked out boxers with abnormal or borderline CT scans had participated in more bouts than younger knocked out boxers with normal CT scans. However, this difference

TABLE 69
Characteristics of 335 active professional boxers according to findings on EEG. (From JORDAN et al. 1992).

Type of EEG	No. of Patients	Mean Age (y)	Mean No. of Wins	Mean No. of Losses	Mean Total No. of Bouts	History of TKO or KO*	Abnormal CT Scan*
Normal	293	25.1	8.1	2.8	11.3	123 (42)	23 (8)
Borderline	24	27.9	14.8	4.0	19.6	13 (54)	1 (4)
Abnormal	18	25.6	13.1	3.4	17.2	6 (33)	1 (6)
All	335	25.3	8.9	3.0	12.3	142 (42)	25 (7)

*Numbers are numbers of patients; numbers in parentheses are percentages.

was according to JORDAN et al., not statistically significant.

The apparent dissociation in JORDAN et al.'s study between having an abnormal CT scan and having an abnormal EEG raises an important issue in the evaluation of potential brain injury in boxers. The CT scan evaluates the cerebral anatomy, the EEG evaluates function. Accordingly, they should not be substituted for one another, because they serve different roles. This point is reinforced, according to JORDAN et al., when one notes that 92% of the boxers with abnormal CT scans had normal EEG's. Therefore, the neurologic evaluation of an active boxer should include a CT or magnetic resonance image for detection of anatomic lesions and an EEG for detection of electrophysiologic dysfunction.

ROCHON (1994) reported on a 68-year old ex-boxer:

> He began boxing at age eight. He had 100 amateur fights. From his 17. to his 26. year of life (10 years) he was a professional and had 100 professional fights as a middleweight.
> He prersented slight memory problems and behavioral change over a period of a year. Neuropsychological examination confirmed slight frontal cognitive damage and damage at the level of the motor signs of the upper left hemicorpus. This symptomatology will, accordimg to the author, raise the possibility of a chronic punch drunk syndrome.

Amateur boxing is faced with criticism over the potential damage the sport inflicts on those who participate. The most sensitive measure of early neurological dysfunction is neuropsychological investigation. Ten studies employing such assessment on 289 amateur boxers are reviewed. The forms of analysis undertaken include controlled comparison with other sportsmen, of both active and former boxers, detailed pre- and post-bout analysis, analysis of the influence of within boxing variables, length of career, level of competition and prospective longitudinal investigation. Amateur boxers were found to exhibit no signs of neuropsychological dysfunction in any analysis. However, some trends emerged suggesting a lengthy career in amateur boxing might induce fine motor reactions, although such findings are within the normal range and did not represent central neuropsychological functioning. Thus some amateur boxing does not appear to expose individuals to neurological dysfunction.

MENDEZ (1995) reviewed the neuropsychiatry of boxing.

This update considers the clinical, neuropsychological, diagnostic, neurobiological, and management aspects of boxing-related brain injury.

Professional boxers with multiple bouts and repeated head blows are prone to chronic traumatic encephalopathy. Repeated head blows produce rotational acceleration of the brain, diffuse axonal injury, and other neuropathological features. Chronic traumatic encephalopathy includes motor changes such as tremor, dysarthria, and Parkinsonism; cognitive changes such as mental slowing and memory deficits; and psychiatric changes such as explosive behavior, morbid jealousy, pathological intoxication, and paranoia. Screening with neuropsychological tests and neuro-imaging may help predict those boxers at risk for chronic traumatic encephalopathy.

Boxing results in a spectrum of chronic traumatic encephalopathy ranging from mild, non-progressive motor changes to dementia pugilistica. Recent emphasis on safety in the ring, rehabilitation techniques, and other interventions do not eliminate the risk for chronic traumatic encephalopathy. For this reason, there is an active movement to ban boxing.

DAVIE et al. (1995) described the clinical and magnetic resonance spectroscopy (MRS) findings in three ex-professional boxers who developed a Parkinsonian syndrome.

Proton magnetic resonance spectroscopy is a non-invasive method that provides information about the chemical patholgy of conditions affecting the CNS. The largest peak visible with MRS is derived principally from N-acetylaspartate (NAA), and amino acid contained almost exclusively within neurons and their processes in adult brain (URENJAK et al. 1993).

Case 1: This boxer was first described in the medical literature by HARVEY and DAVIS in 1974. He is now a 45 year old man, who was noted by his family to have slurring of speech after being involved in 16 professional bouts. The patient attributed his symptoms to a head injury he had received while in police custody. He did not lose consciousness during this alleged assault. Six months later he started to shuffle when walking and his legs appeared stiff. During his last few bouts he was aware of slowness on his feet and difficulty avoiding trouble in the ring. After retiring, his gait deteriorated; he developed lability of mood and became aggressive. When seen in 1974, he was noted to be cooperative but mentally slow. He had a symmetric extrapyramidal disorder with an expressionless face, generalized poverty of movement, cogwheel rigidity in all four limbs, and an extrapyramidal gait. Tendon reflexes were brisk, with a left extensor plantar. Treatment with a levodopa preparation at that time resulted in some improvement in his gait.

Before his current neurological assessment he had undergone treatment with lithium for a bipolar depressive disorder. *On examination* he was easily distractible and poorly cooperative. He had reduced facial expression. There was no rest tremor of his limbs and he displayed reduced arm swing on walking. There was bradykinesia and cogwheel rigidity that was worse on the left side. Power was normal. Reflexes were brisk but symmetric with flexor plantar responses. There were no cerebellar signs, no sensory deficit, and sphincter functions were normal.

Case 2: This 52 year old man fought professionally as a middle and cruiser weight. He was involved in a car accident when 28 and received a head injury for which he required 24 stitches. He did not lose consciousness during this event but his injuries were severe enough to prevent him continuing with his boxing career. At the age of 50 he began to drag his left leg and had trouble with his balance. He began to slow down. His speech became slurred and his memory deteriorated. There was slight improvement in his symptoms with levodopa treatment. On examination he had mild bradykinesia, with some cogwheel rigidity and pronounced blepharospasm. He had some slowness of gait with reduced arm swing. There were no cerebellar signs and no sphincter disturbance.

Case 3: This 69 year old man first began boxing at the age of 20 when he entered the navy. He boxed for about 18 months before turning professional. He denied being knocked out in any of his amateur or professional bouts (Table 70). At the age of 66 he developed rest tremor of the right arm that soon spread to the other side. Levodopa produced only mild improvement.

On examination, he was moderately rigid and bradykinetic with bilateral rest tremor. Posture was stooped and he shuffled with short steps. Postural reflexes were impaired. There were no other neurological signs.

DAVIE et al. studied also six patients, aged 49 to 72 (median 56) years and median duration of disease of seven years with classic, levodopa responsive asymmetric idiopathic Parkinson's disease and six healthy controls with a median age of 59 years.

The study was approved by the joint ethics committee at the *Institute of Neurology* and the *National Hospital for Neurology and Neurosurgery, London.* Informed consent was obtained before each study. *Magnetic resonance imaging* and *MRS* were performed on a 1.5 Tesla GE whole body imager with a standard quadrature head coil. The study commenced with a coronal, T2 weighted, spin echo imaging sequence (TR 2000 ms TE 80 ms) (3 mm slices with 1 mm gap, 256 x 256 matrix, echo train length 8).

Magnetic Resonance Imaging: MRI revealed abnormalities in all three boxers, consisting of discrete periventricular or subcortical white matter high signal lesions. There were no high signal lesions present in the basal ganglia. In one of the three boxers (case 1), there was evidence of a cavum septi pellucidi. In three of the patients with idiopathic Parkinson's disease there were small lesions present on T2 weighted MRI. Similar changes were seen in two of the six controls. There was no evidence of a cavum septi pellucidi in any of the patients with idiopathic Parkinson's disease or controls. In two of the patients with idiopathic Parkinson's disease there was a moderate degree of low signal extending from the globus pallidus into the medial border of the putamen. This was not present in the controls or in the three ex-boxers.

Magnetic Resonance Spectroscopy: Table 70 shows the concentrations of N-acetyl aspartase (NAA) from the lentiform nucleus in the three ex-boxers. These were all significantly lower (median 6.1 mmol/l) than the NAA concentrations from the age matched controls (median 9.94, range 9.26 – 10.77) mmol/l, $P < 0.03$). By contrast, there was no significant difference in the concentration of NAA in the idiopathic Parkinson's group (median 9.99 range 9.0 – 11.0) mmol/l $P > 0.9$) compared with the controls. There was no significant difference in the concentrations of creatine or com-

TABLE 70
Boxing and MRS data. (From DAVIE et al. 1995).

Case No.	Age	Began boxing	Finished boxing	Amateur bouts	Professional bouts	Knocked out	Lentiform [NAA]	Fenestrated septum pellucidum
1	45	14	24	75	25	1	6·1 mmol/l	Present
2	52	15	27	1	24	3	7·7 mmol/l	Absent
3	69	20	25	?	30–40	0	5·14 mmol/l	Absent

Control medium [NAA] 9·94 mmol/l; idiopathic Parkinson's disease, median [NAA] 9·99 mmol/l.

pounds containing choline between the three groups.

The most striking finding in the study of DAVIE et al. was the significant reduction in the concentration of N-acetylaspartase (NAA) from the putamen and globus pallidus in the three ex-boxers with a Parkinsonian syndrome, compared with the patients with the idiopathic Parkinson's disease and the control group. This reduction is likely to reflect neuronal loss occurring in the putamen and globus pallidus and supports the hypothesis that the extrapyramidal syndrome that may occur in ex-boxers is a distinct entity from idiopathic Parkinson's disease.

PORTER and FRICKER (1996) carried out a controlled prospective neuropsychological assessment of active experienced boxers.

The authors used a practical battery of eight neuropsychological tests for the detection of an association between amateur boxing and chronic traumatic encephalopathy. The study was carried out over a 15-18-month period. The setting was Dublin, Ireland, March 1992 to September 1993. Twenty experienced, actively competing, amateur boxers and 20 controls matched for age and socioeconomic status were examined.

At the end of the study period, the boxers performed significantly better than the controls in both the Trial-Making Tests A and B (TMT-A, TMT-B), whereas the control group's scores for the Finger Tapping Tests (FTT) were significantly higher than those of the boxers. The boxer's scores for the dominant-hand FTT showed a significant deterioration, but there was no association between this change and boxing exposure.

The authors concluded that there was no evidence of neuropsychological impairment in the boxers as compared with socio-economically, educationally and age-matched controls, and there was no association between boxing exposure and performances in any of the neuropsychological tests used. There is accumulating evidence that amateur boxing is not usually associated with chronic traumatic encephalopathy, but longer term prospective studies are needed.

JORDAN et al. (1997) published an article on *'Apolipoprotein E ε 4 Associated with Chronic Traumatic Brain Injury in Boxing'.* Due to its importance this paper is reviewed in greater depth than it otherweise would have been. The authors refer to the similarities between Alzheimer's disease (AD) and dementia pugilistica, and evaluated the relationship between apolipoprotein E *(APOE)* genotype and chronic traumatic brain injury (CTBI) in boxers to determine whether there is a genetic susceptibility to the effects of head trauma. JORDAN et al. examined 30 professional boxers aged 23 to 76 years, who underwent neurological and behavioral assessment in conjunction with *APOE* genotyping. The authors write that: "Chronic traumatic brain injury is the most serious public health concern in modern-day boxing. Also known as dementia pugilistica or chronic traumatic encephalopathy, CTBI representsd the long-term and cumulative neurologic consequences of repetitive head trauma. This syndrome has been most typically described in active or retired boxers after a long exposure to the sport."JORDAN et al. noted, that head trauma has been implicated as a possible enviromental trigger for Alzheimer's disease. The current investigation by JORDAN et al. was under-

TABLE 71
Chronic brain injury scale (CBI). (From JORDAN et al., 1997).

Motor (incoordination, dysarthria, parkinsonism, gait disturbance, or pyramidal signs)	
Normal	0
Mild	1
Moderate	2
Severe	3
Cognitive (deficits in mental speed, memory, attention, executive function, language, or visuospatial function)	
Normal (Mini-Mental State Examination [MMSE] = 28-30)	0
Mild (MMSE = 20-27)	1
Moderate (MMSE = 10-19)	2
Severe (MMSE $\leq$ 9)	3
Behavioral (agitation or aggression, delusions, hallucinations, dysphoria, anxiety, euphoria, apathy, disinhibition, irritability or lability, or aberrant motor behavior)	
Normal	0
Mild	1
Moderate	2
Severe	3
Total CBI score (range, 0-9)	

taken to determine whether *APOE* genotype influences the risk of CTBI and its severity among boxers. The authors enrolled boxers from several sources. The majority were local boxers in retirement who volunteered for neurologic examination and *APOE* genotyping. Five professional boxers were referred by the New York State Athletic Commission (NYSAC) for neurologic evaluation, and the five volunteered to undergo *APOE* genotyping. One additional boxer was referred directly for the evaluation of dementia. Boxers underwent a detailed neurologic examination performed by one of three neurologists (B. D. J., N. P. R., A. R. J.). Two examineers were present in 10% of cases to confirm interrater reliability. Impairments were rated using a new scale designed to assess the long-term consequences of brain injury, the *Chronic Brain Injury (CBI)* scale (Table 71). The CBI scale quantifies the clinical findings of motor, cognitive, and psychiatric deficits associated with boxing-related brain injury. Scores on the CBI score range from 0 to 9, with higher scores reflecting greater impairment. JORDAN et al. obtained information about each boxer's past medical history, boxing exposure, family history, and social habits. In cognitively intact boxers, boxing exposure was obtained by self-report. In cases of significant cognitive impairment, boxing exposure was obtained from a surrogate responder or by boxing records. Overall boxing exposure was classified as high if the boxer participated in twelve or more professional bouts. The use of 12-bout criterion of exposure was based, according to the authors, on a prior study of a representative sample of 338 professional boxers, among whom the number of bouts was 12.2. Amateur boxers with no history of professional bouts were classified as low exposure based on the lack of conclusive evidence for long-term neurologic sequelae in asssociation with amateur boxing. Measures of severity of CTBI included, according to the authors, the total CBI score and a categorical classification based on CBI score consisting of normal (CBI = 0), mild (CBI = 1- 2), moderate (CBI = 3 – 4), and severely impaired (CBI >) ratings. Subjects were given the option of receiving their *APOE* genotype test results, contingent on pretest and posttest genetic counceling, following the guidelines of the National Institute of Aging and the Alzheimer's Association Working Group on *APOE* Genotyping in Alzheimer's Disease.

Results

JORDAN et al. presented the characteristics of the 30 boxers who participated in the current investigation in Table 72. Twenty-seven boxers were retired from boxing, and three were active. The average age was 49 years (S D, 16.5 years; 23 – 76 years). There were eleven whites, ten blacks, and nine Hispanics. The boxers in the current investigation had, according to JORDAN et al., a varied exposure to the sport. Among the 30 boxers, 18 (60%) were considered high exposure. The high exposure boxers tended to be older (mean [S D] age, 53.4 [15.7] years; range 29 to 76) compared with those with low exposure (mean [S D] age, 42.6 [16.2] years; range 26 to 68 years; P = .08). The high exposure boxers had significantly higher CBI scores (mean = 2.6, S D = 1.9) compared with low exposure boxers (mean = 0.3, S D = 0.7, P < .001).

TABLE 72
Chararcteristics of the study population. (From JORDAN et al. 1997).

Case	Age, y (Race)	Boxing Exposure	Amateur Bouts, No.	Professional Bouts, No.
1	70 (B)	High	NA	162
2	26 (H)	Low	14	8
3	48 (B)	High	NA	79
4	29 (B)	High	150	13
5	31 (B)	High	NA	34
6	29 (B)	Low	8	3
7	57 (W)	Low	100s	NA
8	57 (W)	Low	200	1
9	36 (W)	Low	32	0
10	59 (H)	High	50	71
11	39 (W)	Low	5	NA
12	70 (W)	High	26	22
13	65 (W)	High	31	31
14	76 (W)	High	21	29
15	69 (B)	High	47	45
16	67 (B)	Low	4	NA
17	64 (H)	High	32	117
18	37 (H)	High	105	58
19	41 (H)	High	33	20
20	35 (W)	High	15	28
21	64 (W)	High	1	40
22	60 (H)	High	41	56
23	68 (W)	Low	48	NA
24	47 (W)	Low	45	4
25	31 (B)	Low	20	2
26	39 (H)	High	110	23
27	30 (B)	Low	55	10
28	64 (B)	High	200	25
29	37 (H)	High	44	20
30	23 (H)	Low	25	0

W indicates white; B, black; H, Hispanic; and NA, not available

According to JORDAN et al. eleven boxers (37%) had normal neurologic findings (CBI = 0), while 19 boxers (63%) exhibited abnormalities. Of the 19 boxers with CBI score greater than 0, 12 boxers were classified as probable CTBI, and seven boxers were classified as possible CTBI. Among the seven possible CTBI cases, four boxers had mild to moderate cognitive impairment with or without minor motor disturbances (e. g. mild dysarthria, difficulty with tandem gait, abnormal Romberg sign) in the setting of regular alcohol usage (i. e. alcohol ingestion at least 4 times per week). A fifth boxer had a history of crack cocaine abuse and exhibited mild psychiatric disturbances along with mild dysarthria and an abnormal Romberg sign. The sixth boxer in the 'possible' group had mild mental status changes and symptoms consistent with parkinsonism. The seventh exhibited essential tremor. JORDAN et al. present the results of the neurodiagnostic evaluation in Table 73. Twelve boxers (40%) had mild CTBI (CBI = 1 – 2). Four boxers (13%) were moderately impaired (CBI = 3 – 4), and three boxers (10%) were severely affected (CBI > 4). Neurologic findings in the boxers with mild CTBI included mild cognitive impairment (seven cases), mild dysarthria (four cases), mild gait disturbances (four cases), positive Romberg (three cases), mild parkinsonism (two cases), and mild pyramidal tract impairments (i. e. hyperreflexia, mildly increased tone (two cases). One boxer with mild CTBI was diagnosed with bipolar disorder during his boxing career and displayed hypermania and increased talkativeness during the examination. In seven boxers with moderate to severe CTBI, the predominent features were dementia, dysarthria, and ataxia. Three boxers with severe CTBI also exhibited psychiatric manifestations.According to JORDAN et al. nineteen boxers (63%) were homozygous for *APOE* ε 3 / ε 3. There were seven boxers (23%) with an ε 3 / ε 4 genotype, two (7%) with ε 2 / ε 4, one (3%) with ε 4 / ε 4, and one (3%) with ε 2 / ε 3. Among the twelve boxers classified as probable CTBI, six (50%) has either *APOE* ε 3 / ε 4 or ε 4 / ε 4 compared with three (11%) of the 18 boxers labeled as normal or possible CTBI (P = .004).

All three boxers (100%) who exhibited severe CTBI had at least one copy of the *APOE* ε 4 allele, as JORDAN et al. report. This is compared with 50% (two of four) of those with moderate CTBI, 25% (three of twelve) with mild CTBI, and 18% (two of eleven) with normal examinations (Figure 234). This trend of higher CBI scores with the ε 4 allele was statistically significant (P = .01). The mean CBI score for the ten boxers with an *APOE* e 4 allele was 2.8 (S D, 2.6), compared with a mean of 1.2 (S D, 1.2) among the twenty boxers without an *APOE* ε 4 allele. This difference did not reach statistical significance (P = .06).

Boxers were stratified according to boxing expo-

TABLE 73
Results of diagnostic evaluations. (From JORDAN et al., 1997).

Case	Neurologic Examination	MMSE Score	CBI Score	CTBI Classification	Comment
1	Severe dementia; severe spasticity; agitation; dysarthria; gait ataxia	3	8	Probable	
2	Normal	28	0	Normal	Referred for evaluation by the New York State Athletic Commission (NYSAC); magnetic resonance imaging (MRI) scan showed a paraventricular cyst, mild atrophy, and cavum septum pellucidum; admitted to use of recreational drugs; normal electroencephalogram (EEG)
3	Posturing on stressed gait; decreased simple and complex attention; mild dysarthria	28	2	Probable	Normal computed tomography (CT) scan and EEG; referred by the NYSAC
4	Normal	29	0	Normal	Experienced a prolonged postconcussion syndrome with 24 h of amnesia; also had unexplained syncopal episodes; normal MRI and EEG
5	Difficulty with Tandem gait	N/A	1	Probable	Complained of difficulties with balance; referred by the NYSAC; 1 episode of head trauma outside the ring
6	Normal	30	0	Normal	Referred by the NYSAC; abnormal EEG; normal CT scan
7	Normal	29	0	Normal	One episode of head trauma outside the ring secondary to a motor vehicle accident (experienced a few seconds of unconsciousness)
8	Normal	28	0	Normal	Magnetic resonance imaging scan showed nonspecific white matter changes consistent with focal ischemia
9	Normal	30	0	Normal	. . .
10	Mild dysarthria	27	2	Possible	Regular alcohol use
11	Normal	30	0	Normal	Occasional recreational drug use; regular alcohol use
12	Normal	27	1	Possible	History of meningitis as a child and history of schizophrenia that preceded boxing; regular alcohol use
13	Difficulty with tandem gait; positive Romberg sign	25	2	Possible	History of alcohol use
14	Mild-moderate memory impairment	24	2	Possible	Regular alcohol abuse
15	Severe dysarthria; slight increased tone in lower extremities; ataxic gait; slight dysmetria on left	22	4	Probable	Hospitalized once after a fight secondary to a concussion
16	Normal	29	0	Normal	. . .
17	Mild hyperreflexia on left; resting tremor and mild cogwheel rigidity	27	2	Possible	. . .
18	Flat affect with pseudobulbar-type laughing outbursts; dysarthria; mild paratonia of right upper extremity; ankle clonus; slowed rapid alternating movements; ataxic gait; memory impairment	20	5	Probable	MRI scan: moderate cortical atrophy and periventricular white matter changes; EEG showed diffuse slowing
19	Mild pseudobulbar affect; dysarthria; memory impairment; dysmetria; ataxic gait	17	5	Probable	. . .
20	Normal	29	0	Normal	Normal EEG and MRI
21	Moderate dysarthria and incoordination; slight increased tone in right lower extremity with wide based ataxic gait; hyperreflexia in right upper extremity	30	2	Probable	CT scan: mild cerebellar atrophy

TABLE 73
Results of diagnostic evaluations. (From JORDAN et al., 1997). (continued)

Case	Neurologic Examination	MMSE Score	CBI Score	CTBI Classification	Comment
22	Difficulty with sharpened Romberg sign	27	2	Probable	. . .
23	Mild tremor with titubation	29	1	Possible	. . .
24	Mild dysarthria and cogwheel rigidity	30	1	Probable	Remote history of polio
25	Mild dysmetria and difficulty with Romberg sign; hypomania and increased talkativeness	28	2	Possible	One episode of head trauma outside of the ring with loss of consciousness; still boxing; history of drug abuse
26	Anisocoria; abducens palsy diplopia on lateral gaze; dysmetria; ataxia; pseudobulbar affect and behavioral disinhibition	29	3	Probable	History of alcohol abuse; cavum septi on CT
27	Normal	29	0	Normal	. . .
28	Mild dysarthria; gait ataxia; resting tremor	25	3	Probable	. . .
29	Slight increased tone in lower extremities; abnormal sharpened Romberg sign; mild rigidity; mild depression	27	3	Probable	Complained of memory loss
30	Normal	29	0	Normal	. . .

*MMSE indicates Mini-Mental State Examination; CBI, chronic brain injury; and CTBI, chronic traumatic brain injury.

sure and presence or absence of the e 4 allele. JORDAN et al. report that Table 74 shows mean, median, and ranges of CBI scores and age according to boxing exposure and *APOE* genotype. Low-exposure boxers with and without the ε 4 allele had a mean CBI score of 0.3 (S D = 0,7). High-exposure boxers with an *APOE* ε 4 allele had a mean CBI score of 3.9 (S D = 2.3) compared with 1.8 (S D = 1.2) in high-exposure boxers without the ε 4 allele. This difference was statistically significant (P = .04) (Table 75). The tendency for greater CTBI symptoms among those with increased exposure and an ε 4 allele was statistically significant (P < .001) (Figure 235).

According to the authors, three boxers had an *APOE* ε 2 allele. All three had high-exposure to boxing (range, 34 –79 professional bouts), and two had the *APOE* ε 4 allele in combination with *APOE* ε 2. However, none showed significant evidence of CTBI (i. e. scores suggesting moderate

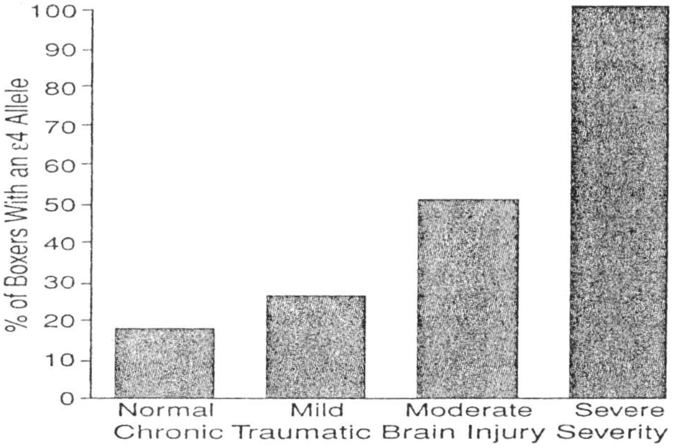

Fig. 234. Frequency of apolipoprotein E epsilon 4 carrier frequency according to severity of chronic traumatic brain injury. (From JORDAN et al., 1997).

TABLE 74
Chronic brain injury (CBI) score stratified by APOE epsilon 4 status and boxing exposure. (From JORDAN et al., 1997).

	Low Exposure		High Exposure		
Variables	ε4 Absent (n=9)	ε4 Present (n=3)	ε4 Absent (n=11)	ε4 Present (n=7)	*P* Value
Age, y					
Mean (SD)	44.0 (18.0)	38.3 (9.0)	59.6 (14.5)	43.6 (12.7)	
Median (range)	36.0 (24–68)	39.0 (29–47)	64.0 (29–76)	39.0 (31–70)	$P > .05$†
CBI					
Mean (SD)	0.3 (0.7)	0.3 (0.6)	1.8 (1.2)	3.9 (2.3)	
Median (range)	0.0 (0–2)	0.0 (0–1)	2.0 (0–4)	3.0 (1–8)	$P < .001$‡

**APOE* ε4 indicates the type 4 allele of apolipoprotein E; and SD, standard deviation.
† *P* value determined using the *t* test.
‡ *P* value determined using the Kruskal-Wallis test.

TABLE 75
Pairwise comparisons of groups: genotype and boxing exposure. (From JORDAN et al. 1997).

Comparison	*P* Value*
Genotype	
ε4 Absent vs ε4 present (low exposure)	> .05
ε4 Absent vs ε4 present (high exposure)	= .04
Exposure	
Low exposure vs high exposure (ε4 absent)	< .001
Low exposure vs high exposure (ε4 present)	= .001

*Mann-Whitney *U* test used to determinate *P* value.

or severe deficits). Although the mean CBI score for the three boxers with the ε 2 allele was lower (1.67) than the CBI score for high-exposure boxers without an *APOE* ε 2 allele (2.8), this difference did not reach statistical significance (P = .30).

JORDAN et al. comment that the results of this preliminary investigation suggest that the *APOE* ε 4 allele may predispose a boxer to developing CTBI, especially in those with high-exposure to the sport. Boxers who were *APOE* ε 4 positive with a high-exposure to boxing had the highest scores on the CBI. Despite their greater overall impairment, the high-exposure boxers with the *APOE* ε 4 allele tended to be younger (average age, 45.2 years) than high-exposure boxers without the *APOE* ε 4 allele (average age, 56.5 years). This age difference approached statistical significance. While it is plausible that the *APOE* ε 4 allele may be associated with a younger age of onset of CTBI, the age difference may reflect a self-selection bias toward younger *APOE* 4-positive boxers in the authors volunteer population. It is posssible that fewer older, high-exposure boxers who carry *APOE* ε 4 were available owing to the greater burden of disease among such individuals.

JORDAN et al. continued, that the frequency of *APOE* ε 4 allele was noted to be highest (67%) in boxers with typical traumatic encephalopathy related to boxing (probable CTBI). The *APOE* ε 4 allele was also overrepresented in those boxers with severe CTBI (100%) and moderate CTBI (50%).

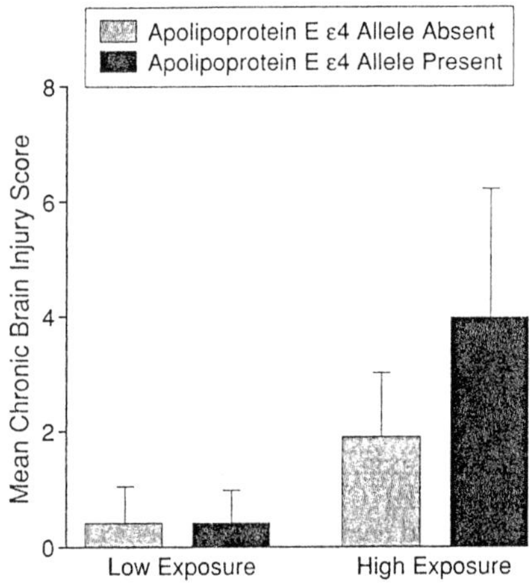

Fig. 235. Chronic traumatic brain injury scores of boxers by apolipoprotein E genotype and exposure. The bars represent the standard deviation. (From JORDAN et al., 1997).

These findings are statistically significant and support the hypothesis that *APOE* ε 4 is associated with CTBI in high-exposure boxers. There may be no association between *APOE* ε 4 and CTBI in low-exposure boxers, or the lack of an observed effect may be due to the small sample size in the authors study.

According to the authors the three boxers who had an *APOE* ε 2 allele had relative mild CTBI scores, despite high boxing exposure. This finding is consistent with the literature, which suggests that the *APOE* ε 2 allele may convey protection against neurodegenerative disorders such as AD.

JORDAN et al. note that the *APOE* carrier frequencies and the odds ratio of developing AD as a function of *APOE* genotype have been reported to vary as a function of national origin or race. The authors study population contained approximately equal numbers of whites, blacks, and Hispanics. It is unlikely that the effects *APOE* ε 4 possession on CTBI that the authors observed are due to the heterogeneity of the racial origin of their study subjects.

JORDAN et al. summarize:

"Although the findings of this preliminary investigation suggests that there may be a genetic susceptibility for development of CTBI in boxers, our results should be interpreted with caution. Since the majority of these boxers were either referred or volunteers, there is a potential self- selection bias. It is conceivable that more symptomatic boxers may come to medical attention or may volunteer to participate. Although the *APOE* allele frequencies in our study population are comparable with those reported in several large population studies, our sample size is relatively small. The confirmation of our findings requires *APOE* genotyping in a large, representative sample of boxers with appropriate attention to other variables such as age, boxing exposure, national origin, and other putative risk factors for CTBI and dementia."

JORDAN et al. continue:

"The hypothesis that APOE ε 4 conveys a genetic predisposition to CTBI potentially has extraordinary ramifications for the regulation of health and safety in boxing and other high-risk sports including American football, soccer, and ice hockey. Boxers who are at high risk could conceivably undergo more detailed neurological evaluations and therefore be better advised regarding future participation. The presence of an *APOE* ε 4 allele in a boxer would not nessessarily preclude a boxer from participation but could be an indication to minimize or more strictly limit exposure to the sport (e. g., excessive number of bouts, long duration of career, and numerous knock outs."

JORDAN et al. state finally:

"These preliminary findings suggest that the *APOE* ε 4 allele may be associated with the pathophysiology of CTBI secondary to boxing. The magnitude of this effect as well as its precise mechanism remain to be elucidated. In light of the established role of *APOE* ε 4 as a risk factor for late-onset forms of AD, the current findings lend support to the hypothesis that both AD and dementia pugilistica are complex disorders arising from the interaction of inherited susceptibilities and environmental exposures, including (but not limited to) *APOE* genotype and head trauma."

McLATCHIE et al. (1987) studied *twenty active amateur boxers* seeking evidence of neurological dysfunction and the best method for detecting it. Seven of the group had an abnormal clinical neurological examination, eight an abnormal EEG and nine of the fifteen examined had abnormal neuropsychometry. The CT scan was abnormal in only one boxer. An abnormal clinical examination correlated significantly ($p < 0.05$) with an increasing number of fights, and an abnormal EEG with decreasing age ($p < 0.05$). In several of the neuropsychometric tests, the boxers were significantly worse than controls ($p < 0.05$). Neuropsychometry was the best method for detecting neurological dysfunction.

BROOKS et al. (1987) carried out neuropsychological examinations *on 29 amateur boxers* and *19 controls matched for age, ethnicity and education.* There was no evidence of significantly impaired performance in the boxers. Within the boxing group, a variety of features of boxing history were examined as possible predictors of cognitive performance (such as of knock outs, duration of boxing). No feature was a significant predictor of lower cognitive performance.

HAGLUND et al. (1990) noted that Sweden had banned professional boxing in 1969 and had also considered banning amateur boxing. The authors therefore analyzed possible chronic brain damage in *47 former amateur boxers* who had started their careers after the introduction of stricter Swedish amateur boxing rules. The boxers were compared with three control groups – 25 soccer players, 25 track and field athletes and 19 conscripts. All athletes were interviewed about their sports career, medical history and social variables. They then underwent a physical and a neurological examination, including a mini-mental state examination. Personality traits were investigated and related to their platelet MAO activity in the athletes as well as in the conscripts. No significant differences were found between the groups in any of the physical or neurological examinations. All had a normal mini-mental state examination. Thus, results from these test methods did not reveal any signs of chronic brain damage from Swedish amateur boxing. Neither were any significant differences found with regard to platelet MAO activity, while significant differences were found in some of the social and personality traits variables.

HAGLUND and PERSSON (1990) wrote that the aim of the present study was to investigate possible chronic brain damage due to *Swedish amateur boxing*. Forty-seven amateur boxers, 22 with many (HM = high matched) and 25 with few matches (LM = low-matched) during their career were examined and compared with two control groups of 25 soccer players and 25 track and field athletes in the same age-range. No severe EEG abnormality was found. There was a somewhat higher incidence of slight or moderate EEG deviations among HM (32%, 7/22) and LM (36%, 9/25) boxers than among soccer players (20%, 5/25) and track and field athletes (12%, 3/25). Brain electric activity mapping (BEAM), brainstem auditory evoked potential (BAEP) and auditory evoked P 300 potential (P 300) did not differ significantly between the groups. No neurophysiological variable was correlated to the number of bouts, number of lost fights or length of boxing career. Thus, no signs of serious chronic brain damage were found among the amateur boxers or the soccer players and the track and field athletes. However, it cannot be excluded that the EEG differences between the groups may be a sign of slight brain dysfunction in some of the amateur boxers.

HEILBRONNER et al. (1991) performed in *23 amateur boxers prefight* and *postfight neuropsychological tests*. Comparisons between their prefight and postfight performance provided the data analyzed in the present study. No boxer had a history of chronic alcohol or drug abuse, hospitalization for head injury sustained outside the ring, and/or other neuropsychiatric disorder. No financial compensation was offered for participation in the study.

HEILBRONNER et al. noted that their results are generally consistent with the results from other studies that have shown only minor changes in cognitive performance in amateur boxers compared to controls (THOMASSEN et al. 1979, BROOKS et al. 1987, LEVIN et al. 1987, McLATCHIE et al. 1987). For the first time, however, these authors could demonstrate mild changes in cognitive functions assessed immediately after a boxing match. Impairments were noted in verbal recall and incidental memory, while improvements were observed in bilateral motor speed and on a task of sequencing and cognitive flexibility. On all other measures there were no significant differences between the prefight and postfight scores nor were there any observed differences between winners and losers.

The reported number of fights revealed that 15 boxers had participated in 0 to 10 fights, three in 11 to 50 fights, and three in more than 50 fights; in two boxers the number of fights was unknown.

HEILBRONNER et al. pointed out that one finding is of particular interest, namely that the four boxers with the most extensive fight histories (subjects 1, 3, 4, and 23) all demonstrated slower dominant hand tapping speed after the fight compared to their prefight speed. Three out of four also showed nondominant hand speed. The authors noted that the likelihood exists that these boxers could have sustained enough head blows and subsequent brain dysfunction so as to affect their psychomotor speed. This would be consistent with reports in the literature of slowed movement and delayed reaction times in boxers with extended fight histories (MAWDSLEY and FERGUSON 1963).

The authors conclude their study with the statement: "Based on the present results, it is difficult to determine whether or not amateur boxers are sustaining significant and irreversible brain damage as a result of engaging in irregular boxing event. This study is limited by the failure to calculate the number of subconcussive head blows and by the lack of a matched control group of boxers who engaged for a period of time equivalent to a three-round boxing match. Boxers should likely be tested longitudinally over an extended period of time or at least serially after repeated concussive and subconcussive head blows in order to monitor changes in their cognitive status before significant and irreversible damage becomes a reality. While it is unlikely that a single boxing match leads to irreversible and permanent cognitive deficits, this question remains unanswered given the limitations acknowledged in this study."

HOLZGRAEFE et al. (1992) reported about the significance of diagnostic imaging in acute and chronic brain damage in boxing, a *prospective study in amateur boxing using magnetic resonance imaging (MRI)*. The authors investigated the significance and possible extent of structural damage to the Central Nervous System (CNS) due to boxing. The authors consider small hemorrhages, especially microhematomas, to be the probable cause of the chronic encephalopathy of boxers.

In addition to neurological and neuropsychological examinations, CCT, EEG, and neurophysiological (olfactometry, nystagmography) tests were conducted. The study was conducted of 13 amateur boxers. The average age was 24 years, and the total number of fights varied betweeen 18 and 203. Weight classes ranged between 67 and 91 kg. (Table 76).

We have pointed out that the in the medical literature again and again mentioned small hemorrhages are not the substrate and possible cause of the chronic encephalopathy in boxers. COURVILLE had coined the term 'diffuse contusions.' These traumatic alterations, erroneously identified by COURVILLE as 'diffuse contusions,' are not true contusions, but hemorrhagic necroses. Such necroses play an important role in acute boxing fatalities, – as in other head injuries too, – but

TABLE 76
Compiled data from the study subjects. (From HOLZGRAEFE et al., 1992).

Boxer No.	Age	Fights won	Fights lost	Lost by knockout	Weight division up to (kg)	Head punches		Neurological symptoms	Interval between last fight and MRI (in days)
						light	severe		
1	23	33	21	1	75	22	7	+	4.30
2	26	40	44	3	81	67	18	–	4.60
3	26	52	52	1	91	60	16	–	4
4	21	16	4	0	67	18	2	–	4
5	26	52	48	1	91	19	3	–	3
6	23	13	5	1	71	42	12	–	3
7	21	34	20	0	71	61	18	+	4
8	25	56	42	0	91	43	13	+	5.60
9	23	45	37	0	71	30	12	+	3.21
10	28	139	64	0	71	32	6	–	3
11	23	114	26	0	75	30	1	–	3
12	26	60	12	0	67	4	–	–	3
13	23	50	8	0	75	12	2	+	3

they are not the morphological substrate of the encephalopathy of the boxer. These small hemorrhages, in reality early hemorrhagic necroses, are again and again quoted in the pertinent literature. They should be buried. The chronic pathomorphologic damage in the boxers' brain is the result of hypoxic-ischemic alterations, which affect the entire cerebrum and cerebellum. This damage consists of loss of neurons with glial reaction and resulting atrophy of cerebrum and cerebellum. Neurons in different layers of the cerebral cortex, neurons in the Ammon's horn formation and the Purkinje cells in the cerebellum are especially affected.

Among the thirteen boxers, five demonstrated focal neurological signs following the fights, without evidence of small hematoma or other structural alterations. The number of head punches did not correlate with the occurrence of neurological signs. These results indicate, according to HOLZGRAEFE et al., that up to now imaging methods cannot clarify the development of chronic encephalopathy.

Again HOLZGRAEFE et al. did not find small hemorrhages in the brain of boxers. This is an exact finding, a correct observation or conclusion, because they do not exist. But one cannot conclude there was no brain damage from boxing simply because no small hemorrhages were found when the brains were examined. Such hemorrhages were never found in boxers, as they do not exist, they are not the substrate of the punch drunk syndrome. Not that the absence of these in reality not existing hemorrhages can be used as an argument that the examined brains do not show any damage as the result of boxing.

HAGLUND and ERICKSSON (1993) examined *50 Swedish amateur boxers* and *compared the findings with two control groups of soccer players and track and field athletes.* All subjects were interviewed regarding their sports career, medical history, and social variables. They underwent a physical and a neurological examination. Personality traits were investigated and related to the platelet monoamine oxidase activity. Cerebral morphologic changes were evaluated using computer tomography and magnetic resonance imaging. Further, clinical neurophysiological tests were made as well neuropsychiatric tests.

The authors continue that no significant differences were found between the groups in any of the physical or neurological examinations or in platelet mono-amino oxidase activity. Socially, the boxers had a lower degree of education and had chosen less intellectual professions, but they were less impulsive and more socialized. The computed tomography images and magnetic resonance imaging studies showed no significant differences between the groups. There was a significantly higher incidence of slight to moderate electroencephalography deviations among the boxers. Neuropsychologically, the boxers had an inferior finger-tapping performance. Thus, no signs of serious chronic brain damage were found among any of the groups studied. However, the electroencephalography and finger-tapping differences between the groups might indicate slight brain dysfunction in some of the amateur boxers.

KEMP et al. (1994) noted that the few scientific studies in the literature purporting to assess the safety of amateur boxing produce conflicting results (THOMASSEN et al. 1979, HAGLUNG and PERSSON (1990). The authors wrote that studies using mixed data from amateur and professional boxers generally show adverse findings in the amateur boxer (KASTE et al. 1982, ROSS et al. 1983, CASSON et al. 1984, JORDAN and ZIMMERMAN 1990). However, the numbers studied were too small to derive significant conclusions, having been fewer than ten in all.

KEMP et al. noted that amateur boxing is encouraged in the armed forces to promote both fitness and self discipline.

The authors invited amateur boxers in the boxing associations of the Royal Navy or Army to participate in this study. At the time of the request, all the amateurs who represented the respective services volunteered and gave informal consent, thus minimizing selection bias. None of the boxers had sustained a relevant injury outside the ring (defined as more than momentary loss of consciousness or concussion lasting more than a minute) or had a history of neurological illness.

Most of the boxers underwent both psychometric tests and Tc-99m HMPAO SPECT brain scanning. However, due to logistical constraints a few boxers were not able to have both sets of tests done. In total, 42 boxers participated in the trial. Forty-one boxers underwent psychometric tests, of which 33 also had Tc-99m HMPAO SPECT scans; one boxer underwent TC-99m HMPAO SPECT without psychometric testing. The boxers varied in age from 8 to 30 years, the number of competitive bouts fought ranged from 0 to 250 (median 40). The duration of their participation in boxing extended from three months to 17 years.
Servicemen undergoing medical assistant training were invited to participate as controls in the study.

Age and School Attainment

Table 77 in the series of KEMP et al. show the summary of the data of the distributions of the ages and number of O levels for the amateur boxing and control groups,

Tc-99 HMPAO SPECT scanning technique and interpretation of image: All subjects had 5500 MBq of Tc-99 m HMPAO injected intravenously into an antecubital vein and were scanned within ten minutes of injection with a Siemens 7500 ZLC tomographic camera, with a low energy, all purpose collimator linked to a Bertec Micas V Sun system. The images produced of the boxers and controls were compared with an 'atlas of normality' to detect focal abnormalities. The atlas of normality was derived from the images of 50 healthy male and female subjects, civilian and military personnel who were aged 18 – 60 years and had no history of head injury or neurological illness.

Psychometric Tests: The battery of tests selected a wide range of cognitive functions (reaction time, psychomotor learning, reasoning, perception and comparison, short term memory, and motor function). The test battery included the following tasks: Four choice reaction time, Code substitution, logical reasoning, simultaneous pattern comparison, Sternberg, and finger tapping.

Tc-99m HMPAO Scanning: KEMP et al. defined abnormality in their scans as "the presence of one or more regions of significant aberration in cerebral perfusion when compared with the atlas of normality." Table 78 shows the summary of the results of Tc-99m HMPAO SPECT scanning for the groups of boxers and controls.

The authors found that 41% of the amateur boxers had abnormal scans, compared with 14% of the controls ($x2 = 5.9$, $df = 1$, $P < 0.02$). In the control group the aberrations in cerebral perfusion were largely singular, whereas in the boxing group the aberrations were generally multiple. In both groups there were no predominant sites of abnor-

TABLE 77
Summary data of the distributions of the ages and number of levels for the amateur boxing and control groups undergoing TC-99M HMPAO SPECT and psychometric testing. (From KEMP et al., 1995).

	No	Mean (SD)	Minimum	Maximum
		Tc-99m HMPAO SPECT		
Age				
Boxers	34	22·0 (2·9)	18	29
Controls	34	22·2 (3·1)	18	29
O levels:				
Boxers	34	1·9 (2·6)	0	8
Controls	34	3·1 (2·8)	0	9
		Psychometric testing		
Age				
Boxers	41	22·0 (2·8)	18	29
Controls	27	22·0 (2·9)	18	29
O levels:				
Boxers	41	2·4 (3·1)	0	8
Controls	27	3·1 (2·6)	0	9

TABLE 78
Results of comparing HMPAO SPECT of amateur boxers and controls with the 'atlas of normality'. (From KEMP et al., 1995).

	HMPAO SPECT		
	Normal	Abnormal	Total
Boxers	20	14	34
Controls	29	5	34

mality; the lesions were scattered throughout the cerebral cortex.

Psychometric testing: KEMP et al. summarized the average psychometric test scores for 41 boxers and 27 controls, without adjustment for co-variates, in Table 79. The logical reasoning task had to be excluded from the analysis as many boxers and several controls had scores around 50% which suggested that those subjects found the task too difficult and were guessing.

KEMP et al. summarize the average psychometric test scores (untransformed) with the adjustments for the covariates, in Table 80. In four out of the five tests there were significant differences between the groups, to the detriment of the boxers. The differences ranged from 10 – 20%.

Comparisons of Boxers with Low and High Number of Bouts: In the study of KEMP et al., test results showed a high correlation between the number of bouts and the number of years of boxing. Consequently either variable could be used as a measure of boxing exposure.

The median number of bouts fought could be used, according to KEMP et al., to divide the boxers into two groups, 21 boxers with a maximum of 40 bouts (low bouts) and 20 boxers with more than 40 bouts (high bouts). The summarized average test scores (untransformed) are shown in Table 81.

Excluding the Sternberg test due to an interaction between groups (low and high bouts) and education, a univariate ANCOVA showed that low bout boxers performed significantly better that the high bout boxers on the four choice reaction time and pattern recognition ($P < 0.05$). There were no significant differences on the remaining two tasks.

KEMP et al. noted that it is not surprising that functional brain imaging with Tc-99m HMPAO MRI scanning is more sensitive than CT or MRI at showing the consequences of closed head injury. The authors write, "Cerebral amyloid protein deposition, immuno-histochemically similar to that seen in Alzheimer's disease, has been noted in the brains of ex-boxers and in people with head trauma. Most importantly, it has also been shown that these regions of amyloid protein deposition are associated with abnormal vasculature, which could

TABLE 79
Summary od average psychometric test scores for 41 boxers and 27 controls adjusted for covariates (age, age^2, number of 0 levels. (From KEMP et al., 1995).

Psychometric test	Group	Mean	* (95% CI for difference of means (boxers – controls)
Four choice reaction time	41 boxers	0·49 s	(0·04 to 0·10)*
	27 controls	0·42 s	
Code substitution	41 boxers	46·6 correct	(−8·0 to −0·9)*
	27 controls	51·0 correct	
Pattern recognition	41 boxers	90·2 correct	(−15·7 to −2·1)*
	27 controls	99·1 correct	
Sternberg reaction time	41 boxers	1·17 s	(0·1 to 0·2)
	27 controls	1·12 s	
Alternate finger tapping	41 boxers	90 taps	(−32 to −12)*
	27 controls	113 taps	

*$P < 0{\cdot}05$

TABLE 80
Average psychometric test scores for 41 boxers and 27 controls without adjuestment for covariates. (From KEMP et al., 1995).

Psychometric test	Group	Mean	(95% CI for difference of means (boxers – controls)
Four choice reaction time	41 boxers	0·49 s	(0·04 to 0·10)*
	27 controls	0·42 s	
Code substitution	41 boxers	46·5 correct	(−8·3 to −1·1)*
	27 controls	51·2 correct	
Pattern recognition	41 boxers	89·9 correct	(−17·6 to −2·0)*
	27 controls	99·7 correct	
Sternberg reaction time	41 boxers	1·18 s	(0·0 to 0·2)
	27 controls	1·11 s	
Alternate finger tapping	41 boxers	90 taps	(−33 to −13)*
	27 controsl	113 taps	

TABLE 81
Summary of average psychometric test scores for 21 low bout boxers and 30 high bout boxers adjusted for covariates (age, age^2, number of 0 levels. (From KEMP et al., 1995).

Psychometric test	Group	Mean	(95% CI for difference of means (low – high bouts)
Four choice reaction time	21 low bout	0·46 s	(−0·10 to −0·01)*
	20 high bout	0·52 s	
Code substitution	21 low bout	48·3 correct	(−1·3 to −8·7)
	20 high bout	44·6 correct	
Pattern recognition	21 low bout	94·2 correct	(0·7 to 17·1)*
	20 high bout	85·3 correct	
Alternate finger tapping	21 low bout	88 taps	(−16·0 to 8·6)
	20 high bout	92 taps	

*P < 0·05.

also account for the abnormalities noted on Tc-99m HMPAO SPECT cerebral perfusion imaging."

KEMP et al. list the results of recent major studies that assessed the safety of amateur boxing in Table 82. According to those authors, their study was relatively more fortunate than most in terms of selection bias, extraneous head injuries, and choice controls in their subjects.

KEMP et al. concluded: In summary, psychometric testing and Tc-99m HMPAO cerebral perfusion imaging were undertaken in a group of service amateur boxers and a control group of sports men. *Psychometric testing showed detrimental findings in the group of boxers, and this was more apparent in those who had the most bouts. Computer analysis of Tc-99m HMPAO SPECT images showed significantly greater aberrations in cerebral perfusion in the group of boxers than in the controls. This study noted subtle impairment in two aspects of neurophysiology in those young men, who were still engaging in amateur boxing and receiving the consequent head trauma at the time of the study. KEMP et al. close their study: "It is hoped that the investigations of the $ 1.5 million study in the United States (STEWART et al. 1994) will follow up on their large cohort of boxers to provide a definitive answer."*

BUTLER (1994) carried out a neuropsychological investigation *of amateur boxers. The authors noted, that: "Amateur boxing is faced with criticism*

TABLE 82
Major studies assessing the safety of amateut boxing. From KEMP et al., 1995).

Investigators	Participants	Findings	Comments
Thomassen *et al.* DENMARK 1979	53 former boxers 53 footballers	No significant differences on clinical examination, EEG, or psychometric testing.	These findings are confounded by the fact that 88% of the controls and 75% of the boxers sustained head injuries outside the ring.
McLatchie *et al.* SCOTLAND 1987	20 active boxers Variable number of controls	Abnormal clinical examination in 35% of boxers, abnormal EEG in 40% of boxers, abnormal psychometric testing in 63% of boxers	No controls had clinical examination or EEG. University staff used as controls for psychometric testing.
Brooks *et al.* SCOTLAND 1987	29 active boxers 19 controls	No significant differences on psychometric testing	Only 39% of invited boxers participated. Controls matched for age, education, and ethnicity.
Haglund *et al.* SWEDEN 1990	50 former boxers 50 athletes	No significant differences on clinical examination, CT, MRI, EEG showed significantly greater abnormalities in boxers.	Although the authors concluded that amateur boxing appeared 'safe' the differences on EEG examination remain unexplained.
Butler *et al.* ENGLAND 1993	86 active boxers 31 water polo and 47 rugby players	Significant differences between boxers and controls on initial psychometric testing to the detriment of the boxers. Improvement noted on longitudinal assessment. No differences on clinical examination, evoked responses, EEG, CKBB isoenzymes, or ophthalmological assessment.	No allowance for the intellectual differences between the controls (some were university students) and the boxers (many early school leavers). Authors concluded no evidence of damage in boxers.
Stewart *et al.* USA 1994	484 amateur boxers	Significant tests of trend between psychometric testing ability and number of bouts. No deterioration on 2 year follow up.	Authors urge caution in the interpretation of the longitudinal findings as a longer latency period may be required before effects are manifest.
Kemp *et al.* ENGLAND 1994 (This study)	(i) 34 active boxers 34 controls (ii) 41 active boxers 27 controls	Boxers had significantly more abnormalities on Tc-99m HMPAO cerebral perfusion imaging. Psychometric testing revealed boxers to perform significantly worse than controls and, in addition, those boxers with the greater number of bouts performed significantly worse than those with few bouts.	All invitees participated thus reducing selection bias. Subjects with extraneous head injuries were not included. Controls matched for age and eduction.

over the potential damage the sport inflicts on those who participate."

According to the author, the most sensitive measure of early neurological dysfunction is neuropsychological investigation. Ten studies employing such assessments on 289 amateur boxers were reviewed. The forms of analysis undertaken include controlled comparison with other sportsmen, of both active and former boxers, detailed pre- and post-bout analysis, analysis of the influence of within-boxing variables, length of career, level of competition, and prospective longitudinal investigation. Amateur boxers were found to exhibit no signs of neuropsychological dysfunction in any analysis. However some trends emerged suggesting that a long career in amateur boxing might reduce fine motor reactions, although such findings are within the normal range and do not represent central neuropsychological functioning. Thus BUTLER concluded that amateur boxing does no appear to expose individuals to neurological dysfunction

STEWART et al. (1994) reported that bouts in amateur boxing are limited to three rounds of three minutes each, gloves are heavier, larger and more absorbent than those used by professional boxers; and, as standard praxis, amateurs are trained to pursue 'points' rather than knock out blows to attain victory *(1989 Official Rules: USA Amateur Boxing Federation).*

Between 1984 and 1986 a number of safety measures were introduced in amateur boxing. Headgear is required, novices (fewer than ten bouts) are now matched with opponents of similar level, and safety measures have been implemented to stop bouts when boxers are at risk of head injury. Mandatory suspension rules are imposed for observable head injuries. In 1986 a mandatory 'passbook' system was implemented to record the outcome of each bout and directly enforce suspension rules. While observable concussive injuries are not frequent in amateur boxing today, one cannot consider the subconcussive blows in amateur boxing to be 'safe'. It is possible that repeated jarring, non concussive or concussive blows, and less frequent concussive blows cause consistent patterns of brain damage, especailly if studied over time.

STEWART et al. quoted earlier studies (KASTE et al. 1992, ROSS et al. 1983, CASSON et al. 1984) and controlled cross-sectional studies (JEDLINSKI et al. (1970), THOMASSEN et al. (1979), McLATCHIE et al. (1987), BROOKS et al. (1987) showing that an amateur boxer who competed primarily during or before the 1970's were at risk of central nervous system function impairment. Thus the regulatory changes listed above were well motivated and justified. STEWART et al. quoted a more recent study of 50 highly competitive, active boxers who competed primarily in the 1980's (MURELIUS et al. 1991), where no differences were found between amateur boxers and soccer players on an extensive battery of standardized neuropsychologic and imaging measures.

STEWART et al. studied active amateur boxers over a relatively short period of time. The authors obtained measures of cerebral nervous system function at initial entry into the study and two years later. Baseline and follow-up data were obtained to measure changes in central nervous system function over a two-year period. As a predictor of change in central neuron system function, according to STEWART et al., boxing-related exposure was evaluated as (1) baseline exposure, defined as the number of bouts and amount of sparring that had occurred up to the baseline visit, and (2) follow-up exposure, defined as the number of bouts and amount of sparring that occurred during the follow-up interval of two years.

STEWART et al. enrolled 484 of 718 eligible boxers. Participants enrolled in the study of baseline were reexamined two years later.
The examination carried out by STEWART et al. consisted of the following:
(1) Interviews administered at the beginning of each visit (questions about socio-demographic characteristics, history of concussive head injury from causes other than competitions, sparring history, and psychosocial factors, including recent use of drugs and alcohol).
(2) A battery of neuropsychological tests, administered in the following order: Grooved pegboard, Rey Complex Figure, Copy, and Rey Complex Finger, Immediate recall, Rey Auditory verbal Learning Test, Digit Span, Forward and Digit Span, backward from the Wechsler Adult Intelligence Scale-Revised (WAIS-R), Raven's Progressive Matrices, Block Design from the WAIS-R), Word Fluency Test, Symbol Digit Modalities, Alphabet Recital, Trail making Test A; Trail-making Test B; Recognition from the Rey Auditory Verbal Learning Test; Rey Complex Figure' delayed Recall; Two-Choice Reaction Time; and Vigilance Test.
(3) Measurements of brainstem auditory evoked potentials,
(4) Electro-encephalography,
(5) ataxia/vestibular test for drug testing,
(6) urine sample for drug testing.
The exposure was defined by STEWART et al. according to number of bouts, sparring years, and sparring with a professional boxer.

PORTER and O'BRIEN (1996) tried to determine the incidence, pattern, and severity of injuries resulting from *participation in amateur boxing.*

The authors carried out a prospective 5-month survey of injuries that occurred during competitive amateur boxing and training. The setting was amateur boxing competitions held in Dublin between November 1992 and March 1993, and the six largest amateur boxing clubs in Dublin. Participants were all competitors in the tournaments and boxing club members over 16 year old.

The incidence of injuries in *competition* was 0.92 injuries per man-hour of play (or 0.7 injuries per boxer per year), while the incidence in *training* was 0.69 injuries per boxer per year. Cerebral injuries were reported as occurring only during competition, most of these being mild concussions. Hand, wrist and facial injuries were related to direct impact and occurred more frequently in competition than training, while injuries to other body parts were predominantly chronic and were considered training-associated. Shoulder and knee injuries were the most debilitating injuries seen.

The authors conclude that the yearly risk of injury resulting from participation in amateur boxing is relatively low when compared with other sports. Cerebral injuries, which occur almost exclusively in competition, are predominantly mild concussions. Had the authors been able to continue following up on their subjects in this study, their conclusions would have had greater applicability to the risks of injury from amateur boxing. As they were only able to follow up for a two year time period, their conclusions must be limited to acute injuries or injuries detectable during two years of amateur boxing.

Electroencephalograpic Findings in Boxers

Electroencephalographic recordings of boxers were carried out by TEMMES and HUHMAR (1952), BUSSE and SILVERMAN (1952), ARDITO (1954), HAJUSEK and MEDVED (1954), RAVINA (1955), PAMPUS (1956), PAMPUS and GROTE (1956) TITECA (1957), BEAUSSART (1959), BEAUSSART et al. (1959, 1960, 1970), DEVOE (1959), NESARAJAH et al. (1961), PATTI (1964), GALOWA and GALA (1965), GEBELT (1965), CHODKIEWICZ (1967), GATARSKI (1969), JOHNSON (1969), ROBERTS (1996), MORIYASU et al. (1969), THOMASSEN et al. (1979).

Here are only those publications mentioned in which EEG recordings of boxers were performed as the main topic of the studies. Papers which deal with the chronic clinical alterations and pathomorphological findings in which EEG recordings were in some or all cases carried out, are presented in summarized form.

Many EEG recordings of boxers have been obtained either as part of neurological examinations or in an attempt to measure acute traumatic injuries of the central nervous system. In the latter case they were recorded immediately before and after boxing bouts. A review of the interpretation of these recordings shows that a large number of authors saw evidence of alterations in the EEG linked to boxing activity, while a smaller group denied that any correlation existed between EEG findings and boxing.

The very first recordings, performed on two boxers, showed alterations in the form of irregularities, diminished amplitude of the potentials, and spikes and waves (GUILLAIN et al. 1948).

SJAARDEMA 1950) examined 200 boxers electroencephalographically. Of the amateur boxers, 85% had normal EEG's, 21% revealed bilateral theta waves and 2.8% slow theta waves. The respective percentages for professional boxers were 64.4%, 18.6%, and 17%. A group of boxers who had participated in severe bouts during the time of the EEG recording showed normal EEG's in only 59.9% of boxers who had taken severe blows against the head. Theta waves were seen in 23.5%.

Compulsory electroencephalographic examinations were made by BUSSE and SILVERMAN (1952) on 24 professional boxers with the cooperation of the *State Athletic Commission of Colorado*. Of these 24 professional boxers who were examined immediately after their fights, nine boxers (more than one-third) showed dysrythmias, with knocked out boxers showing a higher incidence of abnormalities and irregularities. Focal disturbances were more frequent in the younger men than in the older men, who also, generally speaking, had fewer electroencephalographic disturbances. By comparison, in a control group of non-boxers, abnormal EEG's were seen in only one out of eight. The authors were unable to establish a correlation between the degree of electrical abnormalities and the features of the boxers' fighting career.

BUSSE and SILVERMAN presented details of their findings: The regulations of the *State Commission of Colorado* had passed the following regulations requiring periodic electroencephalographic examinations on all professional boxers performing in the State of Colorado: (1) A professional boxer must have an electroencephalogram at least once a year; (2) in the case of a 'knock out,' an electroencephalogram must be done within two weeks of the injury; and (3) frequent, repeated examinations could be done in the case of a boxer with suspicious recording.

Compulsory electroencephalographic examinations were made on 24 professional boxers with the cooperation of the *State Athletic Commission of Colorado*. A statistically significant, increased incidence of dysrhythmic electroencephalograms was found in nine boxers (37.5%) in their series. There was some indication that men who had been knocked out showed more severe electroencephalographic disturbances than those who had not been knocked out. Focal disturbances were more frequent in the younger boxers than in the older ones, who also, generally speaking, had fewer electroencephalographic disturbances.

Of the electroencephalograms done on these men, four showed severe disturbances, five were moderately dysrhythmic. BUSSE and SILVERMAN mentioned that this is well above the 10 to 15% reported by GIBBS et al. (1943) and others in control groups, whose findings are in accord with those in their own series. There were four records showing severe disturbances, three of these showed a temporal slow wave focus, and one revealed diffuse slowing and paroxysms. Five records showing moderate disturbance were noted. Three of these were diffusely slow, and two were indicative of temporal slow-wave foci. Thus, in totaling the severely and moderately disturbed records, five were found to be focal and four showed diffuse findings. In addition, paroxysms were seen in one each of the focal and diffuse records.

BUSSE and SILVERMAN presented the case history of a boxer who was finally suspended from boxing largely on the basis of progressive encephalographic abnormalities.

The patient, a professional boxer, aged 21, was first seen in November, 1950. He had had three professional fights prior to his first electroencephalogram. In July, 1950, he was knocked out, but the following month he won his next fight by a technical knockout in the first round. Three days before a scheduled fight in November, 1950, the patient had his first EEG which revealed a left temporal slow-wave focus. The fight was cancelled. A repeat EEG revealed more extensive temporal focal findings. (The patient, incidentally, showed no abnormal neurological or psychological signs or symptoms at any time). He was temporary suspended from boxing, but continued to train. A third EEG in January, 1951, revealed the persistence of the focus and the addition of paroxysms. This suggested that he had received additional brain injury during the period of training, from which he could not legally be prohibited. It was now believed that this man had suffered some degree of irreversible damage, and he was permanently suspended from boxing in January 16, 1951, for his own good. It was felt that in his case the EEG had picked up intracranial damage before it had become clinically apparent and that late neurological disability or even death had possibly been averted by this decision.

In a *mixed group of eight active and retired amateur boxers who had been boxing for at least four years and having experienced a knock out at least twice and who sought medical advice because of varying neurologic symptoms,* TEMMES and HUHMAR (1952) found seven to have pathological EEG's, such as dysrythmias combined with reduced amplitudes of alpha waves. These irregularities were observed even beyond the active career of the boxer.

The subjective complaints of seven of the patients who consulted a doctor were surprisingly similar: temporary or continuous headache, loss of memory, almost permanent fatigue, tremor of the hands or lowered hearing or buzzing. Moreover, most of the patients had experienced somewhat inexactly described fits of 'weakness' or 'dizziness,' and one of them had clinically obvious epileptic attacks. According to information obtained from the relatives, most of the patients had exhibited distinct signs of a change in personality, the most common complaints being irritability and quarrelsomeness and repeated absenteism. In all subjects the efficiency of their work and their social adjustment in their employment had deteriorated progressively. Among the objective symptoms the following may be mentioned: two were clearly demented, one of them (a former typesetter who was now carrying used typebeds to the printer's apprentices for breaking up) was also suffering from disturbances of speech and from a distinct tendency to perseveration. All had tremor of the hands, and one showed uncertainty in the finger-nosetip and finger-finger tests. The facial expressions had been reduced to a greater or lesser extent in all the patients. Despite this, the appearance of most of the patients was, even ignoring the typical boxer's face, considerably older than their age presupposed.

Those five patients who were in hospitals for an examination or had been sent by neurological outpatients' clinics, and among these were all those showing severe clinical or EEG symptoms, had negative Wassermann reaction in blood and liquor. The patients were sent by private doctors, without any note of the result of a possible Wassermann test. The latter three on enquiry denied having suffered from syphilis.

The eight patient showed a clinical picture and EEG findings differing from those of the others. About 15 minutes after a prize fight he had won, he had become confused, and soon after that, unconscious. He was sent directly to the *local hospital.* He regained consciousness on the following day, but was still confused at times. He complained that his right arm and leg felt paralysed. Urine was passed involuntarily at the beginning of his condition. Within a fortnight the somatic symptoms had abated considerably, but as the patient was temporarily confused, he was sent for further examination to another hospital, where the following observations were made; partial loss of memory, the strength of the right leg was reduced, Babinski on the right +++, on the left -. Romberg: cannot stand upright, falls backwards and to the left. Papilledema on both sides. A marked cerebral atrophy was roentgenologically diagnosed.

The *EEG examination* revealed a diffusely limited area of dysrhythmia (delta waves) in the left temporo-parietal region, with a center in the frontal parts of the temporal region.

The EEG of the remaining seven patients showed the following characteristics: One of the patients had a slightly irregular EEG, still keeping within the normal range, but the remaining six had a distinctly dysrhymic EEG. Typical of the disturbances was an irregularity of the alpha waves; in most cases they were considerably reduced or almost completely lacking, and their frequency was labile. If the alpha waves were dominant, they showed variable irregular damping in different regions. All cases showed fast activity mingled with the dominant frequency. In one patient, the EEG was greatly reduced, and there was only a trace of measurable function; the right hemisphere showed more abnormality than the left. Another patient exhibited only a minimum of alpha waves all over the convexity; there were also irregular fast waves, and slow and medium frequency (delta and 4-7/s) waves, registered more in the frontal parts of the brain than in the occipital region. Medium frequency waves also occurred, in addition to the aforementioned disturbances, in varying proportion in all the other patients examined. No instance of specific dysrhythmia was observed.

ARDITO (1954) described in *ten amateur boxers who had not boxed for a long time dysrythmias in the EEG.*

MARQUARDSEN (1954) reported similar irregularities. TAYLOR (1953) also saw reduced activity in all recordings from a boxer who had switched from amateur to professional status. Five ex-professionals examined by SOEDER and ARNDT (1954) exhibited pathological changes in the over-all picture, and three also showed a focus and labile frequencies.

LARSSON et al. (1954) used the EEG to study effects upon brain functions in individual amateur bouts. *Their recordings were made 15 to 30 minutes after the fights and showed generalized flattening of the record with slow irregular activity in 13%.* Significantly, these findings were recorded in 30% of the boxers who had sustained a knock out blow or technical knock out defeat. Control recordings had been made prior to 3 fights. The members of another group were examined only after fights that were considered particularly rough. The resulting EEG recordings were pathological in 52% whereas the overall percentage of pathological recordings before the bout was 16%.

The paper of LARSSON et al. was quoted before in the chapter discussing groggy stages. In the following four EEG recordings are shown (Figs. 236, 237, 238, 239).

PAMPUS (1956) and PAMPUS and GROTE (1956) *recorded the EEG's of 250 active and 17 retired amateur boxers at the German Amateur Boxing Championship. The boxers were involved, during the period of the week, in up to five bouts, and a total of more than 500 EEG recordings were examined.*

The findings were arranged in *three groups:* (1) *Normal findings* in regard to the age of the participants; (2) *mild, questionably pathological recordings;* and (3) *distinct changes* and *pathological alterations.*

Of 190 boxers, 70 had a normal EEG (36.8%), 39 mild pathological findings (20.5%), and 81

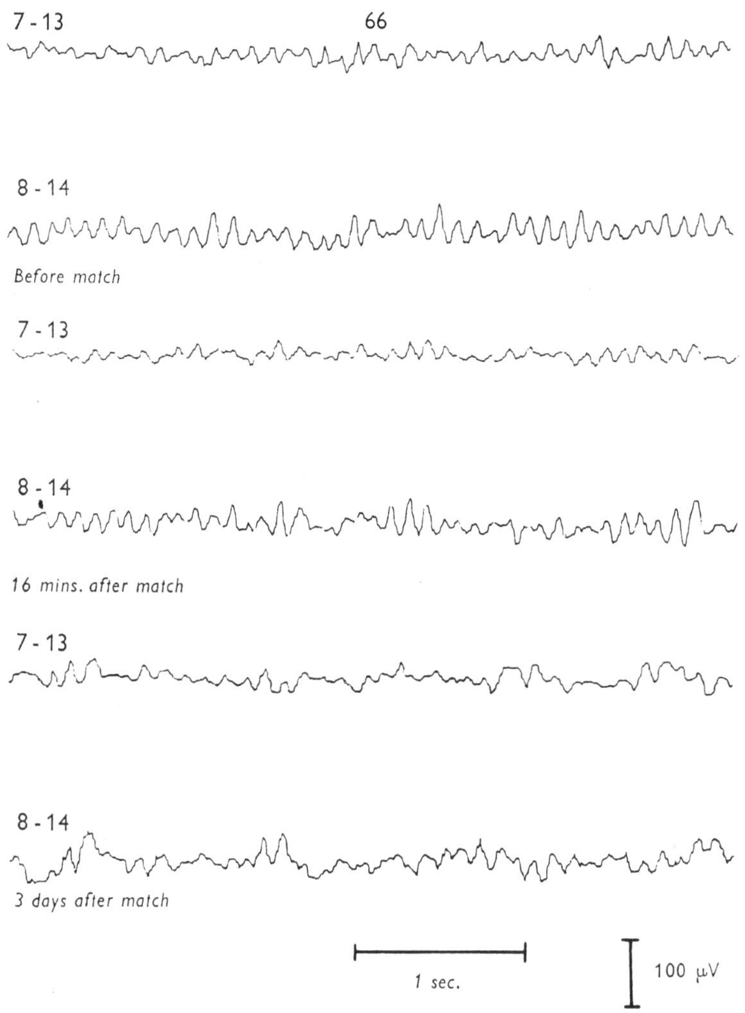

Fig. 236. Knock downs. Confused memory of entire match and does not know definitely that he lost the match. Disorientation in time. Lucid after 20 mins. The EEG 16 mins, after match is practically unchanged but three days later there is an increase of slow activity in the record. (From LARSSON, et al., 1954).

(42.7%) had in at least one recording a pathological curve.

PAMPUS found, out of *17 inactive amateur boxers,* four pathological, three mildly altered and ten normal EEG's.

Of one 116 EEG's, which were recorded *immediately before* or *one to two days before the fights,* 29.3% revealed pathological, 19.8% slightly altered, and 50.8% normal readings. Two of the EEG recordings of the authors are presented (Figs. 240 and 241).

The EEG's taken *after the bouts* were *normally recorded within 5 to 15 minutes after the fight,* although in a few cases this *was extended to 30 to 45 minutes later.*

This methodological shortcoming may in part account for the contrasting results in the evaluation of records obtained by a number of authors. Thus, no significant changes were seen in the post fight recordings of 1.000 professional boxers made by KAPLAN and BROWDER (1954), who also watched the boxers during their fights as to any effect of the blows. The au-

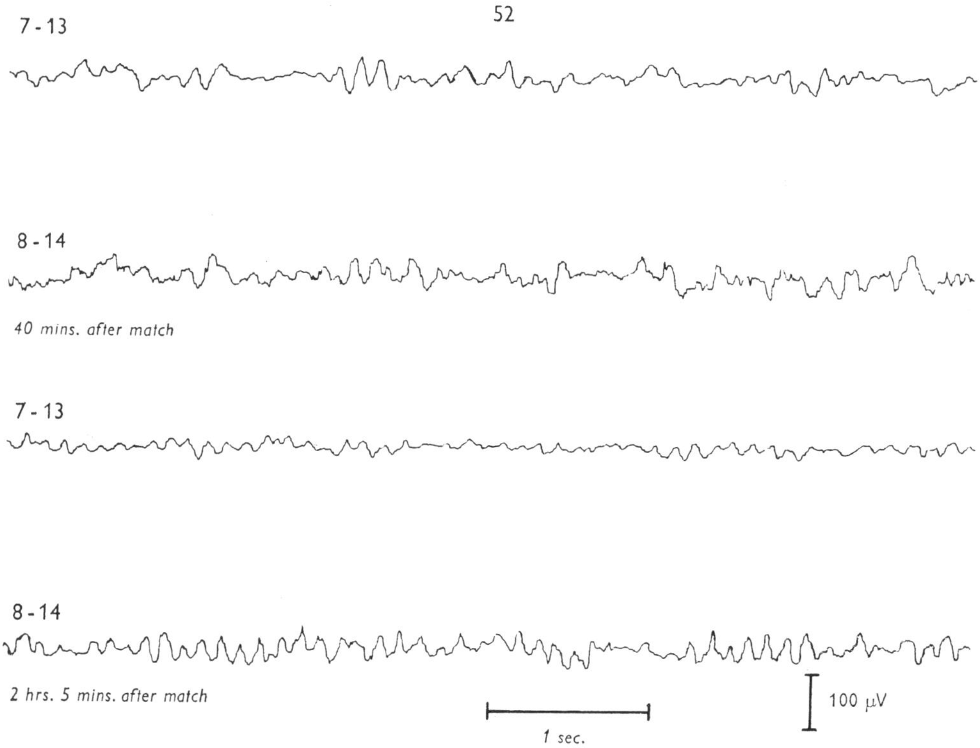

Fig. 237. Mentally hazy, 'cannot see clearly'. Appears to think sluggishly. Uncertain orientation in time. Lucid after half an hour. The EEG 40 mins. after the match shows an irregular slow activity, but is defitively improved after further 85 mins. (From LARSSON, et al., 1954)

thors admitted, however, to a great number of disturbed recordings among the poorly rated fighters.

MEDVED (1955) examined *32 Yugoslav amateur boxers* between 20 and 34 years of age who had been active for several years and had performed in between 15 to 250 bouts. MEDVED wrote of only four abnormal recordings, although 46.8% of the men had been knocked down at last once.

The entirely normal recordings reported by BLONSTEIN and CLARKE (1957) concerned the neurological and electroencephalographic examination of *29 amateur boxers* (0.58% of the *London Amateur Boxing Association)*, who had been knocked out more than once.

In conclusion, it can be said that the majority of investigators noted unspecific irregularities and abnormal recordings in boxers in a far greater proportion than in control groups. For purposes of comparison it should be noted that, according to GIBBS et al. (1943), 10% to 13% abnormal recordings can be found in any group of 'healthy' persons. A small group of authors reported pathological recordings among boxers in nearly equal incidence with this rate.

Generally, the EEG is considered a less reliable indicator in boxer's encephalopathy than neurological and psychopathological examination, and its value is recognized only in the context of a neurologic-psychopathologic evaluation. Also, it must be remembered that a normal EEG finding does not exclude the possible existence of a traumatic encephalopathy.

Alternatively, of ten boxers examined by MAWDSLEY and FERGUSON (1963) five showed excessive theta activity and two showed

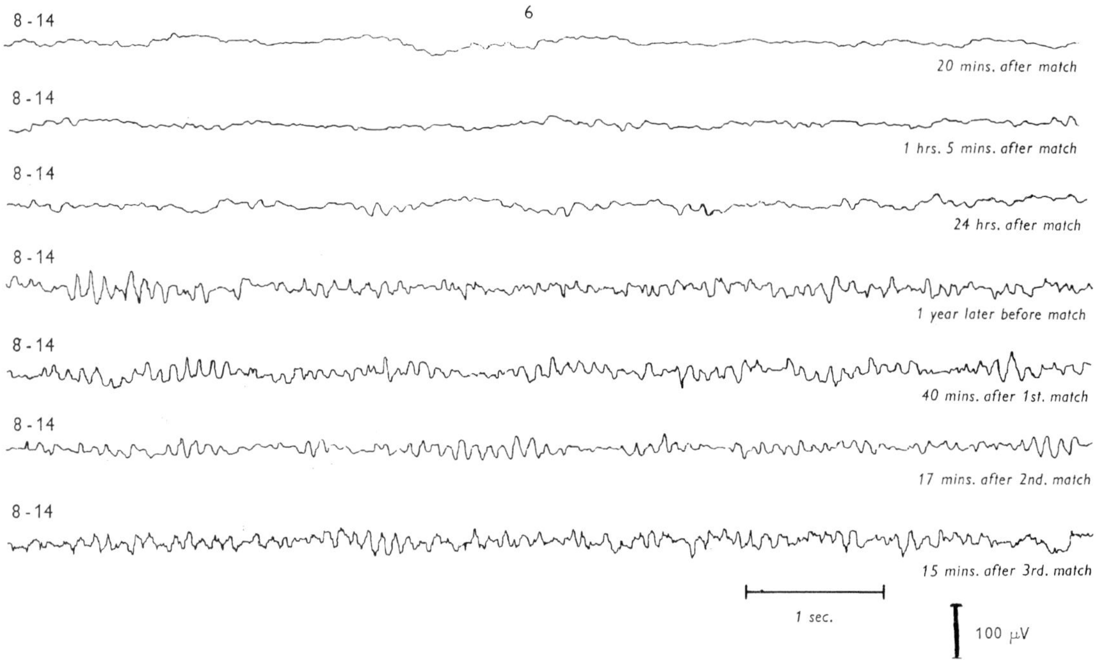

Fig. 238. Knock out. Amnesia for match, disorientation for time. Euphoric. Amnestic period of 20 mins. duration. Headache afterwards. The EEG 20 mins. after match shows an outspoken amplitude decrease still remaining 24 hours later. The next year he fought three matches, winning them all. The EEG was essentially unchanged throughout.(From LARSSON, et al., 1954).

delta waves. They discovered no correlation between EEG and clinical findings. In their opinion, EEG recordings are useful in determining the severity of trauma directly after a fight, but they deny their usefulness to any extent in the diagnosis of chronic disturbances in the boxer. "The EEG may be of value in assessing the severity of cerebral injury immediately after a fight. It is of little help in the diagnosis of chronic lesions in boxers" (MAWDSLEY and FERGUSON)

TITECA (1957) who recorded the EEG's in two professional boxers stressed the point that the EEG may be already pathological before clinical signs become manifest.

CRITCHLEY (1957) remarked that the EEG of the successful boxer tends to reveal minor and non-specific abnormalities, but that these are seen more frequently in those suffering from a punch-drunk state. On the other hand, says CRITCHLEY, the EEG is a less sensitive indicator of boxer's encephalopathy than are clinical signs such as may be noticeable even when the EEG recordings read with in the norm. In his opinion, the return to normal of a series of EEG recordings for an individual patient cannot be viewed as a restitutio ad integrum either physiologically, anatomically, or clinically.

BEAUSSART et al. (1959) reported on EEG recordings of a group of young amateur boxers with few fights. The EEG's were recorded six to ten minutes after the bout. Only one boxer who was knocked out showed a reduced electrical activity, the other EEG's were within the physiological limits.

NESARAJAH et al. (1961) *examined a total of 50 Ceylonese boxers:* they were drawn from the *Ceylon University, Police, Navy, Army, Royal Ceylonese Air Force,* and some of the secondary schools. In this group were included the best boxers of the island, who represented Ceylon at meetings abroad. They had been in training for periods of one to 18 years. As a control group, the authors

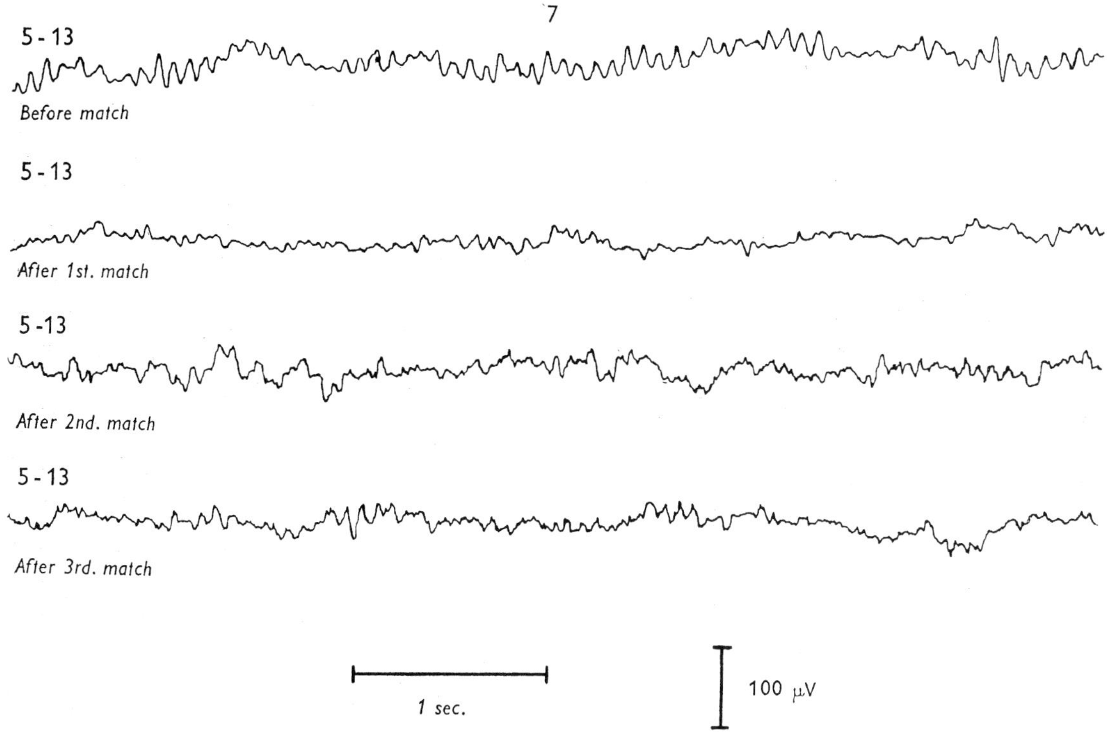

Fig. 239. Winner in first match but received several hard head-blows. No clinical symptoms. The EEG shows an amplitude decrease of the alpha activity and a certain incrase of 15/s waves, not observed in the record before the match. After second match without hard blows to the head and without clinical symptoms, there is an increase in irregular slow activity, presumably an after-effect of the first match. After third match, which was stopped because he got several hard head-blows, he claimed that he was in center of ring when the match was stopped (hung on the ropes). Examination 5 mins. later showed disorientation in time; completely lucid some mins. later. The EEG shows a decrease in slow activity in comparison with the day before. The subject is not included in tables III and IV because it is impossible to relate the EEG-changes to one match only. (From LARSSON, et al., 1954).

had 75 Ceylonese adults, University students, who took part in other sports but not in boxing.

An analysis of the control group showed that six cases (8%) revealed spontaneous bursts of abnormal activity – that is either high amplitude spikes or large slow waves (of 4 – 7 c/s). The rest of the control tracings were within the limits of normal variaton and consisted of 4-12 c/s activity with an amplitude of 30-50 microvolts.

The boxers were examined and a careful history taken to eliminate, as much as possible, head injuries from any other cause. [There were four rugger players and two wrestlers amongst our group, and only one rugger player gave a definite history of being knocked out twice during games.] While recording their histories they questioned them especially for (1) any knockouts during their boxing career, (2) frequent attacks of headaches and (3) loss or impairment of memory.

Seventeen (34%) gave a history of at least one knock out, while six (12%) complained of persistent headaches and eleven (22%) of loss or impairment of memory. It was interesting to note that some boxers actually said, "Doctor, I regret having taken up boxing; I will never allow my sons to box."

A careful analysis of the EEG recordings of these 50 boxers showed that 30 (60%) showed spontaneous bursts of disordered activity. The criteria used to classify a record as abnormal have already been mentioned.

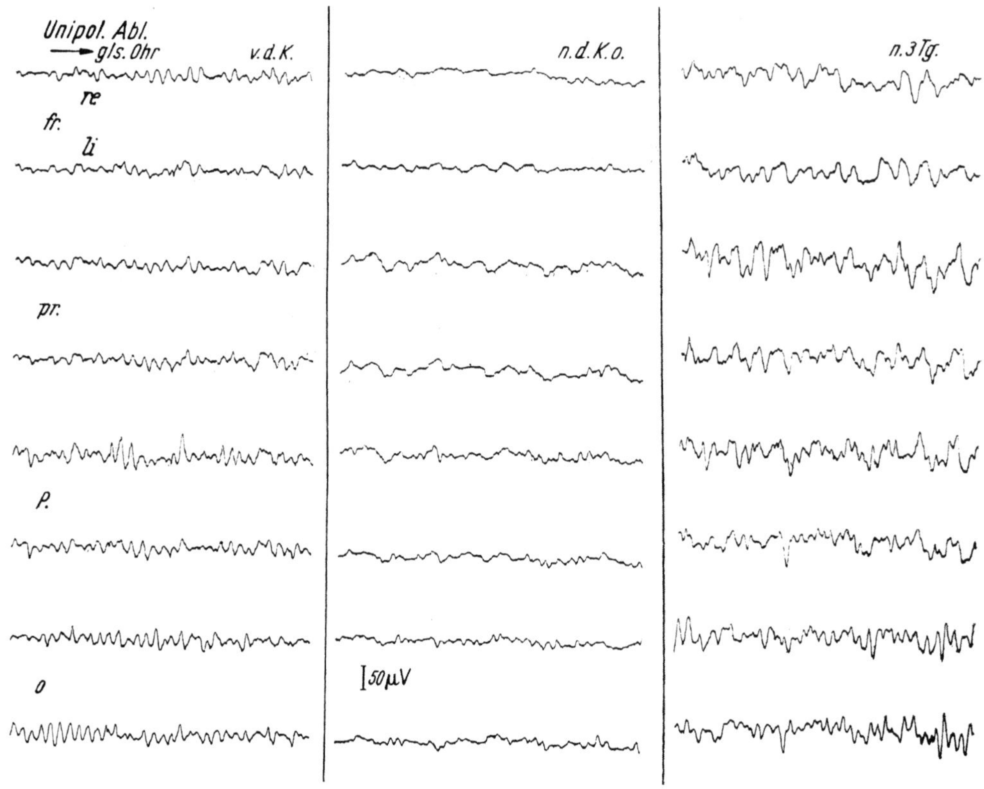

Fig. 240. Comparison curves before and after a knock out (before the match: normal curve, after the match: Reduction of the ground rhythm, precentral dysrhytmia, three days later: general dysrythmia. (From PAMPUS, GROTE, 1956).

The authors presented three case histories with the EEG findings:

Case 1: This man, aged 25 had boxed since 1948. He represented Ceylon in the *1952 Asian Games*, the *1956 Olympics in Melbourne*, the *1957 South East Asiatic Championship Meeting in Burma*, the *1958 Empire Games in Cardiff*, and the *1959 Indo-Ceylon meeting in India*. He gave no history of knock outs but complained of impairment of memory. He has felt 'dazed' after hard blows on the head. *EEG* recordings taken at rest showed sporadic bursts of disordered activity in both 'unipolar' and bipolar records.

Case 2: This man, aged 21, had boxed from 1952 to 1956; he then stopped until 1960. He gave a history of having been struck by a stone on the head in 1954, and was unconscious for about two hours on that occasion. He complained of frequent attacks of headaches and was referred to us for an EEG in May, 1960. The EEG was reported as being within the limits of normal variation. In August 1960, he was sent for an EEG as he had been unconscious for about 20 minutes after a blow on the chin during a university boxing meet. This EEG was taken three days after the 'knock out.' The recording showed spontaneous bursts of disordered activity.

Case 3: This man, aged 27, had boxed for the last five years. He did not complain of headaches or loss of memory, but had had one knock out and was unconscious for eight hours (two years before the EEG recording was made). His EEG showed spontaneous bursts of disordered activity.

NESARAJAH et al. summarized that the variation found in their control group of cases (8%) is well within the expected figures. The presence of disordered activity among 60% of the boxers is statistically very significant. Further, it is noticed that, although only 34% of the boxers gave a history of knock outs, 60% showed disordered activity. Their findings therefore show that EEG recordings of boxers taken under identical conditions (at rest) show a significant change in the wave form in a large percentage of cases.

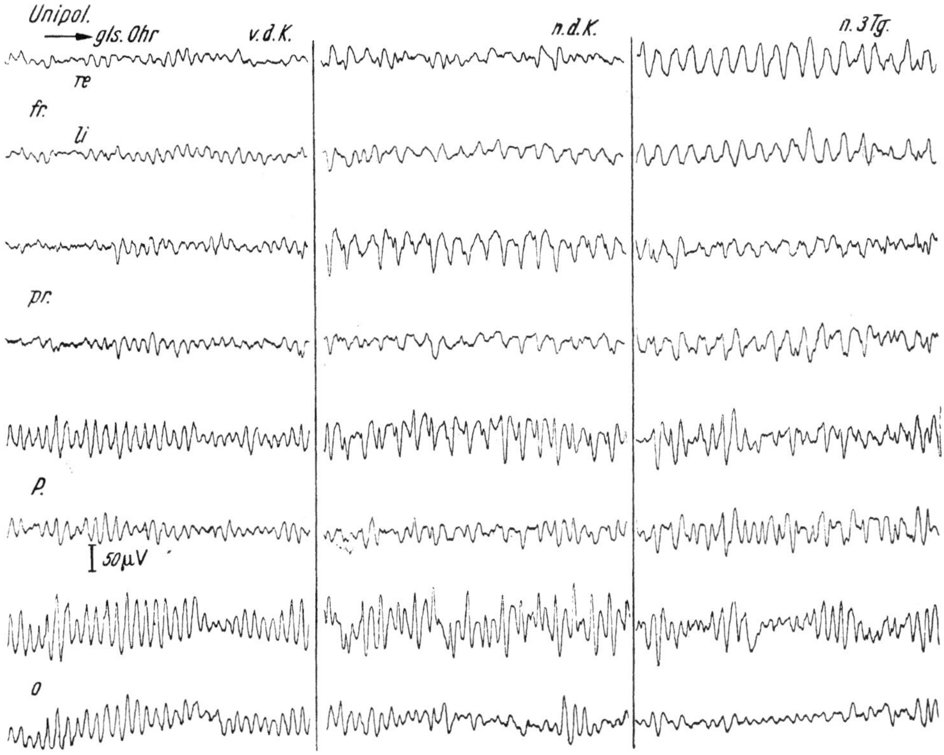

Fig. 241. Unilateral activation of the ground rhythm and unilateral development of paroxysm-like waves above one hemisphere after a knock out. (From PAMPUS, GROTE, 1956).

The psychiatric aspects of the encephalopathy of boxers were described in 17 ex-boxers who were referred to a hospital because they were thought to be suffering from this disorder (JOHNSON 1968). The EEG findings were described in detail by JOHNSON (1969) and an attempt was made to evaluate their diagnostic value in relation to the other forms of investigation, by correlating the degree of abnormality to the severity of the clinical syndrome.

Bipolar, scalp EEG's were carried out by JOHNSON in the 17 cases of the series, abnormal records were present in 10. These were severely abnormal in one case, moderately abnormal in four cases and borderline abnormal in five. Two main types of abnormal records were recognised:

(1) A flat, low voltage record with minimal (below 25 µV) or absent alpha rhythm throughout the record (Fig. 242), was present in three records (1, 3, 7).

(2) An alpha dominat record with randomly occurring mixtures of slow and intermediate slow background activity was present in seven records (Fig. 243) (2, 6, 9, 14, 15, 16, 17).

All the records were examined 'blind' without knowledge of the clinical details of the cases and were assessed numerically along four different parameters, viz., abnormalities in the dominat rhythm, the background rhythms, transient abnormal rhythms and abnormalities produced by activation procedures (hyperventilation and photostimulation). Each parameter was assessed as
0 = no abnormality
1 = mild abnormality
2 = moderate abnormality
3 = severe abnormality

The maximal total score for abnormal record was 12: summated scores are shown in Table 83.

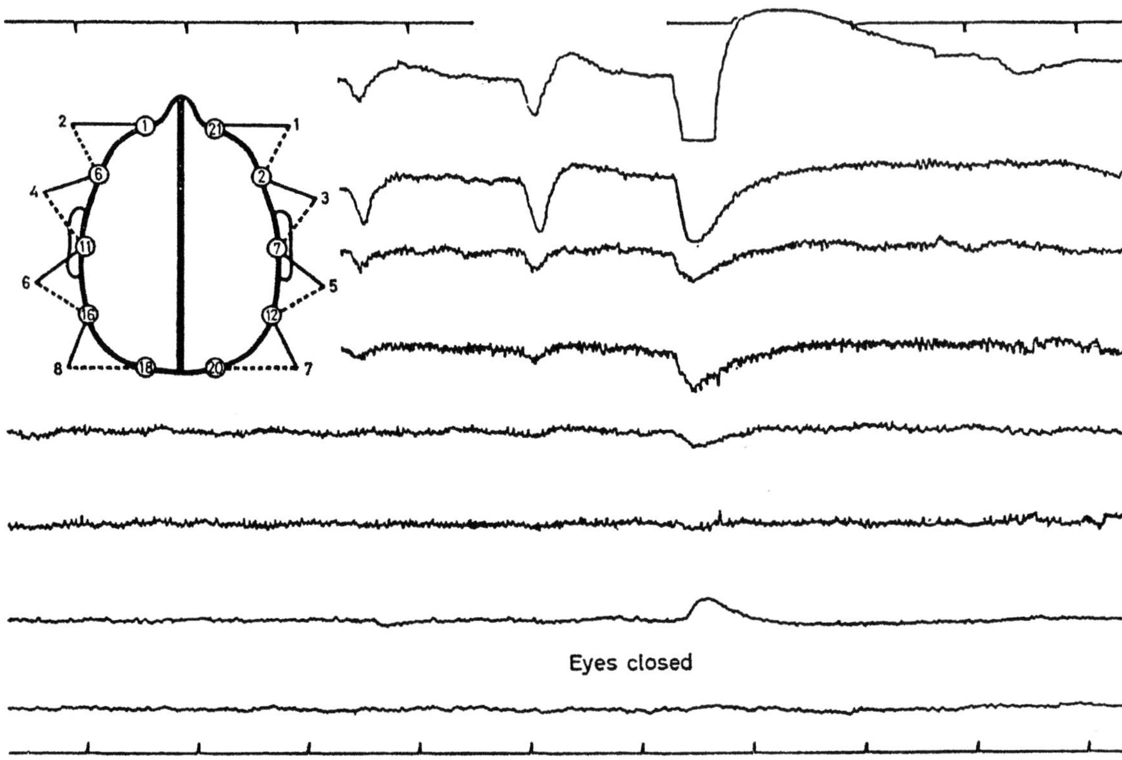

Fig. 242. J. B. 6-3-63. Flat, low voltage record with minimal (below 25 μV) or absent alpha rhythm. (From JOHNSON, 1969).

Neuroradiological Findings

Lumbar air encephalography (AEG) was carried out by JOHNSON in 15 of the 17 cases and evidence of brain damage demonstrated in 13. ISHERWOOD et al. (1966) described the main neuroradiological abnormalities in this traumatic encephalopathy, the specific features of which had first been described by SPILLANE (1962). Three abnormal radiological findings were assessed as present or absent from the neuroradiologist's reports viz. communicating cavum septi pellucidi, cortical or cerebellar atrophy and ventricular dilatation. A maximal rating of 3 was therefore possible if all abnormalities were present in the AEG (91).

Psychometric Findings

Psychometric tests of brain damage (Wechsler Adult Intelligence Scale, Benton Visual Retention Test and an auditory Word Learning Test), were carried out by JOHNSON in 15 cases by a clinical psychologist and wee abnormal in 11. As these tests measure different aspects of brain damage a single point was allocated for evidence of abnormality in any one test (Table 83).

Clinical Findings

The clinical manifestations of the traumatic encephalopathy in the series of JOHNSON varied widely in severity. In some cases they are mild and non-progressive (Cases 5, 7, 14) whilst in others they are severe and progressive (1, 8, 9, 16). Neurological features were numerically rated according

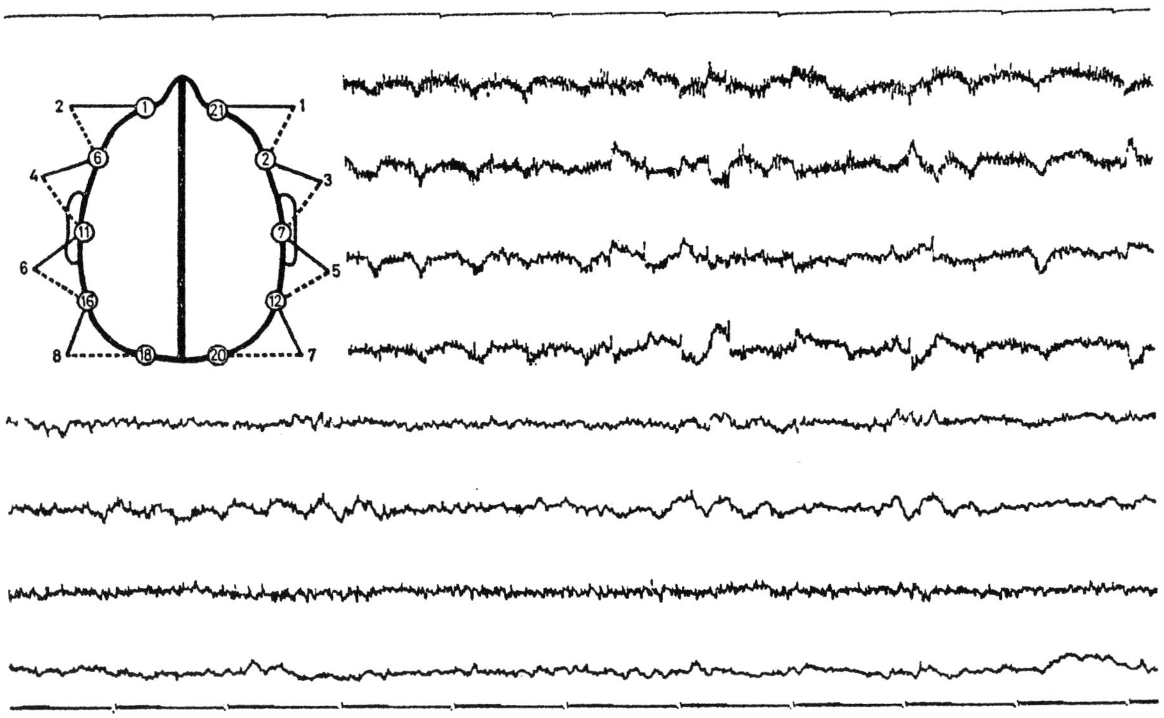

Fig. 243. Th. M. 25-5-65. Alpha dominat record with randomly occurring mixtures of slow and intermediate slow background activity. (From JOHNSON, 1969).

due to the presence of pyramidal, extrapyramidal, cerebellar signs, epilepsy and impotence. Psychiatric features were enumerated according to the presence or absence of an amnestic state, dementia, morbid jelousy syndrome, psychosis, explosive personality reactions. A total clinical score was obtained by summating the psychiatric and neurological abnormalities to give a maximum of 10 possible symptoms (Table 83).

The investigations and the clinical state were then arranged in rank order according to the degree of severity. Each investigation was then correlated in turn with the severity of the clinical state and Spearman's rank correlation coefficient calculated (Table 84).

An EEG abnormality of a non-specific nature was present in 60% of the patients in JOHNSON's series. The low voltage 'flat' record, was present in three cases (1, 3, 7) has been previously described in patients suffering from Huntington's chorea and dementia (MARGERISON and SCOTT 1965). GORDON and SIM (1967) have also noted a high incidence of this type of record in patients with presenile dementia of the Alzheimer type. The EEG abnormality in the remaining seven cases affected the background rhythms and occurred randomly except in one case (14) where there was a localized focal abnormality in one parietal area.

Although the overall incidence of EEG abnormalities in the traumatic encephalopathy is high, according to JOHNSON, the correlation between the severity of the clinical syndrome and the degree of disorganization of the EEG is poor (r = 0.1103). On the other hand evidence of brain damage on the air encephalogram shows a high correlation with the degree of severity in the clinical syndrome (r = 0.5777). The lesions responsible for the encephalopathy are probably located in two main

TABLE 83
Rating scores of severity of clinical state and investigation. (From JOHNSON, 1969).

Case No.	Neurological	Psychiatric	Total clinical rating	AEG (0–3)	Psychometric tests (0–3)	EEG rating (0–12)
1	4	4	8	3	3	3
2	3	1	4	0	0	2
3	0.5	1	1.5	2	1	3
4	0	0	0	0	0	0
5	0	1	1	–	–	0
6	4	3	7	2	2	2
7	1	0	1	2	2	2
8	2.5	3	5.5	2	0	0
9	2	4	6	2	2	2
10	1.5	3	4.5	2	2	0
11	1	1	2	2	–	0
12	3	1	4	2	0	0
13	2	1	3	2	1	0
14	0.5	0	0.5	2	1	7
15	3	2	5	2	3	2
16	3	3	6	–	3	3
17	1	3	4	1	2	4

areas (JOHNSON 1968). Those responsible for the neurological features are, according to JOHNSON, in the upper brain stem whilst those related to the psychiatric features are in the hippocampal-fornical system. As the scalp EEG predominantly reflects neural disorganization in the cerebral cortex it is not surprising that there is a low level of correlation between the EEG abnormalities and the severity of the clinical syndrome since the responsble lesions are largely subcortical.

Three cases (1, 16, 17) showed definite evidence of progressive, intellectual deterioration i. e. dementia, whilst a fourth case (15) was showing evidence of early dementia at follow-up; all had abnormal EEG's. Although the EEG is frequently abnormal in patients suffering from dementia, most investigators have found the level of correlation to be low between the degree of dementia and the severity of the EEG abnormality.

TABLE 84
Correlations between clinical state and investigations. (From JOHNSON, 1969).

	Clinical state		
	AEG	Psychometric	EEG
No. Subjects	15	15	17
Spearman's Rank Correlation Coefficient	0.6187	0.5777	0.1103
't' test (N-2)	2.83	2.5	0.42
Percentage	1–2%	2–5%	>10%

Transient abnormalities in the EEG immediately after a boxing bout (LARSSON et al. 1954) reveal temporary neuronal disorganisation particularly when the boxer has been involved in a 'knock down.' The EEG might be used, according to JOHNSON, as a convenient method of deciding when such disorganization has resolved and when a fighter can safely resume his boxing activities. The use of the routine EEG as a screening procedure for the detection of the incipient traumatic encephalopathy therefore appears to be limited in view of the low level of correlation between the severity of the syndrome and the EEG abnormality. Psychometric testing and air encephalography (where justified) offer more accurate evidence of the severity of the traumatic encephalopathy

The findings of this investigation must, however, be interpreted with reservation. Visual analysis of EEG records by clinicians combined with simple computations of their abnormalities are of limited value. Whilst the changes in a single record as in this series may be normal or only minimally abnormal, the changes observed in serial records may be more significant. The application of wave analysis and computation to serial records on boxers may well reveal early and more significant evidence of an insidious encephalopathy than isolated records can do. Although the EEG is revealed in this investigation as poorly correlated with the severity of the traumatic encephalopathy of boxers, this is probably due to the limitations imposed by present clinical methods of interpretation.

GATARSKI (1969) took *electroencephalographic recordings of 60 amateur boxers, and 30 candidates who were used as controls.* Pathologic EEG changes were found in the EEG's of 40% of the boxers, and in only 6.6% of the controls.

MORIYASU et al. (1969) recorded *EEG's of 78 professional* and *ten amateur boxers before and after boxing fights.* These recordings were compared with those of 300 of general head injury.

After boxing matches the authors saw a decrease in the amplitude of alpha waves, and the appearance of slow waves.

The EEG's which were recorded on the day following the boxing match showed slow waves in 41% of the 53 professional boxers, and in 20% of the ten amateurs. Many of these recordings showed 6-7 Hz/s theta waves. Abnormal EEG's with 4-5 Hz/s theta waves and 3 Hz/s delta waves were found in some of the professional boxers. Half of the cases which exhibited slow waves after the boxing fights diminished after one month, but in the other half the slow waves remained.

When JOHNSON (1969) *examined 17 ex-boxers who had symptoms of traumatic encephalopathy,* his two most remarkable findings were flat low voltage recordings (three cases) and an alpha-dominant record with randomly occurring mixtures of slow and intermediate background activity (seven cases). Correlations between the degree of electroencephalographic abnormalities, psychometric tests and neuroradiological findings showed the EEG to have the lowest correlation. JOHNSON suggested that this might be due to the limits imposed by visual analysis of isolated records, and that serial EEG recordings during the active careers of boxers would probably be of more value in detecting the encephalopathy.

After the first group of 177 recordings, 36% showed pathological findings after the second bout (which took place two or three days later), of 73 boxers, 40% showed pathological findings.

The percentage of pathological findings increased after three bouts from 29% to 42%. Only two boxers had EEG's recorded after a fourth bout, and both revealed pathological findings.

It has to be taken into consideration that in all these EEG's the victorious boxers, in other words the less damaged, were recorded.

The percentage of the single EEG alterations in regard to the entire group of examined boxers was, before and after the first fight: Diffuse irregularities, 15.5%/15.8%; pathological hyperventilations irregularities, 11.2%/12.4%; local dysrhythmias, 8.8%/8%; alterations over an entire hemisphere, 1.7%/5.6%; and local focal alterations, 8.6%/10.8%.

In the study of ROBERTS (1969) there were available for assessment the records of 168 professional boxers and 98 controls. The assessment of each record was termed as *abnormal, borderline* or *normal,* and separately as *low voltage,* has been considered in the case of the boxers in Table 33, with respect to age, professional career, as measured by the number of years he fought professionally, and also with respect to the clinical diagnosis of traumatic encephalopathy.

ROBERTS noted first that there is no previous report of an electroencephalographic study of a random sample of ex-professional boxers with which these results may be compared. ROBERTS confirmed that:

"more relevant to the question of permanent, cumulative structural damage, investigated in the present study, are the small number of electroencephalographic reports in the cases already reviewed and judged to provide adequate details of the boxing career and the neurological syndrome to support a clinical diagnosis of traumatic encephalopathy." There was in all, according to ROBERTS, among these cases, mention of records of twenty-four punch-drunk ex-boxers (GUILLEMAIN et al. 1948, Mc ALPINE and PAGE 1949, TAYLOR 1953, SOEDER and ARNDT 1954, CRITCHLEY 1957, NEUBÜRGER et al. 1959, WOLOWSKA 1960, SPILLANE 1962, MAWDSLEY and FERGUSON 1965, GORALSKI and SYPNIEWSKI 1967, HESE and SIBILIAK 1967). Among these recordings there were, according to ROBERTS, fifteen judged abnormal, and nine normal.

ROBERTS continued:

"The most frequently recorded abnormality was an excess of theta activity, often of low voltage and generalised, but occasionally focal and then most commonly in the temporal areas. There was otherwise no characteristic abnormality noted, nor were the most severe abnormalities constantly related to the areas most severely disabled clinically, and there was no constant distinction between the abnormalities found in those with predominantly cerebellar or predominantly extrapyramidal syndromes. Non-specific abnormalities of the kind described in most of these cases are found in the electroencephalograms of 10 to 15 per cent of the population (WILLIAMS 1941, HILL and WATTERSON 1942). In this small series of highly selected cases of traumatic encephalopathy of boxers the proportion with abnormal recordings is more than double that which might be exspected in the population."

I like to add to the names of scientists who described non-specific abnormalities in electroencephalograms (GIBBS et al. 1943).

ROBERTS stresses:

"It may therefore, be unwise to assume that a single electroencephalographic recording can be used to establish or support a diagnosis of chronic traumatic encephalopathy in an ex-boxer."

Acute Pathomorphological Findings

An editorial in *The Lancet* (1898) brings one of the first recorded medical cases of a fatal glove fight:

"Mr. TROUTBECK, the *coroner for Westminster,* held an *inquiry* on November 14 into *the circumstances attending the death of Thomas TURNER, aged 23 years, who died on November 10 in Charing-cross Hospital after engaging in a glove fight at the National Sporting Club on November 7th.* The fight was 'a match of 20 rounds at 8 st. 8 lb. with 4 oz. gloves.' Actually 5 oz. gloves were used. The referee gave evidence that the fight was in no way a severe one. The men were very evenly matched and during the thirteenth round TURNER received a light blow on the side of the head close to the right eye but some inches away from the 'point.' He then collapsed and the referee stopped the fight. After some brandy had been administered he revived, stood up, shook hands with his opponent, sat down again, and became unconscious."

"Mr. Jackson LANG, who was present as a spectator, said that he attended TURNER, who *showed symptoms of extreme collapse and inequality of his pupils.*"

"Mr. Malcolm SMITH, who made the *post-mortem examination,* said that TURNER's heart was extremely weak and small. On the right side of the brain was a clot of blood. The jury returned a verdict of *'Accidental death,'* with a rider suggesting that the club should instruct a medical man to examine the physical condition of candidates for boxing competitions before allowing them to take part in any fight. TURNER, it appears, had not undergone any medical examination before fighting, but even if he had we are not certain whether an medical man could have certified him as unfit. We doubt whether the diminution in the normal size of the heart could have been diagnosed, and if it had been, a small heart need not necessarily mean an unhealthy heart. However, there can be no harm in the *National Sporting Club* adopting the suggestion of the jury, but this particular accident appears to us to have been unavoidable."

The rider of the jury recommended that a medical man should examine the physical condition of the candidates for boxing competitions before allowing them to take part in any fight is insofar of importance, as candidates whose physical condition is not good can be eliminated to participate in boxing. But that the presence of a physiciam could prevent the occurrence of a subdural hemorrhage due to boxing blows is quite naïve. A physician is able to diagnose an injury, as Mr. Jackson LANG did, who was present as a spectator.

Mr. Jackson LANG, a physician, who was present as a spectator and who attended the injured boxers reported *inequality of his pupils.* It can be concluded from this finding, that with a very high probability an *acute* or *peracute right-sided subdural hemorrhage* existed. The autopsy revealed that on the right side of the brain was a clot of blood. There is no detailed description of the location of this hemorrhage, but one can conclude from the autopsy report that it was subdurally located. There are no data presented about the time of survival. A subdural hemorrhage is generally the result of rotational acceleration of the head of a boxer that leads to rupture of bridging veins. External bruises of the face can be completely missing or can be quite minimal.

It is not probable that the altogether small heart was the cause of death. A medical examination before the bout would have had no effect on the fatal outcome of this fist fight.

ORTON (1901) gave the following editorial comment on the deaths from boxing:

"Sirs, – Surely the faculty ought to be able to help in throwing some sort of light on what appears to be a ghastly mystery in the matter of these last three deaths, after boxing under apparently the safest of conditions, conditions non-existent in other clubs where great numbers of severe contests take place. The mystery has hitherto been treated as a surgical affair, but I cannot help thinking there is a medical aspect to it which may be worth while considering, especially as the mystery remains unsolved by the surgico-pathological theory. Admitting that intracranial hemorrhage from a blow without fracture is 'on the cards,' I contend that, considering the millions of knocks, blows, and punches on the head which occur, it is most rare. In all these cases the boxing has been of such a high-class nature that no really violent blow was received on either side. In the first case, that of TURNER, postmortem examination revealed no lesion whatever, the only peculiarity that could be spoken of was that the heart was smaller than the average. Yet he had won an open competition, meeting the picked men in England, only a week or two before. He was to be married the week after the contest at the *National Sporting Club* and alone supported a widowed mother. *What could he have died from but nerve collapse due to over excitement and anxiety?* Yet public opinion put it down very naturally to boxing, pure and simple – killed by it. In the cases of RILEY and SMITH the necropsy showed intercranial hemorrhage with no external marks of violence, much less fracture. Now I put it in this way: if this 'case' last week had been brought into the hospital, not as one from the National Sporting Club, but one found insensible, sitting on a doorstep, of course the strictest search would be made for signs of outward violence; finding none, post-mortem examination shows intercranial hemorrhage. How would death be certified – from 'violence' or from 'natural causes?'"

"Apoplexy may be uncommon in the young and apparently healthy, but I contend that nerve tension would be much more likely to cause it than any outward violence, though the surgical evidence was that the hemorrhage could not have arisen without external concussion. I have witnessed practically every round of the amateur boxing championships for the last 21 years, where infinitely more severe punishment is given and taken than in these THREE cases. *St. James's Hall* is packed from floor to ceiling, any amount of shouting and chaff being indulged in by the spectators, the din being terrific at times, yet never has anything serious happened to the boxers. Can it be that the very strict rule of 'silence during the boxing' at the National Sporting Club has the effect of making the contest a much more intense mental strain to the competitor and conducing, therefore, towards nerve collapse and apoplexy? In all other clubs spectators are allowed to cheer or otherwise during the boxing, thereby letting off the steam of excitement for onlookers and competitors. The officials of the club must indeed be almost driven crazy at finding their well-intentioned efforts to provide for the safety of the boxers end in such lamentable failure, and to sportsmen who take an interest in boxing these deaths have been very pathetic and heart-breaking. One does not get 'used' to these kind of sad endings to contests fought out, as they almost always are, with the best of good feeling and even generosity."

The first of the three cases reported by ORTON was the same boxer who was already described in the literature by TROUTBECK. Here the author noted that the postmortem examination revealed no lesion whatsoever. The only peculiarity was that the heart was smaller than average. The 'blood clot' on the right side of the brain, mentioned in Mr. TROUTBECK's autopsy report, is not mentioned here. The question asked by ORTON, "what could he have died from but nerve collapse due to over excitement and anciety?" This argumentation must be considered in the historical context; just one hundred years ago, it was an accepted theory in medicine. Yet public opinion, attributing the death very naturally to boxing, 'pure and simple,' and killed by it, was not accepted in the medical community.

Autopsy in the other two cases, RILEY and SMITH, showed intercranial hemorrhage with no external marks of violence, much less fracture. It is highly probable that these were also subdural hemorrhages.

ORTON raised the question whether the *very strict rule of 'silence during boxing'* at the *National Sporting Club* had the effect of making the contest much more intense, a mental strain to the competitor, causing nerve collapse and apoplexy. ORTON noted: "In all other clubs spectators are allowed letting off the steam of excitement for onlookers and competitors. The officials of the club must indeed be almost driven crazy at finding their well-intentioned efforts to provide for the safety of the boxers end in such lamentable failure..." The answer is simple, a blow or blows to the head, producing rotational acceleration as the result of an uppercut or hook, results in ruptured bridging veins with following subdural hemorrhages. Mr. LANG, the first physician to examine the injured boxer immediately after the bout, described "inequality of his pupils." With this observation he came very close to the right diagnosis!

Another editorial in *The Lancet* (1902) dealt with death from fisticuffs:

"The coroner's jury at Hackney which recently inquired into the death of a man named BARNETT and returned a verdict of 'death by misadventure' showed more good sense and discrimination than are always exhibited on such occasions. The deceased and a man named SMITH had a quarrel originating in a discussion over the meaning of the word 'punter,' and the discussion of this somewhat trifling matter – which perhaps makes the prevalence of betting the originating cause of the tragedy – seems to have aroused considerable ill-feeling in the mind of BARNETT. After some preliminary scuffling it was arranged that the TWO men should formally fight later in the day, BARNETT throughout being the aggressor and SMITH showing himself unwilling to continue the quarrel. In the end, however, SMITH proved himself the better man with his fists, and, after what appears to have been a perfectly fair fight in the presence of a number of his fellow workmen, BARNETT received a blow on the jaw which rendered him insensible. Very soon afterwards he died. The medical evidence was to the effect that BARNETT had been severely punished about the head and face and that the blow on the jaw had probably caused concussion of the brain, while the fall had ruptured a blood vessel on the brain. In such circumstances men before now have been committed for trial for manslaughter and indicted for... they did 'feloniously kill and slay' their opponents. The verdict of the jury, found under the direction of Dr. Wynn WESTCROTT, coroner for North-east London, seems to recognize the fact that if rough men choose to settle their disputes with their fists a fatal occurrence is not necessarily chargeable against the survivor as a crime. It is for the pistol, the knife, and the attack by an organised gang as modes of settling quarrels between men or boys that the criminal law should be set in motion. The fist, as the occurrence we are commenting on shows, may, guided by an unhappy chance, be a deadly weapon when wielded against a man possibly not in the best physical condition and not trained to receive or to avoid blows dexterously; but the annals of the prize-ring are sufficient to show the extraordinary amount of battering with the bare fist that the head and body of a man will take without injury when he is in sound health, practiced, perhaps, to avoid the more dangerous strokes, and injured to receive blows as well as skilled to deliver them. It no doubt weighed with the Hackney jury that the man whose fate was in its hands had not been the aggressor but must be taken to have had no choice between submitting to the bullies and fighting in self-defence."

In this case the coroner noted that BARNETT had been severely punished about the head and face and that the blow on the jaw had probably caused concussion of the brain. One could add that unconsciousness occurred and due to the fall a blood vessel of the brain ruptured.

Again anatomical details of the location of the hemorrhage are not presented. Also the author does not mention whether it was a bare knuckle fist fight. A subdural hemorrhage is very probable.

The following case histories with autopsies were published by KOHLRAUSCH (1921):

Case 1: Seventeen year old muscular, healthy looking *amateur boxer* won a bout by points, in which slight blows are apparent on both sides. The opponent, smaller of the TWO, connected in the second round with some light blows. A few minutes after the bout the boxer began to vomit, and became unconscious a few minutes after that. Pulse 70/min and regular. Clinically no abnormal findings of the heart, pupils the same size on both sides and react to light. The corneal reflexes were abolished after one hour and 45 minutes. Five hours and 15 minutes after the bout, dilation of the left pupil was observed. Patient succumbed before a craniotomy could be performed.

Autopsy confirmed a left-sided acute subdural hematoma over the entire cerebral hemisphere of about 100 ccm. The pole of the left temporal lobe revealed a circumscribed subarachnoid hemorrhage. Otherwise unremarkable findings in the brain and body organs.

This amateur boxer had identical pupils that reacted to light after the fight. Five hours and 15 minutes after the bout, dilation of the left pupil was observed. The patient succumbed before a craniotomy could be performed. Here, an acute subdural hemorrhage over the entire left cerebral hemisphere was correctly diagnosed.

Case 2: The boxer participated in a *training bout* and received blows against the stomach region, thorax, and chin. The last-mentioned blow caused a *knock out.* The boxer felt nausea about 15 minutes later and had to sit down. Decreasing consciousness. A physician began cardiopulmonary resuscitation. *Admission to a hospital,* pulse 70/min. Corneal reflexes were absent on both sides, pupils showed no reaction to light. No pareses: later Cheyne-Stoke respiration. Cyanotic face and hands. The pulse could no longer be felt; patient succumbs after 3 hours and 30 minutes.

The *autopsy* revealed an acute right-sided subdural hematoma and a fresh endocarditis.

The boxer had an acute right-sided subdural hematoma and fresh endocarditis. The survival time was 3 hours and 30 minutes. Whether the subdural hematoma was the result of a boxing blow against the head or resulted from the fall due to the knock out cannot conclusively be determined.

Case 3: The boxer was hit by a blow against the chin and was immediately unconscious. He died the same night without regaining consciousness.

Autopsy revealed a cerebral hemorrhage.

In this case the autopsy report stated that there was a cerebral hemorrhage. Data about the boxing match and location of the cerebral hemorrhage are not presented. Without doubt this boxer died due to an cerebral hemorrhage sustained in a boxing match.

FRAENKEL (1922):

Nineteen year old *amateur boxer* who fought a very successful boxer in his first public boxing bout. The boxer tried to cover his face with both arms and gloves while his opponent connected with a series of hard blows against his head. Finally he began to stagger, his face covered with blood, and was knocked out. He had to be carried out of the ring, and died 6 hours 30 minutes later in a hospital without regaining consciousness.
The *autopsy revealed* a left-sided subdural hematoma.

This inexperienced amateur boxer – it was his first public boxing fight – fought a very successful opponent who connected a series of hard blows against the head; he was knocked out. He became unconscious in the ring and had to be carried out; he died 6 hours and 30 minutes later. Undoubtedly, this young amateur boxer was overmatched.

The autopsy revealed a left-sided acute subdural hematoma. One can suspect rupture of bridging veins.

Von SURY (1922):

Twenty-five year old *amateur boxer* participates in a training bout. In the second round he receives a blow against the left mandible and lateral neck region. A few minutes later, after a break, and before the third round begins, sudden collapse and unconsciousness.
Autopsy revealed hemorrhages in both sternocleidomastoid muscles, more on the left than on the right, clotted blood around the left carotid artery, and hemorrhages over the larynx. The vocal cords were constricted and dark red in color.

The *autopsy of the brain* revealed subarachnoid hemorrhages over the left hemisphere and over inferior aspects of the left cerebellar hemisphere. Otherwise unremarkable brain.

This 25 year old amateur received a blow against the left mandible and the lateral neck region; he collapsed and became a few minutes later unconscious. Hemorrhages in both sternocleidomastoid muscles, left more than right, could be seen at autopsy. Clotted blood was visible around he left carotid artery, and hemorrhages over the larynx. We do not know how long this patient survived. The question is whether the hemorrhage around the left carotid artery lead to a thrombotic occlusion of the vessel and a beginning necrosis or hemorrhagic necrosis of the brain developed in the area supported by this artery. Only subarachnoid hemorrhages over the left cerebral hemisphere and the inferior aspects of the left cerebellar hemisphere are described. The question of whether this was an early stage of hemorrhagic necrosis cannot be answered.

MUNCK (1923):

Case 1: In a training bout, a 23 year old boxer was hit against the head and immediately knocked out. Four days later he won a boxing bout in a superior manner. He was not hit by any blows worth mentioning. The next night he boxed again, the first two rounds without incident. In the third round he was hit by a right hook in the area of the left mandibular angle and was knocked down. At the count of '9', he was on his legs again and continued the bout. Shortly after, he was hit again by the same blow and knocked down again to the count of '9'. It is reported that on neither occasion did his head connect with the ring canvas. He continued to fight, then began to stumble and collapsed unconscious.
He was admitted to a hospital one hour later; pulse 40/min. Both pupils same size. Next day jerks of arms and legs and bladder incontinence. Fourteen hours after the beginning of unconsciousness, irregular respiration began, pulse 144/min; patient succumbed three hours later.
Autopsy revealed subdural hematoma over the right cerebral hemisphere.

This boxer received blows against his head in a training bout and was knocked out. Four days later he won a boxing bout in a superior manner. The following night he boxed again. While the first two rounds were unremarkable, he was hit in the third round by a right hook on the left side of the mandibular angle, which knocked him down. After a count of '9' he continued the bout. He was hit with the same blow and knocked down again to a count of '9'. It was reported that on neither occasion did his head connect with the ring canvas. He continued the fight, began to stumble and collapsed unconscious.

It bears repeating that this boxer participated in three boxing matches on five consecutive days. In the first match he was knocked out, in the third he took two hooks on his right mandibular angle, both leading to knock downs. The two hooks certainly must have produced considerable rotational acceleration of his head. On admission to hospital, pupils were allegedly identical. He died 17 hours later.

The autopsy revealed an acute subdural hematoma over the right cerebral hemisphere due to ruptured bridging veins is probable. The case history reveals how dangerous boxing bouts can be which occur in short intervals.

Case 2: Nineteen year old boxer in boxing match that should have extended over six rounds, two minutes each. He was knocked down in the fifth round but was able to get to his feet without help and go to his corner, where he reported fatigue. Shortly after the start of the sixth round he was hit with the same type of blow, against the left side of his chin. The intensity of the blow made him fall backwards and he hit the canvas with his head and neck. He immediately became unconscious, with irregular respiration, and 20 minutes later succumbed due to respiratory arrest.
The *autopsy revealed* subdural hematomas over both cerebral hemispheres. The impact of his occiput at the canvas caused the bilateral subdural hematomas due to ruptured bridging veins.

The boxer probably developed the subdural hemorrhage in his first bout. In the second bout he suffered a recurrence of the hemorrhage despite relatively light blows. He was knocked down in the fifth round, was able to get on his feet without help and went to his corner. Shortly after the sixth round began, he was hit again by the same blow against the left side of his chin. The intensity of the blow made him fall backward and he hit the ring canvas with his head and neck. He became immediately unconscious and succumbed 20 minutes later, due to respiratory arrest. The bilateral subdural hematomas were peracute. It is possible that the blows during the boxing match produced ruptured bridging veins and the subdural hemorrhage, followed by a second severe impact (deceleration trauma) from the impact of his head on the canvas, after falling backward. The second impact was undoubtedly more severe, but it is impossible to say which impact injured the bridging veins. Another possibility is, that the boxer developed a subdural hemorrhage in the first bout and blows in the following bout affected a brain with a subdural hemorrhage.

WOLFF (1928):

An *amateur boxer,* lightweight, is hit (in an elimination bout) by hard blows; subsequent headache. He begins the following bout with headache. He was knocked down in the second round due to blows against the abdominal region, as a newspaper reported, "without any reason." The blows against his head were, according to some knowledgeable spectators, not severe. He gave up the bout in the break after the second round. He had to be carried out of the ring unconscious. A physician transferred him to a hospital where he remained unconscious for the next three days. He developed a left-sided oculomotor paresis. A subdural hemorrhage was diagnosed. The patient was released after four weeks; the oculomotor paresis had disappeared, his sensorium was clear. Lues reactions in blood and cerebrospinal fluid were unremarkable.

WOLFF (1928):

Case 1: A *professional boxer* of unknown age was hit in the tenth round of a comparatively easy fight, by a fairly light blow against the chin and stomach region. The fight continued for a few seconds, until the patient collapsed and became unconscious. He had to be carried out of the ring; face cyanotic. Succumbed shortly thereafter.
Autopsy revealed a severe luetic vascular process of heart and aorta. No injuries to the brain or enveloping structures.

This boxer did not succumb to brain injuries received in a boxing match. A severe luetic vascular process of heart and aorta was responsible for the death. A *physical examination* with serological tests should have revealed the diagnosis.

Case 2: A 21 year old amateur boxer who had previously only fought in training bouts. In his first public fight against a superior boxer (8 oz. gloves), he was hit in the second round by a blow to the stomach region. He went home unassisted but later sought the services of a physician. He complained of abdominal pain; surgery was performed on a ruptured intestine. The patient died the following evening.
Autopsy revealed a 2 cm long rupture of the intestine, about 80 cm distal of the pylorus; peritonitis was present. There was no other organic alteration, such as an ulcer, in the traumatized area.

This 21 year old amateur boxer, who participated

in his first boxing match, was hit in the second round by a blow to the stomach region. He complained about abdominal pain, an operation was performed for a ruptured intestine, he died on the following evening.

The post-operative autopsy revealed a 2 cm long rupture of the intestine, about 80 cm distal of the pylorus, with peritonitis. Rupture of the intestine due to a boxing blow is a very rare complication. This boxing bout took place before sulfonamids and antibiotics were available.

Case 3: A 17 year old *amateur boxer* joined a boxing club and began training. He asked to participate in a real boxing fight and was matched to fight against a boxer who was heavier by 10 lbs. (4,5 kg), using 8 oz. gloves. During the fight, he was hit on the left temple by a powerful hook. He continued the fight. Twenty-four hours later he complained of headache and vomited, which continued for the next three days. Headaches continued for the next three to four weeks. Five weeks later he went back to work as a locksmith's apprentice. In the following weeks he again experienced headaches, especially at night, and mostly in the frontal area. On admission to hospital: unremarkable clinical findings, fully oriented in time and space. During hospitalization, he continually complained of headache. Sudden death followed.
Autopsy revealed a massive subdural hemorrhage, of 1 to 1 1/2 cm thickness, over the entire middle part of the left cerebral hemisphere, not touching the frontal or occipital lobes. This hematoma extended into the base of the skull. A massive shift of the left cerebral hemisphere to the right was seen. The left cerebral ventricle appeared as a slit. The right cerebral hemisphere was compressed, the gyri flattened and the sulci shallow.
Histological examination showed a thickening of the dura mater with a proliferation of capillaries and hemosiderin pigment. The hematoma was located between two membranes of connective tissue.

This 17 year old amateur boxer joined a boxing club and began training. He wanted to participate in a real boxing bout and was matched against a heavier boxer (4,5 kg). His opponent landed a heavy hook on his left temple. He continued the fight. Twenty-four hours later he complained of headache, and he had to vomit. Headaches continued for the next four weeks. Five weeks later he began to work again as a locksmith apprentice. In the following weeks he experienced headaches, especially in the frontal area.

On admission to hospital there were unremarkable clinical findings. He died suddenly.

At autopsy a massive subdural hematoma was found over the entire middle parts of the left cerebral hemisphere, not extending into the frontal and occipital lobes, of 1 to 1 1/2 cm thickness. The hematoma extended into the base of the skull. There was a massive shift of the left cerebral hemisphere over the midline structures to the right. The right cerebral hemisphere was compressed, the left cerebral ventricle appeared as a slit. It was a chronic left-sided subdural hematoma, which was clearly a space-occupying lesion with compression and shift of the brain to the other side. This boxing match took place at the time when Egaz MONIZ (1927) published his arteriography technique.

WEIMANN (1931):

During an *amateur boxing fight* between two teams, in a lightweight bout in the juvenile group, the 15 year old *amateur* showed good defensive action during the first round. In the second round his opponent increased his attacks, especially with left straight blows, and connected at one minute 30 seconds with a right hook against the left side of the chin. The boxer was immediately knocked down – he fell straight backwards and hit the canvas with his occiput. (Canvas covered with 1.5 cm thick felt and tarpaulin). He was counted out and carried unconscious to his corner. Revival was attempted in the usual manner, with cold water. He was carried out of the ring and transferred to hospital 15 minutes later. Shortly after admission to a hospital, about 30 minutes after the knock out, exitus lethalis.
Autopsy showed subgaleate hematoma in occipital region. Massive subdural hematoma over both cerebral hemispheres.

In this case the *amateur boxer* in the lightweight class was hit in the second round with a right hook which connected with the left side of his chin. The force of the blow knocked him down; he fell backwards and hit the canvas with his occiput. He was counted out and carried unconscious to his corner. He was immediately transferred to a hospital where he succumbed 30 minutes after the knock out.

The autopsy revealed a subgaleate hematoma at the occipital region (deceleration trauma), the area of impact of his head at the canvas. There was a *massive peracute subdural hematoma over both cerebral hemispheres.* The possibility that the right hook started the bilateral subdural hematoma before the head hit the canvas cannot be excluded with certainty, but it is highly probable it was caused by the occiput impacting against the canvas. In such a deceleration trauma cortical contusion like injuries can occur at sites opposite the impact point – par contrecoup. Although they are not described in the autopsy report, the canvas cover of felt and tarpaulin probably attenuated the fall, so that the threshold for cortical contusions, par contrecoup, was not reached. But a rotational component in this impact ruptured bridging veins on both sides of the cerebral hemispheres, producing these very massive and deadly peracute subdural hematomas.

Von MARENHOLTZ (1932):

> The boxer was hit in the second round by a severe blow from the right, against his left mandible, in the area of the left foramen mentale. He fell backward like a tree with a completely rigid body, and hit the canvas with his occiput. He immediately became unconscious. There is no data about his survival time.
> *Autopsy revealed* a left-sided subdural hemorrhage. The central cortex revealed bilateral subarachnoid hemorrhages near the midline.

This boxer was hit in the second round by a severe right blow against his left mandible, probably a right hook. The boxer fell backward like a tree with a completely rigid body, and hit the canvas with his occiput. He was immediately unconscious. There are no data about his survival time.

The autopsy showed a left-sided subdural hemorrhage and there were bilateral subarachnoid hemorrhages near the midline. Again, one cannot exclude the possibility that the right hook produced rotational acceleration with rupture of bridging veins, but it is more probable that the backward fall with impact of his occiput on the canvas produced the left-sided subdural hemorrhage. It was a classic deceleration trauma.

The case report said that the boxer fell backward 'like a tree, with a completely rigid body.' Such cases have been reported in the literature. Fig. 226 shows such a boxer falling backward with a completely rigid body. There are no data available how frequently such falls occur.

TYRELL (1932):

> The following is an *excerpt from the file of a public attorney.* " The deceased asked me several times to box with him. I refused several times but finally started a boxing match with him. In general I defended myself and did not connect with a single severe blow. In the third round I connected with a blow to the thorax region – he staggered and slowly collapsed. I helped him up and put him on a chair. He gasped for air, fell from the chair and became cyanotic. I contacted a physician, but after I had done that he was dead. The physician arrived and certified that he was dead."

Autopsy revealed a 2 x 2 cm hematoma in the left pectoral muscle. Otherwise unremarkable.

We do not know whether boxing gloves were used or the match was a bare knuckle fist fight.

The autopsy report mentions only a hematoma in the left pectoral muscle. It is not clear whether a cardiac contusion existed, despite the autopsy report which noted 'otherwise unremarkable.'

FOERSTER (1932):

> The patient was a *'plant' in the audience around a boxing booth,* hired by contract with the owner to fight the boxer in the booth. On this particular day, he had 'challenged' a booth boxer six times. He was knocked out by his opponent and remained unconscious. He died six to seven hours later without regaining consciousness.
> *Autopsy revealed* a marked subdural hematoma over the right cerebral hemisphere. Clotted blood of about 1/2 cm thickness was found over the left cerebral hemisphere, in the area of the central gyrus, and at the base of the skull.

This is a booth boxer, a 'plant' in the audience, who was hired by contract to box, challenging boxers to exhibition fights in the booth. He had already 'challenged,' six booth boxers, making the last match his seventh of day! His opponent knocked him out, and he died six or seven hours later, without regaining consciousness. Details of the fight and his clinical findings are not known.

The autopsy revealed a marked subdural hematoma over the right cerebral hemisphere, extending to the base of the skull. It is possible that the injuries were inflicted during one of the earlier bouts, perhaps a day or several days earlier, since it existed a large subdural hematoma.

WERKGARTNER (1935):

Twenty year old amateur boxer. During a training session the trainer demonstrated to the pupil a boxing combination of feinting and the so-called solar plexus blow. Trainer reported that trainee reacted clumsily and ran the tip of his chin into the second blow after the feinted one. Pupil reacted with a counterblow but then took off gloves and got dressed. He reported nausea, vomited, and became unconscious after a few minutes. Succumbed an hour later.
Autopsy showed a subdural hematoma over the left cerebral hemisphere in the area of the central cortex, also at the base of the brain. Shift of left cerebral hemisphere to the right side. Slit-shaped cerebral ventricles. The tentorium cerebelli ruptured a little to the left of the midline. The straight sinus was also ruptured.

This injury occurred in a training bout. After a combination of feinting and a so-called solar plexus blow, the pupil reacted clumsily and ran the tip of his chin into the second blow after the feinted blow. There was no knock down. The pupil got dressed, reported nausea, vomited, and after a few minutes became unconscious. He died an hour later.

The autopsy revealed a *peracute subdural hematoma* over the left cerebral hemisphere. It had already developed into a space-occupying lesion, extending to the base of the brain. The left cerebral hemisphere had begun to shift to the right side. The intracranial pressure had already compressed the lateral ventricles. The straight sinus and the tentorium cerebelli, a little to the left of the midline, were ruptured. In this case the bleeding came from venous sinuses.

KAPPIS (1938):

A 15 year old *amateur boxer* fought against a 13 year old, who was smaller and weaker. The 15 year old was hit by a blow against the left mandible, staggered and gave up the fight. Immediately complained of headache, sat down, and five minutes later collapsed and became unconscious.
On admission to hospital two hours later, the diagnosis was increased intracranial pressure. Intravenous application of hypertonic solutions. Eight hours later pulse and respiration increased. Two hours later, deterioration with exitus.
Autopsy showed bilateral subdural hematoma greater on the right than on the left.

The blow against the left mandible probably produced a rotational acceleration of the head with tears of bridging veins and an acute subdural hematoma.

SCHEPPACH (1938):

An 18 year old *amateur boxer,* middleweight class, was used as a replacement against an opponent who was known as a good boxer with good technical skills. Eight-ounce gloves were worn for three rounds of three minutes each.
The amateur boxer was leading in the first round by points, but in the second round he was hit by a hook to the chin which showed some effects. He went into a clinch with his opponent, and during this time was hit by a short left hook to the chin. Both boxers fell to the canvas. The patient did not hit the canvas with his head. He was counted out, got to his feet on his own and reported that he could have continued the fight but did not want to because a tooth was knocked out. *Immediate examination* revealed fracture of the mandible on the right side. *Admitted to hospital.* Death five weeks after bout.
Autopsy revealed an operative scar of 9 cm length, in area of right mandibular angle. There was a metallic splint in the area of the right mandible. Sigmoid sinus on the right side showed pus like whipped cream. Tissue around sella turcica was gelatinous and putrified. Muscle below foramen magnum on left side putrified and showed cavities filled with pus. SCHEPPACH concluded that the boxer died of sepsis related to the fractured mandible.

A fractured mandible was found.

Admission to a hospital, he died five weeks after the boxing bout. In the area of the right mandibular angle, a long operative scar (9 cm) was visible. There was a metallic splint in the area of the right mandible. The sigmoid sinus on the right side revealed pus, and the tissue around the sella turcica was putrified. The boxer died of sepsis related to the fractured mandible.

CARR and MOODY (1939):

Case 1: Professional boxer. In the second round, struck apparently on the jaw and knocked down. Rose shortly and finished the round. In the following two rounds, seemed dazed and to have lost some coordination. During these two rounds his head was struck repeatedly by his opponent, rocking it back and forth from side to side. In the fifth round he was hit by another sharp blow, momentarily losing consciousness, and he turned clinging to the ropes. Struck sharply from behind just below the occiput, forcing his chin abruptly against his chest. Slumped on the ropes and hung there, receiving several more blows before the referee disengaged his opponent. Carried from ring, did not regain consciousness. *Taken to hospital,* where he remained unconscious and died, before he could be examined, three hours after the knock out.

Autopsy revealed considerable brain edema, scattered small contusion-like hemorrhages beneath the cortex of the frontal, occipital and parietal lobes, and marked hemorrhages of a similar type in the brain stem extending from the base into the medulla. The cerebellum also contained scattered small hemorrhages. *Diagnosis:* Subpial intracerebral and pontine hemorrhages with characteristics of those occurring as the result of indirect violence.

This professional boxer was knocked down in the second round, he rose and finished the round. He seemed dazed in the next two rounds and seemed to have lost some coordination. In these two rounds his opponent landed several blows which rocked his head backward and forced it from side to side. In the fifth round, he was hit by another sharp blow, and he immediately lost consciousness. Slumping and hanging on the ropes, he received several more blows and was hit by a blow from behind, just below the occiput. It was an illegal, so-called 'rabbit punch' that forced his chin abruptly against his chest. The referee should have disqualified his opponent. He did not regain consciousness and had to be carried from the ring. He was transferred to a hospital and died there before he could be examined, three hours after the knock out.

The autopsy report mentions considerable brain edema, and scattered contusion-like hemorrhages beneath the cortex of the frontal, occipital and parietal lobes and marked hemorrhages of a similar type in the brainstem extending from the base into the medulla oblongata. These were probably no so-called cortical contusions, but a beginning hemorrhagic necrosis. The brain was not examined histologically, so the diagnosis could not be secured.

Case 2: Professional boxer. In the third round he was knocked out for 10 to 15 seconds by a sharp blow to the head. After he was counted out, he walked to his corner and seemed normal.

Two days later, after a few groans, he lost consciousness. He was taken to a hospital immediately, in a deep stupor; on admission showed no response to painful stimuli. Reflexes about equal, possibly a little more active on the left. Bilateral Babinski, bilateral ankle clonus, better sustained on the left than on the right, both pupils dilated and fixed. All superficial reflexes were absent. Alternating periods of restlessness and deep stupor, with irregular, varying respiration; periods of restlessness from 8 to 30 minutes. Pulse regular, but varied in rate with respiration from 20 to 48. Spinal tap released frankly bloody fluid under pressure of 198 mm/H_20. About 15 ccm were removed, after which pressure down to 100 mm /Hg. Patient placed in respirator to stabilize respiration and pulse. Four days after injury, frontal craniotomy was done without anesthesia. No pulse, respiration only by respirator. When dura opened, no pulsation could be seen. Patient was obviously dead; no attempt was made to clean out hematoma.

The *autopsy* revealed a very edematous brain, but no subdural hematoma. Brain symmetrical, convolutions flat and sulci shallow. Fresh hemorrhage from several small areas of contusions in the pons and into medulla. Blood had passed from contusions into the fourth ventricle and into spinal canal.

This professional boxer was knocked out in the second round by a sharp blow to the head. He was counted out, was for about 10 to 15 seconds unconscious and walked after regaining consciousness to his corner and seemed normal.

Two days later, after a few groans, he lost consciousness and was immediately transferred to a hospital. There were pathological pyramid signs, left more that right. Both pupils were dilated and fixed. The patient was placed on a respirator to stabilize respiration and pulse.

The autopsy revealed a very edematous brain, but no subdural hematoma. There were fresh hemorrhages described, allegedly rising from several small areas of contusions in the pons and extending into the medulla oblongata. Localization of the so-called contusions is very unusual. Contusions of the pons and midbrain do not exist; the terminus 'brainstem contusion' is an inadequate term. These traumatic lesions are the result of temporal lobe herniations at the free edge of the tentorium cerebelli, producing incisures at the crus. The term 'boxer's hemorrhage' is not adequate, since these alterations are also seen in patients with head injuries unrelated to boxing. The authors stress the point that they have collected an additional series brain contusion hemorrhage cases. These traumatic lesions are probably hemorrhagic necroses in an early stage. The cases described by COURVILLE, as so-called 'diffuse contusions' are not contusions, but are hemorrhagic necroses. These hemorrhagic necroses were described quite frequently in boxers who were injured in the ring and died later. Also the traumatic alterations described by OSNATO and GILIBERTI (1927) and MARTLAND and BELING (1929) are hemorrhagic necroses.

CARR and MOODY concluded that these cases were the result of severe beating of the boxers, with the individual (at least in the terminal stages of the beating) only partially conscious, or, in the boxing sense, 'out on their feet.'

"Because these first three cases which we encountered have occurred in connection with fisticuffs of some nature, the specific and limited injury is regarded as being peculiar to boxers and fighters – to those people who are involuntarily beaten, or to others where the type of injury causes acute angulation, and pinching of the pons and medulla over the tentorium. The term 'boxer's hemorrhage," therefore, is suggested for these injuries" (CARR and MOODY).

CARR and MOODY quote other authors who described cases of fatal head injury, in which similar hemorrhages were found. OSNATO and GILIBERTI (1927) reported a case in which the period of survival was 36 hours; there was no fracture of the skull. These authors found petechial hemorrhages scattered through the centrum ovale, the corpus callosum and pons. MARTLAND and BELING (1929) reported 23 cases of multiple deep hemorrhages in 309 fatal cases. GONZALES et al. (1937) have found multiple small hemorrhages of petechial size associated with violence producing a sudden jarring of the head, but still not of sufficient moment to cause a fracture of the skull. They indicate that in these cases the brain is made to oscillate, and contuses itself against the skull and the different reflections of the dura like the falx cerebri and tentorium. They believe that such movement causes the pons and cerebral peduncles to be injured against the sharp edges of the tentorium, sometimes to the point of complete laceration. In some instances they believe that the oscillation of the brain inside the skull might give rise to punctate hemorrhages throughout both the gray and white matter of the cerebral hemisphere.

In a discussion of the paper of CARR and MOODY SCHALLER (1939) commented that brainstem hemorrhages, especially punctile hemorrhages following trauma, have been difficult to explain because of the excellent protection of this region of the brain, lying deep-seated and resting on the water cushion of the basal cisterna. According to SCHALLER, these hemorrhages have been regarded as venous in origin, due to stasis and subsequent rupture of an impaired vessel wall. The stasis has been explained by thrombosis of larger vessels, such as the veins of Galen or the venae striae terminalis, to which the ruptured vein is contributory, or by the mechanism of local vasodilatation and stasis.

An important step forward in explaining the mechanogenesis of these lesions were the findings in an paper written by KERNOHAN and WOLTMAN 1929) on *'Incisura of the crus due to contralateral brain tumor.'* The grooves caused by the pressure of the tentorium are often associated with petechial hemorrhages, as described by these authors.

MORRISSEY (1939), in a discussion of the paper by CARR and MOODY, expounded upon the petechial hemorrhages, especially those localized in the brain stem. He belied that they offer an excellent explanation for the variety of symptoms and neurological findings noted in boxers who have survived several severe beatings in the prize ring, often referred to under the illicit but popular term 'punch-drunk.'

MORRISSEY correctly noted that the two day symptom-free interval in case 2 of CARR and MOODY would indicate delayed hemorrhages. We agree with this author fully when he writes, "Although these petechial hemorrhages of the brainstem, as pointed out, are rather characteristic of the cerebral trauma resulting from blows about the head received during fistic encounters, they do occur in other types of head injuries, and therefore the advisability of introducing the term *'boxer's hemorrhage'* might be questioned."

JOKL (1946):

Case 1: This was a young soldier, in an amateur boxing match. In the second round, a blow against the jaw dazed him so that he was counted out on his feet and lost the bout. In the dressing room the patient became disoriented and complained of a severe headache. Condition deteriorated rapidly, and he *arrived at the hospital* unconscious. Cheyne-Stokes breathing, slow pulse (50/min), temperature 100° F. Both pupils were dilated and did not react to light. A lumbar puncture revealed bloodstained CSF.
A middle meningeal hemorrhage was diagnosed; an operation was performed immediately. The dura mater was under considerable tension, and on incision much fresh blood escaped. An underlying laceration of the brain tissue was seen. At this stage of the operation the patient stopped breathing and died shortly afterwards.
Autopsy showed a two inch long, one-half inch deep laceration of the left temporal lobe. Extensive intracranial hemorrhage.

The description of the findings is not clear. A hemorrhage of the middle meningeal artery was diagnosed, so one could expect an epidural hemorrhage. But through an incision of the dura mater escaped fresh blood, therefore is was a subdural hemorrhage. A laceration of the underlying brain tissue was seen. The patient died in tabula.

Case 2: A 20 year old Marine was knocked out, he regained consciousness a few minutes later. There was a chronic subdural hematoma over the right cerebral hemisphere. A right subtemporal craniotomy was performed and the patient was able to return to duty.
Case 3: A Navel rating, 20 years old, was taking part in a Service boxing contest. After he had received a certain amount of head punishment he was knocked down twice for short counts, a decision of t. k. o. was givin against him. The diagnosis was *'early delayed traumatic subarachnoid hemorrhage.'* There are no data, whether he survived, but since no autopsy report is given, he probably did.

KREFFT (1952):

Case 1: A 23 year old boxer was hit in the first round by a hard left against the head and knocked down to a count of '5'. His opponent drove the boxer in front of him and connected with a series of blows to the body and head. An especially severe blow to the liver brought his defense down. His opponent then connected with a second hard blow to the head. The boxer went down, turning as he went down and supporting himself with his gloves on the canvas. After a count of '1' he came to his feet, but as he did not take a position, the fight was stopped.
After the fight the boxer was benumbed. He ate a piece of bread and was driven in the 'buddy seat' of a motorcycle to his home, 25 km away. He immediately lay down, vomited, and complained of a severe headache. Shortly after he became unconscious. *When admitted to hospital,* he was deeply unconscious and displayed Cheyne-Stokes respiration. Both pupils were wide. The following hour the right pupil was wider than the left – an operation was performed on suspicion of an epi- or subdural hemorrhage on the right side. A palm-sized right-sided subdural hematoma of 4 cm thickness was removed. During the operation there was improvement of the clinical status with normal pulse frequency, but no change in consciousness. At four o'clock in the morning sudden increase of pulse to 116-128/minute. Motor excitation, vomiting. Patient succumbed at 5:30.
Autopsy revealed epi- and subdural hematomas over the right cerebral hemisphere. Generalized brain edema especially over the right cerebral hemisphere with smaller ventricles on the right. *Brain weight* 1,560 g. Lentil-sized hemorrhages immediately dorsal to right posterior horn; pin-sized to lentil-sized hemorrhages in pons.

The fight was stopped, in technical terms it was a t. k. o. When admitted to hospital, he was deep unconscious, displayed Cheyne-Stokes respiration and both pupils were wide. The following hour the right pupil was wider than the left. A palm sized right-sided subdural hematoma of 4 cm thickness was removed. The patient succumbed the next morning.

The autopsy revealed epi- and subdural hematomas over the right cerebral hemisphere. It existed a general brain edema especially over the right cerebral hemisphere. The massive acute subdural hematoma of 4 cm thickness could be considered a space-occupying lesion. Due to the increased intracranial pressure there had developed lentil-sized hemorrhages immediately dorsal to the right posterior horn and pin-sized hemorrhages in the pons. The qualitiy of the hemorrhages whether primary or secondary traumatic is not discussed. Furthermore it is not mentioned whether lesions of the uncus gyri hippocampi existed.

Case 2: A 27 year old *amateur middleweight boxer* fought in a county championship bout wearing 8 oz. gloves. Shortly before the end of the first round he was hit by a right hook to the liver that did not show an immediate effect. He continued to box, but then stepped back into the ropes and collapsed without receiving further blows. After the 10-count his second tried to lift him up. He was deeply unconscious, and had a gray-yellow discoloration of the skin, especially of the cornea. He was carried out of the ring, and a physician pronounced him dead.
Autopsy revealed a plumsized hemorrhage in the left pelvis of the kidney. There was a slight swelling of the brain.

KREFFT assumes a shock death to the injury.

Case 3: Boxing fight between two equally matched boxers. The first round was unexceptional. In the second round the boxer was knocked down to a count of '8'. Shortly after he was hit on the tip of the chin and fell backwards, with his occiput and back striking the canvas together. He was immediately unconscious, and had to be carried out of the ring. The examining physician *transferred him to a hospital.* Development of a right-sided hemiparesis with transfer to another hospital. On admission the patient reacted to questions but was not fully oriented as to time and space. Has swelling of galea over left ear and right sided paresis of both extremities and facial nerve. *Lumbar puncture* showed bloody cerebrospinal fluid. Pulse 50/min. No difference in the size of the pupils. Left *temporal craniotomy revealed* rupture of branch of medial meningeal artery which was ligated. No impression of the brain substance, no blood coagula. Patient remained unconscious. Four days later dilatation of left pupil. *Another left-sided epidural hematoma,* which indented the dura mater, removed. Late afternoon rapid deterioration, pulse 176/min, Cheyne-Stokes respiration. *Another left-sided epidural hematoma* was removed, patient died in tabula.
Autopsy revealed *status after removal of left epidural hematoma.* Convolutions flat and sulci shallow. Herniation of cerebellar tonsils. Subdurally in both middle cerebral cavities, a layer of coagulated blood about 3 mm thick. Multiple subarachnoid hemorrhages. Dove-egg sized cortical contusions of the 2nd and 3rd temporal gyri on the right side, additional similar lesions at base of right occipital lobe. General brain edema with compression of lateral ventricles and sanguinolent cerebrospinal fluid.

This patient was knocked down in the second round to a count of '8'. A short time later he was hit on the tip of the chin and fell backwards, striking the canvas with his occiput and back. It was a classic deceleration trauma. The immediately unconscious boxer had to be carried out of the ring and the examining physician transferred him to a hospital. Since a right-sided hemiparesis developed he was transferred to another hospital.

A left temporal craniotomy revealed a tear of the middle meningeal artery which was ligated. The patient remained unconscious. Four days later a dilatation of the left pupil developed and another left-sided epidural hematoma, which indented the dura mater, was removed. At late afternoon, a rapid deterioration developed and a recurrence of a left-sided epidural hematoma was removed, the patient died in tabula. At autopsy a generalized brain edema existed with compression of the lateral ventricles and dove-egg sized cortical contusions of the second and third temporal gyri on the right side, and additional similar lesions at the base of the right occipital lobe, par contrecoup existed.

JOKL (1954) reported a boxing injury allegedly to the midbrain:

Case 1: In the first round of a fight, a boxer received a powerful blow to the head. He finished the contest, winning on points. One hour later he became unconscious and never regained consciousness. During the ensuing three weeks pyramidal, extrapyramidal, diencephalic and mesecephalic symptoms were in evidence and at the 21st day after the fight the patient died.
At the *autopsy* tissue changes and softening were noted throughout the brainstem. No major hemorrhages were present.
Histological studies failed to reveal pathological changes of the blood vessels. The fatal condition had presumably been caused by vasomotor disturbances.

This boxer received a powerful blow to the head in the first round, but he finished the contest, winning on points. One hour later he became unconscious and never regained consciousness. During the ensuing three weeks pyramidal, extrapyramidal, diencephalic and mesencephalic symptoms existed. A detailed description of the clinical findings is, however, not given. At autopsy tissue changes and softening were noted throughout the brainstem. A detailed autopsy report is not presented, probably there was a necrosis of the brain tissue.

BERGLEITER and JOKL (1956):

Case 1: A 20 year old amateur boxer received heavy blows against the head in the first round. He lost the fight by points. As he left the ring to go to the dressing room, his friends noticed that he was confused and disoriented. He collapsed in the dressing room and was deeply unconscious when admitted to hospital. Slow pulse, irregular respiration, corneal reflexes bilaterally negative. Three days later patient became lucid, but soon after was deeply unconscious again. On the sixth day clonus-like movements of the left hand; exitus later that day.
Autopsy revealed a *massive hematoma over the right cerebral hemisphere.*

This 20 year old amateur boxer was severely punished. He collapsed in the dressing room and was deeply unconscious when admitted to hospital, where he died later that day.

At autopsy a massive hematoma over the right cerebral henisphaere was found. It was probably a subdural hematoma. Since the clinical findings were evident for a right-sided lesion of the brain, it is therefore unintelligible why a right-sided craniotomy was not performed.

Case 2: A 23 year old professional boxer was knocked down in the first round and got to his feet at the count of '5'; he was knocked down again in the fourth round, and got to his feet before the count of '10'. A few seconds later he was knocked out by a hook to the chin. He was unconscious for six days. *A subdural hematoma over the right cerebral hemisphere was removed.* At first he showed a clinical recovery, but after one week there was a recurrence of a right-sided subdural hematoma, which was removed, the patient died shortly after the operation.
Autopsy revealed a *recurrence of a massive subdural hematoma over the right cerebral hemisphere.*

This 23 year old professional boxer was knocked down several times in the first and fourth round. In the fourth round he was knocked out by a hook to the chin. Unconsciousness existed for six days. But finally a subdural hematoma over the right cerebral hemisphere was removed. At first the patient showed a clinical recovery, but suffered one week later a recurrence of the massive right-sided subdural hematoma which was removed, but the patient died shortly after the operation. It is unintelligible why the first operation was performed after six days of unconsciousness. An earlier operation was indicated.

Case 3: A 22 year old professional boxer participated in a bout scheduled for 15 rounds. In the fourth round he was knocked down for a count of '8' but continued to fight until the fifteenth round. Shortly before the end of the fight he was knocked down again, and his head hit the canvas. He got to his feet again, was knocked down a third time. He got up and was hanging on the ropes after a blow to the chin during which time he received more blows to the head. The fight was stopped at this point, and his opponent declared the winner by a technical knock out. The boxer became unconscious and was *transferred to hospital.* An *epi- and subdural hematoma over the left temporal cerebral cortex was removed,* but the patient succumbed after the operation.
Autopsy revealed a *marked subdural hemorrhage over the entire right cerebral hemisphere.* The origin of the hemorrhage seemed to be the surface of the brain, where small cortical contusions were visible. Infarctions were visible in the brainstem, in the area of the fourth ventricle.

This 22 year old professional boxer participated in a bout scheduled for 15 rounds. He was knocked down in the fourth round for a count of '8', but continued the fight until the fifteenth round. Shortly before the end of the fight he was knocked down again, and his head hit the canvas. It was a deceleration trauma. He suffered both an acceleration trauma due to the boxing blow and a deceleration trauma through the impact of his head on the canvas. He got to his feet again and was knocked down a third time. He got up and was hanging on the ropes after a blow to the chin. During this time he received more blows to the head. There is no doubt that he was 'groggy' and defenseless. The bout was finally stopped and his opponent was declared the winner by a technical knock out. This case shows convincingly how dangerous situations are, when a boxer is hanging draped over ropes. The referee should have stopped such a bout immediately, especially in a boxer who had previously taken severe punishment and had been knocked down several times. The already 'groggy' boxer became unconscious and was transferred to a hospital. An epi- and a subdural hematoma over the left temporal cerebral cortex was removed, but the patient succumbed after the operation. The autopsy revealed a marked subdural hemorrhage over the entire right cerebral hemisphere which originated from the surface of the brain, where small cortical contusions were visible. There were infarctions in the brainstem, in the area of the fourth ventricle. There are no clinical data given, therefore it cannot be determined on which side the contrecoup lesions were.

Case 4: A 17 year old boxer was hit by a heavy blow against the right temple, in the second round of a fight, but was not knocked down. He collapsed in the third round and became unconscious without receiving further blows against his head. He had a generalized tonic-clonic seizure in the dressing room. The patient died a few minutes after *admission to hospital.*
Autopsy showed a *marked subdural hematoma over the entire left cerebral hemisphere.*

Case: (intracerebral hemorrhage) A 20 year old healthy soldier was hit, in the second round of an *amateur tournament,* against the chin and was knocked out.

He was disoriented in the dressing room and complained of severe headache. Soon after he became deeply unconscious and was *transferred to hospital.* Cheyne-Stokes respiration, bradycardia, raised temperature (39°C). Both pupils dilated and showed no reaction to light. Bilaterally positive Babinski reflexes. *Lumbar puncture* showed sanguinolent cerebrospinal fluid. Hemorrhage of the middle meningeal artery was diagnosed and a trepination was performed. The dura mater was under great tension, and after incision a large amount of fresh blood was removed. Patient died during operation, due to respiratory failure.
Autopsy revealed a 5 cm long, 2.5 cm *deep lesion in the left temporal lobe. Marked intracerebral hemorrhage* without fracture of the bone.

This 20 year old healthy soldier who participated in an amateur tournament received blows in the second round against his chin and was knocked out. He was deeply unconcious and was transferred to a hospital. Both pupils were dilated and showed no reaction to light. A hemorrhage of the middle meningeal artery was diagnosed, and a trepination was performed. The side is not reported. But now it is reported that the dura mater was under great tension and after incision a large amount of fresh blood was removed. The patient died in tabula. It was therefore a subdural hemorrhage, while first a hemorrahge of the middle meningeal artery was mentioned, which produces normally an epidural hemorrhage. This case is identical with case 4 reported by JOKL in 1954.

BENES (1956) reported the case history of an amateur boxer who died of an acute subdural hematoma.

PAMPUS and Norbert MÜLLER (1956):

This 19 year old amateur boxer began boxing at age 15, and participated in 74 bouts from May 1949 to March 1954. He was described as a good boxer who won the majority of his fights. He had allegedly never been knocked out, and physical examinations at irregular intervals revealed no pathological findings.
In January 1954 he was matched against a superior opponent, and lost in the third round by a technical knock out. No detailed report of this boxing match exists. It is interesting that this boxer lost the following four boxing bouts. In his last fight, March 13, 1954, he received a hook against the left side of the head in the first round and was knocked down. He got to his feet at the count of '8' and continued to box. According to his trainer, between rounds he was fully clear and boxed the third round to its end. He lost the fight by points and complained 40 to 60 minutes after the fight of headache. A few hours later he began vomiting, and then became unconscious. The same night he was *admitted to a neurosurgical hospital,* deeply unconscious, with pale skin, superficial respiration, bradycardia and dilatation of the left pupil with reduced reaction to light. Corneal reaction was positive on both sides; otherwise reactions equal on both sides. There were no pathological pyramidal signs, and the patient showed clear cerebrospinal fluid.
A *left-sided epi- and subdural hematoma* was diagnosed, and a burr hole made over the left temporal bone. A *small extradural* and a *minimal subdural hematoma were removed.* During the operation patient showed marked improvement and a few minutes later was conscious and oriented. He asked which hospital he was in. In the following days, he became somnolent again. He was also restless, and tried to get out of his bed. Clinically, minimal neck rigidity existed. The left pupil remained dilated, but reacted slowly to light. The right side of the mouth dropped lower than the left. Reflexes in both extremities were greater on the right than on the left.
A *left-sided carotid angiogram* revealed no abnormalities. An *EEG showed* generalized alterations to the same extent on both sides.
Twenty days after the bout, patient succumbed.
The autopsy revealed a moderate tonsillar herniation but otherwise no signs of increased intracranial pressure. In posterior

parts of the second frontal lobe and outer parts of the first temporal lobe on the left were small subdural blood coagula, easily removed from the dura mater. The leptomeninges in this area were marked with a brownish discoloration and the cerebral cortex showed a trough-shaped indentation.
Histological examination of many slides from all parts of the brain revealed multiple small infarctions in the left amygdala in the cortex of the fourth right temporal gyrus, in the left caudate nucleus, in middle parts of both pallida and putamina and in both lateral thalamic nuclei, in the right paraventricular and supra-optical nucleus and in the reticular zone of the substantia nigra. All these lesions could be described as softening in the second stage, but the amount of macrophages and the extent and distribution of vascular proliferation allowed the conclusion that these necrotic zones developed at different times.
Partial, and in some areas total, necroses could be seen. In the marginal zones of necrotic areas, a massive glial reaction could always be seen. Small diapedetic exudates of erythrocytes could be seen in many necrotic areas. There was a small microscopically recognizable cortical contusion at the tip of the second left temporal gyrus, in the form of a cone, with the pointed end inside the brain and the wide end on the surface. Perifocal edema was seen in the neighborhood of infarcts and of cortical contusions.
The authors noted that patterns in CNS damage appearing after multiple and repeated closed head injuries are not found in injuries from a single head trauma.

PAUL (1957):

On June 21, 1956 at an *amateur boxing meeting* open to boxers who had never won trophies at intermediate or open meetings, two men met for a bantamweight contest (118 lbs. or 53.5 kg). Both contestants were 20 years old. Raymond PERERA weighed 114 lbs. (51.7 kg), his opponent, 116 lbs. (52.6 kg). PERERA was knocked out in the fight. The judge's observations were recorded:
The knock out was delivered shortly after the first minute of the first round. Up to this stage the deceased PERERA, had been leading 'lefts' at his opponent and the blow which knocked him out was the first struck by his opponent. It was a right cross which caught PERERA on the left side of his chin. PERERA appeared to be stunned by the impact, and gradually went dizzier, and slowly slumped down till his bottom reached the canvas. Then he rolled over to one side. He did not appear to be unconscious, and he was raised by his two seconds, who helped him walk to his corner.
The timekeeper was watching this match between rounds. He observed that when PERERA reached his chair he was lying with his limbs fully extended and his jaws clenched, and that he was breathing deeply and audibly like a man sound asleep. To the timekeeper these appearances were unreal and so unlike those of the victim of a knock out that he wondered whether the victim was shamming. He noticed that the victim never spoke from the time of the knock out and that he still remained rigidly extended when being lifted into the stretcher for *transport to hospital.* The knock out occurred at 6:30 p.m. PAUL was on duty at 7 p.m. in the casualty ward of the General Hospital, Colombo, when PERERA was brought in. He lay on his stretcher breathing deeply as if in deep sleep, his lower limbs were rigidly extended at the hips and knees, and his upper limbs lay by the side of the body rigidly extended at the elbows, with the forearms fully pronated and the fists clenched. Passive movements were resisted by the rigidity and if the resistance was overcome by force there was movement of the limb, restoring it to its original position. PERERA was unconscious, and made no response to a pin tickling the inside of the nostril or pricking the skin of an eyelid.
The left pupil was widely dilated and the right pupil was small. There were no external injuries, no bruises, and no hemorrhage from the nostrils, mouth or chin. The temperature was 97.4° F (36.3° C), pulse full and regular 104 and respiration 20.
At 8 p.m. a trephine hole in the left temporal fossa was made and the dura mater bulged with subdural blood. Incision of the dura mater evacuated 2 oz (57 ml) of fluid blood. A second trephine hole in the right temporal fossa showed no extradural or subdural blood. At 11 p.m. the pulse was 88, respiration 38, and temperature 102° F (38.9° C). PERERA's condition continued to deteriorate and he died at 1:30 a.m. on June 22.
PAUL obtained permission of the coroner to make a post mortem examination on June 22 at 11 am. On turning down the scalpel there was a diffuse film of coagulated blood on the sub-epicranical surface of the scalp and on the periocranial surface of the skull extending upwards from the temporal fossae from the surgical incision in the scalp.
When the vault of the skull was removed, the dura mater covering the whole of the cortex surface of the left cerebral hemisphere was blue from subdural blood. When the dura mater was incised, a thin film of fluid subdural blood covering the convex surface of the left cerebral hemisphere flowed away. The blood was pooled posteriorly, where it bathed the left occipital lobe. The brain was not compressed by the pooled blood. There was no subdural blood over the right cerebral hemisphere. There was a film of subarachnoid blood coating the pre- and post-central convolutions of both cerebral hemispheres. It extended a little in front of and behind these convolutions as well.
On the left side, this film of subarachnoid blood continued into the Sylvian fissure, over the island of Reil and over the outer surface of the left temporal lobe. The medial and inferior surfaces of the cerebral hemispheres had no films of subarachnoid blood. There were no cerebral lacerations. There was about 3 oz (85 ml) of dark blood in the posterior fossa, but the cerebellum and the brainstem showed no subarach-

noid films of blood and no lacerations. It is likely that this blood flowed into the posterior fossa from above.
Sections of the cerebral hemispheres showed no intracerebral hemorrhages or lacerations, but the white matter of both hemispheres was studded with the pinhead orifices of dilated vessels. There were no hemorrhages at these orifices. Sections of the brain stem and the cerebellum showed no hemorrhage or laceration at these sites.

PAUL commented that decerebrate rigidity of the victim presented some striking features. That it came on with the receipt of the blow was evident from the observation that the victim appeared to be stunned right away. He became dizzier, and that he then slowly slumped down till his bottom touched the canvas, an unusual fall for a knock out blow, explainable if decerebrate rigidity broke the fall. Then, too, the 'automatic walk' back to the corner, when he had been picked up and supported by his two seconds, is another phenomenon referable to the state of decerebrate rigidity. The poker-back posture when he was laid down was clear evidence of the state of decerebrate rigidity, as was the maintenance of this posture till the time of death.

PAUL noted that: experimental transections of the midbrain have established that decerebrate rigidity is the result of a lesion between the vestibular and the red nuclei. The post-mortem examination showed that there was no visible lesion in the midbrain, but it did establish that there had been an appreciable displacement of the cerebral hemispheres within the brain case. Subarachnoid hemorrhages on the convex surfaces of both cerebral hemispheres, across the waists between their frontal and occipital poles, were clearly the result of both cerebral hemispheres being driven upward, against the cranial vault. The force came from the blow on the chin, as it was transmitted upward, through the condyles of the mandible, up into the cranium. The diffuse subdural blood covering the convex and inferior surfaces of the left cerebral hemispheres must have been from a traction rupture of veins in the subdural space passing from the dura mater to the inferior surface and the lower convex surface of the left cerebral hemisphere. As the punch was delivered on the left of the chin this would account for the greater upward shift of the left cerebral hemisphere rupturing the veins in the subdural space on this side.

The post-mortem findings provide clear evidence that both cerebral hemispheres shifted upward. It would be logical to deduce that the upward shift strained the mid-brain area between the vestibular and red nuclei, causing decerebrate rigidity.

The site of strain in the mid-brain can be postulated precisely in this case, due to the fact that decerebrate rigidity was observed and recorded. The theory that concussion is a result of strains just above the red nuclei in the mid-brain cannot be postulated with the same degree of certainty. However, the fact that in this case the strain occurred at a slightly lower level suggests that a midbrain lesion is a causal factor in concussion.

A 20 year old bantamweight amateur boxer was knocked out in the first round by a right cross, which caught him on the left side of his chin. He appeared stunned by the impact and slowly slumbed down till the bottom reached the canvas. He then rolled over to one side. He did not seem to be unconscious, and he was raised by his two seconds, who helped him to walk to his corner. We have a good description from the timekeeper who observed the boxer closely. His observations are presented in the case history. Already in the ring a decerebrate rigidity developed. A left temporal subdural hemorrhage was removed through a trephine hole. At autopsy a left-sided subdural hemorrhage was fond.

KRAULAND (1961):

Case 8: A 19 year *old amateur boxer in a sparring fight with a professional.* The professional boxer was wearing 12 oz gloves; the amateur did not wear head protection. The amateur was hit two to three times on the left side of his face. After two minutes he was taken, out of the ring with a bloody nose. He complained of nausea, had to lie down, and after a short time became deeply unconscious. Cheyne-Stokes respiration; increased pulse; exitus one hour after the fight.
He had taken up boxing training two years before, and had participated in matches after just one year. He had had seven fights; six he won by knock out, and the seventh he lost on points, 29 days before his death. In that fight he had received

several severe blows against the head in the third round. He had been groggy but did not go down. He complained of headache, vomited in the dressing room, and had to lie down. He was given a painkiller, to good effect. He took his sports friends home in his own car, but drove slowly. At home he complained of headache and felt tired. Ten days later he took up training again, and complained of nausea several times.
Autopsy revealed a *fresh subdural hematoma over the right cerebral hemisphere,* of about 55 ccm. On the *right side were three ruptured bridging veins,* and the older brownish remnants of a hemorrhage in a state of organization were visible on the inner side of the dura mater to the right. Brain edema.

For details see legends of Figs. 247 -251
QUANDT and SOMMER (1965):

A 24 year old *amateur boxer* suffered a cerebral knockout, and was counted out while standing, collapsed, unconscious, in the ring shortly afterward. On *admission to hospital* complete areflexia with marked dilatation of both pupils and slow reaction to light.
During the next five days he slowly regained consciousness, but again lapsed into unconscious on the sixth day. This occurred together with a right-sided spastic hemiparesis. He was transferred to a neurosurgical unit, where a *left-sided carotid angiogram* and *EEG* showed signs indicative of severe diffuse traumatic lesions of the brain. Death 33 days after the fight, due to circulatory arrest.
Autopsy showed no sign of brain edema, but several recent fronto-parietal cerebral contusions. Cerebral ventricles moderately dilated. Multiple symmetrical anemic infarcts in basal parts of both caudate nuclei, the putamina, medial parts of the thalamus, and in the brainstem, in the immediate neighborhood of the substantia nigra.
Microscopically, most lesions showed total necrosis with a loss of neurons and glial cells. Multiple ischemic necroses in the above described areas. Especially remarkable were the nearly symmetrical necroses in both hippocampal gyri with typical segmental loss of neurons in Sommer's sector (h_1) and bands of necrotic tissue parallel to the substantia nigra. Glial and mesodermal tissue reaction was so far advanced that small cystic defects were visible. Many neurons showed ischemic alterations.

This 24 year old amateur boxer was counted out while standing, he collapsed and became unconscious in the ring. Shortly afterwards on admission to hospital a complete areflexia with marked dilatation of both pupils existed. He regained slowly consciousness in the next five days and lapsed into unconsciousness on the sixth day. This occurred together with a right-sided spastic hemiparesis. In a neurosurgical unit a left-sided carotid angiogram showed signs indicative of severe diffuse traumatic lesions of the brain. Death occurred 33 days after the fight due to circulatory arrest.

The autopsy revealed several recent frontoparietal cortical contusions and multiple anemic infarcts in basal parts of both caudate nuclei, the putamina, medial parts of the thalamus, and in the brainstem in the immediate neighborhood of the substantia nigra.

At microscopic examination the lesions showed a total necrosis with loss of neurons and glial cells. Especially remarkable were the nearly symmetrical necroses in both hippocampal gyri with a typical loss of neurons in Sommer's sector (h_1) and and bands of necrotic tissue parallel to the substantia nigra.

The case of QUANDT and SOMMER is exemplary well examined especially histologically.

BURGER and MINAROVJECH (1966) reported the case history of a 20 year *old amateur* boxer who suffered an *acute right-sided subdural hematoma.*

The patient had suffered a cerebral concussion as a toddler. He took up boxing at the age of 16. He trained 4 to 6 times per week. He participated in 32 fights, of which he won 21, lost seven and drew four. He suffered a cerebral knock out in June of 1963.
The *neurological examination* showed unremarkable findings; he was considered a simply structured personality. The *EEG* revealed a generalized lability.
In his next bout the patient received several severe blows against his head, early in the second round; two or three of those blows hit the left temple. He showed some ataxia and numbness but was not knocked down. The referee stopped the fight immediately, over the boxer's protestation. The boxer complained about headache in the dressing room and was upset. A neurologist who examined the boxer reported normal findings, especially symmetrical pupils that reacted to light. The physician watched parts of the following bout, but about five minutes later was called into the dressing room. The patient was lying on a bench. He was pale and perspiring profusely and his speech was inarticulate and confused. The right pupil was wider then the left one. About 15 minutes later he was *transported to a hospital.* The admitting physician diagnosed a state of shock caused by blows against the solar plexus since the patient again had an unimpaired sensorium

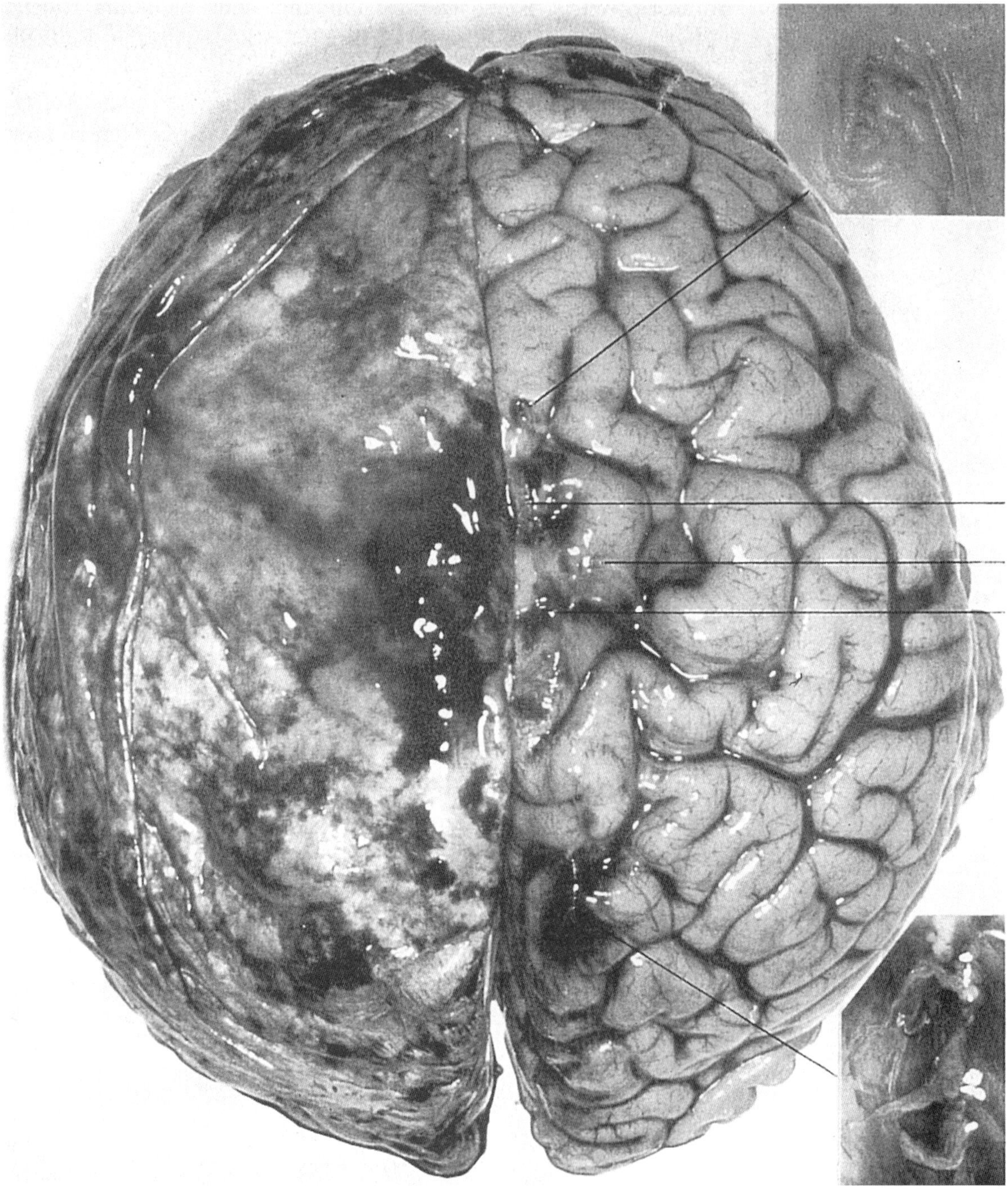

Fig. 247. Case 8. Amateur boxer. Head injury in a boxing match. Acute subdural hemorrhage, 55 ccm, right, survival time one hour. Upper left of the cerebral hemisphere, the dura mater is folded to the left. Fresh hemorrhage removed. The internal plane of the dura mater reveals fresh hemorrhages in the neomembranes.
R_1, R_2 and R_3 show fresh tears of parasagittal bridging veins with a subarachnoid hemorrhage around the peripheral stump.
a_1 and a_2 healed stumps of torn veins. (From KRAULAND, 1961).

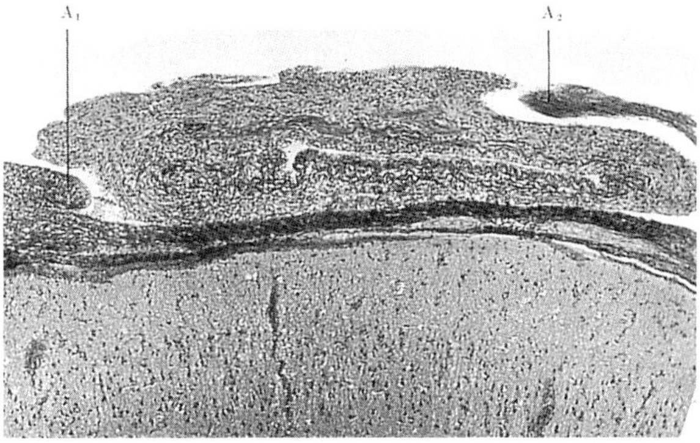

Fig. 248. Case 8. Cross section through the peripheral vein stump a_2 from Fig. 247. The loosened up wall of the vein is densely interspersed with cells, the lumen agglutinated. The vein proper is embedded in granulation tissue.
A_1 and A_2 = margin of the arachnoidea. Van Gieson technique, 52 : 1. (From KRAULAND, 1961).

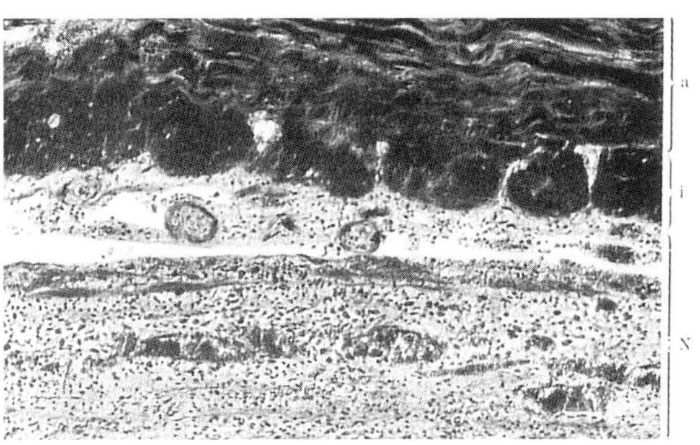

Fig. 249. Case 8. Cross cut of the dura mater.
N = Neomembrane, i = inner, a = external layer of the dura mater.
The neomembranes can be distinguished clearly by their colouring and richness in nuclei from the of the dura mater. In the porous external layers are cross sections of small arteries visible. Van Gieson technique, 78,5 : 1. (From KRAULAND, 1961).

and was fully oriented. But when the physician went into the hospital room, he found the patient deeply comatose; he did not react to stimuli. The right pupil was wider than the left one. A contralateral spasticity with increased tendon reflexes, patellar, and ankle clonus, and a bilateral Babinski were found. In intervals of 15 to 20 seconds, tonic clonic seizures of the extremities with accentuation of the left side occurred. Bradycardia, 52/min.
About four hours after the bout, trepanation was performed. A *massive right-sided subdural hematoma* was removed. Fresh blood gushed from a deeper ruptured vessel, which was very difficult to stem. Left-sided trepanation did not show a subdural hematoma. The boxer succumbed 4 1/2 hours later.

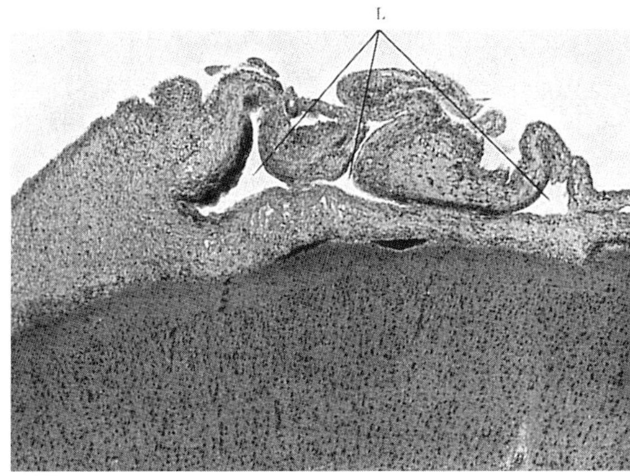

Fig. 250. Case 8. Longitudinal section through the peripheral healed vein stump a_1 from Fig. 247. The walls of the veins were agglutinated, partly interspersed with fresh hemorrhages. The partially preserved remaining lumen (L) without thrombosis. The subarachnoid space is closed by proliferation of connective tissue. Ths irregular spots are giant cells. No loss of nerve cells in the cerebral cortex. Hematoxylin-Eosin technique, 27 : 1. (From KRAULAND, 1961).

The *autopsy* revealed, in addition to the right-sided subdural hematoma, recent hemorrhages in the pons.

This 20 year old amateur boxer trained 4 to 6 times per week. The patient had suffered a cerebral knock out. The neurologic examination was normal and the EEG revealed a generalized lability. In the next bout the patient received several severe blows against his head early in the second round. He showed some effect, the referee stopped the fight immediately over the boxers protestation. A neurologist reported normal findings, especially symmetrical pupils, which reacted to light. While the physician watched parts of the following bout, he was called about 5 minutes later into the dressing room. The patient was lying on a bench, was pale and perspiring profusely and his speech was inarticulate and confused. Now the right pupil was wider than the left. About 15 minutes later he was transferred to a hospital. The admitting physician diagnosed a state of shock caused by blows against the solar plexus. But when the physician went into the hospital room, he found the patient deep comatose, the right pupil was wider than the left one. A

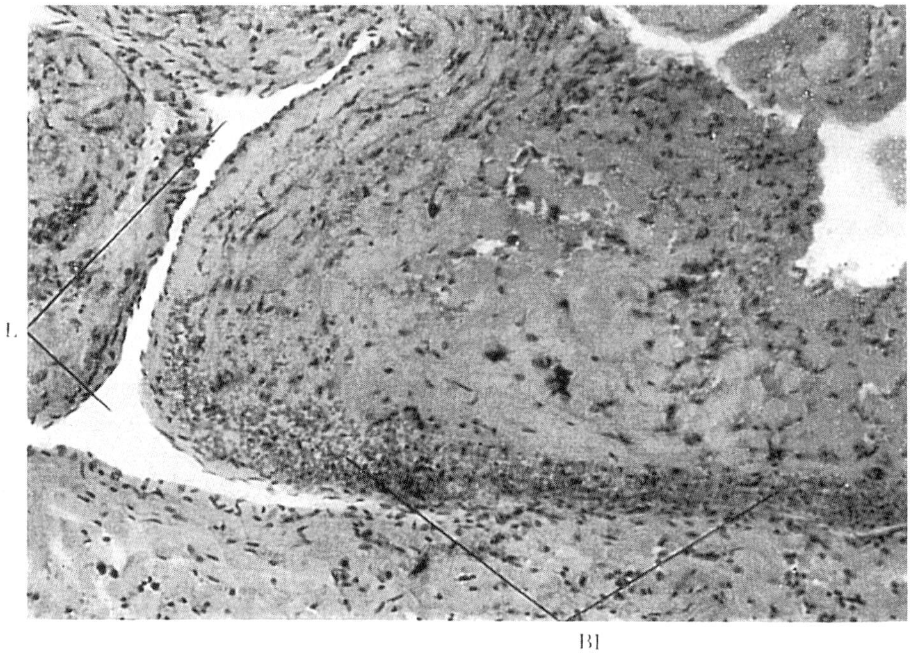

Fig. 251. Case 8. Magnified photo from the right upper left of cerebrum.
L = lumen of a vein. The margin of the endothelial nuclei is clearly visible. The wall of the vein sticks together with the arachnoidea, here are multinucleated cells.
Bl = thin fresh bleeding along the sticked together walls of the veins. Hematoxylin-Eosin technique, 63 : 1. (From KRAULAND, 1961).

contralateral spasticity with increased reflexes and bilateral Babinski were found. In intervals of 15 to 20 seconds tonic-clonic seizures with accentuation of the left side were found.

About four hours after the bout, a massive right-sided subdural hematoma was removed. Fresh blood gushed from a deeper ruptured vessel, which was very difficult to stem. The boxer died 4 1/2 hours later.

The autopsy revealed, in addition to the right-sided subdural hamatoma, recent hemorrhages in the pons.

It existed a peracute right-sided subdural hematoma which had developed into a space-occupying lesion which lead to hemorrhages in the pons, which were responsible for the death of the patient.

SCHMID et al. (1968) reported two acute ring deaths of *amateur boxers:*

Case 1: 25 year old heavyweight boxer whose autopsy revealed a *left-sided subdural hematoma,.*
Case 2: A 19 year old *amateur boxer* received several blows against his neck and thorax. Respiration stopped, followed by *cardiac arrest.* Although direct external heart massage was applied, followed by treatment in an intensive care unit, the boxer died four days later.
The *autopsy* revealed severe *'axonic alterations of the brain,'* whose quality and distribution are not discussed in detail.

STRASSMANN and HELPERN (1968) reported *that 23 autopsies of boxers injured in the ring were carried out between 1918 and 1950,* in the *Office of the Chief Medical Examiners of New York.* In *thirteen cases subdural hemorrhages* were determined to have been the cause of death.

STRASSMANN and HELPERN published four additional cases of subdural hemorrhage, which were found in autopsies after 1951.

Case 1: A 20 old *amateur boxer* was knocked out in the second round by a blow against his chin. He immediately became unconscious and remained unconscious for 54 hours until his death.
A bilateral craniotomy was performed, *with removal of a left-sided subdural hematoma. Tracheotomy.*
Autopsy revealed protrusion of the brain through the burr holes, and the *remnants of a small subdural hematoma on the left side. Brain* weight 1.450 g. Diffuse hemorrhagic infarctions of all lobes at base, basal ganglia, corpus callosum, pons, and crura cerebri. Glottis edema, emphysema of lungs, acute hemorrhagic ulcers of duodenum.
Microscopic examination revealed diffuse perivascular edema, vacuoles and diapedesis of erythrocytes, stasis and thrombi in small vessels, degenerative alterations of neurons and loss of neurons. Perivascular hemosiderin pigment from earlier traumatic lesions.
Case 2: A 19 year old *professional boxer* was knocked out in the tenth round after a series of heavy blows against the head. He sank slowly to the canvas and remained unconscious until his death 54 hours later. A bilateral craniotomy was performed, with the *removal of a subdural hematoma on the right side,* as well as a tracheotomy.
Autopsy revealed *remnants of a right-sided subdural hematoma, ruptured bridging veins, thrombosis of the superior longitudinal sinus,* and *brain swelling* (weight 1.350 g). Herniation of the right uncus gyri hippocampi. Softening, hemorrhages, and duodenal hemorrhages of the mucosa.
Microscopically this case revealed findings similar to those of case number one. In the hemorrhagic infarctions diffuse and perivascular diapedesis of erythrocytes and transudates, thrombi in veins, degenerative alterations and loss of neurons could be seen. There were also the beginnings of vacuole formations and additional perivascular hemosiderin deposits.
Case 3: A 20 year old *professional boxer* was knocked out in the eighth round after a series of blows against the head. He fell backward and hit the canvas with his occiput. He remained unconscious until his death five days later. While he was unconscious, a craniotomy was performed and a left sided subdural hematoma was removed; a tracheotomy was also performed.
The *autopsy* revealed *remnants of a subdural hematoma on the left side,* ruptured bridging veins with thrombus formation, and additional thrombus formation in the longitudinal sinus. *Brain* weight was 1.380 g. Brain edema with protrusion of the brain through the burr holes. Swelling of both temporal lobes with smaller hemorrhages. Ischemic necrosis of globus pallidus (more on the left than on the right), pons and medulla oblongata. Glottis edema, lung emphysema and hemorrhagic infarctions of both lower lobes.
Microscopic examination revealed degenerative alterations and loss of neurons in the lenticular nucleus, pons, medulla oblongata and both temporal lobes. Perivascular edema with transudate, lymphocytes and hemosiderin-laden gitter cells. Proliferation of endothelial cells of vessels in involved areas and in the subarachnoid space.
Case 4: A 25 year old *professional boxer* received a series of heavy blows against his head, in the twelfth round. He sank slowly to the canvas and remained unconscious until his death nine days later. A bilateral craniotomy was performed, with removal of a *right sided subdural hematoma,* as well as a tracheotomy and treatment with antibiotics.
Autopsy revealed brain edema with protrusion of the brain tissue through the burr holes. Over the *right cerebral hemisphere were remnants of a subdural hemorrhage. Brain weight* 1.250 g. *Ruptured right bridging veins with thrombus,* thrombi in other pial veins and in the superior longitudinal sinus. *Cortical contusion of the left occipital lobe. Massive hemorrhagic infarctions of all lobes of the base of the brain, of the basal ganglia, the thalamus, and the crura cerebri.* Severe bronchio-pneumonic alterations in all lobes of both lungs with abscess formation and fibernus pleuritis.
Histological examination showed diffuse and perivascular extravasation of edematous fluid, erythrocytes and lymphocytes, hemosiderin and lipoid-laden gitter cells, thrombosis of pial veins, diffuse loss of neurons in the cerebral cortex and loss of Purkinje cells in the cerebellum. Proliferation of endothelial cells, formation of vacuoles and demyelination in the white substance. Marked lymphocytic accumulations in the subarachnoid space.

Subdural hemorrhages existed in all four cases, and were immediately evacuated in hospitals. Despite the fact that the space occupying lesions were removed, none of the patients regained consciousness. They died between 54 hours and nine days after the knock out. They were all young and healthy persons between 19 and 25 years old; *three were professional and one was an amateur boxer.* The latter was knocked out in the second round by a blow against the chin. He hit the canvas with his occiput. A similar case was observed in one of the professional boxers in the eighth round; he had already taken severe punishment against his head. The other professional boxers were knocked out in the tenth and twelfth rounds, respectively. They slumped to the canvas without landing on the occiput.

The cause of death in all cases was a marked swelling of the brain and hemorrhagic infarctions of large areas. In three cases the *edema had produced a prolaps of brain tissue through the openings of the dura mater at the burr holes.* In one case there was a *uni-*

lateral prolaps of the uncus gyri hippocampi. In another case the infarctions were of an ischemic nature, but all the others were hemorrhagic in nature. Hemorrhages and necroses at the base of the frontal, temporal, and occipital lobes of both hemispheres. Hemorrhages and necroses extended from the cortex into the white substance, the caudate nucleus, the thalamus, corpus callosum, pons, and sometimes into the cerebellum. These alterations were especially pronounced. One boxer with a survival time of nine days after the knock out revealed an ischemic necrosis in the anterior lobe of the pituitary. Microscopically, the histological picture was very similar in all hemorrhagic infarcts after 54 hours.

One remarkable pattern was observed. Thrombi caused the early occlusion of veins and capillaries of the pia and the brain, with extravasation of edematous fluid into the perivascular spaces and the surrounding brain tissue, which showed many lacunas. The neurons in the gray substance showed different forms of degenerative alterations. *Sporadic deposits of hemosiderin pigment were seen around vessels, proof of former brain injuries.* Alterations of the neurons were evident in more distant parts of the brain where there were no hemorrhages, i. e. in the medulla oblongata and cerebellum, where there was a *severe loss of Purkinje cells.*

Reactive proliferation of capillaries and glial elements was seen in hemorrhagic or ischemic areas only in those cases in which the patient had survived his knock out for five or more days. These reactions consisted of massive extravasation of leukocytes (mostly lymphocytes), occurrence of gitter cells, of microglial proliferation and of proliferation of adventitial cell elements, which were filled with debris, lipoid and hemosiderin. There were fresh unaltered erythrocytes.

The frequent secondary traumatic hemorrhages in the brainstem were missing only in case 3, which showed, not hemorrhagic, but ischemic necrosis in the basal ganglia, pons and medulla oblongata.

The autopsy of a 30 year old professional boxer published by STRASSMANN and HELPERN (1968), who was knocked out and remained unconscious for about 30 minutes revealed a thrombotic occlusion of the intracranial part of the right carotid artery and the middle cerebral artery with hemorrhagic necrosis of parts of the right cerebral hemisphere, especially the basal ganglia and the temporal lobe. This case is discussed in detail in the chapter about traumatic vascular lesions of lateral parts of the neck, see p. 281.

IKEDA et al. (1980) published the case of a boxer, aged 26 years, who received a series of five punches, was knocked out, and died after three days.

The boxing match was video-recorded, so the authors thought to be able to study the injury mechanics involved. By comparing the video-record with the autopsy findings they were able to correlate the brain damage they found with specific blows to the head during the fight.
The *postmortem examination* revealed a *slight subdural hematoma on the convex surface of the right cerebral hemisphere* and a *marked cerebral contusion from both frontal bases to the inside of the right temporal lobe and from the left thalamus to the upper part of the mesencephalon.* The contusions occupied the central and anterior part of the brain. The localization of the contusion suggested allegedly that the blows spun the head horizontally, around the cervical column, causing a hasty twist and strain of the brain and brainstem.
Analysis of the five punches revealed that the last two punches made the head rotate quickly from left side to right, around the cervical column in the horizontal plane. These results of the analysis support the inference that this rotating movement of the head brought about the damage.

This 26 year old boxer, who had received a series of five punches, was knocked out. At autopsy a slight subdural hematoma was found on the convex surface of the right cerebral hemisphere and allegedly marked cerebral contusions from both frontal bases to the inside of the right temporal lobe to the inside of the right temporal lobe and from the left thalamus to the upper part of the mesencephalon. To begin with in these areas are cortical contusions normally not found. It is in my opinion more probable that these lesions are infarcts whether anemic and/or hemorrhagic. In my opinion the intensity of boxing blows is sufficient to produce the clinical symptoms of a cerebral concussion or commotio cerebri, but not sufficient to produce cortical contusions. When cortical contusions in boxers are found at an autopsy they have to be considered the result of impacting the head of the boxer on the canvas as the result of a deceleration trauma.

European Championship Fight in Welterweight between Jörg EIPEL and Alain MARION

An account of the clinical and pathomorphological findings of the *professional welterweight boxer* Jörg EIPEL (J. E.) COLMANT and DOTZAUER were reported in 1980. On December 18, 1977 EIPEL lost the European welterweight title fight against the French boxer Alain MARION, due to a technical knock out in the fifteenth round. He died after seven days in a coma.

The two authors discuss *the Report of the Federal Institute of Sport Sciences (Bericht des Bundesinstitutes für Sportwissenschaften),* volume 3, for the *Federal Ministry of the Interior (Bundesministerium des Innern).* Medical and sports athletic views from both the amateur and the professional boxing federations were included. The report summarizes the results of a symposium held in 1975, in Saarbrücken, Germany.

COLMANT and DOTZAUER stress their observation that of all participants, only two expert advisers of the meeting – a clinician and a neuropathologist – expressed serious concerns with regard to boxing. The other authors alluded to certain possible injuries and preventive measures, but they were circumspect about the danger of brain injuries.
The boxer J. E. had allegedly taken during the boxing match around 22:00 h ephedrine nose spray, one teaspoon of Melissengeist (over the counter drug) and one tablet Reactivan. The urin sample from the first 12 hours in the hospital revealed a Pervitin content of a least 40 gamma%. Accordingly J. E. must have taken additional substances before the fight.

J. E. made an impression at the boxing match, which was duly noted by the physicians present. Dr. H. said: "in the first rounds J. E. seemed to be much stronger than in previous boxing bouts. He has taken never before so many blows against his head, without being knocked down. In the sixth round J. E. had to take very severe blows against his head which he could not have taken without doping substances. Both boxers were in the last round completely exhausted. "Both showed a very bad reaction."

The preceding commentaries were made by a physician, in fact, a ring physician. His belief that a boxer was able to withstand many blows to the head without falling down because he was in good physical condition, is, in the best case, naive. Good physical condition does not protect the brain from the physical force of repeated blows against the head. It is difficult to imagine what specific protections the commenting physician had in mind, except to say that the well conditioned boxer may be able to avoid contact by quick and clever evasive moves. Such was not the case in the example of J. E., however, as punches against the head landed time after time, in quick succession. The knock down in the fifth round at the '8' count was a so-called 'reflex knock down', but on consideration and with hindsight, was probably a carotid sinus syndrome. It was not a knock out but a knock down, because after the '8' count the boxer was on his feet again, to continue the boxing match. In the description of the boxing match a blow against the carotid artery is mentioned, a mechanism, which, along with carotid sinus syndrome, must be considered.

The quote above implies that doping substances have the ability to protect the brain against the effects of rapid impact, letting the boxer take more severe blows against his head, "which he could not have taken without doping." This was a novel idea indeed. Apparently Pervitin, the substance later detected in his urine sample, was thought to have protected J. E. as a kind of roboricum or therapeuticum against repeated mechanical impacts?

Finally, the last sentence is unintelligible: "both boxers showed a bad reaction." We do not have the benefit of elaboration on what was meant by this conclusion. Probably he meant to say that the boxers were completely beaten up and were in traumatic shock, both of which appear to have been true: "... "Both boxers were in the last round completely done." Probably he meant that both boxers were completely worn out and were in a groggy state of traumatic shock. Adequate medical nomenclature could have been used profitably in describing the events of the fight.

Description of the Boxing Match: J. E. began fighting with a relative fast tempo. In the sixth round he was knocked down after a blow against lateral parts of his neck (carotid sinus reflex or carotid artery?). The referee counted to '8'. Between the sixth and the seventh round EIPEL seemed confused and disoriented, he went, for example, after the end of the round to the wrong ring corner. In the eighth round a decline in his performance occurred. Hereafter he was 'tired,' his reactivity was decreased, especially in the eighth to the tenth, and the thirteenth to the fifteenth rounds. In the eight and especially in the eleventh round he had to take severe punishment. From the twelfth round he seemed to be rather groggy. In the last three rounds both boxers were completely exhausted.
Due to the testimony of the ring physician, later, in the legal proceedings, MARION fought with an improperly attached glove, open on the left side. He landed many illegal punches, especially at the occiput and the nape of his opponent' neck, so-called 'rabbit punches.' By his own count, in the first six rounds 18 times, in the following rounds 21 times, for a total of 39 times during the match. Compared with this, the court records show a notice, whereupon the ring physician (Dr. H.) counted only 21 (illegal!) 'rabbit punches.'

A 'rabbit punch' is against the rules, illegal, and unnecessarily dangerous. The referee should have warned MARION after the first rabbit punch, and should have disqualified him after a single repetition! Only in the fourteenth round did the referee warn MARION, for the first time, about the 'rabbit punches.' In the fifteenth round J. E. went down on one knee after a blow to the head. While going down, he took another blow on the nape of his neck. This, too, was both illegal and ignored by the referee. As he began the count, J. E. got up after a few seconds and lifted his hand up to surrender, ending the fight to save himself.

Here, a boxer (MARION) fought a match using illegal blows throughout, without any interference from the referee, who should have taken action to prevent many of the blows. Not only did the referee aid and abet an act of manslaughter by not punishing obvious violations, but he inspired similar behavior in the future, by rewarding such conduct with a professional win and the purse. What kind of honor can there be in winning a match by patent, obvious cheating? What can be said of a match where a boxer is mortally pummeled by overly dangerous illegal blows, in front of officials, ring physicians, and the public, in a fully regulated and sanctioned title fight, without objection? This fight tragically demonstrates that legal protections do not assure that boxing in a ring is safer than it is in the streets.

Six to seven minutes after the fight J. E. became unconscious, suddenly, with maximally dilated pupils. Shortly afterwards tonic seizures occurred, with clonism of the feet and a temporary arrest of respiration. J. E. was immediately placed on artificial respiration and transferred to the *Department of Neurosurgery of the University of Cologne.* The admission in the hospital took place just 30 minutes after the end of the boxing match.

In surgery, a *left sided temporal subdural hematoma was removed.* As origin of the hemorrhage tempero-basal veins and a tear of a fronto-polar bridging vein was mentioned, additionally considerable contusions of the left temporal lobe.

External injuries consisted of a small laceration at the root of the nose and periorbital edema.

The treatment in the hospital consisted in a typical manner. Due to the expected brain edema the dura mater was not closed. A dehydrating treatment was given, assisted by an endotracheal lying Portex tube; vegetative suppression and antibiotics were given.

After slight improvement of the responsiveness, at about the sixth day, deterioration began. A coma persisted for seven days. Death occurred under the clinical picture of a cerebral hemorrhage.

Pathologic-anatomical Diagnosis: Seven days survived blunt impacts of the nape of the neck with local hemorrhages in the neck muscles, hemorrhages in dorsal parts of the anulus fibrosus of the discs of C2/C3 and C3/C4, in the latter one extending to the nucleus pulposus.
These morphological findings of the hemorrhages in the neck muscles and the hemorrhages in dorsal parts of the anulus fibrosus of cervical discs, the lower located extending into the nucleus pulposus, therefore deeper into this anatomical structures, are the results of the mentioned so-called 'rabbit punches.' These findings are traumatic tissue alterations that have not been described before as result of boxing.

Retropharyngeal Hematoma: The retropharyngeal hematoma of traumatic origin, the direct result of a boxing blow landed at anterior or lateral parts of the neck. The morphological tissue damage of the *retropharyngeal hematoma has not been described before, as a result of boxing,* so it will be necessary to begin with a description.

After *operative removal of a subdural hematoma from left-frontal tears of bridging veins* and *hemorrhages from extensive cortical contusions of the cerebral cortex,* the status is as follows: *slight subarachnoid hemorrhage near the inter-hemispheric fissure* (Fig. 252). Extensive hemorrhages in the leptomeninges, in the area over the left gyrus olfactorius, and a single focal one on the right. Partially confluent focal hemorrhages over basal parts of the left temporal and occipital lobes of the cerebral hemisphere. Slight mass deviation of the left cerebral hemisphere over the midline structures to the right. *Typical prolapse of the brain in the operative area. Pressure cones of the cerebellar tonsils.* Middle sized cerebral ventricles. *Extensive cortical contusions in both frontal lobes, smaller ones in the cortex of the left temporal lobe and in the island.* Numerous hemorrhages around the longitudinal fissure.

Necroses and hemorrhages in the supply area of the left posterior cerebral artery. Extensive, as secondary-traumatic to be considered hemorrhages in the area of the pons.

Considerable edema of the white substance of the left-sided cerebral hemisphere and *numerous hemorrhagic necroses.* These are located in left orbital and median convolutions bilaterally between the optic fascicles, in the area of the nucleus amygdalae, the Ammon's horn formation and the roof of the lower horn of the ventricles bilaterally, in the area of left-sided tem-

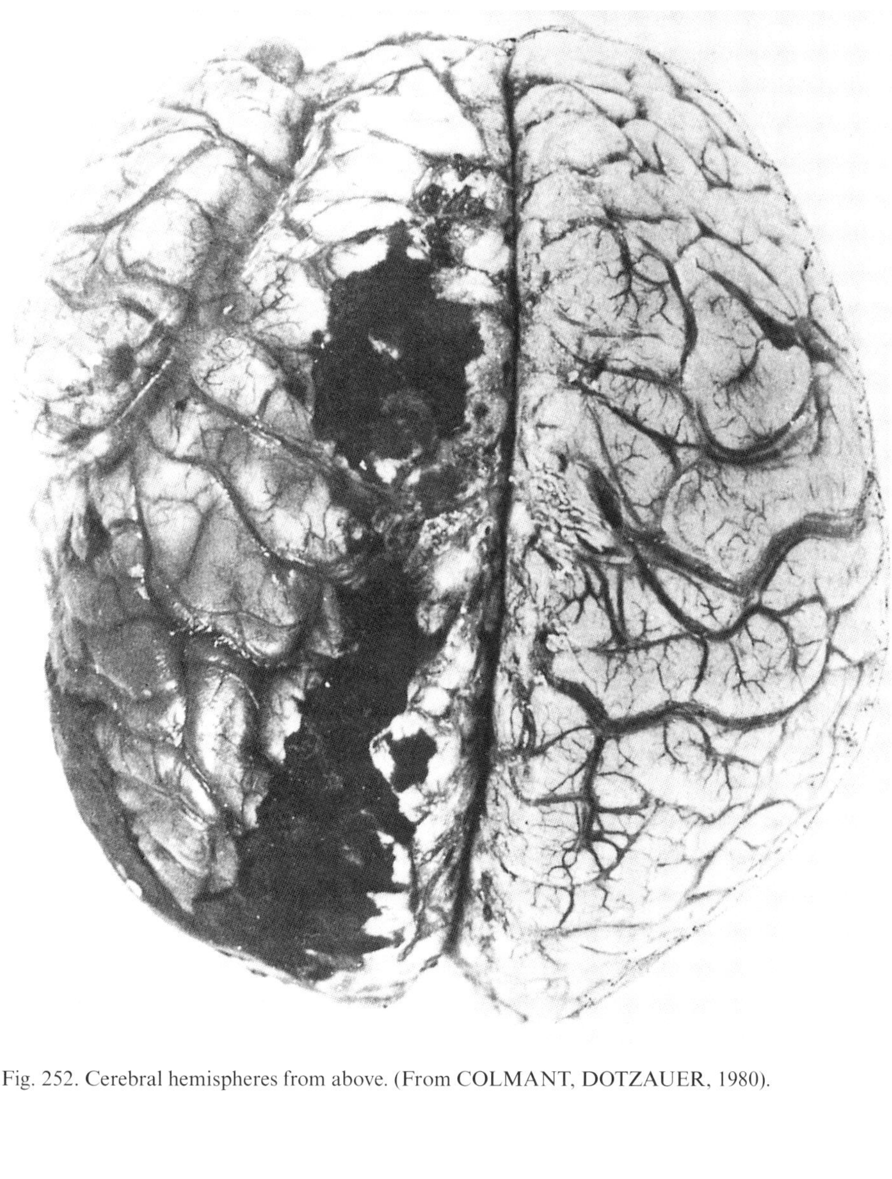

Fig. 252. Cerebral hemispheres from above. (From COLMANT, DOTZAUER, 1980).

poral convolutions, at the incisure of the prolaps of brain tissue, in the hypothalamus, in the thalamus and subthalamus – immediate connection with additional hemorrhages in the midbrain and the pons. Involved are furthermore the posterior gyrus cinguli, the dorsal layer of the corpus callosum, and very widespread median convolutions of the left occipital lobe. Smaller hemorrhages are in the insular cortex, in the white matter beside congested vessels, in the left mammilary body, basal parts of the internal capsule left and in the lateral cortex of the left occipital lobe under a space-occupying subarachnoid hemorrhage (Figs. 253 – 255).

These alterations are to be considered partly as so-called cortical contusions. This is true for the extensive orbital tissue alterations – in which at the left by inspection with a magnifying glass typical wedge-shaped foci can be seen, the alterations in the amygdala bilaterally and the lesions in the first to third temporal convolutions left. The alterations at the incisure of the prolapse and in the left gyrus cinguli can be explained by local disturbances of the blood supply, due to strangulation of vessels in this area. Also the alterations in the Ammon's horn formation bilaterally, in the left occipital lobe and in the tectum and tegmentum of the midbrain, also in the pons, have to be attributed to the well-known strangulation mechanism in the area of the tentorium cerebelli. Presumably the very exceptional extensive and to orally extending hemorrhages in the neighborhood of the third ventricle have to be explained in this way.

Histological Techniques: Tissue staining methods were cresylviolet (Nissl), hematoxylin-eosin, van Gieson, Kluever-Barrera, PAS and iron staining after embedding in paraplast.

Histologial Findings: The alterations are altogether of the same age and can be summarizing described as total and partial tissue necroses with and without hemorrhages. The tissue alterations are according to the size and intensity of the tissue damage in different stages of development. Small tissue necroses are in a stage of disintegration of tissue with abundant macrophages and compound granular cells and proliferating capillaries with a thickened wall. In addition to tissue reactions, others show a distinct wall of proliferating macro- and astroglia, largely active in the center. Sometimes remnants of ischemic, necrotic, in many cases eosinophilic nerve cells can be seen. Widespread focal and laminary parenchyma necroses often surround the described hemorrhagic necroses, but are sometimes found independently as well.

Especially instructive are the discreet alterations in the right cerebral hemisphere. Frontally at the superolateral surface of the longitudinal cerebral fissue a macroscopically invisible, wedge-shaped, so-called cortical contusion can be seen, also as a contusion damage to be interpreted alteration in the nucleus amygdalae. Additionally, many 'white' necroses, or elective parenchyma necroses in cortex and basal ganglia, are present. They appear alongside small necroses in the optic chiasm. White necroses are also present dorsally at the base of the brain, a round focus in the pallidum and a total necrosis of the corpus Luys' necroses in the oral substantia nigra and in parts of the subiculum, also in the Ammon's horn formation (dentate fascia, endleave and dense layer, furthermore in median nuclei of the lateral nuclear complex of the thalamus). The cortex is, with the exception of a cortical edema in the neighborhood of the calcarine fissure not involved. In the central white matter is a perifocal glial reaction with appearance of protoplasmic astrocytes and a ubiquitous reaction of microglial elements evident.

The alterations on the left are more severe than indicated on macroscopic inspection. The lesions extend deeply into the white matter of the orbital lobes, and they resemble or appear to like a colossal cortical contusion, in which lies a distinct, typical wedge-shaped focus (compare macroscopic findings), lateral of the gyrus rectus. In the farther neighborhood of the temporal contusion are differently marked elective parenchyma necroses of the cortex, also at the pole of the temporal lobe, especially in the sulci of the convolutions. The cortex of the island is with inclusion of the claustrum almost completely destroyed. This nucleus reveals a circumscribed softening, ventrally, also putamen and the cauda nuclei caudati, not only orally at the lateral ventricle wall but also caudally in the roof of the lower horn. The pallidum reveals elective parenchyma necroses. The fornix shows a central necrosis and there is a dorsal hemorrhage in the corpus callosum with perifocal edema, which extends deeply into the ventral splenium of the corpus callosum. Remarkable are, especially in the ventral splenium of the corpus callosum, but also at the other areas, round, perivasal located demyelination. In the occipital cortex extravasation of polymorphous leukocytes or debris of nuclei can be seen. In general the leptomeninges reveal hemorrhages, macrophages and gitter cells. In the midbrain exists a total necrosis of the substantia nigra and massive, partially confluent hemorrhages in the periaqueductal gray, further edema and necrosis in the stadium of tissue disintegration in the area of the raphe of the tegmentum. On the contrary the substantia reticularis is preserved. There were small hemorrhages in the pons, in the raphe of the base and bilaterally, in the lateral tegmentum. The medulla oblongata reveals in the examined planes no pathological alterations. The basal ganglia show a marked arteriosclerotic plaque.

From all tissue blocks iron staining was performed. There was pervasive early disintegration of blood (hemosiderin), especially in macrophages of the leptomeninx, can be seen, but barely in the brain substance. There was no indication of older hemorrhages.

Also generally glia fiber staining techniques were

Fig. 253. Coronal sections through the cerebrum. (From COLMANT, DOTZAUER, 1980).

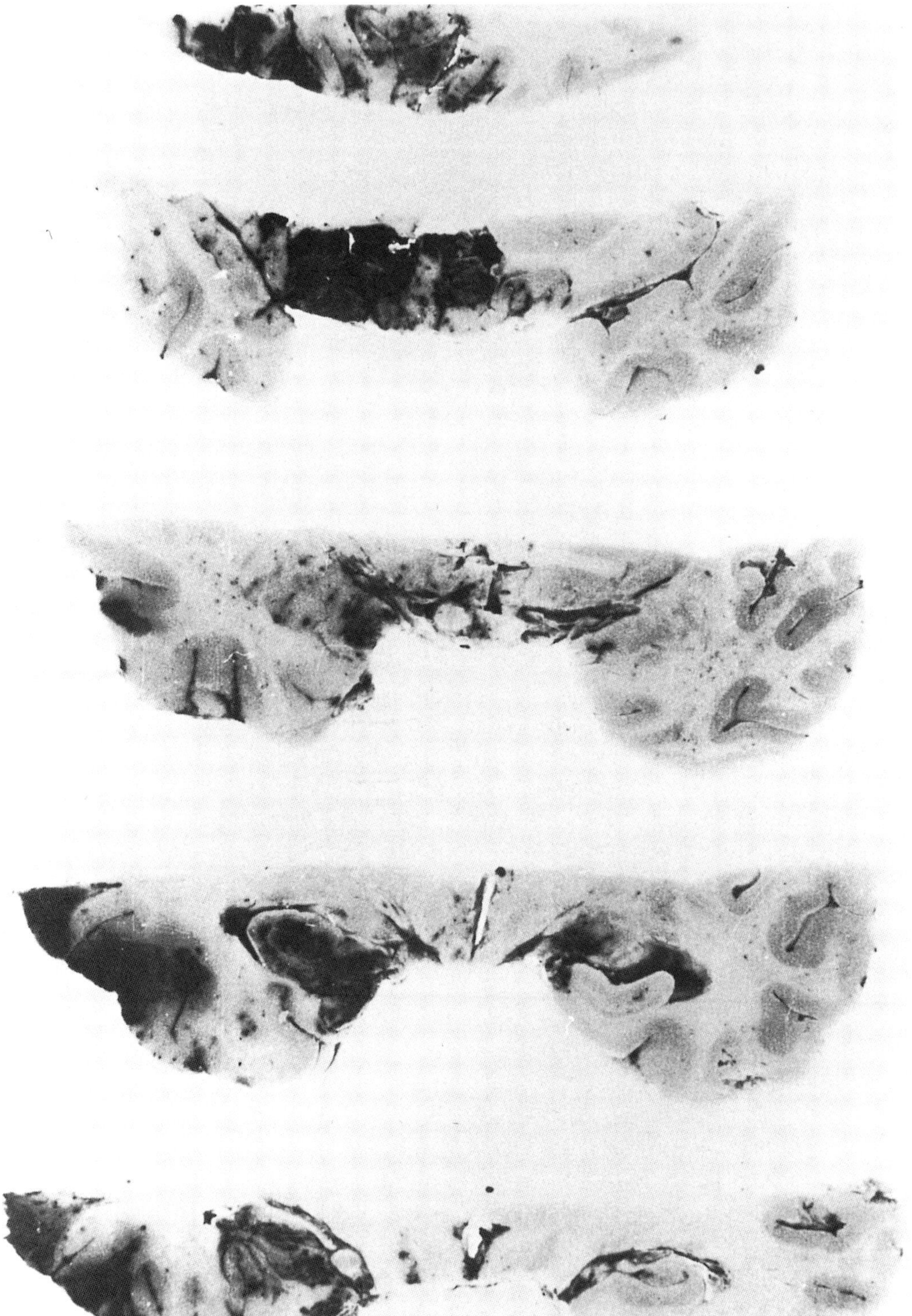

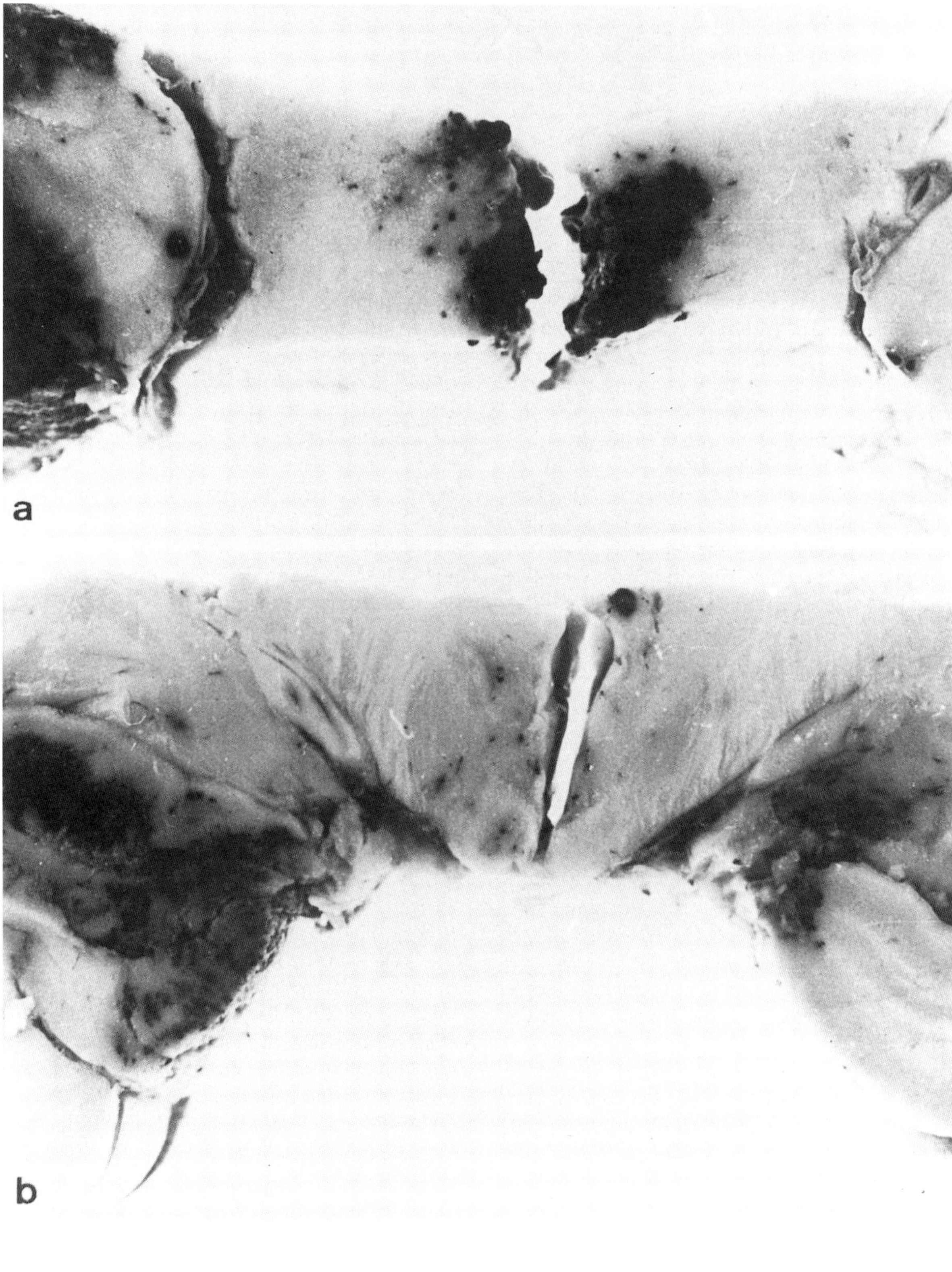
a
b

performed. They reveal a ubiquitous light glial fiber proliferation in the white matter of the cerebral hemispheres.

The authors stated that they are not experienced ring physicians. They report their morphological findings, but with regard to the boxing bout they relied on television recordings and remarks made by the commentator.

Both authors noted, that up until the end of the boxing match, a victory by points in favor of EIPEL was considered a possibility. It appeared that an inexperienced observer would not have detected a clear disturbance in coordination, of a groggy condition or complete disability, especially in the last round. Neither boxing officials nor television commentators took notice of EIPEL's deteriorating condition. The fight was ended by EIPEL himself, after a knock down in the fist part of the fifteenth and last round; no towel was thrown from his corner. The surrender was unexpected, according to the authors. One can only speculate about the subjective reasons for it, that is EIPEL's own observations and incapacitation, as they were not recorded. By the rules, the fight ended in a technical knock out, in favor of MARION. EIPEL broke down and suddenly lost consciousness about six or seven minutes after the referee had stopped the fight.

COLMANT and DOTZAUER report however, that on the tape recording one could hear the commentators' conversation: "Jörg EIPEL does not seem to recognize what is happening around him." In those moments EIPEL experienced an extreme narrowing of the state of consciousness, or a 'traumatic amnesia' ('traumatischer Dämmerzustand'). It is also possible that the changes in consciousness began in an earlier round, not at the end, in the fifteenth round.

After six or seven minutes, EIPEL fell into a state of deep unconsciousness. Both pupils were extremely dilated; respiration temporarily stopped. The neurosurgery commenced after about 30 to 45 minutes in the Department of Neurosurgery of the University of Cologne. *A leftsided subdural hematoma, under high pressure, was removed.* The boxer did not regain consciousness again, and he died one week later.

The authors emphasize that contrary to the typical slowly developing subdural hematoma, here the hemorrhage must have developed very quickly, as massive brain edema was observed.

These incidents in the boxing ring are not rare, as the analysis of the presented casuistics reveals. Sometimes they occur after severe punishment with series of blows against the head or an especially powerful single boxing blow a seamless transition from a syndrome of a cerebral concussion into deep unconsciousness with massive brain edema, a course, which in single cases can be interrupted by a short clearing up of awareness (WOLFF 1928, WEIMANN 1934, FELC 1939, KAPLAN and BROWDER 1954, STRASSMANN and HELPERN, 2 cases, 1968).

In some cases unconsciousness is completely missing immediately after a boxing match. The unconsciousness occurs only minutes up to about an hour later, as in the observations of STEIN (1952), PAMPUS and Norbert MÜLLER (1956) could be demonstrated.

In the case of Jörg EIPEL one must assume, as COLMANT and DOTZAUER correctly noted, that the space-occupying brain edema developed already during the boxing match, that the comparatively small subdural hematoma was lifted over a critical threshold. The severe brain edema of the boxer EIPEL was resistant to therapy, a fact, which the two authors rightfully deduce from the continuing unconsciousness and the severe brain damage.

The presence of large so-called cortical contusions is surprising. Such lesions were described in the brains of some boxers (KREFFT 1952, PAMPUS and Norbert MÜLLER 1956, QUANDT and SOMMER 1967, PAYNE 1968). Up until now, literary descriptions of the so-called

Fig. 254 a and b. Coronal sections through the cerebral hemispheres. (From COLMANT, DOTZAUER, 1980).

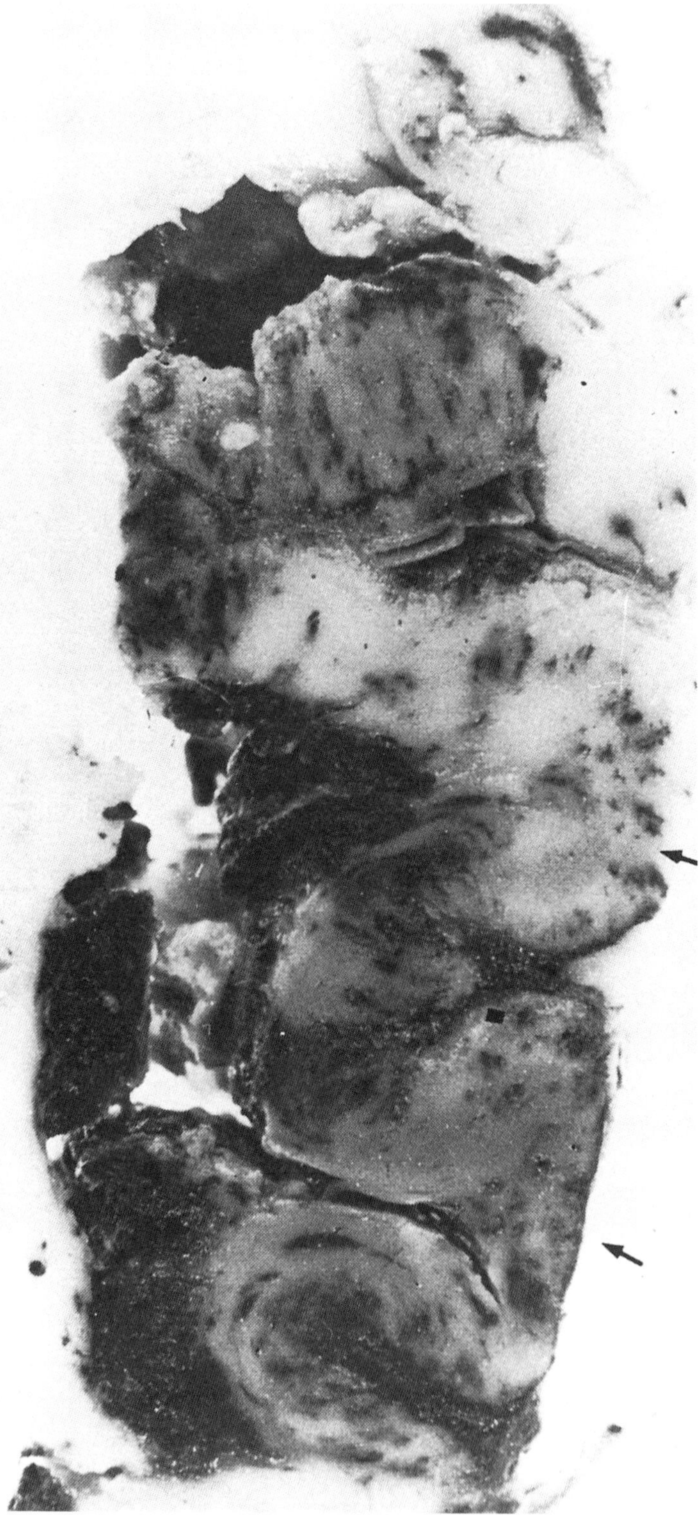

cortical contusions in boxers are, without exception, small. They are not generally explained as the sequel of boxing blows to the head, but due to impact of the head on the ring canvas, as a result of a deceleration trauma. EIPEL did not impact the canvas with his head, so that this attempted explanation here must be eliminated. The very massive and very extensive so-called cortical contusions are in this case an extremely unusual morphological finding, which was up to date not described or considered possible. The question I asked myself immediately was, are the described traumatic lesions really so-called cortical contusions? The injury pattern is from that of a so-called cortical contusion so different, the missing of an impact of the head on the canvas as a deceleration trauma, but especially not only due to its massive size, but especially to the quality of the tissue damage. Also the localization is so unusual, because they are in those areas, were they are normally not be found. Such large, we can use the word gigantic, so-called cortical contusions I have not seen before. If they are not so-called cortical contusions, what are they then? Most likely offers to me an attempted explanation that these are early stages of hemorrhagic necroses, which developed already in the brain of the still boxing EIPEL. The existing severe brain edema, already the result of the repeated blows against the head, the result of the punishment of EIPEL, and not mainly as the result of the subdural hemorrhage, which both lead to strangulation mechanisms of the brainstem against the tentorium cerebelli, which caused secondary-traumatic tissue alterations. Unfortunately I had no change to examine the histological slides of these tissue alterations under a microscope. Summarizing, these as so-called cortical contusions described tissue alterations are in my opinion no cortical contusion, but beginning hemorrhagic necroses. We must remember that such traumatic lesions do not occur in neuropathology, because there exists no syndrome in which such a repetition of physical force, of hammering the head and and brain, can be described.

COLMANT and DOTZAUER mention the very extensive partial and total necroses of the cerebral cortex and the basal ganglia (striatum, caudatum, putamen, pallidum, thalamus and subthalamus). The authors see in these tissue alterations a parallel to those, which were described by UNTERHARNSCHEIDT (1958, 1963, 1993) in an animal model. A detailed presentation and discussion of these tissue alterations was presented by UNTERHARNSCHEIDT (1963). They are located outside of those regions which are affected by strangulation mechanisms especially of those caused by the tentorium cerebelli and by so-called cortical contusions, as COLMANT and DOTZAUER rightly stressed. An explanation offer the in boxing indeed unique situation, in which numerous repeated blows to the head occur in short intervals, an event which can be labeled 'hammering' trauma or 'pummeling knock out' ('Verhämmerungs-knock out').

The existing so-called cortical contusions – we have to discuss here another aspect of the already mentioned tissue alterations – are rightly considered remarkable by COLMANT and DOTZAUER, because they could develop without a cerebral concussion. EIPEL had been knocked down, respectively had gone down on one knee for a few seconds, and was able to get up on his own to continue the boxing match. He did not hit his head on the canvas. During the boxing match he was never unconscious! And this boxer had the largest so-called cortical contusions! COLMANT and DOTZAUER consider this finding of general interest for neurotraumatology. From their own observations they are familiar with cases, in which a so-called cortical contusion without an cerebral concussion occurred, a quite relative rare occurrence, an observation to which I consent from my

Fig. 255. Coronal section through basal parts of cerebral hemispheres. Arrows show so-called cortical contusions. (From COLMANT, DOTZAUER, 1980).

own experience. But that such cortical contusions of such a gigantic size could occur especially in a patient who was not unconscious is not probable.

The above quoted animal models (UNTERHARNSCHEIDT 1958, 1963) clearly showed that in relative low accelerations of the head, a subconcussive syndrome occurred first. After a single application of this force, the animals were not unconscious, or if one should have uneasiness to use this term consciousness in an animal, the animals were responsive; they reacted. Only after repeated application of the force at several of the following one after the other, or days or repetitions in weekly intervals developed after a free interval of weeks a severe permanent brain damage of secondary traumatic nature, without any primary-traumatic lesions. If the intensity of the force was increased the syndrome of a cerebral concussion developed, the animals were immediately after the impact unconscious, respective unresponsive. But after a single impact with a concussive intensity there existed no morphological findings, see p. 76.

At histological examination of these animals there existed no so-called cortical contusions. They occurred for the first time after an additional increase of the impacting force. Then primary-traumatic lesions, among them so-called cortical contusions appeared. In the boxer EIPEL existed here apparently an event in the sense, that here allegedly primary-traumatic tissue alterations developed after higher intensities of the impacting force, and that a syndrome of a cerebral concussion, which should occur at a lower threshold, is missing.

At the following discussion of the pathomorphological alterations of the boxer Jörg EIPEL referred COLMANT and DOTZAUER that in the medical commentaries of the *German Amateur Boxing Confederation (Deutscher Amateur-Boxsportverband)* and the *Confederation of German Professional Boxers (Bund Deutscher Berufsboxer)*, edited by ANDERS et al. (1977) one author (Dr. FUNKE) noted: "Theoretically at a against a wall leaned head even at frequent and repeated impacts, inasmuch the boxer resist, in the brain nothing can happen, because the head is not accelerated, it occurs no change of velocity." This has to be considered an extremely silly comparison. Such an objection has nothing to do with the situation in boxing. The boxing ring is after all not surrounded by a wall, such impact at a fixated, not freely movable head, does not occur in boxing. In boxing the blows are landing always at a freely movable head, also when a boxer stands in his corner or near the ropes. That after all severe traumatic brain lesions, lethal ones, can occur at a fixed and not freely movable head, is sufficiently known in *so-called compression* or *crushing injuries of the head and brain.* A detailed discussion of these prototypes of brain injuries was presented by UNTERHARNSCHEIDT (1993).

COLMANT and DOTZAUER found in the brain of EIPEL a discrete diffuse faser gliosis in the white matter of the cerebral hemispheres, of which the authors could not exclude, that they in regard of the profession of the deceased represent the sequels of passée brain edemas. Here one should point to the young age of the deceased. In that way the finding gets a more considerable importance.

Adolf METZNER *(Nach der Ringtragödie in Köln). (After the ring tragedy in Cologne)* (*Die Zeit, no. 163, Tuesday, July 2, 1968, p.19)* wrote about the death of EIPEL:

> "Comparisons with the danger of automobile racing sport, as they are again and again by advocates of boxing are made, are misleading. Beside boxing there is no sport, at which a decision is possible which rests on damage to the health of the opponent or even causes his death. Boxing is therefore, strictly speaking, not a sport, but rather a duel. It is similar to a students' duel 'mensur') at which it comes to sharp wounds, to bodily injury. The first boxing instructors in England were primarily fencing masters. In substitute of a sword appeared then at arguments and duels the fist: 'The noble art of self-defence.'
>
> "As the mensur is already a half adaptation ('Versportlichung') of the former wild fighting of students, at whom no seldom deaths occurred, since safety goggles, leather apron, and so an was introduced, boxing became by introduction of gloves, which appeared in England already around 1800, but much later in the United States a presumable sport. As we know today, not at last by the studies of neurologists, is box-

ing in the current form only half a sport, since the protection by gloves is insufficient and serious damage is almost the rule. Thereby plays their weight, which is, as in many other sports, termed in an English unit of measure, here in ounces, no important role, alike whether the gloves weigh four ounces, as in professional boxing matches in the United states, or six ounces, as in professional, or more, as at the amateurs,"

"If now after the death of Jörg EIPEL, officials try to lay the blame on the new eight ounces gloves, so is this plainly ridiculous. The guilt lays in it, that boxing is no genuine sport, that in professional boxing, the boxers are nearly unprotected. No doubt the absurd distance of fifteen rounds includes much longer as they are possible at three rounds of amateurs, despite the fact that here death has occurred, too."

"At the fist fight between EIPEL and MARION both boxers were in the last rounds completely exhausted, because they had began in an actually fulminant tempo, that they could not, despite the one minute breaks, endure 45 minutes. Both thrashed at each other, rabbit punches were, as always at the profis, an everyday occurrence. A responsible referee would have had the boxing match beforehand stopped. At he latest the ring physician should have stopped this murderous 'scrap' at which heads became punching balls. But nothing happened, until EIPEL, already dying, lifted his hand up to sign to give up."

"When one tried to make excuses, saying the 'reflexes' had been 'fully existed,' demonstrated this only an awful ignorance. In the boxing parlance under 'reflexes' is the reaction understood, but it depends not so much, whether a boxer still reacts, but whether he is fully conscious. The consciousness was at both boxers temporarily muddy. EIPEL went in a break first into the wrong corner. He boxed only like in a trance."

"Naturally is the consideration for the paying audience or better said the fear these spectators who absolutely want to witness a merciless fight, instead of an elegant action. Concessions to lowest killer instincts are here involved."

"What should now take place? To prohibit boxing? Since 5.000 years has been boxed. The oldest sport historical presentations, which we know, are Sumeric boxers, from the beginning of the third century B.C. Not prohibit, but render boxing harmless and make it human, should be the order of the hour. The knock out as a decision for the fight has to be abolished and at first sign of transitory loss of consciousness the contest has to be stopped by an independent sport physician, who should be trained neurologically. The supervision about the professional boxer must be transferred to amateurs, as it is in professional soccer, too. Also the referees should be, as in soccer, amateurs."

METZNER concludes his report:

"Not the sensationalism of the moloch audience, but alone the worry about the maintenance of health and life of the boxer must determine all measures."

Mc QUILLEN et al. (1988) discuss in a paper *'Trauma, Sport, and Malignant Cerebral Edema'* sudden cerebral swelling and death secundary to cranial trauma.

In case 4 the authors ask the question 'Can encephalitis, even when asymptomatic, lead to sudden cerebral swelling in response to trauma?

The Boxer

S. M. was a 24-year-old schoolteacher who died five days after admission to the hospital in coma following a blow received while boxing. He had entered the ring in a supervised sparring match in which both contestants wore head gear. He received a punch to the head that knocked him down. Although he did not lose consciousness, he did look dazed and commented on the strength of the blow. Approximately 2 – 3 min later while sitting beside the ring, he became pale, his eyes rolled back, and he lost consciousness.

In a nearby emergency room, he was unconscious with bilateral positive Babinski signs and decerebrate posturing. An emergency CT scan revealed a small, right subdural hematoma and right-sided cerebral swelling. An emergency craniotomy revealed cerebral swelling involving the right hemisphere and a small, clinically insignificant subdural hematoma. He never regained consciousness. On the sixth day following admission, a radionuclide brain scan was performed and revealed no cerebral blood flow. The patient was pronounced dead.

The most striking finding on neuropathological examination was the microscopic appearance of the brain. There were prominent lymphocytic infiltrations in the walls of many small vessels including capillaries, venules, and small arterioles. There was no significant parenchymal cellular infiltrate.

Inquiries after autopsy revealed that the patient had suffered from the 'flu' for about two weeks prior to his sparring match, during which time he complained of headache. He had remained at home from his teaching position for two days prior to the match. A reasonable hypothesis to explain this catastrophic event is, accoring to the authors, that there was mild cerebral swelling secondary to encephalitis that dimished intracranial compliance at the time of the lethal injury.

The number of cases by the autors here is small and the experience is anecdotal. However, with one exception (case 1), the cases are based on clinical and autopsy data; and even without autopsy in case 1, the CT scan speaks for itself. Since the two cases of respirator-assisted survival were associated with the changes of respirator brain, the cellular reaction – absent in that condition – must be secondary to an underlying disease process rather than an artifact. These cases represent a unique type of sudden cerebral swelling. They may help explain the perplexing finding of rapid cerebral 'edema' in other cases of sudden death.

Case 1 occurred in a downhill racer, case 2 in a footbal player, and case 3 in a honeymooner who participated in scuba diving.

Chronic Pathomorphological Findings

Case Histories of Histologically Examined Brains of Boxers

Excellent studies about boxing injuries were published in the last decennia. Fatal incidents in the boxing ring have been analyzed, and a large number of clinical studies have shed light on the aftermath of boxing, especially with regard to permanent brain damage from repeated blows against boxers' heads. Regarding the pathomorphology of brain damage in boxers, we refer to the following sources: BRANDENBURG and HALLERVORDEN (1954), GRAHMANN and ULE (1957), the two cases of NEUBÜRGER et al. (1959), the observation of COURVILLE (1962, 1964), case no. 5 of SPILLANE (1962), case 10 of MAWDSLEY and FERGUSON (1963), QUANDT and SOMMER (1965), BURGER and MINAROVJECH (1966), CONSTANTINIDES and TISSOT (1967), the six observations of PAYNE (1968), the 15 cases of CORSELLIS et al. (1973) and the classic paper by DRACHMAN and NEWELL 1999).

Boxer who have been injured in the boxing ring with following death are not included in this chapter, but are discussed in the preceding chapter of 'Acute Pathomorphological Findings.' It is true, these boxer, especially, if they had a longer active career as boxers, reveal chronic tissue alterations in the brain, which can be considered chronic pathomorphological findings but in the pathomorphology of these case histories exist a combination of acute and chronic pathomorphological findings. It was thought to be a better solution to separate these two entities.

The following discussion of permanent brain damage in boxers is presented using histories, clinical, and pathomorphological findings from the sources cited above.

BRANDENBURG and HALLERVORDEN (1954) were the first authors who published a paper on permanent brain damage in a boxer.

They described a patient with normal development and a junior high school diploma, who became a tradesman. He began to box at 18 years old, and at 23 years old became the amateur German middle-weight Champion. He retained the title for a period of six years, and was known as a good tactician. He was knocked out several times.

His wife mentioned hard training bouts involving numerous head blows that had temporary effects without knocking him out. He gave up boxing after eleven years, at the age of 29, and continued to work as a tradesman. His boxing fellows noted that he had become 'soft.'

This case history shows very convincingly how damaging hard training bouts are, involving numerous head blows. At age 29, his sports friends noted that he had become 'soft.'

Ten years later, at 39 years old, the first symptoms of a slow personality change emerged. He became 'nervous,' restless, and anxious. He was frequently depressed, but now and then euphoric. He had difficulty falling asleep, and had nightmares. He experienced increasing forgetfulness, but retained the memories of his youth. He was discharged from the military service due to these cognitive disturbances. The forgetfulness became more severe in his mid-forties, when he even forgot significant events of his boxing career. He became talkative and chatty, and his speech lost clarity until it became indistinct.

A *neurological examination* at age 40 revealed no alterations of the cranial nerves, with the exception of paresis of the oral branch of the facial nerve. The tendon reflexes on both sides were brisk. The Babinsky on the left was suspicious, on the right, positive. Muscle tone was normal, no pareses or ataxia. Mimic rigidity was noteworthy. He appeared muddleheaded, and sometimes disoriented. He vacillated between depression and euphoria.

In the following years, his condition deteriorated; he became unfit for work and completely helpless. He lost orientation in time and location. He was childishly anxious, unable to keep himself occupied. He became irritated and annoyed when left alone. With increasing dementia, Parkinsonism became apparent.

At age 50, he was once again admitted to a hospital. At this time he had remarkable akinesia with a shuffling walk: short stepping, disturbed by propulsions, but without the usual associated movements. Mimic was completely stiffened and his body was bent forward. His fingers were continually engaged in money counting movements. The muscle tone was increased in all extremities, while muscle strength was reduced. There were brisk tendon reflexes bilaterally. Mentally, he was severely slowed down. Patient commented on suspension of thoughts, and that he was unable to remember things. Loquacious, restless, mostly obliging, but sometimes severely irritated. The diagnosis: post-traumatic Parkinsonism and brain atrophy.

Finally, stationary foster care became necessary. Periods of lost drive alternated with phases of restlessness and irritation. Three months after admission, he was found unconscious after a fall; he died three days later.

Cranial Cavity: The skull revealed no signs of injury or other abnormal findings. The dura mater over the right cerebral hemisphere was somewhat stretched; the inner surface was smooth. The sinuses did not show thrombosis.

The right cerebral hemisphere was more voluminous than the left one. The convolutions were moderately flattened, the grooves shallow. The relief of the left cerebral hemisphere revealed broad grooves and small convolutions. These signs of cerebral atrophy were uniformly distributed in all parts of the cerebral cortex, with varying severity. No distinct area of preference could be seen. No cortical contusions were found. The cerebellum, pons, olives and medulla oblongata were unremarkable.

The leptomeninx, especially the anterior parts over the cerebral hemispheres, were severely opaque in appearance. There was fibrous thickening in some areas, while other parts appeared 'milky.' The subarachnoid space was considerably widened and filled in with clear cerebrospinal fluid.

The arteries at the base and in the body of the brain were unremarkable, only solitary atheromatous plaques were visible in the intima. The lumen of the vessels were patent.

Examination of the brain revealed a nearly child fist sized mass hemorrhage in the white matter of the right cerebral hemisphere, centered in the frontal and temporal area. The hemorrhage extended into the orbital lobes, with destruction of claustrum, and putamen in the orbital lobes.

Coronal sections of the brain revealed considerable widening of the lateral ventricle in the left cerebral hemisphere. The third ventricle was less markedly widened. The ependyma had a smooth surface in all ventricles. The plexus choriodeus revealed no peculiarities. In addition to the enlargement of the ventricular system, the cerebral white matter had atrophied. Midbrain, pons and medulla oblongata were unremarkable.

Histological Findings: The leptomeninges, especially over the frontal lobes, were thickened and contained abundant collagenous fibers. The subdural space contained some blood, especially over the convexity and in the fossa Sylvii. Concentrations of lymphocytes and plasma-cellular elements and some macrophages were present.

There were vessel wall alterations in the subarachnoid space. The intima and the subintimal layer were mostly intact. The media, to the contrary, shows partially degenerative alterations, i. e. the nuclei of the smooth muscle cells were not stained. The muscularis seemed homogenized. Many vessels had widened adventitia, with abundant fibers. Some small vessels had thick homogenous walls, without discernable layers. In a larger branch of an artery near the island of Reil, a hemorrhage extended into the vessel wall, destroying the media in a larger area, and generating a development resembling a dissecting aneurysm. Atheromatous alterations were not visible in the meningeal vessels. There were solitary occlusions of small arteries with streak like necrosis of the upper and median layers of the cerebral cortex.

In all areas of the cerebral cortex and in the subcortical white substance, vascular alterations were noted and described. The walls of the small arteries were considerably thickened, without narrowing of the lumen. The intima was well preserved. A subintimal layer and elastica were not discernable, since the intima of the vessel wall became homogenous. There was no muscularis. The homogenous layer was relatively sharply demarcated peripherally, and was clearly distinguishable from a pale disintegrated margin consisting partly of mesenchymal, partly of glial fibers (Fig. 256). *Elastica staining* did not show fibers. The homogenous vessel wall was stained red by the van Gieson technique, blue by the Azan-Mallory technique, and green under Masson-Goldner. Using Kongo-red some vessel walls were tinged reddish. Some were metachromatically tinged with cresylviolet. Sudanophilic substances were barely visible; only single vessel walls showed a pale reddish tone with Sudan III staining. The Bielschowsky staining technique revealed that the apparently homogeneous vessel wall consisted of a threadlike, granulated argentophilic substance. It permeated the entire vessel wall and was sometimes visible in the neighborhood outside of the wall (Fig. 257). Cup- and balloon-like silver stained, druse-like structures adhered to the vessel wall. Apparently it was a substance originating in the lumen and subsequently invading the vessel wall. It saturated the entire vessel wall and the surrounding area. Its precipitation products were argentophilic. This is the substrate SCHOLZ called the 'drusige Entartung' of the brain vessels.

Less significant alterations were also seen. Scattered tiny intramural hemorrhages in small arteries were found, with dissecting aneurysm-like alterations. There was ubiquitous, substantial enlargement of the adventitial space of small vessels. They had very few connective tissue fibers without corpuscular elements between the vessel wall proper and the peripheral

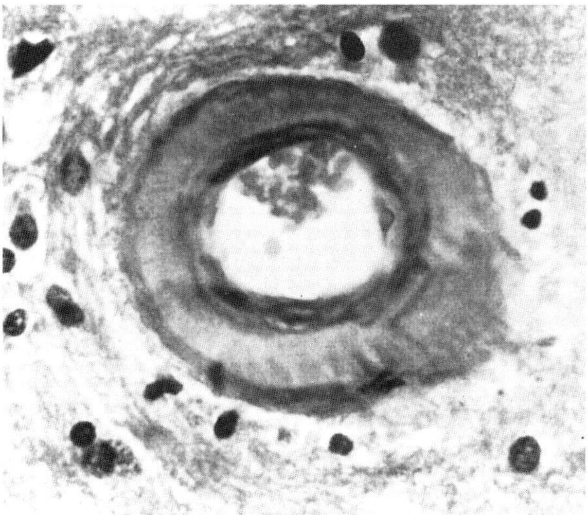

Fig. 256. Small cortical artery with thickened homogenized wall. Hematoxylin-Eosin technique, 800: 1. (From BRANDENBURG W, HALLERVORDEN J, 1954).

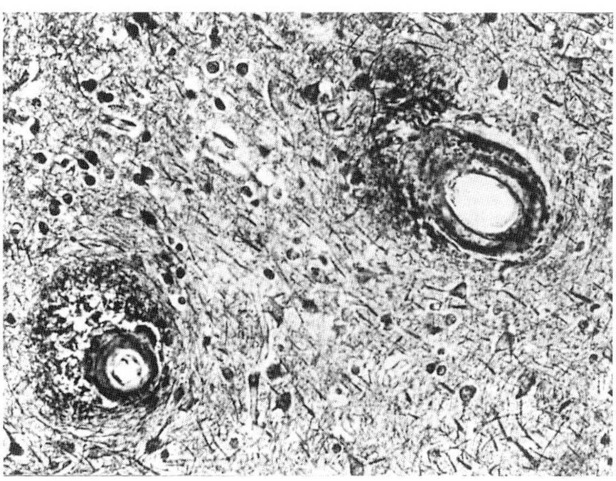

Fig. 257. Small arteries of the cerebral cortex with thickened wall and accumulated plaques. Bielschowsky technique, 200 : 1. (From BRANDENBURG W, HALLERVORDEN J, 1954).

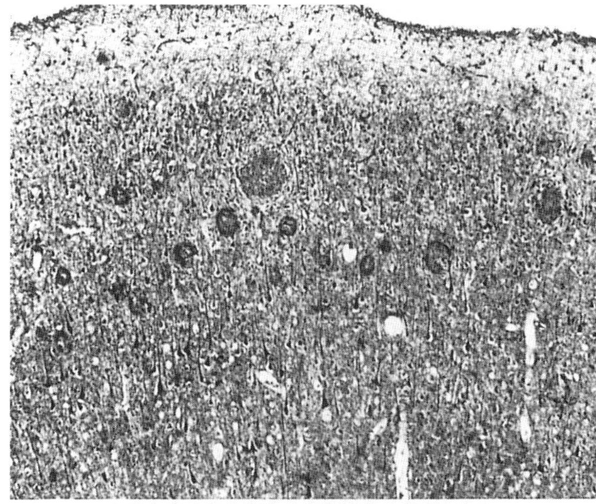

a

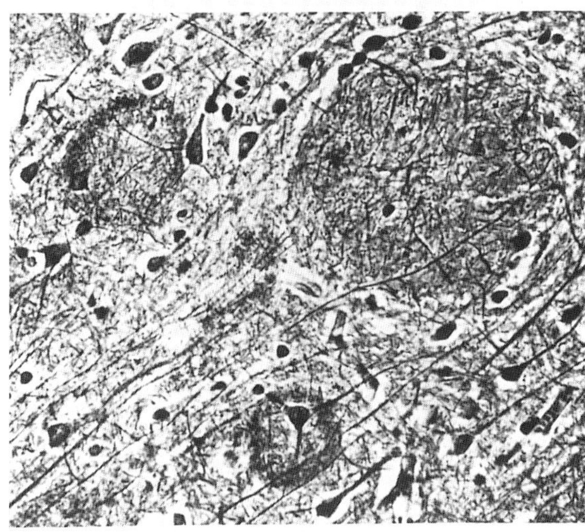

b

Figs. 258 a and b. Plaques in the cerebral cortex. a. Synopsis, 60 : 1. b. Single condensation plaques, 300 : 1. Bielschowsky technique. (From BRANDENBURG W, HALLERVORDEN J, 1954).

glial tissue. The condition of the walls of these small and medium-sized vessels – mostly arteries – was obviously the result of edema. They were found in the cerebral cortex in large numbers, as well as in the cerebral white matter and the basal ganglia.

In the upper cortical layers the tissue was consistently disintegrated and had the appearance of spongiosis. Here, and in deeper layers, the cortex was filled by round or ovoid structures that looked like elastic fiber tangles. They contrasted very well against the porous surrounding area (Figs. 258 a and b). They appeared lightly red or rose-colored by staining with *Hematoxylin-Eosin technique,* while they were less visible when the *thionin staining technique* was applied. They varied in size, but were about the diameter of a medium-sized vessel in the cerebral cortex. They were the so-called *'primitive'* or *'Verdichtungs' ('condensation) plaques').* The tissue elements seemed to be largely displaced or wiped-out, though now and then single undamaged nerve and glial cells were seen. The *Bielschowsky technique* clearly revealed a distinct coiled fiber texture. Fibers were more stain impregnated in some parts than in others. Using the *Azan-Mallory staining technique* they appeared blue-gray, the *Masson-Goldner staining tech-*

nique brought out a green tone. Thus, two additional plaque forms in Alzheimer's disease could be distinguished by using the *Azan-Mallory* and *Masson-Goldner staining techniques along with the silver impregnation technique* (Figs. 259 a, b).

(1) *Drusen with a compact nucleus.* This plaque was found to be *argentophilic.* Nodular, thickened ligamentous structures appeared at the nuclear periphery, surrounded by a clear halo with no impregnable substances. Veiled deposits about half the diameter of a nucleus were seen, and finally a corona, i. e. a broad meshwork of thread-like, granular structures.

(2) *Nuclear plaque without the halo* or *corona.* This plaque was smaller, forming a small, sharp, circumscribed dark spot.

(3) *Condensation plaque.* Along with these typical forms there were small thread-like or felt-like plaque structures, the size of a nerve cell or a little larger. Each was a small hirsuite appearing focus, sometimes isolated, some located on small vessels. They were most often found in the brainstem. Argentophilic material was found with the plaque: short, fiber trucks of nervous or gliogenous origin. The fibers were found in thickened, swollen, or coiled areas, usually as single fibers, though sometimes in multiples, forming a small thicket.

Of the three mentioned main forms, the *condensation plaques predominated,* but groups of nuclear plaques and drusen were found here and there.

Alzheimer's disease does not usually present with plaque in the white matter. However, in the case described by BRANDENBURG and HALLERVORDEN, they were relatively common. Most were small nuclear plaques, while the condensation plaque was rare. These three forms alone filled the entire cerebral cortex. They were more dense in the temporal lobes than in the frontal lobes. They predominated in the Ammon's horn formation (Fig. 260).

Nuclear plaques and *the nuclei* were *stained red with Best-Carmin.* They maintained a red appearance long after fixation of the brain, from which we can conclude that they were glycogen or protein bound. *PAS staining technique* tinged them slightly pinkish. With *Kongo-red,* there was a light red staining of the nuclear plaques and the drusen. BRANDENBURG and HALLERVORDEN concluded that an amyloid substance, or, to be more precise, a high molecular, starch like compound was present. The vessel walls were homogenous, with similar alterations, but to a much greater degree.

The *Azan-Mallory technique* yielded deep blue staining of the altered vessel walls, in the nuclear plaques and in the nuclei of the drusen. However the condensation plaques only turned a light blue-gray tone. The green staining due to Masson-Goldner was the same. One can only conclude that the substance producing homogeneity of the vessels was also present in the druses and nuclear plaques and, to a lesser degree, in the condensation plaques. Using myelin staining preparations, the plaques were spared, so that the lower parts of the cortex looked perforated, like a sieve (Figs. 262 a and b).

The close proximity of the vessels was already pointed out. The capillaries were particularly involved. They appeared both at the periphery of the plaques and penetrating them. The intima was consistently intact. A dense, argentophilic substance was often visible on capillary walls, sometimes wid-

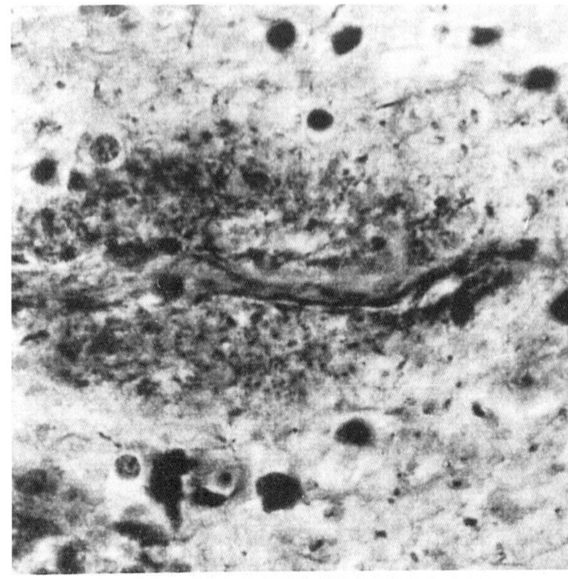

a

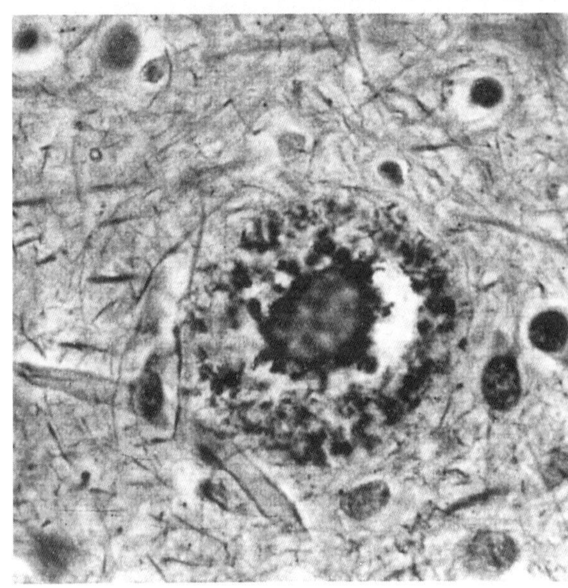

b

Figs. 259 a and b. a. Condensation plaques, of capillaries traversed, cerebral cortex. Bielschowsky technique, 800 : 1. b. Plaque from the cerebral cortex with nucleus, halo and wreath. Bielschowsky technique, 800 : 1. (From BRANDENBURG W, HALLERVORDEN J, 1954).

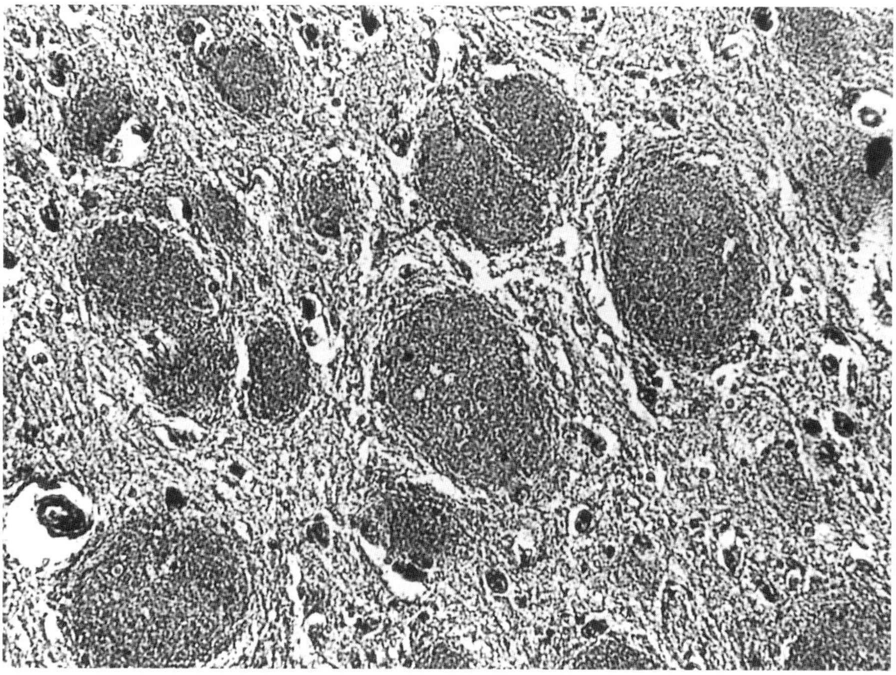

Fig. 260. Numerous condensation plaques in the cortex/white matter zone of the Ammons' horn formation. Azan-Mallory technique 200 : 1. (From BRANDENBURG W, HALLERVORDEN J, 1954).

ening near a plaque nucleus. A halo and corona frequently surrounded them. Due to the *Bielschowsky preparation,* one must presume that the thread-like, granular substance emerged from the capillary walls. There were also capillaries partially covered by thorny, cristalloid masses, as described earlier by SCHOLZ.

Many neurons showed *Alzheimer's fibrillary tangles* (Figs. 261a, b) consisting partly of a thickening of endocellular fibrils, partly in a coil-formed structure inside of the cells. Frequently a tract of the dendrites' axis cylinders had thickened.

The *architecture of the cerebral cortex* was somewhat disturbed by the alterations and interrupted by gaps caused by the plaques. The stratification remained visible, though cresylviolet stained them very little or not at all. In the gaps, single shrunken neurons or glial cells appeared in dense piles, with increased numbers of glial cells. In addition to *neuronophagia, shrunken, hyperchromatic nerve cells* were visible. *Cellular swellings* and *pigment atrophy* were also seen. The lipoids were hardly altered at all.

Findings in the caudate nucleus and the putamen were identical to those seen in the cerebral cortex. Plaques were common in both structures, with small bundles of recognizable myelinated fibers transversing the centers. Surprisingly, considering the condition of the total cell population, many neurons and glial cells were preserved. There was neuronal loss with a proliferation of astroglial elements. *The large neurons were often swollen with Alzheimer's fibrillary tangles.* The vessel alterations resembled those of the cortex, however, to a lesser degree. Many plaques were found in the claustrum.

Most neurons there were shrunken in the pallidum, but plaques were absent. Alterations of the fibrils were rarely visible and plaques too, in the hypothalamus. The thalamus contained numerous plaques and as well as some fibrillary tangles in the gray nuclei.

The mesencephalon contained single plaques and fibrillary tangles. The *black zone of the substantia nigra* revealed *pigment dissemination in the glial cells* and *near the vessels. There was scant loss of neurons.* Some of the ganglion cells were swollen. The pigment was displaced to the side and there were many fibrillary tangles in neurons, about five to ten in one section. The loss was less severe than one would have expected to find in a case of post-encephalitic Parkinsonism. A loss of specific cell groups, as is typical for the paralysis agitans, could not be seen.

The *pons* held single small plaques, especially in the tegmentum, but also in the ganglia at the base of the pons. Fibrillary tangles were visible only in the tegmentum, especially in the locus coeruleus, about five to eight in a section, therefore much more than in the layer of the substantia nigra. There is greater loss of melanin containing neurons here than in the locus coeruleus.

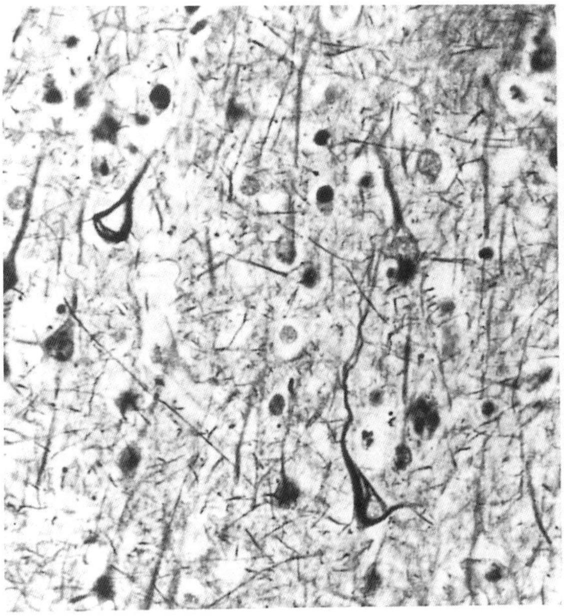

a

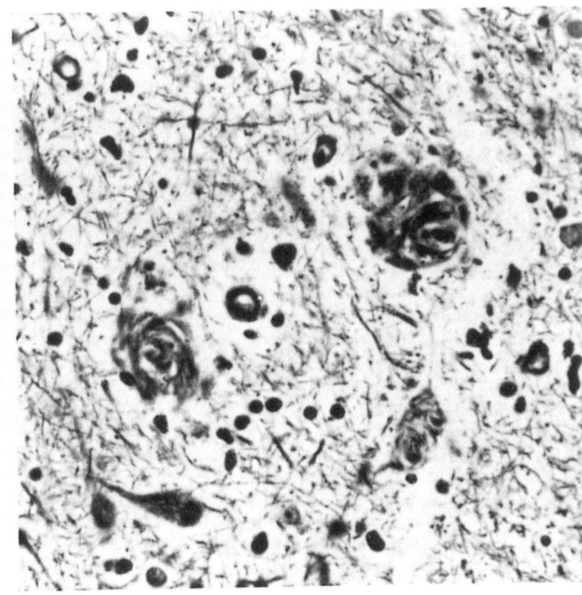

b

Figs. 261 a and b. (a) Myelin technique. (b) Azan technique. One can discern myelin bundles of the putamen and numerous condensation plaques, which in (a) appear blank, in (b) as dull-grey spots. 35 : 1. (From BRANDENBURG W, HALLERVORDEN J, 1954).

Some plaques were found in the tegmentum of the medulla oblongata. The previously described small felt-like structures, with preserved melanin containing cells, were especially numerous in the area of the vegetative nucleus of the vagal nerve. The neurons in the olives were shrunken, but otherwise not altered. The pyramid tract was intact.

The *cerebellum* suffered the *loss of some Purkinje cells.* Other cells there were shrunken, but these alterations were present only to a modest degree. Fibrillary tangles were not found in the cells of the cerebellum. The molecular cell layer and the granular cell layers revealed nuclear plaques throughout, though only rarely between Purkinje cells. The dentate nucleus and its cells were intact, with some isolated plaques, but no fibrillary tangles.

The myelin technique, when applied to the spinal cord, showed some lightening and glial proliferation in the pyramid tracts. The neurons in the anterior horns and the spinal roots were well preserved. No plaques were visible in the spinal cord.

All iron staining techniques were negative. Only a few plaques turned a light bluish tinge. Remnants of older diapedetic or rhectic hemorrhages were not visible at all. No further hemorrhages were visible outside of the vessels around the intramural vessel hemorrhages.

No inflammatory alterations were found. The mass of the white matter in the cerebral hemispheres was reduced, due to brain atrophy. The white matter of the occipital lobes showed a moderate lightening around the posterior horn of the lateral ventricles, probably due to edema. Otherwise, no other alterations were seen.

The 51 year old former amateur boxer, who boxed from the age of 18, became the middle-weight German Amateur Champion for six years, and was knocked out several times during his pugilistic career. At age of 29 years, after a career of eleven years as an amateur, he retired from the ring. At age of 38 years, therefore nine years later, personality changes developed. He became irritated, forgetful, and his speech became indistinct. At the age of 50 years, a Parkinsonism developed, with dementia. The patient died one year later due to 'apoplexy.'

The *autopsy* revealed a mass hemorrhage in anterior parts of the right cerebral hemisphere with moderate brain atrophy and dilatation of the ventricular system, but no cortical contusions. Some vessels in the subarachnoid space and in the basal ganglia revealed intramural dissecting hemorrhages. It is worth noting here that the vessels at the base of the brain were not altered in appearance at all.

Histological examination revealed degenerated drusen, ('drusige Entartung'), of the brain vessels and a great proliferation of senile plaques, among which condensation plaques dominated. They were not only in the entire cerebral cortex, but

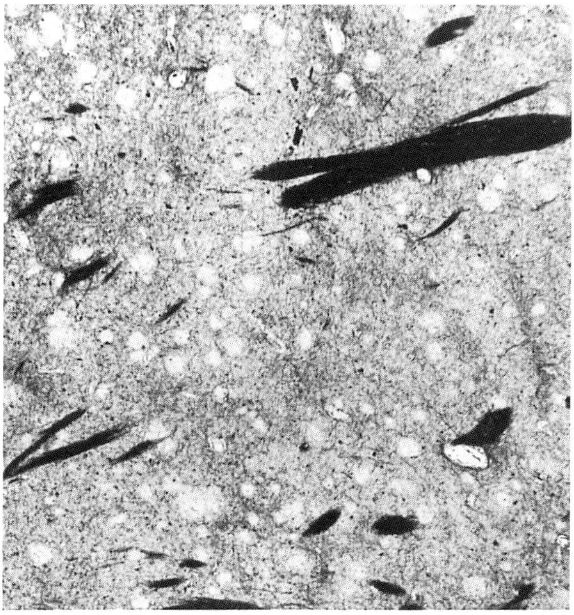

a

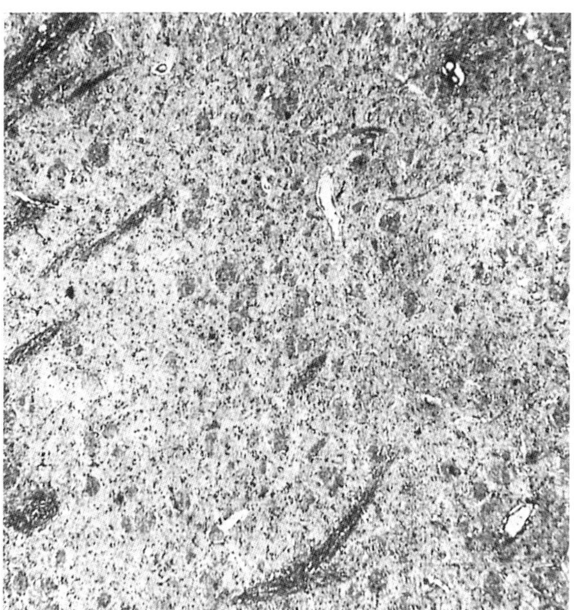

b

Figs. 262 a and b. Alzheimer's fibrillary tangles in nerve cells. (a) Two fibrillary tangles from inner cortical layers. (b) Three tangle like interlaced fibrillary tangles in the melanin containing nerve cells of the locus coeruleus from the pons tegmentum. Bielschowsky technique, 300 : 1. (From BRANDENBURG W, HALLERVORDEN J, 1954).

also in the striatum and thalamus. They were found in fewer numbers throughout the brain stem.

Additionally nuclear plaques were seen in the molecular cell and granular cell layers of the cerebellum. Ample Alzheimer's fibrillary tangles of the neurons in the cerebral cortex and in the striatum were visible, and also in the hypothalamus, the substantia nigra, and in the locus coeruleus. In the spinal cord, a moderate degeneration of the lateral pyramid tracts was found.

The histological picture was dominated by profuse fibrillary tangles of the plaques, and the degenerated druses of brain vessels, as one would expect to find in severe cases of Alzheimer's disease.

Plaques and fibrillary tangles are considered to be characteristic, concomitant symptoms of the aging process, but they can be missed? Interestingly and significantly they are never found in old animals. Because they are found with different diseases in all stages of human life, including juveniles, they cannot be exclusively attributed to the aging process.

The appearance and distribution of each of the three forms, especially using the aforementioned staining techniques help predict future developments.

Suggestions:

Although the precise origin of the three plaque forms is still controversial, it is safe to presume that they are composed of argentophilic substances deposited in the tissue. A discussion of plaque form genesis was published by BIELSCHOWSKY (1910) and SPIELMEYER (1922).

SPIELMEYER (1922) suggested that Alzheimer fibrillary tangles are argentophilic deposits on the intracellular fibrils, with parts of their dendrites and axis cylinders. They are pervasively impregnated and after destruction are visible in the tissue.

BRANDENBURG and HALLERVORDEN pointed out that the same argentophilic products forming the plaques are also deposited in the vessel walls. Their own case showed this, and it is also found in cases of Alzheimer's disease.

These authors stated in 1954 that their published case is evidence of a special form of traumatic Parkinsonism. They also observed that such cases were probably exceptional and rare. The Parkinsonian picture is the result of damage in the substantia nigra or globus pallidus.

The substania nigra alterations described by these authors matches the cell loss known to occur in cases of paralysis agitans.

The conclusion reached by BRANDENBURG and HALLERVORDEN, that they had presented a case of post-traumatic dementia with Parkinsonism, was reached by sound technique and analysis, and has stood the test of time. Those who argue that Parkinsonism cannot be caused by trauma must account for the findings of these and other subsequent authors, with sound alternative explanations, in order to prevail. Since these findings appeared in print, no alternative theories worth debate have been advanced to explain them, and we are personally unaware of any under development at this time.

The paper was important because it described the first full morphological examination of a boxer's brain, in which the clinical picture and specific detailed morphological findings are directly attributable to the numerous head blows received during a professional boxing career. The pathomorphological alterations in the substantia nigra were also discussed.

While we fully agree that the case presented post-traumatic dementia with Parkinsonism, there is doubt about their additional diagnosis of 'delayed apoplexy.' This diagnosis has not been confirmed by subsequent inquiry, and we find it unconvincing.

The authors stressed the serious consequences of repeated blows to the head appropriately, but they placed excessive importance on the effects of the relatively few knock out blows received. It is not possible to identify a set of specific blows as the cause of broad spread alterations such as those described above. Rather, they are the cumulative result of numerous head blows, many in rapid succession. Although a single blow can cause damage, the morphological findings would be different. Subsequent findings have shown that the pervasive damage in this case was not caused by several knock out blows alone, but by those blows combined with thousands, or tens of thousands more head blows received throughout his active career. Professional boxers such as this one sometimes receive hundreds of blows in a single match. Although a single blow is certainly capable of causing injury, the quality and distribution of pathomorphological changes would be different from those described above.

On publication, the work of BRANDENBURG and HALLERVORDEN was decried as a coincidence: the case of a boxer who also happened to have Alzheimer's disease.

The fact that many years had passed between boxing and the recognition of clinical signs raised even more objections. One such critic was PETERS, who discussed the relationship between brain injury, senile plaques, and Alzheimer's fibrillary tangles, in 1962. He quoted NEUBÜRGER and von BRAUNMÜHL, NEUBÜRGER, and GRÜNTHAL, who had published suitable cases following single head injuries.

NEUBÜRGER and von BRAUNMÜHL described the brain of a 56 year old man, who suffered a head injury 34 years before his death and was later diagnosed as having a 'traumatic dementia,' with numerous senile plaques.
NEUBÜRGER found an abundance of senile plaques near alterations that were known to have been the result of a brain injury.
GRÜNTHAL published findings in the case of an 80 year old patient who had strikingly dense senile plaques in the cerebral hemisphere that had been injured. This patient had suffered a brain injury from a cobblestone. The injury caused a cicatrization of scalp, skull and dura mater over the right parietal lobe; the right cerebral hemisphere was shrunken in appearance.

PETERS discussed BRANDENBURG and HALLERVORDEN's 'dementia pugilistica,' i. e. their suggestion that numerous blows against the head cause circulatory disturbances, acidosis, and increased permeability of the blood brain barrier, and that these events trigger a syneretic syndrome.

PETERS expressed doubts about this theory, because, in his opinion, it is possible to have pre-senile dementia without boxing related brain damage. PETERS denied the possibility that additional cases might change his opinion. In particular, he was not persuaded by the case of a 48 year old boxer, published by GRAHMANN and ULE (1959).

PETERS continued to say that the observations of BRANDENBURG and HALLERVORDEN, and GRAHMANN and ULE differed from findings of previous authors. Others had discussed repetitive head trauma occurring over the course of several years, in relatively short intervals. PETERS referred to results of extensive experimental investigations by UNTERHARNSCHEIDT (1963), quoting discussions on secondary traumatic circulatory and tissue alterations in the brain. "But UNTERHARNSCHEIDT did not find senile plaques and Alzheimer fibrillary tangles" (PETERS 1962). In response, it is correct to say that I did not find senile plaques and Alzheimer's fibrillary tangles in my experimental material. However, those alterations have never been found or described in animals. BRANDENBURG and HALLERVORDEN had suggested this point already in their paper. In any case, even if fibrillary tangles do exist in animals, it is clear that they do not appear quickly. The survival time of the subjects in my study was much too short for such changes to develop, so my findings could not have been dispositive of this particular issue.

Finally, PETERS said that it is not possible for single injuries to set the syneretic syndrome into motion prematurely or frequently.

It would be hasty to conclude that multiple blows do not produce the pattern of alterations and plaques described above simply because a single blow cannot produce the same pattern. Multiple blows may produce an even greater variety of morphological changes than single blows. It bears repeating that in a single match, a boxer may take hundreds of blows to the head, within a very short period of time. Over a few short years the number of head blows might well reach into the thousands. The punches are delivered by well trained individuals with the specific intent to cause physical incapacitation. The blows received in training and exhibition over the average boxing career must be a staggering figure, especially considering the intensities with which they are often delivered.

The criticism against BRANDENBURG and HALLERVORDEN and GRAHMANN and ULE may have been well founded initially, but studies during the following years confirmed their observations and silenced the critics and ended the debate in their favor. In particular we refer to the 15 cases of CORSELLIS et al. (1973). Their exhaustive analysis and discussion of the connection between specific alterations and the effects of boxing is still well worth reading. For details see p. 488.

PAMPUS and Norbert MÜLLER (1956) were of the opinion that BRANDENBURG and HALLERVORDEN's observations were not convincing, simply because the first clinical symptoms did not appear until nine years after retirement from boxing. They felt that the difference between morphological findings in the reported case and in pre-senile dementia were insignificant.

GRAHMANN and ULE (1957) published the following observations:

At the age of 15 years, the patient joined an amateur boxing club and traveled some years with a boxing booth. Allegedly, he was never knocked out, but was often benumbed and groggy. After ten years of boxing, he became 'soft,' he gave up boxing at the age of 25 years and worked as a lock-keeper. At thirty-five years old, he had his first attack of convulsions of the right arm and leg, without unconsciousness. Repeated attacks occurred, at irregular intervals; they ceased after one year, without treatment. His wife noticed the first mental changes in the patient, who was now 36 year old. He was slowed down, uninterested, sitting around idle, cried about nothing and became more forgetful. During this time he perspired noticeably, having taken a corresponding amount of additional fluids. The patient had never been seriously ill, with the exception of a nasal fracture from boxing; he did not smoke and drink.

Admission to a hospital: Robustly built, with coarse features, traumatic saddle-nose, deformed ears ('cauliflower ears'). Expression was poor, staring gaze, left corner of his mouth less innervated. Slow, sluggish speech. Patellar tendon reflexes more brisk on the left than the right.

EEG: Slightly abnormal, but with no local disturbances or spikes.

PEG: External and internal hydrocephalus.

Mentally: Lack of impulse, retardation, impeded comprehension, memory decreased.

Could not hold a steady job due to incompetence, discharged from work and unemployable. At this point deterioration was quick; his wife found him "motionless and dull." He was good-natured, in general, but sometimes became irritated and explosive. He required providing for, like a child: dressing, toilet, did not find his way in the apartment, immoderate eating. Libido and potency completely lost.

After one year he was *once again admitted to a hospital,* this time in considerably worse condition. Extremely demented, destitute, filthy, lying in bed without movements, an imbecil simper at his lips. He took no notice of the surroundings, interested only in eating. He stuffed food eagerly into his mouth, until he was no longer able to close it.

Personally and locally oriented, but always gave the wrong date. Memory lapses over the past ten years. Decreased memory, amnesic functions were not as bad as on might expect from his desolate condition. The handwriting, still intelligible just two years earlier, was now a narrow, small scribble. He was yielding, willing, good-natured, could not distinguish between left and right. Associated movements of the left arm were missing, short-stepping, shuffling gait. Rigid face, unmodulated speech.

PEG revealed further enlargement of the ventricular system since the last examination one year before.

After a pneumo-encephalogram, his condition deteriorated. He had Jackson seizures of the left arm and the left side of his face, which changed soon into a status epilepticus. Increasing absence of mind, high temperatures, with profuse perspiration. He died because of circulatory insufficiency.

Morphologically, no old cortical contusions were found. Dilated ventricular system, large cavum septi pellucidi. Moderate loss of neurons in the cerebral cortex. Marked loss of cells in the compact zone of the substantia nigra with abundant melanin, lying free in the tissue. The cells of the tegmentum show alterations of the Alzheimer's fibrillary tangle type. In the cerebellum, Purkinje and granular cells were lost. Fibrillary tangles were found in smaller numbers, also in the substantia nigra and occasionally in cells of the thalamus.

In the cerebral cortex, the gyrus hippocampus, and in the neuronal band of the Ammon's horn formation, fibrillary tangles were numerous. In the frontal lobes, the parietal lobes, and the occipital cortex there were only isolated fibrillary tangles. Senile plaques could not be detected using silver impregnation methods or PAS reactions. The cerebral vessels did not reveal peculiarities using the PAS method.

This former amateur boxer had boxed from his fifteenth to his twenty-fifth year of life, therefore over a space of time of ten years. At 46 years old he developed dementia. Focal neurological findings were not found. On the other hand, disturbances of the extrapyramidal system were present, with a progressive internal and external hydrocephalus. The patient died at age 46, due to a hemorrhagic infarction of the right frontal and parietal lobes, due to a thrombotic occlusion of the dural sinus and veins of the meninges. The brain was atrophic. It revealed no plaques; however, there were neurofibrillary alterations.

The authors referred to the diagnosis 'dementia pugilistica', which they assigned to cases with progressive disturbances occurring after a free interval. Neurological disturbances in boxers should be diagnosed as 'traumatic encephalopathy of the boxer'. Their definition is a good one, and is still appropriately used today.

We have already mentioned that BRANDENBURG and HALLERVORDEN's case was not unconditionally accepted as a condition caused by boxing. Some authors (PETERS, PAMPUS and Norbert MÜLLER) referred rather to the clinical and pathomorphological similarity to Alzheimer's disease. They were troubled by the lengthy symptom-free interval between retirement from boxing and the first clinical findings. GRAHMANN and ULE had already responded to the free interval issue. It is well-known that broad application of the silver impregnation method proves that these alterations are not, as was formerly believed, an indicator of a specific disease. Rather, they are an attendant facultative symptom resulting from many different pathological processes in the brain.

GRAHMANN and ULE are correct in stating that the fundamental conformity of the findings in both boxing cases subjected to careful analysis is no accident. Could one at the observation of BRANDENBURG and HALLERVORDEN still consider an accidental coincidence of two different disease entities, perhaps a traumatic brain injury and an Alzheimer's disease, so is this to reject since the second pathological examined observation offers essential the same findings, so that a doubt is no longer justified.

GRAHMANN and ULE summarized the most important aspects with remarkable clarity, given the current state of knowledge. Individuals who have boxed for many years have suffered many hard head blows, from which different types of brain damage are to be expected:

(1) Chronic progressive syndromes with a more or less distinct Parkinson-like picture occur only after a free interval. Here a mental disintegration is in the forefront, so that one may speak of ' boxer's dementia' or 'dementia pugilistica.'

(2) Observations with stationary deficits, especially in the neurological sector, which develop in a direct temporal connection with the suffered repeated blows against the head, which are based in a generalized destruction of neurons with resulting scar tissue development. GRAHMANN and ULE are of the opinion, that these predominant neurological defect syndromes are best characterized with the terminus 'traumatic boxer encephalopathy' or 'encephalopathia pugilistica.' They consider it quite possible that a stationary defect syndrome does not rule out the appearance of the progressive clinical picture, dementia pugilistica. In such cases the characteristic free interval is filled with neurological defect syndrome, to be exacerbated later by the appearance of dementia pugilistica symptoms.

(3) Clinical syndromes with paranoid and catatonic psychoses, at which at that time the pathogenesis and especially the causal relationship between the peculiarities of boxing were extremely problematic and yet not possible to decide. Here too, were published in the following period many additional observations, that the question of a connection with the suffered repeated head injuries at boxing, apart from some exceptions, could be positively decided.

A span of 15 and 20 years, respectively, separates the end of their active pugilistic careers and the time they came to medical attention in the cases of *two professional boxers* reported by NEUBÜRGER et al. (1959).

Case 1 was a 46 year old man, had boxed from age 14 to age 26 and, as a light heavyweight and heavyweight. He had 140 professional fights, in the course of which he was knocked out 30 times. He then toured the fairs as 'a fighter who would take on all comers.' He claimed he was never knocked down during this period but conceded he was frequently 'out on his feet.' From age 50 to 53, he was observed frequently and complained continually of headaches, occasional blackout spell, and tremulousness. He was unable to work consistently, but did odd jobs.

Neurological examination revealed an intermittent fine, rapid tremor of the head that occasionally spread to the arms, particularly the right one. A slight increase in tone was noted in the right arm, along with some decrease in ability to perform rapid alternation movements. There was a defect of upward gaze. Otherwise, he was neurologically unremarkable.

The *EEG* showed intermittent bursts of 4- to 5- second slow waves, with or without sharp waves, apparently originating in the left temporal area.

Psychic tests revealed loss of immediate recall and of ability to learn new tasks. The full-scale intelligence performance was 84.

Biopsy of the frontal cortex exhibited a slight numerical reduction of nerve cells in the lower layers, where there was also a striking diffuse proliferation of large astrocytes. A suggestion of faintly argentophilic structures in the form of thin, kinked fibrils was observed in the immediate neighborhood of small vessels. They vaguely resembled small 'senile' plaques; however, typical plaques, fibrillary alterations, and vascular lesions were not found.

Case 2 was a 53 year old man who had boxed from age 18 to 24, having had ten bouts before the age of 20. Early newspaper clippings describe him as aggressive, 'always ready to give or take a punch.' He fought as a middleweight but apparently also fought heavier men. Numerous newspaper accounts stress his staying power, i. e. his ability to win or finish a fight after having been knocked down. The first entry in his scrapbook states that he was knocked down three times in the first round, but with his left leg almost paralyzed and barely able to support him, he continued, received numerous blows, and was again knocked down. The fight was stopped in the seventh round. He stopped boxing because of a paralyzed left side, and tried an unsuccessful comeback one year later. He then went into business with some success. His wife stated that at the age of 48 he became forgetful, confused, irritable, and moody.

His *examination* at the *Mayo Clinic* yielded the following observations: he was an affable, alert, restless patient, disoriented as to time and place, able to perform only the simplest calculations, and unable to find his way about Rochester unescorted. He showed an ataxic gait, decreased speed of motion in the left hand, increased tendon reflexes, and extensor plantar reflexes on the left. The impression was 'psychotic reaction – the result of organic brain disease – most likely in the nature of traumatic encephalopathy (punch-drunk).' The patient's condition progressively deteriorated, with several hospitalizations becoming necessary because of confusion, hyperactivity, loquaciousness, and 'nervous breakdown.' He died of progressively pulmonary insufficiency.

The authors concluded that trauma was the only factor that might have played a primary role in the cerebral disorder. The pulmonary insufficiency developed about three years after the onset of the disturbance of mental function, and it did not become severe until 18 months before death.

Severe atrophy was evident in the frontal lobes, more pronounced in the *right cerebral hemisphere* (Fig. 263). *Microscopically,* in conformity with the gross appearance, the most severe changes were found in the cortex of the frontal lobes. The convolutions were thin, the sulci were wide, and the ventricles were slightly dilated. The cortex was thinned and the stratification was partially lost; numerous nerve cells were absent (Fig. 264). A suggestion of status spongiosus was seen in the upper two layers of the occipital cortex. A fairly advanced degree of gliosis was observed, many large fibrillary astrocytes were present throughout the cortex, and a network of glial fibers occupied the cell layers. The adjacent and periventricular white matter likewise showed numerous astrocytes and considerable degree of fibrillary gliosis that was accentuated around vessels (Fig. 265). The hippocampus exhibited loss of nerve cells in the presubiculum, resembling that seen in senile and presenile brain disease. The myelin sheath stain showed advanced loss of myelinated fibers in the diseased cortex but only slight focal diminution of fibers in the white matter. Lipoid droplets were present about blood vessels in the white matter and were abundant in the substantia nigra. The cortical neurons contained greater amounts of lipid material than is usually seen at this age. There were no vascular lesions, senile plaques, fibrillary alterations of nerve cells, or intracellular argentophilic inclusions. The thalamus exhibited a mild plasmatic gliosis, but the number of nerve cells in both thalamus and lenticular nuclei was normal. The pyramids showed severe fibrillary gliosis, more advanced on the right side.

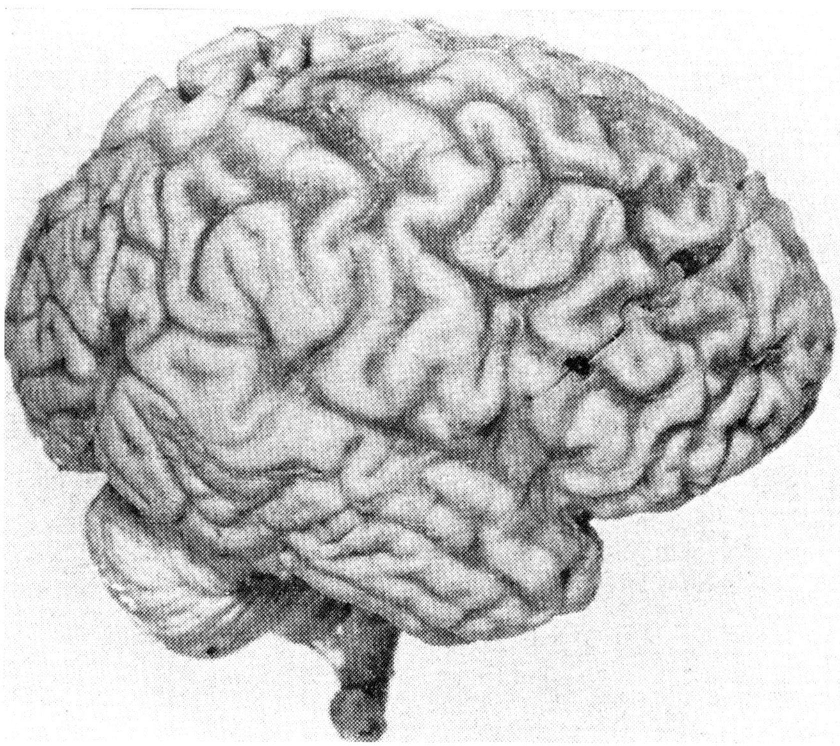

Fig. 263 . Case 2. Right cerebral hemisphere from lateral. Brain atrophy, especially of the frontal lobes. (From NEUBÜRGER K T et al., 1959).

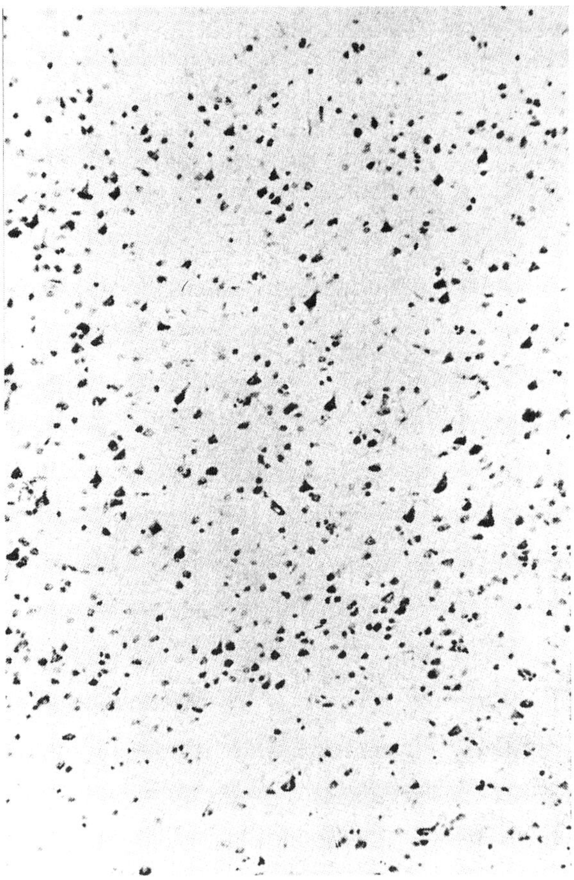

Fig. 264. Case 2. Atrophic right frontal lobe with loss of neurons. Nissl technique, 90 : 1. (From NEUBÜRGER K T et al., 1959).

The cerebellum exhibited a slight reduction in the number of granule cells.

The case published by COURVILLE (1962) is complicated by severe alcoholism and diabetes mellitus.

The clinical picture of the 49 year old *retired professional boxer* with four years of active fighting was complicated by severe alcoholism and diabetes mellitus, and the chronic brain syndrome which he exhibited cannot be conclusively attributed to his boxing activities alone.

This patient was admitted to a local hospital in a confused state, but he was able to say that he had tripped over a rug at the rest home where he was staying (he could not remember the name of the rest home). He had fallen on his out-stretched right hand, injuring the wrist.

Roentgenograms revealed a *Colles type fracture of the distal portion of the radius.* He was a diabetic and was treated with regular insulin, 25 units at 11 p.m., and 10 to 30 units later in the day, depending upon the amount of sugar in the urine. On the fourth day of hospitalization, after one dose of 20 units of insulin, the patient became more listless and stuporous, with depression of his respiratory activities. He failed to respond to the administration of oxygen and died that evening. The immediate cause of death was an *acute hypoglycemic episode* (blood sugar was 2.5 mg%, 90 minutes before death), the cause of which was not understood because he had received only 20 units of insulin that day.

The *postmortem examination* disclosed none of the usual signs of a pugilist, such as cauliflower ears or chronically swollen brows. His body was poorly nourished, and weighed only 85 pounds. The liver showed evidence of diffuse fibrosis. Focal hemorrhages were found in the adrenal cortex, especially along the cortico-medullary junction. The parenchyma of the pancreas was decreased in amount and showed extensive replacement by scar tissue. The lobules were decreased in number and size.

The *brain* weighed 1.120 g. The leptomeninges were considerably thickened and were composed of knot-like masses of collagen material, which was considerably hyalinized. The cerebrum showed atrophy of the frontal convolutions (Fig. 266). The convolutions of the upper dorsolateral surfaces of the frontal, central, and parietal areas showed a definite cortical atrophy. There was meningeal thickening in that area, especially adjacent to the superior longitudinal sinus, as there was over the base of the brain. The cerebral arteries forming the circulus arteriosus cerebri and the cortical vessels showed no evidence of atheromatous degeneration.

Sections through the cerebral hemispheres showed an incidental fifth ventricle; the lateral and third ventricles were somewhat enlarged. The cortical ribbon was narrowed, suggesting a diffuse degree of atrophy of that structure. Otherwise, no gross architectural changes were apparent in the brain stem, cerebellum, or cerebral hemispheres.

Examination of the cortical sections showed a marked patchy loss of the pyramidal cells, chiefly those in the intermediate laminae (Fig. 267 a). The changes in individual cells as seen in the Nissl preparations varied considerably, from a simple loss of tigroid material to structural deterioration of the entire cell. The preservation of cells in clusters was characteristic. In the reduced silver preparations, the variation in cell structure was also apparent. The variation, however, involved a deterioration of the neurofibrillar apparatus, which went through several forms of fine to coarse granular degeneration, or clumping, of accumulative portions of these fibrils (Fig. 267 b).

In the routine Hematoxylin-Eosin sections, the basal ganglia appeared to be sparsely cellular except for the bundles of

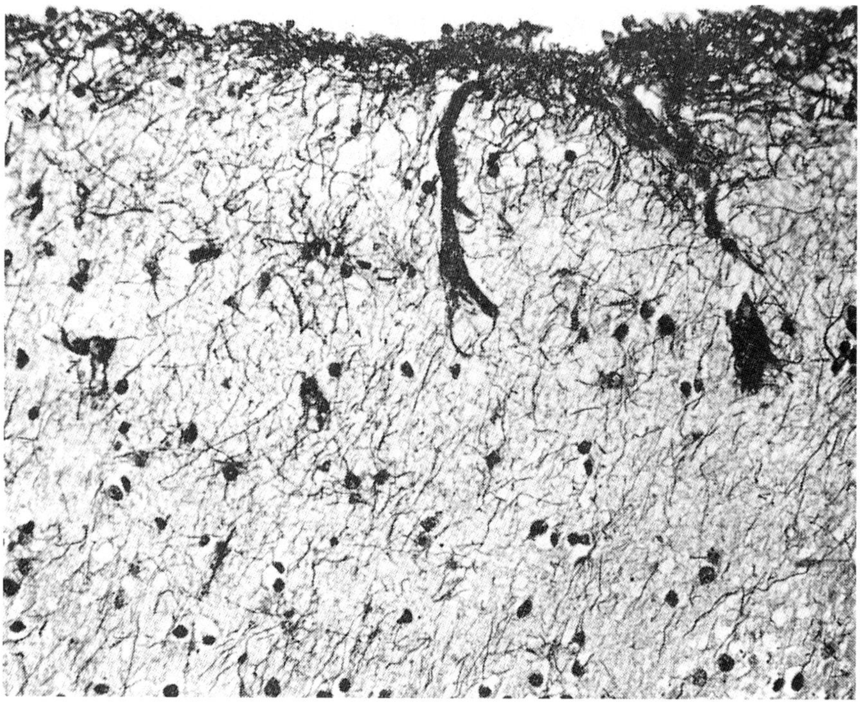

Fig. 265. Fibrillar gliosis and proliferation of astroglia in the frontal cortex. Holzer technique, 200 : 1. (From NEUBÜRGER K T et al., 1959).

white fibers passing through, which seemed to be marked by closely packed bundles of acutely swollen oligodendroglia. The large cells of the corpus striatum were greatly reduced in number and the few that remained showed a variety of friability. They showed total loss of tigroid material, vacuolar degeneration, loss of tinctorial coloring, increased lipoidal degeneration (Fig. 268 a), degeneration of neurofibrillar substance in the form of fine and coarse granulation, and clumping of argentophilic material (Fig. 268 b). On the other hand, the small cells were still present in large numbers, the chief morphological change being pyknosis.

The *reduced silver stains* demonstrated the nerve cells very well, but in addition showed up the oligodendroglia and possibly the microglia as well, so that this preparation appeared to be unusually cellular.

The *gold sublimate preparation* of the lenticular nucleus showed a remarkable degree of deterioration of the astrocytes, which appeared to affect almost all of these elements. It was impossible to find any resemblance to a normal cell of this type. The astrocytes were reduced either to irregular clusters of fragments or small masses, completely devoid of processes (Fig. 269).

The *Mahoney method of silver impregnation* showed an essentially normal pattern of blood vessels, although the scarlet red stain showed some small globules of fat around some blood vessels. The source of this fat was probably in the cells, indication of lipoidal degeneration.

Routine sections of the cerebellum showed *minimal loss of Purkinje cells.* Those remaining were either swollen and devoid of tigroid material or shrunken and deep purple in color (pyknotic change). This layer of cells was enveloped by a zone of fluid. The granule cell layer appeared to be entirely normal throughout, as did the white matter.

From the series of former boxers published by SPILLANE (1962), *case 5* is discussed here:

A laborer aged 45 entered the Department of Neurology in March 1962. He started to box when he was ten, and turned professional at 18. Until he retired at the age of 32 he had approximately 300 fights in various countries. His wife thought that his ill-health began about that time. The first symptom noted was a progressive slurring of speech. He had been drinking heavily for a few years, and there were bouts of depression. However, despite the alcoholism there were long periods of abstention, but his family and friends were puzzled because his speech remained slurred. Two pints of beer would often render it incomprehensible. His wife also noticed that nocturnal urinary incontinence, which she had attributed for-

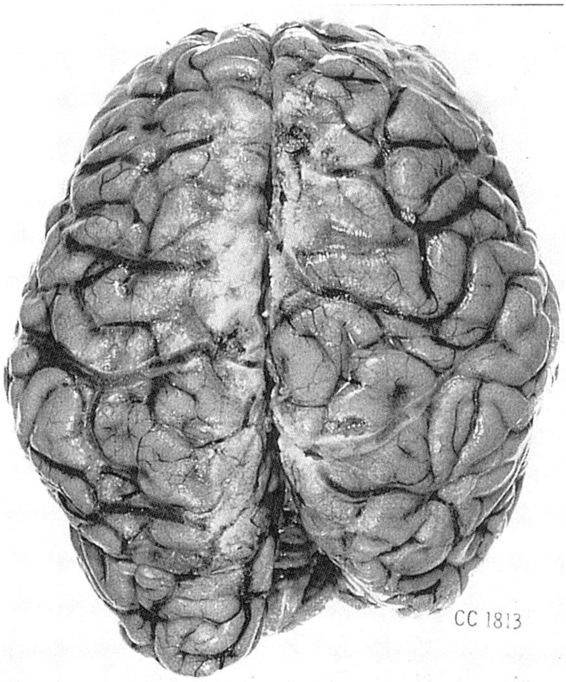

Fig. 266 Brain of a boxer. Diagnosis: 'punch drunk,' showing cortical atrophy and thickening of the leptomeninges. (From COURVILLE C B, 1962).

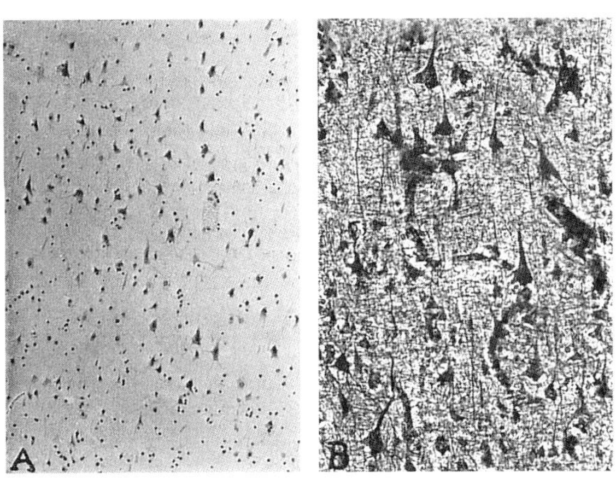

Figs. 267 a and b. Microscopic view of the cortex. (a) Low power showing patchy loss of nerve cells. The stratification of the cerebral cortex is lost. Nissl technique, 125 : 1. (b) High power showing irregular granulation of neurofibrils. Reduced silver technique, 250 : 1. (From COURVILLE C B, 1962).

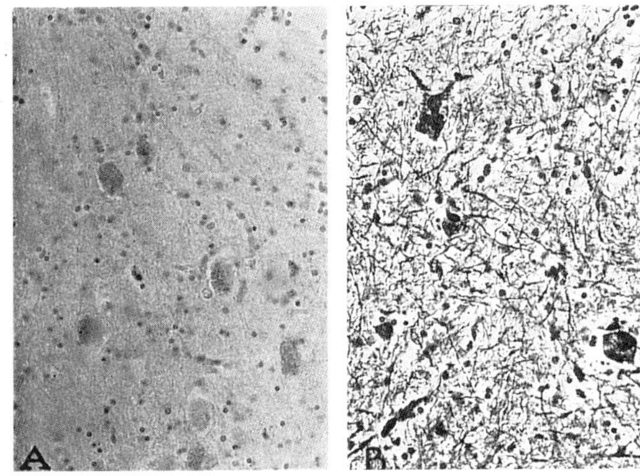

Figs. 268 a and b. Microscopic view of lenticular nucleus. (a) Lipoid deposits in remaining large cells of putamen. Scarlet red, 250 : 1. (b) Fine and coarse granulation of remaining large cells of putamen. Reduced silver technique, 325 : 1. (From COURVILLE C B, 1962).

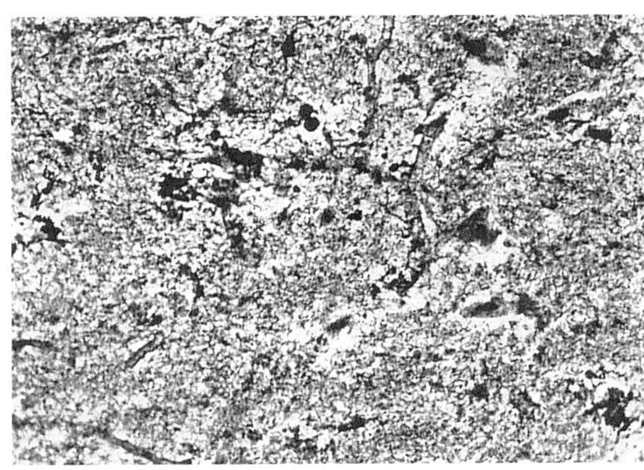

Fig. 269. Degeneration of astrocytes in lenticular nucleus. Gold sublimate technique, 250 : 1. (From COURVILLE C B, 1962).

merly to inebriation, persisted during his periods of sobriety. In the past ten years he has been a patient in a mental hospital on several occasions. During one such admission, about three years ago, it was observed that he was dragging his left leg when he walked and tended to hold his left arm flexed across his body. He had a perforated peptic ulcer in 1951 and a pleural effusion in 1955. In 1957 a partial gastrectomy was performed. At this time an enlarged liver was noted, but liver-function tests were within normal limits. There was no family history of mental or neurological disorder.

Examination – On admission to hospital he was cooperative, and his general physical condition was good. He had been working regularly at a steel mill since August 1961, and his wife informed SPILLANE that he had taken no alcohol since then. He was morose and depressed. He had a broken nose and one cauliflower ear. His speech was thick and slurred and his gait unsteady. His pupils were normal. There was nystagmus on right and left lateral gaze. Power in all limbs was good, reflexes were brisk and symmetrical, and the plantar responses were flexor. There was some diminution of sensation (tactile, pin-prick, and vibration) in the left leg. B.P. 130/80. He was slow, his power of concentration and his memory for recent events were poor, but he was not demented. He was correctly oriented and his behavior was quite normal. A speech therapist formed the view that in addition to his gross dysarthria there was slight executive dysphasia. Psychometry revealed a dull normal level of intelligence (I.Q. 85), with no indication of intellectual deterioration. He attributed his slurred speech, unsteady gait, and alcoholism to boxing, and he thought 'professional boxing should be abolished altogether.' He recalled one episode during his boxing career of 24 hours' amnesia. This included a journey of 50 miles (80 km) to a neighboring town, and eight-round contest, which he won by points, and the journey home.

Blood W.R. was negative.

Radiographs of skull were normal. The cerebrospinal fluid protein content was raised (60 mg/100 ml); cells 4; normal Lange curve; negative W.R.

EEG was normal.

Lumbar air encephalogram was normal. There was no evidence of myxoedema.

Liver-function tests were normal, and his liver was not enlarged.

He died suddenly at work, of acute myocardial infarction, two months after leaving hospital. At necropsy there were foci of softening on the left side in the cortex of the parietal, temporal, and cerebellar lobes. There was a cystic scar in the anterior limb of the left internal capsule. The meninges were thickened at several sites and the cerebral arteries showed slight to moderate atheroma. The pathological picture did not resemble the restricted form of cerebellar cortical degeneration of alcoholics described by VICTOR et al. (1959) in which the main atrophy is largely confined to the vermis. The lesions were probably the result of vascular disturbance, but what part trauma played could not be determined.

The patient started to box when he was ten, and turned professional at 18. Until he retired at the age of 32 he had approximately 300 fights in various countries. His active career as professional boxer extended over a period of 14 years. At retiring from his active career as a first symptom a progressive slurring of speech was noted. There existed a left-sided hemiparesis. Stigmata of his profession were a broken nose and one 'cauliflower ear.' The patient remembered an amnesia of 24 hours duration in which he traveled to his 50 miles (80 km) distant home town, it included the eight round contest, which he had won by points. His power of concentration and his memory for recent events were poor. But there was no dementia. These findings were obtained at age 45. He had insight and attributed his symptoms to boxing.

MAWDSLEY and FERGUSON (1963) published a paper on *'Neurological Disease in Boxers.'* Cases 1 to 9 review the neurological findings of former boxers. *Case 10* died recently at the age of 51.

He began his career as a boy of 14, fighting nightly in local booths. He had 300 bouts as a professional bantam weight. He retired when he was 35. He was never knocked out but often lost on points later in his career. At his peak he was very successful, and between the ages of 22 and 26 fought an average of 14 top-class contests each year. When he retired many people taunted him with being 'punchy.' He had a set, expressionless face, slurring of his speech, and was noticeably clumsy in the use of his hands. His wife noticed a progressive mental impairment from the age of 35. His memory deteriorated, and he lost his last job when he was 42, because he could not remember his duties and was unreliable in keeping to a schedule. During the last ten years of his life he spent most of the day sitting apathetically at home. He became slovenly in his dress and habits, wearing the same clothes and underclothes for weeks on end unless he was badgered to change. He took no interest in his family's welfare. The lethargy was in striking contrast to the liveliness and alertness described by people who knew him in his twenties. Never a heavy drinker, he was easily rendered ataxic by small amounts of alcohol. He was an amiable man and had no record of aggressive or violent behavior. He was partially impotent from the age of 35, and this was absolute after the age of 47. On three occasions he lost consciousness for a few seconds. He died from a squamous-cell carcinoma of the floor of his mouth.

Clinical Examination: He had a broken nose and old fractured metacarpals. There was clinical evidence of dementia. He had pronounced slurring dysarthria and an expressionless Parkinsonian-like face. The tendon reflexes were abnormally brisk. The abdominal reflexes could not be elicited.

Investigations: The *EEG* was probably within normal limits, showing intermittent low and medium voltage theta activity at 5-7 c/s, in all areas, symmetrically.

Macroscopic Examination of the Brain: The *brain* weighed only 1.050 g and there was gross diffuse cortical atrophy, most evident in the frontal lobes. The lateral ventricles were dilated. There was a cavum septi pelucidi. The cavum is seen in Fig. 270, showing a coronal section through the cerebral hemispheres, cutting the anterior horns of the lateral ventricles in front of the interventricular foramina. The two laminae of the septum are thick and intact there. They remain imperforate anteriorly, where they are attached to the genu and ros-

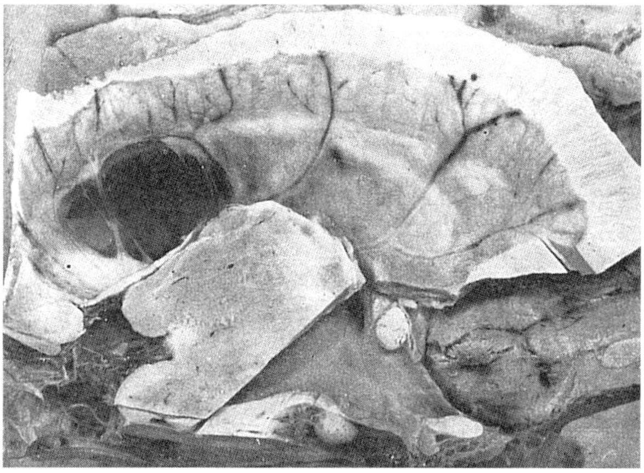

Fig. 270. Section through corpus callosum. Anatomical specimen showing absence of the posterior third of the septum pellucidum. Note remaining strands of tissue.Makrophoto (From SPILLANE J D, 1962).

trum of the corpus callosum. In the central part of the lateral ventricles just above the foramina of Monro the two leaves show multiple perforations along their superior margins, near their attachment to the trunk of the corpus callosum. The lattice-like appearance of these perforations is seen in the oblique view (Fig. 271). Posteriorly, towards the splenium, the lamellae have almost disappeared. All that remains of their backward continuation are a few strands, arranged in two columns (Fig. 272).

This boxer began his pugilistic career at age 14, he

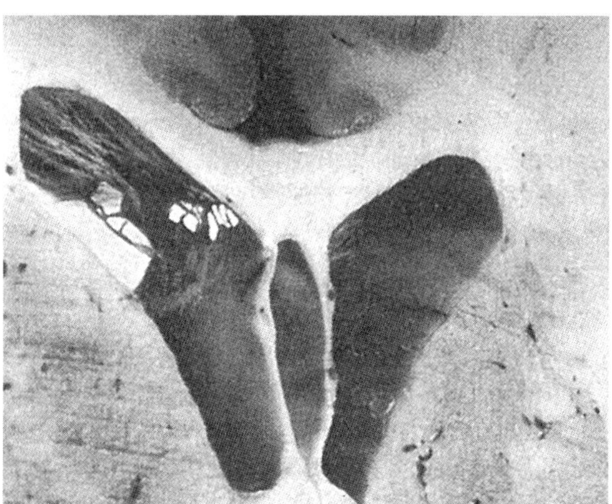

Fig. 271. Case 10. Oblique view of coronal section of cerebrum showing perforation in leaves of cavum septi pellucidi. (From MAWDSLEY C, FERGUSON F R, 1963).

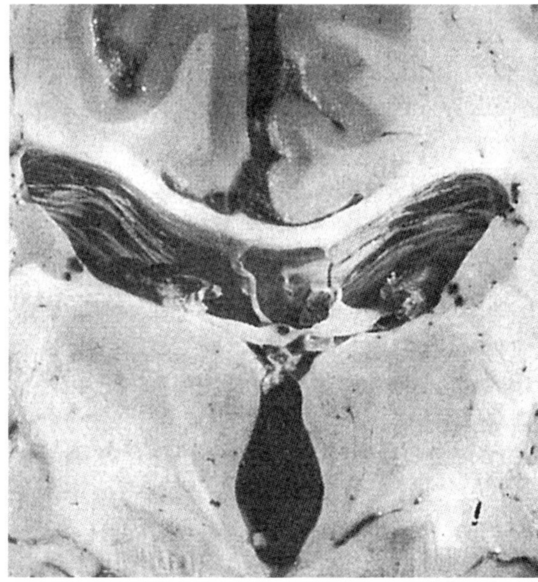

Fig. 272. Case 10. Coronal section through the cerebrum at posterior parts of the lateral ventricles, showing remaining strands of the septum pellucidum. Makrophoto. (From MAWDSLEY C, FERGUSON F R, 1963).

fought nightly in local booth, the exact number of bouts is not known. He had 300 professional bouts in the bantamweight class. He retired from his active career as boxer at age 35. At his peak in his active career he was very successful and between the ages 22 and 26 fought an average of 14 top-class contests every year, that is about one per month. At the time he retired he was by many considered 'punchy.' His progressive mental impairment began at the time of his retirement from active career. At age 42 he lost his job. The last ten years of his life he spent most of the day sitting apathetically at home. A depravation developed. There was a striking contrast to the liveliness ansd alertness of his twenties. Stigmata of his profession were a broken nose and old fractured metacarpals. Clinically there existed a dementia. An articulatory speech disturbance existed. The face was expressionless and parkinson-like.

There was a dilatation of the lateral ventricles and a cavum septi pellucidi was found with many perforations.

QUANDT and SOMMER (1965) published the

case history of a 24 year old amateur boxer who had suffered a knock out during a match, due to a blow against the left eye brow and right mandible.

This boxer was counted out standing in the ring, but collapsed in the ring a short time later. *At admission to a hospital* a short time later, there was complete areflexia with marked dilatation of both pupils. During the next five days the unconsciousness slowly lightened up and the patient became again responsive. At the sixth day the patient became again unconscious, developing right-sided spastic paresis. The patient was transferred to a neurosurgical department. A *left-sided carotid arteriography* and an *EEG* revealed severe, diffuse, traumatic damage of the left cerebral hemisphere. The patient died 33 days after the boxing match, due to circulatory failure.
The *autopsy* revealed hypostatic bronchopneumonia, marked cachexia and considerable decubital necroses.
Autopsy of the brain revealed several second stage cortical contusions in the fronto-parietal area, at the transition to the fissura interhemispherica.
Coronal sections of the brain showed the cortical contusions described above and a slightly dilated ventricular system. Multiple, mostly anemic, infarcts in the subacute stage, were seen at basal parts of the caudate nucleus, the putamen, and in median parts of the thalamus of the temporal poles (Fig. 273), including the gyrus hippocampus and in the midbrain in the immediate area around the substantia nigra. Brain stem and cerebellum are macroscopically unremarkable.
The *histological examination* of sections of different regions of all cerebral lobes revealed typical cortical contusions in the fronto-temporal region (Fig. 274). These contusions were in a stage of total tissue necrosis, with complete loss of neurons and glial elements. Numerous ischemic necroses exist in the above described areas of the cerebrum. Especially impressive were the almost symmetrical necroses in the gyrus hippocampus, with a typical segmental loss of neurons in Sommer's sector (Fig. 275) and a striate shaped necrosis parallel to the substantia nigra (Fig. 276). The gliogenous and mesenchymal tissue alterations were already in such an advanced stage that small cystic formations existed. Some of the cells of the substantia nigra showed ischemic alterations; melanin pigment was deposited in the neighboring tissue.
The *neuropathological examination* of the brain of this amateur boxer revealed two different injury patterns.

(1) There are several cortical contusions in the second stage in the fronto-pariental region. QUANDT and SOMMER attributed the cortical contusions as 'contrecoup' injuries to the boxing blows against the left eyebrow area and the right lateral parts of the mandible. We do not agree with this interpretation. Cortical contusions in boxers are normally the results of falls in the ring, with impact of the head against the canvas, or, less often, in some cases against a ring rope. There is no detailed report about the end of this boxing match, only that the boxer suffered knock out blows against his head and collapsed while still in the ring. Whether he hit his head falling to the canvas is not mentioned. We remain unconvinced that a boxing blow against the head can produce sufficient acceleration to cause a cortical contusion.

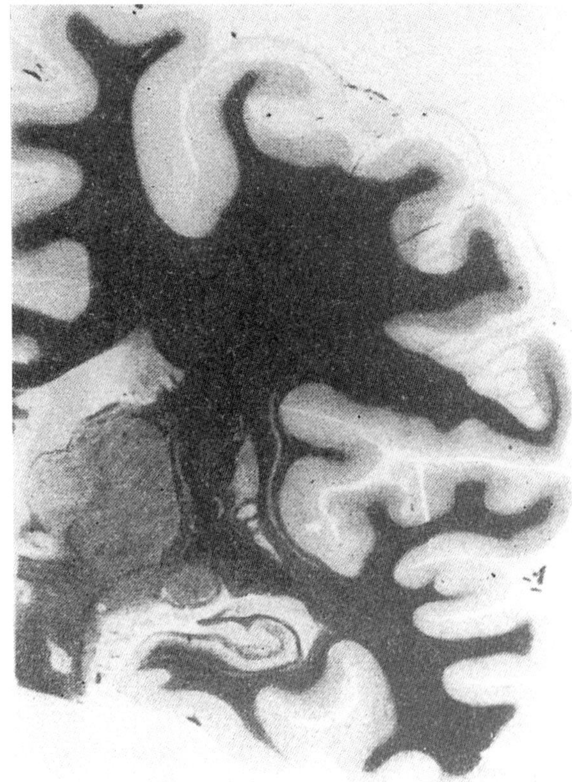

Fig. 273. Coronal section through the cerebral hemisphere in the area of the hippocampal gyrus. Distinctively are recognizable ischemic necroses in the Ammons' horn formation, of the substantia nigra, as well as nucleus ruber and in median parts of the thalamus. Heidenhain-Woelke technique. (From QUANDT J, SOMMER H, 1965).

(2) The *second injury pattern* reveals extensive, almost symmetrical, necroses in basal parts of the cerebral hemispheres and basal ganglia. The midbrain revealed a striated tissue necrosis parallel to the substantia nigra, with ischemic alterations of

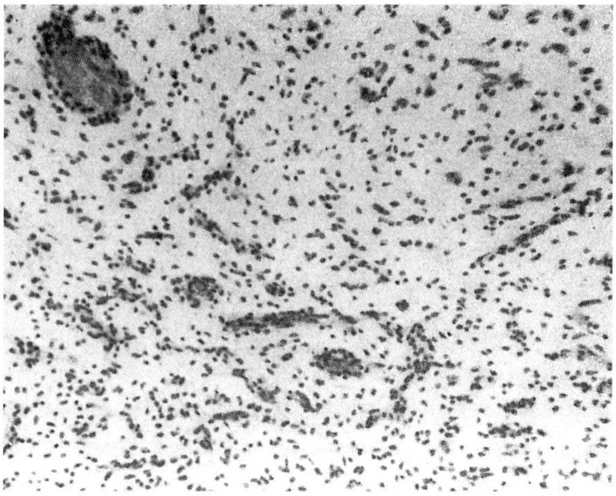

Fig. 274. The authors using the term 'cortical contusion' in the depth of the parital cortex. Nissl technique. (From QUANDT J, SOMMER H, 1965).

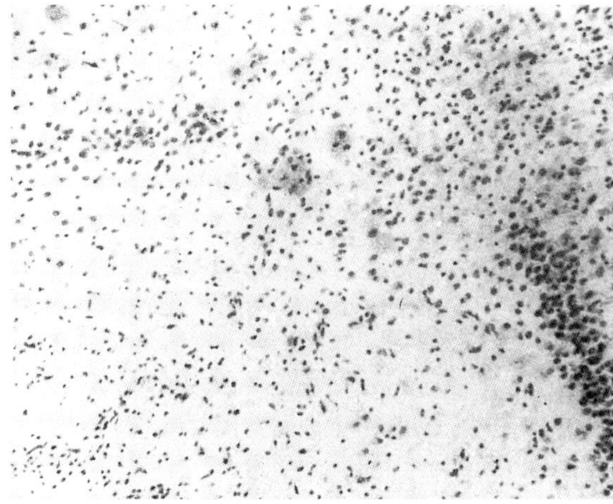

Fig. 276. Ischemic necroses in the Sommer's sector (h_1) with considerable activation of microglial elements. Nissl technique. (From QUANDT J, SOMMER H, 1965).

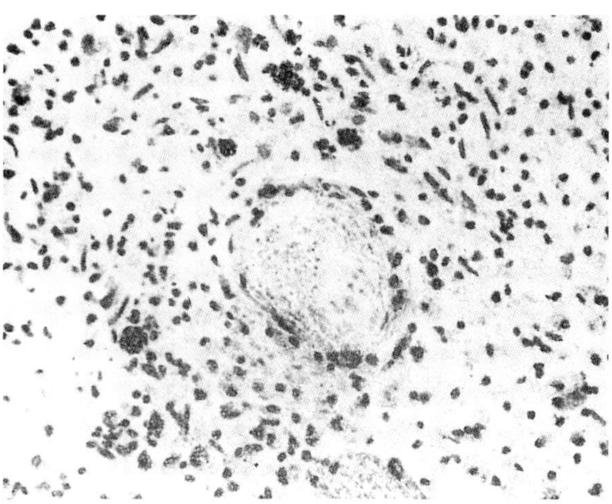

Fig. 275. Necrotic area in the substantia nigra. In the neighborhood is niger pigment deposited. Nissl technique. (From QUANDT J, SOMMER H, 1965).

the neurons. There was melanin pigment free in the neighboring tissue, with a proliferation of local glial elements. The sudden deterioration of the clinical picture, on the sixth day after the knock out, can be explained by gradual brain edema development, eventually resulting in reduced blood supply to arteries supporting the midbrain and basal ganglia.

If this patient had survived for a longer period, one would expect the development of a Parkinson syndrome, due to the involvement of the substantia nigra.

PAYNE (1968) reported on *six professional boxers*. The main pathological features were an enlarged ventricular system, miniature scars and microscars in the grey matter, foci of myelin degeneration, scattered chronic inflammatory cells and phagocytes in the perivascular spaces, and irregular areas of proliferation on the surface of the gyri. These consistent findings indicated that small foci of brain damage had occurred in each brain. PAYNE concluded, however, that it was impossible to completely exclude any other disorders, such as cardiovascular disease, as the cause of the microlesions in his cases. However, the general picture presented by these six cases suggests that the scars are the result of brain trauma. It is necessary to note that at least two of the cases were alcoholics.

Case 1: He was a *well-known bantam-weight professional boxer before World War I.* He was born in 1884, one of eleven children. He attended school infrequently and was not a good scholar. After leaving school at 12 years of age, he took up labouring jobs and drifted into professional boxing. His boxing career extended over a period of about twelve years, during which time he traveled to Australia to take part in boxing contests. There is no record of the number of fights he took part in, or the number of times he was knocked out. For some time he was a sparring partner to a world champion. He drank alcohol, but the amount is unknown. In 1914 he was admitted to hospital with 'Manic Depressive Psychosis' and thereafter he spent many years in mental institutions. He had a guilt complex about his mis-spent youth. His blood pressure was recorded as 150/100 mm/Hg.

Cause of Death: Coronary atheroma.

The *brain* showed slight general atrophy of the frontal lobes with thickening of the leptomeninges at the base. There was moderate cerebral atheroma. The ventricular system was enlarged and there was a *cavum septi pellucidi.* There was a small linear scar 3 mm by 0.5 mm in the cortical ribbon of the right occipital lobe.

Histology: The arterioles in the cerebrum and cerebellum showed thick walls with degenerative changes: the thickening was mainly of the adventitial layer, although subintimal hyaline thickening was fairly common. There were a number of amyloid bodies scattered throughout the brain, with clusters in the hippocampal gyrus. *Silver stains* showed a small number of argyrophilic plaques in the cortical ribbon; the neurofibrils were normal.

Study of the cortical ribbon showed irregular proliferation of the surface layer. This proliferation was most common at the upper corners of the convolutions. The proud tissue on the edge of the gyri produced a step-like appearance where it met the natural level of the gyral surface (Fig. 277). The proliferated tissue was composed of a nondescript kind of connective tissue. Occasional sulci showed partial obliteration. This appeared to be due to fusion of parts of two adjacent gyri. *Cajal stain* revealed small foci of hypertrophy and hyperplasia of astrocytes scattered about in the gray matter of the cortex and at the base (Fig. 278). These microsacs were found scattered about in various areas and layers of the grey matter.

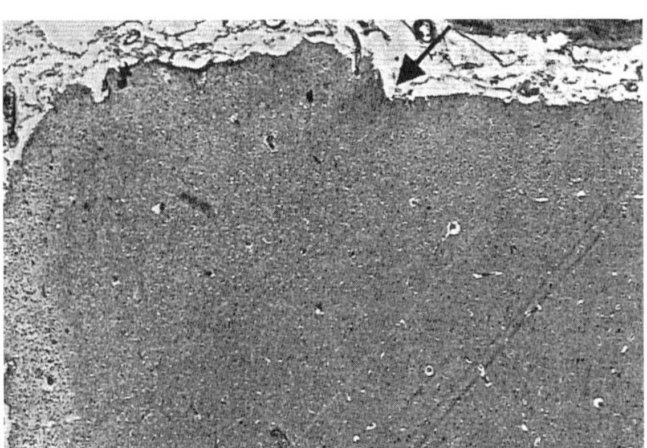

Fig. 277. Case 1. Proliferation of surface layer of cerebrum. The proliferation of tissue at the angle of a gyrus is shown by the step-like appearance on the surface of the cortex at the point arrowed. (From PAYNE E E, 1968).

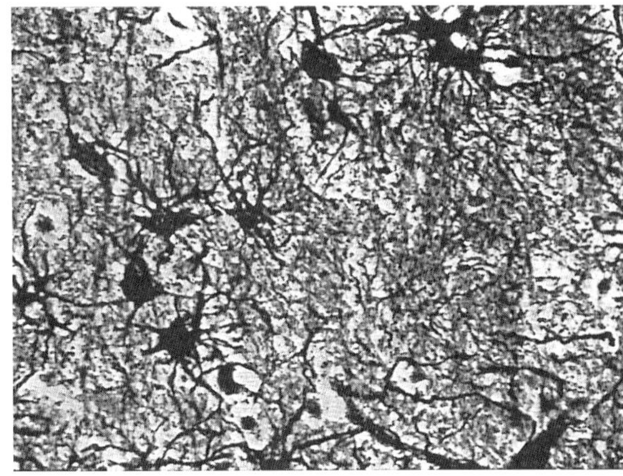

Fig. 278. Case 1. Microscar. A small focus of hypertrophic and hyperplastic astrocytes which form a microscar. Cajal stain, high magnification. (From PAYNE E E, 1968).

There was a fairly diffuse, slight gliosis in the subcortical white matter. Many parts of the perivascular space throughout the brain contained a small number of chronic inflammatory cells, mainly lymphocytes, and a few phagocytes containing products of myelin breakdown. Several small foci of myelin degeneration were found scattered about in the white matter. These foci, sometimes round, sometimes irregular in shape, varied in size up to 150 microns in diameter; the larger ones contained a granular eosinophilic material within ghost-like cell boundaries (Fig. 279).

The *septum pellucidum* contained several large thick walled veins, and the arterioles showed pronounced subintimal hyaline thickening. The laminae of the septum pellucidum showed irregular gliosis and a segment of one lamina was

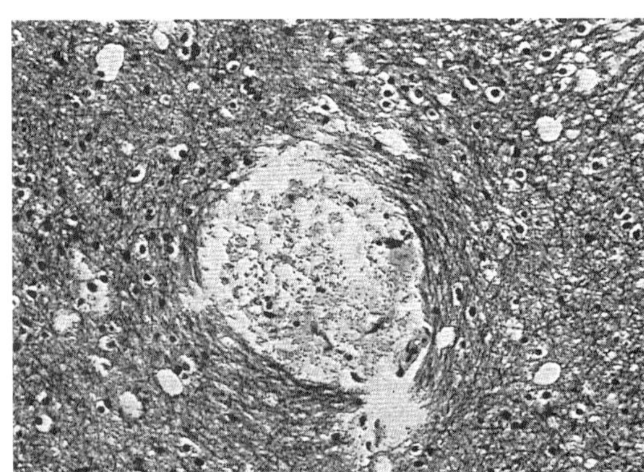

Fig. 279. Case 1. Focus of myelin degeneration. The round focus contains eosinophilic granular material. (From PAYNE E E, 1968).

thinned and attenuated. The corpus callosum adjacent to the attachments of the septal laminae showed foci of myelin degeneration.

Case 2: An *ex-professional lightweight boxer* of 46 years of age. He began boxing in his early teens and became a professional at about 20 years of age. It was said that he had been knocked out only once, although he had taken many beatings in over 300 contests. He had fought in traveling boxing booths before giving up boxing at about 33 years of age; he then went to work in a factory. At about that time, he complained of headaches and of the noise of the machinery in the factory. A year later he fell off a motor cycle and was unconscious for some minutes. On admission to hospital, examination of his central nervous system did not reveal any abnormality.

He took only an occasional drink of alcohol, but he was a compulsive gambler. At 43 years of age he had left ventricular failure due to malignant hypertension. Examination at that time noted that he had slurring of speech and an unsteady gait. His hypertension was treated, but he became depressed, partly because of unemployment, and his dysarthria and ataxia became more severe. Episodes of strife occurred between his mother and himself, and he became violent and paranoid. In the month before his death, one of the attacks led him to an attempt to set his house on fire.

On admission to hospital he had dysarthria, ataxia, papilledema, and a left third nerve palsy. He was emotionally labile and required constant supervision. His blood pressure was 190/130 mm/Hg; he was dyspneic, and he had edema of his legs.

An *EEG* showed an unstable rhythm maximal posteriorly, with focal abnormalities suggesting a right parietal lesion. He died as the result of heart failure from malignant hypertension.

Externally, the *brain* showed areas of congestion in the meninges. The cut surface showed enlarged ventricles. There was *a cavum septi pellucidi with fenestration of the laminae.* There were areas of infarction in the white matter of the frontal lobes. There were a few pin-point scars in the cortex of the parieto-occipital lobes. There were irregular areas of cystic scarring in the pons and basal ganglia. In the subcortical region of the left occipital lobe, there was an irregular area of grey stippling about two centimeters in diameter.

Histology: There was widespread arteriosclerosis and atheroma. There were microscars scattered in the cortex of the cerebrum and parts of the subcortex showed gliosis. In the white matter, there were foci of degeneration of myelinated fibres; there were periventricular glial nodes and irregular areas of gliosis in the centra, the corpus callosum, and cerebellum. There were cystic scars containing compound granular corpuscles in the thalamus, the centra, the cerebellum, and the pons. The perivascular space contained foci of lymphocytes with occasional phagocytes containing hemosiderin and breakdown products of myelin. There was slight irregular proliferation of small parts of the surface layer of the cerebral cortex. There were no senile plaques and the neurofibrils were normal.

Case 3: A man of 46 years. He left school at 14 years of age and went to work in the coal mines. His boxing career began at 19 years of age. While in the Services, he once held a Welterweight Championship. After leaving the army, he was a professional boxer for about two years, before he retired from the ring at 31 years of age. During his boxing career, he had more than 200 contests, and he admitted to having been knocked out on some 50 occasions. He became a heavy drinker of spirits and was treated in several hospitals for alcoholism. He had many court convictions for drunkenness and several for larceny. He worked intermittently as a labourer, but was always work-shy; in 15 years his longest period of continuous work was three months. He was unable to get on with his wife, and he left her. After giving up boxing, his main complaint was of headaches and lack of concentration, although he also suffered with insomnia, enuresis and periods of depression. Eight years after the end of his boxing career, it was first noted that he had slurring of speech. On most of his many admissions to hospital he was pleasant and cooperative, although on a few occasions he was abusive and threatening to the medical staff and to other patients. On several occasions the CSF protein was slightly raised (70-80 mg%); his speech was slightly abnormal, and there was mild impairment of coordination in his left arm. In the last year of his life his memory was poor, he was sometimes confused and vague, and he became dirty and untidy in his dress.

His *EEG* showed a diffuse abnormality, and his *PEG* showed evidence of brain atrophy.

He was found dead in bed after an alcoholic bout: the cause of his death was given as 'acute interstitial pneumonia.'

The *brain* revealed slight thickening of the leptomeninges at the base. The hemispheres appeared normal. There was minimal cerebral atheroma. Coronal section of the cerebrum showed a *small cavum septi pellucidi,* and the upper corners of the lateral ventricles were rounded. There was a thin linear scar in the cortical ribbon in the occipital lobe.

Histology: Cajal stains revealed microscars (Fig. 280) scattered in the cerebral cortex in the hippocampal gyrus, in the base, in the cerebellum and in the medulla. There was mild subcortical gliosis. *Silver stains* showed some thickening or clumping of the neurofibrils, but there were no argentophil plaques.

The *Luxol Fast Blue stain* revealed small foci of myelin degeneration in the white matter and gliosis in the corpus callosum, near the attachments of the septum pellucidum. There were areas of proliferation of the surface layer of the convolutions, partial obliteration of sulci, areas of thickening of the leptomeninges and paraventricular gliosis. The perivascular spaces showed a few lymphocytes and phagocytes containing breakdown products of myelin.

Case 4: A bookmaker's runner and ex-boilerman of 45 years of age. He had been *amateur boxer* for several years and a *pro-*

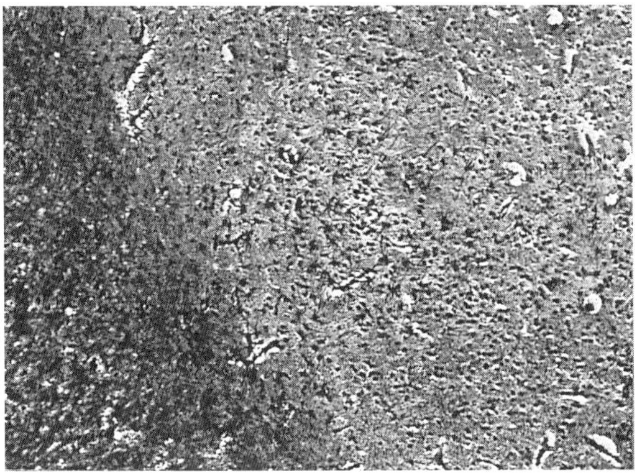

Fig. 280. Case 3. Microscar. A lower-power view of a large microscar near the junction of grey and white matter. (From PAYNE E E, 1968).

fessional for 14 years. He had taken part in about 300 contests, and, at one time he was contender for the British Middleweight Championship. He retired from the ring at 32 years of age and in the ensuing seven years he was admitted to hospital on four occasions for medical or surgical treatment. On these admissions there were no abnormalities of his nervous system noted, but his blood pressure was recorded as 145/95 mm/Hg and 150/100 mm/Hg. One year later he was under care as an alcoholic; his speech was slurred, and his gait shuffling. He had been drinking about 10 pints of beer (4,7 liter) a day for many years, but at that time two pints of beer (0.94 liter) made him impulsively violent.

An *EEG* then showed slight theta activity in the left temporal lobe.

His intemperate habits had resulted in marital disharmony.

In the following five years, he undertook repeated medical treatment for his alcoholism, but in the intervals he was consuming large quantities of beer. During this period he complained of headaches, depression, lack of concentration, insomnia, enuresis of his limbs, and a mild sensory impairment in the left leg. His intelligence quotient was at the lower level of normal; there was no evidence of intellectual deterioration. His *CSF* protein was 60 mg%, *PEG* was normal, and his Wassermann reaction was negative. His blood pressure in his last year was 130/80 mm/Hg. He died suddenly as the result of myocardial infarction.

The *brain* revealed a slight thickening of the leptomeninges at the base, over the right Sylvian fissure and over the posterior part of the left temporal and parietal lobes. The vertebral arteries and basilar artery showed slight atheroma. The intracranial part of the internal carotid arteries showed moderate atheroma, most marked on the left side.

In general, the cerebral convolutions and sulci appeared normal, except in the posterior part of the left parietal and temporal lobes. The upper limb of the posterior parietal gyrus was depressed below the level of the adjacent cortex, and the posterior limb of the transverse temporal gyri showed a ragged yellow area about one centimeter in diameter, a 'plaque jaune.'

Internally: The white matter of the cerebral hemispheres was congested. In the anterior limb of the left internal capsule there was an irregularly scarred area about 1 cm by 0.5 cm in diameter; the scarred area was partly cystic and pigmented, and it extended from the body of the caudate nucleus through the internal capsule to the putamen.

In the posterior part of the left temporal and parietal lobes and in the anterior part of the left occipital lobe there were a number of small scars (Fig. 281). These scars were situated mainly within the cortex; they were yellow in colour and some were tinged orange. Most of the scars were situated about halfway along a sulcus; they were about two millimeters by one millimeter and were disposed at right angles to the surface of the sulci. The miniature scars (miniscars) were found in the parietal gyri, posterior to the superior, middle and inferior gyri, on the lateral surface of the anterior part of the occipital lobe. Fourteen such scars were found.

At the posterior end of the superior surface of the left middle temporal gyrus the cortex was almost completely replaced by a linear scar about 3 cm in length; parts of the adjacent white matter showed small cysts. The inferior temporal gyrus showed a cortical cystic scar one centimetre in diameter with a thin membrane containing yellow necrotic material.

The *left lateral ventricle* was lightly enlarged, with rounding of the apex. *A cavum septi pellucidi was present.* The left cerebellar hemisphere showed four small, irregular, degenerated areas scattered in the folia of the anterior lobe; the lesions were pigmented and partly cystic.

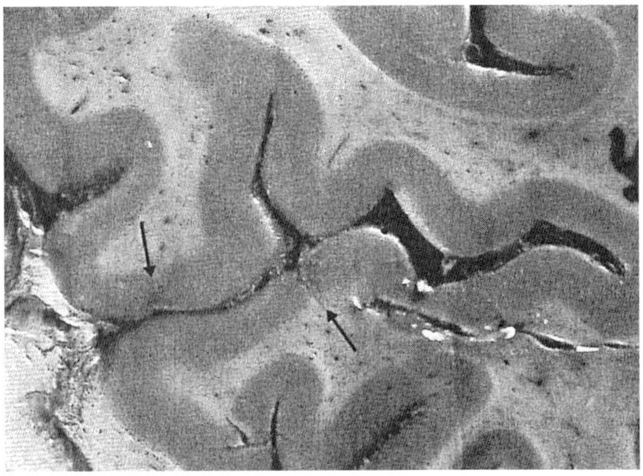

Fig. 281. Case 4. Miniscar. Two miniature scars (arrowed) in the cortical ribbon. (From PAYNE E E, 1968).

Histology: Examination of the histological sections from the temporo-parieto-occipital area showed that the miniscars (Fig. 282) were triangular in shape, with the main axis lying at right angles to the pia; the base of the triangular scar was adjacent and parallel to the sulcus but was often separated from the pia by an intact molecular layer. The large scars in the inferior and middle temporal gyri were mainly orientated along the cortical ribbon; the periphery of these lesions was composed of a dense layer of fibrous astrocytes, while the center was filled by a mass of compound granular corpuscles and tissue debris, with a few cystic spaces. In the adjacent white matter there were irregular cystic areas and gliosis. Some of the extracerebral arterioles in the sulci adjacent to the large scars showed endarteritis. In addition to miniscars and cystic scars with softening, Cajal staining revealed many foci of glial hypertrophy and hyperplasia (microscars). The microscars were scattered in the grey matter of the cortex and at the base. There was diffuse mild gliosis in the subcortical white matter. The scars in the cerebellum were irregular in shape and about 5 mm in diameter. The scars involved both the grey and white matter of the folia and contained granular corpuscles (Fig. 283); occasional adjacent vessels showed endarteritis. Examination of the white matter of the brain revealed a number of small foci of myelin degeneration scattered about in the cerebrum, corpus callosum, and cerebellum. A few perivascular spaces contained a small number of chronic inflammatory cells and occasional and phagocyte with degenerate products of myelin. There was an irregular proliferation over surface parts of the gyri, and parts of the adjacent leptomeninges were thickened.

Case 5: A *former lightweight champion,* who had over 200 professional fights. His boxing career began at the age of 15 years, when he fought in the booths for 7/6d a night. He retired from boxing at 28 years of age. He drank alcohol, but this never appeared to present a social problem. Some of his acquaintances thought that he was quite normal, while others

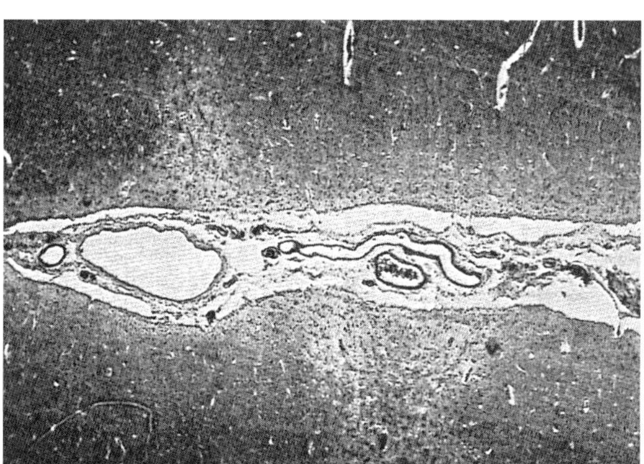

Fig. 282. Case 4. Histological section of miniscars by Luxol Fast Blue stain. (From PAYNE E E, 1968).

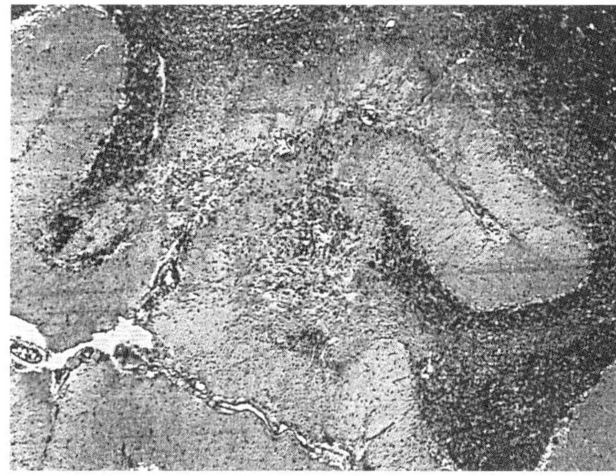

Fig. 283. Case 4. Cerebellum. To show loss of architecture of the folia due to large areas of scarring. (From PAYNE E E, 1968).

considered him to be 'early punchy.' His speech was occasionally indistinct, and he was sometimes unsteady on his feet. He died at the age of 44 years from coronary atheroma thrombosis.

Brain: The external surface of the brain did not show any evidence of atrophy. There was slight cerebral atheroma. Coronal section of the cerebrum revealed slight enlargement of the ventricular system. *There was a cavum septi pellucidi* (Fig. 284) *with small fenestrations in the leaves of the septum.*

Histology: There were a few chronic inflammatory cells scattered about, in the perivascular spaces throughout the brain. In the cerebrum there were a number of microscars in the

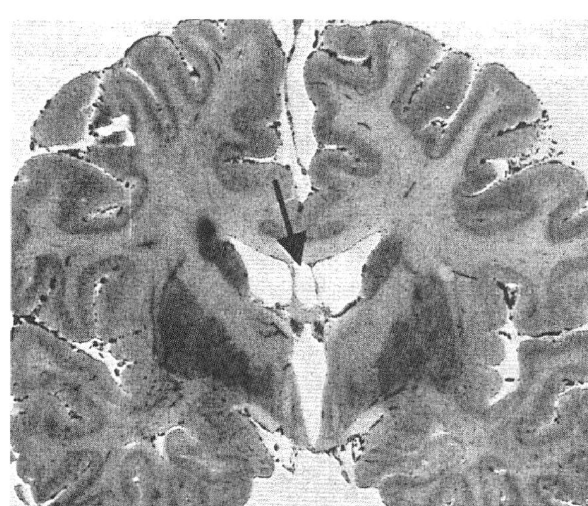

Fig. 284. Case 5. Cavum septi pellucidi. Coronal section of the cerebrum shows the two leaves of the septum pellucidum are separated to form the cavum septi pellucidi (arrowed). (From PAYNE E E, 1968).

grey matter, a few foci of myelin degeneration in the centra and the corpus callosum. There were areas of proliferation on the surface layer of the cerebrum and partial obliteration of a few sulci. There was paraventricular gliosis. In the cerebellum there was an irregular area of scarring in the white matter at the base of a folium. Silver stains showed slight thickening of neurofibrils in some neurons; there were no argentophil plaques.

Case 6: He started boxing at the age of twelve years and *won a schoolboy title.* He became a *British Army light-heavyweight champion before turning to professional boxing.* In his professional career, which lasted for four years, he attained a light-heavyweight championship. The official records show that he had 28 professional contests of which he won seventeen and lost eleven. He was not counted out in any of these contests, although on three occasions the referee stopped the contest, once because he was unable to defend himself. His boxing career ended prematurely because of a prison sentence. He died four years later as the result of a stab wound of the heart.

External examination revealed no evidence of brain atrophy. There was a small (1 cm) circular herniation of the left parietal cortex, through a fault in the dura, over the vertex. The ventricular corners were rounded, and there was a cavum septi pellucidi present. There were no macroscopic scars to be seen.

Histology: There were a few small areas of the glial hyperplasia in the tangential layer of the cerebrum, and the adjacent leptomeninges were slightly thickened. An occasional small vessel within the brain showed a few chronic inflammatory cells and some granules of haemosiderin in the perivascular space (Fig. 285). There was partial fusion of adjacent gyri (Fig. 286). There were no argentophil plaques and neurofibrils appeared normal. The white matter showed small foci of degenerative changes a few fibres.

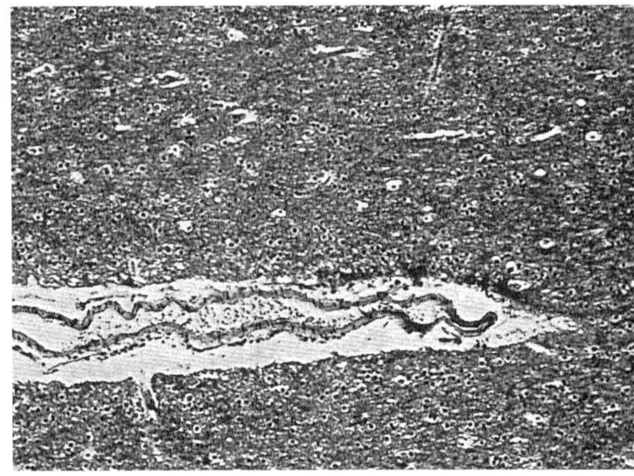

Fig. 285. Case 6. Small vessel in the brain. There are a number of chronic inflammatory cells in the perivascular space. (From PAYNE E E, 1968).

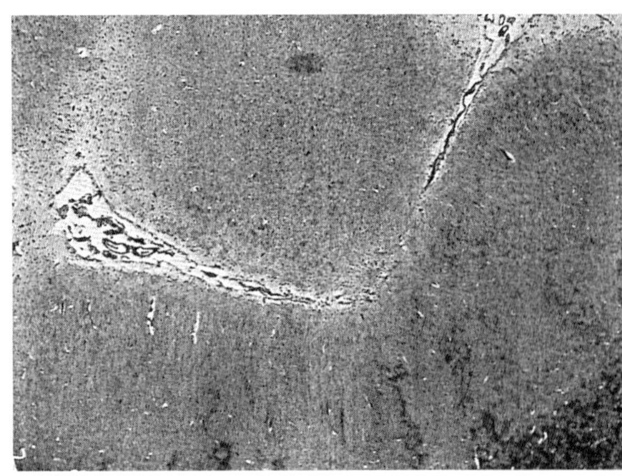

Fig. 286. Case 6. Partial fusion of adjacent gyri with obliteration of the intervening sulcus. (From PAYNE E E, 1968).

PAYNE commented on the morphological findings:

Thickened leptomeninges: Slight thickening of the leptomeninges at the base of the brain is a common finding in routine autopsies of persons over middle age. Four out of six cases presented showed a slight thickening of the leptomeninges at the base.

Macroscopic Atrophy of the Cerebral Convolutions: Macroscopic atrophy of the cerebral convolutions was present in one of the cases examined (case 1). The atrophy mainly affected the frontal lobes and it was slight to moderate in degree. This finding does not appear to have any significance in relation to boxers, as the affected case was a 82 year old patient. Frontal lobe gyral atrophy is a common finding in routine autopsies on the aged.

Ventricular Dilatation: The normal ventricles are little more than narrow slits within the brain substance; they enlarge slightly with increasing age. The ventricular system is seen to be large in many diseases which affect the brain.

Cavum Septi Pellucidi: Separation of the two laminae of the anterior part of the septum pellucidum to form a cavity, 'the cavum septi pellucidi' is not an uncommon finding in the examination of the brain at routine autopsies.

A clinical report of five ex-professionl boxers, of whom four had a chronic cerebral disorder in later life, was given by SPILLANE (1962). Air encephalographic studies showed that in four of these patients there was evidence of either a cavum septi pellucidi, defects in the laminae of the septum pellucidum or cerebral atrophy. MAWDSLEY and FERGUSON (1963) performed air encephalographic studies on eleven ex-boxers with neurological disease and studied the septum pellucidum of one other case after post mortem. They commented that their findings supplemented the earlier clinical accounts of neurological disease in boxers and augmented SPILLANE's observations about changes in the septum pellucidum.

In routine autopsies a very small triangular space is commonly found between the attachments of the laminae of the septum pellucidum to the corpus callosum. When the ventricular system enlarges the corpus callosum becomes stretched and the attachments of the septal laminae diverge with consequent enlargement of the triangular space. The tension on opposed laminae cause them to separate with the formation of the so-called 'cavum septi pellucidi.' The formation of a cavum is dependent upon the initial presence of a small cleft between the laminae and the subsequent divergence of the attachments of the septal laminae that occur as the ventricles dilate. *All the brains examined in this series showed a cavum septi pellucidi.*

Fenestration of the Septum Pellucidum: DYKE and DAVIDOFF (1935) commented on the considerable number of cases of perforation of the septum pellucidum that occur in the literature and how with the exception of these cases all were associated with internal hydrocephalus. General dilatation of the ventricular system inevitably produces a stretching of the septum pellucidum with consequent thinning of the lamina. Injury or disease of the blood vessels of the septum would predispose to or even initiate perforations of the laminae. Fenestrations of the septum pellucidum affecting one or both laminae were seen in three of the six cases presented.

Proliferation of the Surface Layer of the Cerebrum: On microscopic examination the surface of a normal gyrus is fairly smooth and regular. The outermost layer of the cortex, the molecular layer, is fairly regular in depth; it is composed of few cellular bodies and many transverse fibres. In parts of all the specimens examined, there were irregularities and proliferation of the surface layer. The proliferations were most frequently encountered on the gyri of the posterior part of the cerebrum. The proud tissue commonly formed a hump on the surface especially at the lateral edges or corners of the gyri. The excessive tissue was slightly more than the normal molecular layer and here and there a fibrous cell with tangled processes was seen. It seems likely that these areas of prolifeartion represent foci of organisation of subpial hemorrhages.

Partial Obliteration of the Sulci: Localized partial obliteration of the sulci was observed in the histological sections in four of the cases. The obliteration was seen in either the mid or lower part of the sulcus and it was due to fusion of the adjacant surface layers of the cortical ribbon. There was a slight increase in glial cells in the affected part of the molecular and it appeared that the fusion occurs as the result of mild gliosis of the adjacent surface layers.

Focal Scars and Nerve Cell Loss: Macroscopic examination of the brains showed one or more small scars in four of the specimens. The scars were found in the posterior parts of the brain, in the parieto-occipital lobes or cerebellum. These scars were situated in the cortical ribbon, usually about half way along a sulcus. Most of the scars were found only after prolonged search of each brain slice: it was discovered that the scars could be identified more readily after the cut surfaces of the brain were allowed to dry in air for a few hours. The macroscopic scars were about two millimetres in lengst by approximately one millimetre in width and lay at right angles to the cortical ribbon. Case 4 showed the greatest number of macroscopic scars; in the specimen several cerebral arterioles showed a non-specific endarteritis which was probably the cause for the scars. The aetiology of the endarteritis is not known but in the absence of other evidence it may be postulated that it was due to trauma.

It is difficult to evaluate with certainty the presence of slight diffuse cell loss in the cerebrum. It is generally agreed that there must be a relatively large fall-out of the neuron population before cell loss can be discerned in a study of serial brain sections.

Focal loss of neurones is more readily perceived as this results in an alteration of the architecture of the part. The loss of neurones in a small area of brain tissue is replaced by fibrillary astrocytes which form the scar tissue of the brain.

Histological sections stained by routine methods readily reveal recent repair of injured tissue by the attendant cellular reaction; but small scars of long standing may be difficult to distinguish. Special staining techniques for the demonstration of fibrillary astrocytes are necessary in order to demonstrate small foci of astrocyte proliferation which replace areas of nerve call loss and constitute microscars.

Histological study of the brains revealed that in addition to the miniature macroscopic scars a number of small microscopic foci of astrocytic hypertrophy and hyperplasia were found scattered in the cortex, the grey masses at the base of the cerebellum and brain stem. These small glial foci or microscars, seen in Cajal stain, were composed of large fibrous astrocytes. The foci varied in size up to 1.000 microns in diameter. Microscars, in varying numbers were found in each brain, the least numerous were in case 6.

Focal Degeneration of the White Matter: Scattered foci of degererate white matter was seen in all the brains. These foci were usually situated adjacent to a small blood vessel. The smaller foci of about fifty microns in diameter were irregular in shape and the areas of myelin loss was replaced by an homogenous eosinophilic material. Larger foci were less common, they tended to be round in shape and varied in size up to 150 microns in diameter. The foci were composed of clumps of eosinophilic granular material; ghost like outlines of cell boundaries could sometimes be distinguished which suggested that the larger foci were made up of degenerate compound granular corpuscles. Occasionally the foci contained a few glial nuclei and in some a few fine myelinated fibres traversed the focus. The larger foci, stained by Cajal, showed a marked peripheral gliosis.

Perivascular Inflammatory Cells and Phagocytes: In each brain a number of small vessels showed a thin layer of chronic inflammatory cells with an occosional phagocyte. The inflammatory cells were almost all lymphocytes. The phagocytes contained either a brown granular pigment or degenerate products of myelin. The breakdown products of brain damage are slowly removed over a period of years.

Senile Plaques and Neurofibrillary Change: A small number of senile plaques were seen in the cortex of one brain (case 1) and early neurofibrillary changes were observed in two brains (case 3 and 5). The changes indicate non-specific degenerative phenomena.

Discussion: The main features in the pathology of the cases presented by PAYNE are the enlarged ventricular system, the miniature scars and microscars in the grey matter, the foci of myelin degeneration, the scattered chronic inflammatory cells and phagocytes in the perivascular spaces and the irregular areas of proliferation on the surface of the gyri. These findings indicate that small foci of brain damage have occurred in each of these brains. In case 1 this might not be unexpected in view of his age and the moderate cerebral atheroma. Case 2 had malignant hypertension for several years which might be called to account for his lesions and symptoms. Case 3 and 4 were chronic alcoholics who suffered with psychiatric and neurological symptoms. Cases 5 and 6 did not complain of any symptoms, as they did not have any other disorder that their practitioners were aware of but they did have micro-scars in their brains. This evidence indicates that small scars in the brain may occur without producing overt clinical symptomes. The degree, both the size and the number, of brain scars is clearly important: it is of interest to note that at least one macroscopic scar was found in the brains of those cases which had overt clinical symptoms. The presence of numerous micro-lesions must impair the function of the brain. The effects of somewhat lesser number may be never sufficient to produce clinical symptoms. It seems likely that during young manhood the large functional reserve, a normal feature of the young human brain, is capable of masking the effects of small lesions. When the functional reserve begins to wane with increasing age or if the reserve is reduced by some other concomitant disorder then symptoms from the lesions may slowly appear.

It is impossible to exclude completely other disorders such as cardiovascular disease as the cause of the micro-lesions in these six ex-boxers, but the general picture that the cases present suggest that the scars are the result of trauma to the brain. It is here pertinent to emphasize that the cause and type of the trauma cannot be specified, thus head injury from causes other than boxing may well produce similar lesions.

It is interesting to note that at least two of the cases were alcoholics. It is not clear whether or not alcoholism plays a role in the punch drunk state. Several boxing enthusiasts have voiced the opinion that boxers become 'punchy' when they abuse themselves and take to alcohol. It may be that the punch drunk state as the name suggests is a composite syndrome due to brain injury from boxing on which is superimposed a concomitant disorder such as the deleterious effects which follow the regular consumption of large amounts of alcohol. This hypothesis may in part explain why many well known ex-boxers of temperate habits and with a competent vascular system show no signs of this syndrome.

PAYNE summarised the findings in the boxer brains:

The lesions found in the brains of six ex-professional boxers were enlarged ventricles, miniscars and microscars in the grey matter, foci of myelin degeneration in the white matter, areas of proliferation of the surface layer of the cortex, and chronic inflammatory cells and phagocytes in the perivascular spaces. The number of these lesions found in selected blocks from the specimens probably represent only a proportion of the total number present in each brain. It is clear that small scars collectively reduce the competence of the brain to carry out its functions. As yet there is no means of correlating the degree of clinical disturbance with the number of scattered small lesions in the brain. However there seems little doubt that brain disturbances which sometimes occur in ex-boxers are in part, at least, due to these lesions.

BURGER and MINAROVJECH (1966) published the case history of a *former amateur boxer*, who died at age 44, due to massive hemorrhage from an ulcus ventriculi.

He had begun boxing at age 15; in the first seven years of his boxing career he participated in about 70 fights. His boxing career was interrupted for seven years during World War II. He continued to box from 1948 to 1952 and participated in 50 additional fights, so that the total number of his bouts was around 120. In none of his bouts was he knocked out. He had the reputation of being a skilled technical boxer. There were no other head injuries, aside from boxing.

He worked as a heating engineer and was known to be a skilled professional. During the last three years of his life he developed a tremor of his hands, a speech disturbance and emotional irritability.

The *histological examination* of the brain revealed circumscribed and lamillary loss of cerebral cortex neurons and Purkinje cells in the cerebellum. The loss of neurons was attributed to his activity as a boxer.

This former amateur boxer, who died at age 44, participated in around 120 amateur bouts. He suffered allegedly no knock outs. He was considered a skillful technical boxer. Three years before his death, at age 41, tremor of his hands, an articulatory speech disorder and emotional instability developed. The cerebellum revealed a circumscribed and laminary loss of Purkinje cells. This case history reveals that 120 amateur boxing bouts with the typical sparring produced a permanent and irreversible brain damage in an amateur boxer.

BETTI and OTTINO (1969) reported the following case history:

A 38year old boxer, who started boxing at the age of 14 and participated in 380 fights. He allegedly lost one bout by a knock out due to a blow against the solar plexus. In 1957, then 26 years old he felt an acute pain in the cervical spine, since this time cervical rotation was limited, especially to the

left. In 1958, then 27 years old, he suffered a head trauma, falling down stairs, his head hit against the edge of a chair.
A tremor in the upper limbs, particularly on the left, appeared and his gait impairment increased. Posttraumatic parkinsonism was diagnosed and treated with Artane.
The *neurologic examination* reveals a parkinson-like tremor of the upper limbs. Stiffness in all limbs and gait impairment exist with oscillating steps. There is a moderate dysarthria, his speech is poor.
He was aggresssive and misbehaved.
The *EEG* is within normal limits.
A *frontal cerebral biopsy* was performed. The histopathologic examination revealed numerous pyknotic cells and intense gliosis of the white matter, specially around vessels. There was 'disintegration pigment'. The cerebral cortex showewd a severe atrophy of the pyramidal cells, especially in the deeper layers.
Endofibrillary lesions of the pyramidal cells were observed with del RIO HORTEGA's silver technique.
A *stereotaxic operation* was performed. Electrodes were implanted into the ventral lateral nucleus of the right thalamus and into the right pallidus. The operation had, according to the authors, an excellent result, with disappearance of the contralateral tremor and stiffness.
The pallidus biopsy was examined, revealing a severe decrease of cells, with atrophy and neuronophagia.

The author summarize that all lesions may be applied to boxing injuries.

CORSELLIS et al. (1973) studied the brains of fifteen retired boxers. Life histories of the subjects were investigated in retrospect. A characteristic pattern of cerebral damage was identified. The pathological image that emerged from the fifteen case studies appeared not only to be damage from boxing, but it generated an outline of the many features of punch drunk syndrome.

Case 1: This boxer's ex-wife, a sister, and a brother were interviewed. He was a bright, healthy boy who did well at elementary school and excelled at sport. His boxing career began at a charity show when he was eleven years old, and he went on to fight as a professional for the next 14 years. The number of his fights was estimated at 400; he lost a handful, and one contest was stopped. He fought in the United States and became a British and a World Champion.
As a young man he was quiet, generous, and abstemious. He married in his early twenties, when he had already become a social as well as a boxing success. Soon his life became more hectic and 'he changed completely.' He wenched and drank and gambled heavily. His memory began to fail him. He had three car accidents; in one he suffered severe scalp lacerations and an injury to the right eye, followed by three months in hospital. His marriage broke up, and he drifted away from his family, only returning for the occasional embarrassing visit. He had violent outbursts, he was 'knocked out' by only a small amount of alcohol, his behaviour was 'disgusting.' His brother remarked that "his brain was not functioning – he made mistakes in reckoning." He could not settle in a job and he became a vagrant. In his fifties, his jaw was broken in a brawl. At the age of 62, he was found lying neglected and louse-ridden, in the boiler house of a hotel. In hospital he could answer questions to the point, but his speech was slurred and indistinct. He was continent.
A *routine physical examination* revealed a right-sided ptosis and a cataract of the right eye. He was unsteady on his feet. He complained of back pain, and a prostatic carcinoma with skeletal metastases was diagnosed. He died several months later, aged 63. At *necropsy* 'cauliflower ears' and a broken nose were described together with metacarpal exostoses of the right hand. Only the brain could be examined.
Naked-eye appearance of brain: The fixed *brain* weighed 1.310 g. The leptomeninges were moderately thickened over both convexities, but no evidence of cortical scarring was found. Slight atherosclerosis affected the large cerebral vessels. Both hemispheres were a little atrophic. The floor of the hypothalamus was thin and ballooned, with sunken mammillary bodies. Coronal cuts revealed moderately enlarged lateral ventricles, in which only a few strands of a grossly fenestrated septum remained. Anteriorly, the remnants were about 8 mm apart; posteriorly none was left, and the two flattened fornices were totally detached from the undersurface of the corpus callosum (Figs. 287 a and b). The third ventricle was moderately dilated. A cystic softening, 2-3 mm in diameter, was present in each thalamus and in the left putamen. The cortical ribbon generally appeared a little narrow. The substantia nigra had a noticeable lack of pigment, most marked in the lateral two-thirds, on both side (Fig. 288 b). There was a pin-head softening in the nuclei pontis; the cerebellum and medulla appeared normal to the naked eye.
Histological findings: Cerebral hemispheres: The leptomeninges were moderately thickened. The cerebral cortex generally showed a slight diffuse loss of neurons with some increase in astrocytes. This atrophy was most marked in the medial temporal gyri, and the hippocampus, in particular, showed an almost total loss of nerve cells in the Sommer's sector (Figs. 289 a and b). Neurofibrillary tangles were scattered in moderate numbers through the frontal and temporal cortex, but were strikingly intense in the uncus, amygdaloid nucleus, parahippocampal gyrus, and hippocampus (Figs. 290 a and b). Tangles were also seen in small numbers in the diencephalic grey matter. No senile plaques were seen in any area. The bodies of both fornices were thin and elongated. They appeared to be poorly myelinated and were moderately gliosed.

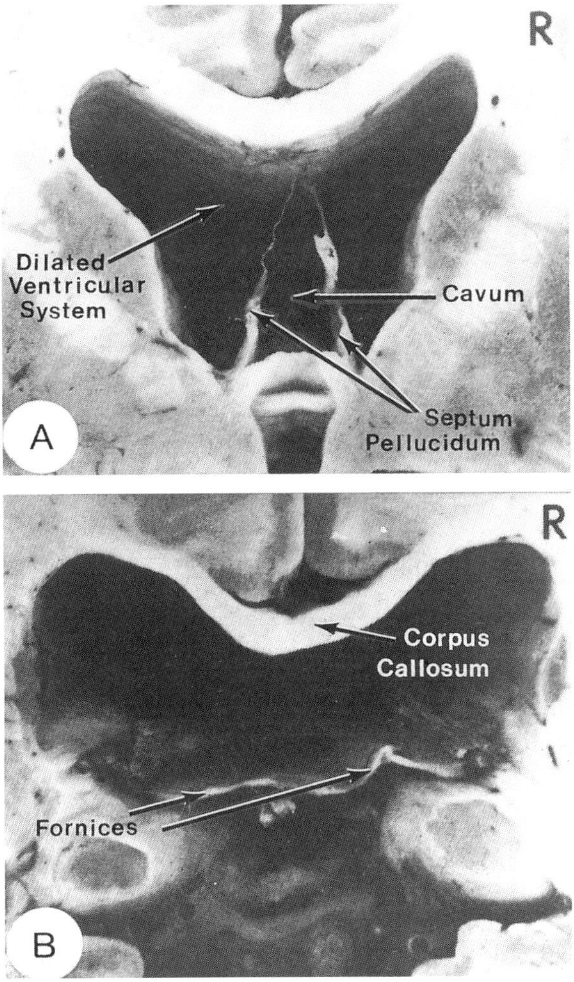

Figs. 287 a and b. Case 1. (a) Coronal cut through the anterior horn of the lateral ventricles. Only a few strands of the septum pellucidum have survived and they are widely separated by an intervening cavum. (b) Case 1. Coronal cut at a more posterior level showing the flattened bodies of the fornices completely detached from the under-surface of the corpus callosum. There is no septal remnant at this level. (From CORSELLIS J A N et al., 1973).

The mammillary bodies were small; many of the neurons were shrunken, and there was a marked increase in astrocytes.
Brainstem: One inferior colliculus was partly destroyed by an old cystic softening; the cause was not clear. The substantia nigra showed a moderate loss of pigmented nerve cells with some free lying pigment and an increase in astrocytes. The medial group, on both sides, was least affected. A moderate number of nigral neurons showed neurofibrillary change; no Lewy bodies were seen. The nuclei pontis contained two 3 mm cystic softenings adjacent to thickened and necrotic vessels. Many cells of the locus coeruleus, and a few other neurons in the medulla, showed neurofibrillary change.
Cerebellum: An area of folial atrophy and gliosis with marked loss of granule and Purkinje cells was present in the inferior surface of both hemispheres in the region of the cerebellar tonsils. The digital white matter was partly demyelinated in this area and there was a marked fibrous gliosis.
Case 2: This man's brother, sister-in-law, and cousin were interviewed. He had boxed from boyhood at first in booths where he often fought several times a day, and later throughout the world as a professional. He was a World Champion for some years and reputedly had more than 700 contests. He retired in his early thirties after a severe battering and was described at that time as 'a little old man... playing the part of an animated punching bag...'
By the age of 50 he staggered slightly when he walked, and his speech was slow and slurred – "In fact, you'd think he was always drunk but he never touched a drop." During the following decade he became childlike, often wanting to be cuddled and reassured. He suffered two head injuries, which were not severe, but they may have aggravated the deterioration in memory and in behaviour; while recovering from the second he was found to be mildly diabetic.
When he was 67, he developed an acute appendicitis and was noticed to be confused, incontinent of urine, and disorientated; his memory was severely impaired. His manner was still childish; he would sulk when left alone and would smile happily when approached. He retained some idea of his boxing success but had forgotten the details. He spoke indistinctly. He was markedly ataxic, walking on a wide base and often stumbling. He could not stand on one leg, and heel-toe walking led him to fall to the right. He had nystagmus on the right and tremor of his upper limbs. Muscle power, tendon reflexes, pupillary responses, and blood pressure were all normal.
The *CSF* was normal and serological tests for syphilis were negative. A *lumbar PEG* showed marked symmetrical dilatation of the lateral ventricles and a well-outlined septal cavum. Cerebral atrophy and punch-drunk syndrome were diagnosed.
He died demented and doubly incontinent, in a psychiatric hospital at the age of 77. Death was attributed to bronchopneumonia, as a consequence of organic dementia.
Naked-eye appearance of brain: The leptomeninges were moderately thickened over both convexities; no focal damage or scarring was identified. The cerebral hemispheres were grossly atrophic (the brain weight after fixation was 960 g), the frontal and temporal poles on both sides being the most shrunken. Slight patchy atheroma and thickening of the large cerebral vessels were seen. The hypothalamic floor was stretched and translucent; the mammillary bodies were unusually small. Coronal cuts revealed marked enlargement of the lateral and third ventricles. Anteriorly the fenestrated remnants of the

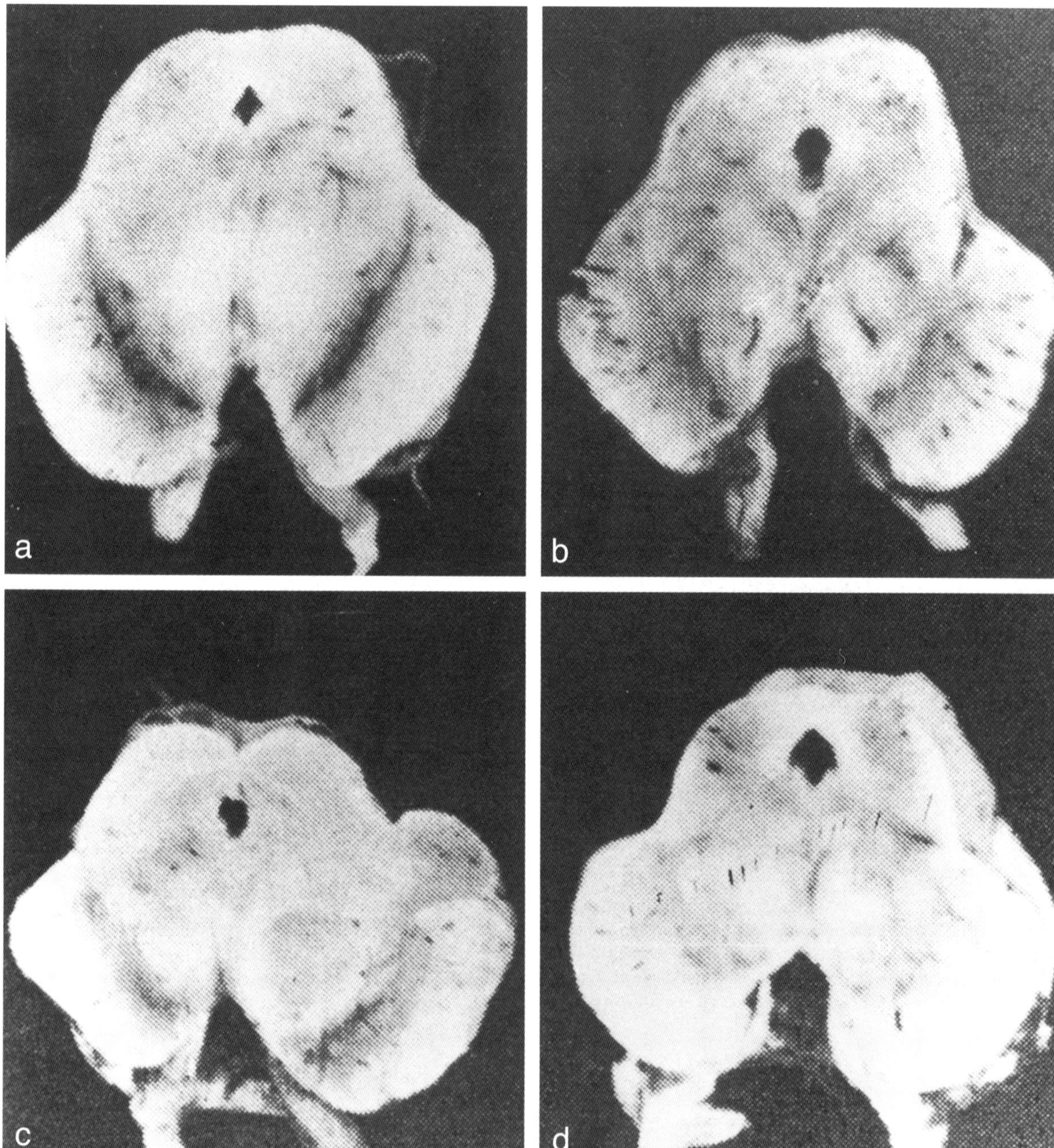

Figs. 288 a – d. (a) Transverse cut through the mid-brain of an elderly non-boxer to show the pigmentation of the normal substantia nigra for comparison with (b) Case 1; (c) Case 6; (d) Case 4. All showing a marked loss of pigment. (From CORSELLIS J A N et al., 1973).

septal leaves were separated by a cavum about 7 mm wide (Figs. 291 a and b). Posteriorly no septum was found and the thin, flattened fornix was separated by a gap of 1.5 cm from the under-surface of the atrophic corpus callosum. The anterior horn of the left lateral ventricle expanded into a rostral diverticulum with smooth walls, about 2.0 cm in diameter. The

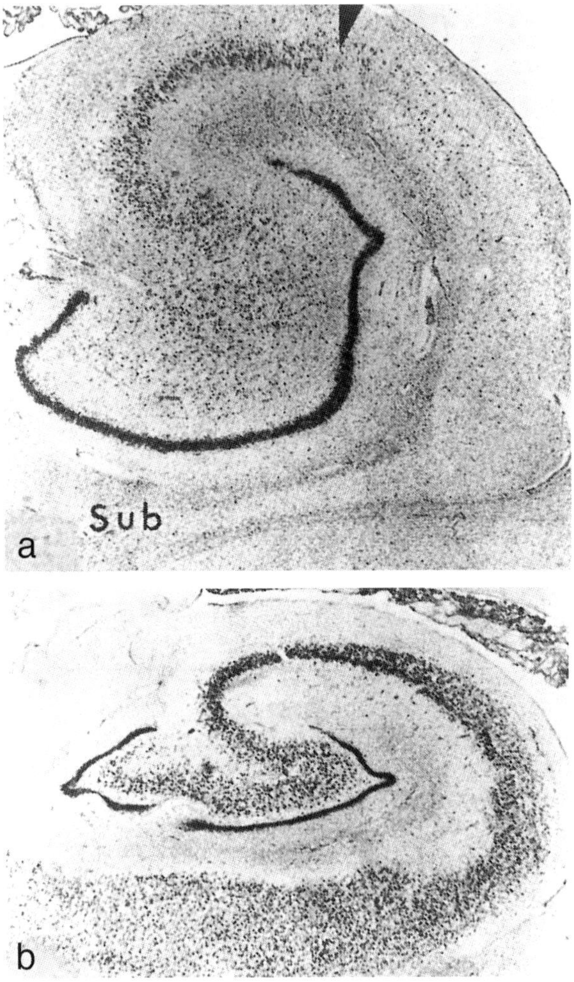

Figs. 289 a and b. (a) Case 1. The degenerated hippocampus showing almost total loss of nerve cells from the arrow head round to the subicular region (sub). Cresyl violet, 12 : 1. (b) The normal cell population at a slightly lower magnification. Cresyl violet, 10 : 1. (From CORSELLIS J A N et al., 1973).

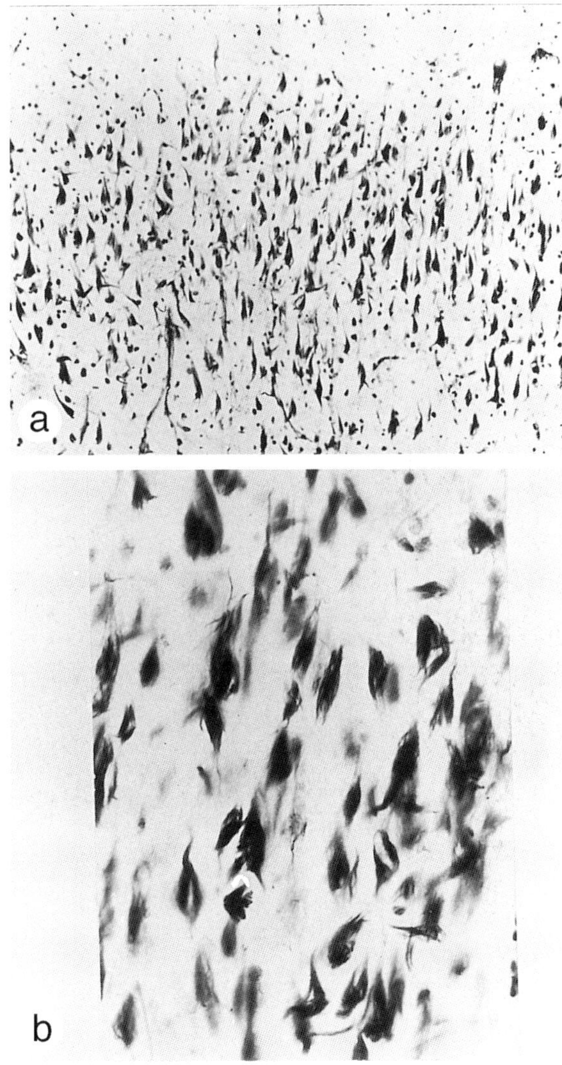

Figs. 290 a and b. Case 1. Alzheimer's neurofibrillary tangles affecting most of the neurons in the cortex of the parahippocampal gyrus. Von Braunmühl, 140 : 1. (b) A higher power view of the tangles in the same case as in (a) Von Braunmühl. 140 : 1. (From CORSELLIS J A N et al., 1973).

mid-brain was grossly atrophied and the aqueduct dilated. The pigmentation of the substantia nigra appeared normal in the medial half but was reduced laterally. The pons, medulla, and cerebellum were mildly atrophic.

Histological findings: Cerebral hemispheres: The leptomeninges over both convexities were greatly thickened. The cortical ribbon was narrowed, particularly in the frontal and temporal lobes. There was severe neuronal loss in all but the first layer (Figs. 292 a and b), and this was accompanied by a dense proliferation of astrocytes and a sub-pial fibrous gliosis. Many cortical nerve cells showed neurofibrillary change with a marked emphasis on the medial temporal grey matter. A few neurofibrillary tangles were seen in the thalamus and hypothalamus. No senile plaques were identified in any area. The white matter of both hemispheres showed a diffuse loss of myelin and a moderate glial increase. The large cerebral arteries showed occasional patches of atheroma; many of the smaller intracerebral vessels were thickened by collagen.

Brainstem: Many pigmented nerve cells in the substantia nigra had disappeared; the remainder were mostly shrunken and some showed neurofibrillary change. The degeneration was most marked in the lateral and intermediate groups and

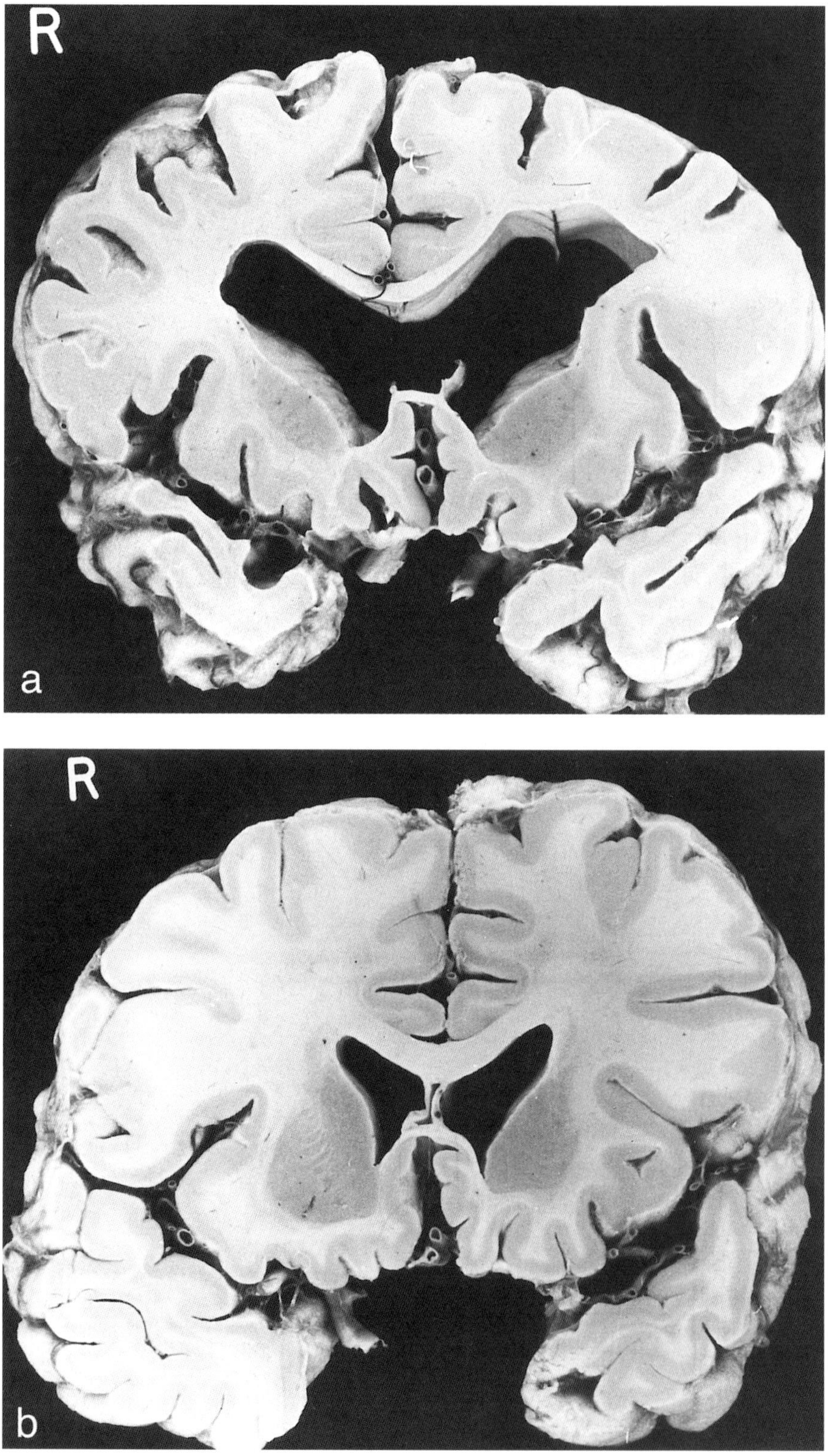
R
a
R
b

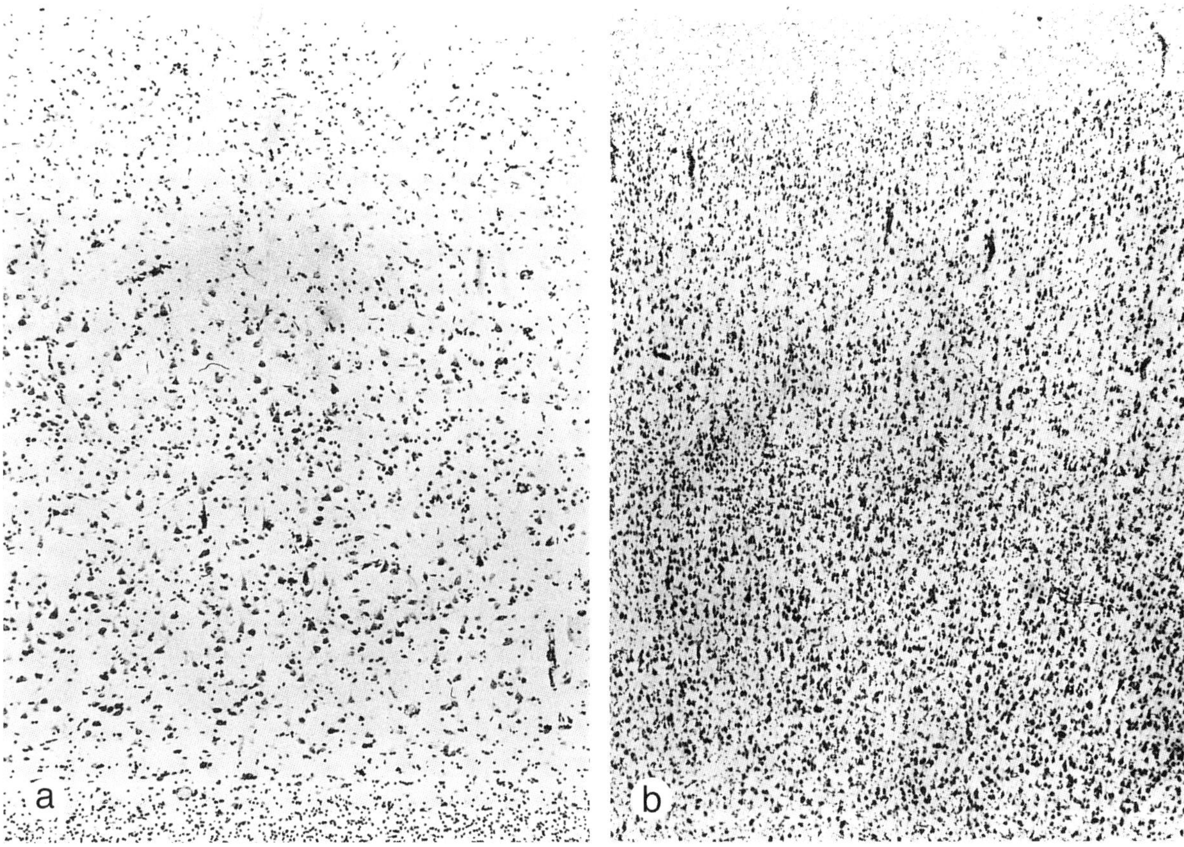

Figs. 292 a and b. (a) Case 2. Marked loss of neurons in the anterior temporal cortex can be seen when compared with b. which shows the architecture of the same area in the brain of a non-boxer, dying at 68. (a) and (b) Cresyl violet, 40 : 1. (From CORSELLIS J A N et al., 1973).

on the right side more than on the left. In the pons, many of the pigmented cells of the locus coeruleus showed neurofibrillary tangles. There was some neuronal loss and fine fibrous gliosis in both inferior olivary nuclei.

Cerebellum: The molecular layer of many folia was unduly cellular and showed patchy fibrous gliosis. In addition, there were two circumscribed foci of more severe folial damage. The first affected the cerebellar tonsils. In this region there was a marked loss of Purkinje cells and to a lesser extent of granule cells, with some demyelination of the digital white matter. The second localized lesion consisted of a small area of recent infarction on the dorsal surface of one hemisphere. The cerebellar white matter, in general, showed a diffuse pallor of the myelin and there was an increase in fibrous astrocytes.

Case 3: This boxer's wife was interviewed. He was born abroad and came to England as a young child. He started boxing at 16 and fought about 300 professional contests during the next 13 years, winning an area championship and travelling abroad. He suffered a neck injury (unrelated to boxing)

Figs. 291 a and b. (a) The patient died demented, aged 77 years. Coronal cut through the anterior horn of the cerebral hemispheres: the anterior horns are considerably enlarged and a ventricular diverticulum is present at the dorsolateral angle of the left lateral ventricle. The septum pellucidum is represented by two small tags which were up to 7 mm apart. There is marked cortical atrophy particularly of the right temporal lobe. (b) Case 12. The patient died, mentally alert, aged 91 years. Coronal cut similar to (a) the lateral ventricles are slightly enlarged and there is a septal cavum 1 mm wide but the brain appears essentially normal. Both specimens have been reduced equally in the illustration and gross loss of cerebral tissue in (a) compared with (b) is clearly seen. (From CORSELLIS J A N et al., 1973).

when about 26 and boxed less after this, becoming a manager when approaching 30. His wife met him shortly after this and she recalled that "he had already begun to get a bit muddled." By the age of 36 he would often fall over backwards. He began to suspect his mother of theft and his wife of infidelity. He was seen in hospital when aged 44, because of back pain. Osteoarthritis of his cervical vertebrae was found on x-ray; a rigid gait was noted; both arms were slightly spastic; early Parkinsonism was queried. At 50, slurring of speech was mentioned, and the term 'punch drunk' was used. Six years later a neurological report included an extrapyramidal tremor of both hands and ataxia of both legs. There was a slight dysarthria. A *PEG* shortly afterwards showed enlarged lateral ventricles with a septal cavum. A further report, when he was 61, stated that his unsteadiness and shaking hands had become worse. He was 'dribbling a bit,' he had a Parkinsonian facies, his speech was difficult to understand, and his memory was poor. His wife recalled that he had a bad temper, 'the scenes were awful,' 'he raved at night' but was 'so kind and nice' afterwards. It was concluded in hospital that he had 'diffuse degenerative brain disease, affecting extrapyramidal and pyramidal systems as well as causing mental impairment.' He could not talk or swallow properly; he had several falls and after the last one he died from bronchitis and bronchopneumonia, when aged 63. He was said to have drunk moderately at times, and an attack of pancreatitis when aged 55 was attributed to alcohol. He appears to have given up the drink since then.

Naked-eye appearance of brain: The weight after fixation was 1.260 g. The leptomeningeal and cortical surfaces were normal and no superficial scarring was seen. There was no atheroma of the large cerebral vessels. The sulci of both hemispheres were slightly widened. Coronal cuts showed moderate enlargement of the anterior horns of the lateral ventricles. The two leaves of the septum pellucidum were about 5 mm apart; they were shredded and consisted only of a few remaining strands. The fenestration ran the sagittal length of the septum. The grey and white matter of both hemispheres was normal apart from the floor of the hypothalamus, which was unusually thin. There was a marked lack of pigment in the substantia nigra, on the right side more than on the left. The pons and medulla appeared normal. The cerebellum looked slightly atrophic.

Histological findings: Cerebral hemispheres: The leptomeninges were moderately thickened over both convexities. The cerebral cortex generally showed slight neuronal loss and a mild fibrous gliosis. The hippocampal neuronal population was considerably depleted. There were many neurofibrillary tangles in the nerve cells of the frontal and temporal cortex. They were particularly marked in the antero-medial temporal grey matter, including the amygdaloid nucleus, the hippocampus, and the parahippocampal gyrus (Figs. 293 a and b), but were scanty in the diencephalic and periventricular grey matter. No senile plaques were found in any area. The centrum

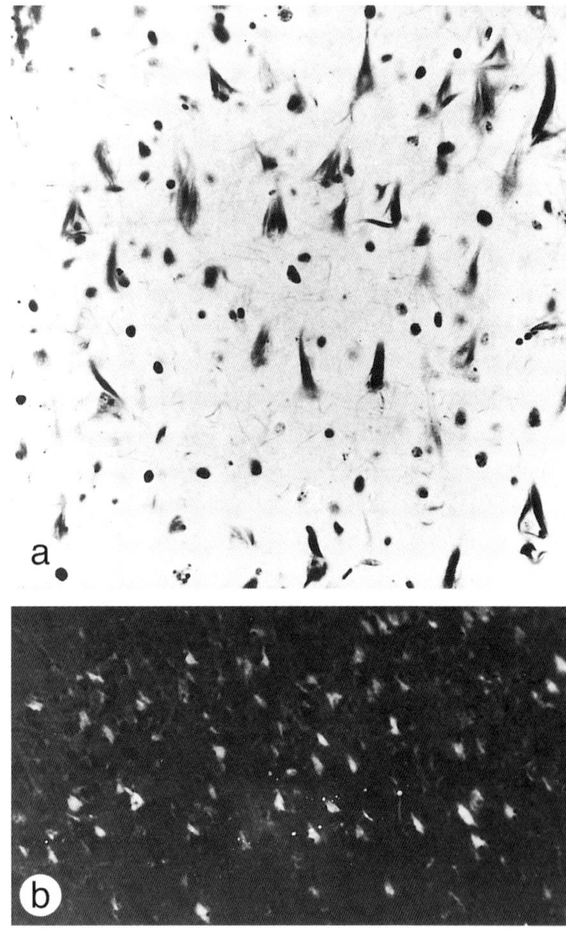

Figs. 293 a and b. (a) Alzheimer's fibrillary tangles affecting large numbers of neurons in the fusiform gyrus. Von Braunmühl technique, 240 : 1. (b) Congo red stain under polarized light showing the intensity of the neurofibrillary change in the medial temporal cortex of the same case. 100 : 1. (From CORSELLIS J A N et al., 1973).

semi-ovale was well myelinated and no glial increase was noted. The bodies of the fornices were partly detached from the roof of the ventricles and appeared slightly atrophic and demyelinated. The mammillary bodies appeared small but otherwise normal. No other abnormality was found in the deep hemispheral grey matter.

Brainstem: The substantia nigra, on the right side more than on the left, was severely degenerated. Most of the pigmented nerve cells had disappeared and the remainder were shrunken and poorly pigmented (Figs. 294 a and b). Much pigment was present in the adjacent tissue. Some neurons showed neurofibrillary change (Figs. 295 a and b); no Lewy bodies were seen. The medial part of the substantia nigra was least af-

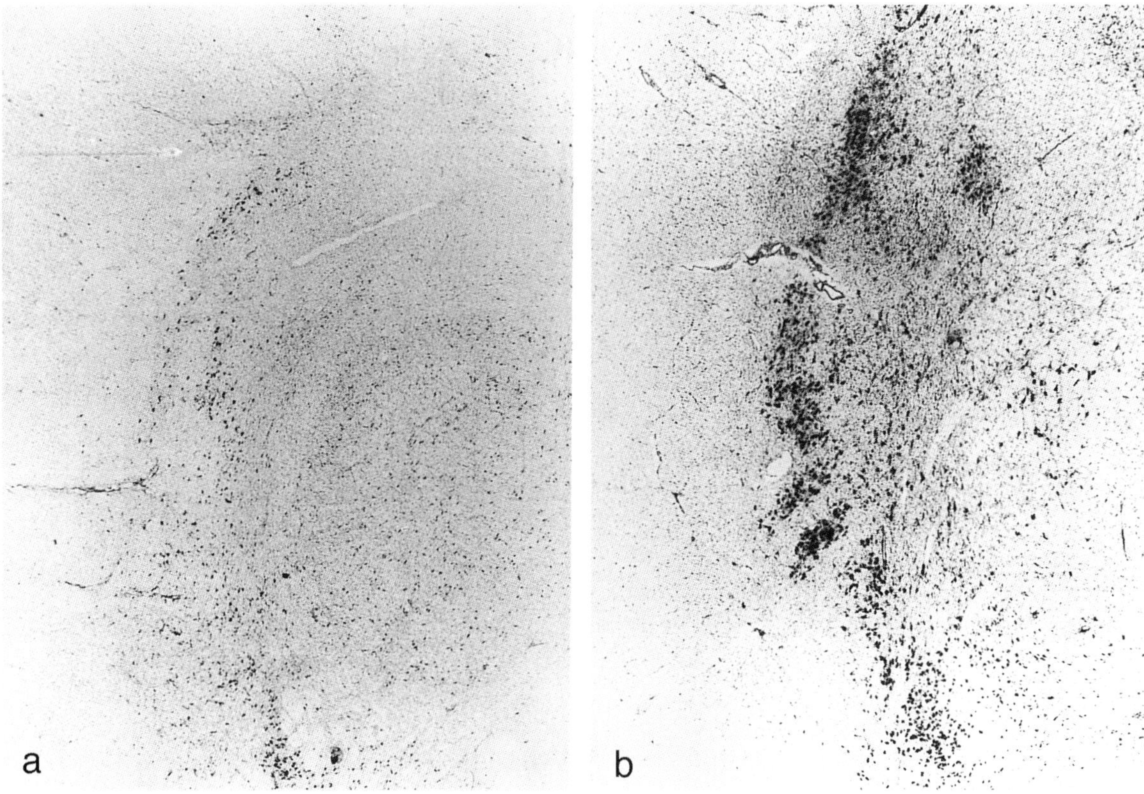

Figs. 294 a and b. (a) Case 3. The considerable depletion of pigmented neurons in the substantia nigra can be seen when compared with the normal picture in (b) (a) and (b) Cresyl violet, 11 : 1. (From CORSELLIS J A N et al., 1973).

fected. The locus coeruleus in the pons and the occasional neuron in the periventricular grey matter of the pons and medulla showed neurofibrillary tangles.

Cerebellum: Throughout the entire cerebellum, the molecular layer showed a fine fibrous gliosis which became considerably more intense in patches. The most affected area lay in the region of the tonsils, in which there was also a moderate loss of Purkinje cells, with some thinning of the granule layer and slight loss of myelin.

Case 4: This man's son and wife were interviewed. The son had also boxed but gave it up when 28 years old because "I was getting hurt and did not want to get like my father."

The patient himself went through elementary school and won, but rejected, a place at high school. He joined the Royal Navy when 15. Because of his talents as an amateur boxer he turned professional, and during the following 25 years he had an estimated 600 fights in many parts of the world. He was a challenger for a British Championship. A boxing journal described how he had once fought a 15 round contest every day of the week with a 20 round contest on Saturday. As a young man he was considered shrewd and not unintelligent. He had a minor car accident in his twenties. In his mid-thirties he became unsteady on his feet, and after the age of 40 he had a shuffling gait. His legs were apt to give way, and he developed a marked tremor of the hands. He had difficulty in dressing himself. He drank heavily for a few years at this time but cut his intake down drastically when he found that he could no longer take it. He rarely smoked. He retired from boxing when he was 40 years old and worked as a carpenter for 15 years. He was then registered as a disabled person and could only manage unskilled work.

When he was 57 he attended hospital for backache and was found to have a spastic gait and slurred speech, both of which were recorded as having been present for some years. At the age of 64 he was reinvestigated in a neurosurgical unit. A history of progressive unsteadiness on his feet and mental deterioration, with impairment of memory, were described by his wife. He lost all sense of date and time and did not recognize his relatives. His speech was dysarthric, he had a wide-based ataxic gait, and Romberg's sign was positive. He was considered mildly demented. No tremor was recorded. Air studies showed general enlargement of the ventricles without dis-

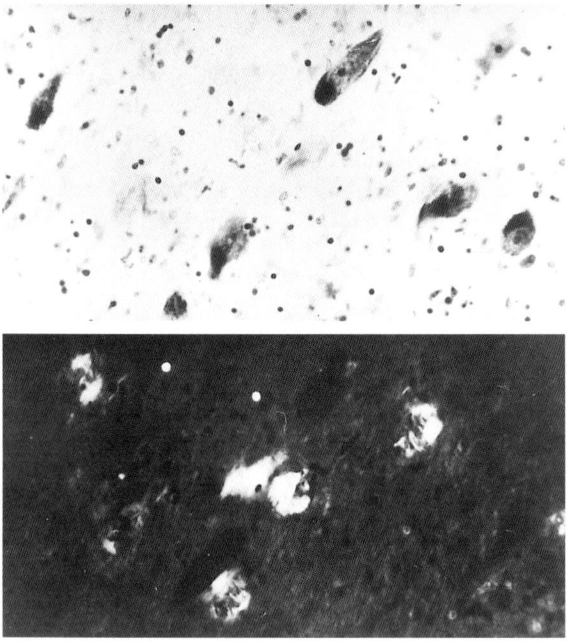

Figs. 295 a and b. (a) Case 3. An area of the substantia nigra stained with Congo red and hematoxylin to show a few shrunken neurons containing pigment and the faint remnants of several more degenerated cell. (b) The same field under polarized light. Doubly refractile neurofibrillary tangles can be seen particularly in the places where the remnants of degenerated neurons can be faintly identified in (a). (a) and (b), 270 : 1. (From CORSELLIS J A N et al., 1973).

placement; the septum was not mentioned, and the punch-drunk syndrome and spinal osteoarthritis were diagnosed. When aged 69, he entered hospital with acute retention of urine. He had a tremor of both legs and a mask-like facies. He developed a chest infection and died three weeks later.

Naked-eye appearance of brain: The *brain* weighed 1.205 g after fixation. The olfactory bulbs and the surrounding orbital cortex showed slight rusty discoloration. The large cerebral vessels were not atheromatous. The leptomeninges over the convexities were slightly thickened. The sulci were a little widened. Coronal cuts showed the lateral and third ventricles to be moderately enlarged. The septum pellucidum was intact and without a cavum anteriorly, but it was grossly fenestrated in its posterior half (Figs. 296 a and b). The corpus callosum showed no scars but was only about 2 mm thick posteriorly. The floor of the hypothalamus was thin and translucent. The mid-brain was unusually small; the substantia nigra was almost without pigment. The cerebellum looked a little shrunken. The pons and medulla were normal.

Histological findings: Cerebral hemispheres. The large cerebral vessels appeared normal but the intracerebral arteries showed

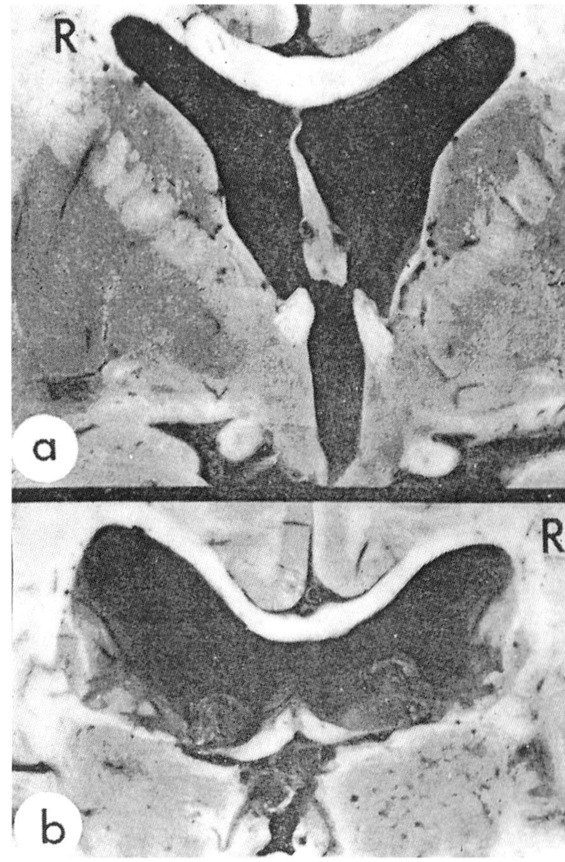

Fig. 296. Case 4. Anterior (a) and (b) posterior coronal views of the septum pellucidum. (a) No cavum is present. This gives the appearance of a normal but slightly elongated septum which may be compared with the fenestrated cava in Figs. 1, 5, and 12. (b) The same case as in (a) The septum has been largely destroyed posteriorly and the bodies of the fornices are nearly detached from the corpus callosum. (From CORSELLIS J A N et al., 1973).

some fibrous thickening and hyaline change. The cortical ribbon was not thinned and no generalized loss of nerve cells was identified. The antero-medial temporal cortex, however, and in particular, the pes hippocampi showed a moderate degree of neuronal loss. A slight generalized astrocytic increase was noted. The white matter of the hemispheres was well myelinated, apart from a few small areas of recent softening lateral to, and in, the basal ganglia.

Neurofibrillary tangles were scattered in small numbers through the cortex of the frontal and parietal lobes and the three lateral temporal gyri. They were moderately intense in the amygdaloid nucleus and uncal cortex and in the hippocampus and the parahippocampal gyri. No senile plaques were found in any area.

The thalamus showed some loss of nerve cells in the anterior nuclei. Both fornices were slightly atrophic and were moderately gliosed in the lateral portions. The mammillary bodies were poorly populated, and many of the remaining neurons were shrunken.

Brainstem: Very few pigmented cells were left in the substantia nigra. This area was severely gliosed and contained much pigment debris, the lateral and intermediate nuclear groups being most affected (Figs. 297 a and b). A few neurofibrillary tangles were seen in the nigral neurons, but no Lewy bodies were found in them. The locus coeruleus showed many tangles.

Cerebellum: There was a circumscribed area of folial scarring on the inferior surface of the cerebellar hemispheres in the tonsillar region. Marked loss of Purkinje cells and some loss in the granule layer was seen, with proliferation of the Bergmann glia. The molecular layer in this area was narrowed and showed marked fibrous gliosis. A similar but less severe patch of folial atrophy affected the antero-dorsal surface.

Case 5: Two of this man's boxing colleagues, one a promoter and the other his second, were interviewed. The patient had been a deaf mute and had boxed as a child. He became a professional when 18 years old, first mainly in the booths and later in promoted contests. He held a Northern championship for two years, during a career which lasted 18 years. He had fought in South Africa.

"He was always aggressive and spiteful and was so unpopular with the crowds that they liked to see him beaten." He never refused a challenge, regardless of weight, and often took 'terrible punishment' to the head. He sometimes boxed in a trance. He drank excessively in his twenties. Towards the end of his career at the age of about 30, he was no longer able to look after his personal affairs. He became enuretic when about 35 and eventually had to be cared for in a home. He was noted to be "irascible, forgetful, but clean and tidy and quite nimble on his feet." When 60 years old he developed a right facial weakness and a right homonymous hemianopia. In the same year, and a few months before he died, he was transferred to a psychiatric hospital. Death was attributed to bronchopneumonia and cardiac ischaemia. The heart weighed 370 g.

Naked-eye appearance of brain: The fixed *brain* was small and weighed 1.095 g. The leptomeninges were moderately thickened. The frontal and, to a lesser extent, the temporal sulci were moderately widened. The large cerebral vessels were atheromatous and the cortical area supplied by the occluded left posterior cerebral had been destroyed by old infarction. Coronal cuts showed moderately enlarged lateral and third ventricles. Anteriorly the two laminae of the septum were fenestrated and were separated by a cavum about 8 mm wide. A few tags of septum remained posteriorly. The corpus callosum was 3 mm thick and contained a small yellowish cystic patch towards its right border. The right putamen and adjacent frontal cortex were softened. The aqueduct was dilated; the pigmentation of the substantia nigra was a little reduced. The pons and medulla appeared intact. The cerebellum looked mildly atrophied around the tonsils.

Histological findings: Cerebral hemispheres: The leptomeninges showed moderate collagenous thickening and the large cerebral arteries were markedly atherosclerotic. There was a moderate degree of generalized cortical atrophy, which was noticeably severe in the frontal lobes. There were several patches of necrosis, the largest measuring 5 mm across, scattered through the centrum semi-ovale on both sides. There was a small softening in the body of the corpus callosum, which contained many fat phagocytes and some Marchi-positive material. There were also several focal areas of necrosis in the striatum and pallidum in addition to the extensive necrosis and scarring of the medial gyri in the right temporal lobe and the striate area. No senile plaques were found in any part of the grey matter, but neurofibrillary tangles were scattered through all lobes of the hemispheres. They were strikingly intense in the neurons of the uncus and the parahippocampal gyrus on both sides.

Brainstem: The substantia nigra showed a slight loss of pigmented neurons, with melanin pigment lying free in the adjacent tissue. A few nerve cells showed neurofibrillary change; no Lewy bodies were seen. The corticospinal tracts on both

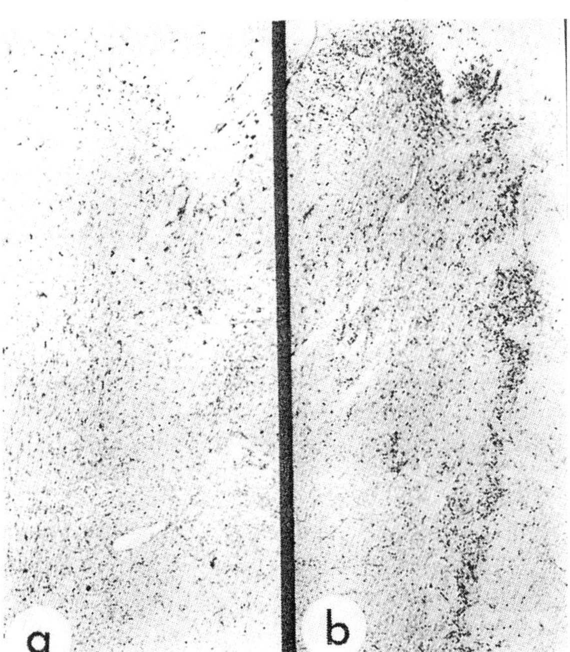

Figs. 297 a and b. (a) case 4. Almost complete loss of pigmented nerve cells in all but the uppermost (or medial) part of the substantia nigra. The neuronal loss may be compared with the normal picture in (b). (a) and (b) Cresyl violet, 12 : 1. (From CORSELLIS J A N et al., 1973).

sides were smaller than usual, and were partially demyelinated.
Cerebellum: The molecular layer of many folia showed a fine fibrous gliosis. There was a small patch of folial necrosis adjacent to several thickened vessels on the dorsal surface of the hemisphere. There was also a circumscribed patch of folial scarring in the tonsillar region of both hemispheres. The affected folia showed a marked fibrous gliosis with patchy Purkinje cell and granule cell loss. The cerebellar white matter was generally poorly myelinated.
Case 6: The man's younger brother and one son were interviewed. The brother had boxed during his teens but said that he had been persuaded to give it up by his wife and by the sight of two punch-drunk boxers.
The patient was an average child. Like his mother, he was fond of a fight. He began boxing at school and left home at 13 to travel with the booths. As a professional in Great Britain and in the United States, he had an estimated 500 contests in a career spanning 25 years. He was a 'Mirror of Life' Champion of England. He sometimes took severe punishment and was knocked out occasionally. He could not always recall the fights that he had had the previous night. He was fit until the age of 32, when his legs 'began to give way,' he developed a 'hopping-dancing walk,' and his speech became slurred. His memory was good. He retired from boxing six years later and became a referee and a promoter. Eventually he took any job he could get. His two marriages broke up, largely because of his violent nature. When 65 years old, he went to live with his son but aggressive outbursts led to his removal to an old people's home. He became more subdued but he still had the occasional scuffle and one of these brought him to a psychiatric hospital.
On *admission* he was orientated but his memory was poor. He thought people were robbing him. He was dysarthric, and ataxic, with a slow shuffling gait. He had a Parkinsonian look and some weakness of his left leg. He was incontinent of urine. His death, at 83, was attributed to bronchopneumonia,
Naked-eye appearance of brain: Brain weight after fixation was 1.300 g. The leptomeninges were moderately thickened; there was slight rusty staining of the olfactory bulbs and of both temporal poles. The large cerebral vessels were tortuous and slightly narrowed by flecks of atheroma. No cortical atrophy was seen. Coronal cuts showed moderate enlargement of the lateral and third ventricles. The septum pellucidum consisted anteriorly of two fragmented leaves about 6 mm apart (Fig. 298). Posteriorly only a few fine strands remained, and the fornix was separated from the corpus callosum by a vertical gap of 1.4 cm. The floor of the hypothalamus was paper-thin and translucent. The cortex and deep grey matter appeared normal. The substantia nigra was strikingly short of pigment (Fig. 288 c); the rest of the brainstem appeared normal. There was slight widening of the main cerebellar sulci suggesting mild generalized atrophy.

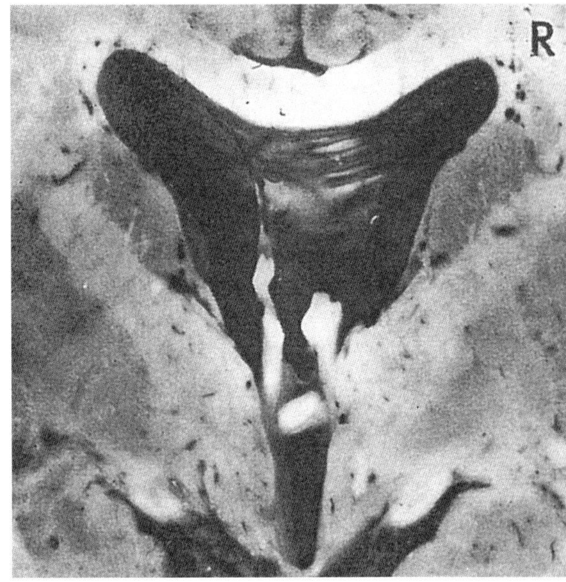

Fig. 298. Case 6. Widely separated strands of a grossly fenestrated cavum septi pellucidi. Macrophoto. (From CORSELLIS J A N et al., 1973).

Histological findings: Cerebral hemispheres: The leptomeninges and large cerebral vessels were moderately thickened. The cortex showed patchy nerve cell loss and gliosis especially in the medial temporal gyri. A few senile plaques were seen in the frontal, parietal, and occipital lobes; many were found in the parahippocampal gyrus and in the hippocampus. Neurofibrillary tangles were considerably more intense and were particularly numerous in the medial temporal grey matter. Both fornices were gliosed; the left one was shrunken. The mammillary bodies were small, some of the neurons were shrunken, and there was a marked increase in glia. The white matter of the centrum semi-ovale was poorly myelinated and gliosed.
Brainstem: The neuronal population of the substantia nigra was depleted, particularly in the lateral nuclear groups. Many remaining cells were shrunken or showed neurofibrillary change (Fig. 299). Pigmented neurons in the locus coeruleus were similarly affected. No Lewy bodies were seen.
Cerebellum: The molecular layer of many folia showed a fine fibrous gliosis. In addition, an area of folial atrophy, gliosis, and demyelination, with considerable Purkinje and granule cell loss was present in the cerebellar tonsil (Figs. 300 a and b). The cerebellar white matter generally was poorly myelinated.
Case 7: In this case the wife was interviewed. Her husband came from a family of boxers. He was not good at school – "he could read and write but never did." He started boxing as a professional at the age of 16 years, and during the next 20 years or so he fought an estimated 400 contests. He was a

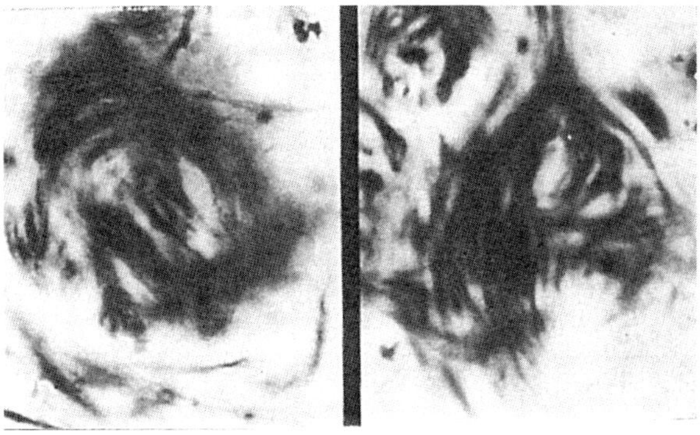

Fig. 299. Case 6. Two pigmented neurons in the substantia nigra showing neurofibrillary change. Von Braunmühl technique, 1000 : 1. (From CORSELLIS J A N et al., 1973).

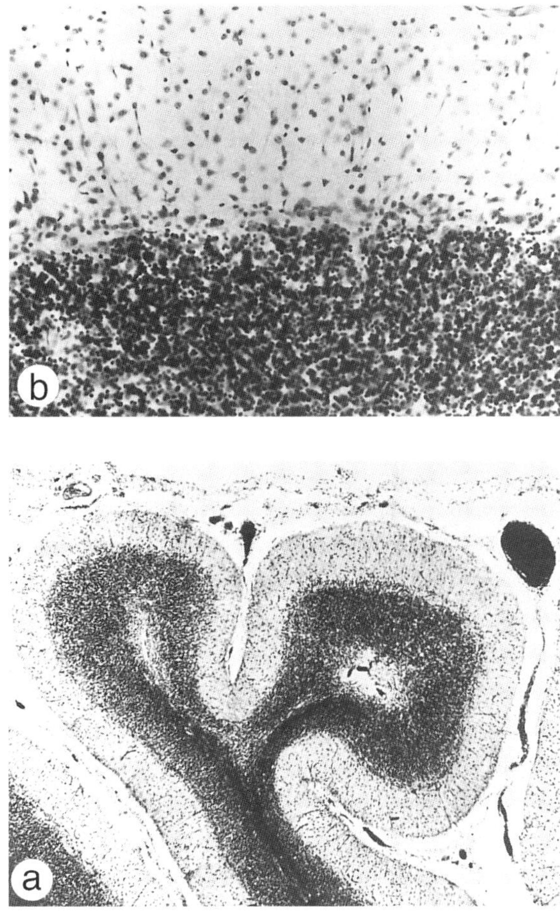

Figs. 300 a and b. (a) Case 6. Myelin stain of the cerebellum showing loss of myelin in the digital white matter in the tonsillar region. Heidenhain-Woelke technique, 36 : 1. (b) A small but representative part of the area in (a) showing the complete loss of Purkinje cells and some proliferation of Bergmann glia. Cresyl violet, 35 : 1. (From CORSELLIS J A N et al., 1973).

popular, successful man and he traveled on many boxing tours around the world He retired from the ring after a particularly damaging fight, and started to teach physical training and boxing in boarding schools. He gave this up after a few years at a time when he was drinking heavily and beginning to become moody and violent. A few years later he lost a job in a printing works, for hiding away to sleep, and he seems not to have worked again. When about 54 he began to be 'very unsteady on his feet' and kept falling about. He had spells of 'going within himself' followed by aggressive attacks on his wife and his home. He complained of violent headaches; his sexual demands gave his wife little rest. He wandered, out, half-dressed, at night and would importune for money. He had 'glassy-looking eyes;' he was doubly incontinent at times. He spoke normally and had no tremor; he could not be left alone. When 59 he was admitted to a psychiatric hospital and found to be grossly demented. The summary stated that "examination of the CNS, including gait, revealed no abnormality. *PEG* showed a somewhat dilated third ventricle with cortical atrophy and dilated lateral ventricles." The dementia deepened and in the last two years of his life "he aged, became like an old man," he developed bronchopneumonia and died at the age of 62. At the post-mortem examination he was noted to have cauliflower ears and a flattened nose.

Naked-eye appearance of brain: The fixed brain was small, weighing 1.090 g. The leptomeninges were moderately milky. The frontal sulci were slightly widened. No scarring or other focal abnormality was seen on the cortical surface. The large cerebral vessels were moderately atheromatous. The lateral and third ventricles were moderately enlarged. A septal cavum was present anteriorly, 4 mm wide. Posteriorly, the separated leaves were grossly fenestrated. The body of the corpus callosum was thinned posteriorly to about 2 mm. The cortical ribbon appeared normal. There were small cystic softenings up to 3 mm across in the thalamus and putamen of both sides. The right internal capsule was stained brown.

The midbrain appeared slightly atrophic but the substantia nigra showed average pigmentation. A few pin-head softenings were present in the nuclei pontis. The left superior and middle cerebellar penduncles showed small areas of yellow pigmentation, similar to those seen in head injury. The medullary pyramids were normal. The cerebellar white matter was diffusely greyish and shrunken.

Histological findings: Cerebral hemispheres: The leptomeninges and small cerebral vessels were greatly thickened. The larger arteries showed moderate atherosclerosis. The cortex was well populated except in the medial temporal areas, which showed patchy nerve cell loss and gliosis. Many neurofibrillary tangles were concentrated in the uncus, amygdaloid nucleus, hippocampus, and parahippocampus gyrus; they oc-

curred sporadically elsewhere. There were no senile plaques in any area. Several small perivascular softenings were present in the corpus striatum and thalamus; the largest of these had destroyed part of the right putamen and extended into the adjacent internal capsule. The white matter of the hemispheres showed a slight loss of myelin and a moderate glial increase.
Brainstem: The substantia nigra and locus coeruleus showed a slight loss of pigmented cells with melanin particles scattered nearby. The nigral damage was more severe in the lateral nuclear group. A few cells showed neurofibrillary change; no Lewy bodies were seen. The tegmentum of the pons had been partially destroyed by cystic softenings that surrounded thickened vessels. The left inferior olivary nucleus showed marked neuronal loss and gliosis.
Cerebellum: The molecular layer of a few folia showed a fibrous gliosis. The cerebellar vessels were thickened and small softenings were scattered throughout the poorly myelinated white matter. The cerebellar tonsils were apparently damaged or scarred.
Case 8: The second wife of this boxer was interviewed. He was born abroad and she knows little of his early life. He came to London when 14 years old. He was said to have been of average intelligence and to have spoken four languages. His professional boxing career started at about 17 years, and he claimed to have had 565 contests. He boxed in many countries, sometimes having three contests a day. He retired at about the age of 40 years. After this he worked until he was 68 as a labourer, although at some time he was registered as disabled.
His wife remembered that he had a cauliflower ear and, because of a broken nose, he could not breathe properly. She said that his only interests were boxing and gambling. "He drank little, but scarcely needed to since two light ales knocked him out."
He began to go downhill when he was 60 years old. He then lost consciousness for two hours, and on recovery he was ataxic and had some weakness of his left side. Several similar episodes occurred during the next ten years; between them he remained ataxic with a tendency to fall. He had a tremor and could not tie a bow. The last attack affected his right side, and his intellect and behaviour then deteriorated. He became violent towards his wife but would afterwards apologize. He burnt all his newspaper cuttings and then asked for them back. He would try to go unclothed into the street. He died at home aged 71.
Naked-eye appearance of brain: The fixed *brain* was small and weighed 1.040 g. The leptomeninges were considerably thickened. The large cerebral vessels were moderately atheromatous; there was an old softening 2 cm in diameter in the lateral orbital part of the right frontal lobe. The frontal and temporal poles were unusually pointed and the sulci over both convexities were widened. Coronal cuts showed moderate ventricular dilatation. The septum was difficult to assess but it was manifestly absent posteriorly, while rostrally a few tags of the septal walls indicated the presence of a fenestrated cavum, about 4 mm wide. Several patches of old softening were scattered through the hemispheres and one lay in the right internal capsule. The cortical ribbon was slightly thinned and the basal ganglia were atrophic. The aqueduct of the mid-brain was dilated. The right cerebral peduncle appeared smaller than the left; the substantia nigra seemed normally pigmented. The right side of the pons and the right pyramid in the medulla were shrunken. The left cerebellar hemisphere was infarcted and partly destroyed on its ventral surface; the right hemisphere appeared unaffected.
Histological findings: Cerebral hemispheres: The leptomeninges and cerebral vessels were moderately thickened. The cortical ribbon was well populated except in the medial temporal region where there was neuronal loss. Neurofibrillary tangles were present in small numbers throughout the cortex and diencephalon; the greatest numbers were found in the parahippocampal gyrus and the hippocampus. A few senile plaques were seen in the frontal and occipital cortex. The mammillary bodies were slightly gliosed; the nerve cells appeared intact. Many cells in the thalamus were shrunken. The hemispheral white matter was poorly myelinated and gliosed.
Brainstem: The substantia nigra and locus coeruleus showed a slight depletion of pigmented cells; free-lying pigment granules were scattered in the adjacent tissue. Some of the remaining neurons showed neurofibrillary change; no Lewy bodies were seen.
Cerebellum: The ventral surface of the left hemisphere, including the tonsillar region, showed widespread folial necrosis. The necrotic areas were adjacent to thickened and narrowed vessels. The dorsal surface of the right hemisphere showed a similar patch of folial infarction. The right tonsillar region appeared normal.
Case 9: This man was a patient in a psychiatric hospital for the last seven years of his life. His younger brother was interviewed. As a young man the patient had taught physical training. He boxed in the Army and had later been 'quite a successful' professional boxer in Great Britain and the United States, having fought at Blackfriars Ring and Madison Square Gardens. On his return to England, when aged 31, his brother noticed that there was a tremor of his left hand and that his speech was slurred and hoarse. The hoarseness had followed a blow to the throat.
He started drinking heavily and drifted from job to job. He tried to kill himself when aged 65 and was admitted to a psychiatric hospital. A Parkinsonian facies was noted, with tremor and rigidity of the left upper limb. The left vocal chords were markedly thickened and he was dysarthric. The blood pressure ranged around 210/130 mmHg, and he was considered to be demented, possibly as a result of cerebral trauma. A transient left-sided hemiparesis was noticed nine months before his death from bronchopneumonia at the age of 72 years.

Naked-eye appearance of brain: The fixed *brain* was normal in size and in weight at 1.435 g. The large cerebral vessels were severely atheromatous. The leptomeninges were only slightly thickened and no evidence of past cortical injury was seen. Coronal cuts showed the lateral ventricles to be slightly enlarged. Anteriorly a well-defined cavum 3 mm wide was present; posteriorly the septum was mildly fenestrated.
No other abnormality was noted in the hemispheres. The substantia nigra was markedly lacking in pigment. A 3 mm ball haemorrhage was present in the nuclei pontis on the left side. The cerebellum appeared normal.
Histological findings: Cerebral hemispheres: The leptomeninges and cerebral vessels were slightly thickened. The cortex showed a diffuse increase in astrocytes with slight neuronal loss. Moderate numbers of neurofibrillary tangles were seen in all regions but were particularly noticeable in the hippocampus and in the parahippocampal gyrus. Senile plaques were not seen. Both fornices were well myelinated; they showed a slight glial increase. The deep grey matter and hemispheral white matter was also gliosed.
Brainstem: The substantia nigra showed a marked loss of pigmented nerve cells especially from the lateral nuclear zones (Fig. 301). Considerable quantities of melanin pigment were lying free in the adjacent tissues or in phagocytes. Many of the remaining cells showed neurofibrillary change. No Lewy bodies were seen. The pons and the medulla were normal.
Cerebellum: There was a circumscribed area of folial atrophy and scarring in the region of the tonsils (Figs. 302 a and b). The rest of the cerebellum appeared normal.
Case 10: This patient had been a professional boxer. He was said to have fought as a champion but this could not be confirmed. His second wife, who had been interviewed while he was still alive, knew little of his early career. She first noticed that his memory was poor when, aged about 40, he was serving in her shop. Later he worked as a labourer, and became unemployed for five years after the loss of an eye (unrelated

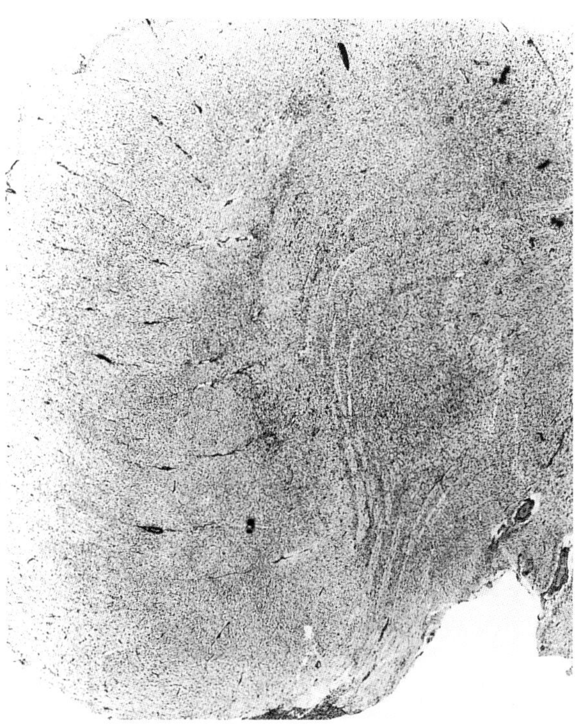

Fig. 301. Case 9. Gross loss of pigmented nerve cells in the substantia nigra. Cresyl violet, 14 : 1. (From CORSELLIS J A N et al., 1973).

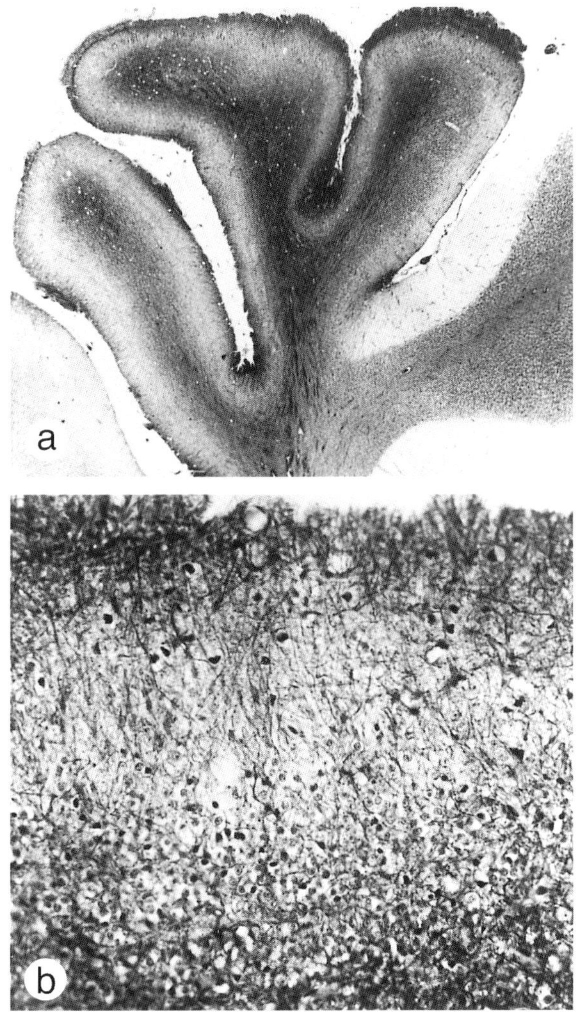

Figs. 302 a and b. (a) Case 9. Atrophy and gliosis of cerebellar folia in the tonsillar region. Mallory PTAH technique, 16 : 1. (b) A higher power view of the molecular cell layer, showing the intense fibrous gliosis. Mallory PTAH technique, 220 : 1. (From CORSELLIS J A N et al., 1973).

to boxing). He returned to work as a road sweeper. He and his wife had drinking bouts and he was known in his locality as a violent man, although "he had been a good husband." At the age of 60 he developed a left hemiparesis and was admitted to a mental hospital three years later. He was disorientated and had a marked loss of recent memory. He was paranoid and deluded, and became confused and aggressive. The Kahn and Price's precipitation reactions were negative. The diagnosis was of cerebral arteriosclerosis and organic dementia, probably post-traumatic in origin. The term 'punch drunk' was used. He gradually deteriorated over the next four years and died aged 67.

Naked-eye appearance of brain: This was a small atrophied brain that weighed 1.095 g. after fixation. The leptomeninges were moderately thickened over the surface. They were particularly dense at the base and concealed an abnormally thin hypothalamic floor with the mammillary bodies scarcely protruding. The large cerebral vessels showed a minimal degree of atheroma; no scarring of the cortical surface was found. Coronal cuts showed severe dilatation of the third and the lateral ventricles. A septal cavum, 8 mm wide, was present and the septal leaves were grossly fenestrated. The right globus pallidus and the right internal capsule had been destroyed by a recent circumscribed haemorrhage. The brainstem and cerebellum were small but otherwise normal.

Histological findings: Cerebral hemispheres: The leptomeninges and cerebral vessels were moderately thickened. The cortical ribbon showed a diffuse neuronal loss and an astrocytic increase. Neurofibrillary tangles were plentiful throughout the cortex; they were especially concentrated in the hippocampus and parahippocampal gyrus and there were several profuse patches in the lateral temporal gyri and the insula. Senile plaques were confined to the medial temporal grey matter. The fornices were thin, poorly myelinated, and gliosed. Part of the right pallidum had been destroyed by haemorrhage from a cavernous haemangioma. The damage had spread to involve the adjacent internal capsule. The thalamic and hypothalamic nuclei were gliosed; some of the neurons showed neurofibrillary change. Many small vessels in the centrum semi-ovale had widened perivascular spaces; the white matter showed a slight glial increase but was well myelinated.

Brainstem: The substantia nigra showed moderate loss of pigmented cells with melanin debris from disintegrated cells scattered nearby. The medial part was less affected than the lateral nuclear groups. Many remaining neurons showed neurofibrillary change; no Lewy bodies were seen. Some neurons in the locus coeruleus contained tangles. The medulla appeared normal.

Cerebellum: The molecular layer of many folia showed a fine fibrous gliosis. In addition, an area of folial atrophy was present of the inferior surface of the hemispheres in the tonsillar region. The affected folia showed a patchy Purkinje cell and granule cell loss with proliferation of the Bergmann glia; the molecular layer showed a dense fibrous gliosis.

Case 11: The patient's wife was interviewed but she knew little of his early history. He was born in the West Indies and had no relatives in this country. He had boxed professionally until about 35 years old and had then been in the Army during the war. His wife believed that he was discharged from the Army, and later from the maritime service, on health grounds. He took small jobs interspersed with National Assistance for many years. "He was a lovely man – never a cross word." When aged 48 his Wassermann reaction had been positive in the blood and in the cerebrospinal fluid and he was treated with procaine penicillin for suspected tabes dorsalis. In the ensuing years he suffered from a papilloma of the bladder. During the last six years of his life he remained bright and alert and worked regularly as a road sweeper. When aged 67 he was admitted to hospital in heart failure with a history of increasing confusion. He was disorientated and aggressive; Argyll Robertson pupils were present. The cerebospinal fluid was normal. The Wassermann reaction was positive in the blood but negative in the cerebrospinal fluid. He died three months later, aged 67.

Naked-eye appearance of brain: The fixed *brain* was small, weighing 1.030 g. The leptomeninges were moderately thickened; no evidence of old cortical scarring was found. The sulci generally were slightly widened. There was marked, confluent atheroma of the large cerebral vessels, the lumen of the basilar artery being reduced by half. The lateral ventricles were moderately enlarged. The septum consisted of a few strands of tissue 6 mm apart. The body of the corpus callosum was 3 mm thick anteriorly and thinned further to 2 mm towards the splenium. No abnormality was seen elsewhere in the white matter or in the deep grey matter. The mid-brain, including the substantia nigra, appeared normal. The cerebellar leptomeninges were unusually milky but the cerebellum appeared otherwise normal, as did the pons and the medulla.

Histological findings: Cerebral hemispheres: The leptomeninges and cerebral vessels were thickened. The cortex generally showed a slight neuronal loss and a marked increase in astrocytes. Many of the remaining nerve cells were shrunken and darkly stained. A few neurofibrillary tangles were visible in the temporal cortex and in the hippocampus; no senile plaques were seen. The fornices and mammillary bodies could not be assessed. The diencephalic grey matter was moderately gliosed; there were several cystic softenings around thickened vessels in the striatum and pallidum. The subcortical white matter was poorly myelinated and gliosed. No histological evidence of a syphilitic or other infection was found.

Brainstem and spinal cord: The mid-brain, pons, medulla, and spinal cord appeared normal; the posterior columns were well myelinated.

Cerebellum: The white matter was poorly myelinated and slightly gliosed. The cerebellar tonsils were not damaged.

Case 12: Nothing could be found out about this man beyond the fact that he was reputed to have boxed professionally in his youth. He spent his last years in a home for the partially disabled as he had become almost blind, for reasons that were never stated. He was solitary and placid but occasionally became aggressive. He was considered active and mentally alert up to the time of his death at 91, from cardiac ischaemia. The matron of the home said that he never talked about boxing.

It was noted at the post-mortem examination that there were no stigmata of his former profession and that he was a "remarkably well-preserved old man." His brain, however, was referred for study because of his reputed profession.

Naked-eye appearance of brain: The brain was of normal size, with no more than slight widening of the sulci generally and slight atheroma of the large cerebral vessels. The leptomeninges were moderately thickened. No focal lesion was seen on the surface. Coronal cuts revealed slight enlargement of the lateral ventricles; the septum pellucidum had a 1 mm wide cavum but the leaves were otherwise intact and no further abnormality was detected at any level (Fig. 291 b). The substantia nigra was well pigmented. The cerebellum appeared normal.

Histological findings: Cerebral hemispheres: The leptomeninges and cerebral vessels were moderately thickened. The cortex and deep grey matter appeared well populated; a few nerve cells in the medial temporal cortex and in the thalamus showed neurofibrillary change but no senile plaques were seen in any area. The hemispheral white matter appeared normal.

Brainstem: The mid-brain, pons, and medulla were all normal.

Cerebellum: The covering leptomeninges were considerably thickened. Two necrotic foci, due to small vessel disease, had destroyed several folia in the lateral lobe of the left hemisphere; the cerebellar tonsils, however, were not damaged. The digital white matter and the dentate nucleus were normal. In summary, the brain appeared particularly well preserved.

It was later found out that this patient had not been a boxer. The matron of the home had already said before that he never talked about boxing.

Case 13: The patient's daughter was interviewed. She had been told by her father that he was poor at school and only interested in sport. He joined the Royal Navy a 14 years and rose to Petty Officer rank. During this time he boxed as an amateur and won a Royal Navy Championship in his weight. He won "several cups and many canteens of cutlery;" he took part in exhibition bouts; he was knocked out several times. He had cauliflower ears and a deformed nose. His wife had described him as good-tempered and popular at that time with a keen sense of responsibility and strict personal standards. After leaving the Navy he worked for several years in the police, followed by a period in the docks. His daughter first remembered him some years later when he was a labourer employed by the local council. She recalled that he was "always having accidents if he rode on a bicycle" and when aged 34 he was concussed after an accident. This needed outpatient treatment; he was delirious at home the following night and needed a week to recover. He rejoined the Navy when aged 43 and trained cadets during the war. After his discharge a few years later he became 'more vague' and 'spoke more slowly.' He began to neglect his appearance and he complained that he could not grip things or get on a bus. His sight and his memory seemed a little faulty. He had never drunk alcohol excessively but he now became violent after two pints of beer, sometimes 'fighting with his wife.' He worked as a milk roundsman, a bus conductor, and later as a machine operator. When he could no longer manage this, the firm transferred him to labouring and caretaking. After two years he was given notice without warning. He wandered away bemused and had to be taken to a mental hospital. He was then 53. *On admission* he was mildly confused and complained of blackouts. His wife, from whom he was separated, but who was still concerned for him, said that his personality had been changing for the worse. A brief neurological report confirmed this and mentioned a deterioration of habits. He could not recall his address or the date. No incoordination was found but a 'little dyspraxia' was queried. During the next year his memory deteriorated further, he was completely disorientated and needed help with dressing. These features were emphasized in a detailed neurological investigation. In addition his speech was slurred, and like his movements was slow and tremulous. He could not undress himself. An *PEG* revealed generalized ventricular dilatation. All other investigations, including serological tests for syphilis, were negative and it was concluded that "the cortical atrophy and dementia may be related to his previous experiences as a boxer."

The deterioration continued. He became incontinent, paranoid, and aggressive. He ate off the table with his fingers; a cigarette packet he called a flower; he undid buttons when asked to put his tongue out. He continued to have blackouts and a few days after a convulsion he died from bronchopneumonia, aged 57 years.

Naked-eye appearance of brain: The *brain* weight after fixation was much reduced (950 g). The large cerebral vessels appeared healthy. The leptomeninges were moderately thickened. No cortical scars were seen, but there was a moderate degree of cerebral atrophy which was most marked fronto-laterally. Coronal cuts showed considerable enlargement of the lateral ventricles and a cavum at least 2 mm wide was present. The septum appeared to have been fenestrated but because of post mortem artifact this could not be established with certainty. The corpus callosum was thinned to about 2 mm above-down throughout its length; a minute, old softening was noted in it anteriorly on the right side. The deep grey matter was atrophic. The substantia nigra was well pigmented. The surface of the cerebellar tonsils appeared scarred and slightly atrophic.

Histological findings: Cerebral hemispheres: The leptomeninges and cerebral vessels were slightly thickened. The cerebral cortex showed a massive diffuse neuronal loss, with proliferation of astrocytes and fibrous gliosis. Many remaining nerve cells showed Alzheimer's neurofibrillary change; large numbers of senile plaques were seen in all regions (Fig. 303). The deep grey matter showed a moderate glial increase; neurofibrillary tangles were present in the thalamus and hypothalamus. The fornices were thin, poorly myelinated, gliosed. The mammillary bodies showed a slight glial increase. The hemispheral white matter was poorly myelinated and gliosed. *Brainstem:* The substantia nigra showed a slight loss of pigmented cells with free-lying melanin deposits nearby. The nigral damage was most marked in the lateral nuclear groups, the medial group being relatively intact. No Lewy bodies were visible but some of the remaining cells showed neurofibrillary tangles. A few neurons in the locus coeruleus and elsewhere in the pons were similarly changed. The medulla and spinal cord were normal.
Cerebellum: The molecular layer of some folia showed a fine fibrous gliosis. In addition there was well-defined circumscribed folial atrophy, gliosis, and demyelination in the tonsillar regions (Fig. 304). The white matter of the cerebellum showed a slight diffuse loss of myelin. The dentate nucleus appeared normal.

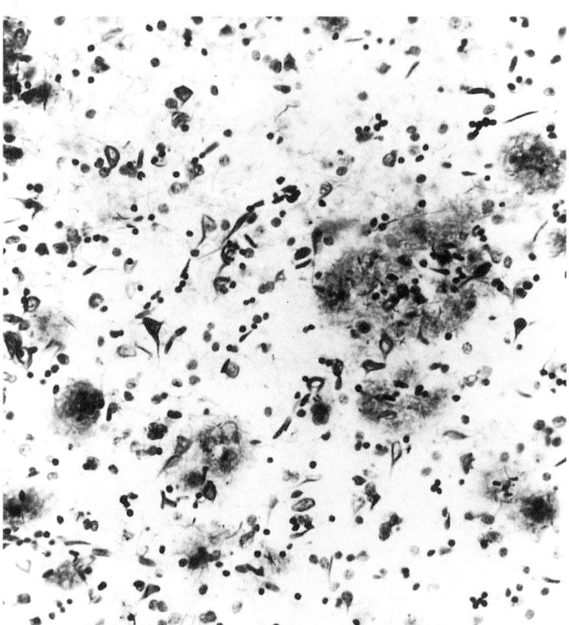

Fig. 303. Case 13. A profusion of senile plaques and of neurofibrillary tangles in the cerebral cortex. This is the typical histological picture of Alzheimer's disease. King's amyloid technique, 140 : 1. (From CORSELLIS J A N et al., 1973).

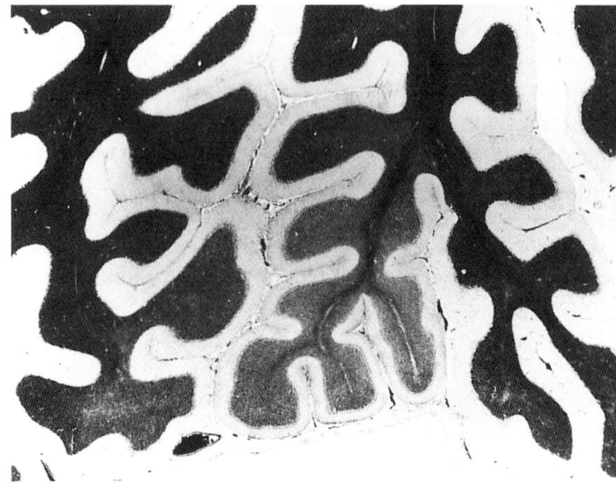

Fig. 304. Case 13. Pallor and demyelination of several cerebellar folia in the tonsillar region. The severest loss of myelin is in the digital white matter. Heidenhain-Woelke technique, 7 : 1. (From CORSELLIS J A N et al., 1973).

Case 14: This man's wife was interviewed at home. He, his father, and his two brothers had all boxed for sport. His childhood and school life were average and uneventful. He joined the Royal Air Force when aged 18 and served in it for 28 years. During this time he was a keen amateur boxer, winning the Wakefield R.A.F. Boxing Trophy and an R.A.F. Group Championship in his weight. After leaving the Service he worked for 15 years as a pipe installer. He was a placid, popular man, happily married with five children. He drank and smoked little. He suffered no injuries during his boxing career and had no serious illnesses. His wife said that his eyesight and hearing, his speech and his memory were normal up to the time of his death, from a cerebral haemorrhage, at the age of 61.
Naked-eye appearance of brain: The *brain* weight after fixation was 1.395 g. The brain was full, the convolutions were flattened, and a film of subarachnoid blood spread up from both lateral fissures. There was marked uncal herniation. Coronal cuts showed all ventricles to be moderately dilated and filled with blood clot. The cerebral vessels appeared healthy. No aneurysm or other possible source of haemorrhage was identified. A septal cavum, about 1 mm wide, had probably been present but the intraventricular haemorrhage prevented any reliable assessment. The mid-brain was swollen; the substantia nigra did not lack pigment. The cerebellum was swollen and there was some grooving of both tonsils.
Histological findings: Cerebral hemispheres: The leptomeninges were normal. The cerebral vessels were congested; moderate quantities of recent haemorrhage surrounded vessels in the white matter and lay within the subarachnoid space over the convexities. The cortex appeared well populated and showed some astrocytic increase; many nerve cells were

shrunken. A very few senile plaques and neurofibrillary tangles were seen; they were confined to the parahippocampal gyrus and to the hippocampus. The fornices were normal The mammillary bodies could not be assessed. Patches of recent haemorrhage had partially destroyed the caudate nucleus. Many shrunken darkly stained nerve cells were present in the thalamus and the hypothalamus. The cavities of the lateral ventricles contained blood clot; the leaves of septum pellucidum were separated and they had been partly destroyed, possibly by haemorrhage.

Brainstem: The cerebral aqueduct was dilated and filled by blood clot. The substantia nigra was normally pigmented. No neurofibrillary tangles or Lewy bodies were seen. The pons and the medulla were normal.

Cerebellum: The hemispheres were swollen and the tonsillar region had been compressed. The molecular layer of many folia showed a fine fibrous gliosis with some proliferation of the Bergmann glia; there was some reduction in Purkinje cells particularly in the tonsillar region.

Case 15: This boxer's sister and brother-in-law were interviewed. The patient was average at school. Later he was an amateur boxer in the Territorial Army and in boy's clubs. He was always healthy and kept physically fit. No neurological or psychological abnormality was noted at any time. He neither drank nor smoked. His work record was satisfactory until he died from a fractured skull and cerebral haemorrhage after a road accident at the age of 58. His brain was sent for neuropathological study because of the history of boxing.

Naked-eye appearance of brain: This was a large, swollen brain, the right hemisphere being the more affected. No atheroma of the large cerebral vessels found. A film of blood clot covered the right temporal convexity, the cortex of which was lacerated in patches. Blood clot filled the left foramen of Luschka and spread across over the undersurface of both cerebellar hemispheres. The cerebellar tonsils appeared a little atrophic. They and the right uncus, were grooved by herniation. Coronal cuts showed that the lateral and third ventricles were small and that they contained blood clot. The septum could not be conclusively assessed owing to the haemorrhage but a fenestrated septal cavum about 3 mm wide was present. No septal leaves could be identified posteriorly. The white matter of both hemispheres was diffusely swollen. The thalamus and hypothalamus on both sides had been largely destroyed by haemorrhage and by infarction which had spread through much of the brainstem. The substantia nigra could not be assessed.

Histological findings: Cerebral hemispheres: The leptomeninges were normal. The cerebral vessels were congested and small amounts of recent subarachnoid haemorrhage were overlying the frontal, temporal, and occipital cortex. The lateral convexity of the right occipital lobe was largely necrotic; the calcarine cortex had been partially spared. The right frontal cortex was less severely damaged, but many nerve cells showed ischaemic change. The remaining cortex was well populated. Neither senile plaques nor neurofibrillary tangles were seen.

Many nerve cells in the striatum and pallidum were shrunken. The right anterior thalamus and hypothalamus had been totally destroyed by haemorrhage. The subcortical white matter stained palely for myelin and contained numerous pin-point haemorrhages.

Brainstem: The mid-brain and pons had been virtually destroyed by haemorrhage, and the substantia nigra could not be assessed.

Cerebellum: Moderate quantities of subarachnoid haemorrhage covered the folia; more extensive bleeding had partially disrupted the inferior surface of the right hemisphere. Beneath this a single folium contained a small patch of fibrous gliosis in the molecular layer. There was no focal tonsillar damage or scarring.

Summary and discussion of the pathomorphological findings of CORSELLIS et al.

Abnormalities of the septum pellucidum: This anatomical structure consists of two thin lamina or leaves of nervous tissue, which extend in the sagittal plane between the lower surface of the corpus callosum and the fornix. If the brain is dissected in coronal sections the lamina are usually fused or contiguous. In some cases, however, they are separated and at anterior parts the space between the two leaves form the 'cavum septi pellucidi.' In posterior parts exists an interseptal space which is termed the cavum Vergae. These cava do not communicate with the ventricular system, but communicate when the laminae are ruptured. We follow CORSELLIS et al., who made no distinction in their study between the cavum septi pellucidi and the cavum Vergae. Both cava are covered by the term 'septal cavum.'

In the observations of CORSELLIS et al. two of the 15 boxers (nos. 14 and 15) died after an intraventricular haemorrhage and in both of them leaves of the septum were torn and separated. As these abnormalities might have been accentuated by the bleeding, these two cases were not considered further in the study of CORSELLIS et al.

A *septal cavum* was identified by CORSELLIS et al. in 12 of the remaining 13 brains, the maximum coronal widths ranging from 1-8 mm with a mean of 5.17 mm.

In the single case (no. case 4) with no cavum anteriorly, the septum was nevertheless fenestrated to destruction posteriorly. One case (no.12) with a caval width of 1 mm showed no fenestration, while in another (no. 13) with a 2 mm wide cavum, the septum could not be properly asserted, and it has been treated as not fenestrated in the tabulation. In the other eleven the septum had been perforated; in ten this was severe and in most only a few thin tags remained. CORSELLIS et al. listed these findings in Table 89. In the opinion of these authors it is clear that septal cava with fenestrated leaves have proved to be a common finding in the brains of boxers.

In the opinion of these authors the significance of these observations depends largely on how often such anomalies are to be found in the brains of those who have not boxed. Since no firm data could be traced, CORSELLIS et al. examined the brains of 500 adults, 200 men and 300 women, in the same way as those of the boxers.

Twenty-five of the brains had come from a general hospital but the rest were from patients who had died in a psychiatric hospital. The cerebral degenerative conditions and the organic dementias associated with middle and old age were therefore represented in large numbers in the study of CORSELLIS et al. The age range is shown by CORSELLIS et al. in Table 85.

The incidence of a septal cavum and its size were compared by CORSELLIS et al. in Fig. 305, with the findings. Both were tabulated according to the age at death and the non-boxers are also divided according to sex.

Table 86 gives the figures on which Fig. 305 is based.

According to CORSELLIS et al. the mean width of the cava, in the 142 cases in the non-boxing sub-groups b and d, whether fenestrated or not, was 1.6 mm. In the twelve boxers in the same sub-groups it was 5.17 mm.

The outstanding difference is in sub-group d, for a fenestrated cavum was present in 77% of the boxers compared with 3% of the non-boxers.

The presence of a septal cavum in the non-boxers was found not to have been influenced significantly, as CORSELLIS et al write, either by the sex of the patient, by the age at death or by the size of the lateral ventricles. Moreover, patients with moderate or severe frontal atrophy – for example, Pick's disease, Huntington's chorea, Alzheimer's disease, senile dementia, and after leucotomy – showed no special liability to develop a cavum.

In summary, the boxers' septa were in most cases of CORSELLIS et al. strikingly different from those of the non-boxers. A cavum was the rule rather than the exception; its average width was three times that of the non-boxers and septal leaves were almost without exception grossly penetrated.

CORSELLIS et al. continue, indeed, the remnants of the old boxer's septum often has a look of its own which is rarely seen in other conditions. "The fornix becomes almost totally detached from the undersurface of the corpus callosum and the two flattened fornical bodies splay out horizontally over the dorsal surface of the thalamus. The body of the corpus callosum is usually markedly thinned as well."

Even when a fenestrated cavum occurred in non-boxer group and there were examples in the 500 cases of CORSELLIS et al. – the width tended to be narrower: a mean of 2.5 mm for the non-boxers against a mean of 5.9 mm for the boxers.

CORSELLIS et al. noted the precise way in which the act of boxing produces a large and fenestrated cavum is unclear. The authors continue that the present findings have shown that it is not, as

TABLE 85
Age range of 500 control adults. (From CORSELLIS J A N et al., 1973).

Age (yr)	Men	Woman	Total
21–50	15	15	30
51–60	22	19	41
61–70	46	48	94
71–80	61	108	169
81–90	54	98	152
91–100	2	12	14
Total	200	300	500

TABLE 86
Comparative incidence of fenestrated and non-fenestrated cava in boxers and non-boxers. (From CORSELLIS J A N et al., 1973).

	Sub-group								
	a		b		c		d		
	Cavum – Fen. –		Cavum + Fen. –		Cavum – Fen. +		Cavum + Fen. +		
	(no.)	(%)	(no.)	(%)	(no.)	(%)	(no.)	(%)	Total
Non-boxers	327	66	127	25	31	6	15	3	500
Boxers	0		2		1		10	77	13

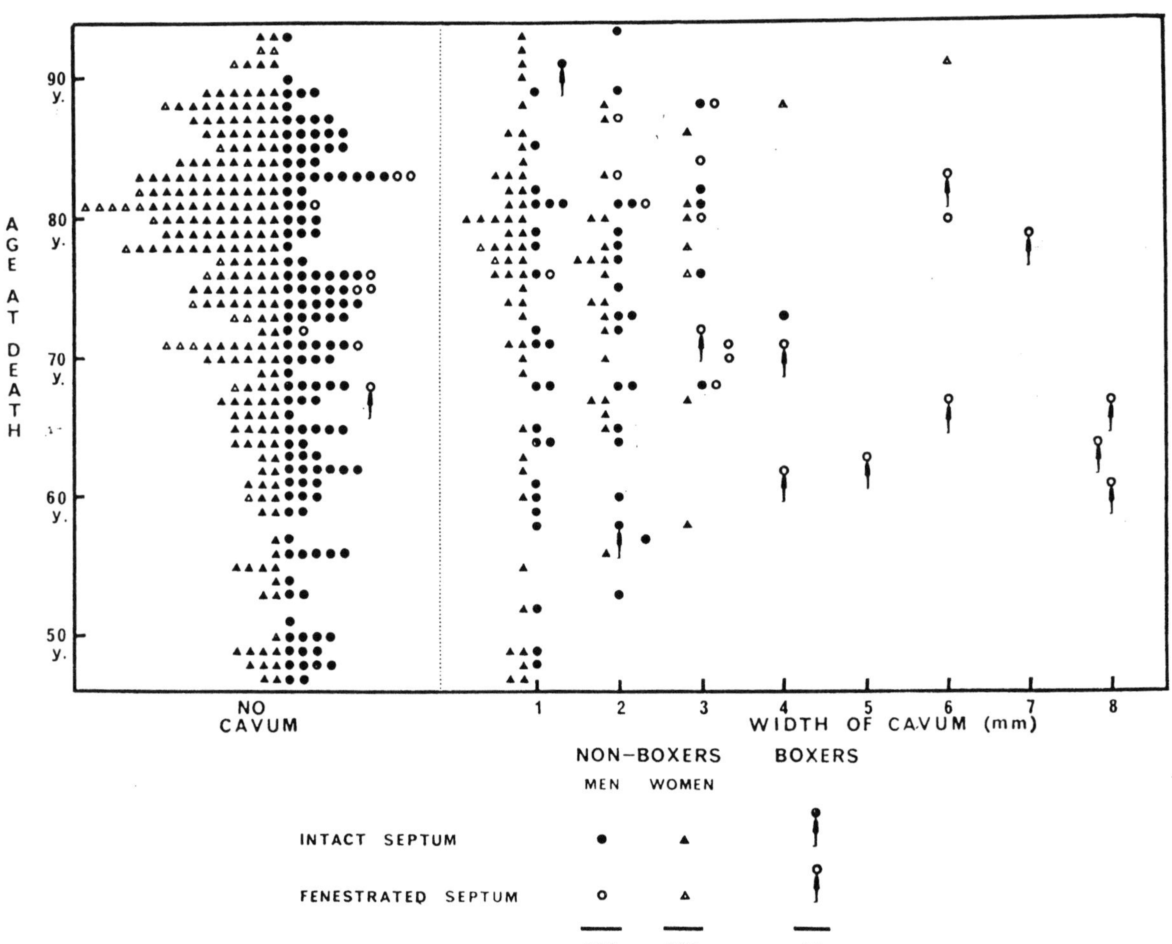

Fig. 305. The incidence of fenestrated and nonfenestrated cava in the brains of 500 nonboxers and of 13 boxers tabulated at death. (From CORSELLIS J A N et al., 1973).

PAYNE (1968) implied, initiated merely by the passive enlargement of the ventricular system, even though some enlargements manifestly takes place. CORSELLIS et al. quote MAWDSLEY and FERGUSON (1963) who preferred to invoke the effect of repeated sudden rises in intraventricular pressure leading to rupture of the septal leaves and they were less inclined to incriminate the tearing effects of rotational and accelerative forces. CORSELLIS et al. noted that these latter, however, may not be unimportant, for not only the corpus callosum, but also the septum and the fornices are particularly vulnerable to movements of the brain during the momentary, catastrophic head injury that may lead to posttraumatic encephalopathy or dementia (STRICH 1961). CORSELLIS et al. stress the point that the septum along its dorsal attachment to the under surface of the corpus callosum is a relatively thin and fragile membrane, whereas it thickens up ventrally into the more solid fornices. It may be, according to CORSELLIS et al., that during the repeated, but less forceful blows to the head in boxing the flimsy dorsal septal attachment is strained and torn by the repeated movements of the corpus callosum on the tethered fornix. Then, as the lateral ventricles enlarge, as they undoubtedly do, the septum stretches, as it must, and the fenestration grow in size."

This percentage (92%) is a vastly higher incidence than the 28% found in their age matched controls. HARVEY (1977), who commented on these findings, concluded that it was apparent that head injury incurred during a long boxing career produced enlargement of the normally small *cavum septi pellucidi (cavum pugilisticum)* in nearly every case, often followed by *septal fenestration (cavum pugilisticum rupturum)*. The fluid found within the cavum is of unknown origin (UNTERHARNSCHEIDT et al. 1968; SHAW and ALVORD 1969). However, it is obvious that the size of the cavum increases when the brain is 'boxed.' HARVEY comments, "the fluid accumulation may be primary, followed by further separation of the leaves of the septum and enlargement of the formerly small cavum. Conversely, the primary event may be that the leaves of the septum are jostled in opposite direction when 'boxed,' with enlargement of the potential cavum followed by fluid accumulation."

Cerebellar and other scarring of the brain: The clinical picture of the encephalopathy of the boxer and the punch drunk syndrome in many cases suggest that the cerebellar system has been affected. One can fully agree with CORSELLIS et al. that the cerebellum itself has received little attention in previous neuropathological studies. BRANDENBURG and HALLERVORDEN (1954) reported scattered loss of Purkinje cells, while GRAHMANN and ULE (1957) noted localized focal atrophy. NEUBÜRGER et al. (1959) mentioned a reduction in the number of cells in the granular cell layer. UNTERHARNSCHEIDT (1963) described severe loss of Purkinje cells in animals that had received repeated blows in a short time interval, simulating the situation in boxing. These pathomorphological alterations are described in detail on p. 76.

Histological sections of the cerebellum were examined by CORSELLIS et al. in all cases in the plane illustrated in Fig. 306. The most striking finding of these authors was the well-defined cortical scarring on the inferior surface of the lateral lobes of the cerebellum. This was, according to CORSELLIS et al. always most marked in the tonsillar region – that is, in the folia around the groove formed by the sloping edge of the foramen magnum. Cortex and adjacent to the area also tended to be patchily affected, but the dorsum of the cerebellum usually appeared to be spared. In the scarred areas the molecular cell layer was narrowed and contained many glial nuclei with a dense fibrous gliosis. Purkinje cell loss was marked, and there was usually some thinning of the granular cell layer. The digital white matter of the affected folia was partially demyelinated. This type of damage could be identified by CORSELLIS et al. in ten of the fifteen brains.

The *loss of Purkinje cells* in the affected cerebella appeared to spread well beyond the limit of frank scarring. In order to quantify this, cell counts were

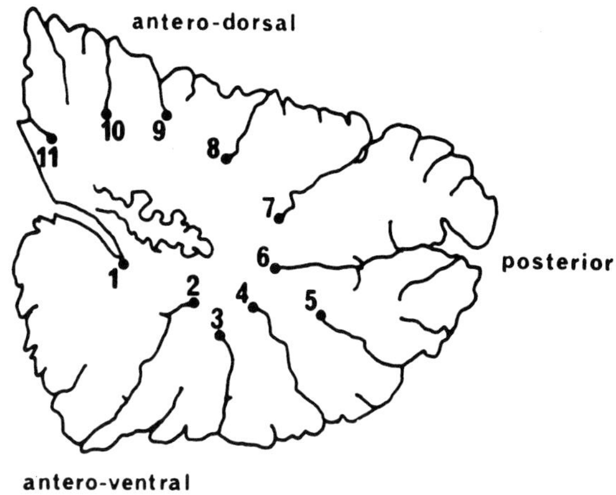

Fig. 306. Drawing of the standard histological section of the cerebellum used in the Purkinje cell counts and measurements. (From CORSELLIS J A N et al., 1973).

made on eleven of the cases by CORSELLIS et al. (1973) and compared with eleven normal specimens obtained from non-boxing patients who had died within the age range of the boxers.

The results were tabulated by CORSELLIS et al. in Table 87. The mean distance (in μ) between Purkinje cells is given for each sector in the sections of each boxer (Table 87) and each of the non-boxers (Table 88). The salient features of these observations are illustrated by CORSELLIS et al. in Fig. 307. The 'normal range,' obtained from the measurements and counts in 'non-boxers,' is the shaded area. The mean reading for each section of the eleven boxers is plotted as a continuous line. The figures of CORSELLIS et al. show that the Purkinje cell population did not vary greatly from one sector to another in any given person to person. In the boxers, on the other hand, the mean distance between Purkinje cells was often two or three times as great on the ventral aspect of the cerebellum as it was on the dorsum, where, apart from the more antero-dorsal folia, the figure usually fell back to within normal limits. This means, according to CORSELLIS et al., that as many as half or even two-thirds of the Purkinje cells had disappeared in the affected parts. The counts were within normal limits in all sections in two of the boxers. One of these was case no. 12, without obvious evidence of any neurological disorder, who died at the age of 91 and about whose boxing career very little was known. It was later established that this patient had not been a boxer.

In the other case (no. 9), a small but indubitable tonsillar scar was present, even though the Purkinje cell loss over the sector as a whole proved to be negligible.

TABLE 87
Purkinje cell counts in boxers. (From CORSELLIS J A N et al., 1973).

Case no.	Age at death (yr)	Mean distance (μ) between Purkinje cells (areas of cerebellum)									
		1/2	2/3	3/4	4/5	5/6	6/7	7/8	8/9	9/10	10/11
1	63	193	459	340	232	173	205	230	211	251	–
2	77	348	381	244	229	238	241	691*	432*	225	300
3	62	218	202	408	258	210	179	181	227	148	150
4	69	151	258	245	208	237	144	240	250	679	–
5	61	176	657	653	362	297	193	184	168	232	181
6	83	167	172	524	375	230	182	194	206	198	168
9	72	174	146	191	195	176	156	153	159	170	171
10	67	187	400	264	192	184	198	185	197	271	–
12	91	229	233	217	299	211	264	307	281	235	189
13	57	173	408	291	232	220	180	168	175	168	191
14	61	100	312	325	252	191	168	193	241	180	–
Mean	69	192	330	337	258	215	192	204	212	251	193
SD		62	151	142	63	36	35	45	39	147	49

*Case 2. Two readings excluded as due to a small recent infarction.

TABLE 88
Purkinje cell counts in non-boxers. (From CORSELLIS J A N et al., 1973).

Case no.	Age at death (yr)	Mean distance (μ) between Purkinje cells (areas of cerebellum)									
		1/2	2/3	3/4	4/5	5/6	6/7	7/8	8/9	9/10	10/11
97/70	67	153	188	197	217	223	229	212	239	205	228
53/60	65	146	181	178	188	191	175	187	177	152	170
104/63	72	161	166	191	188	229	205	182	199	189	–
46/63	60	145	138	143	132	194	162	179	166	158	130
83/58	68	182	243	230	247	242	243	221	221	185	178
83/59	58	175	194	181	193	149	155	101	165	150	136
72/63	56	198	234	218	223	228	194	178	207	208	193
94/70	68	180	229	214	200	208	200	199	185	170	136
28/64	50	146	177	175	170	174	172	158	143	139	125
85/62	69	179	183	153	184	173	177	155	144	156	157
179/69	81	108	146	175	154	165	115	118	120	118	117
Mean	65	161	189	187	191	198	184	172	179	166	157
SD		25	34	27	32	30	36	37	36	28	35

The cerebellar tissue around the foramen magnum seems to be pushed and pressed into this opening, probably due to increased intracranial pressure resulting from the repeated head blows in boxing. We fully agree with CORSELLIS et al. that such a mechanism has been the origin of the glial scarring. The marked loss of Purkinje cells, especially visible in the animal model of

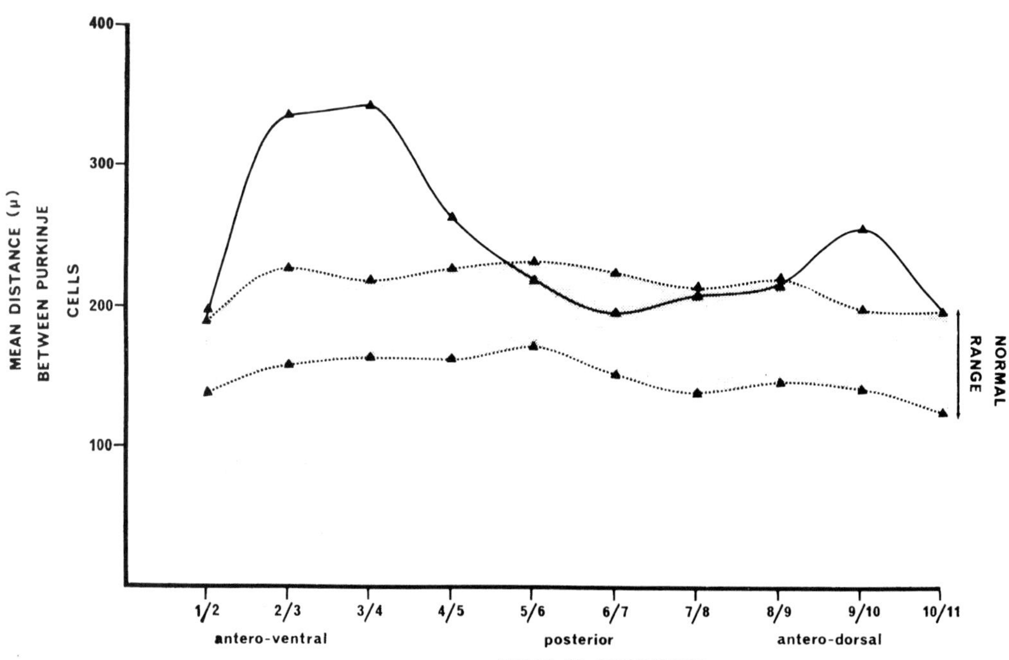

Fig. 307. A comparison of the Purkinje cell counts in 11 boxers and in 11 non-boxers. (From CORSELLIS J A N et al., 1973).

UNTERHARNSCHEIDT (1963), seems to be due to ischemic processes.

There is no doubt that, in most of these authors' cases, not only the trend to a reduced brain weight but also the enlargement of the lateral and the third ventricles, and the thinness of the corpus callosum, pointed to a material reduction in the volume of the cerebral hemisphere.

Degeneration of the substantia nigra: A *Parkinsonian syndrome* was named in the hospital records of four of the cases (3, 4, 6, 9) of CORSELLIS et al. In several more the descriptions of the patients' movements in later life suggested that a degree of the syndrome had developed. On the pathological side, naked eye examination showed gross lack of nigral pigment in the four cases diagnosed clinically, while, although lesser degrees of reduced pigmentation are difficult to assess, seven more of the remaining eleven were materially affected. In its extreme, the histological lesion consisted of almost complete loss of pigmented cells, while many of the remaining ones showed neurofibrillary change. There was a tendency for the medial nuclear group to be spared, according to CORSELLIS et al., whereas the intermediate and lateral ones were more vulnerable.

CORSELLIS et al. summarize, it now seems clear that repeated blows to the head may be followed by degeneration within the extrapyramidal system. The authors continue: "This was shown in the previous section to be true of many neurons in the cerebellar efferent pathway; it is now seen to apply to many of those in the substantia nigra."

Regional occurrence of Alzheimer's neurofibrillary tangles: CORSELLIS et al. noted that Alzheimer's neurofibrillary tangles are found in small numbers of neurons in many elderly people but they are then *virtually always accompanied* by senile plaques. The authors write: "The degree of plaques and tangle formation vary greatly but both end to be at their most extreme in Alzheimer's disease and in senile dementia. Although in these disorders (if indeed they are not the same) vast numbers of nerve cells throughout the cerebral cortex may be 'entangled,' the brunt of the change normally falls on those in the antero-medial temporal grey matter, including the amygdaloid nucleus and the hippocampus (JAMADA and MEHRAEIN 1968, CORSELLIS 1970). The neurons of the diencephalic and the periventricular regions are less affected."

CORSELLIS et al. continued to say that, in contrast, there are other rare conditions in which neurofibrillary change occurs in the *complete, or almost complete absence* of senile plaques. They include, according to CORSELLIS et al. the Parkinson-dementia complex of Guam, some forms of chronic encephalitis, Parkinsonism, particularly when post-encephalitic (HIRANO and ZIMMERMAN 1962) and progressive supranuclear palsy (STEELE et al. 1964).

In the cases of CORSELLIS et al. the distribution of the *neurofibrillary tangles* did not conform to either topographic pattern, for they were spread diffusely through both cerebral cortex and the brainstem. The most noticeable feature, however, was the intensity of the neurofibrillary change in the medial temporal grey matter. Vast numbers of neurons in the uncus, in the cortico medial part of the amygdaloid nucleus, in the hippocampus, the parahippocampal gyrus, and the fusiform gyrus were affected. The most incongruous feature in this material was the *scarcity,* or *in most cases the total absence, of senile plaques,* which had already been noticed by GRAHMANN and ULE (1957) and later by CONSTANTINIDES and TISSOT (1967).

The incidence of the four types of pathological change was summarized by CORSELLIS et al. in Tab. 89. It illustrates the tendency for all types of change to occur together, if they occurred at all.

CORSELLIS et al. finally discussed the extent to which the damage is a consequence of boxing, for it is possible that other factors are more important. They write: "These include head injury received outside the boxing ring, drink, and the conditions such as neurosyphilis and the degenerative diseases of old age."

Head injury outside boxing: According to CORSELLIS et al. there was a history of such trau-

TABLE 89
Incidence of four types of pathological damage. (From CORSELLIS J A N et al., 1973).

Case no.	Age at death (yr)	Boxing status	Cavum Width (mm)	Septum Fenestration	Cerebellum Tonsillar scarring	Substantia Nigra		Cerebral cortex	
						Loss of pigmented cells	Neurofibrillary tangles	Senile plaques	Neurofibrillary tangles
1	63	P Ch	8	+	+	+	+ +	0	+ + +
2	77	P Ch	7	+	+	+ +	+ + +	0	+ + +
3	62	P Ch	5	+	+	+ + +	+ + +	0	+ + +
4	69	P Ch	0	+	+	+ + +	+	0	+ +
5	61	P Ch	8	+	+	+	+ +	0	+ +
6	83	P Ch	6	+	+	+ +	+ +	+	+ +
7	62	P	4	+	0	+	+ +	0	+ + +
8	71	P	4	+	0	+	+	+	+ +
9	72	P	3	+	+	+ +	+ +	0	+ +
10	67	P	8	+	+	+ +	+ +	+	+ +
11	67	P	6	+	0	0	0	0	+
12	91	P?	1	0	0	0	0	0	+
13	57	Am Ch	2	Npa	+	+	+ +	+ + +	+ + +
14	61	Am Ch	?1	Npa	0	0	0	+	+
15	58	Am	?3	Npa	+	Npa	0	0	0

P: professional
Am: amateur
Ch: champion

0: no change
+ change present
Npa: not possible to assess

0: no change
+ slight change

+ + moderate
+ + + severe

ma in four of the 15 cases; a fifth (no. 15) died in a road accident.

Case 1 was involved as a young man in three car accidents, two of which were minor. In the third, no skull fracture was found but he was concussed and spent three months in hospital. Some time in later life he was a publican, and his jaw was broken in a brawl.

Case 2 was attacked and knocked unconscious when in his 60s. A few years later he suffered scalp lacerations in a road accident and was admitted to hospital. There was no radiological evidence of a skull fracture but he was confused and amnesic and he was found mildly diabetic.

Case 4 spent two days in hospital after a car accident but was back in the ring on the third day. The history suggested that he had been concussed for a short time.

Case 13 was once taken to hospital after a fall from a bicycle. He was not admitted but remained confused for 24 hours at home.

No history of head injury could be traced in the remaining ten cases.

Case 3 suffered damage to the first cervical vertebra when a plank fell on his neck. He continued boxing after this for several years; he later wore a collar.

Alcohol intake: Many of the 15 men drank heavily at some time during their lives and several became unusually sensitive to small amounts.

Cases 1, 3, 4, 5, 7, 9, and 10 were reported to have drunk to excess.

Cases 1, 4, 6, 8, 9, and 13 were described as being 'knocked out' or as becoming 'violent' after about two pints of beer.

Case 14 was abstemious, and *case 15* was teetotaller.

Other medical conditions: One patient (no. 11) had been treated ten years before for tabes dorsalis. This treatment had been successful clinically, although the Wassermann reaction in the blood (but not in the cerebrospinal fluid) remained positive. No histological evidence of the condition was found after death.

Case 2 was mildly diabetic.

Case 3 had an attack of acute pancreatitis, possibly precipitated by an excess of alcohol.

The commonest complication was cerebrovascular disease and the clinical picture was manifestly influenced by this in four patients (nos. 5, 8, 9, and 10), all of whom became hemiplegic resulting from infarction. In addition there was a variable degree of atherosclerosis and small vessel disease in other cases, although several of even the older men were remarkably free of it. The possible influence of the senile form of cerebral degeneration is discussed later by CORSELLIS et al. in relation to dementia.

None of these various factors occurred often enough or was severe enough to justify its acceptance as of more than peripheral importance to the central problem. The strongest argument, however, against the incrimination of any of them is the neuropathological fact that none is known to produce the *pattern* of cerebral damage that has been clearly delineated. That is to say, noted CORSELLIS et al., although any single item in the tetrad of pathological changes might conceivably be attributed to a cause other than boxing, it is not possible to account for all four in this way. Thus, according to CORSELLIS et al., cerebellar degeneration may occur as a rarity in chronic alcoholic patients, but the topography of the damage is radically different from that in the published cases (VICTOR et al. 1959). Nor could the septal, the nigral, or the neuronal changes be explained on this basis. Moreover, no other histological evidence of the various forms of brain damage associated with alcoholism was found by CORSELLIS et al. Even though the hypothalamic floor in several cases was unusually thin, with an associated tendency for the mammilary bodies to be small and withdraw, the histological appearance were, according to CORSELLIS et al. never those of a longstanding Wernicke's encephalopathy.

Head injury unrelated to boxing had been recorded according to CORSELLIS et al. in only a minority of the 15 histories and in any case has never been shown to result in the particular combination of lesions that had been found by CORSELLIS et al. Nor would neurosyphilis be held to do so, particularly as this condition had been diagnosed clinically only once and even then no histological evidence of the past infection was found after death by CORSELLIS et al.

The conclusion, therefore, wrote CORSELLIS et al., seems unavoidable; the paramount reason for this insidious neurological and psychological deterioration of so many of the 15 men was the brain damage incurred while boxing. According to CORSELLIS et al. the conclusion is all the more definite, since the concordance of the clinical and the neuropathological findings is considerably closer than might been expected in such a complex situation. Thus the neurological features pointed time and again, as CORSELLIS et al. point out, to the involvement of the cerebellar and the extrapyramidal pathways and both regions have been shown to be particularly prone to undergo structural change.

CORSELLIS et al. noted furthermore: "Perhaps the more intriguing problems, however, are raised by the psychological components of the syndrome. These have been rarely looked at in the detail accorded to the neurological side..."

The authors quoted WINTERSTEIN (1937) who had lucidly summarized his psychiatric impression of about 50 professional boxers. He wrote about 60 years ago: "Most of those badly affected showed an impairment of intelligence. A bad memory for recent events was very obvious. A characteristic sign of mental dullness is the ease with which these men have adjusted themselves to their fall from a comparatively brilliant position to that of poorly paid occasional labour in ordinary life. Some have great difficulty in concentrating and are garrulous; others again may be paranoid and suspicious, other appear over-cheerful..."

In a paper by JOHNSON (1969), who studied one amateur and 16 professional middle-aged ex-boxers with neuropsychiatric symptoms. This author described four 'organic psychosyndromes' that he considered to be a result of boxing. They were a *chronic amnesic state, dementia, rage reactions in a personality disorder,* and a *morbid jealousy syndrome.*

Memory loss and dementia: Disorders of the memory, and particularly of recent memory, are, according to CORSELLIS et al. liable to result from disturbances centered on the limbic areas of the brain. The pathways that appear to be most directly concerned are those in the medial temporal grey matter including the hippocampal formation. These areas project, according to these authors via the fornix to the mammillary bodies and to the thalami and they are also linked to the septal region (CORSELLIS 1970).

Many of the boxers in the series of CORSELLIS et al. had begun to show a degree of memory impairment towards the end of their boxing careers. In some the condition worsened, in a few it was noticed only in later life. In sharp contrast, there were four other boxers in the series of CORSELLIS et al., two professionals (nos. 11, 12) and two amateurs (nos. 14, 15), whose memories were reputedly unaffected at any time during their lives.

CORSELLIS et al. noted that in all the cases in the larger, dysmnestic, group, the neurons of the medial temporal grey matter of both hemispheres were riddled with neurofibrillary tangles to an extent far in excess of that anywhere in the brain. Partly as a consequence of this degeneration, the loss of nerve cells also tended to be most severe in the same areas. In these boxers the bodies of both fornices in the group of forgetful boxers were displaced by the septal damage and usually appeared both atrophic and poorly myelinated. They were also virtually denuded of their dorsal attachment to the corpus callosum. Furthermore, CORSELLIS et al. noted, the mammilary bodies, in which most of the hippocampal efferent fibres end, were often sunken into the hypothalamic floor of the dilated third ventricle and were then small and gliotic.

The findings in these cases may be, as CORSELLIS et al. write, contrasted with the observation made on the four men who had no recorded disorder of memory. In them the degeneration in the temporal components of the limbic pathways was strikingly less marked or even absent.

CORSELLIS et al. summarized:

"When, however, the pathology of the boxer's brain is compared with that of the various organic dementias, material differences become evident. Firstly, it has been shown that the gross abnormalities in the septal areas are as common in boxers as they are rare in any of the known forms of organic dementia. Secondly, the cerebellar damage in the boxers is essentially different from that encountered in any of the various hereditary or other cerebellar atrophies that may be accompanied by dementia. Thirdly, nigral degeneration shows no marked tendency to occur in demented patients, other than in certain rare and specified conditions. On the other hand, there is one neurohistological feature which does clearly link the pathology of the boxer's brain to the organic dementias. This is the intense formation of the neurofibrillary tangles, since these form one of the mayor features of Alzheimer's presenile and senile dementia."

Curiously enough, the authors commented, the probability that a *single* head injury might occasionally precipitate the onset of Alzheimer's disease has been mooted for some years (CLAUDE and CUEL 1939, CORSELLIS and BRIERLEY 1959, HOLLANDER and STRICH 1970), the idea that the *repetitive* trauma inherent in boxing might have the same effect was put forward by BRANDENBURG and HALLERVORDEN (1954).

PETERS (1962), PAMPUS and Norbert MÜLLER (1956) questioned whether such an association was not merely a coincidence, but a second example occurred in the series (no. 13) of CORSELLIS et al. "Alzheimer's disease, fatal before the age of 60, is most uncommon, and the death of two ex-boxers from this condition in their 50s, may not be a chance finding."

Rage reactions: CORSELLIS et al. commented that the issue of whether any of the more prominent disorders of behaviour, such as unprofessional outbursts of rage or of aggression, might be a consequence of brain damage can only be surmised. Certainly, they continue, some of the men in their study seemed to have been preternaturally violent even in their younger days and one, for example, went to prison for throwing his bride's lover out of a first floor bedroom window. "This might still be taken, however, more as a sign of virility than of disease and few boxers would get far without some attachment to the used force." Nevertheless, continue CORSELLIS et al., several men remained addicted to violence long after they had given up boxing. Similar tendencies were presented by JOHNSON (1969).

In their general conclusions CORSELLIS et al. noted:

"There is now a solid foundation for the view that some experienced boxers develop a clinical disorder, the greater part of which has a neuropathological basis. The severity of the condition varies greatly, ranging from a mild clumsiness of speed and movement, with or without some loss of memory, to the ataxic, dysarthric, and perhaps Parkinsonian demented. All these features may be attributed to the novel pattern of cerebral damage or degeneration that is concentrated on the septal regions, in the deep temporal grey matter and on certain neurons along the cerebellar and nigral pathways."

CORSELLIS et al. end their classic contribution with a short chapter *'Boxing: Past and Present.'* They write that the published conclusions have been drawn from the study of men who have boxed in the period between 1900 and 1940. "This was a time," they say, "when the sport came to its peak and it was possible to fight frequently for a long time. Thus over half the boxers in the present series had reportedly taken part in over 300 contests or had careers lasting 15 years and more; some had a history of both. It is scarcely surprising therefore that so many had suffered brain damage, for ROBERTS (1969) found clinical evidence of this in 50% of those with half as much experience in the ring. Conversely, in the small number of cases in the present study with minimal cerebral damage the histories indicated that the men had tended to box less, while two of the four had fought only as amateurs. Again, ROBERTS found the same trend towards fewer fights and less neurological abnormality."

CORSELLIS et al. come to the following conclusion:

"It may well be, therefore, as it is sometimes claimed, that the introduction of stricter medical and administrative control of the number and nature of the contests has considerably reduced the prevalence of cerebral damage. It seems, however, unlikely that all risk can have been eliminated. A single punch, or even many punches, to the head need not visibly alter the structure of the brain but there is still the danger that, at an unpredictable moment and for an unknown reason, one or more blows will leave their mark. The destruction of cerebral tissue will have then begun and, although this will usually be slight enough in the early stages to be undetectable, it may built up, if the boxing continues, until it becomes clinically evident. At this point, however, it could already be too late, for destroyed cerebral tissue can never be replaced, while the further danger exists that the process of the degeneration could smoulder on even after the boxing has stopped."

"When these reports are considered together," CORSELLIS et al. (1973) continue, "pattern begins to emerge. Firstly, septal anomalies have occurred suspiciously often, particularly when the evidence from air encephalographic studies is taken into account. Secondly, Parkinsonism as well as neuronal degeneration with loss of pigment in the substantia nigra have been mentioned several times. Thirdly, a diagnosis of Alzheimer's disease has sometimes been broached, although in this connection the more striking, and the more anomalous feature has been the tendency for neurofibrillary tangles to occur in the absence of senile plaques."

Summarizing their findings, CORSELLIS et al. (1973) conclude, "When the pathology of the boxer's brain is compared with that of the various organic dementias, material differences become evident. Firstly, it has been shown that the gross abnormalities in the septal areas area as common in boxers as they are rare in any of the known forms of organic dementia. Secondly, the cerebellar damage in the boxers is essentially different from that encountered in any of the various hereditary or other cerebellar atrophies that may be accompanied by dementia. Thirdly, nigral degeneration shows no marked tendency to occur in demented patients, other than in certain rare and specified conditions. On the other hand, there is one neurohistological feature which does clearly link the pathology of the boxer's brain to the organic dementias. This is the intense formation of the neurofibrillary tangles, since these form one of the major features of Alzheimer's presenile and senile dementia."

The *brains of 22 ex-boxers were examined histologically* by C.W.M. ADAMS and BRUTON (1990) to determine the frequency of recent and old hemorrhage. Four boxers had died from acute intracerebral bleed, usually soon after a boxing bout. Seven of the 18 showed evidence of previous perivascular hemorrhage, as detected by *Perl's ferrocyanide test for iron,* and a similar number showed minor degrees of *meningeal* or *subpial siderosis,* consistent with previous meningeal bleeding. *Cerebellar siderosis* was present in six cases. Seventeen of the 22 boxers showed *evidence of recent* or *past hemorrhage.* Control material showed an incidence of 11% for perivascular iron deposition and only 4% for minor degrees of meningeal siderosis.

The brains of 22 boxers were studied from the neuropathology collection at Runwell Hospital. These included 13 of the original 15 cases reported by CORSELLIS et al. (1973), ten of the 13 men had boxed professionally, although in one case the information was unreliable; two were at one time *world champions* and six more *had won or fought for a national or regional championship.* All, but one of them (the ex-professional with the unreliable history) showed neuropathological evidence of brain damage of the type described in the 'punch drunk' state (Table 90).

The brains of a further nine boxers were also examined, four of these boxers had become comatose and did not regain consciousness. At necropsy, their brains were found to have naked-eye evidence of recent subdural, subarachnoid or intracerebral hemorrhage or a combination of these complications. The other five boxers (four amateur and one professional) had lived for many years after the end of their boxing careers. The brains of three of the latter five cases showed the typical stigmata of dementia pugilistica (Table 90).

C.W.M. ADAMS and BRUTON *studied the brains from 94 'control' patients, their ages ranged from 54 – 95 years; most had died from cardiovascular disease.*

The authors reexamined histological sections from all 22 boxers' brain. Three of the four cases who had died soon after a boxing bout showed evidence of *recent* subdural hemorrhage combined with subarachnoid or intracerebral hemorrhage (Table 90); the fourth boxer who had remained in a coma for a year before death had evidence of *old* subarachnoid and intracerebral hemorrhage.

Intracerebral Hemorrhage: Seven of the remaining 18 ex-boxers described by C.W. M. ADAMS and BRUTON (1990) *showed intracerebral iron-staining (hemosiderin) within macrophages* or *neu-*

TABLE 90
Cerebrovascular lesions in 22 boxers. (From ADAMS C W M, BRUTON C J, 1989).

Boxin status	No	Mean age (yr, SD)	Dementia pugilistica*	No vascular lesions	Iron around cerebral vessels	Meningeal or subpial iron	Cerebellar siderosis	Recent haemorrhage	Type of recent haemorrhage
Acute death	4	22 (4)	0	0	0	1†	0	4	SD/SA + SD/SA + IC/SA + IC†
Ex-professional	10	66 (8)	10	1	5	5	4	1	ANG
Ex-amateur or doubtful status	8	67 (12)	5	4	2	2	2	2	IC

*Data in Corsellis *et al.* (1973).
† In coma for 1 year after boxing bout.
SD, Subdural haemorrhage; SA, Subarachnoid haemorrhage; ANG, Angiomatous haemorrhage; IC, Intracerebral haemorrhage.

roglia or *free lying around cerebral or cerebellar blood vessels* (Table 90). The authors noted that although perivascular iron and calcium deposits is not uncommonly found in the brains of normal old people (WERTHAM and WERTHAM 1934), the affected boxers were all under 65 years of age and, in general, younger than those boxers who showed no such change. C.W.M. ADAMS and BRUTON concluded that the perivascular iron would therefore seem unlikely to be the result of normal aging.

In total, the authors wrote, *nearly 39% of the ex-boxers showed evidence of past intracerebral hemorrhage;* this figure is nearly four times the 11% recorded in the 94 control brains (Table 90).

Meningeal Hemorrhage: Histological evidence of iron-staining was also present in the meninges, as C.W.M. ADAMS and BRUTON wrote, the subpial region and/or around the cranial nerves of seven of the 18 ex-boxers presumably as a result of past non-fatal bleeding. Indeed, two of the cases of the authors, showed naked-eye evidence of previous hemorrhage in the form of a localized brownish staining of the leptomeninges. Of the seven cases with evidence of past meningeal hemorrhage, five showed a patchy or continuous band of iron staining in the subpial region of the cerebellum or the pons, which resembled that seen in patients with subpial hemosiderosis (McGEE et al. 1962, TOMLINSON and WALTON 1964, BRAHAM and WOLMAN 1965, HUGHES and OPPENHEIMER 1969, KOEPPEN and BARRON 1971, KOEPPEN and DENTINGER 1988).

C.W.M. ADAMS and BRUTON wrote, that the *subpial hemorrhages* seen in the boxers was *sometimes found within glial cells* but often *was distributed more diffusely over glial fibres.*

In all in the authors material 11 of the 22 brains showed evidence of recent or old *meningeal hemorrhage,* four of which proved to be fatal.

Cerebellar Siderosis: C.W.M. ADAMS and BRUTON found in six ex-boxers *evidence of minor cerebellar siderosis,* namely iron deposition in the white matter of the cerebellar folia and three of six had focal iron deposition over Bergmann glia and Purkinje cells. One of the patients showed *ferrugination of the Purkinje cells.* The authors emphasize that these cerebellar changes were but a minor manifestation of those seen in the fully developed superficial siderosis of the CNS. They were unlikely to be the cause of the cerebellar signs of dementia pugilistica, as they were noted in only three of the 13 original cases that C.M.W. ADAMS and BRUTON were able to study.

C.M.W. ADAMS and BRUTON *concluded that no less than 17 of the 22 ex-boxers showed evidence of past cerebral, cerebellar or meningeal hemorrhage which had occurred sometime during the boxer's life,* although the extent of past hemorrhage was less evident in the amateurs than in the ex-professional boxers.

C.W.M. ADAMS and BRUTON summarize: "However, even in the ex-professional boxers the degree of cerebral damage, as measured by the amount of free-lying hemosiderin, was relatively minor for the extensive neuropathology that characterizes the 'punch drunk' state. In fact the damage appeared to be additional to the cerebral pathology normally associated with dementia pugilistica. Nevertheless, it is self-evident that in any of these 17 cases, the hemorrhages could have been fatal. Indeed no less than four were so."

ROBERTS et al. (1990) *examined archival formalin-fixed material from the temporal lobe from 14 of the 15 cases of dementia pugilistica originally reported by CORSELLIS et al. (1973),* one *additional case of a boxer who died during a fight,* three additional cases of *professional boxers* and two *amateur boxers* (Table 91): Each case had previously been investigated to establish career details and had undergone extensive neuropathological examination.

The extent of tangle and plaque formation was established by *silver and Congo red staining* (Table 92). Paraffin embedded 20 μm sections were stained by hematoxylin-eosin and Congo red and by imuno-cytochemical methods (PAP) using a well characterized monoclonal antibody 4 D 12/1/6 raised to a synthetic peptide corresponding to residues 8 – 17 of the beta-protein amyloid. Immunostaining was carried out with

TABLE 91
Biographical and neuropathological details of cases. (From ADAMS C W M, BRUTON C J, 1989).

		Details of cases		Neuropathology[1]		
Case	Age	Boxing career (years)	No fights	Cavum septum	Tangles	Plaques
Professional						
1	22	3	3 (died during fight)	+	−	−
2	53	25	>100	+	+	−
3 (5)*	60	18	>100	+	+ +	−
4	61	15	150	+	+ + +	−
5 (7)	62	20	400	+	+ + +	−
6 (3)	63	13	300	+	+ + +	−
7 (1)	63	14	400	+	+ + +	−
8	65	20	>200	+	+	−
9 (11)	67	20	?	+	+	±
10 (10)	67	15	?	+	+ +	±
11 (4)	69	25	600	+	+ +	−
12 (8)	71	23	565	+	+ +	±
13 (9)	72	10	?	+	+ +	−
14 (2)	77	17	700	+	+ + +	−
15 (6)	83	25	500	+	+ +	+ +
Amateur						
16 (13)	57	15	>100	+	+ + +	+ + +
17 (15)	58	5	> 50	±	−	−
18 (14)	61	20	>100	±	+	+
19	68	2	< 10	±	−	−
20[A]	72	2	< 10	−	+ + +	+
AD cases						
N = 20	55–85	N/A	N/A	−	+ + (MIN)	+ + (MAX)
Controls						
N = 20	58–90	N/A	N/A	−	+ (MAX)	+ + (MAX)

[1] Tangle and plaque assessed by routine silver stains.
[*] Case No. in Corsellis *et al.*
[A] Probable Alzheimer's disease.

or without a fifteen minute pre-incubation 70% formic acid. Sections were incubated with the primary antibody (dilution 1:1000 ascites fluid for one hour) and then with a second layer anti-mouse Ig G peroxidase conjugated antisera (Miles-Yeda 1:100 for 30 minutes). The reaction site was visualized by a final incubation in a 0.05% diaminobenzidine solution.
For comparison, sections from *20 confirmed cases of Alzheimer's disease and 20 age-matched controls* were investigated in a similar fashion.

ROBERTS et al. found that all dementia pugilistica cases with substantial tangle formation showed evidence of extensive beta-protein immuno-reactive deposits (plaques). The diffuse 'plaques' were not visible with Congo red or standard silver stains. The degree of beta-protein deposition was comparable to that seen in Alzheimer's disease. *The data of ROBERTS et al. show that the present neuropathological description of dementia pugilistica (tangles but no plaques) should be altered to acknowledge the presence of substantial beta-protein deposition (plaques). The molecular markers present in the plaques and tangles of dementia pugilistica are the same as those in Alzheimer's disease.* Similarities in clinical symptoms, distribution of pathology and neurochemical deficits also exist. Epidemiological studies have, according to ROBERTS et al., shown that head injury is a risk factor in Alzheimer's disease. "It is probably that dementia pugilistica and Alzheimer's disease share common pathogenic mechanisms leading to tangle and plaque formation."

TABLE 92
Detaction of amyloid plaques. (From ROBERTS G W et al., 1990).

	β-Amyloid immunostaining		
Case	No Formic Acid	Formic Acid	Silver staining[1]
1	–	–	–
2	–	+ +	–
3	–	+ +	–
4	±	+ + +	–
5	–	+ +	–
6	±	+ + +	–
7	±	+ + +	–
8	+	+ +	–
9	+	+ +	±
10	+	+ + +	±
11	–	+ +	–
12	+	+ + +	±
13	±	+ +	–
14	–	+ + +	–
15	+ +	+ + +	+ +
16	+ +	+ + +	+ + +
17	±	+ +	–
18	+	+	+
19	+	+	–
20	+ + +	+ + +	+ + +
AD Cases	+ + + (MIN)	+ + (MIN)	+ + (MIN)
Controls	+ + (MAX)	+ + (MAX)	+ + (MAX)

[1] von Braunmühls or Bielschowsky (from table 1).

TOKUDA et al. (1991) carried out a histopathological study on the *brains of eight ex-boxers* – aged 56 to 83 – using conventional histological staining methods and immuno-cytochemistry with antibodies to amyloid beta-protein and the PHF-related tau protein (Table 93). All cases showed a large number of tau immuno-reactive neurofibrillary tangles and also beta-protein immuno-reactive senile plaques in the cortex. In the areas with many neurofibrillary tangles, neurophil threads with tau-immuno-reactivity were also observed, and some of the senile plaque lesions were surrounded by abnormal neurites with tau-immuno-reactivity. Moreover, three cases revealed beta-protein-type cerebrovascular amyloid deposits on both leptomeningeal and cortical blood vessels. *The present observations indicate that the cerebral pathology of dementia pugilistica is very similar to that of Alzheimer's disease and suggests that these two disorders share some common etiological and pathogenic mechanisms.*

CLINTON et al. (1991) examined cases of post-traumatic Alzheimer's disease: preponderance of a single plaque type.

The cause of Alzheimer's disease is, according to CLINTON et al., unknown. Several factors have

TABLE 93
Semiquantitative estimation of neurofibrillary changes and beta-protein deposits in eight cases with dementia pugilistica. (From TOKUDA T, 1993).

Case	Age/sex (years)	Congo-red			Silver impregnation		Immunostain				
							β-protdin			tau	
		PL	NFT	CAA	PL	NFT	PL	NFT	CAA	NFT	NT
1	56/M	+ +	+ +	+ +	+ + +	+ +	+ + +	–	+ +	+ +	+ +
2	61/M	+ +	+	+ +	+ + +	+	+ + +	–	+ +	–	–
3	61/M	–	+ + +	–	–	+ + +	+	+	–	+ + +	+ + +
4	63/M	+ +	+ +	–	+ +	+	+ +	+	–	+ +	+
5	71/M	–	+ + +	–	+ +	+ + +	+ +	+	–	+ +	+
6	75/M	+ +	+ +	+ +	+ + +	+ +	+ + +	+	+ +	+ +	+ +
7	77/M	–	+ + +	–	–	+ + +	+	+ +	+	+ + +	+ +
8	83/M	+ +	+ + +	–	+ +	+ + +	+ +	+	–	+ + +	+

PL, senile plaque; NFT, neurofibrillary tangle; CAA, cerebral amyloid angiopathy; NT, neuropil thread; M, male
The following four grades of severity was observed under low power magnification; –, not observed; +, a few lesions were seen, but only after extensive search; + +, a substantial number of lesions were observed, but some optic fields were still free of lesions; + + +, many lesions were visible in every field.

been proposed including head trauma. At present, the link between head injury and a subsequent neuro-degenerative process is largely circumferential, except in the case of dementia pugilistica (punch drunk syndrome) found in boxers. Recent studies have shown that the brains of boxers with this syndrome contain large numbers of 'diffuse' beta-protein immuno-reactive plaques. CLINTON et al. supposed that this plaque type might be associated with trauma induced Alzheimer-like degeneration. In order to test this hypothesis the authors re-investigated a previously reported case of posttraumatic premature Alzheimer's disease. Immuno-cytochemistry using antibodies to amyloid beta-protein revealed large numbers of 'diffuse' non-congophilic plaques with little or no neuritic component. A similar preponderance of this plaque type is present in the brains of boxers with dementia pugilistica. *The authors observations support the idea of a trauma induced Alzheimer-like degenerative process and indicate that such a condition is associated with a marked preponderance of 'diffuse' plaques.*

HOF et al. (1992) obtained the *brains of three patients presenting with dementia pugilistica* (case 1: 58 years old; case 2: 63 years old; and case 3: 69 years old) and from *eight patients with Alzheimer's disease* (80.3 +/− 6.5 years old) at autopsy and fixed in a 15% formalin solution.

The clinical and neuropathological data were obtained from the clinical records of the *Department of Psychiatry, University of Geneva School of Medicine, Geneva, Switzerland* (case 1, CONSTANTINIDES and TISSOT 1967), and the Alzheimer's disease cases; and from the *Division of Neuropathology, Mount Sinai Medical Center, New York, USA* (cases 2 and 3). All three dementia pugilistica cases were former professional boxers with careers that extended over 25 years and who fought in excess of 600 bouts. The clinical presentation of all three boxers was consistent with the diagnosis of dementia pugilistica and revealed severe memory disturbances, impairment of judgment and abstract thinking, temporo-spatial disorientation and dysphoria, combined with a constellation of neurological symptoms including to various degrees of extrapyramidal rigidity, resting tremor, gait disturbances and hypomimia and dysarthria.
A clinical and neuropathological summary (case 1) has been previously reported (CONSTANTINIDES and TISSOT 1967) and material from the Alzheimer disease cases have been included in previous analyses (HOF et al. 1990, 1991).

The entire brain was obtained from case 1, while only samples from the superior-frontal, orbito-frontal, and occipital cortex were available from cases 2 and 3. The following areas were analyzed in case 1 (numeration according to BRODMANN's nomenclature): 9, 46, 45 and 11 in the prefrontal cortex; 4 (primary motor cortex); 3, 1 and 2 (primary sensory cortex; 17, 18 (primary and secondary visual cortex) and 19 in the occipital cortex; 7 b in the posterior parietal cortex; 24 and 23 in the cingulate cortex; 20, 21, 22 and 41 (primary auditory cortex) in the temporal cortex and 35/36 (perirhinal cortex), anterorhinal cortex and hippocampus. From each tissue block, 30 µm-thick sections were cut; on a cryostat and stained *with modified thioflavine S* (GUNTERN et al. 1992, VALLET et al. 1992), *Bielschowsky* and *hematoxylin-eosin techniques.* From each block, additional sections were processed with specific antibodies to the microtubule-associated protein tau (MV4) or to the amyloid beta protein A4 (AG 8). Briefly, the authors incubated overnight 30 µm thick sections at 4° C with the anti-tau antibody MV 4 at a dilution of 1:2.000 or at a dilution of 1:4.000 from the anti-amyloid beta protein antibody 4G 8. Following incubation, sections were processed by the avidin-biotin method using a Vecta stain ABC Kit (Vector Laboratories), and diamino-benzidine.

HOF et al. surveyed all the sections systematically and lesions were counted using a computer-assisted image analysis system consisting of a Zeiss Axiophot photomicroscope equipped with a motorized stage, a high sensitivity Dage 666 SIT camera, a DEC 3.100 workstation and Macintosh microcomputer, and custom designed morphometry software. Qualitative analyses were performed on thioflavine S-stained sections. Comparative counts were made on sections from area 9, stained with anti-tau antibody. On each slide, neurofibrillary tangles were counted under five to ten 1mm-wide cortical traverses in layers II – III and V – VI separately in each area, and in the hippocampus. In neocortical areas, traverses were counted in regions were the tissue was cut perpendicular to the pial surface to avoid artifacts resulting from oblique sectioning of the cortical layers. Computer-generated map of neurofibrillary tangle distribution were created by collecting the x – y coordinates of all neurofibrillary tangles present under adjacent cortical traverses through the entire cortical thickness, and were assembled automatically to create the map (Figs. 308 a and b).

HOF et al. reported that *routine neuropathological evaluation of the three cases of dementia pugilistica revealed a cavum septi pellucidi, widespread neurofibrillary tangles, gliosis and neuronal loss in the neocortex and hippocampal formation* (Fig. 309 A, C, E), as well as *neurofibrillary tangle formation* and *severe depigmentation in the substantia nigra, and loss of Purkinje cells in the cerebellum.* In addition, the entire amygdala, the nucleus basalis of Meynert, and several hypothalamic nuclei contained numerous neurofibrillary tangles in case 1. Senile plaques and amyloid deposition were not mentioned in the neuropathological report of all three dementia pugilistica cases based on Bielschowsky- and thioflavine S- stained materials. The eight Alzheimer disease cases presented with high neurofibrillary tangles and senile plaque densities throughout the neocortex and hippocampal formation Figs. 309 B, D, F).

According to HOF et al. in all the neocortical areas investigated, the *distribution of neurofibrillary tangles was qualitatively comparable in the three dementia pugilistica cases,* and exhibited a *preferential distribution in layers II and III,* whereas *layers V and VI were less affected* (Figs. 309 and 310). *This distribution differed,* according to HOF et al., *strikingly from that usually observed in Alzheimer's disease cases,* where neurofibrillary tangles are present in both superficial and deep layers, with a higher density in layers V and VI in most neocortical areas (Figs. 308 C-F, Table 94). In addition, HOF et al. reported, that neurofibrillary tangles were denser in the upper two thirds of layer II, in the dementia pugilistica cases, whereas they tended to be preferentially located in the lower one third of layer II in Alzheimer's disease cases. In the hippocampal formation (case 1) the neurofibrillary tangle distribution was comparatively to that observed in the Alzheimer's disease cases (Figs. 309 C, D), with numerous neurofibrillary tangles in the CA 1 field of the Ammon's horn, subiculum and layers II and V of the entorhinal cortex (Figs. 309 A, B 310, D). The neurofibrillary tangles in three dementia pugilistica cases were strongly labeled by an antibody to the microtulule-associated protein tau. Staining with the anti-tau antibody revealed, according to HOF et al., a neurofibrillary tangle distribution similar to that obtained with thioflavine S in all cortical areas (Figs. 309 D – F), although neurofibrillary tangles were generally lower than with thioflavine S. Finally, staining with an antibody to the amyloid beta protein failed to reveal widespread amyloid deposition or senile plaques in the three dementia pugilistica cases. In fact cases 2 and 3 were, as HOF et al. reported, devoid of amyloid deposition, and case 1 showed a unique nest of senile plaques in the inferior temporal cortex. Interestingly, so continue HOF et al., a few neurofibrillary tangles were labeled by the anti-amyloid beta protein antibody in the vicinity of this senile plaque cluster (Figs. 310 A – C). No neurofibrillary tangle staining was observed when tissue was incubated in absence of primary antibody or with an irrelevant secondary step. In addition, neuritic threads were occasionally encountered in the dementia pugilistica cases. They were more frequent in the hippocampal formation than in neocortical areas in case 1, although their density generally was much lower than in the Alzheimer's disease cases.

A quantitative analysis demonstrated the neurofibrillary tangle density in the neocortex was globally higher in dementia pugilistica than in Alzheimer's disease cases, whereas in the hippocampal formation comparable values were observed in both conditions (Table 94). *Neurofibrillary tangles were absent from the occipital cortex of all three dementia pugilistica cases.* Generally, according to HOF et al., *temporal neocortical regions showed higher neurofibrillary tangle densities than parietal and frontal areas in case 1* (Table 95). The anterior and posterior cingulate cortex (areas 24 and 23), primary sensory, primary auditory (areas 3, 1, 2 and 41), and primary motor (area 4) cortex had relatively low neurofibrillary tangle densities (Table 95). The highest neuro-

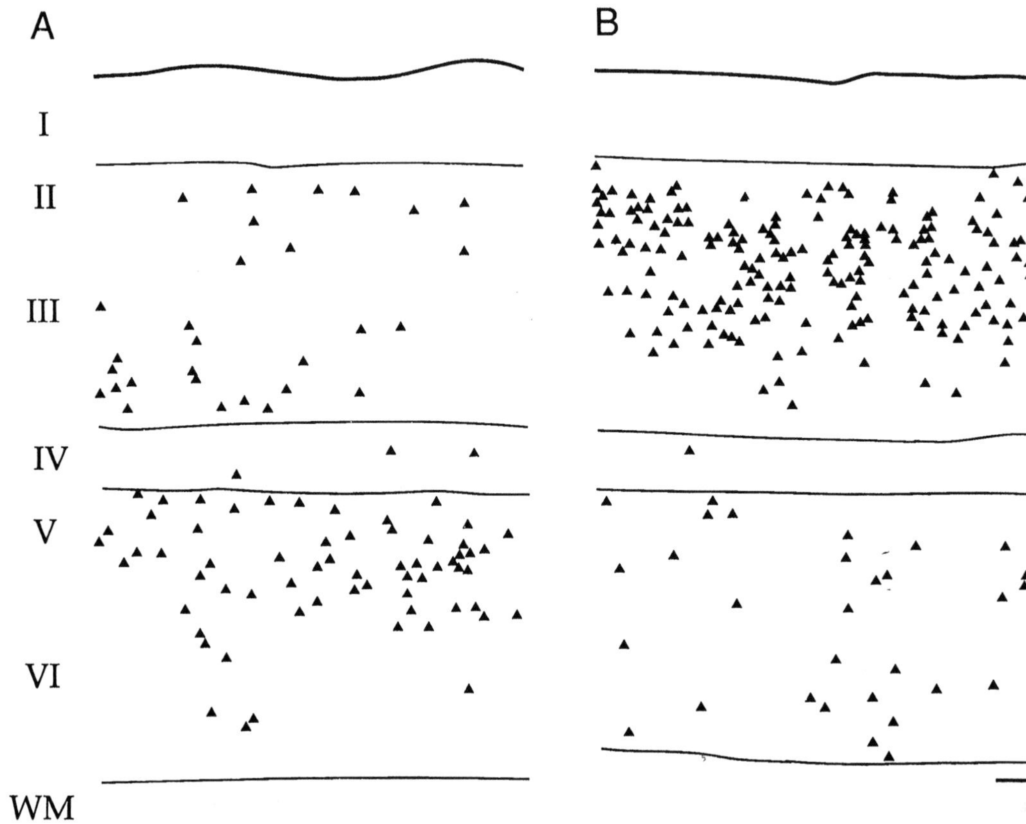

Figs. 308 A and B. Computer generated maps of neurofibrillary tangle (NFT) distribution in the inferior temporal cortex (area 20) of an Alzheimer 's disease (AD) case (A), and of case 1 (B). The map was created at a magnification of 200 x by collecting the x-y coordinates of all the thioflavine S-positive lesions present under adjacent tissue traverses through the entire extent of the section. The tissue traverses were then automatically assembled to create the map. Boundaries were manually edited on the computer screen display. Layers are indicated by Roman numerals. NFT are labeled by *black triangles.* Note the inverse NFT distribution in the dementia pugilistica case as compared to the AD case. Senile plaques in the AD case were omitted. *WM,* White matter. *Scale bar* = 100 μm. (From HOF P R et al., 1992).

fibrillary tangle densities were observed by HOF et al. in the perirhinal and inferior temporal cortex, while the lowest counts were collected in the primary somatosensory region (Tables 94 and 95).

HOF et al. noted that the presence of neurofibrillary tangles in the cerebral cortex of dementia pugilistica cases has been well documented (BRANDENBURG and HALLERVORDEN 1954, GRAHMANN and ULE 1957, QUANDT and SOMMER 1965, UNTERHARNSCHEIDT and SELLIER 1971, CORSELLIS et al. 1973, CORSELLIS 1978, ROBERTS 1988, TOKUDA et al. 1991, TOKUDA 1993). HOF et al. continue: *"However, to our knowledge, the striking difference in laminar distribution of neurofibrillary tangles between dementia pugilistica and Alzheimer's disease has not yet been reported.* Thus, in all neocortical areas investigated in the dementia pugilistica cases, neurofibrillary tangles predominated in the superficial layers, whereas in Alzheimer's' disease they are located both superficially and deep, with a higher density in the deep layers. Interestingly, the laminar pattern of neurofibrillary tangles is comparable to that existing in other neurological illnesses. For instance, in Guamanian amyotrophic lateral sclerosis/Parkinsonism-dementia (ALS/PD)

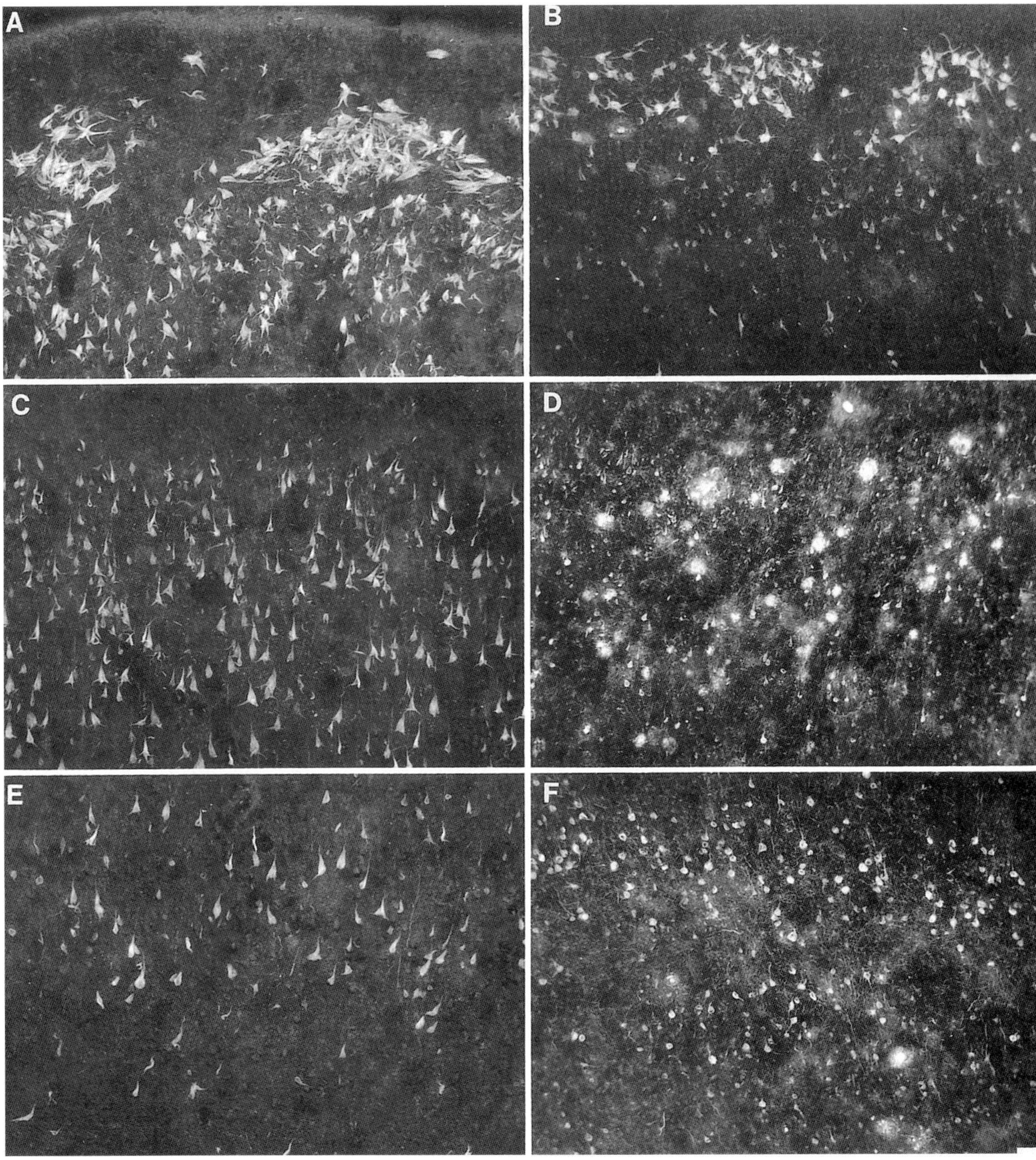

Figs. 309 A – F. Comparison of NFT distribution in the cerebral cortex in dementia pugilistica (A, C, E) and AD (B, D, F). (A, B) Layers II and III of the entorhinal cortex. There is a comparable distribution and density of NFT in both the dementia pugilistica case (A) and the AD case (B). (C, D) Layers II and II of the mid temporal gyrus cortex (area 21). The lesions are preferentially located in the upper part of layer III and in layer II in the dementia pugilistica case (C) and are more numerous than in the AD case (D). (E, F) layers V and VI of area 21. Note the inverse distribution between the two cases and the fact that NFT are denser in the AD case (F). Materials were stained with thioflavine S. *Scale bar* = 100 μm.(From HOF P R et al., 1992).

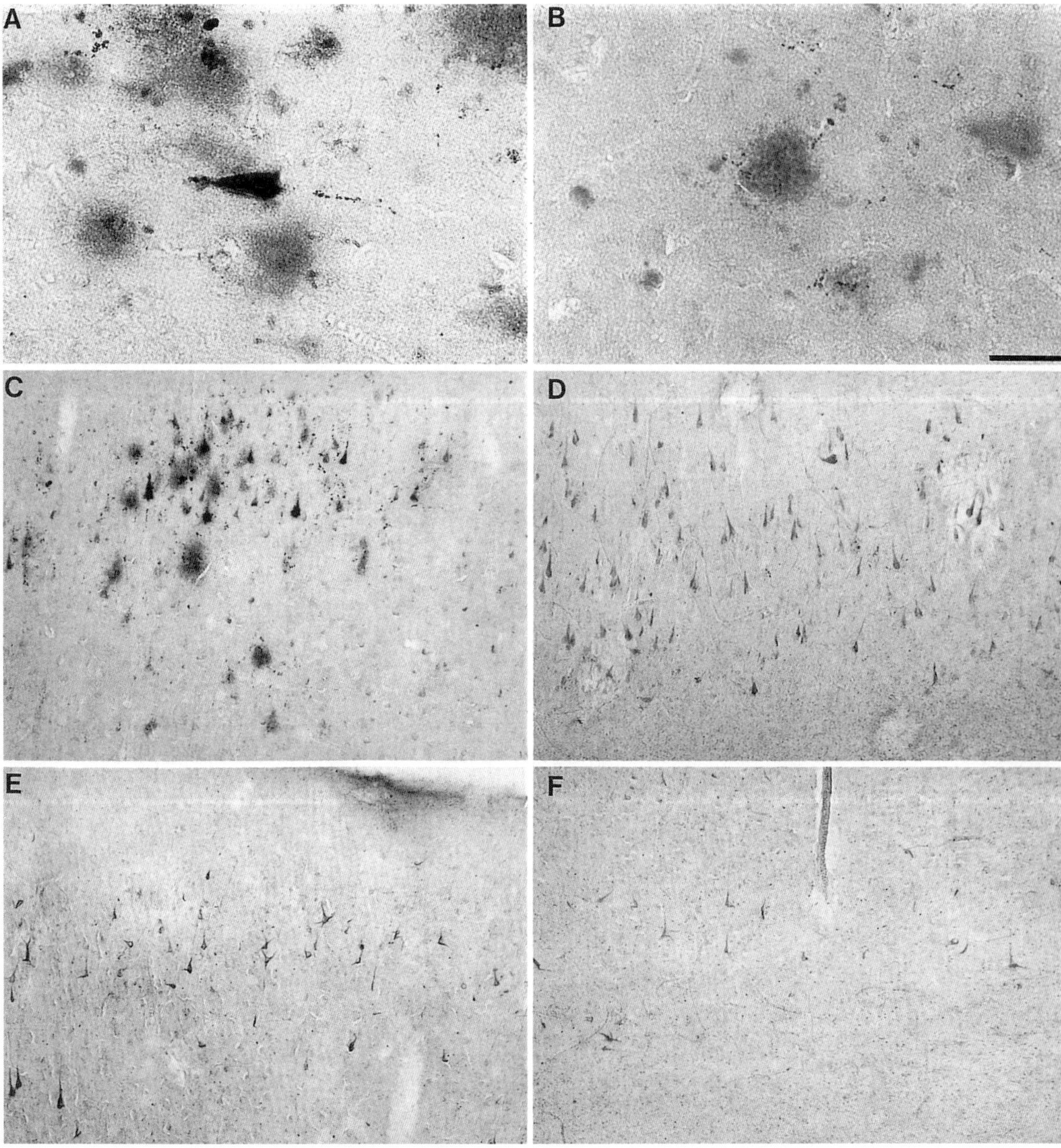

Figs. 310 A – F. High magnification of amyloid beta protein-positive extracellular NFT in layer II of area 20. (B) Amyloid deposits within the cluster, same magnification as in (A). Note the lack of classical amyloid core and that no neurites of threads are labeled. (C) View of the entire cluster of senile plaques stained with the antibody to the amyloid beta protein in the inferior temporal cortex of case 1. Note the number of labeled NFT. This was the only occurrence of amyloid deposition among the three dementia pugilistica cases. (D–F) Staining pattern obtained with the antibody of the microtubule associated protein tau. Note the high NFT density in the CA field of the hippocampus (D). (E, F) tau-positive NFT in layer III (E) and V (F) of area 9. There are only few tau-immunoreactive neuritic threads. The staining pattern is comparable to that observed with thioflavine S. *Scale bars* B, F = 100 μm. (From HOF P R et al., 1992).

TABLE 94
Comparison of neurofibrillary tangles (NFT) counts between a case of dementia pugilistica (case 1) and eight Alzheimer's disease (AD) cases. (From HOF PR et al., 1992).

Area	Dementia pugilistica	AD
CA1	76.5 ± 6.4	70.1 ± 10.2
Entorhinal cortex		
Layers II	119.5 ± 6.4	91.0 ± 5.9
III	86.0 ± 3.7	65.2 ± 6.6
V	114.3 ± 21.0	81.2 ± 8.6
VI	51.5 ± 3.3	37.4 ± 3.7
20		
Layers II–III	217.3 ± 15.5	65.2 ± 6.7
V–VI	71.5 ± 4.4	98.4 ± 6.5
9		
Layers II–III	63.3 ± 3.5	35.4 ± 4.1
V–VI	24.2 ± 5.4	45.7 ± 3.9

The results represent means ± SEM and are expressed as NFT counts/mm of cortical traverse. Data from the AD cases were pooled. Note the inverse NFT distribution in the dementia pugilistica case as compared to AD. Neocortical areas are numbered according to Brodmann's nomenclature. Layers are indicated by Roman numerals. See text for details.

TABLE 95
Regional and laminar distribution of NFT in dementia pugilistica (case 1). (From HOF P R et al. 1992).

Area	*46*	*45*	*11*
Layers II–III	25.0 ± 4.9	87.3 ± 8.0	45.4 ± 4.0
V–VI	14.6 ± 1.6	32.8 ± 2.4	10.8 ± 1.2
Area	*35/36*	*21*	*22*
Layers II–III	224.3 ± 8.4	160.8 ± 15.4	196.0 ± 13.5
V–VI	67.5 ± 5.8	43.3 ± 3.6	66.8 ± 5.5
Area	*7b*	*23*	*24*
Layers II–III	87.6 ± 3.7	15.2 ± 1.4	20.2 ± 1.3
V–VI	10.6 ± 0.7	6.6 ± 1.0	12.0 ± 1.5
Area	*41*	*3, 1, 2*	*4*
Layers II–III	34.2 ± 4.0	13.8 ± 1.7	17.0 ± 1.5
V–VI	12.8 ± 1.6	9.4 ± 1.3	9.8 ± 0.8

The results represents means ± SEM and are expressed as NFT counts/mm of cortical traverse. Note the preferential localization of NFT in layers II and III. A similar NFT distribution was observed in samples from cases 2 and 3. Areas are numbered according to Brodmann's nomenclature. Layers are indicated by Roman numerals. See text for details.

cases, post-encephalitic Parkinsonism (PEP) cases demonstrate a preferential localization of neurofibrillary tangles in layers II and II of the neocortex, while neurofibrillary tangle distribution in the hippocampal formation is similar to that observed in Alzheimer's disease patients.

HOF et al. summarized, the association cortex of brains from dementia pugilistica patients demonstrated an inverse neurofibrillary tangle distribution as compared to that found in Alzheimer's disease. This finding suggests that a more circumscribed population of cortical pyramidal neurons might be affected in dementia pugilistica than in Alzheimer's disease.

GEDDES et al. (1996) found neurofibrillary tangles, but not Alzheimer-type pathology, in a young boxer. The authors noted: "The chronic neurological sequelae of boxing are well described, but there have been few neuropathological studies of boxers dying early in their career."

GEDDES et al. reported the case of a 23-year-old boxer whose brain showed neurofibrillary tangles in all neocortical area. Remarkably, the medial temporal lobe structures were spared. The tangles, assumed to be the result of repetitive head injury, were the only detectable abnormality: None of the other changes found in the brains of the retired boxers described previously were seen. The distribution and features of the neuropathological findings in this case suggest that the mechanism of tangle formation induced by repetitive head trauma may be different from that in Alzheimer's disease.

A classic study was published by DRACHMAN and NEWELL (1999) in the Weekly Clinicopathological Exercises of the Massachusetts General Hospital (Case 12-1999):

"A 67-year old, right handed man was admitted to the hospital because of increasing dementia."

"Three years before admission, the patient's cognitive functions began to decline. One year later, a neurologist diagnosed progressive dementia with parkinsonism."

"During the two months before admission, the patient became agitated, and trazodone was prescribed. His family noted transient periods when he stared into space and did not respond to his surroundings. He fell several times, wtih apparent loss of conciousness but no obvious injury."

"One week before admission, the patient punched his wife because she would not let him drive. Soon thereafter, he began to look under his bed, locked the doors and windows, and became convinced that the neighbors were trying to steal the family's car. Subsequently, he struck then his teenage grandson without provocation and wandered from home. He was referred to his hospital."

"The patient had boxed professionally for ten years in over 100 bouts. He drank alcohol excessively in his youth but had been sober for several decades. There was no family history of psychiatric problems. A sibling died 10 years earlier of probable amyotrophic lateral sclerosis."

"The temperature was 37° C, the pulse was 84, and the respirations were 16. The blood pressure was 145/80 mm Hg. The results of a general physical examination were normal."

"On neurological examination, the patient was alert but was oriented only to himself. He believed that he was in his house in the city and that the year was 1970. He lost concentration when attempting serial tasks. Abstract tasks were performed poorly. He recalled none of the objects at five minutes, even with cues. Remote autobiographical memories were relatively preserved. Left-right confusion was present. The patient's language was intact, and phrases were repeated accurately. His speech was hypophonic. His facial expression was reduced, and he drooled. Cranial-nerve functions were preserved. There was severe atrophy of the shoulder girdles and arms and legs. Diffuse fasciculations and muscle twitches were observed. Motor power was normal. The patient's posture was slightly stooped. His spontaneous gait was normal and tandem walking was unsteady. No resting tremor or cogwheel rigidity was detected. His sense of joint position and perception of vibration were intact. The deep-tendon reflexes were + or ++ and symmetric. No bradykinesia or postural instability was evident, and the glabellar-tap sign and Babinski sighs were absent. The electrographic and laboratory findings were normal."

"A magnetic resonance imaging (MRI) examination of the brain showed severe, diffuse ventricular and cortical sulcal enlargement (Fig. 311). T 2 -weighted images showed slight thinning of the corpus callosum, atrophy of the mammillary body, a moderately reduced signal in the globus pallidus and putamen, and minimal hyperintensity of periventricular white matter. The pars compacta was thinned on the proton-density-weighted images. A cavum septi pellucidi was noted. The major intracranial vessels were patent."

"An electroencephalographic study, performed while the patient was awake, revealed symmetric 8-Hz alpha activity, considerable generalized, symmetric 7-Hz theta slowing of low-to-moderate voltage, in the left temporal lobe. No response to photic stimulation was observed. Electromyographic examination showed widespread spontaneous activity and chronic denervation – reinnervation in the arms and legs, paraspinal muscles, and face, without evidence of a generalised polyneuropathy. The findings were consistent with the presence of a generalized disorder of the motor neurons, their axons, or both."

"Trazodone was continued, and treatment with risperidone was begun. The patient became less agitated. On the 16th hospital day, he was transferred to a nursing home, where he died suddenly five days later."

Differential Diagnosis

DRACHMAN: "Dementia is the 'decline of memory and other cognitive functions in comparison with the patient's previous level of function' (Mc KHANN et al. 1984). Distinguishing among the 70 or more causes of dementia (MAYEUX et al. 1993) can be a formidable task. Dementia may be 'static,' resulting from a single event such as a stroke, severe head trauma, or encephalitis, or it may be progressive, as in this case. Progressive dementias, which account for the vast majority of cases that neurologists see as diagnostic problems, are divided into two groups: symptomatic dementias and degenerative dementias."

DRACHMAN continues:

"Symptomatic dementias are these in which cognitive impairment is caused by a known disease process, such as anatomical lesions, toxic, or metabolitic disorders, or infection. Since for many of these diseases the diagnostic methods are reliable and the treatments are effective, much effort should be devoted to identifying them as either the primary or a contributory cause of the dementias. These diseases include anatomical lesions such as brain tumors and normal-pressure hydrocephalus, metabolic disorders such as hypothyroidism and vitamin 12 deficiency, and infectious processes such as Creutzfeld-Jakob disease in human immunodeficiency virus infection. Only when all the symptomatic dementias have been ruled out can the diagnostic possibilities be narrowed to the category of degenerative dementia."

"Several tests are useful in the differential diagnosis of dementia, and at least one brain imaging study is necessary (COREY-BLOOM et al (1995). May we see the MRI scan?"

Michael H. Lev: " MRI examination of the brain without contrast material showed severe, diffuse ventricular and cortical sulcal enlargement, which was slightly more prominent at the temporal lobes (Fig. 311). The temporal horns and the thinning of the corpus callosum and minimal, nonspecific hy-

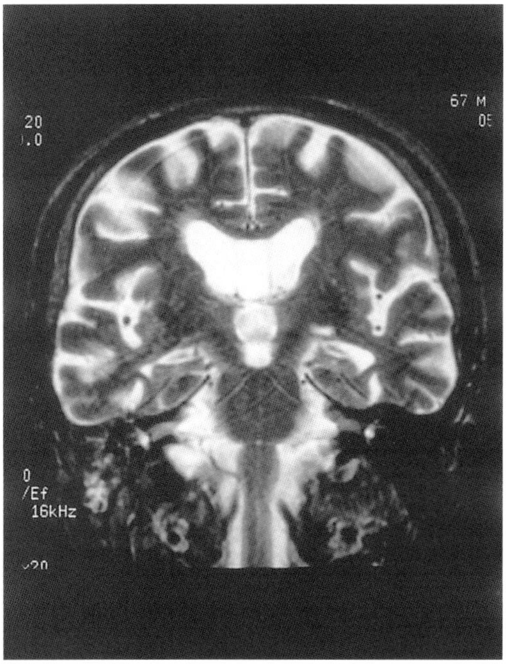

Fig. 311.Coronal T_2 – weightes – MRI scan obtained at the level of the lateral and third ventricles, revealing severe cortical atrophy and cavum septi pellucidi (arrows). (From DRACHMAN DA, NEWLL KL, 1999).

perintensity of periventricular white matter on T2-weighted images. The pars compacta was thinned on the proton-density-weighted images, and the signals in the globus pallidus and putamen were moderately reduced on the T2-weighted images. A cavum septi pellucidi was present.There was no evidence of a stroke, bleeding, or a mass lesion."

DRACHMAN: "The findings on the MRI scan help rule out anatomical lesions among the degenerative dementias. In a few conditions, such as Huntington's disease and advanced Pick's disease, the scan may show distinctive changes. In this case, there is moderately extensive cortical atrophy, but the absence of disproportionate hippocampal atrophy makes a diagnosis of Alzheimer's disease less likely" (JACK et al. 1997). The presence of a cavum septi pellucidi, the thinning of the corpus callosum and the substantia nigra, and the slight atrophy of the mammilary bodies resulted from the patient's long history of boxing."

DRACHMAN continues:

"The common metabolic and infectious diseases that cause symptomatic dementias have been ruled out by the normal laboratory findings. Thoroughness is essential when one is determining the cause of dementia in older patients, which requires as complete an evaluation as is feasible (Table 96). Even when a cause is identified, it is unseful to search for contributing factors, particularly in the elderly (SNOWDON et al. 1997), since two thirds of people over 65 years of age have at least one chronic illness (DRACHMAN 1991)".

This patient appears to have had a progressive and apparently degenerative dementia. Degenerative dementias can be divided into four groups: Alzheimer's disease, dementias with movement disorders, those associated with cortical-lobar abnormalities, and miscellaneous dementias, including cerebellar degenerative disorders and corticobasal ganglionic degeneration. The clinical findings can usually be used to differentiate these groups, with confirmatory help from laboratory studies."

"The nature of the cognitive impairment is often an imperfect clue to the differential diagnosis. In many degenerative dementias, cognitive changes begin with a nonspecific history of impaired memory. Only occasionally is there a more focal manifestation, such as progressive aphasia, which may occur in some patients with frontal-lobe dementia. Detailed cognitive tests reveal more about the severity of the dementia than about its specific cause. Such tests, can also be helpful when the cognitive pattern suggests focal rather than global deficits, acting as a reminder to look again for one of the symptomatic dementias."

"Neurologic clues to the diagnosis of dementia may be found outside the cognitive sphere and are often motor or behavioral. Patients with diffuse Lewy body dementia (also called Lewy body disease), for example, may have early extrapyramidal signs, postural instability, prominent early hallucinations and delusions, and undue sensitivity to neuroleptic drugs (Mc KEITH et al. 1996). In contrast, patients with impaired memory and global dementia, mild extrapyramidal signs, behavioral

TABLE 96
Useful tests in the diagnosis of dementia. (From DRACHMAN D A, NEWELL K L. 1999).

Psychometric evaluation
Imaging studies
  CT or MRI of the head†
  Duplex scanning of carotid or vertebral arteries
  Positron-emission tomography or single-photon-emission CT (optional)
  Magnetic resonance angiography or spiral CT (optional)
  Radioisotope cisternography (optional)
Blood studies
  Complete blood count
  Chemical measurements
  Serum electrolyte measurements
  Thyroid-function studies
  Serologic test for syphilis
  Serum B_{12} and folate measurements (optional)
  Serum protein electrophoresis or immunofixation (optional)
  Lipid profile (optional)
  HIV test (optional)
  Test for paraneoplastic antibodies (optional)
  Apolipoprotein E genotyping (optional)
  Drug screening (optional)
Lumbar puncture (optional)
  Measurement of fluid pressure
  Cell count
  Chemical measurements
  Culture
  Cytologic studies (optional)
  Measurement of amyloid or tau protein (optional)
  Measurement of 14-3-3 protein (optional)
Genetic studies, such as tests for familial Alzheimer's dementia and Huntington's disease (optional)
Other studies
  Electrocardiography
  Chest roentgenography
  Electroencephalography (optional)
  Echocardiography (optional)

* Optional studies are those that are of uncertain diagnostic value or that are performed for special indications. CT denotes computed tomography, and HIV human immunodeficiency virus.
† MRI is the preferred study.

TABLE 97
The age and sex distribution of patients shown by CT to have a 5th ventricle. (From MACPHERSON P, TEASDALE E, 1988).

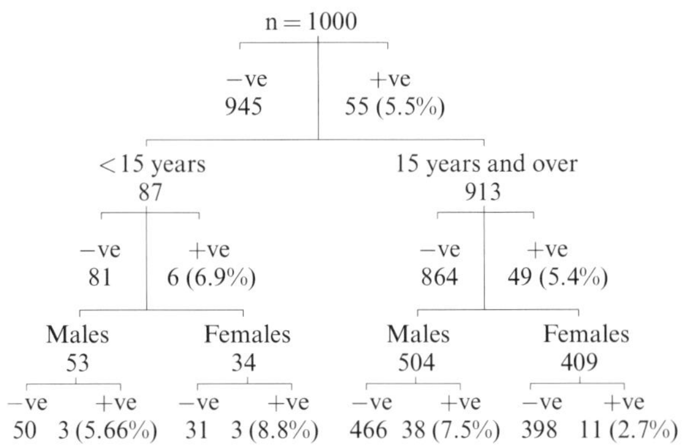

TABLE 98
Fatal incidents in the ring, (From GROH H. 1962).

Author	Year	Number of Fatal Incidents
Flint (England)	1914–1958	124
Knoll (Hamburg)	from World literature	130
Baum	from World literature	111
Ferlaino	1946	108
Ferlaino	1952	17
Pohle	1953	21
German Amateur-Boxsport Association	1945–1956 in Germany	19

disturbances, paranoid ideas, and hallucinations develop only later and variably."

"Much effort has been devoted to developing laboratory tests that will distinguish more reliably among the degenerative dementias. Studies have included blood analyses for the epsilon 4 allele of apolipoprotein E, urinanalysis for neuropil-thread protein, cerebrospinal fluid measurements of tau protein and beta-amyloid, positron-emission and single-photon-emissions computed tomographic scans of cerebral blood flow or glucose metabolism, volumetric MRI studies to detect atrophy in the hippocampal complexes and elsewhere, and molecular studies to identify the genes that produce Huntington's disease or the familial forms of Alzheimer's disease. Of these studies, only the molecular genetic analyses provide definitive diagnostic informations. The other studies, especially MRI, may add to the probability of the diagnosis (JACK et al. 1992), but no tests, singly or in combination, are sufficiently reliable to provide a certain diagnosis. For example, a study of Alzheimer's disease found that, as compared with clinical evaluation, determination of the apolipoprotein E status was only slightly more specific and considerably less sensitive (MAYEUX et al. 1998). Laboratory studies discriminate imperfectly among the degenerative dementias partly because of the long period of latency before dementia develops in Alzheimer's disease (estimated to be decades) and the overlap between degenerative dementias and the normal aging process. The clinical history and the findings on physical examination often provide the most valuable diagnostic information during life. Non-pathological examination is usually considered the gold standard for the diagnosis of degenerative dementias."

"Several clinical and behavioral observations merit closer analysis in this case. Two years before the patient was hospitalized, a neurologist had diagnosed 'progressive dementia with parkinsonism.' Neurologic examination at this hospital revealed only minor extrapyramidal features: some reduction of facial expression, mild hypophonia, and a sligtly stooped posture, but without marked rigidity, a resting tremor, a gait disorder, or a glabellar sign. The combination of dementia and extrapyramidal feature raises the possibility of diffuse Lewy body dementia, but the patient had only minimal postural and extrapyramidal features and no visual hallucinations or fluctuations in mental function (McKEITH et al. 1996, ALA et al. 1997)."

"Abnormal thought and behavior were an important, although not initial, part of the patient's dementia, with paranoid ideas, loss of inhibition, and aggressive acts. Obstreperous behavior is a common and important manifestation of many dementias (DRACHMAN et al. 1992, SWEARER et al. 1988), with the specific type of behavior varying according to the type of dementia. Disinhibited behavior is typical of frontal lobe or frontotemporal dementia (BRUN and PASSANT 1996, MILLER et al. 1997), although it occurs in Alzheimer's disease as well (DRACHMAN et al. 1992, SWEARER et al. 1997). In this case, the patient's prominent behavioral problems were the reason for his prolonged hospitalization and treatment with neuroleptic drugs."

"A striking feature of the case, with important diagnostic implications was the presence of motor neuron disease, with severe atrophy of both shoulder girdles and distal muscles, and diffuse fasciculations. Muscle strength was reported to be normal, and Babinski signs were absent. Nevertheless, an electromyogram, which is not part of the routine workup for dementia, showed chronic denervation and reinnervation in all muscle groups, a finding consisting with the presence of a 'generalized disorder of motor neurons, their axons, or both.'"

"Few conditions cause widespread motor neuron disease. This patient was too old to have spinal muscular atrophy, and there is no evidence of mercury poisoning. The post-polio syndrome occurs in persons who have had poliomyelitis many decades earlier. Since this patient had no such history and had been strong enough to box professionally, we can discard that diagnosis."

"Amyotrophic lateral sclerosis is the most common cause of diffuse motor neuron disease, with an annual incidence of about 1 case per 100.000 population and a prevalence of 4 to 5 cases per 100.000 (HUDSON 1981). The prevalence of dementia is about 1 percent in this patient's age group (HEBERT et al. 1955). The probability that dementia and amyotrophic lateral sclerosis would coexist as unrelated disorders in a previously healthy patient in his 60s is between 1 in a million and 1 in 10 million. These odds suggest that the association of dementia with motor neuron disease in the patient was not random, but that instead he had a specific disorder in which these two clinical features were connected."

"The best known association between dementia and motor neuron disease occurs in the amyotrophic lateral sclerosis-parkinson-dementia complex, seen only in native Chamorro Indians on Guam (CHEN 1955, McGEER et al. 1997). The cause of this disorder, which may be manifested as amyotrophic lateral sclerosis, parkinsonism with dementia, or both, is unknown. No mendelian mechanism of inheritance has been identified, and the idea that the disorder might be due to a toxic effect of a native flour made from the cycad plant has been discarded (CHEN 1995). I have no reason to suspect that diagnosis in the case."

"Outside the Mariana Islands, there are several interesting associations between dementia and motor neuron disease (HUDSON 1981, KIERNAN and HUDSON 1994, LOPEZ et al. 1994, MITSUYAMA 1993, WILHELMSEN 1997, JACKSON et al. 1996, NEARY and SNOWDON 1996, HOOTEN and LYKETSOS 1996, SCHMITT et al. 1995, HEUTINK et al. 1997, TALBOT 1996. MAJOOR-KRAKAUER et al. 1994). Dementia develops in about 5 percent of patients presenting with sporadic amyotrophic lateral sclerosis and in about 15 percent of patients with the familial form of amyotrophic lateral sclerosis (HUDSON 1981). In both forms the dementia is often very mild, develops late in the course of the disease, and is difficult to evaluate when severe bulbar weakness impairs speech. The clinical picture is different in patients who present with progressive dementia and in whom motor neuron disease subsequently develops. At one time, they were thought to have an amyotrophic form of CREUTZFELD-JAKOB disease, but in 1983, SALAZAR et al. (1983) reported that spongiform encephalopathy did not develop in primates inoculated with brain tissue from such patients. Many cases of dementia with motor neuron disease, with or without extrapyramidal features, have been reported since then (KIERNAN and HUDSON 1994, MITSUYAMA 1993. The variety of names applied to this clinical syndrome has been confusing. Most authors with an interest in neuromuscular diseases consider it to be a variant of amyotrophic lateral sclerosis. Whereas those who approach the disease behaviorally refer to it as the frontotemporal dementia syndrome."

"The patient had many features of extended frontotemporal syndrome. His symptoms began with dementia, he later had mild parkinsonism, his behavior became troublesome and disinhibited, and he had considerable evidence of motor neuron disease. The early death of a sibling from probable amyotrophic lateral sclerosis supports a connection between familial amyotrophic lateral sclerosis and the frontotemporal syndrome (MAJOOR-KRAKAUER et al. 1994)."

"Frontotemporal dementia, however, refers to several clinical and pathological pictures, reflecting variations in neurologic classifications according to whether etiologic, clinical, or neuropathological criteria are used. The cause of frontotemporal dementia is largely unknown, but there are both sporadic and genetic forms (LYNCH et al. 1999). The signs and symptoms vary – patients may present with primary progressive aphasia, personality change with delusions, hallucinations, and disinhibition; or dementia with motor neuron disease, sometimes combined with the parkinsonian syndrome (MILLER et al. 1997, MENDEZ et al. 1996, NEARY 1995). Even the gold standard of pathological examination is mixed with baser metals, since frontotemporal dementia may take several forms: typical Pick's disease, frontotemporal dementia without Pick's bodies, or frontotemporal dementia with motor neuron disease (MAJOOR-KRAKAUER et al. 1994). Although the issue has been debated (NEARY 1997, KERTESZ 1997), frontotemporal dementia is probably not due to a single cause. The defining features of this type of dementia are a loss of neurons in the frontal and temporal cortex, with gliosis and often a spongiform change, without findings that are characteristic of Alzheimer's disease. In the form associated with motor neuron disease, there is a loss af anterior horn cells. Cortical atrophy is typically mild, unlike that seen on this patient's MRI scan."

"Finally, repeated, severe head injuries in professional boxers may result in chronic traumatic encephalopathy, leading to the 'punch drunk syndrome,' or dementia pugilistica. Brain damage is especially due to acceleration of the head, usually rotational, the force of a single blow by a professional boxer has been estimated at 0,63 ton delivered in 100 ms (MENDEZ 1995). Repeated blows produce much more damage than a single punch, especially when the fighter is groggy, since the neck is less rigid at such times. Neurologic impairment usually begins near the end of the fighter's career or shortly after retirement (MENDEZ 1995) and evolves over a period of a few years, although its onset may be delayed for a considerably longer period (JORDAN et al. 1995). Up to 87 percent of professional boxers have some neurologic impairment (CASSON et al. 1984), with changes on imaging, electroencephalographic, and neuropsychological studies in more that half of them. An estimated 17 percent of boxers have clinically important neurologic deficits (CASSON et al. 1984), with severe dementia in about 6 percent, most often those who have boxed in more than a dozen bouts (ROBERTS 1969). The patient's history of more that 100 fights, together with the MRI changes, suggests that he must have considerable brain damage as a result of his boxing career."

"Was dementia pugilistica the primary cause of the patient's dementia? The long period of apparently normal functioning after his retirement from boxing is somewhat unusual. Motor neuron disease is not part of the punch drunk syndrome, and the role of trauma in amyotrophic lateral sclerosis is unclear (NELSON 1995, JORDAN et al.1997). Nevertheless, the one world champion boxer in my clinical practice had amyotrophic lateral sclerosis. The patient under discussion had a sibling with amyotrophic lateral sclerosis, which points to the familial syndrome of frontotemporal dementia and amyotrophic lateral sclerosis."

"The possible combination of dementia pugilistica and frontotemporal dementia could present a neuropathological dilemma, however. The changes in the brain that are associated with dementia pugilistica somewhat resemble those seen in Alzheimer's disease, which many cortical neurofibrillary tangles, amyloid angiopathy, and diffuse (but not mature) amyloid plaques. But since frontotemporal dementia is characterized, in part, by an absence of the changes that are characteristic of Alzheimer's disease, it may be difficult to identify pathologically this type of dementia in the brain of a boxer."

"In this case, the cause of the episodes of decreased responsivenss and occasional falls is obscure. They could have been the result of transient cardiac arrhythmias or partial complex seizures. We have no information from electrocardiographic monitoring, which would have indicated the presence or absence of an arrhythmia. The electroencephalographic findings were nondiagnostic, but the study was performed while the patient was awake, and seizure activity may therefore have been missed. Prizefighters often have post-traumatic cortical scars and 'plaques jaunes,' which may cause seizures."

"The cause of the patient's sudden death is speculative. Overall, arrhythmia is the most frequent cause of sudden death, but the absence of clinical or electrocardiographic evidence of cardiac disease makes it less likely. In patients with motor neuron disease, aspiration with secondary cardiorespiratory arrest may cause sudden death, but the absence of bulbar signs only a week before this patient's death is evidence against this cause. In boxers, however, large or small cerebral hemorrhages may occur. Dementia pugilistica is more likely to develop in boxers with the apolipoprotein E epsilon 4 allele than in those without it (ROBERTS 1996) but boxers with the allele are also more susceptible to cerebral hemorrhage because of the presence of cerebral amyloid angiopathy. Since this patient had an extensive history of head trauma, the possibility of hemorrhage from cerebral amyloid angiopathy is worth considering."

Bruce H. PRICE: "We considered several possible diagnoses. Because of the patient's fasciculations and electromyographic abnormalities, we considered the possibility of amyotrophic lateral sclerosis with concomitant frontotemporal dementia or Alzheimer's disease, but bulbar signs were not prominent, there was no atrophy of the intrinsic head musculature, his language was relatively spared, and he had a rather severe amnesia. Because of the muscle findings, however, we believed that an atypical variant of amyotrophic lateral sclerosis with dementia could not be ruled out."

"We learned that the patient had been knocked down 13 times in one fight. We favored the diagnosis of dementia pugilistica, although the 40 year period of latency between his retirement from boxing and the emergence of his disease was not consistent with this diagnosis."

Clinical Diagnoses

? Dementia pugilistica.

Possible atypical amyotrophic lateral sclerosis with dementia.

Dr. David A. DRACHMAN's Diagnoses

Dementia pugilistica.

Dementia with motor neuron disease (frontotemporal dementia).

Cerebral amyloid angiopathy, ? cerebral hemorrhage.

Pathological Discussion

Dr. Kathy L. NEWELL: "At autopsy, the frontal and anterior temporal lobes were slightly atrophic, with a slight reduction of white matter in the frontal lobes and a slight enlargement of the lateral ventricles. Thickening and fenestration of the left septal leaf of an anterior cavum septi pellucidi (Fig. 311) were seen, and the substantia nigra (Fig. 312) and locus ceruleus were pale."

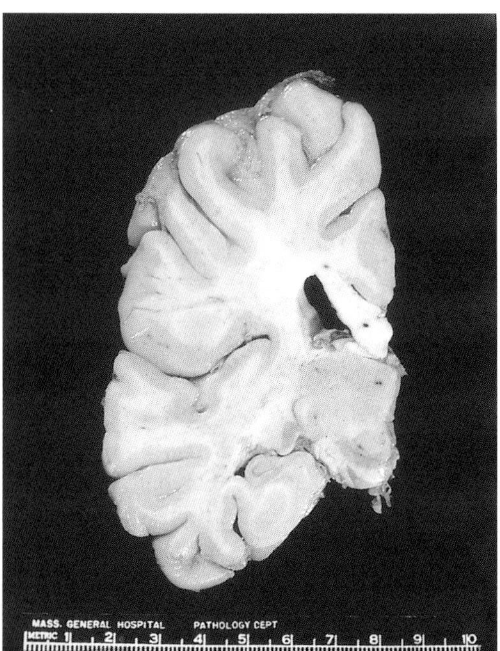

Fig. 312. Coronal section of left half of the brain through the lateral geniculate nucleus. The substantia nigra (arrows), exhibits marked pallor. (From DRACHMAN DA, NEWELL KL, 1999).

"Microscopical examination of multiple cortical sections revealed neuronal loss and many scattered neurofibrillary tangles that were immunoreactive to a polyclonal antibody against phosphorylated tau; the tangles were most numerous in the superficial cortical neuronal layers of the inferior frontal and temporal lobes (Fig. 313). Abundant tangles in the hippocampus were associated with marked neuronal loss and gliosis. Occasional tangles were present in the deep layers of the primary motor cortex, but Betz'a cells were well preserved. Modified Bielschowsky's silver staining and beta-amyloid immunohistochemical staining revealed rare senile plaques. No cerebral amyloid angiopathy or acute hemorrhage was detected."

"Severe cell loss with numerous neurofibrillary tangles (Fig. 314) and gliosis was correlated with the pallor of the substantia nigra. No Lewy bodies were found. Other nuclei in the brain stem, cerebellum, and cerebral hemispheres (including the locus ceruleus, red nucleus, pontine and raphe nuclei, motor nucleus of the fifth cranial nerve, hypoglossal nucleus, arcuate nucleus, inferior olivary nucleus, thalamus, dentate nucleus, caudate nucleus, putamen, globus pallidus, subthalamic nucleus, and amygdala) showed neurofibrillary degeneration."

"Examination of the spinal cord revealed neuronal loss, gliosis, and axonal spheroids in the anterior horns, and neurofibrillary tangles in the anterior and posterior horns. In contrast to their usual appearance in motor neuron disease, the corticospinal tracts in the lateral and anterior funiculi were well preserved. The anterior spinal roots were characterized by the loss of myelinatsd fibers (Fig. 315). Peripheral muscle contained scattered small, angulated fibers with occasional groups of end-stage muscle fibers – findings consistent with the presence of denervation related atrophy (Fig. 316)."

"Although the degenerative findings in the spinal cord gray matter may be the pathological basis for the clinical manifestations of motor neuron disease in this patient, the presence of neurofibrillary tangles in the spinal cord and other sites is not typical of motor neuron disease. Similarly, the frontal and temporal lobe changes may explain the dementia, but they are not diagnostic of Alzheimer's disease or frontotemporal dementia. The pattern of neurofibrillary changes and the small number of cortical plaques suggest several rare conditions, including the amyotrophic lateral sclerosis-parkinson-dementia complex of Guam and postencephalitic parkinsonism (HIRANO et al. 1961, GEDDES et al. 1993), but both disorders are ruled out by the clinical history."

"Another condition in which neurofibrillary tan-

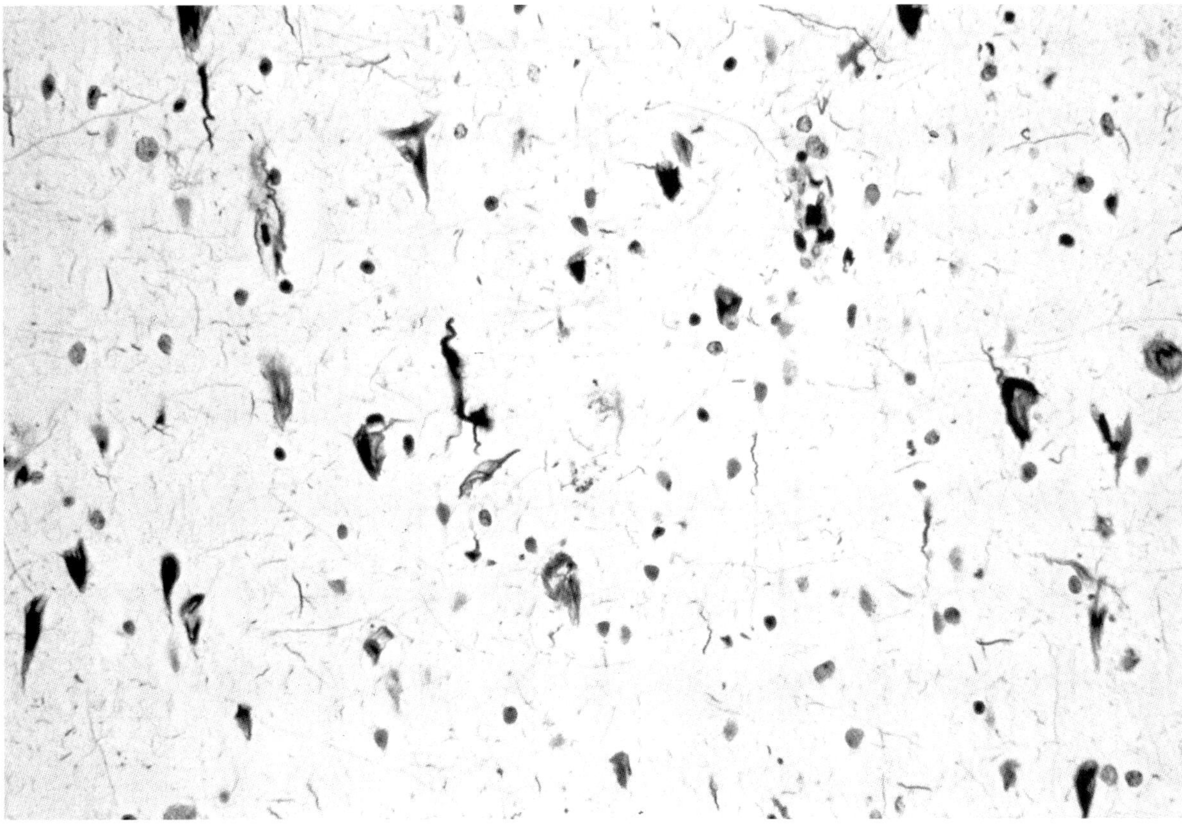

Fig. 313. Numerous neurofibrillary tangles in the temporal cortex (arrows). Modified Bielschowsky's sivler stain, 425:1. No senile plaques are present. (From DRACHMAN DA, NEWELL KL, 1999).

gles may occur with few or no amyloid plaques is dementia pugilistica (CORSELLIS et al. 1973). Since this observations was reported, however, immunohistochemical studies have identified numerous diffuse amyloid plaques in the brain of many patients with this disorder (ROBERTS et al.1990, TOKUDA et al. 1991). In a study of 15 ex-boxer, CORSELLIS et al. 1973) found common features, septal cava with fenestrated leaves, cerebellar and cerebral hemispheric scarring, degeneration of the substantia nigra with tangles but without Lewy bodies, and cortical neurofibrillary tangles with no or few senile plaques. Septal cava are frequent incidental findings at autopsy, with a mean width of 1.6 mm; in the 15 ex-boxers, however, the mean width was 5,17 mm (CORSELLIS et al. 1973). In this case, the width of the septal cavum was 5 mm. Except for the cerebellar scarring, the other common features were also present in this case. Dr. Edward P. RICHARDSON, Jr., who examined the slides in this case without knowing the history, suggested the diagnosis of dementia pugilistica, supporting our interpretation."

"Dementia pugilistica is especially likely to occur in boxers with a particular style of fighting, such as slugging (MARTLAND 1928), which was reportedly the style of this patient. The risk of the disease is also correlated with the length of the boxer's career and the number of bouts fought (ROBERTS 1969, CRITCHLEY 1957)."

"Clinical and neuropathological descriptions of neurologic disease in boxers have rarely included motor neuron disease. PARKER diagnosed lateral sclerosis in the spinal cord in a 28 year-old boxer with reduced memory but without marked weakness (PARKER 1934). Unlike the symptoms in

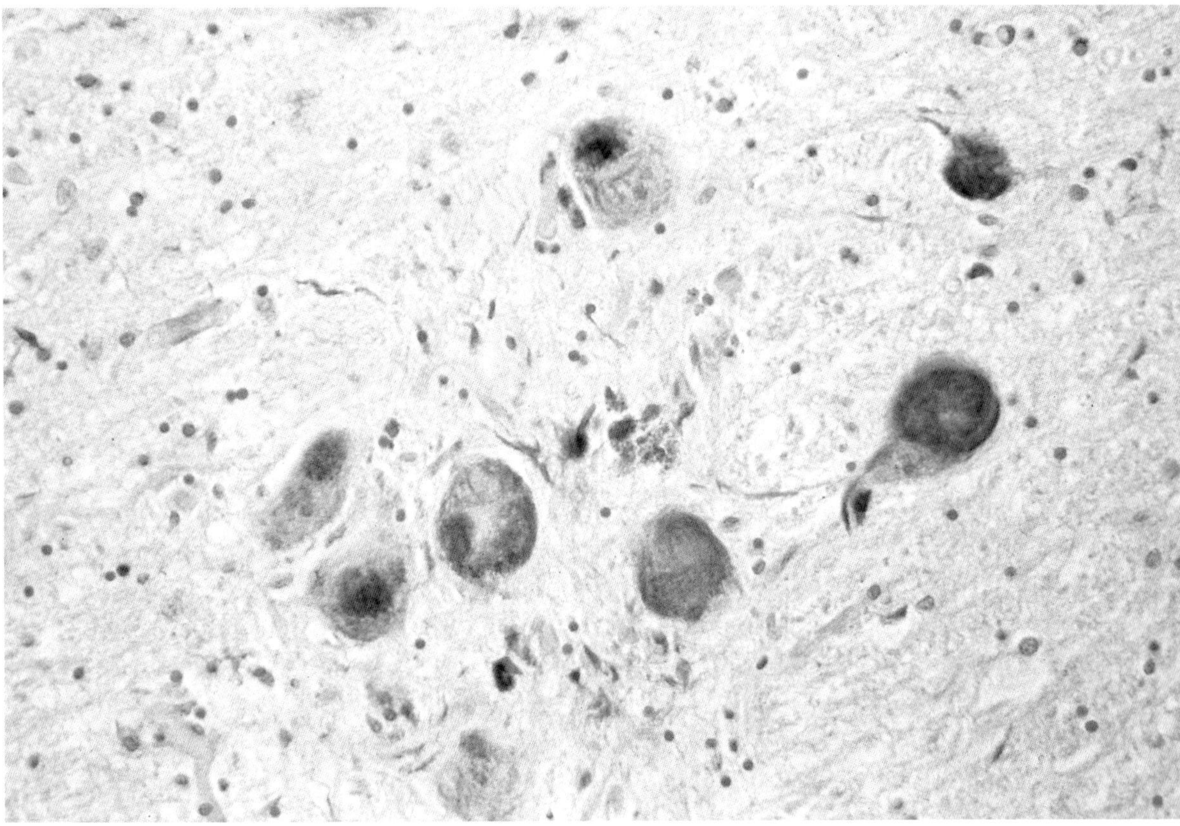

Fig. 314. Neurofibrillary tangles in remaining neurons of the substantia nigra pars compacta. Immunochemical stain for phosphorylated tau with Hematoxylin-Eosin stain, 550:1.

our patient, that patient's symptoms developed during the busiest period of his 10-year fighting career. Also, at follow-up examination 11 years later, the patient had the same disturbance of speech and gait as before, without further physical or mental degeneration. In one of the ex-boxers described by CORSELLIS et al., the corticospinal tracts were 'small and partially demyelinated (CORSELLIS et al. 1973)."

"This patient's sudden death was due to aspiration pneumonia, and vegetable material was found in multiple sites in the lungs."

Dr. PRICE: " One constructive approach is the problem of dementia pugilistica is to identify boxers at high risk for the disease, caution them about their susceptibility, and if they decide to continue their boxing careers, recommend that they undergo serial neuropsychological examinations. This patient was shown to have the apolipoprotein E epsilon 3 and epsilon 4 alleles on postmortem examination."

Dr. DRACHMAN: " This case raises the question of why neurofibrillary tangles, amyloid deposits, and amyloid angiopathy develop after head trauma. The notion that Alzheimer's disease and its characteristic lesions are due to highly specificc enzymatic defects or metabolic disorder is called into question by the fact that blunt head trauma may cause disorders with many of the clinical and pathological features of Alzheimer's disease. Understanding the mechanism by which blunt injury to the brain can cause neurofibrillary tangles and amyloid deposits may provide further insights into the process underlying Alzheimer's disease and other degenerative dementias."

Anatomical Diagnosis

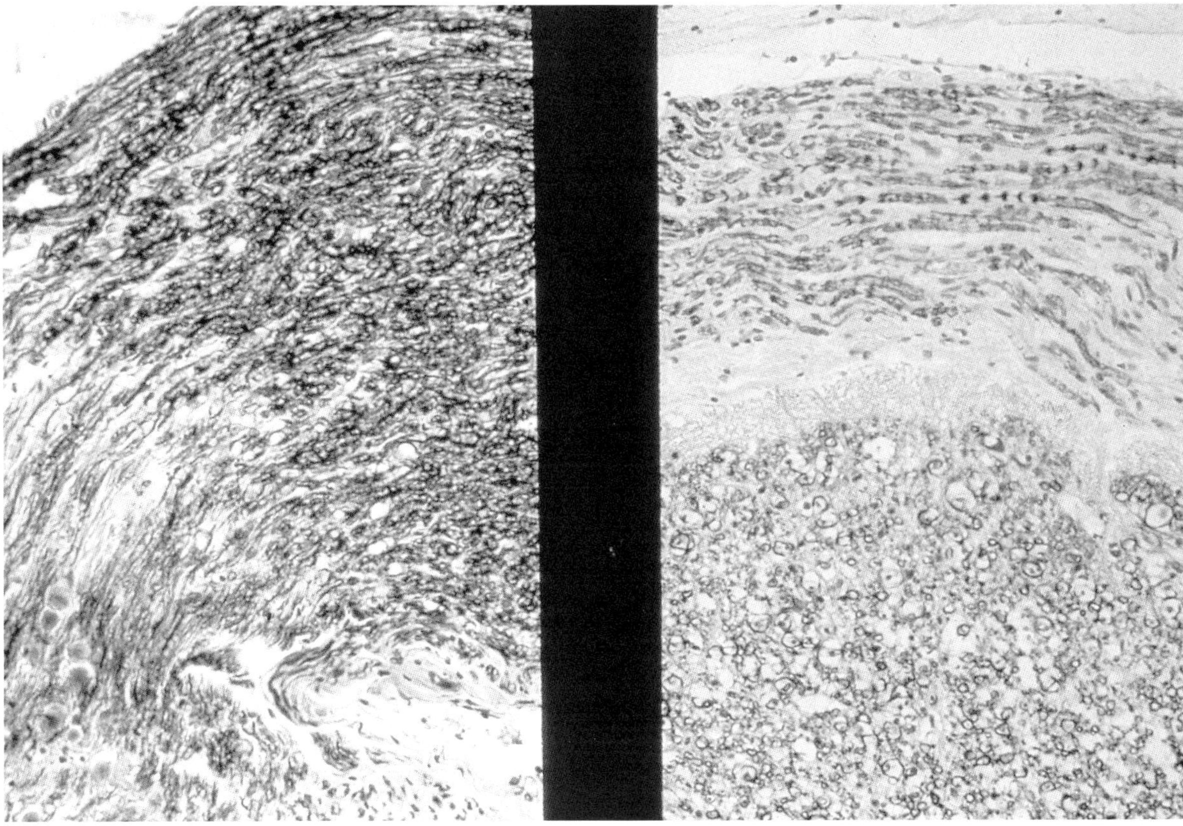

Fig. 315. Loss of myelinated fibers in the anterior spinal roots (right), with myelinated fibers preserved in the posterior roots (left). Luxol Fast Blue-Hematyoxylin and Eosin technique, 125:1. (From DRACHMAN DA, NEWELL KL, 1999).

Multisystem neurodegenerative disease (characterized by neurofibrillary changes and few plaques), findings consistent with dementia pugilistica.

The paper by DRACHMAN and NEWELL should be considered a classic. The authors provide a comprehensive history with detailed neurological findings, including all noninvasive modern diagnostic scans. The culmination is an exemplary differential diagnosis by DRACHMAN. The pathological discussion, presented by NEWELL, is thorough and demonstrates all pathomorphological findings with a series of outstanding histological slides. It would be fortunate if other papers were of similar quality and clarity.

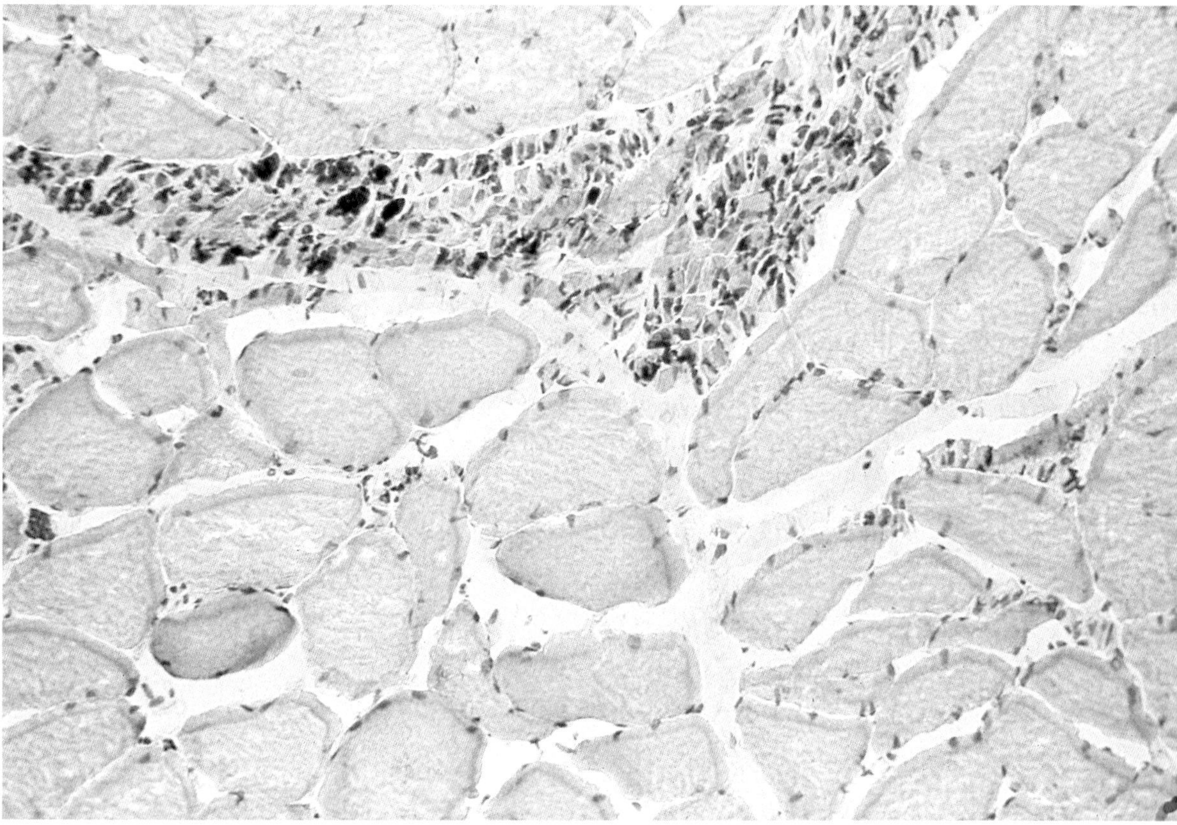

Fig. 316. Deltoid muscle containing a group of very small, denervated myofibers. Hematoxylin and Eosin technique, 320:1. (From DRACHMAN DA, NEWELL KL, 1999).

Cavum Septi Pellucidi in the population and in boxers

The size of a cavum septi pellucidi varies from a narrow slit to a wide cystic enlargement. The statements about its frequency varies considerably, since some authors take into account small slit-shaped cavities, while others do not.

The cavum can be more or less completely closed, with near fusion of both septal laminae or leaves. Alternatively, there can be a nearly complete fusion of the laminae with loose connective tissue in the median part and a slit shaped cavum, or a relatively large cavum septi pellucidi.

The cavum is usually in anterior parts, less often at the median or dorsal.

The enlarged cavum septi pellucidi and is not lined by ependyma. It is filled with a clear fluid.

Although VERGA claimed to have discovered the cavum septi pellucidi before FERRARIO with his description in 1851, it was in fact discussed in the literature much earlier. DUNCAN called it *'la petite cavité'* in 1678, after which time it was also called *'Duncan's cavum.'* It was also mentioned by Vicq d'AZYR, and Frédéric COUVIER termed the cavum *"ventriculus quintus."*

Ancient authors, such as Vicq d'AZYR, SANTORINI, and VERGA and, more recently, DANDY considered the openings to be artificial and not congenital.

Some other names for the cavity are: *VI. ventricle, fornix ventricle, ventricle of Strambio, ventriculus fornicis, ventriculus triangularis,* and *Tricone medium.*

Communication with the lateral ventricles or the third ventricle may be accomplished by one or several openings, which are termed *as Foramen Vieussenii* or *Foramen Mihalowski.*

In our experience, at least a small slit shaped cavity can be found in every autopsy, provided that it is conducted with care. A small cavum can easily be missed if the coronal sections are too big, since the cavities are only 2–3 mm wide and the two lamellae or leaves are in close contact with each other. This is in agreement with the findings of SHAW and ALVORD (1969) and HARVEY (1971) who found a small slip-shaped cavum between the septal leaves in practical 100% of the examined brains. COTTINI (1939) observed a slit-shaped cavum between the septal leaves in 95% of the brains examined.

Because there is no agreed standard to say when an opening is large enough to be considered a cavum septi pellucidi, there is some confusion in the literature on this point. DANDY (1931) found a cavum septi pellucidi in 5%, SCHWIDDE (1952) in 20%, von WAGENEN and AIRD (1934) in 66%, and HUGHES et al. (1955) in 85%. SCHUNK (1963), who examined 1.000 pneumencephalograms found a cavum septi pellucidi of 5 mm or more in only 6 cases (0.6%). SPILIANE (1965) reported that he found no cavum septi pellucidi in 100 consecutive cases of cerebral atrophy in pneumencephalograms. If there were clear standards for the minimum size of the cavity for a cavum septi pellucidi, the statistical frequency figures would no doubt be in better alignment.

The eminent anatomist Johannes LANG (1985) is of the opinion that the term cavum septi pellucidi should only be used when a communication of the cavity with the ventricular system exists. He has suggested that a secluded, fluid-filled cavity between the septal leaves could be termed a *'septal cyst'.*

Apart from the size of the cavity, other variations are also found. Posterior parts of the septal leaves or laminae are sometimes perforated or reveal fenestrations. In such a case only single strands with vessels, branches of the venae septi pellucidi exist as remnants. Johannes LANG (1972) found defects of the septum pellucidum in three of 250 brains examined. The openings provided a connection between both lateral ventricles. These defects are a known consequence of hydrocephalus. Of special importance to our discussion, they are frequently found in the examined brains of boxers.

Presence of Enlarged Cava Septi Pellucidi in Ex-boxers

Evaluation of the clinical and morphological literature about brain damage in ex-boxers reveals unequivocally that repeated and numerous blows against the head of *amateur* and *professional boxers* produces a dilatation of the normally tiny cavum septi pellucidi. The dorsal parts of the cavum septi pellucidi are especially dilated and enlarged. In other words, the distance between the two leaves or laminae of the two septa of this anatomical structure is considerably increased. Another significant finding is added to this: dilatation is frequently combined with a fenestration of the two laminae or leaves, especially in dorsal parts. The so-called *cavum pugilisticum* becomes a *cavum pugilisticum rupturum.*

By which mechanism can the dilatation and fenestration of this structure be explained? One can argue with some certainty, each boxing blow against the head causes the leaves of the cavum septi pellucidi to fluctuate sidewise, perhaps vibrate, or oscillate. The frequent, continuing disruption over a boxing career leads to an enlargement or dilatation of the cavum septi pellucidi. It would be fair to characterize the repeated action as a 'hammering' of the two lamellae. The septal walls of the cavum septi pellucidi rarefy during the process of dilatation, become thinner, and finally rupture. They form 'windows.' Eventually only small strands of tissue remain, until finally these too have disappeared, so that there is a direct communication between the two lateral ventricles. The etiology, with which we agree, was discussed by HARVEY (1977). His alternate interpretation was that the fluid accumulation in the cavum between the two septal leaves would be the primary process, after which the two laminae of the septum deviate from each other. However, since the fluid between the septal leaves acts like a hammer on the septa, we do not find these to be mutually exclusive explanations.

MAWDSLEY and FERGUSON (1963) noted that in six air-encephalograms of ex-boxers a cavum septi pellucidi was seen. The degree of separation of the two septal leaves and their radiological definition varied. In cases 1 and 8 the outlines of the leaves were so faint, that at first sight the septum appeared to be absent. In case 5 the leaves were well defined but thin, and in case 2, though two lamellae were clearly visible, they were interrupted along their length in the anteroposterior view. Case 6 showed relatively thick leaves which were not widely separated, and seemed to be intact. In the remaining five cases the leaves, as well as being separated, were probably perforated, since air passed freely from one lateral ventricle to the other and the 'comma sign' of DYKE and DAVIDOFF (1935) could be seen in lateral views.

ISHERWOOD et al. (1966) reported that in their present series of 16 cases, nine patients had evidence of a communicating cavum septi pellucidi, an incidence of 56 %. If the case histories described by SPILLANE (1962) are considered, the recorded incidence of abnormal communication between the lateral ventricles, with or without a cavum, in ex-boxers, is 13 out of 21 patients, i. e., 62 %.

The natural incidence of a communicating cavum septi pellucidi has never previously been determined accurately, though TIEFFREY et al. (1958) suggested 1 %. In the last 300 unselected encephalographies, carried out in the *Manchester Royal Infirmary* three such cases have been recorded, giving an incidence similar to that suggested by THIEFFREY et al.

Tomography demonstrated the walls of the cavum to be intact anteriorly, but to have multiple perforations behind the Foramina of Monro, increasing in number posteriorly. These observations have been confirmed by an autopsy study of the brain of an ex-boxer, where it was noted that the fenestrations increased posteriorly to a point where only a few residual strands of the lamellae remained. A free communication between the posterior ends of the lateral venricles was thus allowed.

ISHERWOOD et al. concluded, without doubt, that the changes observed in the septum pellucidum in ex-boxers are a significant feature, and traumatic causation has been conclusively proven. The changes may be produced by a sudden elevation in ventricular pressure, with thinning and finally rupture of the floor or walls of the potential cavum septi pellucidi. Later pressure rises might cause dilatation of the cavum and thinning and perforation of its wall.

In his series of thirteen qualifying observations, CORSELLIS et al. (1973) found a cavum septi pellucidi in twelve ex- boxers (92 %). This is a much higher incidence than the 28 % found in the corresponding age-matched controls of the non-boxers. In addition, the cava septi pellucidi averaged larger in the boxers than in the controls.

Ex-boxers have enlarged cava septi pellucidi and related anomalous tissue variations so much more frequently than the general population that the relationship is undeniable. Yet there are still a few lone voices insisting that the cavum septi pellucidi in boxers is a congenital anomaly unrelated to boxing. For example, BLONSTEIN (1974) wrote:

"The cavum septum pellucidum (sic) is not pathognomonic of the punch drunk syndrome. This also occurs in women who have never boxed."

I call special attention to this statement, it is the 'finest' differential diagnosis I ever read!

Unfortunately the statement above was not only made by a licensed physician, but one with authority, since BLONSTEIN was also the *leading official of the British Amateur Boxing Association.* The good health and welfare of thousands of young amateur boxers were entrusted to him, presuming his authority on medical issues related to boxing. The statement is true in part, in that a non boxing women can present with an altered cavum septi pellucidi. But because boxing is not the only cause of these alterations, offering such a statement as proof is inappropriate and misguided, unless perhaps BLONSTEIN has data suggesting that boxers are genetically predisposed to cavum septi pellucidi, which of course, would be the subject of another entire book, to be sure.

Punch drunk Syndrome

There is no doubt that the 'punch drunk syndrome' was seen and recognized by lay people, boxing managers, trainers, boxers, referees, many years before the medical profession recognized and used the term. The story is that the boxer, after a moderately successful earlier career in the ring, begins to slow up, he is knocked out more often, to win fewer contests, and to be seedy for progressively longer periods after each affray. That punch drunkenness was known to a large extend by the laity, can also be concluded from the fact, that punch drunk boxers were imitated by comedians.

The comedian Red SKELTON imitated a punch drunk boxer, 'Mr. Pug,' who lived in Queer Street, who heard bells and whistles and recognized sudden flights of birds over his head. The comedian imitated the short, tripping steps of a punch drunk boxer and let his hands and head tremble. The spectators laughed again and again, about this, his most popular comedy routine. A punch drunk boxer is a severe brain damaged person, a patient, one should not laugh about him!

CRITCHLEY (1942) noted:

"Perhaps the most characteristic feature of all is the fact that the boxer admits, to direct questioning, that he finally gave up the ring... 'because the wife disapproved.'"

The punch drunk syndrome includes impaired mental functions, slowed speech, ataxic gait and it simulates an alcoholic intoxicated person.

The pathologist Harrison MARTLAND (1928) — seventy years ago — in a classic study presented the first clinical description of the psychopathological and neurological features of the punch drunk syndrome. He reported that of 23 punch drunk boxers known to him, four were permanently in mental institutions. MARTIAND assumed that the condition afflicted, in mild or severe form, at least 50% of pugilists of a certain type who had kept at their profession over a certain period.

We quoted BARNES who wrote in 1958:

"Among the most tragic figures are the boxers who have finished their lives in mental institutions. The list stretches back to the early 1900's. One of the first prominent examples was Terry McGOVERN, a featherweigt champion. Over the years such great fighters as Joe GRIM, Ad WOLGAST and 'Battling' NELSON ended up in asylums. Then there was Frankie CARBONE, a former middleweight, who was picked up one night in front of the Roxy Theater in New York City after shouting, 'People are trying to take me out to the middle of the ocean, right into the waves!'"

"There was also Bob MARTIN, champion of the American Expeditionary Forces in World War I, who once fought Gene TUNNEY. MARTIN was taken to Walter REED Hospital in Washington suffering from what was described as 'boxer's waltz.' Physicians said his disability was the result of cerebral lesions directly due to the many blows he had taken on the head. He was classified as incurable."

"This unhappy list continues to grow. Johnny BRATTON, former N.B.C. welterweight champion, has been in several Illinois state hospitals since May 1956. Another sad recent case was that of Jimmy FLOOD a onetime 'pleasing' middleweight contender, as he was often introduced by the ring announcer. He was sent to Bellevue Hospital for mental observation after being arrested on a wife-beating charge. His wife, Dorothy, told the magistrate that her husband had suffered brain damage as a result of his ring career."

MILLSPAUGH (1937) Introduced the term *'dementia pugulistica.'* This term should be used only for those ex-boxers who reveal a dementia, however, many ex-boxers show neurological findings, with no or only very mild psychopathological deficits. The diagnosis *'chronic progressive traumatic encephalopathy of boxers,'* introduced by CRITCHLEY in *1957,* is here acceptable. Here should be objected that evidently not all cases seem to be 'progressive.' It was SYMONDS (1960) who noted that in some cases the condition may be slight and nonprogressive. BOWMAN and BLAU (1960) asserted that the syndrome is only progressive for a few years and that the condition then becomes stationary. The evaluation of the literature reveals that the syndrome is progressive in the majority of cases but there are other cases were it remains stationary over many years.

The evaluation of the literature reveals that the neurological features may assume predominance over the psychopathological features and vice versa, as had already expressed by JOHNSON (1969). Both features may progress independently of each other which would indicate that different areas in different parts of the brain are affected.

There exists an enormous polemic which surrounds the relationship of boxing to the punch drunk syndrome. Defenders of boxing are often arguing that the brain damage in dementia pugilistica is not caused by boxing but is the result of alcoholism, venereal diseases and other degenerative processes. Such an opinion cannot be maintained today. The presented case histories show convincingly that a solid differential diagnosis in each case was performed. Venereal diseases were excluded using serologicaol tests. Alcoholism alone does not produce clinical features seen in the punch drunk syndrome. If alcoholism existed in periods of a boxer's life it is in each case discussed by the authors who published the case histories.

In boxing circles are many who deny that there exists such a syndrome. Mr. Jack SOLOMONS, *British boxing promoter,* has allegedly never seen such a case.

Mr. SOLOMONS was for some decades the leading British boxing promoter. He must have met with many punch drunk ex-boxers and must have heard from boxing managers about such cases, many of whom must have been in asylums. This statement of Mr. SOLOMONS must be regarded as unbelievable and incredibile.

There are ring physicians who deny the existence of the punch drunk syndrome in boxers.

KAPLAN and BROWDER (1954) published their findings of a four-year study of 1.043 professional boxers in New York. Ringside observations – one of the writers watched the bouts at ringside once per week for three years. Their conclusions are that clinical observations "failed to reveal any abnormal neurological features, even in those contenders who lost their bouts by a knock out." Slow-motion cinematography "confirmed one major impression gained at the ringside, namely, that most blows to the head were short of their mark or deflected by their opponent." Electroencephalography "primarily serves the purpose in professional fighters in the detection of grossly disorganized electroencephalograms." The authors stated that all effects of cerebral concussion are reversibel. *The authors allegedly did not find any cases of the punch-drunk syndrome.*

The authors write that all of cerebral concussion are reversible, there were no cases of punch drunk syndrome? Such statements are incomprehensible and unconceivable. They studied 1,043 professional boxers in a period of four years in the early fifties. One of the writers watched the bouts at ringside once per week for three years. He must have watched many of the professional fights in the Madison Square Garden. These authors must have have contact with many ex-pugilists and boxing managers and they must have recognised many punch drunk ex-boxers and have heard of them from managers.

GILMER (1957) wrote:

"These people, who usually have done no boxing themselves, attract attention by using the expression 'punch drunk,' which was an imaginative phrase coined by a New York sports writer 30-odd years ago. They picture the boxer as a fellow with glassy eyes, stuttering speech, and appearing just a little drunk. After 33 years of association with boxers ranging from nine of age to upward of forty, the only one I have seen who answers this description is the fictitious character portrayed by Red SKELTON on TV. Mr. STEINHAUS says he has talked to many of such people. Its strange that although I have been both an amateur and a professional boxer, heve taught, officiated, and finally supervised the sport of boxing, I have not met any who answers the 'punch drunk' description.

GILMER made this statement in the year 1957. He was associated with boxers, as he mentions, for 33 years, with other words from 1924 on, in a period in which the 'punch drunk syndrome' was seen frequently. MARTLAND's published his classic paper in the year 1928, he as a pathologist, was not directly connected with the boxing fraternity, reported that 23 punch drunk boxers were known to him.

McCOWN (1959), *Medical Director of the New York State Athletic Commission* wrote that between 1952 and 1957 among 9.871 professional boxers there were 259 knock outs and only nine serious head injuries requiring admission to a hospital. 138 boxers retired or were denied a license – McCOWN does not say why. McCOWN wrote in regard to punch drunkenness:

"It has never been proved to be a neurological syndrome peculiar to boxers and produced by boxing... It is a slick medical cliché."

McCOWN's statement that the punch drunkness syndrome has never been proved to be a neurological syndrome peculiar to boxers and produced by boxing... that it is a slick medical cliché. First a man in such an official position in New York must have seen numerous punch drunk ex-boxers, and secondly he must have found numerous cases published in the medical literature.

SPILLANE (1962) commented on those statements:

"But surely inquiries such as these do not tell us what happens to the brains of professional boxers in later years. One might just as well say that regular medical and radiological examination of young cigarette smokers for four or five years, failing to reveal any examples of pulmonary carcinoma, means that cigarette smoking does not cause that disease. The problem is the delay in the appearance of the symptoms.

This is an important observation. But the previous quoted persons were all closely connected with boxing, they must have seen dozens or hundredth of ex-boxers showing all stages of the punch drunk syndrome. There remains only one conclusion, namely, that their statements were deliberately false.

PARKER (1934) wrote:

"The result, immediate or remote, of repeated injuries to the brain of a professional boxer forms a study all in itself, and contrasts with the more usual sequelae seen when patients have received only one injury in the course of industrial or other pursuits."

The eminent British neurologist Mcdonald CRITCHLEY noted already 1942:

"The punch drank condition is usually a progessive one, with both psychological and neurological features."

CRITCHLEY noted that punch drunkenness is found more often in professionals than in amateurs, in the second- or third rate performer rather than in the scientific exponent with a county or provincial title. As fighters, they are rarely quick on their feet, but are better known as able to 'take it.' CRITCHLEY noted that there is a *direct ratio between the degree of facial scarring (cauliflower ears, broken nose, hypertrophied superciliary ridges) and the intensity of the punch drunk symptoms.* The sum-total is important. We agree fully with CR1TCHLEY that being 'knocked out on their feet' is more significant than being floored.

Of special etiological importance is that humbler branch of boxing whereby the contestant travels around in fairs in *'boxing booth'* or *('blood-tubs')* taking on all comers at any weight at half a crown a time.

Dailan THOMAS portrayed such a boxer:

"Outside his booth stood a bitten-eared and barndoor chested pug with a nose like a twisted swede and hair that started from his eyebrows and three teeth yellow as a camel's inviting any sportsman to a sudden and sickening basting in the sandy ring or a quid if he lasted a round, and, wiry, cocky, bow-legged, coal-seared, boozed, sportsmen by the dozen strutted in and reeled out; and still those three teeth remained chipped and camel-yellow in the bored, teak face."

Almost as characteristic is the boxer who retires from the ring having failed 'to make the grade,' and later becomes a *'punch-bag,'* that is a term of sparring partner to a first class heavyweight.

In the boxing world other synonyms have been applied – more picturesque than scientific -and etymologists rather than neurologists might be intrigued with such expressions as *'slap happy,' 'goofy,' 'slug nutty,' 'cutting paper dolls,' 'cuckoo.'* German pugilists used the term, 'weiche Birne,' ('soft pear'), 'er ist matchig gehauen'; ('he is punched slushy'). In Italian it is called *'suonati,' 'he hears the bells ringing.'*

CRITCHLEY reported"

"Up to date, my own neurological experience has comprised a series of over 100 cases of frank neurological disease in boxers. Many of these – perhaps even the majority – should be looked upon as examples of punch drunkenness, either early or well established."

Why are Woodpeckers not Punch Drunk?

At different times I was asked in discussions after I had spoken about the severe permanent brain damage in ex-boxers due to the repeated blows against their heads, why woodpecker, whose lifelong activity consist to cleave cavities with their beaks into trees, are not 'punch drunk,' too? To begin with, I was not able to answer this earnest question. In preserved parks in the States of Texas and Louisiana and in my garden I was able to watch woodpecker in their natural habitat, they impressed me by all means not as 'punch drunk' or brain damaged. Subsequently some aspects of the biomechanics of cleaving or hacking with threir beaks, as well as anatomical particularities of the head and brain of the woodpecker will be discussed.

MAY et al. (1975) examined the trajectories of hacking of woodpeckers with the aid of highspeed photography. These authors could show: (1) that the trajectories of woodpeckers are essentially linear, with only a slight rotation of the head, (2) that after impact at the target only insignificant movements exist, (3) that the impact velocity is in the order of 600 — 700 cm/s, and (4) that the deceleration at impact of the beak is in the order of an average of 982 g. It must be pointed out that the eyelid of the woodpecker immediately before the impact of the beak closes and thereafter opens again and remains open at the backward trajectory. SPRING (1965) and MAYER et al. (1979) put forward the theory that the closing of the eyelid immediately before the impact at the target is a kind of 'security guarantee,' namely to hinder the eye bulbs to leap out at the impact. The authors asked whether a natural selection favored the survival of wookpeckers with linear trajectories.

About the anatomy of the woodpecker: It has very strong neck muscles, a very narrow cerebrospinal space with relatively little cerebrospinal fluid, the brain weight is between 1.25 and 3.95 g and is due to its little size and weight about 100 times less sensible to impacts than that of a human. Between skull and beak is an area of cartilage, which works as a shock absorber. Interesting is the fact that the woodpecker before he begins his frequency of hacking, first he taps slightly one or two times the target as to 'take measurement.'

These findings are not only of a pure academic interest but its analysis can provide essential findings for construction of protective helmets.

The conclusion of MAY et al. (1979), however, that, because woodpeckers have a linear trajectory and are not brain damaged, therefore a rotational acceleration must be more dangerous than a linear acceleration, is not justified. In another chapter I have shown that the injury patterns due to linear and angular accelerations are different (UNTERHARNSCHEIDT 1963, 1993).

Examples of Brain Damaged Boxers

Ad Wolgast

A well known and often quoted example for a punch-drunk ex profi-boxer is Ad WOLGAST *(Literary Digest, 123, 1937, pp. 39–40):*

"Most famous of the world champions who didn't retire in time is Ad WOLGAST, who in 1910 won the lightweight title from Bat NELSON, and lost it in 1912 to Willie RITCHIE. WOLGAST is now in a asylum in California 'getting ready for his fight with Bat NELSON.' The 'fight' is always scheduled for 'to-morrow.'"

A classic description of the punch drunk Ad WOLGAST was presented by McGEEHAN *(News of a champion, New York Herald Tribune).*

"Ad WOLGAST, once lightweight champion of the world, is in the news again briefly. Friends who have been caring for him in California have made application to have him committed to an asylum as hopelessly insane."

"For more than ten years WOLGAST has been living in a phantom world populated by old prize-ring ghosts. His mind failed him shortly after he lost his championship, and he labored under the hallucination that he was to meet Joe GANS (long since dead) for the lightweight championship. Jack DOYLE, a Los Angeles promoter, with more heart than most of the men who make their money out of the manly art of modified murder, took pity on him and assumed full charge of him."

"Through DOYLE's generosity Ad WOLGAST was provided with a little gymnasium, where he did his training. He seldom used to miss a morning on the road. It was his hallucination that he was to meet Joe GANS, dead even before WOLGAST's mind went into the fog, in a championship bout. For years he settled down into this routine with only the idea that he must be in condition for the championship bout."

"Sometimes they would take him to boxing matches. He always used to say, 'I could whip either of them.' But he never insisted on meeting any of the fighters he saw. He was concentrating on the phantom bout with GANS. Physically WOLGAST seemed to be all that he ever had been, but his mind was gone forever."

"I saw WOLGAST win the lightweight championship from Battling NELSON in a ring pitched in the adobe mud near Port Richmond, California. For concentrated viciousness, prolonged past fourty rounds, that was the most savage bout I ever have seen. Both men were badly enough battered, for in giving NELSON a beating, WOLGAST was forced to take almost as much as he gave. It was inevitable that the effects would tell."

"Somewhere around the thirtieth round I think it was, WOLGAST dropped by a body blow and it looked like the end. But he was up in an instant, snarling and lashing at NELSON. After that it was NELSON, the Durable Dane, who showed signs of weakening and whose face began to look like a raw slab of steak. The features were obliterated and only the slit of one eye remained open."

"In the forty-second round NELSON was pressing feebly forward while WOLGAST's gloves were hurling crimson splashes around the ring every time they struck the battered face. The Dane would not yield an inch, but it had become so cruel that the most hardened ringsiders were calling upon the referee to stop it."

"Finally Eddie SMITH stepped between the men and pushed NELSON to his corner. The Dane snarled at him, then tried to protest through the twisted and battered mouth. The only sound that came was one such as might have been made by an exhausted and terribly wounded wild animal. NELSON's seconds caught him and pushed him onto the stool in the corner. The referee raised

WOLGAST's hand. He had become the lightweight champion of the world."

"Not so long after I saw him lose his title to Willie RITCHIE of California, then a young graduate from the amateur ranks. It started out as though WOLGAST would rush RITCHIE off his feet. The 'Cadillac Bearcat' was beating the Californian from the start, when RITCHIE landed a wild swing that caught WOLGAST flush on the jaw. WOLGAST dropped to his knees, all but out. As he was about to collapse he drove two foul punches upward at the Californian. He had the rattlesnake's instinct to strike, even when mortally hurt."

"It was WOLGAST's fate to have won the championship in one of the hardest fights ever staged. Only to have lost it through a chance blow – if there is such a thing as a chance blow."

"Said one of WOLGAST's former managers, 'He was one of the greatest fighters I ever knew, and it was a pleasure to manage him. He never cared whom he fought or how often. He would fight anybody I signed him up with."

"'Once I tried him out for fun. I knew that he was in his hotel room so I rang him up and told him I had signed him as a substitute boxer for some bouts that night. The schedule called for him to appear in three hours, 'All right,' he said, 'Wake me up in a few hours, and I'll get ready to go into the ring.'"

"But you haven't asked who it is that you are going to fight,' I said, 'No,' said WOLGAST, 'and I don't care. Just wake me up so that I won't be late getting into the ring.' He was that way all the time. He would say, 'Get me anybody for any time, and I'll fight him any time you say."

"You do not meet that kind of fighter these days, the fellow you can have hop into a ring on a few hours' notice. The boys are very careful, particularly the topnotchers, as to the kind of matches you make for them. They never want to take a chance with any of the rough ones. Can you picture any of the later lightweights being careless about whom they are signed up with?"

"From that point of view of the manager there is no doubt that WOLGAST was the ideal prize fighter. He always was willing to step into the ring at the command of the mastermind, to take a beating, and to listen to the voice of the manager shouting, 'Go on in. He can't hurt us.'"

"Nearly all of WOLGAST's former managers have their health and are in no danger of the almhouse or the insane asylum. Oh yes. There is no doubt that Ad WOLGAST will be taken to a state insane asylum shortly. The blows that the managers did not feel seem to have had effect upon WOLGAST. Perhaps the fighter was just a trifle more sensitive than the manager."

Muhammad ALI

There has, unfortunately, been some evidence of a very famous case recently. In *Sports Illustrated, April 11, 1983)* BOYLE and AMES spoke of a recent visit to England by MUHAMMAD ALI, ex-heavyweight champion and renowned showman.

"Two weeks later (1981) ALI visited England. Interviewed on BBC radio, he slurred his speech, and when he recited a poem on how he would beat HOLMES in a rematch, listeners found most of it incomprehensible. The *BBC* canceled the broadcast of a taped interview for another program because ALI's speech was too slurred to be understood. A *BBC spokesman* said, 'It was very sad that so much of what history's most celebrated fighter said was unintelligible.' When a London reporter asked him if he could possible be punch-drunk, ALI replied, 'I have heard about people being punch drunk, but I do not feel drunk.' He added, 'When you get as great as me, people always look for some type of downfall.'"

A little later in the article was an inset entitled *'A Conversation With the* Greatest, 'which is quoted in toto:

"'I'd been trying to reach him for weeks, but I was still surprised – unprepared, in fact – when I picked up the ringing phone one night and there was MUHAMMAD ALI. He was in fine fettle, charged up, and what ensued was not an interview so much as a vintage ALI monologue.'"

"'Why do you want to check my brain? How

about white boxers – Jerry QUARRY, George CHUVALO, are you going to show their brains? No, it's me, a leader of the black people, they want to say I have a brain injury, that I'm crazy. I won't be no guinea pig, it's an embarrassment, a humiliation. What about my reputation, my family, my children – they're worth more to me than anything. For me to take that gamble, even to show nothing's wrong, I'd have to have $ 10 million...'"

"'Why don't you do a test on Richard NIXON's brain? Get the top whites and let them submit to it. I'll tell you what. You get the wisest of the whites, you get William F. BUCKLEY and some others, the brainiest. I'll pay'em $ 25,000 to go on TV with me and let'em try and trip me up, ask me any questions they like, any subject they want. Then we'll do the brain show afterwards, for everyone, on TV, I want an hour and a half...'"

"'Do you think, talking to you, that I sound crazy? Do I? Do you have this on tape? You don't? Well there's somethimg wrong with your brain... Why are they picking on boxing? It's because the black men are so superior in boxing that they want to stop it. Well, the black man's so hungry, he's got to fight. Yes, I think it't racial, the whole thing..."

"ALI talked for half an hour. I got in a few sheepish words edgewise. But as he went on, he began to wind down and repeat himself. He was not interested in my questions. It has always been his pattern, even when he was in his early twenties, to grow tired of declaiming. Now his fatigue was obvious. He began to mumble and speak more and more softly. The angry tape in his head, what he wanted to say, was running out, or going onto a loop. At the end he was talking as if in a trance of self-hypnosis. He didn't hang up he just dropped the phone."

As if they were not a strong enough indication, Howard COSELL recently remarked, "The other day, Ken NORTON walked into my office. My God, he's as thick tongued now as ALI..."

"There seems to be a famous case of boxer's encephalopathy presently in our midst. It is more tragic because of ALI's personality and showmanship which so pleased those who followed his lead. it is difficult to maintain charisma when people cannot understand one's speech."

One of the great problems with detecting dementia pugilistica is that it is so difficult to do before the typical symptoms appear.

The sports journalist Dave ANDERSON *(For Ali, what price the Thrilla in Manila? The New York Times, Sunday, September 23, 1984, p. 20)* presented a comprehensive and wellinformed description of the boxing career of MUHAMMAD ALI, who had began his career with the name Cassius CLAY. This presentation analyses three phases in his career as a boxer and tries to discover, in which fist fights MUHAMMAD ALI had to take the hard punching boxing blows. which lead to his encephalopathia pugilistica.

ANDERSON wrote:

"When that epic of brutality ended in the Phillippine Coliseum nearly nine years ago, Joe FRAZIER had lost. His manager, Eddie FUTCH, had surrendered for him after the 14th round.

MUHAMMAD ALI had retained his world heavyweight title. About 15 minutes later, Joe FRAZIER, his left eye closed below the towel over his battered head, trudged into the interview room to answer a few quick questions. But even after Joe FRAZIER departed, MUHAMMAD ALI did not appear."

"'The champ's too tired," one of his aides explained. 'He's just layin' on a couch in his dressing room.'"

"More than half an hour later, MUHAMMAD ALI finally arrived, worn and weary. For once, he had little to say, even as the victor in what will be remembered as perhaps his noblest hour as a gladiator."

"What you saw,' he said slowly, 'was next to death."

"During his fourteen fights in the six years that followed, MUHAMMAD ALI was never the same after that Thrilla in Manila, as he had titled it. And in retrospect now, with ALI on medication for what one of his doctors calls 'Parkinson's syndrome, very possibly' as the result of blows to the head, the question is obvious – for ALI, what was the eventual price of the Thrilla in Manila?"

"No answer is possible, but to someone who was with him during those years, the memories linger of ALI in the hours after that Manila fight and also of him in his later fights."

"Over his 21-year career, MUHAMMAD ALI had a 61–5 record, with 37 knock outs. But that career had three phases – one, from when he turned pro in 1960 to when his three-and-a halfyear exile began in 1967 for refusing to be inducted into the Army; two, from his comeback in 1970 to when he regained the title in Zaire; three, from 1975 until his eventual surrender to himself in 1981."

"In examining his total of 552 rounds, not counting exhibitions and training sessions, ALI took most punishment when his body could least afford it – in phase three."

"Over twenty bouts in *phase one,* ALI fought only 110 rounds – an average of 5.5 rounds. In his 27 bouts in *phase two,* he had 266 rounds, a 9.9 average. But during his final fourteen bouts in *phase three,* he went 176 rounds for a 12.6 average."

"Of all of ALI's talents, one of his best was his ability to take a punch, an ability that may have betrayed him. Beginning in Manila, the punches multiplied quickly."

"To appreciate the extent of ALI's weariness in Manila, it is necessary to know that he usually recovered quickly after a hard fight. The year before Manila, for example., ALI had dethroned George FOREMAN in Zaire with an eight-round knock out after his body had absorbed dozens of hard punches with his 'rope-a-dope' style. Less than five hours later, ALI was twirling an ivory-inlaid mahogany cane as he clowned outside his villa along the Congo River."

"For weeks I kept hollerin' 'Be ready to dance,' he said, his eyes wide. 'But I didn't dance. That was the surprise, that was the trick."

"About eight hours after the Manila fight, in contrast, ALI appeared at a reception arranged by President Ferdinand MARCOS, but he resembled someone who had been mugged. The skin on his face was drawn tightly, as if it were a mask. His narrowed eyes appeared to be underlined in purple crayon. He had lumps on his forehead. His nose was scraped pink. He moved stiffly, almost in a limp."

"When he shook hands with a softly folded right fist, he winced. When he sat, he was hunched in soreness."

"Death is so near,' he wrote in a visitor book for Mrs. MARCOS, 'and time for friendly action is so limited. Love and peace always.'"

"Perhaps an hour later, ALI was more like himself, joking and laughing. But not when he talked about the fight."

"This'll kill you,' he said. 'This is next to death. I'm a super human. So when I'm that tired, it's dangerous.'"

"He hinted that 'you may have seen that last of ALI' and if he had it to do over now, he might agree that he should have retired from boxing then. But early in 1975 he resumed his career against Pierre COOPMAN, a harmless Belgian who disintegrated with the first good punch. ALI didn't even bother to take a shower after that appearance but two months later, he struggled through a disputed 15-round decision over Jimmy YOUNG."

ALI's trainer. Angelo DUNDEE, acknowledged, 'This was my guy's worst fight.' His boxing physician, Dr. Ferdie PACHECO, who literally had been in his corner for years, was more explicit."

"ALI was missing, he had no timing, although the other guy was making him miss,' Dr. PACHECO said. 'He was getting tired a lot sooner than usual. His reflexes were only 25 to 30 percent of what they should be.'"

"After a fifth-round knock out of Richard DUNN in Munich, Germany, another gift payday, ALI struggled by Ken NORTON in another disputed 15-round decision. But in May of 1977, he was in the ring again, ths time with Alfredo EVANGELISTA."

"'ALI is now at a dangerous mental point where his heart and his mind are no longer in it,' Dr. Ferdie PACHECO said even before that fight. 'In his last fight with NORTON, that's the night I thought

ALI stepped onto the last downhill phase of his career.'"

"Physically, he does not look that much different than three or four years ago. His reflexes you can't really measure except in a laboratory. But at his age, repetitive punishment usually has taken its toll on a boxer. His speech is slurred. He walks with a shuffle. But not him."

"Even so, Dr. PACHECO believed that MUHAMMAD ALI should box no longer. For that honest medical opinion he was banned from the entourage by Herbert MUHAMMAD, the champion's manager."

"ALI kept fighting, surviving a 15-round pounding from Earnie SHAVERS but then, slow and soft, he lost his title to Leon SPINKS on February 15, 1975. Seven months later, he regained the crown that, perhaps more than anything, was too difficult for Leon SPINKS to wear. By now ALI's words occasionally were a little slurred. Not much, but to those who remembered the speeches of his youth, it was there."

"Perhaps he heard it, too. On June 26, 1979, he announced he would box no more."

"'My toughest fight?' he said a few days later. 'The toughest was Manila, feeling so dead, so tired, a hard brutal fight, taking a lot of punishment from Joe FRAZIER, giving a lot, an endurance test.'"

"But 15 months later, ALI returned, only to be embarrassed by Larry HOLMES in a charade that was stopped after ten rounds. The next morning, his eyes shielded by rose-colored glasses, he tried to pretent he wasn't 38 years old."

More that a year later, ALI would fight for the last time, losing sadly to Trevor BERBICK in another 10-round bout that finally convinced him of his age. Asked the next day what it was like to know that his reflexes may have gone, ALI looked up quickly"

"'They have gone, not my have gone,' he said. 'They have gone. I could tell I was 40.'"

"For more that two years now, ALI has been talking like a man who was 80, his words clinging together like cobwebs of dust. After saying for too long that 'there's nothin' wrong with me, I'm still the greatest.' he has agreed to medical supervision, prompting the inevitable question – should boxing be banned?"

"Ideally, yes. Practically, no."

"Even if boxing were banned in the United States and other large nations, its popularity would maintain it in some other countries. Boxers would travel wherever they were accepted, just as ALI did. When he couldn't get a license to fight Trevor BERBICK except in South Carolina, he went to the Bahamas to learn he was 40.

"Boxing needs to be governed better, much better. But boxing is like sin. It's to popular to be abolished."

In the second year after MUHAMMAD ALI had retired from the ring – he was 38 years old – his body weight increased to 225 pounds, he 'was out of shape.' On March 5 1980, Bob ARUM announced that MUHAMMAD ALI would fight John TATE, the *heavyweight champion of the World Boxing Association.* Since TATE had been knocked out by Mike WEAVER on March 31, 1980, Larry HOLMES, who had been a sparring partner from 1973 to 1975, was taken into consideration, and formally a contract was signed, that MUHAMMAD ALI should receive a purse of $ 8 millions and HOLMES of $ 2.3 millions.

On July 23, 1980, MUHAMMAD ALI was admitted at the *Mayo Clinic* for a two day, especially nephrologic and neurologic examination. MUHAMMAD ALI *needed a physical examination to get a licence for the planned boxing match;* the medical findings were handed over to the *Nevada State Athletic Commission.* Two reports were delivered to the *Commission,* one from Dr. John MITCHELL. of the Department of Nephrology and Internal Medicine and the other one from Dr. Frank HOWARD of the Department of Neurology.

Dr. MITCHELL wrote:

"Peliminary examinations would indicate that the patient is in excellent general medical health with no evidence of renal impairment or chronic or acute medical illness" (quoted after HAUSER 1991).

Dr. HOWARD's report (quoted after HAUSER 1991) reads:

"I saw MUHAMMAD ALI for a neurological evaluation on 7–23–80. He was concerned about some comments that he might have brain damage, because following a filmed TV interview in London, it was commented that he staggered and that his speech was slurred. The patient attributes this to excessive fatigue from an all-night flight to London, and claims that he has always had some mild slurring of his speech for the past ten to twelve years. He stated that he was tired on the day of the examination here and that he had gotten little sleep. He denied any problems in coordination as far as jogging, sparring, or skipping rope. He also says that his memory is excellent and that he can deliver five forty-five minute lectures without notes."

He has never been knocked out, even though he had his jaw broken in a fight. He has been dazed on two occasions – once in a 1964 fight with Sonny LISTON where his mentation was fuzzy for a considerable length of time, and in 1972 in another fight he was briefly dazed. He has never been in any car accidents where he sustained an injury, he has not had any infections with any high fever, and he takes no medication. Other than occasional tingling of the hands in the morning when he awakens which clears promptly with movement of the hands, he denied any other neurologic symptoms."

On *neurologic examination,* he seems to have a mild ataxic dysarthria. The remainder of his examination is normal except that he does not quite hop with the agility that one might anticipate and on finger to nose testing there is a slight degree of missing the target. Both of these tests could be significantly influenced by fatigue."

A CT scan of the head was performed and showed only a *congenital variation in the form of a small cavum septum pellucidum* (cavum septi pellucidi, author). The remainder of the examination was normal, and *the above mentioned structure is a congenital abnormality and not related in any way to any head trauma.* "

"On extensive psychometric testing, he showed a minimal decrease in memory that was more pronounced when he was fatigued, but all other intellectual functions appeared to be intact."

In summary, there is no specific finding that would prohibit him from engaging in further prize fights. There is minimal evidence of some difficulty with his speech and memory and perhaps to a very slight degree with his coordination. All of these are more noticeable when he is fatigued."

At this time already existed a cavum septi pellucidi, which was considered as a congenital abnormality, despite at this time there were reports in the medical literature, which described and considered such a finding as for a boxer typical permanent brain damage.

MUHAMMAD ALI was concerned about some commentaries, that he suffered a permanent brain damage following a filmed television interview in London, it was commented that he staggered and that his speech was slurred. He said that slurred speech had been existing since about ten to twelve years. At that time had been in the medical literature already called special attention to the fact, whether a knocked out defeat had existed or not, was not significant, since the repeated, numerous blows against the head in a boxer were responsible for the severe permanent brain damage. It existed, too, *a mild ataxic dysarthria* and a *decrease of the memory functions.* The finger-nose test was slightly disturbed.

The summary reads:

"In summary, there is no specific finding that would prohibit him from engaging in further prize fights."

The *neuroradiological findings* in the report of the *Mayo Clinic* reveal a beginning for a boxer typical permanent brain damage; this report contained enough pathological findings to decline a license for participation in a further boxing match. Interestingly, the *Nevada State Athletic Commission* refused to provide copies of this report to sports journalists.

MUHAMMAD ALI was at this time already a growing old and an already brain damaged boxer, who had to fight against a good boxer.

The boxing match took place on October 2, 1980 in an arena at the parking space at *Caesar's Palace* in Las Vegas. 24.740 spectators paid $ 5.766.125.-

HAUSER (1991) wrote:

"It wasn't a fight: it was an execution." Ritchie GIADEET! (quoted after HAUSER 1991) said: "I thought they should have stopped it in the sixth round. After that, there was no point in going on."

In the ninth round HOLMES landed an uppercut which hurled MUHAMMAD ALI into the ring ropes; he held his gloves in front of his face, as HOLMES landed a blow in his kidney region. Sylvester STALLONE (quoted after HAUSER 1991) said later: "Like watching an autopsy on a man who's still alive." After the tenth round Angelo DUNDEE signaled the referee Richard GREENE: "That's all."

Four days after this boxing match MUHAMMAD ALI was admitted at the *UCLA Medical Center.*

On December 11, 1981 the last boxing match of MUHAMMAD ALI took place, his opponent was Trevor BERBICK. BERBICK was a not very experienced boxer, but very aggressive, who was able to land his body blows again and again at MUHAMMAD ALI. All three judges declared BERBICK an unanimous victor.

MUHAMMAD ALI's state of health deteriorated. September 1984 he was admitted and examined at *Columbia-Presbytarian Medical center in New York.* He told reporters, that he had trembling in his hands, that his speech were slurred and uneven, and that he was almost always exhausted. He was discharged after an eight day clinic stay. A short comment of his attending physician Dr. Stanley FAHN reads:

"To whom it may concern:"

"With regard to my patient, MUHAMMAD ALI, no definite diagnosis can be given at this time. However, the following has been determined:"

"(1) MUHAMMAD ALI does not have Parkinson's disease."

"(2) ALI exhibits some mild symptoms of Parkinson's syndrome."

"(3) ALI is responding well to medication. It is anticipated that his symptoms can be reversed with medication, and that he can lead a normal life."

(4) There is strong reason to believe that ALI does not have a progressive degenerative condition. It is anticipated that with medication his condition will stabilize with ALI doing very well."

"(5) ALI does not suffer from dementia pugilistica, commonly referred to as 'punch-drunkenness' ALI's mind is impressively alert and well-oriented."

"(6) ALI's life is not treatened by his condition. His life expectancy is in no way diminished by this condition."

"(7) ALI's condition is not contagious."

"(8) ALI should be fully able to effectively carry out personal appearances and other business activities indefinitely."

A comprehensive point of view of Peter ALFANO, in which some managers, boxers and sports journalists had their say, follows:

"Teddy BRENNER, the boxing promotor, who arranged MUHAMMAD ALI's first professional boxing match in *Madison Square Garden*, on February 10, 1962 against Sonny BANKS, said: 'I am that one, who told him 1977 after the boxing match against Earnie SHAVERS, that he was hit too often and too hard. I had the impression that he had achieved enough. He had been world champion three times, he had earned millions of dollars and that he had earned a niche in the history of boxing, that I told him. From here on, every young, who is not able to carry your water bucket, to beat you. He was an important heavyweight, great in each era, and one of the friendliest and charming athletes, whom I ever had known."

"It is sad," said Howard COSELL, 'to see the problems, who ALI has now. I have never dreamed, that it would end like that. The physicians do not know what the matter is with him, but it must be connected with boxing. He took too many blows! He had an unusual ability to take punishment."

"ALI sometimes took delight, to take blows.

MURAD MUHAMMAD and BRENNER spoke about it, how the former world champion allowed his sparring partners punish him at discretion and tried to knock him down. It was in training and in the ring, that ALI took too many blows."

BRENNER continued: "He was his worst enemy. If he did not wanted it, to take blows, there was no chance to hit him."

"ALI's first defeat came after 31 victorious boxing matches on March 8, 1971, when he in an extremely hard 15 round boxing match in the *Madison Square Garden* was defeated by the at this time world champion Joe FRAZIER by points."

"Two just as extremely hard boxing matches against Ken NORTON, Larry HOLMES and finally Trevor BERBICK in his last fight on December 11, 1981. 'You know,' said Murad MUHAMMAD, 'until he finally was beaten, I thought ALI were not a human. Many thought, that nobody would hit him."

"In his last time, remarked ALFANO, 'ALI's abilities allowed him, to move in the ring without being hit. He had so fast hitting hands, despite PATTERSON remarks, that his were faster. Of his 56 victories in 61 professional boxing matches, 37 were achieved by a knock out."

"He changed boxing,' said Angelo DUNDEE, Ali's trainer, 'well sort of trainer!' He told me once, you do not train me. He ruined a lot of boxers, who wanted to be as him.":

COSELL continued:

"Perhaps this is an occasion, that congress will be induced to act

Abigail TRAFFORD *(Is the gold worth the price? The Washington Post, August 6, 1996)* wrote:

"After the cheering stops and the crowds go home, I am left with two enduring images of the Centennial Olympic Games. The first is of MUHAMMAD ALI, a champion and a gentleman, appearing as a ghost at the opening ceremonies, dressed in white, shaking terribly, seemingly confused as he lit the cauldron, setting off a roar of adoration from the crowd. The president, teary-eyed with emotion, later said: 'He was great, wasn't he?' Yes, The Greatest."

Abigail TRAFFORD continued:

"But the sport that made him great may have left him a terrible medical legacy: a Parkinson's syndrome that is likely caused by brain damage from repeated blows to the head in a long boxing career. The sport is considered so dangerous that the American Medical Association has called for it to be banned at the Olympics."

Abigail TRAFFORD asks finally:

"What price athletic glory?"

Bubi SCHOLZ

The *former German professional middleweight boxer Bubi SCHOLZ* - in 1997 67 years old – *is now a patient in a closed ward in the Max BÜRGER Hospital in Berlin -Patricia BARTELS, 'Bubi SCHOLZ verlor das Gedächtnis' ('Bubi SCHOLZ lost his* memory'), *Bildzeitung, December 30, 1969, and Bild, December 12, 1997).* SCHOLZ has extreme difficulties with his visual acuity and does not recognise his friends. It exists a severe mnemory defect. He is suffering on dementia pugilistica, he is punch drunk.

Jerry QUARRY

Jerry QUARRY is a *professional heavyweight* who fought MUHAMMAD ALI and Floyd PATTERSOIN. Among the highlights of QUARRY's career were two fights against PATTERSON, the heavyweight champion. The first fight ended in a draw and he won the second one on a controversial split decision.

QUARRY, a 6 foot, 195 pound heavyweight, who seemed to be easy to cut, had his biggest payday, $ 338.000, by fighting MUHAMMAD ALI when ALI returned from his banishmant in 1970. Early in this fight in Atlanta, the two boxers butted head and a gash opened above QUARRY's left eye. ALI peppered the spot with jabs, spraying blood and knocked out QUARRY in the third round.

ALI won the rematch again, and knocked him out in round seven. QUARRY was severely punished and beaten up by Joe FRAZIER in the second of their two fights, winning in the fifth round at *Madison Square Garden* in 1974.

Jerry QUARRY finished his career as a professional heavyweight with a 53–9-4 record after having fought more than 200 bouts as an amateur.

A *neurological examination* in 1982 – QUARRY was then 37 years old – revealed early sign of dementia pugilistica. His short-term memory loss and motor skills deteriorated already before his last three fights.

QUARRY fought in 1992 – then 47 years old – for one last time. He thought to be able to make a comeback as George FOREMAN had, he participated in a boxing match in Colorado, where no boxing license was required. He was battered and punished in six rounds by a club fighter.

QUARRY, who earned $ 2.1 million in purses as a top contender in the 1960s and 70s, was *living on Social Security.* At age 50, the pounding and punishment he had taken in the ring had turned him into a confused, childlike man whose relatives had to take care of him. A *dementia pugilistica* was diagnosed, a brain damage caused by repeated blows to his head.

PLUMMER and SCHNAUFER reported on Jerry QUARRY *(A life on the ropes, Jerry QUARRY fought ALI and FRAZIER bravely, with no regard for his safety; at 50 he is paying the price. People, February 19, 1996, pp. 66–70):*

"whose home is in the hills north of Los Angeles where he lives with his mother."

"His memory is lost in the haze caused by too many punches and punishment of his opponents from the likes of MUHAMMAD ALI and Joe FRAZIER. QUARRY is suffering from dementia pugilistica, a progressive condition due to numerous blows to the head, which produced a very severe permanent brain damage. 'QUARRY has the mentality, says neuropsychologist Peter RUSSELL of a person with *advanced Alzheimer's disease.* If he lives another ten years he'll be lucky,' says RUSSELL, who has likened QUARRY's dying brain cells to sugar dissolving in water."

QUARRY has lost nearly all the money he made, the more than two million dollars in purses is gone, along with his three wives.

QUARRY started boxing at age twelve. He beat ex-heavyweight champion Floyd PATTERSON and 1973 knocked out Earnie SHAVERS. He lost his bout for the title 1968 to Jimmy ELLIS.

The fight against MUHAMMAD ALI in 1970 gave QUARRY his biggest purse $ 338.000. In a fight against FRAZIER in 1974 QUARRY took a terrible beating. He promised to retire but had nine months later but he a fight with Ken NORTON where he received more punishment. A catscan of his brain revealed in 1983 severe traumatic alterations, but QUARRY continued to box. In 1987 his short term memory began to fail. QUARRY tried a comeback in 1992 – now 47 years old – he boxed for $ 1050 in a six round bout against a boxer with no name. He was severely beaten, when he returned from this fight in Colorado, remembers his brother Jonny, 'he couldn't remember the night before."

Dr. RUSSELL reports, that a recent *MRI* indicates that the damage in QUARRY's frontal lobes is increasing.

QUARRY was hospitalized December 28, 1998 with pneumonia and then suffered a cardiac arrest. He died at 3:52 p.m. after family members directed doctors to remove life support, according to Claude SUTHERLAND, a longtime family friend *(Battered Jerry QUARRY dies at 53, Lexington Herald Leader, Monday, January 4, 1999, C 3).*

Fatal Incidents in the Boxing Ring

Fatal events in the boxing ring are usually well reported in the media, so it is often possible to find at least several accounts of each incident in the press. They describe the observations of witnesses, but do not include clinical or morphological findings. There are so many fatal incidents that they could easily fill a book of their own. Space considerations have limited the discussion to brief summaries of a few well reported events taken from the world press. Statistics on fatalities in the boxing ring are rarely found in the literature. However individual cases are described and will be discussed below.

Fatal boxing incidents are listed in Table 98 below, as reported by several authors. Table 99 reveals the causese of 77 fatalties in the ring.

Professional Boxing Fight between Primo CARNERA and Ernie SCHAAF

One of the most often discussed fatal incidents in the boxing ring occurred in the 1930's. It concerned a professional heavyweight match between Primo CARNERA and Ernie SCHAAF on February 10, 1933, in the Madison Square Garden, in New York.

Ernie SCHAAF was beaten by heavyweight Max BAER, just six months before his fight against Primo CARNERA. It was well noted in the world press that SCHAAF had taken a series of hard blows against his head from BAER during the fight. The press recorded it as a 'bad beating,' and it was generally believed that SCHAAF was severely punished by BAER.

TABLE 99
Causes of 77 fatal incidents due to POHLE. (From GROH H, 1962).

Death due to Head Injury	52	67,4%
Death due to Heart Injury	13	16,8%
Death due to Abdominal Injury	6	7,8%
Death due to Fracture of Cervical Spine .	2	2,6%
Death due to Neck Injury (Larynx, Carotid Sinus)	2	2,6%

The following compilation is taken from contemporaneous press reports *(Death among the heavyweight; CARNERA-SCHAAF fight. Literary Digest, 115, pp. 26-29, February 25, 1933)*. Several contemporary press comments are quoted in it, so it is worth reading to give some sense of the observations at the time :

"The gigantic Primo CARNERA 'thrust his ponderous left arm out in the deliberate fashion of a man pushing his way through a revolving door.'"

"Ernie SCHAAF, of Boston, who was fighting him in Madison Square Garden for a moderate purse plus the glory of being one important step nearer to a fight for the championship, literally 'took it on the chin.'"

"He sank to his haunches (Fig. 321)," relates Jimmy POWERS in the *New York Daily News,* "apparently in full possession of his senses. His eyes wandered casually about the ring, focusing first on the timekeeper, then on the white-jerseyed seconds in his corner; at '9' he started to rise, but went into reverse and sank slowly until his cheek lay flush against the grimy canvas. Counted out, he was carried to his corner. He was not to recover consciousness for seventy minutes. He was to die within three days. But the crowd shrieked "Fake!" His handlers, including champion Jack SHARKEY, tried in vain to restore him to consciousness, and meanwhile, there were more boos from the gallery."

Mr. POWERS continued:

"... and one red-faced fan snatched at Jack SHARKEY's shoulder and shrieked above the bedlam: 'Why don't you take him to a hospital to make the act look good?'" We read on: "SHARKEY savagely pushed the intruder aside and he hollered at Manager BUCKLEY: "Hey! Maybe we better take him to a hospital." And off the procession went to Policlinic, with SCHAAF limp and apparently unaware either of the time of night or his destination."

"Ernie SCHAAF, a sturdy, high-ranking heavyweight from Boston, died today as a result of injuries suffered last Friday in one of the dullest yet most sensational prize-fights in the recent fistic history."

Fig. 321. Heavyweight bout between Primo CARNERA (left) and Ernie SCHAAF (right) on February 10, 1933, at Madison Square Garden, New York. The photo shows the final knock down and knock out of Ernie SCHAAF by CARNERA.

"The clever and hard-hitting SCHAAF was knocked out, and suffered an intracranial hemorrhage at the hands of the towering, awkward Primo CARNERA, giant boxer from Italy, who was taken to the Districts Attorney's office late this morning for questioning about the fight."

"His managers also were questioned. There had been much speculation among boxing experts on the fight, which many fans claimed was 'fixt,' but which boxing officials of New York State defended vigorously."

"All persons connected with the staging of the Primo CARNERA – Ernie SCHAAF prize-fight were directed to appear at the District Attorney's office to-day, P. Francis MARRO, Assistent District Attorney, told the United Press."

"They will be questioned about the fight in which SCHAAF received injuries resulting in his death."

"'We must show there was intent to kill before we can seek an indictment,' MARRO said."

"Through the morning, waiting for his expected arrest, Primo CARNERA sat in his room at the Hotel Victoria, 'his mountainous form slumped in a chair, the picture of complete dejection,' relates Jack CUDDY, of the United Press staff, continuing."

"This, he explained, was the first tragedy of his career – a career that has carried him from circus life as a 'strong man,' through vicious wrestling, and into fighting."

"'I was over in New Jersey last night with some friends,' he said. 'Someone telephoned about Ernie's death. I rushed back as fast as I could. I had hoped so much he would live. He was a good young man.'"

"Under the questioning of reporters, CARNERA said that it must have been the constant pounding, 'those left jabs' that injured SCHAAF so seriously."

"'I remember that last blow,' he said."

"It was started as a left jab, and I turned it into a sort of hook, I didn't think I hurt him, I didn't notice any pained expression on his face, perhaps that was because I was so anxious to win that I couldn't notice.'"

"'You know,' he continued reflectively, 'I hit Ted SANDWINA with that punch six times and knocked him down six times, in Tampa, with that same blow, and it didn't hurt him.'"

"'Will you fight again?' he was asked."

"'I shall have' he started to say when one of his advisers interrupted to remark, 'This is no time to say anything about that.'"

"'But you know I have to fight Jack SHARKEY four months from now,' CARNERA continued."

"'I have never had a thing like this before. Never has any one died like this after being in the ring with me. And I am twenty-six years old.'"

"'There will be probably a big gate for your fight with SHARKEY,' a hanger-on suggested."

"CARNERA waved his big hands in the air."

"'Please,' he said. 'Please don't talk about money at a time like this. This is no question of money. This is the question of the death of a fellow fighter.'"

"'My mother,' he added a moment later, 'my mother has called me up to-day from the little town of Sequale, near Venice. My mother Giovanna. She called me to tell me I must do my best to console Mrs. SCHAAF, Ernie's mother, because she knows how she feels. She told me I must send flowers, but of course I had sent them already. But she was thinking about this thing.'"

"'Then, too,' he continued, 'I could not do more than tell Mrs. SCHAAF how I felt. So I sent her a telegram.'"

"According to Ed FRAYNE, *sports editor* of the *New York American,* there had been no hint of the dramatic climax in the clumsy rounds that preceded the finish, and he proceeds to give the high-lights of the fight:"

"For one round, SCHAAF had managed to outscore the towering Pisan, and then the contest resolved itself in a boxing exhibition, with CARNERA doing all the exhibiting."

"Round after round, CARNERA stood off, thudding ponderous left jabs in the rushing blond face. Now and then SCHAAF would bring a yell from the crowd with a left dig to CARNERA's body, but the blows bounced off like horsehide off a bat. CARNERA's stomach was muscled like a washboard."

"Round after round, SCHAAF rushed head-long into that blistering jab. At times, CARNERA stabbed seven and eight times without a return. The monotony began to bore the folks in the top shelf, and they booed as the eighth and ninth rounds ended."

"CARNERA bled from the nose in the fourth, but Mannie SEAMAN and Billy DUFFY know their business. They stopped the flow before he came out for the fifth, and SCHAAF could not tab it again."

"SCHAAF made a great rally to win the tenth with a left-handed body attack, but he blew up from his exertions as the round ended."

"The big red glove on CARNERA's left hand resumed its pounding in the eleventh. CARNERA took that round and the twelth. The constant pounding wore away the steel chin of the Boston blond. The vertebrae can stand such a jolting only so long. It was the accumulative effect of the punching that preceded the thirteenth as much as the slingshot jab that numbed SCHAAF's mind in the thirteenth. Nature took all she could stand."

Professional Fist Fight between Harry LANE and Lem FRANKLIN

Heavyweight Lem FRANKLIN won 19 consecutive boxing fights and was finally knocked out by Bob PASTOR, in 1942. After the knock out he fought unsuccessfully, and took so much punishment in the ring that he retired from boxing. He later tried a come back, and was defeated in a knock out match with Harry LANE. He died eleven days later, in July 1944.

Professional Fist Fight between Kid POINT and Roland PRAIRIE

Roland PRAIRIE won a ten round fight against Kid POINT, in April 1946 in Quebec. Afterward he died from a traumatic brain hemorrhage.

Fatal Outcome of a Professional Fist Fight in Providence, Rhode Island

Professional boxer Bobby BURTON, from Pittsburgh, participated in a boxing match on October 30, 1964. He took a right blow against his body in the fifth round that produced a knock out. The unconscious boxer died shortly afterward, in the boxing ring (Fig. 322). Right in the photo is Dr. James FLANAGAN, member of the State Boxing Commission, who had tried life restoring measures. A police lieutenant Charles POTTER (in white shirt) tried likewise to revive the unconscious boxer with a resuscitator. A catholic priest, father CAVALLERO, gave the last rites to the dying boxer.

Fig. 322. Lieutenant Charles POTTER (white shirt) tries to revive the knocked out boxer Bobby BURTON (Pittsburgh) who died a short time after he was knocked out on October 30, 1946 in Providence, Rhode Island. The Reverend CAVALLERO (left) administers the last rites of the Catholic Church. At right Dr. James FLANAGAN of the State Boxing Commission tried to revive BURTON after he was floored by a right hand punch to the body in the fifth round. Photo: U.P.I.

Fatal Outcome of an Amateur Boxing Match for Francesco VELASQUEZ

The 19 year old Francesco VELASQUEZ, an *amateur boxer* and *soldier,* was knocked out in Carbondale, Pennsylvania, at a charity benefit. An ambulance unit tried in vain to resuscitate the boxer in the ring (Fig. 323). He was transferred to a local hospital but was dead on arrival (d. o. a) as the result of a blow to the head.

Professional Boxing Match between Manfred NAUKE and Karl Heinz BICK

Professional boxing match between Manfred NAUKE (left) and Karl Heinz BICK (right) on June 15, 1957, in the Westphalen Halle in Dortmund, Germany. BICK took a severe beating against his head, as the press universally reported; he was 'severely groggy.' Not till the tenth round did BICK lift his hand up as signal to give up the fight (Figs. 324, 325). BICK died in a hospital in Dortmund, due to the sequelae of brain hemorrhage.

BICK had undergone a *tonsillectomy* some weeks before the fight and had been released from the hospital. The *Association of German Professional Boxers (Verband Deutscher Berufsboxer)* attributed the ring death to the after-effects of the oper-

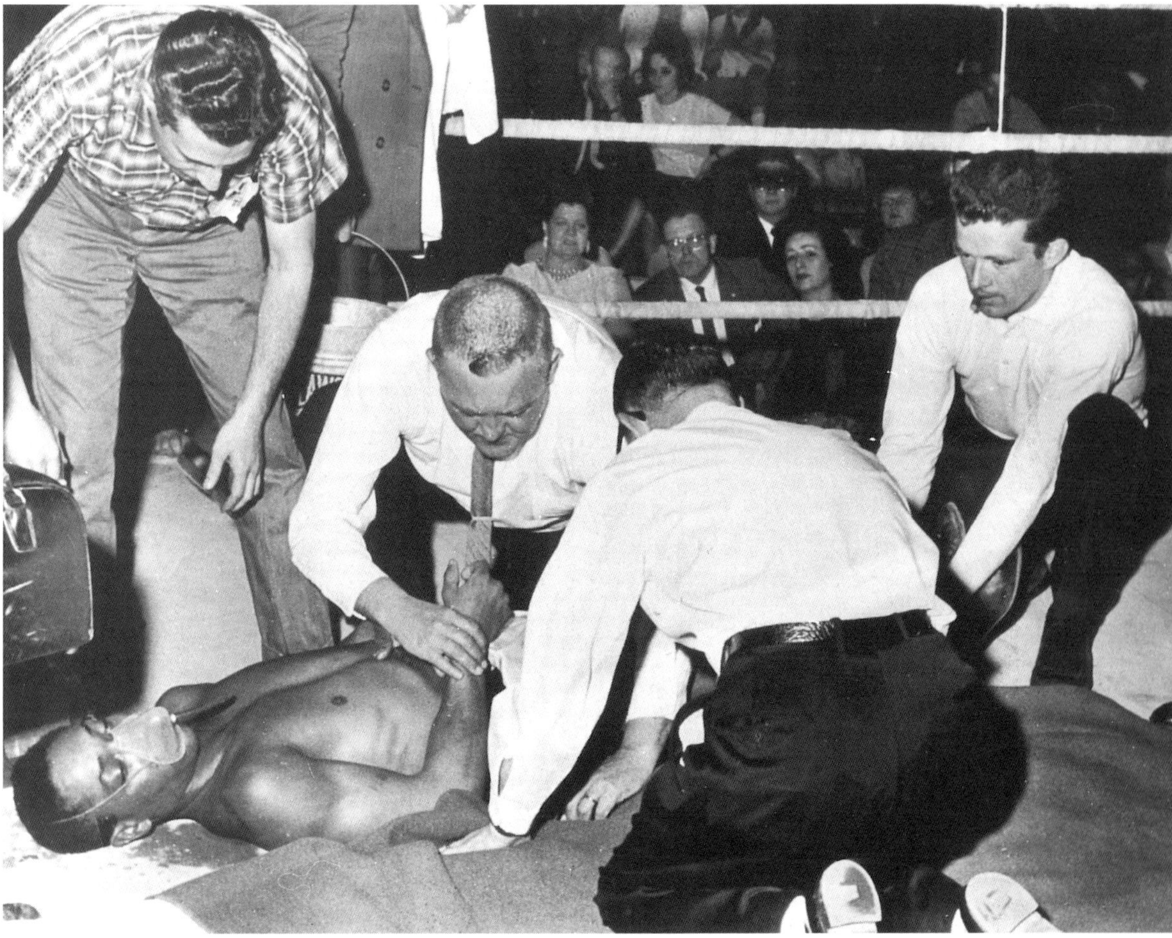

Fig. 323. Ambulance squad members work over 19 year-old Francesco VELASQUEZ, an amateur boxer, in a ring at Carbondale, PA, after he was knocked out and fatally injured in a charity bout. He was moved from the ring to a hospital where he was pronounced dead after being felled by blows to the head in a middleweight bout. The amateur boxer was stationed at nearby Tobyhanna Army Depot. Photo: A.P.

ation. If the tonsillectomy had created any risks, the association should not have allowed BICK to participate in the match. In the end it was proven that BICK died because of a brain hemorrhage, which was clearly the sequel of the many severe head blows during his last ten round fight.

Professional Boxing Match between Rodrigo CONTRERAS and Sonny NUNES

Professional boxing match between Rodrigo CONTRERAS from Tucson, Arizona who fought the 22 year old Sonny NUNES from Glendale, Arizona, in a professional match. In the photograph, NUNES is lying unconscious on the ring canvas. He died 40 minutes later because of brain hemorrhages (Fig. 326). Both boxers were participating in their first professional boxing match.

Professional Boxing Match between 'Sugar' Ray ROBINSON and Jimmy DOYLE

In a muggy June night on June 24, 1947, 'Sugar' Ray ROBINSON defended his welterweight title successfully in Cleveland. He knocked out Jimmy DOYLE in the eighth round (Figs. 327, 328). DOYLE died the next morning due to a traumatic brain hemorrhage.

Fig. 325. This photo shows the dying boxer BICK as he is being carried out of the ring. Photo: KEYSTONE.

Fig. 324. Professional lightweight bout between Manfred NAUKE (left) and Karl Heinz BICK on June 15, 1957 in Dortmund, Germany. The photo shows the last phase of the fight. Despite the fact that he had taken severe punishment, BICK waited until the tenth round before he gave up the fight. BICK died in a Dortmund hospital of a cerebral hemorrhage. One should note the unnatural expression in BICK's eyes, the referee obviously did not recognise it. Photo: d.p.a.

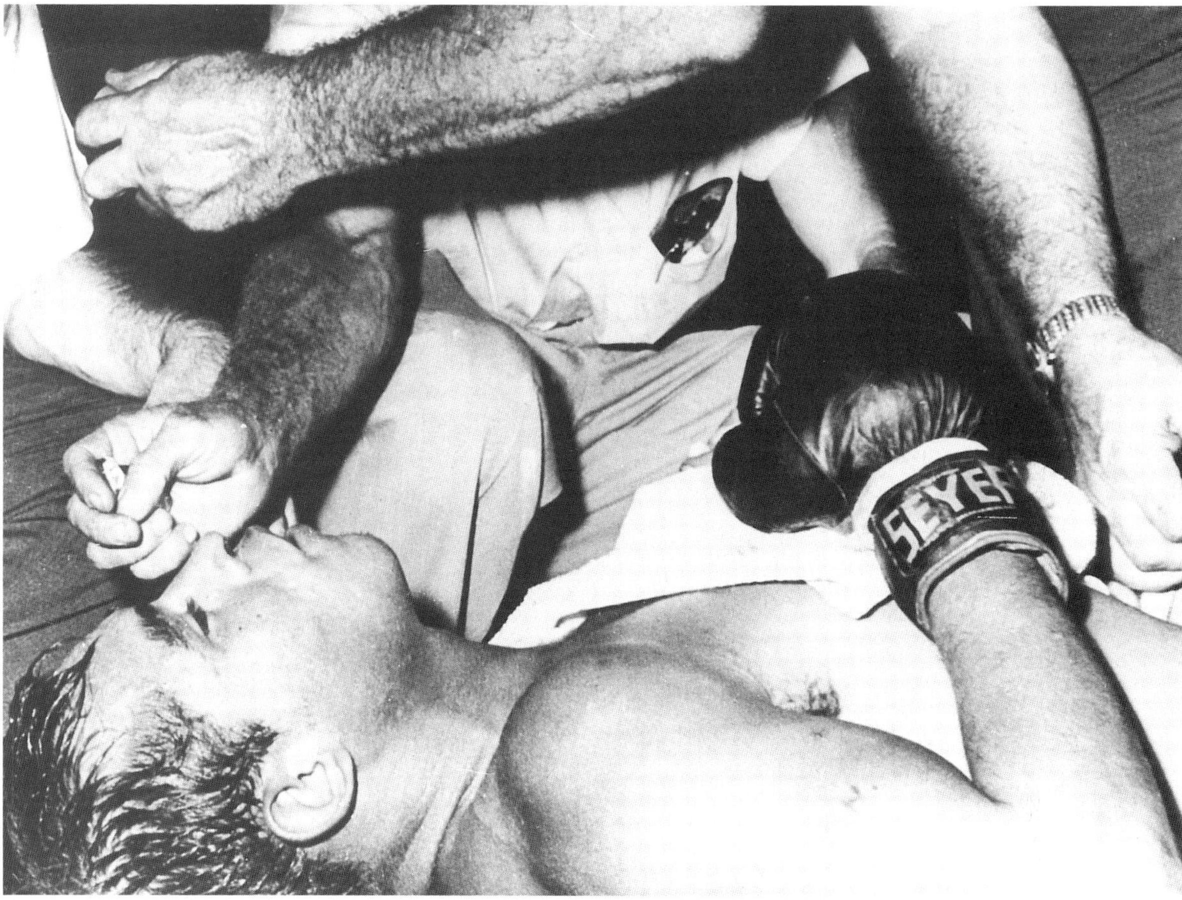

Fig. 326. Professional boxing fight between the 22 year old Sonny NUNEZ and Rodrigo CONTRERAS. NUNEZ, who was participating in his first professional boxing match, lies unconscious on the canvas and died 40 minutes later of cerebral hemorrhage. CONTRERAS also made his debut as a professional boxer in this fight. Photo: U.P.I.

A journalist asked ROBINSON: "Could you tell he was in trouble?" "That's my business," replied ROBINSON with utmost sang-froid: "Getting people in trouble is my business."
A short, summarizing description of the pugilistic careers of Jimmy DOYLE and 'Sugar' Ray ROBINSON follows:
Jimmy DOYLE fought as *professional boxer* from 1941 until 1974. He participated in *52 professional fights;* he won 42 and lost seven. Three fights ended in a draw. He achieved 14 knock out victories and was knocked out two times. He challenged 'Sugar' Ray ROBINSON for the championship in the welterweight, which took place on June 24, 1947.
'Sugar' Ray ROBINSON (Walker SMITH jr.) *fought 202 professional fights,* he won 175, lost 19, six fights ended in a draw and participated in two 'no decision' contests. He achieved 110 knock outs and was knocked out one time. He won the championship in the welterweight division by a point decision, on December 20, 1964, after a 15 round fight against Tommy BELL. He then defended his title successfully four times and gave up his welterweight title and fought then as a middleweight on February 14, 1951 by a knock out victory in the thirteenth round over Jack LA MOTTA, he lost his title in a decision on points after a 15 round fight against Randy TURPIN on March 3, 1951 and won the title back from TURPIN by a knock out victory in the tenth round, he had two successful title defenses and retired December 18, 1952 from the ring. He tried a 'comeback' on January 4, 1955 and won the middleweight championship on December 9, 1955 by a k. o. victory in the second round. He defended his title successfully one time and lost it by a decision on points in a fight against Gene FULMER on January 2, 1957. He regained the championship after a knock out victory over FULMER in the fifth round, on September 23, 1957. He

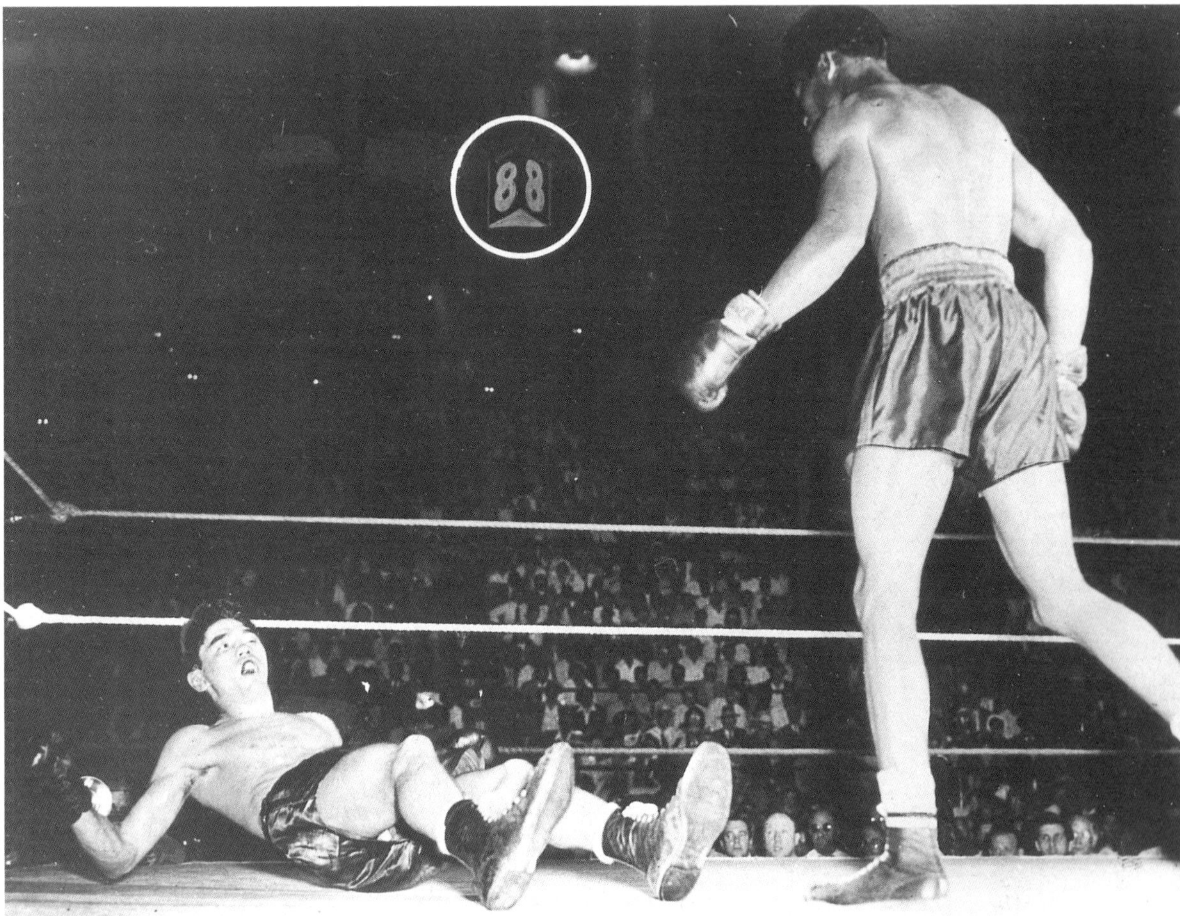

Fig. 327. Professional title fight in the welterweight between 'Sugar' Ray ROBINSON and Jimmy DOYLE on June 24, 1947 in Cleveland, Ohio. DOYLE was knocked out in the eighth round by a left hook to his head by ROBINSON. Photo: d.p.a.

then lost it again in a decision on points, after a 15 round match against Carmen BASILIO. He later lost in a point decision after a 15 round bout against Paul PECKER on January 22, 1960.

About 15 months before Doyle's last boxing match, middleweight Artie LEVINE knocked him out in the Cleveland Arena on March 12, 1946. At DOYLE's autopsy, his brother Edward DELANEY said that his brother had been a 'changed man' after losing the fight against LEVINE. While before this fight DOYLE was an active person and a good dancer, after the match he sat at home, idle, and had mentioned never again to go back into the boxing ring. However, after nine months DOYLE took up boxing again, and his manager arranged a title bout against ROBINSON, who was at the zenith of his boxing career.

In an article *(Death by permission, Newsweek, Vol. 30, July 7, 1947, p. 30)* the following words appeared:

"How quickly and how easily ROBINSON wins depends entirely upon the size of a soft spot in the noggin (slang for head, author) of DOYLE... It is possible – quite possible, I fear – that DOYLE could be seriously harmed by another thorough lacing," so had Franklin LEWIS in his report warned (*Cleveland Press,* June 24, 1947).

From press reports follows that DOYLE was hopelessly inferior and was never able to land an effective blow at the champion ROBINSON.

James P. DAWSON, who described the boxing match (*The New York Times*) wrote:

Fig. 328. Professional boxing match between 'Sugar' Ray ROBINSON and Jimmy DOYLE on June 14, 1947 in Cleveland, Ohio. DOYLE was knocked down in the eighth round with a left hook to his head by ROBINSON and had to be carried from the ring on a stretcher. DOYLE died later in a Cleveland hospital despite an operative intervention.

"Only for his courage under fire could DOYLE be credited. For the doughty little Californian, though buffeted about a sea of blows against a champion who is a master boxer and a paralyzing puncher as well, never stopped coming."

DOYLE had to take in the first seven rounds severest punishment against head and body; in the eighth round ROBINSON landed a short left hook at DOYLE's chin, DOYLE fell in a heap and hit with his head the canvas; it could be heard a resounding thud. The knocked out boxer had to be carried on a stretcher out of the ring (Fig. 328) and not regaining consciousness again died 17 hours later, an neurosurgical operation was not successful. DOYLE had already in the boxing match against LEVINE hit his head at the canvas.

In an article *(Death by permission, Newsweek, 30, No. 80, July 7, 1947)* it reads:

"Following the death last week there was the customary mushroom crop of editorials suggesting ways to reduce boxing's fatalities and its steady output of damaged minds. As usual, two Cleveland investigations were quickly and vigorously launched. At the week end nothing had come of them."

Professional welterweight title match between Emile GRIFFITH and Benny 'Kid' PARET

One of the most controversialof boxing legends has to do with the death of Benny 'Kid' PARET at the hands of Emile GRIFFITH in a welterweight match in Madison Square Garden in New York, on March 24, 1962. According to the story PARET provoked GRIFFITH in their weight-in by calling him 'maricon' ('fagger'), and was in effect killed by GRIFFITH in the ring that night. Recalling the events years later GRIFFITH said he was only following his trainer's instructions to hit PARET, to hurt PARET to keep punching PARET until the referee made him stop. By which time, as it turned out, PARET was virtually dead. (He died about ten days later in a New York hospital).

OATES (1987) wrote:

"... In the infamous Benny PARET – Emile GRIFFITH fight of March 1962 the referee Ruby GOLDSTEIN was said to have stood paralyzed as GRIFFITH trapped PARET in the ropes, striking as many as eighteen times in the head."

The final moment of the begin of the end of the fight between Emile GRIFFITH (right) and Benny 'Kid' PARET (left). GRIFFITH lands a right hand which hurls PARET into the ropes. PARET's eyes are open.

Benny 'Kid' PARET of Cuba is just seconds away from losing his welterweight title to Ermile GRIFFITH of New York. PARET staggers at the ropes in the twelfth round of the championship fight. PARET is trapped in a corner between the middle and upper ring rope.

The end of the fatal boxing match is shown in Figs. 329–334.

Fig. 329. The final moment, the begin of the end of the professional welterweight fight between Emile GRIFFITH (right) and Benny 'Kid' PARET (left). GRIFFITH lands a right hand and hurls PARET into the ropes. His eyes are open. Photo: A.P.

ROUND
12
CHICAGO

Fig. 331. This photo shows the end of the bout of another line of sight. PARET is trapped between the ropes and his eyes are closed.

GRIFFITH who had landed 25 hard blows against PARET's head, in the twelfth round, without that the blows were answered by the defenseless PARET, is restrained by referee Ruby GOLDSTEIN, after PARET had sank unconscious to the canvas.

Benny 'Kid' PARET is carried on a stretcher from the ring after being knocked out in the twelveth round. GRIFFITH regained the title by a TKO victory.

The unconscious Benny PARET in New York's Roosevelt Hospital. In the foreground is Mrs. PARET, in the background his manager Manuel ALFARO.

Fig. 330. Benny 'Kid' PARET of Cuba is just seconds away from losing his welterweight title to Emile GRÎFFITH of New York. PARET staggers at the ropes in the 12th round of their championship fight on March 24, 1962, in New York's Madison Square Garden. Referee Ruby GOLDSTEIN stopped the fight after 2 minutes, 9 seconds of the round. PARET knocked unconscious, was hospitalized. Photo: A.P.

Fig. 332. GRIFFITH who had landed about 25 hard blows at PARET's head, without returning a blow, is finally retained by referee Ruby GOLDSTEIN. PARET had dropped unconscious in the ring corner to the canvas. PARET died due to his brain injuries. Photo: A.P.

Professional Boxing Match between 'Sugar' RAMOS and Davey MOORE

A little more than a year later the world champion in the featherweight Davey MOORE suffered in a boxing match against 'Sugar' RAMOS fatal injuries, that television had again to expand for a limited time to other objects.

The 29 year old *professional boxer Davey MOORE* retained the championship in the lightweight and fought in a lightweight championship fight on March 21, 1963 in Hollywood in an at ten rounds arranged boxing match against 'Sugar' Ramos. He was considered the favorite, generally was agreed that he would win this boxing match in an early round.

The fist fight was fierce, but carried out by both boxers fair. MOORE began fast and landed on RAMOS in the second round series of tough blows, which made him confused and let him stagger. Every time, when RAMOS was hit, his hanger-on shouted 'RAMOS – RAMOS – RAMOS,' and he answered this with attacks. In the third round RAMOS changed his tactic. He kept distance and began to bring MOORE with a hard rotated jab out of his balance. In the fifth round a jab let MOORE whirl through the rope, a push brought him down. Referee LATKA called it stumble. MOORE got up and landed at RAMOS with a right, the Cuban countered with five straight jabs and a right cross, a combination of punches which let MOORE's mouthpiece fly out and break. MOORE used this mouthpiece furthermore, despite it caused wounds in his mouth and made him swallow blood. His manager KETCHUM had another mouthpiece, which did not fit over MOORE's loosened teeths. In the eigth round a sharp

Fig. 333. Benny 'Kid' PARET is carried on a stretcher from the ring after being knocked out by Emile GRIFFITH in the 12th round of their scheduled 15 round welterweight championship fight. GRIFFITH regained the title on a TKO victory. Photo: A.P.

right of MOORE closed the right eye of RAMOS to a slit, but RAMOS used his left, which landed on MOORE. RAMOS connected at the end of the ninth round with a hard right, which shook him. At the begin of the tenth round MOORE jumped out of his corner and landed two solid rights. RAMOS countered these blows and MOORE went into a clinch. The fanatic spectators shouted 'Arriba! Arriba' and RAMOS hit MOORE five uppercuts, which whirled MOORE stumbling through the ring, after it hard left jabs, which were followed by a right. MOORE fell with such a force on his buttocks that his occiput hit the lowest rope. MOORE was at '3' again at his legs. The referee cleaned MOORE's gloves and let the fight go on after the mandatory '8' count.

Later referee LATKA said: "MOORE's eyes looked o. k., despite I recognised, that Davey had to take some hard blows. His arms moved and his 'reflexes' seemed to be in order. He seemed to be very, very exhausted, but his eyes were really clear, sharp and he was really alert." But strange to say, LATKA added, that he was apprehensive since the begin of the fight about the footwork of MOORE. He did not move as he had done as before. He was over the entire time confused. From the first round his footwork was not right.

MOORE stumbled as the fight continued. He staggered defenceless around the ring while RAMOS punished him severely. At the end RAMOS was pushed by a powerful right through the ropes and remained lying with his back at the

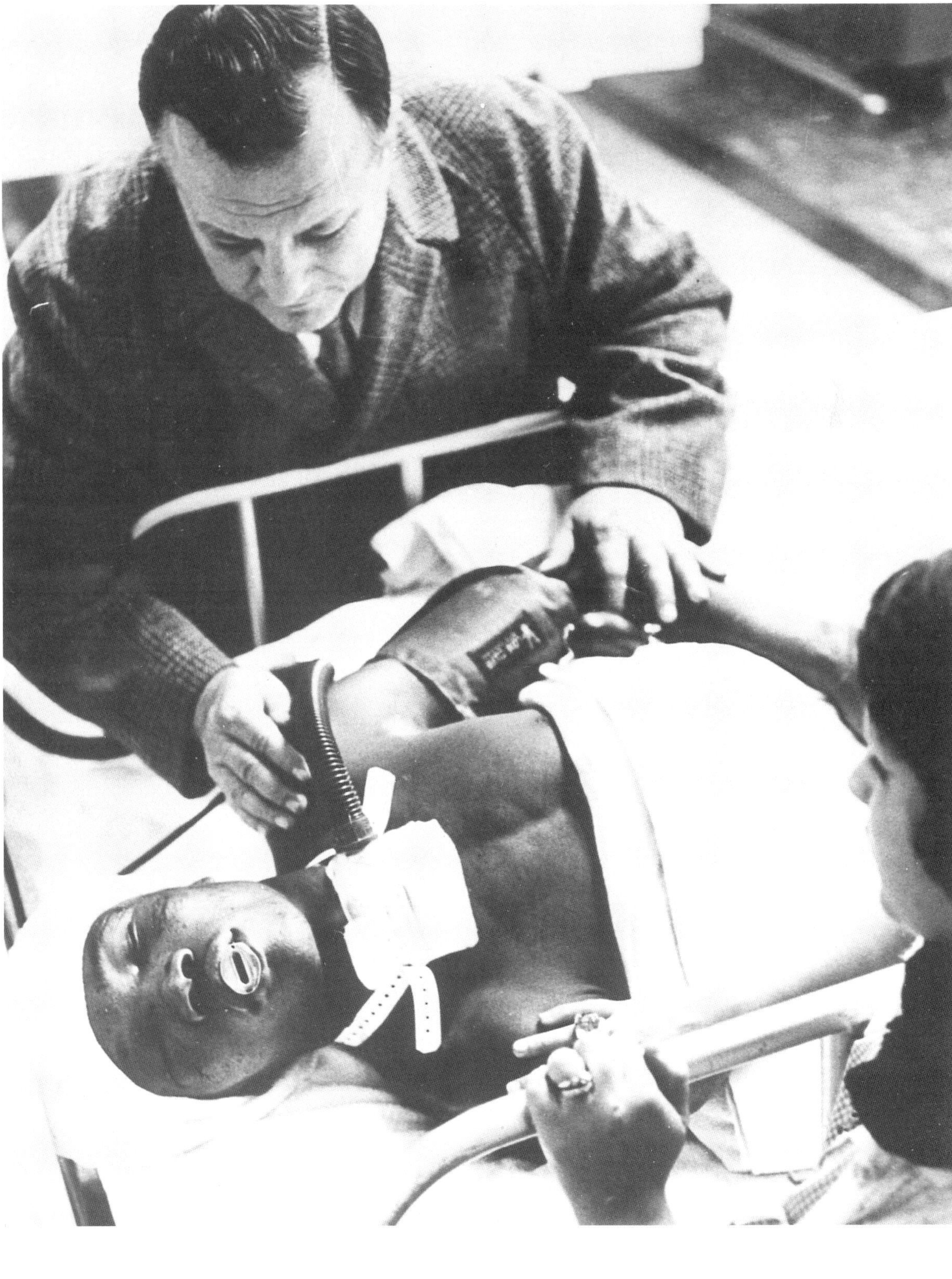

middle rope. Even RAMOS seemed to have enough, because he stepped to the side and watched. LATKA remarked: "I took RAMOS hand and began to give MOORE a mandatory '8' count, despite the fact that he was not lying at the ropes. But then indicated the gong the end of the round and I pulled MOORE up. I wrote my scoring for this round down, and I was in the act of looking at MOORE, as Willie [KETCHUM] came to me and said, he wished to interrupt the fight. I for myself had decided to stop the fight, if MOORE showed no changes in the next 30 seconds."

"I did not want to harm MOORE" said RAMOS, "the boxers in the ring are partners. They carry out the fight, not to harm or kill, but to show talent, and win the challenge. After the fight my opponent is my brother. But this tragedy is a thing with which all fighters have to reckon. I myself could have been the person who got severely injured. The knowledge, that it can happen, I have to accept, and perhaps MOORE did it, too. Perhaps was yesterday his destiny, mine will be on another day."

Summarising, in the ninth round MOORE was hit by a straight blow in his face, a blow, which let him stagger. In the same round he was hit by many blows at his head. In the last ten seconds of the tenth round MOORE was pushed into the ring ropes (Figs. 335 a – c, 336, 337), allegedly he hit with his occiput the lowest rope, but was able before being counted out to get on his feet. Since it was obvious, that he was defenceless, his manager interrupted the fight.

MOORE at first did not show significant after effects. He was able, to speak shortly into the television microphone, to announce a revanche bout. He walked without assistance into the dressing room and seemed unremarkable. About 45 minutes after the end of the boxing match he became in short time unconscious and was transfered immediately into a neurosurgical department.

MOORE was admitted in the hospital deeply comatous with extremely wide pupils and a tetraparesis. During the survival time of 72 hours no improvement of his condition occurred. A lumbar puncture revealed clear cerebrospinal fluid.

Shortly before his death the blood pressure increased to 200/130 mm/Hg, the pulse rate to 130/min. A neurosurgical intervention was not considered necessary.

The autopsy revealed a slight subdural hemorrhage over dorsolateral parts of the right cerebral hemisphere (it was, according to COURVILLE, too small, to be clinically recognisable, but could have developed with a longer survival time into a hematoma).

There is a congestion of vessels at the surface of the cerebral hemispheres. The convolutions are flattened, especially at the left side. The base of the frontal and temporal lobes shows subpial hemorrhages. Both unci gyri hippocampi show incisures in this structure.

The brainstem is cut from the cerebellum by the method introduced by SPATZ. The aqueductus cerebri is surrounded by a small zone of hemorrhagic softening, of about 1-2 mm diameter (Figs. 338 a and b).

The left peduncle shows a focal hemorrhage which extends into the substantia nigra, additionally there are extensive so-called contusions in the uncus and gyrus hippocampi. They extend allegedly from the right side into the left subfrontal region. The extent of this marked lesion can especially be seen at inspection of median aspects of the right cerebral hemisphere (Figs. 339 a and b). Also the corpus callosum and the occipital region are involved, COURVILLE speaks of a 'diffuse contusion.'

These by COURVILLE described processes are obviously no contusions, but hemorrhagic necroses due to the increased intracranial pressure. Also the morphological alterations in the uncus gyri hippocampi are no contusions, but these structures were due to the increased intracranial pressure pressed against the margin of the tentorium cerebelli, which acts as a blunt knive, it is a 'pressure necrosis.'

Midbrain and upper pons reveal focal hemorrhages in the tegmentum, smaller one are seen in the basal parts, too.

A macrophoto and a drawing of the base of the brain reveal the extent of the hemorrhagic necroses, which COURVILLE erraneously termed 'diffuse contusions' (Figs. 340 a and b).

COURVILLE describes the sequence of the end of the boxing match as follows:

" A heavy blow against the head of MOORE produced a knock down, the boxer falls first at his buttocks. He falls then backward and hits with his occiput the lowest ring rope. The vibrating rope impacts allegedly several times against MOORE's head (Fig. 341)."

This mechanism produced according to COURVILLE a contrecoup injury. COURVILLE continues that this event was clearly visible at the television film. A drawing shows the last phase of the knock out (Fig. 342).

It seemed to me much more probable that the powerful series of blows against MOORE's head in the last rounds are responsible for the extensive hemorhagic necroses in the brain of MORRE which led to MOORE's death.

The interval of 45 minutes between the technical knock out and the beginning of the clinical symptoms can be so ex-

Fig. 334. Mrs. PARET sits at bedside of her husband in New York's Roosevelt Hospital. In the backgroundbd is PARET's manager Manuel ALFARO. Photo: A.P.

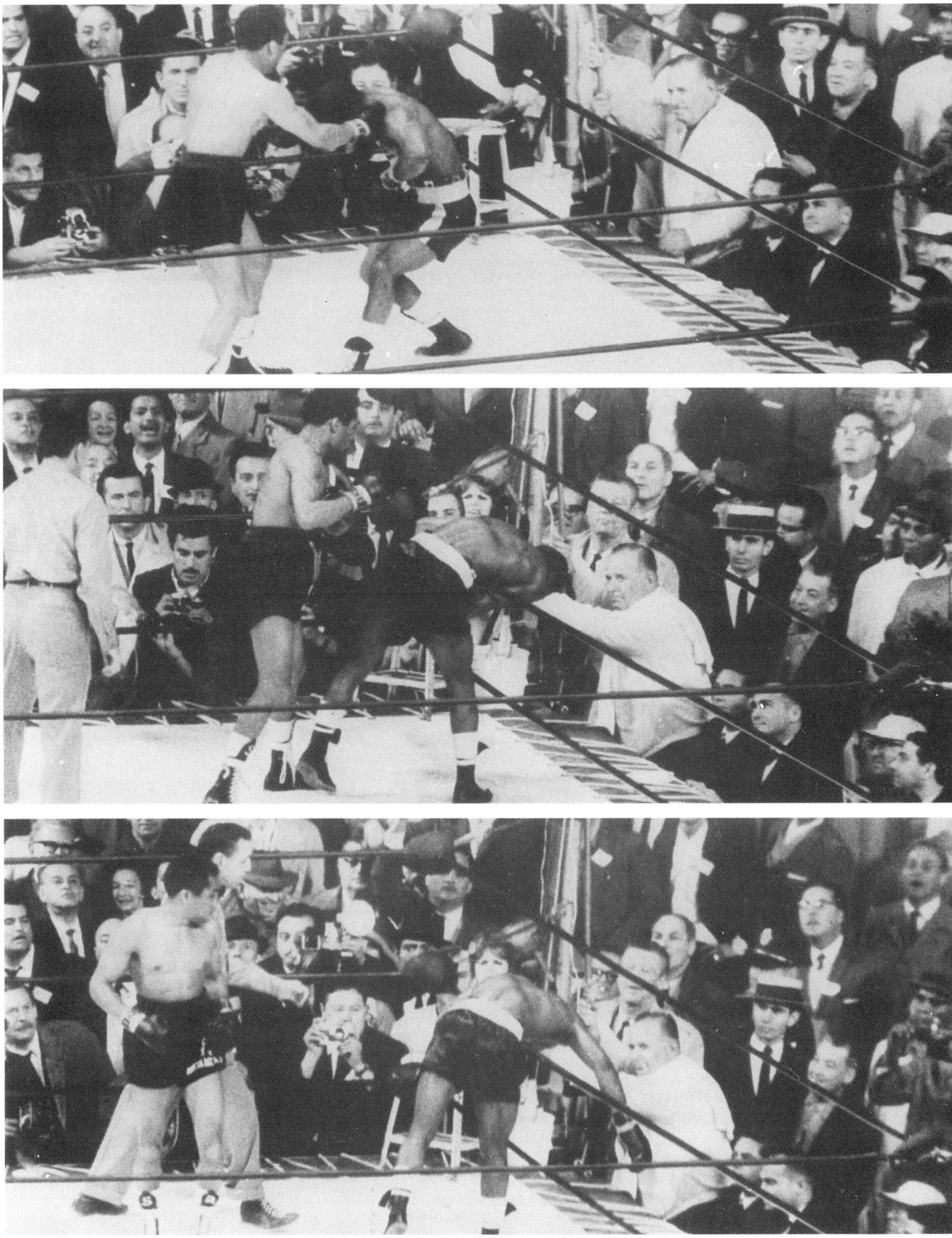

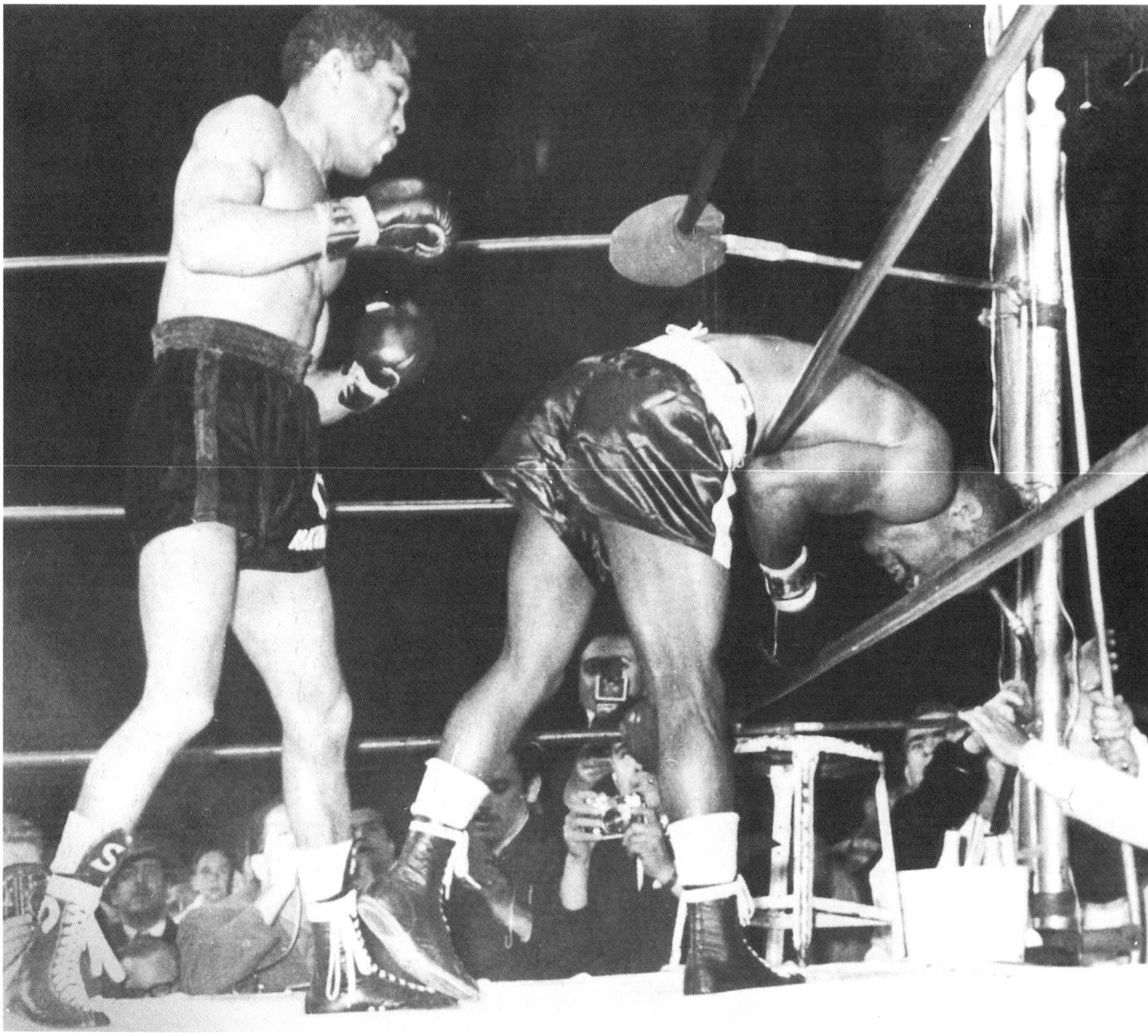

Fig. 336. Davey MOORE is draped over the ropes, beaten and helpless, at the end of the tenth round of his unsuccessful defense of his world's featherweight championship. Standing behind and over him is the new champion 'Sugar' RAMOS, a Cuban refugee, now living in Mexico City. Photo: A.P.

plained, that this time was necessary until the developing brain edema exerted a pressure at the brainstem.

Figs. 335 a – c. Featherweight title bout between 'Sugar' RAMOS and Davey MOORE on March 21, 1963 in Dodger Stadium, Los Angeles, California. As MOORE's manager, Willie KETCHUM (white shirt at right), looks on, 'Sugar' RAMOS connects with a right hook to MOORE's head (top). MOORE reels and hangs on the ropes (center) and referee George LATKA (bottom), walks in to call a halt to the fight and award a technical knock out victory to RAMOS. Photo: U.P.I.

Fig. 337. Unable to defend himself, featherweight champion Davey MORE sags on the ropes at the end of the tenth round by the beating given him by 'Sugar' RAMOS to the canvas. Photo: A.P.

Professional Boxing Match in the Middleweight between Jackie DARTHARD and Bert LYTELL in Milwaukee

A fatal collaps of the boxer Jackie DARTHARD occurred in Milwaukee between two rounds of a boxing match in Milwaukee on April 23, 1948, who died 12 hours later.

A description of this fight was published under the title *'The kill' (Time, Vol. 51, May 3, 1948, p. 43):*

"In Milwaukee's smoky Auditorium the rumbling shouts rose to a roar. The crowd was clamoring for the kill. Jackie DARTHARD, a promising 18 year old Negro middleweight – rated No. 5 in the United States by Ring Magazine – was down twice in the third round. At the bell, he stumbled woozily to his corner. To test his reactions, the manager threw questions at him quickly, 'What ya name... What town we in... What round it is?' DARTHARD muttered: 'Cut out the jive. I'll get this guy.'"

"The guy he had to get was 160 lb. Bert LYTELL, also a Negro and more noted for shiftness than for a punch. But after the sixth round, DARTHARD looked in terrible shape. His manager tried questions again and got only a thick-tongued reply: 'All I know is I'm fighting in Milwaukee.' Then

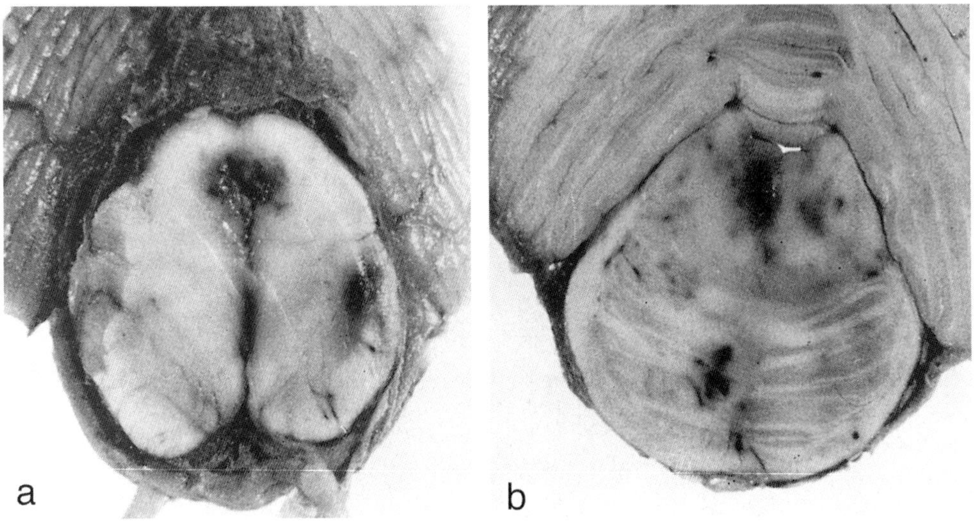

Figs. 338 a and b show midbrain and pons with periaqueductal hemorrhages, as well those in the raphe, the substantia nigra, and in the pons. Makrophoto. (From COURVILLE C B, 1964).

DARTHARD began tossing crazily on his stool and complaining that his head ached. He slumped over, mumbling incoherently. They put him on a stretcher and carried him, unconscious, into the dressing room. A few hours later he was dead of a brain hemorrhage."

Professional Boxing Matches of the Heavyweight Boxer Alejandro LAVORANTE against Archie MOORE, Cassius CLAY and finally against Johnny RIGGINS, in which LAVORANTE was fatally injured

Alejandro LAVORANTE, a *professional heavyweight boxer* from Argentina was severely beaten by Archie MOORE and by Cassius CLAY. After the boxing match against MOORE, who had knocked him out, he had to be carried on a stretcher out of the ring (Fig. 343). Four months later he was knocked out by Cassius CLAY in the fifth round. Nine months later he fought against an outsider, Johnny RIGGINS, wo knocked him out in the sixth round. He was treated in a hospital for a time of 558 days, before he died. Three neurosurgical operations had to be carried out (Figs. 344 and 345).

Professional Boxing Match between Ernie KNOX and Wayne BETHEA

"Ernie KNOX, was 'a sleepy-eyed negro who would trade leather with anyone for the prize of a train ticket and a night on the town. Once the *Ring Magazine* picked him as its *'Promising Fighter of the Month'"* – but that was in 1958, and the promise was mostly unfullfilled. He lost almost as often as he won (ten wins, seven losses, three draws), and it was not long before Ernie KNOX, 26, was eking out an uncertain living from part-time jobs and unemployment checks. But always there was that dream of the big time *(Prize fighting. The tenth death. Time, 82, October 25, 1963)."*

"One night last week, Ernie climbed into the ring at the *Baltimore Coliseum* to fight New York's Wayne BETHEA, 31. A hulking 205-pounder, BETHEA had fought Sonny LISTON and Ezzard CHARLES, had a record of 32 – 19 – 3. Ernie weighed 184 lbs. on the boxing commissioners scales and he had not fought in nearly a year. 'You have to prove yourself,' shrugged his manager. 'You have to be ready to take a chance.'"

"Some chance: In the ninth round, BETHEA unloaded a series of bombs and Ernie toppled to the canvas. Rushed to a hospital for 'observation,' he asked for a drink of water, sank into a deep coma, and died of a brain hemorrhage – boxing's tenth fatality of 1963."

"A Baltimore coroner started investigating the circumstances of Ernie' death. By week's end it had turned up two sickening bits of information. Ernie's cut of the $ 1620.- purse was hardly enough to pay his burial expenses. It came to $ 243.-

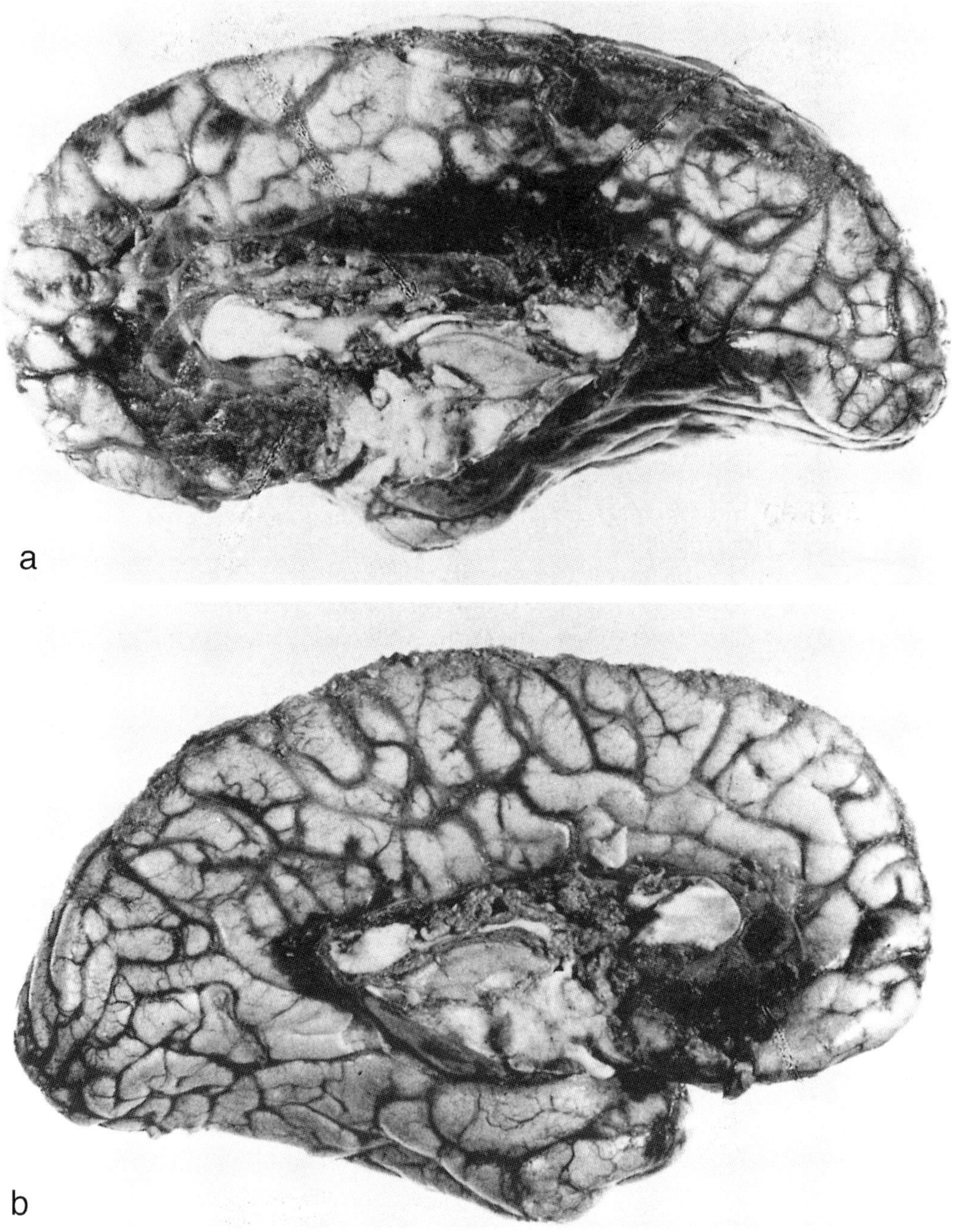

Figs. 339 a and b. Makrophoto of both median planes of the cerebral hemispheres, which show the hemorrhagic necrosis of the corpus callosum, as well as those of median parts of the cerebral hemispheres. (From COURVILLE C B, 1964).

and at the city morgue, Ernie KNOX's body weighed only 151 lbs. The boxing commission sheepishly admitted that KNOX had been permitted to weigh in with his clothes on. Said the autopsy surgeon: 'All you had to do was look at this kid's body to know he did not weigh 184 lbs."

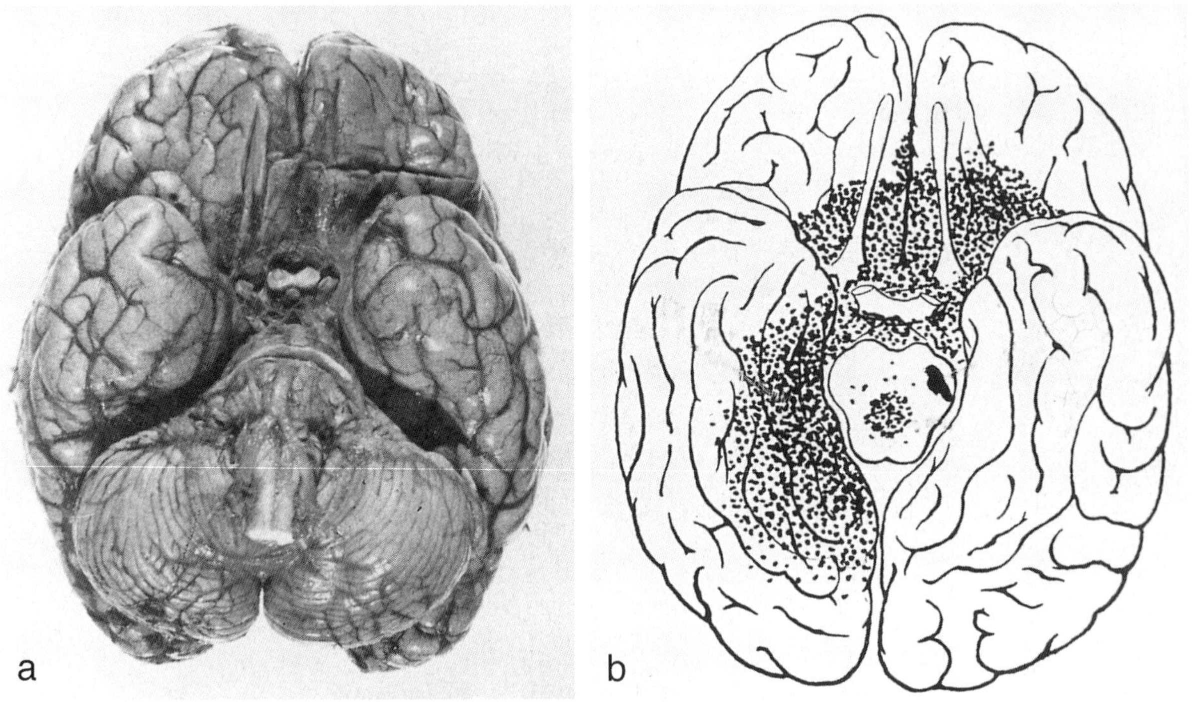

Figs. 340 a and b. (a) Makrophoto of the base of the brain, and (b) drawing of the hemorrhagic necrosis of the cerebrum, right more pronounced than left, also around the aqueduct and in the left substantia nigra. The tissue alterations are wrongly termed, using inadequate nomenclature 'diffuse contusions.' (From COURVILLE C B, 1964).

Professional Boxing Fight between Juan Jesus MELERO and Francisco RODRIGUEZ

On the occasion of the ring death of the *Spanish professional middleweight boxer Juan Jesus MELERO (Wieder Tod nach k.o. Spaniens Presse klagt an). (Again death after a k.o. Spains press accuses), Münchener Merkur, No. 45, February 23, 1978, p. 30)* it reads:

"MELERO had been knocked out in his eighth professional boxing match against the Spanish champion in the middleweight Francisco RODRIGUEZ (Figs. 346 and 347). The 23 year old MELERO had suffered hemorrhages in the brain, despite an operation in a neurosurgical unit of a Madrid hospital he died four days later. It developed a brain death syndrome, a zero-line EEG existed."

The Madrilenean newspaper *"Ya"* practised severe criticism at the Spanish Boxing Federation:

"The Federation says after all nothing, one closes the eyes and smiles like narcotized." Especially was denounced, that not till 20 minutes after the knock out of MELERO an oxygen-apparatus and an ambulance were available.

Professional Boxing Match between Angelo JACOPUCCI and Alan MINTER

About the ring death of the *Italian professional boxer Angelo JACOPUCCI (MÜLLER-MEININGEN, Tod im Ring) (Death in the ring) Süddeutsche Zeitung, No. 166, Saturday/Sunday, July 22/23, 1978)* the following commentary was given by MÜLLER-MEININGEN:

"The television spectators was offered at Wednesday evening the execution – one can it hardly otherwise term – of the Italian professional boxer JACOPUCCI. He is now a patient after a neurosurgical operation in a quite hopeless condition in a hospital in Bologna. On an another day died in Valencia another Spanish boxer. After such incidents begins at times a fresh debate, whether at least professional boxing should not be prohibited. In circles of experts the argumentation moves between two poles: the boxing has a smaller injury than that of other types of sport (soccer, wrestling, handball, skiing), so say some, boxing is legalized murder, the other."

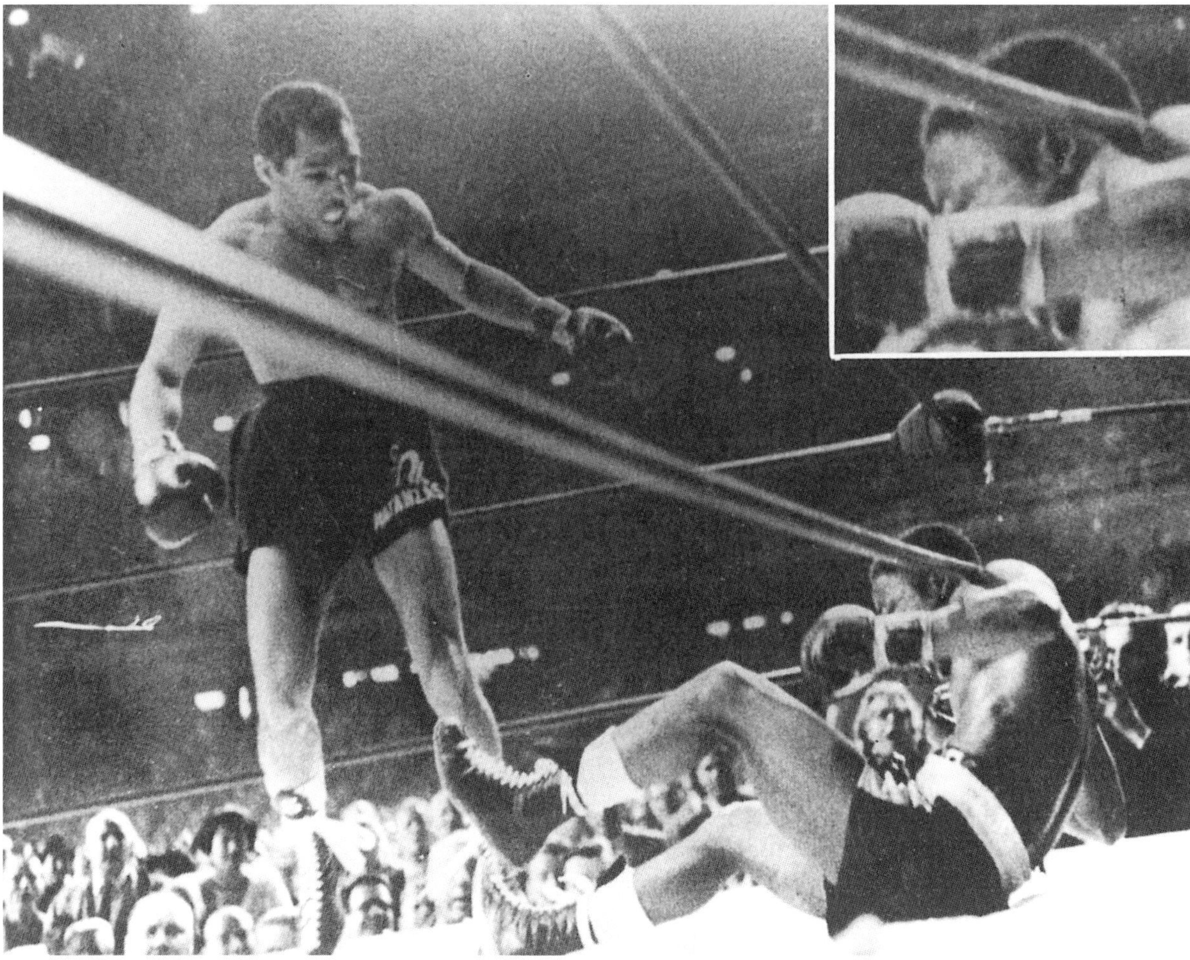

Fig. 341. Photo of the knock dow at the end of the bout. Inset enlarged to show the mechanism of impact of back of head against lowermost rope of the ring. (From COURVILLE C B, 1964).

"One could object: professional boxing is without doubt a dangerous matter, everybody knows that and everybody has to know it, who has to do with it; he puts up with danger. By the way it is a show business, in which in certain cases much money may be earned, also for the boxer self. Top competitive sport is indeed in general not really wholesome. Nobody is forced to box; the professional boxer is protected by the fundamental law for free choice of a profession, the moreover by more severe protective measurements protected amateur has his fun on it – shortly: one should let him then alone."

MÜLLER-MEININGEN continued:

"In the Federal Republic of Germany was at motion of the member of the Lower House of Parliament (Bundestag) MÜLLER-EMMERT (1973) even an own expert commission troubled, which after many years epistemologically divulged, casulties as maximal events occur in all types of sport again and again and about chronic brain damage – the syndrome of the 'weiche Birne' ('soft in the head' or 'batty') there exists no thrustworthy material. Organizer, referees, ring physicians – so it seems to us – had to be drawd into greater responsibility."

An additional commentary *(Hans Joachim LEYENBERG. Boxen im Widerstreit: Bis zum nächsten Mal?) (Boxing in the conflict of opinions: up to the next time?) (Frankfurter Allgemeine Zeitung, July 25, 1978)* follows:

"Angelo JACOPUCCI was on Sunday led to rest in his hometown Tarquinia. At the grave his family, his closest friends, in secured distance all those who keep faith to the box sport. The fascination which can come out from box sport, is such indivisible as the harm. When we imagine Max

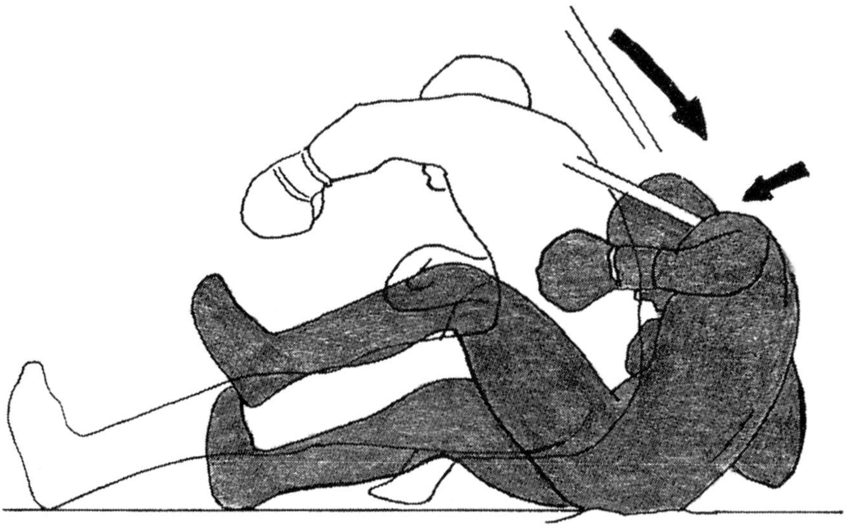

Fig. 342. Mechanism of head injury. Diagram of final positions of boxer with head against the lowermost rope at the end of the tenth round (see Fig. 341), superimposed on outline of position taken only a second before. Taken from photograph reproduced by *Life Magazine,* 1963, from COURVILLE C B, 1963).

SCHMELING, MUHAMMAD ALI, but Angelo JACOPUCCI, that is the grimace of this sport, which lets its friends become unsettled."

"What is to do? More protective measures, education of the tune of combat, more responsibilty for referees and ring physicians? That all is in amateur boxing already materialized, and despite of that there are no guaranties against fatal risikos – perhaps head gear and heavier gloves? But the summation of many blows is sealing the fate of a boxer, damages his brain 'Boxing is not a dangerous sport,' states the German Federal Labor Council (Bundesarbeitsgericht) in Kassel in January of this year in a leading decision."

"Conflict of opinions? About which there can be no discussion, is the uneasiness behind the scenes, things about we hear in the acute later – when at all."

"The shocked Welshman Alan MINTER, said immediately after his return to London: 'JACOPUCCI seemed to be completely o. k. Nobody told me, that he had have this type of difficulties before.' And MINTER emphasized, he had nothing known about it, that there had been hesitations against this fight."

"Hesitation is also necessary, when the *Federation of German Professional Boxer* a man as Conny VELENSEC after a two year abstinence grants a new boxing licence, as happened last week end. Supported by two medical examinations returns the today 36 year old innkeeper from the village Schöningen at the demarcation line (between East and West Germany) back into the boxing ring. The uneasiness grows."

"Into this portrayal of customs fits the news from Berlin, according to which Jupp ELZE after his European Championship fight in Paris 25 days being in a coma, wants to box again. His manager ZELLER has declared, that it would be no comeback under him. Where are the physicians, the backers in the two with each other rivalising German professional boxing organisations, who want to pass today their hand to ELZE for a suicide mission?"

"On Friday ervening Georg STEINHERR, whom they call 'Hammer Schorsch' knocked out his collegue INDRISSA from the Senegal already in the third round. Some between 2.500 spectators whisled, because INDRISSA went down too early. Inasmuch the conduct of some consumers."

"And what are the officials doing? Elucidations facilitate the preparations for the *Congress of the World Federation (WBA) of Professional Boxers*, which should take place in October in Johannesburg, but now has to be postponed. The *South-African Boxing Federation* had first promised to cover half of the costs for air journey and the per diem for the delegation of the World Boxing Association and their wives. Since most of the delegates wanted to stay twelve instead of the five for the meeting necessary days in Johannesburg, the Federation withdrew its offerte. Evidently for most of the delegates tourism is at the termin scheduling."

"The death of Angelo JACOPUCCI, son of a scavenger, has revived an old discussion. It was the 'seventh' case this year. But since the cameras do not make them so publicly as in the television broadcast on Thursday, so long is the death a more or less anonymous affair. How did it come that for the six other cases the report of execution was not worthwhile?"

"'The joy at the European championship title has died with Angelo,' declared the new European champion MINTER,

Fig. 344. Professional heavyweight fight between Johnny RIGGINS and Alejandro LAVORANTE on September 21, 1962 in Las Vegas. LAVORANTE, once a contender for the world's heavyweight crown, sits crouched down, dazed and beaten in his ring corner after being knocked out by Johnny RIGGINS. His manager, Pinky GEORGE, hurries on to help him. LAVORANTE had to be carried on a stretcher from the ring. Photo: U.P.I.

and his manager Doug BIDWELL added, that we do not know it for sure. This is one of the things where we have to sit down and talk about it. A honorable intention, which coincides with the deliberations of many afficinados of the box sport. Deliberation due to piety – up to the next time.''

Professional Boxing Match between Ray 'Boom Boom' MANCINI and DUK KOO KIM

In a defending champion title match Ray 'Boom Boom' MANCINI fought against DUK KO KIM, an in the United States completely unknown Korean boxer, on November 3, 1982, who was termed by the *World Boxing Association (WBA)* as no. 1 challenger. Before this boxing match, so was reported, KIM wrote allegedly at a lampshade in a hotel room in Las Vegas 'kill or be killed.'

Fig. 343. Professional heavyweight fight between Archie MOORE (right) and Alejandro LAVORANTE on March 30, 1962 in Los Angeles. LAVORANTE sags in the tenth and final round to the canvas. Referee Tommy HART stopped the fight after 38 seconds of the tenth round after he had taken a severe beating from MOORE. LAVORANTE was groggy and no longer able to defend himself. On the right is Archie MOORE. Photo: A.P.

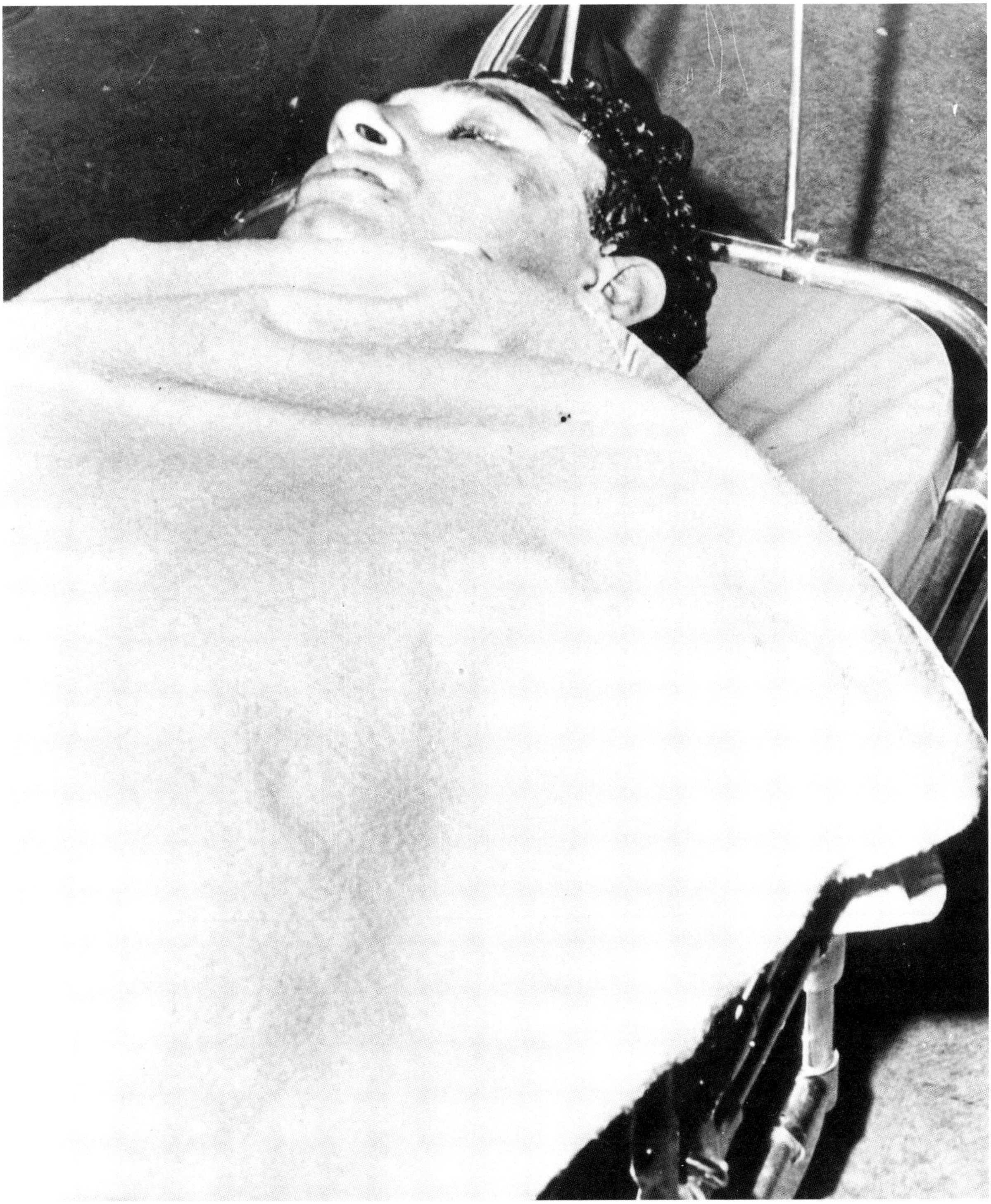

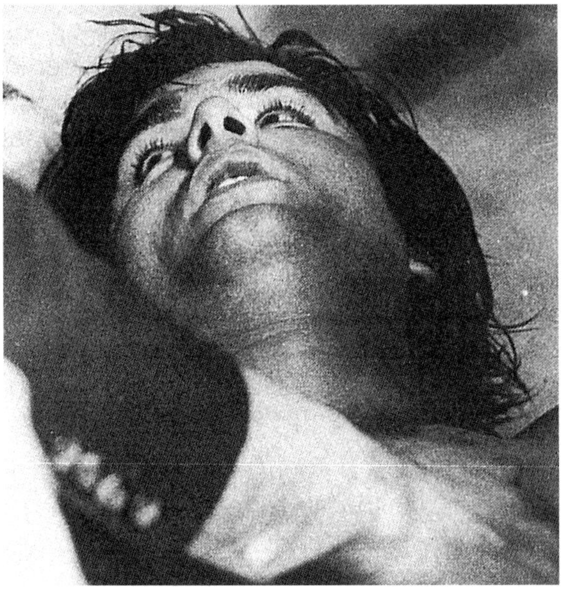

Fig. 346. Juan Rubio MELERO was knocked out by the Spanish middleweight champion Francisco RODRIGUEZ. He is attended by the ring physician. Photo: KEYSTONE.

Short summarising description of the pugilistic careers of DUK KOO KIM and Ray "Boom Boom" MANCINI:

DUK KO KIM fought as *professional boxer* from 1978 to 1982, he participated in *20 professional boxing fights,* he won 17, lost two and one bout ended in a draw, he achieved eight victories by a knock out and was knocked out one time. He challenged MANCINI for the WBA lightweight title and lost this boxing match in the fourteenth round by a knock out.
Ray 'Boom Boom' MANCINI (Raynold Michael) MANCINI fought as *professional boxer* from 1979 till 1988, he participated in 33 professional fights, he won 29, lost 4, and he achieved 23 knock out victories and won the WBA lightweight title by a knock out victory in the first round against Art FRIAS, he defended his title four times successfully, and lost the championship title to Livingstone BRAMBLE by a knock out defeat in the fourteenth round on June 1, 1984.

The boxing match between Ray 'Boom Boom' MANCINI and the Korean DUK KOO KIM took place on November 13, 1982 and was transmitted 'live' by a national television broadcast corporation.

NOONAN, an eyewitnss of this boxing bout had observed that DUK KOO KIM had in his last fist fight passed through several stages. During the first seven rounds he had to take many blows, but he dealed out just as many. He fought a man against man fist fight, in a style, which was similar to that of his opponent MANCINI. In the eigth round KIM changed his manner to fight. For the first time KIM slowed down his actions. Despite the fact that he in the following rounds again returned to attack, it seemed, as if KIM had problems with his coordination and had a 'reduced reflex ability.'
In the next three rounds gave MANCINI's better physical condition him advantages and symptoms of fatigue were now more distinctly visible at his opponent. In the eleventh round KIM's attacks became wilder, he missed his target wide. In the twelfth round he stumbled sporadic and had to take series of blows against his head. In the thirteenth round MANCINI hit him about forty times successively. The majority of these blows, which remained unanswered, were directed against the head. They were series of blows, which usually lead to 'groggy states' in a boxer. He staggered and tottered and seemed no longer be able to defend himself. A commentator remarked, that his legs looked like spaghetti. In the fourteenth round KIM had to take two especially powerful blows against his head. At the end of this round KIM attacked again, for the last time. The 'groggy' KIM collapsed and was in short time deeply comatous. The last right of MANCINI against KIM's head thrust the challenger staggering through the entire ring and at the canvas. He was immediately deep unconscious and was a short time later braindead. KIM was treated with a heart-lung maschine. After the district judge of the State of Nevada, Paul GOLDMAN, had declared KIM legally dead, the maschine was disconnected.

Millions of horrified spectators were able to see the entire fight 'live' and watched this execution. It was up to this time the largest television audience, who watched and was present at a ring death.

The boxing elite made some cosmetic changes. The decisive change, which the *World Boxing Council (WBC)* carried out, was a decrease of the number of rounds from 15 to 12. One reasoned that shorter boxing matches would reduce injuries, which were allegedly the result of fatigue. The

Fig. 345. The Argentinian heavyweight LAVORANTE is carried on a stretcher from the ring after his knock out by Archie MOORE. He was transferred unconscious by an ambulance to a Los Angeles hospital. Photo: A.P.

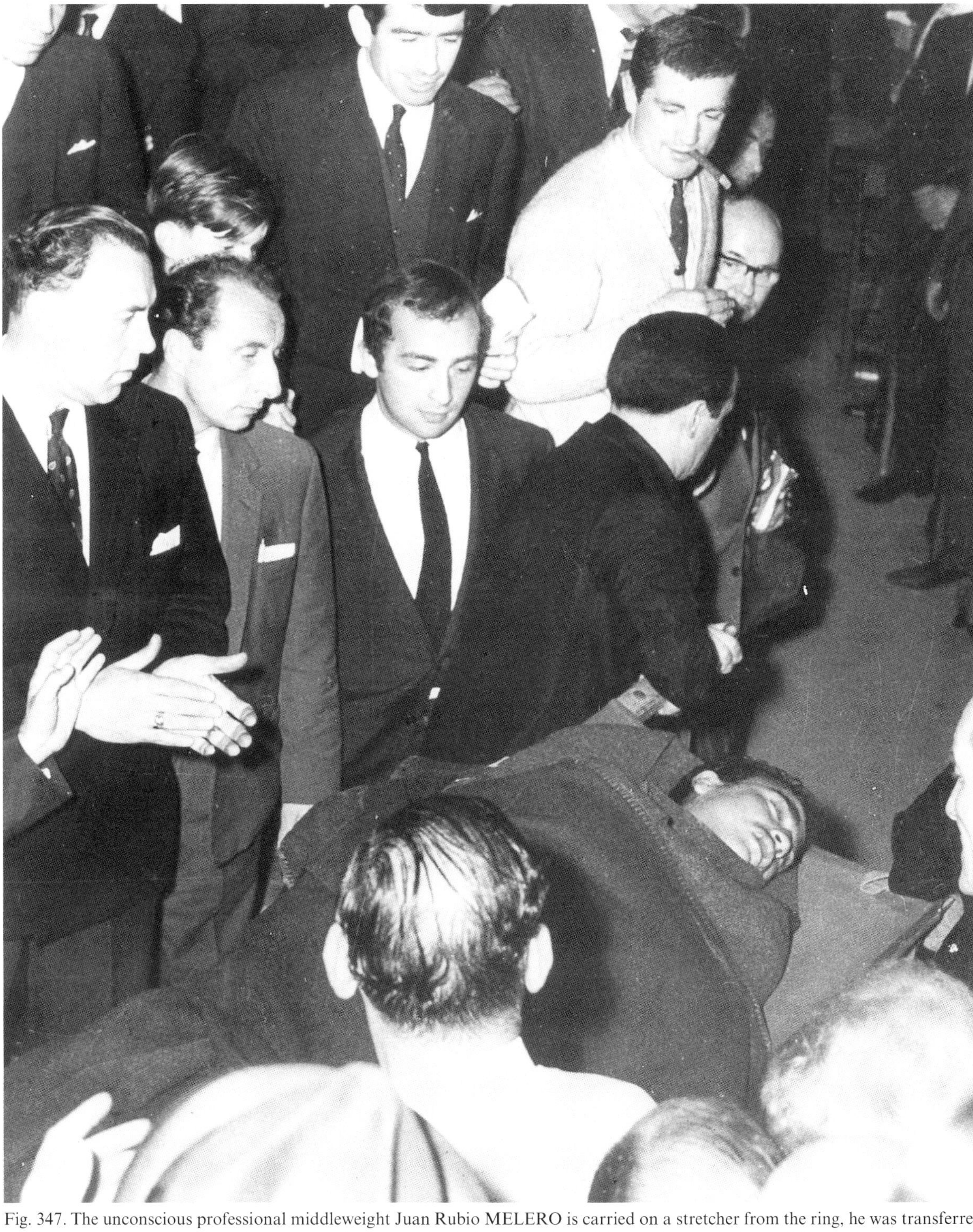

Fig. 347. The unconscious professional middleweight Juan Rubio MELERO is carried on a stretcher from the ring, he was transferred to a hospital, where he died four days later on the sequels of a brain hemorrhage. The reaction on the faces of the spectators is different. Photo: KEYSTONE.

problem at KIM's death was, however, not so much exhaustion but his death was the result of series of hammer hard blows against his head, it was a classic example of a *'hammering or pummeling knock out' ('Verhämmerungs-knock out').*

Professional Boxing Match in which Rickie SANDOVAL was knocked out

The American writer Carol Joyce OATES, who had published in 1987 a noteworthy book with the title *'On Boxing'* remarked as follows:

"More justified perhaps, was my sickened sense that boxing is, simply, wrong, a mistake, an outlaw activity for some reason under the protectorate of the law, when, a few weeks ago, in March 1986, I sat in the midst of a suddenly very quiet closed-circuit television audience in a suburban Trenton hall watching bantamweight Richie SANDOVAL as he lay flat and unmoving on his back... very likely dead of a savage beating the referee had not, for some reason, stopped in time. My conviction was that anything was preferable to boxing, anything was preferable to seeing another minute of it, for instance standing outside in the parking lot for the remainder of the evening and staring at the stained asphalt..."

Professional Boxing Fight between Lupe PINTOR and Johnny OWEN

The tragically ending *professional boxing match* between Lupe PINTOR and Johnny OWEN was described by McILVANNEY (1988):

"The first round was probably about even. OWEN might have edged the second on a British scorecard and he certainly took the third but already PINTOR's right hooks and uppercuts were making occasional dramatic interventions, sending nervous chills through the challenger's friends around the ring."

"It was in the fourth round that PINTOR's right hand first struck with a hint of the force that was to be so overwhelming subsequently, but this time it was thrown overarm and long and OWEN weathered it readily enough. He was seen to be bleeding from the inside of his lip in the fifth (the injury may have been inflicted earlier) but, since both PINTOR's eyebrows were receiving attention from his seconds by then, the bloodshed seemed to be reasonably shared. In fact the laceration in the mouth was serious and soon the challenger was swallowing blood. He was being caught with more shots to the head, too, but refused to be discouraged, and an American voice behind the press seats said incredulously 'I dont believe this guy.'"

"PINTOR was heaving for breath at the end of the fifth but in the sixth he mounted a surge, punished OWEN, and began to take control of the contest. The official doctor, Bernhard SCHWARTZ, checked the lip for the second time before the start for the seventh. PINTOR dominated that one but OWEN revived heroically in the eighth, which made the abrupt disaster of the ninth all the more painful."

"PINTOR smashed in damaging hooks early in the ninth but their threat appeared to have passed as the round moved to its close. Then, without a trace of warning, PINTOR chopped a smattering right hook over OWEN's bony left shoulder. The blow hurled him to the floor, and it was here that his courage began to be a double-edged virtue. He rose after a couple of seconds although clearly in a bad condition. There was a mandatory '8' count but even at the end of it he was hopelessly vulnerable to more hooks to the head and it took the bell to save him."

"By the tenth there was unmistakable evidence that the strength had drained out of every part of OWEN's body except his heart. He was too tired and weak now to stay really close to PINTOR skin against skin, denying the puncher leverage. As that weariness gradually created a space between them, PINTOR filled it with cruel, stiff-armed hooks. Every time OWEN was hit solidly in the eleventh the thin body shuddered. We knew the end had to be near but could not foresee how awful it would be."

"There were just 40 seconds of the twelth round left when the horror story started to take shape, OWEN was trying to press in on PINTOR near the ropes, failed to prevent that deadly space from developing again and was dropped on his knees by a short right. After rising at '3' and taking another mandatory count, he was moved by the action to the other side of the ring and it was there that a ferocious right hook threw him on to his back (Fig. 348). He was unconscious before he hit the canvas and his relaxed neck muscles allowed his head to thud against the boards. GARDINER and his father were in the ring long before the count could be completed and they were quickly joined by SCHWARTZ who called for oxygen. Perhaps the oxygen might come rather more swiftly that it did but only if it had been on hand at the ringside. Obviously that would be a sensible precaution. Just as it might be sensible to have a stretcher immediately available. It is no easy job to bring such equipment through the crowd of spectators at an arena like the Auditorium where PINTOR's supporters were mainly concerned about cheering the arrival as a symbol of how comprehensive their man's victory had been. The outward journey to the dressing room with poor Johnny OWEN deep in a sinister unconsciousness was no simple task and the indifference of many of the crowd was emphasized, when one of the stretcher bearer had his pocket picked."

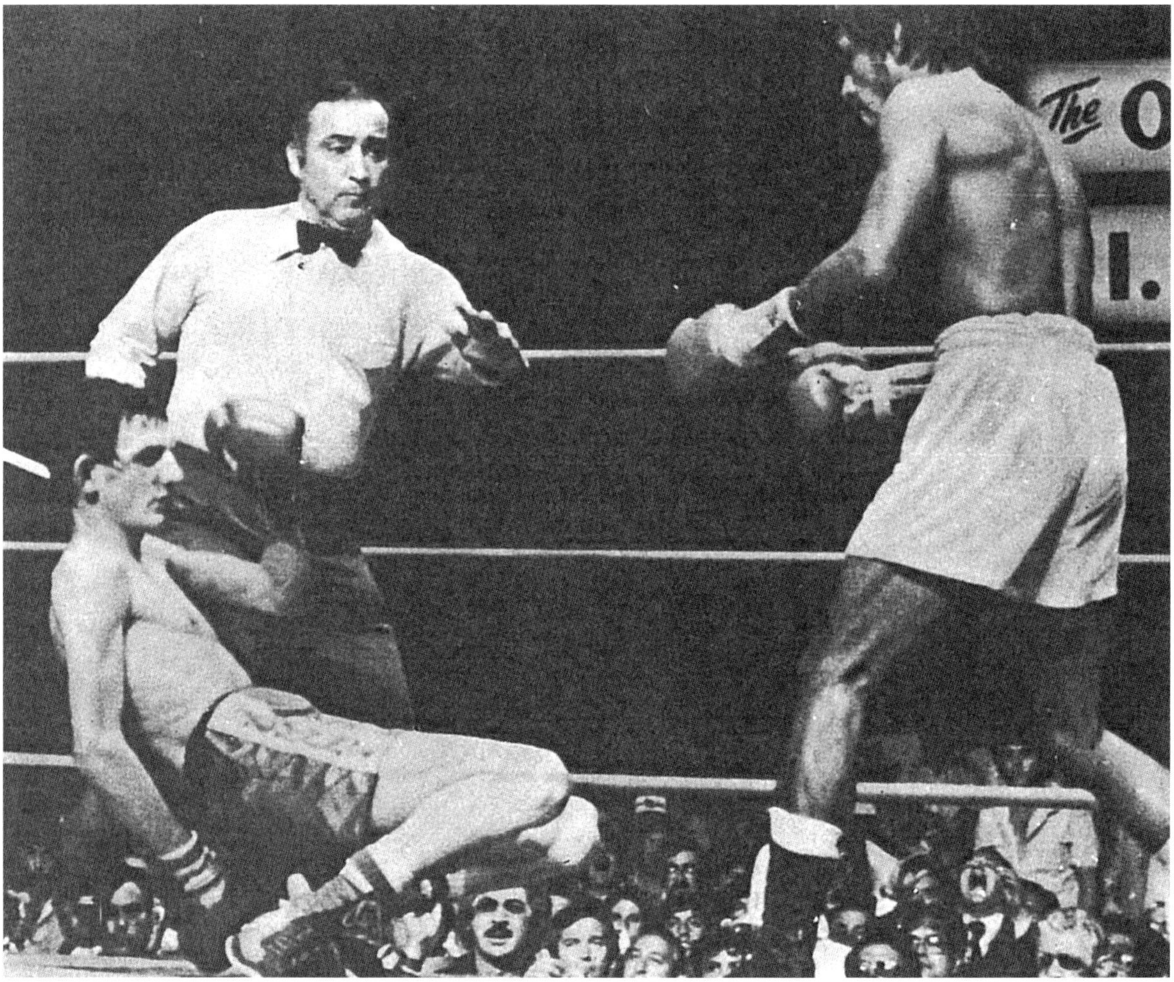

Fig. 348. Professional boxing match between Lupe PINTOR and Johnny OWEN on September 19, 1980, in Los Angeles, California. The photo shows the knock out of OWEN in the twelth round. OWEN died in a hospital.

"OWEN died in deep coma in the first week of December 1980."

"PINTOR comes out of a gym in Mexico City were more than a hundred work out regularly and others queue for a chance to show that what they can do in the alleys they can do in the ring."

OATES (1997) noted about the outcome of the fight:

"Spectators at public games derive much of their pleasure from reliving the communal emotions of childhood but spectators at boxing matches relive the murderous infamy of the race. Hence the occasional savagery of boxing crowds – the crowd, largely Hispanic, that cheered as the Welchman Johnny OWEN was pounded into insensibility by the Mexican bantamweight champion Lupe PINTOR, for instance – and the excitement when a man begins to seriously bleed."

Fatal Outcome of a Professional Fist Fight between Ken KATAGIRI and Fusaaki TAKANAGA

The *Japanese bantamweight* Ken KATAGIRI died due to a brain injury, two weeks after he lost a fight by a knock out against Fusaaki TAKENAGA. KATAGIRI was knocked down by Fusaaki TAKENAGA in the ninth round of a scheduled 10-round boxing match at the *Korakuen Boxing Hall in Tokyo* on October 12, 1998 by TAKENAGA. He got up and

tried to continue the fight, but had to take severe punishment to his head and was knocked down again. KATAGIRI became unconscious and died in a coma, neurosurgery failed to safe him. KATAGIRI began 1991 to fight as a professional, he had a record of five victories, six defeats and two draws.
Since the foundation of the *Japanese Boxing Commission* in 1952 he is the *28th Japanese boxer who died due to a fatal incident in the boxing ring.*

Professional Boxing Match between Jimmy GARCIA and Gabriel RUELAS

The 23 year old *Columbian professional featherweight boxer* Jimmy GARCIA fought on May 6, 1995 in Las Vegas for the championship title of the *WBO.* GARCIA died despite a neurosurgical operation 13 days later. The attending physician interrupted at this time the artificial respiration of the brain-dead GARCIA.
Frank BLACKMAN *(Boxing, like other sports, risks death, Lexington Herald Leader, Tuesday, June 6, 1995)* commented:
"So now it begins again. A boxer dies, and there is a clamor to abolish the sport. How predictable. How absurd."
"Jimmy GARCIA's death in the ring was a tragedy. Only a ghoul would want to see someone injured – or worse."
"But was GARCIA's death any less of a tragedy than the horrific crash during the Indianapolis 500 last week that sent Stan FOX to the hospital in critical condition? Please explain what intrinsic value there is in strapping a man into a metal shell, installing a humongous engine on the frame, and allowing to drive in traffic at more that 200 miles an hour."
"Still, despite stomach-turning film of FOX unconscious, his exposed legs dangling from the sheared off end of the racecar, there wasn't a spontaneous call for a referendum on the validity of auto racing in a civilized society."
"Was GARCIA's death any less of a tragedy than the fall from a horse that has paralysed actor Christopher REEVE and put him on a life support system? Should show jumping be banned, too?"
"Isn't there one thing all three tragedies have in common? Each man was exercising his free will to pursue a sport that was potentially dangerous."
"And of the three, GARCIA probably had the most motivation and was the most grateful for the opportunity. A poor kid from Latin-America, he decides this was his way to earn a living for himself and his family. If he could take a punch and deliver one, he might make more money in one night than during an entire lifetime cratching out an existence back home."
"Would you risk your life for a dream of wealth and fame? Why is that ignoble? GARCIA's heart was bigger than his body.
"He took a punch too well, absorbing so many blows from Gabriel RUELAS that it eventually killed him. But GARCIA died doing something he wanted to do. He died young, but with honour. Why is that shameful?"
"Of course there is violence in boxing. Men hit men. Damage is inevitable. And there is an atavistic response from spectators that can't be ignored. There is something primal in watching one man hit another. It resonates deep within us – at least within some of us. Maybe it's why some of us are confirmed carnivores and others only eat leafy green vegetables."
"But there is also rare skill involved, the beauty a boxer displays nimbly avoiding harm while inflicting it. There was nothing brutish about MUHAMMAD ALI at work. Or Sugar Ray ROBINSON. Or Sugar Ray LEONARD. They were unforgettable"
"Don't forget Benny Kid PARET, either. Or DUK KOO KIM. Or GARCIA. Don't forget any of the other fighters who climbed into the ring and had to be carried out of it. They are a reminder that this can be deadly."
"So make sure that the referee is alert and that a doctor is in attendance. Try to limit risks as much as humanly possible."
If there is anything about GARCIA's death to be abhorred, it is that he was allowed into the ring by advisers either too stupid or greedy to have his best interests at heart,"
"GARCIA reportedly had to loose 30 pounds to make the weight for the 130-pound title fight. Yet no one in his camp apparently considered the implications of that. With his weight went his power. He could barely fend off RUELAS and was pawing at him in an effort to stop a relentless attack that ultimately wore him down and killed him."
"Maybe GARCIA's father, who was his trainer, didn't know any better."
"There are fine people involved in boxing. Spend time talking to the men and women who referee and judge matches and you begin to understand that their love of the sport is pure."
"But boxing also attracts creatures who crawl out of sewers. Their only goal is to make a buck. And they'll exploit whomever they have to to do it. It was the slimeballs who doomed GARCIA."
"So outlaw them. Jail them. And, then get the hell out of the way and let the men fight."

Human Immunodeficiency Virus (HIV) Exposure

Concern for the spread of HIV through loose or flying blood in the boxing ring has steadily risen in recent years. In comparison with response to other known medical risks and concerns in boxing, enactment of regulations to prevent HIV exposure have been effectuated with lightening speed, though the actual risk of HIV exposure in the boxing ring is very limited. Initially it appears that boxing regulators have acted to protect boxers and officials from medical risks of boxing, but on reflection, these regulations may better protect the bloody nature of boxing.

Some boxers have admitted engaging in activities such as IV drug use, increasing the risk of HIV infection. The initial fear was that their bloody injuries in the ring would expose referees, seconds, physicians, or others to HIV. Thus universal precautions were mandated by many local regulators, and the use of gloves by seconds and officials came into common use.

The issue of boxers being infected by their opponents was later raised in a letter to the editor of the *New York State Journal of Medicine* (Valiere ALCENA, 88:392, 1988) where it was suggested that admitted drug users be given confidential HIV tests, and if positive, they should be required to observe precautions to prevent their exposing others to the disease. Head protection was suggested as an example of one such precaution, presumably to protect areas that most commonly produce bloody wounds. Although it would be difficult to keep HIV test results confidential if extra equipment were required, the point was made that boxers might be more likely to risk exposure than referees and seconds.

Copious amounts of blood are common in the ring and are not considered sufficient reason to stop a fight. HIV exposure in the ring would require co-mingling of the infected blood with that of another person. Spattering or smearing of HIV infected blood could only infect the other person if it were to enter the body, for example, through an open wound. As a practical matter, the only individuals in or near a boxing ring who might have open wounds would be the two fighters themselves. Even between the fighters, the odds of becoming HIV positive as a result of ring activities is low. At present there are no known cases of HIV infection from boxing matches.

HIV testing is now required as a matter of local boxing regulations. In the United States boxing regulations are primarily enacted by state governments. At present a number of states have added HIV testing to their statutes or regulations, including Nevada, California, New Jersey, Louisiana.

Current HIV regulations consider prevention of HIV exposure for boxers and officials alike. The HIV regulations require that all boxers be tested for HIV prior to a match, and if the results are positive, they are banned from the ring. There has been at least one case of a boxer who was banned because he was found to be HIV positive. Of course these rules would be difficult or impossible to enforce in sparring or in boxing gyms, where similar contact occurs on a daily basis.

A letter to the editor of Journal of the American Medical Association compared HIV regulations in boxing with those in other sports. JAMA, D. Peter DROTMAN, MD MPH. Peter DROTMAN, MD MPH 276:193 1996. Dr. DROTMAN, of the CDC, suggests that nearly all major athletic organizations, at all levels, have adopted some form of the HIV precautionary rule that when a bleeding injury occurs, play must stop, only to resume when all injured players have stopped bleeding and their wounds have been dressed. Athletes are to be educated on the risks of HIV, but are not tested for the purpose of excluding HIV positive players. He adds that there are no known cases of HIV infection from sports activities and that in the only sport where the HIV risk factor has been estimated, professional American football, the risk of transmission was found to be miniscule. He concludes that the only reason for HIV testing of boxers is so that they can continue to box regardless of how badly they are bleeding during a match, saying "If the health of the boxers is paramount, professional boxing should do what other major sports do: stop bouts when a fighter is bleeding and tend to the wound."

Amateur Boxing in the Armed Forces and Police Academies

BRENNAN and O'CONNOR reported 1968 on *'Incidence of Boxing Injuries in the Royal Force in the United Kingdom 1953 – 1966.'*

The report begins: "In recent years the injuries suffered by boxers have been the subject of world-wide interest. The evidence presented by those who regard boxing as an unacceptably dangerous sport and those who disagree with this view has been conflicting and has come from so many sources that objective assessment of the hazards of this sport has not hitherto been possible. The divergence of opinion which exists was shown at a symposium arranged by the *British Association of Sport and Medicine in London* in November 1963 (BASS, BLONSTEIN, JAMES, and WILLIAMS (eds) 1965)."

In their paper the authors describe the injuries suffered by *Royal Air Force amateur boxers in the United Kingdom during the 14 years 1953 – 1966*. They listed the boxing injuries which caused absence from work for 48 hours or more, the nature of these injuries, the number of days under treatment, and the permanent damage caused. They also estimated the incidence of injuries per 'man-bout' boxed during the seven years, as the *Royal Air Force* began strict control of boxing in 1959. The authors use the term *'man-bout'* to count the number of occasions on which a *Royal Air Force boxer* entered the ring. The incidence of injuries per man-bout is a measure of how often accidents occur to boxers.

BRENNAN and O'CONNOR continue:

"Boxing in the *Royal Air Force* is controlled by the *Royal Air Force Boxing Association (RAF BA)* under the rules of the *Combined Services Boxing Association*. The *RAF BA* organizes and controls its own medical scheme, which was introduced in October 1959. This scheme depends for its success on *three factors:* first, the issue of a *medical record card to each registered RAF boxer,* without which he is not permitted to take part in competitive boxing under *CSBA rules; second,* a system of medical examinations which is intended to ensure that a boxer is physically fit before he enters the ring; and *third,* the duty of the Association officials supervising a tournament (the official in charge, the referee, and the tournament medical officer) to protect boxers from excessive punishment by stopping contests which are ill-matched and ensuring adequate time for recovery from injury by periods of suspension.

Detailed recording of all bouts was begun in 1959 as an integral part of the *RAF BA* medical scheme. These records describe the method of win, particulars of any injuries suffered, and details of any suspension on medical grounds.

Method: An important fact is, according to BRENNAN and O'CONNOR that details of all airmen who are off duty for 48 hours or more because of illness or injury are recorded in full on Royal Air Force documents. For this reason the sickness statistics in the closed community of the Royal Air Force are extremely accurate.

Results: In the period from *1953 – 1966 240 Royal Air Force amateur boxers* sustained injuries in the ring, which resulted in the loss of 2.627 working days (Table 101). These accidents included 142 injuries of the head and neck; 139 of these men returned to duty after a mean period off work of nine days. There were 39 injuries to the upper

TABLE 100
Details of three men who did not return to duty after boxing injuries. (From BRENNAN, O'CONNOR, 1968).

Year	Casualty Detail	Disposal
1953	Head injury	Died: cause, laceration of brain
1956	Head injury	Invalided from RAF: cause, post-concussional state
1959	Head injury	Died: cause, laceration of brain

limbs, which resulted in a mean period off duty of 12 days. All other injuries due to boxing totaled 59. These men returned to work after an average stay in hospital of 16 days. In the period 1953 – 1966 there were two fatal head injuries due to boxing and one injury which resulted in invaliding (Table 100).

Table 102 shows how often *Royal Air Force boxers* have been injured since the *RAF BA* tightened their safety precautions in 1959. The injury rate varied widely from year to year, but on average out of every 1.000 men entering the ring, 6.2 sustained an injury which made them unfit for work for some days. Of the 6.2, 3.9 were injured in the head or neck. There have been no fatal injuries and no invalidings as a result of boxing in the *Royal Air Force* since the new regulations were introduced in the 1959/1960 season.

TABLE 101
Boxing injuries sustained by RAF contestants 1953-66, showing number of injuries and site. (From BRENNAN, O'CONNOR, 1968).

Year	Site of Injury							
	Head and Neck		Upper Limbs and Shoulders		Other		Total	
1953	15 (+ 1 died)	(12)	7	(14)	3	(27)	26	(14)
1954	20	(9)	5	(22)	9	(35)	34	(18)
1955	20	(9)	4	(15)	10	(10)	34	(10)
1956	13 (+ 1 invalided)	(7)	2	(3)	7	(15)	23	(9)
1957	13	(6)	2	(29)	7	(13)	22	(11)
1958	5	(15)	2	(6)	6	(13)	13	(12)
1959	7 (+ 1 died)	(7)	4	(5)	3	(6)	15	(6)
1960	20	(8)	6	(8)	3	(3)	29	(8)
1961	1	(5)	2	(7)	6	(20)	9	(16)
1962	7	(7)	1	(13)	3	(6)	11	(7)
1963	4	(5)	0	(–)	0	(–)	4	(5)
1964	8	(9)	1	(22)	1	(6)	10	(10)
1965	4	(5)	3	(9)	0	(–)	7	(7)
1966	2	(14)	0	(–)	1	(14)	3	(14)
Total	139 (+ 2 deaths) (+ 1 invalided)	(9)	39	(12)	59	(16)	240	(11)

The mean number of working days lost is given in parentheses.
See Table 100 for details of three men who did not return to work.

Table 103 shows that out of 11.820 man-bouts, 2.706 (837 plus 1.869), or less than one quarter, ended prematurely either because a contestant was counted out or because the referee stopped the bout. Contests were stopped because the boxers were uneven matched or if one contestant risked further injury by continuing. Medical suspension from boxing for a period varying from a few weeks to a whole season followed on 952 occasions. Medical suspension was invoked when the *RAF BA officials* considered that a man had sustained punishment that results of which may or may not require medical treatment but could have an adverse effect on his health, especially if closely followed by further boxing injuries. In the opinion of the medical advisor to the *RAF BA* about one third of the 952 medical suspensions followed concussive blows to the head. This figure of about 300 minor head injuries in 11.820 man-bouts may be, according to the authors, the most important statistic in measuring the dangers of boxing as a sport.

Table 103 also shows that the ratio of medical suspensions to man-bouts for the first three years of detailed recording (1960 – 1962) was 6.6%, whereas the four-year period 1963 – 1966 the ratio was 10%. While there has been an increase in medical suspensions, the number of recorded injuries sustained in the ring has remained low, and it is suggested that the increase in medical suspensions reflects a more rigid application of safeguards by referees and tournament medical officers. It is considered of interest to note that the proportion of bouts which were stopped or ended in knock outs has remained fairly constant throughout the seven years (Table 103).

Discussion: BRENNAN and O'CONNOR noted that many doctors would accept that the acute injuries from carefully supervised amateur boxing are not excessive. They continued that the important question is whether chronic traumatic encephalopathy (the punch drunk syndrome) occurs after amateur boxing. The authors quote CRITCHLEY, who stated that he had seen the syndrome in an amateur boxer. During the 14 years 1953 – 1966 no case of the punch-drunk syndrome

TABLE 102
Number of contests which ended prematurely (counted out plus referee stopped bout) and resulting suspensions. (From BRENNAN, O'CONNOR, 1968).

Year	Man-bouts	Counted Out	Referee Stopped Bout	Bouts Stopped or Counted Out/100	Medical Suspensions
1960	2.190	141	361	22.9	182
1961	2.582	190	382	22.2	155
1962	1.964	158	280	22.2	109
1963	1.626	125	262	23.8	154
1964	1.186	72	208	23.6	125
1965	1.384	97	238	24.2	161
1966	888	54	138	21.6	66
Total	11.820	837	1.869	–	952

TABLE 103
Number of man-bouts boxed by RAF contestants 1960–66. Injuries sustained in these bouts, and incidence of injuries per thousand man-bouts. (From BRENNAN, O'CONNOR, 1968).

Year	No. of Man-bouts	Head and Neck		Upper Limbs and Shoulders		Other		Total	
		Cases	Rate per 1.000	Cases	Rate per 1.000	Cases	Rate per 1.000	Cases	Rate per 1.000
1960	2.190	20	9.1	6	2.7	3	1.4	29	13.2
1961	2.582	1	0.4	2	0.8	6	2.3	9	3.5
1962	1.964	7	3.6	1	0.5	3	1.5	11	5.6
1963	1.626	4	2.5	—	—	—	—	4	2.5
1964	1.186	8	6.7	1	0.8	1	0.8	10	8.4
1965	1.384	4	2.9	3	2.2	—	—	7	5.1
1966	888	2	2.3	—	—	1	1.1	3	3.4
Total	11.820	46	3.9	13	1.2	14	1.2	73	6.2

was diagnosed in the *Royal Air Force* and neither, during that time, did any man claim that boxing had impaired his mind. The authors mention that it may be that their period of observation was too short as most of the men leave the *Royal Air Force* before the age of 56. The authors note: "If chronic traumatic encephalopathy does occur in amateur boxing and it is due to repeated cerebral concussion, then the incidence of concussion in Royal Air Force boxers may be the most important statistic."

The authors continue to say that there have been two deaths from amateur boxing in the *Royal Air Force* during the 14 years under review, showing that boxing carries some risk to life. Apart from the two fatal accidents, known permanent damage was rare. One out of 240 casualties in 14 years was left with disabilities necessitating discharge from the Service. According to BRENNAN and O'CONNOR, this figure is much smaller than they had expected, in the light of criticism of the sport. They continue: "It may be that neurological or other sequelae may yet develop because the period of observation has been too short."

OELMAN et al. published 1983 a paper *on 'Boxing in the [British] Army."* The authors begin their contribution by remarking that the recent death of a junior soldier following a civilian boxing match raised questions about the safety of boxing as a sport.

Methods: Records of all admissions to Hospitals and Medical Reception Stations of Service Personnel for two days or more are held in an archive. The authors retrieved details of all Army personnel admitted because of boxing injuries during the years 1969 – 1980 or who died or were medically discharged between 1952 and 1980.

During the 12 year (1969 – 1980) there were 437 admissions. Table 104 shows the distribution of the injuries by year and site. There is, according to the authors, no obvious trend over time for the number of injuries or between types of injury but, as might be expected, head injuries predominated and accounted for 67.7% of all admission. Table 105 gives the total cases and inpatient days for selected diagnoses.

There were 296 head injuries, made up of 83 (28%) *fractures of the face* (ICI code 802), 115 (19%) cases of *concussion* (code 850), 69 (23%) cases of *'intracranial injury of other and unspecified nature'* (code 854), 13 (14.4%) cases of *contusion of face, scalp, and neck* (code 920), 5 (1.7%) cases of *contusions of the eye* and *orbit* (code 921), two cases (0.68%) of *cerebral laceration* and *contusion* (code 851) of which one resulted in a medical discharge, and 3 (1%) cases of *subdural haemorrhage* (code 852) all of which resulted in medical discharges. The others were made up of a miscellaneous collection ranging from two cases of open wound of ear (code 872) to *one crushing injury of face* and *neck* (code 925).

The *average length of stay for head injuries* was 8.77 days. The highest average length of stay was 242 days for *cerebral laceration* and *contusion* but one of the cases was in hospital for 471 days before being medically discharged from the Army. The *fractured face bone* cases stayed for 8.88 days and the *concussion* patients for 3.91 days on average. The other *intracranial injuries* averaged 8.58 days with a maximum of 21 days. The other high average of this group was 59.3 days for three cases of *subdural haemorrhage,* all resulting in medical discharge.

Of the 59 *upper limb injury* admissions 28 (47.5%) were for *fractured metacarpal bones* (code 815) and these cases averaged 12.1 days in hospital. The other cases were for various *minor fractures, dislocations, sprains, strains,* and *contusions* and the only groups of any note were eight (13.6%) cases of *fractured fingers* or *thumbs* (code 816) and 9 (15.5%) cases of *dislocated shoulders* (code 831). Of the two cases of *open wounds of the upper limb* (code 881) one case of *laceration* of the *right wrist* was in hospital for 174 days before being medically discharged, which accounts for the high average stay (12.5 days) for arm injuries.

"There were only 21 *injuries to the lower limbs* and no large group, however, three cases of *fractured tibia* and *fibula* (code 823) involved a total of 136 days in hospital and seven cases of *dislocated knee* (code 836) totaled 248 days. One of these cases was medically discharged after 82 days in hospital and another required extensive rehabilitation with a total stay of 84 days. There were three cases of *sprains* and *strains* to the *knee* and *leg* (code 844) involving 152 days in hospital but this was inflated by one of the cases which required an operation on the patella ligament and joint capsule

TABLE 104
Admissions to medical unit from boxing injuries Army personnel, 1969-1980. (From OELMAN et al., 1983).

Year	Site Total	Head	Upper Limb	Lower Limb	Trunk	Internal	Other
1969	53	35	9	2	2	1	4
1970	34	20	6	1	3	1	3
1971	18	11	4	0	0	0	3
1972	40	28	5	1	1	1	4
1973	27	18	4	2	0	1	2
1974	24	11	3	3	6	0	1
1975	34	26	4	1	1	2	0
1976	26	18	4	0	2	0	3
1977	43	32	2	5	0	2	2
1978	47	30	7	2	2	3	3
1979	36	29	2	2	1	0	2
1980	54	38	9	2	2	2	1
Totals	437	296	59	21	20	13	28
(%)		(67.7)	(13.5)	(4.8)	(4.6)	(3.0)	(6.4)

TABLE 105
Number of admissions and inpatient days for selected diagnosis for boxing injuries. Army personnel, 1969-1980. (From OELMAN et al., 1983).

Diagnosis	ICD Code Ninth Edition	Number of Admissions	Total Inpatient Days	Average Inpatient Days
Total	—	437	4540	10.4
Head Injuries	—	296	2595	8.8
Fracture of Face Bones	802	83	737	8.9
Concussion	850	115	450	3.9
Cerebral laceration and contusion	851	2	484	242.0
Subdural haemorrhage	852	3	178	59.3
Other Intracranial Injury	854	69	592	8.6
Contusion of Face Scalp and Neck	920	13	104	8.0
Other	—	11	50	4.5
Upper Limb Injuries	—	59	740	12.5
Fracture of humerus	812	1	21	21.0
Fracture of radius and ulna	813	3	48	16.0
Fracture of metacarpal bones	815	28	342	12.2
Open wound of elbow forearm and wrist	881	2	177	88.5
Other	—	25	152	6.1
Lower Limb Injuries	—	21	580	27.6
Fracture and tibia and fibula	823	3	136	45.3
Dislocation of knee	836	7	248	35.4
Sprains and strains of knee and leg	844	3	152	50.7
Other	—	8	44	5.5
Trunk Injuries		20	265	18.3
Spondylosis	713	1	153	153.0
Fracture or fracture dislocation of the vertebral column without spinal lesion	805	2	45	22.5
Fracture of the ribs or larynx	807	3	40	13.3
Spinal cord lesion	958	1	33	33.0
Other	—	13	94	7.2
Internal Injuries	—	13	81	6.2
'Other' Injuries	—	28	179	6.4
Injuries other and unspecified	959	21	125	6.0

Note: Conditions for inclusion of specific diagnosis – twelve or more cases during the period or an average of ten or more a day in hospital.

followed by physiotherapy, the total stay being 143 days.

There were twenty admissions for *injuries to the trunk with contusions* (code 922) accounting for seven of them. Of the remainder two *fractures of the vertebral column* (code 805), one case of *spondylosis* (code 952) were the more serious cases. The latter two cases resulted in medical discharges.

Internal injuries accounted, according to OELMAN et al., for only thirteen of the cases and there were no outstanding types either for number or length of stay. The injuries varied between *injuries to kidneys* (4) to those of the *pelvic organs* (3). Average length of stay was 6.23 days. The 'other' injuries group consists mainly of those injuries where sufficient information was not given on the records received for us to determine the location and nature of the injury. All the cases were of short duration.

Figure 349 shows number of cases with each length of stay. It can be seen that 173 out of the 437 cases (40%) only stayed in hospital for three days (note that (1) both the day of discharge and the day of admission are counted, thus three days

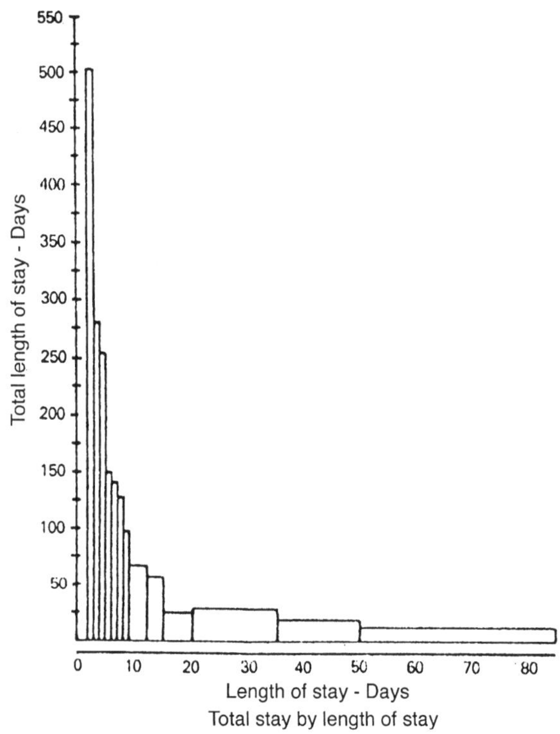

Fig.349. Number of cases by length of stay. From OELMAN et al., 1983).

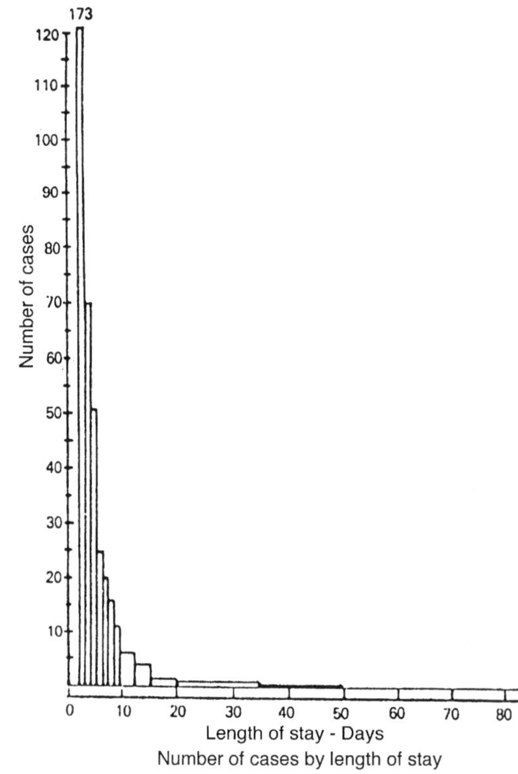

Fig. 350. Total stay by length of stay. (From OELMAN et al., 1983).

stay represent only two nights in hospital and (2) no cases of two days or less are included in the study). A further 16% of the cases were in for only four days and 90% of cases were discharged within two weeks of admission.

Figure 350 shows the same data but highlighting the total length of stay in each group. Thus although only ten (2%) cases had lengths of stay of more than 50 days they acconuted for 39% of the total days in hospital.

Medical Discharges and Deaths: During the 29 year period from 1952 to 1981 no deaths occurred in the Army as a result of injuries received while boxing. There were, however, *29 injuries serious enough to result in the patient being medically discharged from the Service.* Details of these cases are given in Table 106. The cases were evenly spread over the period and there is no evidence to suggest either a rising or falling trend. There also seems to be no trend in the type of injury for which people are being discharged.

The *majority of discharges* (16 out of 29, 55%) were in patients suffering from *injuries to the head* and seven of these were for *intracranial haemorrhage.* The remaining cases were mainly for *fractures, dislocations* and *sprains.* Most of the discharges occurred after long hospital stays. Conversely many of the long stay cases resulted in medical discharges. 67% of discharges occurred in a different year from the original injury.

Discussion: OELMAN et al. note that one of the problems with trying to assess how dangerous an activity is lies in estimating the population at risk. ROSENBAUM (1967) found in 1962, that of a random sample of 3.185 Army personnel 113 boxed and that they spent on average 4.9 hours a month each boxing (or training).

If the above ratios hold true, continue the au-

TABLE 106
Medical discharges 1952-1980. (From OELMAN et al., 1983).

Year of Discharge	Year of Injury (1*)	Age at time of Discharge	Princ Dis at Discharge ICD Code (2*)	Sec Dis at Discharge ICD Code (2*)	Princ Dis at Discharge If Different ICD Code (2*)	
1952	—	—	8141	—	—	Fracture of Carpal bones
1953	—	—	8540	—	—	Head injury of other and unspecified nature
1955	—	—	8340	—	—	Dislocation of finger
1956	—	21	8530	3100	—	Intracranial haemorrhage, borderline intelligence
1957	—	20	8530	9968	—	Intracranial haemorrhage injury of other and unspecified nature
1958	—	19	8141	—	—	Fracture of carpal bones
1958	—	19	8520	—	—	Subarachnoid subdural and extradural haemorrhage
1958	—	40	2962	3007	—	Depressive, Psychoneurotic disorders
1959	—	22	8300	—	—	Dislocation of Jaw
1961	—	18	8540	—	—	Head injury of other and unspecified nature
1962	—	18	376	8700	—	Detachment of retina, open wound of eye and orbit
1963	—	22	8120	—	—	Fracture of humerus
1964	—	20	8130	—	—	Fracture of radius and ulna
1964	—	23	8460	—	—	Sprains and strains of Sacro-ilias region
1964	—	22	3899	—	—	Other deafness
1964	—	50	8540	—	—	Head injury of other and unspecified nature
1964	—	23	8130	—	—	Fracture of radius and ulna
1968	1965	23	8720	—	—	Traumatic perforation of ear
1968	1967	24	9500	—	9210	Injury to optic nerves, contusion of eye and orbit (sub total blindness)
1969	1968	20	8540	—	8530	Intracranial haemorrhage
1970	1969	25	8812	—	8819	Laceration of wrist
1972	1972	16	8360	—	—	Dislocation of knee
1973	1973	23	8520	2250	—	Subdural haemorrhage, benign neoplasm of brain
1975	1973	18	8519	—	8510	Cerebral laceration and contusion
1976	1974	20	7131	9954	—	Spondylolysys-adhesions and nerve root irritation
1977	1976	22	8529	7810	—	Subdural haemorrhage, disturbance of vision
1978	1978	17	8122	—	—	Fracture of humerus
1979	1979	19	9580	—	—	Cervical spine injury with nerve root contusion
1980	1979	21	8520	3451	—	Subdural haemorrhage, epilepsy

*Note: 1 Not available prior to 1968 2 ICD Codes refer to the eighth edition throughout.

thors, then through the period 1969 – 1980 there would have been about 5.700 boxers in the Army each spending about 59 hours a year boxing. On average there were 35 injuries a year serious enough for admission to hospital and one medical discharge a year so that about one in every 150 boxers is admitted per year, or one injury occurs after every 9.000 man hours of boxing. These figures are very approximate and the rates are somewhat less than found by ROSENBAUM.

With an average length of stay per injury of 10.4 days there will, on average, be one person in hospital at any time because of a boxing injury.

The authors noted that the aim of their paper is not to compare boxing with other sports, but to outline the types of injury received and to highlight the problems arising from them. As would be expected, they found that head injuries predominated, and concussion was the most common single diagnosis, with 115 cases, 39% of all admissions. Most cases were fairly minor, 57% of them being of less than five days duration, but many injuries to the head have secondary and late effects (which the authors were not able to quantify) and a number of the medical discharges did not occur until well after the original injury.

Table 107 shows the distribution of injuries by injury group amongst the ages 15 – 26 and over 26. The percentages are given in two ways; first as a percentage of the injuries within an injury group and secondly as a percentage of the total injuries at a given age. Also shown is the distribution of the average strength of the Army seen from 1969 – 1980.

It can be seen, continue OELMAN et al. that while only 8.6% of the Army population was less than 18 years of age, 31.1% of the injuries occurred to people in that age group. It is not known what proportions of soldiers within each age group participate in boxing, therefore it is impossible to say whether this is significant. However it appears that even amongst the younger members of the Army (up to age 26), who might be expected to be equally at risk, the injury rate decreased with age and thus experience. This is borne out by the fact that the proportion of injuries involving the head is considerably higher in the younger age groups than in the older groups. Injuries to the limbs and trunk appear to increase with age while internal injuries remain fairly constant.

The authors summarize that in absolute terms boxing injuries are costing the Army one medical discharge per year and one person in hospital at any one time. In addition there must be many minor injuries that do not result in admission but which may have long term effects on the health of the patient. It is not for us to say, conclude OELMAN et al. in their paper, whether this price is too high for the 3.5% of the Army who participate in the sport.

In an editorial *'Boxing in the Army'* (1984) is noted that boxing is a popular sport in the Services. In 1984 there were 5.260 registered boxers in the Army.

The Combined Services Boxing Association is affiliated through the British Amateur Boxing Association to the International Boxing Association. The medical control of boxers and boxing is as detailed in Army General Administrative Instructions is at least as strict as that recommended by these bodies. During the period 1981-1983, in addition to 60 knock outs, 500 bouts were stopped by the Army boxing referees which suggests that they, too, conscientiously play their part in preventing injury.
And yet, as OELMAN et al. (1983) have pointed out in this journal, each year about 35 boxers require hospital admission and one is medically discharged, as a result of about 60 hours boxing and training, per boxer, per year. TOINTEN (1984) in this issue puts this in perspective. Table VIII shows that boxing accounts for a mere 2% of all admissions due to sports injuries in the three Services. Of the boxing injuries, 70% were to the head. After football, which accounted for 55% of all head injury admissions, boxing was the principal cause of head injury. So nagging doubts remain.
NEEDHAM (1984) considers that 'it is incumbent on us to consider the arguments and the evidence and be prepared to inform participants of the risks they run in the ring, and to be prepared to provide this information for authority.' Here are some arguments, evidence and recommendations, reflecting the Editorial, though not necessarily the Medical Directorate's, view.

Whether the adversary is an icy mountain, a goalkeeper, or a fencing swordsman, many sports consist in a sort of duel which, despite all precautions, may have destructive and even deadly consequences for the contestant. In bullfighting and boxing, however, the skilful infliction of damage to the opponent is the prime objective of the antagonists, and not a secondary consequence.

In order to win a fight conclusively a boxer delivers hard blows to the opponent's head so that, apart from accumulating points, he can, under the well-controlled condition of amateur boxing, get the fight stopped by the referee. This may happen if the loser sustains orbital laceration, is dazed and unable to defend himself, or is concussed. To say that the object is not to damage the brain but merely to interfere with its function is like saying, when one strikes a similarly delicate but well protected mechanism – a watch – with a hammer, in order to slow it down or stop it, one does not intend the mechanical damage which will inevitably result.

One of the criticisms by sceptics of some of the studies of brain damage in boxers, such as those of ROBERTS' in 1969

TABLE 107
Number and percentage of cases by age and injury group. (From OELMAN et al., 1983).

	Age												Over	
	15	16	17	18	19	20	21	22	23	24	25	26	26	Total
Injury Group *Number of Cases*														
Head	7	61	35	38	29	30	20	18	17	11	10	8	12	296
Upper Limb		1	7	10	8	8	8	4	6	3	3	1		59
Lower Limb	1	2	2	4	1	2	1	2	1	1	2	1	1	21
Trunk	1	1		1	4	2	3	2	1	1			4	20
Internal	1	1	1	1	2	2	1		1	1			2	13
Other	6	5	4	2		2	2	2		2	2		1	28
TOTAL	16	71	49	56	44	46	35	28	26	19	17	10	20	437
Average Strength 1969–80 (1000's)		13.8		9.2	10.9	11.5	10.3	9.4	8.9	8.3	7.9	7.4	62.8	160.4
Percentage of Cases *in each Age Group*														
Head	2.4	20.6	11.8	12.8	9.8	10.1	6.8	6.1	5.7	3.7	3.4	2.7	4.1	100
Upper Limb		1.7	11.9	16.9	13.6	13.6	13.6	6.8	10.2	5.1	5.1	1.7		100
Lower Limb	4.8	9.5	9.5	19.0	4.8	9.5	4.8	9.5	4.8	4.8	9.5	4.8	4.8	100
Trunk	5.0	5.0		5.0	20.0	10.0	15.0	10.0	5.0	5.0			20.0	100
Internal	7.7	7.7	7.7	7.7	15.4	15.4	7.7		7.7	7.7			15.4	100
Other	21.4	17.8	14.3	7.1		7.1	7.1	7.1		7.1	7.1		3.6	100
TOTAL	3.7	16.2	11.2	12.8	10.1	10.5	8.1	6.4	5.9	4.3	3.9	2.3	4.6	100
Strength 1969–80 (+)		8.6		5.7	6.8	7.2	6.4	5.8	5.6	5.2	4.9	4.6	39.2	100
Percentage of Cases *in each Injury Group*														
Head	43.8	85.9	71.4	67.9	65.9	65.2	57.1	64.2	65.4	57.9	58.8	80.0	60.0	67.7
Upper Limb		1.4	14.3	17.9	18.2	17.4	22.9	14.3	23.1	15.8	17.6	10.0		13.5
Lower Limb	6.3	2.8	4.1	7.1	2.3	4.3	2.9	7.1	3.8	5.3	11.8	10.0	5.0	4.8
Trunk	6.3	1.4		1.8	9.1	4.3	8.6	7.1	3.8	5.3			20.0	4.6
Internal	6.3	1.4	2.0	1.8	4.5	4.3	2.9		3.8	5.3			10.0	3.0
Other	37.5	7.0	8.2	3.6		4.3	5.7	7.1		10.5	11.8		5.0	6.4
Total	100	100	100	100	100	100	100	100	100	100	100	100	100	100

based on men who had been fighting between 1930 and 1955, was that these boxers were damaged under less well-controlled fighting conditions than prevail in amateur boxing today. But what these studies did do was to establish beyond question the essential link between boxing and brain damage.
SPILLANE in 1962 and MAWDSLEY and FERGUSON the following year, had made their important radiological observations on the separation and tearing of the septal leaves in the depths of the brain. ROBERTS found that 17% of 200 randomly selected professional ex-boxers had developed neurological evidence of brain damage.

CORSELLIS and his colleagues in 1973, in describing 12 punch drunk out of 15 ex-boxers, defined the typical neurological picture varying from 'mild clumsiness of movement and of speech, with or without some loss of memory, to the ataxic dysarthric and perhaps parkinsonian dement.' They also linked these to the 'novel pattern of cerebral damage and degeneration that is concentrated on the septal regions, on the deep temporal grey matter and on certain neurones along the cerebellar and nigral pathways.'

As the result of the sudden acceleration and deceleration of the jelly-like contents of the skull the characteristic boxing injuries are produced:

a. Septal tears.
b. Focal scarring of the cerebral and cerebellar hemispheres.
c. Degeneration of the substantia nigra.
d. Widespread neurofibrillaery tangles within the substance of the brain.

Hence the ataxia, parkinsonism and memory loss.

Now what is most alarming is not so much the 20 sad boxing dements now vegetating in Rumwell Hospital, relics of the old days, but persistent reports of neurological, EEG and CAT scan evidence of in 1982 reported on 14 boxers, a 1/5 with septal cava and 1/3 with cerebral atrophy. Similar pathological findings led ROSS and others in the USA last year to 'support the hypothesis that the more one boxes the more likely he is to have evidence of cerebral atrophy.' CASSON and co-workers in New York in 1982 came to a similar conclusion having demonstrated cerebral atrophy in a study of 10 boxers none of whom had been concussed for more than 10 seconds.

The remarkable powers of recovery of function (though not of regeneration of brain cells), as a result both of healing and redundancy in the architecture of the brain, obscure the fact that every heavy blow sustained is accelerating the degeneration of nerve cells, tearing their connections, using up the reserves, and eventually impoverishing thought, feeling and action. The problem is that of insidious cumulation.

What can we do to limit this process? Firstly the present Army regulations allow a man who is dazed and unable to defend himself to box again in one week, and if he is knocked out or the fight is stopped due to a head blow he may box again in four weeks. That is about the time it takes for a brain clot to liquefy and the scarring process to *begin*. The turnover of protein in nerve cells takes about six weeks, regeneration of boutons terminaux (in the spinal cord) three to four months, gliosis months and recovery of function after head injury (in those who escape post-traumatic epilepsy) years. It is submitted that if, as in the conditions quoted, there is ipso facto demonstrable evidence of brain damage, the boxer should be protected willy-nilly from further deliberate damage to his brain for a minimum of six months.

Secondly, given that the principal damage, apart from the occasional catastrophic haemorrhage, is acquired by accumulation of not nessessarily fight-stopping head blows, an Army boxer should not box for longer than ten years. This would permit the development of a vigorous amateur sport amongst the vast majority of soldiers without allowing any soldier to drift into pseudo-professionalism. The only way to ensure this restriction of the span of an Army boxers' career is to declare an upper age limit of 28.

Given that 25% of the oxygen consumption and 30% of the glucose used by the human body is metabolished by the brain, an engineer, thinking merely in terms of fuel consumption, might suppose that this small lump of jelly was an important part of the organism. But we know more than that. We know that every squaremillimetre of cerebral cortex contains 50.000 cells and each cell may have up to 50.000 connections to other cells by means of dendrites and boutons termineaux. Whether or not a cell discharges depends upon the subtle interplay of a myriad excitatory and inhibitory influences. The nature of every cerebral event is determined by the integrated pattern of activity of all the countless cells involved. It is this organ of infinitive complexity and delicacy, in which the mind and personality are manifest, that we seek to protect in the interests both of the individual and the Service. All of us who have this knowledge, and who are in a position to advice, cannot in conscience do other than insist that a man's brain must be allowed to heal before he fights again and that the process of degeneration should not be allowed to continue beyond the age of 28.

ENZENAUER and ENZENAUER (1987) published an editorial called *'Ban Military Boxing.'* The authors began their article with some general remarks about the dangers of this activity:

> "Boxing is a controversial sport, and numerous reports have been published about boxing related injuries. Neurologic and ocular injuries can be particularly devastating sequelae of competitive boxing. Evidence that boxing produces irreversible brain damage is now as indisputable as the link between cigarette smoking and lung cancer."

The authors continue that currently, competitive boxing is promoted throughout the *U.S. military,* with inter-unit matches common on most large installations, and inter-service competition typical at many oversees locations. Boxing is also a part of the physical conditioning programs at the *U.S. service academies.* The advisability of promoting boxing by the U.S. military, write ENZENAUER and ENZENAUER, based on medical grounds and the historical experience of military boxing, has not yet been addressed.

ENZENAUER and ENZENAUER noted:

"The most worrisome injury in boxing, boxer's encephalopathy, or the 'punch drunk' syndrome, has been recognized of over a half-century. MUHAMMAD ALI's Parkinsonism has brought the subject of chronic brain damage to the attention of the general public. A military surgeon called the condition 'dementia pugilistica' (MILLSPAUGH 1937). It is estimated that chronic brain damage, characterized by dementia, memory loss, slurred speech, tremor, and abnormal gait, is seen, according to the authors, in up to 10 – 15% of professional boxers. Growing evidence incriminates repeated head trauma as being responsible for measurable alterations in brain structure and function. Gross examinations of brains from punch-drunk boxers reveal enlarged ventricles, widened sulci, and narrowed gyri ... Cerebral atrophy has been documented by head computerized tomography scan and is more common in the boxers with the most bouts" (CASSON et al. 1982, ROSS et al. 1983).

The authors continue that additionally, vision-threatening eye injuries are not uncommon in boxing. Vitreous hemorrhage, retinal detachment, retinal disruption, and various other ocular trauma can result in serious visual impairment (DOGGART 1955, MAGUIRE and BENSON 1986). The authors noted, that 'Sugar' Ray LEONARD's recent traumatic retinal detachment brought this boxing injury to the attention of the world.

ENZENAUER and ENZENAUER pointed out that numerous professional medical organizations have made official statements calling for significant reforms to make boxing safer or for a total ban on boxing. The earliest official statement came in 1962 when the *Committee on the Medical Aspects of Sport of the American Medical Association (AMA)* concluded that "the advisability of boxing is a debatable and unsettled issue." The authors continue that *The Journal of the American Medical Association* has been sounding board for much of the popular medical sentiment calling for the abolition of boxing. In December 1984, the *AMA House of Delegates* adopted a resolution calling for a ban on all boxing, amateur and professional, in the United States. In addition to the *AMA, the British, Canadian,* and *Australian Medical Association,* have also recently recommended that boxing be banned. The *American Academy of Pediatrics* opposes boxing in any sports program for children and young adults, concluding that they "should be encouraged to participate in sports in which intentional head injury is not the primary objective of the sport." The *American Neurological Association* and the *American Academy of Neurology* called for a ban on boxing. The *American Association of Neu-*

rological Surgeons has also spoken out against boxing.

The authors continue to say that the extent of military boxing injuries has been evaluated only once. In a *British Army* study, OELMAN et al. (1983) noted that about 35 boxers require hospitalization and one is medically discharged annually, as a result of about 60 hours of boxing and training, per boxer, per year. Recently, in a study of the instructional boxing program at the *U.S. Military Academy at West Point,* WELCH et al. (1986) reported an injury rate of less than 4% over a two-year period, 1983 – 1985. ENZENAUER and ENZENAUER continue: "Personal experience in the spartan environment of the U.S. service academies (R.W.E., *U.S. Military Academy, 1975,* R.J.E., *U.S. Naval Academy, 1975)* would suggest that any 'official' injury survey would probably be a gross underestimate. Additionally, a more extensive review will reveal at least one death attributable to the 'instructional' boxing at West Point."

The National Collegiate Athletic Association no longer recognizes boxing as an intercollegiate sport. Proponents of boxing contend that it builds character and self-confidence, while teaching self-defense and self-discipline. They argue that it is relatively safe, especially with different gloves and protective headgear. However, write the authors, it is unlikely that protective headgear might actually increase cerebral injury. Proponents of boxing will argue that 'serious' injury is much more common in other professional sports, such as football or rugby, than it is in boxing. They continue: "However, the main difference between boxing and any other 'sport' is the intent to do bodily harm, and specifically harm to the central nervous system."

ENZENAUER and ENZENAUER note that boxing has been described as a "sport in which two fighters battle each other with their fists... The boxers throw powerful punches as each tries to win the bout, render his opponent unconscious, or force him to give up the fight" (SULLIVAN 1984). The authors continue: "Most serious medical authors have reminded us that the 'principal purpose of a boxing match is for one opponent to render the other injured, defenseless, incapacitated, unconscious" (RICHARDS 1984)... The ultimate goal of a boxing match is, whether amateur or professional, is the knock out. A knock out necessarily connotes significant neurologic injury. We agree with our more eloquent colleagues that however boxing is classified, the goals and ideals of boxing are inconsistent with those of other sports. Boxing has been compared to cockfighting and to 'privately staged dogfights' (Van ALLEN 1983), and these comparisons are not totally unwarranted. In other sports, injury is considered a consequence of the athletic encounter. In boxing, direct injury is the intent of the encounter, and the 'intent is wrong'" (RICHARDS 1985).

ENZENAUER and ENZENAUER summarize: "War is obviously a dangerous encounter. Preparing for war can necessarily be a dangerous occupation, many military greats have stressed the importance of athletic competition in the development of military leaders. General of the Armies Douglas McARTHUR said that it is 'upon the field of friendly strife that are sewn the seeds' that will on future battlefields 'bear the fruits of victory.' Physical fitness and conditioning are stressed in all echelons of the military. However, physical fitness, strength, endurance, self-confidence, self assurance, and self-discipline can all be achieved through athletic endeavors other than boxing, which offer less probability of acute and chronic brain injury."

The authors finally write that a *recent editorial* in *the Journal of the Royal Army Medical Corps* (130: 140-142, 1984) sought to reduce the extent of central nervous system injury among military boxers by limiting an Army boxer's career to ten years and by extending the limit between bouts. The authors continue that the intent here is admirable. However, it simply does not extend the discussion to its logical conclusion. The proposal does not go far enough. "It is a needless risk to the mind and body of our nation's youth to actively promote competitive boxing in the *U.S. Armed Forces.* Boxing should be eliminated from the physical conditioning programs of *the U.S. service academies.* Additionally, boxing should not be promoted by the *U.S. Armed Forces* and should be totally prohibited on *U.S. military installations.*"

ENZENAUER et al. (1989) reported on *'Boxing-Related Injuries in the US Army, 1980 through 1985'* The authors elected to retrospectively evaluate the extent of morbidity and mortality directly attributable to participation in military boxing.

Records of hospitalizations for boxing-related injuries in the *US Army Medical Treatment Facilities worldwide* between Jan.1, 1980, and Dec. 31, 1985 were obtained through the *US Army Patient Administration Systems and Biostatistics Activity at Fort Sam Houston, Texas.* The accessible data for each patient included all diagnoses made, duration of hospitalization, and procedures performed, along with demographic data such as the patient's age, sex, race, etc. Further details concerning an individual patient's hospital course were not readily accessible.

For purposes of analysis, according to the authors, the boxing-related injuries among *US Military Academy (USMA)* cadets were included and also evaluated as a separate subset of all the hospitalizations.

Results: During the six-year period under consideration (1980 through 1985), there were 410 admissions for boxing-related injuries (Table 108). The nine civilians and dependents were excluded from the study. Although the total military active duty admissions remained relatively constant, there was generally a trend of increasing number of admissions among the *USAM cadets* and decreasing admissions among other active duty soldiers.

Figure 351 graphically depicts the age distribution of the hospitalized boxers. Not surprisingly, predominately men (99% [396 /401 admissions]) were hospitalized for boxing-related injuries, with only five women admitted. This is consistent with the obvious fact that boxing is largely a male-dominated sport, as well as the requirement that male and not female cadets participate in the instructional boxing program at the *USMA.*

Figures 352 and 353 show the number of cases by *length of hospitalization 'bed days'* and by *length of disability 'sick days.'* The median length of hospital stay was two days, as was the median length of disability for both the *USMA cadets* and the *Army at large.* The boxing-related hospitalizations both for *USMA cadets* and for the *US Army at large,* are summarised in Table 109. The average *length of hospitalization (bed days)* was 5.1 days *US Army-wide,* while the average hospitalization of the injured *USMA cadets* was only 2.0 days. The average *length of disability (sick days)* was 8.9 days *Army-wide* and 2.0 days among *USMA cadets.*

ENZENAUER er al. found that *head injuries* accounted for 68% (272/401) of all boxing-related hospitalizations and 81% (140/173) of injuries to *USMA cadets.* Head injury admissions included the following diagnoses: *intracranial injury* (not

TABLE 108
Admissions to US Army medical treatment facilities from boxing injuries, 1980 through 1985). (From ENZENAUER et al., 1989).

Year	Total Admissions	Total Admissions of Active Duty Personnel	Admissions of USMA* Cadets	Admissions of Other Active Duty Personnel	Admissions of Dependents/Civilians
1980	62	59	21	38	3
1981	77	74	25	49	3
1982	74	72	22	50	2
1983	61	61	28	33	0
1984	67	67	39	28	0
1985	69	68	38	30	1
Total	410	401	173	228	9

*USMA indicates US Military Academy.

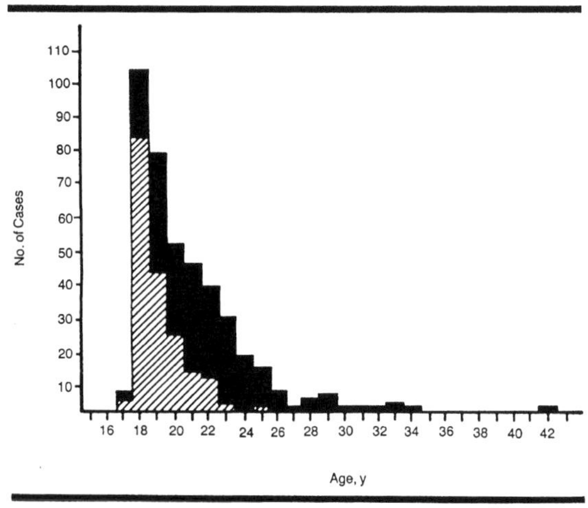

Number of cases by age, 1980 through 1985. Black area indicates total number of cases US Army–wide (401 cases); slashed area, number of cases among US Military Academy cadets (173 cases).

Fig. 351. Number of cases by age, 1980 through 1985. Black area indicates total number of cases US Army-wide (401 cases) slashed area, number of cases among US Military Academy cadets (173). (From ENZENAUER, ENZENAUER, 1989).

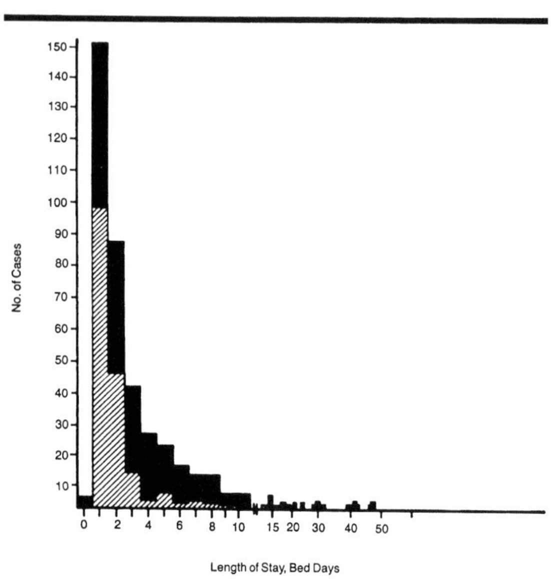

Number of cases by length of stay, "bed days." Black area indicates total number of cases US Army–wide; slashed area, number of cases among US Military Academy cadets.

Fig. 352. Number of cases by length of stay, 'bed days.' Black area indicates total number of cases US Army-wide, slashed area, number of cases among US Military Academy cadets. (From ENZENAUER, ENZENAUER, 1989).

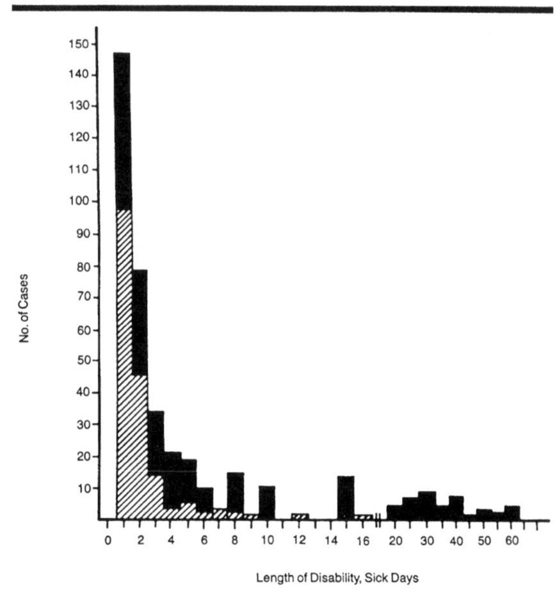

Number of cases by length of disability, "sick days." Black area indicates total number of cases US Army-wide; slashed area, number of cases among US Military Academy cadets.

Fig. 353. Number of cases by length of disability, 'sick days.' Black area indicates total number of cases US Army-wide, slashed area, number of cases among US Military Academy cadets. (From ENZENAUER, ENZENAUER, 1989).

elsewhere classified), *concussion/cerebral contusions, intracranial hemorrhage, skull fracture, fracture of the facial bones, ocular injuries, head/neck contusion, and 'other.'* Included in the *US Army-wide admissions for head injuries* were one soldier who died from complications of intracranial hemorrhage and one soldier blinded in one eye after a ruptured globe.

Upper extremity trauma accounted for 17% (67/401) of all hospitalizations and 95 (16/173) of the cadet injuries. *Upper extremity injuries* included the following diagnoses: *dislocation of shoulder, dislocation of elbow/wrist/finger, fracture of humerus, fracture of radius/ulna, fracture of hand bones, sprain/strain of upper extremity, open wounds/infections, and other.*

Injuries to the trunk and *torso* accounted for 7% (29/401) of admissions *US Army-wide* and 6% (19/173) of *cadet injuries. Injuries to the trunk* and *torso* included the following diagnoses: *sprain/strain of neck, sprain/strain of back, contusion of*

TABLE 109
Number of admissions, inpatient bed days, and disability sick days from boxing injuries, US Army-wide/USMA cadets, 1980 through 1985). (From ENZENAUER et al., 1989).

Injury	No. of Admissions	Total Bed Days	Average Bed Days	Total Sick Days	Average Sick Days
Total injuries	401/173	2056/349	5.1/2.0	3581/349	8.9/2.0
Head injuries	272/140	1252/267	4.6/1.9	2200/267	8.1/1.9
Upper extremity injuries	67/16	418/27	6.2/1.7	767/27	11.4/1.7
Trunk/torso injuries	29/10	217/28	7.5/2.8	323/28	11.1/2.8
Lower extremity injuries	25/7	147/27	5.9/3.9	269/27	10.8/3.9
Other injuries	8/0	22/0	2.8/0	22/0	2.8/0

USMA indicates US Military Academy.

chest/abdomen/genitalia, injury to kidney/intra-abdominal organs, fractured ribs, traumatic pneumothorax, and other.

Lower extremity injuries were involved in 6% (25/401) of *all admissions* and 4% (7/173) of *cadet admissions. Lower extremity admissions* included the following diagnoses: *Fracture of femur, fracture of ankle/metatarsals, tear of knee cartilage/meniscus, sprain/strain of hip/knee/ankle, and other.*

Table 110 summarizes the relationship of age with type of injury. Head injuries predominated in the younger and presumably less experienced boxers.

Head injuries predominate as the leading cause of serious injury to competitive boxers. This should be a surprise to no one, since the goal of a boxing match, whether amateur or professional, is the knock out. And the knock out, or head injury resulting in unconsciousness, connotes significant neurological injury. Additionally, concussion is a serious head injury. Boxers who appear 'dazed' or 'out on their feet' all represent cases of concussion, at the very least.

Boxer's encephalopathy, or the *punch-drunk boxer* is, as ENZENAUER et al. noted, the most obvious of the neurological sequelae of a career in boxing. It was first reported a half century ago (MARTLAND 1928) and was described by a military physician as 'dementia pugilistica.' MUHAMMAD ALI Parkinsonian-like symptoms have brought the subject to the attention of the general public. The authors noted, that "Growing evidence incriminates repeated head trauma to be responsible for measurable alterations in the brain structure and function. It is estimated that chronic brain damage, characterized by dementia, memory loss, slurred speech, tremor, and abnormal gait, is seen in up to 10% to 15% of professional boxers.

ENZENAUER et al. noted that the actual incidence of boxing injuries is impossible to determine

TABLE 110
Number of injuries by age and injury type. (From ENZENAUER, et al., 1989).

	Age, y											
Injury Group	17	18	19	20	21	22	23	24	25	26-29	30-42	Total Injuries
Head	3	80	54	34	29	29	18	11	7	5	2	272
Upper extremity	2	8	14	7	9	3	6	3	3	7	5	67
Trunk	1	5	7	4	1	1	0	2	3	2	3	29
Lower extremity	0	8	1	5	4	2	2	0	0	2	1	25
Other	0	1	1	0	1	2	2	0	0	0	1	8
Total	6	102	77	50	44	37	28	16	13	16	12	401

from their data, since boxers are not registered in the US Army, but the authors conclude that their study demonstrates that, however, that military-supported boxing is the case of significant morbidity and occasional mortality. The cost of 2.056 days of hospitalization and 3.581 days of lost work is certainly considerable. Additionally, without a prospective study, it would be impossible to ascertain how many soldiers were discharged from the hospital only to be unable to continue in their previous military occupational specialty because of disability. The authors continue: "Although we personally do not condone boxing, our study definitely confirms that the instructional boxing program at the *USMA* is characterized by less serious injury, compared with injuries in the *US Army as a whole,* as evidenced by the markedly shorter duration of hospitalization and disability."

ENZENAUER et al. summarize their findings and opinion: "The advisability of continued military promotion of competitive boxing needs to be addressed. Proponents of boxing at the *US service academies* argue that boxing is unexcelled in improving the self-concept of cadets, helping them develop physical courage, and teaching them how to deal with fear. These alleged benefits of boxing need to be considered against the definite risks associated with participation. Only serious, prospective studies of military boxing can adequately answer these questions. A controlled prospective study of mild and moderate head injury in *Army boxers* and its impact on military performance has been drafted by the *director of the Army Head Injury Unit at Walter Reed Army Medical Center;* however, adequate funding for this rather extensive study is in doubt. A proposal for a prospective study of *boxing-related ocular injuries* has been submitted to the *USMA*, however, eventual approval through the multiple echelons of the military hierarchy is also unlikely... The significant morbidity associated with military boxing reported in our study makes the continued promotion of competitive boxing in the military a controversial question. Proponents of professional boxing argue that, although there is the risk of brain damage and death, the boxer accepts the risk of injury just as other athletes do. *Military service, however, is not a democracy.* A military commander may restrict the soldiers in his command from activities that are deemed unreasonably dangerous. Similarly, participation in military boxing is not entirely voluntary. *Instructional boxing is a mandatory requirement at the US service academies.* Additionally, participation in unit boxing matches is often characterized by subtle coercion to encourage soldiers to box for 'the good of the unit,' troop morale, esprit de corps, to win a streamer for the unit guidon, etc."

The final paragraph in the study of ENZENAUER et al. reads: *"When it was deemed that cigarette smoking offered a significant risk to health, cigarettes were no longer included in soldiers' C rations. When physical hazing was determined inappropriate behavior for future officers, it was discontinued at the US service academies. Perhaps military boxing, with its definite risks and arguable benefits, will suffer a similar fate."*

Three replies were send to the editor to the article of ENZENAUR et al:

To the Editor: "The article titled *'Boxing Related Injuries in the US Army, 1980 through 1985'* contains important information about the cause of injury. From the information they obtained, the authors made a giant leap from being scientists to being experts in military preparedness and questioned the advisability of promoting boxing in the US Army. This was done in such a way that one can only conclude that they believe it should be banned. In doing so they may have meant well but the action is misguided or naive."

"In matters of prevention, diagnosis, and treatment of illness and injuries, the options of physicians have special status. The authors seem to believe they also have special competence to judge whether boxing does more harm than good as a tool to develop soldiers. In other words, they address the broader question of the appropriateness of boxing as a military training tool."

"Taking a young man or woman from our society and adequately training him or her for the rigors of war is a formidable task. We cannot divorce ourselves from reality and deny this is necessary. There will be wars in the future, and our soldiers must be prepared. The military persons appointed to transform civilians into soldiers are most knowledgeable about how to do this best. Those who lead them in peacetime and in war are in the best position to know how to keep them ready to carry out their duties at all times. If our military leaders believe boxing is an important tool in the making of good soldiers, then it should be continued in the military services. Let us not make the mistake of substituting the opinions of physicians for those of professional soldiers about an item of soldiering. The background and motivation of physicians are generally not the kinds to make one expect their opinions on military preparedness to be informed enough to change established policy."

"Besides, to ban boxing in military services of the level of risk would not withstand the scrutiny of logic and ordinary common sense. Parachuting is dangerous. So is training to use military tanks, assault weapons, and artillery pieces. Surely no one would advocate that the US Army ban them. And who could keep a straight face while arguing that competitive boxing is unreasonably dangerous for someone who must know how to fight and kill, or protect himself and others from being killed, by a determined enemy in combat?"

Jack I. FINNEY, MD, retired, US Army, Los Angeles, Calif.

To the Editor: "In the March 10 issue of *JAMA*, ENZENAUER et al. report boxing-related injuries in the US Army that were serious enough to require hospitalization. In view of the medical, political, and social implications of their study, there are significant methodological issues that need to be addressed. The delineations of these methodological limitations will ensure the proper interpretation of their data and will prevent biased and incorrect extrapolation."

"Using records of hospitalization, the authors reported that on average there were 67 hospitalizations annually, with the injured spending an average of 5.1 days in bed and 8.9 days disabled, i. e., unfit for duty. To assess accurately the magnitude of such a finding, one needs to know several factors. First, what percentage of average annual hospitalizations comprised 67 hospitalizations for all military personnel? In fact, an average of 67 hospitalizations per year may be a small number of hospitalizations when one considers all military personnel worldwide. Second, the reader should be informed of the average length of hospitalization and disability for all hospitalizations. To establish the magnitude of the burden of boxing to the US Army, it would be of interest to compare boxing-related hospitalization with those related to football or ice hockey."

"Another methodological limitation of the study is the failure to delineate the level and supervision of the boxing. There are essentially three types of military boxing. One is the relative noncompetitive boxing that occurs as part of instructional programs at the US Military Academy in West Point, NY. This type of program is associated with a low injury rate. The second type of military boxing is that associated with interservice competitions, the Olympic Games, the Pam American Games, and National American Championships. The third level of military boxing comprises what often is referred to as 'smokers' or open competitions between various military units. This third level of boxing is potentially dangerous because many of the participants lack the proper training, supervision, and medical surveillance afforded to the other two types of military boxing. Accordingly, to assess the magnitude of the burden of military boxing, the reader should be told what percentage of injuries were related to instructional boxing, Olympic-style boxing (i. e., boxing sanctioned by the US Amateur Boxing Federation), or the less structured smokers. Intuitively, one would suspect that the majority would be associated with the smokers."

Barry D. JORDAN, MD, Hospital for Special Surgery, Cornell Medical Center, New York, NY

To the Editor: " I read the article on boxing injuries in the army with interest, as well as the associated editorial on the American Medical Association's position."

"As far as training for self-defense, the data are of little help. Certain group of individuals need to learn to defend themselves because their jobs require exposure to personal hazards."

"I believe combat arms soldiers, whether enlisted or officers, should learn self-defense. The same training is needed by law enforcement personnel. Any self-defense skill has dangers."

" The data presented in the accompanying article refer only to numerators. We should compare the rate of permanent injuries per 10.000 persons engaged in various types of training and body contact sports and then determine the likelihood of injury. Many other sports now require compulsory wearing of protective equipment to prevent many injuries. Boxing injuries can be minimized while training soldiers for personnel combat."

" I do not believe we can train soldiers to engage in personal combat without risks. We need to be sure the training is optimal to achieve its goal and with the least risk consistent with proper training. We need better studies than the one reported to determine which sports provide the best self-defense training and ensure minimum likelihood of individual survival with the least risk of training injury."

C.M.G. BUTTERY M.D., MPH, Midlothian, VA.

ENZENAUER et al. wrote in reply:

"We anticipated the controversy our article might engender. The army's medical mission is 'to preserve the fighting strength.' We object to Dr. FINNEY's suggestion that we made a 'gigantic leap' to question boxing. Our concern is the health of the soldiers we are sworn to protect."

"We do possess the 'special competence' Dr. FINNEY describes to assess boxing as military training. Two of us graduated from service academies (R. W. E. *USMA;* R. J. E., *USNA)* and experienced boxing's 'training benefits' first hand. Our opinions come from more that merely a 'medical background'."

"A military boxing ban would withstand Dr. FINNEY's 'scrutiny of logic.' Medical criteria should be tempered by training considerations when evaluating any activity. However, that military boxing injuries can be disabling or fatal is indisputable."

"War is a dangerous encounter; preparing for war is a dangerous occupation. Military training is often dangerous. However, valuable training does not have to be hazardous. Soldiers should learn self-defense and how to deal with fear. Other activities, however, can teach them just as well under conditions more realistic than the boxing ring: for example, the martial arts, Airborne/Ranger School, and the standard army 'confidence course.' The association between boxing and military skill is unproved and irrelevant on today's 'high-tech battlefield of the future.' Quoting a military physician, 'The skill of hand-to-hand combat may not prove as useful as a sharp mind and a steady eye on the highly technical battlefield of the future. Unfortunately, boxing destroys both.'"

"Military leaders might be in the 'best position' to decide proper basic training methods, but an 'informed' decision requires all pertinent information. The presumed training benefit of boxing has been considered without substantiation of its safety. Except for instructional boxing at the service academies, military boxing is rarely 'training.'"

"We accept that many other athletic activities have higher acute injury rates than boxing. However, this does not constitute evidence that boxing is safe."

The 'methodological limitations' described by Dr. JORDAN do not prevent 'proper interpretation.' The percentage of military hospitalizations attributable to boxing is not as important as the true incidence rate of boxing injuries. However, until the services 'register' all military boxers, such data are not obtainable."

"Our study does delineate two types of boxing: the *US Military Academy instructional program* and the *larger US Army-wide experience. Academy graduates* would object to describing their boxing experience as 'relatively noncompetitive'. Cadets and midshipmen are fiercely competitive. Also many soldier-boxers 'do it all' – box at the local boxing club, in Golden Gloves, and in inter-unit smokers. Therefore, further delineation as Dr. JORDAN suggests, would add little to the military boxing discussion."

"It is our opinion that it is probably in the army's best interest to get out of the 'boxing business.' In speaking of military boxing, ALGEO (1988) says succinctly: 'Our commissions as medical officers obligate us to conserve our most precious resource: the soldier. As Transportations Officers would not permit demolition derbies with their vehicles, can we tolerate this needless abuse of a vital 'weapons system?'"

Edward W. ENZENAUER, MD, MPH,
Raymond J. ENZENAUER. MD,
Fitzsimons Army Medical Center, Aurora, Colo.

R.W. ENZENAUER et al. reported on *'Boxing-Related Ocular Injuries in the United States Army, 1980 to 1985'*:

"Boxing is a controversial sport in which neurological and ocular injuries can be particularly devastating. Currently, competitive boxing is promoted throughout the military. Ocular injuries attributable to military boxing were studied as a subset of a larger study of boxing-related trauma. Ocular injuries occurred in 5% (22/401) of soldiers hospitalized for boxing-related trauma. The 22 patients with eye injuries spent an average of 8.1 days in the hospital and 17.8 days unfit for duty. Ocular injuries, on average, resulted in a longer hospitalization and disability than all other boxing injuries. One soldier was blind in one eye after complications from a ruptured globe."

Marines Suspend Boxing Training (Lexington Herald Leader, Wednesday, June 4, 1997, A 9)

The Marines have suspended boxing at boot camp after one recruit from Kentucky died of head injury and another recruit was seriously injured.

Recruits will continue to learn fighting techniques but will not punch each other until a seven member committee reviews the training, base spokesman Maj. Rick LONG said yesterday."

The punching part of 'combat hitting skills' consists of recruits fighting three or four 15-second rounds while wearing boxing gloves, mouthpieces and head and groin protection. The training lasts two days, and most recruits show the necessary skills in two rounds, LONG said.

Recruit Michael J. CECIL, 19, of New Haven, Ky., died in February of brain injuries he suffered during boxing training four days earlier.

Another recruit, Eugene J. KENNEDY III, 27, was in serious condition after being injured Friday.

Lt. Col. WRIGHT noted in a letter to the editor of the Journal of the Royal Army Medical Corps about 'Boxing in the Army' in 1990:

"Sir, I was saddened to read in the Army Medical Services Magazine of an RAMC success in boxing. The medical profession has long been aware of the dangers of boxing and the effects on the brain. An editorial in this journal drew attention to the problems and made recommendations."

"It is often said that Army boxing is so well controlled that no harm can ensue. In this context the wording of the report would indicate otherwise... under a ferocious barrage from the strong and aggressive... boxer."

Statements of Medical Societies and their Perspectives on Boxing

The Royal College of Physiscians of London

"The Royal College of Physicians of London has always concerned itself with the health of the community. In recent years there has been increasing public interest in the possible long-term hazards to health from boxing, and as a result of a debate in *The House of Lords* in 1962 the question of the medical aspects of boxing was referred to the *College* by the late Lord BRAIN. His suggestion for an independent and scientific investigation by the College, which would provide the facts needed for a more accurate assessment of the dangers was accepted and the *College* set up a *Committee 'to report on the medical aspects of boxing from the data at present available.'"*

"The Committee on Boxing submitted an *interim report to the College in 1964.* It had sought and received evidence on injuries observed personally by their members from a number of medical societies and associations with specialised knowledge and experience in various medical and surgical fields. These had included *The Association of British Neurologists, The Society of British Neurological Surgeons, The Ophthalmological Society of the United Kingdom, The Royal Medico-Psychological Association, The British Cardiac Society, The Society of Thoracic Surgeons, The British Neuropathological Society, The British Association of Urological Surgeons, The Coroners' Society, The Association of Police Surgeons,* and *The Medical Officers of Schools Association.* Information had also been received from the *Medical Directors-General of the Armed Services.* Medical Officers in charge of professional and amateur boxing had given evidence to the Committee."

"The *Committee* had reviewed in detail the reports it had received of individual cases in which acute and chronic injury had been caused, or was thought to have been caused, by boxing. These suggested that fatalities and severe acute injury to the brain were an uncommon occurrence in both amateur and professional boxing. Only in the case of chronic injury to the brain and of injury to the eyes were there indications that serious permanent damage to appreciable numbers of boxers might have occurred. The *Committee* decided that firm conclusions and recommendations about these particular dangers of boxing could not be drawn from the available information, and recommended an assessement of the frequency with which they occurred by a carefully planned study of a random sample of boxers whose careers were adequately authenticated. In this way, the possible hazards of boxing in two classes of boxers, amateur and professional, who boxed under different conditions, might be related to the populations at risk."

"Records were made available by the *British Boxing Board of Control* and by the *Amateur Boxing Association.* It was found that only the records of professional boxers were adequate for the proposed study in providing an unselected population of boxers, over a long enough period, from which a suitable sample might be drawn. The investigation was, therefore, confined to the population of professional boxers registered with the *Board of Control between 1929 and 1955.* The *College* is greatly indebted to the *British Board of Control* and to the two hundred and twenty-four boxers whose generous help has made this study possible."

"The clinical investigation was carried out at the *National Hospital, Queen Square,* by Dr. A.H. ROBERTS, the Research Scholar appointed by the *Committee* in January 1967. The electroencephalographic study was undertaken by Dr. W.A. COBB, *Director of Electrophysiology at the National Hospital.* Officers of the *Metropolitan Police* generously provided the control population for this part of the survey."

"The full account of this epidemiological survey, which the *Committe* recommended should be published at the same time as this report *(Brain Damage in Boxers: A Study of the Prevalence of Traumatic Encephalopathy Among Ex-Professional Boxers, by A.H. ROBERTS, Pitman Medical & Scientific)."*

"The present report of the *Committee* includes a summary of these findings, and reviews them in relation to the present day supervision of professional, amateur and Services' boxing."

The Clinical and Pathological Picture of Traumatic Encephalopathy

"Boxers, by the nature of their sport, are liable to injury. In rare accidents this may result in acute severe brain damage, haemorrhage and even death. It has been suggested until recently that what seemed to be a comparatively mild injury to the head, causing only a shortlived disturbance of function, might lead to cumulative and permanent damage to the delicate tissues of the nervous system if often repeated."

"Many professional boxers know of colleagues who, as a result of their careers, have developed a chronic disabling condition which they call *'punch drunk.'* The first medical account of this condition was published in 1928. *(Ref.: MARTLAND. H.S. (1928), J Am Med Ass, 91: 1103).*

"Since then the *'punch drunk' boxer* has been commented on extensively in the lay and medical press. The clinical picture which has been described has usually been reminiscent, though not altogether typical, of commonly occurring etiological disease, but few cases have been reported with sufficient detail to provide an adequate idea of the relationship between the boxers' career and the subsequent disorder of his nervous system. In a number of these well documented cases the condition appeared to have progressed despite the retirement of the boxer from the ring."

"It was not until 1962, when Dr. J. D. SPILLANE published a report on five former professional boxers, that there was definite evidence of a specific pathological change in the brains of boxers that was almost certainly due to boxing. In four of these boxers, three of whom had clinical evidence of a chronic brain disorder, defects were demonstrated by air-encephalography in the septum pellucidum (a thin membraneous partition that separates the lateral ventricles of the brain). Damage to this structure is very rarely found when this X-ray investigation is used to study other types of neurological disease in adults *(Ref.: SPILLANE, J. D., (1962), Brit Med J II 1205)"*

"The *Committee* received evidence from Dr. C. MAWDSLEY and Professor P O. YATES. Dr. MAWDSLEY had made a study of thirty-two retired boxers (thiry-one professional, one amateur), all of whom had clinical evidence of neurological lesions. All but one of the men had fought more than 100 bouts. Dr. MAWDSLEY had so far published only a brief account of his findings *(Ref.: MAWDSLEY, C. and FERGUSON, F.R., (1963), Lancet, II: 795; FERGUSON, F. R. and MAWDSLEY, C., (1965), 8th International Congress of Neurology, Vienna, Proceedings, 1: 81; ISHERWOOD I., MAWDSLEY. C. and FERGUSON, F. R., (1966), Acta Radiol, 5: 654)."*

"The cinical picture he found was similar to that previously described. Most of the men dated the onset of their symptoms to the end of their boxing careers, when they were relatively young. Many thought that their disabilities had gradually progressed. Slurred speech, slowness and clumsiness of movement, unsteadiness of gait and tremors were the commonest features. Dementia was apparent in about half the group, and six of them gave a history of probable epilepsy. In twenty-three out of the twenty-eight boxers in whom air-encephalography was performed the septim pellucidum was defective, conforming the earlier observations. This particular lesion was not considered in itself to be responsible for the clinical picture of chronic brain damage in boxers, but it did confirm the presence of permanent structural damage, almost certainly due to boxing, in the brains of a large proportion of retired boxers who had a characteristic neurological disorder."

"In the brains of four boxers examined post mortem, cerebral atrophy and similar ragged holes in the septum pellucidum were found and could be seen with the nacked eye. Microscopically, it was apparent that there was extensive loss of cells in the cerebral cortex, and neurofibrillary degeneration of many cells of the kind seen more usually in older people, and also in the brains of patients with a relatively uncommon disease causing progressive dementia in middle life. These particular cell changes were found in great profusion in a area of the brain stem especially concerned in the overall control of voluntary movement; an unusual finding in pathological examinations of the brain, but not confined to the brains of boxers."

"This evidence, however, provides no information as to the frequency of these conditions in former boxers. The epidemiological survey was designed to provide the facts which had not, so far, been obtained about the prevalence of brain damage among ex-boxers and to relate these to the details of their boxing careers."

The Prevalence of Traumatic Encephalopathy in Ex-Professional Boxers

(a) Method of Investigation

"In 1929 the *British Boxing Board of Control* was established in order to supervise the interests of the professional sport in this country and, since then, the *Board* has required all professional boxers to register with it and to re-license themselves annually. A total of 16.781 boxers were registered and first licensed between 1929 and 1955. The investigation was based on a random sample of two hundred and fifty boxers who had been licensed for three years or more and who were nationals resident in England, Scotland or Wales. This sample, although only 1 1/2 per cent. of the total population of registered boxers represented 6 per cent. of the estimated 4.300 nationals resident and professionally active for three years or more."

"Every one of the two hundred and fifty ex-professional boxers included in the sample was traced. Sixteen had died, nine had emigrated and only one refused to be examined. The available two hundred and twenty-four were questioned and examined, and the details of their careers were checked as far as possible with official records and press records. The examination took the form of a detailed interrogation and physical examination by an experienced neurologist and included some psychological testing. Additional investigation by electroencephalography was made in over three-quarters of the boxers, and in a small number certain blood tests were also done. Because the number of boxers sampled for the study has been kept proportional to the numbers registering each year, the majority of those examined had done their boxing before the Second World War, when the sport was at the height of its popularity, and there was a far larger number of professional boxers."

(b) The Clinical Evidence of Brain Damage

"Those boxers who were found to be most severely affected suffered from slurring of speech to the point of being almost unintelligible, defective memory and slow thought. In some the limbs were stiff, the face expressionless, their walk a slow shuffle and their body and hands affected by a tremor at rest, the picture very like paralysis agitans. In others, who walked with a stagger, purposeful movements brought on a marked tremor of the limbs and trunk which, together with other signs, produced a clinical picture superficially resembling disseminated sclerosis. The milder forms of these disabilities were evident only as a slight thickening of speech or some unsteadiness on walking, of which the boxer was often but not invariably aware himself. These signs suggested that those parts of the brain had been damaged which are particularly involved in the control of balance and the co-ordination of posture and movement, the cerebellar and extra-pyramidal systems; there was often also evidence a asymmetrical damage to the pyramidal system, another nervous pathway concerned in articulation and other voluntary movement."

"Most of those affected had developed their disabilities towards the end of their boxing career. In a few cases the condition appeared to have become more obvious, especially to the wives, as the boxers had grown older, and in four cases it had advanced more rapidly and earlier than it seemed could have been accounted for by normal ageing processes alone. The condition of most those affected appeared to have remained static when once they had stopped boxing, and there were some instances of improvement after retirement."

"The whole clinical picture, although varying in degree, was remarkably consistent from one case to another. It was not typical of the clinical picture of the commoner diseases involving the nervous system, and the more severe examples resembled closely a number of the previously reported cases of traumatic encephalopathy in boxers."

"The eyes were another part of the nervous system in which there was definite evidence that damage had been caused by boxing. Retinal detachment causing virtual blindness in one eye was present in two cases. There were some indications that boxing had contributed to the slightly impaired vision found in a small number of other older boxers. Double vision due to damage to a muscle concerned in eye movement was found in five boxers, and in four of these it was severe enough to cause a permanent compensatory tilting of the head to one side."

"There was nothing to suggest that boxing had contributed to, or aggravated, any other systemic disease found."

"The electroencephalographic examination of the boxers revealed no specific changes that could be related either to their clinical condition or to their career, and the freqency of epilepsy was no greater than that found in the general population. It, therefore, seems unwise to rely upon a single electroencephalographic record either to establish or refute a diagnosis of traumatic encephalopathy in boxers."

(c) The Prevalence of Brain Damage

"Of the two hundred and twenty-four ex-professional boxers examined thirty-seven showed evidence of brain damage which was disabling in thirteen. This prevalence rate of 17 per cent. indicates that one in six of the boxers studied showed evidence of some damage to their nervous systems attributable to boxing, but probably only a third of these were affected severely enough to be recognised by a layman as 'punch drunk.' The others had evidence of similar damage to their nervous system but this was mild and likely to constitute little more than an inconvenience to the individual."

"Assuming that the random sample studied is representative it can be estimated that in the country as a whole there may be about two hundred ex-professional boxers who are handicapped by brain damage sufficiently to be described as 'punch drunk,' and about another four hundred to five hundred who are likely to have some evidence of the same kind of damage which would be apparent to a neurologist but possibly not so a layman."

"The evidence available did not suggest any direct relationship between boxing and the cause of death in the 16 exboxers who had died."

(d) Correlation of Brain Damage with Occupational Exposure

"The very close association between the present age of the exboxer and the period in which he was professionally active, that is pre- or post-war, prevented any clear demonstration of the relative importance of these two factors. In general, the prevalence of brain damage was higher among the older men professionally active in the 1930s, than among their younger colleagues with post-war careers. It was striking, however, that when both the age of the boxer and the period in which he had boxed were held constant, the prevalence of brain damage increased with increasing 'occupational exposure' measured by the length of the professional career. Among ex-boxers aged 50 years and over, almost half (seventeen out of thirty-six, or 47 per cent.) of those who had boxed professionally, licensed or un-licensed, for more that ten years showed evidence of brain damage, a proportion that declined to 17 per cent. (five out of thirty) among boxers with a career lasting six to nine years, and to 13 per cent. (three out of twenty-four) among those who had boxed no longer than five years. Among younger boxers (thirty to forty-nine years) the percentage with brain damage was lower but the trend with career duration was even more marked, declining from 25 per cent (five out of twenty) to 14 per cent (six out of forty-three) and to 1 per cent. (one out of seventy-two) respectively. This direct association between exposure to boxing and brain damage was also apparent when clinical findings were related to other measures of boxing experience such as the number of professional fights or age at retirement from the ring."

"Fairground boxing booths and work as a paid sparring partner have always provided employment for boxers whose abilities are declining. There was some evidence that individual boxers who had been actively engaged over long periods in these activities, neither requiring a license, were more liable to have evidence of brain damage. On the other hand, almost a third of those with this condition, including about a third of those who were most severely affected, had never fought in booths. There was also some evidence that traumatic encephalopathy had occurred more frequently among those boxing in heavier (welterweight and above) than in the lighter classes (lightweight and below). But with each of these weight groupings it was still notable that the important finding remained, the longer the exposure in boxing the higher the frequency of brain damage."

"This interpretation has, therefore, established that the prevalence of the clinical signs of brain damage of the kind that are found in ex-boxers increases with increasing exposure to boxing."

Discussion

(a) Professional Boxing

"While the clinical picture of the chronic encephalopathy that may occur in ex-professional boxers has been well described in the past, the frequency with which it occurred appears not to have been determined until now. The prevalence found among the group of ex-boxers studied in the epidemiological survey, and the relationship between this prevalence and the amount of exposure to boxing are striking."

"In assessing these findings it must be remembered that they relate to professional boxers licensed between 1929 and 1935, and that professional boxing has altered in many ways since the mid-war depression of the thirties. In the earlier years many started their professional careers when they were younger, fought more frequently and over a longer period, and more often without the benefits of professional and medical supervision of the sport. The prevalence of the encephalopathy is greatest among the oldest boxers whose careers took place before the Second World War, although it is notable that among the younger, post-war boxers, as among the older, prevalence is related to boxing exposure."

"The *Committee* has had discussions with members, including *Medical Officers of the British Boxing Board of Control* and appreciate that professional boxing under their auspices is, today, closely and carefully controlled. It is, therefore, not possible to draw conclusions about the risk of brain damage in professonal boxing today."

(b) Amateur Boxing

"*The Committee* has had discussions with representatives, including *Medical Officers, of the Amateur Boxing Association,* and received detailed information about the care of individual boxers and about the general control of bouts exercised by the Association. The *Committee* appreciates that every effort is made to safeguard the amateur boxer. The *Association* co-operated most fully in making their records available to the *Committee,* but these were not adequate for an investigation of amateur boxers comparable with that of professional boxers."

"The *Committee* also received information about amateur boxing in the Services from the *Directors-General of the Medical Services in the Armed Forces.* The *Royal Air Force* supplied figures of injuries sustained in boxing which have since been published *(Ref.: BRENNAN, T. N. N., and O'CONNOR, P. J., (1968), Br J Industr Med, 25: 326).*"

"The *Armed Forces* are uniquely able to exercise a strict control over boxing, and it appeared to the *Committee* that they will avail themselves fully of their opportunities in this respect."

Conclusions and Recommandations

"The evidence examined by the *Committee* suggests that severe acute injury to the brain is an uncommon occurrence in both amateur and professional boxing, though a high mortality in young boxers has been associated with the rare accidents which have caused bleeding into the membranes covering the brain. "

"The *Committee* is satisfied, on the evidence it examined, that there is a danger of chronic brain damage occurring in boxers as a result of their careers. It draws attention to the findings of an enquiry into the manifestations of traumatic encephalopathy and its prevalence among a group of former professional boxers, the majority of whom retired from boxing over twenty years ago. "

"There were few boxers who were severely disabled by this disorder among those examined in the epidemiological survey, but evidence was found of lesser degrees of the same disorder in an alarmingly high proportion of older boxers who had pursued their careers for over ten years."

"The completeness of the personal records available for each boxer varies among the different bodies supervising the sport, and professional booth boxing may take place without adequate supervision. The *Committee* recommends that all forms of competitive boxing should be supervised by organisations able to undertake responsibility for the medical welfare of the boxer, and that continuing personal records of all boxers engaged in competitive boxing should be maintained. The control of the sport, both amateur and professional, appears in general to be close at the present time. If adequate records are maintained the *College* would be able to carry out further surveys."

"It may be that careful professional and medical supervision, which has become increasingly a feature of the sport in recent years, has reduced and will continue to reduce the chances of chronic traumatic encephalopathy developing in boxers. The hazards inherent in the sport will continue to carry with it the likelihood of some degree of permanent injury to the brain."

American Medical Association, Council of Scientific Affairs. Brain Injury in Boxing *(J Am Med Ass 249: 254-257, 1983)*

"The resolution regarding brain injury in boxing was introduced to the *House of Delegates of the American Medical Association* by *the American Association of Neurological Surgeons.* The sponsor of the resolution noted that brain damage, as evidenced by dementia, memory loss, slurred speech, tremor and abnormal gait, is seen in perhaps 15% of professional boxers. The sponsor further noted that death is an occasional consequence of the sport in which the ultimate goal is to reduce the oppponent to a state of total and complete helplessness."

"The resolution was referred to the *Board of Trustees* with the request that the board 'study the matter of brain injury in boxing and report the results of the study, along with such remedies as may be appropriate.' The *Board of Trustees* referred the matter to the *Council on Scientific Affairs."*

"The *Council on Scientific Affairs* concurs with the findings in the following report, which was prepared by its expert *Advisory Panel on Brain Injury in Boxing."*

Introduction

"Boxing is a collision sport that stimulates extreme emotion from both proponents and opponents. Proponents claim that the sport requires rigorous training, strict discipline, tolerance to pain, resolution, alertness, courage, and endurance. These are desirable qualities that may contribute to the physical and social development of youth. Furthermore, boxing has been proposed as a controlled outlet for aggressive human instincts. Under properly controlled and supervised circumstances, proponents claim that the risk of serious head injuries is no greater than for other collision sports."

"Opponents of boxing use the adjectives 'suicidal,' 'brutal,' and 'murderous' to describe action in the sport, and further point out that it is the only sport in which, within the rules, each contestant deliberately tries to inflict severe physical injury on his opponent and render him senseless through a 'knockout' blow. Accordingly, they claim boxing to be morally indefensible. They further state that no present equipment can completely protect the brain from short-term or long-term injury. The benefits and intangible values attributed to boxing can be derived as well from other less dangerous individual sports."

"The A*MA* has previously addressed the issue of the dangers of boxing through the statement of the *AMA Committee on the Medical Aspects of Sports, 'Statement on Boxing' (J Am Med Ass, 181: 242, 1962).*

Boxing Activity in the United States

Amateur Boxing

"Approximately 15.000 boxers between the ages of 10 and 15 years are registered with the *National Amateur Athletic Union (AAU) Junior Olympic boxing program.* Each boxer will average between ten and thirty bouts a year for a period of five to six years. The incidence of knock outs is calculated at approximately 5% for the 14- to 15-year age group and is extremely rare among the younger age groups."

"Current estimates indicate that an additional 12.500 amateur boxers participate in the *Golden Gloves Association of America* boxing program. Each boxer averages approximately 20 bouts per year for three to five years. Fewer than 5% of these amateur boxers sustain a knock out blow during their careers."

"In recents years, 'tough man' and women boxing contests have been promoted and staged in various sections of the counry. The contests usually involve poorly conditioned, unlicensed amateurs, and are not sanctioned by appropriate state boxing commissions. The potential for serious injury is high in any such unlicensed boxing bouts. In at least three reported instances a 'tough man' contest has resulted in brain injury and, in two instances, death."

"The *National Collegiate Athletic Association* does not recognize boxing as a varsity sport. There are several colleges where boxing is a club sport. However, intramural boxing contests continue to be held under close supervision at the service academies."

Professional Boxing

"Boxing is regulated by state or local boxing commissions established under law in 46 states, five territories, and the District of Columbia. The states of Georgia, Oklahoma, South Carolina and Wyoming have no boxing statutes. There may be as many as 5.000 licensed professional boxers in ths country, although the precise number is unknown. Licensed boxers frequently move between states for scheduled bouts. There are reports that some boxers perform under different names in several states. The various boxing commissions provide medical standards for the physical examination and licensure of boxers, and medical supervision of boxing bouts. However, there appears to be incomplete and fragmentary exchange of information between many state boxing commissions regarding the routine identification and medical conditions of injured boxers."

"A report from the *New York State Athletic Commission* indicated that 856 boxers with an average age of 22.3 years were licensed in *New York between 1976-1981*. These boxers participated in approximately 677 bouts per year, fighting an average 2.907.4 rounds of boxing. During the five-year period, there were 544 suspensions of licensed boxers."

"The principal professional boxing governing bodies are the *World Boxing Council (WBC)* and the *World Boxing Association (WBA)*. Each organization provides ratings for individual boxers under its control and sponsors championship matches in the various weight divisions. Neither organization routinely accepts the ring ratings of the other. Other active boxing associations with ring ratings and individul champions include the *United States Boxing Association (USBA)* and the *North American Boxing Federation (NABF)*. A majority of the state and local boxing commissions are voluntary members of one or more of the various boxing associations. The large sums of money involved in professional boxing encourages accusations of fraud and corruption, which has been the subject recently of several articles published in major sports magazines."

Boxing Deaths

"According to MOORE, 335 deaths occurred among amateur and professional boxers worldwide during the 35-year period between 1945 and 1979. Calculation of mortality rates for the sport is inprecise, since the exact number in the world is unknown. However, the fatality rate for boxing has been calculated as 0.13 death per 1,000 participants. The following are calculated fatality rates per 1.000 participants for other sports during the same period: college football, 0.3; motorcycle racing, 0.7; scuba diving, 1.1; mounteneering 5.1; hang gliding, 5.6; sky diving, 12.3; and horse racing (jockeys and sulky drivers), 12.8. The *advisory panel* had no information on how these statistics were compiled, and cannot attest to their validity or reliability. The studies of REFSHAUGE and GONZALES suggest low mortality rates in boxing compared with other sports."

"A study by PAYNE indicated also that the usual cause of boxing fatality was subdural hemorrhage and its complications. The mechanism of the fatal injury is related to rotational acceleration of the head from forceful blow with rupture of bridging or connecting veins. Such hemorrhage results from the direct effects of the blow or from subsequent impact of the head against the floor or ringposts."

Cerebral Concussion

"A 'knock out' and, in many instances, a technical knock out (TKO), indicates that the boxer has suffered a cerebral concussion. The recent incidence of knock outs in professional fights in *New York State* has been reported to be 43/677 bouts *(Frank S. FOLK, M.D., written communication, September 1961)*. More than 90% of the knock outs occur during the first three rounds. BLONSTEIN and CLARKE assessed boxing injuries encountered in 3.000 *London Amateur Boxing Association contests* during 1955-1956. Only 29 boxers (0.58%) had severe concussions or were knocked out more that once. The results of neurological and EEG examinations were normal in all boxers examined."

"In their report of the brain-wave patterns of professional boxers, KAPLAN and BROWDER mention that the EEG examinations performed on 40 boxers within ten minutes after losing the fight by a decision or TKO (a number of them had been dazed by a blow to the head, which in many instances lead to a TKO) showed no constant change from their prefight tracing. These authors also mention also that 'detailed examinations' after the fight failed to disclose any abnormal neurological signs, even in those boxers who were knocked out."

"BUSSE and SILVERMAN carried out routine EEG studies on boxers who had been knocked out in the ring and compared them with those of a control group. Although they reported abnormalities (37%), they were unable to correlate the degree of electrical abnormality with the knock out."

"A cerebral concussion may be associated with other brain injuries, such as a subdural hematoma. The number of nonfatal acute intracranial hemorrhages sustained in boxing is not known. Two cases in young amateur boxers were recently reported by CRUIKSHANK et al."

Longterm Neurological Sequelae of Boxing

"The majority of the literature on the chronic neurological sequelae of boxing has dealt with the 'punch-drunk syndrome.' This term, first introducd by MARTLAND in 1928, describes boxers who were ataxic, had a broad-based gait, slurred speech, and dementia. ROBERTS performed examinations on 224 of 16.731 professional boxers registered with the *British Boxing Board of Control* between 1929 and 1955. Boxers were given physical examinations, psychiatric interviews, and a number of memory tests. The study indicated that 17% of those wo had boxed for six to nine years displayed brain damage, and one third showed signs of punch-drunk syndrome. The extent of the damage was related to the number of bouts fought."

"SERCL and JAROS reported the incidence of chronic boxer's encephalopathy to be 9%. JOHNSON reviewed the cases of 15 professional fighters about 22 years after their careers ended. Each boxer had experienced from 200 to 300 fights. Half of this group were found to have neurological signs compatible with traumatic encephalopathy. All but three boxers showed damage to the brain evident by air-encephalography. Eleven boxers had diffuse abnormalities. MAWDSLEY and FERGUSON studied ten retired boxers who displayed neurological abnormalities. The neurological dysfunctions included dementia and extrapyramidal and cerebellar signs, with pyramidal dysfunction being less frequent. Air encephalography disclosed evidence of cerebral and cerebellar atrophy and cavum septi pellucidi was common. They concluded that boxing is sometimes the cause of progressive neurological disorder."

"There are other isolated case reports describing similar neurological abnormalities in boxers. HARVEY and DAVIS reported the syndrome in a 25-year old middleweight boxer."

"CORSELLIS et al. examined the brains of 15 retired boxers and interviewed relatives and friends about the boxers' lives. From these investigations, they concluded that in some boxers a neurological disorder developed that correlated with 'cerebral damage or degeneration that is concentrated on the septal regions, and on certain neurons along the cerebellar and nigral pathways.'"

"Many boxing physicians feel that the incidence of the punch drunk syndrome has been sharply reduced because of increased medical supervision. This is perhaps substantiated by THOMASSEN et al. who performed neurological examinations on a total of 53 former champion amateur boxers. Only minor neuropsychological disturbances were found in the boxers, and were most pronounced in impaired motor functions of the left hand. The EEG study of KAPLAN and BROWDER as well as those of BUSSE and SILVERMAN show an increased incidence of abnormalities in boxers, however, a cause-and-effect relationship has been questioned."

"Other potential long-term neurological sequelae from boxing, such as post-traumatic epilepsy and traumatic cranial nerve palsies have not been adequately documented."

Comment

"The *advisory panel* has carefully reviewed studies of deaths from boxing injuries and reported instances of brain injury in boxing. It has been reasonably well established in the medical literature that there was a punch-drunk syndrome that had adversely affected boxers in past years and may be a specific long-term occupational hazard of the sport. A traumatic encephalopathy resulting from repeated blows to the head has affected unspecified numbers of boxers. This progressive and disabling disorder may be prevented by sound medical and administrative measures in the sport."

"Evaluation for the possible confounding effects of excessive alcohol abuse, sexually transmitted disease, and the aging process on the brain has been lacking in previous investigations, and reports frequently have not included suitable control populations. In addition, there has not been a comprehensive medical study of boxers since many medical surveillance measures were instituted in the 1950s. Earlier investigations included boxers who had fought in the 1920-1940 period when medical supervision was infrequent, and when boxers had a greater number of bouts and generally longer careers in the ring."

"In recent years, medical supervision of boxing has improved in some boxing jurisdictions. Improvements have been noted in the following areas: (1) basic standards and requirements for the medical examination of boxers; (2) adequate medical supervision and equipment for the boxer at ringside; and (3) periodic examination and evaluation of poor medical and boxing risks. Such precautions minimize and may even prevent acute intracranial hemorrhages and death."

"The activities of the *New York State Athletic Commission* provide one example of improved medical supervision of professional boxers. Following the death of a boxer in 1979, the *Athletic Commission* made a series of changes in its boxing rules and practices. A boxer must have a 'passport' to fight in New York, and all medical and previous boxing data are checked by computer before he receives clearance for a fight. The state assigns two physicians at ringside before the start of any boxing program; each physician can enter the ring at any time and stop the fight for medical reasons. A boxing match is never started until an ambulance and emergency equipment are on the premises. If he sustains lacerations of the face, a boxer may receive a suspension of 60 days, while TKO's for minor injuries entail a suspension of 30 days and those for head injuries receive a 45-day suspension. Boxers with immediate response to a knock out receive a 60-day suspension, while those with a slow response receive a 90-day suspension plus a computed axial tomography scan and EEG within 24 hours."

Conclusions

"Boxing is a dangerous sport and can result in death or long-term brain injury. However, other sports may also result in additional death or brain injury for participants."

"Amateur boxing is fairly well supervised in this country through several national organizations. Professional is less well controlled since the supervision of the sport is carried out worldwide through numerous uncoordinated national, state, and local boxing commissions. Therefore, it is dfficult to determine the medical chronology of injuries in boxers."

"No reliable test exists to identify boxers at risk for sudden death or impending brain injury. To reduce this risk, central administrative supervision should be required for the sport of boxing."

Recommendations

"Some would favor banning boxing completely, but this is not realistic solution to the problem of brain injury in boxing. Moreover, the sport does not seem any more dangerous than other sports presently accepted by society. The *Advisory Panel on Brain Injury in Boxing* does not see a need for specific improvements in administrative and medical standards, and the *Council on Scientific Affairs* recommends that the *AMA* implement the following measures:"

"(1) Encourage the establishment of a *'National Registry of Boxers'* for all amateur and professional boxers, including 'sparring mates,' in the country. The proposed functions of a computer-based central registry would be to record the results of all licensed bouts, including TKOs, knock outs, and other boxing injuries, and to compile injury and win-lose records for individual boxers."

"(2) Plan and conduct a conference with representatives of the *Ameican Association of Ringside Physicians,* medical representatives, medical representatives of the various state and local boxing commissions, and representatives of organized professional and amateur boxing organisations, to review criteria for the physical examination of boxers, to determine other comprehensive medical measures necessary for the prevention of brain injury in the sport, and develop specific criteria for the discontinuance of a bout for medical reason."

" (3) Recommend to all boxing jurisdictions that the ring physician should be authorized to stop any bout in progress, at any time, to examine a contestant and, when indicated, to terminate a bout that might, in his opinion, result in serious injury for either contestant."

"(4) Urge state and local commissions to conduct frequent medical training seminars for all ring personnel."

"(5) Recommend to all boxing jurisdictions that no amateur or professional boxing bout should be permitted unless (1) the contest is held in an area where adequate neurosurgical facilities are immediately available for skilled emergency treatment of an injured boxer; (2) advanced life-support systems are available at ring-side, and (3) a comprehensive evacuation plan for the removal of any seriously injured boxer to hospital facilities is ready."

"(6) Inform state legislatures that unsupervised boxing competition between unlicensed boxers in 'tough man' contests is a most dangerous practice that may result in serious injury or death to contestants, and should be condemned."

"(7) Urge state and local boxing commissions to mandate the use of safety equipment, such as plastic safety mats and padded cornerposts, and encourage continued development of safety equipment."

"(8) Urge state and local boxing commissions to extend all safety measures to sparring partners."

"(9) Urge state and local boxing commissions to upgrade, standardize, and strictly enforce medical evalutaion for boxers."

The *Canadian Medical Association* resolved in 1983 that "the *CMA* recommend to appropriate government authorities that all boxing in Canada be banned."

The *Canadian Psychiatric Association* supported this action.

"MURRAY, *past president* of the *Canadian Neurological Association* and past vice-president of the *American Academy of Neurology,* who was previously a boxer, supported the ban.

The *World Medical Association* recommended in October 1983, that boxing be banned.

The *British Medical Association* published in February 1984 a *'Report of the Board of Science and Education Working Party on Boxing.'*

"The enquiry on which this report is based resulted from a resolution passed by the *BMA's* policy making body, the Annual Representative Meeting, in June 1982:"

"That in view of the proven ocular and brain damage resulting from professional boxing, the *Association* should campaign for its abolition."

"The *Annual Representative Meeting* delegates its power during the year to the *Council of the Association* who requested that the *Board of Science* should consider the *ARM* resolution on boxing."

"The *Board of Science* considered the resolution at its meeting in October 1982. Two important factors were immediately apparent to the *Board.* The first was that a definitive report on boxing had been published as long ago as 1969 by the *Royal College of Physicians.* This report was based on investigation carried out on professional boxers who had been fighting between 1929 and 1955. There are those who question the relevance of the *RCP report* to boxing under the more stringent regulations imposed since then."

"The second point noted by the *Board of Science* was that the *Annual Representative Meetings's resolution* referred only to professional boxing. This raised the obvious question as to whether the adverse effects of boxing were confined solely to those who undertook the professional sport."

"It was clear to the *Board of Science* that a review of the recent medical evidence should be undertaken. Accordingly, the Board set up a *Working Party* under the *Chairmanship of its Deputy Chairman,* Professor J. P. PAYNE. The following terms of reference were agreed:"

"To consider the medical effects of boxing and to make recommendations in relation to the *ARM* resolution of June 1982."

"The membership of the Working Party was as follows:"

Professor James F. PAYNE – Research Professor of Anaethetics at the Royal College of Surgeons

Professor J. A. N. CORSELLIS – Emeritus Professor of Neuropathology of the University of London

Dr. Jeffrey M. CUNDY – Consultant Anaesthetist at Mewisham Hospital

Mr. Andrw R. ELKINGTON – Consulant Ophthalmologist at Southampton Eye Hospital

Sir Ronald GIBSON – General Practitioner, School Medical Officer

Dr. William GOODDY – Former Senior Physician at the National Hospital for Nervous Diseases, Queens Square

Professor Bryan JENNETT – Professor of Neurosurgery, Institute of Neurological Sciences, Glasgow

Secretary – Miss Maryse BARWOOD

"The *Working Party* began its task by issuing a press statement calling for evidence on boxing injuries and by conducting an extensive review of the recent medical literature. During the course of the Working Party's early discussions, it became clear that further information was needed in certain areas. Mr. ELKINGTON agreed to make further enquiries into the incidence of eye injuries in boxing. Dr. Adrian WHITESON, *Chief Medical Officer of the British Boxing Board of Control*, at an informal meeting with Mr. ELKINGTON said that he would be willing to release the necessary data but unfortunately the information has not been forthcoming. Meanwhile Professor JENNETT's Unit in Glasgow had already started a study of amateur boxers in the Glasgow area. Each volunteer gave details of his boxing career, was examined by a neurologist, a psychologist, had a physical examination and a CT scan. The study relies on volunteers and the numbers are only slowly accumulating; it will be some time before results are available."

The *Working Party* was grateful for the co-operation it received from many experts and Associations. In particular it would like to mention the *Amateur Boxing Association* and the *Armed Services*. However, the Working Party was disappointed at the lack of co-operation received from the *British Boxing Board of Control (BBBC)*. In all communications with the *British Boxing Board of Control* care was taken to emphasise the independence of the *Working Party* and its potential importance as an influence in amending *BMA* policy. The *Working Party* regrets the refusal of the *BBBC* to provide information or even to send representatives to meet the working party."

Since World War II there has been an increasing medical interest in injuries related to boxing; and the improvement on the medical supervision of the health of boxers both professional and amateur has been due to some extent to the concern of the medical profession."

"The medical interest in boxing has been international and in 1982 important reports on boxing were published by the *American Medical Association*, the *World Medical Association*, the *Australian Medical Association*, the *Swedish Government* and the *Swedish Academy of Medical Sciences*. Professional boxing has been banned in Sweden since 1969 and in Norway since 1982."

"There has been a much greater interest on the part of the public in the damage caused by boxing so that the term 'punch drunk' is now widely used. The regular coverage by television of boxing matches has been enabled an enormous number of people to witness the damaging blows sustained in the ring and it may be the influence of this vast number of spectators which has caused the promoters, managers and especially referees to be more careful in not allowing contests to proceed beyond what they think the public will accept. A doctor is now required to be in attendance at both professional and amateur contests. However, he has no power to stop a fight although he can indicate to a trainer if he thinks a fight should be stopped."

"In this country from the historical point of view the celebrated fights of the 18th century, sometimes going to more than one hundred rounds (and the disablement of one or possibly both the contestants) faded away after the *Queensberry Rules* were introduced in 1860. The *Amateur Boxing Association (ABA)* was formed in 1880."

"One form of boxing, which has now almost vanished, is that of the *'booth boxer'* who was to be found in circuses and fairs. He was rightly known as a *'bruiser'* and often was fit enough to knock down a local lad. It has been suspected that the booth boxer may often have been the subject of punch drunkenness, having taken part in innumerable fights. It may have been because of his failing health that he was obliged to undertake such a crude form of boxing."

"In the modern era amateur boxing is, as a sport, related to athletics. Those who support amateur boxing encourage it as a means of helping young men to improve their physical health to a high level and also, by learning the rules of boxing to establish their own discipline and self-control. In a Boxing Club the extremes of personality, from very aggressive to inadequate are recognised and to some extent understood and assisted. Boxing is considered as an acceptable activity in some branches of the three Services where to have some experience of boxing is regarded as a great advantage. Training for most sports (including boxing) discurages drinking, smoking and drug taking."

"In general the amateur starts boxing as an adolescent and until the age of sixteen the signed consent of his parents is required before he is allowed to box. Amateur boxing becomes less common after the age of twenty. The demands of a career make it difficult for the boxer to keep up with his training. It is possible for a successful amateur boxer to become a professional boxer if he thinks he has reasonable career ahead of him and most professional boxers have had an amateur career."

"The approach to professional boxing is different from that to amateur boxing. Whereas the latter is usually regarded as a personality-developing form of sport related to good physical health, the former is 'big business.' The professional boxer is an asset to his manager. The objective is to promote him so that he will graduate from national to international contests and earn increasing fees. A series of wins is needed for the records so that the boxer may be deemed a serious contender for the British Championship. From these he progresses to the European Championships and he will get automatic world ranking if he wins. On average a professional boxer will fight eight times per year. At the very top of the ladder some of the rewards are in excess of one million dollars or pounds for a single fight. The loser in these contests also receives a very substantial fee. The commercial aspects of 'top class' bouts extend to television fees and the sponsorship of commercial goods. The 'big business' aspects of professional boxing are obviously tempting to a young man and the prospect of financial gain may cause him to ignore the risk to his health. There have been instances of the most successful professional boxers playing a kind of political or religious role because of their fame or notoriety. At the end of their boxing careers they continue as popular sporting figures, sometimes running Clubs or Bars and continuing to appear on television. On at least one significant occasion the appearance of serious neurological problems which so far have not been publicly acknowledged or documentad in this country. However, the April 1983 edition of *Sports Illustrated* gives a detailed account of well-known and named boxers who are brain damaged. The account refers to their CT scans."

"The medical literature on injuries associated with boxing is limited whereas there is much data on head injuries in general. This includes experimental studies of mechanisms and pathology of injury, on the effects of repeated concussion and on the insidious development of progressive mental dilapidation and unco-ordiantion."

"The definitive work on clinical evidence of brain damage in living boxers is that of Dr. A. H. ROBERTS, FRCP, who was appointed in 1967 by the *Royal College of Physicans in London* under the chairmanship of the late Lord BRAIN 'to report on the medical aspects of boxing.' The *BBBC* co-operated in that study and the statistics were supervised by the *Medical Research Councils's Statistical Resarch Unit.* Two hundred and tweny-four ex-professional boxers were selected at random from the population of 16.781 boxers registered between 1929 and 1955. This investigation led to the recognition of a syndrome including dysarthria (slurred speech), lack of balance (unsteady gait), rigidity and tremor and varying degrees of dementia."

"Those who support boxing point out that the number of injuries, serious or otherwise, is relatively small in comparison to the total number of contests and that there are other sports, both amateur and professional, which also have serious hazards. It is now recognized that rugby footballers and jockeys, especially 'jump jockeys' may suffer repeated and sometimes serious injuries. It also has been suggested that sports like parachuting and hang-gliding have their dangers. It should be recognised that the rules of these sports are generally designed to prevent injury to the participants and that most of the injuries sustained are accidental in a way that deliberate blows to the head are not."

"The problem is one of preventable injury. The pressure of 'big business' is the most formidable obstacle to the banning of boxing. It remains to the medical profession to ensure that the risks are widely known."

The Pathology of Brain Damage due to Boxing Introduction

"The bony skull. with its tough fibrous lining, contains and protects the brain. The brain fills the space within the skull and is a soft and jelly-like organ, covered by two thin membranes between which is a thin cushion of fluid. The brain is actually an enormous collection of nerve cells (called neurons) and the human nervous system contains a total of some 15-20 million of these cells. Within the skull the brain consists of two hemispheres which sit together. The hemispheres are joined at the base and large numbers of nerves cross from one side to the other for a number of purposes."

"As an embryo grows the brain develops around the upper end of a long hollow tube and as it develops a number of recognizable structures become prominent. Several of these structures form what is known as the old brain and these are found in the brain stem between the top of the spinal cord and the surfaces of the two hemispheres. Another organ, the cerebellum, develops and is attached to the brain stem by three groups of nerve fibres on each side. The cerebellum lies below the rear half of the two hemispheres and is situated low down at the back of the head. The surface of each of the hemispheres is covered by the cerebral cortex which is a layer of grey matter a few millimetres thick. The deeper parts of the brain lying at the base consists mainly of grey matter and the intervening space between the deep grey matter and the cerebral cortex is filled by white matter. The grey matter consists mainly of nerve cells with thin branching connections which make up the mass of the white matter. A number of connections (synapses) in contact with a nerve cell of any type is very large and may be up to 10.000 in some instances."

"Each of the millions of nerve cells in the brain is specialized for the conduction of impulses, and it sacrifices may other biological functions to this end. A nerve cell cannot reproduce itself and its metabolism is so simple and immediate that it cannot live for more than a few minutes without oxygen. If the blood supply to the head is interrupted for a period as little as 15 seconds a person may become unconscious because elements of the brain cease to function."

The Nature of Cerebral Damage

"There are two main ways in which boxing can lead to structural damage to the brain. The first occurs as an acute episode in which one or more severe blows during a single fight lead to loss of consciousness and, very occasionally, death. The second kind of damage tends to develop over a much longer period and changes begin to take place as the number of fights mounts up."

Acute Damage

"Death in the ring or in the days or weeks after the contest is usually attributable to one or other form of bleeding within the skull. This haemorrhage is usually under the thick membrane (dura) which lines the inside of the skull and the blood clot usually lies on the surface of the brain. With prompt and expert treatment the clot may be evacuated successfully. Not all such collections of blood or haematomas however, are followed immediately by loss of consciousness and mortality remains around 45% even with all modern neurosurgical facilities. In some instances bleeding and destruction of brain substance may occur and the brain may swell dangerously (UNTERHARNSCHEIDT 1975)."

Insidious or Chronic Damage

"The term 'punch drunk' was introduced in 1928 by MARTLAND to describe the tendency for some experienced boxers to become unsteady on their feet and to move and think more slowly. MARTLAND described lucidly the many clinical features of the syndrome, but he could only conjecture on the underlying cerebral damage. Neither he nor anyone else at that time had direct evidence of the nature of cerebral damage or even of its existence. In the 1950s clinico-pathological case reports began to appear. The first of these by BRANDENBURG and HALLERVORDEN (1954) concerned a man who had become demented, tremulous and unable to co-ordinate his movements after eleven years of amateur boxing. The brain showed most of the features of the common severe degenerative disorder known as Alzheimer's disease."

"GRAHMANN and ULE (1957) examined the brain of one of four ex-boxers whose illnesses they had studied clinically. In this case the man had become 'soft' after ten years of amateur and professional boxing and had stopped fighting. However, he deteriorated physically and mentally for a further twenty years until he died demented and parkinsonian at the age of 46. The examination showed that the brain was atrophied and as part of this atrophy the fluid-containing cavities inside the brain, that is the lateral ventricles, were enlarged. There was a sizeable septal cavum but they did not comment on this fact. The cortical atrophy was confirmed with the microscope and many nerve cells in the cerebral cortex, and to a greater extent in the brainstem, were found to be grossly abnormal with thickened neurofibrils (Alzheimer's neurofibrillary tangles). No senile plaques were found. There are minute foci of degeneration in those areas of the brain which contain nerve cells – that is the grey matter including the cerebral cortex. These plaques are often found in the brains of old people together with neurofibrillary changes."

"In 1959 NEUBURGER et al. found the brain of a 53 year old, mildly demented, ex-boxer to be severely atrophied with marked loss of nerve cells and gliosis, or scarring, in the cerebral cortex but without the tangles described above. A cortical biopsy from a second patient showed a similar histological picture."

In 1962 SPILLLANE emphasised the frequency of chronic neurological disorders in professional boxers and made important radiological observations of abnormalities of the deep mid-line structures, the septum pellucidum, in four of the five boxers, and FERGUSON and MAWDSLEY (1965) elaborated on SPILLANE's discovery and described how they had found a cavum septi pellucidi with fenestrated walls in each of the brains of four ex-boxers. The main histological abnormality was the constant presence of Alzheimer's neurofibrillary change in some cortical neurons and specially in those of the midbrain."

"In 1967 CONSTANTINIDES and TISSOT examined the brain of a Swiss ex-boxing champion who had developed a hemiparesis in his twenties, had later deteriorated mentally and had become parkinsonian. He died aged 58. The brain was moderately atrophied and a fenestrated septal cavum was noted. The substantia nigra had lost its pigment. Many neurofibrillary tangles were scattered predominantly through the medial temporal grey matter and in the pigmented cells of the brainstem."

"In 1968 PAYNE reported briefly on the brains of six professional boxers. Four and possibly five had died before reaching the age of 50. Enlarged ventricles and a septal cavum were found in all six of the boxers while the septal leaves were fenestrated in three. The only other prominent features were occasional degenerative foci scattered in the grey and white matter. Neurofibrillary change was identified in three patients, one being that of an octogenarian in whose brain scattered senile plaques were also seen."

In 1973 CORSELLIS et al. investigated the past histories and the brains of 15 ex-boxers. Twelve had boxed professionally; three had been amateurs. Two were world champions and six were reported to have won a national or regional championship. Their findings confirmed the pattern of change in the brain of the boxer which was to be noted in earlier studies.

Within a general setting of cerebral atrophy, four main component are liable to occur:"
"(1) Abnormalities of the septum pellucidum and the adjacent periventricular grey matter."
"(2) Cerebellar and other scarring of the brain."
"(3) Degeneration of the substantia nigra."
"(4) The regional occurrence of neurofibrillary tangles."

(1) Septal and Related Abnormalities

"The normal septum consists of two thin, imperforate sheets or leaves of nervous tissue which lie vertically in the mid-line of the brain since they represent one part of the medial section (i.e. a cut made vertically from 'ear to ear') the septum is seen to be stretched between the corpus callosum above, which carries the two fornices below, each of which is another major fibre pathway but which in this case links different parts of the same hemisphere (i.e. hippocampal region to hypothalamic septal areas). In this coronal section the two leaves are usually fused or contiguous but a narrow space or cavum normally separates them in about a fifth in unselected human brains (SCHWIDDE 1952). However, CORSELLIS et al. noted that in twelve of the thirteen ex-boxers's brains suitable for examination a cavum was found with a mean width of about 5 mm compared with the normal mean of 1.4 mm. In the single case with no cavum anteriorly, the septum had been destroyed posteriorly. In ten of the boxers the septum was so severely fenestrated that it was represented by only a few tags of remaining tissue. Thus the septa of these brains was strikingly different from those of the 500 non-boxes with which they were compared. A cavum with an average width of three times of the non-boxers was the rule while the septal leaves were almost without exception grossly torn."

"CORSELLIS commented in a later study (1978) that the remnants of the old boxer's septum often has a look almost of its own. The fornix becomes detached and grossly atrophic, while the mammillary bodies, in which the fornix terminates, and even the walls of the hypothalamus appear shrunken and scarred. It may well be relevant, along with the medial temporal grey matter, to be associated with disturbances of memory."

"The way in which the act of boxing produces a large and fenestrated cavum is unclear. It is merely the result of enlarging lateral ventricles for when this occurs it ends to result in fenestration without a cavum. MAWDSLEY and FERGUSON (1963) invoked the effect of repeated sudden rises in intraventricular pressure and were less inclined to consider the possible distortion and tearing produced by rotational acceleration or accelerative forces. However, these zones may be important as the deep mid-line regions are known to be particularly vulnerable in acute head injury (STRICH 1961). The septum changes from a thin fragile membrane below the corpus callosum to merge ventrally into the thickening of the fornices. It seems possible that repeated blows to the head strain the dorsal attachment by the movement of the corpus callosum on the tethered fornix. Then, as the lateral ventricle enlarge, as they undoubtedly do, the septum stretches, ruptures and the fenestration appears. Whatever the prime mechanism may be, the practical importance of the lesion is the proof it provides that the brain has been damaged and that nervous tissue lies along the limbic pathways of the brain has been lost."

(2) Cerebellar Scarring

" The clinical picture in the punch-drunk syndrome often indicates that the cerebellar pathways have been affected but earlier neuropathological studies have paid little attention to the possibility. The most clear-cut finding has been a well defined area of cortical scarring on the under-surface of the cerebellum, and particularly those parts of it (i. e. the cerebellar tonsils) which lie against the bony rim of the whole in the base of the skull (foramen magnum) through which pass the mass of nervous connections to and from the rest of the body (i.e. the lower part of the brainstem and spinal cord). In this scarred area, in addition to a dense fibrous gliosis spreading through the molecular area, a marked loss of nerve cells occurs. This loss of these large neurons, that is the Purkinje cells, which are of major importance to the control of movement, was cearly demonstrated when comparisons were made with cell counts on the lateral lobes of the cerebellum. This loss reached at times a half or even two-thirds of the expected population.... Although the lower brainstem itself in the boxer's brains showed no changes comparable to those in the cerebellum it is conceivable that a considerable strain might be placed on the neural pathways or running through it and that this might be an important, if over-simplified factor, in the pathogenesis of the 'knock out.'"

Scarring in the Cerebral Hemispheres

"Although he had no histological evidence MARTLAND (1928) believed that 'in the punch-drunk there is a very definite brain injury due to single or repeated blows to the head or jaw which cause multiple concussion haemorrhages.' However by the time an old boxer dies the residue of any small bleeds will either have disappeared or will show up only as focal or multifocal scarring in the grey or in the white matter. None of the earlier studies with the exception of PAYNE (1968) identified more than the occasional minor scar. Much the same is true of the fifteen cases in the study of CORSELLIS et al. (1973). Rusty staining of the undersurface of the frontal parts of the brain (which are particularly vulnerable in severe blows to the head) was found only twice while micro-scars were not identified either in the cerebral cortex or the deeper grey and white matter, unless vascular degenerating blood vessels were present to a degree that would at least throw doubt on the nature of the source. In other words, MARTLAND's hypothesis about the occurrence of haemorrhages in the acute stages may well be correct. This has been confirmed by CORSELLIS in the brain of an amateur boxer who died in the ring. The brain in the chronic state, however, is not a suitable vehicle in which to study the possibility."

"For similar reasons it was not possible to assess the relevance to boxing of STRICH's important observations (1956, 1961) on the pathogenesis of cerebral degeneration following a single head injury. STRICH showed, using a special (Marchi) technique, that many foci of myelin breakdown were disseminated through the white matter of the cerebral hemispheres of patients 'de-mented' by sudden trauma. By its nature the presence of this degenerating myelin gives strong evidence of damaged nerve fibres and therefore of loss of connections between nerve cells. The damage was attributed to the shearing of nerve fibres due to the sudden distortion of brain tissue by a blow to the unsupported head. A similar mechanism is more than likely to be important to the boxer. The difficulty is to prove this morphologically, for most boxers die long after they have given up organized contests and the longer the survival the more difficult it is to demonstrate myelin breakdown by the Marchi method. Furthermore, the older the patient the more likely it is that degenerative changes in the blood vessels of the brain have had a part to play. Nevertheless there is no doubt that in most cases not only the trend to reduced brain weight but also the enlargement of the lateral and third ventricles and the thinness of the corpus callosum pointed to a material reduction in the amount of tissue in the cerebral hemispheres, even if the precise mechanism of tissue loss cannot be identified. Even in the younger patients this loss of tissue affected the deep periventricular region more than the cerebral mantle. This indicated that the damage was due to the effects of boxing and not merely to those of ageing."

(3) Degeneration of the Substantia nigra

"Degeneration and loss of a major category of nerve cells in the substantia nigra (which gets its name from the dark pigment in these cells) are the classical and most recognised neuropathological accompaniment of the disorder of movement implied in the term 'parkinsonian'. Clinical evidence of such a disorder was identified by ROBERTS (1969) in 20 of the 52 case reports abstracted from the past literature on boxing. Three parkinsonian boxers, whose brains had been examined, were reported to have shown the typical loss of the pigmented nigral cells mentioned above accompanied by reactive scar tissue (BRANDENBURG and HALLERVORDEN 1954, GRAHMANN and ULE 1957, CONSTANTINIDES and TISSOT). In the study of CORSELLIS et al. (1973) a parkinsonian disorder had been diagnosed in four patients and in several more the clinical evidence indicated that the syndrome had been developing. At post-mortem examination the changes in the midbrain were often obvious to the naked eye while the histological feature were also typical. Of the fourteen cases that could be assessed eleven were found to have been materially affected. Many of the remaining nigral cells showed neurofibrillary tangles and since Lewy bodies were absent the picture approximated more to that seen in the form of parkinsonian that can be late sequel to some forms of encephalitis than that in the spontaneously occurring paralysis agitans (GREENFIELD and BOSANQUET 1953). The relation between trauma and nigral degeneration has been repeatedly and inconclusively discussed over the years (i.e. GUILLAIN et al. 1948) but it now seems clear that repeated blows to the head may precipitate a degenerative process within the neuronal and nerve fibre pathways concerned with the control of movement."

(4) Regional Occurrence of Alzheimer's Neurofibrillary Tangles

"GRAHMANN and ULE (1957) first drew attention to the presence of 'entangled' neurons in the brain of the boxer which they had examined, including that by CONSTANTINIDES and TISSOT (1967). These latter authors emphasized that neurofibrillary tangles, which are a form of neuronal degeneration, usually virtually occur in relation to advancing age but are then, already emphasized, virtually always accompanied by the other form of senile degeneration known as the senile plaques. Tangles without plaques have been identified but this has nearly always been in certain cerebral disorders such as post-encephalitic parkinsonism or some other rare forms of encephalitis (HIRANO and ZIMMERMAN 1962)."

"The striking feature in so many of the boxers, which was confirmed by CORSELLIS et al. (1973) is the massive number of altered cells spread through both the cerebral cortex and the brainstem. The most noticeable fact was the intensity of the brain change within the medial temporal grey matter. Vast numbers of neurons in the uncus, in the corticomedial part of the amygdaloid nucleus, in the hippocampus, the parahippocampal gyrus anf the fusiform gyrus were affected, the more laterally placed temporal, insular and frontal cortex was also heavilty if patchily, involved while the parietal and occipital cortex was largely spared."

"Interestingly enough, these affected areas, which lie deep in the medial part of each hemisphere, together make up a major sector of the limbic pathways, and as already briefly mentioned in section 1, have along other things, an important role in memory function."

"The cause for this neurofibrillary change is not known. It is generally looked on as a degenerative change loosely linked to age but also influenced by the presence of certain disease processes. The evidence from the studies on boxing strongly suggests that the repeated cerebral trauma of boxing should be added to the list of precipitating, if not causal, agents. Certainly the intense tangle formation seen in some boxers brains cannot be attributed to the effect of age."

"When the different categories of change, under the four headings described above, are taken together they are seen to form an unusual and relatively stereotyped pattern of cerebral damage. The immediate question which arises from this assertion is whether such damage is truly a consequence of the boxing. It has been suggested that other factors could be of greater importance. Those most often put forward are head injury outside the boxing ring, alcohol in excess, and conditions such as neurosyphilis or the degenerative processes of old age. There are good reasons to reject these suggestions. Although their occurrence was deliberately sought none of them occurred often enough or to a severe enough degree to justify its acceptance as of more than peripheral importance. The strongest argument against the incrimination of any of them is the neuropathological fact that none is known to produce the pattern of cerebral damage that has been described. That is to say, although any single item in the tetrad of pathological alterations might conceivably be attributed to a cause other than repeated blows to the head, it is not possible to account for all four having developed simultaneously in so many of the boxers studied. For example cerebellar degeneration occurs rarely in chronic alcoholic patients but the topography of the damage is radically differnt from that seen in the boxers studied. Moreover, no other histological evidence of the varous kinds of cerebral damage associated with alcoholism was found. Even though the hypothalamic floor and the mammillary bodies were unusually atrophic, the histological appearances were not those of Wernicke's encephalopathy, a condition commonly associated with drinking."

"A head injury, unrelated to boxing, had occurred in a few patients but it could never be linked to the combination of lesions found while neither clinical nor pathological evidence of a syphilitis infection was found apart from a clinical suspicion in one case."

"The conclusion that the paramount reason for the insidious neurological and psychological deterioration of so many of the group of ex-boxers was the boxing itself seems unavoidable. This conclusion appears all the more definite as the concordance of the clinical and neuropathological findings was markedly close when the complexity of the situation is realised. The neurological features point repeatedly to the involvement of the cerebellum and the extra-pyramidal pathways, while the prominence of behavioural and particularly of memory difficulties, could well be related to the relative intensity of neuronal and other changes in the temporal and diencephalic components of the limbic areas. The evidence suggests that all the changes that have been described would appear particularly when taken together, to be restricted to the punch-drunk syndrome. This condition seems largely to have been confined to boxers (Lancet 1976) and because of its insidious nature it can prove especially devastating. Both the clinical and the neuropathological disintegration may progress over many years and long after the boxing has stopped."

"This means that in any study of the late effects of damage to the brain many years must pass before they can be fully investigated. Any up-to-data autopsy and histological data are inevitably based on the examination of the histories and the brains of a previous generation of boxers. Even a clinical study, such as that published by ROBERTS in 1969, was based on men who had been fighting between 1929 and 1955. Therefore the sceptic usualy maintains that such dated observations have become irrelevant because the many improvements in the modern medical and lay supervision of boxing have reduced immeasurably the risk of any cerebral damage. There was also a tendency in the past for some boxers to continue fighting in booths and fairs after retiring from the professional ring. It may be claimed that the chances of damage are now so slight as to be acceptable, particularly in view of the advantages which accrue both to amateur and professional boxers, the former gaining from a healthier and more disciplined youth, and the latter, along with their entourages, from the prospect if not the reality of money and fame. Until the last few years this superannuation of the older studies has not been easy either to justify or refute."

"Nevertheless, it seems unlikely that all risk can be eliminated. A single punch, or even a number of them, need not alter the structure of the brain as seen under the microscope. There is still, however, the danger that one or more blows will leave their mark. The destruction of cerebral tissue will have then begun, and although this may well be undetectable in the early stages, it may build until it becomes clinically evident. At this point, it is already too late for, once lost, nerve cells can never be replaced while the further danger exists that the degenerative process may smoulder on even after the boxing has stopped."

Recent Medical Developments in Understanding and Detecting Brain Damage

"Medical advances in the last decade have led to better understanding of the nature of the brain damage that results from concussion, of the type of injury needed to produce it, and of psychological and radiological techniques for detecting it during life. Some of these studies are experimental and some related to head injuries due to other causes. From these studies, however, it is possible to draw conclusions about the brain damage caused by boxing. Direct evidence from CT scanning is now accumulating that shows beyond doubt that permanent brain damage still commonly occurs in men who box whether professionally or as amateurs."

Experimental Evidence

" For the first time an experiment has been successfully devised which allows engineers, neurosurgeons and neuropathologists to produce at will a whole range of lesions that occur naturally in man soon after accidental injury due to road accident, fights and falls. These experiments show that even brief concussion is associated with tearing of fibres in the white matter in many areas of the brain. Moreover such lesions are known to occur in man after mild injuries that result only in concussion. Repair processes to the brain are very limited and it is now widely accepted that even after mild concussion a brain is left with a small amount of permanent structural brain damage. The experiments also show that the whole range of lesions seen in man can be produced without the head ever being struck. They result from the acceleration of the head and the damage is most severe when there is an asymmetric rotational acceleration of the head – exactly what happens with most blows to the head in boxing. Experiments elsewhere have calibrated the force of a boxer's punch on a dummy head and have shown that even a single blow can generate enough force to produce permanent brain damage."

Psychological Evidence

"GRONWALL and WRIGHTSON (1975) and WRIGHTSON and GRONWALL (1983) have been studying for many years patients who have recovered from mild concussion, often not serious enough to have warranted even over night admission to hospital. They find marked impairment of psychological function for two to three weeks in most cases and for six weeks or longer in others. These signs are more marked and more prolonged in patients who have suffered a previous concussion, even though it was years before. This suggests that although recovery from the first concussion appeared to have been complete, permanent brain damage had occurred to which the second concussion was added – hence the term cumulative concussion."

"These workers performed an experiment which yielded further evidence that people who seemed to have completely recovered from concussion in fact have permanent brain damage. University students who had had a concussion during the previous three years but who had successfully returned to their studies underwent psychological testing after they had been at a simulated altitude of 3.800 metres for 30 minutes. Under these conditions of mild hypoxia the previously concussed students performed much less on these tests than did the control group of students."

Radiological Evidence

"Until recently the only readily available tests of brain function were the clinical neurological examinations, psychological tests and the EEG. Now that CT scanning can show so much about the brain in a matter of minutes and with no risk or discomfort to the patient it is being realised just how insensitive and inadequate are these three established methods of examination. Even with marked pathology of various kinds in the brain that is obvious on the CT scan all three of these other tests can be normal."

"CT scanning is an X-ray technique that examines several transverse slices of the brain each 1 cm thick. Dilated ventricles show up clearly as does atrophy of the cerebral cortex; the characteristic tearing of the septum pellucidum can also be seen. Progressive changes can be recognised when scans are repeated at intervals over years. Only a limited number of CT scan studies in boxers have so far been reported, but they are disturbing. They show what clinicians had begun to expect from the experimental and psychological investigations – that boxers can have marked brain damage that goes completely undetected as long as the examination is limited to the traditional tests. These studies have such serious implications for boxing that they are summarised individually:"

" (i) Typical marked traumatic encephalopathy with classical clinical and radiological signs (CT scan + air study) noted in a man of 25 who boxed 72 amateur bouts and 25 professional. He lost only ten fights as an amateur and four as a professional and was knocked out only once and stopped once, both when he was a professional. He had nine further fights after he had begun to show slurring of speech and stiff and shuffling gait; in spite of this he won his last five bouts before retiring in 1972. The authors note that this throws serious doubt on BLONSTEIN's claim in 1966 that 'the punch drunk syndrome has not been seen during the last 20 years in professional boxers' (HARVEY and DAVIS 1974)".

"(ii) Two amateurs aged 17 and 22 were found on CT scan to have intracranial haematomas that required surgical removal. One had had headache after a sparring bout but he worked the next day. The following day the headache was worse and on the next day again he became drowsy and had a fit. The other suffered a one minute knock out but returned to work after four days. A day or so later he had headache and then hemiparesis and went into a coma (CRUIKSHANK et al. 1980)."

"(iii) Five of ten American professionals aged 20 to 31 had marked cerebral atrophy on CT can. The boxers with this abnormality were those who had had most bouts. The neurological examination was normal in all five boxers with radiological atrophy and the EEG was normal in the four in whom it was done (CASSON et al. 1982)."

"(iv) Some signs of brain damage was found in five out of six professional and four out of eight amateurs who were all previous Scandinavian champions (mean age 31). CT scan was abnormal in five, of whom four had normal neurological examination. The title of this 1982 paper was *"Is chronic brain damage in boxing a hazard of the past?* (KASTE et al. 1982)."

"(v) CT scans were abnormal on 53% of 38 American ex-boxers. This abnormality was correlated with the number of bouts but not with the number of knock out (actual or technical); it did not correlate with clinical neurological examination (ROSS et al. 1983)."

"These five studies using CT scans indicate that:"

"(i) Brain damage of the type previously associated with the punch drunk syndrome is now being detected before obvious clinical signs have developed."

"(ii) Such cerebral atrophy has been found in comparatively young boxers including amateurs, and in those whose careers have been successful."

"(iii) Acute intracranial haematomas may be more common than has hitherto been realised, and can occur in boxers who do not appear soon after a bout to have sustained any unusual degree of damage."

" (iv) Normal neurological examination does not exclude brain damage."

"It is therefore doubtful if the Danish study (THOMASSEN et al. 1979) which did not show neurological, psychological and EEG abnormalities in 53 amateur ex-boxers was justified in concluding that 'modern medical supervision and safety precautions protect amateur boxers against the risk of serious or permanent brain damage."

"Brain damage is a likely consequence of boxing whether in amateur or professional fighting. It may also be sustained during sparring which is controlled only by the manager. The risk to sparring partners is often ignored. It may be aggravated in some boxers by bouts fought on private circuits after they have retired and at which no doctor is present."

"Occasional deaths will continue from acute intracranial haematomas. Whilst the outcome of some incidence in the past might have been averted by more expeditious medical care, the mortality for this condition in the best neurosurgical hands is still 40 – 50% and many of the survivors are permanently disabled."

"An unpublished survey of 165 British Neurologists in 1974 about their encounters with the punch drunk syndrome associated with various sports, yielded: twelve jockeys, five soccer players, two rugby, two professional wrestling and one parachutist and 290 boxers."

"The frequency with which cerebral atrophy occurs and its exact relation to a boxer's fighting career will be known only when boxing authorities agree to a systematic study of boxers and ex-boxers by CT scanning. The importance of CT scanning must be emphasized as the other tests made on boxers are often normal, whilst the scan is abnormal. It remains to be seen how many of the younger boxers who already have atrophy on their scan will go on to develop clinical features or whether the condition will remain static if they stop boxing now. It is difficult to envisage any benefit from routine CT scans before and/or after each fight which any event would not be feasible for practical or financial reasons, except as a research project in a limited number of boxers."

"Many blows that do not cause a knock out, actual or technical can cause brain damage. It is the movement of the head in the air that appears to cause the damage and seldom is the head hitting the floor an important component; film sequences of fights show that the head is usually the last part of the collapsing boxer's body to reach the canvas. Head guards are therefore unlikely to reduce the amount of brain damage although they may protect the eye. In 1960 a professional champion in the USA died, in defence of his title, of brain haemorrhage despite wearing a head guard during a bout. This has led to the banning of boxing as an intercollegiate sport in the USA. Boxing is a contest in which the winner seems often to be the one who produces more brain damage in his opponent that he himself sustains. CT scanning can be helpful in the supervision of a boxer over a period of years in that as soon as brain damage is evident a boxer can be advised to retire. However, by the time this damage is visible it has already occurred and the brain has very limited capacity for repair. The most that can be done has already occurred and the brain has very limited capacity for repair. The most that can be done is to prevent further insult to the brain and to hope that degeneration will not be progressive – in any event such boxers should remain under medical follow-up."

"It might be worth considering following the lead of Canadian neurosurgeons, who have recommended to their Health Minister that boxers should sign a form of informed consent similar to that given to patients prior to major brain surgery. It would spell out the risks of acute and chronic brain damage. Only then could it be truly said that boxers were choosing to continue to fight in the full understanding of ths risks that they were running."

Boxing and the Eye

"Before discussing the literature on damage to boxers' eyes it may be helpful to review briefly how the eye works and to explain some of the technical terms that are frequently used to describe it."

"The eye can usefully be likened to a camera. Rays of light enter the eye through the transparent cornea and pass through a small lens to be brought to a focus at the back of the eye onto a light sensitive membrane – the retina. The central, and most sensitive part of the retina, is known as the macula."

"The space between the cornea and the lens is known as the anterior chamber. This is filled with a fluid which normally escapes from the eye from the periphery or angle, of the anterior chamber. When he fluid is prevented from excaping then the pressure within the eye increases. This in turn damages the retina and the condition glaucoma results."

"The space between the lens and the retina is filled with a jelly, known as the vitreous. Outside the retina is a vascular layer the choroid, which nourishes the retinal light receptors. Nerve fibres from these light receptors converge to a point at the back of the eye (the optic disc) before piercing the outermost coat of the eye to pass back to the brain as the optic nerve. One sequel to a blow on the eye is that the retina separates from the underlying nutritive layer and such a retinal detachment may lead to all sight being lost."

"The function of the eye is conveniently assessed by comparing the vision to that of a normal individual. For instance 6/24 means that the eye in question can read at 6 feet what at normal eye can read at 24 feet. The movement of the eyes is normally perfectly co-ordinated. If the eyes become even only very slightly malaligned then double vision results. This may occurs either when the tiny muscles responsible for moving the eyes (the extraocular muscles) are damaged or when the brain's control of eye movements is defective. Nystagmus refers to an oscillation of the eyes which may stem from brain damage."

"There is much anecdotal evidence that boxing damages the eye. This evidence has been widely reported over many years in medical literature. In a report from South Africa BOSHOFF and JOKL (1948) described three eye injuries sustained by ten boxers. Three had retinal detachments, two had choroidal tears involving the macular and another had a vitreous haemorrhage. DOGGART (1951) in his paper *"Fisticuffs and the Eye"* described many of the ocular complications which are known to stem from blunt trauma to the eye. These ranged from avulsion of the lids and damage to the cornea in front to 'cauliflower choroids' and retinal detachments behind. In the same year FAVORY and SEDAN wrote a complementary paper of their experience that over a five year period of the 9.871 participating boxers, sixteen had suffered eye damage; eleven had corneal injuries and five had retinal detachments. In a further report from the United States of America, PALMER (1972) found that of fifty-five retired boxers, nine had evidence of past contusion of the angle of the anterior chamber. However, despite an average follow-up period of thirty-three years, not one had developed glaucoma – a well-recognised complication of such angle recession. SPILLANE (1962) described five ex-professional boxers, four of whom were 'punch drunk' and the fifth exhibited bouts of agression. One patient had a partially atrophic disc, and another had nystagmus on lateral gaze. HRUBY (1979) reported that he had examined five boxers who had six detached retinas between them. However, WHITESON, writing in 1981 as the *Senior Medical Officer of the British Boxing Board of Control,* stated that he could recall since 1949 'only ten cases' of retinal detachment occurring amongst professional boxers."

"The most comprehensive report from the United Kingdom is contained in ROBERT's work. In this epidemiological study designed to assess the neurological status of 224 ex-professional boxers, five boxers were found to have double vision due to damage to an extraocular muscle. Four of these boxers were noted to have permanently adopted a compensatory head posture. There were two cases of retinal detachment reducing the vision to counting fingers in the affected eye in each case. However, one of these two boxers said the detachment stemmed from a blow on the eye in childhood. Another boxer had a pigmentary disturbance of the macula which was thought to be due to trauma. The precise cause of impaired vision of several other individuals with defective vision was not determined. When assessing ROBERT's work on eye injuries account should be taken of the fact the subjects were not examined by an ophthalmologist and although the posterior structures of the eye were routinely examined the pupil was dilated prior to examination in two cases."

"RAKUSIN (1970) examined two boxers in a survey of 370 patients seen with bleeding into the anterior chamber of the eye as a consequence of trauma. It is interesting to note that both said they had been hit by either the oppponent's thumb or the lace of the glove. During the three-year period of this study five cases of traumatic retinal detachment amongst boxers were seen."

"Boxing injuries in the Army for the period 1969-1980 were reported by OELMAN et al. (1983). There were no reported deaths but ten men were discharged from the service for medical reasons which stemmed from damage sustained in the ring. Two had damaged eyes; one having a detached retina and the other an injury to an optic nerve."

"All of the studies published so far on eye injuries in boxing note a number of retinal detachments. In this connection it is interesting to note that the incidence of retinal detachment in the general population is 1 per 10.000 patients per year. Most of these patients would be elderly people. Many would be either shortsighed (myopic) or have had a cataract removed (aphakic). It is well recognised that retinal detachments are more likely to occur in myopic or aphakic eyes."

"In contrast boxers who develop retinal detachment are relatively young. Moreover the causative tears in the retina are frequently of a type (dialysis) that is characteristically seen after trauma to the eye. Several boxers have presented with giant retinal tears, which are seen only very rarely amongst patients with detachments from the general population."

The reports outlined above have been corroborated by information given in oral evidence. However, to date the profession lacks information derived from an epidemiological study on damage to the eye due to trauma sustained in the ring. Any study undertaken would have to take into account the fact that damage may become manifest only many years after the initiating injury."

"Members of the Working Party had the opportunity of learning about many aspects of Amateur Boxing when four *Medical Officers of the Amateur Boxing Association* gave both written and oral evidence. They explained that under the rules of the *ABA* a contestant must have an acuity of at least 6/12 and 6/24. This standard is not as stringent as that for profes-

sional boxers which appears to be currently 6/9 and 6/18 for each eye respectively."

"One aim of recording the unaided vision of each eye is to exclude from boxing those with myopia of more than a slight degree. However, an impaired unaided acuity may stem from other forms of refractive error (e.g. astigmatism) or from a squint. There is no evidence to suggests that either the astigmatic or squinting eye is more prone to suffer damage in the ring than is the normal eye. The acuity criteria are therefore arbitrary and could be reassessed.

"The decision as to whether a boxer should return to the ring after treatment for a retinal detachment is often a contentious issue. Mr. Mc LEOD said that after discussions with other Eye Surgeons experienced in sports injuries he has adopted the rule 'that a boxer may return to the ring if he satisfies the Board's visual standards and if his ocular status is such that there is no 'significantly increased risk' of further injury beyond that which pertains to any boxer engaged in this activity.' Our survey (which is discussed in detail later) indicates that most eye surgeons feel strongly that if a boxer has had a retinal detachment repaired he should not box again."

"Many boxers feel certain that their eye injuries stem from 'thumbing.' There is currently a great deal of interest in designing a 'thumbless' glove but available models are not proving popular largely because the wearer feels that his grip on the glove is inadequate. If an acceptable model could be produced then there might be fewer eye injuries."

"Although the *Working Party* was unable to mount a full epidemiological study of eye injuries in boxing due to difficulties in obtaining access to record and appropriate information, a questionaire was circulated to the 717 members of the *Ophthalmological Society of the United Kingdom (OSUK)* asking about their experience, if any, of boxers' eye injuries."

"By February 1984 314 ophthalmologists (43.79%) had replied to the questionnaire. 170 of them (54.14%) did not recall seeing an injured boxer whereas 144 (45.85%) recalled seeing a boxers with an eye injury. Of the ophthalmologists who recalled seeing an injured boxer 15 (10.41%) remembered seing them at a centre other than in which they now work. This accounted for 17 of the injured boxers. Of the 144 surgeons who had seen injured boxers 108 (75%) gave details of only one boxer each and 36 (25%) gave details of two or more boxers. These surgeons estimated that they had seen an additional 77 boxers between them but were unable to give specific details."

"The many limitations of a study based on a questionnaire are well recognised. The response rate was satisfactory. However, there is inevitably a problem of 'over reporting' (i. e. several surgeons reporting one patient). This might occur particularly at a specialised centre with many postgraduates *(e. g. Moorfields Eye Hospital)*. However the fact that only 10.4% of responding surgeons reported seeing boxers at a centre other than that at which they now work suggests that 'over reporting' is not a major factor in this study.' Under reporting' is probable a more important factor because the survey depends upon surgeons' recollections; the most commonly reported injury, detachment of the retina, may develop a long time after the causative injury and so not to be ascribed to boxing; as stated above surgeons remembered seeing 77 additional cases about whom they could not give details."

"Retinal detachment appeared more often in professional boxers but the amateur boxer also sustained the injury."

"The proportion of total retinal detachments reported as occurring in the last five years is almost identical to the split for injuries in general. There is thus no evidence that the present rules of boxing protect the boxer from serious eye damage."

American Academy of Pediatrics

Committee on Sports Medicine, 1983 – 1984 (Pediatrics 74: 311-312, 1984)

"The *American Academy of Pediatrics* opposes boxing in any sport program for children and young adults. Amateur boxing is potentially dangerous and yet youngsters are involved in boxing at ages 3 to 4 years. Approximately 15.000 boxers between the ages ten and fifteen are registered with the *National Amateur Athletic Union (AAU) Junior Olympics boxing program.* There may be an even larger number of young boxers in community organizations. Impoverished youths view boxing as a means of financial gain with the potential of providing a new life. Unfortunately, for many, it is a means of improving their physical condition at the risk of slow progressive brain injury, with occasional or no financial rewards. Other sports offer the same conditioning opportunity with minimal or no risk of brain injury."

"In contrast to professional boxing, amateur boxing is apparently for prestige, recognition, and the enjoyment of winning. Proponents of boxing suggest that it teaches self-defense and discipline, that it builds character and confidence, and that it is relatively safe. Opponents of boxing stress that the principal goal is to render the opponent senseless."

"Ironically, protective headgear may actually increase brain injury. The degree of of physical injury in boxing correlates with the physical strength and activity of the participants; the gretest risk exists when participants are obviously mismatched. The fatality rate in boxing is reported to be low. However, the freqency of chronic brain damage is an increasing concern in the medical community."

"Recent studies using *computed tomography (CT) scanning* have revealed brain injury in young boxers. Detection of such brain injury was previously missed by EEG, neurologic testing, and other standard prefight medical examination procedures. Studies have shown that neuropathologic changes occur in human beings or animals knocked unconscious from a blunt blow to the head. 'Accumulated destructive (brain) effects of repeated blows even when consciousness and posture are not lost are well known and accepted.'"

"It is unlikely that boxing will ever be abolished in the United States; therefore, it is crucial for pediatricians to become vigorous opponents of boxing as a sport for any child or young adult. Simple changes in the rules and medical supervision, and increased awareness of the dangers of boxing are not enough. Our opposition to boxing should be expressed at the time of health maintenance or preparticipation examination; opposition should be expressed in public whenever the opportunity presents itself; and our opposition should be expressed as a printed recommendation in brochures available in pediatric waiting rooms. Children and young adults should be encouraged to participate in sports in which intentional head injury is not the primary objective of the sport."

"At a recent interim meeting of the *AMA (1985)* delegates took a stand against boxing following testimony which pointed out that boxing is the only sport in which the purpose is to inflict injury and that 87 percent of all boxers show some signs of the sport, the... public is best served by a strong *AMA stand in opposition to boxing."*

The American Academy of Pediatrics, Committee on Sports Medicine and Fitness expressed *(Pediatrics: 134-135, 1997)* its Opinion on Participation in Boxing by Children, Adolescents, and Young Adults:

"The *American Academy of Pediatrics* opposes the sport of boxing for children, adolescents, and young adults. Amateur boxing is a collison sport in which winning is based on the number and force of punches successfully landed on an opponent's head and/or body. This deliberately exposes boxing participants to potentially devastating neurologic and ocular injuries (ENZENAUER and MAULDIN 1989, CASSON et al. 1984). Despite these potential dangers, thousands of boys and girls continue to participate in amateur boxing."

"Supporters of amateur boxing suggest that it teaches self-defense, discipline, strength, and agility while building self-confidence, character, and courage (ENZENAUER et al. 1985). For some youth, boxing may provide a supervised, structured, goal-oriented alternative to the streets. Impoverished youth may see professional boxing as a means of financial gain-without regard to attendant medical risks (CASSON et al. 1984)."

"The overall riks of injury in amateur boxing is actually lower than in some other collision sports such as football, rugby, and ice hockey. However, as opponents of boxing have emphasized, boxing is the only sport where direct blows to the head are rewarded and the ultimate victory may be to render the opponent senseless. Participants in boxing are at risk for dementia pugilistica, a chronic encephalopathy caused by the cumulative effects of multiple subconcussive blows to the head. Numerous studies of professional boxers document this hazard and its potentially devastating consequences in long-term health (CASSON et al. 1984, DREW et al. 1986, KASTE et al. 1982, JORDAN et al. 1992). Because amateur fights last only three rounds, compared to as many as 15 rounds in professional boxing, it is reasonable to expect that amateur boxers receive fewer blows to the head and thus suffer fewer brain injuries. However, recent studies have shown that amateur boxers still are at risk for acquiring cognitive abnormalities and /or focal neurologic deficits (HAGLUND and PERSSON 1990, HEILBRONNER et al. 1991, HAGLUND and ERIKSSON 1993, HOLZGRAEFE et al. 1992, Mc LATCHIE et al. 1987). Although other studies have not confirmed these findings, the 'safety' of amateur boxing remains unproven (HAGLUND et al. 1990, JORDAN and ZIMMERMAN 1988)."

"Prophylactic measures with helmets, unlimited lengths of hand bandage, and heavier gloves have not decreased the frequency of matches that are stopped for neurologic reasons (KASTE et al. 1982, SCHMIDT-OLSEN et al. 1990). Ocular injuries are a risk, GIOVINAZZO et al. 1987, WEDRICH et al. 1993, JONES 1989) even for amateur boxers, and may account for more hospitalization that neurological injuries (ENZENAUER et al. 1989)."

"Despite abundant evidence of the medial risks of boxing and clear opposition from medical associations in the United States, boxing is likely to remain as a sporting option for interested youth. Pediatricians can help young people make more informed choices about participation in boxing and hopefully direct their patients toward safer activities."

"*The American Academy of Pediatrics recommends that pediatricians:"*

"(1) Vigorously oppose boxing as a sport for any child, adolescent, or young adult;"

"(2) Educate 'at risk' patients about the medical risks of boxing and provide information that supports the Academy's opposition to the sport; and

"(3) Encorage young athletes to participate in sports in which intentional head injury is not the primary objective."

Dana SWARTZBERG *(Air Force Academy ends mandatory boxing activity, J Am Med Ass 274: 784, 1995)* writes:

"In respose to mounting pressure from the medical community, the *US Air Force Academy* has eliminated boxing as a mandatory activity."

"Until now, boxing was offered at the *Army, Navy,* and *Air Force academies* as an intramural sport and a mandatory combative training course. But new *Air Force cadets* will not be required to box as freshmen; they will have the option of taking the course as sophomores and may still participate intramurally. 'If any cadet boxes at the [Air Force] academy, it will be because he wants to,' says Capt. Will LANE, executive officer at the athletic department."

"Events leading to this decision stem in part from an editorial by George D. LUNDBERG, MD, editor-in-chief of the *Journal of the American Medical Association,* published June 8, 1994. A long-time opponent of boxing, LUNDBERG called on the *surgeons general of the US Army, Navy, and Air Force* to influence the elimination of boxing at the military academies (*JAMA,* 1994, 271: 1790)."

"In response to LUNDBERTG's request, the *Air Force surgeon general's medical operations agency* issued a statement last spring: 'The risks of participation [in boxing] far outweigh any potential benefits.' As a result, the *Air Force Academy* eliminated mandatory boxing."

"Members of the boxing community argue that the sport is not only safe but also critical to the proper training of military men.' I have nothing to say about boxing of a negative nature,' says Edward J. WEICHERS, Jr, *head boxing coach* and *assistant professor of physical education at the Air Force Academy.'* Boxing creates an environment of controlled stress that teaches men to deal with pressure, fear, and anxiety. When done properly and with common sense, it is a productive sport."

"Since the early 1980s, there has been a movement on many fronts to improve safety for boxers. In 1983, the *American Medical Association* collected data on the high frequency of chronic brain damage in boxers who have fought many bouts. A broad international medial consensus to ban boxing now exists."

"However, the *Army* and *Navy acedemies* still require it in their combat training programs for men. 'It appears that the Air Force values the brain power of its officers more highly than the Army or the Navy;' says LUNDBERG."

"Jim McNALLY, *head boxing coach* and *associate professor of physical education at the US Naval Academy* does not consider the elimination of mandatory boxing at the *Air Force Academy* a threat to his program. In fact, he says plans are being made to allow women to box in their own division this fall."

"McNALLY and WEICHERS argue that a distinction be made between professional and amateur participants. McNALLY says that amateurs participate to shorter rounds and must abide by more stringent safety rules. He says that since better gloves and headgear were introduced in 1984 'the ratio of knock outs has lessened considerably, and virtually no signs of neurological disorder have surfaced.' He criticizes studies published in *JAMA* as poorly designed and says they do not prove the dangers of amateur boxing."

"Nevertheless, boxing has been singled out from all other sports because blows to the head in boxing can injure the brain. Studies show that chronic brain damage results from repetitive subconcussive blows during training sessions and matches. The frequency of chronic brain damage in boxers who fight often is estimated at 60% to 80%."

"When you see what boxing has done for the many men who have gone on to become great leaders, you see how important boxing can be, says WEICHERS. 'The medical community is looking out for the safety of mankind. But let's be fair and not give a tainted picture.'"

Discussion about Boxing initiated by Dr. Lundberg

In an editorial entitled *'Boxing should be banned in Civilized Countries,'* LUNDBERG (1983), the editor of the *Journal of the American Medical Association (JAMA),* expounded upon professional and amateur boxing:

"The principal purpose of a boxing match is for one opponent to render the other injured, defenseless, incapacitated, unconscious. No caring person could have observed the events in professional prizefighting in the past few months and not have been revolted. No prudent physician could have watched the most recent debacle/mismatch on Nov 26, 1982, between Larry HOLMES and Randall 'Tex' COBB and believe that the current boxing control system is functioning. The fact that this massacre came on the immediate heels of even more tragic fights serves to accentuate the uncontrolled situation."

"The *American Medical Association* recognized this problem some time ago, and its *Council on Scientific Affairs* commissioned a panel to study the problem and to make recommendations. The report, presented on page 254 in this issue, is the *official AMA position.* It is solid, balanced, and reasonable. It operates with the assumption that boxing cannot be stopped, so it recommends ways in which it should be controlled better. To continue its interest in the safety and medical care of boxers, the *AMA* is cosponsoring a conference with the *Association of Ringside Physicians* on *'Medical Aspects of Boxing"* at Caesar's Palace Hotel in Las Vegas on Feb 18, 1983. Eleven faculty members will discuss the duties and responsibilities of the ring physician, emergency medical procedures in the management of the severely injured boxer, and several other important topics."

"Since the *Council report* was approved by the *House of Delegates* in 1982, two other major studies have appeared. KASTE et al. (1982), writing in a recent issue of *The Lancet,* studied eleven boxers who had been national champions in Finland and who had been carefully screened and found not to have other known reasons for brain atrophy. They report computed tomographic (CT) evidence of brain injury in four of six professional and one of eight amateur boxers. Also, two of the professionals and eight of the amateurs had EEG abnormalities that may have been caused by brain injury. KASTE and colleagues state, 'The most predictable and permanent reward ... is chronic brain damage,' and 'The only way to prevent brain injuries is to disqualify blows to the head.' Appearing on page 211 of this issue, ROSS et al. (1983) report a stuly of 38 boxers with CT scans, 24 of whom had a complete neurological cxamination and EEG as well. They report a significant relationship between the number of bouts fought and brain damage detected by C'T scan and demonstrate no significant relationship with neurological symptoms or findings or number of knock outs or technical knock outs. This is additional strong evidence of chronic brain damage with cerebral atrophy in many fighters."

"Some have argued that boxing has a redeeming social value in that it allows a few disadvantaged or minority individuals an opportunity to rise to spectacular wealth and fame. This does occur, but at what price? The price in this country includes chronic brain damage for them and the thousands of others who do not achieve wealth, fame, or even a decent living from the ring. Others argue that man must fight and that surreptitious fights will occur if boxing is outlawed, producing an even worse situation. I suggest that such is equivalent to arguing that gunfighter duels should be instituted, tickets sold, and betting promoted since, after all, homicide by gunshot is also common in our society."

"This editor believes personally that boxing is wrong at its base. In contrast to boxing, in all other recognized sport, injury is an undesired by-product of the activity. Boxing seems to me to be less sport than is cockfighting; boxing is an obscenity. Uncivilized man may have been bloodthirsty. Boxing, as a throwback to uncivilized man, should not be sanctioned by any civilized society."

In the same issue of the *Journal of the American Medical Association* a commentary by VAN ALLEN about *'The Deadly Degrading Sport'* appeared:

"How strange that, in this climate of preoccupation with health and physical fitness and with near-hysterical concern for every conceivable deleterious factor in the environments, so few raise their voices against boxing. How strange, when strident voices urge equality for all and make capital of support for equal rights, that poor and minority youth are recruited and rewarded for sacrificing themselves to a spectacle for the more favored of whatever ethnic or fiscal group."

"What factors contribute to this continued public spectacle of brutality, and the literal sacrifice of minority youth for the profit and delectation of self styled sportsmen?"

"In fairness to the boxing game and its proponents, let us review the widely shared ignorance about the effects of trauma on the brain and the implications of being knocked unconscious."

"Head injury from falls and blows is a common incident in the animated cartoons of children's shows. The hero or villain whether animal or human, is often momentarily stopped in his action by a blow to the head – the circumstances are entertaining, and the victim quickly recovers and is as fast and effective as before. This may happen repeatedly to the same character with no harmful effect. Children can grow up with the belief that head injury is amusing, recoverable, and of little consequence. Novels and television shows bludgeon their private eyes, heroes, and villains with never a suggestion of post-traumatic symptoms lasting more than a few minutes."

"The football player who is stunned or senseless has bad his 'bell rung' or is 'shaken up on the play.' That he can walk off the field with help or even reenter the game is ample evidence of the triviality of the incident and its apparent short-lived consequences. No matter that memory of the incident may be lost and confusion be present for several days afterward. The sports commentators, whose own fortunes are invested heavily in the game, never allude to and are probably only vaguely aware of the implications of these brain injuries – hence, their prattle of euphemisms to cover only vague discomfort."

"With this kind of folklore about brain injury, small wonder that those who enjoy and profit from regulated brawling and violence easily convince themselves that little harm is done in boxing. The 'punch drunk' fighter is an amusing oddity, seldom the object of pity and not, it seems, a catalyst of guilt."

"The fight game provides an opportunity for ambitious youths to climb from scandalous social circumstances through a disgraceful 'sports' opportunity to some kind of fame or hero: status. We are told this in different terms by those who justify boxing and who find the bashing of others to be financially and emotionally rewarding."

"In boxing, we are reassured by the concern of the announcers for facial cuts and by the referee, who will stop a fight when superficial hemorrhage may obscure the fighter's vision, or perhaps offend some in the audience and remind them of their involvement in the guilt of promoting a vicious and deadly game. At the end, some functionary will appear in formal evening attire to announce the winners. His ruffled shirt and black tie attest to the dignity of the proceedings, and to the gentlemanly way in which they are conducted."

"Perhaps you will say that, with human nature as it is, some important societal needs are served by this vicarious outlet of violence for the viciousness hidden in all of us and that a good fight by others relieves tensions and lubricates communal living. Others, in defending the recruitment of children to the fight game, will point to the advantages of the discipline that comes from preparation for fighting and to the moral benefits of fighting within a set of rules. They will emphasize the opportunity for the otherwise hopeless to achieve fame, no matter what the price to the brains of the unsuccessful and successful alike. No matter the basic degradation of those who fight for the entertainment of others even when victorious. The owners and managers of a 'stable' of fighters recall those who solved the energy crisis created by the cotton gin."

"We are assured by the television networks responsible for bringing us a boxing spectacle that an ambulance will be available throughout the bout, and of course that physicians are present at ringside – a flattering faith in the ability of modern medicine to repair irreversible damage to the nervous system. We are not so reassured when clearly mismatched fighters are paired in the ring and one game but less-talented gladiator is finally unmercifully beaten while the referee, for reasons of his own, allows a bout to proceed when the outcome is clear to all. The physical and mental consequences are smothered in euphemisms and suppressed by announcers, promoters, and audience."

"We are reassured again, when we reflect on the respectability given to the sport by the *Olympic Committee,* since these self-appointed guardians of sportsmans virtues endorse fighting (under careful jurisdiction), suggesting that well-regulated sin is perhaps not very sinful after all."

"Heroes usually arise from sacrifice, often in hazardous circumstances and at high cost. The high cost that is paid by the fighters in boxing is buried in emotional bookkeeping."

"When a human or animal is struck on the head so that consciousness is lost, pathological changes – minute or larger hemorrhages – contusions often at the base, and tearing of nerve fibers that may not be easily identified, are all consequences of a blunt blow of sufficient force to render the subject unconscious. Detectable symptoms of a beating may not be apparent to a victim preoccupied by the pursuit that caused the injury, but have been admitted by the more introspective who go on to other occupations."

"The accumulative destructive effects of repeated blows, even when consciousness and posture are not lost, are well known and accepted."

"Is now not the time to suppress exposure of this fragment of our savagery by the mass media and leave boxing to those who enjoy privately staged dogfights?"

VINCENT (1985), a physician in Nashville, Tennessee, commented on LUNDBERG's (1983) editorial:

"To the Editor.- This letter concerns the editorial by Dr. LUNDBERG advocating the banning of boxing in civilized countries. Dr LUNDBERG and others have made the personal judgment that boxing is immoral because the participants risk death or injury. If the medical profession accepts this type of judgement about boxing, then we should be consistent and institute measures to ban all sports that involve personal risk for participants. Why not ban football, auto racing, ice hockey, mountain climbing, hang gliding, spelunking, and sky jumping, since all of these sports put the participants at personal risk?'

"Furthermore, let us not confine our efforts only to organized sports. After all, many of us have seen good friends and even medical colleagues sustain serious injuries while on skiing vacations. Surely the social conscience of our profession should be such that we throw the full weight of our organization behind legislation to ban the dangerous practice of skiing vacations. After this has been accomplished, we could lower the boom on roller rink parlors."

"After reading Dr. LUNDBERG's editorial, I realized that the 1984 envisioned by George ORWELL is closer to reality than ever before."

LUNDBERG responded:

"In Reply.—Unfortunately, Dr. VINCENT missed the point. I believe that boxing is immoral not because the participants risk death or injury but because the intention of the sport is to produce injury as a principal way to win the contest."

OLIVER (1985) entered the discussion with the following comments on LUNDBERG's statements:

"To the Editor.- While frankly I follow few sports with enthusiasm and follow boxing not at all, I read THE JOURNAL's recent editorial calling again for a ban of the sport with mixed feelings. I applaud the efforts of my colleagues to reveal the morbidity inherent in the sport and to alert the American public to these dangers. The threat of permanent brain damage is frightening and should be recognized by participants and afficinados of the sport. In contrast, the departure from an objective description of these dangers into the realm of polemic as found in the editorial was disturbing. Not, mind you, because I do not appreciate a rousing and well-written polemic as much as any American with strong opinions on politics, morality, or the commonwealth, but because of the coercive aim of this particular editorial."

"Boxing is described as 'organized brutality,' 'slaughter,' and 'carnage,' with the implication that those who support boxing are, in turn, barbaric atavists who are neither moral nor intellectual peers of those who wish the practice banned. It is with this sense of superiority that we in the American Medical Association claim the right to force those who stand for the game to their knees, if not with our own hands then with the arm of law. It has been my perception that in a free society it is not the function of the state to impose moral or behavioral restrictions except in those cases where the civil liberties of others in turn suffer infringement. This is not the question addressed in the editorial but is instead an effort by those with presumably higher values to impose these on a dissident segment of society."

"That it is neither practical nor advisable for a segment of society with such a self-defined higher morality to force their will on another should have been the lesson learned from the 18th Amendment. Instead, the same arguments of protecting the citizen from himself, the social cost of such freedom, and the sense of righteous indignation of those who would impose restrictive laws are being pressed on the public again. Even given that the suppression of boxing is feasible in a practical sense. I suggest that the social cost of such suppression in terms of the continuing erosion of civil liberties and intrusion of the state in our everyday lives may be as great as the social cost of allowing participation in a dangerous sport. That a young man should be discouraged from a pugilistic career may be laudable. To subject him to the threat of imprisonment, fine, and surveillance, as is the case in other banned sports such as gambling, may be a bit much."

"In a free society, it is the responsibility of the individual to make his or her own decisions about life. With that responsibility comes the risk of making a bad decision. It is the responsibility of the physician to provide education and advice so that an informed decision can be made. It is not his or her place to take such freedom away from the citizen on the basis of educational, moral, or social superiority. The American dream of utopia is that of a society built by common consent achieved through the exercise of personal freedoms and mutual respect. Such a dream will never be achieved through coercion. We should, therefore, strive to make boxing unpopular, not illegal."

LUNDBERG answered with the following:

In Reply – As Dr OLIVER suggests, we are striving to make boxing as unpopular as drunk driving, street gunfights, heroin addiction, and public tobacco cigarette smoking (some other great freedoms) ought to be."

An additional comment to LUNDBERG's editorial was presented by CORCUM *(1985)*, The John Hopkins University School of Medicine, Baltimore, MD:

"To the Editor. – Amid the barrage of disparaging and admonitory media coverage, I would like to offer another point of view concerning the sport of boxing. As a former world-ranked professional boxer and current medical student, I think that I have a unique vantage."

"There are many aspects of the fight game that the average spectator and, most assuredly, the avid critic cannot appreciate. Through rigorous training and diligent practice, individuality and a sense of self are nurtured. How better can one realize one's capabilities and limitations than through controlled, physical, one-on-one competition? Motivating him with the desire to win and - the threat of defeat, the situation of the fight brings forth the finest in the athlete, psychologically as well as physically. The costs and rewards of hard work are quickly realized. Determination and self-discipline are unquestionably admirable qualities. Developed through necessity, they become a part of the fighter's person. Are not these qualities valued throughout society and applicable to every aspect of life?"

"One of the major criticisms of the sport is its barbaric nature. Is it the fighter who is instinctively barbaric or is it the sellout crowd that eagerly anticipates the sight of blood or the knock out blow? First and foremost, the fighter is an athlete. He strives for fluidity of movement, strength, precision of timing, and endurance. These attributes are central to all of athletics. Through the rigors of training, the fighter views his superb physical conditioning and definition with pride. Because they share common emotional and physical hurdles, it is not difficult to comprehend the mutual respect between fighters. Outside of the ring, this virtue is directed toward all individuals who strive for a goal."

"I cannot describe to you the feeling of being face to face with an opponent of superior caliber and having no alternative but to stand up in the face of seemingly insurmountable odds. Nor can I ever begin to relate what itis like to be the center of attention in an arena full of thousands of spectators with your career and longevity as a fighter on the line. But imagine if you will – and I emphasize the word imagine – the pressure. Speaking from experience, I can truthfully say that situations such as these develop emotional stability and psychological maturity. And probably most admirable is the refusal to concede in the wake of apparent defeat. Aren't those qualities of universal value?"

"Without a doubt, there are certain risks with blows to the head. I am much in favor of mandating measures that would increase the safety and diminish the risk of injury to the individual fighter. Congressmen, representatives, and sportscasters on the ever-growing publicity bandwagon have lost sight of their initial objective. They have become obsessed with the corruption and deviant business dealings of the game, instead of concern for the athlete. All of professional sports is riddled with overtones of venality. This does not justify banning the sport of boxing."

"Throughout the thousands of boxing gyms across the country, very few fighters progress to the level of professional, but the sport serves an important social function. It takes wayward youths, who are victims of the public educational system, off the street. It teaches them self-discipline and control and reinforces the adage that nothing of value is acquired without hard work and sacrifices. Self-confidence can only be promoted through an individual sport where the athlete must rely on his own talent and believe in his own ability. Only through conflict can hidden resources surface. In a controlled situation, with the safety of each boxer as the primary concern, the sport of boxing can be justified socially as well as personally."

LUNDBERG (1995) replied to the comment of CORCUM:

"In Reply — How would Mr. CORCUM propose to prevent brain damage from blows to the head?"

The mother of an US Army Military Academy, West Point, cadet, Mrs. Marolyn Caldwell WILSON (1985), whose son died of a head injury suffered in a boxing match which was part of the physical fitness training program responded as follows:

"To the Editor. – I dreamed about my son Richard Mull last night and woke up crying. It has been seven years since he entered a boxing ring at West Point for his first graded bout, came out apparently uninjured to learn he had flunked the bout, and was left alone apparently because he did not feel well enough to return to the locker room. Someone found him a short time later. He was quickly slipping into the coma from which he never returned. It was his 19th birthday."

"Today, I read a newspaper article summarizing your editorial. A few moments later the publication itself appeared in my employer's in-box."

"If there is any way I can add my voice to that of yours and others who wish to end the brutal 'sport' of boxing in this country, please feel free to call me. A few years ago I put gentle pressure on West Point to end mandatory boxing, to no avail, unfortunately. I wonder how many West Point cadets who have gone on to responsible positions in the Army have brain damage that could affect crucial decisions in times of great stress. It is worth thinking about, I believe."

BERNSTEIN (1985), a physician in Elmwood Park, Illinois, wrote to the editor:

"To the editor. – I congratulate you for your strong stand against boxing and would like to add my own voice to the debate. I went to a military high school where I boxed in amateur tournaments for three years. During that time, I sustained a dental injury, as well as a fracture of the first metacarpal bone in my dominant hand. At the age of 43 years, I have a resultant basilar arthritis of the right thumb. What bothers me the most about my own injuries is that I was told recently by an ophthalmologist that I have a 'retinal scar' that he believes is a result of fighting."

"For several years in early practice I did some moonlighting as a boxing doctor. I performed physical examinations and stood by for many of the local tournaments, as well as the Golden Glove Tournament of Champions in Chicago. It was during this period that I became opposed to boxing. Nothing was done at that time to protect the fighters, and, indeed, many of the regulations were completely ignored. Youngsters who were not physically fit were allowed into the ring despite my protests. They were also allowed to wear certain articles that I felt were objectable, such as a headband, which later fell into the fighter's eyes and occluded his vision. There are many anecdotes that could be related highlighting the inhumanity of fighting, but the ultimate insanity of fighting was revealed to me on the last evening of the Golden Glove Tournament of Champions, when I attempted to stop a fight because a fighter had sustained cuts above his eyes that were bleeding into his eyes and he was unable to see. I was told by the referee that 'You can't stop this fight Doc, it's a championship!"

"Boxing could easily be replaced by any number of highly athletic sports that teach self-reliance, courage, and endurance."

PATTERSON (1986), a neurosurgeon, Department of Neurosurgery, The New York Hospital – Cornell Medical Center, New York, stressed the following point in his contribution *'On Boxing and Liberty:*

"SHOULD a sport as brutal as boxing be allowed to exist in civilized society? Is a ban on boxing an excessive restraint of individual freedom? We perhaps could all agree that there should exist a certain minimal area of personal freedom that must not be violated or else the individual will be so confined that he is unable to develop his individuality even to the slightest degree. This means that a boundary or frontier must be defined between private life and public authority. The debate is on where that line should be drawn" (AVEEN et al. 1984).

"John Stuart MILLS wrote in his essay, *'On Liberty,'* that there was only one circumstance under which society was entitled to use physical force, legal penalties, or moral coercion. He wrote: 'That the only principle that is the sole end for which mankind are warranted, individually or collectively, in interfering with the liberty of action of any of their number, is self-protection. That the only purpose for which power can rightfully be exercised over any member of a civilized community against his will is to prevent harm to others. His own good, either physical or moral, is not a sufficient warrant. He cannot rightfully be compelled to do or forbear because it will be better for him to do so, because it will make him happier, because, in the opinion of others to do so would be wiser or even right. These are good reasons for remonstrating with him, or reasoning with him, or persuading him, or entreating him, but not for compelling him or visiting him with any evil in case he did otherwise. To justify that, the conduct from which it is desired to deter him must be calculated to produce evil to someone else...."

"But neither one person nor any number of persons, is warranted in saying to another human creature of ripe years that he shall not do with his life for its own benefit what he chooses to do with it" (MILLS 1947).

"Why was MILL so rigid on drawing the line on public authority only at the point of self-protection? Certainly he understood that men are interdependent and that no one is so completely isolated as not to ever obstruct the lives of others in any way. Liberty for the pike is death for the minnows."

"Isaiah BERLIN reasoned in his *'Four Essays on Liberty,'* that none of us is truly free because we are slaves to our passions and our habits, if not to another man. Besides our habits and pleasures, we also tend to pursue long-range goals of a higher level because we believe them to be for our best. Such goals might vary from honesty and charity to a set of religious beliefs. Often we extend these beliefs to something wider than ourselves, perhaps to our family, our tribe, our church, our country, or even to the rest of mankind. If our children or our friends seem unwilling to assume our beliefs as their own, we try to educate them, argue with them, or otherwise convert them. We believe them to be blind or ignorant. If only they understood themselves as we understand them, they would not resist. The use of logic and education to alter the ideas of others, we would all accept, still leaves the individual free to make the final decision for himself."

"However, it is but a small step to reason that because others fail to understand their own best interests and because they are not as rational and wise as we are, it is our duty to coerce them for their own sake and not for our own self-interest, If they understood their own inner spirit they would no longer resist. It is only their hidden inner self that we take into account. We are now in a position to ignore the actual wishes of men or society and to bully, or to oppress, or to torture them in the name and on behalf of their hidden, real self."

"Therefore, the concept of liberty or freedom changes from a freedom that an individual actually feels he wishes and needs to a freedom buried within the individual and not dreamed of by his empirical self. This kind of freedom is then easily translated into some superpersonal entity, a state or a class or a nation or even to the march of history. The lessons of the last 50 years show us that this issue is more than academic" (BERLIN 1969).

"KANT wrote that 'no one can tell me to be happy in his own way.' Paternalism is the greatest despot imaginable. This is so because it is to treat men as if they, were not free but human material for one man, the 'benevolent protector,' to mold in accordance with his own purpose. This manipulation of men to propel them toward goals that only the social reformer sees and they do not is to deny them their human essence, to treat them as objects without wills of their own, and therefore to degrade them. To drive men toward a goal not conceived by them, even if it is for their own benefit, is in effect to treat them as subhuman, to behave as if their ends are less sacred than one's own. It is not enough to say that it is in the name of some value higher than themselves, because as KANT pointed out all values are the creation of men and consequently there is no value higher than the individual. All forms of tampering with human beings, shaping them against their will, all thought control denies them their freedom and deprives them of their most basic right."

"Boxing lasted 1.000 years at the time of the Greek and Roman empires, when Western Civilization as we know it was thriving. It ended in about 400 A.D., probably because it had become thoroughly corrupt and professionalized and also because it was banned by a Christian emperor who had a dislike for heathen spectacles. The Greek temple to Zeus at Olympia was razed, the magnificent statue of Zeus was moved to Constantinople, and Christianity became dominant. Western Civilization entered the Dark Ages. Man's spirits soared again during the Italian Renaissance, and again boxing briefly flowered. But society – became more restrictive, boxing and also learning disappeared, and the Renaissance faded. In the first part of the eighteenth century, with the industrial revolution and the rise of individual freedom, boxing again began to flourish. Art and science flowered during the first part of the twenteeth century, and boxing prospered as well."

"The association between man's highest intellectual accomplishments and boxing is no accident as I see it. Arts, letters, and science only thrive in a time when the government and the church remove the hobbles from individual creativity. If the government becomes an arbiter of moral standards and personal safety, it does so at the price of individual freedom. Whether the curbs are on boxing or on the scholar. How much better to educate, cajole, inform, entreat, then let the individual make his own decision. I personally have great concerns about any restriction of an individual and his freedom. The banning of boxing could be looked on as telling boxers that their desires and their values do not count. They are not competent to decide for themselves, and the choice is better left in the hands of other individuals who are more educated and know better what they need. This modest tyranny, to me, is an early sign of repressive authority. I am against it."

LUNDBERG (1986) commented again about his contribution:

"I am going to punish him. .. I am going to beat him that he'll need a shoehorn to get his hat on again." Cassius CLAY prior to World Heavyweight Championship fight with Floyd PATTERSON *(Time, Dec. 3, 1956, p. 73).*"

"The organized brutality of boxing has become widely recognised for what it is since we published our pioneering articles on Jan 14, 1983 (ROSS et al. 1983, LUNDBERG 1983, VAN ALLEN 1983). Most noteworthy have been the changes in the policies of organized medicine. Four years ago the *American Medical Association (AMA)* had no policy on boxing. As the new information became widely known the *AMA* went from calling for many safety measures recommended by its *Council on Scientific Affairs* and expert panel in December 1982 (Brain injury in boxing, 1983) to a much more aggressive posture in June 1983 (LUNDBERG 1984). Data from four research groups in three countries have documented the high frequency of chronic brain damage in boxers who have had many fights, and a series of strong editorials has argued that boxing is wrong medically and morally (LUNDBERG 1983, VAN ALLEN 1983, LUNDBERG 1984, RICHARDS 1984). We have seen the development of a broad international medical consensus that boxing should be abolished. Among the many medical groups that have taken official positions are the *national medical associations of Britain, Canada, Australia, and the United States, the World Medical Association,* specialty societies such as the *American Academy of Pediatrics, the Canadian Psychiatric Association, and the American Academy of Neurology* and the *state medical associations of California and New York.*"

"In December 1984 the *AMA House of Delegates,* reflecting the new scientific data and clinical experience extended its June 1983 opposition to boxing. After substantial debate, the *House* passed Substitute Resolution 26 which stated as follows:"

"RESOLVED. That the *American Medical Association:* 1 . Encourage the elimination of both amateur and professional boxing, a sport in which the primary objective is to inflict injury. 2. Communicate its opposition to boxing as a sport to appropriate regulating bodies. 3. Assist state medical societies to work with their state legislatures to enact laws to eliminate boxing in their jurisdictions, and 4. Educate the American public, especially children and young adults about the dangerous effects of boxing on the health of participants."

"Twelve months later, in December 1983, in Washington, DC, after much less debate and with little dissent, the *House of Delegates* reaffirmed the December 1984 resolution."

Opposing Viewpoints

"Three opposing viewpoints consistently appear, two represented in this boxing theme issue. One is that amateur boxing is not as bad as professional boxing, can be monitored and controlled, and has a low likelihood of harm. A description of such an approach to boxing in Sweden is contained in the commentary by LUDWIG. Another is the libertarian view that a person should be able to do anything he or she wishes as long as it doesn't hurt others or society. Such a view is presented by PATTERSON. The third view is that poor and minority children have only one way to escape the ghetto in this and other countries and that way is fighting. Obviously, we disagree with these three views and have presented our reasons for disagreement many times. in addition, we have included in this issue an ORIGINAL CONTRIBUTION on eye damage in boxing (MAGUIRE and BENSON 1986) and a thorough review article on medical and public health aspects of boxing" (MORRISON 1986)."

"Many people question why boxing (along with full-contact karate) has been singled out from all other sports for such major opposition by physicians. The data are clear blows to the head in boxing damage the brain. While a few severe blows may kill acutely by cerebral contusion and edema and subdural hematoma, chronic brain damage results from repetitive subconcussive blows over multiple training sessions and matches. It is the high frequency of chronic brain damage (60 % to 87 %) among boxers who have had many fights that sets boxing apart medically, in comparing risk sports in regard to odds of death per unit of time of participation, mountain climbing, parachuting, and hang gliding are considered 'high risk' while professional boxing and scuba diving are 'medium risk.' (REIF 1981). Figures from 1983 for US injuries by sport showed boxing leading the list at 50 % of participants experiencing injury, followed by American football at *30 %, baseball 3 %, basketball 29 %, and the others trailing" (Accidental Facts, National Safety Councel, Chicago, 1984).*

"American football is violent. It results in many serious injuries, including cerebral concussions and fractured cervical vertebrae with paralysis. There may be a substantial prevalence of chronic brain damage in football players, but at this time no one seems to know. The peer review literature is apparently mute. One senses that the football enthusiasts, including the sports medicine establishment, may not want to know, since one does not hear of many people planning to do proper brain studies on football players. Can the football sports medicine community examine scientifically the brains of current or former football players prospectively, retrospectively, or epidemiologically cross-sectionally and send the reports of such studies for peer review and subsequent publication. We would welcome such at *JAMA*."

"The other major difference between boxing (and full-contact karate) and all other contact or collision sports is the intent to win by deliberately harming the opponent. It is morally wrong for: one human being to attempt intentionally to harm the brain of another. A major purpose of a sport event is to win. When the surest way to win is by damaging the opponent's brain and this becomes standard procedure, the sport is morally wrong. Many say that it is often the intent of American football players to harm their opponent's, thereby removing them from the game, especially defensive linemen and linebackers going after the quarterback. This may be true, but such harm is not the intent ot the game. Each football game continues to its conclusion, no matter how many injuries occur, by the use of substitutes."

Does Violence Beget Violence?

"Yet another aspect or boxing's influence on society has recently been noted. Many people have suggested that boxing gives some men who may be naturally violent an opportunity to work off their aggressions in the ring. Others say that observers of boxing may rid themselves of the need to be violent by vicariously living such an experience. To my knowledge, no evidence exists to support such views. On the contrary, many believe that people who participate in and observe violence over the long haul become tolerant of it, as if violence were normal, and cease to find it abhorrent. There are articles in the psychology and sociology literature tying exposure of angered laboratory subjects to a filmed prize fight scene to an increase in aggressive behavior (BERKOWITZ and RAWLINGS 1963, BERKOWITZ and ALUOTO 1973). PHILLIPS (1983) has presented evidence that there is a sharp, nontrivial increase in US homicides after strongly publicized heavyweight championship prize fights. These and related observations that support the hypothesis that mass media portrayals of violence incite imitative responses among the public have recently been disputed on methodological grounds (BATON and REISS 1985, PHILLIPS and BOLLEN 1985), and clarification of any cause-effect relationship awaits future research. The evidence supporting the view that violence shown on television produces additional actual violence was strong enough for the *House of Delegates of the AMA* in 1976 to pass a resolution on this subject, which reads as follows:"

"RESOLVED, That the *AMA:* 1. Declare its recognition of the fact that TV violence is a risk factor threatening the health and welfare of young Americans, indeed our future society. 2. Commit itself to remedial action in concert with industry, government and other interested parties. 3. Encourage all physicians, their families and their patients to actively oppose TV programs containing violence, as well as products and/or services sponsoring such programs."

Support from Wide-Ranging Sources

"There are many people out there who believe as we do about boxing. Support may come from surprising places. From the *Roman Catholic Church,* Jesuit theologian John CONNERY of Loyola University of Chicago makes a strong case against boxing on moral grounds. He points out that 'there is no way you can justify doing damage to another unless you can show some compensating good.' He says that the risk of cumulative brain damage rules out any blows above the collar bone as being moral (BANK 1986). Television sports reporter Howard COSELL, once one of the most influentiai spokesmen for boxing, ended many years of covering professional boxing in 1982 and has recently announced that he agrees with our position that professional boxing should be abolished in

America and that he also has grave doubts about amateur boxing (COSELL 1986). Of course, in our 'age of litigation,' the lawyers have also gotten into the act. The case of a young amateur fighter in West Virginia who died of brain damage shortly after his third bout was settled in favor of the plaintiffs (the boy's parents) after three days of a jury trial in 1986. This was apparentiy the first such court case in the United States"

"Where do we go from here? As scientists, we must continue to be responsive to new data, especially prospective and controlled, that will shed additional light on boxing's dangers and their possible elimination. As long as the data are there, we must resolutely stand by the position of the international medical community and do everything possible to abolish boxing? Until it is banned, we must support all reasonable efforts to improve safety in the ring. Many states, as well as the federal government, have moved toward improved safety, but none has yet abolished boxing."

"Is there hope for amateur boxing? Generally speaking, the people who have come forth to defend amateur boxing seem well intentioned and responsibly motivated. It may be possible to develop rules that would allow this sport to continue to be one of speed, skill, enthusiasm, and discipline by simply abolishing blows to the head and having boxers win by points."

Unfortunately, blows to the head damage the brain, whether or nor the assailant and the recipient are paid for their efforts. If the highly vocal proponents of amateur boxing truly do not wish their fighters to win or lose by virtue of brain damage, they should be able to devise rules to prevent this. Society no longer sanctions sword duels in which one participant runs the other through; we now see skillful swordsmanship in the ritualized form of fencing. Amateur boxing could also be ritualized into safety if the responsible leaders so choose."

"There is little support for the continuation of professional boxing, except from some of those with an obvious financial interest – some fighters and their entourages, promoters, the media, and various state officials who depend on regulating boxing for their livelihood. The final report of the *State of New Jersey Commission of investigation* of Dec 16, 1985, entitled *'Organized Crime in Boxing,'* concluded that boxing should be abolished, that 'the inherent problems of professional boxing and most particularly its constant threat of bodily destruction, mentally and physically – cannot be effectively resolved at any governmental level,' that the sport is 'marred by official misconduct, promotional greed and matchmaking barbarism,' and that 'not even the sturdiest of statutory controls will reduce the brutality of the sport to any significant degree (PATERSON et al. 1985). In response to this report, *The New York Times , Dec. 16, 1985, p. 24)* called on New Jersey to become the first state in the Unired States to outlaw boxing."

"Professional boxing is hopeless and doomed to meet the same fate in the United States that it has met in Sweden and Norway: extinction. Some state will ban it in this decade. Most will ban it by the end of this century. Who will be first? Will it be a populous state with many citizen fighters to protect, such as New Jersey, Florida, New York, or California? Or will it be a state where boxing has little monetary influence, but where traditional morality runs high, such as Nebraska, Vermont, Arkansas, or North Carolina? Or might Oregon and Washington ban it regionally? Who will be first to abolish and who will be last to sanction this tragic obscenity?"

"We will see that hard scientijic data, medical concern for the brains of all members of humanity, and the power of the pen are mightier than greed, human exploitation, and the fist" (LUNDBERG 1985).

Dr. VOY (1989), *USA Amateur Boxing Federation, Colorado Springs, Colorado,* gave the following commentary:

"To the Editor. – I was disappointed and dismayed by the misleading article on boxing in the January 6 issue of *JAMA* (COWART 1989). Juan Antonio SAMARANCH, *president of the International Olympic Committee,* publicly stated his support for including boxing in the Olympic program. The committee's investigation of boxing was directed toward reviewing the scoring system."

"It is unfair to lump all types of boxing together. Professional and amateur boxing have vastly different rules, objectives, and requirements. No matter what television commentators might say, the style and technique are not the same. To combine events and actions of the two types of boxing in one article totally misrepresents the facts. Moreover, the boxing programs for police and military training programs in no way resemble amateur or professional boxing. They are not organized under the rules of either sport and are not subject to the supervision or regulations of either. Frankly, they do not teach boxing; self-defense perhaps, but not sparring or training as are found in amateur boxing."

"Participants in amateur boxing are taught offensive and defensive techniques and are matched by weight and experience. Both practices and competitions are supervised, and multiple knock downs are not tolerated. If a boxer is stunned by a blow or knocked to the mat three times in a round or four times in a match the bout is halted. The boxer is suspended from boxing or sparring for 30 days, and he must pass a medical and neurological examination before he can return. In fact, amateur boxing is one of the most medically controlled sports, not only in the Olympic movement but in sports in general."

"The *International Olympic Committee 's Medical Commission* changed its stance on boxing after learning of the medical studies that have been and are being conducted. The commission was especially impressed by the four-year study now being done at *The Johns Hopkins University,* Baltimore, MD."

"With regard to injuries, the current medical condition of MUHAMMAD ALI truly is sad, but that cannot be laid at

the door of amateur boxing. His amateur career was brief, and as a professional he continued to fight against medical advice. He is visible because he is a celebrity. What about the more than 200 young men who have become quadriplegic because of playing football? Yet, there is no move to abolish football."
"The article in *JAMA* is a reflection of a negative mentality toward boxing that exists in a portion of the medical community. I challenge the medical community to look at amateur boxing objectively and to join with the amateur boxing community in its efforts to make the sport even safer.

Virginia S. COWART, Chicago, Illinois, answered him:

"In Reply.- The people who run amateur boxing seem to be genuinely concerned with the safety of the sport, and there is no doubt that it is well regulated. However, most amateur boxers themselves, at the Olympic level anyway, do not seem to see amateur boxing as an end in itself but rather as a means to go into professional boxing and perhaps become rich and famous. When boxers are betting among themselves, as they did in Seoul, South Korea, on who gets the quickest knock out, it does not sound as if they value style and technique over a convincing demonstration of physical superiority. With regard to professional boxing, it is a paradox of our society to condemn violence and brutality on one hand while paying professional boxers millions of dollars to batter each other in a sport whose main point seems to be to inflict pain and suffering."

David C. FIORE (1996), MD, *University of Nevada School of Medicine, Reno, Nevada,* presented the following comments on LUNDBERG, under the title: *'Boxing: Does the Size of the Prize affect the Drain of the Brain?'*

"To the Editor. – Dr. LUNDBERG's fifth, and most recent, Editorial (1996) calling for a ban of boxing compels me to reply to two major areas of concern: factual and philosophical."
"I understand and share Dr. LUNDBERG's abhorrence of violence and of needless injury and suffering inflicted on any members of our society. I also believe, however, that as physicians we must be extremely careful not to let our own prejudices and moral stances lead us to unfounded or poorly supported actions. One only needs to recall the abject failure of Prohibition to understand how a 'morally right' stance can lead to a 'pragmatically wrong' solution. Furthermore, asking physicians to call for the ban of boxing because it is 'morally wrong' may be setting a dangerous precedent. What is the next morally wrong activity that we should try to ban? Perhaps abortion or gay lifestyles, or birth control, or the possession of firearms. Who writes the list, and where do we draw the line?"

"I believe that LUNDBERG's argument that boxing ought to be banned because it is harmful to the participants is one that physicians may rally behind, if it could be supported. When talking about the current state of professional boxing, I cannot help but agree with LUNDBERG's conclusion: it is a corrupt and dangerous sport. Unfortunately, LUNDBERG and other opponents of boxing confuse the issue by failing to recognize amateur and professional boxing as the two very distinct sports that they are."
"If we focus on injury and fatality rates, rather than the morality of sports, we find that boxing is actually safer than many other sports. In a study involving boxing during its heyday of unregulated matches (1918 to 1950), GONZALES (1951) found that boxing 'produced fewer deaths, in proportion to the number of participants, than occur in baseball or football.' More recently, this finding has been supported by the study of boxing by the *American Medical Association's Council on Scientific Affairs (Report on Brain Injury in Boxing, 1983),* which reported a fatality rate of boxing (amateur and professional) to be one-third that of college football."
"The issue of brain damage due to boxing has now become the rallying cry of those who wish to ban the sport. In citing the studies that have convincingly demonstrated the longterm brain damage trom professional boxing, LUNDBERG and others have failed to give any notice to the growing body of literature that demonstrates that amateur boxing does not lead to similar consequences. There are recent studies from the *United Kingdom* (BUTLER 1994), *Ireland* (PORTER and PRICKER 1996), and *Sweden* (HAGLUND and PERSSON 1990) that support the conclusion of PORTER and FRICKER: 'There is an obvious dichotomy between the business of professional boxing and the sport of amateur boxing, and these differences significantly alter the relative risks of chronic cerebral atrophy. All levels of amateur boxing have been investigated, and there is no evidence of an association between participation in amateur boxing and chronic traumatic encephalopathy."
"Perhaps it is time for us to step away from moral rhetoric and focus on the issue of brain injury in boxing, primarily professional boxing, and help institute changes that would make the sport safer for those who chose to participate. Indeed, I do not think we could do much better than to start with the guidelines recommended by the *1983 report of the American Medical Association's Council on Scient fic Affairs."*

LUNDBERG (1996) gave the following answer:

"In Reply.- Blows to the head damage the brain, whether or not the person who first delivers the blow is paid to deliver it. I hope the ongoing studies of brain damage from amateur boxing include boxers like Dale FOREMAN of Cincinnati, Ohio, who died of head injuries suffered on June 29, 1996, during a three-round Golden Gloves bout in Richmond, KY."

Pros and Cons of Boxing. Statements by Physicians, Officials, Writers, Moral Theologians, Politicians, etc.

A more complete presentation of the far ranging public debate on boxing will be found in the historical volume of this work, however, the medical volume would not be complete without a discussion of the boxing debate in medical circles. Prominent physicians and medical journals have participated on both sides, arguing both for and against boxing as a sport and public spectacle. Arguments have been based on injuries found or not found, public policy, the appropriate position for the medical community to take, the nature and extent of the effect on spectators, or ethical and moral questions about what is in the best interest of the general public. Contributions to medical discussions have been international, many national medical associations voicing a firm position on whether boxing should be allowed to continue. The debate has been fervent, as the comments below will reveal, strong feelings are the consistent theme throughout the literature.

In order to present some historical background for the current debate on boxing, the following extracts and comments, from a variety of sources, are quoted below. They are representative of opinions that have appeared internationally in periodicals, newspapers, and books, concerning the value and costs of boxing, both amateur and professional. The material below is presented in temporal sequence.

The condemnation of boxing is not a recent phenomenon. It has been around almost as long as the sport, as witness this dialogue by LUCIAN (AD 125-180) between the Greek SOLON and the legendary ANACHARSIS of Scythia (Southern Russia), whose reported visit to Greece in quest of wisdom gripped the imagination of both ancient and modern writers. [The dialogue is quoted only in part; the original form being quite lengthy].

LUKIANOS from Samosata lived ca. from 120 – 180 AD. He wrote „*Anacharsis e peri gymnasion*' It is a written dialogue in Greek between the Scytian ANACHARSIS and the Athenian SOLON about the value of athletics and sports.

ANACHARSIS: "And why are your young men doing this, SOLON? Some of them are tripping one another up... Others, standing upright, themselves covered with dust, are attacking each other with blows and kicks. This one here looks as if he were going to spew out his teeth, unlucky man, his mouth is so full of blood and sand; he has had a blow on the jaw, as you see. But even the official there does not separate them and break up the fight – I assume from his purple cloak that he is one of the officials; on the contrary, he urges them on and praises the one who struck the blow."

"I want to know, therefore, what good it can be to do all this, because to me at least the thing looks more like insanity than anything else, and nobody can easily convince me that men who act in that way are not out of their mind."

SOLON: "It is only natural, ANACHARSIS, that what they are doing should have that appearance to you, since it is unfamiliar and very much in contrast with Scythian customs. But have no fear, my dear sir; it is not insanity, and it is not out of brutality that they strike one another and tumble in the mud... the thing has a certain usefulness, not unattended by pleasure, and it gives much strength to their bodies. As a matter of fact, if you stop for some time... in Greece, before long you yourself will be one of the muddy and dusty set; so delightful and at the same time so profitable will the thing seem to you."

ANACHARSIS: "Get out with you, SOLON! You Greeks may have those benefits and pleasures. For my part, if one of you should treat me like that, he will find out that we do not carry these daggers at our belts for nothing! But tell me, what name do you give to those performances? What are we to say they are doing?"

SOLON: "... when they stand up and strike one another, we call it pankratium [an early combined form of boxing and wrestling, author]. We have other such athletic exercises, too – boxing, throwing the discus, and jumping – in all of which

we hold contests, and the winner is considered best in his class and carries off the prizes."
ANACHARSIS: "And these prizes of yours, what are they?"
SOLON: "At the Olympic games, a wreath made of wild olive, at the Isthmian one of pine, and at the Nemean one of parsley, at the Pythian some of the apples sacred to APOLLO, and with us at the Panathenaea, the oil from the holy olive. What made you laugh, ANACHARSIS? Because you think these prizes trivial?"
ANACHARSIS: "No, the prizes that you have told of are absolutely imposing, SOLON; they may well cause those who have offered them to glory in their munificence and the contestants themselves to be tremendously eager to carry off such guerdons, so that they will go through all these preliminary hardships and risks, getting choked and broken in two by one another, for apples and parsley, as if it were not possible for anyone who wants them to get plenty of apples without any trouble, or to wear a wreath or parsley or of pine without having his face begaubed with mud or letting himself be kicked in the belly by his opponent!"
SOLON: "You are still unacquainted with our ways, I tell you. After a little you will think differently about them, when you go to the games and see that great throng of people gathering to look at such spectacles, and amphitheaters filling that will hold thousands, and the contestants applauded, and the one among them who succeeds in winning counted equal to the gods."
ANACHARSIS: "That is precisely the most pitiable part of it, SOLON, if they undergo this treatment not before just a few but in the presence of so many spectators and witnesses of the brutality who no doubt felicitate them on seeing them streaming with blood or getting strangled by their opponents; for these are the extreme felicities that go with their victory! With us Scythians, SOLON, if anyone strikes a citizen, or assaults him and throws him down, or tears his clothing, the elders impose severe penalties upon him, even if the offense takes place before just a few witnesses, not to speak of such great assemblies as that at the Isthmus and that at Olympia which you describe. I assure, I cannot help pitying the contestants for what they go through, and I am absolutely amazed at the spectators, the prominent men who come, you say, from all sides to these games, if they neglect their urgent business and fritter their time away in such matters. I cannot yet conceive what pleasure it is to them to see men struck, pummeled, dashed on the ground, and crushed by one another."

The tone of this dialogue is a familiar one (although strictly speaking ANACHARSIS was ridiculing the entire Athenian system and not just boxing). Those who oppose the 'sport' do so usually because they do not see why a fight between two men which will end with both of them injured to some extent should be considered sporting, athletic, or in any way beneficial to contestants or spectators.

In 1898, REMONDINO wrote the following somewhat philosophical piece *'Athletics and Pugilism in their Relation to Longevity'*, of which a part appears below:

"The blows upon the head and on the chin, which often leave the vanquished unconscious cannot help but be classed as factors in any subsequent breakdown, and the injury inflicted upon the real breathing organ – the nose – not only upon its outer or cosmetic parts, but upon the inner and more complicated anatomy and physiology are also injuries which must be tend to favor the subsequent productions of the many chest ailments or diseases of the respiratory organs, to which so many pugilists subsequently fall victims. As remarked, we cannot altogether accuse wild debauchery for the consumption that takes off so many of the calling in early live; were is such a main factor, many of the rioting and wildly hilarious and even drunk nobility would most certainly die of the same disease. The severe training to which many are subjected – long, excessive and continuous hard exercise – is probably much more of a factor in inducing diseases of the heart and lungs than the intemperance."
"The fighting term of a modern pugilist's life is short..."
"Here we have another factor that tends to shorten the life of the professional pugilist. Without being a philosopher, or without taking any stock in the saying that we begin to die the moment that we begin to live, he nevertheless, has its future limits – of usefulness and activity – much more sharply defined than those of any other occupation. He has the lives and exploits of those gone before to warn him of the shortness of his own career. He knows that age and physical degeneration will overtake him early in the race, and that the Chinese meaning attached to the age of fifty will come upon him before he has reached the half of those years. In pugilism it is only the ascending star that shines – the descending star quickly sinks into obscurity and neglect – of this he is well aware. When the time arrives that he feels the zenith is reached, it becomes a time of anxiety for him, and he begins to look among the younger and ascending aspirants, for the one that will hurt him – bruised, bleeding and possibly unconscious – from his pinnacle of glory and reality of life. A true philosopher – one possessing all the required moral stamina to well live over a defeat and its subsequent depressions, miseries and neglects – never would adopt the life of a pugilist, and we find but very few of the class who are philosophical enough to stand up under the overcoming moral depression consequent upon their downfall. This anxiety and their depression are no uncertain mortality factors to the class."
"As a rule, a pugilist does all in his power to use up or injure his opponent in either his resistance or aggressiveness, regard-

less of immediate or future ill results. The number of pugilists that have of late met death in the ring – although not so frequent an occurrence as death from football – is sufficient evidence that any accidental blow may kill an opponent. Thus to place life unnecessarily in jeopardy requires considerable blunting of conscience or rather an education into the idea that death is a desirable condition. When the pugilist systematically hammers over his opponent's liver or heart so as to make him cave in, he cannot be ignorant of the probable future results of this hammering."

"The idea that a man must hammer each other up to determine which is the best man, is, or should be, a relic of primitive days. At any rate it should have ended with Homeric civilization. Any intelligent physician can easily determine all that to the certainty of a mathematical fraction, with the assistance of improved diagnostic instruments. This would apply to disputed questions of endurance, vitality, or of strength. As to the question of scientific attainments, these could be easily determined by the boxer alternatively taking the offensive or defensive singly. JACKSON felt himself so much a master of the science, that he defied any one to reach him with a blow. 'Of course, JACKSON would observe, 'this challenge does not apply at times when I am trying to hit someone else.' In other words, JACKSON felt that the moment one entered into a contest 'give and take,' that no known science might prevent your antagonist reaching you at times. It then ceases to be pure science. Then again, as long, as observed by Jonathan HUTCHINSON, as between many men, it is only a question of accident or chance – something altogether foreign to science, skill or any precision and even outside of pluck or courage – as to who is to come off victorious, it seems that such decisions might as well be left to the chance toss of a coin. To settle disputes –as many do – by a resort to the street or backyard, is simply reverting back to the times when a man who had spoken wrongly or viciously of a woman was condemned to do her battle partly sunk in a hole in the ground and both armed with cudgels. If she won she was virtuous and if not her character was lost; it is but a revival of the wager of battle; something, which in our time is a reproach to our civilization or to the inefficiency of our social laws, which fail to give character or property and person the proper protection. It is simply going back to the principle of might being right of the old feudal or robber barons, which was a cheerful and conscious quieting way of looking at things until some stronger baron came around and made you change your mind."

"In the present, advances in war and battle science, where a decrepit old man weighing no more than a hundred and thirty or forty pounds can sit in an office and with telephone and telegraph make his army move as if playing a game of chess, and rout a Charles MARTEL, or a Richard. The muscular soldiery of former days are at a discount. The days of cavalry charges and heavy massing of troops began to decline with the Crimean war where with the Russians were seen the last of the great square solid formations. With the present machine and rapid fire guns and rifles, where troops begin to exchange fire a mile or over distance, and where the new artillery explosives will lay waste a whole brigade, muscular soldiers are no longer of use, and the old training of the chasseur, jager, bersaglieri or voltigeurs is now useless, unless possibly in a mountain warfare. The study Swiss pikeman of Morat and Morgarten, with their serried and compact formation backed by stout muscles and stouter hearts, would count but for little in this beginning age of BULWER's 'Coming Race' with its wholesale and silent destructive powers. It is useless for a man, now to cultivate the physique beyond the requirements needed to maintain excellent health, as it is useless to put horses to trails of long distance military rides, as has lately been done in Europe, where railroads propelling their cars by steam or electricity will soon gridiron the continent. NAPOLEON'a great dash with the bulk of his army, by diligence and coach and forced marches, from the shores of Boulogne sur Mer to the surrender of MACK at Ulm – in the beginning of this century – could now be duplicated in one-fifth of the time without being a subject of wonderment. The chemist and the electrician with the inventor are fast replacing muscle and pluck as well as numbers on the battle field, just as mechanics are replacing them in agriculture or taking the place of the nimble fingers of the seamstress..."

"Rough and violent exhausting sports of all kinds are a positive physical injury. If there is any temporary benefit to be derived from them, it is like borrowing money at three per cent a month. No exercise should be indulged in that cannot be taken regularly – and at the age of forty as well as at twenty."

"Athletics do not seem to render the system disease proof, a noted pugilist died from drinking a glass of ice water after some fatigue and overheating, the cold water using him as unceremoniously as it would have done a New England school marm. If you are a philosopher you will not need any great fighting ability and therefore avoid anything that will prevent you living to the age of the early saints or of many French philosophers; never attack a man you might easily whip, as that would be cowardly and mean and in every way discreditable to you; nor go the indiscretion of attacking someone who can easily whip you, for that would be like being demented; and as to attacking your equal, it is a foolish waste of time and a foolish waste of strength and exertion; it is better in that case to toss for heads and tails as to who shall consider himself the beaten man, – as a piece of banana peel might otherwise decide it for or against you. As the day of not firing 'until you see the whites of their eyes,' as at Bunker Hill, or fighting at close quarters a la NELSON's are things of the past, and as in all likelihood the days of leading a boarding party with a cutlass or leading a forlorn party into a breach or of wild cavalry charges, are also things of the past, our midshipmen and cadets need no longer cultivate athletics to that degree that they have in the past; the necessity no longer exists; the most that they will be called upon to do in the future in the physical way, will be to go down suddenly with the ironclad or to go into space through the means of a well directed dynamite shell fired from a gun several miles away, or of being bored through by a small ball that makes a rapid gimlet hole through the anatomy, that comes from a rifle a mile or two away, and whose report you have not heard, nor seen the smoke. Our embryo generals and admirals can therefore dispense with football and other athletics and husband their physical powers and stamina so as to live as old as von MOLTKE and not to lose their chances of promotion. If you wish to cultivate something that will stand by you at all the times cultivate a stoical philosophy, be an optimistic philosopher – even if you have to live like the lamented PANGLOSS of VOLTAIRE's 'Candide' – avoid sadness and cultivate cheerfulness, and don't do anything that will bring about a disease or injured condition of your circulatory organs – something, which any athletics, carried out beyond certain limits of moderation, are sure to do."

In fact, it may be seen that throughout this century (at least) there have been voices raised against boxing from what amounts to largely the same reasons. In 1919, the following article appeared in the *Literary Digest.*

"Why do peaceable, intelligent, respectable average Americans become enthusiastic over the spectacle of two big men pummeling each other until one or both arrive at a state of disfigured helplessness? The enthusiasm has been present in large measure on every such occasion in the long history of championship boxing. It was responsible for a tremendous pilgrimage to Toledo, Ohio, when WILLARD fought DEMPSEY on July 4. Airplanes, wires, and newspapers cooperated in rushing to the four corners of the country the tremendous news that Mr. WILLARD had collapsed, suffering from an extreme fistic indisposition, in the third act, or round. Mr. DEMPSEY, whose courage some of the soldiers' papers are questioning because although unmarried and within draft age, he did not appear in France, is hailed as the 'world's champion fighter.' However, the *New York Tribune* refers to a 'fine distinction' made by Mr. Grantland RICE, who opines that DEMPSEY is 'a marvel in the ring, the greatest hitting machine, or what you will,' but he is not 'the world's champion fighter.' As Mr. RICE puts it:
"Not by a margin of 50.000 men who either stood, or were ready to stand, the test of cold steel and exploding shell for anything from 6 cents to a dollar a day. It would be an insult to every young American who sleeps today from Flanders to Lorraine, from the Somme to the Argonne, to crown DEMPSEY with any laurels built of fighting courage."
"The *Tribune* comments that 'the call of combat is loud, and human beings, the most orderly and peaceable of them, find a thrill in the dramatic values of any duel between giants... But the heroes of the prize ring are something else again. The war gave us new values, new standards of courage and moral value, new contrasts, and a new meaning for 'champion fighters.' Despise the men and the methods as we will, a so-called championship fight 'holds a curious and intense fascination for us all,' in the words of the *Macon (GA) Daily Telegraph,* and has held a similar fascination wherever men had red blood in their veins. It is an 'instinct' remarks in the *New York Times,* 'and there is no quarreling with an instinct.'"

Obviously, this author did 'despise the men and the methods.' And seemingly put forward the idea of fighting being 'an instinct' as an attempt to explain the interest in prize fighting.

Max BERBOOM *('Yet again.' Knopf, New York, 1923)* wrote:

"If two men are angry with each other, let them fight it out (provided I be not one of them) in the good old English fashion, by all means. But prize-fighting is to be deplored as an offence against the soul of man. And this offence is committed, not by the fighters themselves, but by us, soft and sedentary gentlemen who set them on to fight. Looking back at ancient Rome, no one blames the poor gladiators in the arena. Everyone reserves his pious horror for the citizens in the amphitheatre. Yet how are we superior to them? Are we not even as they – suspended at exactly their point between barbarism and civilization? In the course of time, doubtless, 'the ring' will die out. For either we shall become so civilized that we shall not rejoice in the sight of painful violence, or we shall relapse into barbarism and go on into the mauling business on our own account. Our present stage – the stage of our transition – is not pretty, I think."

Don C. SEITZ (1926) wrote in the '*Millionaire Art of Self-Defense.' (Outlook, 144, October 6, 1926, pp. 179-180)* that boxing was:

"... a mere throwback to the Roman coliseum where the populace gathered for gore. The cries from the usual ringside, as repeated over the radio, are fit to chill the blood. There is something thrilling in the roar of lions; even the long note of the wolf and the barking of the hyena are not so bad. But this concentration of the human voices, expressing ferocity and utter lack of mercy, will long ring in the ears of those who have listened in. Probably those who join in the savage chorus are unconscious of the horrifying sounds produced in their ecstasy. That there was, at Philadelphia, a smaller manifestation of this sort than common was probably because the rain chilled the vocalism."

SEITZ continued:

"In the noble process of raising the standards of American immorality no uplift has been greater than that of the prize-ring. Four governors, cabinet officers, public functionaries of all degrees of eminence, and not a few clergymen were among those present. Millionaires were there in vast plenty, in just admiration of one of their class, Mr. DEMPSEY's modest earnings having long passed the million mark, which presents a little less than half of the gate receipts. All this was different from the days of Tom CRIBB, and like celebrities, who lived in Britain in the long ago. Prize-fighting is a British sport. It is an English trait to take 'punishment' with pleasure."

The following commentary was published *('Fighting invalid, The dirtiest game. Benny LEONARD tells why he isn't coming back to the ring. An interview by KENNEDY J B' (Colliers, 78, 134, September 4, 1926, p. 13 +):*

"He said more, much more. It all proved that he was wholly, permanently through; that the gloves he hung up two years ago, after sending Pal MORAN into the gaunt, grim hall of the defeated, will never be used again."

"We went back to his first beginning, one of eight children, five boys and three girls. In a club that housed horse auctions by day, Benny LEONARD got his first fight."

"A burly announcer with a four-wheel-brake voice couldn't solve his right moniker of Benjamin LEINER. He called him Benny LEONARD. What meant the etiquette of a right name to a four-round preliminary kid?"

"Benny got $20 for the go, plus a cracked nose bridge that he wears to this day; plus, also, an indigo eye. Within three years he got a crack at the title that had wandered from America. He brought it back."

"It's the most valuable title in the grimy game of commercialized fisticuffs. The heavyweight champion can command his $ 300.000 for a 'big shot.' A purse like that may come once every two or three years. The lightweight champion can fight every week or so and pile up purses averaging from twenty to thirty thousand dollars, without risking his diadem. It's a million-dollar cinch to an active man."

"LEONARD met some 200 contenders. There are three fights he'll never forget. He had a tough time with Willie RITCHIE of San Francisco, quodam champion, and after an inconclusive rough-house with Sailor BARTFIELD he let the middleweights and near-heavyweights alone."

"But his most sensational fight came in the first ring affair to be promoted by women in America. Miss Anne MORGAN wanted money for devastated France, so the purses of the sports were devastated to the tune of $25 to see Mr. LEONARD engage with Mr. MITCHELL, of Milwaukee. "

"HOMER might have described the third round in that fight. I'm afraid I can't. And I was there."

"MITCHELL took a straight left to the point of the chin and went down fifteen seconds after the round started. He took a count of six and got up but went down for nine. MITCHELL grasped the ropes and rolled along one side of the ring, LEONARD chasing him. The champion dropped him again."

"Women screamed, men roared! The champion calmly waited. At eight, MITCHELL tottered to his feet."

"Then lightning struck. From somewhere the groggy MITCHELL whipped a haymaker! For the first time the lightweight champion of the world heard a referee doing his sums. One, two, three, four! In the uproar LEONARD planted his gloves in the resin, and stared in a trance toward his corner, where Billy GIBSON, his manager, shrieked and tore the part out of his hair."

"Five, six, seven. LEONARD remained on his knees. MITCHELL, goggle-eyed, staggered towards him. *Eight.* LEONARD didn't move. The referee's swishing arm stirred damp hair screening his eyes. *Nine.* LEONARD heaved himself up and nearly dropped again."

"He clung to the ropes, MITCHELL wallowing after him. LEONARD turned. His eyes saw nothing. His hands guarded his jaw, stalling by his corner, GIBSON yelled, 'Forty seconds for the bell.'"

"LEONARD didn't hear. By blind instinct he defended himself. The bell!"

"LEONARD was out on his feet. Ammonia, ice water, and savage calf pounding brought him to. In deathly silence he staked from his stool for the next round, and within thirty seconds MITCHELL was a bloody wreck."

"Such is the way of champions."

"'I'll never forget it as long. I seemed as though MITCHELL ripped my head off with one punch. I don't know what happened after that until I saw him sprawled at my feet in the following round.'"

"A close call, and I swore to myself that night that it would never be repeated."

"TENDLER smacked me hard in Philadelphia. He tore off an eyebrow that had to be sewn back and the scar will be a permanent memento. But I've come through two hundred battles with my ears and eyes normal. I value them more that any money the ring can give me to stage a comeback."

"'Boxing,' he continued, 'made me – physically, mentally, in every way. But professional fighting is a nasty game. See a modern champion riding around in a limousine, and you don't think so. But get into the slums of the business, where gunmen and hijackers thrive on poor little kids trying to make an easy dollar. Go to the cheap gymnasiums and the cheaper clubs and see how the youngsters have to earn their money, not by the sweat of their brow, but by their very life blood – and you'll realize that it holds nothing further for any sane man who has got all it can give."

"Yes. In New York two gunmen sat in the ring-side seats ready to shoot a boxer in his teens who had refused to lie down so that a gambling outfit could play heavy odds and win at his expense. In Chicago a gang riddled to death a ham-and beaner who 'accidentally' double-crossed a boy he had agreed to dive for. Even in the august alley-ways of the New York Boxing Commission itself the rat-faced leeches of the game make their boasts and ply their bets, using dumb athletes as their pawns."

"The manly art of self-defense!"

"'I'm not knocking the game that made me rich,' says LEONARD, 'but I'm through with it. I want to see it continue, strictly regulated, with the harpies and hard-boiled birds driven out. The game can be kept clean only by jailing any hijacker who threatens a boxer and by blacklisting every grafter who holds him up for a part of his end. The youngster who gets into the ring quickly should get what he can quickly – and get out.'"

"Unless you're a champion or a near-champion, it's the dirtiest game in the world."

Robert LYND *('Brutality of the prize ring.' Living Age, 333, September 1, 1927, pp. 427-439)* comments on professional boxing:

"It is easy enough to see why many people should dislike prize fighting as it is at present carried on. Prize fighting has become professionalized and commercialized to such a degree that its proper place seems to be in the music halls rather than among the sports. The rewards paid to prize fighters, both in fame and in money, are ludicrously out of proportion to any service they render the community. Great cricketeers and footballers frequently become professionals, and make a living out of playing games; but they are content with moderate rewards, like poets and clergymen. Cricket remains a game – a game that deceives us into the feeling that it is played for pleasure, not for profit – in spite of its having been professionalized. There is no haggling over huge purses before championship matches can be arranged."

"Perhaps it is absurd to object to one kind of sportsman being able to make more money than another; but, at the same time, it is a sound instinct that dislikes seeing money playing a predominated part in sport. We like to see a holiday spirit maintained in games, and we do not like to see them made the preserve of those who have far more money than ourselves. Possibly our objection to professional boxing is fundamentally an objection to the prices charged for admission. It is, we realize, of as little use to criticize professionalism at this time of day, as it would be to criticize an eclipse of the sun; but we resent its triumphant spread. We should like to see more people boxing and fewer people making money out of it."

"To dislike some of the concomitants of a modern prize fight, however, is one thing; to denounce prize fighting as though it were a savage and bestial orgy is another. There has been an extraordinary outpouring of violent and vituperative speech over the fight that took place the other day between WALKER and MILLIGAN – speech scarcely less sanguinary than the fight itself. Sir Hall CAINE led the chorus with a denunciation of the fight as a 'debauch of brutality.' Sir James CRITCHTON-BROWNE followed with an onslaught on the audience as 'vultures at their carrion,' and declared that, when he read the report of the fight, he exclaimed to his wife: 'Oh, for a CARLILE, fearless and arresting, to denounce with righteous indignation this sanguinary iniquity!' The fight, he said, had been 'pushed to the verge of homicide,' and he called sarcastically for the institution of a new kind of fighting between electric lay figures with arrangements for periodical spurts of blood from the figures 'so as to give *vraisemblance* to the scene.' Another writer, who was apparently not present at the fight, declared that 'a Spanish bullfight would have been sweet in comparison.' A lady called it 'hideous.' Another said it was 'a terrible reflection of our much-vaunted civilization and professed Christianity' that 'our so-called Christian leaders and humanitarians' had made no protest against the fight. In vain did a sportsman on the other side cry out that these attacks on boxing were 'insulting' to the Prince of WALES and the Royal Family, since 'the Prince of WALES is himself a fine boxer, and on more than one occasion has honored a championship contest with his presence.' The critics of the fight saw red, and were far beyond the reach of such reasonable arguments."

"Probably it was the blood that did it. There is something peculiarly horrible in the sight of blood. There have been people who fainted on seeing it flow from a cut finger. I myself, I confess, shrink from the sight of blood. If I were present at a fight, and blood began to flow, I should cease to enjoy the fighting. It is very unpleasant to see a man daubing his opponent with red stains all over his body with a glove that has been drenched by a bleeding nose. If you did not know that nobody was seriously hurt, you would imagine that you were witnessing some heathen and homicidal orgy. If you were honest with yourself, however, you would realize that it was not the brutality of the scene, but the blood, that had shocked you. To the boxer, the loss of blood as a rule means nothing. He does not remember it among the dark and painful episodes of his career. It is only a minor discomfort, and probably hurts some of the spectators more than it hurts the boxer."

"The bloodshed that took place at the WALKER – MILLIGAN fight was, it must be admitted, a more serious matter than a bleeding nose. With his last terrific blow WALKER split MILLIGAN's lip in such a fashion that, when MILLIGAN fell and rolled over, blood seemed to be gushing from his mouth in torrents, and some who were present were afraid that he had received a mortal injury. The scene was undoubtedly one that would have horrified most of us, and, if such scenes were common in prize fights, most of us would be reluctant to witness them. We have no more love for serious accidents in games than in the streets. No one but a Sadist could take pleasure in seeing human flesh gashed and bleeding. Yet the denunciation of the fight as a 'debauch of brutality' seems to me to rest on the assumption that the audience was largely composed of Sadists, who took a particular pleasure in seeing MILLIGAN's lip gashed. There can surely be no brutality in being spectators of an accident which no one could have foreseen."

"Boxing, no doubt, is a more primitive sport than cricket, and rouses passions of the brute. But if we condemn all sports in which the primitive passions are displayed, we shall find ourselves condemning, not only boxing, but Rugby football. No one can deny that Rugby football is rough. No one can deny that men constantly get hurt when playing it. I have seen games that had to be stopped three or four times while a player obviously suffering excruciating pain was rubbed and tugged back into something like normal life. A Rugby football field is a battlefield on which nobody is supposed to hurt anybody, but on which people inevitably get hurt. If you judged the game by its worst accidents, you would sign petitions to the Government to suppress it. What must Sir Hall CAINE have thought, when he read the description of the last international match between Ireland and Scotland – a match that was played in the mud during an icy blizzard, and, what with the rigor of the game and the rigor of the weather, resulted in two players being carried unconscious off the field and, if I remember right, in the swooning of the referee? Compared to such a game, an ordinary prize fight would seem like a Sunday-school entertainment. And the truth is, Rugby football cannot be defended except on the assumption that it is good for young men, within limits, to treat one another violently and to take part in contests of strength and skill that tax human endurance to the utmost."

"There may be moral heights from which all contests for supremacy seem odious, but we do not live on them. We match team against team, man against man, horse against horse, and dog against dog, and we rejoice in the beauty of the strength and skill that triumph in the contest. It would, no doubt, be better for the race if we could devote ourselves entirely to intellectual and spiritual pursuits, but that for most of us is not the alternative to strenuous pastimes. If boys and young men give up Rugby football, it will scarcely be in order to study *Hours with the Mystics,* or to lead the lives of holy hermits. Games are the alternative, not to a better kind of employment, but to a worse kind of idleness. They are among the most innocent means of enjoying life and of making it dramatic. It is true the contest of HAMLET with destiny affords us infinitely richer pleasure than a contest between WALKER and MILLIGAN; but we cannot always be reading or seeing *HAMLET,* and we do not enjoy SHAKESPEARE the less for admiring the power of WALKER or the courage of MILLIGAN. Moreover, if you are going to call any contest brutal, surely the contest in *HAMLET* and in *King LEAR* is a bloodier affair even than a prize fight. It would be easy to attack all tragedies as 'degrading exhibitions' on the ground that the audience enjoys watching the struggles of human beings, knowing that they are doomed to end in death and pain. Everybody realizes, however, that the spectators at a tragedy enjoy, not the homicides and suicides that are so essential a part of it, but the spectacle of the spirit of man heroic even in defeat."

"And I am sure that it is in the splendor of the body and spirit of man that makes all the world excited about a great boxing contest. No one would pay sixpence merely in order to see one man 'punishing' another, as the phrase is, unless there were skill and courage to make the punishment given and the punishment received beautiful. One may convincingly be mistaken in thinking the muscles and frame of a boxer beautiful, but magnificently used, they represent an ideal of strength and power that all men desire. Here is something human disciplined till it has become something almost more than human, and the race of man seems nobler when we see it. Boxing and football bring the elation of poetry – not the best poetry, but still quite tolerable poetry – into the ordinary man's life. He enjoys them with all the better conscience because he knows that they are both far less dangerous than they appear to be, and that the ordinary boxer, far from being butchered to make a spectator's holiday, runs considerably less risk of injury than a railway worker or a motorist. Boxing can no more be justly condemned because of MILLIGAN's battering and his gashed lip than horse racing can be condemned because HUMORIST died as the result of his desperate effort to win the Derby. (If we condemned everything in which horrifying accidents occur, we should never allow a cathedral to be built, and we should have to put an end to motoring, mountaineering, and flying."

An editorial (*Literary Digest, 94: 32, September 10, 1927)* reads:

"The *Church Federation* stands unequivocally against commercialized prize-fighting in all forms as a sordid perversion of true sportsmanship, and as blighting in its influence upon the young as open gambling, vice and drunkenness. It brutalizes the instincts of youth and dulls their finer sensibilities. For officials of the city and State not only to lend their endorsement but also to enthusiastically espouse prize-fighting and openly court this TUNNEY – DEMPSEY bout, for Chicago is a lamentable display of low-mindedness, which can only bring about harm. The declaration that Chicago would commercially profit by this battle is only a smoke-screen to hide the real purpose of getting the prize-fight to be held and to lend it an air of respectability. Chicago is rich enough that it does not need more wealth at such a price. Those who make this argument had better begin to take stock of that challenging question which still holds good: 'What doth it profit a man if he gain the whole world and lose his own soul? It applies to cities also.'"

"We would call to mind not only to the hosts of our own faith but also to the millions of the other great communion, that this prize-fight is proposed for the same spot which not so long ago was one vast outdoor cathedral, where the millions in the Eucharistic Congress honored God and partook of the sacrament, in memory of the broken body of Christ. We are sure that this proposed declaration of ground dedicated to a high purpose will call forth a mighty wrath. It has its warrant in the declaration just made in the *Osservatore Romano*, official organ of the Vatican, that 'the passion of the people for these spectacles (referring to the DEMPSEY – TUNNEY fight, the subject of the editorial) is the indication of the resurrection of paganism and barbarism in its most vulgar and ferocious form!"

"Prize-fighting has only recently succeeded in staging a return, having previously been outlawed except in some frontier communities. But under the guise of boxing contests and sportsmanship, it has returned in all its viciousness, with brutality, gambling, and foul play rampant. It is a sordid perversion of lower instincts for the purpose of money, both on the part of contestants and promoters. It is distinctly a moral issue, summoning the churches to the crusader's spirit in doing battle to the death against it in all its forms.'"

"Therefore the *Chicago Church Federation enters* its protest first against the fight being held in Chicago, and secondly against it being held in Soldier's Field, and calls upon churches of each and every community, church members, and other citizens, to register their emphatic opposition to the TUNNEY – DEMPSEY battle."

A well-written article appeared in (*Literary Digest, 1928),* comparing the views of Bertrand RUSSELL and Benny LEONARD. Unfortunately, he did not bring the two men together for a debate, but interviewed Mr. RUSSELL and Mr. LEONARD separately. The editor/author was not obviously of one opinion or the other:

"Two captains of the cauliflower industry,' as W. O. McGEEHAN has renamed it in the *New York Herald Tribune,* trying to out-hammer, out-think each other to a knock out in the prize-ring, cheered on by thousands of other men raised emotionally to a state of frantic enthusiasm, make a spectacle susceptible of strikingly different interpretations. 'Sadism,' says Bertrand RUSSELL, British philosopher, in an interview by Edna B. MANN in the *New York World,* admitting that he has never seen a prize-fight because 'it just never came my way.' Mr. RUSSELL proves himself as pronounced a pacifist concerning the prize-ring as his reading public finds him in regard to war. But his diagnosis of fight-fan excitement does not go undisputed. 'The normal outbursts of normal people, excited by exciting events,' contradicts Benny LEONARD, undefeated lightweight champion of the world, writing in the same paper. In the interview Miss MANN asked why men love fighting and prefer a 'he-man prize-fighter like DEMPSEY with his knock out blows.' It might be, she suggested, mentioning the rough play of children and puppies, pure delight in the activity. 'Doesn't the world need this delight?' she asked, and we read Mr. RUSSELL's reply:

"There is a difference between playing and fighting,' said RUSSELL. 'You watch two puppies, What they are doing when they pretend fighting is not fighting. It is playing. Although they seem to be going to bite, they never do bite. They just rehearse movements. It is the same with the children you describe. Real fighting is different. There is a sentiment of anger, a flow of adrenalin, and so on.'"

"Prize-fighting, I think, comes near to the real thing. It gets too close to it. I think it arouses brutal emotions. It appeals to Sadism in man, his lust for cruelty. The ordinary person has far more cruelty in his nature than he realizes. We go about looking for legitimate opportunities to satisfy it."

"This, then, is why the average fight fan complains he does not get his money's worth when there is no knock out or heavy hitting; it explains why the fighter with a paralyzing punch – a killer – is so much more popular than a clever boxer who lacks the knack of a knock out; it explains the crowd's savage roar of excitement when a hitter connects and his unfortunate opponent staggers, groggy from the blow. The basic appeal of prize-fighting is to man's primitive Sadism."

To this charge Mr. LEONARD speedily made the following replay:

"In the first place, Dr. RUSSELL admitted he had never seen a boxing exhibition – he calls it a prize-fight. Hence, what the philosopher knows of boxing is merely what others told him."

"It may seem strange to Dr. RUSSELL when I state that boxing is just the opposite of what he pictures it to be. I wish he could come with me some day and see what boxing has done for thousands of boys in every section of the city. It has curbed the gang spirit; it has helped to refine countless boys, who, because of the conditions that surrounded them, ever eager for a brawl, when the boot was used as quickly as the fist."

"It has curbed the headlong lad ever eager to settle a quarrel with his fists. Instead, boxing has taught him restraint. It has taught him that the battle is never won by a lad who loses his head; that an argument does not have to be decided by the fists, with possible injury to his opponent, but is frequently won by the lad who keeps his wits about him."

Dr. RUSSELL dwells at comfortable length on the cruelty of the prize-ring as a form of Sadism. The very fact that he never saw a bout discounts his argument.

"There is nothing in the ring that shows the slightest suspicion of Sadism. The boxers have no malice in their hearts. They know that a blackened eye, a sore side, a broken nose, while it may hurt, seldom causes permanent injury. When they measure their opponent for the blow that deprives the rival temporarily of his senses, there is not the slightest bit of malice, and, in 99 out of 100 cases, not the least bit of anger."
"Sadism does not exist in the ring any more than it does in other rough sports, such as football, lacrosse and hockey."

For Bertrand RUSSELL had given a clean bill of health to certain other sports. The interviewer had asked him about football, baseball and other popular diversions in which the fight element is strong, and he had replied:

"There you have genuine play. Games satisfy in imagination what you would otherwise satisfy in reality. We mustn't be cynical about any human impulse. We should instead find the least harmful way of indulging it. Football athletic contest, debates, work, offer abundant useful outlets for the fighting impulse."

Before Miss MANN had come to interview RUSSELL, she had been talking with a he-man editor of vigorous bulldog proportions, who, we learn as we read on, now stood revealed in the light of RUSSEL's analysis:

"'We want no pacifistic stuff,' he said to me. It's fighting that men like. Why, when I look back over my whole life (he is now over 45 and is an important figure in the magazine world), I get most satisfaction recalling a fist fight I had with two sailors out on the snow in front of a saloon. For an hour and a half we battled, rolling over in the ice and snow. My face was cut and bleeding from forehead to chin. It was so black and bruised I was hardly recognizable... Then there was another fight, "he licked his chops with relish, as evident as it was sincere, over other fist fights in, before, behind saloons, cafes, and dives, all the way up and down the scale."
"I like to win,' he said, "But even when I don't win, it's the fighting I like. Damned if I know why."
"Do you mean to say," I asked, "that nothing in your life, no other achievement, say in your work, or with women, has given you satisfaction equal to fighting?" 'Women?' he stumbled over the word as though an inner resistance blocked him. 'Women? I never got much out of women. Never had much to do with them – but it was too much for my own good. Never want much to do with them, either. But don't think I've stopped fighting because I don't want to fight. It's just because I'm too old and fat. It's fighting I get the kick out of.'"
"Miss MANN told Bertrand RUSSELL this story, and he replied: 'The way to deal with Sadism is to make men more satisfied, happier, to find out the sources of happiness and try to realize or legalize them. I am constantly struck with the fact that there is a great deal of Sadism in America.'"
But to return to Mr. LEONARD. The former ring star, and present sports writer, takes up Mr. RUSSELL's assertion that Sadism does not enter into other sports, and retorts:
"Certainly it is not Sadism when we shout with joy as the fullback of Notre Dame plows down the field, and several of the opposing line are stretched out on the field, their wits temporarily knocked out of them."
"Not at all; that's not Sadism; that's red-blooded American men and women venting their excitement in shouts at a great exhibition of sport."
"Perhaps the good doctor has seen a lacrosse game, where there are more injuries in one clash than in a dozen battles in the ring. Certainly the philosopher would not call a lacrosse player a Sadist as he bashes in the nose of an opposing player."
"Dr. RUSSELL's interview was most interesting reading, but I fear that few of us, especially those who like red-blooded sports, will agree with him that our delight in these games or lines of athletic endeavor are exhibitions of Sadism."

And, to conclude, Mr. LEONARD makes an offer, thus:

"I wish Dr. RUSSELL would be my guest at some big bout in Madison Square Garden. I know he is no Sadist. But I venture the prediction that it would require one or two special police to keep him in his seat when the old cry of 'there he goes' rings out as one of the boxers starts to sag and his opponent measures him for the finishing punch."
"Sadism on the part of Dr. RUSSELL? Not at all; just a red-blooded man doing what nine out of ten who never had training in boxing, will do – losing his head."

"It is probable," says CARROLL (1936) "that no head blow is taken with impunity, and that each knock out causes definitive and irreparable damage. If such trauma is repeated for a long period it is inevitable that nerve-cell insufficiency will develop ultimately, and the victim will become punch drunk."

JOHNSTON in his book *'Ten and Out,'* published in 1937, writes:

"Fighting with the fists, whether 'the raw 'uns' or gloves, is not a gentle sport. Blood will always be shed, eyes will be blackened, and there may be accidents of a more serious nature, but the rough old game serves a place in our life, helping to stave off the effeminacy that is one of the dangers of nations that grow old and soft and unwilling to endure hardships."

It reads further:

"Twenty-five centuries earlier, in the epics of HOMER, THEOCRITUS describes AMYCUS and POLYDEUCES, meeting in midring and breathing slaughter against each other. In between AMYCUS and Sugar Ray came Christianity, knighthood, the eighth Marquis of QUEENSBERRY, the padded glove, Gentleman Jim CORBETT, Tex RICKARD, and Jack DEMPSEY. But the homicidal theme has never changed."

Finally we read:

"If the referee is considered third man in the ring, death has always been the fourth presence. And the fifth member of boxing's personage is the attending doctor who must stand by and watch the primeval urges are being loosened, who must hear the spat of leather against flesh and the thud as a boxer's head hits the ring apron. Those in the employ of boxing commissions deal routinely with 'hemorrhaged ears, swollen noses, detached retinas."

"Boxers do die regularly – at a rate of about ten a year for the last twelve years in both amateur and professional ranks, but it is not necessarily the number who die – dozens die of football injuries each year – that sets doctor against doctor in bitter polemic. Rather, it is the way they die – by calculated violence – that troubles the conscience of some doctors."

"When a boxer dies in the ring, the medical cause of death may never be established. Only about half of boxing fatalities are autopsied."

WINTERSTEIN (1927) wrote:

"Boxing is one of the popular sports, but it has rarely been realized that punches and their consequences may provide the medical investigator with interesting material for the study of head injury. Unlike accidents these injuries are caused by trauma always of the same kind and acting with almost laboratory exactness. Everything occurs in the presence of experienced spectators, such as the trainer, manager, and the referee. Film records are taken. No police inquiries follow to interfere with psychological observation; there is no question of compensation, so that the possibility of a compensation neurosis is excluded. These points make boxing a model experiment for the study of traumatic head injury."

JOKL (1941) wrote:

"... of all major sport, boxing occupies a special position since its aim is that of producing injuries, more particularly to the brain. As the more dramatic manifestations of such injuries are colloquially referred to by such terms as 'knock out,' 'grogginess,' 'punch drunk,' etc., it is not usually appreciated that these conditions indicate the presence of serious cerebral disorders. It is true that similar injuries occur in sports other than boxing, e. g., in football or wrestling. But here they are accidents rather than sequelae of intentional acts. Only in boxing are traumatic injuries unavoidable even if the rules are adhered to."

JOKL noted:

"To say that boxing is 'a character-forming sport' and that it engenders 'courage, self-reliance, physical fitness, and chivalry' is an assertion not based on evidence. There seem in fact, to be good reasons to justify qualified statements to the contrary... Mr. LESLIE points out that the head is not the only 'recommended target' in boxing, implying, I take it, that the rest of the body is not vulnerable to blows. A sufficient number of medical reports are proving the fallacy of such a pronouncement... No categorical difference exists between the dangers of amateur and professional pugilism... In its reaction to injury, living tissue is not influenced by the monetary arrangements between the boxer's managers."

The Roman moral theologian Father DAMEN wrote 1947, in a later edition AERTYNS-DAMEN):

" It is sinful to reconcile prize fighting as we have it today with Catholic principles of morality. Boxing is often accompanied by injury that it may easily assume sinful features."

Boxing is explicitly mentioned in the *Theologia moralis,* edited by AERTYNS-DAMEN (1947). The pertinent section (Vol. I, p. 476, lib. III, Tract. V, Caput V) reads as follows in Latin:

"Nihilominus hujusmodi certamen facile illicitum fieri potest, leviter vel graviter pro gradu perieuli magis vel minus probabilis laesionis vel etiam mortis; ut si debitae cautiones non observentur, vel si certamen eo tendat ut unus ex certantibus sensibus destituatur (to knock out).

A free translation from Latin into English reads as follows: "

This type of contest can easily become unlawful. Either venially or mortally so, in accordance with the degree of the more or less probable danger of inflicting injury or even of death, for example, if the due cautions are not observed, or if the contest tends to the 'knock out' of one of the fighters."

An article appeared in the *Reader's Digest (1947)* entitled *'Facts on the manly sport.'* The author was largely concerned with corruption in the *'boxing racket:'*

"The boxing business is booming as never before. There are some 10.000 professional boxers in the United States, and about 100 fight clubs to keep them busy. To support this industry, fans pay out some $ 16.000.000 a year."
"On the surface, boxing is a manly, exciting sport. Unfortunately, there are strong undercurrents of physical tragedy and sordid, crooked dealing. 'Deaths occur from time to time in every red-blooded sport', says Andy O'BRIEN in *Sportsfolio,* 'but no other sport produces anything to match boxing's list of 'permanently injured.' Last year eleven boxers in the United States and Canada died from hammerings they took from opponents. The most publicized fatality in boxing history was the death of heavyweight Ernie SCHAAF after a knock out by Primo CARNERA in Madison Square Garden in 1933."
"The business is infested with racketeers and hoodlums who operate behind the scenes."
"At least one third of the professional boxers," write John FIELD and Earl BROWN in *Life,* "are owned or controlled by crooked managers who fix fights in order to pull off gambling coups. One of the most feared men in boxing today is Frankie CARBO, alias Jimmy the Wop. In the last 25 years he has been arrested eleven times and served one term in Sing Sing. In 1941, he beat a murder rap. He controls a string of good fighters."
"Sometimes the hoods 'cut themselves in' on the earnings of promising fighters by threatening the boxers' managers with bodily harm if they refuse to make a deal. Not long ago, manager Freddie SOMMERS brought a good lightweight, Cleo SHANS, to the East. The hoodlums talked to SOMMERS. He turned them down. A few days later he was found beaten almost to death."
"Early this year New York District Attorney Frank S. HOGAN started a grand jury investigation of the boxing racket. The grand jury has called many witnesses, including CARBO and Jimmy PLUMERI another ex-convict. One manager testified 'the mob' told him that if he wanted a certain match his fighter would have to agree to loose. Testimony also showed that boxers are sometimes victimized by managers who, in violation of *New York State Athletic Commission's* rules, deduct unwarranted 'expenses' from the fighters purses. The boxers themselves hesitate to report such abuses, fearing either that they will be beaten up or that they will be barred from future matches."

Father Eugene HILLMAN of Norwalk, Connecticut, wrote 1951 about *'The Morality of Boxing.'* Father HILLMAN wrote:

"The moral implications are obvious. A boxer sins when he attempts to land severe blows on the head of his opponent. The matter is grave because of the importance and delicacy of the organ which is attacked. "

He continues to say that:

"... by his enthusiasm the spectator positively sustains, fosters, and encourages the contestants in their immoral actions can hardly be considered anything less than formal cooperation in sin."
"The moral implications are obvious. A boxer sins when he attempts to land severe blows on the head of his opponent. The matter is grave because of the importance and delicacy of the organ which is attacked. "

He continues to say that:

... "by his enthusiasm the spectator positively encourages the boxers. This concurrence which the spectator positively sustains, fosters and encourages the contestants in their immoral actions can hardly be considered anything less than formal cooperation in sin."
"The moral implications are obvious. A boxer sins when he attempts to land blows on the head of his opponent. The matter is grave because of the importance and delicacy of the organ which is attacked."

He continues to say that:

"... by his enthusiasm the spectator positively encourages the boxers. This concurrence which the spectator positively sustatins, fosters and encourages the contestants in their immoral actions can hardly be considered anything less than formal cooperation in sin."
"For the most part, discussions on the morality of boxing have remained on the emotional level. Those who like rugged sports are apt to think that boxing is all right; those who do not are apt to think it is all-wrong. Very few theologians have discussed the matter as theologians. And among these few, none has offered a really scientific treatment. They have mentioned principles, expressed feelings, and stated some very definite conclusions for both sides. These opinions have been

well summarized by Fr. Gerald KELLY in the March issue of *'Theological Studies.'*"

"Here we shall offer several serious reasons why boxing, as we have it today, must be regarded as morally reprehensible. For the sake of an orderly treatment we shall present all the pertinent material in three separate arguments under the principles of the Fifth Commandment. The boxing with which we are primarily concerned is our modern professional prizefighting, and all fighting which is after the manner of our famous title bouts. Secondarily, our conclusions touch amateur boxing, to the extent that it is like professional boxing. In other words, we are here passing judgment on all boxing which holds the knock out and technical knock out as goals, gives credit for punching power, and caters to the brutishness of the spectators."

"It might be wondered that the morality of boxing should even be called into question, since it has been for so long accepted as a wholesome form of recreation, and since it has received so much sponsorship from organizations of noble purpose. The reasons for this are not hard to find. The application of moral principles to particular cases depends upon certain facts in the cases. The morality of a case may be wholly changed by the discovery of some previously unknown physical fact. On the basis of newly discovered medical data it sometimes becomes necessary to reconsider moral principles in their application to particular cases. Thus it sometimes happens that practices which were previously considered licit may no longer be permitted. This is the situation today with regard to boxing. Medical findings have compelled us to reconsideration of our moral principles in respect to boxing as it is carried on at present."

First Argument

"Our first argument is based on the principle that only a proportionately grave reason can justify a man's placing an action which is calculated to deprive another of his consciousness violently. Although this principle has not been fully thought out by moralists in respect to boxing, it has been well analysed in its application to drunkenness which, in the judgment of medical men, is a state closely resembling that of a boxer who has been dazed by repeated blows on the head, even short of what is referred to as a knock out."

"To place a man violently in such a state is to deprive him of his specifically human powers; it is to dehumanize him. For it is man's most precious good which is attacked, the good by which he is distinguished from the brute, his ability to think rightly and choose freely and to control his body as a responsible human being. Even the desire or the intention to attack such a good is sinful if there is not present some proportionately grave reason, such as the good of integral bodily health or the repelling of an unjust aggressor. There is hardly another reason which would justify the violation of so fundamental a human good. Certainly, no sum of money, no amount of fame or pleasure can be considered equal to the good of human consciousness. Indeed, if we may not deprive our neighbor of his material goods unless we are in a state of extreme need, then surely we are not justified in depriving him of his spiritual powers for any reason which is less than extreme need. Even in extreme need we may deprive our neighbor of his material goods only when there is no other means of alleviating the situation."

"The malice and the gravity of an unreasonable intention or attempt to deprive another of consciousness may be seen through a consideration of the parallel state of perfect drunkenness which is gravely sinful, although admitting of parvitude of matter. VERMEERSCH, PRÜMMER, and others tell us that the essential gravity of an unreasonable deprivation of consciousness precinct from any duration of time, and consists precisely in the violent deprivation of the ability to think normally and judge rightly, without a proportionately grave justifying reason."

"Lest it be thought that the knock out suffered by boxers is really not a deprivation of consciousness in the proper sense, but merely a state of temporary grogginess, let us consider the statements of some competent physicians who have made a serious study of the medical aspects of boxing. Dr. Ernst JOKL, who is regarded as the leading medical authority on boxing injuries, has this to say about the knock out:"

"'... if a blow is sufficiently strong the opponent may be rendered unconscious, irrespective of the region where the blow is landed. The most frequent form of knock out is that from a punch on the chin. There occurs a sudden loss of consciousness. The duration of the loss of consciousness is quite varied; as a rule it is less than ten seconds. The consciousness returns slowly. Frequently the boxer may be capable of continuing the fight, even before the complete return of consciousness. In some cases the loss of consciousness may persist for a few minutes and occasionally for a few hours. It leads not only to complete loss of memory, but occasionally even to a retrograde amnesia extending to the entire fight.'"

"The parallel between the knock out and the state of perfect drunkenness is clear not only at the point of complete loss of consciousness and in the period of partial consciousness following the knock out, but also in those cases where a series of blows on the head has brought a boxer to a verge of unconsciousness. This state is described by Dr. G. W. WILL (1939):"

"'Such a man appears dazed, his resistance to his opponent is more or less automatic, and there may be amnesia for all or part of the contest. This confused state may follow a single knock out blow where there has been a period of unconsciousness, but it may also follow a series of blows none of which has been serious enough to produce unconsciousness.'"

Father HILLMAN continues:

"In this state of confusion or grogginess the specifically animal functions of man continue to operate, but he is unable to think and control his body as a responsible human being. For there occurs, according to Dr. JOKL, 'a degree of cloudiness of the consciousness which leaves intact the automatic motoric functions, but interferes with the conception ability, and decreases the reaction rate.' A man in such a state hardly seems capable of moral responsibility. From a moral viewpoint, one in such a state is lacking consciousness no less than one in a state of perfect drunkenness. With this problem of the knock out in mind, it might be well to read pp. 123-126 of Father John C. FORD's *'Depth Psychology,* Morality, *and Alcoholism'* (1950)."

"Because of the moral principles concerning the violent and reasonable deprivation of consciousness, and because of the nature of the knock out as explained by those who have seriously studied the matter, it must be said that boxing as we have it today involves serious sin, at least materially, on the part of the contestants who attempt or intend to win by a knock out."

"That most boxers, at least on the professional level, make such attempts and have such intentions can be shown from the testimony of boxers themselves, and from the nature of the contest, as it is today."

"Every boxer wants to win, and he knows that today the most desirable means to victory is the knock out. If a fighter wants to be popular and successful, he knows that it is knock outs that count. It is a fact that the majority of title bouts end in knock outs; and young boxers look to such bouts as the point of success in their profession. It is exceptional when a promising young boxer does not have on his record a high percentage of knock outs. Even technical knock outs are professionally more desirable on a boxer's record that victories by decision. This information has been gathered from the knock out records of title bouts since 1946, from the records of individual boxers, and from consultation with MURNANE's statement:"

"'Most fighters want to win by knock outs, if possible, for knock outs pay off. They build up a fighter's reputation, making it possible for him to demand more money and to advance in division ratings... When a fighter sees his opponent in a bad way, he doesn't let up. On the contrary, he renews his efforts to score a knock out... The head is the main target and most knock out blows are delivered to the head. Even when a fighter directs his attack to other parts of the body, it is generally a diversionary measure to get an opening to the head.'"

"The nature of the contest today, at least on the professional level, holds the knock out to be the most desirable means of victory. This is what the spectators want, so the boxer who would be successful must also want it and produce it. Judging from statistics in professional boxing, this type of victory is very frequently attained."

"So, from boxing statistics on the number of knock outs, from the nature of the contest today, and from the testimony of boxers we conclude that the boxers who want to get ahead, probably most boxers, have the intention of knocking out their opponents whenever this is possible. Is it even conceivable that one could engage in such a profession without ever intending or attempting to deprive another of his consciousness?"

"We have seen above that the duration of time has nothing to do with the essential gravity of an unreasonable and violent deprivation of consciousness. So, the mere intention of rendering a man unconscious for a moment is a sin, if there is no proportionate reason. And, as there is question here of man's most precious good, there is grave matter from the very nature of the case."

"The boxer's knock out is certainly a more violent deprivation of consciousness than that effected by an excess of alcohol; it is also more calculated and deliberate. In both cases the dehumanizing effect is essentially the same, regardless of the time element. So, we may conclude that, while the full knock out has at least the same malice as perfect drunkenness, the partial knock out or the state of grogginess has at least the same malice as imperfect drunkenness. It is, therefore, sinful for a boxer to intend or attempt to inflict a knock out on his opponent. And, since the proximate occasion of sin must be avoided, one may not engage in a contest which has as a desirable and frequently attained objective an action which is sinful from its very nature."

Second Argument

"In virtue of the Fifth Commandment it is forbidden, except for reasons of integral body health or of just punishment by civil authority for serious crimes, to harm a man in his rights to bodily integrity or well-being by inflicting wounds, acts of violence which impair the strength or beauty, or break the continuity, of his body. Under this principle fall all acts of mutilation. The application of this principle to boxing will be clearly seen through an examination of the studies of MARTLAND, JOKL, CARROLL, and others whose findings have been confirmed by recent medical discoveries and experiments of the nature, the causes, and the effects of brain injury. It need hardly be demonstrated that only a most grave reason could justify one who would inflict a permanent wound, or place an action which by its very nature will very probably produce a permanent wound, upon a man's most delicate vital organ. Where such an action is placed, without a proportionally grave reason, there is mortal sin – at least material."

"Obviously, no amount of money or fame can be considered a good effect proportioned to the evil effect of certain, or even very probable, permanent brain injury. Again, if we may deprive our neighbor of his material goods only when we are in a state of extreme need with no other means of escape, then certainly we may not violate his bodily rights merely for the sake of a large sum of money and a bit of glory. A fighter has, then, no just reason for harming a neighbor in his rights to bodily integrity or well-being by inflicting wounds which certainly impair the strength and beauty, and break the continuity, of his body."

"God alone has direct dominion over man's body; no man has the authority to dispose of his own body just as he wishes. Just as one may not, without a proportionately grave reason, intend to harm another, so no one has the right to allow another to harm him without such a reason. Indeed, one may not permit another to inflict upon him even moderate blows except for reasons of just correction, discipline, or punishment (MERKELBACH 1938). Boxers, then, may not implicitly grant mutual permission to punish one another."

"Boxing may not be regarded as a licit dangerous occupation such as the work of a test pilot, a miner, or an acrobat. Between such occupations and boxing there is a parallel in respect to the good effect (money), but not in respect to the evil effect. For in the dangerous occupations, which are permitted for the sake of money, the evil effect (the danger of physical harm) may not be held in any way as a goal, or as a desirable and most efficient means of obtaining the good effect. Yet, boxing today holds the knock out to be a most desirable way of obtaining the good effect; and the very intention to win a fight, since this is almost always done by a series of head blows, implies the intention of directly inflicting physical harm to the extent that the opponent is unable to defend himself. Moreover, to engage without sin in so-called dangerous occupations, it is required that at least ordinary means be taken to make the danger remote. In boxing it is precisely the hitting on the head that makes proximate the danger of permanent brain injury. Let this proximate danger be removed by ordinary means (e. g., a second foul line at the shoulder level), and the whole nature of modern boxing will be changed."

"The above principles may be readily granted, but the facts remain to be seen. Does a boxer really attempt to inflict serious wounds, and does he frequently succeed? An affirmative answer to this question follows from an examination of (a) the nature of modern pugilism; (b) the nature of the brain structure, and recent data on the causes and effects of brain injuries; and (c) a consideration of the common opinion of the competent medical authorities who have studied boxing injuries."

The Nature of Modern Pugilism

"Modern pugilism, as we have already seen, has as a principal and most desirable objective to inflict such punishment on the opponent's brain that he is no longer able to act as a human being. The objective is frequently attained. This was not the case forty years ago. In seeking the reason for an 'increased frequency and severity of cerebral traumas among pugilists," a writer offers two very plausible suggestions:"

"' ... technical progress in boxing, leading to more and rapid blows; boxers nowadays know that facial and chin blows are more effective than any other, and they concentrate on these. Together with the technical development, a change of the public taste has led to a significant modification of the character of boxing; ...years ago the science of evasion and parry was stressed, nowadays the public wants offensive action to the utmost, without regard to blows received... This technique... exposes the fighter to more traumatisms'" (RAVINA 1937).

Father HILLMAN continues:

"A boxing promoter wants a crowd. So the thing he likes better than a puncher is two punchers, says a writer in a popular boxing magazine. Of a *'grudge fight'* he says: 'There's nothing like it to pack an arena.' *(BUCK A, 'How to pack them in.', Ring, April 1951, p. 19).* Invariably the most hard-hitting fights are regarded as the best. If fights stall or slow up, the crowd boos and urges them to *'mix it up,'* to *'cut out the exhibition,' and* to *'fight.'* Boxing magazines, which naturally try to keep people interested in the sport, are a good indication of the boxing fans' tastes and the current of the sport itself. These magazines are filled with gruesome pictures of the most popular fights in their most bloody moments. Why emphasize this aspect of the contest, unless it is what the fans want? This is the aspect emphasized by sport writers with all their talk of *'fierce,' 'hard-hitting,' 'smashing,' 'slaughtering,' 'savage' blows."*

"It is a standard technique among prizefighters today to try for a technical knock out whenever the opportunity presents itself. This is done by concentrating as many severe blows as possible upon any wound. Often it is an eye cut, a weak or broken nose which is concentrated upon, until a doctor or the referee stops the fight. Frequently this enrages the crowd. What amount of money or what degree of fame can justify these deliberate and direct attempts to aggravate an open wound or to displace an oft-broken nose? Yet they say: 'It's all part of the game!' Although statistics are hard to find on the point, an examination of countless pictures of boxers' faces has revealed only one who does not seem to have a broken nose; and he is very young in the sport. Blindness resulting from a detached or torn retina is all too common a fate of boxers. The nature of this profession seems to make it almost impossible for a man to retain the normal beauty of the human face."

"The fighter today is forced by the desires of the spectators to take the maximum beating. If he confesses that he is hurt and retires, he will be derided by the mob, as is the referee when he stops a fight because there is too much bloodshed or because there is too much 'glassiness' in the eyes of one of the contestants who has already been knocked about sufficiently. Frequently full knock outs occur because a boxer has been so dazed by previous blows that he leaves himself open to a hard and calculated final punch."

"It is a fact that medical attention and state regulations do not protect the boxer from taking a severe beating. For only a severe beating can render a man defenseless; and this happens to be a commonly attained objective of the contest. There are recorded incidents of fights in which examinations and regulations did nothing to prevent men from fighting in spite of severe injuries suffered previously. It will be seen presently that the most serious wounds suffered in a fight are hidden from the best medical examination. Television's boxing fans know this already. Not long ago they saw a physician, after looking into the eyes of a groggy fighter, permit that man to fight on to his death. Nor is the X-ray of any help, aside from cases of bone injuries, in aiding one to estimate the effects of a previous head injury (WILL 1939). Recent experiments on brain-wave alterations resulting from concussions, and the electro-encephalographic findings in the cases of two hundred boxers studied, cast doubt upon the efficiency of electro-encephalograms as a means of making boxing safe."

"It is clear then that a boxer intends to hit his opponent on the head as often and as severely as his skill will permit him. He does this because he wishes to render the man defenseless and he knows that unconsciousness is a most effective means. Now what is the significance of all these severe blows to the head?"

Causes and Effects of Brain Injuries

"The weight behind a boxer's punch is very great. In a test at the *University of Wisconsin* it was shown that a one hundred and forty-five pound amateur is able to deposit an impact of six hundred pounds of pressure on his target (STEINHAUS 1950). In a fight, the actual weight of a punch may be less, or it may be more, depending upon many conditions. However, aside from weight of the blows, it is the very fact of the blows that does the real harm. According to Gene TUNNEY, the real danger in boxing is the aggregate effect of the minor head blows received over the years, which beat a man into a mental wreck. The truth of this is consideration of the nature of the brain which, in the judgment of a brain specialist consulted on this point, is so constructed that it cannot suffer a series of such blows without certainty, or at least very probably, incurring thereby some permanent injury."

"... The cause-and-effect experiments on animals and human brains, the latest deductions from post-mortem findings in cases of brain injury, all go to support the theories of the doctors who have studied the matter of boxing injuries during the past twenty years. Among those who have studied this problem there exist no significant contradictions and no opinions which oppose our contention that boxing, in which the head is the object of severe blows, involves serious permanent wounds to the brain."

"Depending upon the number of these wounds which progressively extend as long as a man continues to fight, and depending upon the exact location of these wounds, there will be perceptible bad effects, especially such ones as impairment of speech, gait, thinking powers, emotional stability, and other forms of 'slowing up.' So, even though there are immediately perceptible bad effects, the fact is certain that some permanent harm has been done to the brain..."

"Hence, it seems almost impossible for a boxer to avoid suffering, in the course of varying periods of time, a degenerative progressive lesion or series of scars in the brain tissue. This may eventually result in the condition known as traumatic encephalitis (punch drunkenness) or it may cause premature death; for there is evidence from autopsies that death may occur from a cranial injury in which there are no other lesions but multiple punctate hemorrhages in the deeper structures of the brain, without the presence of skull fracture or scalp laceration."

The Common Opinion of the Experts

"The competent medical authorities, who have studied the significance of cerebral traumas among pugilists have offered opinions which must be considered seriously by anyone forming a moral judgment on boxing..."

"... We have seen the significance of head punches from the medical viewpoint. Real and serious wounds are inflicted on man's most delicate vital organ, and the natural beauty of the human face is directly attacked by the boxer who aims blows at the head of another in an attempt to win a fight for the sake of money and fame. In such an action there is no just proportion between the bad effect (permanent wounds on the brain with progressive impairment of functions and loss of mental power) and the good effect (a sum of money and a measure of 'popularity'). The moral implications are obvious. A boxer sins when he attempts to land severe blows on the head of his opponent. The matter is grave because of the importance and the delicacy of the organ which is attacked."

Third Argument

"As a corollary of all that has preceded, a third argument presents itself against boxing as we have it today. A study of the physiological facts involved in such boxing has led to the conclusion that modern boxing cannot be reconciled with the principles proceeding from the Fifth Commandment. When a contest or spectacle of such a kind is held primarily for the spectators, something must be said about the morality of being one of the spectators. Boxing fans, by their presence and their enthusiasm, give to the spectacle its reason for existing and its peculiar character. Is it possible to contribute in this way to such an institution without participating to some extent in the sin."

"By his presence at a fight the spectator contributes financially to the contest; for the boxer engages in this business primarily for money which is supplied by the spectators. And by his enthusiasm the spectator positively encourages the boxer in the sordid aspects of his profession; for only the matches which promise 'a good hard fight' will draw 'a good crowd.' This occurrence which the spectator positively sustains, fosters, and encourages the contestants in their immoral actions can hardly be considered anything less than formal moral cooperation in sin."

Father HILLMAN continues:

"And, since the desires of the spectators set the tone of the contest, it is very difficult to see how those who manifest approval of modern pugilism by their presence and their enthusiasm can be from the sin of scandal. Such is the situation of the spectator from the viewpoint of objective sin. In each case, however, the degree of cooperation and scandal must be judged from the circumstances. And, as in the case of the boxers themselves, it does not seem likely that the sins are ordinarily formal."

"Still, something must be said about the morality of being a spectator at a modern prizefight. Aside from the physiological facts, which have been the basis of our considerations up to now, there is something in modern boxing which might be called the psychological element. The sinfulness involved in this element might be treated under the virtue of temperance; but, for our purpose, it may be considered as it touches the Fifth Commandment. For this law of God forbids man to cooperate and take pleasure in the unjust and unnecessary sufferings of His creatures."

"'A man who is angry without being hurt, or with one who has not offended him' is said by Saint THOMAS 'not to be cruel, but to be brutal or savage.' It belongs to his vice of brutality or savagery to take pleasure in the unnecessary sufferings of man. Besides being directly opposed to the gift of piety, this form of pleasure is an emotional perversion, which is directly opposed to the respectful and tender emotions we are obliged to cultivate towards one another. In explaining the Old Testament prohibitions against cruelty to animals Saint THOMAS tells us that the object of such a prohibition is to turn men's minds away from the practice of inflicting needless suffering on animals, lest anyone by exercising cruelty towards animals should also become cruel toward man... God's purpose, often recommending and commending kind treatment of brute creation, is to dispose men to pity and tenderness for one another. Any contest which fosters dispositions exactly the opposite of pity and tenderness for one another is directly opposed to God's purpose."

"Moreover, scholastic theologians rest their condemnation of cruelty to animals primarily on its demoralizing influence. Cock fights, for example, are condemned by some moralists because of their unwholesome influence upon the spectators. Now if the causing of needless animal suffering is evil for this reason, it may be said a fortiori that the spectacle of human beings pummeling one another, needlessly and unjustly, is even more evil for the same reason."

"The fact that the most violent boxing matches bring the largest crowds seems to indicate that for most of the spectators, it is the 'pleasure of a good fight' which principally motivates their presence. This is not to say that the display of skill and the element of human drama may not be motive forces for those who attend fights. The point is that for most fans the unwholesome motives seem dominant; otherwise, the character of boxing would be different. The fights are calculated by the promoter to please the crowd."

Father HILLMAN summarizes:

"Anyone who has attended fights will admit that the animal emotions are dominant; that dispositions of tenderness are opposed; that the spectators are given to a general emotional instability, and are frequently stimulated to the point of frenzy; that the contest easily occasions in the spectators expressions which are akin to hatred and cruelty. For the pleasurable feeling which is fostered by the spectacle of one man's beating another into an inhuman state seems to be an emotion of beastliness which ordinarily proceeds from the vice of savagery or brutality."

"If anything has the power of releasing the beast in man, it is a fight. For this reason it seems inevitable that an unhealthy enthusiasm should be generated among the spectators at a 'good fight.' And if the spectator at such a contest does not actually take pleasure in the needless suffering and the crude atmosphere, he does at least cooperate in it and he places himself in a free absolute proximate occasion of taking sinful pleasure."

"A contest which has as a direct purpose the violation of the human body, and which fosters such crude emotionalism among the spectators, should appear particularly reprehensible to the Christian conscience in times like ours. For when in history has the body of man suffered violation on a larger scale than it does today? And when have human emotions been more wanting in the cultivation of tenderness and pity?"

"The brutishness fostered by boxing appears especially sinister when we consider that there is always some degree of identification of the spectator with the characteristics in a drama; and a fight is perhaps the most powerful means of bringing out the animal in man. In a world so filled with suffering and crudeness what place is there for a form of recreation which deliberately places acts of brutality as a means of pleasure, and positively fosters a perverse emotionalism? Boxing, as we have it today, is badly in need of an apologist."

Editorial *'The Nations most dangerous trade – Professional boxing,' (Industrial Medicine Surgery: 20: 571-572, 1951):*

"Professional boxing is an industry. Professional boxers are players of a trade. The prime objective of professional boxing is money – for the pugilist, for the manager, for the promoter, for advertisers, for taxes. Only euphemistically is boxing a sport. That term 'sport' ever must be preserved. Eliminate the implications of 'sport' and no customer stands in front of the box office. No customers – no boxing industry. Only the naive belief that boxing is even flavored with the forthright, untainted by the sordid, the conniving, back-ring machinations. The stalwart athlete becomes a frail reed in the objectives of ring promoters – and the devil take the hindmost."

"In no other trade is the product palpably 'injury.' The successful bout is always attended by physical damage. The customer pays for, and demands injury. With injury the lot of both participants, the winner is he who best survives injury. Human injury is the sorriest product ever turned out by any industry."

"This Journal continually dedicated to health conservation and promotion in every industry, exercises its avowed function in earmarking professional boxing as the nation's most dangerous trade. The accusing eye is focused on five deaths in New York State alone within the past three years. Consider the total number of professional athletes fighting in New York in that period. That number of deaths, percentage-wise – not in a single mishap such as is the misfortune of some industries, but in five distinct incidents – would have closed many an industry until due safeguards provided some promise of no further catastrophes. Refusing to ignore its sponsorships, this Journal joins in righteous indignation, protesting further blindness of promoters and authoritative controllers of professional boxing, to the obvious that they are engaged in a hazardous trade – attested by the not too pretty sight of shiny, unnecessary, coffins."

"Professional boxing, however distasteful to the many, will continue. Never do those engaged in industrial health protection demand the destruction of any industry solely because it is perilous. Always the objective is to rid any dangerous trade of its disastrous threats. That becomes the objective in this present condemnation of professional boxing as it now performs. The boxer is fully entitled to due safeguards in his trade, whether their absence spring from the boxer himself, the matchmakers, the promoters, unskilled or unscrupulous physicians, or governmental officials. The actuality of hazard permits of no debate. What is wanted is the remedy."

"So long as boxing continues, injury will continue – that's the product. But unnecessary, predictable and anticipatable injury may not be condoned. At least in New York State, legal enactments made provisions for skilled medical protection for boxers with prospective engagements. No medical commission, even with near scientific omnipotence, may foresee every disaster. But long before some recent fatalities, medical foresight warned against undesirable eventualities. Their omens were ignored, their advice belittled – the fight must go on. Disheartened, conscientious physicians tossed in their own sponges – not in futility but in protestation over the manifest neglect of rights of a group of industrial tradesmen to safeguarded work conditions, appropriated to the performance of their work."

"The skilled physician, clothed with authority, is the boxer's best friend."

"What constitutes an adequate medical program of protection? Already that is fairly well known to those physicians guided by long experience in the pugilistic world. Give them a chance to apply that program – to save the boxer for another bout, next quarter, or next year. Further deaths in the squared ring will not be passed over as only the natural risks of the trade – or the will of God. Maybe God is mad, too."

Both of these papers concentrate on professional boxing, as it seems to have been more corrupt than the amateur form of the sport. However, those physicians who have made their feelings known have made it plain that there are very few differences.

Arthur DALEY published an article: *'Is boxing on the ropes. Yes says a veteran critic of the manly art who can remember Tony CANZONIERI as a preliminary boy – and he blames it all on TV. New York Times Magazine, January 31, 1954, p. 19).*

DALEY writes:

"Tex RICKARD was the promotional genius who gave boxing its first million-dollar gate. The cold-eyed gambling man from the Klondyke strove earnestly to attract a better class of people to his fisticuffing productions, preferably folks with blue blood or gold dollars. Those cold eyes glistened at the sight of bejeweled women in evening dress at the ringside."

"RICKARD didn't just seek class because he had no class himself. In his practical, down-to-earth fashion he realized it was better for business that way. Gentlemen in dinner jackets gave more tone to the Busted Beak industry than crum bums in turtle-neck sweaters ever could. 'I never seed so many nice people,' drawled Tex one night about thirty years ago. It was to become his favorite expression, and he was to say it with ever-increasing frequency until the day he died.

"Boxing has never drawn so vast a number of 'nice people' as it draws today. Its steady spectators are in the countless millions. Is boxing therefore thriving as it never thrived before? Nope."

"The television revolution has dislocated all sports in one way or another. But boxing has become its prize captive. No form of athletics reproduces on the video screen with the faithful exactness of boxing – unless you insist on including wrestling, which isn't a sport but a dramatic burlesque of consummate fakery. The TV cameras can't begin to cover all the fringe happenings in baseball or football or basketball or hockey or any other sport. But boxing has no fringes. The ring is small enough to be always in focus. The contestants are the absolute minimum of two. It's the ideal arrangement because every screen is a ringside seat. And the price is perfect – free."

DALEY continues:

"The sponsors don't give away their foaming glasses of beer or the sharpest razor edges ever honed. But they've made boxing a gigantic give-away program. There's a boxing show on TV practically every night in the week and it has created millions of new fans. But none of them ever got in the habit of digging into his jeans for the price of a ticket at the box office. Those free samples more than satisfy their wants. Amateur boxing is deader than MARLEY's ghost. The smaller fight clubs are dying off fast. Only a few big ones are left, most of them controlled by that pugilistic monopoly, the International Boxing Club."

"By the virtually indiscriminate use of television, the major league baseball teams have been killing off the minor leagues. This is akin to eating one's young. But boxing devours its young even faster than baseball does. Once upon a time the Amateur Athletic Union and its various district associations received a high proportion of the revenue from amateur boxing. With considerable regularity, the simon purses drew capacity or near capacity crowds to Madison Square Garden. Even the 'bootleg amateurs' (who were semi-pro fighters) did well in outlying sections of the country."

"But when boxing fans could see genuine professionals for free in the comfort of their homes, they resolutely refused to inconvenience themselves and pay money to watch amateurs. So that phase of the sport is dead. Or if it's alive at all, it would take an expert coroner to detect the flicker of a pulse beat. Amateur boxing was the kindergarten for professional boxing. Then the small clubs were the grade school and the high school, graduating only the brightest, cleverest and best-equipped students to college – Madison Square Garden, the Boston Garden, the Detroit Olympia, Chicago Stadium the San Francisco Cow Palace and other such showplaces."

"All the stars of yesteryear followed the same routine – amateurs, small clubs and the big time. And they had many small clubs were they could learn their trade. New York City, for instance, was loaded with them. There were the Broadway Arena, Ridgewood Grove, Dexter Park, Fort Hamilton, Pioneer, Coney Island Stadium, Henderson Bowl, the Coliseum, Star Casino, Rockland Palace, Manhattan Casino, Lenox, Sunnyside, Queensboro, Eastern Parkway and St. Nick's, among others."

"How many of those are left? Eastern Parkway is still in operation. St. Nick's is used by the Garden when the big arena has to yield time to the circus, rodeo, ice shows or some other major production. All others have closed down. This is not a situation indigenous to New York. It's the same the country over. The little fellow just can't compete with the big fellow any more, especially since the big guy is giving his away on nation-wide networks."

DALEY continues:

"This reporter still can remember one of his earliest assignments for The New York Times more than a quarter of a century ago. He went to the Broadway Arena in Brooklyn and found himself fascinated by a kid from New Orleans, a youngster just up from the amateur ranks. The kid could box, could hit and could take it. So the reporter assumed a proprietary interest in him, watching him move up to Garden preliminary bouts, Garden semi-final and eventually Garden final. There was a vicarious thrill for an awfully young and awfully impressionable reporter when his new hero, Tony CANZONERI, went on to win both featherweight and lightweight championships of the world. CANZONERI was a great little fighter, one of the best. He had been properly schooled in the small fight clubs."

"In boxing, as in life, there are few short cuts to success. Maybe there are fewer short cuts. Hard work is still the only sure road to the mountaintop and only by toiling indefatigably at his job can a man learn it properly. Television has done a double disservice to boxing. It has helped kill off the kindergartens and the secondary schools. So they are uneducated folks who suddenly reach the pugilistic college these days. They've skipped so many grades that they flunk the big test as soon as it comes."

"The story of Chuck DAVEY, television's darling, is the perfect illustration of what happened to boxing. DAVEY had been on the boxing team at Michigan State, a top-flight intercollegiate performer. Then he turned pro. DAVEY was photogenic and he had flash in his style. He was an instant hit on TV. The guys and dolls went gaga over him and thought him the greatest thing the ring had seen since John L. SULLIVAN."

"The dolls liked him, because he looked like the All-American boy. The guys loved him for the superficial class he showed. Not being genuine fight experts themselves, but only the synthetic TV brand, they couldn't recognize superficiality when they saw it. The DAVEY build-up was shrewd and well handled. He was artfully matched against men with styles that made him look better than he was. The big coup was matching him with Rocky GRAZIANO, the one-time dread killer of the ring, but then an over-the-hill washed-up fighter. DAVEY outdanced the lumbering rock."

"So he was matched with Kid GAVILAN for the welterweight championship. The kid from Havana is probably the best all-round ringman in the game today – which may be damning with faint praise. He started fighting in the Cuban canebrakes as a tot and he learned his trade soundly. If you'll pardon a personal allusion. I wrote a column before the fight in which I said the television experts expected DAVEY to beat GAVILAN. 'With what?' I asked. And they went on to say why I thought they didn't have a chance."

"Oh, brother, how the mail filled in! Most of the letters were of the Dear-Sir-you-can. I was a dope, a rock-head, an ignoramus and so on. I wasn't being uncannily prescient. I was merely stating an obvious fact, one that anyone who knew the slightest about boxing couldn't fail to see. GAVILAN, of course, punched DAVEY's ears off."

"When Jimmy CARTER, the lightweight champion bounced Tommy COLLINS off the floor, as though COLLINS were a rubber ball, indignant TV fans almost demanded a Congressional investigation. So violent were their outcries that this actually became front page news. COLLINS, by the way, was just another of those synthetic TV fighters."

"Things used to be different. When Tex RICKARD was in charge of Madison Square Garden's fistic promotions, he didn't break his back to have a show every week. He'd rather wait until a good match came along. When radio took over Friday nights at the Garden, a weekly show was a must. Not all of them were good, but it was no undue drain. Then came television with Wednesday for suds and Friday for blades, both on a coast-to-coast basis. There aren't that many good fights or fighters any more. Boys are brought up before they're ready for quick. They go down as fast as they rise."

"Nor are there as many of them any more. One reason is the disappearance of the small clubs. Another is general economics. Boxing's most overwhelming appeal was to the underprivileged. It offered a chance to leap quickly over racial and financial barriers. Joe LOUIS, a cotton picker, earned millions of dollars. So did Jack DEMPSEY, once a hobo. So did Gene TUNNEY, once a shipping clerk."

"But the nation's economy and way of life are such these days that there are not many underprivileged left. If a kid can earn more than $ 100 a week as a truck driver or bricklayer or something better, boxing has little appeal for him. Why should he train endlessly and then fight once a month for a $ 50 purse he has to split down the middle with his manager?"

DALEY continues:

"Once upon a time, a good manager was the key figure in the development of every fighter. Most of them had been fighters themselves and they knew every trick of the trade. What's more, they could teach it expertly. The few who couldn't, latched on to clever trainers."

"They taught their pupils proper techniques and proper strategies. They taught them how to throw a punch with maximum power, how to roll with the blow, how to sidestep, how to maneuver in the clinches, how to stay in balance, how to piece their punches into telling sequences. There were a thousand and one things they taught them. Some even taught extra refinements – if that's the correct word – such as jabbing a thumb in the other guy's eye or giving him the laces across the face."

"The main point is that they taught them. They might have been crude about it at times but they were effective. For instance, Jack DEMPSEY never used his left hand until Doc KEARNS, his manager, angrily tied the Mauler's right hand to his side and made him practice fighting only with his left. That's how DEMPSEY developed the most murderous left hook the game has ever known. Joe LOUIS had the big punch but little finesse in the beginning. But the wily Jack BLACKBURN taught him well. Before the bomber's second fight with Billy CONN the experts wrote that it was a match between a slugger and a consummate boxer. That nettled the Jolter, 'I kin box pretty good, too,' drawled LOUIS."

"However, the great teachers have almost all faded from the scene. The manager of today is anything but an ex-fighter. Some of them are merely fronts for far more sinister characters whose police records wouldn't permit them to manage fighters openly. And most of them are linked together in a tight cabal, which almost amounts to a syndicate operation of sorts. They give nothing to boxing. They only take from it. Least of all do they give any instruction to their charges. Perhaps that's the most alarming aspect of the situation."

"At the same time, American youth these days is more inclined to seek security that pot of gold at the rainbow's end. They also realize – they're better educated than yesteryears fighters – odds are prohibitively against their winning a championship. Suppose they win a title? Will it pay off?"

"Champions reach such an economic saturation point these days that they can lose money once their income tilts over a certain figure. Please, remember, though, that there are only eight champions of the world and even that is stretching matters as far as importance is concerned. The big one, of course, is the heavyweight title and a colorful performer can make it pay (although taxes nowadays take a big bite) fairly handsomely. But dull champions like Jersey Joe WALCOTT and Ezzard CHARLES couldn't. The middleweight crown can be a valuable asset if the right man wears it. So can welterweight and lightweight laurels. But the other four are not revenue producers of any particular consequence."

DALEY continues:

"The lure of the ring just isn't as compelling for the youth of America as it once was. This, perhaps, is reflected in the roster of the champions of the present. MARCIANO's place in history has yet to be determined. So terrific a hitter is he that he may yet rank with the great ones, although his crudeness as a boxer and his advanced age of 30 militate against him. Most of the other champions are second-rate. Maybe GAVILAN would rank the top ten in his weight division on an all-time basis, but none of the others would. Yet most of these champions dominate their divisions so utterly that there aren't even logical contenders in sight."

"Although it may seem at the moment as though boxing is dying of pernicious anemia, that's a wrong conclusion. It's too primitive, too fundamental, too tough ever to die."

"Perhaps it will get the infusion it needs as pay-as-you-go television should replace the give-away television of the present. The shoddy match-making of today would have to yield to competitive match-making. The small clubs might even get back into business, each shooting for the customer's dollar or quarter or whatever it is he'd put in the coin machine."

"It could be a throwback of sorts, to the past. The little clubs built up the fighters and matched them shrewdly with an eye toward the gate receipts. A hot attraction would draw a packed house. A dud would leave it empty. So there rarely was a dud. Pay-as-you-go television would be the same thing, except on a larger scale."

"Rare were the nights when the Garden drew an attendance of less that 10.000 spectators. But the Garden has become a ghost arena on fight night. It's now an oversized television studio. Not long ago, for instance, a Friday night fight attracted – if that's the word – 2.991 spectators."

"And folks wonder what's happened to boxing? This was merely a fight to fill the provisions of the contract with the television sponsor. But if the Garden had to compete for the customer's dollar on a pay-as-you-go basis, there'd be no clinkers. With more fights in more clubs for more money, interest in ring careers might be revived among the youth of the land."

"The blight of give-away television is stunting boxing's growth and making it wither. The promoters, ostrichlike, refuse to admit what's happening. But there can be some significance in the fact that there is a new trend among promoters. Whenever there's an out-of-the-ordinary fight, the city where the show originates is blacked out to TV."

"The professional football people discovered this first and made it work. Baseball is edging until the day may come when all home baseball games will be blacked out. Only the away games will be televised, a policy that the Cleveland Indians will inaugurate next season."

"It seems utterly ironical that there should be more fans and more interest in boxing than ever before in history while there is a shrinkage in sites, competitors and caliber of performance. Maybe pay-as-you-go television will be the ring's salvation. This give-away method of operation simply doesn't work."

The eminent English ophthalmologist DOGGART, an eloquent, witty, outspoken and vigorous opponent of boxing, has published numerous papers on the effects of boxing on the eye. One of these *'Fisticuffs and the Visual Organs'* (1955) is a beautifully written condemnation.

DOGGART began:

"So deeply implanted is the fighting instinct that if a school boy were to inveigh against boxing, he would probably withered by the sidelong glances of his more pugnacious contemporaries, who cling to the belief that some disputes lie too deep to allow settlement otherwise than by fist. They even maintain, until experience has supplied innumerable instances to the contrary, that right is stronger than might in the realm of fisticuffs. If a boy is to qualify as a popular hero, then, as W. S. GILBERT *(H.M.S. Pinaforte) says*, 'His bosom should heave and his heart should glow, and his fist ever ready for a knock down blow.'"

"Literature has raised the status of boxing by vivid descriptions in which cads and bullies are usually overcome by fair haired striplings, even though the later weigh several stone less than their aggressors. Conan DOYLE, who was himself an enthusiastic wielder of gloves, enriched the fiction of boxing. Millions of readers have been thrilled by Jack LONDON..."

A little later in the paper, DOGGART noted:

"Most boxers either enjoy their pastime, or else look upon it mainly as a ladder to fame and wealth. The average spectator would claim to be chiefly preoccupied with boxing technique, but it seems probably that many are more motivated by blood-lust. At any rate they adopt the human device of rationalizing their own brand of pleasure. Boxing, we are told, promotes pluck and endurance. It qualifies a man to protect the weak and stand up for himself. These devotees will proceed to explain, with the aid of statistics, that fists are scarcely more dangerous than a ping pong ball. Slight cuts on the forehead, occasionally the chipping of a tooth, bruised eyelids that soon revert to normal, are the only things to happen (sic), except perhaps a broken nose, but then flat profiles are as good as medals to the initiated."

His conclusion:

"In spite of all assertions to the contrary by gladiatorially minded sentimentalists, facts will out. Medical literature abounds with records of havoc wrought by boxing, and it seems clear that sight, sanity and life itself are endangered when a man pursues this sport over a period of years, especially in the professional ring. We know that spectators at a boxing contest emit howls of joy when the thud of the gloves upon flesh sends them into an ecstasy. Let us hope that their jubilation would be tempered if only they were to think of the hazards inseparable from a boxing career."

"Boxing devotees are no exception to the rule that men can always evolve a set of reasons to justify their favorite pleasure; and even if their arguments be exposed as fallacies, those who love a fight will readily produce alternative pleas to support their craze. Before we look at some of the conventional arguments for boxing, let it be emphasized that many of us possess firsthand evidence against HAISLET's contention that the sport is not dangerous. 'The injuries that do occur,' he says. 'are minor and are not serious'. According to him, there are 'no permanent scars or injuries.' Anyone who can believe those statements in defiance of the facts has no right to despise an ostrich for burying his head in the sand. The literature contains many records of eyes wrecked by boxing, and we must also remember that the eyes and their adnexa, though necessary for vision, are nevertheless incapable of appreciating images or making visual judgments. The act of seeing is mediated by almost inconceivable complicated cells and fibers extending through that organ which every boxer's opponent strives to knock out of action."

He went on to present a fairly comprehensive argument:

"When people offer reasons in justification of boxing, they usually assume, or pretend to assume, that the risk of serious mental and physical damage is negligible. Some of them carry wishful thinking a stage further and persuade themselves that the fair-headed stripling with right on his side will usually beat the bully ruffian even if the latter resorts to foul methods. Many people also proclaim the usefulness of boxing knowledge as a means of defense against blackguards and skubberdegullions, but unfortunately, the efficient thug does not stand facing his intended victim in a good light, waiting to be hit on the side of the chin. If his fists come into the picture at all, they will be buttressed with knuckle dusters; but probably he will use a knife, a revolver, or at least a spanner. Moreover, jiu-jitsu experts and trained wrestlers without adventitious weapons can soon put a boxer out of action. Perhaps the most curious argument ever advanced in favor of fisticuffs is that of George BORROW. In his gypsy book *'The Romany 'ye'* he calls it a fine, manly, English art, and a great defense against Popery,' but of course he lived in the days when almost every dispute was assessed in relation to theology."

"Extremists have maintained that a certain amount of ocular damage does not matter, because defective sight is not incompatible with prowess in the ring. Did not Slugger WHITE's artificial eye pop out on the floor while he was being medically examined before a fight in San Francisco? And what about the Boston Tar Baby Sam LANGFORD? Vision in his right eye was far below normal when the other eye went blind in 1917, but he went on fighting until one day in 1924, when he was observed wandering vaguely about the ring looking for the opponent he had knocked down."

"Humor is a useful brush for whitewashing the ring. Boxing fans gleefully fasten upon items of damage arising in the course of more sedate pastimes. S'pose that a boxer is flipped on the eye by a Ping-pong ball. 'Oh yes,' they will say, wagging their heads in mock solemnity, 'he'd much better have stuck to boxing. It's so much safer.' Then they collect amusing examples of players being accidentally hit on the head by a croquet mallet or knocked unconscious on the tennis court. Football supplies them with more solid ammunition, because there is undoubtedly a long list of injuries produced by this game, even if we exclude damage arising from foul play. In this connection, however, we should remember three things. First, the number of footballers vastly exceeds the combined total of amateur and professional boxers. Secondly, the rules of football forbid any deliberate infliction of injury, whereas the prime object of boxing is to knock your opponent unconscious. Thirdly, the damage sustained at football, even if it be permanent, is far more likely to be located in the limbs than in the head. From time to time footballers are concussed, and very rarely there has been a death; but this game creates no genuine counterpart to the slap-happy veterans of the ring."

"Clearly, there is some truth in the contention that boxing promotes endurance, quick footedness, and muscular development. On the other hand, many other forms of sport offer the same or greater benefits without the disadvantages inherent in pugilism. One objectionable feature of boxing is that it tends to develop the upper half of the body out of proportion to the lower. Furthermore, the intensive bouts of exercise, together with drastic rapid alterations in weight necessitated by an impending fight, often drive boxers to fling up the sponge and abandon asceticism as soon as they go out to grass. Although there are several noteworthy examples to the contrary, it is evident that retired boxers are only too liable to sag."

"We have already seen that the art of boxing does not qualify a man to protect others or himself from jiu-jitsu exponents or efficient thugs. Nor should boxing be countenanced as a form of exercise if in fact it entails unreasonable risks."

DOGGART (1955) in his *'Main Conclusions About Boxing Amateur and Professional'* concluded that:

"... many people suppose that the major hazards of boxing are confined to professionals. They admit that the careerist boxer is in grave danger, especially if he continues to fight after the age of 30, but they maintain that amateur boxing is practically safe."

"If all boxers before every contest were submitted to thorough clinical examination, together with elaborate laboratory, radiographic, ophthalmoscopic, and other special investigations, and if all these precautionary tests were to yield negative results, the boxer might still be damaged for life, or fatally injured in the first round. Admittedly, medical supervision can pick out various unequivocal defects, and exclude from championship fighting the manifest cripples and morons, but it cannot alter the anatomy of the human skull or lessen the delicacy of those organs which subserve our highest faculties. All medically qualified people have had the opportunity to dissect the head and neck. These are not only targets for disabling blows, together with a most fragile set of sense organs, a sequence of delicate nerves, and a number of other structures nourished by richly anastomosing blood vessels. The very thought of setting out to smash all this artistry with fists is sacrilege, not sport, and it behooves us to speak out against such barbarism."

Listen also to the testimony of LAURIE (1955), who has attended more than 2.000 bouts as medical officer to a prominent boxing club in Glasgow:

"Far from being a useful and justifiable sport, amateur boxing is a dangerous, and often brutal, happening."

LAURIE goes on to ask whether his nation is in a position to afford this type of disability production; he continued:

"It may be contended that the total damage produced by the boxing in a whole decade could not match the immediate devastation from one small hydrogen bomb, but such an argument ignores the baneful influence of boxing upon fans and spectators. For every onlooker who takes a genuine interest in the technique of the game, there must be many who consciously or subconsciously utilize boxing to foster their sadistic impulses. Otherwise, how can we explain the bloodthirsty howls with which a championship fight is so often punctuated? How else can we account for the specious arguments put forward to bolster pastime which will soon, we hope, be as obsolete as galley slaves or the chain gang."

The well-known sport journalist Paul GALLICO (*The Daily Mail, September 6, 1955*) spent many years organizing, promoting, watching, popularizing, and writing about boxing. He summarizes his opinion about boxing, as follows:

"If I were a dictator I would abolish prize-fighting in my country by decree. I would scrap all rings, burn all boxing gloves, and never let a youth be taught to strike another with his fists, for prize-fighting and boxing are stupid, senseless, unappetizing, inefficient, and 100 percent useless."

"Most of the things said or written about boxing – that it is healthful, manly, sporting, virile, scientific, and good for the development of character and self-reliance – are untrue."

"I have covered boxing, promoted boxing, watched it, thought about it, and after long reflection I cannot find a single thing that is good about it from the point of view of either participant or spectator."

"It brings out the worst in human beings."

"As entertainment it is uncivilized; as exercise it is contradictory and ridiculous; as a means of self-defence it is wholly absurd..."

"It is known that serious lesions in the brain, leading to eventual paralysis and even death, as well as damage to the optic nerves causing blindness, can and do result from these blows.

"It is further known that a light blow delivered low can render a man as quickly hors de combat, flooring him and causing him to lose all further interest in fighting, but without doing him any permanent injury leading to darkness, imbecility, or the grave."

"Yet the low punch is barred and called foul and when used causes the sportsmen among the spectators to seethe with indignation and cry out in anger, demanding punishment for the guilty party while every wallop to the head and jaw, eyes, nose, and ears, all of the delicate organs, is hailed with delight and cheered, particularly when those blows bruise, maim, cut and tear."

"I reproach myself when I recall that I invented and started the Golden Gloves Amateur Boxing Tournament and, by glamorizing and publicizing an useless, dangerous sport, am responsible for hundreds of young men becoming professional prize fighters. If there is a chapter in my life I would care to erase, it is this one."

"Prize-fighting, amateur or professional, was, is, and always will be a low, tawdry, greedy game."

"There is no magic about prize-fighting that rescues the lost soul. Kids can be just as much interested in football, hockey, tennis, cricket, and athletics. What they want is the sense of belonging, of being wanted and needed."

"Any team game is immediately the superior, in social work, of boxing."

"Boxing is the most selfish and self-centered sport there is."

"Beating a man into insensibility with a padded fist takes time. The first rule of the sport is to weaken the opponent by belaboring him in painful and vulnerable spots such as the kidneys, the stomach, the solar plexus, and the heart, as well as the head. If a cut or abrasion appears on the face, this must be enlarged. A broken nose is to be hammered, a torn ear further lacerated."

"If all this is fair and above board, consider what the shadier methods must be and are."

"To sum up: prize-fighting and boxing have no value whatsoever in our civilization, ethically, morally, or spiritually."

Billy GRAHAM (1955), a former welterweight contender, now a sales promotion executive for a liquor distributing firm had 131 professional fights and was never knocked out:

"When you go the distance that many times, you absorb more head blows than the fighter who takes one on the button in the early rounds and goes down."

"He estimated that he had received an average of 15 punches to the head in every three minutes of fighting. "I'm not pretty," he says, '"but I think I function pretty well."

"Fitness is not the only long-range concern. With the leveling of the educational process, authorities close to the sport believe the reservoir of hungry, footloose, and educated young men necessary to maintain the professional sport is rapidly being depleted. Certainly, amateur boxing appears to be hitting the canvas. Thomas B. QUIGLEY, Harvard University's medical director, notes that leading British schools – Charterhouse, Gordonstown, and Eton no longer teach the sport and interest is waning at Oxford."

CARROLL (1956) wrote:

"It is probable that no head blow is taken with impunity and that each knock out causes definite and irreparable damage."

MACKAY (1956) noted:

"To defend boxing is to defend concussion and brain injury, and to say that it does no harm is to blind ourselves to what is quite obviously a fact. We shouldn't need a lot of statistics to prove to us that to knock a man out several times in the course of a few years of his life is apt to cause some harm. Certainly we don't think concussion is a good thing for people in other walks of life, and I can't see why we should feel obliged to defend concussion when it takes place in a sport."

DeVOE (1959) wrote:

"In boxing immediate death due to injury is rare. Abolition of this sport would constitute an infringement of individual liberty... "

The well-known neuropathologist NEUBÜRGER (1960), who published pathomorphological findings on two boxers, asked:

"Should we, then, favor abolition of boxing? This is a question of philosophy and, of course, of business interests. Leaning aside the latter, of which we don't pretend to understand anything, we feel that the responsible parties should realize that professional boxing is a potentially very dangerous endeavor... Nevertheless, we are convinced that boxing fans, promoters, referees, and advertisers would be shocked if they would check on the mental condition of some of their pugilists 15 or 20 years after the termination of their boxing careers... the characteristic history of repeated battering, at first with no after effects, but later with progressive blunting of everything that matters most in the human mind, and the final mental wreck."

Recently *Sports Illustrated ('Is professional boxing immoral?')* invited Father Richard A. McCORMICK, S.J., the eminent Catholic moral theologian and teacher, to discuss the moral aspects of professional boxing. Here is his considered judgment:

"Professional boxing is a part of us. Yet every now and then a tragedy (such as the recent death of Benny PARET) shocks us into enquiry. It revises and reveals the morality of professional boxing as a legitimate question. This is in some ways unfortunate. The outbursts surrounding tragedy tend to obscure the real issue by focusing exclusively on fatalities. They also provoke us to continue to think with our hearts rather than our heads. Rarely has morality been clarified in such an atmosphere."

"Boxing can be and has been defined as a giving and parrying of light blows with no intention of striking the opponent severely. If no one has ever questioned the morality of this type of thing, neither has anyone ever thought it a realistic definition of modern professional boxing. Recent moral theologians who have reflected on the matter wisely restrict their considerations to 'professional boxing as it is today.' When the theologian says 'as it is today,' he is trying to highlight an existing situation, perhaps not an inevitable one. Some, possibly many, elements of professional boxing could be radically altered, in which case it is quite conceivable that a different moral evaluation of the sport would have to be made."

"By using the phrase boxing as it is today the theologian does not mean to concentrate on the fight-for-pay element which distinguishes amateur from professional; his intention is to emphasize characteristics of professional boxing once the distinction has been made. He is trying to paint a picture in a single phrase. Among these characteristics there is the element of a career involving a whole series of fights with cumulative effects. There is the admitted effort of most professionals to win by a k. o. – or at least a t. k. o. – rather that by decision. There is the medical report of injury, particularly to the brain. There is the synthetic notion of courage, wherein confession of injury followed by retirement from a fight invites decision by a crowd that enjoys a beating, clamors for the kill, and lustily boos evasive tactics. There are the undeniable benefits that boxing has brought to the lives of many individuals. There are television contracts which create severe scheduling demands; there are boxing commissions and control groups. Finally, there is a specific set of rules. Professional boxing involves more and longer rounds, lighter gloves and sometimes different scoring criteria. These are the things the moralist attempts to evoke with the phrase professional boxing as it is today. It is not an individual fight that is his immediate concern. Individual fights may not contain the elements widely present in the sport as a whole. Nor is his concern boxing at the level of the Golden Gloves, the CYO and the private club. Still less is it a judgment of the individual fighter and his motives. It is a whole institution as it touches human conduct."

"The defenders of professional boxing regard boxing as a science demanding skill, strength and discipline. In boxing there is splendid opportunity for physical development, alertness, poise, confidence, sportsmansship, initiative and character-building in general. Statistically professional boxing is, they point out, far less dangerous than auto racing, college football and several other sports. Furthermore, the game has given underprivileged youngsters a chance to better themselves. In summary, the advantages outweigh the disadvantages."

"With an eye to these claims, some earlier moral evaluations of professional boxing were at times relatively tolerant. In fairness to these earlier views, it must be pointed out that they were formulated before widespread publication of pertinent medical findings. In fairness to professional boxing, however, it should be said that even those who now regard the sport as immoral concede the above advantages. Their objections are elsewhere."

"The application of immutable moral principle will vary with the variation of concrete fact or its understanding. Thus in the past twenty years or so there has been a growing consensus among theologians that the sport will not survive moral scrutiny. The three most recent American studies (HILLMAN 1951, BERNARD, LAFORET 1962) conclude that the current version of professional boxing is immoral. Most moral theologians would endorse and defend this position, not as the official position of the Catholic Church (the Church has never spoken officially on the matter) but as their own conviction after thoughtful application of their principles to the facts as they see them. If they have been less than enthusiastic about publicizing their conviction it is not because of reluctance to take publicly an unpopular stand. That would be cowardice. Rather it is because the conviction has matured slowly and painfully and because even now some uncertainties still cling to it. But as the subject receives intensified study, it is increasingly difficult to find defenders of the sport among theologians."

"Professional boxing is unique among the sports. It is admittedly the only sport whose primary objective toward victory is to batter and damage an opponent into helplessness and the incapacity to continue. In a sport where the infliction of damage is rewarded, one would expect a wide variety of injuries."

Father McCORMICK quoted that ophthalmic injury is far from unknown even to the extent of actual blindness.

"Maxillofacial and aural trauma, including damage to the jaw, teeth, nose, and hearing apparatus, are more common. Boxer's nose and cauliflower ear are commonplace."

Father McCORMICK mentioned here that there is the possibility of brain damage.

"While these and other types of injury do occur, it is cranio-cerebral injury that recently has engrossed the attention of the medical world. Because of the premium placed on the k. o. and the t. k. o., the head has always been the prime target in professional boxing. Blows directed to the head or face comprise about 85% of all blows delivered in the ordinary bout. Body blows are principally diversionary tactics to lay open this prime target. The injuries caused by head blows have provided excellent opportunity for medical investigation because, as noted in *'The Lancet'* (1937), 'unlike accidents these injuries are caused by traumas almost always of the same kind and acting with almost laboratory exactness.'"

"Scientists indicate that the human brain weighs about three pounds. It is fluid-packed but not secured within the skull. A blow to the head causes it to wobble, slide and bounce back and forth inside its cranial container. If a moderate blow can bang the brain against its sidewall, a more severe blow can bring it into contact with the bony sphenoidal ridge, to produce selective damage to the frontal lobes, either bleeding or bruising. Where there is destruction of nerve cells the damage is permanent and, when repeated, cumulative."

"Medical scientists also call attention to another injury not infrequently suffered by boxers: the punctate (small) hemorrhages in the pons and medulla, probably caused by the jamming of cerebrospinal fluid. Again, where such hemorrhages destroy nerve tissue the damage is permanent, though this need not imply that malfunction of the brain ensues. Such symptom would be a matter of extent and degree. The possibilities for brain damage appear to be as multiple as the organ is delicate."

"What are the noticeable results of brain injury? The most sensational, if not the most tragic, is death, generally associated with hemorrhage. Depending on how one reads statistics, will one conclude with Dr. Arthur H. STEINHAUS, former chief of the *Division of Physical Education and Health Activities of the U.S. Office of Education,* that 'Professional boxing is 83 times more deadly than high school football and 50 times more deadly than college football?'"

"Or with T. A. GONZALES, that '32 years of boxing competitions... have produced fewer deaths in proportion to the number of participants than occur in baseball or football? The point is not clear."

"But if death is a relative rarity, the same does not seem to be true of brain damage. In 1928 H. S. MARTLAND concluded his discussion of the punchdrunk syndrome with the statement that 50% of fighters will develop the condition in mild or severe form if they stay in the game long enough, and that this 'seems to be good evidence that some special brain injury due to their occupation exists'."

The paper of Harrison MARTLAND is a classic.

Father McCORMICK quoted CARROLL (1936):

"...who came to know fighters intimately through professional and nonprofessional contacts, estimated that after five years of boxing 60% of the boxers will develop mental and emotional changes which are obvious to people who know them. He stated that 'no head blow is taken with impunity and ... each knock out causes definite and irreparable damage'."

The brilliantly written contribution of CARROLL is quoted and discussed in detail on p. 301.

"If such trauma is repeated for a long period, it is inevitable that nerve cell insufficiency will develop ultimately... The recent work of LA CAVA in Italy and PAMPUS in Bonn tends to substantiate these claims."
"Findings such as these received fresh emphasis by sparring partner Ben SKELTON's report (*Sports Illustrated, September 24)* that LISTON's left jab is so hard 'that for a week after being hit with it, I was taking pills to kill the pain.'"

Father McCORMICK once again quoted STEINHAUS (1957) who had been:

"so impressed with the medical evidence concerning brain damage in boxing that he feels a second foul line must be created at the shoulder'. He cites a noted brain surgeon with wide experience with boxers as contending that every head-pummeling is likely to leave some small portions of the brain tissue permanently damaged, even though this may not be noticed for some time. The treacherous aspect of such injury is that it apparently does not manifest itself clinically until rather late in the degenerative process. Furthermore, there are obvious reasons why professional fighters would be reluctant to report symptoms of brain damage."

Father McCORMICK continues:

"When one reads these statements – and there are many more of the same – one has an indefinable sense of uneasiness, of inconclusiveness. There is almost the sense of being in the presence of a crusader. Is it really this bad? Could it be that the admirable tendency of the doctor to regard any disease or injury as too much has expanded these statements?"

"With this in mind, I submitted the following statement, attributed to a prominent brain specialist, to ten of the top neurosurgeons of the U.S. and Canada: 'The brain is so constructed that I cannot suffer a series of head blows over the years in boxing without certainly or at least very probably incurring thereby some permanent injury.' These experts agreed that the statement could be endorsed as a general statement, while it is probably correct, is poorly written. He could not accept the inference in it that malfunction of the brain follows brain damage. Such a symptom would be a matter of degree."
"If these specialists are incorrect in their estimate about brain damage, then the moral theologian would desire to reexamine certain aspects or emphases of his argumentation, as we shall see. But it is this type of evidence that makes one take a long look at the words of Abe SIMON, former heavyweight contender: '... jarring of the brain. That what causes the trouble – my headaches and those of every fighter who has taken punishment. It's not a single punch; it's the constant jarring... [The fighter] is always soothed by the falsehood that he will be just as good as new after a short rest. He never is, and no fighter living today who has had fifty or more reasonably hard fights can honestly make the claim.'"
"Such a medical review was necessary preparation for a moral estimate. Since everyone familiar with the sport concedes its advantages, the moral discussion boils down to this: Are the arguments against professional boxing conclusive? Of the many moral objections one hears, the most serious are reducible to three."
(1) *"The knock out.* It is simple unrealistic to deny that most professional fighters aim for a knock out. This is regarded as the most decisive and impressive way to win a fight. It is what the fighter wants and very often what the fans want. The long climb to contender status or the comeback often hinges on it."

Father McCORMICK cited Nat FLEISCHER who wrote in the 'Heavyweight Championship of Joe LOUIS' comeback tour after his 1936 loss to Max SCHMELING:

"There was only one way to do that – to roll up victory after victory over the knock out route."
"Not a few moral theologians find it difficult to admit that the knock out is justifiable. They frequently formulate this as follows: directly and violently to deprive oneself or another of the use of reason is morally reprehensible except for a sufficient cause. It is the rational faculties, intellect and free will that distinguish man from the brute. Directly to deprive man of these faculties, without sufficient reason, is to dehumanize. These theologians are reluctant to admit that sport, money, fame qualify as sufficient reason. If such violent deprivation of higher controls is reprehensible, then the intent to do so is

equally reprehensible. Hence a sport in which this intent plays such an integral role must be condemned."

"Is the argument convincing? I do not believe it is. First of all, the knock out is understood in the rather limited sense or 'rendering unconscious'. This is not a necessary sense of the word. A knock out is, more realistically, beating a fighter to the point where he is physically incapable of continuing. This is what the ordinary professional desires. Deprivation of the use of reason is not essential to this. Hence, practically, it is hard to show that the knock out in this limited sense is an essential aim of most fighters. Second, even if it were the fighter's aim, it would be difficult to show how the knock out of itself (independent of injury) is sufficient to condemn the sport. It can be argued that, generally, deprivation of the use of reason lasts only a few seconds at most (eight to ten) and that this is so little as to be negligible. If this were the only thing at stake, it is highly doubtful that there would be as much objection to boxing as there seems to be."

(2) *"The intent of injury.* If the argument concerning the knock out is not satisfying, the objection from injury is much more arresting. Professional boxing is the only sport where the immediate objective is to damage the opponent. A puffed or cut eye, a lacerated cheek, a bleeding nose – these are signals for an intensified attack on the vulnerable area. When Jimmy DOYLE died after being knocked out by Sugar Ray ROBINSON, ROBINSON was asked if he noticed that DOYLE was in trouble. He is widely quoted as answering, 'Getting him into trouble is my business.' In all other sports the immediate objective is to cross a goal line, tip in a basket, throw a strike. Injury and incapacity to continue are incidental. A knee to the groin, a fist to the face in football, bean balls and deliberate spiking in baseball are penalized and would unhesitatingly be branded as immoral by the theologian."

Father McCORMICK mentioned PATTERSON, who described the unique character of professional boxing when he wrote in his book *'Victory over Myself:'*

"[after] the first JOHANSSON knock out, of his desire never to be vicious again: 'At the same time I know that I must be, because I am in a business of violence.' If direct damage to the opponent is immoral in all other sports, why not in this business of violence?"

"It is here that the medical evidence assumes some importance. Were the injury passing and negligible, theologians might perhaps mitigate their judgment. But if the specialists are right in their claims about injury, particularly brain injury, this must give us pause. The sport as now practiced tends directly to inflict this damage. When injury to the cranium and its contents occur, it is, as BLONSTEIN and CLARKE (1954) note, 'a direct product of boxing and not an accident as in all other sports.' Since this is true, then the fighter's explicit desire is to maim or cause lasting damage. Few would be that inhuman. As a rule, the fighter's only explicit desire is to win as decisively as possible. But the means he chooses are means that are dangerous. Hence he implicitly intends this damage as a means. How could he disown it? The point might seem a bit fine, but can one choose to pound and sink a nail and yet disown the hole in the wood?"

"At this point professional boxing encounters the disapproving frown of many a moralist. Man, they argue, does not possess the right directly to inflict damage on himself or another in this way. He is not the absolute master of his person with the power to destroy or mutilate as he wishes. Absolute dominion over man's integrity is possessed by God alone. As a creature, man is an administrator charged with the duty and privilege of reasonable administration. His ability directly to mutilate himself is severely limited."

"This is a cardinal principle of sound moral thought. If there is indecisiveness here, there will inevitably be ambiguity or error in the evaluation of many aspects of modern living. Once the limit on man's ability to mutilate himself is obscured, the condemnations of suicide, euthanasia, reckless medical experiment, useless surgery and so on tends to lose rational defense. The novelist knows that the first chapters profoundly affect the outcome of the final chapters of his book. Similarly moral theology is jealous of her basic principles because they contain the germ of practical conclusions."

"Applying these principles, theologians believe that when a man pounds another into helplessness, scars his face, smashes his nose, jars his brain and exposes it to lasting damage, or when he enters a contest where this could happen to him, he has surpassed the bounds of reasonable stewardship of the human person. Surely there are equally – or more effective ways for men to learn the art of self-defense."

"Does the fact that this is done for money affect the moral analysis? Certain medical experiments on the human body, even if done for money, would remain objectionable. In fact, is there not a legitimate sense in which it is true to say that the greater the spoils, the more objectionable the whole business? For as the cash at stake increases, so does the danger of viewing the integrity of the human person as salable at a price. Many can be overvalued. When it is, something else is undervalued. If this something else happens to be the integrity of the human person, have we not made a wrong turn somewhere?"

(3) *"Fostering of brutish instincts.* Man is a delicate combination, midway between animal and angel, with a bit of both in him. His characteristic balance is achieved when he harmonizes these elements. When he fosters one to the neglect of the other, he tends to become either a disengaged dreamer or a savage. Thomas AQUINAS knew nothing of professional boxing, but with an unerring knowledge of human nature he pointed out that to take pleasure in the unnecessary sufferings of another man is brutish."

"Anyone who has watched professional fights will know what AQUINAS was talking about. The crowd too often has come for blood and the knock out. The knock out is the touchdown pass, the home run of boxing. The nearer it is, the more frenzied the howling of the crowd. As Nat FLEISCHER said simply of the first PATTERSON – JOHANSSON fight: 'The crowd, sensing the kill, went wild.' We occasionally hear the referee urge the boys to mix it up, give the fans their money's worth. When a boy is being mauled around the ring, the arena comes alive and emotions run high. The fighter is goaded by the crowd; his own fury further stimulates them. The brutish instinct is in command. At this pitch that finest moves in boxing are missed – or worse – greeted by a chorus of hissing and booing. TUNNEY was so disgusted with this type of thing in one of his fights that he created the phrase, 'the bloodthirsty yap of the mob.' The modern prizefight is increasingly the canonization of brute force – and that at a time when we are struggling with all our might to understand the meaning of force in the world."

"Is not man too weak a creature to unleash and give free play to these forces with impunity? Does he not tend to grow in the image of that which he cheers? If this is true, how long can he cheer these exhibitions without acting at variance with the demands dictated by his own rational nature? To many, this is the strongest indictment of professional boxing, an objection sufficient in itself as a stricture of the game."

"These arguments are not frivolous. Any discussion from natural law; whatever validity they have would surly be intensified by the Christian revelation through which man becomes conscious of an even more starting personal dignity. It was probably arguments such as these that led the Vatican Radio to announce its conviction that professional boxing is objectively immoral. *L'Oservatore della Domenica* insinuated the same thing. Informed Catholics, however, are rightly distressed at the implication in the assertion that these views are 'semiofficial' ecclesiastical positions."

"The Catholic Church has not condemned professional boxing. Many have wondered why not. The eager expectation of ecclesiastical intervention could easily contain a distorted notion of the functions of the Church. While she jealously guards the purity of morals, it does not follow that condemnations do or should issue from her at the slightest provocation – if for no other reason than that this discourages intellectual effort in the ranks by seeming to render it unnecessary. We stand to learn a great deal from this controversy."

"Many reasons suggest themselves in explanation of the Church's official silence on the matter. First of all, and most important perhaps, the matter simply is not clear to her. While the Church speaks frequently on changeless moral principles, she is generally quite content to leave the application of these to her theologians. But theological opinion has been, possible still is, somewhat divided, or at least hesitant. Most of the serious writing has been unfavorable to boxing, but there are many voices yet to be heard. "

"Second, professional boxing is largely, but not exclusively, a local American problem. The U.S. champion is the world champion, the big gate is here and the big fights are generally here. If public statements are called for, it is reasonable to think that this would be left to local bishops."

"Finally, even should the Church desire to take a strong stand, there is the difficulty of formulating a statement which will avoid the impression that all boxing is being censored. It is foolish to lump the pillow fights of the sixth-grader with the hard smashes of the professional. And even at the professional level the differences between individual fighters are tremendous. There are those in superb condition, who fight once or twice a year to defend a title – and these are the champions who are hit least. Then there are those who all but drag themselves into the ring to have their brains rattled on a month-to-month basis. To group these together in a single sweeping rejection would be unrealistic and hazardous."

"Not only is the central moral issue challenging: there are also many fringe problems no less tantalizing. One that is increasingly aired: Is the victorious fighter guilty when his opponent dies as a result of blows received in the ring? Though boxing is different from other sports in its direct aim (the infliction of damage), death is such a departure from the average that its occurrence should be regarded as an undesired byproduct of the sport. Morally it is an accident."

"Problematic too are the possibilities involved in allowing a man with a past to contend for the crown (*Sports Illustrated, February 12).* The issue is scarcely one of Christian forgiveness or rehabilitation. Surely we can hope that we are both humble enough and large enough for this. The problem is rather the defenselessness of our children against their own hero-worshipping simplicity. On the other hand, a clean break with an unfortunate past might actually provide a very helpful example to youngsters. Whatever the answer may be, there is a moral dimension even here."

"The question of professional boxing is a vexing one. The issues seem clear. Defenders of the sport agree that the advantages outweigh the disadvantages. Those who censure it, while admitting the advantages, believe that the moral discussion must begin with the sport itself, not only the circumstances. They see the sport as directly injurious and as one which tends unduly to foster the instinct of brutality in all concerned. Perhaps this is not necessarily true: there are many laudable attempts being made to supervise the sport more thoroughly (*Sports Illustrated, April 23*). Nor is it factually true of all professional fights; but it is too generally true of the sport today. Thus the majority vote among those who have written on the moral questions is unfavorable."

Father McCORMICK summarized:

"Unless the arguments leveled at professional boxing as it is today can be answered, I believe the sport would have to be labeled immoral. I realize that other theologians may take a different point of view. It could be that not all the facts are in. Perhaps, too, we have a great deal to learn about our own principles. Premature conviction slams the door to enlightenment as effectively as refusal to face the moral issue."

"If there remain some uncertainties to haunt us, the general implications of our sincere and honest interest are clear. For, regardless of what answer we come up with, it is both a sign and guarantee of our biding spiritual health to face issues at their moral root. It is never easy to question the moral character of our own pleasure and entertainment. Since, however, moral issues are not defined by the convenient and inconvenient, the pleasant and the annoying, but reach to the division between good and evil, they are too important to receive less than an earnest, but calm and dispassionate, treatment. Failure to this would be a collective shrug-of-the-shoulder at moral values and, as such, a thread to the spiritual goods upon which we have built our dignity and freedom."

James Stewart GORDON *('Abolish professional boxing.' Readers Digest, 76, April 1960, pp. 80-85)* wrote:

"When you watch a fight on TV, the odds are overwhelming that you see two muscular young men who are nothing but puppets of the mobsters who control the business. Often the fighters are subnormal mentally, either because they were born that way or because they have practiced their track too long. They show who is to win and who is to lose – and when and how. And at the end of their ring career they are usually both battered and broke, robbed by the men who have used them... Considered coldly and on the basis of the evidence, professional boxing has absolutely no excuse for continuing to exist. It is true we were civilized enough to abolish this shabby game, on both medical and moral grounds."

GORDON continued:

"On moral grounds the truth is that professional boxing is not a sport at all. It is a sordid business designed to yield maximum profit for the men who control the fighters, the arenas and the TV outlets. The fact that men can be permanently injured or killed in its practice is of little concern to the behind-the-scenes manipulators. Through their monopoly, known as the 'Octopus,' they decide for virtually every boxer fighting today – regardless of his skill when he will win or lose, when and whom he will fight next."

GORDON summarized:

"John BONOMI, *assistant district attorney of New York and assistant counsel to the KEFAUVER Committee,* which recently announced plans to expose some of boxing's dirty laundry, says: 'In six years of investigation I have found few fighters who are not wholly or partially controlled by the mob.'"

"Middleweight titleholder (in New York and Massachusetts) Paul PENDER has spoken in less restrained fashion: 'Boxing is rotten clear through, infested by gangsters and thieves."

GORDON noted:

" There is no place in the United States today for a 'sport' that is as brutal and crooked as the boxing racket. Harold BARNES, *retired official of the New York State Athletic Commission,* a boxing judge for 38 years, says, 'I consider boxing legalized murder. I would outlaw it.'"

"Isn't it high time the United States banned this public butchery?" asks GORDON!

The English boxing promoter Jack SOLOMONS noted:

"Today boxing is business, big business. As such it has given me a few headaches over the years, but I love every moment of it."

Lady Edith SUMMERSKILL wrote 1956 an essay entitled *'The Ignoble Art'* from which I will quote some of her thoughts.

The first chapter deals with *"The Aggressive Instinct."*

Lady SUMMERSKILL writes:

"A child has certain destructive impulses which, if directed towards some objects in the home, distress and anger a parent. Nevertheless, the same parent is prepared to bring into the living room, a television screen, a picture of two men battering each other, accompanied by a commentary which emphasizes the pain, the dangerous blows, and the extent of the bleeding suffered by the fighters."

Lady SUMMERSKILL continues:

"Youngsters are allowed to watch and listen to this in the presence of a father who obviously favours this kind of behaviour. The next day the same father may box his son's ears for ill-treating his little sister."

"It is argued that the individuals watching fights which are accompanied by hysterical yells, as the pace quickens and the contestants show bloody signs of wear and tear, are worthy men and ordinary citizens, who should be allowed to enjoy their leisure time as they think fit."

"Would those who argue that they see no harm in these exhibitions like to see one of these yelling, over-excited men in a place of authority in time of war? Would they like to see one of them as a sergeant-major in the Army in control of young recruits? If there is some hesitation in answering these two questions, the reason is clearly that the circumstances of the fight unleash sadistic impulses, which are revealed in the ugly behaviour so often accompanying a prize-fight. You would not like to see the same impulses given rein in a position of authority where a word might determine *your* fate. Neither would you like to see them freed to make your son's life a misery in the Army."

"Should we, in a world still full of cruelty and inhumanity, permit a public see them freed to make your son's life a misery in the Army, display of men fighting each other, when all who are honest with themselves know that such a display caters to the basest instincts?"

"Civilized man has come to recognize a moral self which is concerned with the control of his instincts; to contend that, but reject what he values and has slowly acquired, he would achieve a balanced personality and happiness, is to ignore that slow evolution towards a civilized society which in itself affords him satisfaction."

"While certain things are instinctive and natural, that does not necessarily make them desirable. Man is polygamous by nature, but nevertheless he accepts a code of behaviour which permits only one wife. The father, who would protest loudly that his son must be allowed to use his fist because the aggressive instinct is natural, would protest equally loudly if he were commanded to take a second wife because his procreative instinct was sufficiently strong to warrant it. In this matter he would be guided by a social code which he has accepted, approves and understands. Moreover, while he might privately wish that the marital customs of this country were a little more flexible, he knows that the marriage law serves to protect the unity of the family and ensures the stability of the society in which he lives."

"It is reasoning of this nature which will eventually persuade men that wars are the end product of a certain course of behaviour which man himself can control and direct."

"The average boy is taught to believe with W. S. GILBERT *(H.M.S. Pinaforte) that* to be a hero:"

"His bosom should heave and his heart should glow,
And his fist be ever ready for a knock down blow...
His energetic fist should be ready to resist
A dictatorial word.'"

"Schoolboy literature is crammed with stories of the handsome head of the school knocking down the cad of the fifth form. Reared as I was on the *Gem* and the *Magnet,* I still have memories of Harry WHARTON and Tom MERRY dealing in this manner with 'the rotter' who was caught smoking."

"Having sat as the only woman on innumerable committees and in the course of the years ranged over all kinds of subjects, with all due respect to some of my male colleagues I must say that on occasions men ten and twenty years my senior have roused my maternal instinct when the subject of fisticuffs has inadvertently been touched upon. As I have listened to some of these aging gentlemen defending the case of pugilism, the shades of Harry WHARTON and Tom MERRY seem to have joined the committee."

"Women are fortunate in this matter. We are taught as little girls that it is possible to scream if you are frightened. Indeed some mammas of the Victorian period taught their daughters to scream on the slightest provocation if there happened to be a desirable young man in the offing, for this would appeal to his protective impulses."

"On the other hand the boy, however sensitive he may be, is warned that to show fear is unmanly and that therefore this emotion, which is common to us all, must be suppressed."

"And so fighting on boxing rings, in street gangs, in wars, has been glamorized and the boy child is taught to believe that there is some special virtue in fighting, even though he or his opponent may be maimed or killed in the process."

"With this conditions of the young male from his earliest years, the fight promoter finds no dearth of material. Most professionals are recruited from the best of the amateurs, therefore the clubs which provide boxing facilities for boys are the places which are regarded as the nurseries of the boxing world."

"The temptations for a promising boy from a poor home to yield to an offer, which might lead to the earning of a sum of money which exceeds his wildest dreams, is very great. Of course most succumb, little realizing the risk of impairing their health in the space of few years. Their parents, dazzled by the prospect of having a champion in the family, seldom stop them. There is only this to be said in defence of parents who willingly consent to their son entering the ranks of the professional boxers: they are generally ignorant of the dangers involved."

"In our schools and our clubs boys are encouraged to box, while others are compelled to provide an audience. The perfectly nice people who are responsible for the boxing clubs know little about primitive instincts; or if they are assailed by some doubt as to the advisability of teaching young humans to punch each other on the nose, they quickly smother the thought."

"As one extremely proper and pleasant woman teacher, who helps in a boys' club in the East End one night a week, said to me, 'If it were not for the boxing we could not keep the Club together. The boys like fighting each other.'"

"That was a confession of failure, of which this highly intelligent woman was entirely unaware."

"Of course the aggressive boy likes fighting. There are lots of other things human beings would like to do unchecked."

The second chapter of Lady SUMMERSKILL's essay deals with *"Medical Evidence."*

"I am well aware that the usual attack from those opposed to my views takes the form of ridicule and a charge of crankiness."

"Therefore I should like to emphasize that I am supported by some of the most brilliant and respected men and women, who have earned a reputation not only in their own countries but throughout the world."

"Dr. Karl EVANG, M.D., F.R.S.M., Hon. F.A.P.H.A., is not only the *Director-General of the Public Health Services of Norway,* but has established a world-wide reputation, which led to the *World Health* organization of the *United Nation* using his services as the director and, adviser of the health services in other countries."

"In a paper *on 'The Responsibilities of Health Authorities in Relation to Sport,'* which he read at the international conference on sport and health in Oslo, he wrote:"

"'We are faced with a number of physical injuries, traumas and diseases caused by sporting activities, and the number has assumed such proportions that many people naturally start to question whether the whole thing is worth while. In boxing, for example, a certain number of people are killed outright every year; a much larger number get 'punch drunk' or have to be registered as 'ring wrecks' for years or for a lifetime. How many boxers reduce their intelligence permanently and irrevocably nobody knows, but it is generally accepted that the risk exists.' He goes on to enumerate the main responsibilities of health authorities in relation to sports, with special reference to boxing. 'Branches of sport which are directly dangerous should be stopped. Personally I feel that boxing should definitely be eliminated.'"

"A man of *Dr. EVANG's* medical standing does not make statements of this nature without the most careful consideration of the medical and social aspects of the matter."

"Those who contend that the brain is not permanently damaged by successive fights are today very much on the defensive. Doctors and others of the highest repute, who have no interest in the business other than to protect the innocent lad who risks permanent damage to his health, are publicly stating that boxing is harmful and should be prohibited."

"I have never advanced my own medical views on the matter, but have always sought the advice of surgeons and doctors and other professional people who I have reason to believe are fully competent to express an opinion."

"Dr. A. H. STEINHAUS, *Professor of Physiology at George Williams College, Chicago,* who was formerly *Chief* of the *Division of Physical Education and Health of the United States Office of Education*, summarized the findings in the cases of 200 boxers suffering from concussion who had their brain waves recorded on an electro-encephalograph. He found that there was no conclusive evidence that an examination of this nature could ensure the safety of the boxer. It is, I think, generally agreed that no instrument has yet been invented which can detect the small haemorrhages of the brain which a boxer sustains in the course of a fight and which if repeated over the years produces the condition of 'punch drunkenness.'"

"The weight of a boxer's punch can be very great; it has been estimated that a ten-stone amateur boxer can deliver a punch equal to an impact of six hundred pounds pressure. The degree of damage to the brain cannot be measured solely by a few heavy punches, but must be assessed in relation to the minor head blows received over the years. A few skilled experts may escape, but the ordinary fighter, after about fifty fights in a few years, begins to show the unmistakable signs of deterioration. He fails to time his blows properly because the damage to his brain has caused cerebral atrophy, his defence is inadequate, and soon he is a back number, his promoters have quite deserted him, and with them his good-time friends. He is forgotten by his public and he finds himself a bitter and disillusioned man."

Lady SUMMERSKILL continues:

"His great tragedy is, however, that he has lost his mental health, his capacity to concentrate, very often to co-ordinate his movements and to hold down a responsible job."

"A death in the ring hits the headlines, and while still fresh in the memory causes adverse comment, but the injuries inflicted on a man's brain over the years, reducing him to a sub-human mental condition without necessarily making him certifiable, are only fully appreciated by those in direct contact with the unfortunate individual."

"When the structure of the brain is considered it must surely be obvious that this sensible organ cannot possibly fail to be damaged following the impact of a succession of blows over years."

"The brain weighs about three pounds; it is of the soft consistency of cold porridge; it is not tied down, and rests in a fluid in the bony skull. A blow causes it to wobble from side to side, even a moderate blow will cause it to bang against the side of the skull, and a more severe blow may cause such a sudden forceful movement against the bony parts of the skull as to bring on bleeding or bruises."

"These injuries do not heal and the destruction of the brain cells is permanent. The frontal lobes which are damaged in this way are those parts which control man's highest functions: namely the power to co-ordinate his movements, to restrain his impulses and to exercise his power of self-control."

"Another injury to the brain can be caused by small haemorrhages in the pons and medulla. These haemorrhages are apt to occur during a severe beating in which the boxer continues to fight although he is only partially conscious, a condition called 'out on his feet.'"

"This condition generally precedes the knock out. These small haemorrhages are not revealed in an X-ray and therefore any medical examination in this respect is useless; they are only revealed on post-mortem examination."

"Gene TUNNEY once said, 'When I see a young fellow being quickly knocked out I feel relieved in his interest. I think he might be saved from the real danger; from being hit on the head for years by minor blows and thereby being beaten into a mental wreck.'"

"Dr. Edwin J. CARROLL in a paper *'Punch Drunk,'* published in the *American Journal of Medical Sciences,* says that a boxer begins to 'soften up' after he has engaged in 30-60 professional bouts. This 'softening up' is one of the first manifestations of 'punch drunkenness' (traumatic encephalitis), which comes on slowly from many small wounds in the brain. It is difficult to secure the cooperation of this kind of patient because in a larger number of cases nervous and mental symptoms exist together with a good general condition of health, and consequently the patients like to think that they are perfectly healthy and object to being medically examined. Dr. CARROLL is of the opinion that in that period of five years of professional boxing there occur about five cases of manifest traumatic encephalitis out of every hundred fighters. Out of the same number, sixty will develop mental and emotional changes which are obvious to their friends and relatives."

Dr. CARROLL, who spent two years in the haunts of boxers, in order fully to understand the medical hazards to which they were subjected, writes:

"'It is probable that no head blow is taken with impunity and that each knock out causes definite and irreparable damage. He describes the gradual changes which take place in the young boxer who finally becomes an obvious punch drunk as follows:'"

"'A young sturdy man trains in a gymnasium as an amateur boxer, in order that he may develop himself physically and master the finer points of pugilism. He becomes a fairly tough opponent in the ring, but he tends to fight rather than to box, and so receives some degree of punishment in each of his thirty to forty short amateur matches. Later, he becomes a professional, meets stiffer opposition, engages in longer fights; but he is young, strong and making good money. After a period of approximately four years as a professional boxer, during which time he is engaged in about thirty to sixty bouts, he begins to 'soften up.' His tolerance to all head blows is reduced. After being struck on the jaw he remains dazed for a longer period than formerly and is more likely to be knocked out *('glass jaw').* After a blow on the head his legs will be a little shaky and feel numb. His timing begins to fail and the fighter is no longer as dangerous on the offensive as he had been. His defence, which had been steadily improving with each encounter, becomes less effective, and the pugilist, who previously had been adroit enough to go through a career unmarked, now begins to develop a flat nose and cauliflower ears. Later, his knees tend to give way after a head blow, and a slight dragging of the feet may be noticed, although often only as the contestant is walking to his corner at the end of a round. The fighter still boasts of feeling fine and capable, but now loses engagements which, formerly, he could have won with ease.'"

"'Meanwhile the boxer's manager and associates begin to notice changes in him. They find that he has developed deterioration in attention, concentration and memory. In the midst of a conversation he may go into a reverie, suddenly change the subject or ask the same question several times. He is apt to become too sociable and voluble. Usually an impediment of speech will develop and the eyes may become slightly glazed or have a staring expression *('glass eye').* In short, *he may continuously simulate a person who is just a little drunk.'"*

"'When the symptoms have once appeared, they will not only persist, but may progress for a period of a year or so, when they will usually become stationary. This 'punch drunk' is a self-limited rather than a progressive encephalopathy.'"

"The old boxer who excitedly, illogically and incoherently with defective enunciation seeks to defend prize fighting is manifesting punch drunk symptoms which bear out Dr. CARROLL's statement."

The paper of CARROLL is discussed in detail on p. 301.

"Dr. H. S. MARTLAND in a paper *'Punch Drunk,'* published in *the Journal of the American Medical Association* in 1928 talks of, 'nearly one half of the fighters who have stayed in the game long enough to develop this condition, either in a mild form or as severe and progressive form which often necessitates commitment to an asylum.' This,' he says, 'seems to be good evidence that some special brain injury due to their occupation exists.'"

The classic study of MARTLAND is discussed in detail on p. 288.

"Mr. J. H. DOGGART, M.D., F.R.C.S., *Surgeon to Moorfields Eye Hospital,* having on many occasions operated on the eyes of injured boxers, denounces boxing without any reservations. Recently in a speech in the U.S.A. he drew attention to the tragedy of the ex-boxer with the 'cauliflower eye.' He has said in an address to the *British Ophthalmological Society:"*

"'The contention that serious and permanent injuries are exceptional cannot be maintained by anyone who takes the trouble to consult medical literature. Here we find a long tale of disaster, especially to the eyes and their adnexa. It must be remembered that gloved and ungloved fists are not the only instruments of destruction. Boxers' heads colliding with each other in the heat of combat can inflict fearful injury. So can the hard floor coming into contact with a stricken man's temple. Complete severance of an eyelid is not unknown, and ocular palsies, notably of the oblique muscles, have often occurred.'"

"'Among possible lesions of the anterior segment may be mentioned ruptured sclera, corneal abrasions, partial or total dislocation of lens and cataract... Cauliflower ear is the hallmark of old pugilists, but the term 'cauliflower choroid' might well be applied to some of their fundi, in which we find irregularly furrowed regions of greyish-yellow discoloration, presumably the result of former choroidal haemorrhages....'"

"'Severe and repeated damage to the brain obviously explains the phenomenon of 'punch drunkenness.' We have already seen that multiple blows on the head lead to haemorrhages in the brain followed by irregular patches of cerebral atrophy.'"

"His general conclusion should be brought to the notice of all boxing promoters and the friends and relatives of those physically fit lads who are contemplating the ring as an easy step to fame and fortune."

"Mr. DOGGART says:"

"'In spite of all assertions to the contrary by gladiatorially-minded sentimentalists, facts will out. Medical literature abounds with records of havoc wrought by boxing, and it seems clear that sight, sanity and life itself are endangered when a man pursues this sport over a period of years, especially in the professional ring. We know that spectators at a boxing contest emit howls of joy when the thud of the gloves upon flesh sends them into an ecstasy. Let us hope that their jubilation would be tempered if only they were to think of the hazards inseparable from a boxing career.'"

"After this address the President of the Society agreed with Mr. DOGGART that ophthalmic surgeons, other surgeons also physicians were not so much in love with pugilism as were the general public, who appeared to be so ignorant of the dangers attendant on a boxing bout. He said that at one time or another he had had a number of pugilists as patients. One he had seen at *Moorfields* curiously enough had the name of Bill SOFTLY. Bill had on one occasion suffered an injury to the superior oblique muscle. When trouble had recurred following subsequent injury, he had pointed out to Bill that seeing double must be rather awkward for him, whereupon Bill had replied, 'Yes, I see two chaps in the ring and I hit the one that isn't there and the one that is there hits me.'"

"That was an amusing and not so tragic case. The tragic cases come later, *Mr. DOGGART* said, and he cited the case of a professional boxer who had a double detachment and was completely blind. He has seen also amateur boxers suffering from the same disability."

The papers of DOGGART are quoted at length.

"There seems to be agreement among the experts as to the causes and effects of head injuries. There have been experiments upon animals and model human brains, and these together with post-mortem findings support the view which have been advanced by experts in this field during the past twenty years, that boxing in which the head is subjected to severe blows involves serious permanent damage."

"Now those in the boxing business these days, apprehensive as to the future of their own vested interest, tend to brush aside the opinions of the experts and even attempt to ridicule them. I should like to quote again the evidence of one of the most famous heavy-weight champions of his day, Gene TUNNEY, which cannot be so lightly disregarded."

"In 1932 Gene TUNNEY wrote in his autobiography, '*A Man Must Fight,*' of an incident that occurred between his two DEMPSEY fights."

The incident – in the words of TUNNEY – is quoted.

TUNNEY ends his descriptions with the following comment:

"'...And let me say here, probably for the first time, that a fighter suffers just as much from punches during training as he does from wallops taken in actual combat. He wears a headgear, yes, and the training gloves are pillow-like in structure, but the protection cannot forestall jarring of the brain.'"

Lady SUMMERSKILL continues:

"However, it is possible to find among medical men those who are prepared to encourage youths to enter the ring despite the hazards to health and life itself."

"Professional men in this category are always found attached to the different *Boxing Boards of Control* which, in this and other countries, are concerned with ensuring that the regulations which govern boxing are observed."

"There are others also who, despite the most damning medical evidence, are reluctant to condemn boxing. With all due respect to the few who adopt this attitude, may I suggest that here is a clear case of emotion defying reason? As I have already said the small boy, already aggressive, is encouraged from his earliest years by his schoolmasters and parents to regard fighting with the fists as a manifestation of the finest male attributes."

"Later, in University and Medical School the same code of behaviour is encouraged. So powerful is the effect of public opinion that even the certain knowledge that blows to the head are injurious would not persuade most young men to dissociate themselves from a code of behaviour which might imply that they were lacking in courage."

"In the First World War this primitive training and its consequences were exploited in every recruiting speech. The implication that you were unmanly and a coward if you did not respond to the call to arms was used, often with tragic results in the field."

"Women, by handing white feathers to those in civilian uniform, exploited both their sex and the secret fear of young men of being charged with cowardice."

"With this background training it would be remarkable if male doctors could completely re-educate themselves emotionally, and allow only their reason to dictate their attitude towards boxing."

"However, the *British Medical Association,* which is reluctant to pronounce on social questions until it is fully informed, made a well-balanced comment on the medical aspects of boxing in an article, *'Brain Injury in Boxing'* (*British Medical Journal,* December 25th, 1954)."

"It summed up the position as follows:"

"'While many of the athletic activities of young men are associated with risk of injury, and rarely of death, their value in developing physical fitness and character more than justifies the risk involved.'"

"'Boxing is unique among the sports of civilised communities in that the object of the boxer is not only to hurt his opponent, but if possible to render him unconscious. This has led to differences of opinion on the merits of boxing as a sport and a spectacle...'"

"'While injuries to the face, teeth and hands are common, the most serious is head injury. This is to be expected, since the head is the most important target and knock outs are usually achieved by head blows producing cerebral concussion.'"

"'Concussion is regarded as a disturbance of brain function without anatomical change, and appears to be particularly transient resulting from a head blow in the ring. Nevertheless, out of twenty-four amateur boxers investigated electro-encephalographically after being knocked out, by Dr. J. L. BLONSTEIN and E. CLARKE *(British Medical Journal,* December 25th, 1954) sixteen per cent showed a diffuse slowing of the rhythm; and in one boxer who had been concussed on four occasions, the abnormality persisted for four months.'"

"'Abnormalities of this duration suggest organic brain damage. Dr. LARSSON and his colleagues (*Acta Psychiatr Neurol Scand,* 1954) record that thirty-six out of seventy-five amateur boxers showed either clinical signs of brain injury, E.E.G. changes, or both.'"

"'When E.E.G. records were taken both before and after a fight, these workers found abnormalities in thirteen per cent of all of those who were knocked out, knocked down, or who lost on technical knock outs owing to hard head blows.'"

"'Drs. KAPLAN and BROWDER (*Journal of the American Medical Association,* 1954), whose investigation was confined to professional boxers, state that electro-encephalography failed to give statistically significant results, except for the fact that the highest proportion of disorganised E.E.G. records was found among the individuals with the lowest ring ratings (who doubtless lost more fights).'"

"'It would seem, too, that these authors regard many different types of record as coming within the range of normal, and of the twelve varieties of E.E.G. record they regard as normal some would consider more than fifty per cent abnormal.'"

"'Useful information could be obtained by further electro-encephalographic study – records being taken at the beginning of each boxer's career and at intervals while he continued to box.'"

"Since trivial head injuries which produce no concussion may at times be followed by subdural hemorrhage it is not surprising that intra-cranial bleeding, often with fatal results, occurs as a result of boxing injuries."

"Drs. BLONSTEIN and CLARKE record six fatalities from this cause, but make no reference to any cases of intracranial haemorrhage treated successfully after a boxing injury."

"Drs. KAPLAN and BROWDER have seen five brains at post-mortems on boxers in the New York district over a period of twenty-five years and consider that in every case the deaths were due to cerebral contusions and lacerations in the fronto-temporal regions as a result of a fall on to the occiput resulting from concussion."

"They, too, make no mention of any case of intracranial haemorrhage treated successfully by surgery – though such cases undoubtedly occur, and it would be useful to know what proportions of knocked out boxers are sent to hospital because of evidence that the injury is more serious than a mild cerebral concussion.'"

"Dr. Ernst JOKL in the *'Medical Aspects of Boxing'* sums it up thus:"

"'Of all major sports, boxing occupies a special position since its aim is that of producing injuries, more particularly to the brain. As the more dramatic manifestations of such injuries are colloquially referred to by such terms as 'knock out,' 'grogginess,' 'punch drunk,' etc., it is not usually appreciated that these conditions indicate the presence of serious cerebral disorder. It is true that similar injuries occur in sports other than boxing, e. g., in football and wrestling. But there they are accidents rather than consequences of intentional acts.'"

"To compare boxing with any other sport is to ignore these facts. Professional boxing is the only sport in which a participant seeks to knock his opponent out in the shortest possible space of time."

"It is the only sport in which to confess injury and retire is to risk a hostile demonstration from the spectators."

"It is the only sport in which wounds inflicted and blood drawn gives colour, zest and a sadistic thrill to the whole performance."

"It is a business in which the fighter himself may not realize that he has been damaged and that his mental processes are gradually deteriorating. Or if he does, he may not admit it for fear of being barred from further fights."

"Supporters of boxing argue that Rugby football is equally dangerous. This argument cannot survive examination. In the first place the primary objective in football is to score goals and not to render your opponent insensible. An injured man is immediately removed from the football field with the sympathetic approval of the crowd. Furthermore I have never yet heard of a 'punch drunk' Rugby football player."

The next chapter in Lady SUMMERSKILL's essay deals with *'The Old Prize Fighters.'*

"As the study of the brain is comparatively recent we have no statistics or records which can provide a medical picture of the condition of the old bare fist prize fighters. Some stories, however, have been passed down which make interesting reading."

"In 1833 Jem BURKE, a prize fighter from the East End popularly known as 'The deaf 'Un,' fought and defeated Simon BYRNE. The fight lasted ninety-nine rounds, three hours, sixteen minutes, and BYRNE died as a result of it. Later BURKE was advertised in the music halls:"

"The Champion Boxer of England, the celebrated and Herculean Deaf BURKE"

"BURKE had two fights in the U.S.A. and earned large sums for those days. He returned to England, turned to a life of dissipation, and finally died penniless in a lane off the Waterloo Road."

"No doubt BURKE's deafness was caused from a blow, and his end was typical of the 'punch drunk.'

"In 1889 Jem SMITH fought 109 rounds with KILRAIN on an island in the Seine. Later a man named SULLIVAN beat KILRAIN and the fight lasted for seventy-five rounds."

"To those who are accustomed to glove fighting it seems difficult to believe that men fought with bare fists for over a hundred rounds lasting two and perhaps four hours."

"The great difference between fighting with 'the raw'uns' and with the gloves is that the gloves provide not a protection for the face of the person hit by them, but for the hands of his opponent."

"The whole tempo of the bare fist fight was slower, for otherwise the fighters would have damaged their hands. Moreover wrestling was allowed in the ring and in the early days."

"The standard size of the ring was twenty-four square feet, much larger than the ring used by the glove fighters later on, and this offered greater opportunities for manoeuvre."

"The training methods of the day are described as 'raw eggs to improve the wind and raw beef to make them savage."

"The idea that the comparatively modern practice of wearing gloves ensures safety for the person who is punched is quite fallacious. On the subject, Bill McGOWRAN of the *Evening News,* (June 17th, 1955) wrote:

"Once again we have endured the agonizing spectacle of a gallant boxer fighting a battle with blood streaming from cuts over both eyes.'"

"'Latest victim of this pestilential injury, which is becoming increasingly common, was Wally THOM who lost his European welter-weight title under the handicap of two deep eye cuts, both of which had to be stitched after the contest....'"

"'Damaged eyes have ended the careers of many great fighters – Percy BASSETT, the 'interim' holder of the world featherweight title, Archie SEXTON, Andy NEWTON, George ANGELO, Bruce WOODCOCK and Jack PETERSEN, to name only a few...'"

"'I have held the opinion that the real culprit is the modern boxing glove... The modern glove is a tough, near-lethal weapon, which merely protects the fist and allows the striker to deliver a harder blow without injury to his hand. In all other sports where there is a risk of permanent injury, protection is allowed. Soccer players wear their guards, rugger men may use scrum caps to guard their ears, cricketers have their pads, fencers wear masks and even horses are allowed to run in bandages!'"

"'In America, of course, footballers turn out in armoured suits that turn them into square-shouldered spacemen. Only the poor boxer is compelled to take it on the naked flesh. Even beards are regarded as 'protection!' When Canadian George ROSS weighted in for his fight with Albert FINCH a few years ago, he wore a few days' growth of stubble on his chin. A *Board of Control official* ordered him to shave before the fight.'"

"It is encouraging today to find sports commentators, whose livelihood depends in part presenting boxing to a section of the public eager to have its appetite whetted with horrifying descriptions, having the courage to criticise the business."

The next chapter in Lady SUMMERSKILL's essay is entitled *'The Noble Art.'*

"The spectacle of the prize fight has become so distasteful and repellent nowadays that it is a source of surprise that apparently respectable and reputable organisations continue to sponsor this form of entertainment."

"The reason why public opinion is slowly hardening against prize fighting can be found in the advance of scientific knowledge, which has revealed to the medical profession and the general public the physical danger to which a boxer is exposed. Furthermore, the sight of a human being seeking to render another unconscious in the shortest possible space of time has become repugnant to a people who during the last half-century have raised their standard of living and advanced a vastly improved code of behaviour."

"I do not pretend to be an authority on boxing technique, therefore I sought experts' opinion in what I presume in the boxing world is looked upon as a classic – Vol. XI on *Boxing* in the *Lonsdale Library of Sports, Games and Pastimes,* where Viscount KNEBWORTH contributes a *Guide to Modern Methods,* which is supplemented by W. CHILDS, *coach of the Cambridge University Boxing Club* (1937). This is what Lord

KNEBWORTH advises in Chapter 12, under *'Ring Craft; Seize Opportunities'."*

"'Last of all it is good ring craft never to miss an opportunity. Perhaps the hardest thing in all boxing, certainly the thing which is most often made a mess of, is to finish a man off once he has been shaken. There is a golden line somewhere between hasty enthusiasm and careless calm, which seems very difficult to find. It is certain that a rush madly after a shaken opponent is to lose all sense of distance and coolness of judgment, and to give him, unwittingly, a chance to recover. Equally often are opportunities missed the other way.'"

"'A man sometimes tends to be too cool, to wait his time too relentlessly, and so the golden seconds are allowed to slip by. The best way is to step back first of all, make sure just what the situation is, measure the man, and then keep after him relentlessly until the gong goes.'"

"'A great secret is when an opponent is groggy, to go for the body. He will feel blows there much more than when he was himself, and in all probability he will drop his hands to defend his body. That is the chance a boxer has been waiting for. Let him have it hard on the point of the jaw.'"

"'Above all, avoid the practice, much too common in the amateur ring, of standing aside and allowing a man to recover deliberately... To let a man recover once he has been shaken is not good sportsmanship, is poor ring craft and absolutely rotten boxing.'"

"I must confess that until I read this I was under the impression that there were those in this field of sport who regarded punching a man suffering form mild concussion as unsportsmanlike, like hitting a man when he is down. A man staggering round the ring suffering from blows which have had an impact on his brain is often as incapable of defending himself as if he were on his knees. Apparently it is not 'good sportsmanship' and is 'rotten boxing' to let a man come to his senses. Knock him out before he is fully conscious, is the advice of the best authorities, who no doubt echo the advice of trainers who know nothing about the *Lonsdale Library* and have never read a chapter on *'Ring Craft'* in their lives."

"In the light of this advice by the experts, can anyone talk of the 'noble art' without his tongue in his cheek? The man who is dizzy and reeling and then receives the final knock out blow has little chance of saving his head against the boards. The advice to 'let him have it hard on the point of the jaw' in order to render him insensible, is medically accurate, because unconsciousness and sometimes death supervenes more quickly after this particular blow than after any other. Boxers have learned this through experience, not through consulting medical textbooks. Indeed, the advice to hit the groggy man on the body is based on the medical fact that his brain injury befuddles him; in consequence he forgets to protect his head, and his hands automatically go down to defend his body. Then is the time, apparently, to give the already slightly damaged brain another tremendous blow in order to concuss it and to induce complete unconsciousness. When this so-called sport is analysed in this way, even its devotees must feel just a little dubious about those rules which, we are told, ensure that high standards of sportsmanship are observed in the boxing world."

"The following description of T. H. MURNAME *in 'Let's Face the Facts About Boxing'* must even disconcert those who consistently try to protect the 'noble art' from criticism: 'Mostly fighters want to win by knock outs, if possible, for knock outs pay off. They build up a fighter's reputation, making it possible for him to demand more money and to advance in division ratings... When a fighter sees his opponent in a bad way, he doesn't let up. On the contrary, he renews his efforts to score a knock out... The head is the main target and most knock out blows are delivered to the head. Even when a fighter directs his attack to other parts of the body, it is usually a diversionary measure to get an opening to the head.'"

"Our more detailed knowledge of the brain and its structure has taught the modern prize fighter that his most important objective should be to inflict injury on the brain, if he is to finish the fight in the shortest possible space of time. This point was emphasized by Dr. A. RAVINA *(Presse Médicale)* in an article on *'Traumatic Encephalitis or Punch Drunkenness':"*

"'Boxers nowadays know that facial and chin blows are more effective than any others and they concentrate on these. Together with technical development, a change of the public taste has led to a significant modification of the character of boxing. Years ago the science of evasion and passing was stressed, nowadays the public wants offensive action to the utmost, without regard to blows received.'"

"Fight promoters usually assume an air of complete innocence and mystification when these exhibitions are described as appalling to the sadistic impulses, and consequently having a brutalising effect on the spectators. However, on May 13th, 1954, Tom PHILLIPS wrote in the *Daily Herald* under '*What Do the Fans Want?"*

"'I get puzzled at times, maybe I am dumb or blind. I saw Dai DOWER win a great ten-round fight against the Dutch champion at the Empress Hall, and woke up yesterday to a chorus of 'He wasn't so hot, was he?' What do the critics want? Bear-baiting – with men standing-in for animals? Not once during the last four rounds could you tell that DOWER's right foot was searing hot with pain. True he slowed up. Everyone put it down to the fact that he was getting tired because he had never gone ten rounds before. I like boxing. But when the day comes that most fans want grunt, snort and gore, let them go elsewhere for it and I shall stay at home.'"

"The boxer generally tries to please the customers. If he fails to do so or ignores them and concentrates on style, then he risks ridicule and boos."

"Randolph TURPIN, in his last appearance at the *Harringay Stadium,* was the victim of a running commentary of taunts from those who had come to see him batter his opponent into insensibility and were disappointed. Although TURPIN was the winner he was stung into retaliation and shouted as he left the ring, 'You don't want boxing, you want murder!'"

"Yet we are told by the middle men that the sole object in view is to give the spectators a display of a sport regulated by rules calculated to ensure the maximum enjoyment of boxing technique. If this is so, then why are these bouts accompanied, in the newspapers and on the radio, by a commentary which is deliberately phrased to emphasize the bestial element?"

"We are familiar with the radio commentator who gives a graphic description of freshly-opened cuts, particularly those over an eye which bleeds profusely and so blurs the vision of the fighter. Blows over the heart and kidneys are mentioned with special relish and the words 'savage,' 'slaughter,' and 'smashing' represent the verbal currency of those whose duty it is to give an accurate picture to the listening public of the 'noble art.'"

"The following reports reveal that one experienced sports commentator knows that a cold descriptive account of the technique of boxing would not have the same appeal as a highly coloured picture."

"Peter WILSON wrote in the *Daily Mirror* in September 1953 on the fight between Tommy McGOVERN and Joe LUCY:"

"'Fighting from behind a red blindfold of his own blood, twenty-nine-year-old Tommy McGOVERN from Bermondsey should get an even greater red badge – the red badge of courage.'"

"'That persistent right, probing as a dentist's drill, finally unleashed a red trickle from underneath McGOVERN's left eye. This seemed the encouragement which LUCY needed. He piled on punches...'"

"'LUCY still refused to use his left but increasingly played on the damaged area of McGOVERN's face.'"

"'Although he could see LUCY's rather crude blows coming his way, McGOVERN could do little or nothing to avoid them, so he took them and took them, and took them, and because the human face is not as tough as the leather of a boxing glove another cut opened under his right eye... '"

"'In the eleventh round LUCY finally let go with his left hand. For now it seemed McGOVERN was having more difficulty in sighting with his right eye than with his left. In those last four agonising rounds some of the crowd shouted, 'Stop it! Stop it!...'"

"'Tommy from Bermondsey, his face a bloated india-rubbery mask, with the cuts making him look as though he had three mouths all gaping redly, fought on his guts and his pride...' "

"' I fear LUCY did not add to his reputation by being unable to stop an opponent who was fighting from behind blinkers of blood.'"

"Would anybody seriously suggest that the screaming crowd round a boxing ring are having implanted the fine qualities of pluck, endurance and restraint?"

"Do the pathetic punch drunk boxers bear this out as they amble uncomprehendingly through the rest of their miserable existences?"

"If these fine qualities are inculcated in the boxing ring, why do we not see boxing promoters putting their young sons in the ring for, say, thirty to forty fights, just to toughen them up and strengthen their moral fibre so that they can face life as better and finer men?"

"Those commentators who relish giving the most lurid details were in their element in reporting the MOORE – MARCIANO fight. This exhibition was held on September 21st, 1955, in the *Yankee Stadium, New York,* within easy reach of the *National Assembly of the United Nations,* where representatives of many countries are seeking to establish a form of society in which men will settle disputes without resort to force."

"MARCIANO, heralded long before the fight as a 'killer,' was opposed by Archie MOORE, a man long past his physical prime, who gave his age as thirty-eight but was believed to be some years older. The fight, which lasted for nine rounds, shocked even the hard-bitten reporters. It was headlined as a massacre in which MOORE apparently was sacrificed for the sake of satisfying the appetite of the sensation-seeking crowd. MOORE knew the part for which he had been cast. He had to struggle on, even though he was in a semi-concussed condition, long enough to ensure that the crowd had had their money's worth of sadistic pleasure. His price was twenty per cent of the takings, and he meticulously kept to his part of the bargain."

"MARCIANO delivered twenty-seven smashing blows, most of them to the head, before MOORE was beaten into partial insensibility. One writer said, 'There was less blood about and the corners looked less like clearing stations than we have grown accustomed to in these days of big blood demand.' This did not mean that the fight was conducted in a matter calculated to demonstrate the finer points of boxing. It happened in this case that, apart from MARCIANO's bleeding nose, most of the blows resulted in a gross disfigurement of MOORE's face, especially his right eye. Unfortunately, nobody could see the shaking and bruising of the sensitive brain each time a blow was landed on the skull. The Marquis of QUEENSBERRY's Rules – for what they are worth – were forgotten. Said one eyewitness, 'Just call it a berserk bash – twenty-five minutes, nineteen seconds of savage, no quarter thumping, with the puffed and swollen face of Methuselah MOORE on the consumer end of the pain and punishment.'"

"MOORE knew only too well that if a fighter shows that he has had enough and drags his feet or retires prematurely, then he is the target of abuse."

The fight between MARCIANO and MOORE is described in detail on p. 119.

"The same treatment is meted out to the referee who may in all good faith stop a fight because he thinks that enough damage has been inflicted on one of the contestants. The referee cannot be solely concerned with the conduct of the fight and the condition of the fighters unless he is a very high-minded person with a private income on which he can fall back. Time and again the referee's ruling has been challenged by a hostile crowd afraid of being deprived of a sordid entertainment."

"There was the well-known case of Eugene HENDERSON, the referee who stopped the first Freddie MILLS v. Gus LESNEVICH contest for the world cruiser-weight title in 1946. He stopped the fight in LESNEVICH' favour in the closing seconds of the tenth round, when MILLS had reached what was, in his opinion, the utmost state of physical exhaustion to which a boxer should be allowed to go. MILLS's manager, Ted BROADRIBB, fully concurred in this action, and, subsequently, MILLS – in writing of the incident – expressed approval of Eugene HENDERSON's decision. Nevertheless, the wisdom of his action was called in question, and subsequently Eugene HENDERSON resigned from refereeing for a period."

"The knock out blow which so often takes the crowd by surprise, and even the man who delivers it, is often decisive because a boxer has been so dazed by previous blows that he leaves himself exposed to the final punch. Of course to all this the inevitable reply is forthcoming from those with a monetary interest in the business that the referee is there to see fair play, and to protect the fighters from suffering an injury which might prove to be of serious nature. There is, of course, no real protection for the fighters; the crowd sees to that."

The next chapter in Lady SUMMERSKILL's essay is titled *'Champion for a Day'*

"It has been argued by the more enlightened in the prize fight world that there is some truth in my contention that the spectacle of two men stripped to the waist battering each other for the entertainment of the crowd is calculated to have a degrading effect on the spectators. This, it is pleaded in mitigation, is only the case in those contests where two heavyweights are engaged; in all other classes the contestants depend upon technique to achieve victory, and the fights tend to resemble boxing displays given in well-conducted gymnasia."

"Unfortunately the pictures and reports in the newspapers and the commentaries on television do not support this."

"The nature of the Dai DOWER – Young MARTIN flyweight fight on October 3rd, 1955 sickened even the most seasoned reporters."

"In the *Daily Mail*, October 4th, Frank McGHEE, gives this account:"

"'Dai DOWER, the pride of Wales, was battered and bludgeoned into a sensational twelfth-round defeat by Young MARTIN, of Spain, at Nottingham last night.'"

"'He lost his European fly-weight title. He lost his unbeaten record. And a great dream exploded. For he has lost – for a long, long time – the chance of bringing a world championship to Britain.'"

"'Dai absorbed so much punishment that it is doubtful whether he will ever be as good a fighter again as he was before this knock out night.'"

"'He was knocked down TEN times in the fight – SIX times in one round, the tenth.'"

"'That tenth round was three minutes of sheer brutality that is burned into the memory of every man who was there.'"

"'He took three counts of eight, and one of seven.'"

"'I can see it now. I shall always see it.'"

"'It started with a left hook from the southpaw MARTIN, digging deep into DOWER's body, unhinging his legs and sending him DOWN on his knees for the second time in the fifth – he was DOWN for eight from a similar punch in the fifth.'"

"'The tiny, slim-shouldered, curly-haired Welshman was half erect on unsteady legs again at eight, pawing feebly with the straight left that was one his most useful weapon.'"

"'MARTIN just brushed it aside, ignored it, almost laughed at it... and crashed a left to the head. DOWN went DOWER again and we thought it was all over.'"

"'I wish it had been all over. This wasn't pretty to watch. It isn't a pleasant story to tell.'"

"'But DOWER dragged himself up again at eight and was slammed DOWN again by another rib-bending hook.'"

"'He was up at nine, then DOWN again.'"

"'Surely this was the time for the French referee René VAISBERG, to call a halt. Surely even a European title wasn't this important.'"

"'But impassive-faced, he let it go on and on, while DOWER got up and was smashed DOWN, got up and was smashed down.'"

"'The end was in sight and everyone in the place longed for it.'"

"'But we had to wait through one whole round, the eleventh, forty seconds of the twelfth, and THREE MORE NIGHTMARE KNOCKDOWNS before it was all over.'"

"'The punch that finished it was a left to the head... and even then it wasn't a clean-cut merciful K. O.'"

DOWER, that brave little man, was still trying to force his body erect when the count reached ten... He couldn't make it... and we were all so glad.'"

"'For this fight was more than a tragedy for Wales and for DOWER. It was a tragedy for boxing.'"

"'It made a mockery of what used to be the noble art of self-defence.'"

"'MARTIN is no more than a slugger – a miniature MARCIANO, tremendously strong, tremendously fit, immune to punishment.'"

"'He crowded DOWER right out of the fight, bulling his way in...'"

"'And afterwards DOWER looked as if he had been in a road accident.'"
"'Twenty minutes after the fight he was still flat on his back, hardly breathing, while his young wife, Evelyn, sat beside him, sobbing.'"
"'THAT, TOO, IS SOMETHING I'D LIKE TO FORGET.'"
"Young Mrs. DOWER, aged nineteen, certainly voiced the views of every sensible women in the country when she went running to the dressing-room of her boy husband after the encounter, crying and sobbing.'"
"'Are you all right, Dai? I'm not worried about you losing silly titles. All I'm worried is – are you alright yourself?'" *(Tom PHILLIPS, Daily Herald,* October *4th, 1955).*
"Only the future can answer that question."
"After each fight which seems to mark the end of the hitherto uninterrupted progress of a boxer, the sports commentators express astonishment and shocked surprise to discover that suddenly the champion fails to deliver his punches accurately, that there is an elementary misjudgment in placing his blows and a lack of co-ordination which make him exceedingly vulnerable. Are these experienced men, who have spent years at the ringside, genuinely surprised, or are they compelled as part of the 'game' to make the appropriate sympathetic noises in order not to focus the attention of the public, and other aspirants to ring fame, on the tragic end of yet another young boxer? The sports commentators have seen one young man after another heralded as a future champion, seen him box for a few years, acclaimed by the crowd, surrounded by good-time friends and then after one night's failure sink into obscurity.'"
"The memory of the public is short and the names which have made boxing news are quickly forgotten, to be replaced by those of other simple, ignorant young men with sound brains. The pattern of life will be repeated. A few short years of glamour and excitement and then the most sensitive part of the human being, the delicate brain, will fail to deliver its messages accurately to the fists, and the promising young boxer, however desperately he may train, is doomed to mediocrity or worse."
"Names such as Randolph TURPIN have been headline boxing news within the last two years. Why do such men all fail to make the grade in their latest fights? Is it coincidence, lack of training or something over which they have no control?"
"Randolph TURPIN for a short period was the great British hope in the boxing world. Yet on October 18th, 1955, he was knocked out in the fourth round, at *Harringay,* by Gordon WALLACE, a comparatively unknown boxer from Canada. This fight was planned to be a comeback for TURPIN preliminary to his meeting with Archie MOORE, and his opponent's only claim to fame was that he beat Johnnie SULLIVAN of Preston when he was a novice in Canada."
"Yet TURPIN suffered a crushing defeat."
"This man, whose powerful, well directed blows gave him a victory over 'Sugar' Ray ROBINSON a comparatively short time before, had deteriorated in a dramatic fashion. It was reported that his timing and distance were so poor, his punches so wild, that he missed his target badly."
"Two rights to TURPIN's head put him down for a count of nine. He got up groggy, reeling on the ropes. WALLACE crashed in another right, TURPIN's head went bang on the canvas. Desperately he struggled to his knees at eight, but failed to beat the count" *(Tom PHILLIPS, Daily Herald, October 19th, 1955).*
"After the fight he said in his dressing-room:"
"'I'm finished! I'm retiring from the ring. I did my best but it just wasn't there. I'm sorry.'"
"TURPIN's manager, George MIDDLETON, said, 'This is the end. He has worked hard in training to get back to the top and we have done our best. You can see the result. We are disappointed but our decision is final. Randolph TURPIN is finished. He will fight no more" *(Tom PHILLIPS, Daily Mail, October 19, 1955).*
"Those who had an opportunity of observing Randolph TURPIN since he was defeated in a return fight by 'Sugar' Ray ROBINSON in New York, in the course of which he was subjected to the most violent blows to the head, say that this encounter appeared to affect all his subsequent performances. The match with 'Sugar' Ray ROBINSON was stopped in the tenth round because TURPIN was unable to defend himself. He was knocked down in that fight by a terrific blow to the temple and he was dazed and helpless when the referee stopped the contest. Subsequently he was beaten by a mediocre middle-weight, Carl Bobo OLSEN, for the world title in New York."
"Then he was knocked out in sixty-eight seconds in Rome by Tiberio MITRI."
"And so the young boxer, only twenty-seven years old, tasted the bitterness of complete defeat."
"His courage, his tenacity, and his capacity for hard training were not in question, but the rapid timing and the power of co-ordination which derives from the sensitive brain were lacking."
"If he does retire from the ring his absence will not be noticed; there are plenty more of young willing to take his place."
"While certain names are still fresh in our memory there are others now faded which were as famous in their day. Bruce WOODCOCK had a world reputation but he was knocked down by Joe BAKSI, the American, at *Harringay on April* 15th, 1947, after one minute of the first round. The blow fractured his jaw and damaged his eye, yet he fought on for another six rounds. After eighteen months enforced rest, he tried to return and fought Jack GARDNER who defeated him in the tenth round at *Earl's Court* on November 14th, 1950. He then announced that he must retire owing to the double vision in one eye from which he had suffered since the BAKSI fight."
"Another fighter who it was believed might become a world champion was Danny O'SULLIVAN, a British bantamweight champion."

"He fought Luis ROMERO of Spain, European Champion, at *Harringay,* April 25th, 1950. He took eleven counts before the referee stopped the fight in the thirteenth round. He suffered a terrible beating and after a few more fights retired."

"Jack CASEY was another up-and-coming young fighter advertised as the 'Sunderland Assassin.' He fought Len HARVEY, the British middle-weight champion, at *Newcastle-on-Tyne,* and according to Len HARVEY, 'It was like hitting a tank. In that first fight he just came forward taking everything. I couldn't stop him smiling. My gloves nearly burst because my hands were swelling, badly bruised through hitting him so much. Yet it was impossible for me to knock him out!"

"That terrible fight left his mark. One more terrific but unsuccessful encounter with HARVEY altered the career of the 'Sunderland Assassin."

The following chapter of Lady SUMMERSKILL's essay is entitled *'The Audience and the Law'*

"There is only one reason why two mature men are prepared to enter a small roped enclosure to batter each other for a certain period of time, and that is of financial benefit to a promoter."

"The boxer's part in the business is to earn money, not to demonstrate boxing technique. The spectator provides him with that money via the promoter and expects to get his money's worth. The big crowds are attracted to the heavy-weight championships because they provide 'a good hard fight.'"

"To listen to the noise and to watch the faces of the spectators at a heavy-weight championship bout, is to observe human beings at their worst – uncontrolled, and indulging in an orgy of cruelty. Of course the spectator always claims that he is only interested in technique, and this may be true of a few. At least this can be said for the average spectator at a boxing bout; he knows little about the psychology of man, nor has he given any serious thought to the primitive impulses involved."

"The fact is that a pleasurable feeling is induced by the sight of one man beating another into insensibility, and this primitive emotion, which in civilized people should be disciplined, is fostered by those responsible for staging these bouts. The brutishness of man is no secret. It is revealed every day in our criminal courts. We have yet to evolve into a higher society."

"Meanwhile, let us recognize, without being mealy-mouthed about it that, however much man may protest that it is the technique of boxing which attracts them, they, the promoter, the commentators, and the principal performers, the boxers themselves, know that a 'good fight' means catering to the worst instincts of man."

"Cock-fighting was condemned not solely in the interest of the cock, but because of the unwholesome effect upon the spectators. One could think that if the sight of cocks fighting each other to the death was once calculated to raise the blood-lust in human beings, then the sight of their own species battering each other's faces into pulpy masses would offer an equally, if not more degrading, spectacle."

"The position of the spectator of a prize fight in the eyes of the law raises some interesting legal points. Is he aiding and abetting an assault by his shouts and gestures?"

"It was as long ago as 1882 that Mr. Justice HAWKINS said in R. v. CONEY:"

"'Every fight in which the object and intent of each of the combatants is to subdue the other by violent blows is a breach of the peace; and it matters not, in my opinion, whether such fight be a hostile fight begun in anger, or a prize fight for money or other advantage. In each case the object is the same, and in each case some amount of personal injury to one or both of the combatants is a probable consequence; and although a prize fight may not commence in anger, it is unquestionable calculated to rouse angry feelings of both before its conclusion. I have no doubt then, that every such fight is illegal and the parties to it may be prosecuted for assaults upon each other. Many authorities support this view.'"

"This ruling made every prize fight unlawful and each active participant guilty of assault. While an individual could give permission to a surgeon to perform an operation, according to the letter of the law in 1882, he could not give permission for an assault upon himself for the entertainment of others."

"Apparently the Marquis of QUEENSBERRY was too preoccupied with studying anatomy to concern himself with the finer points of the law which governed the display of fisticuffs."

"Mr. Justice HAWKINS adopted a lofty and judicial attitude towards the spectators. He differentiated between the passive and the active devotee of the noble art."

"In his opinion an aider or abettor must take 'some active steps by word or action to instigate the principals.' He held that:"

"'Encouraging does not necessarily amount to aiding and abetting. It may be intentional or unintentional. A man may unwittingly encourage another by his presence, by misinterpreted words by gestures, or by his silence or non-interference; or he may encourage intentionally by expressions, gestures, or actions intended to signify approval. In the latter case, he aids and abets, in the former case he does not.'"

"Judging from the yells of the crowd to-day, when they have been roused to a pitch of frenzy it would, I should think, have been difficult to find one who at the end of a series of bouts could not be labeled as an aider and abettor and liable to be indicted for assault. However, today the legal position is a little obscure. From various judicial pronouncements it would seem that provided the contestants wear gloves they are not committing any offence."

The next chapter in Lady SUMMERSKILL's essay has the title *'Amateurs and Professionals.'*

"There are many who condemn prize fights, but when asked about their attitude towards amateur boxing are inclined to hedge. This is understandable because the sordid business interests which surround the professional ring, for which amateurs have the utmost contempt, are lacking in the amateur world."

"As an observer of the boxing world, I find it difficult to distinguish between amateur and professional boxing. The rules are the same apparently and the objective is the same, namely to knock your opponent out in the shortest possible space of time."

"The amateur generally covers his chest with a little vest. Is this a concession to modesty or to make some visible distinction between the amateur and the professional?"

"As boxing to some extent is concerned with the glamorization of the male body, it is a pity that the amateur should decide to hide what might be an excellent chest development."

"It is generally thought by men that women are the worse snobs. I am well aware that women waste much physical and mental energy and precious house-keeping money trying to live up to the JONESES, but the most distasteful kind of snobbery is practised by men in the field of sport, although sport is supposed to offer an opportunity to demonstrate the finest human qualities."

"For years cricketers were expected to use two doors, one for the 'gentleman' and one for the 'Players,' in order to distinguish those who received a monetary reward. A member of the 'gentleman's' team might be engaged in all kinds of shady transactions during the week as a result of which he made a large income, transactions which a 'player' with his moderate income would not consider fair and sportsmanslike. When the week-end arrived the 'player' was relegated to such an inferior social position that he had to use what might be compared to the tradesman's entrance."

"Similarly in boxing, apparently it is right and socially proper to box for a canteen of cutlery or a gold watch, but not for the equivalent in terms of hard cash."

"Again, the amateur who does a little boxing at the weekend may be far inferior in skill to the professional, but as he does not accept payment he is superior socially, whatever means he adopts to earn his livelihood in the course of the week."

"The sheer humbug of the professional versus the amateur distinction in sport, whether it be in boxing, ice skating, or lawn tennis, is revealed when an individual who has been a stickler for all the proper forms of etiquette being observed in relation to the amateur, turns professional overnight. This transformation is invariably accompanied by a contract carrying some little compensation for the rapid metamorphosis."

"Now it is undoubtedly true that to-day professional boxing carries a social stigma. Why has this happened? Lord KNEBWORTH in the volume of *'The Lonsdale Library'* gives this explanation:"

"'When there entered the professional ring a number of ill-made, ugly, unworthy men who could no more box than they could appreciate fresh air, but who understood the subtle arts of finance and showmanship and 'boost,' the degradation of the professional ring began...'"

"'As a lure of money took the place of pride in victory, as clean fighting was superseded by clever manipulation, so the mad lunacy of twentieth-century sport burst upon a bewildered and little understanding world."

"In my condemnation of boxing I confess that I have never become quite so offensive as this authority on the business. He goes on to say:"

"'An age which counted money as more important than everything else in the world; an age of insincerity and folly; an age of small men and huge half-educated crowds seized upon the professional ring like harpy and scattered its fine traditions and splendid ideas to the four quarters of the world...'"

"' A championship became not a well won and worthy possession, but a negotiable financial asset. A boxer became not a fine specimen of physical man engaged upon the greatest of known physical endeavour, but either a shrewd hard-headed business man or a commodity for the manipulation of other and more competent business men.'"

"The public which attended boxing contests was no longer either interested, well informed, or enthusiastic so much as a great herd of ignorant and befuzzled people, potential ticket buyers, fooled by the press, fooled by the promoter, fooled by the boxer, and fooled by every book-maker who troubled to relieve them of their cash."

"If all this is true – and the evidence is such that most honest people believe that it is true – I can understand why the amateur dons his little vest to distinguish himself from those who frequent the professional world."

"The money making aspect of prize fighting in the last century certainly played an important part, but it was in the hands of a few 'gentlemen backers' who arranged fights between their pet pugilists."

"When boxing gloves were introduced the 'backers' disappeared, and the middle men in the shape of the manager and promoter came into the picture."

"It is the manager's job to find the youth who shows exceptional promise, train him, and groom him for the attention of the promoter."

"The promoter must understand the public as well as the good points of a fighter, for his success will depend on pleasing the crowd, whose assessment of a boxer is not solely related to his boxing ability. Personality, whether it be on the stage, in Parliament, or in boxing, always attracts the interest of a crowd. Boxers like Billy WELLS may not have had great ability, but the crowd liked him and this was good for the box office."

"The promoter's dream is to find first-class boxers with highly attractive personalities; but in order to fill a large hall these personalities must be carefully exploited. Before a fight the public are not interested in reading about the fine footwork of a fighter or his own special technique in the ring. They like to hear about his domestic life, his wife, mother, children or girl friend. They like to know what he eats, what he wears and the time he goes to bed. The attention of the public before the VALDES – COCKELL fight was concentrated, not on boxing technique, but on the all-absorbing question whether VALDES would wear white pants or not."

"This build-up calls for good relations with the press and the co-operation of the boxer with the sports commentators."

"Boxers are not generally business men, and fights conducted on a mammoth scale as they are to-day, with large sums at stake, attract men who view a simple youth who shows boxing promise as a good financial proposition. The boxer and his family do not fail to understand the reason for the interest which potential managers take in their affairs. The consequence is that the boxer-manager relationship is not always harmonious."

"When the contract between them expires, the boxer whose reputation has grown is often in a hurry to get rid of a manager who he feels has had more than his fair share of his earnings. He may be right."

"On the other hand the boxer may have developed a swelled head and consequently minimise what has been done for him by a good manager."

"Then there is the manager, anxious to make as much money in a short period of time, who overworked a boxer to the latter's detriment."

"The manager-boxer relationship is certainly a delicate one in a world akin in some respects to the theatrical world. The long weeks of preparation for 'the night' can be likened to rehearsals, the press interest is similar, and the boxer, surrounded by admirers, flatterers, sycophants and the rest, is inclined to bouts of temperament."

"It has been known for a manager to sell his contract, and to walk out in the middle of training, or on the eve of the fight."

"Against this background fights take place for enormous purses and before huge crowds. When TUNNEY fought DEMPSEY in Chicago, £ 600.000 was taken for seats. Subsequently, when they met again, DEMPSEY's share of the purse was about £ 147.000."

"This was a little different from the days when Deaf BURKE commented on receiving his share. 'Twenty bob wasn't bad for twenty minutes' easy work."

"The cost of organising a big fight is tremendous and the promoter has no guarantee that he will make a profit, and therefore the whole thing is in the nature of a huge gamble, at least so the promoter always assures us."

"For this reason the press is of primary importance and the sporting journalist plays an important part in the boxing world. Although the part he plays to-day is in his working capacity, the sporting journalists and other writers were, in the early years of boxing, among those who encouraged boxers at places like the *Pelican Club* and the *Blue Anchor, Shoreditch,* and finally welcomed the fraternity to the *National Sporting Club* and later even the *Albert Hall."*

"In these days the sporting journalist concentrated on a boxer's form and technique, but lately he has been compelled to bow to the great advertising interests which finance his paper."

"To-day every newspaper manager is haunted by the fear of a falling circulation. If advertisers are to be attracted, then the circulation must be increased. This frantic competition to increase circulation has had a disastrous effect upon the press of today. Sensational journalism appears to be the sole objective of many newspapers; the aesthetic, cultural and intellectual needs of the public are ignored and every news incident which can be presented in a manner which may appeal to the lowest impulses is magnified and elaborated."

"The public, so the circulation manager argues, is not interested in the scientific approach to boxing – it wants gossip about the fighter before the event and a bloodthirsty description of the fight afterwards. Many of the sporting journalists are deserving of sympathy. Their technical ability has to be sacrificed to this mad rush to build circulation through an appeal to the stupid and brutish. The journalist must earn his living and he knows only one way of doing this. If he is to feed his family he must please the editor, and the editor, is to hold his job, has to listen to the business side of the paper, which pins all its hopes of greater prosperity on an increase in circulation."

"Of course there are newspapers which still attach great importance to moral values, but a boxer does not look to these to give him the publicity."

The next chapter in Lady SUMMERSKILL's essay is entitled *'Boxing in the U.S.'*

"The conditions in the boxing world here are only a pale reflection of what happens in America. However, as those engaged in the business on both sides of the Atlantic are so closely associated, the degrading scenes may well multiply and worsen."

"The conditions under which fights are staged in the U.S.A. are so disgraceful that the *National Boxing Association of America* asked Congress on August 16th, 1955, to use its powers, as the *Association* was unable to put its own house in order."

"Heavy gambling on fights causes frequent scandals which are exposed in the papers, Tom PHILLIPS describes a crime of which he had personal experience. Ray ARDL, a friend of his, was the victim of a murderous attack in Boston in 1953. ARDL had been a manager and trainer and decided to become a promoter of television boxing without the protection of a gang. For this demonstration of independence his skull was fractured with an iron bar. Apparently this was just the beginning of a brutal persecution, which culminated in a further beating up."

"It is said that on previous occasions when inquiries have been instituted the gangsters have treated the whole matter as a joke and finally no action has been taken."

"If the United States Government took strong action now it would have a salutary effect on the business of prize fighting throughout the world."

"While the conduct of professional boxing in Britain is slowly rousing people to a sense of their own responsibility in this matter, the so-called 'fight business' in the U.S.A. has provoked many highly responsible individuals to make strong protests."

"If one is to judge from the uproar which followed the COCKELL v. MARCIANO fight, one must assume that the American boxing rules and regulations are different from ours and not rightly adhered to. I have no intention of venturing into the technicalities of the business. I come to this conclusion after reading the press reports of the enthusiasts who watched the fight and subsequently commented upon it in this country."

"Certainly the fights in the U.S.A. seem to be publicised weeks before the event, with details of the boxers' private lives, their daily condition and reactions, and forecasts alternating from week to week as to the outcome."

"Apparently some boxers, who are accustomed to this fierce publicity, enjoy it or are indifferent to the hubbub; this is just as well, for others, who do not respond well to the persistent pressure of the reporters, fare badly. They are pictured as being bad-tempered and moody, and gradually an unattractive personality is built up which reflects on the training quarters and which, I presume, affects the attitude of the crowd on the night."

"During the preliminary period, while this artificial excitement is being whipped up, bets are laid and, of course, the odds vary according to the fluctuating news from the two 'stables.'"

"The bookmaker seems to be an indispensable part of this sport and his activities both before and during the fight contribute to the rather sordid atmosphere. The excitement of a fight, whether viewed from the arena or on television, can certainly be intensified if the viewer also has a financial interest in the result, however small."

"In America boxing in the past has tended to symbolise the American approach to life; it has emphasised the importance of aggressive qualities, of strength and endurance and the ability to take punishment."

"During the last twenty years a slow but noticeable change has taken place in the attitude of many responsible people. Scientists, doctors and physical instructors who have studied the effects of these bouts on the contestants and the public, are now openly opposing these exhibitions."

" A few boxers also have expressed adverse opinions based on their own personal experience. One of the most courageous writers and speakers on the subject is Dr. Arthur H. STEINHAUS, *now Professor of Physiology at George William College, Chicago,* and a *past president of the American Academy of Physical Education."*

"Dr. STEINHAUS stirred the conscience of the public and infuriated the boxing fraternity by an article in *Look (January 1950)* called *'Boxing – Legalised Murder?'"*

"He reminded the American public that forty-two boxers had died in the previous years; for every death there are hundreds of other tragedies – 'the young men who have had their brains knocked out. Their cases get no publicity. They are the living dead of pugilism, the victims of its occupational disease: punch drunkenness.'"

"He drew attention to the investigations of Dr. Ward C. HALSTEAD of the *University of Chicago,* who examined 147 cases of brain injury in which no trace of injury was visible outside the skull. Brain injuries, he pointed out, do not heal. A broken leg has to set. A black eye clears up. But the destruction of brain cells is permanent. Impairment of thinking powers, speech, gait, and emotional stability results."

"Dr. STEINHAUS urges that boxing should be abolished in the high schools and colleges and for amateurs generally. The view is supported by the recorded vote of half the *College Directors of Physical Education* in the country; by virtually all State offices of public instruction; by the *National Association of Physical Education* and by every doctor who has given the subject serious study."

"On a radio programme sponsored by the *American Medical Association*, it was stated, 'Heavy gloves or head-gear may reduce external injuries, but their effectiveness against brain injury is open to question."

"Shortly after Dr. STEINHAUS expressed his views forcibly to the American public in *Look,* Larme ROACH, a twenty-four-year-old professional boxer died of a brain injury received in a fight at *St. Nicholas Arena* in New York. Carmine VINGO, twenty-year-old heavy-weight, was luckier. He was released from the same hospital in which ROACH died, after six weeks' treatment of a near-fatal brain injury, which left him with a semi-paralysed left leg."

"In order to focus attention of the American public on the urgency of the problem, Dr. STEINHAUS wrote an article for the *Journal of the American Association for Health, Physical Education and Recreation*, entitled, *'Boxers Brains Swapped for Medals.'"*

"He said, 'Much is written of death in the ring. But the forty-seven professional and at least twenty-two amateurs killed since the beginning of 1946 are only exclamation points in the records of boxing. Of greater significance is the much larger number that is beaten into various stages of mental incapacity, from Bowery 'stumble bum' to institutionalised insanity.'"

"A noted brain surgeon, who had seen hundreds of amateur and professional boxers with marks of brain injury, expressed the view to Dr. STEINHAUS that, whether it resulted in a knock-down or not, he felt certain that every head pommeling was likely to leave some small portion of brain tissue permanently damaged. This, he said, might go unnoticed for a time because there is an abundance of cortical tissue, but where permanent damage was involved repeated injuries added up."

"Dr. JOKL put it this way: 'Headaches on the day following a bout are consequences of brain swelling, which though viewed as serious symptoms should they appear following an auto accident, for example, are generally ignored among people in the world of sport."

"The American commanding officer of a hospital unit in Australia decided to introduce boxing for the purpose of introducing variety into the lives of those working in a place far removed from home. Approximately 50 of the 500 men attached to the unit took part in this new activity. Within a month the attending brain surgeon hospitalised four of them for serious headaches, ringing in their ears, and dizzy spells."

"Within a year three of the four were court-martialled, although previously they had merited regular promotion."

"Dr. STEINHAUS says:"

"'Some educators, though ready to condemn professional boxing, are convinced that amateur boxing, especially as conducted in high school and college circles, is virtually without danger.'"

"The ideals of those who promote college boxing are stated in the 1950 *Official Boxing Guide* of the *National Collegiate Athletic Association (U.S.A.)* as follows:"

"'Skill is the criterion, not brute power of force; points, not knock downs; clean hits, not knock outs. Intelligent boxing de-emphasises power and emphasises skill. A well-conceived skillful attack that builds up points is indicative of college boxing. A knock down or knock out, should it occur, is considered incidental to the bout, not the objective.'"

"College boxing has profited much from such safe-guarding regulations, restrictions and philosophy. But when the safety of America's highly selected University students is at stake it is necessary to ask, 'Is this enough?' One naturally turns to the records to see how many have taken beatings of the kind that damage the brain."

"Reporting is not uniform from year to year and usually only the number of bouts are given, with no indication of total entries. Still some calculations are possible."

" The *N.C.A.A.* year books report six annual tournaments of the *Eastern Collegiate Boxing Association* for the years 1945-1950 inclusive. In all these are reported the outcome of 172 bouts. Of these four ended in knock outs, thirty-three in technical knock outs, two were forfeited, one defaulted and three were stopped because of cuts and other injuries."

"Generally, the total number of entries in each weight class exceeds the number of bouts in that class by one or possibly two. Calculated on this assumption, there were about 196 entries in the six tournaments. If thirty-seven bouts ended in either K. O's or T. K. O's we are led to conclude that nineteen per cent of the boys who engaged in the *Eastern Intercollegiate Boxing Tournaments* between 1945 and 1950 suffered blows that probably left them with some permanent brain damage."

"An investigation in the U.S.A. in 1949, of the attitude of the authorities in eighty-four colleges and universities, conducted by Messrs. KENNEDY, THAKER and GEBHART of the *University of Illinois,* revealed that slightly more than fifty per cent of the *health and athletic directors* did not favour the sport for intercollegiate competition and did not consider it a healthy activity. Nearly twenty per cent of those who had varsity teams opposed it and of sixty-two *health service directors only* fifteen per cent regarded boxing as an appropriate sport for high schools. Another eighteen per cent favoured it with, certain qualifications, whereas fifty-eight per cent outspokenly opposed it, and nine per cent were doubtful. Twelve universities reported cases of punch drunkenness resulting from their boxing programme and four reported cases of amnesia. Several death were also reported."

"Dr. A. H. STEINHAUS in a statement on boxing in *High School,* prepared at the request of the *District of Columbia Board of Education* while it was considering whether boxing should be introduced into the schools of the District, expressed the following opinion:

"'It is often stated that the teaching of boxing with heavy gloves to boys can in no sense be compared with prize fighting, and therefore the consequences of prize fighting should not enter the picture in argument against 'controlled' boxing instruction with adequate protection. Those who defend this position must face the following questions:"

"1. Can the full advantages credited to boxing be obtained short of genuine 'all-out' boxing?

2. Can an average high school coach or teacher control a large class of boxers to ensure 'safe' boxing?

3. Is there available and on hand a thoroughly safe head-gear to protect against not head injury, but *brain* injury?

4. Is the average boy who becomes good in the sport willing to use his skills only in self-defence and unlikely to use the sport as an offence in overcoming opponents both in and out of athletic contest?

5. Can a high school teach boxing without at once giving *implied* approval to amateur boxing tournaments and to the professional ring?

6. Would you favour a 'Golden Glove tournament' in which head-gear was used that would absolutely eliminate the possibility of brain injury and knock outs?"

"A single 'no' answer to any of the above questions is, I believe, evidence that boxing has no place in an educational system that desires to align its own practices and its influence on American sport in accord with the best available knowledge."

"Dr. KENNEY and his team, who conducted investigations into intercollegiate boxing, aptly remark:"

"'Instruction classes in boxing are not particularly dangerous, but it is impossible to justify teaching an activity and the same time to advice students against participation in the activity taught.'"

"'Much has been done to clean up boxing and make it safe. Such efforts should not be disparaged, neither should they be valued above their real worth. The shortening of bouts to three rounds reduces the likelihood of contestants fatigued, but it does not prevent the hardest blows that are dealt when boxers are fresh. The death, on February 5th, 1951, of sixteen-year-old *Golden Gloves contender,* John SHODDY, at Fort Wayne in the first minute of the first round; that of thirteen-year-old David ZIMMERMAN after three one-minute rounds in Kent, Minnesota, on February 14th, 1951; that of Dixon WALKER of the University of Maryland, in his first round in a match with *Catholic University,* March 3rd, 1946, must embarrass the three-round argument."

"Careful and expert medical services before the bout and at the ringside, even electro-encephalographs, are frequently entrusted with making boxing a safe sport. The best a physician can do is to keep an unfit man from boxing or stop a bout when a fit man has been hurt."

"Where else, except in war, does medical talent permit itself to be associated in like manner, with a practice that is calculated to produce damage to human flesh?"

"A doctor peering into the eyes of a boy and then permitting him to fight one more round to his death has recently exposed to millions of America's television audience the futility of medical supervision."

"In two years, twenty-six *University of Wisconsin boxers* reported being temporarily dazed thirty-eight times in supervised matches; but the attending physicians were able to identify this condition by their objective criteria in only nine instances."

"If this can and did happen under the most careful and extant medical supervision ever given to boxing, need more be said?"

"Mrs. MATZ of Lake Crystal, Minnesota, whose seventeen-year-old son was killed in an amateur bout, allowed Dr. STEINHAUS to publish her letters to him. She said:"

"'I did not know the danger in boxing, and I'm sure my son never did either. It would be much fairer for men who put these boxing bouts on to explain to these young lads the danger in Boxing... Can't this sport be stopped?'"

"In 1944 Abe GREENE, commissioner and Chairman of the *National Boxing Association* is quoted in the *New Orleans States Item* as follows, 'A constant beating about the head induces in a fighter the stage generally known as punch drunkenness. Oftimes a fighter may be in great shape outwardly, but he may be suffering any one of a number of impairments."

"His sight might be in danger. His head may be damaged by heavy blows in the past, which have caused a series of smaller haemorrhages and concussions. His nerve system may be jolted by past knock outs. This impairment may not show in a superficial examination."

"J. H. HASANG, a successful boxer, manager and referee, about 1915 wrote as follows, 'I once wrote a letter to the paper about thirty-five years ago, defending boxing, because I like it, like many other immature minds, and I have been sorry ever since... Abe GREEN says boxing needs a doctor, what it needs is an undertaker.'"

"Francis J. CONNELL (1950) on *'Prize Fighting and Boxing'* in the *American Ecclesiastical Review* quoting DAMEN from *Theologia Moralis,* Rome 1947, concludes that even boxing is often accompanied by injury that it may easily assume sinful features. He says, 'It is difficult to reconcile prize fighting, as we have it today, with Catholic principles of morality... This opinion may seem somewhat severe in view of the widespread belief of the American people that prize fighting is a 'good clean sport.' Yet it is difficult to see how any other interpretation of the Sixth Commandment can be given.'"

The next chapter in Lady SUMMERSKILL's essay is entitled *'Nigeria and Belgium.'*

"While European supporters of boxing may vigorously assert that exhibitions of men fighting each other inculcate in the contestants, and those who watch, some of the finest attributes of the human race, are their consciences equally easy about the promotion of boxing in other countries?"

"This century is witnessing the emergence of small countries into full nationhood. The people of some of these countries are still illiterate, superstitious and desperately striving towards a civilised way of life. Evolution is slow and difficult for many who are still subject to tribal customs, traditions and prejudices. They are endeavouring to master their primitive instincts, and it is to their great credit that people who are only a generation removed from the primitive life of wild and desolate countryside are learning the arts of self-government. Codes of behaviour in these countries are generally based on Western standards, and to shirk our responsibilities to *these people is to betray our trust.*"

"As I write I have before me the *Sunday Times* of Nigeria, the brain child of an English newspaper office."

"The editor has considered it desirable to give a page to news on boxing, illustrated by men in fighting attitudes. Apparently the Nigerians have not taken up the pastime of 'civilized' fighting seriously and it is therefore considered necessary to give it more publicity."

"On July 3rd, 1955, the Sports Editor writes, 'The men of ideas in the *Twentieth -Century Promoters' Association* are always out for novelty to see to it that boxing returns to its own in Nigeria. For the current month they hope to stage what will be known and remembered for a long time as the 'boxing match of the year.' It is a long time since Lagos boxing fans saw an instructional boxing tournament. The matches staged in recent years were rather disappointing."

"Boxing in the Gold Coast has not been properly organised as it is in Nigeria, and this was confirmed by Kop BADIR, Sports Editor of the *Daily Graphic*, Accra, when last he came down to Lagos for the international athletic meeting."

"At least it is good news that the people of the Gold Coast, who are striving to establish a well-ordered society, have not thought it fit and proper to stage prize fights. But no doubt some *Promoters' Association* whose only interest in the fighting business is the amount of profit which it can make, and not the orderly progress of a young state or a high standard of human behaviour, will come along and see to it that the Gold Coast enjoys to the full the brutish spectacles associated with heavy-weight fights."

"The *Daily Times*, another paper which has close English associations, gives the fighting business a helping hand by goading the young Nigerian male into action. Under the heading:"

"Five Boxing Titles Are Vacant"

"Five boxing titles out of the eight open to Nigerian boxers are still vacant – heavy-weight, middle-weight, welter-weight, light-weight, and feather-weight."

"The would-be managers and promoters will be pleased to learn that the source of supply of the above is also receiving attention."

"The boxing sub-committee of the *Ibadan Boys' Clubs* plans a series of boxing tournaments 'to keep the boys in fighting form for their forthcoming league, inter-district and individual championships. It will also enable the committee to select boxers to represent Ibadan against other provincial boxers."

"It seems that the European managers and promoters who have been bewailing the fact that fewer people are watching their fights need no despair. African boys are being trained so that in the years to come they may provide some of the most bloodthirsty exhibitions in the fight business."

"It is, however, encouraging to learn that just across the Channel, public opinion is changing in its attitude towards prize fighting. This was reflected in the *Belgian Parliament* when members of all parties supported a Bill which was introduced into the *House of Representatives* in May 1954. This action was taken following a statement issued by the *Belgian Royal Academy of Medicine* in February 1954."

"The text of a Resolution passed unanimously by the *Belgian Royal Academy of Medicine* on February 17th, 1954, was follows:"

"The *Belgian Royal Academy of Medicine,* after having examined the consequences of repeated blows to the head to which boxers are exposed, which can lead to a state of irremediable cerebral injury, and considering that public boxing matches often constitute a degrading spectacle capable of exercising a bad influence on the moral development of the young, draws the attention of the Public Authorities to the need of regulating very strictly the practice of boxing."

"The Bill was short but uncompromising. It contained four clauses which made the following provisions:"

"Article 1"

"The organisation of boxing and 'catch' matches is forbidden."

"Article 2"

"Persons letting or leading their premises and equipment to this end; persons organising boxing fights and 'catch' matches; the fighters and umpires; will be punished by imprisonment of from eight days to six months and a fine of from a hundred to a thousand francs."

"The King, after consultation with competent sporting organisations on the national level and with *the High Council of Physical Education on Sports and Outdoor Work,* determines the conditions on which demonstrations or bouts of boxing, and 'catch' trial are to be considered as exhibitions."

"Article 3"

"The penalties will be doubled if one of the fighters is injured and is incapacitated for work. They will be tripled in the case of the death of one of the fighters as the result of blows received."

" The country responsible for taking this action has been called *Brave Little Belgium*. Her conduct in two world wars in standing up to powerful aggression evoked the admiration of the whole world."

"No one would charge Belgian politicians with being cowardly or unrealistic or failing to strengthen the moral fibre of the nation, yet they have considered the whole question of the effects of boxing on the population and decided to promote legislation to ban it as demoralising pastime."

The next chapter of Lady SUMMERSKILL's essay is entitled *'My Debate with Jack SOLOMONS* (Fig. 354).'

"I have already said that it is difficult to persuade a man, however honest he may be in other matters, publicly to repudiate prize fighting. He has been taught, quite wrongly, that to fight is to demonstrate virility. Virility is strictly associated with the procreative power, and large biceps have nothing to do with a man's capacity to reproduce his species. A heavy-weight boxer may lack virility in this respect, while an undersized sedentary worker may process it in a marked degree. However, this aspect of masculinity is apt to be overlooked. It is accepted by a male public that to be muscular tough and prepared to fight when provoked is highly desirable, and all advertisements for physical culture are illustrated with pic-

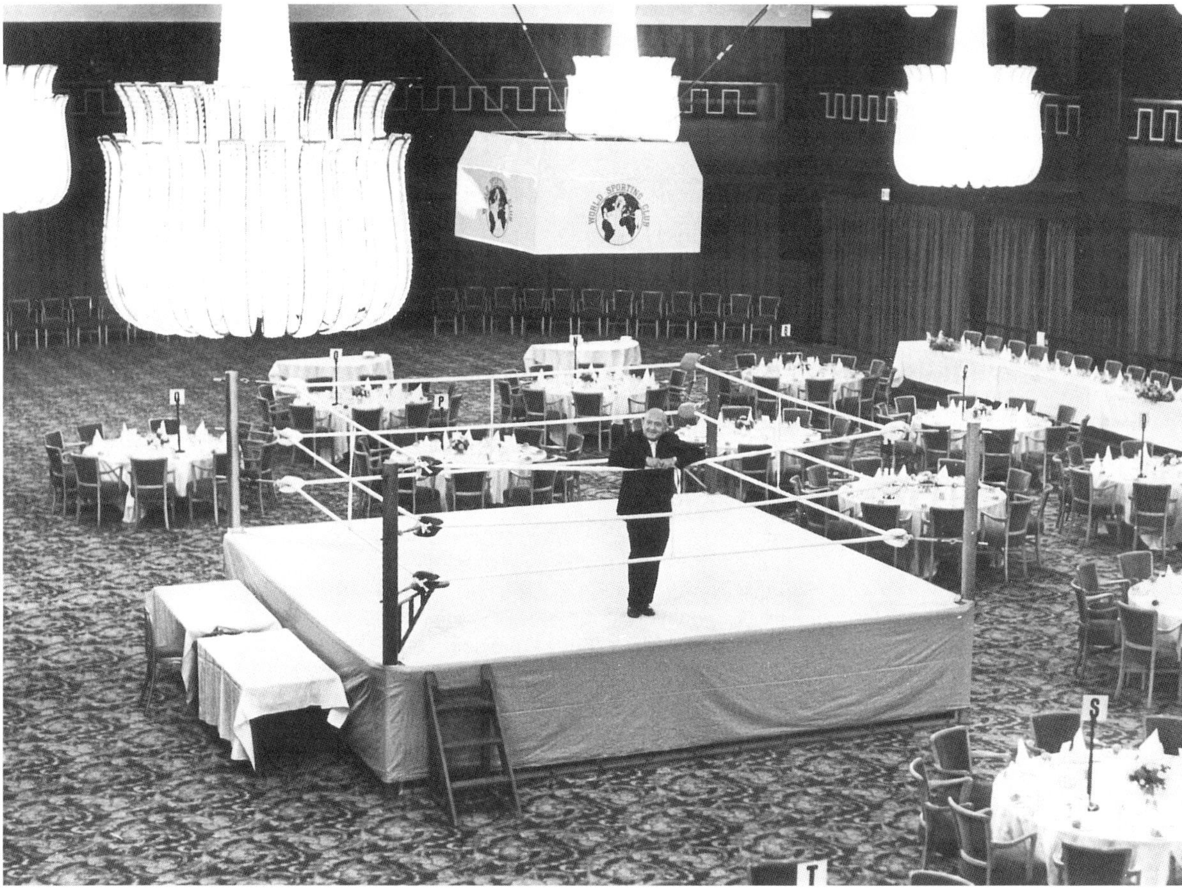

Fig. 354. Promoter Jack SOLOMONS opened the *'New Home'* of the *'World Sporting Club'* of the British box sport. The club will organize each year ten sporting events in the luxurious Great Room of the Grosvenor Hotel in Park Lane in London. The *Great Room Arena,* which can seat 1.200 members and guests, is illuminated by eight colossal cristal chandeliers, in which 832 candles burn. Photo KEYSTONE.

tures of outsize men in shorts demonstrating their muscular prowess."

"Furthermore, it is always assumed that any woman, whether her intellectual powers are nil or highly developed, is solely interested in the he-man, even though he may be an intellectual pigmy. This myth is assiduously propagated in novels, plays, the cinema and wherever else the relationship of the sexes is portrayed."

"While women, at least those with a modicum of intelligence, know that there are male attributes which are infinitely preferable to mere brawn, most men never outgrow the boyish idea that everybody, particularly women, must admire the tough man. Consequently, although we have reached a stage of civilization when the cultivation of the mind is regarded as highly desirable, it is possible to find cultured men who cannot bring themselves to denounce prize fighting publicly."

"It was with some interest therefore and with little hope of success that I accepted an invitation from the *Hardwicke Society,* a debating society of lawyers to which all barristers and students of the *Four Inns of Court* are eligible."

"I was asked to propose the motion, *'That this House wishes professional boxing could be banned.'"* I was told that Mr. Jack SOLOMONS, the boxing promoter, would oppose the motion, supported by Mr. Peter WILSON, the sports writer, and that a member of the *Hardwicke Society* would support me, as I had some difficulty in finding somebody to second the motion."

"The meeting was held in December 1953 at *Niblett Hall, Kings' Bench Walk, Temple,* and I suppose never had such a curious assortment of people entered that Hall, which seemed to smell of musty law tomes and reflect the austerity of the legal world in which it is embedded."

"According to the minutes there were sixty-three people present. As I contemplated the young male President, Treasurer, Secretary and Assistant Secretary and the lawyers in the audience, I knew that for the most part their training had been very similar. The public school and university had left their mark and neither of these institutions are characterised by any new approach to what have always been regarded as strictly male activities. Indeed, even in medical schools which generally form integral parts of a university, the male medical students are as conservative, in is broadest sense, as their grandfathers. They had helped to hold the doors of medical schools shut against women in London until they were compelled to admit a few by the *University Grants Committee,* which made a financial grant dependent upon the admission of women. It seems difficult to believe, but such is the prejudice that one famous medical school, even to-day, refuses to let women students in the Christmas play; the men play the women's parts rather than allow the women to participate. With all this in mind and the knowledge that in the field of law women had yet to make their mark, I anticipated a resounding defeat. The audience was certainly mixed. Apparently those opposing the motion had brought a few friends from the boxing world. Mr. Freddie MILLS was very much in evidence and a few others who appeared to have had some knowledge of the ring if one could judge from the condition of the bony structure of the nose."

"There were a few women present, for which I was thankful; at least I thought that they might not be prejudiced from the outset. In view of the preponderance of visitors it was decided that only the tweny-four members of the Club present should vote."

"Mr. SOLOMONS and I were given twenty minutes each to put the case. I was armed with a skull, which I had borrowed from my daughter – a medical student."

"I methodically explained the anatomy and structure of the brain and bony skull and described the impact of the soft brain on the hard bone following a blow to the point of the jaw – the most damaging form of attack – and explained that even light blows caused the brain to bounce about in its bony case, I then supported my case by quoting medical authorities in other parts of the world. Altogether I took a little more than my twenty minutes."

"Then Mr. SOLOMONS followed, to speak on a subject on which he surely should be an authority. To my amazement and that of the audience, he spoke for only a few minutes."

"Mr. SOLOMONS case for Prize fighters was chiefly based on the fact that he knew of prize fighters who on retirement had bought public houses and doing very well and others who had proposed to rich women – one had acquired a vast sum of dollars. 'What's wrong with the business if they can do that?' he asked."

"Mr. Peter WILSON, who must have written thousands of words on prize fighting, also spoke only for a very few minutes and subsided. If you have a good case, surely twenty minutes is certainly not too long in which to put it?"

"In the course of the discussion Mr. Freddie MILLS drew our attention to the suit he was wearing, assuring us that it cost a lot of money and was made by one of the best tailors. He also told us that he had a very nice house and garden. 'That only goes to show what there is in the fight business,' he observed sagely. He certainly brought the house down when he inquired of the learned lawyers whether they thought he was punch drunk."

"The debate from the floor renewed my faith in my fellow-men. Although it was evident that early training had been responsible for prejudicing many against my views, it was also clear that a legal training had helped others to overcome their prejudices and to listen to me impartially. I was glad that I had plodded along in a pain-staking fashion and explained the medical aspects in detail, for it was now clear that I had had an attentive audience."

"When the vote was taken I had a narrow majority; it was thirteen for the motion and eleven against."

"This was indeed a victory in such a selective audience, and I would recommend anyone who is in accord with me on this subject to try their hand with any audience. They may discover, as I did, that the British public is well aware of the harmful and degrading influence."

"It was shortly after this debate that Mr. SOLOMONS announced that he was going to contest West Fulham – the seat I then represented in Parliament – as an independent candidate at the next Election. He told reporters that he had been taking lessons privately in public speaking and learning the tricks of the trade."

"Having already taken the measure of Mr. SOLOMONS in debate, I looked forward to a full-scale political contest and the opportunity of delivering a decisive verbal knock out, Mr. SOLOMONS, however, discreetly withdrew before the General Election."

The next chapter in Lady SUMMERSKILL's essay is entitled *'Boxing in Parliament'*

"Another place in which I have endeavoured to influence public opinion on this question is the *House of Commons."*

"In 1953 and 1954 amendments were put down to the Budget, the purpose of which was to reduce the Entertainment Duty on boxing."

"In 1953, in speaking in opposition to the amendment rather late in the evening, such was my reception that I found it difficult to make myself heard. In 1954, only one year later, again I rose, anticipating the same kind of treatment. I was delighted to find that I spoke to a quiet and attentive House. Indeed the difference was so striking that I commented upon it. I said, 'You will recall, Mr. HOY, (the Chairman) that last year a similar amendment was moved and that I opposed it.

It was late at night and my speech was accompanied by jeers, cheers and interjections of all kind. I must confess that it is very pleasant this afternoon to rise in this peaceful and, I hope, slightly sympathetic atmosphere."

"I wish to present a rational case... "

"I think that it is clear that public opinion is changing and I believe that even since last year, a volume of public opinion has been created in this country of such a degree that there are many people who will read my speech to-morrow with much greater sympathy than that with which they read my speech last year... "

"In 1953 Lieutenant-Colonel W. H. BROMLEY-DAVENPORT (Knutsford) moved the amendment to reduce the tax. He argued that if the tax was not reduced prize fighting would become extinct."

"He said, 'There has been a vast decrease in the number of professional tournaments held all over the country. Promoters find it almost impossible to make a profit but – and I am sure that I shall carry the Committee with me on this point – they find it very easy to make a loss and therefore it is not worth taking the risk...'"

"The gross takings at a recent Midlands promotion amounted to £ 7.200. After all expenses and taxes had been paid, the promoter was left with a profit of £ 200."

"He argued that unless the duty was reduced, young fighters to-day would not get the great chances that their predecessors enjoyed."

"Among others he instanced Jack DEMPSEY. 'My last example is Jack DEMPSEY. Again on the way up the ladder, in the year 1918, before he won the heavy-weight championship of the world he had twenty-two fights. Seventeen of them were knock outs and twelve of them were in the first round. There is practice for you, Sir Charles (the Chairman). That is the way to get your eye in. Knock 'em bow-legged in the first round and bring home the bacon..."

"Apparently then, the amount of profits to be made determines the future of this 'sport.'"

"I find this a curious argument in the light of the frequent pronouncements of an altogether different nature that are made by those who have an interest in boxing. We are told that it is essential that youth should be taught to fight if they are to develop all the most desirable male attributes, that boys want to fight, and that nothing will stop them cultivating the 'noble art.'"

"Now what are the facts?"

"The latest figures I have, which were sent to me by the *British Boxing Board of Control,* refer to profits made in 1953."

"Gross Takings
Large Tournaments – £ 347.100
Small Tournament – £ 183.400
"Entertainment Duty
Large Tournaments – £ 87.100
Small Tournaments – £ 35.900"

"That still leaves a very considerable sum to be divided between the various participants. It is a little difficult to believe that the promoters are suffering any great hardship or that they will cease in the very near future to organise fights. The attempt to remove the Entertainment Tax on professional boxing by those who have a commercial interest in the business is conducted in a highly efficient manner, and there is no evidence of any shortage of money in the conduct of the campaign. All Members of Parliament are circularised with such meticulous thoroughness that even I received a communication asking for my consideration of the unhappy financial plight of the fight promoters."

"Deputations headed by the most vocal advocates of the prize fighting are taken to the Treasury, here they are given a polite hearing. But fortunately their advice is not always acted upon."

"While it is quite impossible for the comparatively few people who support my view to compete with the huge financial resources of the fight promoters, I felt that the strength of our moral case was such that it should also be put to the Treasury."

"Accordingly, in March 1954, I arranged a deputation of men representing various aspects of our social life, whose standing and authority in their own fields could carry weight."

"I should like to pay a tribute to those men for having the moral courage to express their views."

"The deputation consisted of Mr. J. H. DOGGART, F.R.C.S., *Surgeon to Moorfields Eye Hospital,* Mr. Cyril RALPH, *Organiser of Physical Education for the Gloucestershire County Council,* Dr. Fred ROBERTS, M. A., *Headmaster of Willesden Grammar School,* and Mr. Llewellyn WILLIAMS, a *Nonconformist minister and Member of Parliament."*

"I have mentioned these in alphabetical order, for to me each one made a contribution to the deputation different, but equal in knowledge, understanding and persuasiveness."

"As a *Member of Parliament* for seventeen years, who has taken deputations to Ministers, and as a former Minister who has received many, I can say quite sincerely that never have I been so impressed as I was by this deputation of fine, public-spirited men."

"Mr. DOGGART, an eminent ophthalmic surgeon, described with the aid of diagrams the exact nature of injuries to the eye and the brain to which a boxer is subjected."

"I doubt whether Mr. BOYD-CARPENTER, the then *Financial Secretary to the Treasury and now Minister of Pensions and National Insurance,* who received us, had ever heard such a lucid anatomical and surgical exposition of a case."

"Mr. RALPH, a former R.A.F. officer, said that all his efforts were directed against the teaching of boxing as part of the Physical Education programme in school, and against all forms of boxing competitions for youths, because in the R.A.F. he had learned of the danger to which the brain was exposed even in what appeared to be a harmless amateur bout. While boxing is not prohibited by the R.A.F., wise commanders warn their men of the risks of a pastime that involves violent blows to the head and damage to the delicate brain structure. The R.A.F. need men whose faculties are not impaired even in the smallest degree, and to expose the most delicate instrument in an aircraft – the pilot's brain – to the slightest chance of injury would be the most stupid policy."

"Mr. RALPH certainly disposed of the theory that fighting with the fists was an indispensable part of the training of lads who might be called upon to display conspicuous courage."

"Mr. ROBERTS, headmaster in a part of London where youth from poor homes were exposed to all kinds of temptations, strongly condemned fighting with the fists as an entertainment. He said it appealed to the very worst elements in his neighborhood. As I listened to Mr. ROBERTS's pleasant, quietly spoken, persuasive argument, I wondered what effect he had upon his school. A year later I had the pleasure of presenting the prizes at his school and found there a happy, friendly atmosphere which reflected the personality of the head."

"Then Mr. WILLIAMS spoke on the moral aspects of prize fighting and the bad influence it undoubtedly had on susceptible youth."

"I think that deputation, which left a lasting impression on my mind, received its reward when the Chancellor of the Exchequer, in the Budget which he presented the following month, did not reduce the Entertainment Tax on prize fighting."

"Now I do not believe that men will stop playing football or cricket if profits are removed altogether. These are healthy sports which men enjoy for the game alone and for the team spirit which they engender. On the other hand I certainly cannot imagine one man knocking another unconscious in a ringed enclosure just for the game alone. There must be some other motive, and the agitation of the promoter against the Entertainment Tax reveals it in all its crudity. Promotions – the delightful euphemisms for prize fights – would stop unless there were substantial profits in it for the middle men. That is the plain truth."

"The conditions in which some of these fights are conducted was well presented by Tom PHILLIPS, the journalist, in an article in the *Daily Herald* in 1954:"

'Boxing Board Urge Clean-Up at Ringside'

"'The *British Boxing Board of Control* are to warn promoters to take strict measures against ringside betting boys.'"

"'This is welcome news. Every programme carries the notice, 'Betting is strictly prohibited.' Yet recently at two prominent arenas – the *Albert Hall* and the *Empress Hall* – a certain section has behaved in a most unpleasant and undesirable manner; and that is putting it mildly.'"

"'It is growing into open intimidation. There was one instance of this at the *Albert Hall* on Tuesday.'"

"'After Mr. Frank AMOR, the referee, had disqualified George HAPPE, of Bethnal Green, for an alleged low blow in the second round, he returned to collect his coat from a seat near me.'"

"'He was immediately insulted by shouts from men nearby. 'What did you want to stop the fight for? We pay to come in here to see fights, not to see you stopping them!' Similar remarks from the noisy betting section were later aimed at Mr. AMOR. *Then he wanted quiet!*'"

"'Shouts like, 'I'll take 30 to 20 on the one in that corner!' were all too frequently heard, Yet one had the nerve to tell me to get a silencer fitted to my typewriter. I replied 'There are some people here I'd like to fit silencers to!"

"'Betting boys frequently try to influence referees – rarely with success – during fights.'"

"'When the boxer they want to lose gets a slight graze over an eye, the cry goes up, 'Look at that eye, ref! Stop it, ref!' If the man they have betted against is a body-puncher, they shout, 'Watch him, ref! He's hitting low!'"

"Apparently ringside betting is now an accepted part of the business. Does this also help to improve the character of our youth? Or does it on the contrary introduce them for the first time to the 'betting boy'?"

"While Tom PHILLIPS's revelations shock, Peter WILSON's comments on fights in Britain in the *Daily Express* of January 5th, 1953, must alarm the nation. He said:"

"'There are fights in Britain that are rigged. Big boxing more than any other sport is big business and when has big business ever had a lily-white reputation? Furthermore, it is so very much easier to fix a fight than it is, say, a football match. Make an 'arrangement' with two fighters, their two managers and a total of four men is all you need to carry out the swindle. Even that number can be reduced. In the fight which above all others I am convinced was crooked. I am very nearly certain that one boxer – the winner – did not know that the whole thing was rigged... "

"For years I have been campaigning against the managers who are allowed to take as much as twenty-five per cent of their boxer's earnings – lots of boxers."

"There are many honourable men in the management of professionals, but there are a number whom I once referred to as licensed pickpockets and then had to tender an apology, because Scotland Yard told me they weren't licensed:"

"These men are variety booking agents *peddling flesh in a new slave trade.*"

"Would anybody say that British morale would suffer if every boxing promoter was taxed out of existence; that British youth would deteriorate physically and morally if it did not see, in the flesh or on television, two men pounding each other's faces and bodies until one is overcome?"

"In an atomic age when productive processes and wars will be conducted with forces invented by scientists who have never been near the ringside, can we really believe that the boxing promoter and his entourage are indispensable to Britain?"

"Is there any evidence that the boys who do not learn boxing or attend prize fights are of an inferior calibre to those who do?"

"Similarly, has any evidence been produced that in those towns where fights are held regularly the crime record is lower because youth is thus given the opportunity to acquire some of the finer qualities?'

"It seems that when the case has to be put for boxing it is reduced to a plea for the continuance of a sensational entertainment of a dubious character, which makes a special appeal to the baser instincts, and all the other arguments fall."

The next chapter in Lady SUMMERSKILL's essay reads *'Criticism from the Inside.'*

"Now, having tried to put the physical and the moral case against prize fighting, I hope it will be read by those physically fit youngsters who in their spare time do a little sparring and dream of world fame. However, I am sufficiently realistic to know that if they do glance at this book the very fact that it has been written by a woman doctor and Member of Parliament will be sufficient for some of them to dismiss it with the comment, 'What does she know about it?'"

"I agree that my life and the lives of potential young fighters and their families are totally different, and perhaps my standards of behaviour with regard to prize fighting may be regarded as too academic."

"It is possible for a healthy young man to dismiss the medical views of doctors, because he firmly believes that ill health only affects the elderly, and furthermore that he was born under a lucky star and will be able to avoid the injuries which other young fighters, more stupid than he, have sustained. It won't happen to me' is the secret belief not only of boxers but of many others engaged in some hazardous job."

"I am, I hope, sufficiently tolerant of the point of view of youth not to be dismayed by the rejection of my own opinions and those of others who agree with me. Let us then seek the views of men who have spent part of their lives at the ringside, looking at fights and talking to and observing those connected with the business. There is nobody more likely to be familiar with this world than the sports writer, who is in close contact with promoters, managers and boxers and all the smaller fry whose livelihood depends indirectly on prize fights."

"It is significant that to-day well-known sports writers are expressing a certain uneasiness about the conduct of fights and the whole atmosphere which surrounds the business."

"A sports commentator has a vested interest, in so far as part of his livelihood depends upon reports of boxing bouts. Therefore he must be considerably disturbed before he feels himself bound to criticise adversely some part of the source of his income. Moreover, a man who is prepared to do this has the moral courage to defy the boxing promoters, whom presumably he meets socially and who no doubt for their own reasons dispense generous hospitality."

"Those of us who in the campaign are dubbed as cranks, killjoys and prudes, and are subject to ridicule from the more generous and abuse from the less generous devotees of boxing owe a debt of gratitude to the commentator who has the moral courage to make a protest against the commercial exploitation of the boxer."

"I presume the young fighters who are unconcerned with my views on the matter will be prepared at least to consider seriously the opinions of two well-known sport writers."

"The first, Paul GALLICO, was for years a columnist and sports commentator. A time came, when he could not endure the so-called sport of prize fighting any longer and felt compelled to tell the world what he had been privately thinking for years."

"He was a man who originated and organised the first Golden Gloves tournament, and who, it is said, is the writer responsible for more memorable pieces on the great fights of the century than anyone."

Paul GALLICO would ban boxing. This is what he says in *Esquire* of September 1955. This is quoted already..

Lady SUMMERSKILL discusses GALLICO's comments:

"Now what has the simple youngster who has been told about the glamorous world of prize fighting to say about that? It was written by a man who knew the fighting world, and who has not one good word to say about it. He ridicules those who call it a sport, destroys the rather frayed argument that it is the noble art of self-defence indispensable to every boy, and floodlights the corrupt and rotten methods employed in the business."

"Another writer, A. J. LIEBLING, who is still an active journalist of the *New Yorker* said."

"He wrote in *The Observer* of September 4, 1955, an article more moderate, but showing a supreme contempt for the fighting business and all its unpleasant associations."

He wrote:

"'I was therefore delighted on Thursday to receive an invitation from Jack SOLOMONS, a self-effacing British boxing promoter, to attend a quiet luncheon he was giving at a restaurant he has a piece of in Wardour Street, to introduce VALDES and GLEASON to the resident and visiting press. Mr. SOLOMONS, handsomely acknowledging that he had imported VALDES to box ten rounds with Don COCKELL at White City on September 13th, intimated that, since news of the affair was bound to leak out sooner or later, he was sacrificing his personal feelings and letting us in on the secret.'''

'''Before attending this formal opening of the social season, I attended a still more intimate reception for Mr. GLEASON in his bedroom at the Cumberland Hotel, where he had just arrived from the International Airport with VALDES and Luis MOLINAO, the fighter's Cuban trainer. GLEASON had sent VALDES and the trainer to sleep in an adjoining bedroom. He, a vigorous, Napoleonic figure in exiguous under-shorts, was selecting from his wardrobe bag the elements of his first London ensemble. The Cumberland, at Marble Arch, is handy to Rotten Row, where the fighter will do his roadwork.'''

'''Dignity is da keynote here. Joe 'GLEASON said, waving me to the arm-chair, as he chose a black Italian-silk suit with a large superimposed hairline diagram in white and a tie somewhat between ashes-of-roses and burnt-orange, infused with mulberry. I'm glad to get away from dem Philistines around *I B.C.'* The *I.B.C, or International Boxing Club,* is the corporation that, through its bonds with commercial television, has a firm grip on professional boxing in the United States.'''

'''What is a Philistine, Bobby?' I asked him, and he answered without hesitation, 'Al WEILL' WEILL is the manager of the world's heavy-weight champion Rocky MARCIANO, whom he has not allowed to assault VALDES. 'Da way I figure, he'll never fight me while he got the title,' Mr. GLEASON said. (This requires a bit of interpretation – he meant that Mr. WEILL would never let MARCIANO fight VALDES).'''

'''So what of it'' I can make as much with ten fights a year for ten thousand as one big shot for a hundred grand. I like Europe anyway.' Ten fights, at ten thousand dollars apiece, entails ten times as much fighting as one at one hundred thousand, but GLEASON wasn't going to do the physical fighting, which is, after all, a mechanical detail. It is why I think of him as an economist.''

"This picture of the business is even more damning in its presentation than Paul GALLICO's onslaught. No doubt those mentioned will think it highly complimentary."

"VALDES, this tall, simple boxer, is a piece of merchandise being hawked around to the highest bidder. For just a little while the limelight will be on VALDES and white pants, and then what? Another strong, ignorant, innocent, white, brown or black, will take his place."

"While the professional writers are now boldly challenging a business disguised as sport, it is encouraging to find men quite unknown to the public writing in their works magazines in the vein. This calls for a high degree of moral courage, for men like G. KEELING, writing in *The Review News*, a newspaper for the employees of Messrs. STEWART and LLOYDS, a great engineering works, must face his workmates every day and no doubt be subject to the ridicule of those small men who may privately believe he is right, but lack the courage to support him."

"This is how he put the case under 'Seconds Out.''

'''Seconds out of the ring and leave it to the Editor to call 'time.'''

'''The British are noted as a sporting nation. Soccer, cricket, rugger – bustling, boisterous games, watched by thousands cheering themselves hoarse. Sometimes, in spite of rules framed to avoid it, someone gets hurt, and a hush of sympathy falls on the crowd. Very rarely does one player intentionally hurt another, and if he does the crowd roars its disapproval – 'Send him off, ref!'''

'''Good, clean sport.'''

'''But let us consider another branch of 'sport.' The crowd again roars its approval or condemnation – approval when one player succeeds in hurting another; condemnation if he fails. The sight of blood whips the spectators into a frenzy. The spectacle of a semi- conscious man, with bruise-closed eyes and breath whistling from a chest red from pummeling, struggling to withstand or inflict more punishment – this brings forth delighted roars.'''

'''In a word – BOXING'.''

'''Boxing has been called 'the noble art of self-defence.' It is, however, far from noble and far from art. I have heard that if it is not so much inflicting blows as defending against them that is the sport. Hooey! Read the sports writers in the daily press and you will see such expressions as 'so and so has stamina but lacks a killer punch;' 'landed a beautiful left over the heart;' —— 'eye was cut and his opponent kept leading for it.'''

'''Sport? Again – hooey!'''

'''A contest in which those taking part do their best to hurt each other as much as possible is not sport. If the referee thinks the contestants are not trying hard enough (to hurt each other) he orders them to put more life into it. Permit me to tell you that the best self-defence is for both boxers to sit down in the centre of the ring and say to each other. 'Why should we knock each other senseless for the profit of the promoters and to provide a spectacle for the sadists lining the ringside?'''

'''Finally, man is blessed with certain senses. Sight is provided by that wonderful instrument the eye; yet in this 'sport' blow after blow is landed on just this spot. Fools! Would you allow your child to bounce a ball against your television screen – replaceable though it is? The ear, too, is a delicate mechanism. Grandpa never lends his hearing aid for baby to play with. Also in the region of the ears is the mechanism which controls the sense of balance and direction. The captain does not allow the cabin boy to tamper with the instruments on the bridge. And what of the irreparable damage caused to the sense of smell?'''

'''Men have been killed in this sport' and were you and I to fall out over what I have written and fight about it, inflicting on each other such injury as boxers do in the ring, we should dispute all this, and if the Editor will allow the space, please write to him and say so. I will try to defend myself.'''

"It gave me great satisfaction to follow Mr. KEELING in the next issue of *The Review News,* giving him my unstinging support and expressing my admiration for his moral courage. I only hope that it helped others in that engineering works who feel equally strongly, but who are slower to express themselves."

The next chapter in Lady SUMMERSKILL's essay is entitled *'The Boxer's Wife.'*

"British women, as represented by the wives of professional boxers, were severely criticised after the fight between Don COCKELL and Nino VALDES, in September 1955, which ended in COCKELL receiving a severe head blow in the third round which compelled him to retire. Mr. SOLOMONS, the promoter, said, 'I'm calling off the British heavy-weight title fight between COCKELL and GARDNER unless both men agree to leave home and go into strict training (Tom PHILLIPS, *Daily Herald, September 15th, 1955).*"

"Tom PHILLIPS wrote, 'Yet boxers' wives and sweethearts seem to dominate them. These so-called tough guys go as soft as cuddly dolls when their old woman or their girl friends begins to rib them. Poor little darlings. Dey get so lonely when their man is away. Not earning a crust, as you and I do for our sweat, but huge wads of white money. They like the money boxers earn – but they don't want him to get hurt. My wife wouldn't worry if I had a broken nose – if I retired at forty with a hundred grand. Nor would I.'"

"'Now, if Don COCKELL had not been slightly concussed and suffering from profuse bleeding from a cut which required two stitches, he might have continued. In the event of his winning, no doubt the boxing fraternity would have pointed with pride to the prowess of men in the ring and particularly outside it in the managerial and promotion side of the business. Apparently, in the eyes of the organisers, COCKELL failed and reason had to be sought."

"It is not unusual to hear those responsible for organising elections which prove successful for their party, boasting afterwards of the good organisation, of the persuasive speeches, and the oustanding calibre of the leader, all of which it is understood is in the main done by men."

"If the party fails, then inevitably the woman are charged with having voted the wrong way."

"This story is as old as the Garden of Eden itself, but it is rather amusing to find that boxing promoters have to fall back on the old excuse. Now, apparently, if a heavy-weight champion fails to defeat his opponent it is some woman's fault."

"I rather suspect that Don COCKELL's real crime in the eyes of big business was that he did not allow himself to be knocked about for a few more rounds for the benefit of the crowd. The plain truth was that the audience felt that they did not get their money's worth. They went to the fight to be entertained. Instead, the fight was all over in the third round and the frustrated crowd had to disperse without having given full vent to its emotions."

"Mrs. COCKELL and Mrs. GARDNER, who was also the target for attack because her husband prefers to live with his wife and family while training rather that in the uncongenial atmosphere of a training camp, must not be influenced by this criticism. They and their husbands should be proud of the fact that the only charge against them is that they love each other and consequently want to be near each other."

"Furthermore they need not worry about the headlines in the *Daily Herald* of September 15th, 1955. *'British Boxers are too Soft to Win Titles.'*"

"The British know just what the British can do when forced to defend themselves in battle, and we are not very concerned with the spleen of those who are only interested in the profits made out of setting one man against another in a roped enclosure. Many of the most courageous men despise everything connected with commercialised prize fighting and feel that the conduct of the business is a reflection on British sport and does not enhance its reputation."

"I urge the COCKELL and the GARDNER families to conduct their lives as they think fit and not be influenced by those who are completely regardless of their personal happiness."

"Every decent woman would agree that it is better to manage on a smaller income than jeopardise the happiness of a lifetime by allowing your husband and the father of your children to expose himself to the possibility of becoming unstable, irritable, moody punch drunk."

"I have received confidential letters from the wives of professional boxers which have endorsed everything I say. The lives of these women have been made almost unendurable by husbands whose brains, in the course of the years, have suffered permanent damage affecting their whole pattern of behaviour. Moodiness, bad temper, bouts of drunkenness, combined with brutal assaults on his wife, often mark the behaviour of the old pugilist."

"This apparently has been the experience of the wife of a prize fighter who felt impelled to write to the *Daily Mail* to make her protest."

"'I have never seen a boxing match, and yet I agree wholeheartedly with much that Paul GALLICO has said. I have had personal experience of the fight game, you see. I was a boxer's wife."

"'Boxing, I believe wrecked my marriage and brought years of unhappiness to me and my son. This is why I am writing this letter – as a warning to any woman who may be thinking of marrying a professional boxer. What happened to me could happen to her.'"

"'Jack, my husband, was twenty-three when I met him, but he had already been fighting for nine years, first as a boy amateur, and later as a professional.'"

"'Outwardly, he was a fine, athletic man, a steeplejack by trade. But quickly I learned the truth – that boxing had affected his mind. I believe he was punch drunk.'"

"'He would be moody, scarcely speaking to me for days at a time. Worse were his bouts of drinking, his periodic fits of violence. Sometimes I had to call the police for protection.'"
"'At times he was fight-crazed. He would pick a quarrel with a stranger in the street and beat him unmercifully.'"
"'When I saw what boxing was doing to him, I begged Jack repeatedly to give it up. But he never would; it was as though something in his blood made him carry on.'"
"'After his fights the drunken bouts usually followed, and the fits of violence towards me. Next day he would break down and cry like a baby and tell me he was sorry. Soon life with him became unbearable, and I left him and brought up Dennis alone.'"
"'In 1940 he volunteered for the R.A.F. and left the district. A year later he was badly injured in a crash and invalided out, and in 1946 he died after a long illness, aged thirty-seven.'"
"'But for boxing I believe Jack would have been a different man. He had his good qualities.'"
"'Boxing may have its good points, but after my experiences I know that once if affects a fighter's mind it can bring only misery to his family.'"
Lilian LIPTROT
"'Bradley Street, Newhey, Near Rochdale.'"

The *'Conclusions' in* Lady SUMMERSKILL's essay read:

"Boys in all countries should be taught from earliest childhood that fighting in all its forms is a primitive expression of the aggressive instinct which civilised people should have progressed further towards establishing a peaceful world by all the regional pacts which have ever been devised.'"
"If the new pattern of behaviour were accepted then the occasional bellicose demagogue would-be dictator could be defeated by ridicule alone.'"
"The two big world wars were responsible for an appalling waste of human life, chiefly that of young healthy men in their prime. What was the result?"
"France was stripped of a generation of her finest young men and now after two world wars in which she was declared the victor she sees the vanquished neighbor, Germany, prosperous with a balanced budget and a high standard of living. No feeling of hate, jealousy, or revenge motivated me when I mention Germany in this context. I use her simply as an illustration of the futility of war, and the appalling waste of life which it involves."
"The boys of Germany idealised HITLER, who glamourised war and fostered their aggressive instincts by every means in his power. Women, who HITLER regarded as a peace-loving element in the community, were consigned to the kitchen."
"The young men of Germany were sacrificed, and so carefully had they been trained that when the time came they marched towards certain death, arm in arm, singing songs glorifying HITLER."
"This is what can be done by suggestion, daily indoctrination of the child, and the glorification of brute force."
"In this matter of education of boys in social values the important role of the mother is often overlooked. Indeed, mother unfortunately tends to overlook it herself. She – if she only exercised it – has tremendous power for good or evil in the home, for she is closely associated with the child during its most important formative years. It is the hand that rocks the cradle which could rule the world, if only that hand were a little more sure of itself."
"Unfortunately, many women are inclined to accept other people's values without carefully examining them to see if they are genuine or bogus. A woman may instinctively know that something is wrong, but she is not prepared to defend in the home what may be an unpopular viewpoint. The mother regards harmony in the home as one of her main objectives. Quarreling, wrangling, even controversy on current problems disturb her. She knows without having read books on child psychology that a child needs a peaceful and happy environment if it is to develop into a well-balanced adult from morbid fear of insecurity. She knows that political and social questions are apt to provoke heated arguments between members of the family, and therefore she finds it more prudent not to take sides."
"Male values on social questions are often accepted without question and therefore for the sake of peace father and sons are permitted to strike certain attitudes and lay down certain canons of behaviour, although they do not bear examination."
"Now a mother must not fail to do what the father, because of his sex bias, carefully avoids. She must teach her son that violence in all forms is harmful, physically, morally and socially, and she will very often have her reward in finding an intelligent boy agreeing with her wholeheartedly."
"Then there is the part that teachers can play in helping the boy to think for himself. In the fact that it is a school tradition that boys should learn to punch each other sufficient reason for continuing a pastime which belongs to an earlier phase in our evolution? Here again we find – although there are many outstanding exceptions – teachers lacking the moral courage to say what they believe is the truth – that boxing is as outmoded as bear baiting and cock fighting – for fear of losing face in the school, or of upsetting some fathers or their less enlightened colleagues."
"To-day, all our media of propaganda – the cinema, the theatre, television, the boxing ring – present scenes of violence or sex, or both, which appeal to the undisciplined individual whose love of sensationalism makes him an easy target for the unscrupulous purveyor of this form of entertainment."
"War, when it comes, is a kind of collective insanity. There is no time to think coherently, and the killing goes on until one side is exhausted."

"Now is the time to think and to try to understand why human beings, who one day are leading quite orderly lives in their respective countries, should, at a signal, turn on one another and blow each other and their families to pieces."

"Propaganda, which forments hate once war is started, is not sufficiently powerful to rouse the aggressive instincts of a disciplined people in peace-time. Nevertheless, it can have been encouraged to believe that fighting in any form is praiseworthy."

"Now is the time, while we have a breathing space, for us to ask ourselves whether we are doing everything in our power to educate our young people to hate violence and to control their aggressive instincts for their own sakes, for the sake of the community in which they live and for the peace of the world."

In a short piece, Karel HENNER, the *late Chairman of the Department of Neurology and Psychiatry of the University of Prague, Czechoslovakia,* in 1956 discusses moral damage, not to the contestants, but to the spectators:

"It must be borne in mind that traumatically induced unconsciousness is an absolutely normal and legitimate feature of this form of sport. Both criminal and civil law punish even small injuries to the head and provide for compensation for the victim, and rightly so. In boxing, the contestant who succeeds in 'putting his opponent to sleep' is held in esteem, feted, and set up an example. This cynical euphemism means that the aim of boxing is to pummel your opponent's head so cleverly and brutally that he falls unconscious. If he falls and rises again, he is given a series of terrible blows until he finally remains motionless. Then there is the moral damage to the spectators. This leads to over-admiration of brute force and exerts an unhealthy influence on young, immature onlookers and may even lead to sadism... Boxing with the purpose of training and conditioning, without blows to the head – yes; boxing contests in their present form- certainly no."

Karel HENNER recommended in a later paper (1963):

(1) Absolute prohibition of blows to the head; (2) counting to five during knock outs, instead of ten as hitherto; and (3) electro-encephalographic examinations of juniors and then compulsory EEG examination every year for all active boxers, even if no knock out has occurred. EEG recordings with spikes and sharp waves as well as other obviously pathological findings should be taken as sufficient cause for a prohibition from boxing.

Karel HENNER wrote this in 1963. At this time an EEG was the most advanced diagnostic tool available for neurologists examining patients with brain injuries. Today, 40 years later, new diagnostic techniques have been developed, as computerized axial tomography (CAT scans), magnetic resonance imaging (MRI), which have revolutionized diagnostic testing of brain injured patients. HENNER would have recommended the use of these modern techniques available today for the examination of the permanent brain damage in boxers.

A debate on television between Dr. STEINHAUS and Dr. GILMER took place on December 16, 1956, in the *NBC program 'Debate.'* Dr. STEINHAUS expressed his opinion for a prohibition of boxing.

Dr. GILMER, the Chairman of the *Illinois Athletic Commission* emphasized the allegedly positive aspects of this activity.

STEINHAUS 'Should the sport of boxing be discontinued? Yes!' *(Physical Education, Phi Epsilon Kappa 14: 131-132, 1957)* wrote:

"The sport of boxing as now practiced should be discontinued:"

"1) The head is not constructed to take the blows delivered in boxing."

"2) It is not a true sport when measured by modern standards of sports and sportmanship."

"3) It is immoral when judged by the principles of Christianity and the standards of an enlightened social conscience."

STEINHAUS continues:

"Man's head is not constructed to take the blows of boxing... When I see a young fellow being quickly knocked out, I feel relieved in his interest. I think he might be saved from real danger, from being hit on the head for years, though by minor blows, and thereby from being beaten into a mental wreck."

"Every head pummeling, with or without a knock out, is likely to leave a small portion of brain tissue permanently damaged. It is very probable that from fifteen to twenty-five percent of the men who enter a national or Olympic tournament come out with some permanent damage. In time these small injuries add up to produce serious mental defects."

"Boxing is not a true sport when measured by modern standards of sports and sportsmanship."

"First there is no sport except boxing in which the primary purpose is to inflict bodily injury on an opponent, and in which victory goes to a man because he has successfully beaten his opponent into temporary unconsciousness."
"Second, in no sport except boxing do spectators expect and ever demand that bodily damage be inflicted."
"Third, in every sport except boxing everyone feels badly when someone is injured, and the injured player is respected, even applauded by both sides."
"Fourth, in every American sport except boxing there is provision for time out and for substitution when there is bodily injury. The injured contestant may leave the field without disgrace and without loss of self respect."
"Fifth, in no sport except boxing is a man's competence or courage judged by his readiness to continue in combat after suffering bodily injury."
"Sixth, in every sport except boxing a bloody nose, a lacerated face, or temporary unconsciousness are recorded as accidents. In boxing, these are produced on purpose and not reported as accidents. Thus boxing has a fictitiously (sic) low accident record."
"Boxing is immoral when judged by the principles of Christianity and the standards of an enlightened social conscience."

GILMER 'Should be the sport of boxing be discontinued?" No!' *(Physical Education Phi Epsilon Kappa 14: 123-124, 1957)* wrote a panegyric companion article to STEINHAUS:

" The sport of boxing is a valuable asset and should not be discontinued."
"Boxing is the art of hitting with gloved fists without getting hit – a science of attack and defense. The value of boxing is primarily in its training the reflexes and develops technique, confidence, aggressiveness, endurance, physical courage, individual thinking, poise and sportsmanship. Boxing is a vigorous, rough, body contact sport suitable for those physically sound boys of all ages, weights and sizes who want and enjoy such competition. Boxing is not a sport which should be forced upon anyone."
"The key to successful boxing is intelligent supervision and instruction. Only physically sound boys should box. Proper instruction, equipment, conditioning, and matching according to weight, age, and ability are essential. Boxing must be conducted according to its rules. Proper rest must be taught. Careful referees must protect the contestants from mismatches and fouls. Given these common-sense precautions, and teaching boxing as a game of skill, speed and condition, it is a body-contact sport which involves very little danger."
"Through training in boxing boys learn coordination, gain physical strength, learn to stand up under fire, and learn through experience never to quit."
"During World War II, Secretary of the Navy, Frank KNOX, said on January 11, 1944 in a Touchdown Club address:
'"We in the navy definitely believe in the type of physical exercise and sports which involves bodily contact with your opponent... We, as a nation have had a frightful shock as we've come to realize that fifty per cent of the men who were called for service with the Colors could not qualify for physical or mental reasons. That means a nation, one half of which physically could not defend his own freedom... One of the lessons we have learned in the war ought to be that the building up of the physical side of young Americans is one of the most important and critical tasks in the days after the war is over. The price of freedom is the ability to defend it.'"
"Those who contended that the sport of boxing should be discontinued say that the danger of head injuries which may be sustained in the sport outweighs the benefits to be derived from it. These people, who usually have done no boxing themselves, attract attention by using the expression 'punch drunk' which was an imaginative phrase coined by a New York sports writer 30-odd years ago. They picture the boxer as a fellow with glassy eyes, dragging feet, stuttering speech and appearing just a little drunk. After 33 years of association with boxers ranging from nine years of age to upwards of forty, the only one I have seen who answers this description is the fictitious character portrayed by Red SKELTON on TV. Mr. STEINHAUS says he has talked to many of such people. Its strange that although I have been both an amateur and a professional boxer, have taught, officiated, and finally supervised the sport of boxing, I have not met any who answers the 'punch drunk' description."
"Injuries and deaths have been suffered by participants in every other sport. Most of the injuries and deaths occurring to boxers could have been prevented by better supervision. Some were unavoidable accidents."

GILMER (1957) continues:

"Some people, like STEINHAUS, consider boxing brutal. Boxing is not as rough or dangerous as football; does not require as exhausting application as wrestling, ice hockey, crew, lacrosse, cross-country running or basketball. The University of Wisconsin made an objective study of college boxing during the four seasons 1949 and 1952. All boxers were tested, examined and observed, and were personally known as individuals. Electroencephalograms testing brain waves were obtained, and all known medical literature, including writings on the so-called 'punch drunk' state, were reviewed by the University doctors in charge of the study. No evidence that a 'punch drunk' state had occurred in anyone as the result of college boxing was discovered. Even papers on electroencephalograms made by others did not indicate any significant changes among amateur contestants after their boxing experience. The objective 4-year University of Wisconsin study reported no ill effects from college boxing. That study demolished the unfounded and unchecked assertions so glibly made by non-medics that all boxing is dangerous and undesirable as a recognized sport, regardless of supervision. Such assertions must be classed as mere sensational writing for profit."

"Mr. STEINHAUS has erroneously stated that the primary purpose of boxing is to inflict bodily injury on your opponent. That is about as fair as saying that the primary purpose of baseball is for the pitcher to throw a beanball at the batter, or is like saying that the primary purpose of defensive line man in football is the slug and foul the star back of the other team. A boxer, baseball player or football player whose primary purpose is to inflict bodily injury on an opponent should be barred from the sport. Any boxing coach who coached his boxers to concentrate on knock outs would be out of a job in no time. His boxers would continually miss haymakers and lose decisions."

"Mr. STEINHAUS also says that victory goes to the one (in boxing) who has successfully beaten his opponent into temporary unconsciousness. Most boxing bouts are decided on points. Knock outs sometimes do occur and occasionally the boxer who is unable to continue is temporarily unconscious. Usually the referee will stop an uneven bout and avoid pointless continuation of such a match. Spectators applaud such action."

"Mr. STEINHAUS complains that there are no 'time outs' in boxing but forgets that after every short round there is a mandatory rest period which is devoted to the well-being of the contestants. Proper supervision is the key to boxing's success or failure as a worthwhile sport."

"Finally. I have this to say in connection with Mr. STEINHAUS' theory that the brain wobble, slide and twists in the skull when a head blow is received, and that permanent damage necessarily results. This is a medical problem. Neither Dr. STEINHAUS nor I are doctors of medicine. But upon conferring with Dr. Irving E. SLOTT, *Chief of the Medical Staff of the Athletic Commission of Illinois,* who has specialized in such matters, I am advised that the 'wobble, slide and twist' theory of Mr. STEINHAUS is regarded as the least likely of brain theories concerned with the effect of blows on the head. According to Dr. SLOTT, the latest expressions by leading neurologists in recent years on this subject do not support Mr. STEINHAUS' theory at all."

This debate in television had taken place on December 16, 1956 in the *NBC program 'Debate.'* Dr. STEINHAUS expressed his opinion for a prohibition of boxing.

Dr. GILMER, the *Chairman* of the *Illinois Athletic Commission* emphasized the allegedly positive aspects of this activity.

The debate was judged by a Deputy Chief of Police, a President of the League of Women's Voters, and a Professor of Speech of the Northwestern University in Chicago. Dr. STEINHAUS won the debate, to use a boxing term by a split decision on points, thus 2: 1.

GILMER makes boxing sound safe, controlled and healthy, but he makes one very sweeping (and revealing) statement:

"Proper supervision is the key to boxing's success or failure as a worthwhile sport."

According to a great number of authors, both boxers and non-boxers 'proper supervision' is an idealistic dream that has little or nothing to do with professional boxing at least. As one author said, the only problem with amateur boxers is that they become professionals.

Harold BARNES *'Let's abolish boxing.' (Saturday Evening Post, 231, 42 +, October 25, 1958)* was a ring judge for the last 37 years. In the time, he wrote his report he was the *Senior Official of the New York State Athletic Commission;* he estimates that he was a ring judge at 6.000 boxing matches, of whom were 2.000 main event and 100 world championship fights.

BARNES wrote:

"There is one other statistic. I'm not at all proud of. I've seen almost as much bloodshed in the rings as if I had been working in a Chicago slaughterhouse. I'm 63 years old now, and I've seen enough. If I had the power to do so, I would outlaw boxing. To put it blandly, I consider boxing legalized murder."

"Since 1945 some 156 fighters have died from injuries inflicted in the ring. This total takes in boxers all over the world – both amateur and professional – and may not sound too staggering. However, when you consider that there are more death per number of participants in boxing than in any other sport, you can get a better idea of the situation."

"In any event, it isn't just the deaths that disturbs me. Only a small percentage of prize fighters lose their lives in the ring. But a large percentage lose something else in their years of fighting. Way back in the 1920's a doctor in Newark, New Jersey, (MARTLAND, author) made an exhaustive study of fighters. He found that 50% of those he examined were punch drunk to some degree."

"Whatever the exact percentage, it is too high for my peace of mind – all those men walking the street on wobbly legs, their minds not quite so sharp as before they started fighting; their reasoning, memory, self-control, speech and coordination all impaired by repeated head blows."

"TUNNEY was lucky to get out in time. Other fighters have been less fortunate. Lavern ROACH, Sonny Boy WEST, Jimmy DOYLE, Ed SANDERS, 1952 Olympic heavyweight champion, and George FLORES are among those who, have died in recent years in professional fights. I officiated at one time or another in bouts featuring most of these boys. In the case of George FLORES, I was one of the judges at the fight that resulted in his death – a bout with Roger DONOGHUE in Madison Square Garden on August 29, 1951."

"Going into the eighth round I had FLORES losing the fight on points, but he hadn't seemed hopelessly outclassed. Nor did he seem an unfit opponent for DONOGHUE. Then, in the eighth, I was taken by surprise when FLORES was knocked down by a strong left hook to the jaw. He fell directly in front of me, not more than two or three feet away. I heard his head strike the ring floor with a resounding thud. He tried to get up, but couldn't. Finally he was able to walk out of the ring under his own power."

"Ten minutes later in the dressing room he went into a coma. An emergency operation was performed almost immediately and a blood clot removed from the left side of the brain. But it was too late. The autopsy showed that FLORES had been injured in such a way as to tear a small vein in the head and cause clots which cut off circulation to the brain."

The death, naturally, created a furore. Sports columnist Bill CORUM, of the *New York Journal-American,* wrote, 'if the killing or near killing of lads as such as this, in the name of sport, can't be stopped, and isn't stopped, then boxing itself must be."

"But boxing goes on today. A wry postscript to that fatal fight is that FLORES' widow, who was originally awarded $ 80.000 in a damage suit against the state because of her husband's death, ultimately got nothing. A higher court unanimously set aside the damage award, saying, in effect, that boxing was such a dangerous sport that the state could not be expected to insure the participants against death."

"Twenty-seven years earlier I witnessed my first ring fatality. It happened at the old Madison Square Garden on January 11, 1924, in a semifinal match between Bud TAYLOR and Frankie JEROME."

"That was one of the bloodiest fights I ever saw. TAYLOR completely outclassed JEROME. On my card he took every round but the fourth and tenth. In the eleventh he brought JEROME down with a right to the chin. Frankie took a count of '8' and lurched to his corner at the bell."

"At that moment I noticed that JEROME's eyes were glassy. I stood up and called in the referee, Jack O'SULLIVAN, to stop the fight. I felt the referee hadn't seen the condition of JEROME's eyes. But he paid no attention to me, and one of the New York Boxing Commissioners, Walter HOOKE, hollered, 'You're out of order, BARNES! Sit down before you start a riot.'"

"JEROME staggered out for the twelfth round. In the first mix up, TAYLOR floored his opponent. JEROME took a count of '9' and again staggered to his feet, blind and groggy. He moved in instinctively, but referee O'SULLIVAN finally stepped between them and stopped."

"JEROME fainted as he was being assisted to his corner. He was carried to his dressing room unconscious. He died the next day."

"Afterwards I made it a point to tell Commissioner HOOKE that JEROME might not have died if the referee had ended the fight when I hollered to him. The commissioner didn't answer me."

"I don't contend that boxing is as brutal a sport today as it was in the days of Frankie JEROME. Refinements have been instituted to protect fighters. Padding in the ring is much heavier, gloves are thicker, boxing commissioners are diligent public officials. The fighters also are helped by more stringent regulations and stricter medical supervision."

"After FLORES died in 1951, a Medical Advisory Board was established to aid the New York State Athletic Commission. This medical board, headed today by Dr. Mal STEVENS, a former college football coach, has done a fine job. There hasn't been a death in a New York ring since that of FLORES."

"But these safeguards are still not enough to guarantee the welfare of young men who participate in boxing. There is only one sure way to prevent them – by abolishing the sport altogether."

"Last May 24th, in Detroit, a former leading heavyweight contender, Johnny SUMMERLIN, was knocked out by Nino VALDES. Several days after the fight it was revealed that SUMMERLIN had passed the prefight medical exam even though his left side was numb from head to toe. His condition was eventually diagnosed as hypesthesia, 'a diminished capacity for sensation,' caused by pinpoint hemorrhages in the brain."

"Authorities in Michigan were so shocked by this disclosure that they immediately appointed a medical-study committee headed by Dr. Joseph CAHALAN. Doctor CAHALAN took the job, although he advocated abolishing the sport, saying, "As long as there is boxing, there will be injuries and possible deaths.""

"The head isn't the only vulnerable part of the body. The bones in the hand are extremely fragile, and were never meant to deliver forceful blows. The eyes are also extremely vulnerable. As an optometrist of some forty year standing – I taught optometry at Columbia University for thirty-five years – I know something about the eye structure. The frequency of retinal detachments and hemorrhages among fighters is appalling. I would estimate that 10 per cent of the 'catchers' in boxing – the fellows who absorb unusual amounts of punishment

– eventually suffer retinal detachments. Such detachment results in loss of vision in that eye."

"Not long ago Pat VALENTINO, a former heavyweight contender, revealed that he had undergone an eye operation nine months before his Ezzard CHARLES fight in 1949, and was partially blind at that time. He also admitted that when he lost to Joe LOUIS two months later he was completely blind in the right eye."

"The worst thing about ring damage, as far as I am concerned, is that it is premeditated. That's why I call boxing legalized murder. It is the only sport in which the definite objective is to injure an opponent sufficiently to knock him out. So, together with an admittedly long, glorious record of champion and challengers and great, exciting fights, you also have a long, inglorious roll of boxers who have died in the ring or sustained permanent injury."

BARNES continues:

Among the most tragic figures are the boxers who have finished their lives in mental institutions. The list stretches back to the early 1900's. One of the first prominent examples was Terry McGOVERN, a featherweight champion. Over the years such great fighters as Joe GRIM, Ad WOLGAST and Battling NELSON ended up in asylums. Then there was Frankie CARBONE, a former middleweight, who was picked up one night in front of the Roxy Theater in New York City after shouting, 'People are trying to take me out to the middle of the ocean, right into the waves!'"

"There also was Bob MARTIN, champion of the American Expeditionary Forces in World War I, who once fought Gene TUNNEY. MARTIN was taken to Walter REED Hospital in Washington suffering from what was described as 'boxer's waltz.' Physicians said his disability was the result of cerebral lesions directly due to the many blows he had taken on the head. He was classified as incurable."

"This unhappy list continues to grow. Johnny BRATTON, former N.B.C. welterweight champion, has been in several Illinois state hospitals since May 1956. Another sad recent case was that of Jimmy FLOOD a onetime 'pleasing' middleweight contender, as he was often introduced by the ring announcer. He was sent to Bellevue Hospital for mental observation after being arrested on a wife-beating charge. His wife, Dorothy, told the magistrate that her husband had suffered brain damage as a result of his ring career. Indeed, FLOOD had been retired by the New York Medical Advisory Board after an electro-encephalogram revealed such brain damage. Too many present-day fighters face the same kind of future."

"It may seem hypocritical for me to be sounding off about boxing when I've stayed in it as a fight judge for thirty-seven years. There have been times when I've come home after a particularly vicious and one-sided fight and told my wife, 'That was terrible. I'm going to get out of it.' And she'd said to me, 'You can't, Harold. It's in your blood.'"

"That may be true to some degree. At its best, the sport can be a most moving experience for those who watch it. However, there are other reasons why I've remained connected with it. For one thing, the work of judging fights presents a tremendous challenge. But more important, as long as boxing is legalized, somebody must officiate. I like to feel I am contributing to keeping the sport as decent as it can be. It was in this spirit that I applied for renewal of my judges license this fall."

"Lately, though, I 'm beginning to wonder how clean boxing really is. I may seem naive, but until recently I never thought the boxing world was the corrupt jungle that some critics made it out to be. Oh, I've officiated at certain fights – I'll talk about them later – that were distinctly suspicious. But there weren't many of them. And never in my life has anyone approached me and tried to bribe me to vote a certain way."

"Nobody has ever seen me in the company of, or in conversation with a fighter or manager. I don't think that I would recognize a manager if he walked into my back yard. I have never set foot in the usual haunts of the fight crowd in New York City. I don't go to fights as a spectator. The only exceptions since I became a judge were the second Max SCHMELING – Jack SHARKEY bout in 1932 and the second Rocky MARCIANO – Ezzard CHARLES fight in 1950. In each instance, I had officiated at the first fight and was very much interested in seeing the rematch."

"I don't mean to imply by all this that I regard fighters and managers as outcasts of society. I simply believe that a fight judge, to insure his impartiality, shouldn't associate with these people."

"The indictment last July of one of my fellow judges, Bert GRANT, shocked me profoundly. No New York State boxing official ever had been indicted before. It was alleged in the indictment that GRANT accepted bribes totaling $ 400 to influence his decisions in five specified fights."

"I don't know GRANT personally. I know him only as a reserved, introverted type, who talked little when in the official's room. On the basis of his decisions, he seemed very capable. The handing down of an indictment doesn't mean a man is guilty. Unless he is proven guilty, one should not pass judgment."

"If any fight judge or referee does take part in a 'fix,' it amazed me that he would jeopardize his career and his family's reputation for a mere pittance. In GRANT's case the alleged bribes ranged from $50 to $100."

"Most of the officials I have worked with have been business and professional men of the highest caliber. They got into officiating primarily because of a love for the sport and a desire to contribute something to it. They certainly didn't take up officiating for the money involved. Fees to judges today ran from anywhere from $10 to $40 a night at a small club, depending whether upon a bout draws well, and from $75 to $150 for a big fight with an exceptionally large gate. The club that stages the fight pays the officials. In the old days of the million-dollar gate I earned up to $250 for judging a fight. The last such an occasion was the second Joe LOUIS – Billy CONN fight, in 1946. Most years, after license fee, income tax and other miscellaneous expenses, there has been very little left of my officiating fees. "

"One of the fights for which Bert GRANT allegedly accepted a bribe was the Alex MITEFF – Nino VALDES battle of last February. According to the New York indictment, GRANT was paid $100 for that one in the expectation that he would be designated a judge. But he wasn't selected – I was. Thus I feel personally involved – especially since I voted for MITEFF, Hymie WALLMAN's fighter. WALLMAN is the manager who alleged in the indictment to have paid off GRANT."

"That was one of the most savage fights I've witnessed recently. VALDES' white trunks were spattered with MITEFF's blood. Referee Teddy MARTIN's hands were covered with blood. The canvas was crimson too. It was a sickening sight."

"Despite all the blood that was shed by MITEFF, I awarded him the decision, four for VALDES and one even. Referee MARTIN agreed with me, giving it to MITEFF, six-four. Judge Nick GAMBULI called it for VALDES, six-three-one. It was a close fight, but MITEFF gained an early lead. As I saw it, he won the first three rounds by a definite margin. Thereafter VALDES came along better. But I felt that he hadn't shown effective aggressiveness, that he hadn't sustained the action by the greater number of skillful attacks. MITEFF continually forced the action despite his cuts. I think the majority of the newspapermen at ringside agreed with my decision."

"After the fight there was a disgraceful exhibition in the ring by VALDES, who went around crying that he had been robbed. His manager Bobby GLEASON was just as bad. GLEASON later apologized to the boxing commission for his behavior. He was simply following the path of many other fight managers. Hymie WALLMAN, for instance, has always been a chronic squawker."

"A boxing judge is subject to all sorts of abuse. Everyone seems to take a personal pleasure in questioning his decisions. I have often been awakened at three in the morning by bums who would curse me over the phone saying something like, 'You dirty crook, what made you think you could judge a fight?' I never fail to be disturbed by this of thing."

"Under the New York system of judging we use a round and point system, with rounds prevailing unless the rounds are even. The points are added up. The winner of each round is awarded anywhere from one to four points, based on such factors as: (1) strong, effective punching; (2) effective aggressiveness; (3) skillful defense; (4) all-round ring generalship."

"I think rounds should be eliminated altogether in judging, and the points simply added up. The fallacy of the rounds-won system is that a fighter can do more damage to an opponent throughout a bout, yet still lose because the other won a number of rounds by a mere eyelash."

"I became a boxing judge in 1921, when I was twenty-six years old. I was quite a fight fan in those days, although I never did any boxing myself, and I used to attend bouts at the Y.M.C.A. and the Broadway Sporting Club."

"One night I watched Marty CROOS knock Soldier BARTFIELD all over the ring during a bout. Between rounds some fellows kept going up and down the aisles offering to bet on BARTFIELD. Since he was taking such a beating, nobody could understand why anybody would want to bet on him. However, BARTFIELD managed to survive until the end of the fight, and then was declared winner by a decision."

"I felt that even without any experience I could judge a fight at least as well as the men who gave BARTFIELD that decision. So I applied for a license. I became a judge in April, 1921."

"My first important fight was in March, 1921 – a flyweight fight between the champion, Pancho VILLA, and challenger Frankie GENARO. I scored that fight fourteen-one for VILLA. Nearly every newspaperman at ringside concurred with my decision, but the two other officials gave it to GENARO and took the championship away from VILLA."

"I was stunned. I remember yelling at the time, 'I'm through! I won't be a goat. I'll submit my resignation tomorrow morning.' The reason I was so angry was before the fight Commissioner Walter HOOKE had taken me aside and said, 'This is a very important fight, your first championship. You've got to remember that just because a man is champion, that's no reason why he should necessarily receive the decision. It didn't enter my mind until after the fight that HOOKE might have been trying to influence my decision."

"Perhaps the most suspicious-looking bout I ever worked was a lightweight bout between Tony CANZONERI and Al SINGER in 1930. SINGER was 'knocked out' by CANZONERI in one minute, six seconds of the first round. Only three punches landed, and none seemed sufficient to break an egg."

"Year later when SINGER was an official, he told me that he never recalled being hit by CANZONERI. 'You know, Harold,' he said, 'I think I might have been doped before the fight.'"

"The questionable fight that caused me the most trouble personally was the Robert VILLEMAIN – La MOTTA battle of March 25, 1949. Two months before, VILLEMAIN, a courageous Frenchman, had fought Steve BELLOISE in a gory battle at which I officiated. BELLOISE won the fight, and VILLEMAIN won the glory by lasting the distance. He was a bloody mess. Four stitches had to be taken above of VILLEMAIN's eyes, and eight more on his cheek."

"VILLEMAIN's brave performance that night endeared him to the Madison Square Garden crowd. In his next outing there, the La MOTTA fight, he was a strong sentimental favorite. Not many people liked La MOTTA anyway. In fact, he was looked upon as an undesirable champion. He was a surly individual. Also, in a number of his fights the betting odds had fluctuated so wildly as to cause some suspicion."

"There had been a lot of rumors that La MOTTA was going into the VILLEMAIN fight to lose. I don't know how true those rumors were, but to this day I think Jake La MOTTA wanted to lose that fight. It is not true, as a sports columnist wrote recently, that La MOTTA screamed at the officials after being awarded the fight. 'How could you do it to me?' But I must say he didn't look very happy about winning."

"Most of the 11.491 people at the Garden that night, including all of the newspaperman, thought VILLEMAIN had earned the win. I gave it to La MOTTA, six-five- one. Harry EBBETTS, the referee, backed me up with a six-four-two verdict. Judge Charlie SHORTELL saw VILLEMAIN the winner."

"The way I saw it, while I thought La MOTTA tried to lose, he couldn't make it bad enough. VILLEMAIN's punches, when they landed were powder-puffy. It would have been easy to award VILLEMAIN the night, but I couldn't in good conscience do it."

"Commissioner Eddie EAGAN was quite unhappy over the decision. He held an immediate investigation, and decided EBBETTS and I should be suspended indefinitely. One of the things that sustained me during that period was a letter I got from Jim FARLEY, a former boxing commissioner and a wonderful man. FARLEY wrote, 'When I was a member of the boxing commission, there was never anyone I regarded more highly than yourself, not only from the point of view of integrity and character, but also efficiency... I don't think anything has been gained by the act of the commission. Don't let it bother you because your friends, who know you, will still have the same degree of confidence in you.' Two months later I was reinstated."

"Among the boxing commissioners I worked under, I felt that EAGAN was the weakest – and not because of my suspension. He always seemed to be more in what other people thought than in standing by his officials. He treated his officials shabbily. Before the second LOUIS -WALCOTT fight, for instance, he herded a group of us to a room at the Concourse Plaza Hotel near Yankee Stadium. We were kept there for hours, until it was time to go to the stadium. I always regarded it as a silly act, and one that displayed a complete lack of confidence in his officials."

"The decision of mine that created the greatest clamor came in the Max SCHMELING – Jack SHARKEY heavyweight championship bout of 1930, before 80.000 fans at Yankee Stadium. This was the fight that led to the present rule that a man cannot win on a foul."

"I was the only official in a position to see the controversial blow in the fourth round. The boxers were directly in front of me. I could almost touch them. SHARKEY landed a sweeping low left. The punch seemed to me about eight inches below fair territory. SCHMELING fell to the floor, apparently in great pain. Referee CROWLEY looked at me, and I nodded to signal that it had been a foul. Then the bell rang, and all was confusion."

"CROWLEY asked me four times if I had seen the foul. Each time I said yes. 'Finally I got disgusted,' I said 'How many times do you want me to tell you it's a foul.'

"CROWLEY finally awarded the fight – and the heavyweight championship – to SCHMELING. Afterwards, the influential newspaper columnist Arthur BRISBANE, leaned over to me and said, 'Sonny, you've just saved boxing in New York State.'"

"Today I wonder if boxing was worth the saving. Looking back over the years, I'm convinced the answer is no. I've seen Harry GREB, perhaps the most vicious of all fighters, heel and gouge an opponent's eye until it was closed tight. I've seen Beau JACK bleeding from half a dozen wounds and hobbling about the ring on a bad leg. I've seen Ezzard CHARLES after Rocky MARCIANO got through with him in their first fight – unable to talk because of a smash on the Adam's apple, and with his right eye puffed, his left cheek ballooned by a blood clot, his lower lip swollen. I've seen Frankie RYFF after a fight when he needed twenty-three stitches. I've seen Hurricane JACKSON take terrific punishment in the ring. His second fight with Floyd PATTERSON, which I'm thankful I saw only on television, was one of the most brutal, one-sided affairs in my memory."

"Last July, JACKSON was finally retired by the medical board in New York. The board declared that he was 'deteriorating' and unable to defend himself. However, he turned up in action again in a fight in Canada, in which he decisioned an obscure heavyweight named Johnny VICK."

"Will JACKSON quit in time? And what about the other fighters who today are choking on their own blood in the ring"

"I say, let's become civilized. Let's do away with boxing."

McCOWN (1959) wrote:

"Its seems incredible that the many scientific groups and organizations which have pioneered and made so many valuable contributions to trauma in other hazardous occupations and industries, have not become more interested in athletic injury. Certainly, there is no more challenging area in traumatology than sports injuries."

McCOWN continues in his general statement:

"Until recently, accurate information and reliable statistics in competitive sports have been difficult to obtain. This is largely due to the lack of any medical program for the supervision and regulation of the sports. This applies particularly to amateur and professional sports. There are exceptions in intercollegiate sports as exemplified by the Harvard group, under the direction of THORNDYKE, in which a well organized medical program has been in existence for years and has provided excellent medical facilities for the diagnosis, evaluation and definitive care of athletic trauma."

McCOWN continues:

"Medical interest in boxing, until recently, had been very lax and little thought or concern has been given to the safety and welfare of the boxer. Many factors obviously have contributed to this apathy, the most important of which is that the very nature and concept of boxing is contrary to medical regulation and supervision. Unfortunately, there has been little interest shown by physicians and sports officials, or by the medical profession generally, in the medical aspects of professional boxing. Until recently there were no reliable statistics or data sufficiently accurate or informative enough to be of value in determining the medical status of boxing. Furthermore, there are factors inherent in boxing which have made it difficult to plan or establish any long range or satisfactory medical program for this sports."

"The second point of view is well represented by the quotation by ROCKE (1967) at the beginning of this section. Such physicians see nothing than dangerous or less than wholesome about boxing."

McCOWN comments on boxers' encephalopathy:

"The so-called punch drunk syndrome described by MARTLAND over 30 years ago, has become symbolic of head injury. It has never been proved to be a neurological syndrome peculiar to boxers and produced by boxing. It has, unfortunately, become a slick medical cliche with which to label any boxer whose performance and behavior in or out of the ring is unsatisfactory or abnormal."

"During the past seven years, over 11.000 boxers have been examined and licensed in New York State. Twenty year old, retired boxers were examined and 148 boxers out of the total number were retired or refused a license because of medical disability. Not one of the boxers examined revealed a typical punch drunk syndrome. We have no evidence at hand to prove that such a syndrome exists in boxers."

"No clinical conclusions or laboratory evidence was found which would substantiate the so-called punch drunk syndrome that has so often been erroneously identified with boxers.' (We would refer the reader to p. 540 on Late Permanent Brain Damage. The evidence there put forth for the existence of the traumatic encephalopathy of boxers is beyond refutation).

Elsewhere in his article, McCOWN mentions two observations made with the help of cinematic photography of numerous bouts – the first, less than earth shattering observation was that "performance in the ring closely parallels the physical fitness and experience of the fighter."

Indeed.

The second was more interesting, McCOWN stated that "relatively few blows in a single fight are effective, most of them falling short of their mark or striking their opponent tangentially. The majority of the blows to the head are dissipated on the surface with no obvious evidence of the force being transmitted to the structures within the cranial cavity. In an average ten round bout, approximately 1.000 punches are struck to various parts of the body but relatively few of them are effective, as exemplified by the fact that on 100 consecutive fights, only four knock outs resulted. The entire list of 32 knock outs represents only approximately 3 percent of the entire group of those participated in boxing over the seven year period."

MCCOWN continues:

"The effectiveness of the blow is in direct proportion to the speed and manner in which it is delivered. A straight or short punch delivered from the shoulder is far more effective than a wild so-called haymaker swing. Blows which land on the chin or side of the head in the anterior temporal region are the most effective and produce the majority of knock outs. An effective blow to the head or chin causes a fighter to pause momentarily, lower his guard or reach for his opponent, clinch, drop to one knee or fall completely prostrate. A dazed fighter is an easy target for a knock down blow. The knock out blow is one which renders the boxer temporarily unconscious for a period of ten seconds, during which he is counted out. Soon thereafter, within 30 seconds, he rises under his own power, returns to his corner and is able to leave the ring unaided. Complete consciousness has returned and the neurological examination is negative for abnormalities. He has no retrograde amnesia, headaches or other associated neurological signs or symptoms. This is the routine pattern which is generally observed when an opponent is knocked out in the ring."

DeVOE (1959) wrote:

"In boxing immediate death due to boxing is rare. Abolition of this sport would constitute an infringement of individual liberty..."

PÖSCHL (1960) noted:

"The best prophylaxis is in my opinion the prohibition of boxing."

STONE (1960) wrote:

"Every state in the Union has at one time outlawed 'prize fights,' and it was not until 1917, when boxing became an integral part of the U.S. Army's physical training program, that legal bans were significantly relaxed."

James Stewart GORDON *(Abolish professional boxing, Readers Digest, 76, April 1960, pp. 80-85)* wrote:

"On March 5, 1959, Johnny SAXTON, 28 years old, a former welterweight champion stood before a New York City judge, charged with housebreaking. The proceeds of the burglary: $ 5.20. SAXTON, who had reportedly earned more than a quarter of a million dollars during eight years in the ring, was destitute, he said, and owed the government some $ 16.000 in back taxes."

"The judge looked down at SAXTON, 'Johnny,' he said, 'where did your money go?' SAXTON answered in a slurred, barely audible voice, 'I didn't get much of it.' 'Why did you give up fighting, Johnny?' the judge went on."

"SAXTON dropped his gaze to the floor, 'They didn't need me no more.'"

"That's also what boxing did for Johnny BRATTON, another former welterweight champion, who at 30 years of age had become an old man in an Illinois mental hospital."

"The frightening truth is this: professional boxing is the down-at-the-heels prostitute of organized sport. The cases of Johnny SAXTON and Johnny BRATTON are no rarities. They serve merely to illustrate the racket's outstanding characteristics: brutality, and the venality and crookedness of boxing's managers and promoters."

"When you watch a fight on TV, the odds are overwhelming that you see two muscular young men who are nothing but puppets of the mobsters who control the business. Often the fighters are subnormal mentally, either because they were born that way or because they have practiced their trade too long. They know who is to win and who is to lose – and when and how. And at the end of their ring careers they are usually both battered and broke, robbed by the men who have used them."

"Considered coldly and on the basis of the evidence, professional boxing has absolutely no excuse for continuing to exist. It is time we were civilized enough to abolish this shabby game, on both medical and moral grounds."

"On moral grounds, the truth is that professional boxing is not a sport at all. It is a sordid business designed to yield maximum profit to the men who control the fighters, the arenas and the TV outlets. The fact that men can be permanently injured or killed in its practice is of little concern to the behind-the-scene manipulators. Through their monopoly, known as the 'Octopus,' they decide for virtually every boxer fighting today – regardless of his skill – when he will win or lose, when and whom he will fight next."

"Apparently the network people were not satisfied with the results. The National Broadcasting Co., recently decided to discontinue its Friday-night fights as of September 1960. According to an editorial in *Sports Illustrated,* the sponsor was happy because the fights were selling razor blades. The advertising agency was happy because the sponsor was. But the network was unhappy – presumably, as columnist Red SMITH surmises, 'because it is not possible to offer 'exciting' matches with brand-new performers every Friday."

"John BONOMI, assistant district attorney of New York and assistant counsel to the KEFAUER Committee, which recently announced plans to expose some of boxing's dirty laundry, says: 'In six years investigation I have found few fighters who are not wholly or partially controlled by the mob. Middleweight titleholder (in New York and Massachusetts) Paul PENDER has spoken in less restrained fashion: 'Boxing is rotten clear through, infested by gangsters and thieves.'"

"There is no place in the United States today for a 'sport' that is as brutal and crooked as the boxing racket. Harold BARNES, retired official of the *New York State Athletic Commission,* a boxing judge for 38 years, says, 'I consider boxing legalized murder. I would outlaw it.'"

"Isn't it high time the United States banned this public butchery?"

Six years after the statement of Karel HENNER, LAFORET *'Medico-legal aspects,* boxing.' *(Carney Hospital Journal 4: 54-56, 1962)* makes essentially the same points, but at greater length in greater detail:

"Since boxing as a sport and profession already enjoys legal sanction throughout most of what we are pleased to call the civilized world, any discussion of its medico-legal aspects suggests the quixotic..."

"Any final judgment concerning the propriety of boxing must rest neither on its legality (which is a fait accompli) nor on its medical aspects (which indicate that it is, like football, a potentially hazardous contact sport) but squarely on its morality."

"Among all extant forms of athletic endeavor pitting man against man, boxing is unique in that its prime and direct object is the physical injury of the participants."

"As in other sports, it is the desire of the contestants in boxing that victory be both convincing and obvious. And the most convincing and obvious way of winning is to make it physically impossible for an opponent to continue, either by producing various degrees of cerebral concussion ('knock out') or other injuries ('technical knock out'). When neither competitor is patently disabled, the decision may be awarded, on a 'point' basis, to the one who has sustained the fewer and less severe blows, although this method of winning is not considered ideal. Thus, injury directly intended and deliberately procured, forms an essential part of boxing, while in all other sports it is secondary, accidental, and not directly related to the object of the contest. Apologists for boxing are fond of citing statistics that indicate it to be probably less hazardous than football or baseball, and certainly less dangerous than driving a car. Because there is insufficient evidence to contest the validity of such statements they are usually made in the secure knowledge that contradiction is impossible. Nevertheless, true or not, they do not alter the premise that only in boxing is injury directly intended and produced. With this in mind, it is difficult to conclude that the potential hazards and actual injuries intrinsically associated with boxing can be justified morally."

"But is this moral stricture one of the theory only, or is there medical evidence that *significant* trauma may occur? Quite the latter. Acute cranio-cerebral injury is seen frequently, and ranges from the molecular disturbance of the simple 'knock out' to the gross vascular insults that may prove immediately lethal. Chronic cranio-cerebral injury, less dramatic but more pathetic, results in the punch drunk syndrome. The causal relationship of boxing to such cumulative chronic brain damage has been challenged in some circles but most data suggest a definite link. Documentation is usually retrospective and hence difficult, particularly when the defects are more psychological than neurological. Ocular, maxillo-facial, and aural trauma occur frequently. A special note is due renal injury, which an extensive and careful investigation has recently shown to be common. Thus, there is nothing theoretical about a subdural hematoma, or a cauliflower ear, or total hematuria."

"At this juncture it may be desirable to indicate that, despite somewhat varying details, amateur boxing partakes of the essence of boxing as such, and that only accidental differences exist between the professional and nonprofessional forms of the sport. The moral aspects already presented apply with equal force to both varieties."

"Since the absolute number of tragedies in boxing is probably quite small when compared, for example, with the carnage on the highways, it may properly be asked why one should be so overly concerned with a pastime that is entertaining to the viewers and voluntary entered upon by the participants. The reason is simple. Trauma and violence are a sad part of every day life, and no one is more aware of this than the physician. However, they are recognized for what they are, evils – unfortunate, perhaps inevitable, but evils nonetheless. In boxing, on the other hand, these same evils have been surrounded with an aura of respectability and have been placed in a context that is socially acceptable."

LAFORET continues:

"Of particular interest is the pre-fight medical examination, which is always cited as evidence both of protection afforded the boxer and of concern for his welfare. In some respects this can be considered a perversion of preventive medicine. The analogy to the exhibitionist preparing to go over the Niagara Falls in a barrel and clutching in his hand a physician's statement that he is physically capable of doing so is not too finely drawn."

" An accepted part of the ritual associated with boxing, and frequently prescribed by law, is the attendance of a physician at ringside. Since antiquity the physician has been a foremost and, at times, the sole defender of the sanctity of human life; the incongruity of his presence at such bouts need not been stressed. Nevertheless, the legal requirement that a bout not be permitted without his attendance has interesting implications. If, as recently, a group of physicians can refuse cooperation for politico-economic reasons, what can be said of a similar move from the motive of *true* preventive medicine?"

SAVASTANO (1962) also felt that boxing was too dangerous to continue, at least in its present form. He wrote of the opponents of boxing that:

" They further argue that it is the only sport in which a man is declared a winner simple because he has dealt more punishment to his opponent than he himself has received; and that in many cases this punishment consists of rendering one of the contestants unconscious, or even causing his death. Those who oppose boxing emphasize that it is the only sport in which both contestants and spectators may at times forget, or lack, that wonderful quality known as 'sportsmanship.' In every sport except boxing, the spectators feel sorry when a player injured, and even applaud him when he is carried off the field. In every other sport there is a provision for 'time out' when a player is injured. However, in boxing the injured contestant is put at a disadvantage when injured because his opponent will concentrate then on landing heavier blows with greater rapidity. In every sport except boxing a player who is unable to continue because of injuries leaves the contest with dignity and self respect. The boxer who is being beaten and cannot fight back effectively is on occasion booed by the spectators. One writer reports that boxing is 83 times more deadly than high school football, 50 times more deadly than college football, and many, many times more deadly than professional football."

"The chief argument against boxing is that the contents of the skull (chiefly the brain and its appendages) are the chief target. In other words, the objectives of most boxers is to inflict trauma, and put out of action the brain which is the most important and one of the most vulnerable organs in the human body. It is pointed out further that repeated blows to the brain have led and can lead to blindness, imbecility, paralysis, psychotic disorders and even death. In the United States alone, approximately, ten fighters lose their lives yearly as a result of brain injuries sustained in boxing bouts."

"The objectors to boxing maintain that, if a person becomes even momentarily unconscious as a result of an automobile accident, an industrial accident or as a result of engaging in any sport other than boxing, he is invariably admitted to a hospital where he remains under observation and treatment for lengthy periods of time. In addition to the detailed clinical examination which is conducted by competent neurosurgeons and neurologists such laboratory tests as X-ray films of the skull, encephalograms, and spinal fluid examinations are also done. These patients are seldom if ever allowed to leave the hospital until the chemistry of the spinal fluid has returned practically to normal. The average boxer who has been knocked out is not afforded this type of treatment. In many cases, after he regains consciousness, he receives a short quick inadequate examination while still in the ring, and is then allowed to go home on his own. Only those who fail to regain consciousness in the ring are hospitalized. This practice is defended by some physicians on the basis that observations at the ringside and more detailed examinations in the dressing room after the fight fail to reveal any abnormal neurological findings even in those participants who have lost their bouts by a knock out."

"It should be emphasized that for every boxer who becomes definitely punch drunk or who dies, many others become forgetful, irresponsible, moody, irritable, thick in speech, and subject to chronic headaches or dizziness."

"It has been previously established that the head is the primary target in practically every boxing match. However, secondary targets such as the back of the neck, the heart, the loins, and the solar plexus are not overlooked. If an abrasion or laceration is noticed, an effort will immediately be made to make it larger. If a swelling is noted over the eye or cheek, this then become the immediate primary target. In other words any visible contusion, abrasion, or laceration on a boxer's face during a bout immediately becomes the pinpoint target."

"The elimination of boxing gloves and the establishment of an 'upper foul line' should appease those who propose the banning of boxing. Perhaps some of the promoters will say that if the following rules were adopted, boxing would die a slow natural death in spite of the fact that the number of injuries would undoubtedly diminish."

(1)*"The elimination of the boxing glove.* I believe that this suggestion was originally made by Doctor LA CAVA of Rome, Italy. It is unreasonable to expect the unprotected hand repeatedly to strike the head of an opponent without its becoming injured and painful. The more the striking hand is injured, the less the force of each succeeding blow. I would anticipate that more contusions, lacerations, sprains, dislocations, and fractures of the hand would result; however, it is better to have an injury to the hand than an injury to the brain. It is therefore suggested that boxing gloves and bandaging of the hands be eliminated. It is recommended that the boxer be allowed to wear only an ordinary leather glove (not a boxing glove) without any padding whatever under the glove."

(2) *"Establishment of a second foul line.* As boxing rules now stand a boxer commits a foul when he strikes his opponent below the waistline. It is obvious that while being struck below the beltline may be painful, there are no recorded deaths from being struck below the beltline. It is also to be noted that when a boxer strikes his opponent below the designated beltline, officials warn the offender not to do it again, and the boxing fans become incensed against him. Objectives are raised to the so-called 'low blow,' even though the receiver of the low blow has well developed lower abdominal muscles and is probably wearing an adequate genital protective device. On the other hand, when a boxer scores with a heavy solid blow to his opponent's head, he is cheered even though his blow may have caused a permanent injury to its recipient. Since a low foul line is respected by contestants and enforced by the referees, why cannot a 'high foul line' also be established? It is recommended that the line be placed at the junction of the neck and the chest in front, and encircle the entire neck at this level. Purposely striking an opponent above the upper line or below the lower line should carry an automatic disqualification of the offender. Warnings should not be allowed."

This is an idea that has been put forward time and again. But SAVASTANO said, despite the fact that the number of injuries would be diminished, boxing would die a slow, natural death if head blows were not allowed.

The eminent British neurologist McDonald CRITCHLEY (1963) noted about the brain damage in boxing:

"Despite assurances to the contrary, there is little doubt that many boxers enter the ring with the intention of securing an unequivocal verdict by rendering the opponent unconscious, usually by a blow to the head. Inevitably, therefore, boxers are liable to suffer neurological damage, which may at time be severe, although it must be admitted it is usually apparently slight and transient. However, it does appear that there is a cumulative effect of multiple neurological disturbances received in the boxing ring leading either, from a short-term point of view, to the 'groggy state' or, from a long-term point of view, to the punch-drunk syndrome."

"It must be said that the mechanism of the cerebral knock out remains unclear. Sometimes loss of consciousness is acute following a single heavy punch to the head or the concussion of the head on the floor matting, ring stanchion or some hard part of the opponent's body. In the case of the acute loss of consciousness it is possible that damage to the neural tissue itself is slight, although exceptionally it may be severe. A blow of the severity necessary to cause instantaneous loss of consciousness may be sufficient to damage other intra-cranial structures with more or less disastrous consequences. The cumulative effects of blows not in themselves sufficiently powerful to render the patient immediately unconscious may lead to the so-called 'groggy state.' In this condition the patient, although remaining on his feet, is disoriented, hypotonic and helpless, becoming little more than a passive target for his opponent. It seems likely that if the bout is allowed to continue when one of the boxers is suffering from the 'groggy state,' severe damage will follow. It is therefore necessary for the referee to be constantly on the watch for the first signs of this condition, stopping the fight immediately they appear."

"Probably the most important neurological consequence of boxing, however, is the so-called punch drunk syndrome or dementia pugilistica. I prefer to call this the chronic progressive traumatic encephalopathy of boxing a somewhat lengthy term that is both descriptive and explicit. This condition tends to be a later sequel to boxing and its manifestations may not show themselves until after the boxer's career is concluded. Once it appears it is slowly but steadily progressive. The presenting features are diverse, often making for some difficulty in diagnosis although it is, of course, to be suspected when the stigmata of the ring (cauliflower ear, thickened supra-orbital margins) are visible. Differential diagnosis is between Parkinson's disease, General Paralysis of the Insane, and frontal tumour. The patient usually complains of headache, dizziness and disturbances of vision. Intellectual impairment is seldom obvious to the patient, although relatives will notice changes in personality with euphoria, fatuousness and defective emotional control. There is frequently a correlation between the degree to which the stigmata are manifest and the severity of the dementia. On clinical examination the usual findings are extra-pyramidal signs."

"This condition is much more common in professionals than in amateurs (though it has been described in amateur boxers) and is more common in Whites than Negroes. The class of the boxer will frequently determine his liability to this condition since it is much more commonly seen in second-raters who make up for lack of skill by aggressiveness and a tendency to 'slug it out'".

"The great problem of this condition is that it is extremely difficult to diagnose in the early stages; the commonest early signs include failure to win bouts which the boxer would normally be expected to win and an inordinate recovery period required after the fight. The ability of the 'punch drunk boxer' to absorb further punishment in the ring is disputed; some seem to have a raised threshold, being able to take progressively more and more punishment before going down, while in others the opposite occurs. These latter are sometimes said to be 'glassy jawed.'"

"The morbid anatomy of the punch-drunk syndrome is uncertain. The changes described include progressive cortical atrophy, internal hydrocephalus and gliosis. In some cases rupture of the septum pellucidum has been described. These changes are, of course, similar to those in extreme senility. It appears that the progress of the disease is relentless. It is not yet certain whether it can be arrested if boxing is discontinued at the appearance of the first signs. Certainly we must regard this condition, albeit becoming less common, as extremely serious and a very heavy price to pay for a career in the ring."

"Boxing is not without danger, least of all danger to the brain. Fatalities are regularly recorded, the causes of death being remarkable varied. It is not so long since three fatalities were recorded in teen-age boys. In some of these cases the cause has been intra-cranial hemorrhage and it is sometimes discovered that the skulls of these boxers are abnormally thin. So serious is the risk of injury, and so peculiar is boxing (in that it is the only sport in which competitors deliberately set out to inflict injury upon their opponents) that there is ample justification on medical grounds for its discontinuation."

CRITCHLEY's view has been proven correct by many more scientific publications concerning head injuries due to boxing.

In 1964, Jack DEMPSEY wrote an article *(What about boxing today, Western Medicine 5:267-268, 1964)* from essentially the same viewpoint.
DEMPSEY wrote:

"*Give up boxing?* Hell, no. Maybe the sport is a little sick today, but if you doctors have made such wonderful progress in curing diseases with magic drugs and operations, I think we boxers can cure the disease in boxing today."

"You know, if you want a watch fixed you must take it to a watch repair man. If you want a car fixed, you had better take it to the garage. If something goes wrong with the stomach or brain, you had better let a doctor fix it. And if anything goes wrong with boxing, why not let the boxers fix it? All professions, trades, and businesses need consultation at times, but there are altogether too many people trying to cure diseases that they don't know anything about... whether the disease is in a car, watch, stomach, or in a sport."

"Since I have boxed a few times in my life, I think I can help solve some of the problems of the sport. We'll need some changes to catch up with other things today, but we don't want to completely emasculate the sport, though. Like you doctors who practice what you call 'preventive medicine,' I think we can take the bull by the horns and practice a little preventive treatment in the boxing game, too – and clean it up! I think it is possible to correct the major ills of the sport if we begin *now!*"

"First of all, boxing is an occupation just like any other occupation. A kid has a right to choose whether he wants to become a bricklayer, taxi-driver, doctor, lawyer, boxer, or engineer. And when he makes up his mind, let's encourage him in his chosen field so that no matter what it is, he will strive to become the very best in that undertaking. We need all types of workers, so it really doesn't matter what he chooses – as long as what he wants to do is an honorable profession. Boxing has always been an honorable profession, and the temporary setback it is receiving these past years can be corrected easily."

"If a youngster chooses boxing for his livelihood, let's give him the very best of everything to become a champion – just like we would like any kid to be the very best surgeon, lawyer, bricklayer, or pilot – or whatever he likes... you name it, *but help him. Don't hinder him.*"

"One of the best apprenticeships a boxer can have is the training he gets at the colleges with boxing teams. The colleges have fine instructors and coaches and equipment. They have excellent doctors who watch over them, too. But in the past couple of years, many colleges have removed boxing from their list of sports because of a fatality that occurred after a bout. This was very unfortunate, for the young boxer was an excellent man. The coach was one of the most careful and best liked in the business – but the accident did happen. However, dozens of kids have died after football, basketball, and other sports but no one ever cries to the high heavens to get rid of these sports, too. I wouldn't want them to get rid of any of the sports, but I do think that boxing should be returned to the sports card of every college. You know, I knew a guy who dies walking down the street; I knew another who died in a merry-go-round; and another who died swimming. Should we eliminate all these activities, too? As long as we continue to do what is best for each athlete, and if he is willing to participate in that sport, then we should do everything to make it safe for him to compete... no matter what the sport is!'

"In the State of Nevada, where the boxing teams still reign as powerful teams of the country in boxing, a committee of the University of Nevada almost ruled the team out this year. However, the vote was put to the student body for their final opinion they voted to keep boxing on their schedule by a great majority. It seems as though the female student population was as anxious to keep it as the males were. This is a great credit to Dr. Art BROTEN, who is head of the Department of the University of Nevada, and a tremendous tribute to the coach, Jimmy OLIVAS – who has been given high praise a number of times at the meetings of the sports of the American Medical Association, and by many physicians who understand the sport of boxing. OLIVAS an outstanding boxer of some 300 fights in his day, is one of the most prominent boxing coaches in the country today."

"We don't have many colleges like the University of Nevada left who have boxing in their schedules so the next place the kids can get some knowledge, training and experience is at the gyms and local boxing clubs. TV has practically closed all the private clubs and gyms, but they can be instituted again with very little trouble. And the quicker we do it, the better it is going to control a good element of kids who are getting mixed up in juvenile delinquency problems. Let's save the kids, make good sports out of them, and watch them progress financially and every other way."

"When boxing is efficiently taught, guided, and controlled, the problems are minimized considerably for it is a wholesome sport for the keen, alert athlete with extremely rapid reflexes who combines strength with science and know-how to mould sportmanship. It demands strict training and concentration in addition to a very healthy body. Those with less than the maximum of the health should not participate in the sport on a championship basis."

"It should not be considered more dangerous than many sports that are participated in all strata. For example, my good friend, Dr. Joe ELIA who is very familiar with boxing medical problems, learned while he was doing some research on the subject, that the most dangerous sport was softball, with soccer running a close second... because these two sports in particular were participated in by the 'occasional' athlete who generally is not in good physical condition. If you ask me my choice of a boxing ring, or a hot rod on a highway, I'll choose the ring for a couple of kids anytime to becoming a 'statistic' on the freeway."

"If we can institute the rebirth of local clubs; and gyms, we can save many kids, and at the same time give them an apprenticeship they are being denied in the colleges – and at the same time give them a chance at the 'big time' when they become good enough. Can't you just see the picture now? A hundred or more clubs all around the country for kids to train, boy, wouldn't this boxing picture of 1964 change? I's love to see that change, believe me."

"If we can get a good start for the kids, we can have just about the finest group of boxers this world has seen – and also, there will be less accidents. The medical profession has gone far in promoting good ideas to make the sport safer. Those doctors who know the sport can help tremendously, and many who do not understand it can get tips from those who do know it quickly. With the very best of testing equipment, we can almost predict what a man will do in the game. For a lot depends upon his own personal physical make-up. Good examinations and good physicians to watch over the fights are an excellent contribution to the sport – before and after the fight."

"The kids must be covered by workman's compensation while engaged in the occupation of boxing. There is no reason in the world why a man who get injured while participating in his occupation – whether it is shoemaking, bricklaying, engineering, or boxing should be covered by the compensation laws as pertains to any labor in that state. Also, the kids' purses must be protected from the vultures by law. This is where the National Czar is of much value. I think the National Czar, like in baseball, would organize the sport once and for all... and boy, watch the hoodlums disappear. Once there is a National Czar, there will be complete control of the sport so that each individual will not be chiseled by 'handlers' and parasites. Boxers who are not apt to par will not be thrown into the ring too quickly, or against competition which they are not capable of competing with, or whom gamblers are interested in. As a matter of fact, a Czar would give the great heave to anyone associated with illegal gamblers. Gangland would be eliminated if there was some federal or 'central' control, especially if they were given the treatment that organized baseball gave the Black Sox some years ago. Rules and regulations must be made uniform some day – all states must have the same rules and regulations – if boxing is to attain any degree of sanity. It's silly the way it is now. A boxer can be a champ in a few states and not in others. Everything must come under one control so that all things will be equal in boxing. We've got to get rid of the 'return bout clause' in these ridiculous contracts, too. When TUNNEY took me the first time, I didn't even ask for a return bout without working my way up to the top while TUNNEY went on fighting others. Eventually, I got the contender position again. That's the way it should be all the time. Federal control, or a Czar to control, would really make a difference."

"The medical men have a good number of suggestions to help the sport but first we must build good strong healthy bodies – these bodies must be healthy physically, mentally, and morally – then we will have the sport back to where it belongs."

DEMPSEY seems more than a little confused, at times. His opening sentence, referring to 'magic drugs and operations,' implies that there is some 'magical' solution to boxing's ills.

His attempted analogy fails miserably. The examples are all objects, and are all 'taken' to those trained specifically to repair or cure them. Letting boxers 'fix' boxing's problems may be a good idea – after all, John BROUGHTON set up the first rules for the sport in England and he was a pugilist – but whether professional prize fighters will ever break out of the manager-trainer-promoter circle seems doubtful. The amount of money involved to the business (and it is most definitely a business) is very great, and most boxers, sadly, are not intelligent enough to know the first thing about major business deals.

To call boxing 'an occupation just like any other occupation' is ludicrous, as are the statements that 'we need all types of workers, so it really doesn't matter what he choose,' and 'boxing has always been an honorably profession.' The last is patently untrue, as history will disclose; and as for boxing being like 'any other occupation,' let it be remembered that even soldiers (the only occupation that is even close) do not fight amongst themselves for an occupation.

As for helping, not hindering a young man in his pursuit of boxing, it would be contrary to a physician's essential philosophy to help a youth along the road to certain physical (and very probably neurological) injury. Luckily, since 1964, boxing has ceased to be taught in most universities and is not sanctioned by the NCAA.

It is unfortunate, like many others, Mr. DEMPSEY misses the point behind one of his statements. He wrote '... dozens of kids have died after football, basketball and other sports but no one ever cries to the high heavens to get rid of those sports...' He does not ask why although no one attempts to abolish the other sports, boxing is constantly battered by cries of abolition. There is a very good reason. Other sports death are caused by *accidents; deaths* in boxing result from the fists of one man striking the head and body of another man, for money or 'sport.' Only 'accidental deaths' in boxing result from the accident of the dead man's body falling when perhaps the body of another wouldn't.

To suggest that the availability of boxing training in local clubs will reduce juvenile delinquency and/or 'make *good* sports' out of them (DEMPSEY caps) seems a bit far-fetched at best. Those young people who are naturally inclined toward delinquency to a large extent will not be at all influenced by the chance to learn boxing. These juveniles are interested in power and luck, not in facing a chance to be hit and being expected to stand there and take it (or dish it out). In fact,

DEMPSEY's description of boxing – 'a wholesome sport for the keen, alert athlete with extremely rapid reflexes who combines strength with science and know-how to mould sportmanship' – apply much better to the martial arts of judo, karate, jiu-jitsu etc., or perhaps fencing, than to boxing.

When he defends boxing by comparing it to 'a hot-rod on a highway,' DEMPSEY writes 'I'll choose the ring for a couple of kids anytime to becoming a 'statistic' on the freeway,' he seems to miss a vital point. Cars are means of transportation ('hot-rodding' being a fad in the 1960's) while boxing is a 'sport.' Something that is good for you.

Although DEMPSEY recommends testing and examinations, he never makes the final step suggesting that physicians be given the power to stop fights. 'Before and after' says DEMPSEY, essentially implying that what goes on *during* the fight is no business of a doctor's. Most would disagree.

Does DEMPSEY *really* wonder why boxers are not covered by workman's compensation? What insurance company in it right collective mind would conceive of covering a man whose profession is fighting other people who are intent upon incapacitating him?

It is some consolation that, in his penultimate paragraph, DEMPSEY recommends national control of boxing (as did the 1983 report of the American Medical Association). In this, DEMPSEY was ahead of his time, but the many misconceptions that accompany that recommendation are the building blocks of his confused philosophy.

DEMPSEY published another article *(The Boxing Mess – And What to Do About It. Readers Digest 83:85-90, 1963).*
DEMPSEY wrote:

"When Luis Angel FIRPO's big right fist connected high on my face in the first round of our New York fight in 1923, I drew a blank. If the punch had landed a little lower, it would have been curtains for me. As it was, I never knew what had happened during that first round until I saw the movies. I was out on my feet."

"Being out on your feet doesn't hurt. You don't know where you are; you couldn't, if someone should ask you, add two and two. But you punch, feint and clinch by instinct, as the subconscious takes charge. You are even abnormally dangerous to your opponent: you are uninhibited by caution; you are bent on tearing apart that foggy figure before you."

"That's the way I was against FIRPO in that first round. I knocked him down half a dozen times. And then, *he* knocked me through the ropes and out of the ring."

"When that wild round was over, Jack KEARNS, my manager, gave me a whiff of smelling salts strong enough to wake a dead man. My first conscious thought was, 'I've been knocked out!' I asked KEARNS what had happened. He told me I was doing great, the second round was coming up, and then I could take FIRPO out any time now."

"When we came out for the for the second round, FIRPO was in even worse shape than I was, and I did knock him out."

"Immediately after that fight I went to a doctor and had a brain examination. Everything was all right. That's the point I want to make: one of the main things wrong with boxing today is the lack of thorough examinations before and after a bout. This, plus other abuses, has resulted in the death of a number of fighters."

"Look what happened to the Argentine heavyweight Alejandro LAVORANTE, fighting in the United States. In March of 1962 LAVORANTE was knocked out by Archie MOORE. He had to be carried from the ring on a stretcher. Less that four months later he was allowed to fight again and was knocked out in five rounds by Cassius CLAY. Yet two months later he was permitted to fight Johnny RIGGINS – and without taking a brain test. That time he was knocked out in six rounds, and remained in a coma. For months he lay unconscious in a hospital. Today, after 13 months and three brain operations, he is still semi-conscious. When things like this are permitted, no wonder people say, 'Abolish boxing!'"

"Because of such abuses, the boxing game has, in recent years taken a dive such as no fighter ever took, in a depressing series of deaths, scandals and investigations of hoodlum control. Unless there is thorough reform, I will have to agree with those critics who demand the abolition of the game."

"But boxing *can* be saved, and should be. On the basis of my own considerable experience inside and outside the ring, I'd like to propose a six-point program to restore it as an honest and virile sport."

"1. *Protect the fighter with proper physical examinations.* This means a thorough medical examination one week before each fight, another the day of the fight, and a brain test after if the boxer has taken excessive punishment or been knocked out. As things stand now, most boxing commission doctors give a fighter a quick examination – more ceremony than examination – on the day of the fight. That's all."

"More often than not, it is the fight before the final one that does the fatal damage. For instance heavyweight Max BAER gave Ernie SCHAAF a bad beating in 1932. Then, six months later, SCHAAF, without a brain test, met Primo

CARNERA – and was knocked out. The blow appeared to be light that the Madison Square Garden onlookers jeered SCHAAF as he lay on the canvas, and people hinted openly of a 'fix.' But SCHAAF died three days later. Similarly, welterweight Benny PARET took a bad beating from rugged middleweight champion Gene FULLMER about three months before a knock out by Emile GRIFFITH. As you can see, the final fight only caps an already aggravated condition."

"Another thing: In football, when a player is out on his feet, the coach takes him out of the game, pronto. But in New York a boxer's manager – the man who knows the fighter best – is not allowed to throw in the towel to stop the fight. The towel might have saved Benny PARET. In the 12th round Emile GRIFFITH pinned him in a corner with a flurry of blows, PARET would surely have fallen but for being tangled in the ropes. Then the referee stopped the fight. By then it was too late. If PARET's seconds had been able to throw in the towel when it was clear that he was helpless, Benny might be alive today."

"When I was a young boxer on the way up, I made the mistake of taking on Fireman Jim FLYNN, one of the ranking contenders with a record of 38 knock outs. He knocked me down five times in the first two minutes of the opening round. My brother Bernie threw in the towel and stopped the fight. I was raging mad at Bernie at the time, but I am thankful to him today. If I had taken a few more punches, my career could have ended right there."

"2. *Change the judging system.* When there is no knock out, the outcome of each fight is today determined in most states by the so-called 'point system,' under which two judges and the referee award specified points for performance in each round. This scheme should be replaced by having the referee alone select the winner, and I'll tell you why."

"The only sensible way to judge a fight is to award the decision to the man who 'forced' the fight, did the most damage and *finished stronger.* If the point system existed in track meets, the man who led a race all the way but folded at the finish and lost could be adjusted the winner."

"Over the years thousands of people have told me they thought I won my second fight with Gene TUNNEY. TUNNEY, they insist, was given the benefit of a long count. Well, after several pretty good rounds, I did catch Gene with a couple of good punches in the seventh, and he went down. Even he will tell you he was hurt. But nobody in this world except Gene can tell you whether he could have gotten up sooner. Gene says he could have. That's good enough for me."

"What many people seem to forget is that Gene came back strongly in the eighth round, and that he knocked *me* down in the ninth. For most of the rest of the way he got in the better punches, and he was stronger than I was at the finish. Don't let anyone kid you – Gene was entitled to that decision and the championship. He did more damage, and he finished stronger."

"Under the point system, the 'winner' can be hanging on the ropes at the finish. Many a time I have seen a ring announcer point to a bleeding, groggy wreck as the winner, while the loser was unmarked and fresh as a daisy."

"Along with the point system, we should get rid of the ringside judges themselves. They add nothing to boxing. A good referee – and there are many of them – should be the complete boss, as he used to be in the great days of the game, and as he still is in England and a number of other countries. He is the person who can judge a fight better than anyone else, because he is in there with the fighters and is the only one who can get a clear view of everything, including the in-fighting. That's why sportwriters always ask the referee questions after a big fight, about things they couldn't see. Where were the sportwriters sitting? Right alongside the judges. I have refereed hundreds of fights, and the scoring of the judges has often made me wonder whether we had been watching the same contest. The trouble is that most boxing judges are politically appointed and sometimes know little or nothing about boxing."

"3. *Oust the hoodlums.* Everyone connected with boxing should be carefully screened. For so long much of boxing has been controlled by the underworld. At last, Frankie CARBO, called 'the underworld czar of boxing,' is serving 25 years in prison, for trying to muscle in on the earnings of former welterweight champion Don JORDAN. And three CARBO associates – Joseph SICA, Frank 'Blinkie' PALERMO and Truman GIBSON – were convicted on the same charge and received varying sentences. (All are appealing to the Supreme Court)."

"But these convictions are only a start in the right direction. No mobster or racketeer should be licensed as promoters or managers, where they are in a position to control fighters or to fix fights. And when they operate through stooges they should be ferreted out and banned. "

"I was myself approached for what, to my mind, would have amounted to a fixed fight. John REISLER, a promoter, tried to make himself my self-appointed manager with a phony contract. He wanted me to fight the great Sam LANGFORD, whom he was also managing – me, a kid just 21 years old and weighing 165 pounds soaking wet! I had sense enough to know I wasn't ready for a man of LANGFORD's class. Nevertheless, the promoter kept after me, trying to show me how to beat the man who was his own meal ticket. I walked out on him."

"Today's would-be fixers – gangsters, hoodlums and their hanger-on – have no place in the sport of boxing. When they are swept out, there will be a lot less talk about fixed fights."

"4. *Make fighters earn their chance at a championship.* The automatic re-match, like last July's Floyd PATTERSON – Sonny LISTON two-minutes-and-ten-seconds fiasco, which commissioners happily sanctioned – after LISTON had knocked out PATTERSON in two minutes and six seconds of their first fight – must be outlawed for the good of all concerned. This kind of thing is making boxing a laughingstock. There was no rapid rematching during my time. If you lost then, you had to prove yourself all over again. In my case, I lost the championship to Gene TUNNEY. Before I met TUNNEY again I fought Jack SHARKEY. Had I been beaten, SHARKEY would have fought TUNNEY for the championship and I would have been out of luck. It was a tough fight. SHARKEY was never better. He could box, and he could punch. The fans got their money's worth."

"In those days you had to really earn your crack at a big fight. I fought for ten years before I got to meet Jess WILLARD for the heavyweight title. Yet Pete RADEMACHER fought Floyd PATTERSON for the heavyweight crown in his first professional fight! No wonder that most of the fighters around today are no great shakes. Sonny LISTON is strong and tough, but that kind of stand-up fighter can be beaten. I'd like to see a boxer like Gene TUNNEY or a puncher like Joe LOUIS in the ring with LISTON. I fought several big, tough fighters like LISTON – men like FIRPO, Fred FULTON, Jess WILLARD. They ranged in size from six-feet-four to six-feet-six and from 215 to 250 pounds, and I knocked them out. I weighted 180, but in those days you had to learn how to fight."

"5. *Give fighters economic protection.* Boxers, most of them just youngsters, many of them uneducated, should be protected by law against having an excessive percentage of their earnings chewed up by those around them. Heavyweight champion Joe LOUIS made millions in the ring; he earned big sums for service charities during World War II; yet he wound up broke and owing Uncle Sam more than a million dollars. Why? One reason is that, by contract, his managers received 50 percent of his earnings."

"Due to lack of protection, many other top boxers have seen their winning dissipated in various ways. For example, Johnny SAXTON, former welterweight champion made more than a quarter of a million dollars – and wound up with nothing. And Beau Jack, who made many a fat purse, in now shining shoes at a Miami hotel."

"I was no better than most at protecting myself in the financial clinches during my fighting days. When Doc KEARNS was managing me, I could never get him to put anything on paper. He helped me a lot, but he got half of *everything* – movies, articles, real estate, as well as the fighting. In addition to this, he gradually borrowed $ 150.000 from me. I finally decided to do something. After my fight with FIRPO I went to the promoter's office and collected my purse – a half million dollars. Then I gave Doc his half – *minus* what he owed me. He was fit to be tied. That was the beginning of the end of our relationship."

"Fighters also need protection of a pension fund. Too many men, trained for fighting and nothing else, have no money to fall back on when their careers are finished. The baseball player's pension program (fed partly by a percentage of the proceeds of the All-Star Game and the World Series) provides retired baseball players with as much as $ 725 a month for life. A cut from the profits of both closed-circuit and regular TV could provide the same for fighters. TV certainly owes boxing a great debt for the way it has damaged the sport, through cutting down attendance at the 'big' fights and almost eliminating the small fight clubs. "

"6. *Put boxing under national regulation.* Congress should pass a federal law regulating boxing and setting up a boxing commissioner to rule over every aspect of the game, just as the commissioners of baseball and football do now. Today our 50 states have varying boxing laws; some states recognize one boxer as champion, while others recognize someone else. Sonny LISTON is barred in New York State, so he fought Floyd PATTERSON in Nevada. Uniform nationwide regulation is the only way to put into effects the reforms I have suggested."

"Major-league baseball was under a heavy cloud once; even a World Series was fixed. So the baseball leaders named Judge Kenesaw Mountain LANDIS as commissioner. He used his absolute power to enforce the rules passed by the clubs. Baseball has been healthy ever since."

"Last year two pro football players were found to have been betting on games. Pro football's commissioner acted quickly, ousted the players and aired the scandal. The public's faith in pro football went undimmed. That's what we need in boxing."

"This is not an idea original with me. The late Senator Estes KEFAUVER tried to get a bill through Congress last year that would have established a national boxing commissioner and would have imposed federal regulation upon boxing. For all its obvious merit, the bill didn't get to the floor."

"If boxing continues to be as mismanaged as it is today, it will be abolished by popular demand. But, if boxing is reformed and regulated, it can become again a great sport which, because it puts the very highest premium on physical fitness, courage and manliness, no one but an extremist would ever think of trying to abolish."

Alfred BALK published an article *'Should boxing be abolished?' Today Health, 43: 18-34,* August, 1965*)*. Following a few excerpts are brought:

"Boxing cannot be made safe – only be safer. It's the only sport in which, within the rules, each man deliberately acts out to hurt his opponent."

"There are also numerous cases of permanent injuries suffered by boxers in the ring. Two former welterweight champions, Johnny SAXTON and Johnny BRATTON, for instance, incurred such extensive brain damage during their boxing careers that both were in mental hospitals before they reached age 30."

"Among many fighters who have suffered partial or total blindness as a result of boxing are ex-champions 'Kid'

GAVILAN, Henry ARMSTRONG, Jimmy CARTER, Frankie GENERO, Speedy DUDO, and Vince DUNDEE."

"Because of this and a series of revelations of gangster infiltration of the sport, an increasing number of critics have demanded that boxing be outlawed as inherently unsafe."

"... Here are a large number who have an extremely low I.Q., even to the point of being unable to read or write. Should a person with a low I.Q., but with tremendous physical attributes, be refused the opportunity of applying them to earn a comfortable living?" (KAPLAN).

"Opponents of boxing point out that it is the only sport in which, within the rules, the contestants deliberately sets out to injure his opponent. This, to say, is morally indefensible in a civilized society."

"Various champions, when questioned as to whether this 'intent to injure' or mere skill in strategy and maneuver was paramount in their success, have corroborated the estimate:"

"Says Sonny LISTON: 'It's like war. Either I'm gonna hurt him or he'd gonna hurt me.'"

"Jack DEMPSEY: ' To protect yourself with your fists, you must become a knock out fighter.'"

Sugar Ray ROBINSON: '"Hurting people is my business."'

No discussion of the pros and cons of boxing from a medical standpoint could be complete without an account of the vigorous debate held in 1963 at the Conference of *The British Association of Sports and Medicine*

The following is a review of the *Panel Discussion* of the Conference of the *British Association of Sport and Medicine on the Medical Aspects of Boxing,* held in London in November of 1963, and printed under the title of *Medical Aspects of Boxing,* edited by A. L. BASS, J. L. BLONSTEIN, R. D. JAMES, and J. G. P. WILLIAMS (1965).

The chairman was Sir Arthur PORRIT *(Chairman of the British Association of Sport and Medicine).* The members of the panel were: The Rt. Hon. Baroness SUMMERSKILL of Ken Wood, P.C., M.R.C.S., L.R.C.P; Mr. A. Dixon WRIGHT, M.S., F.R.C.S., D.T.M. & H.; Dr. Philip KAPLIN, M.B., B.S., M.R.C.S., L.R.C.P.; Dr. J. A. WAYCOTT, M.B., B.Chir., M.R.C.P, J.P., of the Middle Temple, Barrister-at-Law; and Dr. Lindsay NEUSTATTER, M.D., B.Sc., M.R.C.P

The highly colored responses, to the questions asked by the Chairman and to the views of the other members of the panel, typify the emotionally charged atmosphere that usually accompanies any meeting of the proponents and opponents of boxing.

The first question: "Should boxing be rationalized by omitting the head from the target area and allowing hits below the belt on the abdominal muscles?" – evoked the following responses:

WRIGHT: "It would make an end of boxing. All kinds of armor would be worn to defeat the object of it, and take out all its interest... Guarding your chin and your face is where the most skill comes in – the best form of defense is attack."

KAPLIN: "This is another form of saying that we should have head-guards, and then, I expect, pantaloons and go back to the days when we were like gladiators, wearing armor. Is not this absolutely ludicrous? We are losing sight of the noble art of self-defense. This is an art or a science – there is no slogging."

Lady SUMMERSKILL: "This would be an excellent first step, because it would protect the most vulnerable part of the body, the human brain."

WAYCOTT: "If one removed the head as part of the target, this would be a first step. I think the whole thing is so impracticable that we should not waste any further time talking about it."

NEUSTATTER: "It might save a head, but what about the future of the human race?" (Laughter).

Opinions were generally divided into two groups. There were those whose main concern was preserving the essential nature of boxing as a competitive sport and spectacle versus those concerned primarily with the wellbeing of the participants.

WRIGHT and KAPLAN suggested that head blows are so critical that without them boxing would be destroyed. NEUSTATTER admitted that heads might be saved though it would be detrimental to human society. WAYCOTT agreed, stating that the whole discussion was a waste of time. They ventured no comments on medical aspects of the alternatives.

The lone voice in defense of the proposed no head blow rule was Lady SUMMERSKILL, who expressed concern for brain injuries in participants. If anyone agreed with her, their opinions were not recorded.

The meeting was intended to be a medical discussion, but no medical issues were mentioned or debated, aside from the generalized comment of

Lady SUMMERSKILL. The focus of debate was muddled and jocular in nature, even to the point of ridicule. In their most serious moments participants commented on social benefits of boxing in its present form. From the tone of their comments, it is clear that there was no earnest discussion of medical issues or any possible changes that might render boxing more safe. There is still a tendency for medical discussions on boxing to go awry, as can be seen in later excerpts and comments from the literature. Possibly the most helpful adjustment in boxing discussion would be discipline in the debate format, with medical issues considered separately from broader social or moral issues. The most powerful contributions to public debate that physicians can make are those based on medical fact and experience, whether for or against boxing as a sport. In representing the view of medical societies, our persuasive impact is diluted when we drift into areas of political, moral, or societal interest. Such topics are better left to personal individual opinion expressed in the proper forum and those experts called upon for public statements.

The *second question* was whether a bout should be stopped as soon as an injury occurred in a boxing match, with victory awarded to the boxer who is ahead on points.

The response to this question was mixed and slightly confused. KAPLIN called repeatedly for a definition of injury. WRIGHT said that such should be left to the discretion of the referee, and Lady SUMMERSKILL said the advice of the 'experts' should be followed and proceeded to misinterpret that advice.

The third question – 'Of all the aspects of boxing, which do you most deplore?' – raised heated response from most of those present. The more-or-less direct responses were as follows:

Lady SUMMERSKILL: "If you had asked me that question 20 years ago, I would have said the injuries which the individual boxers suffered and the hazards involved. Today I must combine it with the spectacle – there is a television set which brings to the people of the household a spectacle every week, of men fighting each other and a commentary giving a graphic description of what is happening. Now he has a cut eye, how he is nearly down, and so on... A boy who is not very bright watching these pictures which glamorize violence is encouraged to think that this kind of behaviour is something to be respected and if possible emulated. In this days and age the worst thing is for the spectacles we have had described to us to be seen by unintelligent youths throughout the country. Remember that I am talking of members of a captive audience... and I believe that we as responsible citizens should recognize that this is a danger to our way of life."

KAPLIN: "We are not arbiters of public morals... We are dealing today with the medical aspects of boxing... Provided there is professional boxing allowed in this country, so long is it absolutely essential to have control... I only deplore the fact that if there is an error in control – that is wrong."

WAYCOTT: "The chief objection to the sport is that it is the only sport in which the rules are designed in such a way that the infliction of injury is a deliberate and necessary part of the game. There are sports more dangerous, but injury when it occurs is an accident and the crowd is duly sympathetic. When injury occurs in boxing it is because it is the intention of the fighter to render his opponent unconscious by a blow to the head. If that is not deliberate infliction of injury, what is?"

WRIGHT: "In answer to Lady SUMMERSKILL, I think the violence in this country is due to violence in Western and crime films and not due to watching boxing."

Question four – "Should headgear be compulsory?"

WAYCOTT: "Yes, certainly, for all... The head is the most vulnerable part of the body in boxing. Although a head-guard cannot give total protection, at least it can give some... I think that this is something that ought to come in this country."

"The sooner the medical objections to the sport and the degree of control of the sport will come more in to line with enlightened medical opinion, the better for all."

By implication, the medical objections to boxing are far removed from 'enlightened medical opinion.' One might wish here more specific definition of terms.

KAPLIN's concern is for appropriate protective floor-coverings, and he dismisses head-guards as 'looking ludicrous' and 'not 100% safe,' citing a record of a man being killed while wearing a head guard. He also denies that any record has been made of a "man being killed where appropriate floor matting is used." One might be inclined to question the need for a form of protection being 100% safe – unless the man wearing the head guard was killed because of it, there would seem little to his claim against protection. To dismiss a protective measure on cosmetic grounds is ludicrous – as if the injuries prevented by the gear would be more aesthetically or emotionally appealing!

WRIGHT: "A head guard only saves you from a thick ear. It does not really safeguard you from anything else but cosmetic damage, and I am against it."

(Lady SUMMERSKILL and NEUSTATTER did not speak on this topic.)

Question five - "Is it true that the vast majority of amateur boxers come from the same socio-economic group? If so, why?"

Lady SUMMERSKILL states that "It is the simple, ignorant boy who boxes," citing various 'public' (American 'private') schools at which boxing is not allowed. She wonders why, "if it is considered unfit for these boys to learn boxing why should the boys in Cranwell learn it today?"

WRIGHT: "There is enough public school boxing that have contests every year."
NEUSTATTER: "In overcrowded districts there is gross lack of recreational facilities and boxing is a recreation which can be comparatively easily obtained, unlike cricket where you need a large field and a whole day to occupy only 22 men."
Lady SUMMERSKILL: "... surely there are other sports these boys could indulge in, not only boxing."
NEUSTATTER: "I agree that boxing is not the only sport, but it is a very difficult thing to find a necessary space for others."
WRIGHT: "Boys' clubs would be lost without their boxing."
WAYCOTT (referring to the *Royal Air Force Base at Cranwell, England*) – "As far as Cranwell is concerned we have young men who are taught to land jet aeroplanes at high speeds when the utmost skill, co-ordination and balance is required. The Commandant of R.A.F. Cranwell, because he has been worried about impairment of co-ordination resulting from boxing injuries, has banned boxing."
BLONSTEIN: "My son was a medical officer at Cranwell. Pilots not allowed to box were allowed to play rugby football and every weekend they had at least four fractured noses in pilots as a result of rugby."
WAYCOTT: "You do not fly an aeroplane with your nose but with your brain."
Lady SUMMERSKILL: "Boxing is more dangerous than rugby. The R.A.F. refused to let them box because of the danger."

This exchange speaks very clearly for itself.

Question six – "Protagonists of boxing claim that boxing is a safe sport because in absolute terms there are fewer injuries. Is it not dishonest to employ absolute figures instead of relative figures, which give a more accurate and honest comparison?"

WRIGHT: "What we do know is that every sport has a risk and we should not give up sports because of risks. In rugby there are many injuries such as spinal injuries which are severe but do not stop people playing. I agree we should consider relative risk, but you would find it jolly hard to do so."
Lady SUMMERSKILL: "I am told that in this country are only 400 registered boxers and not 4.000 as before – this is a terrific decline. To compare the accident rate of these with the hundreds of thousands playing football this afternoon is quite unreasonable..."
Chairman: "I try to remain impartial, but in my opinion this is a tiny problem altogether and we want to keep that in perspective. It really is not of great importance compared to numerous other things which do produce serious injuries."
Lady SUMMERSKILL: "We have to realize we are not talking about obvious injuries only, but also syndromes such as punch drunkenness where the initial signs cannot be detected and when they are detected they cannot be cured and the condition may be permanent... These statistics are completely fallacious."
KAPLIN: "The first important point is to discuss this question of punch-drunkenness. The early signs of this can only be diagnosed at the ringside, and those who do not go boxing can have no possible idea what may occur in the early stages... I have taken boxers' licenses away for 15 years and more for these specific signs and I am proud to say that none of them developed punch drunkenness. It is, of course, difficult to make an exact prognosis. At the risk of repeating, myself, I wish to say that you must know what you are talking about when you say that statistics are fallacious and talk a lot of spurious sophistry of these points."
Audience: "These statistics are available in insurance companies. The percentage accident rate is about half that for football. That discounts any question of numbers. The rate to cover boxers is much lower that for footballers."

WRIGHT's statement that "it would be jolly hard to do" hardly deserves commentary, and cannot be taken as a serious argument for either side."

The Chairman's remark is completely uncalled

for, dismissing in two sentences the central question of determining the dangers of boxing. He views this as a 'tiny problem' that 'really is not of great importance.' He was clearly losing his struggle for impartiality.

Lady SUMMERSKILL's later comment on punch drunkenness is largely that such syndromes do not appear until late in a career, and when detected cannot be cured, a point which is true. KAPLIN replies that such can only be detected at ring side, and once the boxers stop fighting, the symptoms disappear. The progressive traumatic encephalopathy of the boxer, as defined by CRITCHLEY (who is quoted by KAPLIN in his reply) is just that – a generalized loss of neurons, and restitution of the parenchyma by glial scar tissue. Damage of this nature does not disappear. His charge of deceitful, deceptive subtle argumentation seems a harsh judgment, against extrapolations that were essentially correct. A call for additional statistical information would have been more appropriate under the circumstances.

The comment of the unnamed member of the audience is a classic example of the argumentation immortalized in the book *'How to Lie with Statistics.'* He does not mention where 'the insurance companies" get their figures, nor does he take into account those data insurance companies as a class utilize to arrive at a premium rate, i. e. the relative number of participants in the sport in question compared to the number of claims, which is the initial problem.

Quotations *seven, eight,* and *nine* were answered by only a single panel member (or the editor allowed space for only one response).

Question Seven – "Why does the governing body of professional boxing not permit the medical officer greater control, including allowing him to stop the fight if he thinks this necessary on medical grounds?"

KAPLIN: "Every referee, from the moment he is given an area, is given grounding in the medical aspects. The referee is the final arbiter in the ring, and rightly so. No medical officer should be put in the position of making a diagnosis during a fight. If there was a trickle of blood, can you possibly imagine what would happen in the hall if the medical officer said: "You have a fractured septum," and stopped the fight? Let the referee do it, that is what he is there for."

Question Eight – "When an amateur boxer turns professional, no notice is taken of his past record. Should there not be liaison in this matter?"

KAPLIN: "The *British Board of Control* today have a very happy liaison with the *Amateur Boxing Association.* When a certain boxer turned professional, it came to my knowledge behind the scenes that he had been in a mental hospital, because one of his colleagues wrote me a letter. We will continue to have the greatest liaison between amateur bodies and ourselves."

Question Nine – "Can boxing or boxing training have any benefit for other sports?"

WAYCOTT: "Boxing means a rigorous training for co-ordination and balance, These things can be turned to good account in other sports."

The text has been quoted at length to illustrate a point. Over four pages were spent discussing personal prejudices (Question three), and nearly three on the socio-economic background of boxers. Such topics, though interesting and of themselves important, pale into insignificance beside such questions as a boxer's past fight to the medical officers at ringside. Excellent though KAPLIN's contacts are, there is still no official, required liaison between the amateur and professional boxing authorities; and KAPLIN's later comment verges on the ridiculous. That a comparison should be made between the minimal knowledge of the referee, who has been 'grounded' in the medical aspects of the sport, and that of the physician seated a few feet away with years of training behind him, is almost unforgivable.

That boxing encourages co-ordination and balance is obvious – most sports do, and ballet training is much better than any sport.

The *final question* – 'Should boxing be banned for medical reasons?'

Lady SUMMERSKILL answers in the affirmative, reminding the audience that her opinion is held in a number of countries, citing Denmark, Belgium, and Norway, each of which to some extent was working against the sport. Iceland had in fact prohibited it.

> KAPLIN answers that, if it should not be banned, it must be perfectly controlled. "We need every possible safeguard in boxing."
> NEUSTATTER replied "You may dislike boxing because of the alleged aggression or because of the physical dangers, but there are dangers in every sport. I am not satisfied that boxing has many more than any other sport."
> WAYCOTT says, "I should like to see greater control in both amateur and professional boxing." He advocates the use of headgear and speaks of the transformation of dueling several centuries ago," ... This dueling was prohibited by law and dueling has been transformed into fencing and fencing is one of the most elegant and skillful sports you may ever hope to see. This sort of protection may be made available to boxers."

The discussion ends here.

The suggestion put forward by WAYCOTT has certain merit. Fencers are occasionally injured, of course, but the sport is organized specifically to avoid such occurrences. Fencers wear padding over their entire bodies, and have metal face-guards on their helmets. For competition purposes, these outfits are wired electrically. One touch of the end of the foil/saber/epee sets off a signal (light or buzzer) and first touch wins the round. Thus no advantage is ensured by excessive force.

The profession will, of course, never accept this idea – but that very fact reveals something about the real concern with skill and technique, and how important human suffering is to the boxing 'game.'

One last question arises from this great sea of controversy, of abolition versus protection.

Why?

Why does the question of abolition always come up?

Why, when doctors meet to discuss this activity, is there always a strong faction favoring the abolition of boxing altogether?

To our knowledge, only one other sport ever had an abolitionist faction – it is said that Queen VICTORIA sought to stop mountain climbing! Aside from this short episode of history, the overwhelming question must be answered. When doctors seek to protect participants in other sports – by designing better running shoes, quick-release mechanisms for skis, or better landing pads for the high jump – they report the injuries that occur regularly and attempt to solve the problems which cause them. No one insists – or ever suggests – that the sports themselves be abolished.

Why, then, have so many doctors of impeccable repute been willing to hold forth at impassioned length on the side of abolition?

The answer is, unfortunately, simple.

The object and main goal of boxing is the incapacitation of one's opponent.

For this to be lauded as a 'sport' requiring great skill, or worse yet the 'noble art of self-defense' (originally applied to fencing, and attached to boxing only by association) is an insult to those athletes whose sports reflect a desire for excellence, and not a desire to see their opponent stretched unconscious on the ring floor before them.

The neurosurgeon Dixon WRIGHT (1965) wrote:

> "The brain is a very wonderful organ, and it has an equally marvelous system for its protection against injury. First of all, it lies within a hard casing – the skull – which protects its delicate substance from all but the hardest knocks. Within the skull, between it and the brain itself, are the three membranes or meninges, between which are two cushions of fluid – the cerebro-spinal fluids. The brain floats in these fluids like a sponge in a bath, and the fluid is constantly renewed and flows from one part to another through small openings resembling thus the shock absorbers of a car. In addition, between the lining of the skull and the surface of the brain run delicate filaments which lightly tether the brain in place and prevent it from swirling about too much when the skull is struck with a blow. The brain itself is thus buffered in the bath of fluid in which it lies, and although it may move a little under the shock of impact it is really very well protected. Injury to the brain itself is thus not very likely under the force of a boxer's blow. Even though the blow tends to rotate the skull as well displace it off centre. What does happen sometimes is that some of the delicate little filaments may be stretched and torn, and as they often have blood-vessels running near them, this may lead to bleeding. This bleeding can be a very serious matter, but this is not an injury to the brain itself. In these cases the blood may leak out into the cerebro-spinal fluid and mix freely with it, but if in greater quantity, will not

do so and a clot forms. As a result, increased pressure inside the skull does harm by displacing the brain inside the skull."

WRIGHT continues:

"Another form of intra-cranial haemorrhage which may follow injury, usually a more severe injury such as that due to striking the head on unprotected corner stanchions or the floor of an unpadded ring, is one where the blood leaks out between the skull itself and the membranes. This type of haemorrhage is caused by damage to one of the blood-vessels on the inside of the skull and is well localized, and if dealt with quickly is easily remedied."

"These injuries are rare in boxing but they are very important. If they are properly treated they can be dealt with satisfactorily. The reason why some people may die from such injuries is because they may not have been treated properly. I well remember a case of a boxer who sustained one of these injuries – an intra-cranial haemorrhage. He was treated by a young doctor of limited experience who operated on him, but could find no bleeding. After the operation the patient got worse and I was called to see him. It was too late and the patient died. He had got some bleeding but the doctor was not experienced enough to find it. The reason why these injuries are often lethal is because the best individual care is not always available. These patients should only be treated by people of great experience."

"One of the objections to boxing raised by many people that it leads to 'punch drunkenness.' We will be hearing a lot about this. The great difficulty about the punch drunk syndrome is that it comes on late in life, after the boxer's active career is over, and it might not be due to the boxing itself but to the other risks in the boxer's life. Punch-drunkenness used to be fairly common in the old professional boxers and everybody said it was due to the boxing, but you must remember that many of the boxers in those days came from conditions of poverty – they had to fight in order to make a living, and there were, and I should think still are, many temptations for the boxer, particularly the successful ones. These temptations include wine and women. I remember attending to a very well-known boxer once who had neuro-syphilis, which can produce a clinical picture very like the punch-drunk syndrome. You all know that many boxers when they retire become publicans, and certainly many in the old days at least, drank to excess and became chronic alcoholics. The chronic alcoholic may become a person very like the punch drunk person, and very often it is difficult to say which is which. There are also many harmless things which people do which cause brain damage. The heavy smoker with chronic bronchitis who wakes the whole house with his coughing every morning may be causing little tiny brain haemorrhages of the kind said to occur in the punch drunk, every time he coughs. There are many other ways in which the pressure in the blood-vessels of the brain may be raised causing then little haemorrhages. In these days, when there are fewer professional fights, punch drunkenness is almost unknown. Even in cases which do occur it is difficult to be sure that they are due to boxing and not to some other aspect of the boxer's life."

"Of course there is some danger in boxing, but then there is danger in every worthwhile sport. You cannot go through life without an element of risk of some sort. The element of risk in boxing is not sufficiently great to justify doing away with the sport. Every sport could be eliminated on this basis. I have even known a player die from injuries received in billiards!"

Now, at least we know that wonderful organ, the brain, is so marvelously protected against boxing blows! And if the infrequent lesions receive proper care, then all is well. Those poor unfortunate few who died of boxing injuries simply were not treated properly; they were in the care of someone inexperienced, like the young doctor who could not find the hemorrhage. Is it not generally known even under the finest care, boxers often die of severe secondary traumatic lesions without ever having a detectable and operable hemorrhage?

Of course, punch drunkenness is the outcome of many temptations, like women and wine, to which especially the successful boxer succumbs. This absurd statement in untenable: the literature shows that a solid differential diagnosis excludes a luetic disease by using serologic tests. Many case histories deal with real punch drunkenness – by boxing punches, and not by drinking [alcoholic] punch.

And if wine, women and song are not enough, now it seems that coughing produces small hemorrhages in the suddenly delicate, unprotected brain. Boxers must be careful of their respiratory exertions in their singing careers! The case of the poor billiard player merely shows us that no harmless amusement is without an occasional risk.

In an editorial *'Mayhem in the ring puts doctors on the spot." (Medical World News, 1966):*

"One humid June night in 1947, 'Sugar' Ray ROBINSON defended his welterweight title in Cleveland by knocking out Jimmy DOYLE in eight rounds. The next morning DOYLE was dead of a brain hemorrhage and a reporter asked

ROBINSON: 'Could you tell he was in trouble?' 'That's my business' retorted ROBINSON, with utter sangfroid, 'Getting fighters in trouble is my business.'"

Although the controversy over boxing normally rages from two extremes, there have occasionally been those who have at least attempted to present an unbiased view. This may take the form of simple reporting which this first text exemplifies, or it may focus on one of the more important but less debatable aspects of the ring (money, for example).

This first text is almost a sketch of manners. The club described seems almost Victorian in aspect and substance. The article *'Bash Him Again, With Feeling,'* appeared in *Sports Illustrated in October 1966.*

"Having long since tired of discourse on the density of Harold WILSON's pipe smoke, the height of Julie CHRISTIE's miniskirts and the length of the Rolling Stones hair, affluent British businessmen are looking elsewhere for idle-hour entertainment. They are turning, in increasing numbers, to private boxing clubs – elegant establishments serving food, drink and prizefights. They may represent London's last fling at a sport that can now count only 500 professional fighters in all of Britain. There are eleven of these houses of posh and pow, but the grandest of all has always been the *National Sporting Club,* founded in 1891 as a middle-class and Bohemian once bet on bareknuckle fights and shouted, 'More gore,' whenever a fight was not exciting enough."

"As bloodthirsty as the *Pelican* at first, the *National Sporting Club* gradually reformed, and by 1929 had brought professional boxing within the bounds of legality. The QUEENSBERRY Rules for Endurance Contests were the lever, which meant rounds were limited to three minutes, with one-minute intervals and not more that 20 rounds to a contest. The referee halted the bout if one man was unfit to continue and, most important, the club rejected barefist fighting."

"Down through the years the *National Sporting Club became* the place for the boxing in England. 'What the *Covent Garden Opera House is* to the world of music, the *National Sporting Club is* to pugilism. No artist can say he has arrived until he has proved himself on the stage of one or in the ring of the other,' said the Marquess of QUEENSBERRY. This marquuess was Marquess No. 10 – twice removed from Marquess No. 8, who wrote the rules. Translated into *Madison Square Garden terms*, Marquess 10 was saying simply, 'The club has all the big ones.'"

"Yet is was not until 1938 that the club actually had a clubhouse – the first and second floors of the *Hotel Splendide on Piccadilly.* After one move, the club settled in 1955 into its present quarters in the *Café Royal.* Just off Piccadilly Circus, where much of its past glory has been successfully brought to life and the blend of brandy and cigars is as perfect as ever."

"For his yearly subscription of £ 25 ($ 70) a member can watch 21 boxing tournaments promoted by the club between September and June. *National Sporting Club nights*, strictly black-tie events, with women rarely permitted, follow an orderly, gentlemanly ritual. Members and their guests meet around 7 for drinks in the DERBY, QUEENSBERRY and LONSDALE rooms. Dinner follows in the DuBARRY and DAUPHINE suites (one recent menu: iced melon, minestrone, sirloin of Aberdeen Angus or steak-and-kidney pie with French beans and potato in a jacket, followed by fresh pears in kirsch, lemon water ice or sponge fingers), and by 9 sharp everybody is seated at ringside in the NAPOLEON Room, surrounded by sparkling chandeliers and grandiose mirrors. Waiters move quietly down the rows of tables and chairs distributing brandy, whiskey and champagne and lighting cigars. Betting is quiet and subdued, and it is a rule of the club, usually adhered to, that applause may occur only between rounds."

Some data about the history of the National Sporting Club follow:

"Recently 650 members and guests gathered on a fight night to watch four matches, the final one of which was described by *London's Boxing News as* the 'bonniest punchup so far of this club season. Signifying their approval of all that had gone before, members rose up in the traditional manner to shower the canvas with 'nobbins' – silver coins and roller-up bills."

"After it was over, while the boxers changed in what are normally ladies' powder rooms and the members rolled down the great curve of Regents Street in their limousines, a club official explained how modern sensitivities had changed and, with them, the ways of the *National Sporting Club.* 'The old Corinthians encouraged an atmosphere that was more like a cockfight,' he said. 'Boxers at the club in the past were fighting for their lives. We have a more sophisticated membership today. We're a little more appreciative of people's feelings.'"

Some data about the history of the National Sporting Club follow:

The *Horton Law allowed* boxing in New York State. It was repealed at the beginning of the 20th century. The main reason was a boxing match between James J. CORBETT and 'Kid' McCOY in August 1900 at Madison Square Garden, New York. CORBETT won the fight by a knock out in the fifth round. It was generally assumed that McCOY had 'thrown the fight.' Public boxing matches were banned by the repeal of the *Horton Law,* which was replaced by the *Lewis Law,* which allowed staging of boxing matches only for members of constituted clubs.

At a boxing fight at the National Sporting Club in London, on April 4, 1901, a featherweight, Jack ROBERTS, knocked out his opponent Billy SMITH, who struck his head while falling and died of a head injury.

ROBERTS and nine leading members of the National Sporting Club were charged at the *Old Bailey* with 'feloniously killing and slaying' Murray LIVINGSTONE, the real name of SMITH.

The prosecution admitted that the objective was to declare boxing illegal rather than to seek punishment on the defendants. The judge, Mr. Justice GRANTHAM, upheld the desirability of differences in daily life settled with fists in preference to weapons, and the desirability of organised boxing being organised with proper rules.

The jury found that the death of the boxer occurred in a properly regulated boxing contest, was accidental, and that the defendants were therefore 'not guilty.'

As result the police concerned themselves only with keeping order at boxing matches, they did not question the legality of properly organised boxing fights.

The *National Sporting Club prospered* in the early part of the century. The most important event was the introduction of weight classes. In 1909 the *National Sporting Club* announced poundages for eight classes as follows:

Flyweight	8 st (112 lb – 50.4 kg)
Bantamweight	8 st 6 lb (118 lb – 53.1 kg)
Featherweight	9 st (126 lb – 56.7 kg)
Lightweight	9 st 9 lb (135 lb – 60.75 kg)
Welterweight	10 st 7 lb (147 lb – 66.15 kg)
Middleweight	11 st 6 lb (160 lb – 72 kg)
Cruiserweight	12 st 7 lb (175 lb – 78.75 kg)
Heavyweight	over 12 st 7 lb.

This was accepted throughout the boxing world, except for the term 'cruiserweight' in the United States, where there existed already a 175 lb class called 'light-heavyweight.' The light-heavyweight class was a suggestion of Lou HOUSEMAN, a Chicago newspaperman.

The British eventually changed their name for the class from 'cruiser' to 'light-heavyweight' in 1937 and in 1979 a new 'cruiserweight' title was introduced by the *World Boxing Council (WBC),* with a limit of 13 st 13 lb (195 lb – 87.75 kg). Other new weight classes had been introduced since the 1920', so that now 15 weight classes exist.

The *National Sporting Club persuaded* their President, Lord LONSDALE, to give his name to the famous series of belts, to be given the victorious boxer in each class. This was done to give authority to the Club's rule that British championship matches could only be held on their premises. It established a monopoly for the *National Sporting Club,* the holder of the LONSDALE belt was considered the champion. The Fifth Earl of LONSDALE provided the first belt himself.

"Another approach quite different from this colorful account involves the use of statistics and at least pseudo-scientific data in an attempt to inform. Again, the author limits himself to the more concrete, less arguable aspects of boxing.

"In millions of American homes today, fist fighting has become an integral part of the domestic scene. I'm speaking now about professional fighting – not the amateur, or party, variety. Last year the four major networks carried into the nation's homes 197 main-event boxing bouts. Some were watched by as many as 35 million people in the comfort of their living rooms. It's likely, then, that questions which once bothered only the fight fans who crowded smoke-filled arenas or converted ball parks now interest the whole family."

"As these millions of new boxing enthusiasts look on, two fighters may belabor each other for ten rounds – and walk away from the scene slightly short of wind but otherwise unaffected. On another night, one man may drop as if shot after a blow that did not appear greatly different from a 1.000 others."

"Forty-nine – about one in four – of the 197 bouts televised nationally by NBC, CBS, ABC and DuMONT during 1953 ended in knock outs (not counting those stopped because of cuts or arm or leg injuries). These statistics indicate that only one fighter out of every eight, is on the average, successful in any single attempt to knock out his opponent."

"Just how much power is behind a punch? The maximum thrust in a boxer's blow was measured three years ago by the *Cornell Aeronautical Laboratory, in* Buffalo, New York, for the *New York State Athletic Commission.* While wearing 8-ounce gloves – the weight worn in all but championship contests – heavyweights (more than 175 pounds) averaged 17.4 foot pounds of kinetic energy per punch. On the other hand, flyweights (112 pounds or less) managed only 5.6 foot-pounds. When 6-ounce gloves were used, the heavyweights averaged 26.2 foot-pounds and the flyweights 8.7."

"Multiply these figures by the total number of punches thrown in a single fight and you get some impressive totals. In clocking dozens of fights, I've found that two slow-moving heavyweights, in a fight so dull that it induces anesthesia only among the spectators, will land, between 750 punches in 10 rounds. At the other extreme, two light hitting but fast flyweights, flapping like moths around a flame, may land as many as 1.000 over the same number of rounds.

The same sort of approach was used in this short piece from *'Business Week'* in 1962.

"This week's brief encounter at *Chicago's Cominsky Park,* in which Sonny LISTON, boxing's Neanderthal man, dethroned Floyd PATTERSON as heavyweight champion of the world, came close to fulfilling JACOBS' prophecy. A mere 18.894 souls paid $ 665.420 to glimpse 2,1 minute of action. The live gate accounted for 15% of the estimated 5 million dollars the fight grossed. Nearly 2 million dollars of it will go to PATTERSON, and about $ 250.000 to the new champion."

The other way of attempting to remain neutral is to do what this chapter had tried to do – present both sides of the argument. The editor's bias will almost certainly be evident, but material is presented from both sides.

The Swedish government banned professional boxing in 1969 for ethical, medical, and moral reasons. In 1985 the Swedish Parliament moved to forbid amateur boxing, too.

In an article by Lester KINSOLVING *(Muhammad Ali and the manly art. Should a minister be exempted from military service while he makes a fortune in boxing. The Houston Post, February 28, 1971).*

KINSOLVING stated:

"Despite more that four hundred deaths in the boxing ring since 1900 – plus the unnumbered army of brain-damaged ('punch drunk') ex-boxers – the expressed concern of organized religion about boxing has been minimal and isolated."

KINSOLVING quoted several other representatives of various churches who have addressed the issue. Nearly two decades ago, Jesuit Father Alfredo BOSCHI wrote that

"Boxing cannot be justified from a moral viewpoint, but must be condemned as something gravely illicit in itself. It not only produces, but aims to produce serious injuries which can become permanent and lead to death... It makes a beast of a man... adoration of brute strength, of the fist which can pulverize the brain."

In 1963, the Right Rev. Nelson BURROUGHT, Episcopal Bishop of Ohio, noted that:

"Five men have been killed in the boxing ring since the first of January."

Bishop BURROUGHT, now retired, went on to assert:

"To encourage their potential violation of the Sixth Commandment and to pit man against man under the guise of American entertainment is in my judgment a denial of our Lord's emphasis on the sacredness and value of human personality."

Yet such ecclesiastical concern has been rare – despite organized religion's historic (if initially unpopular) opposition to a considerable number of blood 'sports,' from gladiatorial games to bear-baiting.

KINSOLVING continues with the main focus of his article – the use of a 'religious' objection to war by MUHAMMAD ALI (for the purpose of exemption).

"Currently religion is a point at issue in boxing regarding America's most celebrated pugilist – and ham actor. MUHAMMAD ALI contends that he should be exempt from military service because he is a Black Muslim minister."

"Just how 'pacifistic' is this theologically fantastic hate group is well documented in the autobiography of the late Malcolm X, who expressed little doubt as to the identity of those seek-

ing after his life for his having dared to object to the unofficial harem of Black Muslim leader Elijah MUHAMMED."
"The entire concept of clergy draft exemption has been substantially challenged by Father RIGA of California's Saint Mary's College. But for the Rev. Mr. ALI to demand that he be exempted from military service so that he may make several fortunes by brain-bashing is as grotesque as the military chaplain who recently prayed for a large body count of the enemy in Vietnam."
"Perhaps the ultimate irony in this case is in the thousands of his fellow blacks who have either been without the funds to prolong draft resistance in the courts – or who have died, often very bravely, such as the black medic who saved the lives of his comrades by falling on a live grenade."
"And if the parents, wives or children of such men have ever expressed any outrage at this irony, it has scarcely been heard. This of course, in striking contrast to the strident pronouncements of the fighting millionaire minister."

KINSOLVING's views are an echo of those expressed almost fifty years before, when a contributor to the *Literary Digest* was appalled at Chicago's plans to hold the TUNNEY – DEMPSEY fight at the site of a war memorial.

"Regarding the holding of this battle at Soldier's Field, what patriotic citizen is thrilled by the prospect? If Chicago permits this desecration of its memorial to the heroes of the World War, who gave themselves in death without the dollar as the motive, why should not the city be consistent and set forth on a pilgrimage to lay a wreath on Benedict ARNOLD's grave, or honor the achievements of Aaron BURR? There is a lot of shouting now, about Americanism and the principles of George WASHINGTON, Soldier's Field is sacred ground, or its dedicatory exercises were only a mockery. How can it be honored by the brutal spectacle of prize-fighting, especially where the patriotic record of any concerned is under a cloud?"

Norman MAILER (1975) wrote in *'The Fight:'*

"Prizefighter do not, of course, train to kill people at large. To the contrary, prizefighting offers a profession of men who might otherwise commit murder in the street. Nonetheless, the violence capable of being generated in a champion like FOREMAN is staggering to contemplate when brought to focus against another fighter. This violence converted to a special skill, had won him the Championship by his thirty-eighth fight. FOREMAN had never been defeated. On the night he won the Championship, he had accumulated no less than thirty-five knock outs, the fights stopped on an average before the third round. What an unbelievable record that is! Ten knock outs in the first, eleven in the second, eleven in the third and fourth. No need to think of him as psychotic, rather, as a physical genius who employed the methods of catatonia (silence, concentration and immobility). Since ALI was a genius in wholly separate ways, one could anticipate the rarest war of all – a collision between different embodiments of divine inspiration."

An editorial *(Australian New Zealand Journal of Psychiatry, 'Boxing', 5:131-132, 1971)* reads:

"Boxing has been a well recognized human interest and activity for at least 3.000 years. The battered features of the boxer were as well known to the accient Greeks as they are to us, and may be found depicted on vases and statues from that era, as may the equipment of the boxer. The Greeks bound their knuckles with thongs of leather, soft or sharp, according to the fashion of the time. The Romans added metal spikes and other similar mutilating devices, and it is therefore fortunate that boxing fell into oblivion for many centuries, until its revival in England and America in the last 200 years. For years, it took the form of prolonged, bare-knuckle bouts, until more civilized standards were composed. The last bare-knuckle championship in America took place in 1889, when SULLIVAN knoccked out KILRAIN after 2 1/4 hours in the 75th round (UNTERHARNSCHEIDT 1970). It is interesting to note that Oscar WILDE's 'screaming scarlet Marquis', the Marquis of QUEENSBERRY, one of the most uncontrollable violent men of his day who pursued the Prime Minister with a dog whip, played a major part in the civilizing of boxing and in the introduction of the rules which bear his name."
"There is a growing body of opinion in favour of the complete abolition of professional boxing, not only because of the incidence of injuries, but also because of the allegedly brutalizing effect of this so-called sport on its participants, its camp followers, and its spectators. It must be admitted that the crowd at a professional bout, roaring encouragement to two blood bespattered heavyweights endeavouring to batter each other into insensibility, is as close to the Roman crowd at a gladiatorial contest as we can get today. However, we probably have no specifically psychiatric opinion to offer on this aspect of boxing which could not be contributed by any other intelligent and well-informed section of the community."
"As medical practitioners, however, we can and should concern ourselves with the injuries, sometimes leading to death, which boxer suffer in the course of their profession. Injuries to face and jaws, eyes, nose, ears and kidneys are well recognized, but it is injury to the brain which directly concerns us as psychiatrists. Let us bear in mind that, while similar injuries can and do occur in other sports, these are more or less accidental, whereas 'in boxing the aim and object – explicit or implied – is to render the opponent *hors de combat*'" (CRITCHLEY 1937).

"As CRITCHLEY (1957) has pointed out, there is a variety of knock out blows – the 'solar-plexus knock out, the 'heart knock out', the 'carotid sinus knock out', and so on. The variety with which we are concerned results from direct violence to the head, produced by a blow with the fist, and often complicated by a second severe blow when the head hits the floor. The duration of unconsciousness following such a blow is usually quite brief, but longer periods of unconsciousness are known, and in a small minority of cases death may occur without recovery of consciousness" PAUL 1957). The period of amnesia associated with the 'knock out' is also variable. Once again, it is usually brief, but can be much longer, Gene TUNNEY's famous amnesia of 48 hours seems to be well known both to followers of boxing and to those who have not the slightest interest in the sport. UNTERHARNSCHEIDT (1970) points out that all varieties of intracranial haemorrhage which can result from blows to the head have been found in boxers."

"The first description of the punch drunk syndrome, a clinical entity resulting from oft-repeated blows to the head, is usually ascribed to H. S. MARTLAND in 1928, and many cases have been reportetd since. CRITCHLEY (1957), for example, reported that he had collected a series of 69 cases of chronic neurological diseases in boxers, the greatest majority of whom could be regarded as examples of the punch drunk state, which seems a more suitable descriptive label than his suggested alternative, 'chronic progressive traumatic encephalopathy of boxers'. The victims of this syndrome are usually the second- or third-rate performers, and are professionals rather than amateurs. CRITCHLEY noted that there is a correlation between the degree of cranio-facial marking and the severity of the punch drunk syndrome. The characteristic features of the syndrome – gradually progressive dementia with emotional lability, slowness of speech and thought, memory defects, mask-like face, slurred speech, infrequency and slowness of movement, tremor of arms and hand – are, or should be, familiar to every psychiatrist. MAWDSLEY and FERGUSON (1963) reported a further series, and noted that epilepsy may also occur. On investigation, they found cerebral and cerebellar atrophy on pneumencephalography, and also the cavum septi pellucidi was common."

"Evidence of brain damage has been demonstrated by means of E.E.G. in cases where gross damage was not discernible clinically. KAPLAN and BROWDER (1954), in a study which left some things to be desired, studied 1.000 E.E.G.'s from 1.043 boxers. Their study did not permit many detailed conclusions, but it was clear from their observations that disorganized E.E.G.'s came from poor ring performers. i. e., those likely to receive blows to the head. PAMPUS and GROTE 1956) studied 190 active and 17 retired boxers. E.E.G.'s were taken before, immediately after, and days after bouts. Abnormal trainings were found in 41 per cent of cases, and the incidence of abnormal E.E.G.'s increased with each successive fight. Boxers who had been knocked out had more E.E.G. abnormalities than those who had not. NESARAJAH et al. (1961) found an incidence of E.E.G. abnormalities of 60 per cent in boxers, many of them amateurs, even though only 34 per cent had a history of having been knocked out."

"In the light of this evidence, it is clear that a move for the abolition of professional boxing need no longer be based on emotion or on individual or group preconceptions. There is a case for abolition as a matter of psychiatric primary prevention."

A commentary in the *Medical Journal of Australia* (1975) reads:

"It has been suggested that boxing is unique in that self-control has to be exercised in the face of pain; in fact, pain may be inflicted in any body-contact sport, and the mediaeval notion that pain is somehow 'ennobling' is surely now discarded, even in British public schools. Indeed, self-control is developed in other competitive sports, for example, in the face of a disputed umpiring decision. It is wisely considered undesirable for children to view adult boxing contests, and in Victoria boxing is banned from correctional youth training centres, as it is considered that boxing is not a suitable outlet for aggressive tendencies in a correctional setting... unlike other sports, the intention in boxing is to render the opponent ineffective, and this is achieved by blows deliberately aimed at the head, with consequent injury to the brainstem. Since none of the existing rules of boxing leads to avoidance of the knock out, the ideal would be to ban the sport altogether."

"The *Australian Medical Association* has voiced its opposition in principle to boxing because of the dangers to which all contestants are exposed and in particular the risk of brain damage."

An editorial *(Lancet, November 27, 1982, p. 1186) reads:*

"... it is an allusion to suppose that any kind of medical supervision could prevent brain injury while boxer go on hitting each other on the head. If nothing else is changed in boxing, the dangerous illusion of safety provided by modern medical control should be abandoned and honest information given to boxers, their families and the public. The only way to prevent brain damage is to disqualify blows to the head."

CASSON (1982) wrote:

"Boxing is a violent sport in which the participants intentionally attempt to knock out their opponent. Although it has been claimed that very few effective head blows are delivered in professional fights (KAPLAN and BROWDER 1954), the alarming incidence of serious head injuries in boxers provides clear cut evidence of the dangers of the sport."

A somewhat more recent reaction against boxing in the media is the much publicized case of Howard COSELL. The following is an article by Dave KINDRED entitled COSELL *'Giving Boxing Silent Treatment,' and* appeared in the *Washington Post* (quoted from the *Times Picayune/States Item, December 3, 1982):*

Monday night, someone telephoned Howard COSELL in the ABC-TV truck outside Tampa Stadium. Did COSELL want to do the blow-by-blow of two important fights the network would carry soon? This is in his contract. He gets first call because, across four decades, COSELL's masterly work on televised fights has made him as celebrated as the men he covered: MUHAMMAD ALI and Joe FRAZIER, Floyd PATTERSON and Ray LEONARD.

So his answer Monday night is important at a time when Alexis ARGUELLO is punished terribly, when Duk Koo KIM is dead, when Larry HOLMES could have killed the pathetic Tex COBB with ABC-TV and COSELL at ringside for Friday night's heavyweight championship fight from the Houston Astrodome.

"They asked me if – in the wake of that terrible, terrible scene in Houston – I wanted to do these fights coming up," COSELL said from his New York office the other day. "I told them no."

Not only that, COSELL said, he has made another decision after what he calls boxing's 'two weeks of disgust.' He won't work another professional fight:

"I'm not doing any more boxing," he said. "Let me qualify that only a little. I love amateur boxing, and if the company wants me to, I would do Olympic boxing. But professional, no. I have walked away from it. I am past the point where I want to be part of it. I don't want to be party to the hypocrisy, the sleaziness... I'm worn out by it."

The blame for boxing's sickness today falls on three groups, COSELL said:

"First, the networks who make possible the continued existence of organizations such as the WBA and WBC, which exist only for the purpose of creating championships that in turn create fights that the networks, including ABC-TV, can sell for top dollar."

"Second to blame are the ruthless, sleazy promoters. And third is the vast majority of the print-media members who are apologists for boxing."

COSELL's break with boxing is not unexpected. He hadn't worked a fight since HOLMES – Gerry COONEY last summer. His dissatisfaction with, even contempt for, boxing promoters and other parasites is so profound that, in his words:

"I went to Larry HOLMES' room for a set of gin rummy so I could avoid the sleaze in the hotel lobby."

COSELL's voice revealed the melancholy of a man who has passed from love to anger to resignation. Once he railed at authority, asking that fighters be protected from injury and exploitation. In 1972, he testified before the U.S. Senate. He once came to boxing's defense before the New York State Legislature.

But on Friday night in Houston – ten years after the U.S. Senate did nothing, ten years in which boxing has done nothing to help itself – there in the Astrodome, COSELL saw revealed in its ugliness the sport he once loved as art, that moved him to song one night when Ali won. 'Forever young,' COSELL sang, using DYLAN's lyrics.

From the fifth round to the fifteenth, in the HOLMES – COBB fight, COSELL saw it for what it was. This was not sport. This was savagery licensed for profit. At the microphone, COSELL for a half hour cried out for someone to stop the fight.

"Doesn't he know," COSELL said of the referee, "that he is constructing an advertisement for the abolition of boxing?"

As ALI moved COSELL to song, Friday night's fight moved him to fury. The only touch of saving grace, he said, was HOLMES' resolve to keep COBB alive. However, he came by the dubious ability to 'take a punch,' COBB stood up under HOLMES' heaviest punches. HOLMES did

COBB a favor, however, by throwing those punches only often enough to win a round. Had HOLMES mounted an animalistic assault, even a granite chin would not have kept COBB upright.

"I admired HOLMES because he comported himself with dignity," COSELL said.

COSELL loves fighters, hates fighting:

"I have a deep feeling for the men who fight..."

"The other day, Ken NORTON walked into my office. My God, He's as thick-tongued now as ALI... "

"Somewhere along the road, Tex COBB will pay for the punishment he was allowed to take."

From the ninth round on, COSELL made no comment on the fight strategy or tactics. "It was," he said, "as measured a performance as I've ever given."

From COSELL, we heard again and again that the fight should be stopped. He repeatedly named the referee, fixing blame. He spoke of the pall over boxing and said HOLMES – COBB could not have come at a worse time.

Good for him.

We didn't need how-tough-Tex-is anecdotes.

We didn't need a shill telling us to stay tuned because you never know, Tex might win the championship with one punch.

Right now, we need a polemic on boxing. We need someone to shout at the top of his voice that the game is terminally sick.

"If I were the networks," COSELL said, "I would declare a moratorium on boxing until some form of legislation was passed to: (1) Protect the men who fight with the strictest rules for safety and medical examinations; (2) create an honest system of ratings and records; and (3) create one federal government group to administer boxing.

"The print media should support such legislation instead of, by saying the federal government has no business in sports, vilifying those who suggest it. There should be licensing of promoters, and there should be an accounting of money on every fight:

"It has to be in all 50 states under the administration of one group. Either that or abolish the quote – sport unquote."

This from a man who, for almost 40 years has supported boxers and boxing. His experience and his ultimate decision are powerful arguments against the 'sport' and its exposure on television.

A week or so later, Pete AXTHELM published the following article, *'Don't Count Boxing Out,' Newsweek, in* response to COSELL's actions:

"The late Jimmy CANNON, a columnist who wrote as memorably out of the side of his mouth as anyone in the trade, once speculated that if Howard COSELL were a sport, he would be a roller derby. The line drew laughs, but it was off the mark. If Howard were a sport, he would be the one he loves and hates so noisily – boxing."

"Like Howard, boxing can produce simple, powerful statements about the human condition. It can be entertaining, even uplifting at its best. Also like COSELL, it can be a cruel bully in its treatment of the weak. And it can be self-serving and hypocritical. Because the assets and flaws of both COSELL and the sport tend to be magnified by those who observe them, it is not surprising that Howard's public divorce from pro boxing has become a focal point in the latest debate about whether the sport should be banned. Sure, heavy thinkers like George F. WILL and James J. KILPATRICK have weighed in with arguments against the game. Even the less lofty types who run the World Boxing Council preached reform last week: they called for the reduction of title fights from 15 rounds to 12 and the imposition of a standing eight-count when a boxer appears defenseless. But the furor keeps swirling back to the man who legitimized boxing for millions of viewers – and now has washed his hands of the dirty business."

"On balance, I think that COSELL and his fellow critics of the sport are wrong. I was as horrified as anyone by the death last month of the Korean lightweight Duk Koo KIM – a fatality that could not have been prevented by anything short of canceling his fight. I also share in the general dismay over one referee's reluctance to halt Aaron PRYOR's savage beating of the great champion Alex ARGUELLO, another ref's mysteriously quick interruption of Michael DUKE's fight against Mike WEAVER – and a promoter's brazen staging of a mismatch between heavyweight champ Larry HOLMES and white target Randall (Tex) COBB."

"Banquet: But a troubling arrogance lurks in the anti-boxing position. Do those of us who are well paid to cover fights – or well educated enough to choose not to watch them – have the right to cut off the career choices of the unschooled kids who have traditionally used the ring as a step out of hopelessness and poverty? Without boxing, 'Sugar' Ray LEONARD would have still made a delightful after-dinner speaker. But who would have invited him to the banquet? Without his fists, would Cassius CLAY ever have become MUHAMMAD ALI, consummate showman and conscience of many youths of his generation? Or would COSELL have had to bear those roles all by himself?"

"As he reminds us with sometimes cloying frequency, COSELL is a giant of his profession. For all his ear-assaulting faults, he has brought wit, courage and journalism into a medium that did not always welcome the intrusions. He has also been the visible symbol of television's vastly expanded involvement with boxing. As such, he is one of the few figures who could conceivably force more sense and safety into the game."

"I'm not referring to COSELL's occasional grandstanding appearances before congressional committees, calling for federal regulation of the game. Government at various levels has long since proved it can botch up sports as thoroughly as do the most benighted private citizens. The power and purse strings that could achieve real reform lie in television. All three networks have hired experts in recent years to protect against the mismatches and deceptions that create much of the danger in boxing. Obviously, they still have far to go. But fits of pique by leaders like COSELL are hardly a step forward."

"Boxing's most pressing need may be for protection from its own spokesmen. Champion HOLMES, ever graceless, greeted COSELL's recent statements with a strikingly crude and stupid diatribe. But the tactlessness trophy was immediately seized by Wilfred BENITEZ who filled the aftermath of KIM's death with claims that he would 'kill' Tommy HEARNS in their fight. And in a display of chutzpah that was a throwback to his promotion of Evel KNIEVEL's non-jump over the Snake River Canyon, mogul Bob ARUM solemnly called for a two-month moratorium in the wake of the KIM fight that he promoted. It happened that all the major fights in that two-month period belonged to ARUM's enemy, promoter Don KING.

"Logic: But history reveals that this debate has never risen much above the present level. Two decades ago when Emile GRIFFITH killed Benny PARET in a ring, a brilliant South Carolina legislator called for the installation of round boxing rings. That way, he figured, no fighter would ever get trapped in a corner. Even the erudite apologist for the 'Sweet Science,' A. J. LIEBLING, once executed this astounding leap of logic: 'If a boxer ever went as batty as NIJINSKY, all the wowsers in the world would be screaming, 'Punch drunk'. Well, who hit NIJINSKY? And why isn't there a campaign against ballet? It gives girls thick legs.'"

"Against such a backdrop, the situation cries for hard facts. Here is one. In an American Medical Association study of sports fatalities, boxers placed eighth. Jockeys die most frequently, and scuba divers and college football players had a higher mortality rate than boxers. Critics retort that boxing is the only sport in which the harm comes by design rather than accident. But boxing's advocates can argue that intent to injure need not be part of the art. It could certainly be diminished by more padding in the gloves, protective headgear, and better supervision of matchmaking and medical standards."

"The ruling promoters and organizations have shown few flashes of the vision that might produce such changes. But there is something vaguely noble and enduring about boxing that makes me reluctant to abandon hope. Years ago, a middleweight champion named Stanley KETCHEL died of a health hazard only peripherally linked to his occupation. He was shot by the common-law husband of the woman with whom he was having breakfast. When Broadway character Wilson MIZNER heard about the courageous fighters' demise, he said, 'Start counting ten over him. He'll get up.' The whole sport may prove to have that kind of imagined resiliency, as long as its good and honest people – like COSELL – don't all stop counting or being counted."

A few points will suffice as answer to this article (letters in response to *Newsweek's* article follow):

First, although the article's subtitle mentions 'the sport's very real assets,' nowhere does the author even attempt to describe any assets, beyond saying that "there is something vaguely noble and enduring about boxing that makes me reluctant to abandon hope." He links this to the sport's 'imagined resiliency.' His other defenses of boxing – and they are defenses, not statements of its inherent goodness or quality – consist of (1) mentioning boxing's allegedly low fatality rate and (2) saying that "boxing advocates can argue that intent to injure need not be part of the art."

To take the latter point first – true, intent to injure does not need to be part of boxing – but every fighter who speaks truly regarding boxing will admit that the incapacitation of the opponent by injury – as in the CLAY – COOPER fight – is the whole point of the game. That is how a fighter wins a boxing match – by hitting his opponent, or better yet by rendering him unconscious.

We have found only one or two original sources claiming to have found that boxing is less dangerous, although it would be difficult to estimate the frequency with which the conclusions are quoted as true. The inadequacies of the statistical analysis and methodology of these studies have been discussed elsewhere in this text. However, it is worthwhile to note that boxing apologists and enthusiasts adopt the findings, without question, and quote the conclusion as dogma, ad nauseum. It is ironic/symptomatic that such weak claims are accepted and repeated without scrutiny, while the bulk of medical literature is summarily dismissed as being statistically inadequate, simply because it does not speak for the safety of boxing as a sport. Regardless of one's position on boxing, it is difficult not to wince at the dogged attachment to weak and irrelevant arguments by boxing enthusiasts. It is no good to say that those opposed to boxing are more experienced in constructing arguments, or have better access to the press or more resources; historically speaking, boxing enthusiasts have been an eloquent and prolific voice in the press. Perhaps threats to boxing have not been viewed as sufficiently serious to generate substantial argument. That is, as the status quo is satisfactory, it remains for the opposition to come in with the disabling arguments. Regardless of the reasons, it would be useful to hear.

As for the first point, although we do not know the derivation of the statistic (whether it relates the absolute number of deaths, or is based on number of deaths per number of participants), the death rate of boxing is, admittedly, not the most important aspect. The fact is that, according to the best available data, boxers are almost assured of some form of permanent, irreversible brain damage, which will be worse the longer their careers last.

One gets the impression that Mr. AXTHELM's "something vaguely noble and enduring" is the "imagined resiliency" he mentions in his last paragraph. He seems to be lauding boxing for its trait of banging out a fighter's (thus, a fellow-man's) 'gameness' or 'resiliency' in the face of adversity or against an opponent. He seems to have forgotten (or overlooked) the fact that the opponent the adversity are falsely created by the essence of the sport.

Mr. AXTHELM's article evoked several responses from *Newsweek* readers:

(1) "I thoroughly enjoyed Pete AXTHELM's article *'Don't Count Boxing Out' (December, 20, 1982)*. As the referee of the recent Larry HOLMES – Randy COBB fight, I am naturally in agreement with his comment that "there is something vaguely noble and enduring about boxing." The World Boxing Council has embarked upon a far-reaching program to improve the sport and make it safer for boxers. It is unfortunate that most people are unaware of these efforts" (Stephen T. CROSSON, Dallas, Texas).

(2) "Anyone who finds boxing entertaining needs to have his head examined. Here is a sport in which the ultimate achievement is to knock one's opponent unconscious. This is uplifting? As for AXTHELM's reasoning that boxing is a way out of the ghetto for unschooled kids plagued by 'hopelessness and poverty,' is boxing all we have to offer these children?" (Lynn FORD, Salem, S.C.).

(3) "For every 'Sugar' Ray LEONARD, there are countless aspiring young men who fall by the wayside suffering from various stages of brain damage after their usefulness to mercenary promoters has been exhausted. AXTHELM succeeded in jabbing Howard COSELL for unkind remarks against a sport he's paid to admire. But COSELL at least, has seen boxing from all sides, and has made a seasoned, well-reasoned choice – one which I too made in my early boxing days – a choice that deserves respect from those who are far from the ropes" (John M. BOSAK, Harrisburg, PA.).

It is somehow not surprising that the letter in favor of Mr. AXTHELM's article is from the referee of the fight that finally sent Howard COSELL away from pro-boxing. As for his statement that "the WB Council has embarked upon a far reaching program to improve the sport," one wonders why a representative of the Council did not mention these efforts and exactly what they entail, to someone. Unless Mr. CROSSON is a member.

The overall reactions of those on both sides of the media seem to be: (1) That boxing is overplayed to a great extent, and (2) that exposure of this sport on a national level is neither edifying nor enjoyable to those who concern is for the fighters – as witness Mr. COSELL's remark about the referee (CROSSON) – "Doesn't he know that he is constructing an advertisement for the abolition of boxing?"

Indeed!

An editorial *'Boxing dangerous to your health.' (The Washington Post, January 16, 1983, B 6)* reads:

"Each time a fighter is killed in the ring, there is an outpouring of sympathy for the victim, a short and earnest argument about the ethics of what some have called 'the manly art of modified murder' and then a return to business as usual. Even fight fans accept that fact that boxing is the only 'sport' in which the objective is to inflict as much physical damage on one's opponent as possible. In fact, that's what fills arenas and gets the crowd to its feet. And when anyone is killed in the course of this display of skill, power, and endurance, well, accidents happen in every sport, right?"

"Now the American Medical Association, in its current journal, has forced sports fans to come to terms with more than the occasional death in the ring. Although there have been 335 of these in the last 30 years. The doctors have provided evidence of serious brain damage in many fighters simply because they have been hit in the head so often over a period of years. Their information demonstrates that 'boxing is deleterious to the human brain' and that 'the correlation between cerebral atrophy and the number of bouts is significant.' The 'punch drunk' fighter, so often the subject of amusement and ridicule, is, of course, suffering from intentionally inflicted brain damage. His slurred speech, strange walk and lack of coordination are the direct result of an accumulation of beatings suffered over many years. These symptoms are especially sad in men who fought their way to the top and were an inspiration to youngsters hoping to achieve wealth and fame in the ring. The exploitation of poor and minority boys in this sport is particularly obnoxious, for even if a very small number rise to the top, they and thousands of others, some of whom do not even make a decent living in the ring, suffer permanent serious injury."

"While the AMA journal editorializes that 'boxing should be banned in civilized countries,' the association's Council on Scientific Affairs accepts the likelihood that this is not a realistic hope. Instead, the council proposes nine steps that could reduce the incidence of brain damage in the ring. Included are suggestions for a national, computer-based registry for all fights that would include all injuries suffered, mandatory use of safety equipment, not only on the boxer but also on the ring floor and corner posts and authority for fight physicians to stop contests. A conference of AMA specialists and ringside physicians will consider these proposals next month."

"The recommendations seem like small steps, but perhaps they are as much as that public will accept, even in the face of powerful evidence of the harm that is being done. It is as difficult to outlaw boxing as it is to control the possession and use of handguns. Both, nevertheless, remain worthy goals of a civilized society."

George J. ANNAS (1983), possessor of both JD and MPH degrees, wrote an article about *'Boxing: Atavistic Spectacle or Artistic Sport?'*

ANNAS wrote:

"Abraham LINCOLN reportedly told a group who wanted to outlaw gamecock fighting: ' As long as the Almighty permits intelligent men, created in His image and likeness, to fight in public and kill each other while the world looks on approvingly, it is not for me to deprive the chickens of the same privilege.' In the century and a half since, game cocking has been legally abolished in all but a handful of states, but boxing has flourished and boxers continue to be killed and seriously injured. The November of 1982 death of Duk Koo KIM following a bout with Ray 'Boom Boom' MANCINI, and mismatches like the one between Larry HOLMES and Randall 'Tex' COBB *(November 26, 1982),* have led to a chorus of cries to outlaw or regulate boxing (NEWFIELD 1982, LUNDBERG 1983, VAN ALLEN 1983, The American Medical Association 1983).

The Abolitionist Argument

The abolitionist argument was summarised in two editorials in the *Journal of the American Medical Association (JAMA* in January 1983). Dr. George LUNDBERG, the editor, argued that boxing should be banned because 'the principal purpose of a boxing match is for one opponent to render the other injured, defenseless, incapacitated, unconscious.' In contrast, 'in all other organised sports, injury is an undesired by-product of the activity.' A companion guest editorial described boxing as 'a vicious and deadly game,' and a 'public spectacle of brutality.' Injury to boxers was given as the major reason for abolition of this form of 'entertainment.' Others, such as columnist George WILL, have argued that 'a sufficient reason [to outlaw boxing] is the quality of the pleasure boxing often gives to spectators;' it is 'coarsening' and undermines the values upon 'good government and good life depend."

The Reformer Argument

"Reformers agree that there are abuses, but argue that boxing as a sport is no more 'bloodthirsty' than football, auto racing, or hunting, is a voluntary endeavor among adults, and has safety rules and procedures that can be improved.' The ambivalence of philosophers like Albert CAMUS, a boxing fan who opposed both violence and capital punishment, is alluded to, and the potential artistry of boxing praised. Investigative reporter Jack NEWFIELD says of himself: 'I still find a film of 'Sugar' Ray ROBINSON as exhilarating as reading a poem of KEATS, listening to a spiritual sung by Mahalia JACKSON, or hearing a soliloquy composed by SHAKESPEARE. Artistry, regardless of the calling."

"The *American Medical Association's Council on Scientific Affairs* has adopted the report of its *Advisory Panel on Brain Injury in Boxing.* The panel found that the mortality rate for boxers seems to be lower than that for football players, motorcycle racers, scuba divers, mountain climbers, hang gliders, sky divers, and jockeys, although the panel had no information on 'how these statistics were compiled.' The panel's literature review disclosed numerous studies of boxing's 'punch drunk syndrome' related to traumatic encephalopathy from repeated blows to the head. They noted, however, that almost all these studies were poorly designed, with no evaluation for possible confounding variables and no suitable control populations. They concluded that although a dangerous sport, boxing 'does not seem any more dangerous than other sports presently accepted by society.' The panel recommended a series of steps for state and local commissions to take to improve safety equipment, safety measures, and medical supervision of boxing."

Should Boxing be Outlawed? The Abolitionist View

"There is no dispute that the individual states could, if they wanted to, outlaw boxing within their borders. The states' public health 'police powers' are broad enough to prohibit any activity that has as many disabling injuries as boxing. But since boxing appears no more dangerous to the participants than any other sports, JAMA's editor seems correct in focusing the abolitionist argument on the intent of the boxers themselves. The relevant question then becomes: Is boxing the only socially accepted sport where the intention is to injure or incapacitate the opponent? Superficially the answer seems to be yes, since in other sports the objective is usually either territorial or a specific goal. The distinction, however, is probably illusory. The violence of football collisions cannot be persuasively distinguished from the blows delivered in the ring."

"In a lawsuit filed by professional football player Dale HACKBART against the Cincinnati Bengals and one of their players, Charles CLARK (who struck him in the back of the head), US District Court Judge MATCH noted that the most 'obvious characteristic [of football] is that all of the players engage in violent physical behavior.' He summarised the testimony of former Denver Broncos coach, John RALSTON. RALSTON testified that before a game he tries to generate the emotional equivalent in its players of the rage a father, whose family had just been endangered by another driver who attempted to force the family car of the edge of a mountain road, would feel after he overtook the offender and opened the offender's door to take revenge on him. In the judge's words, 'The NFL [National Football League] has substituted the morality of the battlefield for that of the playing field, and the 'restraints of civilization' have been left on the sidelines.' Because of this, the judge ruled that one player could not sue another or injure since there was a knowledgeable assumption of risk. The Appeals Court, however, disagreed, noting: 'It is difficult to understand why social policy should dictate judicial blindness to the violence of professional football. The threat of proper judicial intervention can serve the useful function of discouraging uncontrolled and unnecessary violence.'"

"Football also has injury and regulatory parallels. In one two-year period (1973-1974), for example, of 1.040 total players and their substitutes in the NFL, there was an annual average of 1.101 injuries so severe that the player had to be removed from the game, and 136 players required surgery. In the late 1889s, football was much more dangerous than either it or boxing is today. Players wore no protective padding or helmets and, because of the frequency of death and injury, the public demanded changes in the game. After the 1905 season, in which there were 18 fatalities and 159 serious injuries, President Theodore ROOSEVELT called together representatives of various colleges and threatened to abolish football altogether unless safeguards were adopted."

"What Stance should the Law take?"

"Although theoretically regulated by state and local boxing commissions in 64 states, boxing is the least regulated major sport in the United States, and probably the sport most vulnerable to corruption. The major boxing associations – the World Boxing Association (WBA), and the World Boxing Council (WBC) – are located in Panama and Mexico City, respectively. They recognize different champions in almost every division so television can truthfully sell up to 32 different boxers in 16 different weight classifications as 'world champions.' Neither group has any legal standing in the United States. In the days when boxing was an event only for a local audience, state and local boxing commissions may have made sense. But with television networks and cable stations, boxing is no longer properly viewed as a local public health issue to be handled by exercise of the state police powers. It is – like football, baseball, basketball, and hockey – big-time commerce, and is more appropriately regulated by the federal government under the interstate commerce clause."

"Society presumably tolerates injury in boxing and other violent sports only because it assumes that players do not have a malicious talent to injure, and because it views the likelihood of injury as slight. Neither of these presumptions has yet been proven false in boxing. Nevertheless, federal regulation is needed in boxing because it is basically an 'unstructured, anarchistic sport' with no regular schedules or leagues, standings that are 'political', and matches that are 'improvised and arbitrary.' A Federal Boxing Commission could do what individual state boxing commissions can never do: develop and implement uniform safety standards for the sport. Since most of the money in boxing is from American TV networks, the rules promulgated by a US Commission would stand a reasonable chance of becoming worldwide standards."

"The road to regulatory reform will not be an easy one. Individual state commissions will not want to see their power taken away, and sports promoters will worry about profits. Although the federal government must regulate, it need not preempt the area; states should be permitted to have stricter standards than those imposed by the federal boxing commission, and state boxing commissions should probably retain authority over local 'tough man' contests, which should either be heavily regulated or outlawed."

"The Federal Boxing Commission 's job will be to license and maintain a central registry of all boxers and fights; sanction one set of ratings and champions, grade all boxers to avoid flagrant mismatches; develop reasonable medical standards regarding pre-fight physicals; require the presence of a ringside physician with the ability to terminate the fight or extend rest periods between rounds; require after-fight neurological examinations; enforce mandatory suspensions after knock outs or injuries; develop and enforce safety standards regarding mats and corner-post padding, etc. Its job will be to ensure the sport is as safe as it can be, and that boxers who are likely either to be injured themselves or to maliciously injure others are not permitted to fight. Reasonable regulations could replace an increasingly atavistic spectacle with an artistic sport: We owe it to the boxers and ourselves to try."

In September 1983, we spoke with the *Professor of Moral Theology at Notre Dame Seminary in New Orleans,* Father John B. BALSAM, O.P., S.T.D. He offered these insights into the moral aspects of boxing, from the standpoint of a priest of the Catholic Church and a professor of Moral Theology:

"The purpose of sports in general is, among other things, to benefit one's health, and to contribute to the physical, emotional, and mental well-being of those engaging in them. They also serve to promote friendly competition and recreation both in team versus team (as in baseball or basketball) and person versus person (as in swimming or tennis). Sport may also provide personal and social entertainment for both participants and spectators alike. Thus, the grace and beauty of some sports (e. g., gymnastics or ice skating) may be a source of aesthetic pleasure for all. Finally, sports should develop one's own total prowess, using one's physical and intellectual abilities and gifts. I do not believe that it is possible to become a good gymnast, for instance, without calling into play all one's capabilities. Sports can and should contribute even to one's moral growth; they are often a proving ground for the ideas and practice of fair versus unfair play, of courtesy as against discourtesy, and of such virtues as courage, patience, perseverance and goodwill."

"I would like to look at the principles of ethics which are involved in boxing, whether amateur or professional. There is a medical principle I would like to mention first: 'non nocere', the general principle 'do no harm' to anyone. This most fundamental of medical-moral principles is reflected also in a more general Christian moral principle: *It is wrong to kill or to harm or injure, directly and deliberately, an innocent person.* If harm is inflicted in the process of warding off the unprovoked and unjust attack of another, in self-defense, for instance, such defensive action is justifiable. However, if warding off the attack goes too far, and exceeds the bounds of due proportion adequate to ward off the attack, it begins to be vengeance and cruelty rather than justifiable self-defense. Such an excessive reaction to the attack both offends against the virtue of justice and is a breach of the virtue of clemency by resort to cruelty."

"In other words, to harm another directly and deliberately, one must have justifying and counterbalancing reasons, one of these being true self-defense. This defense, however, must use proportionate means. If a small child approaches me with a stick and threatens to harm me with it, I would take away the stick, not shoot the child or knock him unconscious to protect myself. That would be totally uncalled for and a means of defense completely disproportionate to the seriousness of the attack."

"There are, then, four elements to this problem. There must be *harm involved;* the harm must be *direct and deliberate;* it must be done to an *innocent person,* that is, done without a counterbalancing reason; and the means must *disproportionate to* the violence inflicted."

"Applying this to the type of boxing under consideration, we see that actual harm may not always result in the boxing ring, but the *risk of serious bodily harm is always present.* It is wrong to submit oneself *either to receive or to inflict unnecessary harm or injury,* or to the risk of receiving or inflicting harm or injury, without justifying and proportionate reasons. Taking the risk of being harmed is the first step towards submitting oneself to harm. Thus, the harm (or rather the risk of harm) is always present."

"As for the next two elements, boxers step into the ring by agreement. The professionals sign a contract stating that they will fight one another for a given amount of money. This is a direct, deliberate, and willful action. They are in the ring to pile up points or to knock out the opponent. No fighter has ever won a match by dancing around the ring without inflicting effective punches. A boxer can win a fight in only one way – by out-pointing (which is judged by landing more, and more telling, blows on, and hence out-harming) his opponent, sometimes to the extent of knocking him unconscious by a complete knock out or technical knock out; even at times by killing the opponent. No agreement can give any contestant the moral right to inflict physical injury or harm (often serious and in brutal fashion) on one's opponent, even though it may be legal. Legality and morality are not necessarily always coincident. Neither contestant can be classified in any moral sense as an unjust aggressor, and there is no excusing, justifying, or proportionate cause for the physical injury or risk of it – certainly not the prize money involved."

"I would maintain that I see no counterbalancing reason for submitting oneself to the real risk of harm, or to being knocked unconscious, or killed, or any of the spectrum of injuries and blows in between, simply because of the prize money."

Father BALSAM added:

"We are living in a violent age. When millions watch a fight, this violent sport can readily (and may) dispose people towards a mentality of violence. One of the proofs of this is the behavior of the spectators at a boxing match. The closer one of the fighters comes to a knock out, the more the crowd cries out for the 'kill', sometimes even shouting, 'put him away', 'knock him out.' A mentality of violence can easily be built up, supported, and fed by this violent sport."

When asked what he would do to change the sport, Father BALSAM replied,

"If it could be made a friendly and non-injurious competitive sport – with protective restrictions and equipment and no violence or bodily harm allowed, like Olympic fencing – then I would have no difficulty with such boxing."

As for the suggestion that boxing is a way for 'poor kids to make good,' Father BALSAM replied:

"The price is too high. To say that a child can only better himself in such a costly way is firstly not true – there are many ways, beginning with education – and secondly, it is putting the dollar against the health and well-being of a human life, and there is absolutely no comparison. This attitude brings up a large question of *misplaced values and priorities*, and surely all these years of furthering human rights have taken us farther than that argument suggests."

Father BALSAM continued:

"Learning how to box is fine – the training may be good for one – but not in a violent fashion against an opponent. If the boxers do not come to violent blows – some sort of touche method – I see no harm in that. Light taps are one thing, staggering pounding of an opponent to the point of knocking him unconscious, causing irreparable injuries to organs, bones or brain, or even killing him, is nothing short of an act of unconscionable violence, which cannot be justified by any proportionate or excusing reason, certainly not by mere money from the bout."

Father BALSAM concluded:

"Interestingly enough, a goodly number of moral theologians, if not the majority of them, oppose the violence of prize fighting on moral grounds as being dehumanizing, brutal and unjustifiable by whatever criteria."

"Is the practice manly? I doubt it. There can be no manliness in brutality, yet those of English blood preferred it to the crossing of blades, a Latin amusement, more refined, but much more deadly. So we shall probably keep on building bigger stadiums and charging higher admission to bigger crowds, despite the prayers of churches and the reprehensions of the good."

David NOONAN *(Boxing and the brain. The sweet science constitutes the most purposeful, relentless and sustained assault on the human brain in all sport. The New York Times Magazine, June 12, 1983)* NOONAN (1983) wrote:

"The knock out is the heart of boxing. It is what every boxer wants to do to his opponent in every fight – as soon as possible. While boxing's defenders have often claimed that knock outs are relatively infrequent, an examination of recent records indicates otherwise. The top ten ranked heavy weight contenders in 1981, for instance, had 149 knock outs listed among their 206 ring victories, according to *'The 1982 Ring Record Book.'* Larry HOLMES has won 43 fights as a professional and 30 of them are listed as knock outs. And 'Sugar' Ray LEONARD won 23 of his 33 pro victories by knock outs."

In an article entitled *'Brain damage: will supervision help? Journal of the American Medical Association January 14, 1983, pp. 250, 255)* the author asks, is chronic brain damage in boxing a hazard of the past?

In a recent issue of JAMA a study of forty former boxers yielded "a significant relation between the number of bouts fought and CT changes indicating cerebral atrophy. JAMA's editor George D. LUNDBERG, M.D., asserted, in that issue, "that boxing is wrong and should not be sanctioned by any civilized society."

But the *AMA Advisory Panel on Brain Injury in Boxing*, noting, that: "boxing does not seem any more dangerous than other sports presently accepted by society," has called for admistrative regulation of boxing.

The Lancet *(November 27, 1982, p. 1186)* believes that "it is an illusion to suppose that any kind of medical supervision could prevent brain injury while boxers go on hitting each other on the head. If anything else is changed in boxing, the dangerous illusion of safety provided by modern medical control should be abandoned and honest information given to boxers, their families and the public. The only way to prevent brain damage is to disqualify blows to the head."

LAMPERT and HARTMAN (1984) noted:

"... there can be little doubt that boxing can induce permanent damage of the brain, which may lead to typical clinical and structural alterations known as dementia pugilistica... The public enjoys paying for the spectacle, and for this reason boxers will continue battering each other with damaging force."

In a letter to the editor *'Should boxing be banned?'* by Stratis IOANNOU, *(Canadian Medical Association Journal, 131: 10, 1984)* the author wrote:

"In Lynne COHEN's article on boxing *(Canadian Medical Association Journal 130: 767-768, 1984)* Dr. William GHENT, chairman of the *CMA's council on health care*, states that boxing be banned because its main objective is to injure the opponent. One can quibble and state that the main objective is to get more points than the opponent, not necessarily injure him. Injuring appears to be but one means to an end. In other contact sports, such a football, hockey and wrestling, the purpose is also to win by outscoring the opponent, but no 'deliberate' injury is allowed. The problem is the interpreting 'deliberate.'"

"The *CMA* cites studies that has indicated that boxing causes brain damage. It would be interesting to see how much brain damage is suffered by football players of different ages, at different levels of competition."

"I am not extolling the virtues of boxing. But, while on the surface boxing appears to be the most violent and brutal of sports, other sports, such as football, wrestling and hockey, are just as violent and brutal. The only problem with these sports is trying to determine whether an injury is deliberate, whereas in boxing the intent is explicit."

James J. KILPATRICK *'An argument against boxing.' (The Times Picayune/The States Item, New Orleans, Thursday, December 13, 1984, A 13)* wrote:

"My brother columnist Carl ROWAN, a dedicated sports enthusiast, recently devoted his column to making a case in favor of boxing. Let me respond by offering a case against it."

"Carl's first point is that boxing offers poor kids 'an exit visa from their world of violence, crime, drugs, vice.' He might have mentioned the examples of Joe LOUIS and 'Sugar' Ray LEONARD, in addition to the story of Cassius CLAY. But the problem with this argument is that it falls short of its own infinitesimal weight."

"For every poor kid who gains a measure of fame and fortune in the prize ring, there must be a hundred – more likely, 500 – who never get beyond preliminary bouts in second rate clubs. Many of these wind up as punch drunk stumblebums, their occasional purses diverted by promoters and managers. The prize ring may offer riches for a few; it offers less than a minimum wage in the many."

"Carl's second point is that many professionals and many sports contain elements of risk that are well-known to those who enter these fields."

"The point is interesting, it is also irrelevant. Injuries occur in every sport. Jockeys get hurt; race drivers are killed. But boxing is fundamentally different from these other sports. The object of racing is to get a horse or car first across the finish line. The object of football is to score touchdowns. The object of hockey is to get a puck in the net."

"The object of boxing is to batter one's opponent into unconsciousness. The whole purpose of boxing is to inflict pain."

"For me that is the crucial point. The injuries that occur on a football field are incurred accidentally. They may be unavoidable, but they are not intentional. We have referees to impose penalties for personal fouls and unnecessary roughness. In the prize ring, by contrast, knock outs are the name of the game."

"Carl's third point is perhaps his best point. He takes the libertarian view that nobody's business but the boxer's if a boxer wants to try a career in the ring. It's his own life and his own brain that he risks. What right does society have to prevent a man from earning a living in the only way he knows how?'

"To this I would respond by saying that it is not the injury to prize fighters that concerns me. To be sure, it always is saddening when an athlete dies young. Those who knew MOHAMMAD ALI in his prime, when he was as graceful and as dangerous as a leopard, must sorrow at the image of the man today. But to say that boxers bring their brain damage upon themselves is to duck the main issue. It is not the boxers who are responsible. It is the society that puts up the purses."

"How far have we come from the Roman Coliseum? It was for the amusement of emperors, apart of their policy of bread and circuses, that gladiators met in mortal combat. Today's boxing crowds are not significantly different from the crowds at 2.000 years ago when howling for blood on the sands."

"Oh, it is said that true fans are more interested in skill than in brutality; they come to admire the jab, the hook, the counter punch, the dazzling footwork. This is sanctimonious humbug. The spectator's pleasure depends entirely on the punishment one man inflicts upon another."

"I think society reasonably can say that such 'sport' has no place in our lives. We have all kinds of laws to protect animals from inhumane treatment. In times past we have had all kinds of laws to prohibit prize fighting. It wasn't until 1920 that New York made professional boxing legal."

"In urging state legislatures to prohibit these bloody and sometimes fatal amusements, the AMA is concerned primarily with the damage that prize fighting does to the brain. My own concern, as we nudge in imperceptible steps toward a higher order of age that damage that prize fighting does to us."

Nelson G. RICHARDS published an article entitled *'Ban Boxing.' (Neurology 34: 1485-1486, 1984):*

"Boxing is defined as 'a sport in which two fighters battle each other with their fists... The boxers throw powerful punches as each tries to win the bouts on points, render his opponent unconscious, or force him to give up the fight' (SULLIVAN 1984). The ultimate result of a boxing match is a knock out, which, by definition, is a concussion or more severe brain injury. Lesser degrees of winning may be achieved by an unfair match-up – a 'technical knock out' or the accumulation of enough blows to cause overt confusion without loss of consciousness (and posture)."

RICHARDS summarised:

"After due consideration, it is now my own personal belief that we cannot continue to call an activity a 'sport' if the goal is to injure the brain of the opponent. After reading and listening to proponents of the two sides, it is my conviction that, as a neurologist, I can no longer support boxing in any fashion and I believe that the American Academy of Neurology should clearly state and follow through with the actions necessary to ban boxing as a sport in any form. The matter will be considered again in the 1985 annual meeting."

ROSS et al. (1985) wrote:

"Boxing is the only sport in which the intent is to inflict physical injury. Particularly in professional boxing, the fighter who injures and ultimately causes his opponent to become helpless earns, for some atavistic reason, the plaudits of the fans and therefore the highest monetary recompense."

The responses to ROSS et al.'s paper were rapid and vehement. The following series of letters and replies reflect the still raging controversy.

According to BREGEAT (1984):

"boxing is a natural exercise which belongs to prehistorical eras. The author noted that today three types of exercises which we call the 'noble art' are educational boxing, amateur boxing and professional boxing. The latter is according to the author a real profession and it is there that there are the most numerous accidents. The most serious eye injuries can lead to blindness and cerebral injuries may cause death in the ring or leave aftereffects with the ex-boxers physically and mentally for the remainder of his life."

The author asks:

"Should the 'noble art' be banned? He notes that boxing has its supporters, its reformers, and its abolitionists!"

CROMPTON (1985) noted:

"Indeed, it does seem unfortunate and unnecessary that a brain on which a young man has to depend for the rest of his life, should be addled at an early age for reasons which are hardly necessary or convincing. Surely, his manhood would be expressed and encouraged in ways other than by permanently damaging his own brain and those of others of tender age? The effect might not show itself until the tail end of life, when senility of the mind becomes manifestly obvious one or more years before it would otherwise have appeared."

Johannes KEEL wrote in *'Grenzen der stafrechtlichen Zulässigkeit des Boxsports'* (*'Borderlines of Penal Permissibility of Boxing)' (Magglingen, 10: 16-18, 1986):*

"Despite its rich traditional past, at the present time boxing belongs among the most controversial types of sport. This is undoubtedly connected to the fact that the intentional conquering of an opponent with boxing blows is a behavioral pattern that is entirely prohibited in sport and in daily life."

KEEL continues discussing the penal permissibility of boxing:

"The borders of impermissible behavior in boxing, with respect to the law, are drawn as follows:"

" (1) Boxers, who injure the opponent by intentional or gross negligent fouls and,

(2) Boxers who continue to strike a severely punished, groggy boxer with (rule conforming) blows, are to be criminally sanctioned, because in such a situation, ordinarily, there is no consent from the injured boxer. Moreover, in such an event an ex-

isting consent could not have a lawful effect because the dangers of severe injury or death for the groggy boxer are increased to such an extent that they are disproportionate with the positive aspects of the athletic contest. Likewise, for those indirectly participating, such as trainers and officials (referee, ring physician), the groggy state has special importance, since by their rank they are obligated to interfere. Trainers and officials must follow the higher commandment to protect the health of the athlete. In such a groggy state the boxer is no longer in a situation to look after his interests for himself. In such a situation the officials have to turn in. When inactive or unresponsive they are to be called legally to account. Due to the voluntary or agreed acceptance of their function, trainers and officials have a guarantor position, which binds them to ward off breach of legal duties or obligations."

Allen A. BARTHOLOMEW *'Doctors and boxing contests.'* (*Medical Journal of Australia, 147: 212, 1987*) wrote:

"To the Editor: It is a most extraordinary happening to see a man who is engaged in a boxing contest and who has sustained 'cuts' to the face – above the eyes – being taken by the referee to a medical practitioner for an opinion as to the severity of the 'cuts,' and as to whether or not the fight should continue, and to hear the doctor say 'Yes, you are injured but not sufficiently seriously, carry on.'"

"To support that a seriously-injured man (the referee considered that he needed a medical opinion) continue to fight when the injury could only become worse is surely a most odd manner for a medical practitioner to practice his discipline."

" I would suggest that once an injury (a 'cut') is of such a degree of severity that the referee feels that he needs a medical opinion is when the fight should be stopped by the referee. No doctor should be in a position to have to say, in effect, 'You're not yet sufficiently badly injured to stop the fight: carry on.' The fight will then continue. A few more blows will be rained on the not-as-yet sufficiently serious injury, which then make it sufficiently severe to permit the doctor to say 'Stop the fight.'"

"This odd, indeed unacceptable, scenario occurred during a supporting bout to the FENECH – RICHARDSON fight at the Sidney Entertainment Centre on July 10, 1987."

TOPEROFF (1987) noted:

"Professional boxing is a throwback, a vestige of our dark irrational past. That's one reason it is usually under sharp attack in a society that likes to believe it has evolved very different and superior values."

ENZENAUER and ENZENAUER (1987) wrote:

"Evidence that boxing produces irreversible brain damage is now as indisputable as the link between cigarette smoking and lung cancer."

Jeffrey T. SAMMONS published an article: *(Let's knock prizefighting out of the ring. It brutalizes the spectators and makes false promises to its participants. Newsday, Friday, June 24, 1988).*

SAMMONS (1988) noted:

"As the opening bell for the TYSON – SPINKS heavyweight championship fight, Monday approaches, fans are fed a steady diet of news stories to whet their pugilistic appetites. Yet when two men enter the ring the question that comes to my mind is: Should this fight – or any fight – take place?"

"My answer is no."

"I do not offer this view lightly or ignorantly. I used to be a fan. So, allow me to trace my evolution."

"Although I rarely attend prizefights in person, I used to follow them intensely. I had been conditioned since childhood by frequent television exposure to this accepted entertainment form. I looked forward to hearing that catchy Gillette jingle about looking sharp and feeling sharp because I knew rapid-fire action and a stream of heroes would fill the screen."

"'Hey, Mom! Mom, there's a Sugar, Beau Jack and Jersey Joe,' I hollered as these champions and former champions climbed into the ring, waved to their fans and visited the boxers corners. PAVLOV could not have manipulated me more effectively. "

"Boxing meant more to me than physical combat. This was drama! Boxers were bigger than life. Joe LOUIS had retired before I was old enough to appreciate his greatness, but the stories my folks told convinced me that he was more than a fighter and that boxing's influence extended beyond the ring. I did watch Sugar Ray ROBINSON, the next best thing to LOUIS. His lifestyle fascinated me as much as his fighting style and I knew that boxing had yet another dimension – that of glamour, celebrity."

"Yet Sugar belonged to another generation and I needed a hero of my own. That honor went to a brash young Kentuckian, Cassius CLAY, whose outrageous behavior and sharp wit was just right for me. Before the 'Louisville Lip' had completely cast his spell over me, however, a tragic event opened my eyes to an ugly and sorrowful side of the 'sweet science.' As my grandmother and I looked on in horror, Emile GRIFFITH battered a helpless Benny 'Kid' PARET into permanent unconsciousness as a mesmerized referee, Ruby GOLDSTEIN, looked on. Within a few days, PARET died."

"But even his death – a legal homicide – did not turn me against boxing. The press and officials heaped much of the blame on GOLDSTEIN and I, like others, accepted the explanations and excused the death as a freak accident. When featherweight champion Davey MOORE died a year later, we were told that the fall and not the punches caused his death."

"Moreover, the artistry of Cassius CLAY diverted my attention and society's from the harsh reality of prizefighting. With the exception of Joe LOUIS, no one did more to disguise the bitter truths of the fight game than ALI. When he became a member of the Nation of Islam, his actions outside the ring drew more attention than those inside. By defying the government and challenging the sporting establishment, he became the people's champion – a hero of the dispossessed and the anti-establishment crowd."

"However, ALI stayed in the ring too long, following the bad example of so many other great champions. By the time of his retirement, I had begun my research on boxing and American society. Ironically, the man who had riveted me to the sport eventually helped me to condemn it. For ALI, the man who claimed to have the world, now has Parkinson's syndrome, which many physicians believe was caused by blows to his head, and seems socially disoriented. He is the most notable and noticeable victim of the destructive qualities of boxing, which is set apart from other sports by the quantity of the injuries. "

"A study showed that in 1983 boxing had more injuries than any other contact sport, with 50 percent of the participants suffering an injury, compared with 30 percent for football. Moreover, a study of 1970-78 showed 3.8 death per 1.000 participants for boxing, while college football had only 0.3 and high school football, 0.1."

"Yet these figures and ALI's example have done little to dissuade any number of young men from the ring. A chief reason is that they believe it offers a ladder to success, particularly for ghetto youth. It is a mirage."

"First of all, few boxers make much money. Few of those who fight preliminaries even make a living of boxing, fewer still get a chance at the big purses; hardly any gain fame and fortune. And although the publicity about these big purses gives boxing the appearance of being lucrative, few fighters keep much after taxes, expenses, parasitical followers, purchases on whims, and losses from lax financial controls. Joe LOUIS, Ray ROBINSON, Alexis ARGUELLO, Roberto DURAN, Wilfred BENITEZ, Tommy LOUGHRAN, Beau Jack and Primo CARNERA – champions all – left the sport with little or no money."

"Furthermore, for many in this profession of individuals, retirement brings more emotional difficulties and antisocial behavior than in other sports. Former boxers have no pension plans, almost no chance to earn a living in the game, almost no health benefits or compensation for career-ending injuries or drug counseling."

"And this says nothing of the physical damage suffered by these men. ALI is living proof that even the best risk permanent and disabling injury. Another is Sugar Ray ROBINSON, who has Alzheimer's disease, which many physicians think was, like ALI's injuries, caused by blows to the head. And though other sports produce injury and sometimes death, boxing alone seeks to hurt the opponent. Indeed, injuring him is the surest and quickest way to victory. The courts and boxers have admitted this repeatedly. After Jimmy DOYLE died in a fight with Sugar Ray ROBINSON, investigators asked ROBINSON if he knew he had DOYLE in trouble. 'They pay me to get them in trouble, ' ROBINSON coolly responded."

"In 1984, the American Medical Association called for a ban on boxing because it believed only abolition could prevent nervous injury and death. Still the sport goes on with little protest."

"One defense is that boxing is cathartic, relieving tension and hostility among spectators. This is baloney. Boxing brutalizes. Spectators scream for blood. They shout, "Kill him.' This is not how we want our children to be. Scientific studies show that witnessing violent activities can heighten tension and hostility. Gene TUNNEY and Jack DEMPSEY frequently commented on the bloodlust they observed among fans. A recent sociological study concluded that the murder rate increased after highly publicized championship fights."

"We no longer have gladiatorial combats, as the Romans did, in which fighters killed one another. Civilization has advanced beyond that. So let's turn thumbs down on a sport that demeans us all and which even the winner lose."

"Boxing is one of the supreme anomalies of our time."

The eminent British neuropathologist J. A. N. CORSELLIS, to whom we are indebted for a classic study about the permanent brain damage in boxers, wrote 1989:

"Nevertheless, hitting people on the head must carry some risk to the brain, and as long as the sport is encouraged as a noble activity by some its existence is bound to be questioned by others."

"Secondly, and perhaps more importantly, there is the risk of unremitting disease in those who survive a career in boxing, and it is this kind of damage that is least discussed."

"One of the most interesting paradoxes is the increasing social condemnation of violence and brutality while money and effort are poured into encouraging men to batter one another into a state of confusion if not unconsciousness."

"For what it is, my opposition to the sport derives, in a most prosaic way, from having spent many years examining human brains, some of which were intact but most of which had been damaged by disease or trauma. The irreparable scarring and degeneration in the brains of boxers was unnecessary and the effect catastrophic... The virility, or any other virtue, shown in one man hitting another on the head in a boxing ring cannot transform the act into an accident; the blows are intended. Similarly, an agreement to fight, however amiably, cannot alter the fact hitting one another is deliberate. This point however trite, distinguishes boxing from all other sports. Racing car drivers do not drive at each other in order to collide, nor is the primary aim of soccer or rugby to score hits rather than goals or points."

"To estimate the risks of neurological change in other sports a postal inquiry designed to find punch drunk sports men and women was sent to 162 British neurologists in 1976 and produced five soccer players (including one 'much given to heading and able to do so even if the ball were blasted at him from about six yards') twelve steeplechasers, two amateur rugby players, possible two wrestlers, and a parachute jumper.' Boxer, but not necessary different ones, were mentioned 290 times *(Editorial. Brain damage in sport. Lancet I:401-402, 1976)."*

"My opinion is that the brain should not be a target in any sport, and no amount of juggling with the regulations can take away the risk. Experienced and outstanding boxers are themselves aware of this, difficult as it may be for them eventually to call a halt."

Barry McGUIGAN said: 'Boxing damages your brain. Don't let anybody tell you different;' while Terry MARCH remarked drily: 'I don't need the British Medical Association to tell me getting hit on the head can't do me any good... another might be to recall the remorse of Garry WILLS *'Blood sport.' (New York Review of Books, 2: 5-7, 1980)*, a former boxing fan: 'I stood talking to MUHAMMAD ALI, embarrassed by his inarticulateness, and deeply ashamed, as it was not his own superb body that had bore this terrible thing to his superb mind. I had done it too, as part of the crowds urging him on, applauding the blood. I have not watched a boxing match since then! SAMMONS (1988) merely wrote: 'In boxing parlance, ALI was punch drunk!'"

RICHARDS writes in an editorial *'Boxing and the neurosurgeon', British Journal of Neurosurgery, 7:595-596, 1993):*

"Man is an aggressive animal and this aggression may be channeled in a variety of ways, including war, assault, manipulation in business or medical committies, or, at its best, sporting endeavour. Sport allows aggression to be directed to winning by peaceful means, but depending on the sport a degreee of violence is likely to occur and may even be encouraged. The rugby player making a tackle will be aware that his tackle may injure his opponent, but that is part of the game. The racing driver out-braking his opponent may force the other car off line and into a crash which may be fatal, but this is an accepted risk. Boxing however, is different. In boxing, the aim of the sport is to render your opponent unconscious. Thus, not only is causing brain damage an accepted risk of the sport, it is the reason for the sport."

"The boxer wins his fight by knocking his opponent unconscious for at least 10 seconds. If this is not achieved within the alotted length of the bout the winner is decided on points. These are awarded by ringside judges who allocate them on the likwelihood of a knock out, i. e. the head is the best target. The force that can be landed by a professional glove is rarely sufficient to render coma directly, but invariably it is an angle which twists the head and introduces shearing injury."

"Injuries occur to the soft tissues and bones of the face, the eyes and the brain. Two aspects of brain damage are of concern. First, there is acute damage which may have fatal or disabling effects. This is usually an acute subdural haematoma caused by tearing of a bridging vein or, more rarely, an extradural haematoma. The other aspects is the cumulative effects of multiple blows to the head. The punch drunk syndrome is well known in retired boxers and CORSELLIS had demonstrated the changes in the brains of boxers who have developed presenile dementia. While a bout may only last up to 45 min. boxers will train in the gym and spar for longer periods which may contribute more to chronic damage. Head guards tend to be worn in sparring, but although they may soften the blow to the head, they will not eliminate the effect of shearing."

"What can be done? Sadly nothing can be done to eliminate the risk without fundamental changing what is an immensely popular sport. Any idea about no longer having the aim of the sport to be a knock out and not giving points to head punches are impractical as they would sanitize the sport to a level the public would find unacceptable. Safety factors, such as head guards, will not eliminate risk and will make the target for causing shearing injuries larger. Reducing the padding or eliminating boxing gloves altogether so that the striking force is less will increase soft tissue and eye damage. If these factors cannot be altered what can the medical profession do? The British Medical Association is campaigning for a ban, but unless the public wants the ban they will not be successful. They can attempt to influence the public to turn their back on the sport, but Frank BRUNO and Mike TYSON tend to influence the public far more that the British Medical Association, and the chances of it happening are slim. If lobbying is to achieve anything it should be aimed at the boxing authorities, and could include encouraging them to reduce the number of rounds, increase the time intervals between rounds and control the amount of sparring a boxer can do."

"The other improvement that can be offered is to medical services in the event of an acute injury. The article being written during a Motorcycle Grand Prix where the author is Medical Director. Any rider who falls will be attended by a doctor who has the equipment and ability to begin resuscitation within 30 s. of the rider coming to rest. Present in the Medical Centre is an anaesthetist, general surgeon, traumatologist, thoracic surgeon, neurosurgeon and plastic surgeon who aim to stabilize an injured rider before transferring him to a University Hospital where duty specialists are aware of the event, and their names and direct line telephone numbers are available at the circuit. Thus, whatever trauma a rider receives will be rapidly and effectively dealt with. With an unconscious boxer the situation is simpler. His airway must be secured by a ringside doctor who is skilled in intubation and has the equipment to do so. He must then be transported directly to the nearest neurosurgical unit, and the telephone number and name of the duty neurosurgeon must be available to the ringside doctor. Following a meeting between interested neurosurgeons and the boxing authorities new rules have been introduced which should ensure that adequate facilities for resuscitation and rapid transport are present at the ringside. Neurosurgeons should co-operate with this initiative both to offer a service to injured boxers, and to gain the confidence and trust of the boxing world, so that the latter will accept sensible suggestions to make the sport safer."

PARKINSON (1992) wrote:

"Boxing provides the best human model for the study of concussion."

He continues:

"To many, boxing appears to be a violent, senseless sport. Yet as a sport it ranks a feeble seventh, tied with skating in the per participant incidence of serious head injuries (golf is first)."

Paul HAYWARD *'Regulation needed for primeval urge.' (The Daily Telegraph, October 26, 1995, p. 31)* wrote:

"Two of the vices humanity has been unable to extinguish over the centuries are prostitution and pugilism. The best we can expect is that they will be made safer for both participants. In boxing the conspiracy of embarrassed silence over death and injury has finally come to an end. It would be easy to parody the men who run the sport here as vultures with attaché cases who assist in the exploitation of dirty-faced kids from the back streets. Such a simplification will not do. Britain currently leads the field in ring safety standards and is a haven of common sense and compassion compared with parts of America, where boxing is in no more danger of being abolished that Coca-Cola or apple pie. The unprecedented surge of grief and confusion following the death of Jim MURRAY suggests that the British are more attuned to the moral complexities of this brutal sport. As MURRAY lay on the canvas, twitching like a steeplechaser with a broken neck, even the stoutest defenders of the rights of young men to box will have felt a rush of disgust and remorse. The trouble is that elegies and grief save no lives. If the government and public ignore the abolitionists, what is there left to do? Accept the reality that the fighting will go on and round up the medical experts to ensure that boxing is as safe as it can be. In this, many of us feel, the British Medical Association could spend less time on the soap box and more in the laboratory gathering the hard evidence they lack. The British Boxing Board of Control was forced to accept the need for reform long before Jim MURRAY's funeral cortege set off. In yesterday's 12-point statement there were many abstractions and much that is old. Many of the recommendations of the board's independent panel are familiar from the aftermath of the Michael WATSON tragedy in 1991. Boxers need statutes, not good intentions. An example. The board have stated that weigh-ins should be brought forward 'at least 24 hours,' but 'ideally two or three days.' Which is it? The promise that 'post-contest medical checks should be tightened' is much too vague to reassure the waverers. But at least the debate has been started. A personal view formulated at ringside is that boxing is a highly ritualized expression of man's primeval urge to fight. That urge is better controlled and codified than driven into disused warehouses or the back rooms of pubs. Prostitution is illegal but it exists. Heroin is illegal but people use it. It implies a more rational view of the world to improve and regulate boxing than to wish it away with legislation. Before we congratulate the board too heartily, we should acknowledge the role the boxers themselves have played in forcing change. Any fighter who benefits from having a Magnetic Resonance Imaging brain scan or is saved from the debilitating process of pre-fight wasting will be in debt to two persistent campaigners and colleagues: Barry McGUIGAN and Nicky PIPER, of the Professional Boxer's Association. When a ferry goes down, we hold an inquest. When a plane crashes we send in a QC in order to find out exactly what happened. Yesterday's news conference belonged in the same sad lineage of men in suits endeavouring to be wise after an event and reflecting grimly on a random loss of life. Jim MURRAY could not be present. Nor could Steve WATT or Bradley STONE or even Gerald McCLELLAN, who was maimed in a British ring this year. But if these reforms – and the ones that must follow – save one life or mind, some good will have been salvaged from the graves."

Eileen BATTERSBY *(The Irish Times, October 30, 1995, p. 10)* wrote:

"So yet again outraged demands for the banning of boxing, the human equivalent of cock-fighting, sound put, of solely momentarily, with the recent death of two more fighters. Of course, the obvious medical arguments run counter to those intent on justifying the continuance of the sport."
"But the problem with boxing is not just the high risk element. Many sports are dangerous, aside from tennis and golf; any

sports person, from the schoolgirl hockey player or show-jumper to the rugby player or gymnast, faces possible injury."

"With boxing, it is more a question of ethics, or rather, the abuse of ethics. Boxing is the victim of its own grotesque contradictions. 'The Noble Art of Self Defence,' as its supporters insist on calling it, is an obscenity in itself. There is nothing noble about boxing – nor is there anything particularly 'noble' about any sport whether it be motor racing, golf, Olympic gymnastics or rugby. Sport is sport, something to enjoy, glory in, be passionate about. It is theatre. "

"But is boxing really a sport? Is it not more a licence to engage willingly in legalised combat? Skill and physical conditioning are essential elements of boxing as they are of any sport. But art?"

"The perfect vault is art; 400 metres hurdling is art; world class sprinting is art; but where is the art of setting out to inflict damage? As for self defence – there is nothing self defensive about knowingly climbing into a boxing ring with the express purpose of hurting your opponent. Self defence is swinging an iron frying pan over the head of the intruder who appears in your kitchen at 3 a. m., having gained entrance through forcing a window or door. The irony here is that if you seriously injure your burglar, you might find yourself charged with assault, or worse, should he die. The winning boxer, however, is free to walk away and if his opponent is brain damaged or dies, well, it's the luck of the game isn't it?"

"Interviewed on Channel 4 News last week, a former world welterweight champion Colin McMILLAN, rejected a suggestion made by a brain surgeon that blows above the shoulder be banned, as strikes below the belt are. McMILLAN said: "I'm concerned with winning the fight, regardless how you win it.' This is a typical example of yet another of the dilemmas facing boxing; most of the arguments made in its defence are inarticulate, incoherent and nonsensical. On that programme, the brain surgeon's exasperated logic was overwhelmingly convincing. McMILLAN refused to accept any of the points made about the dangers facing that delicate, jellylike floating mass known as the brain from repeated head blows. McMILLAN did not make sense."

"There is also a social aspect. For every man who boxed at university, there are legions who become involved through the local youth club. Boxing is traditionally regarded as an outlet for working class, disadvantaged youths. "

"Many of its defenders have been busily reiterating lately that 'boxing is good it keeps young lads off the streets and away from crime.' But surely remarks such as that only place a question mark over the heads of all boxers. They imply that all boxers have suspect personalities and display potential for violence and crime. Mike TYSON, a disgrace to any sport, typifies boxing's misfortunes. But insisting that all boxers have anti social tendencies is as dangerously sweeping as to suggest that all world class sprinters take performance enhancing drugs."

"Boxing is seen as exploitative in that many aggressive young boys are drawn to it as a route to a better life, by its promise of high earnings. These expectations are not confined to professional boxing, nor its aggression. Aggression is part of being competitive."

"As for corruption, many sports have been corrupted by professionalism, but perhaps some sports such as boxing have been more corrupted than others. Examine the crazed expression of the faces of spectators at bouts. Then perhaps you will have some insight into the mentality of those who watch other forms of barbarism such as bull and cock fighting."

"Another of the many weaknesses of the pro boxing campaigners is that they repeatedly point to the virtues of amateur boxing as if this excuses the excesses of the professional sport. When will the pro professional boxing lobby accept that professional boxing and amateur boxing are two different sports?"

"Ask boxing fans to explain its attraction and they will say it teaches self defence... This is another of the contradictions. A trained and/or professional boxer is morally obliged not to use his fists outside the ring. The professional fighter's hand outside the ring are illegal weapons. Besides this, anyone who ever witnessed a street brawl must have noticed that the participants are not too concerned with style and as for rules, anything goes – kicking, punching and biting."

"STILL, it is always back to money. You cannot ban boxing. Probably the most understandable appeal of boxing is that it is inexpensive, sustainable by even the poorest youth club. One might say running is also cheap. And it is true that the aspiring middle or long distance runner, like the fighter, needs very little equipment. But sprinters and hurdlers need running tracks; high jumpers and pole vaulters need expensive landing areas. Hurdlers and jumpers and discus and javelin throwers all need facilities and coaching. Swimmers need pools. Gymnasts need facilities and coaches. Equestrian sports, with the exception of racing, are as much associated with the spoilt and pampered middle classes as boxing is with the working class."

"If the government funding of sports was directed at making all sports available to all classes, particular sports would not become so tightly bound by social class and income."

"Probably the best aspect of boxing is the training and the high levels of fitness reached. But if you want to prove you are tough, try running a sub 28 minute 10.000 metres. Should you need to prove your strength, become a 20 metres shot putter. Take up the javelin. Beauty, grace, art, strength, explosive power and discipline are all vital factors of sport which the underlining brutality of boxing somehow manages to obscure. No. Boxing cannot be banned. It can and should be ignored."

Norm FRAUENHEIM (1995) published an article *'Head injury renews heat from critics,' The Arizona Republic, Monday, March 6, 1995, p. D 2)*

"Gerald McCLELLAN's tragic head injury in London ten days ago has re-energized boxing's abolitionists."
"Once again, they're demanding that the sport be outlawed. Of course, this well-intentioned crowd often asks for laws against other undertakings."
"Boxing is on a list that includes war, holes in the ozone layer, Charles BARKLEY's remarks, auto accidents and few hundred other items that are provocative, politically incorrect or downright dangerous."
"Their predictable argument against boxing is becoming a cliché. The sport isn't going to disappear anymore than BARKLEY's profanity."
"Above all, the abolitionists appear naive and oh so elite. Fighters come from tough, gritty neighborhoods. The abolitionists don't. There's a cultural gap as wide as the Grand Canyon between those who fight and those who want to ban the business."
"Dr. Robert VOY, onetime physician for the United States Olympic Committee, summed it up when he was asked about the American Medical Association's longtime opposition to boxing."
"'My colleagues' sons don't have to box," VOY said. 'Doctors' sons don't have to fight their way off those golf courses and tennis courts.'"
"Opposition to boxing is rooted in its naked violence. Unlike football or auto racing, boxing's intent is undisguised. The point is to take out the other guy, plain and simple. Dangerous? You bet."
"But, for the most part, the ring is no more or less dangerous for the fighters than the streets they grow up on. It's nice if those abolitionists could do something about those streets, too. But laws aren't going to turn Michael CARBAJAL's Ninth Street neighborhood into Paradise Valley."
"Strict adherence to ring rules, however, can make the sport a little safer. Over the next several weeks, there will be some discussion on safety. Bigger gloves? Smaller gloves? Head gear? More extensive physical exams before bouts? Fewer rounds? Shorter rounds?"
"For the sake of the fighters, those are the questions that need to be answered. It doesn't matter what the abolitionists say, because fighters will fight, laws or no laws."

At ringside in the telecast of McCLELLAN's injury during the loss to Nigel BENN: Michael WATSON.

Four years ago, WATSON was in a coma for a couple of months after he suffered brain damage in a middleweight loss to Chris EUBANK.

WATSON, crippled, watched from a wheelchair as McCLELLAN collapsed in the tenth round.

Short description of the boxing match between Gerald McCLELLAN and Nigel BENN:

Gerald McCLELLAN turned pro in 1988, quickly he moved up the ranks. In 1991 McCLELLAN knocked out John 'The Beast' MUGABI in the first round and won the World Boxing Organization (WBO) middleweight title. In 1993 he won the WBC title from Julian JACKSON.
McCLELLAN signed for a 12-round title bout with Nigel BENN. His record then was 31-2, with 29 knock outs, 21 of them in the first three rounds. He lost two decisions early in his career.
In the fight with Nigel BENN in the London Arena he knocked BENN within 38 seconds through the ropes, BENN was able to come back into the ring before the referee counted him out. Some ringsiders, including Dr. Ferdie PACHECO, felt that BENN took 14 seconds to be in the ring again.
BENN revived and the savage fight continued. In the eighth round McCLELLAN knocked down BENN again throwing a vicious right. But BENN was able to recover. In the ninth round BENN tried a counter punch to a McCLELLAN blow, he missed with a right hook as he lunged forward. Doing so he apparently head-butted McCLELLAN. In the next round McCLELLAN reacted differently, when BENN hit him with a straight blow, he dropped to one knee, where he was counted out. He then staggered back to his corner and before he could sit down, slid down near the ropes, and collapsed onto the ring canvas.
BENN who was behind on two of the three judges' cards, was declared the winner by a knock out.
McCLELLAN was brought by an ambulance to the Royal London Hospital. His opponent Nigel BENN with a battered face, was admitted to the same hospital. McCLELLAN, then 27, underwent an operation to remove a blood clot in the brain. The surgery saved his life, but after two months in a coma, he was left blind, hearing impaired, brain damaged and unable to walk. His short term memory is severely limited, and his long-term memory is selective.
It is generally agreed that the head butt in the ninth round created the brain hemorrhage. A stinging straight right blow of BENN ended the bout.
McCLELLAN is able to use his hands for every day functions, but he is not able to walk alone. He walks with help of two people.

Edward F. LOVELL, Roanoke, send a letter to the editor, concerning an article *'One punch too many.' Roanoke Times and World News, October 28, 1995)* which was published under the title *'Its time to outlaw boxing,' in* the *Roanoke Times and World News, November 11, 1995, p. A 9):*

"An article in your Oct. 28 Sports section *('One punch too many')* dealt with the life of former pugilist, Jerry QUARRY:"

"What a vivid indictment of boxing. Why do we allow two men, even in controlled circumstances, to try to beat each other into unconsciousness or render him unable to continue the contest? Brain damage is inevitable."
"Fighting in public is illegal. Boxing should be placed under the same prohibition. This form of human brutality should no longer be given public sanction."

Mrs. MOTZ, a mother from Crystal, Minnesota, after her son was fatally injured in an amateur boxing match, wrote:

"I did not know the dangers of boxing, and I am sure that my son never did either. It would be much fairer for men who put these bouts on to explain to these young lads the real dangers of boxing. Can't this sport be stopped?"
"Any traffic that exposes our youth, who are often innocent of the true facts, to condition in which the dangers of permanent brain damage and even death are as certain as they are in boxing, is immoral when judged by the principles of Christianity and the standards of enlightened conscience."

According to NOVICH:

"Boxing is one of the least dangerous sports. And it is high time we stopped listening to the cries of hysterical old women on boxing safety... Most of them have never stepped out of a library in their lives and are simply unaware that relatively few blows in a bout are effective."

NOVICH insists that most fatal injuries occur when the head strikes the floor or ring post producing contrecoup injury, with secondary subarachnoid or subdural hemorrhage. It is rare to have a fatality when the head is freely movable.

TICHY (1997), who reminded us of Karel HENNER's legacy on his one-hundredth birthday. [Karel HENNER was Chairman of the Department of Neurology and Psychiatry of the University of Prag, Czechoslovakia, see p. 689, author]. At the opportunity of the 100th anniversary of HENNER's birth and the fiftieth anniversary of the Clinic he founded in the premises of Saint Catherine TICHY discusses contemporary views on boxing, the risk of repeated injuries for the development of encephalopathia pugilistica and their relationship to Alzheimer's disease.

TICHY noted:

"Contemporary boxing is a spectacle which is a continuation of historical fights of gladiators. If the main objective is to hurt the adversary, this activity does not deserve the name of 'sport.' Sport should promote and maintain physical and mental health which is not the case in professional boxing."

Dr. Bill O'NEILL, science and ethics adviser to the *British Medical Association (Newcastle Chronicle and Journal, October 2, 1997, p. 16)* wrote:

"Brain damage can occur from any blow or cumulative damage over time. Headgear does not protect the brain. We object to boxing in any form – whether it is men, women or children."

There have always been spokesmen for the boxing fraternity, among them George Bernard SHAW. In *'Cashel BYRON's Profession,'* the protagonist is a prizefighter who has been rejected by his lady who is revolted by his profession, he defends himself in beautiful Shawian style, beginning by saying that 'People run my profession down, not because there is a bad one or two in it – there's plenty of bad bishops, if you come to that – but because they're afraid of us."

When the object of his affection, Lydia comments that

"... it seems to me that your art is wholly antisocial and retrograde," he replies:
"I don't know whether it's antisocial or not. But I think it hard that I should be put out of decent society when fellows that do far worse than I are let in. Who did I see here... the most honoured of your guests? Why, that Frenchman with the gold spectacles. What do you think I was told when I asked what his little game was? Baking dogs in ovens to see how long a dog could live red hot! I'd like to catch him doing it to a dog of mine... Do you think I'd have shaken hands with that chap?... And he's to be received and made much of, while I am kicked out!"

This is the tone of SHAW's character – a man who believes that a:

"... good milling does a man more good."
"I was afraid of everybody; afraid that my cowardice would be found out; and as angry and cruel in my ill-tempers as cowards always are. Now you'll hardly believe this; but what saved me from going to be bad altogether was my finding out that I was a good one to fight. The bigger boys were like

grown-up people in respect of liking to see other people fight; and they used to set us young ones at it, whether we liked it or not... at first, when they made me fight, I shut my eyes and cried; but for all that I managed to catch the other fellow tight round the waist and throw him. After that, it became a regular job to make me fight, for I always cried. But the end of it was that I learned to keep my eyes open and hit straight. I had no trouble fighting then. Somehow I could tell by instinct when the other fellow was going to hit and I always hit him first... The power this gave me civilized me. In the end it made me cock of the school; and, as cock, I couldn't be mean or childish. There would be nothing than fighting for licking boys into shape if every one could be cock; but everyone can't; so I suppose it does more harm than good."

SHAW saw boxing, through BYRON's eyes, as a way for a disadvantaged youth to better himself. This view is still shared by a number of people, a century after.

SHAW, however, learned most (if not all) that he knew of boxing by reading other accounts. He had little contact with the sport, and used it more as the vehicle for making a sociological point than as either praiseworthy or damnable in itself.

Ray HELING, a retired Milwaukee trainer, who has been involved in every facet of boxing for 65 years *'Boxing mustn't get a bum rap.' (Milwaukee Journal Sentinel, Thursday, July 25, 1996, p. 17)* wrote:

"Someone I know had his face rearranged by a mugger's but at the Indianapolis 500. But the fact that it happened at an auto race didn't make the incident a blot on the sport. Nor did the Louisville Slugger used in the incident give baseball and my friend a black eye. Likewise, when people were seriously injured in what amounted to a student riot after the Wisconsin Badgers beat Ohio State at Camp Randall Stadium a couple of years ago, there were no calls for the end of intercollegiate football."

"Apparently, it is only boxing that is to blame when some members of its audience go nuts. That's what happened at Madison Square Garden in New York City after the recent heavyweight match between Andrew GOLOTA and Riddick BOWE."

"'Boxing itself took it on the chin and in the process revealed something very ugly about itself,' editorialized the *Journal Sentinal* on July 18. The newspaper cited the outlawry as proof 'once again that boxing is a stupid, violent and dangerous sport if it can be called a sport at all.'"

"Since it cannot claim that boxing is more lethal that sports of which it approves, the *Journal Sentinel,* in its editorial titled 'Boxing, in all its ugliness,' lamely parrots the American Medical Association canard that boxing is the only sport in which the aim is to 'render one's opponent helpless.'"

"Of course, that's also the aim of the 350-pound football linebacker getting ready to flatten the 190-pound quarterback often from behind. And the same can be said for the fastball pitcher throwing at the batter's head. But not only are these sports deemed wholesome and acceptable for public consumption, but taxpayers are muscled to build new stadiums for them as well."

"Denying that boxing is a violent sport would be like denying that grass is green. The violence of the ring is right there in the open, more honest and up-front than in other sport. Granted, getting hit in the head is not designed to add I Q points. Anybody who requires a footnoted, Latin-laced treatise from the American Medical Association to know that fact probably didn't go to a Catholic grade school or have a big brother."

"Hurts, doesn't it? If you didn't enjoy hitting your head, don't go into boxing. No one will force you. But if you think you wouldn't mind putting up with something like that for money, you could be a future heavyweight champion of the world. It's your risk."

"The point is that boxing is a form of expression, like slam dunking a basketball or writing newspaper editorials. And it's strictly voluntary. It's voluntary for both participants and spectators who should know, by the way, that when violence breaks out outside the ropes, the safest place to go is under the ring."

"Bad things happen in boxing. And nobody is more saddened by the plight of MUHAMMAD ALI than those of us who know and appreciate all he did for boxing. But I've never heard him say he would have done anything differently."

HELING concluded:

"In recognition of the dangers the ring entails, most states (including Wisconsin) require professional boxers to have a CT scan of their brains performed at least every two years."

"The fundamental illogic of the latest *Journal Sentinel* editorial against boxing makes me wonder if there shouldn't be a similar requirement for editorial writers."

Nat HENTOFF *(Las Vegas Review-Journal, October 24, 1996, p. 15 B)* wrote:

"For nearly 10 years in Boston, I broadcast boxing bouts sometimes doing blow-by-blow and sometimes color between rounds. I saw a young, extraordinary graceful boxer who seemed choreographed. His name was Cassius CLAY. And I remember such equally and precise combatants as 'Sugar' Ray ROBINSON. They were boxers, not brutal machines like the latter-day Mike TYSON. Most of the bouts I broadcast, however, were between club fighters, professionals who could put on a good, often bloody show for the fans but would never make the big time. One of the last bouts I broadcast was between two fighters who clobbered each other so hard in the early rounds that, neither being able to deliver a knock out blow, they more or less clung to each other. The disgusted crowd, cheated of the chance to see one of them fall unconscious, mockingly started to sing, 'I'll Always be in Love with You' in waltz time.' Watching the deprived sports fans, a venerable sportswriter said to me, 'Once you leave this place, kid, you'll never go to see another boxing bout.' He was right. I don't go to the fights. I don't watch them on television. Friends of mine journalists, lawyers, nearly all of them men make festive evenings of boxing events. They too are disappointed if all that happens is bloodless boxing with nobody on the canvas with his lights out. I have heard defenders of the death penalty claim that each execution creates a catharsis in the surrounding community. Why not then, televise executions with Don KING, Mike TYSON's mentor and promoter, deciding which 'dead man walking' merits prime time? There would be a national catharsis from this entertainment.... When the surest way to win is by damaging the opponent's brain, and this becomes standard procedure, the sport is morally wrong."

Pete HAMILL, who has covered boxing for 40 years *(The Wall Street Journal, Wednesday, March 17, 1999, A 25)* wrote:

"The uproar over the decision in the Evander HOLYFIELD – Lennox LEWIS heavyweight championship fight is ludicrous. This fight was about as entertaining as a morning at the World Trade Organization. Mr. HOLYFIELD is finished as a fighter, while poor Mr. LEWIS is now the most famous pacifist on the planet. From the reaction to the decision – a draw – one would have imagined that the shade of 'Sugar' Ray ROBINSON had arrived in Madison Square Garden in the person of Lennox LEWIS, where he wove a fistic masterpiece, and then was robbed. It was nothing like that. As the late Whitey BIMSTEIN once said, 'They stink the joint out.'"

"If anything, the fight and the decision were emblematic of a sport in full decline. There are no heavyweights worth watching. The greatest box-office attraction, Mike TYSON, is in jail and unlikely ever to recover the form that made him the most feared man in his division. Mr. HOLYFIELD is a shell, his legs gone, the reflexes shot, his balding head a pathetically vulnerable stationary target for other man's punches. Mr. LEWIS enters every fight looking like a man condemned to death. George FOREMAN is a vaudeville act who would be overmatched against most of the nuns of my childhood. The younger heavyweights are almost all immense sides of beef, their records fattened in mismatches with assorted deputy sheriffs and short-order cooks."

"Only three current fighters have the combination of skill and charisma that brings out the fans: Oscar de la HOYA, Felix TRINIDAD and Roy JONES. In his last fight, Mr. JONES defended his light-heavyweight championship against a 40-year-old cop. It was, by all definitions, a fixed fight. That is to say, that outcome was known before the fighters climbed through the ropes. I don't mean that some hoodlum in a pearl-grey hat came into a dressing room and said, as Rod STEIGER told Marlon BRANDO in *'On the Waterfront:'* Kid, this ain't your night – we're going for the price on WILSON.'"

"It doesn't happen that way any more. Now fights are fixed through matchmaking. Champions are put in with opponents who have almost no chance of winning. A sport whose essence is the lack of a script is becoming scripted entertainment, as worthless as wrestling."

"This matchmaking is supported by a ratings system that is subjective at its best and corrupt at its worst. Compounding the problem is an endless array of bogus champions, scattered through fragmented weight divisions, and sponsored by an alphabet soup of boxing organizations. Don't bother trying to decode the differences between the WBA, the WBC, the IBF and the WBU. They are all engaged in a great racket, collecting money from fighters in order to 'sanction' championship fights. The IBF takes 3% of the fighter's purse – off the top. The WBA and the WBC get up to 3.5% – off the top. The purses for last Saturday's fight added up to $ 30 million. You do the math."

"This money is taken by people who do not get punched, who never get hurt. They sit at ringside in the arenas. They do not provide pension or health plans to the fighters. They do not supervise the gymnasiums, where so much damage is done to fighters. They do not protect the fighters from managers (who are often the same people). They just take the money."

"If the LEWIS – HOLYFIELD fiasco is to have any meaning, this mess should be brought to an end. The alphabet organizations should be driven out of business. The television companies, including those that provide pay-per-view services, should be regulated by the federal government, since a fixed fight is a form of consumer fraud. The fighters themselves should form a union, modeled on the Screen Actors Guild, that provides every professional fighter with minimum fees, residuals for all video-rights, and a pension and health care after his boxing days are finished. No manager should be allowed to serve as a promoter, since there is an obvious conflict of interest. Rating fighters should be left to boxing writers, emulating the way college football and basketball teams are rated."

"Most of all, fighters must be protected from themselves. As long ago as 1992, Evander HOLYFIELD was ready to retire. He had taken two savage beatings from Riddick BOWE.

There were problems with his heart. He laid off a while, saying he was through, but he heard whisperings from God and returned. His greatest triumph was over Mike TYSON in their first fight. His courage has never been questioned. But he has looked listless, and uncoordinated in his most recent fights, and in the first round Saturday a soft LEWIS jab made him jerk awkwardly, like someone struck by an electric current. That's usually a sign that a fighter's brain is on its way to becoming oatmeal."

"The solution to Saturday night's fiasco is not a rematch. The first step should be Mr. HOLYFIELD's retirement. Most medical studies indicate that the worst damage is done to fighters after age 30 and that heavyweight fighters, who are hit with the most powerful punches, suffer the most damage. We've seen what his last three fights did to the great MUHAMMAD ALI. We've witnessed what such late career fights did to the brave, skillful, foolhardy Jerry QUARRY, who spent his last years sitting in the dark and is now dead. We've seen what such fights did to the elegant boxer Wilfredo BENITEZ, who was hand-fed by his mother in the final months of his brain-vacant life. This should not happen to Mr. HOLYFIELD, or to any other fighter."

Boxer 'Nicknames'

Several boxers have adopted new names when beginning in a ring career. Many boxers have been given nicknames during their active careers.

Before 1915 were many Americans of Irish descent boxers, they put Irish before their names. Of those boxers was said, they would give a good *'Donnybrook' show*. This word is taken from the village Donnybrook, near Dublin in Ireland, whose yearly annual fair *(Donnybrook Fair)* , which existed until 1885, was famous or infamous of its many brawls. In a figurative sense the term signified a wild fight or brawl, or an unruly scuffle.

OATES who wrote 1987 a remarkable book *'On Boxing'* noted:

"Here we find ourselves through the looking-glass. Values are reversed, evaginated: a boxer is valued not for his humanity but for being a 'killer,' a 'mauler' a 'hit-man,' an 'animal,' for being 'savage,' 'merciless,' 'devastating,' 'ferocious,' 'vicious,' 'murderous.' Opponents are not merely defeated as in a game but are 'decked,' 'stiffed,' 'starched,' 'iced,' 'destroyed,' 'annihilated'."

OATES continued:

"The names of boxers! Machismo as sheer poetry. Though we had in another era, 'Gentleman' Jim or 'Pompadour' Jim CORBETT (world heavyweight champion, 1892-1897), and the first black heavyweight champion Jack JOHNSON (1908-1915) called himself ('Li'l Arthur' as a way of commenting playfully on his powerful physique and savage style. (JOHNSON was a white man's nightmare: the black man who mocked his white opponents as he humiliated them with his fists). He was called 'The Galveston Giant' or 'The Black Bear.' In more recent times we had 'Sugar Ray' ROBINSON and his younger namesake 'Sugar Ray' LEONARD. And Tyrone CRAWLEY, a thinking man's boxer, calls himself 'The Butterfly.' But for the most part a boxer's ring name is chosen to suggest something more ferocious.

In the following selection boxer names are presented. By it, I have used compilations of GOLESWORTHY (1969), OATES (1987) and MARACIN (1992).

Jack DEMPSEY, from Manassa, Colorado, was 'The Manassa MAULER'; the formidable Harry GREB was 'The Human Windmill'; Joe LOUIS was of course, 'The Brown Bomber'; Rocky MARCIANO 'The Brockton Blockbuster'; Jack LA MOTTA 'The Bronx Bull'; Tommy JACKSON 'Hurricane JACKSON'; Roberto DURAN 'Hands of Stone' and 'The Little Killer' variously. More recent are 'Boom Boom' MANCINI'; Thomas 'Hit-Man HEARNS'; James 'Hardrock GREEN'; Al 'Earthquake CARTER'; Frank 'The Animal FLETCHER'; Donald 'The Cobra' Curry; Aaron 'The Hawk' PRYOR; 'Terrible' Tim WITHERSPOON; James 'Bonecrusher' SMITH; Johnny 'Bump City' BUMPHUS; Lonnie 'Lightning' SMITH' Barry 'The Clones Cyclone' Mc GUYGAN; Gene 'Mad Dog' HATCHER; Livingstone 'Pit Bull' BRAMBLE; Hector 'Macho Man' CAMACHO, 'Marvelous' Marvin HAGLER changed his name legally in 'Marvellous' Marvin HAGLER before his fight with Thomas HEARNS; Jessy 'Hook 'em' HADFIELD; John 'Hollywood' MATERI; Jim 'Bang Bang' HARRISON; 'Rocku' Robin BLAKE; Ricky 'The Mongoose' CLEMENTS; Gregg 'Zion Lion' EDELMAN; 'Tiger' Jack FOX; Rick 'King Kong' KELLAR; Johnny 'Honeyboy' BRATTON; Alvin 'Too Sweet' HAYES; Milton 'The Iceman' McCRORY, Joey 'Butcher Boy' ANCONA; John 'Death Wish' GARNER; Eusebio 'Scorpion' PEDROSA; Eduardo 'Tarzan' LOPEZ; Georg 'Hammer Schorsch' STEINHERR; 'Terrible' Terry NORRIS; 'The Toy Bulldog' Mickey WALKER; 'Hammerin' Henry ARMSTRONG; 'Battling' NELSON; 'Tiger' FLORES; 'Gorilla' JONES; 'Gunboat' SMITH; 'Midget' WOLGAST; 'The Astoria Asassin' Paul BERLENBACH; 'The Michigan Assassin' Stanley KETCHEL; 'The Bronx Bull' Jake LA MOTTA; 'The Fargo Express' Billy Petrolle; 'The Nebraska Wildcat' Ace HUDKINS; 'The Boston Strong Boy' John L. SULLIVAN; 'The Ambling Alp' Primo CARNERA; 'The Cinderella Man' James J. BRADDOCK; 'Wild Bull of the Pampas' Luis FIRPO; 'Punch 'em Paul' or 'Paralysing Paul' or the 'Astoria Assassin' BERLENBACH'; 'The Great Gun of Windsor' Tom CANNON; 'The Bristol

Bull' Bill Neat; The 'Bull's Head Terror' John FLOOD; 'The Troy Giant' Paddy RYAN; 'The Iron Man' or 'Grizzly Bear' Jim JEFFRIES; 'The Big Cat' Cleveland WILLIAMS; Roy 'Cut 'n Shoot' HARRIS; 'The Cincinnati Cobra' Ezzard CHARLES; 'Terrible Terry' McGOVERN; 'The Whitechapel Whirlwind' Jack 'Kid' BERG; 'The Boston Gob' or 'The Sobbing Sailor' Jack SHARKEY; 'The Livermore Larruper' Max BAER; 'Homicide Hank' Henry ARMSTRONG; 'The Jewel of the Ghetto' Ruby GOLDSTEIN; 'Smokin' Joe FRAZIER; 'Apollo of the Ring' Jem BELCHER; 'Baby Face' Jimmy Mc LARNIN; 'Barbados Demon' Joe WALCOTT; 'Basque Wood Chopper' Paulino UZCUDUN; 'Belfast Spider' Ike WEIR; 'Benicia Boy' John C. HEENAN; 'Black Panther' Harry WILLS'; 'Black Ulan' Max SCHMELING; 'Boilermaker' Jim JEFFRIES; 'Boston Tar Baby' Sam LANGFORD; 'Boxing Marvel' Jack BRITTON; 'Box o' Tricks' Pedlar PALMER; 'Chicago Spider' Johnny COULON; 'Cleveland Rubber Man' Johnny RISKO; 'Cuban Bon Bon' Kid Chocolate; 'Deaf Un' James BURKE; 'Durable Dane' Battling NELSON; 'Fighting Marine' Gene TUNNEY; 'Flying Dutchman' Harry GREB; 'Game Chicken' Hen PIERCE; 'Georgia Deacon' Tiger Flowers; 'Ghost with a Hammer in his Hand' Jimmy WILDE; 'Hard Rock from Down Under' Tom HEENEY; 'Herkiner Hurricane; Lou AMBERS; 'Homicide Hank' Henry ARMSTRONG; 'Human Punching Bag' Joe GRIM; 'Human Skyscraper' Henry JOHNSON; 'Idol of Fistiana' Jack DEMPSEY; 'Illinois Thunderbolt' Billy PAPKE; 'Kansas Rube' Jim FERNS; 'Kentucky Rosebud' Walter EDGERTON; 'Knob Hill Terror' Monte ATTELL; 'Little Chocolade' George DIXON; 'Madcap Maxie' Max BAER; 'Mighty Atom' Jimmy WILDE; 'Old Master' Joe GANS; 'Old Smoke' John MORRISSEY; 'Orchid Man' Georges CARPENTIER; 'Peerless Jim' Jim DISCOLL; 'Phaintin Phil' Phil SCOTT; 'Pittsburgh Kid' Billy CONN; 'Pottawatomie Giant' Jess WILLARD; 'Rochdale Thunderbolt' Jock McAVOY; 'Saginaw Kid' George LAVIGNE, 'Slapsie Maxie' Maxie ROSENBLOOM; 'Sidney Cornstalk' Frank SLAVIN; 'Tipton Slasher' William PERRY; 'Toy Bull Dog' Mickey WALKER; 'Two-ton' Tony GALENTO; 'Tylerstown Terror' Jimmy WILDE; 'Whitechapel Whirlwind' Jack 'Kid' BERG.

Boxing Spectators

FÜRST (1973) reported on the result of an empiric study made in New York on the basis of interviews with boxers, coaches and spectators, using the method of observation through participation.

FÜRST noted:

DESMOND (1911) observed that the boxing crowd is "a thing with terrible potentialities. Something elusive, baffling, uncertain. Something that requires a mind of hypersensitivity to realize even faintly."

While generally agreeing with DESMOND upon the complexity of the boxing audience, the following factors and analyses challenge the idea that it is baffling. A comprehension of the notions of (1) expectation of conduct and anonymity of the spectator, (2) size of the arena, (3) perceptual contagion and partnership, (4) cognitive tension, and (5) boxing knowledge and ability will constitute toward a greater understanding of the boxing show and its audience.

FÜRST wrote:

"The manner in which boxing has been depicted in novels, movies, plays, and television productions does much to form the *expectations of combat* of those attending a boxing show. Part of these expectations also inhere in the traditional conduct of the sport to be observed. Reactions at a tennis match are markedly different than those at a boxing match. The history and development of these two sports differ radically, thus, what is regarded as proper behavior at one is deplored at another. Publicity pointing out the strengths and weaknesses of the boxers also contributes to the expectations of the audience. If a boxer has a proportionally large number of knock outs, and this fact is publicized, the knowledgeable segments of the audience will anticipate a knock out. These expectations will add to the suspenseful nature of the boxing match."

"The anonymity of a spectator in a sport audience helps to establish a permissive atmosphere, especially when specific regulations or conduct have not been made explicit. Fear of reprisals therefore is diminished, adding to the likelihood of impulsive behavior. Notwithstanding the importance of expectations and tradition, the size of the boxing arena is another factor to be considered."

"The size of the arena has a direct relationship to the kinds of responses emanating from the audience. The small boxing arena facilitates a communal atmosphere especially where there is a consistent number of boxing shows over the year. The regularity increases the probability in a small arena of seeing familiar faces and re-establishing passing friendships."

"In a small arena, a greater proportion of the total audience is near the ring than in a large arena, thereby facilitating greater visibility of the boxers' movements. This proximity increases the opportunity for large numbers of the audience to perceive gestures and movements of the boxers and their supporting cast that might not be observed from a greater distance. This closeness also tacitly indicates that there is greater equality amongst those present. That there is a greater proportion of those present that have paid the same admission price and have comparable views of the match. Although the size of the arena affects the responses of the boxing audience, there are stimuli that generate responses transending the size of the arena and are generalizeable to all boxing arenas. The oblong and tier shape of the arena aids visibility of those sitting across from and below each other. Their actions are immediately perceivable, adding to the excitement wich is *perceptually contagious.* Visibility of other members of the audience also helps to identify partisan members within the audience. Furthermore, it often calls forth counter partisan reactions within the audience."

Fig. 355. Professional fight between Brian LONDON and Dick RICHARDSON at Porthoalwn, Glamorgan. Rioting and brawling break out in the ring at the end of this professional fight. It began at the end of the eight round, when the referee awarded the fight to RICHARDSON after LONDON had sustained a badly cut eye, although he was ahead on points. Immediately, angry members from both corners and enraged spectators burst into the ring and a free-for-all fight broke out. Twenty police officers had to jump through the ropes to break up the fighting and restore order. Photo: KEYSTONE.

"The boxing audience is popularly described as a barbarous collectivity (Fig. 355), reverberating with bloodthirsty cries that seek the brutal destruction of one of the contestants. Spontaneous shouting, where spectators are heard to cry: "Kill him!" or "Finish him!" can be attributed to a tropismatic reaction triggered by blows (action) that signal a resolution or possible termination of the conflict.. These reactions are aided by the concentrated attention of the audience and the high level of expectancy intrinsic to their attentiveness. The format of the boxing match promotes constant expectancy of denouement, especially when action is present. The fact that the audience is not informed of the round by round voting of the judges and referee adds to the suspenseful nature of the contest. Thus the audience is predisposed to spontaneous releases of tension built upon the uncertainty of the outcome of the match. This *cognitive tension* is further intensified when a match provides consistent and sustained periods of action."

"Because of the amount or lack of action present, the response of the audience overflows the boundaries of *partisanship.* Spectators react to the imminent possibility of a resolution. Their reactions are tampered by the vacillating nature of the match. The need to quell suspense overwhelms factionalism, therefore, the audience supports the boxer whose ongoing activity represents a possible outcome."

"At times, the audience will react spontaneously as an unanimous mass to well-known cues. At other time, a majority or segments of the audience will react discriminately to cues that are not generally known. In either case, they are responding to the kind and level of activity in the ring."

"But these responses are also inextricably related to the *boxing knowledgeability* of the spectators. Approximately one third of those members of the audience interviewed demonstrated a knowledge of the strategic nuances involved in boxing. The majority of the audience is ignorant or dimly aware of the fine distinctions in boxing. They are prone to misconstrue activities in the ring and react to them accordingly. Many of the spontaneous responses from a majority or segments of the audience can be attributed to misinterpretations. On the other hand, it is more likely that deliberate responses from segments of the audience are accurate interperetations of ongoing activity in the ring."

"The boxing audience is both the sum of its parts and at the same time greater than its parts, depending on the stimuli present. High levels of *actions* will transcend the possibility of more discriminatory responses and povoke unanimity of response. Situated activities which evoke segmented responses.'

"The stimuli are present in the boxing match episodically as the rhythm of each match unfolds. They are also cumulatively present as the tempo of the boxing show increases"

"The kinds and degrees of audience response are related to the five factors previously enunciated. These factors act as a context by which the dynamic interaction between the activity in the ring and the audience is influenced."

Riots and brawling break out in the ring at the end of the professional fight between Dick RICHARDSON and Brian LONDON at Porthoawl, Glamorgan (Fig. 355). It began at the end of the eighth round, when the referee awarded the fight to RICHARDSON after LONDON had sustained a badly cut eye, although he was ahead on points. Immediately, angry members from both corners and enraged spectators burst into the ring and a free-for –all fight broke out. Twenty police officers had to jump through the ropes to break up the fighting and restore order.

So-called 'Opponents' and Mismatches

"Aside from the champions of the boxing world, of whom we hear most often, another group is usually referred to as the *'opponents.'* These are the countless faces who most often lose their matches. Reliable boxers who lose are useful as opponents to fight against those who are building up their win/loss records. Opponents are also referred to with less flattering terms, such as *'windfall' ('Fallobst'), 'stiffs', 'bums'* or a member of the *'turkey list.'* The use of such terms demonstrates that those who describe boxing as heroic bravery and talk about the courage, strength, and tenacity in being able to take a punch are referring primarily to the winners, for clearly, little respect is afforded the losers, upon whose bodies the champions climb to the top." (Quote from *On Boxing,* J. C. OATES, p. 33-34)

The promising young boxer is able to find financial support in the business, with which to develop his talent. He is matched with an opponent who will deliver an acceptable fight. He will probably come out of the first round, and he will not, this is certain, absolutely not ruin the record of a successful boxer. The opponent may have dreams of winning a 'title', but his value for the fraternity is in helping to build up the record of other boxers. His name is always unknown – probably he has used many different names and aliases. His career is determined, he has no chance of winning. He makes a livelihood from boxing, in that he is a foil against which the glory of younger upstarts will be well reflected, a 'live human punching bag.'

Without them, professional boxers could not exist, according to statements of those who know the most about it. Due to their insufficient boxing abilities and their practical inability to defend themselves well, they are exposed to repeated brain injuries. Many of them have never enjoyed victory in a boxing match. Pragmatic opponents, who know their role, see no problem to lose, as long as they not injured. The romantics among them are convinced that one day they will be recognized and will at last fight for a championship.

Boxing purists complain that their sport would not be the same, what it had been many years ago, as in each city boxing clubs existed, which made the ascent of new attractive boxers with financial help easier.

The boxing historian Hank KAPLAN noted that opponents have always been known to exist in boxing. The only way to build up a boxer would be to find somebody that he could beat up.

SHAPIRO wrote 1988:

"When boxers are known to lose more often than they win, they become useful only as opponents. Often they are called flattering names, but without them, the sport, according to those who understand it best, could not exist. Pragmatic opponents aware of their roles, see no harm in fighting and losing so long as they are not damaged. The romantics among them are sure that someday their time worth will be recognized and they will fight for championships. Boxing purist lament that their sport is not as it was decade ago when there was a club to watch boxing in most every town they visited. But, they add, there has always been a call for opponents to smooth the ascents of new, attractive prizefighters who have the financial backing opponents never know."

"This is boxing as it has been known since the beginning of time," says Hank KAPLAN, a boxing historian. "There have always been opponents. Ever since the dawning someday discovered that the way to build a fighter up was to get him some one he can beat up."

"There are good opponents. A guy who don't get knocked out but who'll always lose a decision," says Chris DUNDEE, the Miami Beach promoter. "An opponent is a fella that is always dependable, that can give a good account of himself and lose."

"There are bad opponents. If a guy goes out in the first round the fans know you got a stiff in there," says Ernie TERRELL, the former heavyweight champion and now a Chicago club-fight promoter who does not like a hard hitter as an opponent because 'punchers are always dangerous.'"

"There are many opponents who, if not necessarily gifted boxers, are still remembered long after they fought: 'I'll never forget the 'Gorilla,' says Jerry WHITE, a long time Miami trainer of his fighter 'Gorilla' GOULD. 'They called me up and needed a heavyweight to fight. They were looking for an opponent. The 'Gorilla' owed me a few bucks. I took 4 to 1 odds that the 'Gorilla' would go the limit. I gave him a couple of cloves of garlic and gave him instructions that whenever you get in a clinch, blow in the other guy's face. He blew and the poor guy stepped back like he was punched.'"

"Jose TORRES, a member of the New York State Athletic Commission, says that boxing is entertainment and that is why opponents are necessary. 'Promoters get people who can excite the public,' he says. 'Opponents exist and I'm sure that promoters will try to get them and good managers will always look for them for their fighters – these 'Tomato cans' who are just in there to lose.'"

"Still, transgressions exist. Although New York State will revoke the license of a fighter who has been knocked out six times,' TORRES says, 'enforcement is difficult. They change their names,' he says."

"TORRES once the world light-heavyweight champion, recalls how his manager, Cus D'AMATO, was judicious in selecting the men he fought. In his first fight he knocked out Gene HAMILTON in the first round. HAMILTON has lost ten of his fifteen fights."

"If I was a promoter and I had a fighter to bring up I'd do the same thing,' says David CONTEH, who has lost fifteen of his nineteen fights. 'Look, I've been around. I'm thirty- four and I'm still pugging. I won't be no champ at my age. If I make a few bucks I'll be lucky. I try to take what I can. A lot of people say I had potential but it never materialized. Everybody wants to use you as an opponent. They say, 'He was a good man when he was younger. What can you do? What can you say? There's nothing you can do.'"

"But paying boxers to 'take dives' is seldom necessary, boxing elders say.

'There's enough bad ones out there to make the really bad ones look terrible,' says Joe MOONEY, a manager and gymnasium operator in Savannah!'"

"MOONEY is a purveyor of opponents. He is not alone in his vocation and maintains that his service is much in demand, 'The main reason they call me,' he says, 'is because I'm dependable.'"

Opponents, however, almost always lose on the road; and even a knock out does not assure victory. TERRELL tells of one especially successful opponent – ' never in my life saw him win a fight' – who was shipped to Memphis from Chicago to battle a local hero. The opponent, who was offered few fights because of his thunderous punch, surprised every one when he quickly knocked out the local man senseless with a mighty blow. Then the arena lights went out. TERRELL says with a laugh, until the local pug was returned to life. This took perhaps twenty minutes, somewhat longer than the traditional ten-count. Upon his revival, the bout continued to its scheduled conclusion, at which point the arm of the victorious local fellow was raised by the referee.'"

Another typical case of a mismatch was the professional boxing match between Ernie KNOX and Wayne BETHEA.

Ernie KNOX had been described as 'a sleepy-eyed negro' who would trade leather with anyone for the price of a train ticket and a night on the town. Once, *The Ring Magazine,* picked him as its 'Promising Fighter of the Month' – but that was in 1958, and the promise was mostly unfulfilled. He lost almost as often as he won (10 wins, 7 losses, 3 draws), and it was not long before Ernie KNOX, 26, was eking out an uncertain living from part-time jobs and unemployment checks. But always there was that dream of the big time *(Prize Fighting. The tenth death, Time, 82, October 25, 1963).*

"One night last week, Ernie climbed into the ring at the *Baltimore Coliseum* to fight New York's Wayne BETHEA, 31. A hulking 205-pounder, BETHEA had fought Sonny LISTON and Ezzard CHARLES, had a record of 32-19-3, Ernie weighted 184 English pounds on the boxing commissials scales and he had not fought in nearly a year. 'You know to prove yourself,' shrugged his manager, 'You have to be ready to take a chance.'"

"Some chance: In the 9th round, BETHEA unloaded a series of bombs and Ernie toppled to the canvas. Rushed to a hospital for observation, he asked for a drink of water, sank into a deep coma, and died of a brain hemorrhage – boxing's 10th fatality of 1963."

"A Baltimore commission started investigating the circumstances of Ernie's death. By week's end it had turned up two shocking bits of information, Ernie's cut of the $ 1.620 purse was hardly enough to pay for his burial expenses. And at the city morgue, Ernie KNOX's body weighted only 151 English pounds. The boxing commission sheepishly admitted that KNOX had been permitted to weigh in with his clothes on. Said the autopsy surgeon: 'All you had to do was to look at this Kid's body to know he didn't weigh 184 English pounds."

An additional case follows *(Charles LEERHSEN,*

HOLMES, The legs didn't go, Newsweek, March 25, 1985, p. 92):

"The boxing cognoscenti differ from other kinds of cognoscenti in that they often don't know very much."

LEERHSEN brings the description of a professional heavyweight fight between the 35-year old professional heavyweight Larry HOLMES and the unknown 25-year old David BEY in Las Vegas in 1985.

"It was just a funny feeling many veterans observers had – based, of course, on the scientific principle that with boxers and blondes, the legs always go first."

"On the other hand, BEY may not be the most reliable witness to what went down in the parking lot of the Riviera hotel – namely himself, twice before the referee stopped the bout with two seconds remaining in round 10. Asked afterwards about hitting the canvas the first time, the challenger looked truly horrified. 'Huh? I only got knocked down once, right?' he said. Then, hearing the truth, he groaned, 'Oh that's terrible.' But in fact BEY himself was anything but, at least during the first four rounds when he muscled HOLMES around the ring like Gerry COONEY never could – and was actually ahead in the point totals on the judges' scorecards. Thanks to BEY, in fact, what promoter Don KING ballyhooed as 'Perseverance – The Grand Finale' turned out to be at least a well-made drama with a beginning, middle and end."

"The turning point came in round 5. That's when HOLMES famous left jab, which had been flicking out tentatively at first, started landing all over BEY's likable, Lonely Guy face. In the parlance of the sweet science, the challenger was being measured, as in 'the undertaker measured the man for a suit.'"

"The fifth round was also when the somewhat blubbery BEY began realizing why HOLMES fights the way he does – with his hands held high and his forearm jutting forward, so that they hover in front of his upper body like the vertical bars of a cage. 'Larry lets you in there,' BEY said, 'but then he keeps hooking you with his elbows, until your muscles get so sore and tired that it's hard to lift your arms' 'It's all perfectly legal – and understandable coming, as it does, from a man who is a link to what may yet go down as the game's golden age. 'You have to remember that I learned to use my elbows and all those other tricks of the trade sparring with ALI, FRAZIER and Ernie SHAVERS, said HOLMES. 'In the end, you know, it's a hard business, and we do what we can to win."

"No doubt. But, some nasty red facial welts notwithstanding, there was nothing cruel about what happened to BEY, perhaps because he dismantled so neatly and quickly. By the sixth round, he was smiling at HOLMES, as if to say, 'You're not hurting me' – which is almost always a sign that a fighter is being hurt badly. And by the eighth, the tide had turned so dramatically that HOLMES could actually be seen shooing away a fly with his left while looking for an opportunity to throw a fight-ending right. About two minutes later the opening came, and, to the accompaniment of flashbulbs, BEY went down like the hunks of lumber he works with on his carpenter's job back in Philadelphia. After that, the challenger did little more than stand in the corner with his own right had cocked while HOLMES jabbed him into oblivion – or wherever it is that all pretenders to the throne have gone during HOLMES's seven-year reign."

"But the even the cognoscenti knew in their hearts that BEY, after just 14 pro fights, would probably go the way of trivia answers like Scott ('Anything's Possible') FRANK and James ('Bonecrusher' – at least he's got a good nickname) SMITH. The larger and more interesting question in this fight was: would HOLMES become that rare champ who walks away while he can still walk straight. The answer unfortunately, is still 'maybe.' In the postfight press conference – where he was at times downright presidential in his bearing, even to the point of addressing the boxing writers by name – HOLMES also managed to look like just another puncher who couldn't bear to say goodbye. 'This is my last fight,' he insisted, only to add, after barely missing a beat, 'unless they come up with a whole lot of money for COONEY or something."

"The frightening thing is, somebody might. Anything is possible in a sport where athletes, desirous of leading a monastic existence, hole up along the Las Vegas Strip. And never before has the game suffered the kind of turmoil that seems inevitable when there are more heavyweight champs around than good cut men."

A classic description of a mismatch is presented in the following article *(The Debut, Sports Illustrated, Vol. 61, No. 27, December 17, 1984, pp. 7 + 10:*

"A few days before Willie de WIT, the Canadian heavyweight boxer who got the silver medal at the L.A. Olympics, was to make his pro debut on Dec. 1 in Edmonton, his scheduled opponent, Al DEGRATE of Phoenix, bowed out with a broken finger. Rocky YOUNG, a Trinidadian living in Mexico, was lined up as a substitute, both ran into passport snags, possibly because he'd also fought as Jerry WILIAMS, John WILLIAMS, James WILLIAMS, Rocky YOUNG and – having just about exhausted the permutations – Young WILLIAMS. Another possible opponent was found, one Abdul MUHAMMAD of Dallas, who was said to have a 17-10 record, but de WIT's manager, Harry SNATIC, nixed that match, because 'we checked him out and his record really was 1-17.' Then a midnight call was made to Ed MORGAN, who'd just gone to bed after his shift as an auto mechanic in

Lake Charles, LA. Hours later MORGAN was 'terrified' to be making his first airplane trip."

"He should have had some qualms about meeting de WIT, too. On the mistaken assumption that MORGAN was a native of Meridian, Miss, Louisiana Boxing Commission secretary-treasurer Mike CUSIMANO told an Edmonton reporter who called to check on the fighter, I 'don't have much confidence in anybody from Meridian. All they bring over here are 'tomato cans,' guys who can't fight.' Actually, MORGAN was born in Hattiesburg, Miss. Also, he looked more like a tomato than a can, packing 247 roundish pounds on his 5' 11" frame. He was billed as having an 8-1 record, but Edmonton boxing authorities could verify only one result – a knockout loss in September in Houston, where MORGAN fought under the name of Walter MORRIS. They were unable to confirm information that he'd also lost at least one fight as Al MORGAN. To add to the confusion, there was a published report apparently untrue – that he might also be Robert MORGAN, who had an 0-4 record, including three first round knockouts."

"MORGAN got to the second round against de WIT before being sent reeling into the ropes with a left to the flab and decked with a right to the head. De WIT then put him away for good with a flurry of punches. Spectators hooted at the mismatch, which prompted MORGAN's trainer, Harold BURCHAM, to shout at them, 'You try getting in here.' MORGAN blew a kiss to one heckler, and later, in his dressing room, he grabbed a roll of fat on his stomach and said, 'This is a spare tire. This comes from eating a lot of steaks. This is out of shape to be an opponent for a guy's professional debut.'"

"De WIT said, 'I'm just glad to get the first one out of the way.'"

Purses in Boxing

We begin this chapter quoting some authors about the financial earnings of boxers and their employment after their retirement from boxing.

OATES (1985) noted:

"There is something peculiarly American in the fact that, while boxing is our most controversial sport, it is also the sport that pays its top athletes the most money. In spite of the controversy, boxing has never been healthier financially (Table 111)."

OATES continued:

"The world's consistently highest paid athletes are American boxers, but it does not follow that boxers as a class are the highest paid athletes. The very opposite is the case. Impoverished people prostitute themselves in ways available to them, and boxing on the lowest levels offers an opportunity fo men to make a living of a kind... In any case boxer's passion for money – for the prodigious sums earned by a very few champions – does not account for the fact that a public is willing, if not eager, to pay them these sums. Private and public obsessions mimic one another but are not identical."

Jack DEMPSEY, former heavyweight champion of the world noted:

"When you're fighting you're fighting for one thing: money"

HALPERN wrote 1988:

"Sure it's a rough sport. Perhaps the roughest. And if, out of the hundred best-paid athletes in America in 1983 the first four are boxers, only five boxers appear in the next 96."

GORN noted:

Far from deflating the myth of social mobility, the fact that most fighters processed limited ability, suffered serious injuries, and left boxing as poor and obscure as when they began, only served to highlight the gaudy success of the few at the pinnacle."

There is not only the danger to suffer brain damage in boxing, however, the income of the average boxer is small and fleeting.

WEINBERG and AROND (1952) (Table 112) examined 127 former boxer, whose careers were traced from 1938 on. They found that only 7.1% had reached national reputation and recognition. Their financial takings obtained under great risks were considerably reduced by expenses during training, payments to managers and their silent partners from the underworld and taxes. The share, which the boxer receives, is often carelessly squandered. 84.2% remained in the local preliminary or

TABLE 111
Prizefighting's million-dollar gates.

In appraising a bout, one criterion for many fans and the fight mob alike is the million-dollar gate.
In the long history of boxing, there have been only 15 such matches – all involving heavyweights:

1921	Jack Dempsey *v.* Georges Carpentier	$ 1.789.238
1923	Jack Dempsey *v.* Luis Firpo	1.188.603
1926	Gene Tunney *v.* Jack Dempsey	1.895.733
1927	Jack Dempsey *v.* Jack Sharkey	1.083.530
1927	Gene Tunney *v.* Jack Dempsey	2.658.660
1935	Joe Louis *v.* Max Baer	1.000.832
1938	Joe Louis *v.* Max Schmeling	1.015.012
1946	Joe Louis *v.* Billy Conn	1.925.564
1955	Rocky Marciano *v.* Archie Moore	2.248.117*
1960	Floyd Patterson *v.* Ingemar Johansson	2.468.278
1962	Sonny Liston *v.* Floyd Patterson	4.665.420
1963	Sonny Liston *v.* Floyd Patterson	4.747.690
1965	Muhammad Ali *v.* Sonny Liston	1.602.190
1965	Muhammad Ali *v.* Floyd Patterson	3.570.000
1966	Muhammad Ali *v.* Henry Cooper	1.200.000

* Receipts from this year on include totals from radio and closed-circuit TV coverage of fight.

TABLE 112
Rank order of number of prominent boxers of various ethnic groups for certain years. (From WEINBERG K S, AREND H 1952).

Year	Rank		
	1	2	3
1909	Irish	German	English
1916	Irish	German	Italian
1928	Jewish	Italian	Irish
1936	Italian	Irish	Jewish
1948	Negro	Italian	Mexican

Data tabulated from *World's Annual Sporting Record* (1910 and 1917); *Everlast Boxing Record* (1929); *Boxing News Record* (1938); and *Ring* (1948 and 1949). The numbers in the succeeding years are: 103, 118, 300, 201, and 149. There may be biases in the listings, but the predominance of two or three ethnic groups is marked in all the years. The Irish were very much above others in 1909 and 1916 (about 40 per cent of all boxers listed); in 1948 nearly half of all boxers listed were Negro. The Jews and Italians did not have so marked a predominance.

semiwindup category; 8.7% became local headliners. "It does not seem unreasonable, according to WEINBERG and ARNOLD, to surmise that similar barriers confront other professional athletes in their quest for ultimate success."

"Eventually the brain is irreversibly damaged by the blows, the seemingly – limitless earnings are plundered by the payoffs, and the fighter stands alone, a hollow hulk. In the end, there is the institution erected by those who cheered, where their discarded creation can be hidden. The damage is done, the brain cannot be restored, and the only issue is custody. Perhaps the Roman crowd which turned thumb down on a 'maimed gladiator' was more merciful."

STONE (1960) noted that boxers find further that, despite their success in sport, their careers terminate at a relatively early age.

He noted:

"Since boxing has been the vocational medium of status attainment and since they have no other skills to retain that status, many boxers experience a sharp decline in status in their post-boxing career."

STONE continued:

As an illustration, of 90 former professional boxers, champions and ranking contenders, each of whom had earned more than $ 100.000 were after retirement from their active careers employed in the following professions: 26 worked in pubs, 18 were trainer or trainer/manager. 18 unskilled worker, 6 worked in movie houses, 3 were taxi cab driver, 3 salesmen at a newspaper kiosk at a race track, 2 wrestler, 2 salesmen in alcoholic beverages, 2 were petrol pump attendant, 2 were doorkeeper and 2 were businessmen.

The earnings of the professional boxer in the first half of the twenteeth century were low. Jim SAVAGE *(Old Fighter, The Saturday Evening Post, 208, March 7, 1936, p. 40 +)* was himself a professional heavyweight boxer, who ended his professional career in 1918. He wrote:

"All a fighter means to a crooked fight manager is a meal-ticket – a dumb, willing, trusting animal out of which he can make money."

Now, after 1936, exists laws, that a fight manager may not receive more than 33 1/3% of the purses of a professional boxer.

SAVAGE continued:

"In my twenty years as a fighter I fought 124 fights, for which I received approximately $ 10.000, or a little more than eighty dollars a fight, or about $ 500 a year. From what I have learned since, some of my various managers received from ten to twelve times that much from my fights. Mind you, I don't accuse them of any dishonesty. A fighter was often lucky to get as much as 10 percent of his ring earnings."

Million dollar gates began in 1921. The following compilation reaches to the year 1966. In an article *(Bull vs. Butterfly. A Clash of Champions. Time, March 8, 1971, pp. 63-69)* are prizefighting's Million Dollar Gates presented. It reads:

"In appraising a bout, one criterion for many fans and the fight mob alike is the million dollar gate. In the long history of boxing, there have been only fifteen such matches – all involving heavyweights (Table 111). But in this table is the last entrance was for year 1966.

In this compilation there is only one bout in which the receipt from radio and closed circuit TV coverage of a fight is included.

The receipts of TV coverage of fights, now called 'pay-per-view,' in the following years reached spectacular sums.

An additional contribution at the same time by Lee WINFREY *(TYSON, HBO talking knock out deal, Times Picayune, New Orleans, Monday, July 17, 1989, C1 and C5)* occupied itself with a similar project:

Home Box Office (HBO) is negotiating with heavyweight boxing champion Mike TYSON on a television contract that would cover the rest of his boxing career – a deal that would pay him more that $ 100 million.

The career-length contract would be the first of it kind between an athlete and a TV network. Even the network agreements with sports leagues, such as the NFL, and the NBA, usually expire every few years.

Seth ABRAHAM, HBO's vice president for sports, said on Thursday that he had been negotiating for about two months with TYSON, promoter Don KING and TYSON's manager, Bill CAYTON.

This is how ABRAHAM said HBO's offer would work:

"TYSON would fight on HBO about four times a year until he retires. Every 18 months, TYSON would have the freedom to fight in a pay-per-view bout not affiliated with HBO."

"TYSON's salary for each bout from HBO would depend on the quality of his opponent. HBO would prepare an 'A' list. 'B' list, and 'B-minus' list, from which his opponent would be chosen by mutual agreement."

"The most HBO has paid TYSON for TV rights was $ 5 million for his knock out of former champion Larry HOLMES on January 22, 1988, in Atlantic City, N.J. Taking that as HBO's payment for an 'A' list fight, a 'B' list bout against lesser opponents would pay 4 million, and a 'B-minus' fight would pay $ 3 million, according to sources close to the negotiations."

"Thus the contract, depending upon the quality of TYSON's opponents, could pay him $ 12 million to $ 20 million a year. Because TYSON is 23, and heavyweight champions usually fight into their 30s, he could punch his way to more than $ 100 million from HBO before he retires."

"It was not clear how the contract would be affected if TYSON were to lose his championship."

"ABRAHAM, who said he is optimistic that the unique deal can be worked out, will have another chance to talk it over with TYSON, CAYTON and KING when TYSON defends his title Friday against Carl 'The Truth' WILLIAMS in Atlantic City."

Neither TYSON, KING nor CAYTON could be reached for comment

"Asked how he could be optimistic when KING and CAYTON had been feuding for more than a year, ABRAHAM said only one word: 'Money.'"

"Under the current terms of his manager's contract, which runs until 1992, CAYTON gets 20 percent of TYSON's earnings. KING gets another cut, which varies in size, for promoting TYSON's fights."

"During the months of bickering between KING and CAYTON, which began during TYSON's tumultuous one-year marriage to TV actress Robin GIVENS, ABRAHAM has remained a middle man trusted by both men."

"ABRAHAM has no illusions about the difficulties of negotiating fight contracts, since he observed years ago that ' the state of boxing is as close to 16th-century buccaneering as anything on the planet today."

"Both the rewards of having TYSON under contract are compelling. Of all the movies and other forms of entertainment that HBO showed during the first six months of 1989, the program that drew the highest NIELSEN rating was TYSON's February knock out of Frank BRUNO in Las Vegas, Nev."

"HBO, whose 17 million subscribers make it the nation's largest pay-cable service, has been smitten with TYSON, even though he often turns in a very short night's work. ABRAHAM said the 12 TYSON fights that HBO had carried to date added up to only 214 minutes of ring action, less that six full rounds per fight. Thirty-two of TYSON's 36 pro fights have ended in knock outs."

"In becoming the unbeaten undisputed champion, TYSON's ferocious punching power has made him an increasingly dazzling but expensive TV star."

"TYSON made his TV debut on February 16, 1986, when he knocked out Jesse FERGUSON on ABC. It was after decking FERGUSON that he spoke one of the most chilling sentences of his career, one that has probably given pause to many opponents since: 'I always try to catch them on the tip of the nose because I try to push the bone into the brain.'"

"After fighting five times on ABC and earning a total of $ 1 million, TYSON signed his first contract with HBO: a total of $ 1,35 million for three fights. This progression from form a broadcast network to a pay-cable service is typical of successful boxers, because ABC, CBS and NBC can't afford to pay as much as HBO can for a heavyweight fight."

"After these contracts expired, HBO paid TYSON a total of $ 5,5 million in participating in three of the 10 bouts the pay-cable service conducted in a series to unify the heavyweight title. TYSON took the World Boxing Council title from Trevor BERBICK in 1986. In 1987, he took the World Boxing Association title from James 'Bonecrusher' SMITH and the International Boxing Federation title form Tony TUCKER."
"TYSON is in the middle of a seven-bout contract under which HBO will pay him a total of $ 26,5 million. This pact includes the WILLIAMS fight, which HBO will present at 10 p.m. Friday from Atlantic City's Convention Hall."
"The current contract and HBO's proposed career-long deal share one important point: freedom for TYSON to step outside periodically and fight on pay-per view, which pays much more than a broadcast network or a pay-cable service can afford."

"TYSON's biggest payday to date was $ 20 million for his only pay-per-view appearance – his knockout of Michael SPINKS in Atlantic City on June 27, 1988."

An article *(The Times Picayune, New Orleans, Tuesday, August 15, 1989, F 4, based on a report of The Associated Press)* reports that heavyweight champion Mike TYSON was offered $ 100 million Monday by a British sports promotion firm to make a year-long world tour that would include a title fight each month.

Winners Worldwide Ltd., which promotes snoobe and soccer as well as British rock bands, said the tour would start in the United States and move into Europe and Japan before finishing in Brockton, Mass., birthplace of the late Rocky MARCIANO, who also was an unbeaten heavyweight champion.

Stephen JONES, a spokesman for *Winners Worldwide,* said the firm has never promoted a world title fight, but "we've done a couple of local shows which have been very successful."

JONES said the company had sent information about its offer to TYSON's associate, Don KING, and was seeking to interest large American corporations such as Mc Donalds and Burger King to sponsor the tour.

"If we can pull it off and get some publicity out of it, then that's no problem,' JONES said, 'We are not just out for the publicity. We are serious about this.'"

A news release from *Winners Worldwide* said:

"There is no reason why a world champion boxer cannot be taken on tour in the same way as a pop group... As the current opposition shouldn't take up too much of his time,' the press release said, 'we feel he can eliminate one a month, putting his awesome talents on display across the world."

At this time the record of TYSON was 37 – 0, with 33 knock outs. His last fight was July 21, when he knocked out Carl "The Truth" WILLIAM, in 93 seconds.

In an article Peter FINNEY *(Boxers shake the money tree, New Orleans Times Picayune, June 7, 1989, C1 and C4)* wrote:

"It took Jack NICKLAUS more than 25 years to win $ 5 million in prize money for hitting a golf ball."
"Martina NAVRATILOVA is tennis all-time leading winner with more than $ 13 million and this took Martina 15 years."
"If *Sunday Silence*, the 3-year-old colt, wins the *Belmont* on Saturday, his owners pick up a $ 5 million bonus, the biggest in racing history."
"Since *Sunday Silence* would earn this for little more than six minutes of running in Triple Crown events, we're getting closer to the point of today's essay."
"And the point is, when you're talking sports and money, everything and everyone – Michael JORDAN, Magic JOHNSON, Joe MONTANA, A. J. FOYT, Wayne GRETSKY, Orel HERSHISER – pales before a boxing match with box-office pull."
"When we say *Sunday Silence* can pick up a cool 45 million for running a fraction over six minutes in the Kentucky Derby, Preakness and Belmont, it should be noted Mike TYSON haul for a 90-second job against Michael SPINKS was 21 million."
"And, for his 90 seconds as a target, SPINKS grossed $ 12 million."
"All of which brings us to LEONARD – HEARNS II on Monday in Las Vegas."
"Sugar Ray LEONARD, who got $ 15 million for his 1987 fight with Marvin HAGLER (HAGLER's take was a then-record $ 18 million), will receive $ 14 million for his rematch with Tommy HEARNS."
"Even if HEARNS, brooding for more than seven years, loses a second time to his longtime nemesis, he will have $ 11 million to ease the pain."
"What other comparison?"
When Mike SCHMIDT of the Phillies retired the other day after 18 years in the majors, it was noted his total salaries over that period – 17.076.738 – were second on the list of baseball players, a fraction behind Dave WINFIELD's."

"Closed-circuit telecasts – piping fights into arenas and theaters – used to shake the sports money tree. Now it's something called Pay-Per-View and the leading players in this stay-home-and- have-a-party era are TYSON and LEONARD."
"Cox Cable is offering Monday's fight to its subscribers in Orleans and Jefferson – $ 29.95 if you order now, $ 34.95 the day of the fight."
"Cox is shooting at an available audience of 195.000 in Jefferson, 80.000 in Orleans. For TYSON – SPINKS, Cox had its biggest audience for any Pay-Per-View event – 17.000."
"So how big is boxing?"
"Consider this: When the same audience was offered a show featuring Wayne NEWTON, the Las Vegas-based entertainer, 79 homes turned in to Pay-Per-View."
"That's seventy-nine!"
"While it might be bruising to NEWTON's ego, there is something to be said for the enduring appeal of LEONARD."
"LEONARD is 32, two years older than HEARNS. This has not stopped the early Pay-Per-View customers from signing up, doing so at a rate that could make LEONARD – HEARNS II one of the biggest money fights of all time, up there with HAGLER – LEONARD and TYSON -SPINKS."
"The Pay-Per-View rights to Southern California alone sold for $ 2.4 million. At the moment, it looks like the gross in that area will exceed $ 5 million."
"What Monday's fight has going for it is the memory of LEONARD's 14-round TKO victory, the night he left HEARNS helpless on the ropes when The Hit Man was leading on points."
"As an entity, there was more action in that fight than in the 12-round decision LEONARD took from HAGLER."
"Because he showed he can take an extended layoff and still fight, Sugar Ray remains a 3-1 favorite, down from 4-1."

COWART (1989) contributed an interesting thought:

"When a young man can lose a fight and still earn as much as $ 13 million for 91 seconds being punched, it may be unrealistic to expect that potentially grave health risks will deter the continuation of professional boxing. Even though few professional boxers ever stand to reap the financial rewards of current heavyweight champion Mike TYSON or challenger Michael SPINKS (whom TYSON knocked out in about a minute and a half) their success seems to keep the stream alive for some youngsters."

James MICHENER (1976) reported that Mike DOUGLAS earns $ 2.200.000 for speaking in television, Charles BRONSON gets $ 1.500.000 for a role in a sadistic movie, Mike JAGGER earns $ 3.000.000 for his songs which can be understood only by a few over the age group of 22 years. Erica de JONG earns more than $ 1.000.000 for her compositions of sexual recollections of an empty-headed young woman. Sally STRUTHERS received $ 156.000 for her part in *'All in the Family.'* A Harvard graduate in philosophy, with an M.A. from Yale in theology is to praise happy when he can earn $ 10.000 per year.

MICHENER quotes Dr. Hervey R. RUTHSTEN, who analyzed this problem in a letter to the *New York Times.* It reads:

"Many women and men study, at significant financial investment, ten to twenty years to attain a doctoral degree. Recent developments and economic setbacks have decreased demand for these intellectually trained people. They now drive cabs, tend bars, sell insurance. Athletes, many of whom were high school drop outs, are building $ 200.000. homes."
"The answer is depressing but inordinately simple. There are few 'gray' zones in sports. A puck goes in the net or misses, the ball through the basket or out, and clearly there is always a winner and a loser. Soul-searching is not required, a weighing of dismal alternatives obviated. We have escaped to the black and white of sports from the frustrations and confusions of such gray zones as the Vietnam War, police corruption, morality of abortion, bounding inflation, dismal political leadership alternatives, safety on the streets."

The following analysis by MICHENER, which made me reflect, reads:

"In ancient Rome, citizens befuddled by the complexities of the waning empire found in the battles waged by the gladiators precisely this simplistic relief form the governmental problem of their day. Cynics are always citing the downfall of Rome as a gloomy precedent; the relevant fact is that after the introduction of those gigantic festivals in the Coliseum, Rome survived successfully for another four centuries, and I suppose we can survive our excessive sports salaries."

MICHENER continued:

"But then the moral problem arises. Should a well-run society divert so high a percentage of its gross national product into sports, when there are so many other aspects of our national life which cry for attention. A recent study has shown that we spend about one hundred billion dollars a year on sports and recreation, considerably more than we spend for national defense. If this vast sum went for the improvement of the national health, fine. If it encouraged our citizens to become active participants and to avoid physical and mental deterioration, fine. And if it encouraged young people to built patterns of living that would serve them throughout life, fine."

"But if a great deal of this investment is merely subtracted from the creative processes of the nation, and siphoned into the hands of a few essentially brash young men and women, the consequences can only be destructive. Our nation may be generating deep resentments in establishing and promulgating such unbalanced priorities."

Karate

Physics of Karate

FELD et al. (1975) investigated in detail how the bare hand can break wood and concrete blocks (and by implication do similar damage to other targets) without being itself broken or injured. The key finding is that the hand of the karateka, or practitioner of karate, can develop a peak velocity of 10 to 14 meters per secomd and exert a force of more than 3.000 newtons, a wallop of 675 pounds. If the hand is positioned properly, it can easily withstand the resulting counterforce.

The maneuvers of the Japanese style karate, that is practiced today were, according to FELD et al., developed on the island of Okinawa. When the Japanese conquered the island in the 17th century, they stripped it of all weapons and banned their making and importation. To defend themselves the Okinawans developed karate, a sytem of empty-hand combat based on the weaponless methods of ancient Chinese monks, warriors and physicians.

The *karateka* concentrates his blows, according to FELD et al., on a small area of the target and seeks to terminate them about a centimeter inside it, without the long deliveries and follow-through of the punches in Western boxing. "Whereas the Western boxer imparts a large amount of momentum to the entire mass of his opponent, pushing him back, the karateka imparts a large amount of momentum to a small area of his opponent's body, an momentum that is capable of breaking tissue and bone" (FELD et al.).

FELD et al. used wood blocks in their experiments which were standard pieces of dry white pine, that weighed .28 kilogram and were 28 centimeters long, 15 centimeters wide and 1.9 centimeters thick. They were cut so that the grain was parallel to the width. The concrete slabs were 'patio blocks' that weighed 6.5 kilograms and were 40 centimeters long, 19 centimeters wide and four centimeters thick. The 'patio blocks' were dried in an oven for several hours to remove excess water, thereby making them uniform in content.

In the block-breaking demonstrations, carried out by FELD et al., the karateka strikes the top of a block in a range of points near the center.

To understand the physics of the breaking process, picture a fist moving forward a block supported by its ends. The impact will cause the block to bend in the direction in which the fist is moving... As the block is bent it is deformed: the upper half of the block is compressed and the lower half is stretched. The top surface is compressed the most and the bottom surface is stretched the most. "Because wood and concrete are weaker under tension than they are under compression the block starts to crack at the bottom surface. The crack spreads rapidly upward as the fist continues to force the block downward" (FELD et al.).

The peak speeds (in meters/second) for several karate maneuvers are listed in Table 113.

TABLE 113
Peak speed of karate maneuvers are listed first for empty-hand blows and then for kicks. The knife-hand stroke is the 'karate chop'. In the roundhouse kick the foot moves in a clockwise circle as the opposite hand moves counterclockwise in order to keep the body balanced by minimizing net angular momentum. In the wheel kick the body rotates with the leg extended. (From FELD MS et al., 1975).

Maneuver	Peak speed (meters/second)
Front forward punch	5.7 – 9.8
Downward hammer-fist strike	10 – 14
Downward knife-hand strike	10 – 14
Roundhouse kick	9.5 – 11
Wheel kick	7.3 – 10
Front kick	9.9 – 14.4
Side kick	9.9 – 14.4

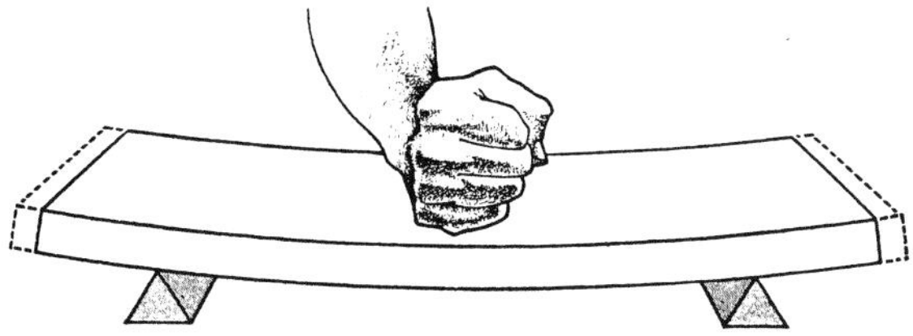

Fig. 356. Hammer Fist Strike breaking a wood block is shown in a highly schematic diagram. In a well-executed strike the fist hits the block along a range of points near the center. The block acts like a small coil spring: it pushes back on the hand with a force proportional to its deflection. The top of the block is compressed; the bottom stretched. Because wood is weaker under tension than it is under compression the block cracks first on the bottom surface (near the center). The crack propagates rapidly upward as the block continues to be pushed downward. (From FELD M S et al., 1975).

Fig. 356 shows a *Hammer Fist Strike* breaking a wood block.

Hydraulic-Press Measurements which were made of the deflection of a wood block (left) and a concrete block (right) as a function of the applied force (Table 114 a and b).

For details we refer to the excellent paper of FELD et al.

Full-contact Karate

An interesting paper about *'Full-contact karate (kick-boxing)'* was published by SCHWARTZ et al. (1986). This activity, which is relatively new in North America, is a marriage of traditional boxing

TABLE 114 A AND B
Hydraulic-Press Measurements were made of the deflecetion of a wood block (left) and a concrete block (right) as a function of the applied force. Each of the plots yielded for the elastic modulus, which for a given material is a constant that indicates its springiness, and for the modules of rupture, which is the minimum stress that causes a block to snap. It is useful to know these moduli in order to find the energy needed to break each kind of block. (From FELD MS et al., 1975).

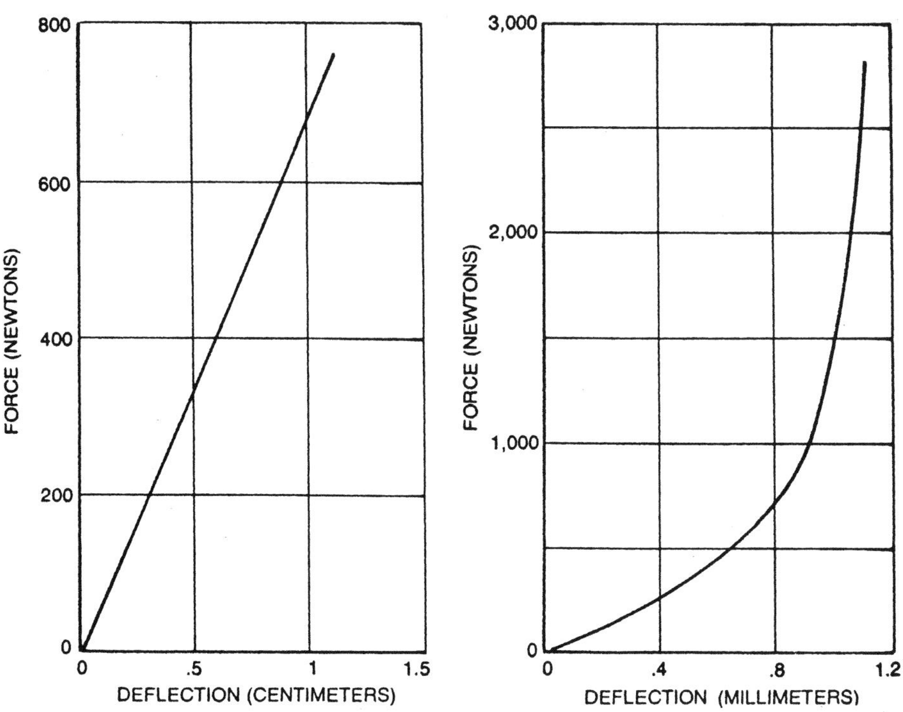

and karate-style kicks. It differs, according to SCHWARTZ et al. (1986), from traditional karate in that contestants fight in a boxing ring and wear boxing gloves and safety-kick foot padding. In full-contact karate, punches and kicks are actually landed rather than being focused to culminate just short of an opponent as practices in traditional karate. "Although a karateka (combatant) can win on points, the surest means of victory is a knock out. Consequently fighters strife to land blows to the head" (SCHWARTZ et al. 1986). A knock down or a knock out can be the result of a blow or a kick. There are full-karate combatants reputed to be experts at winning by knock outs from punches, as well as those known to win by kicking knock outs.

Biomechanical Aspects

SCHWARTZ et al. (1986) mounted a 50th-percentile Hybrid dummy head on a universal joint permitting motion about three orthogonal axes. The two horizontal axes were located 15cm below the attachment of the 'skull,' corresponding approximately to the cervico-thoracic junction. The flexion extension stiffness in the sagittal plane was 0.81 Nm/degree. Side-to-side stiffness to center the head in the coronal plane was 0.54 Nm/degree. Torsional stiffness to center of the head about the vertical axis was 0.54 Nm/degree. The 'neck' mechanism was contained within a stiff steel column that could be bolted directly through the concrete floor for measurements at a lower height (125 cm from the floor to the top of the head) to simulate a 50th-percentile man in a crouched position. The unit was mounted on a pedestal for a higher level (175 cm from the floor to the vertex), the height of a 50th-percentile man standing erect. The steel column was contained within a 30-cm diameter cardboard tube, and the 'neck' mechanism was enclosed within a tube of polyurethane foam to prevent injury to the karate experts who punched and kicked the dummy.

Since the Hybrid dummy does not simulate the visco-elastic energy-absorbing properties of soft tissue, it was fitted with a mask made of 5-cm thick Plastizote to simulate the frictional properties of skin. Accelerometers were mounted at the center of mass of the head, aligned along the sagittal and coronal axes. These were supplied at 10 V direct current. The outputs were fed through differential amplifiers, and a digital storage oscilloscope stored and displayed the two traces in the chopped mode. The time base was triggered from one of the accelerometer signals using a module that enabled data preceding the trigger point to be captured. The sampling rate was 12.8 kHz. Data were stored and analyzed using a microcomputer interfaced to the oscilloscope on the IEEE Bus.

Subject Selection

Fourteen subjects with black belts in karate were involved in the study, each with experience in both kicking and punching. The average height was 173 cm (range 153 cm to 179 cm, standard deviation 8 cm). The average weight was 76 kg (range 49 kg to 95 kg, standard deviation 12 kg).

Each karateka was required to strike the dummy in a testing sequence involving 20 different combinations of style, protective wear, and dummy height (Tab. 115). The combinations were tested in random order. The subjects were encouraged to practice with light blows before attempting to hit the dummy as hard as possible. Following each attempt, the subject was asked if he was satisfied that the attempt represented his maximum force or whether he thought he could do better. If the subject considered that his result was his best effort, then it was stored on magnetic tape for later analysis. If the subject was dissatisfied with his efforts, the result was erased and he was asked to repeat the blow. This sequence was repeated as many times as the subject desired until he was satisfied that the result represented his best effort.

Protective Equipment

All safety equipment used in the testing was new. The 10-ounce boxing gloves were Model IC670, manufactured by International Cougar Equipment Canada, Inc. The karate-style hand protectors and foot protectors were Macho Punch and Macho Kick, respectively, manufactured by Macho Products.

Test Parameters

Tab. 115 provides a summary of the blows and equipment tested. Punches were delivered using bare hands, karate-style 'safety- chop' head protectors, and 10-ounce boxing gloves. The kicks were performed with bare feet and when wearing 'safety-kicks.' Each punch was delivered to both the side and front of the 'head.' The combination of six punches and four kicks was repeated with the 'head' positioned at the lower and the higher height to simulate striking a crouching and erect opponent, respectively.

TABLE 115
Test series of style, protective wear, and dummy height. (From SCHWARTZ M L, et al., 1986).

Type of Blow	Hand or Foot Padding	Front of 'Head'	Side of 'Head'
175 cm (opponent standing erect)			
punch	bare fist	R	B
	safety-chop	R	B
	10-oz boxing glove	U	H
kick	bare foot	S	RB
	safety-kick	S	RI
125 cm (opponent crouching)			
punch	bare fist	R	B
	safety-chop	R	B
	10-oz boxing glove	U	H
kick	bare foot	S	RB
	safety-kick	S	RI

* Punches: R = reverse; U = uppercut; B = back fist; H = hook; Kicks: S = sidethrust; RB = roundhouse; striking with ball of foot; RI = roundhouse, striking with instep.

The sagittal and coronal accelerometer outputs were summed to provide the resultant acceleration. The maximum resultant acceleration was determined and used in the statistical analysis. The generalized linear model (GLIM) statistical package was employed.

Results

Eleven karateka produced twenty 'best effort' tracings each and three produced sixteen tracings each, providing 268 oscilloscope tracings for analysis. Fig. 357 shows a typical (rather than maximal) acceleration curve caused by a punch to the side of the 'head.' Peak acceleration is indicated on the vertical axis with time in milliseconds on the horizontal axis. Peak acceleration in this example is 44 g. Fig. 358 shows acceleration of the dummy, generated by a kick to the side of the 'head' administered by the same subject. Peak acceleration is lower in this case for the kick but acceleration continues for a greater length of time.

When subjects were able to kick the dummy at both the higher and the lower heights, the height of the dummy had no effect on the force of blows landed as there was no significant difference between the maximum acceleration generated at the 'head' in the lower and in the higher positions ($p > 0.25$), regardless of the type of blow (kick or punch). Three karateka were unable to kick the dummy in the higher position. Punches to the side of the 'head' using the back of the fist (with or without safety-chop gloves) produced greater peak acceleration on the average than kicks to either the side or the front of the 'head.' Punches to the side of the 'head' produced greater peak accelerations than punches to the front ($p < 0.005$). There was no significant difference between kicks to the side and front of the 'head' ($p < 0.025$). In summary, on the average, for bare or karate-style equipment-protected (safety-chop and safety-kick) hand and foot blows, punches to the side of the 'head' produced greater peak accelerations than did kicks to the front and side, which in turn produced greater

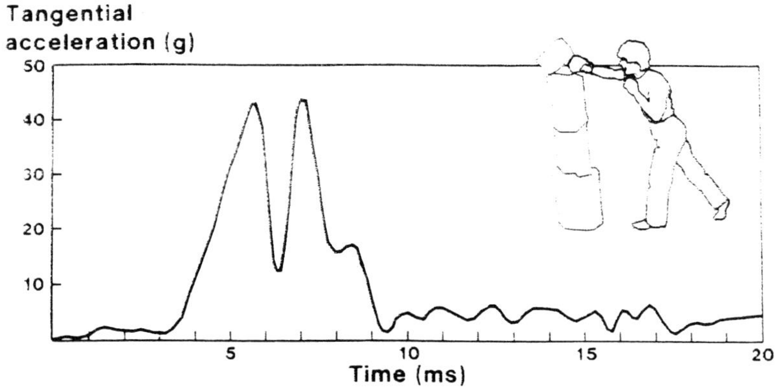

Fig. 357. Acceleration curve of Blow 4 from Subject 1: a back hand punch to the side of the 'head' of he dummy positioned at standing height (175 cm). (From SCHWARTZ ML et al., 1986).

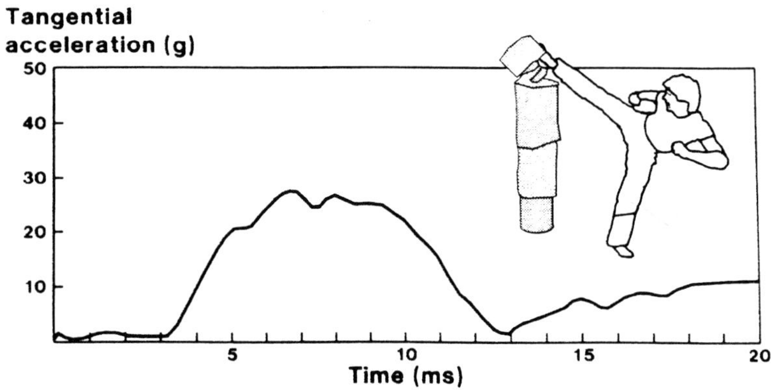

Fig. 358. Acceleration curve of Blow 9 from Subject 1: a bare-foot kick to the side of the 'head' of the dummy at standing height (175 cm). (From SCHWARTZ ML et al., 1986).

acceleration than did punches to the front of the 'head.'

Safety-chop karate-style protective hand-wear did not significantly alter the peak acceleration of the 'head' ($p > 0.1$) but 10-ounce boxing gloves did significantly reduce the peak acceleration of punches ($p < 0.01$). In contrast, safety-kick protective footwear resulted in increased peak acceleration values ($p < 0.01$). It should be noted that there was a difference in the style of kicks to the side of the 'head' with and without the protective footwear, as barefoot blows were struck with the ball of the foot and padded kicks werde struck with the instep.

The analysis of the data in the study of SCHWARTZ et al. (1986) was limited to a consideration for peak accelerations. UNTERHARNSCHEIDT and SELLIER (1971) had bandaged accelerometers to the head of physical education students, unskilled in boxing, and accelerations up to 100 g have been measured when their opponents wore 6-ounce boxing gloves. Sixteen-ounce boxing gloves reduced peak accelerations to 50 g or less. In the study of SCHWARTZ et al. (1986) accelerations of 90 g were recorded on several occasions, and one value of 120 g was recorded.

The findings of SCHWARTZ et al. (1986) show that karate-style safety-chops and safety-kick protectors failed to mitigate peak accelerations. "With the caveat that kicks to the side of the 'head' with and without padding differed slightly in style, it would appear that the karate-style protective hand- and footwear was perceived by the subjects in the study as protection for the wearer rather than for the opponent." The stiffness of the dummy's 'neck' was designed by the authors to simulate that of a conscious, resisting subject, but UNTERHARNSCHEIDT and SELLIER (1971) had described how a boxer often becomes groggy after a series of blows with the result that there is a decrease in muscle tone of the neck permitting the head to move more like a pendulum, with greater rates of acceleration.

An analysis of the kicking force of karate experts was developed by Feld et al. (1979). When kicking a ball, kicks at levels closer to the ground were more powerful. "This, kicking the head of a falling opponent might be expected to be more dangerous than striking him at the crouching or standing height, especially if he has momentarily lost muscle tone in the neck muscles." SCHWARTZ et al. (1986) correctly concluded that if full-contact karate is widely practiced, inevitable cases of kickboxer's encephalopathy will soon be reported.

Afterword

It is not for us to theorize about whether boxing should exist as it is, or not at all. No matter what happens in the future of boxing, all decisions should be made with benefit of the medical facts. We have merely placed the full body of current medical evidence between two covers, for the benefit of those who are interested to know about the medical ramifications of boxing.

Decades of scientific investigation and reflection have lead us to inescapable conclusion that significant head blows and impacts cause pathological alterations in brain tissue. Even a cursory glance at the vast body of medical knowledge shows trauma, especially repeated trauma, to the brain or its enveloping structures is well known to cause various types and degrees of brain injury. There is little we can do to change this well established fact. Protective devices are very limited in their ability to ameliorate the effects of brain trauma. Once trauma has been inflicted, medical practitioners can only mitigate the effects of the injury, often with unsatisfactory results. The damage to cellular structure in the brain and to its circulatory system cannot be undone. From a medical point of view, the only way to prevent brain damage is to prevent the trauma that causes it.

Those who participate in boxing should at least be familiar with the potential kinds of injuries, so that they can make decisions accordingly. Those who are responsible for boxers should also be familiar with the first signs of injury, so that appropriate medical care can be administered as quickly as possible. Without acknowledging the medical facts and making them known, the many tragic deaths and catastrophic injuries recounted in these pages will needlessly continue to occur.

Bibliography

Ababarnel J, Benet AE, Lask D, Kimche D: Sports hematuria. J Urol 143:887-890, 1990

Adams CWM, Bruton CJ: The cerebral vasculature in dementia pugilistica. J Neurol Neurosurg Psychiatr 52:600-604, 1989

Aichmair H: Verletzungen am Auge beim Boxen. In Arzt und Boxsport. Vorträge der Enquete anläßlich der Generalversammlung. Selbstverlag, Wien, November, 1956, pp. 11-13

Aichmair H: Augenverletzungen beim Sport. Münch Med Wochenschr 99:1350, 1957

Ala TA, Yang KH, Sung JH, Frey WH: Hallucinations and signs of Parkinsonism help distinguish patients with dementia and cortical Lewy bodies from patients with Alzheimers's disease at presentation: a clinico-pathological study. J Neurol Neurosurg Psychiatr 62:16-21, 1997

Albaugh CM: Eye problems in boxing. J Internat Coll Surg 17:191-194, 1952

Alcena V: Boxing and the transmission of HIV (letter). New York State J Med 88:392, 1988

Algeo JH: Letter to the editor. Silent serious neurological sequelae among competitive boxers. Mil Med 153:538-539, 1988

Aljamo A: Considerazioni cliniche su alcuni aspetti di traumatologie oculare nel boxer. Arch Ottalmol 61:73-80, 1957

Allegre GE, Wertheimer P: Investigación de las lesiones encefálicas del boxeador. Bol Soc Cir (Uruguay) 29:216-217, 1958

Alonso MR: Traumatic telescoping of nasal bones into the base of the skull. Ann Otol 79: 378-381, 1970

Amelar RD, Solomon C: Acute renal trauma in boxers. J Urol 72:145-148, 1954

Anders G, Felten R, Kirsch A (eds): Boxen und Gesundheit. Zur Frage von Langzeitschäden und ihrer Verhütung. Bundesinstitut für Sportwissenschaft, Schriftenreihe Medizin. Bd 3. Deutscher Ärzte Verlag, Köln, 1977

Annas GJ: Boxing: Atavistic spectacle or atavistic sports? Am J Publ Health 73:811-812, 1983

Aoki N: Brain damage from boxing (letter). J Neurosurg 64:829, 1996

Aoki N: Brain damage from boxing (letter). J Neurosurg 66:153-154, 1987

Ardito, R: Osservatione elettroencefalografiche e cliniche in pugili dilettanti. Riv Neurol 24: 731-738, 1954

Ardito R: Le alterazioni cranio-cerebrali nei pugili dilettanti. Riv Neurobiol 1:79-105, 1955

Ardito R, Fontanari D, Mazzanti L: Traumatische Amnesie bei Boxern. Ärztl Praxis 26: 2111-2114, 1974

Argento M: La medicina in soccorso del pugilato. Riforma Med 49:348, 1933

Arp LC: Boxing, discussion remark. J Am Med Ass 249:2779, 1983

Arthur KE: Tomografia computarizada des cerebro en boxeadores profesionales. (Spanish). Rev Med Panama 7:196-203, 1982

Artiola i Fortuny, Briggs M, Newcombe F, Ratcliff G, Thomas: Measuring the duration of post-traumatic amnesia. J Neurol Neurosurg Psychiatr 43:377-379, 1980

Aschner B: Ueber einen bisher noch nicht beschriebenen Reflex vom Auge auf Kreislauf und Atmung: Verschwinden des Radialispulses bei Druck auf das Auge. Wien Klin Wochenschr 21:1529-1530, 1908

Ask-Upmark E: Carotid sinus and the cerebral circulation: Anatomical, experimental and clinical investigation, including observation on rete mirabile caroticum. Acta Psychiat (Suppl) 5:1-374, 1935

Atha J, Yeadon MR, Sandover J, Parsons KC: The damaging punch. Br Med J 291:1756-1757, 1985

Aubineau: Injections sous-conjunctivales et réflexe oculo-cardiaque. Rev Neurol 36:1156, 1920

Axenfeld H: Zur Frage der zentralen Komponente des hypersensitiven Sinus caroticus. Z Kreislaufforsch 42 55-464, 1953

Axthelm Pete: Don't count boxing out. Newsweek, December 20, 1982, p. 60

Ayella AS: Trauma in boxing. Fact or fable? Pennsylvania Med J 66:15-21, 1963

Baetzner W: Zur Verhütung der Boxschäden der Jugend. Z Ärztl Fortbild (Jena) 36:54-55, 1939

Bailey BN: Maxillo-facial Injuries. In: Bass AL, Blonstein JL, James RD, Williams JGP (eds) Medical Aspects of Boxing. Pergamon Press, London New York, 1965, pp. 29-34.

Bailey JH: The oculocardiac reflex: Report of a case exhibiting a marked reaction following enucleation of the eyeball. Am J Ophthalmol 18: 22-25, 1935

Balabane JM: Contribution a l'etude de épilepsie dite cardiaque

(réflexe oculocardiaque) en qualità de provocateur d' une crise convulsive. Rev Neurol 64:976-981, 1935
Balice G: Il pugilato professionistico sotto accusa. Rass Clin Sci 40:173-183, 1964
Balk A: Should boxing be abolished? Todays Health 43: 18-23, 1965
Bank J: Death and damage in the ring. The case against boxing. St. Anthony Messenger, March 1986, pp. 9-1
Baracu D: Ocular injuries in sport. Rev Chir Oncol Radiol 28:153-157, 1984
Barnes Harold, edit. by Silverman Al: Let's abolish boxing. Saturday Evening Post, 231, October 25, 1958, p. 42
Barnes L: Boxing: more death, more precautions (brief reports) Phys Sportsmedicine 8:20, 1980
Barré JA: Appareil pour la recherche du réflexe oculo-cardiaque. L' oculo-compresseur à ressorts. Paris Méd 42:174-175, 1921
Barré JA, Crusem L: Nouvelles recherches sur le réflexe oculocardiaque normal. Ann Méd 10:303-316, 1921
Barré JA, Crusem L: Etat du réflexe oculo-cardiaque chez les bulbaires et pseudobulbaires. Presse Méd II:1033, 1923
Bartholomew AA: Doctors and boxing contests (letter). Med J Austral 147:212, 1987
Bartsch (1914) quoted after Pöschl M, Krieger G (1963)
Bass AL, Blonstein JL, James RD, Williams JGP (eds): Medical Aspects of Boxing. Pergamon Press, Oxford London New York, 1965
Battalia JE, Aschenbrenner CA, Bennett DR, Friedlander IR, Gado MH, Novich MH, Patterson RH, Bell JA (eds): Brain injury in boxing. Council on Scientific Affairs. American Medical Association. J Am Med Ass 249:254-256, 1983
Baumann JA: L'innervation de l'articulation temporo-mandibulaire et de son ménisque, et les zones réflexogenes du knockout. Bull Acad Nat Med 139:55-56, 1955
Bay E: Traumatische Dämmerzustände. Med Welt :2571-2573, 1968
Beaussart M: Etude électroencéphalographique systématique d'un groupe homogène de 100 jeunes boxeurs amateurs. Rev Neurol 100:416-422, 1959
Beaussart M, Beaussart-Boulenge L: Etude 'expérimentale' de la commotion cérébrale par examens clinique et EEG, avant et immédiatement après combat, chez 123 boxeurs amateurs. Rev Neurol 122:397, 1959
Beaussart M, Beaussart-Boulenge L: Enregistrement électroencéphalographique effectué immédiatement apres traumatisme cranien chez 123 boxers examinés avant et immédiatement après combat. Ann Med Leg 41:426-431, 1961
Beaussart M, Beaussart-Boulenge L: 'Experimental' study of cerebral concussion in 123 amateur boxers, by clinical examination and EEG before and immediately after fights. Electroencephal Clin Neurophysiol 29:529-530, 1970
Beaussart M, Decoulx J, Beausbart L: Study of commotio cerebri by EEG and clinical examination in young amateur boxers. Electroencephal Clin Neurophysiol 14:789, 1962
Beaussart M, Niquet G, Beaussart-Boulenge L: Modifications de l'EEG enregistré immédiatement après combat chez les boxeurs. Rev Neurol 100:392-393, 1959
Beaussart M, Niquet G, Gaudier E, Guislain F: L'EEG des boxeurs examinés immédiatement après combat. Etude comparative avec l'EEG enregistré avant combat dans 52 cas. Rev Neurol 101:422-427, 1959
Beaussart M, Niquet G, Gaudier E, Guislain F: L 'électroencéphalogramme enrigistré immédiatement apres traumatisme crânien dit benin. Etude comparative avec l'examen, effectué avant combat, chez 52 boxeurs amateurs. Presse Méd 68:913-914, 1960
Becher F: Untersuchungen an Spechten zur Frage der funktionellen Anpassung an die mechanische Belastung. Z Naturforsch 8:192-203, 1953
Beck O: Veränderungen an den Hand- und Ellenbogengelenken von Leistungsboxern. Beitr Orthop Traumatol 12:656-658, 1965
Becker B: Das stumpfe Schädeltrauma als Sportunfall. Monatschr Unfallheilk 62:179-186, 1959
Beckett M: Shadow boxing (interview by Daniel Payne). Nurs Times 91:14-15, 1995
Behler Aloys: Schon einmal starb Olympia. Die Welt der Literatur 8: No. 7, 1971, p. 3
Behler Aloys: Hiebe nach allen Regeln. Die Zeit, Nr. 2, 5. Januar 1979
Behrman S, Knight G: Carotid sinus epilepsy. Brit Med J 29:522-1524, 1956
Bellotti P: Fights to live, fights to die (editorial). Recenti Progr Med 87:575, 1996
Bellotti P, Chiarella F, Domenicucci S, Lupi G, Vecchio C: Myocardial contusion after professional boxing match. Am J Cardiol 69:709-710, 1992
Belham GJ, Adler M: Pneumothorx in a boxer. Brit J Sports Med 19:45, 1985
Benda C: Ältere Stadien von Gehirn- und Rückenmarksverletzungen. In: Aschoff L (ed) Pathologische Anatomie. Bd. 8. In: Schjerning O von (ed) Handbuch der ärztlichen Erfahrungen im Weltkrieg 1914/1918. Barth, Leipzig, 1921, pp. 404-414
Bender MB: Syndrome of isolated episode of confusion with amnesia. J Hillside Hosp 5: 212, 1956
Benedict JV, Harris EH, Rosenberg DU von: An analytical investigation of the cavitation hypothesis of brain damage. ASME publication 70 BHF – 3, 1970
Benes V: Death as a result of boxing. (Czech with Engl. summary). Csl Neurol (Prag) 19: 167-170, 1956
Bengtson Hermann: Die Olympischen Spiele in der Antike. Artemis, Zürich, 1971
Benon R: Les troubles 'nerveux' chez les professionnels de la boxe. Prog Med (Paris) 78: 583, 1950
Benson DF, Geschwind N: Shrinking retrograde amnesia. J Neurol Neurosurg Psychiatr 30:539-544, 1967
Berg K: Ein Faustschlag bewirkt Schädelbruch und tödliche Meningealblutung. Deutsch Z Ges Gerichtl Med 21:226-230, 1933
Bergleiter R, Fekas L: Das Cavum septi pellucidi und Cavum Vergae in Klinik und Röntgenbild. Fortschr Neurol Psychiatr 32:361-399, 1954

Bergleiter R, Jokl E: Hirnschädigungen durch Boxsport. Zentralbl Neurochir 16:28-43, 1956
Bernards JA: The orbitocirculatory reflex. I. Effects of pressure on the eyes on the circulation in the dog. Arch Intern Physiol 68:761-773, 1960
Bernstein H: More on boxing (letter). J Am Med Ass 253:2830, 1985
Berlin I: Four Essays on Liberty. In: Two Concepts of Liberty. Oxford University Press, London, 1969, pp. 118-172
Berthold H: Untersuchungen über die Auswirkungen des Boxtraumas auf den Vestibulärapparat. Med Dissertation, Universität Leipzig, 1957
Betkowski A: Hearing loss in boxers and pneumatization of the mastoid bones. (Polish). Otolaryngol Pol 22:99-300, 1968
Betti CO, Ottino CA: Pugilistic encephalopathy. Acta Neurol Latino Amer 15:47-51, 1969
Betti CO, Franke E, Ottino CA, Mendefa NB, Hana VJ: Traumatic encephalopathy. Prensa Med Argent 57:976-979, 1970
Biernakiewicz T: Boks w swietle medycyny. (Polish). Kult Fiz 15:821-825, 1962
Birkmayer W: Über die Gehirnschädigung beim Boxen. In: Arzt und Boxsport. Vorträge der Enquete anläßlich der Generalversammlung. Selbstverlag, Wien, 1956, pp. 9-10
Blagojevich M, Erak D: Eye injuries in boxers. (Serbo-Croatian). Med Glas 20:334-337, 1966
Blitman HM: I know what it means to be punch-drunk. Saturday Evening Post, 227, March 5, 1955, p. 30
Blonstein JL: Boxing injuries. Practitioner 182:515-517, 1959
Blonstein JL: Medical aspects of boxing. Transact Med Soc (London) 76:17-26, 1960
Blonstein JL: The use of 'buccal varidase' in boxing injuries. Practitioner 185:78-79, 1960
Blonstein JL: Enquiry on medical aspects of amateur boxing. J Sport Med 2:180-184, 1962
Blonstein JL: Neurological disease in boxers. Lancet II:937, 1963
Blonstein JL: Some psychological observations on amateur boxers. J Sport Med 4:247, 1964
Blonstein JL: Control of swelling in boxing injuries. Practitioner 193:334, 1964
Blonstein JL: Der Schutz für den Boxsportler. Medizin und Sport 5:192-193, 1965
Blonstein JL: Medical Care of the Boxer. In: Bass AL, Blonstein JL, James RD, Williams JGP (eds) Medical Aspects of Boxing. Pergamon Press, Oxford London New York, 1965, pp. 93-98
Blonstein JL: Sport and medicine. Medical aspects of amateur boxing. Proc Royal Soc Med 59:649-652, 1966
Blonstein JL: Boxing injuries. J Royal Coll Gener Pract 18:100-103, 1969
Blonstein JL: Letter: Traumatic encephalopathy in a young boxer. Lancet II:1213, 1974
Blonstein JL, Clarke E: The medical aspects of amateur boxing. Br Med J II:1523-1525, 1954
Blonstein JL, Clarke E: Further observations on the medical aspects of amateur boxing. Br Med J I:362-364, 1957
Bochnia M: Injuries of the ear in boxers. (Polish). Otolaryngol Pol 39:415-420, 1985
Bock WJ: Kinetics of the avian skull. J Morphol 114:1-41, 1964
Bodechtel G, Spatz R: Amnestische Episoden – traumatische Dämmerzustände – psychomotorische Dämmerattacken. Zur Differentialdiagnose kurzdauernder Bewußtseinsstörungen, Münch Med Wochenschr 113:1333-1338, 1971
Bodensteiner JB, Schaefer GB: Dementia pugilistica and cavum septi pellucidi: Born to box? Sports Med 24:361-365, 1997
Böhm E: Boxunfälle und Schutzbestimmungen. Sportmedizin 8:330-334, 1957
Boehmig AG: Über Boxverletzungen am Auge. Sportmedizin 5:215-216, 1954
Bogdanoff B, Natter HM: Incidence of cavum septi pellucidi in adults: A sign of boxers encephalopathy. Neurology 39:991-992, 1989
Boigey M: Lésions de la face chez les boxeurs. Presse Méd 28 (annexe):444-446, 1920
Boigey M: Lésions et traumatismes sportifs. La boxe. Chap IX. Masson, Paris, 1938, pp. 87-116
Boitano M: Cervical spine in boxing. Unusual ligamentous injuries in Olympic boxer and review of the literature. Hung Rev Sports Med 27:175-189, 1986
Boje O: Chronic brain lesions due to boxing. Ugesk Laeger 101:807-809, 1939
Bolanos Cacho G: Fracturas nasales en los boxeadores. Cir y Cir 32:138-139, 1964
Bollmann F, Clasing D, Wannenmacher M: Entwicklung eines individuell geformten Boxermundschutzes. Deutsch Zahnärztl Z 28:80-86, 1973
Borgogne E, Re F, Re G, Tortini S: Maxillofacial injuries in sports: boxing. Minerva Stomatol 33:203-207, 1984
Boshoff PH, Jokl E: Boxing injuries of the eyes. Arch Ophthalmol 39:643-644, 1948
Bouras C, Giannakopoulos P, Good PF, Hsu A, Hof PR, Perl DP: A laser microprobe mass analysis of brain aluminium and iron in dementia pugilistica: comparison with Alzheimer's disease. Eur Neurol 38:53-58, 1997
Bourguignon A: La cranioencéphalopathie traumatique des boxeurs (Données nouvelles). Semaine Hop (Paris) 26:301-302, 1950
Bourrat L, Micolier R: Remarques sur l'encéphalite des boxeurs et l'évolution de certaines traumatismes crâniens. J Med (Lyon) 25:583-590, 1944
Bousseljot W: Zum Problem der Schädigung des Zentralnervensystems durch den Boxsport. Z Militärmed 10:22-27, 1969
Bousseljot W: Die ärztliche Verantwortung bei der Betreuung von Boxsportlern. Ein kasuistischer Beitrag aus neurologisch-psychiatrischer Sicht. Medizin und Sport 10:45-49, 1970
Bowden SC, Walsh KW : Boxing: time for action (letter). Med J Austral 142:282, 1985
Boyle RB, Ames W: Boxing: A special report. Too many punches, too many little concern. Sports Illustrated, 58:44-82, 1983
Braham J, Wolman M: Subpial siderosis of the central nervous system. Acta Neuropathol 74:559-562, 1965

Brain R: Discussion remark to McAlpine D, Page F. Midbrain syndrome in professional boxers. Proc Royal Soc Med 42:793, 1949
Brain WR: The cerebral basis of consciousness. Brain 73:465-479, 1950
Brain WR: The physiological basis of consciousness. A critical review. Brain 81:426-455, 1958
Braine J, Ravina A: Les lésions traumatiques des boxeurs. Presse Méd 31:849-854, 1923
Bramann C von: Lebensgefahr im Kampfsport. Münch Med Wochenschr 7:634-636, 1927
Brandenburg W, Hallervorden J: Dementia pugilistica mit anatomischem Befund. Virchows Arch Pathol Anat 325:680-709, 1954
Brayne CEG, Dow L, Calloway SP, Thompson RJ: Blood creatine kinase isoenzyme BB in boxers. Lancet II:1308-1309, 1982
Brayne H, Sargeant L, Brayne C: Could boxing be banned? A legal and epidemiological perspective. Br Med J 316:1813, 1998
Bregeat P: Dangers encourous par la pratique de la box. (French with English summary). Bull Acad Nat Med 168:112-119, 1984
Breig A: Biomechanics of the Central Nervous System. Some Basic Normal and Pathologic Phenomena. Almquist & Wiksell, Stockholm, 1960
Breig A: Overstretching of and circumscribed pathological tension in the spinal cord as a basic cause of symptoms in cord disorders. J Biomech 3:7, 1970
Breitner B: Sportschäden und Sportverletzungen. Der Faustkampf. Kap 7. Neue Deutsche Chirurgie. Bd. 58. Enke, Stuttgart, 1953, pp. 71-79
Brennan TNN, O'Connor PJ: Incidence of boxing injuries in the Royal Air Force in the United Kingdom, 1953-66. Br J Industr Med 25:326-329, 1968
Breton F, Pincemaille Y, Tarriere C, Renault B: Event-related potential assessment of attention and the orienting reaction in boxers before and after a fight. Biol Psychol 31:57-71, 1991
Britt J, Dalsi : La boxe anglaise et francaise par les champions de ring. Lafitt, Paris, 1912
Brkovich I: Die Organisation des Gesundheitsdienstes im Boxen in Jugoslawien. Medizin und Sport 5:175-176, 1965
Brock J: Ein neuartiger Boxerschutz zur Prophylaxe von Verletzungen im Zahn-, Kiefer- und Kiefergelenkbereich. Deutsch Med Wochenschr 78:982-983, 1953
Brock J: Boxsportverletzungen im Bereich der Zähne und ihre Verhütung durch ein neues Boxerschutzgerät für das Gebiß. Zahnärztl Welt (Konstanz) 9:97-104, 1954
Bronte RM: Boxer's death in the ring. J Am Med Ass 84:1284, 1925
Brooks DN: Long and short term memory in head injured patients. Cortex II:329-340, 1975
Brooks N, Kupshik G, Wilson L, Galbraith S, Ward R: A neuropsychological study of active amateur boxers. J Neurol Neurosurg Psychiatr 50:997-1000, 1987
Brown JW: Report to the Faculty on the Study of Intercollegiate Boxing at the University of Wisconsin. Document 959, Mimeographed, October, 1950
Bruce DA, Schut L, Sutton LN: Brain and cervical spine injuries occurring during organized sports activities in children and adolescents. Clin Sports Med 1:495-516, 1982
Buck A: How to pack them in. Ring, April 1951, p. 19
Brun A, Passant U: Frontal lobe degeneration of non-Alzheimer type: structural characteristics, diagnostic criteria and relation to other fronto-temporal dementias. Acta Neurol Scand Suppl 168:28-30, 1996
Bucik M, Majdevac Z, Mitrovic R: Fracture of the hyoid bone with laryngeal changes by boxing injuries. (Serbo-Croatian). Srp Arh Celok Lek 114:1075-1081, 1986
Burger H, Minarovjech V: Ergebnisse komplexer sportmedizinischer Untersuchungen an Boxern. Medizin und Sport 5:69-74, 1965
Burger H, Minarovjech V: Sind die Regeln der AIBA für den Gesundheitsschutz der Boxer ausreichend? Medizin und Sport 6:12-15, 1966
Burger H, Welsch G: Über Schocktodesfälle im Boxsport - vermeidbar durch moderne Wiederbelebung. Medizin und Sport 5:186-188, 1965
Burns RJ: Boxing and the brain. New Zealand J Med 16:439-440, 1986
Burri R: Ein Bildbericht. Boxen in Thailand. Atlantis 26:826-837, 1966
Busch KT: Augenärztliche Gesichtspunkte in der augenärztlichen Betreuung von Boxsportlern. Medizin und Sport 6:180-182, 1965
Busse EW, Silverman AJ: Electroencephalographic changes in professional boxers. J Am Med Ass 149:1522-1525, 1952
Butler RJ: Neuropsychological investigation of amateur boxers. Br J Sports Med 28:187-190, 1994
Butterworth JS, Poindexter CA: An electrocardiographic study of the effects of boxing. Am Heart J 23:59-63, 1942
Buttery CMG: Boxing in the Army. To the editor. J Am Med Ass 262:2089-2090, 1989
Byrne A: Neurological injuries in boxers: Lessons from one family. J Am Med Ass 265: 2671-2672, 1991
Byrom F: Head Injuries. In: Bass AL, Blonstein JL, James RD, Williams JGP (eds) Medical Aspects of Boxing. Pergamon Press, Oxford London New York, 1965, p. 73
Cabanis EA, Perez G, Tamraz JC, Iba-Zinzen MT, Roger B, Alfonso JM, Rougemont D: Cephalic magnetic resonance imaging of boxers. Preliminary results. Acta Radiol (Suppl) 369:365-366, 1986
Caldwell M: Letter to the editor. J Am Med Ass 253:199, 1985
Candian FL: Sulle lesioni oculari a carettere professionale dei pugilatori. Arch Antropol Crim 57:647-652, 1937
Canetti E: Crowds and Power. New York, 1966, p. 28
Cantu RC: When to return to contact sports after a cerebral concussion. Sports Med Dig 10: 1-2, 1988
Cantu RC: Head Injury in Sports. In: Grana WA, Lombardo JA (eds) Advances in Sports Medicine and Fitness. Vol 2. Mosby Year Book, St. Louis, 1988
Cantu RC: Second impact syndrome: immediate management. Phys Sportsmed 20:55-66, 1992
Cantu RC (ed): Boxing and Medicine. Human Kinetics, Champaign, ILL, 1995

Cantu RC: Medicine's Stand on Boxing in the 20th Century. Chap. 1. In: Cantu RC (ed) Boxing and Medicine. Human Kinetics, Champaign, ILL, 1995, pp. 3-8
Cantu RC: Brain Injuries. Chap. 2. In: Cantu RC (ed) Boxing and Medicine. Human Kinetics, Champaign, ILL, 1995, pp. 19-32
Cantu RC: Guidelines for Return to Contact Sports after Cerebral Concussion. Phys Sportsmed 14:76, 1986
Cantu RC: How to make Professional Boxing Safe? Chap. 11. In: Cantu RC (ed) Boxing and Medicine. Human Kinetics, Champaign, ILL, 1995, pp. 187-195
Carnall D, Warden J: Tighter medical control proposed for boxing (news). Br Med J 311:1183, 1995
Carp G: Boxing - the hazards and their prevention. Nurs Times 70:1738-1740, 1974
Carr JL, Moody AM: Boxer's hemorrhage. California West Med 51:227-230, 1939
Carroll EJ: Punch-drunk. Am J Med Sci 191:706-711, 1936
Carter JB, Parke DW: Unusual retinal tears in an amateur boxer. Arch Ophthalmol 105:1138, 1987
Cassasa CB: Multiple traumatic cerebral hemorrhages. Proc New York Pathol Soc 24:101, 1924
Casson IR, Sham R, Campbell EA, Tarlau M, Di Domenico A: Neurological and CT evaluation of knocked-out boxers. J Neurol Neurosurg Psychiatr 45:170-174, 1982
Casson IR, Siegel O, Sham R, Campbell EA, Tarlau M, Di Domenico A: Brain damage in modern boxers. J Am Med Ass 251:2663-2667, 1984
Castello R: Criteri e limiti dell audiologia nell accertamento del danno auditivo prodotto da trauma. Zentralbl Hals- Nasen-Ohrenheilk 53:247, 1955
Cavalazzi D: Le lesioni da boxe sotto il rispetto medico legale. Arch Antropol Criminol 57: 683-688, 1937
Cavanagh S: The true 'boxers' fracture? Injury 23:204-205, 1992
Cermak J, Bilek F: The importance of radiographic examination of the hand and wrist in young pugilists. (Czech with Engl. a. French summary). Cas Lek Cesk 104:151-154, 1965
Cerretti A, Montanardo M: Importance dello studio audiometrico nei pugili. Medicina Sportiva (Torino) 9:538-548, 1967
Chamouard F, Trosseille X, Pincemaille Y, Tarriere C: Methodological Aspects of an Experimental Research on Cerebral Tolerance on the Basis of Boxers' Training Fights. Proc 31th Stapp Car Conf, Soc Automot Engin, Warrendale, PA, 1987, pp. 15-28
Chapman PJ: Orofacial injuries in the use of mouthguards by the 1984 Great Britain Rugby League touring team. N J Sports Med 19:34-36, 1985
Charnas L, Pyeritz RE: Neurologic injuries in boxers. Hosp Pract 21:30-31 and 34-39, 1986
Charnas L, Pyeritz RE: Neurological injuries in boxers: The use of tests. J Am Med Ass 264:1532-1533, 1990
Chaudhuri KR, Lemmens G, Williams SC, Leigh PN: Magnetic resonance spectroscopic study of parkinsonism related to boxing (letter), J Neurol Neurosurg Psychiatr 59:561-562, 1995
Chen KM: Disappearence of ALS from Guam: implications for exogenous causes. (Japanese). Rinsho Shinkeigaku 35:1549-1553, 1995
Chenoweth LB: Medical supervision of boxing. Ohio State Med J 47:913-914, 1951
Chodkiewicz JP: La surveillance electro-encephalographique des boxeurs. Med Educ Phys Sport 41:40-42, 1967
Christenberry RK: Correspondence. Boxing in New York. J Am Med Ass 151:317-318, 1953
Civiero G, De Novellis M, Lubich T, Randi V: Cranio-cerebral rheography (RGC) in periodic evaluation of the pugilist and in early ascertainment of post traumatic encephalopathy. Medicina Sportiva (Torino) 2:99-109, 1967
Clarke P: Boxing or judo? J Royal Nav Med Serv 53:109-112, 1967
Clarke TD, Gragg CD, Sprouffske JF, Trout EM, Zimmerman RM, Muzzy WH: Human Head Linear and Angular Acceleration During Impact. In: Backaitis H (ed) Biomechanics of Impact Injury and Injury Tolerances of the Head-Neck Complex. PT-43. Soc Autom Engin, Warrendale, PA, 1993, pp. 87-104
Claude H. Cuel J: Démence présénile post-traumatiques apres fracture du crane. Considérations médico-legales. Ann Méd Leg Criminol 19:173-184, 1939
Cohen L: Should the sport of boxing be banned in Canada? Can Med Ass J 130:767-768, 1984
Cohen RR, Eliezri YD, Silvers DN: Sports Med 10:198-203, 1990
Colla C: Lesivita nello sport del pugilato. Arch Antropol Crim 57:689-693, 1937
Colmant HJ, Dotzauer G: Analyse eines tödlich ausgegangenen Boxkampfes mit ungewöhnlich schweren cerebralen Schäden. Z Rechtsmed 84:263-278, 1980
Colosi HJ, Yanoff M: Intrusion of scleral implant associated with conjunctival epithelial ingrowth. Am J Ophthalmol 83:504-507, 1977
Conrad H: Sportärztliche Tätigkeit beim Boxkampf in besonderer Berücksichtigung des Berufsboxsportes. Deutsch Med Wochenschr (Beiheft) II:50-51, 1935
Constantinidis J, Tissot R: Lésions neurofibrillaires d'Alzheimer généralisées sans plaques séniles (Présentation d'une observation anatomo-clinique). Schweiz Arch Neurol Psychiatr 100:117-130, 1967
Cope RL, Engelhardt T, Baird VC: The carotid sinus syndrome: its importance in clinical medicine. Section on Medicine. State Medical Association of Texas. Annual Session, Dallas, May 7, 1947, pp. 30-34
Corcum JM: Boxing, reply to a discussion remark of Lundberg GD. J Am Med Ass 253: 198-199, 1985
Corsellis JAN: The pathological anatomy of the temporal lobe with special reference to the limbic areas. In: Price JH (ed) Modern Trends in Psychological Medicine. Vol. 2. Butterworths, London, 1970, pp. 296-325
Corsellis JAN: Brain damage in sport. Lancet I:401-402, 1976
Corsellis JAN: Alzheimer's disease: Senile Dementia and related Disorders. In: Katzman R, Terry RD, Bick KL (eds) Posttraumatic Dementia. Raven, New York, NY, 1978, pp. 125-133

Corsellis JAN: Boxing and the brain. Br Med J 298:105-109, 1989
Corsellis JAN: Boxing and the brain - Correction. Br Med J 298:247, 1989
Corsellis JAN, Brierley JB: Observations on the pathology of insidious dementia following head injury. J Ment Sci 105:714-720, 1959
Corsellis JAN, Bruton CJ, Freeman-Browne D: The aftermath of boxing. Psychol Med 3: 270-303, 1973
Coubertin Pierre de: Der olympische Gedanke. Stuttgart, 1967
Courville CB: Punch-drunk. Its pathogenesis and pathology on the basis of a verified case. Bull Los Angeles Neurol Soc 27:160-168, 1962
Courville CB: The mechanism of boxing fatalities. Report of an unusual case with severe brain lesions incident to impact of boxer's head against the ropes. Bull Los Angeles Neurol Soc 29:59-69, 1964
Cousteau JP: Myocardial contusion during boxing (letter, comment). Am J Cardiol 71:493, 1993
Cowart VS: Boxing makes more headlines than usual and a lot of the news hasn't been good. J Am Med Ass 261:14-16, 1989
Cowart VS: Reply to the letter of Voy RS. J Am Med Ass 262:500, 1989
Creutz W: Tödliche Verletzungen bei Sportwettkämpfen in gerichtsärztlicher Beziehung. Deutsch Z Ges Gerichtl Med 15:433-454, 1930
Creutzfeld HG: Discussionsbemerkungen zu Reisch O, Der Boxkampf und seine Auswirkungen auf das Nervensystem. Berliner Gesellschaft für Neurologie und Psychiatrie, Sitzung vom 12, April, 1937. Zentralbl Neurol Psychiatr 90: 704, 1938
Critchley McD: Nervous disease in boxers. Medical Annual 283-284, 1935
Critchley McD: Nervous disorders in boxers. Medical Annual 318-320, 1937
Critchley McD: Medical aspects of boxing, particularly from a neurological standpoint. Br Med J I:357-362, 1957
Critchley McD: Medical aspects of boxing. Transact Med Soc (London) 76:12-17, 1960
Critchley McD: Head Injuries. In: Bass AL, Blonstein JL, James RD, Williams JGP (eds) Medical Aspects of Boxing. Pergamon Press, Oxford London New York, 1965, pp. 65-67
Croce G , Croce A: I traumi nasali del pugilato: patogenesi e trattamento. (Italian). Acta Otorhinolaryngol Ital 7:185-194, 1987
Cruikshank JK, Higgens CS, Gray JR: Two cases of acute intracranial haemorrhage in young amateur boxers. Lancet I:626-627, 1980
Czekalowa L: Complex traumatic fracture of the fifth cervical vertebra with late spinal complications. Report of a case. (Polish). Pol Przegl Radiol 33:737-740, 1969
Czekalowa L: A case of compound fracture of the 5th cervical vertebra after a boxing injury with late spinal cord complications (Polish with Engl. summary). Pol Rev Radiol Nucl Med 33:650-653, 19
Czermak J: Über mechanische Vagusreizung beim Menschen. Jenaer Z Med Naturwissensch 2:384, 1866
Czerner G: Untersuchungen über die Auswirkungen des Boxtraumas auf das Hörorgan. Med Dissertation, Universität Leipzig, 1957
Dabkowski M: Psychiatric consequences of boxing in light of the case of encephalopathy in a boxer. (Polish). Psychiatr Pol 18:69-72, 1984
Dagnini G: Intorno ad un riflesso provocato in alcuni emiplegici collo stimolo dell cornea e colla pressione sul bulbo oculare. Boll Sci Med 8:380-381, 1908
Dale GE, Leigh PN, Luthert P, et al.: Neurofibrillary tangles in dementia pugilistica are ubiquinated. J Neurol Neurosurg Psychiatr 54:116-118, 1991
Dandy WE: Congenital cerebral cysts of the cavum septi pellucidi (fifth ventricle) and cavum Vergae (sixth ventricle). Arch Neurol Psychiatr 25:44-46, 1931
Danielopulo D, Marcou J, Proca GG: Sur la mécanisme de production des variations du reflex sino-cardidieu a l'etat pathogenique. CR Soc Biol 109:767, 1932
Davie CA, Pirtosek Z, Barber GJ, Kingsley DP, Miller PH, Lees AJ: Magnetic resonance spectroscopic study of parkinsonism related to boxing, J Neurol Neurosurg Psychiatr 58:688-691, 1995
Davis PM, McKelvey MK: Medicolegal aspects of athletic head injury. Clin Sports Med 17:71-82, 1998
Dawson James P: New York Times, December 30, 1949
Dax EC: Prevention of brain damage in boxers. Med J Austr 167:502, 1997
De Castro F: Sur la structure et l'innervation du sinus carotidien de l'homme et des mammiferes. Nouveaus faits sur l'innervation et la fonction du glomus caroticum. Trav Lab Recherch 25:331, 1928
De Coubertin P: The Olympic games of 1896. Century Magazine, 53:39-53
Dedola G: Research concerning reflex activity (Dagnini-Aschner) in relation to cranio-cerebral trauma. Rass Ital Chir Med 6:153-166, 1957
De Gispert Gruz I: Sobre la encefalopatia cronica de los boxeadores. Revista Clinica Espan 11:270-273, 1943
De Haven KE, Lintner DM: Athletic injuries: Comparison by age, sport, and gender. Am J Sports Med 14:218-224, 1986
De Judicibus C: Il meccanismo e gli aspetti audiometrici delle lesioni labyrinthiche da trauma cranio senza frattura. Zentralbl Hals Nasen Ohrenheilk 57:148, 1957
De Kleyn A: Proc Koninglike Nederlande Akademie van Wetenshappen 44:787, 1941
Delapchier R: Essai sur les lesions traumatiques observees chez les boxeurs et plus particulierement sur les fractures et luxations du metacarpe (boxe anglaise). Paris, 1909
Delori F, Pomerantzeff O, Cox MS: Deformation of the globe under high speed impact: Its relation to contusion injuries. Invest Ophthalmol 8:290-301, 1969
De Michelis F: Le alterazioni a carico dell' apparato masticatorio nei pugilatori in attivita, negli ex pugilatori e nei pugilatori 'suonati.' Nova Acta Stomat (Parma) 5:21-28, 19
De Nayer PP: Sollte das Boxen unter Studenten verboten werden? Internat Med Kurier (Berlin) 1:2, 1962
De Nayer PP: A propos du problème médical de la boxe et des

dispositions légales en la mátiere. Hermes (Louvain) 1:9-11, 1967
De Nayer PP: 'K.o-Schlag' für den Boxsport durch Skandinaviens Ärzte? Medical Tribune (Frankfurt) :36-37, 1968
Denny-Brown D: Delayed collapse after head injury. Lancet I :371-375, 1941
Denny-Brown D, Russell WR: Experimental cerebral concussion. Brain 64:93, 1941
De Rougemont RR: Boxing – related injuries in the U.S. Army (letter). Mil Med 154:581, 1989
Desmond S: The Soul of the Crowd. Fry's Magazine, Vol. 16, 1911
Despons J: Le syndrome cochléo-vestibulaire dans le traumatismes crâniens. Zentralbl Hals Nasen Ohrenheilk 54:353, 1955/56
De Treigny M: Observations sur le réflexe oculo-cardiaque spontané. Bull Soc Ophthalmol (Paris) 9:725-728, 1937
De Treigny M: Observations sur le réflexe oculo-cardiaque spontané en ophthalmologie courante. Presse Méd 48:1079-1080, 1940
Deutsch F: Sekundenherztod im Boxkampf durch Commotio cordis (Ein Beitrag zur Frage des Sportherzens). Wien Arch Inn Med 20:279-286, 1928
De Voe AG: Injuries to the eye. Am J Surg 98:384-389, 1959
Diamond GR, Quinn GE, Pashby TJ, Easterbrock WM: Ophthalmologic injury. Clin Sports Med 1: 469, 1982
Dimsdale H: Discussion remark to McAlpine D, Page F: Midbrain syndrome in professional boxers. Proc Royal Soc Med 42:793, 1949
Dixon KC: The amnesia of cerebral concussion. Lancet II:1359-1360, 1962
Doggart JH: Fisticuffs and the visual organs. Transact Ophthalmol Soc (United Kingdom) 71:53-59, 1951
Doggart JH: Blessures à la vision par la boxe. Boll d' Oculist 44:273-284, 1954
Doggart JH: The impact of boxing upon the visual apparatus. Arch Ophthalmol 54:161-169, 1955
Doggart JH: Eye Injuries. In: Bass AL, Blonstein JL, James RD, Williams JGP (eds) Medical Aspects of Boxing. Pergamon Press, Oxford London New York, 1965, pp. 3-7
Doleman FT: Delayed traumatic subarachnoid hemorrhage. J Royal Nav Med Serv 27: 309, 1941
Dombrovskii AA: Significance of the design of oral protectors in the prophylaxis of stomatological trauma in sportsmen engaged in boxing. (Russian with Engl. summary). Stomatologia (Moscow) 53:38-41, 1974
Donner M, Kaste, Kuurne: Impossibility of preventing brain damage in boxing. Prohibit head blows, tighen physical control. (Danish). Laekartidningen 78:2234-2235, 1981
Doret M: Manifestations oculaires du syndrome sino-carotidienne fighting arts. Société Médicale de Genève. Séance du 30. juin, 1949. Referat: Praxis :691, 1949
Doris PE, Johnston CC: Large calcified pelvic mass in a boxer. Am J Sports Med 10:117-121, 1982
Drachman DA: Section of geriatric neurology. Neurology 41:1333-1334, 1991 [Erratum, Neurology 41:1828, 1991
Drachman DA, Newell KL: Presentation of a case (12-1999) in the weekly clinicopathological exercises of the Massachusetts General Hospital. The New Engl J Med 340: 1294–1277
Drachman DA, Swearer JM, O'Donnell BF, Mitchell AL, Maloon A: The Caretaker-Obstreporous-Behavior Rating Assessment (COBRA) Scale. J Am Geriatr Soc 40:463-470, 1992
Drew RW, Templer DI, Schuyler BA, Newell TG, Cannon WG: Neuropsychological deficits in active licensed professional boxers. J Clin Psychol 42:520-525, 1986
Drotman DP: Professional boxing, bleeding, and HIV testing (letter). J Am Med Ass 276:193, 1996
Ducroquet quoted from Iselin M (1960)
Durandy : Etude de trois cas de diplopie consecutifs a la pratique de la boxe de combat. These Med, Universität Paris, 1935
Dyke CG, Davidoff LM: Congenital absence of the septum pellucidum; its diagnosis by encephalography. Am J Roentgenol 34:573-578, 1935
Eade GG: Preventing cauliflower ears. Northwest Med 63:99, 1964
Egan DJ: Boxing can be safe. Med Techn Bull 5:102-104, 1954
Eiselt E: Effects of boxing on physical and mental health. (Czech with Engl. summary). Cas Lek Cesk 87:766-769, 1948
Eisenführ G: Stellungnahme des BDB für den Schutzbestimmungen. In: Anders G, Felten R, Kirsch A (eds) Boxen und Gesundheit. Zur Frage von Langzeitschäden und ihrer Verhütung. Bundesinstitut für Sportwissenschaft. Schriftenreihe Medizin. Bd. 3. Deutscher Ärzte Verlag, Köln, 1977, pp. 37-41
Eisenführ G, Schmidt P, Wittenbrink T: Medizinische Stellungnahme des BDB. In: Anders G, Felten R, Kirsch A (eds) Boxen und Gesundheit. Zur Frage von Langzeitschäden und ihrer Verhütung. Bundesinstitut für Sportwissenschaft. Schriftenreihe Medizin. Bd. 3. Deutscher Ärzte Verlag, Köln, 1977, pp. 41-43
Elia JC: Traumatic headache associated with the profession of boxing. Headache 2:138-146, 1962
Elia JC: Editorial comment. A boxing dilemma. Western Med 4:365, 1963
Elia JC: Progress in boxing safety. Western Med 5:98-100, 1964
Elia JC: Head injuries in boxing. Internat Surg (Chicago) 56:334-343, 1971
Elia JC: Physicians and boxing (letter). West J Med 139:717, 1983
Elkington AR: Boxing and the eye: results of a questionnaire. Transact Ophthalmol UK 104:898-902, 1985
Ellis M: Maxillo-facial Injuries. In: Bass AL, Blonstein JL, James RD, Williams JGP (eds) Medical Aspects of Boxing. Pergamon Press, Oxford London New York, 1965, pp. 23-28
Engel GL: Fainting: Physiological and psychological considerations. Thomas, Springfield, 1950
Engel GL: On the existence of the cerebral type of carotid sinus syncope. Neurology 9: 565-568, 1959
Engel GL, Romano J, Mc Lin TR: Vasodepressor and carotid sinus syncope: Clinical, electrocardiographic and electroencephalographic observations. Arch Intern Med 74:100-119, 1944

Engel H: Über einige Verletzungen beim Boxen. Z Ärztl Fortbild (Jena) 19:655-657, 1922
Engel WK: Abolish boxing (letter). New Engl J Med 307:761, 1982
Engelberg AL: Boxing, discussion remark. J Am Med Ass 249:2777, 1983
Enzenauer RW, Enzenauer RJ: Ban military boxing. Mil Med 152:536-537, 1987
Enzenauer RW, Enzenauer RJ: Boxing in the Army. In reply to Finney JL. J Am Med Ass 262:2090, 1989
Enzenauer RW: The borrowed imagery of boxing (letter). J Am Med Ass 264:1531-1532, 1990
Enzenauer RW: Let's stop boxing in the Olympics and the US military (letter, comment). J Am Med Ass 272:1821, 1994
Enzenauer RW, Mauldin WM: Boxing-related ocular injuries in the United States Army, 1980 to 1985. Southern Med J 82:547-549, 1989
Enzenauer RW, Montrey JS, Enzenauer RJ, Mauldin WM: Boxing-related injuries in the US Army, 1980 through 1985. J Am Med Ass 261:1463-1466, 1989
Enzenauer RW: Let's stop boxing in the Olympics and the US Military, in reply to Lundberg GD. J Am Med Ass 272:1821, 1994
Estwanik JJ: Professional and Amateur Rules and Regulations. Chap. 2. In: Cantu RC (ed) Boxing and Medicine. Human Kinetics, Champaign, ILL, 1995, pp. 9-16
Estwanik JJ: Injuries to the Extremities, Trunk, and Head. Chap. 6. In: Cantu RC (ed) Boxing and Medicine. Human Kinetics, Champaign, ILL, 1995, pp. 79-87
Estwanik JJ, O'Connor FG: The Ringside Physician's Role. Chap. 7. In: Cantu RC (ed) Boxing and Medicine. Human Kinetics, Champaign, ILL, 1995, pp. 89-114
Estwanik JJ, Boitano M, Ari N: Amatur boxing injuries at the 1981 and 1982 USA'ABF Nationnal Championships. Phys Sportsmed 12:123-128, 1984
Evans RW: The postconcussion syndrome and the sequelae of mild head injury. Neurol Clin 10:815-847, 1992
Evered DF: Fencing or boxing. Lancet I:857-858, 1964
Ewing CL, Unterharnscheidt F: Neuropathology and Cause of Death in US Naval Aircraft Accidents. Paper presented at the Aerospace Medical Panel Specialists Meeting held in Copenhagen Denmark, April 5-9, 1976. AGARD Conference Proceedings No. 190, Recent Experience/Advances in Aviation Pathology. December 1976, B 16-1-6
Fandre M: Encephalite traumatique des boxeurs. Punch-drunk. Med. Dissertation, Universität Paris, 1945
Farrow R: Hand Injuries. In: Bass AL, Blonstein JL, James RD, Williams JGP (eds) Medical Aspects of Boxing. Pergamon Press, Oxford London New York, 1965, pp. 43-50
Favory A: Les lésions oculaires dues a la boxe de combat. Presse Méd 45:254-257, 1937
Favory A, Sedan J: Traumatologie oculaire du boxeur. Arch Ophthalmol 11:429-456, 1952
Felc W: Zwei neue Todesfälle beim Boxkampf. (Polish). Ref Zentralbl Neurol 94:595, 1939
Feld MS, Mc Nair RE, Wilk SR: The physics of karate. Scientific Am 240:150-158, 1975
Ferguson FR, Mawdsley C: Chronic encephalopathy in boxers. Proc Vol 1, 8th Internat Congr Neurol, Vienna, September, 5-10, 1965, pp. 81-84
Ferlaino FR: Medical program for the boxing industry. J Am Med Ass 150:651-654, 1952
Ferraro S: Su un case di morte violenta fulminea da trauma sul seno carotico (Contributo alla casietica). Policlinico (Sez Prat) 44:36, 1937
Ferris EB, Capps RB, Weiss S: Carotid sinus syncope and its bearing on the mechanisms of unconscious state and convulsions, study of 32 additional cases. Medicine 14:77-456, 1935
Fiedler H: Entwicklungstendenzen des Boxsports. Medizin und Sport 10:33-35, 1970
Finney LA, Reynolds DH, Yates BM: Comminuted subfrontal fractures. J Trauma 4:711-721, 1964
Finney JL: Boxing in the Army. Letter to the editor. J Am Med Ass 262: 2089, 1989
Fiore DC: Boxing: does the size of the prize affect the drain on the brain? J Am Med Ass 276:954, 1996
Fischer H, Spann W: Pathologie des Traumas. Bergmann, München, 1967, pp. 230-233
Fishbein M: Hazards of boxing. Medical World News, January 23, 1970
Fisher CM: Concussion amnesia. Neurology 165:826-830, 1966
Fleischer N, Andre S: A Pictorial History of Boxing, Citadell Press, Secaucus, NY, 1975
Flood FB: Albuminuria and hematuria in boxers. J Am Med Ass 171:1678-1679, 1960
Födisch HJ, Kloss K: Thrombotische Verschlüsse im Stromgebiet der Arteria carotis nach stumpfen Schädel-Hals-Traumen. Hefte zur Unfallheilkunde, Heft 88. Springer, Berlin Heidelberg New York, 1966
Foerster A: Plötzlicher Tod beim Boxkampf. Monatsschr Unfallheilk 39:441-445, 1932
Fog M: Unconsciousness. A clinicopathological study. Lancet I:443-449, 1958
Foltz EL, Jenkner FL, Ward AA: Experimental cerebral concussion. J Neurosurg 10:342-352, 1953
Ford JC father: Depth Psychology, Morality and Alcoholism. 1950, pp. 123-126
Foster JB, Leiguarda R, Tilley PJB: Brain damage in national hunt jockeys. Lancet I:981-983, 1976
Franceschetti A , Duret M: Symptomatologie oculaire du syndrome sino-carotidien. Rev Med 69 603-612, 1949
Frackelton WH: Facial injuries in sports. Am J Surg 98:390-393, 1959
Fraenkel P: Tod im Boxkampf. Deutsch Z Ges Gerichtl Med 1:481-486, 1922
Franke H: Beitrag zur Klinik und Pathogenese der kardialen Form des gesteigerten Sinus-caroticus-Reflexes. Arch Kreislaufforsch 15:198-223, 1948
Franke H: Zur Pathogenese der kardialen Form des hypersensitiven Sinus-caroticus-Reflexes. Verhandl Deutsch Gesellsch Inn Med 55:595-604, 1949
Franke H: Die Auswirkung des hypersensitiven Karotissinus-Syndroms kardialen Typs mit längeren Herzpausen auf das

Hirnstrombild (EEG) des Menschen. Verhandl Deutsch Gesellsch Kreislaufforsch 19:205-210, 1953
Franke H: Das hypersensitive Carotissinus-Syndrom. Versicherungsmedizin 10:5-6, 1958
Franke H: Über das Karotissinus-Syndrom und den sogenannten hyperaktiven Karotissinusreflex. Schattauer, Stuttgart, 1963
Franke K: Mittelhandbruch nach Faustkämpfen. Med Sport 5:74-77, 1965
Franscarelli M, Rocchi L, Oppido P et al.: Analisi delle variazioni dell' attivita elettroencefalografica durante combattimenti fra pugili dilettanti. Ital J Sports Trauma 11: 263-272, 1989
Freedberg AS, Sloan RA: Association of carotid sinus reflexes with syncope and convulsions. Report of four cases. Arch Neurol Psychiatr 38:761-774, 1937
Frester R: Einige psychologische Probleme des moderen Boxens. Medizin und Sport 10: 119-122, 1970
Friedman JH: Progressive parkinsonism in boxers. Southern Med J 82:543-546, 1989
Fritsche P: Ein dynamographisches Informationssystem zur Messung der Schlagkraft beim Boxen. Leistungssport 8:151-156, 1978
Funke W: Diskussionsbemerkung zu dem Vortrag von Professor Unterharnscheidt. In: Stucke K (eds) Verhandlungen, Deutscher Sportärztebund. 24. Tagung, Würzburg. Gräfelfing, Demeter, October 14-17, 1971, p. 122
Fürst T: The Boxer and his Audience. In: Grupe O, Kurtz D, Teipel JM (eds) Sports in the Modern World - Chances and Problems. Springer, Berlin Heidelberg New York, 1973, pp. 421-422
Furtado D: Traumatologia cranio-cerebral dos 'boxeurs.' Gaz Med Portug 10:296-305, 1957
Galowa J, Gala J: Results of EEG investigations in boxers. (Polish with Engl. summary). Wiad Led (Suppl) 18:41-45, 1965
Garcia-Albea E: Boxeo y cerebro. Neurologia 5:113-116, 1990
Gatarski J: Electroencephalographic changes in boxers. (Polish with Engl. summary). Psychiat Pol 3:451-455, 1969
Gebelt H: Die Elektroencephalographie als Untersuchungsmethode bei der medizinischen Eignungsprüfung von Boxsportlern und bei der Früherkennung von sportbedingten Schäden des Zentralnervensystems. Medizin und Sport 5:178-180, 1965
Geddes JF, Hughes AJ, Lees AJ, Daniel SE: Pathological overlap in cases of parkinsonism associated with neurofibrillary tangles: a study of recent cases of postencephalic parkinsonism and comparison with progressive supranuclear palsy and Guamanian parkinsonism-dementia complex. Brain 116:281-302, 1993
Geddes JF, Vowles GH, Robinson SF, Sutcliffe JC: Neurofibrillary tangles, but not Alzheimer-type pathology, in a young boxer. Neuropathol Appl Neurobiol 22:12-16, 1996
Geller W: Paranoide Psychose nach Boxschädigungen. Nervenarzt 24:69-71, 1953
Gibson JK: A perforated septum pellucidum. Anat Rec 28:103-104, 1924
Gibbs FA, Gibbs EL, Lennox WG: Electroencephalographic classification of epileptic patients and control subjects. Arch Neurol Psychiat 50:111-128, 1943
Gillespie RD, Richter CP, Wang G: The oculo-cardiac reflex; its clincal significance. J Ment Sci 72:321-325, 1926
Gilmer FB: Should the sport of boxing be discontinued? No! Phys Educ Phi Epsilon Kappa 14:133-134, 1957
Gimon J: Reglement de la boxe de combat. Med Educ Phys Sport 41:28-31, 1967
Giovinazzo VJ, Smith DJ, Vinger PF: Eye Injuries. Chap. 5. In: Cantu RC (ed) Boxing and Medicine. Human Kinetics, Champaign, ILL, 1995, pp. 65-78
Giovinazzo VJ, Yannuzzi LA. Sorensen JA, Delrowe DJ, Campbell EA: The ocular complications of boxing. Ophthalmology 94:587-596, 1987
Giroud M, Vincent MC, Thiery A, Binnert D, Marin A, Dumas R: Syndrome parkinsonien par hematomes traumatiques des noyaux gris centraux. Neurochirurgie 34: 61-63, 1988
Givre A, Primavesi J: Head injuries in boxers. 2nd Internat Congr Neurolog Surg, October 14, 1961, Washington, DC
Gladden JA: Boxer's knuckle. A preliminary report. Am J Surg 93:388-397, 1957
Glasenapp B: Hörstörungen nach Schädeltraumen. HNO (Berlin) 10:41, 1962
Gleason G: All's fair in Thai boxing. Clipper Magazine, pp. 44-45, 1966
Glees M: Über Boxverletzungen der Augen. Klin Monatsbl Augenheilk 124:101-103, 1954
Göhle S: Der Zahn-und Mundschutz im Boxsport. Deutsch Stomatol 3:340-342, 1953
Göllnitz G: Das klinische Bild bei erweitertem Cavum septi pellucidi. Deutsch Z Nervenheilk 163:1, 1949
Goldsmith MF: Physicians aim to k o. boxers' injuries; focus on eyes as title bout nears. J Am Med Ass 257:1697-1698, 1987
Goltz F: Vagus und Herz. Virchows Arch Pathol Pathol Anat 26:1-33, 1863
Gonzales TA: Fatal injury in competitive sports. J Am Med Ass 146:1506-1511, 1951
Goralski H, Sypniewski J: Boxing encephalopathy complicated by psychosis. (Polish). Neurol Neurochir Polska 1:639-641, 1961
Gordon D: Woodpeckers, gannets, and head injury. Lancet I:801-802, 1972
Gordon EB, Sim M: The EEG in presenile démentia. J Neural Neurosurg Psychiatr 30: 285, 1967
Gorman T: Death in the ring. Hygeia 27:384-385,1949
Gottlieb I, Gottlieb O: Fractura mandibulae opstaet ved Boxning. (Danish with Engl. summary). Ugeskr Laeger 113:1268-1270, 1951
Govons SR: Brain concussion and posture. The knock-down blow of the boxing ring. Confin Neurol 30:77-84, 1968
Graham J W: Professional boxing and the doctor. Br Med J I:219-221, 1955
Graham JW: Medical Care of the Boxer. In: Bass AL, Blonstein JL, James RD, Williams JGP (eds) Medical Aspects of Boxing. Pergamon Press, Oxford London New York, 1965, pp. 85-91
Grahmann H, Ule G: Beitrag zur Kenntnis der chronischen

cerebralen Krankheitsbilder bei Boxern (Dementia pugilistica und traumatische Boxer-Encephalopathie). Psychiatr Neurol (Basel) 134:261-283, 1957
Grasso-Biondi I: Contributo alla conoscenza delle lesioni mortali da pugiliato e dell' assurdo nella ricerca della concausa. Rassegna Previd Sociale 26:33-60, 193
Greco A: Variazioni dinamometriche nei pugili dopo combattimento agonistico. Arch Med Sport (Roma) 5:337-341, 1965
Greenfield JG, Bosanquet FD: Brain-stem lesions in Parkinsonism. J Neurol Neurosurg Psychiatr 16:213-226, 1953
Grewel F: Encephalopathia traumatica bei Boxern. (Dutch with German a. Engl. summary). Nederl T Geneesk 85:154-160, 1941
Groat RA, Magoun HW, Dey FL, Windle WF: Functional alterations on motor and supranuclear mechanisms in experimental concussion. Am J Physiol 141:117-127, 1944
Groat RA, Simmons JA: Loss of nerve cells in experimental cerebral concussion. J Neuropathol Exp Neurol 9:150-163, 1950
Groat RA, Windle WF, Magounn HW: Functional and structural changes in the monkey's brain during and after concussion. J Neurosurg 2:26-35, 1945
Groh G: Should boxing be banned? MD Magazine 27:164-185, 1983
Groh H: Boxsport und Arzt. In: Groh H (ed) Sportmedizin. Biologische und medizinische Grundlagen. Stuttgart, 1962, pp. 149-152
Groh H, Groh P: Boxsportverletzungen. In: Sportverletzungen und Sportschäden. Luitpold-Werk, München, 1975, pp. 60-65
Gronwall D. Wrightson P: Delayed recovery of intellectual function after minor head injury. Lancet II:605-509, 1974
Gronwall D, Wrightson P: Cumulative effect of concussion. Lancet II:995-997, 1975
Gronwall D, Wrightson P: Duration of post-traumatic amnesia after mild head injury. J Clin Neurophysiol 2:51-60, 1980
Gronwall D, Wrightson P: Memory and information processing after closed head injury. J Neurol Neurosurg Psychiatr 44:889-895, 198
Gross AG: Impact thresholds of brain concussion. J Aviat Med 29:723-732, 1958
Gross AG: A new theory on the dynamics of brain concussion and brain injury. J Neurosurg 15:548-561, 1958
Grünthal E (1926) quoted from Peters G (1962)
Grünthal E: Die pathologische Anatomie der senilen Demenz und die Alzheimerscher Krankheit. In: Bumke O (ed) Handbuch der Geisteskrankheiten. Vol. II. Springer, Berlin, 1930
Grupe O, Kurtz D, Teipel JM (eds) Sport in the Modern World – Chances and Problems. Springer, Berlin Heidelberg New York, 1973
Gürtler H: Die Aufgaben des Arztes am Boxring. Medizin und Sport 10:64-66, 1970
Gürtler H, Bousseljot W, Kuhlmey J: Erfahrungen der komplexen Dispensairebetreuung von Boxsportlern. Medizin und Sport 10:61-63, 1970
Guillain G, Sevileano E, Fandre M: L'encéphalite traumatique et les syndromes parkinsoniens chez les boxeurs professionells. Bull Acad Nat Med 132:394-406, 1948
Gunby P: Epidemiologic study to examine amateur boxers' potential risks. J Am Med Ass 255:2397-2399, 1986
Guntern R, Bouras C, Hof PR, Vallet PG: An improved thioflavine S method for neurofibrillary tangles and senile plaques in Alzheimer's disease. Experientia 48:8-10, 1992
Gurdjian ES, Webster JE: Digital carotid artery compression with occlusion of the anterior cerebral artery. Neurology 7:635-640, 1957
Gurdjian ES, Webster JE: Head Injuries in Sports. Chap. XI. In: Head Injuries. Mechanics, Diagnosis and Management. Little Brown, Boston Toronto, 1958, pp. 348-356
Gurdjian ES, Webster JE: Head Injuries. Mechanisms, Diagnosis and Management. Little Brown, Boston Toronto, 1958
Gurdjian ES, Webster JE, Hardy WG, Lindner DW: Nonexistence of the so-called cerebral form of carotid sinus syncope. Neurology 8:818-826, 1958
Gurdjian ES, Webster JE, Martin FA: Carotis-sinus- and allied syndromes. Transact Am Neurol Ass 81:167-168, 1956
Gurdjian ES, Webster JE, Martin JE, Hardy WG: Carotid compression in the neck, results and significance on carotid ligation. J Am Med Ass 163:1030-1036, 1957
Guterman A, Smith RW: Neurological sequelae of boxing. Sports Med (New Zealand) 4: 194-210, 1987
Guttmann E, Winterstein CE: Disturbances of consciousness after head injuries. Observations on boxers. J Med Sci 84:347-351, 1938
Habighorst G: Chile: 84 von 100 Berufsboxern sind körperlich geschädigt. Landarzt 40: 161, 1964
Hage P: To ban or not to ban? Physician Sportsmed 11:143, 1983
Hagler WS, North AW: Retinal dialyses and retinal detachment. Arch Ophthalmol 79:376-388, 1968
Haglund Y, Bergstrand G: Does Swedish amateur boxing lead to chronic brain damage? 2. A retrospective study with cerbral computed tomographyy and magnetic resonance imaging. Acta Neurol Scand 82:297-302, 1990
Haglund Y, Edman G, Murelius O et al.: Does Swedish amateur boxing lead to chronic brain damage? 1. A retrospective medical, neurological and personality trait study. Acta Neurol Scand 82:245-252, 1990
Haglund Y, Eriksson E: Does amateur boxing lead to chronic brain damage? A review of some recent investigations. Am J Sports Med 21:97-109, 1993
Haglund Y, Persson HE: Does Swedish amateur boxing lead to chronic brain damage? A retrospective neurophysiological study. Acta Neurol Scand 82:353-360, 1990
Haislet quoted from Doggart JH (1955)
Haines JD: Let's stop boxing in the Olympics and the US Military, an reply to Lundberg GD. J Am Med Ass 272:1821, 1994
Hall CW, Eckart DE, Nunemaker ME, Crawford RA: Acute traumatic subdural hematoma: Report of a case with slight trauma from boxing. J Kansas Med Soc 53:377-378, 1952
Hallervorden J: Zur Pathogenese des postenzephalitischen Parkinsonismus. Klin Wochenschr 12:692-695, 1933
Hansen M: Letter: Punch-drunk syndrome. Ugeskr Laeger 137:393-394, 1975
Harris C: Correspondence. Dangers for boxers. J Am Med Ass 157:612, 1955

Harvey FH: An explanation for the high incidence of cavum septi pellucidi as an aftermath of boxing. Ann Neurol 2:176, 1977
Harvey PKP, Davis JN: Traumatic encephalopathy in a young boxer. Lancet II:928-929, 1974
Hattingberg von: Diskussionsbemerkungen zu Reisch O: Der Boxkampf und seine Auswirkungen auf das Nervensystem. Berliner Gesellschaft für Neurologie und Psychiatrie, Sitzung von 12. April 1937. Zentralbl Neurol Psychiatr 90:703-704, 1938
Hauke W: Der Gesundheitschutz der Amateurboxer. In: Arzt und Boxsport. Vorträge der Enquete anläßich der Generalversammlung. Selbstverlag, Wien, November, 1956, pp. 17-21
Havener WH: Medicine vs. boxing, Ohio Med 86:427, 1990
Haynes JD: Lets stop boxing in the Olympics and the US military. J Am Med Ass 272:1821, 1994
Heald CB: Injuries and Sport. Oxford University Press, 1931
Hebert LE, Scherr PA, Beckett LA, et al.: Age specific incidence of Alzheimer's disease in a community population. J Am Med Ass 273:1354-1359, 1995
Heilbronner RL, Henry GK, Carson-Brewer M: Neuropsychologic test performance in amateur boxers. Am J Sports Med 19:376-380, 1991
Heinz WC: Knockout. Colliers, 134, September 17, 1954, pp. 94-97
Heiss E: Unfallkunde und Therapie der Sportverletzungen und Sportschäden. In: Arnold A (ed) Lehrbuch der Sportmedizin. Leipzig, 1956, pp. 552-624
Hellgren EG: A previously not described injury to the wrist sustained during boxing. Acta Orthop Scand 2:226-272, 1933
Henneberg: Diskussionsbemerkungen zu Reisch O: Der Boxkampf und seine Auswirkungen auf das Nervensystem. Berliner Gesellschaft für Neurologie und Psychiatrie, Sitzung vom 12. April 1937. Zentralbl Neurol Psychiat 90: 703-704, 1938
Henner K: Why must the neurologist be against the present form of boxing? (Czech with French summary). Cas Lek Cesk 94:859-865, 1955
Henner K: Preventive care in boxing. Rev Czechoslovak Med II:1-8, 1956
Henner K: Why neurologists cannot agree with the contemporary concept and rules of boxing? (Czech with Engl. a. French summary). Cas Lek Cesk 102:417-421, 1963
Henner K: Neurologové k dnesnîm pravidlum boxu. (Czech). Prack Lekar 46:254-256, 1966
Hering HE: Die Karotissinus-Reflexe auf Herz und Gefäße, vom normal-physiologischen, pathologisch-physiologischen und klinischen Standpunkt. Steinkopff, Leipzig, 1927
Herrenschwand F von: Die Schädigungen und Verletzungen des Auges beim Sport. Med Welt 15:34-39, 1941
Herzog K: Neurologische und psychiatrische Veränderungen bei Boxern. Med. Dissertation, Universität Hamburg, 1938
Hese R, Sibiliak J: Encephalopathia pugilistica, a source of diagnostic error. (Polish). Psychiat Pol 1:489-493, 1967
Heutink P, Stevens M, Rizzu P, et al. : Hereditary frontotemporal dementia is linked to chromosome 17q21-q22: a genetic and clinicopathological study of three Dutch families. Ann Neurol 41:150-159, 1997
Hey F: Subdurales Hämatom als Sportverletzung. Deutsch Z Ges Gerichtl Med 5:12-16, 1925
Heymans C: Le sinus carotidien et les autres zones vasosensibles réflexogenes. Presses Universitaires de France, Paris, 1929
Heymans C: Le syndrome sino-carotidien. Bull Acad Royal Méd Belg 1:119-124, 1961
Heymans C, Bouckaert JJ: Le sinus carotidien, zone réflexogene regulatrice du tonus des vaisseaux cephaliques. CR Soc Biol 100:202, 1929
Heymans C, Bouckaert JJ, Regniers P: Le sinus carotidien et la zone homologe cardio-aortique. Doin, Paris, 1933
Heymans C, Delaunois AL, van den Heuvel-Heymans G: Tension and distensibility of carotid sinus wall, pressoceptors and blood pressure regulation. Circul Res 1:3-7, 1953
Hickey JC, Morris AL, Carlson LD, Seward TE: The relation of mouth protectors to cranial pressure and deformation. J Am Dent Ass 74:735-740, 1967
Higgins LS, Schmall RA: A Device for the Investigation of Head Injury Effected by Non-deforming Head Accelerations. Proc 11th Stapp Car Crash Conf, Soc Automot Engin, New York, NY, 1967, pp. 35-46
Higgins LS, Schmall RA, Cain CP, Kielpinski PE, Primiano P, Barber TW, Brockway JA: The Investigation of the Parameters of Head Injury Related to Acceleration and Deceleration. Report TI-118-67-1. Technology Incorporated, Life Sciences Division, San Antonio, TX, 1967
Hill, Watterson, 1942, quoted after Blonstein JL, Clarke E (1964)
Hillman E, father: The Morality of Boxing. Theological Studies 12: No. 3, September 1951, pp; 301-319
Hillman R: Boxing. Resuscitation 8:211-215, 1980
Hirano A, Malamud N, Kurland LT: Parkinsonism-dementia complex, an endemic disease on the island of Guam. II. Pathological features. Brain 84:662-678, 1961
Hirano A, Zimmermann HM: Alzheimers neurofibrillary changes. Arch Neurol 7:227-242, 1962
Hirschfeld K: Ist Boxen gefährlich? Sportmedizin 8:334-335, 1957
Hockaday TDR: Traumatic thrombosis of the internal carotid artery. J Neurol Neurosurg Psychiatr 22:229-231, 1959
Hodgson VR, Thomas LM, Gurdjian ES, Fernando OU, Greenberg SW, Chason J: Advances in Understanding of Experimental Concussion Mechanisms. In: Backaitis H (ed) Biomechanics of Impact Injury and Injury Tolerances of the Head-Neck Complex. PT-43. Soc Autom Engin, Warrendale, PA, 1993, pp. 387-406
Hof PR, Bouras C, Buee L, Delacourte A, Perl DP, Morrison JH: Differential distribution of neurofibrillary tangles in the cerebral cortex of dementia pugilistica and Alzheimer's disease cases. Acta Neuropathol 85:23-30, 1992
Hofmann E von: Schläge mit der Hand ins Gesicht, Tod durch Pachymeningitis haemorrhagica. Wien Klin Presse 17:1437-1440, 1876
Hollander D, Strich S: Atypical Alzheimer's disease with congophile angiopathy presenting with dementia of acute onset. In:

Wolstenholme GEW, O'Connor M (eds) Alzheimer's Disease and Related Conditions. Ciba Symposium, 1970, pp. 105-135
Hollnagel P: The 'punch-drunk' syndrome. Report of three cases of chronic encephalopathy in young boxers. (Danish with Engl. summary) Ugeskr Laeger 136:1871-1874, 1974
Hollwich F, Brandt HP, Zintl F: Der okulo-kardiale Reflex in augenärztlicher Sicht. Med Klin 60:170-172, 1965
Holmes C: Mouth protection in sport in Scotland – a review. Br Dent J 188:473-474, 2000
Holzgraefe M, Lemme W, Funke W, Felix R, Felten R: The significance of diagnostic imaging in acute and chronic brain damage in boxing. A prospective study in amateur boxing using magnetic resonance imaging (MRI). Intern J Sports Med 13:616-620, 1992
HootenWM, Lyketsos CG: Frontotemporal dementia: a clinicopathological review of four postmortem studies. J Neuropsychiatr Clin Neurosci 8:10-19, 1996
Hoppes DA, Olson K, Jenkinson SA: Renal response to boxing: An investigation of changes in the urine in amateur boxers. J Am Osteopath Ass 91:461-464, 1991
Horak J: Srovnani funkcni zdatnosti krevniho obehu a dychani u boxeru s lyzari a plavci. (Czech). Cas Lek Cesk 102:430-432, 1963
Horak J: Circulatory and respiratory response in boxers during adequate laboratory loading. (Czech with Engl. a. French summary). Cas Lek Cesk 102:433-435, 1963
Hornof Z: Beitrag zur Frage der Herzgröße bei Boxern. Med Welt 31:1564-1566, 1960
Horowitz D: Verletzungs- und Todesfälle beim Boxen und ihre gerichtsärztliche Begutachtung. (Polish). Pan Wilenskiego Tow Lek 9:1-20, 1933. Ref. Deutsch Z Ges Gerichtl Med 22:82, 1933
Hromada quoted from Sercl M and Jaros O (1962)
Hruby K: Netzhautablösung beim Boxsport. Klin Monatsbl Augenheilk 174:314-316, 1979
Hudson A: Amyotrophic lateral sclerosis and its association with dementia, parkinsonism and other neurological disorders: a review. Brain 104:217-247, 1981
Hudson CJ: Physicians and boxing (letter). New England J Med 303:1308, 1980
Hughes JT, Oppenheimer DR: Superficial siderosis of the central nervous system. Acta Neuropathol 74:559-562, 1965
Hughes RA:, Kernohan JM, Craig WMC: Caves and cysts of the septum pellucidum. Arch Neurol Psychiatr 74:259-266, 1955
Hugonnier R: Les accidents oculaires de la boxe. Med Educ Phys Sport 38:233-235, 1964
Hulst JPL: Schädelbruch, Gehirnzertrümmerung, subdurales Hämatom, Schlag mit der blossen Faust. Vierteljahresschr Gerichtl Med 62:76-83, 1921
Hussey HH Punch drunk (editorial), J Am Med Ass 236:485, 1976
Huszar I, Környey E: Über neuro-psychiatrische Aspekte des Boxens. Psychiatr Neurol Med Psychol 17:335-338, 1965
Ikeda S, Sato T, Komatsu S, Namiki T, Wada T: Consideration of the mechanism of head injury in boxing – case report. (Japanese). No Shinkei Geka 8:99-103, 1980
Imachi K: Weitere experimentelle Forschungen über die Reizübertragung im sog. Oculo-Vago-Reflex. Über die zentrifugale Bahn des Reflexes im Rückenmark. Jahresber Kurashiki Zentralhosp 14:16, 1939
Imachi K: Weitere experimentelle Forschungen über die Reizübertragung im sog. Oculo-Vago-Reflex. Über die Reizübertragung durch chemische Mittel bei demselben Reflex. Jahresber Kurashiki Zentralhosp 14:31, 1939
Ioannou S: Should boxing be banned? Can Med Ass J 131:10, 1984
Iselin M: Importance de l'arthrose dans le syndrome 'main fragile' des boxeurs. Rev Rhumatol 27:242-243, 1960
Iselin M, Ducroquet J: Traumatismes de la main chez les boxeurs. Mem Acad Chir 74: 803-809, 1948
Iselin M, Ducroquet J: Traumatismes de la main chez les boxeurs. Presse Méd 57:554-555, 1949
Isenberg MT: John L. Sullivan and His America. University of Illinois Press, Urbana Chicago, 1988
Isherwood I, Mawdsley C, Ferguson FR: Pneumoenephalographic changes in boxers. Acta Radiol 5:654-661, 1966
Izeki T, Yano T, Suzuki H, Siota K: Some Findings of the Prevention of Motor Disturbances in Boxers. In: Abstracts, Internat Congr Sport Sci, Tokyo, October 3-8, 1964, p. 110
Jack C, Petersen RC, O'Brien PC, Tangalos EG: MR-based hippocampal volumetry in the diagnosis of Alzheimer's disease. Neurology 42:183-188, 1992
Jackson M, Lennox G, Lowe J: Motor neuron disease inclusion dementia. Neurodegeneration 5:339-350, 1996
Jaillard: Traumas of boxers. Concours Med 83:1743-1744, 1961
Jako P: La rehabilitation des boxeurs. In: Confer Nationale de la Societe Hongroise de Medecine Sportive. Resume des Communications, Budapest, September 6-8, 1966, p. 44
Jakob A: Experimentelle Untersuchungen über die Schädigungen des Zentralnervensystems (mit besonderer Berücksichtigung der Commotio cerebri und Commotionsneurose). In: Nissl F, Alzheimer A (eds) Histologische und histopathologische Arbeiten über die Großhirnrinde. Bd. 5. Jena, pp. 182-341
James W: Texbook of Psychology. London, 1893
Jankovitch L: Suites mortelles d'un combat de boxe. Ann Med Leg 15:795-799, 1935
Jedlinski J, Gatarski J, Szymusik A: Chronic post-traumatic changes in the central nervous system of boxers. (Polish with Engl. summary). Psychiat Pol 3:443-450, 1969
Jedlinski J, Gatarski J, Szymusik A: Chronic posttraumatic changes in the central nervous system in pugilists. Pol Med J 9:743-752, 1970
Jedlinski J, Gatarski J, Szymusik A: Encephalopathia pugilistica (Punch drunkenness). Acta Med Polon 12:443-451, 1971
Jenkins TH: Changes in Ethnic and Racial Representation among Professional Boxers: A Study in Ethnic Succession. MA Thesis, University of Chicago, 1951
Jensen J: Intraoral mouth guards in a group of athletes (boxers). (Danish). Tandlaegebladet 88:681-686, 1984
Jetchev E: Le lourds en boxe et le probleme des arbitres. Bull Inform (Sofia) 12:10-11, 1967
Jimenez Espinosa L, Espinosa Iborra J: Knock-out and syncope

in professional boxing: An EEG study. Electroencephal Clin Neurophysiol 72:196-197, 1960
Joch W, Fritsche P, Krause I: Biomechanical Analysis of Punching in Boxing. In: Morecki A, Fidelus K, Kedzior K, et al. (eds) Biomechanics VII-B. Proc Internat Congr Biomechanics, Warsaw, Poland, 1979, University Press, Baltimore, 1981, pp. 343-349
Joch W, Krause I, Fritsche P: Schlagkraft und Bewegungsschnelligkeit des Boxers. Leistungssport 1:40-46, 1982
Johansen O: Idrett og Skader. Oslo, 1965
Johnson J: The EEG in the traumatic encephalopathy of boxers. Psychiat Clin (Basel) 2: 204-211, 1969
Johnson J: Organic psychosyndromes due to boxing. Br J Psychiatr 115:45-53, 1969
Johnson J, Skorecki J, Wells RP: Peak accelerations of the head experienced in boxing. Med Biol Engineer 13:396-404, 1975
Johnson W, Reid SR, Mamalis AG: Human Body Dynamics, University Press, Oxford, 1982, pp. 524-542
Jokl E: The Medical Aspects of Boxing. Van Schaik, Pretoria, 1941
Jokl E: Medical research in physical education. VI. Boxing injuries. South African Med J 20:209-210, 1946
Jokl E: Brain injuries to boxers. Transact Am Neurol Ass 81:125-128, 1956
Jokl E: Boxing. J Am Med Ass 162:618, 1956
Jokl E: Death of a boxer. Lancet II:422, 1958
Jokl E: Boxing. Lancet II:64, 1959
Jokl E: Über den Boxsport. Landarzt 38:789-791, 1962
Jokl E: Ziele und Aufgaben der Sportmedizin. Medizin und Sport 3:1-7, 1963
Jokl E, Guttman E: Neurologisch-psychiatrische Untersuchungen an Boxern. Münch Med Wochenschr 80:560-562, 1933
Jordan BD: Neurologic aspects of boxing. Arch Neurol 44:453-459, 1987
Jordan BD: Medical and safety reforms in boxing. J Nat Med Ass 80:407-412, 1988
Jordan BD: Boxing in the Army. Letter to the editor. J Am Med Ass 262:2089, 1989
Jordan BD: Neurologic Injuries in Boxing. Chap. 17. In: Jordan BD (ed) Sports Neurology. Aspen Press, 1989
Jordan BD: Amateur boxing in Denmark (letter, comment). Am J Sports Med 18:561, 1990
Jordan BD: Chronic neurologic injuries in boxing. In: Jordan BD (ed) Medical Aspects of Boxing. CRC Press, Boca Raton, FLA, 1993, pp. 177-185
Jordan BD: Epidemiology of brain injury in boxing. In: Jordan BD (ed) CRC Press, Boca Raton, FLA, 1993, pp. 147-168
Jordan BD: Professional Boxing. Experience of the New York State Athletic Commission. Chap. 10. In: Cantu RC (ed) Boxing and Medicine. Human Kinetics, Champaign, ILL, 1995, pp. 177-185
Jordan BD, Campbell EA: Acute injuries among professional boxers in New York State: a two year survey. Physician Sportsmed 16:87-91, 1988
Jordan BD, Jahre C, Hauser WA, Zimmerman RD, Zarrelli M, Lipsitz EC, Johnson V, Tsairis P, Folk FS: CT of 338 active professional boxers. Radiology 185:509-512, 1992
Jordan BD, Kanik AB, Horwich MS, et al.: Apolipoprotein Ee4 and fatal cerebral amyloid angiopathy associated with dementia pugilistica. Ann Neurol 38:693-699, 1995
Jordan BD, Relkin NR, Ravdin LD, Jacobs AR, Bennett A, Gandy S: Apolipoprotein E 4 associated with chronic traumatic brain injury in boxing. J Am Med Ass 278:136-140, 1997
Jordan BD, Zimmerman RD: Magnetic resonance imaging in amateur boxers. Arch Neurol 45:1207-1208, 1988
Jordan BD, Zimmerman RD: Computed tomography and magnetic resonance imaging comparisons in boxers. J Am Med Ass 263:1670-1674, 1990
Jordan BD, Zimmerman RD, Devinsky O, Gamache FW, Folk FS, Campbell EA: Brain contusion and cervical fracture in a professional boxer. Physician Sportmed 16:87-91, 1988
Jordan BD, Voy RO, Stone J: Amateur boxing injuries at the US Olympic Training Center. Physician Sportsmed 18:12, 1990
Juhasz P: Neurologische Probleme des Faustkampfes. Deutsch Gesundheitswes 20:1453, 1965
Kanerva L: Knuckle pads from boxing. Eur J Dermatol 8:359-361, 1998
Kaplan HA, Browder J: Observations of the clinical and brain wave patterns of professional boxers. J Am Med Ass 156:1138-1144, 1954
Kappis M: Über die tödlichen Kopfverletzungen beim Boxkampf. Zentralbl Chir 65:934-938, 1938
Karpov SP: Neglected subluxations of the 3rd metacarpal bone in boxers. (Russian). Ortop Travmatol Protez 7:37-39, 1985
Kaste M, Vilkki J, Katenuo K, Sainio K, Meurala H: Is chronic brain damage in boxing a hazard of the past? Lancet II:1186-1188, 1982
Kasiske K, Schwanewede H von: Aus den Grenzgebieten der prothetischen Stomatologie - Mundschutz für Boxer und Ohrpaßstücke für Hörhilfen. Zahntechnik (Berlin) 14:461-465, 1973
Kaufmann E: Lehrbuch der speziellen pathologischen Anatomie. De Gruyter, Berlin, 1924
Kawamura S: Clinical observation on so-called commotio labyrinthi. Zentralbl Hals Nasen Ohrenheilk 61:326, 1958
Kelly JP: In reply to: The borrowed imagery of boxing. J Am Med Ass 264:1532, 1990
Kemp PM, Houston AS, Mac:Leod MA, Pethybridge RJ: Cerebral perfusion and psychometric testing in military amateur boxers and controls. J Neurol Neurosurg Psychiatr 59:368-374, 1995
Kempe LG: Operative Neurosurgery. Volume 1. Cranial, Cerebral, and Intracranial Vascular Disease. Fourth Printing. Springer, Berlin Heidelberg New York Tokyo, 1985
Kenney HE, Thacker EA, Gebhardt HG: The evaluation of boxing as a college activity. Am Ass Health Phys Educat Recr Quart 11:80, 1940
Kernohan JW, Woltman HW: Incisura of crus due to contralateral brain tumor. Arch Neurol Psychiatr 21:274-287, 1929
Kewalramani LS, Orth MS, Krauss JF, et al.: Cervical spine injuries resulting from collision sports. Paraplegia 19:303-312, 1981

Khosla T, Hitchins RA: Johnny Owen's ill-fated fight (letter). Lancet I:1254-1255, 1980
Kiernan JA, Hudson AJ: Frontal lobe atrophy in motor neuron diseases. Brain 117:747-757, 1994
Kilpatrick James J: An argument against boxing. The Times Picayune / The States Item, December 13, 1984
Kindle J: Isolierte Handwurzelknochenverletzungen. Beitr Klin Chir 67:549, 1910
Kirsch RE, Samt P, Axelrod S, Kugel V: Electrocardiographic changes during ocular surgery and their prevention by retrobulbar injection. Arch Ophthalmol 58:348-356, 1957
Klages U: Thrombose der Arteria basilaris bei einem jugendlichen Amateurboxer. Deutsch Z Ges Gerichtl Med 66:75-85, 1969
Klar B: Boxschäden am Auge. Med. Dissertation, Humboldt Universität, Berlin, 1966
Klaus EJ: Über Boxverletzungen der Hand, ihre Entstehung und Verhütung. Münch Med Wochenschr 87:265-267, 1940
Klaus EJ: Traumatische Zusammenbrüche beim Sport. Berichte 'Bayer' 35: 49-53, 1963
Kleeberg WD: Warnende Stimmen. Leibesübungen 7:16-18, 1956
Kleiman AH: Hematuria in boxers. J Am Med Ass 168:1633-1640, 1958
Klenka L: Zraneni oka pri boxu. (Czech). Cesk Ofthalmol (Prag) 11:333-335, 1955
Kluge G: Boxsportverletzungen und ihre Verhütung aus zahnärtlicher Sicht. Med Sport 6:15-20, 1966
Knoll W: Verhütung von Boxschädigungen, besonders bei Jugendlichen. Sportmedizin 34:1125-1129, 1938
Knoll W, Stille G, Herzog K: Boxschädigungen und ihre Verhütung. Arch Klin Chir 191: 36-42, 1938
Kobayashi S: Experimentelle Untersuchungen über Verletzungen des Gehörorgans. Zentralbl Hals- Nasen- Ohrenheilk 6:336, 1925
Koch J: Studien über die Veränderungen des Gehörorgans. Arch Ohr Nase Kehlkopfkrankh 137:105, 1933
König E: Über einen Fall von Boxsportverletzung des Kopfes. Trepanation, Heilung. Zentralbl Chir 65:204-2206, 1938
Koeppen AH, Barron KD: Superficial siderosis of the central nervous system. J Neuropathol Exp Neurol 30:448-469, 1971
Koeppen AH, Dentinger MP: Brain hemosiderin and superficial siderosis of the central nervous system. J Neuropathol Exp Neurol 47:249-270, 1988
Kolew K: Über die medizinische Kontrolle und Prophylaxe beim Boxen und die Rechte der Ärzte bei der Durchführung von Wettkämpfen. Medizin und Sport 5:191-192, 1965
Kohlrausch W: Boxunfälle mit tödlichem Ausgang. Arch Klin Chir 118:902-907, 1921
Kopecky JA: Pressure distribution developed within the skull during dynamic loading. PhD Dissertation, University of Texas, Austin, TX, 1968
Kopecky JA, Ripperger EA: Closed brain injuries. An engineering analysis. J Biomech 2:29-34, 1969
Koupernik C: Requiem pour un boxeur défunt. Concours Med 84:4191-4192, 1962
Krajina Z, Milojevic B: Resultate der Forschungen der statoakustischen Apparate bei Boxern. Radovi Med Fak (Zagreb) 3:228-236, 1955
Krauland W: Über die Quellen des akuten und chronischen subduralen Hämatoms. Zwangl Abhandl Norm Pathol Anat, Heft 10. Thieme, Stuttgart, 1961
Kraus H: Über eine seltene Sportverletzung. Wien Klin Wochenschr 47:1322, 1934
Kraus H: Über Hirnverletzungen beim Sport. Sportärztliche Praxis, Heft 4. Dezember, 1958
Kraus JF, Conroy C: Mortality and morbidity from injuries in sports and recreation. Ann Rev Publ Health 5:163-192, 1984
Krayenbühl H, Noto GG: Das intracranielle subdurale Hämatom. Huber, Bern, 1949
Krefft F: Über Todesfälle beim Boxen. Deutsch Gesundheitsw 7:1559-1564, 1952
Krejcova H, Czerny R: Vestibular abnormalities in encephalopathia pugilistica. Acta Otolaryngol (suppl) 468:209-210, 1989
Kroeber-Keneth L: Der Tod im Boxkampf und der Arzt am Ring. Med Welt 6:1280-1281, 1932
Külbs F: Commotio cordis durch Boxen entstanden. Med Klin 33:337-338, 1937
Kurland H, Hanson J: Death in the ring; how much contact can the human brain stand? Black Belt 18:58-63, 1980
Kussmann E: Boxkämpfe mit tödlichem Ausgang. Med Welt 1:1380-1381, 1927
Kussmaul, Tenner (1857) quoted from Franke H (1963)
Laane CL: Boxing and boxing injuries. T Norsk Laegeforen 87:677-679, 1967
Labignette P: Le laboratoire dans la surveillance médicale des boxeurs. Concours Med (Paris) 84:4204, 1954
La Cava G: La cranio-encéphalopathie traumatique des boxeurs. Bruxelles Med 29:3233-3245, 1949
La Cava G: La cranio-encéphalopathie traumatique des boxeurs. Bruxelles Med 29:3304-3314, 1949
La Cava G: La cranio-encéphalopathie traumatique des boxeurs. Trav Soc Med Belg Educ Phys 5:3-25, 1950/51
La Cava G: Die Verletzungen beim Boxsport. Deutscher Sportärzte Kongress, Berlin, 1952. Limpert, Frankfurt, 1953, pp. 47-59
La Cava G: Typische Verletzungen beim Boxen. Deutsch Med Wochenschr 79:817-819, 1954
La Cava G: Boxer's encephalopathy. J Sport Med 3:87-92, 1963
La Cava G: A Statistical Review of Boxing Injuries. In: Jokl E, Simon E (eds) International Research in Sport and Physical Education. Thomas, Springfield, 1964, pp. 629-640
Lafon MMR, Muivielle AJ: Le syndrome d'hyperreflectivite sino-carotidienne. Montpellier Med 51:23-53, 1957
Laforet EG: Medico-legal abstracts. Boxing. Carney Hosp J 4:54-56, 1962
Laignel-Lavastine G, Schapira: Le reflexe oculocardiaque unilatéral chez les tuberculeux pulmonaires. Bull Soc Med Hop (Paris) 55:1189-1193, 1939
Lami G: Knockout et morte improvvisa nelle boxe. Riforma Med 50:191-192, 1934
Lamonantara G, De Santi F, Accentura V: Onde lente posteriori nell' EEG di giovani pugili. Med Sport 37:369-372, 1984

Lampert PW, Hardman JM: Morphological changes in brains of boxers. J Am Med Ass 251 2676-2679, 1984
Lang C, Stefan H: Sportverletzungen des Nervenssystems. Fortschr Neurol Psychiatr 67:373-384, 1999
Larose JH, Sik KD: Knuckle fracture. A mechanism of injury. J Am Med Ass 68:893-894, 1968
Larsson LE, Melin KA, Nordström-Öhrberg BP, Silfverskiöld BP, Öhrberg K: Acute head injuries in boxers. Acta Psychiatr Neurol Scand (Suppl) 95:1-42, 1954
Larsson LE, Melin KA, Silfverskiöld B, Öhrberg G, Öhrberg K: Klinische und elektroencephalographische Studien über leichte akute Kopftraumata. 2. Tagung Deutsche EEG Gesellschaft, 1952, Hamburg. Nervenarzt 24:477, 1953
Lasagna L: Death in the ring. Science 23:61-62, 1983
Laurie JC: Medical aspects of amateur boxing. Brit Med J I:227, 1955
Lavorgna G, Russo A, Ambrosio A, Alterio F, Dell Aquila A: Injuries of the nasal bone in boxers. (Italian). Minerva Chir 50:895-899, 1995
Leach A, McGalliard J, Dwyer H, Wong D: Ocular injuries from boxing (letter, comment). Br Med J 304:839-840, 1992
Leach RE: What price glory? Am J Sports Med II:157, 1998
Lealy : Fractures of the metacarpal bones as the result of boxing or fighting. J Royal Army Med Corps 1909
Lechowski S, Jedlinski J: Subdural hematoma in pugilists. (Polish with Engl. summary). Pol Tyg Lek 20:185-186, 1965
Leclerc S, Herrera CD: Sport medicine and the ethics of boxing. Br J Sports Med 33:426-429, 1999
Lee NC: Boxing – what should doctors do? (editorial). S Afr Med J 74:1, 1988
Leerhsen C, Katz S: The AMA tries to k o boxing. Newsweek, December 17, 1984, p. 67
Leger L, Coste M: Syncopes à répétition liées à une hyperréflectivité sinusal. Guérison par énervation sinu-carotidienne. Presse Méd 55:174-175, 1947
Legwold G: Few head injuries found in academy boxing study. Phys Sportsmed 10:43-44, 1982
Lenoir O: La fracture des boxeurs. Fracture de l'extrémité inférieure du premier métacarpien. Presse Med 25:218-220, 1909. Ref. Zentralbl Chir 25:903, 1909
Leonard B: The dirtiest game. Benny Leonard tell why he isn't coming back to the ring. An interview by Kennedy John B. Colliers 78: 13 +, 1926, p. 13
Leoncini F: Lesioni mortali in scontri di boxe. Gass Internaz Med Chir 38:5-18, 1930. Ref. Deutsch Z Ges Gerichtl Med 16:91, 1930
Lerner LH: Painful heel in a professional boxer - a case report. J Am Podiat Ass 51: 425-426, 1961
Levin HS, Lippold SC, Goldman A, Handel S, High WM, Eisenberg HM, Zerlitt D: Neurobehavioral functioning and magnetic resonance imaging findings in young boxers. J Neurosurg 67:657-667, 1987
Leyenberg HJ: Boxen im Widerstreit der Meinungen: Bis zum nächsten Mal. Frankfurter Allgemeine Zeitung, 25. Juli 1978
Leyenberg HJ: Wenn die Elefanten im Ring den Sandmännchenwalzer tanzen. Die sensiblen Kolosse und die Angst vor dem Schlag - das Knallbonbon der Box-Europameisterschaften. Frankfurter Allgemeine Zeitung, Samstag, 11. Mai 1991, Nr. 108, p. 22
Lhermitte MJ: Discussion remark. Bull Acad Nat Méd (Paris) 132:405-406, 1948
Liber AF: Dilated cavum septi pellucidi and juxtaventricular cavities. Acta Neerl Morphol 2: 4-19, 1938
Lietzberg H: Boxen - ein gefährlicher Sport. Z Ärztl Fortbild (Jena) 51:87-90, 1962
Lindau A: Vad anser Ni om boxning? Med Fören Tidskr 25:70-71, 1947
Lindsay KW, Mc Latchie G, Jennett B: Serious head injury in sport. Br Med J 29:789-791, 1980
Littlepage BN, Perry HH: The dangers of boxing. Criminalization would make it worse (letter). Br Med J 307:379, 1993
Livet M: Knock-out et psychoses de guerre. (Considérations médico-sportives 'a propos d'un knock-out). Bull Acad Med (Paris) 87:501-503, 1922
Lobzin VS: On closed injuries of the brain in boxers. (Russian). Zh Nevropat Psikliatr 60: 562-567, 1960
Loebell H: Sport - und Unfallfragen aus der Ohren-, Nasen- und Halsheilkunde. Z Ärztl Fortbild (Jena) 30:490-491, 1933
Löbsack T: Um Kopf und Kragen. Die Zeit, Nr. 45, 6. November, 1964
Löfgren L: A description of the new pneumatic boxing-glove. Medicina Sportiva 11:449-454, 1957
Löfgren L: Prohibition of the use of local anaesthesia in injuries sustained at boxing. J Sports Med Phys Fitness 11:99-100, 1961
Löfgren L: Traumata-preventing under-covers to the floor of boxing rings. Comparative oscillographic study of some new constructions for boxing ring under-covers. J Sports Med Phys Fitn 2:17-19, 1962
Löfgren L: The Pneumatic Boxing Glove. In: Jokl E, Simon E (eds) International Research in Sport and Physical Education. Thomas, Springfield, 1964, pp. 650-655
Löfgren L: A description of the new pneumatic boxing glove. Nordisk Udredningsserie 16:56-60, 1967
Löfgren L: Traumata-preventing under-covers to the floor of boxing rings. Nordisk Udredningsserie 16:52-55, 1967
Löhlein W: Zur Prognose der Boxverletzungen des Auges. Klin Monatsbl Augenheilk 124:570-573, 1954
Loew F: Die gedeckte Hirnschädigung als anatomisches und klinisches Problem. Zentralbl Neurochir 10:132-146, 1950
Lopez OL, Becker JT, De Kosky ST: Dementia accompanying motor neuron disease. Dementia 5:42-47, 1994
Lovas M: Ruptured spleen in a boxer with infectious mononucleosis. Minnesota Med 64:461-462, 1981
Loyarte HF, Mazal A: Necrosis aséptica del semilunar. Resecion. Medicina Atlant 3:658, 1963
Lubell A: Chronic brain injury in boxers: Is it avoidable? Physician Sportmed 17:126-132, 1989
Luckhurst J: Progressive axonopathy in boxers (letter). Vet Rec 107:406, 1980
Ludorf W: Sportmedizinische Probleme des Amateurboxsportes. Sportarzt 9:285-290, 1964
Ludwig R: Making boxing safer: The Swedish model. J Am Med Ass 255:2482, 1986

Lund E, Espersen GT: Pattern of injuries in Danish amateur boxers. (Danish). Ugeskr Laeger 150:1159-1161, 1988
Lundberg GD: Brain injury in boxing. Am J Forens Med Pathol 6:192-198, 1985
Lundberg GD: Boxing should be banned in civilized countries. J Am Med Ass 249: 250, 1983
Lundberg GD: Boxing should be banned in civilized countries - Round 2. J Am Med Ass 251:2696-2698, 1984
Lundberg GD: Boxing, in reply to a discusssion remark of Oliver WR: J Am Med Ass 253:198, 1985
Lundberg GD: Boxing, in reply to a discussion remark of Vincent JL. J Am Med Ass 253:198, 1985
Lundberg GD: Boxing, in reply to a discussion remark of Corcum JM. J Am Med Ass 253:199, 1985
Lundberg GD: Brain injury in boxing. Am J Forens Med Pathol 6:192-198, 1985
Lundberg GD: Boxing should be banned in civilized countries. (Round 3). J Am Med Ass 255:2483-2485, 1986
Lundberg GD: Let's stop boxing in the Olympics and the United States military. J Am Med Ass 271:1790, 1994
Lundberg GD: Blunt force violence in America – shades of grey or red. Ultimate/extreme fighting (editorial). J Am Med Ass 275:1684-1685, 1996
Lydston GF: A case of 'pugilist's ear.' Surg Gynecol Obstet 5:439-441, 1907
Lyle K: In: Bass AL, Blonstein JL, James RD, Williams JGD (eds) Medical Aspects of Boxing. Pergamon Press, Oxford London, 1965, p. 13
Lynch T, Sano M, Marder KS, et al.: Clinical characteristics of a family with chromosome 17-linked disinhibition-dementia-parkinsonism-amyotrophy complex. Neurology 44:1878-1884, 1994
Machalewski J: Boxen als Weltsport. Kosmopolith, Berlin, 1927
Mackay RP: Discussion remark to Jokl E: Injuries to boxers. Transact Am Neurol Ass 81:127, 1956
Mackey-Dick J: Letter to the editor. Boxing. J Royal Army Med Corp 110:121, 1964
MacPherson P, Teasdale E: CT demonstration of a 5th ventricle - a finding to k.o. boxers? Neuroradiology 30:506-510, 1988
Märker K: Die gesundheitlich-soziale Situation ehemaliger Jugendboxer. Medizin und Sport 2:201-204, 1962
Märker K: Prophylaxe im Boxsport aus der Sicht ehemaliger Amateurboxer. Medizin und Sport 5:188-191, 1965
Maguire JI, Benson WE: Retinal injury and detachment in boxers. J Am Med Ass 255: 2451-2453, 1986
Mailer N: The Fight. Little Brown, Boston Toronto, 1975
Maitra A, Sen B: Displaced boxers' fractures: a simple and effective method of external splintage. Brit J Clin Pract 44:348-351, 1990
Maitra A, Sen B: The boxers fracture: a simple and effective method of external splintage. Arch Emerg Med 7:52-54, 1990
Majoor-Krakauer D, Ottman R, Johnson WG, Rowland LP: Familial aggregation of amyotrophic lateral sclerosis, dementia, and Parkinson's disease: evidence of shared genetic susceptibility. Neurology 44:1872-1877, 1994
Mandl F: Chirurgie der Sportunfälle. Boxen. Chap. 9. Urban & Schwarzenberg, Wien, 1925, p. 195-215
Mansell PWA, Hunt AC: Major head injuries in fights and brawls. Med Sci Law 8:181-187, 1968
Maracin PR: A sports fan names the enemy. The Wall Street Journal, Monday, November 30, 1992, A 12
March T quoted from Corsellis JAN (1989)
Marek Z, Jedlinski J, Szymusik A, Gatarski J, Baran E: Problems of medical evidence in boxer's disease. (Polish with Engl. summary). Arch Med Sadowej 22:93-96, 1972
Marenholtz von: Tod im Boxkampf. Med Welt 6:556-557, 1932
Margerison JH, Scott DF: Huntington's chorea: EEG and neuropathological findings. Electroencephal Clin Neurophysiol 19:314, 1965.
Marloth: Wie kommt der k.o. zustande? Z Ärztl Fortbild (Jena) 35:622, 1938
Martin P: Consciousness and its disturbances. Lancet I:1-31, 1949
Martini S, Ricciardi F, Francone A: Encefalopatia cronica progressiva nei pugili: confronto TC-RM. Radiologica :803-806, 1988
Martland HS: Punch-drunk. J Am Med Ass 91:1103-1107, 1928
Martland HS, Beling CC: Traumatic cerebral hemorrhage. Meet, Am Neurol Ass, May 1927, Atlantic City, NJ
Martland HS, Beling CC: Traumatic cerebral hemorrhage. Arch Neurol Psychiatr 22:1001-1023, 1929
Martinez JL, Garcia DJ: A model for whiplash. J Biomech 1:23-32, 1968
Marwick C: Protecting the boxer who won't quit. J Am Med Ass 269:1083, 1993
Mawdsley C, Ferguson FR: Neurological disease in boxer. Lancet II:795-801, 1963
May PRA, Fuster JM, Newman P, et al.: Woodpeckers and head injury. Lancet I:1347-1348, 1976
May PRA, Fuster JM, Haber J, Hirschman A: Woodpecker drilling behavior. An endorsement of the rotational theory of impact brain injury. Arch Neurol 36:370-373,1979
Mayeux R, Foster NL, Rossor M, Whitehouse PJ: The clinical evaluation of patients with dementia. In: Whitehouse PJ (ed) Dementia. Vol. 40. Contemporary Neurology Series. Davis, Philadelphia, 1993, pp. 92-129
Mayeux R, Saunders AM, Shea S, et al.: Utility of the apolipoprotein E genotype in the diagnosis of Alzheimer's disease. New Engl J Med 338:506-511, 1998 {Erratum, New Engl J Med 338:1325, 1998]
McAlpine D, Page F: Midbrain syndrome in a professional boxer. Proc Royal Soc Med 42:792-793, 1949
McCown IA: Boxing injuries. Am J Surg 98:509-516, 1959
McCown IA: Protecting the boxer. J Am Med Ass 169:1409-1413, 1959
McCown IA, Stevens MA: Considerations in the management of traumatic injury in professional boxers. J Sport Med 2:99-102, 1962
McCunney RJ, Russo PK: Brain injuries in boxers. Phys Sports Med 12:52-67, 1984
McGuigan quoted from Corsellis JAN (1989)
McDougall A: Head Injuries. In: Bass AL, Blonstein JL, James RD, Williams JGD (eds) Medical Aspects of Boxing. Pergamon Press, Oxford London, 1965, pp. 80-81

McGee DA, Van Patter HJ, Morotta J, Olszewski J: Subpial cerebral siderosis. Neurology 12:108-113, 1962
McGeer PL, Schwab C, McGeer EG, Haddock RL, Steele JC: Familial nature and continuing morbidity of the amyotrophic lateral sclerosis-parkinsonism dementia complex of Guam. Neurology 49:400-409, 1997
McKeith I, Galasko D, Kosaka K, et al.: Consensus guidelines for the clinical and pathologic diagnosis of dementia with Lewy bodies (DLB): report of the Consortium on DLB international workshop. Neurology 47:1113-1114, 1996
McKenzie JE, Roberts GW, Royston MC: Comparative investigation of neurofibrillary damage in the temporal lobe in Alzheimer's disease, Down's syndrome and dementia pugilistica. Neurodegeneration 5:259-264, 1996
Mc Khann G, Drachman D, Folstein M, Katzman R, Price D, Stadlan E: Clinical diagnosis of Alzheimer's disease: report of the NINCDS-ADRDA Work Group under the auspices of Department of Health and Human Services Task Force on Alzheimer's Diaease. Neurology 34:939-944, 1984
McLatchie G, Brooks N, Galbraith S, Hutchinson JS, Wilson L, Melville I, Teasdale E: Clinical neurological examination, neuropsychology, electroencephalography and computed tomographic head scanning in active amateur boxers. J Neurol Neurosurg Psychiatr 50:96-99, 19087
McLeod D: Ocular injuries from boxing (editorial). Br Med J 304:197, 1992
McNally JJ: Ban military boxing (letter). Mil Med 154:102-103, 1989
McQuillen JB, McQuillen EN, Morrow P: Trauma, sport, and malignant cerebral edema. Am J Forens Med Pathol 9: 12-15, 1988
Medved R: Über Methoden zur medizinischen Kontrolle der Boxer. Deutsch Med Wochenschr 80:191-192, 1955
Mendelblatt FI, Kirsch RE, Lemberg L: A study of comparative methods on preventing the oculocardiac reflex. Am J Ophthalmol 53:506-512, 1962
Mendez MF: The neuropsychiatric aspects of boxing. Int J Psychiatr Med 25:249-262, 1995
Mendez MF, Cherrier M, Perryman KM, Pachana N, Miller BL, Cummings JL: Frontotemporal dementia versus Alzheimers' disease: differential cognitive features. Neurology 47:1189-1194, 1996
Merkelbach (1938) quoted from Hillman E (1951)
Merz B: Is boxing a risk factor for Alzheimer's? J Am Med Ass 261:2597-2598, 1989
Metzel E: Neurologische Aspekte von Boxerverletzungen. In: Anders G, Felten R, Kirsch A (eds) Boxen und Gesundheit. Zur Frage von Langzeitschäden und ihrer Verhütung. Bundesinstitut für Sportwissenschaft. Schriftenreihe Medizin. Bd. 3. Deutscher Ärzte Verlag, Köln, 1977, pp. 184-188
Metzel E, Sellier K: Medizinisches Fachgutachten Boxen und Hirnschäden. In: Anders G, Felten R, Kirsch A (eds) Boxen und Gesundheit. Zur Frage von Langzeitschäden und ihrer Verhütung. Bundesinstitut für Sportwissenschaft. Schriftenreihe Medizin. Bd. 3. Deutscher Ärzte Verlag, Köln, 1977, pp. 17-28
Metzner A: Nach der Ringtragödie in Köln. Die Zeit, no. 163, Tuesday, July 2, 1968, p. 19
Meyran H: Seltene Boxverletzung der Streckaponeurose im Grundgelenkbereich des Mittelfingers. Monatsschr Unfallheilk 54:54-56, 1951
Michaud P: Réflexe oculo-cardiaque observé après une operation de décollement de la rétine par diathermo-coagulation. Bull Soc Ophthalmol (Paris) 50:136-137, 1938
Micheli E, Stoppani F: Lesioni traumatiche nello sport del pugilato. Boll Mem Soc Piemontese Chir 3:1082-1089, 1933
Micolier R: Contribution a l'étude des commotions cérébrales, le punch drunk. Encéphalite des boxeurs. These Med, Universität Lyon, 1942
Mikoleit H: Luftdurchlässiger Mundschutz für Boxer. Quintessenz 16:94, 1965
Mill JS: On liberty. In: Utilitarianism, Liberty and Representative Government. Dutton, New York, 1947, pp. 65-70
Millard RD: Closure of boxing lacerations between rounds. Arch Surg 86:295-298, 1963
Miller AP: Boxing – time for fresh action (editorial). Med J Austral 141:624-626, 1984
Miller BL, Ikonte C, Ponton M, et al.: A study of the Lund-Manchester research criteria for fronto-temporal dementia: clinical and single-photon emission CT correlations. Neurology 48:937-942, 1997
Miller GG: Cerebral concussion. Arch Surg 14:891-916, 1927
Miller OW quoted after Kenney HE, et al. (1940)
Miller OW: Should instruction in boxing be included in the school physical education curriculum? J School Health 34:87-89, 1964
Millspaugh JA: Dementia pugilistica (punch-drunk). US Naval Med Bull 35:297-303, 1937
Mitsuyama Y: Presenile dementia with motor neuron disease. Dementia 4:37-42, 1993
Mittelbach M: Über erworbene Spaltbildungen der Großhirnrinde. Zugleich ein Beitrag zur Frage der pathologischen Markfasergeflechte (Plaques fibromyeliniques) der Rinde. Beitr Pathol Anat 83:445-470, 1929
Mittelbach M, Spatz H: Über eine eigenartige Spaltbildung der Großhirnwindungen ('Schizogyrie'). Zentralbl Ges Neurol Psychiatr 47:700, 1927
Molitoris H: Ob ein Faustschlag für den Tod ursächlich war? Deutsch Z Ges Gerichtl Med 15:320, 1930
Molke W: Some remarks on the traumas in the football players and boxers. (Polish with Engl. summary). Polski Tygod Lek 11:1111-1114, 1956
Molke W: Zlamanie pierwszego zebra u boksera. (Polish). Chir Narz Ruchu 21:167-168, 1956
Molle : Traitement de la perte de connaissance (knock out) consécutive a un chock violent sur le creux épigastrique, par la compression des pneumogastrique, a niveau de cou. Lyon Méd 124:242-244, 1915
Monette S: Canada's health department approves boxing (letter). Lancet I:1138 1980
Monsell RM: Sports medicine. Anaesthesist attends boxing contest (letter, comment). Br Med J 309:474, 1994

Montanaro M: Boxen soll gesund sein! Münch Med Wochenschr 32: Beilage, 1963
Montanaro M: Bedeutung der audiometrischen Untersuchung der Boxer. Medizin und Sport 6:172-174, 1965
Montgomery EG: Boxing trauma. Marquette Med Rev 24:134, 1959
Mooij JJ: Boxing and brain damage. (Dutch). Ned Tijdschr Geneeskd 127:1104-1105, 1983
Moore JR: The medical aspects of boxing. New Zealand Med J 62:335-336, 1963
Moore M: The challenge of boxing: Bringing safety into the ring. Physician Sports Medicine 8:101-105, 1980
Morissey quoted after Carr CJ, Moody AM (1939)
Moriyasu N, Sato K, Miyagami M, Yoshida Y: Two case reports of successful treatment on acute subdural hematoma caused by boxing (Japanese with Engl. summary) Brain Nerve (Tokyo) 21 877-880, 1969
Moriyasu N, Sato K, Nirei M: Electroencephalography in boxers - comparison with cases of general head injury. (Japanese with Engl. summary). No To Shinkai (Tokyo) 21:757-765, 1969
Morris D: Boxing, discussion remark. J Am Med Ass 249:2779, 1983
Morrison RG: Medical and public health aspects of boxing. J Am Med Ass. 255:2475-2480, 1986
Mortimer JA: Epidemiology of post-traumatic encephalopathy in boxers. Minnesota Med 68:299-300, 1985
Moseley IF: The neuroimaging evidence for chronic brain damage due to boxing. Neuroradiology 42:1-8, 2000
Moser M, Schmid P, Wolf W: Zentrale Gleichgewichtsstörungen bei Boxern. Laryngol Rhinol Otol 59:4 67-471, 1980
Mougeot A, Duverger: Bradycardie et réflexe oculo-cardiaque dans les traumatismes oculaires, crâniens, encéphaliques. Presse Méd 25:730-732, 1917
Moulonguet M: Discussion remark. Bull Acad Nat Méd (Paris) 132:406, 1948
Moutoussis M: On the barbaric nature of boxing: thoughts following the panorama debate on banning boxing after yet another death. Med Confl Surv 12:62-63, 1996
Moxon J: Boxing injuries (letter). Br Med J 305:1438, 1992
Müller E: Diagnose und Beurteilung der Boxerencephalopathie. Monatsschr Unfallheilk 61:117-123, 1958
Müller E: Probleme des cerebralen Boxschadens. Med Welt 46:2323-2328, 1963
Müller N: Das subdurale Hämatom als Todesursache nach Boxkampf. Deutsch Z Ges Gerichtl Med 44:763-770, 1956
Müller W: Tötung durch Faustschläge auf den Kopf. Med. Dissertation, Universität München, 1932
Müller-Meiningen M: Tod im Ring. Süddeutsche Zeitung, No. 166, 22.-23. Juli, 1978, p. 4
Muller M: Au sujet de combats de box mortels. Ann Med Leg 11:144-147, 1931
Munck W: Zwei Todesfälle beim Boxen. Ugeskr Laeger 85:848-850, 1923
Munro D: Cranio-cerebral Injuries. Their Diagnosis and Treatment. University Press, New York Oxford, 1938
Munscheck H: Plötzlicher Sporttod bei einem jugendlichen Boxer infolge akuter Coronarinsuffizienz. Sportarzt 15:104-108, 1964
Murelius O, Haglund Y: Does Swedish amateur boxing lead to chronic brain damage? 4. A retrospective neuropsychological study. Acta Neurol Scand 83:9-13, 1991
Murnane: quoted from Hillman (1951)
Murphy P, Rogers CG: Medical groups take positions on boxing. Physician Sportsmed 12:23, 1984
Muskat D: The problem of death in boxers from cerebral injury. Med Proc South Africa 2: 472-485, 1956
Mutch AJ: Boxing. J Royal Army Med Corp 110:65, 1964
Nardiello VA: Twenty-five years at ringside in the recognition and evaluation of medical emergencies in boxers, in New York State Athletic Commission. Proc, July 1952, quoted after Gurdjian ES, Webster JE, 1958
Naviaux RK: Boxing, discussion remark. J Am Med Ass 249:2777-2778, 1983
Neale HW: Deaths due to participation in sports. Medicoleg Bull 137:1-5, 1964
Neary D: Neuropsychological aspects of frontotemporeal degeneration. Ann NY Acad Sci 769:15-22, 1995
Neary D: Frontotemporal degeneration, Pick disease, and corticobasal degeneration: one entity or 3? Arch Neurol 54:1427-1429, 1997
Neary D, Snowdon J: Fronto-temporal atrophy in motor neuron diseases. Brain Cogn 31:176-187, 1996
Nelson LM: Epidemiology of ALS. Clin Neurol 3:327-331, 1995
Nesarajah MS, Seneviratue KN, Watson RS: Electroencephalographic changes in Ceylonese boxers. Brit Med J I: 866-868, 1961
Neubuerger KT: Some neuropathologic aspects of boxing. Industr Med Surg 9:440-441, 1960.
Neubuerger KT quoted from Peters G (1962)
Neubuerger KT, Braunmühl A von: Hirnverletzung. In: Bumke O (ed) Handbuch der Geisteskrankheiten, Bd. 11, Teil 7. Springer, Berlin, 1930, quoted from Peters G (1962)
Neubuerger KT, Sinton DW, Denst J: Cerebral atrophy associated with boxing. Arch Neurol Psychiatr 81:403-408, 1959
Neumann H: Über einen Epileptiker mit klinisch berichteten 6.022 großen Krampfanfällen. Arch Psychiatr Nervenkrankh 199:472-476, 1959
Neustatter WL: Discussion. In: Bass AL, Blonstein JL, James RD, Williams JGP (eds) Medical Aspects of Boxing. Pergamon Press, Oxford London, 1965, pp. 75-76
Newman JA: Tissue stress tolerances and injury indices. Proc 9th Internat Congr of Biomechanics. Symposium on Biomechanical Assessment of Sports Protective Equipment. Internat Congr Biomechanics, Canada, August 1983, pp. 11-21
Nikolic V, Rukavina V: Anatomic variations of the temporomandibular joint and the possibility of ear and base-of-skull injuries in boxers. (Kroatian with Engl. summary). Chir Maxillofac Plast (Zagreb) 7:33-40, 1969
Noble C: Hand injuries in boxing. Am J Sport Med 15:342-346, 1987
Noonan: Boxing and the brain. The New York Times Magazine, June 12, !983
Norman D: Boxing and the brain. The sweet science contributes

the most purposeful, relentless and sustained assault on the human brain in all the sport. The New York Times Magazine, June 12, 1983
Novich MM: Medicine in sports. J Newark Beth Israel Hosp 8:33-42, 1957
Novich MM: Comments on boxing and especially professional boxing. J Sports Med Phys Fitness 4:46-48, 1964
Novich MM: Professional boxing needs a face lifting. J Sports Med 5:82-87, 1965
Novich MM: Boxing program for youngsters. New York State J Med 70:1300-1305, 1970
Novich MM: What really happens in boxing. Phys Sportsmedicin 2:28-32, 1974
Oates CJ: On boxing. The New York Times Magazine, June 16, 1985, pp. 29-38
Oates CJ: On Boxing. Dolphin Doubleday, Garden City, NY, 1987
Oates CJ: On Boxing. In: Oates CJ, Halpern D (eds) Reading the Fights. Holt, New York, 1988, pp. 286-305
Oelker CE, Primiano FD, Wooding HC: The Investigation of the Parameters of Head Injury Related to Acceleration and Deceleration. Technology Incorporated, Report TI 2610-60-4, 1966
Oelman BJ, Rose CME, Arlow KJ: Boxing injuries in the Army. J Royal Army Med Corps 129:32-37, 1983
Oliver WR: Boxing, reply to an discussion remark of Lundberg GD. J Am Med Ass 253: 198, 1985
O'Neil G: Boxing and medicine. Br J Sports Med 33:153, 1999
Orton I: The deaths from (?) boxing. Lancet II:1366, 1901
Osnato M, Giliberti V: Concussion of the brain. Arch Neurol Psychiatr 16:667-669, 1926
Ostertag B: Die Pathologie der Boxverletzungen im Rahmen der Sporttraumen. Deutscher Sportärztekongress Berlin, 1952. Limpert, Frankfurt, 1953, p. 60-69
Ostojic B, Nagulic I, Markovic V: Fatal accidents in boxing. (Serbokroatian with Engl summary). Acta Chir Jugosl 11:97-107, 1964
Ottlitz KH: Cochlearis- und Vestibularisuntersuchung bei jugendlichen Boxern. Medizin und Sport 2:204-206, 1962
Pagno A, Perella F: Caratteri audiologici della' sordita da trauma craneo. Arch Ital Laringol 64:183, 1956
Pakovski K: Nose injuries in boxers. (Polish with Engl. summary). Otolargyngol Pol 18: 271-276, 1964
Pakovski K: Investigation of the organ of hearing in boxers. (Polish with Engl. summary) Vesn Otorinolaringol 27:33-35, 1965
Palazzi S: Cranio-faciale Traumen bei Boxern und Kieferschutzapparat. Z Stomatol 23: 873-882, 1925
Palinkasev A: Injuries in boxing from the forensic point of view. A historical review. (Serbokroatian with Engl. summary) Med Art (Serajevo) 25:61-73, 1971
Palitz L, Frist TF, Kocour E: The effects of pressure on the carotid sinus at various altitudes. Case reports. J Aviat Med 14:346, 1943
Palm E, Strömblad R: Respiratory and circulatory responses to manipulations of the eye. Arch Ophthalmol 32:15-629, 1954
Palmer E, Lieberman TW, Burns S: Contusion angle deformity in prizefighters. Arch Ophthalmol 94:225-228, 1976
Pampus F: Elektroencephalographische Befunde bei wiederholten Kopfverletzungen. Verhandl Deutsch Gesellsch Unfallheilk, 19. Tagung, Goslar, 1955. Hefte Unfallheilk 52:144-149, 1956
Pampus F: Die Boxschäden des Nervensystems. Sportarzt und Sportmedizin, Heft 9:275-285, 1964
Pampus F, Grote W: Electrencephalographische und klinische Befunde bei Boxern und ihre Bedeutung für die Pathophysiologie der traumatischen Hirnschädigung. Arch Psychiat Z Neurol 194:52-178, 1956
Pampus F, Müller N: Über einen Todesfall nach Boxkampf. Deutsch Z Nervenheilk 174: 177-188, 1956
Panov IE, Naumenko VG: Closed head injury in boxers (clinical and forensic medical data). (Russian). Sud Med Ekspert 22:12-16, 1979
Parker AJ: Angular Accelerations of the Head. Hymatic Report PTM: 163, July, 1965
Parker WL: Traumatic encephalopathy ('punch-drunk') of professional pugilists. J Neurol Psychopathol 15:20-28, 1934
Parker WL, Trifunov J: The sport of amateur boxing - good or evil? Can Med Ass J 83: 432-435, 1960
Parkinson D: Concussion is completely reversible; an hypothesis. Med Hypotheses 37:37-39, 1992
Parkinson D, West M, Pathiraja T: Concussion: Comparison of humans and rat. Neurosurgery 3:176-180, 1978
Parkinson D: The biomechanics of concussion. Clin Neurosurg 29:131-145, 1982
Parry CBD: Hand Injuries. In: Bass AL, Blonstein JL, James RD, Williams JGP (eds) Medical Aspects of Boxing. Pergamon Press, Oxford London New York, NY, 1965, pp. 51-53
Parry CH: An inquiry into the symptoms and causes of the syncope anginosa, commonly called angina pectoris. Castell & Davis, London, 1799
Parsons-Smith G, Williams D: Cerebral embolism following contusion of the heart (in boxing). Brit Med J I:10-12, 1949
Pashby TJ, Pashby RC: Treatment of Sports Eye Injuries. Chap. 30. In. Schneider RC, Kennedy JC, Plant ML (eds) Sports Injuries. Mechanisms, Prevention, and Treatment. Williams & Wilkins, Baltimore London Sydney, 1985, pp. 572-591
Patterson RH: On boxing and liberty. J Am Med Ass 255:2481-2482, 1986
Paul M: A fatal injury at boxing (traumatic decerebrate rigidity). Br Med J I:364-366, 1957
Paul W: Über Augenverletzungen beim Boxsport. Deutsch Gesundheitswesen 21:2002-2007, 1966
Paulsen K, Hundhausen T: Hörschäden durch Boxen. Z Laryngol Rhinol Otol Grenzgeb 50:297-324, 1971
Payne EE: Brains of boxers. Neurochirurgia 11:173-188, 1968
Pearce JM: Boxers' brains. Br Med J 288:933-934, 1984
Pearn J: Reducing brain damage in boxers. Med J Austral 168:418, 1998
Pearn J: WMA, boxing and the brain. Med J Austral 141:68, 1984
Perez G: Traumatismas cerebraux de la boxe. Ann Chir 43:703-706, 1989

Petal M: Boxing is murder. Phys Sportsmed 2:67-68, 1974
Peterman AE: Boxing, discussion remark. J Am Med Ass 249:2779, 1983
Peters G: Über gedeckte Gehirnverletzungen (Rindenkontusion) im Tierversuch. Zentralbl Neurochir 8:172-208, 1943
Peters G: Die gedeckten Gehirn- und Rückenmarksverletzungen. In: Scholz W (ed) Handbuch der speziellen pathologischen Anatomie und Histologie. Bd. XIII/3. Springer, Berlin Göttingen Heidelberg, 1955, pp. 84-143
Peters G: Ergebnisse vergleichender anatomisch-pathologischer und klinischer Untersuchungen an Hirngeschädigten. In: Paetzold F, Dierkes C, Goetz F (eds) Arbeit und Gesundheit, Neue Folge. Heft 74. Thieme, Stuttgart, 1962
Petrovcic F, Medved R, Katunaric D: Radiographical changes of boxers skull bones. (Serbocroatian). Zagreb, Izdanja Zavoda Za Fizicki Odgoj :1-11, 1954
Petzetakis M: La réflexe oculo-cardiaque chez les sujets normaux non bradycardiques. Bull Soc Méd Hôp (Paris) 37:562-570, 1914
Playfor SD: Taking a swing at boxing. J Paediatr Child Health 35:153, 1999
Phillips (1954) quoted after Blonstein JL, Clarke E (1954)
Phillips JP, Jones HM, Hitchcock R, Adams N, Thompson RJ: Radioimmunoassay of serum creatine kinase BB as index of brain damage after head injury. Br Med J 281:777-779, 1980
Pia HW: Die Bewußtseinsstörungen bei Hirnverletzungen. Langenbecks Arch Deutsch Z Chir 281:652-661, 1956
Pierini LD: El boxeo como deporte. Incursion clinico neurologica y deontologica. Sem Med (Buenos Aires) 120:209-210, 1962
Pijade R: Alterations radiologiques de la structure des os de la main des boxeurs. In: Prokop K, Rinner L (eds) Sportmedizin. Bericht des 13. Internat Kongr Sportmed, Wien. Hollinek, Wien, 1961, pp. 211-212
Pincemaille Y, Trosseille X, Mack P, Breton F, Renault B: Some New Data Related to Human Tolerance Obtained From Volunteer Boxers. In: Backaitis H (ed) Biomechanics of Impact Injury and Injury Tolerances of the Head-Neck Complex. PT-43. Soc Autom Engin, Warrendale, Pa, 1993, pp. 233-246
Pincemaille Y, Trosseille X, Mack P, Brun-Cassan F, Tarriere C, Renault B, Breton F: Investigation of Relationships between Physical Parameters and Neurophysiological Response to Head Impact. DOT-Kontract, 57-86-C-00037, June 1988
Pincemaille Y, Trosseille X, Mack P, Tarriere C, Breton F, Renault B: Some New Data Related to Human Tolerance Obtained from Volunteer Boxers. Proc 33rd Stapp Car Crash Conf, Soc Autom Engin, Warrendale PA, 1989, pp. 177-190
Piquet J: Hörstörungen nach Schädeltraumen ohne erkennbare Beteiligung des Ohres. HNO Wegweiser 4:256, 1953/1954
Piquet J, Decrois G: La surdite par trauma craniens. Sa place en pathologie auditive. Zentralbl Hals Nasen Ohrenheilk 50:395, 1954
Piston RE: Boxing, personal freedom and the right of lions to Christians (letter). J Am Med Ass 256:1895, 1986
Place HM, Ecklund JM, Enzenauer RJ: Cervical spine injury in a boxers: should mandatory screening be instituted? J Spinal Disord 9:64-67, 1996
Plant JR, Butt JC: Laceration of vertebral artery. A historic boxing death. Am J Forens Med Pathol 14:61-64, 1993
Planten JT: Oculocardiac reflexes in operations for strabismus. Ophthalmologica 135:287-291, 1958
Planten JT: Clinical observation of the oculo-cardiac reflex. Ophthalmologica 146:300-302, 1963
Plimpton G: Shadow Box. Putnam's, New York, 1977
Plum F, Posner J: The Diagnosis of Stupor and Coma. Davis, Philadelphia, 1980
Pöschl M: Sportarzt und Tod im Boxring. Sportmedizin 8:325-329, 1957
Pöschl M: Schäden beim Boxsport. Fortschr Med 78:75-76, 1960
Pöschl M, Krieger G: Todesfälle beim Sport und medizinische Fragen ihrer Prophylaxe. Münch Med Wochenschr 105:2205-2216, 1963
Pohle quoted after Pöschl M, Krieger G (1963)
Polson CJ, Gee DJ: The Essentials of Forensic Medicine. Boxing Injuries. Pergamon Press, Oxford, 1973, pp. 182-185
Popielski B: Todesfälle beim Boxen. (Polish). Polska Gaz Lek 13: 293-295, 1934. Ref Deutsch Z Ges Gerichtl Med 5:48, 1935
Porter M, O'Brien M: Incidence and severity of injuries resulting from amateur boxing in Ireland. Clin J Sports Med 6:97-101, 1996
Porter MD, Fricker PA: Controlled prospective neuropsychological assessment of active experienced amateur boxers. Clin J Sports Med 6:90-96, 1996
Porter ML, Hodgkinson JP, Hirst P, Wharton MR, Cunliffe M: The boxers' fracture: A prospective study of functional recovery. Arch Emerg Med 5:212-215, 1988
Posner MA, Ambrose L: Boxer's knuckle: Dorsal capsular rupture of the metacarpophalangeal joint of a finger. Joint Hand Surg 14A:229-236, 1989
Pothe H: Chronische Subduralhaematome bei Boxern. Beitr Neurochir 8:232-237, 1964
Potter JM: Letter: Boxers' brain damage. Lancet II:1270, 1974
Prakash R: Cockfighting and boxing: a parallel and paradox. J Tennessee Med Ass 78:210, 1985
Primavesi L: El médico frente a la práctica del box. Dia Med (Buenos Aires) 27:147-148, 1955
Prohic H: A fatal accident in boxing. (Czech with German summary). Med Archiv 13:95-98, 1959
Prokop L: Die Boxschäden. In: Arzt und Boxsport. Vorträge der Enquete anläßlich der Generalversammlung, Selbstverlag, Wien, November, 1956, pp. 1-8
Prokop L: Vorschläge zur Entschärfung des Boxsportes. In: Anders G, Felten R, Kirsch A (eds) Boxen und Gesundheit. Zur Frage der Langzeitschäden und ihrer Verhütung. Bundesinstitut für Sportwissenschaft, Schriftenreihe Medizin. Bd. 3. Deutscher Ärzte Verlag, Köln, 1977, pp. 169-170
Pruche A; Le refléxe oculo-cardiaque; etude électrocardiagraphique. Arch Mal Coeur 26:724-751, 1933
Prümmer, quoted from Hillman (1951)
Puxeddu P, Gabrielli L, Orru A: Indagini audiometriche sui pugili. Medicina Sportiva (Torino) 18:812-819, 1964
Quandt J, Sommer H: Beitrag zur Pathogenese der Encephalo-

pathia pugilistica (Boxerencephalopathie). Psychiatr Neurol Psychol (Leipzig) 17:448-451, 1965
Raabe K: Über Mundschutzgeräte für Boxsportler und Eishockeyspieler, ihre Aufgabe und zweckmässige Konstruktion. Deutsch Gesundheitswes 13:572-575, 1958
Raevuori-Nallinmaa S: Brain injuries attributable to boxing. Acta Psychiatr Neurol (Suppl) 60:51-56, 1951
Rafalowska J, Hausmanova-Petrusewicz I: 3 Fälle von traumatischer Encephalopathie bei Boxern. (Polish with Engl summary). Polski Tydn Lek 14:2177-2181, 1959
Rakusin W: Traumatic hyphaema - a clinical study. Ph. D. Dissertation, University of Witwaterstrand, 1970
Randall RD: Prospective Study of Central Nervous System Function in Amateur Boxers in the United States. J Epidemiol 139:573-588, 1994
Raschka C, Parzeller M, Banzer W: 15 years insurance statistics of incidents and accident types of combat sports in Rhineland-Pfalz Federal Sports Club. Sportverl Sportschaden 13:17-21, 1999
Rauchfleisch U, Radu E: Zur Persönlichkeit von Amateurboxern - Resultate einer testpsychologischen Untersuchung. Z Psychosom Med Psychoanal 29:276-285, 1983
Ravina A: L'encéphalite traumatique ou punch drunk. Presse Méd 45:1362-1364, 1937
Ravina A: Les séquelles mentales des traumatismes craniens. Presse Méd 49:1065-1068, 1941
Ravina A: Les accidents mortels provoqués par les sports de compétition. Presse Méd 60: 1136, 1952
Ravina A: Resultats fournis par l'examen des électroencéphalogrammes et des clichés radiographiques du crane de boxeurs. Presse Méd 63:419-420, 1955
Ravina A: La pratique de la boxe et l'organisation des manifestations publique de ce sport. Presse Méd 71:1629-1631, 1963
Refshauge JG: The medical aspects of boxing. Med J Austral 1:611-613, 1963
Reisch O: Der Boxkampf und seine Auswirkungen auf des Nervensystem. Berliner Gesellschaft für Neurologie und Psychiatrie, Sitzung vom 12, April, 1937. Zentralbl Neurol Psychiat 90: 703-704, 1938
Reisner H: Bericht der Kommission des Obersten Sanitätsrates Österreich 1974: Medizinische Aspekte des Boxens. In: Anders G, Felten R, Kirsch A (eds) Boxen und Gesundheit. Zur Frage von Langzeitschäden und deren Verhütung. Bundesinstitut für Sportwissenschaft, Schriftenreihe Medizin, Bd. 3. Deutscher Ärzte-Verlag, Köln, 1977, pp. 166-169
Rhode J, GromE, Bojares AC, Anselmino A, Capriles MA, Rivas C: Am J Ophthalmol 46:367, 1958
Richards NG: Ban boxing: The intent is wrong. Neurology 34:1485-1486, 1984
Richards NG: Boxing: The intent is wrong. Virginia Med 112:122-1231 1985
Richards PG, Hastfield R, Grant HC: Brain damage from boxing (letter). J Neurosurg 65:723, 1986
Richards P: Boxing and the neurosurgeon (editorial). Br J Neurosurg 7:595-596, 1993
Ricker G, Döring G: Commotio cerebri. In: Lubarsch O, Henke E, Rössle R, Uehlinger E (eds) Handbuch der speziellen pathologischen Anatomie und Histologie. Bd.13/3. Scholz W (ed) Nervensystem Springer, Berlin Göttingen Heidelberg, 1955, pp. 177-239
Riederer von Paar V: Ein neuer Beitrag zur Frage der Schizogyrie (Windungsspaltbildung) und ihrer traumatischen Entstehung. Arch Psychiatr 106:71-105, 1936
Riga quoted from Kinsolving (1971)
Ripperger EA, Unterharnscheidt F: Prototypes of Head Injuries: Application to Animal Experiments. Internat Confer on the Biokinetics of Impact, Amsterdam, 26-27 June 1973, pp. 259-270
Roberts AH: Brain Damage in Boxers. A Study of the Traumatic Encephalopathy among Professional Boxers. Pitman Med Scientific Publ., London, 1969
Roberts GW: Immunocytochemistry of neurofibrillary tangles in dementia pugilistica and Alzheimer's disease: evidence for common genesis. Lancet II:1456-1458, 1988
Roberts GW, Allsop D, Bruton C: The occult aftermath of boxing. J Neurol Neurosurg Psychiatr 53:373-378, 1990
Rochon M: Présentation d' un cas: l'encephalopathie des boxeurs. Can J Psychiatr 39:211-214, 1994
Rocke RS: Ring deaths believed to be no cause to abolish boxing. Medical News World Wide Report, 1967
Rockwood CA, Green DP (eds) Fractures in Adults. 2nd ed. Lippincott, Philadelphia Toronto, 1984, p. 492
Rodriguez G, Ferrillo F, Montano V, Rosadini G, Sannita WG: Regional blood flow in boxers (letter). Lancet II:858, 1983
Rodriguez del Valle J: Evoked potentials in boxers before and after a fight. (Spanish). Bol Assoc Med P R 77:19-20, 1985
Rogers T: Canadian boxer dies after fight injury. The New York Times, July 8, 1980, B 10
Rohrschneider W: Sportschädigungen und -verletzungen der Augen. Z Ärztl Fortbild (Jena) 31:464-465, 1933
Romino JD: Boxing and intracranial injuries. West Virginia Med J (Charleston) 45:19, 1949
Rosegay H: Limited value of carotid pulse in diagnosis of internal carotid thrombosis. Neurology 6: 143-145, 1956
Rosenbaum MA: A survey of sport played in the Army. J R Army Med Corps 113:65-74, 1967
Roskam J: Un syndrome nerveuau: syncope cardiaques graves et syncopes répétees par hyperréflectivité sino-carotidien. Press Md :590, 1930
Ross RJ, Casson IR, Siegel O, Cole M: Boxing injuries: Neurologic, radiologic, and neuropsychologic evaluation. Clin Sports Med 6:41-51, 1987
Ross RJ, Cole M, Thompson JS, Kim KH: Boxers-computed tomography, EEG, and neurological evaluation. J Am Med Ass 249:211-213, 1983
Ross RJ, Cole M, Thompson JS, Kim KH: Boxing, reply to a discussion remark of Engelsberg AL. J Am Med Ass 249:2777, 1983
Ross RJ, Cole M, Thompson JS, Kim KH: Boxing, reply to a discussion remark of Naviaux RK. J Am Med Ass 249:2778, 1983
Ross RJ, Cole M, Thompson JS, Kim KH: Boxing, reply to discussion remark of Sukoff MH. J Am Med Assoc 249:779, 1983

Ross WH: Traumatic retinal dialysis. Arch Ophthalmol 91:1371-1374, 1981
Rossier PH, Dressler M, Simmen R: Der zenrale Typ des Karotissinus-Syndroms. Schweiz Med Wochenschr 70:563-570, 1940
Röttgen P: Sobre las lesiones posttraumaticas de las irrigacion sanguinea del cerebro, especialmente despues les lesiones producidas en el boxeo. Med Espan 34:35-40, 1955
Rouvillois, Bercher: Lésions traumatique du maxillair inférior chez les boxers. Rev d' Orthop 33:381, 1926
Rowbotham GF: Acute Injuries of the Head. 4th ed. Livingstone, Edinburgh London, 1964
Roy B, Bernier-Cardou M, Cardou A, Plamondon A: Influence des bandages sur la d'Impact des coups de poing de la boxe. Can J Appl Sport Sci 9:181-187, 1984
Rozmaric A: Tod beim Boxen. (Czech). Cas Lek Cesk 63:1693-1696, 1924. Ref. Deutsch Z Ges Gerichtl Med 5:448, 1925
Rudiengaard P: Boxer's encephalopathy. (Danish with Engl. summary). Ugeskr Laeger 136:2874-2875, 1974
Rugg-Gunn A: Eye Injuries. In: Bass AL, Blonstein JL, James RD, Williams JGP (eds) Medical Aspects of Boxing. Pergamon Press, Oxford London New York, 1965, pp. 9-11
Russell WR: Amnesia following head injuries. Lancet II:762-763, 1935
Russell WR: The Traumatic Amnesias. In: Vinken PJ, Bruyn GW (eds) Handbook of Clinical Neurology. Vol. 3. North-Holland Publ, Amsterdam, 1969, pp. 293-295
Russell WR: The Traumatic Amnesias. Oxford University Press, London, 1971
Russell WR, Nathan PW: Traumatic amnesia. Brain 69:280-300, 1946
Russell WR, Smith A: Post-traumatic amnesia in closed head injury. Arch Neurol 5:4-17, 1961
Ryan AJ: Hit in the head. Phys Sportsmedicine 49:51, 1979
Ryan AJ: Death in the ring. Phys Sportsmedicine 8:47-48. 1980
Ryan AJ: Intracranial injuries resulting from boxing: A review. Clin Sports Med (US) 6:31-40, 1987
Ryan AJ: Eliminate boxing gloves. Physician Sportsmedicine 11:49, 1983
Ryan AJ: Intracranial injuries resulting from boxing (1918-1989). Clin Sports Med 6:31-39, 1987
Ryan HJ: The 'sport' of boxing (letter, comment) Am J Sports Med 21:764, 1993
Sabharwal RK, Sanchetee PC, Sethi PK. Dhamija RM: Chronic traumatic encephalopathy in boxers. J Ass Physicians India 35:571-573, 1987
Salazar AM, Masters CL, Gajdusek DC, Gibbs CJ: Syndromes of amyotrophic lateral sclerosis and dementia: relation to transmissible Creutzfeld-Jakob disease. Ann Neurol 14:17-26, 1983
Sammons JT: Let's knock prizefighting out of the ring. Newsday, Friday, June 24, 1988, p. 93
Sammons JT: Why physicians should oppose boxing: An interdisciplinary history perspective. J Am Med Ass 261:1484-1486, 1989
Sances A, Myklebust J, Kostreva D, Cusick JF, Weber R, Houterman C, Larson SJ, Maiman D, Walsh P, Colbert M, Unterharnscheidt F, Ewing CL, Thomas DJ, Sigesmund K, Ho K, Salzberg B: Pathophysiology of Cervical Injuries. Proc 26th Stapp Car Crash Conf, Soc Autom Engin, Warrendale, PA, pp. 113-142.
Sangrado : Boxe et médecine. Scalpel (Bruxelles) 115:670-671, 1962
Sarno MT: The nature of verbal impairment after closed head injury. J Nerv Ment Dis 168: 685-692, 1980
Saunders RL, Harbaugh RE: The second impact in catastrophic contact-sports head trauma. J Am Med Ass 252 538-539, 1984
Savastano AA: Boxing - its pros and cons. Rhode Island Med J 45:617-620, 1962
Saville PR In: Bass AL, Blonstein JL, James RD, Williams JGD (eds) Medical Aspects of Boxing. Pergamon Press, Oxford London New York, 1965, pp. 15-16
Sawauchi S, Murakami S, Tani S, Ogawa T, Suzuki T, Abe T: Acute subdural hematoma caused by professional boxing. (Japanese with Engl. summary). No Shinkei Geka 24:905-911, 1996
Schacter DL, Crovitz HF: Memory function after closed head injury: A review of the quantitative research. Cortex 13:150-176, 1977
Schaller quoted after Carr CJ, Moody AM (1939)
Scheppach A: Tödliche Folgen von Faustschlägen und ihre forensische Beurteilung (einschliesslich Folgen von Boxverletzungen). Med. Dissertation, Universität Hamburg, 1938
Scherf D, Boyd LJ: Herzkrankheiten und Gefäßerkrankungen. 6. Aufl. Springer, Wien, 1955
Schmid L: Sports medicine and boxing. (Czech with Engl. a. French summary). Cas Lek Cesk 102:427-429, 1963
Schmid L: Die Bedeutung des Boxhandschuhs für die Prophylaxe von Verletzungen. Medizin und Sport 10:117-119, 1970
Schmid L: Kidney lesions in boxing. J Sports Med Phys Fitness 10:265-268, 1970
Schmid L, Hajek E, Votipka F, Teprik O, Blonstein JL: Experiences with headgear in boxing. J Sports Med 8:171-176, 1968
Schmid L, Votipka E, Teprik O: Assessing the intensity of boxing gloves of different weight. Cas Lek Ces 107:40-942, 1968
Schmidt-Olsen S, Kaalund Jensen S, Mortensen V: Amateur boxing in Denmark. Am J Sports Med 18:98-100, 1990
Schmitt HP, Yang Y, Forst H: Frontal lobe degeneration of non-Alzheimer type and Pick's atrophy: lumping or splitting? Eur Arch Psychiatr Clin Neurosci 245:299-305, 1998
Schneider RC, Kahn EA, Crosby EC, Taren JA (eds) Correlative Neurosurgery. 3rd ed. Thomas, Springfield, 1982
Scholz W: Die nicht zur Erweichung führenden unvollständigen Gewebsnekrosen (elektive Parenchymnekrose). In: Lubsrch O, Henke F, Rössle R (eds) Handbuch der speziellen pathologischen Anatomie und Histologie. Bd. 13/1, Scholz W (ed) Nervensystem. Springer, Berlin Göttingen Heidelberg, 1957, pp. 1284-1325
Schönbauer L, Brunner H: Commotio cerebri. In: Alexander G, Marburg O (eds) Handbuch der Neurologie des Ohres. Bd. 2. Urban & Schwarzenberg, Wien, 1928, pp. 273-304
Schunk H: Congenital dilatations of the septum pellucidum. Radiology 81:610-618, 1963

Schwartz ML, Hudson AR, Fernie GR, Hayashi K, Coleclough AA: Biomechanical study of full-contact karate contrasted with boxing. J Neurosurg 64:248-252, 1986
Schwarz B: Chronische Schäden des Zentralnervensystems bei Boxern. Deutsch Gesundheitswesen 8:845-847, 1953
Schwarz B: Veränderungen des Nervensystems durch Sportschädigungen bei Boxern. Theorie und Praxis der Körperkultur, Heft 6:549-557, 1954
Schwarz B: Über Schädel-Hirn-Traumen beim Boxen und ihre Verhütung. Deutsch Gesundheitswesen 10:1319-1324, 1955
Schwarz B: Sportschäden und Sportverletzungen des Zentralnervensystems bei Boxsportlern. Medizin und Sport 5:168-170, 1965
Schwarz B: Über psychopathologische Syndrome bei Boxsportlern auf Grund von Verlaufsuntersuchungen. Medizin und Sport 10:55-60, 1970
Schwarz B, Gebelt H: Verlaufsuntersuchungen bei Boxsportlern unter besonderer Berücksichtigung neurologisch-psychiatrischer und hirnelektrischer Befunde. Medizin und Sport 2:197-201, 1962
Schwidde JT: Incidence of cavum septi pellucidi and cavum Vergae in 1032 human brains. Arch Neurol Psychiatr 67:625-632, 1952
Scott I: Youth boxing ban in some Australian jurisdictions. Inj Prev 5:77, 1999
Scott J: Oral communication to Maguire JI, Benson WE (1986)
Sedan J: Exophorie persistante chez un boxeur après commotion cranienne. Rev Oto-Neuro-Opthalmol 10:79-681, 1932
Segal P: The problem of traumatibility of the organ of vision in boxing. (Polish with Engl. summary). Klin Oczna 29:363-376, 1959
Sellier K: Biomechanische Aspekte des Boxens. In: Stucke K (Hrsg) Verhandlungen, Deutscher Sportärztebund. 24. Tagung, Würzburg, 14-17. Oktober, 1971. Demeter, Gräfelfing, 1971, pp. 113-115
Sellier K: Muß das Boxen verboten werden? Verblödung als Mitgift. Deutsche Zeitung, Nr. 34, 24. August 1973, p. 24
Sellier K: Physikalische Grundlagen von Boxschäden. In: Anders G, Felten R, Kirsch A (eds) Boxen und Gesundheit. Zur Frage von Langzeitschäden und ihrer Verhütung. Bundesinstitut für Sportwissenschaft. Schriftenreihe Medizin. Bd. 3. Deutscher Ärzte Verlag, Köln, 1977, pp. 171-173
Sellier K, Metzel E: Boxen und Hirnschäden. Gutachtliche Stellungnahme, 1975
Sellier K, Unterharnscheidt F: Experimental Studies on the Mechanism of Nonpenetrating Brain Injuries. In: IV Internat Congr Neuropathol, München, September 4-8, 1961, Abstracts, Stuttgart, Thieme, 1961, p. 82
Sellier K, Unterharnscheidt F: Untersuchungen zur Mechanik der gedeckten Schadelhirnverletzungen. In: IV. Internat. Congr. Neuropathol., München, September 4-8, 1961, Proc. Thieme, Stuttgart, 1962, pp. 226-230
Sellier K, Unterharnscheidt F: Mechanik und Pathomorphologie der Hirnschäden nach stumpfer Gewalteinwirkung auf den Schädel. Hefte Unfallheilkunde, Heft 76. Springer, Berlin Göttingen Heidelberg, 1963
Sellier K, Unterharnscheidt F: Mechanik der Gewalteinwirkung auf den Schädel. Abstracts, Internat. Congr. Neurol. Surg., Copenhagen, August 23-29, 1964. In: Excerpta Medica Internat. Congr Ser 93:55-61, 1965
Sellier K, Unterharnscheidt F: The Mechanics of the Impact of Violence on the Skull. Proc. Internat. Congr. Neurol. Surg., Copenhagen, August 23-28, 1965. In: Excerpta Medica Internat. Congr. Ser. 110: 87-92, 1965
Seltzer AP: Surgery of the pugilist nose. Ann Otol Rhinol Laryngol 59:924-930, 1950
Seltzer AP: What should we do about boxing? J Nat Med Ass 59:64-65, 1967
Sercl M: Neurological findings in boxers. (Czech with Engl. summary). Sbornik Lek (Praha) 57:134-144, 1955
Sercl M: On spinal cord injuries of boxers in the ring. Bratisl Lek Listy 51:341-344, 1969
Sercl M, Jaros O: Brain injury in boxing. (Czech). Sborn Praci Vla Jevp C 3: 38, 1956
Sercl M, Jaros O: Klinische Bilder der Beschädigung des ZNS bei Boxern unter Berücksichtigung der häufigsten Verletzungsmechanismen. Sportmedizin 8: 69-74, 1957
Sercl M, Jaros O: The mechanisms of cerebral concussion in boxing and their consequences. World Neurology 3:351-358, 1962
Sercl M, Jaros O: The mechanisms of closed head injuries in boxers and their after- effects. (Czech with Engl. summary). Rozhl Chir 41:597-600, 1962
Sercl M, Jaros O: Neurological findings in boxers. (Czech with Engl. a. French summary). Cas Lek Cesk 102:421-427, 1963
Sercl M, Jaros O: Les troubles tardifs posttraumatiques des blessures fermees du crane des boxers. In Proc 8th Internat . Congr. Neurol., Vol. 1. Vienna, September 5-10, 1965, pp. 253-255
Sercl M, Jaros O: Boxing and the Damage to the Nervous System. Supplement to Collection of Scientific Works of the Faculty of Medicine, Charles University of Hradec Kralove, 1968
Sercl M, Jaros O: On spinal cord injuries arising in the ring on boxers. (Czech with Engl. summary). Bratisl Lek Listy 51:341-344, 1969
Sercl M, Jaros O: Zur Rückenmarksläsion der Boxer im Ring. Sportarzt 22:158, 1974
Shaik F: Physician challenges safety of youth boxing. Phys Sportsmed 9:27, 1981
Shaw CM, Alvord EC: Cava septi pellucidi et Vergae: their normal and pathological states. Brain 92:213-224, 1969
Shaw JH: Health in physical education and athletics. J School Health 34:87-89, 1964
Sheldon T: Dutch examine testing boxers for brain damage (news). Br Med J 317:370, 1998
Sherrington CS: The Integrative Action of the Nervous System. 2nd ed. Cambridge University Press, 1947
Shively RA, Sundaram M: Ununited fractures of the scaphoid in boxers. A therapeutic dilemma. Am J Sports Med 8:40-442, 1980
Sigler LH: Clinical observations on the carotid sinus reflex. The response to cardiac sinus pressure at various age and heart rates and rhythms. Am J Med Sci 186:118-124, 1933

Silfverskiöld BP: Riskerna for Hjarnskador vid boxning. (Swedish). Nordisk Udredningsserie 16: 45-51, 1967
Silfverskiöld BP: Inget förbud nu men nya Säkerhetsföreskrifter. (Swedish). Särtryk Läkart 65:583-587, 1968
Silfverskiöld BP: Die Gefahren einer Hirnschädigung beim Boxen. In: Anders G, Felten R, Kirsch A (eds) Boxen und Gesundheit. Zur Frage der Langzeitschäden und ihrer Verhütung. Schriftenreihe Medizin. Bundesinstitut für Sportwissenschaft. Bd. 3. Deutscher Ärzte-Verlag, Köln, 1977, pp. 155-160
Silverskiöld BP: Protection against blows must be put in the boxers' gloves! (Swedish). Lakartidningen 87:1036, 1990
Silverstein A, Doniger D, Bender MB: Manual compression of the carotid vessels, carotid sinus hypersensitivity and carotid artery occlusion. Ann Intern Med 52:172-181, 1960
Sim E: Die Daumensattelgelenkverrenkung als eine seltene Sportverletzung. Sportverl Sportschaden 4:92-95, 1990
Sime quoted from Giovinazzo VC et al. (1995)
Simonson J: Death associated with amateur boxing. Ugeskr Laeger 142:895 1980
Sironi VA, Ravagnati L: Brain damage in boxers. Lancet I:244, 1983
Sironi VA, Scotti G, Ravagnati L, Franzini A, Marossero F: CT-scan and EEG findings in professional pugilists. Early detection of cerebral atrophy in young boxers. J Neurosurg Sci 26:165-168, 1982
Skehan EM: Rocky Marciano. Biography of a First Son. Kangaroo Book, New York, 1977
Slapak L: Schädigung des Herzens beim Boxen. In: Arzt und Boxsport. Vorträge der Enquete anläßlich der Generalversammlung, 1 November, 1956. Selbstverlag, Wien, 1956, pp. 14-16
Smith DJ: Ocular injuries in boxers (1986) quoted after Enzenauer RW, Mauldin (1989)
Smith EJR: Head Injuries. In: Bass AL, Blonstein JL, James RD, Williams JGP (eds) Medical Aspects of Boxing. Pergamon Press, Oxford London New York, 1965, pp. 69-71
Snowdon DA, Greiner LH, Mortimer JA, Riley KP, Greiner PA, Markesberry WR: Brain infarction and the clinical expression of Alzheimer disease: in the Nun Study. J Am Med Ass 277:813-817, 1997
Sobezuk P: Die Faust als tödliche Waffe. Med. Dissertation, Universität Düsseldorf, 1936
Soeder M, Arndt T: Affektive Störungen und Veränderungen des Hirnstrombildes bei Boxern. Deutsch Med Wochenschr 79:1792-1795, 1954
Söderbergh G: Sur les moyens actuels d'exploration du systeme sympathique en clinique et leur valeur Rev Neurol :721-756, 1926
Somen H: Le mécanisme physiologique du 'knock-out.' Paris Med 13:54-57, 1913/14
Sorensen EJ, Gilmore JE: Cardiac arrest during strabismus surgery; preliminary report. Am J Ophthalmol 41:748-752, 1956
Soutsu M: Auditory brainstem responses of knocked-out boxers. (Japanese). No To Shinkei 9:883-887, 1988
Spatz H: Pathologische Anatomie der gedeckten Hirnverletzungen mit besonderer Berücksichtigung der Rindenkontusion. Zentralbl Ges Neurol Psychiatr 78:6125-616, 1936
Spatz H: Die traumatischen Hirnschädigungen. Zentralbl Neurochir 10:350-351, 1950
Spielmeyer W: Histopathologie des Nervensystems. Springer, Berlin, 1922
Spillane JD: Five boxers. Br Med J II:1205-1210, 1962
Spillane JD: The septum pellucidum in the 'punch-drunk' syndrome. Proc Vol 1, 8th Internat Cong Neurol, Vienna, September 5-10, 1967, pp. 249-251
Spring LW: Climbing and pecking adaptations in some North American woodpeckers. Condor 67: 457-488, 1965
Steele JC, Richardson JC, Olszewski J: Progressive supranuclear palsy. Arch Neurol 10:333-359, 1964
Stein K: Können Schäden beim Wettkampfboxen vermieden werden? Sportmedizin 8: 339-343, 1957
Steinhaus AH: Boxing - legalized murder? Look 14, January 3, 1950
Steinhaus AH: Boxer's brains swapped for medals. J Am Ass Health Phys Educ Recreat 22:12-14, 1951
Steinhaus AH: Statement of boxing in high schools. J Health Phys Educ 15:384-385, 1944
Steinhaus AH: Should the sport of boxing be discontinued? Yes! Phy Educ Phi Epsilon Kappa 14:131-132, 1957
Stern MD: Head trauma as a risk factor for Parkinson's disease. Nov Disord 6:95-97, 1991
Stewart EF, Gordon B: Amateur boxing: Is there a Risk of Brain Injury? Chap. 4. In: Cantu RC (ed) Boxing and Medicine. Human Kinetics, Champaign, ILL, 1995, pp. 33-64
Stewart WF, Gordon B, Selnes O, Bandeen-Roche K, Zeger S, Tusa RF, Celentano DD, Shechter A, Liberman J, Hall C, et al.: Prospective study of central nervous system function in amateur boxers in the United States. Am J Epidemiol 139:573-588, 1994
Stewart-Gordon J: Abolish professional boxing. Reader's Digest, April 1960, pp. 81-82
Stille G: Todesfälle durch Boxschlag. Med. Dissertation, Universität Hamburg, 1938
Stiller JW, Weinberger DR: Boxing and brain damage. Psychiatr Clin North Am 8:339-356, 1985
Störtebecker TP: Posttraumatic oculocardiac syndrome from a neurosurgical point of view. Report of a case. J Neurosurg 10:682-686, 1953
Strano SA, Marais AD: A cervical spine fracture in a boxer - a rare but important sporting injury: A case report. South African Med J 63:328-330, 1983
Strassmann G, Helpern M: Tödliche Hirnverletzungen im Boxkampf. Deutsch Z Ges Gericht Med 63:70-83, 1968
Straube W: Über 'primäre geordnete Dämmerzustände' nach Schädelhirntraumen. Nervenarzt 34:452-456, 1963
Streiff EB: Troubles circulatoires aigus cerebraux et oculaires par hyperexcitabilite sino-carotidienne et cardio-aortique. Bull Schweiz Akad Med Wissensch 5:49-54, 1949
Strich SJ: Diffuse degeneration of the cerebral white matter in severe dementia following head injury. J Neurol Neurosurg Psychiatr 19:163-184, 1956

Strich S: Shearing of nerve fibres as a cause of brain damage due to head injury. Lancet II:443-448, 1961

Strömbäck P: Vad anser Ni om Boxning? Medic Foren Tidskr 25:71-72, 1947.

Strohal R: Boxsportverletzungen der Wirbelsäule. Medizin und Sport 5:177-178, 1965

Struck K: Unterkieferfrakturen beim Boxen. Med. Dissertation, Universität Hamburg, 1939

Sukoff MH: Boxing discussion remarks. J Am Med Ass 249:2778, 1983

Sulat H: Changes in central nervous system in boxers, following organium injuries. (Polish). Wychow Fiz Sport :403-407, 1962

Sullivan quoted from Richards NG (1984)

Summerskill Edith: The Ignoble Art. Heineman, London, 1956

Sunder-Plassmann P: Untersuchungen über den Bulbus caroticus bei Mensch und Tier im Hinblick auf die 'Sinusreflexe' nach Hering HE. Z Ges Anat Z Anat Entwicklungsgesch 93:567-622, 1930

Sury K von: Boxtodesfall infolge akuten Larynxödems. Deutsch Z Ges Gerichtl Med 1: 695-696, 1922

Swearer JM, Drachman DA, O'Donnell BF, Mitchell AL: Troublesome and disruptive behaviors in dementia: relationships to diagnosis and disease severity. J Am Geriatr Soc 36:784-790, 1988

Symonds C: Concussion and its sequelae. Lancet I:1-5, 1962

Szymusik A, Jedlinski J, Gatarski J: Clinical symptoms in boxers' disease. (Polish). Psychiat Pol 4:401-406, 1970

Talbot PR: Frontal lobe dementia and motor neuron disease. J Neural Transm (Suppl) 47:125-132, 1996

Tardieu G, Tardieu C: Le systéme nerveux végétatif. Masson, Paris, 1948

Tarle Y: Sur le mécanisme, la fréquence et la gravité des traumatismes craniens a répétition dans la boxe. Rev Oto Neuro-Ophtalmol 31:60-64, 1959

Tarriere C: Head Injury Mechanisms: Relationship between Experimental Measuring Techniques and Real World Injuries. AAAM/Volvo Symposium on Head Injury Mechanisms. New Orleans, LA, September 1987

Taylor RB: Traumatic encephalopathy from boxing. Br Med J I:200-201, 1953

Temmes Y, Huhmar E: Electroencephalographic changes in boxers. Acta Psychiatr 27: 175-180, 1952

Templer J, Renner GJ: Injuries of the external ear. Otolaryngol Clin North Am 23:1003-1018, 1990

Terziani A: Rilievi elettroencefalografici su un gruppo di pugili. Riv Med Leg 4:443-460, 1962. Ref, Deutsch Z Ges Gerichtl Med 55:225, 1964

Thomes Jay C 'Champ': Boxing Dirty Tricks and Outlaw Killer Punches. Loompanics Unlimited, Port Townsend, WASH, 1979

Thomassen A, Juul-Jensen P, de Fine Olivarius B, Braemer J, Christensen AL: Neurological, electroencephalographic and neuropsychological examination of 53 former amateur boxers. Acta Neurol Scand 60:352-362, 1979

Tomison JB: Man and mayhem (editorial). Southern Med J 76:1081-1082, 1983

Thorndyke A: Athletic Injuries. Prevention, Diagnosis, and Treatment. 4th ed. Lea & Febiger, New York, 1956

Tichy J: Present views on boxing – Henner's legacy. (Czech). Cas Lek Cesk 136:352-355, 1997

Timperley WR: Banning boxing. Br Med J 284: 285-289, 1982

Titeca J: L'examen électro-encéphalographique des boxers. Bull Acad Méd Belg 17:507-526, 1957

Titze K: Hirntraumatologie und Boxsport. Sportmedizin 8:74-77, 1957

Toczolowski J, Gerkowicz M, Jankowska I, Misztal S, Kowalewski J: Examination of the visual system in boxers. (Polish). Klin Oczna 93:63-64, 1991

Tokuda T: Boxers brain and the other types of post-traumatic dementia. (Japanese). No To Shinkei 45:1109-1118, 1993

Tokuda T, Ikeda S, Yanagisawa N, Ihara Y, Glenner GG: Re-examination of ex-boxers' brains using immunohistochemistry with antibodies to amyloid beta-protein and tau protein. Acta Neuropathol 82:280-285, 1991

Tomlison BE, Walton JN: Superficial haemosiderosis of the central nervous system. J Neurol Neurosurg Psychiatr 27:332-229, 1964

Totten J, Buxton R: Were you knocked out? Lancet I: 369-370, 1979

Trullenque W, Schürmann R: Aortitis sifilitica con infarto miocardio en boxeador de 36 anos. Medicina (Conception) 2:99-104, 1974

Tunney GJJ: A Man Must Fight. Houghton Mifflin, Boston New York, 1932

Tunney GJJ: Colliers, March 26, 1932

Tunney GJJ: New York Daily News, August 3, 1969

Tyrell R: Schocktod beim Boxkampf. Deutsch Z Ges Gerichtl Med 19:415-421, 1932

Uhl GR, McKinney M, Hedreen J et al.: Dementia pugilistica. Loss of basal forebrain cholinergic neurons and cortical cholinergic markers. Ann Neurol 12:99, 1982

Ulbrich P. Kibittel W, Marker K: Sportärztliche Aufgaben am Boxring. Z Ärztliche Fortbildung 84:597-600, 1990

Ule G: Zur Klinik und pathologischen Anatomie der Boxerdemenz. Klin Wochenschr 34: 1048, 1956

Ullmark R: Über die Arbeit der Ärztekommission der AIBA während der letzten 15 Jahre. Medizin und Sport 5: 165-167, 1965

Umminger Walter: Helden, Götter, Übermenschen.

Unterharnscheidt F: Experimentelle Untersuchungen über die Schädigungen des ZNS durch gehäufte stumpfe Schädeltraumen. Z Ges Neurol 147:14, 1958

Unterharnscheidt F: Experimentelle Untersuchungen über gedeckte Schäden des Gehirns nach einmaliger und wiederholter stumpfer Gewalteinwirkung auf den Schädel. Fortschr Med 80:369-378, 1972

Unterharnscheidt F: Mecanismo y anatomia pathologica de las lesiones traumaticas cerradas des cerebro producidas por la accion unica, repetida y seriada de agentes contundentes sobre el craneo. Medicina Clinica (Barcelona) 39:200-215, 1962

Unterharnscheidt F: Die gedeckten Schäden des Gehirns. Experimentelle Untersuchungen mit einmaliger, wiederholter und gehäufter stumpfer Gewalteinwirkung auf den Schädel.

Monographien aus dem Gesamtgebiet der Neurologie und Psychiatrie. Heft 103. Springer, Berlin Göttingen Heidelberg, 1963
Unterharnscheidt F: Boxen ist noch gefährlicher. Die Zeit, 25, 1966.
Unterharnscheidt F: Mechanims of Head Injury. Discussion. In: Gurdjian ES, Lange WL, Patrick LM, Thomas LM (eds) Mechanics and Pathomorphology of Closed Head Injuries. Thomas, Springfield, 1970, pp. 43-62
Unterharnscheidt F: About boxing. Review of historical and medical aspects. Texas Rep Biol Med 28:421-495, 1970
Unterharnscheidt F: Traumatische Schäden des Zentralnervensystem bei Boxern. In: Stucke K (ed) Verhandlungen Deutscher Sportärztebund, 24. Tagung, Würzburg. 14-17. October, 1971. Demeter, Gräfelfing, 1971, pp. 116-121
Unterharnscheidt F: Translational versus Rotational Acceleration - Animal Experiments with Measured Input. Proc. 5th Stapp Car Crash Conf., Soc Autom Engin, Warrendale, PA, 1971, pp. 767-770
Unterharnscheidt F: Schädelhirntrauma - Mechanogenese und Pathomorphologie. Gesellschaft für Unfallchirurgie Österreich, Meet October 8-10, 1971, Salzburg, Austria. Hefte Unfallheilk 111:1-16, 1972
Unterharnscheidt F: Mechanogenese und Pathomorphologie der traumatischen Hirnschäden. Referat, 50. Jahrestag Deutsche Gesellschaft für Rechtsmedizin, Cologne, Germany, 1971, Kurzfassung. In: Beitr Gerichtl Med 29:73-76, 1972
Unterharnscheidt F: Head Injury after Boxing. Internat Sympos Rehabilitation in Head Injury, Göteborg, Sweden, September 13-15, 1971, Scand J Rehabil Med 4:77-84, 1972
Unterharnscheidt F: Translational Versus Rotational Acceleration (Animal Experiments with Measured Input). Internat Sympos Rehabilitation in Head Injury. Göteborg, Sweden, September 13-15, 1971
Unterharnscheidt F: Boxing and head injury. Scand Neurosurg Soc, Lund, Sweden, September 16-18, 1971. Acta Neurolog Scand 48:262-263, 1972
Unterharnscheidt F: Mechanogenese und Pathomorphologie der traumatischen Hirnschäden. Z Rechtsmed / J Leg Med 71:153-221, 1972
Unterharnscheidt F: Traumatische Hirnschäden bei Boxern Eine Übersicht. Schweiz Z Sportmed 20:131-175, 1972
Unterharnscheidt F: Hirndauerschäden durch das Boxen. Deutsch Ärzteblatt 70:336-342, 1973
Unterharnscheidt F: Boxen aus nervenärztlicher Sicht. Deutsch Ärzteblatt 71:1244-1248, 1974
Unterharnscheidt F: Injuries due to Boxing and Other Sports. Chap. 26. In: Braakman R, Bruyn GW (eds) Injuries of the Brain and Skull. Handbook of Clinical Neurology. Vol. 23. Amsterdam, North-Holland Publ, 1975, pp. 527-593
Unterharnscheidt F: Biomechanik, Pathomorphologie und Pathophysiologie der Schädelhirntraumen. In: Wieck HH (ed) Neurotraumatologie, Derzeitige Schwerpunkte. 8. Internat Sympos, Erlangen, Germany, June 28-30, Thieme, Stuttgart New York, 1979, pp. 2-14
Unterharnscheidt F: Biomechanik, Pathomorphologie und Klinik der traumatischen Schäden von Gehirn und Rückenmark. In: Brickenstein R, Wedel KW (eds) Symposium 79, Wehrdienst und Gesundheit, Bd. 20, Wehr und Wissen, München, 1979, pp. 80-101
Unterharnscheidt F: Neuropathology of the Injury Caused by Linear Acceleration: Mechanisms, Prevention and Cost. Aerospace Medical Panel Specialists' Meeting, Cologne, West Germany, April 26-29, 1982, pp. 17-1-34
Unterharnscheidt F: Neuropathology of the Rhesus Monkey Undergoing -Gx Impact Acceleration. Chap. 4. In: Ewing CL, Thomas DJ, Sances A, Larson SJ (eds) Impact Injury of the Head and Spine. Thomas, Springfield, 1983, pp. 94-176
Unterharnscheidt F: Traumatic Alterations in the Rhesus Monkey Undergoing -Gx Impact Acceleration. In: Kitamura K (ed) Proc. 6th Conf Japan Soc Neurotraumatol, Tokyo, Japan, February 14, 1983, Tokyo, 1984, pp. 151-167
Unterharnscheidt F: Die traumatischen Schäden des Gehirns: Nosologie. In: Jochheim KA, Rauschelbach HH (eds) Die neurologische Begutachtung. Thieme, Stuttgart, 1984, pp. 118-147
Unterharnscheidt F: Boxing Injuries. Chap. 25. In: Schneider RC, Kennedy JC, Plant M (eds) Sports Injuries: Mechanisms, Prevention, and Teatment. Williams & Wilkins, Baltimore, 1985, pp. 462-495
Unterharnscheidt F: Morphological Findings in Rhesus Monkeys Undergoing -Gx Impact Acceleration. Eleventh Annual Meeting of the Cervical Spine Research Society, The Breakers, Palm Beach, Florida, December 7-10, 1983.
Unterharnscheidt F: Pathological and Neuropathological Findings in Rhesus Monkeys Subjected to –Gx and +Gx Indirect Impact Acceleration. Chap. 21. In: Sances A, Thomas DJ, Ewing CL, Larson SJ, Unterharnscheidt F (eds) Mechanisms of Head and Spine Trauma. Aloray, Goshen, NY, 1986, pp. 565-663
Unterharnscheidt F: Pathologie des Nervensystems 13/VII. Traumatologie von Hirn- und Rückenmark. Traumatische Schäden von Rückenmark und Wirbelsäule (forensische Pathologie). In: Doerr W, Seifert G, Uehlinger E (eds) Spezielle pathologische Anatomie. Bd.13/VII. Springer, Berlin Heidelberg New York, 1992
Unterharnscheidt F: Die indirekten Verletzungen von Kopf und Hals/Nacken (Whiplash-, Schleuder- oder Peitschenschlagverletzungen bei verschiedenen Vektorrichtungen der einwirkenden Gewalt. In: Unterharnscheidt F. Die traumatischen Schäden des Rückenmarks und seiner Hüllen mit Einschluß der Wirbelsäule unter besonderer Berücksichtigung der forensischen Traumatologie. Bd. 13/VII. In: Doerr W, Seifert G, Uehlinger E (eds) Spezielle pathologische Anatomie. Springer, Berlin Heidelberg New York, 1992, pp. 227-314
Unterharnscheidt F: Pathologie des Nervensystems 13/VI.A. Traumatologie von Hirn- und Rückenmark. Traumatische Schäden des Gehirns (forensische Pathologie). In: Doerr W, Seifert G, Uehlinger E (eds) Spezielle pathologische Anatomie. Bd. 13/VI/A. Springer, Berlin Heidelberg New York, 1993
Unterharnscheidt F: Pathologie des Nervensystems 13/VI.B. Traumatologie von Hirn- und Rückenmark. Traumatische Schäden des Gehirns (forensische Pathologie). In: Doerr W,

Seifert G, Uehlinger E (eds) Spezielle pathologische Anatomie. Bd. 13/VI.B. Springer, Berlin Heidelberg New York, 1993

Unterharnscheidt F: Translational versus Rotational Acceleration - Animal Experiments with Measured Input. In: Backaities H (ed) Biomechanics of Impact Injury and Injury Tolerances of the Head-Neck Complex. PT-43. Soc Autom Engin, Soc Autom Engin, Warrendale, PA, 1993, pp. 407-409

Unterharnscheidt F: Pathologie des Nervensystems 13/VI.C. Traumatologie von Hirn- und Rückenmark. Traumatische Schäden des Gehirns (forensische Pathologie). In: Doerr W, Seifert G, Uehlinger E (eds) Spezielle pathologische Anatomie. Bd.13/VI.C. Springer, Berlin Heidelberg New York, 1994

Unterharnscheidt F: A neurologist's reflection on boxing. I. Impact mechanics in boxing and injuries other than central nervous system damage. Revista Neurol 23:661-174, 1995

Unterharnscheidt F: A neurologist's reflection on boxing. II. Acute and chronic clinical findings secondary to central nervous system damage. Revista Neurol 23:833-346, 1995

Unterharnscheidt F; A neurologist's reflection on boxing. III. Vascular injuries. Revista Neurol 23:847-855, 1995

Unterharnscheidt F: A neurologist's reflection on boxing. IV. Late and permanent brain damage. Revista Neurol 23:1013-1026, 1995

Unterharnscheidt F: A neurologist's reflection on boxing. V. Conclude remarks. Revista Neurol 23:1027-1032, 1995

Unterharnscheidt F, Ewing C: Potential Relationship between Human Central Nervous System Injury and Impact Forces based on Primate Studies. Paper presented at the Aerospace Medical Panel Specialists Meeting, held in Paris, France, AGARD, Performance and Protection, Paris, France, November 6-10, 1978, pp. A 18 - 1-6

Unterharnscheidt F, Ewing CL, Thomas DJ, Jessop ME, Rogers JE, Willems G: Preliminary Report on the Neuropathological Findings in Rhesus Monkeys Undergoing -Gx Acceleration. Abstracts No. 201, 6th Internat Congr Neurol Surg, Sao Paulo, Brazil, 19-25 June 1977, p. 80

Unterharnscheidt F, Higgins LS: Traumatic lesions of brain and spinal cord due to non-deforming angular acceleration of the head. Texas Rep Biol Med 27:127-166, 1969

Unterharnscheidt F, Higgins LS: Pathomorphology of experimental head injury due to rotational acceleration. Acta Neuropathol 12:200-204, 1969

Unterharnscheidt F, Higgins LS: Neuropathologic Effects of Translational and Rotational Acceleration of the Head in Animal Experiments. In: Walker AE, Caveness WF, Critchley McD (eds) The Late Effects of Head Injury. Thomas, Springfield, 1969, pp. 158-167

Unterharnscheidt F, Jachnik D, Gött H: Der Balkenmangel. Bericht über Klinik, Pathomorphologie und Pathophysiologie der bisher mitgeteilten sowie von 33 eigenen Fällen von Balkenmangel and ihre differential-diagnostische Abgrenzung. Monographien aus dem Gesamtgebiet der Neurologie und Psychiatrie, Heft 128. Springer, Berlin Heidelberg New York, 1968

Unterharnscheidt F, Ripperger EA: Mechanics and Pathomorphology of Impact-related Closed Brain Injuries. In: Perrone N (ed) Dynamic Response of Biomechanical Systems. Am Soc Mechan Engin Monograph. Am Soc Mechan Engin, New York 1970, pp. 46-83

Unterharnscheidt F, Sellier K: Experimental Studies on the Pathomorphology of Nonpenetrating Brain Injuries due to Single and Repeated Application of Blunt Violence to the Head. In: IV Internat Congr Neuropathol, München, September 4-8, 1961, Abstracts, Stuttgart, Thieme, 1961, pp. 82-83

Unterharnscheidt F, Sellier K: Experimentelle Untersuchungen zur Pathomorphologie der gedeckten Schädelhirnverletzungen nach einmaliger und wiederholter stumpfer Gewalteinwirkung auf den Schädel. In: IV Internat Congr Neuropathol, München, September 4-8, 1961. In: Proc. Thieme, Stuttgart, 1962, pp. 231-233

Unterharnscheidt F, Sellier K: Mechanik und Pathomorphologie der gedeckten Schäden des Gehirns nach einmaliger, wiederholter und gehäufter stumpfer Gewalteinwirkung auf den Schädel. Arch Jap Chir 31:687-713, 1962

Unterharnscheidt F, Sellier K: Pathomorphologie der gedeckten Schädelhirnverletzungen. Abstracts: Third Internat Congr Neurol Surg, Copenhagen, August 23-28, 1965. In: Excerpta Medica Internat Congr Ser 93: 62-73, 1965

Unterharnscheidt F, Sellier K: Mechanik und Pathomorphologie der Schäden des ZNS beim Boxen. In 8th Internat Congr Neurol, Vienna, September 5-10, 1964. Proc, Vol. S, 1965, pp. 119-120

Unterharnscheidt F, Sellier K: Pathomorphology of Nonpenetrating Brain Injuries. In: Proc Internat Congr Neurolog Surg, Copenhagen, August 23-28, 1965. In: Excerpta Medica Internat Congr Ser 110:93-103, 1965

Unterharnscheidt F, Sellier K: Mechanics and Pathomorphology of Closed Brain Injuries. Chap. 26. In: Caveness WF, Walker AE (eds) Head Injury, Confer Proc. Lippincott, Philadelphia, 1966, pp. 321-341

Unterharnscheidt F, Sellier K: Traumatische Schäden des Zentralnervensystems bei Boxern. 30. Jahrestag Deutsch Gesellsch Unfallheilk. Hefte zur Unfallheilkunde 91: 162-168. Springer, Berlin Heidelberg New York, 1967

Unterharnscheidt F, Sellier K: Mechanik und Pathomorphologie der Schäden des ZNS beim Boxen. In: 8. Internat Kongr Neurol, Wien, 5-10 September 1965, Proc, Vol. S, 1966, pp. 119-120

Unterharnscheidt F, Sellier K: Vom Boxen. I. Mitteilung. Mechanik der traumatischen Schäden des ZNS bei Boxern. Medizin und Sport 10:35-45, 1970

Unterharnscheidt F, Sellier K: Vom Boxen. 2. Mitteilung. Pathomorphologie und Klinik der traumatischen Schäden des ZNS bei Boxern. Medizin und Sport 10:11-117, 1970

Unterharnscheidt F, Sellier K: Vom Boxen. Mechanik, Pathomorphologie und Klinik der traumatischen Schäden des ZNS bei Boxern. Fortschr Neurol Psychiatr 39:109-151, 1971

Unterharnscheidt F, Unterharnscheidt JT: Boxing. Historical Aspects. Elsevier, Amsterdam, in preparation

Unterharnscheidt JT, Unterharnscheidt F: Boxing. Legal Aspects. Elsevier, Amsterdam, in preparation.

Vadala G, Mello M, Roberto S, Fea A: Boxing and the eyes; morphological aspects of the ocular system in boxers. Europ J Ophthalmol 7:174-180, 1997
Vajda G: Clinical and electrocardiographic observations in post-traumatic cardiopathy. Acta Med Acad Sci Hung 16:291-301, 1960
Vallet PG, Guntern R, Hof PR, Golaz J, Delacourte A, Robakis NK, Bouras C: A comparative study of the histological tangles and immunohistochemical methods for neurofibrillary tangles and senile plaques in Alzheimer's disease. Acta Neuropathol 83:170-178, 1992
Van Allan MW: The deadly degrading sport. J Am Med Ass 249: 250, 1983
Van Allen MW: Boxing, reply to discussion remark of Morris D. J Am Med Ass 249: 2779, 1983
Van den Bergh T: Letter: Boxers' brain damage. Lancet II:1270, 1974
Vanderfield G: Head Injuries in Sport. In: Grupe O, Kurz D, Teipel JM (eds) Sport in the Modern World - Chances and Problems. Springer, Berlin Heidelberg New York, 1973, pp. 518-519
Van Wagenen WP, Aird RB: Dilatations of the cavity of the septum pellucidi and cavum Vergae. Am J Cancer 20:539-557, 1934
Venco L, Rigemonti L, Boschetti G: Considerations on the ocular injuries in boxing and a case of detachment of the retina due to a fight. Minerva Oftalmol 8:161-164, 1966
Verga A: Sul ventriculo della volta a tre pilastri. Gaz Med Lombarda 7: 1851
Verjaal A: Amnesie nach Trauma capitis. Z Ges Neurol Psychiatr 166:221-235, 1939
Vermeersch, quoted from Hillman E (1951)
Victor M, Adams RD, Mancall EL: A restricted form of cerebellar cortical degeneration occurring in alcoholic patients. Arch Neurol 1:579-588, 1959
Villavicencio C, Arriagada C, Corvalan V: Psychological, neurological and electroencephalographic study of 92 professional boxers. Proc 8th Internat Congr Neurol, Vol. S, September 5-10, 1965, Vienna, pp. 247-248
Villavicencio C, Bórquez Montero M: Influencia del boxeo en la patologia encefalocraneana. Bol Soc Cir (Uruguay) 29:146-160, 1958
Vincent JL: Boxing. Letter to the editor. J Am Med Ass 253:198, 1985
Vincent JL: Boxing, reply to a discussion remark of Lundberg GD. J Am Med Ass 253: 198, 1985
Vinger PF: Sports-related eye injury. A preventable problem. Surv Ophthalmol 25:47-51, 1980
Vinger PF: The incidence of eye injuries in sports. Internat Ophthalmol Clin 21:21-46, 1981
Vinger PF: Sports eye injuries a preventable disease. Ophthalmology 88:108-113, 1981
Vingo Carmen, as told to Shubin Seymour: I was slaughtered to please a crowd. Saturday Evening Post 224: January 12, 1952, pp. 15-17
Völkel G: Über Sportschäden des Auges. Med. Dissertation, Universität, Freiburg. 1943
Voigt GE, Löwenhielm CGR: 'Gliding Contusionen' des Großhirns. Hefte Unfallheilk 117:329-335, 1974
Voigt GE, Löwenhielm CGP, Ljung CBA: Rotational cerebral injuries near the superior margin of the brain. Acta Neuropathol 39:201-209, 1977
Volkner J: Die Schutzbestimmungen im DABV - Bereich. Medizin und Sport 5:170-172, 1965
Voss R: Das psychische Commotionssyndrom. Nervenarzt 24:155-161, 1953
Voy RO: Amateur boxing. J Am Med Ass 262:499-500, 1989
Wacquant LJD: Through the Fighters Eyes: Boxing as a Moral and Sensual World. Chap. 9. In: Cantu RC (ed) Boxing and Medicine. Human Kinetics, Champaign, ILL, 1995, pp. 129-173
Wagemann W: Audiologische Begutachtungen des stumpfen Schädeltraumas. HNO (Berlin) 6:60, 1956/1958
Walczynski JZ, Wosko G, Kulczycki J: Einige gerichtsmedizinische Aspekte von Todesfällen beim Boxen. Aktuelle Fragen Gerichtl Med 3:290-296, 1968
Waller A: Experimental researches on the function of the vagus and the cervical sympathetic nerves in man. Proc Royal Soc Med 11:302, 1862
Warburton N: Freedom to box. J Med Ethics 24:56-60, 1998
Ward RD:, Gallagher JP: Scrappy former boxer fights to maintain physicians' voices. Michigan Med 95:40-41, 1996
Ward SD: Spectator sensitivities (editorial). W V Med J 85:61, 1989
Warden J: MP's would ban all that harms – except boxing (news). Br Med J 313:1168
Waring AJ: Case report. Periorbital pneumatocele after facial trauma in a boxer. Br J Sports Med 14:231, 1980
Warter JJ, Weill JP, Bloch R: Episodes comateux et dissociation auriculo-ventriculaire par hyperréflectivite sinu-carotidienne. Strasbourgh Méd 10:361-373, 1959
Waxman A, Geshelm H: Boxer's bursitis. California Med 69:203-204, 1948
Weber HW: Wenn das Boxen humanisiert würde. Deutsch Ärztebl Heft 1:37-39, 1970
Webster JE, Gurdjian ES, Martin FA: Mechanism of syncope due to unilateral compression of carotid bifurcation. Arch Neurol Psychiat 74: 556-558, 1955
Wedrich A, Velikay M, Binder S, Radax U, Stolba U, Datlinger P: Ocular findings in asymptomatic amateur boxers. Retina 13:114-119, 1993
Wegner W: Boxers' encephalopathy. J Sportsmed 3:170-173, 1963
Wegner W, Pumplun: Über das Aschnersche Bulbusdruckphänomen. Deutsch Med Wochenschr 54:953-954, 1928
Weidenthal DT, Schepens CL: Peripheral fundus changes associated with ocular contusion. Am J Ophthalmol 62:465-477, 1966
Weigelin E, Lobstein A: Ophthalmodynamometrie. Karger, Basel New York, 1962
Weimann W: Zum Tod im Boxkampf. Deutsch Z Ges Gerichtl Med 16:341-344, 1931
Weinberg SK, Arond H: The occupational culture of the boxer. Am J Sociol 57:460-469, 1952

Weisman HF: Myocardial reperfusion injury: the case against shadow boxing. J Thromb Thrombolysis 4:133-135, 1997
Weiss S, Baker J: The carotid sinus reflex in health and disease: Its role in the causation of fainting and convulsions. Medicine 12:297-354, 1933
Weiss S, Ferris EB: Adams-Stokes syndrome with complete heart block of vagovagal reflex origin: Mechanism and treatment. Arch Intern Med 54:931-951, 1934
Welch MJ, Sitler M, Kroeten H: Boxing injuries from an instructional program. Physician Sportsmed 14 81-89, 1986
Welte E: Über die Zusammenhänge zwischen anatomischem Befund und klinischem Bild bei Rindenprellungsherden nach stumpfem Schädeltrauma. Arch Psychiatr Nervenkrankh 179:243-315, 1948
Werkgartner A: Gezelteriß durch Boxhieb. Z Gerichtl Med 25:41-44, 1935
Wertham F, Wertham FP: Brain as a Organ. Its Postmortem Study and Interpretation. Macmillan, New York, 1934
Weve quoted from Doggart JH (1965)
Whicker HW: Cauliflower ears. North Am Review 234:19-20, 1932
Whiteson AL: Injuries in professional boxing: Their treatment and prevention. Practitioner 225:1053-1057, 1981
Whiteson AL: Ocular injuries from boxing (letter, comment). Br Med J 304:574, 1992
Whiting WC, Gregor RJ, Finerman GA: Kinematic analysis of human upper extremity movements in boxing. Am J Sports Med 16:130-136, 1988
Wildes KW: Is Boxing ethically supportable? Chap. 8. In: Cantu RC (ed) Boxing and Medicine. Human Kinetics, Champaign, ILL, 1995, pp. 117-128
Wilhelmsen KC: Disinhibition-dementia-parkinsonism-amyotrophy complex (DDPAC) is a non-Alzheimer's frontotemporal dementia. J Neuol Transm Suppl 49:269-275, 1997
Wilke J: Der Einfluß des Boxens auf das cochleovestibuläre System. Medizin und Sport 5:183-185, 1965
Wilke J, Oeken FW: Zur Ätiologie des Kopf-k.o. Medizin und Sport 4:1-7, 1964
Will GW: 'Punch drunk. J Royal Army Med Corps 72:389-392, 1939
Will GW, quoted from Hillman E (1951)
Williams DJ, Tannenberg AE: Dementia pugilistica in an alcoholic achondroplastic dwarf. Pathology 28:102-104, 1996
Wilson MC: Boxing. Letter to the editor. J Am Med Ass 253:199, 1985
Wilson MC: Boxing, reply to a discussion remark of Lundberg GD. J Am Med Ass 253: 199, 1985
Winterstein CE: Head injuries attributable to boxers. Lancet II:719-720, 1937
Wolff K: Boxsport und Boxverletzungen. Eine kritische Studie unter Mitteilung eigener Beobachtungen. Deutsch Z Chir 208:379-408, 1928
Wolff K: Todesfälle durch Boxkampf. Deutsch Z Ges Gerichtl Med 12:392-401, 1928
Wolff SM, Zimmerman LE: Chronic secondary glaucoma associated with retrodisplacement of iris root and deepening of the anterior angle secondary to contusion. Am J Ophthalmol 54:547-563, 1962
Wolowska J: Encephalopathia pugilistica (Choroba Bokserska). (Polish with Engl. summary). Neurol Neurochir Psychiatr Pol 10:787-793, 1960
Wolter JR: Coup-contrecoup mechanism of ocular injury. Am J Ophthalmol 56:785-796, 1963
Woods RS: Primary dislocation of the left shoulder joint in the boxing-ring. Br Med J I: 387, 1957
Wright AD: Head Injuries. In: Bass AL, Blonstein JL, James RD, Williams JGP (eds) Medical Aspects of Boxing. Pergamon Press, Oxford London New York, 1965, pp. 61-63
Wright DH: Boxing in the Army (letter). J Royal Army Med Corps 136:123, 1990
Wyrsch RB, Spindler KP, Stricker PR: Scapula fracture in a professiomal boxer. J Shoulder Elbow Surg 4:395-398, 1995
Yovanovitch R, Milanovitch JB: Recherches electrocardiographique sur les effects de la compression oculaire bilatérale chez l' homme du vague par l'atropine. CS Soc Biol 127: 1367-1369, 1938
Zange J: Organische Schädigung des nervösen Ohrapparates im Krieg. Münch Med Wochenschr 32:1091, 1915
Zangwill OL: The Amnesia Syndrome. In: Whitty CWM, Zangwill OL (eds) Amnesias. Butterworths, London, 1966, p. 77
Zemina NV: Effect of systematic participation in boxing on the maturation of hand bones in young boxers. (Russian). Arch Anat Gistol Embriol 78:14-18, 1980
Zenneck I: Sportverletzungen des Auges. Med. Dissertation, Universität München, 1938
Ziegan J: Innenschichtriß einer Hirnbasisschlagader mit nachfolgender tödlicher Thrombose als Folge eines Faustschlages. Monatsschr Unfallheilk 72:282-294, 1969
Zorzoli E: La traumatologia del naso nei pugilatori. Arch Ital Otol 62:99-119, 1951
Zurita Serrano I, Gomez Llata S, Moreno Arenas G: Craneocerebrale Läsionen bei Boxern. Mechanismen und gerichtsmedizinischer Aspekt. In: Prokop L (eds) Sportmedizin. Bericht des 13. Internat Kongr Sportmed, 1960, Wien, Hollinek, 1961, p. 220
Editorial: Death from fisticuff. Lancet I:538, 1902
Punchy: Prize-fighters walk on their heels after getting numerous hooks to chin. Literary Digest, 123, 39-40, April 1937
Editorial: Brain injury from high school boxing. J Am Med Ass 136:907, 1948
Editorial: New medical regulations for boxing and wrestling, medical news. J Am Med Ass 142:1304-1305, 1950
Editorial: Safeguards in boxing. J Am Med Ass 142:1298, 1950
Editorial: Towards safety in boxing. Lancet I:1006-1007, 1950
Editorial: The nation's most dangerous trade - professional boxing. Industrial Med Surg 20:571-572, 1951
Editorial: Knock-out. Collier's 134: 94-97, September 17, 1954
Editorial: New York Times Magazines, January 31, 1954, p. 31
Editorial: Boxing. Lancet I: 1185-1186, 1959
Editorial: Ring deaths believed to be no cause to abolish boxing. Medical News, 1962, p. 21
Statement on Boxing. American Medical Association, Commit-

tee on the Medical Aspects of Boxing. J Am Med Ass 181:242, 1962
Editorial: Ban Boxing? Newsweek 61: April 8, 1963, p. 84
Editorial: Prizefighting. Murder on the BBC. Time Magazine 81:1963, p. 66
Editorial: Les méfaits de la boxe émeuvent l'opinion mondiale. Sem Hop Inform 20:8-9, 1963
Editorial: Boxing kills three more. Christian Century 80: April 24, 1963, pp. 515-516
Editorial: Boxser en geneesheer. South African Med J 38:879-880, 1964
Annotations: Boxing: Risks and remedies. J Sport Med 4:53-55, 1964
Editorial: AMA proposes methods for safe boxing matches. Science News. 88:98, 196
Editorial: Should pro boxing be banned? Medical World News, 1966
Editorial: Mayhem in the ring puts doctors on the spot. Medical World News 74-97, 1966
Editorial: Ring deaths believed to be no cause to abolish boxing. Medical News, 1967, p. 21
Editorial: Medical aspects of boxing. Lancet II:1114, 1969
Hazards of boxing. Lancet I:764, 1969
Brawls and fights, lagalized or otherwise. Can Med Ass J 100:1104-1105, 1969
Editorial: (Ex?) dementia pugilistica in ex-boxers. J Am Med Ass 210:2272, 1969
Editorial: Report on the Medical Aspects of Boxing. Committee on Boxing; Royal College of Physicians of London, October, 1969
Brain damage in boxers. Med J Austral II:709-710, 1970
W.H.: Boxen ist gefährlich, sagen die Sportärzte. Med Monatsschr 25:537-538, 1971
Boxing. Austral New Zealand J Psychiatr 5:131-132, 1971
Amateur Boxing Association. The Medical Aspect of Boxing. London, 1972
Editorial: Boxing and the brain. Brit Med J IV:439-440, 1973
Editorial: Boxing brains. Lancet II:1064-1065, 1973
Editorial: Mouth protectors: 11 years later. Bureau of Dental Health Education, Council on Dental Materials and Devices. Am Dent Ass 86:1365-1367, 1973
Editorial: Die Hohenpriester mit dem 'Killerinstinkt.' Frankfurter Allgemeine Zeitung, January 15, 1974, p. 10
Editorial: Letter: Boxers' brain damage. Lancet II:1270, 1974
Boxer's brain damage (letter). Lancet II:1270, 1974
Editorial: Health hazards of boxing. Med J Austral 163-164, 1974
Editorial: Brain damage in sport. Lancet I:401-402, 1976
Urteil des Bundesarbeitsgerichts vom 1.12.1976 zur Frage der Gefährlichkeit des Amateurboxsports. In: Anders G, Felten R, Kirsch A (eds) Boxen und Gesundheit. Zur Frage von Langzeitschäden und ihrer Verhütung. Bundesinstitut für Sportwissenschaft, Schriftenreihe Medizin. Bd. 3. Deutscher Ärzte Verlag, Köln, 1977, pp. 213-215
Jörg Eipel, kaum versichert. General Anzeiger (Bonn), 22. Dezember 1977
Boxkämpfe nur mit Kopfschutz. Süddeutsche Zeitung, No. 168, 25. Juli 1978, p. 4
Reformen im Berufsboxen. Süddeutsche Zeitung, No. 172, 29./30.Juli 1978, p. 46
Wieder Tod nach k.o. Spaniens Presse klagt an. Münchener Merkur, Nr. 45, 23. Februar, 1978, p. 30
Boxen - Totschlag nach sportlichen Regeln? 'Das Gehirn ist nicht fester als Brei!' Sport Illustrierte, Nr 4, April 1979, pp. 32-40
Kampf dem Tod. Der Spiegel, 34, Nr 16, 14. April 1980, p. 224-226
Some high risk sports. Sporting News, August 16, 1980
Impossibility of preventing brain damage in boxing. Prohibit head blows, thighten physical control. A Scandinavian Expert Committee. Lakartidningen 78:2234-2235, 1981
Boxing, Australian Medical Association, 1982
Working Paper on Boxing. World Medical Association, Fernay-Voltaire, 1982
Report on Boxing. American Medical Association, Chicago, 1982
Editorial: Council on Scientific Affairs, American Medical Association, 1983, Brain Injury in Boxing. J Am Med Ass 249:254-257, 1983
Reflessioni medico-legali sulla boxe. Riv Ital Med Leg 5:295-305, 1983
Boxing (letter). J Am Med Ass 249:2777-2779. 1983
Brain injury in boxing. J Am Med Ass 249:254-257, 1983
Review Committee on Kickboxing and Full Contact Karate in Province of Ontario, Canada. Toronto, 1983
New York group calls for ban on boxing. Am Med News, November 2, 1984, p. 7
Proceedings of the House of Delegates, 38th Interim Meeting, December 2-5, 1984, American Medical Association, Chicago, 1984, p. 371
American Academy of Pediatrics, Committee on Sports Medicine: Participation in boxing among children and young adults. Pediatrics 74:311-312, 1984
Report of the Board of Science and Education working party on boxing. Annual Report of council. Br Med J 289:130-131, 1984
New York State Department of State. Laws and Rules Regulating Boxing and Wrestling Matches., 1984
Editorial: Boxing in the Army. J Royal Army Med Corps 130:141-143, 1984
The AMA tries to k.o. boxing. Newsweek, December 17, 1984, p. 67
Interim Report: Inadequate Regulation of Boxing. Commission of Investigation, State of New Jersey, Trenton, NJ, No. 45, 1985
Boxing (letter). J Am Med Ass 253:198-199, 1985
Boxing should be banned in civilized countries - Round 3. J Am Med Ass 255:2483-2485, 1986
Neurological aspects of boxing (letter). Arch Neurol 44:1110, 1987
AIDS antibody testing required for boxers in Nevada. Northwest Dent 67: 10, 1988

Prevention of neurologic injuries in boxing (letter). Arch Neurol 45:713-714, 1988

Hearings before the Subcommittee on Commerce, Consumer Protection, and Competitiveness of the House Committee on Energy and Commerce, 100th Congress, 2nd Sess, 1988. (Testimony of Gerry Cooney and Robert Lee).

Banning military boxing (letter). Mil Med 153:270-273, 1988

Boxing in the Army (letter, comment). J Am Med Ass 262:2089-2090. 1989

American Academy of Ophthalmology. Recommendations for the Prevention of Eye Injuries in Boxing, 1990

Participation in boxing by children, adolescents, and young adults. Americam Academy of Pediatrics Committee on Sports Medicine and Fitness. Pediatrics. 99:134-135, 1997

Editorial: Wettkampfbestimmungen Boxen. Selbstverlag des Deutschen Amateur-Box-Verbandes, e. V., no year

Index